21st Century Nanoscience – A Handbook

21st Century Nanoscience – A Handbook: Nanophysics Sourcebook (Volume One)

21st Century Nanoscience – A Handbook: Design Strategies for Synthesis and Fabrication (Volume Two)

21st Century Nanoscience – A Handbook: Advanced Analytic Methods and Instrumentation (Volume Three)

21st Century Nanoscience – A Handbook: Low-Dimensional Materials and Morphologies (Volume Four)

21st Century Nanoscience – A Handbook: Exotic Nanostructures and Quantum Systems (Volume Five)

21st Century Nanoscience – A Handbook: Nanophotonics, Nanoelectronics, and Nanoplasmonics (Volume Six)

21st Century Nanoscience – A Handbook: Bioinspired Systems and Methods (Volume Seven)

21st Century Nanoscience – A Handbook: Nanopharmaceuticals, Nanomedicine, and Food Nanoscience (Volume Eight)

21st Century Nanoscience – A Handbook: Industrial Applications (Volume Nine)

21st Century Nanoscience – A Handbook: Public Policy, Education, and Global Trends (Volume Ten)

21st Century Nanoscience – A Handbook

Low-Dimensional Materials and Morphologies (Volume Four)

Edited by

Klaus D. Sattler

CRC Press is an imprint of the
Taylor & Francis Group, an **informa** business

CRC Press
Taylor & Francis Group
6000 Broken Sound Parkway NW, Suite 300
Boca Raton, FL 33487-2742

First issued in paperback 2022

ISBN-13: 978-0-815-35528-1 (hbk)
ISBN-13: 978-1-03-233590-2 (pbk)
DOI: 10.1201/9780429347290

Publisher's Note

The publisher has gone to great lengths to ensure the quality of this reprint but points out that some imperfections in the original copies may be apparent.

Library of Congress Cataloging-in-Publication Data

Names: Sattler, Klaus D., editor.
Title: 21st century nanoscience : a handbook / edited by Klaus D. Sattler.
Description: Boca Raton, Florida : CRC Press, [2020] | Includes bibliographical references and index. | Contents: volume 1. Nanophysics sourcebook—volume 2. Design strategies for synthesis and fabrication—volume 3. Advanced analytic methods and instrumentation—volume 5. Exotic nanostructures and quantum systems—volume 6. Nanophotonics, nanoelectronics, and nanoplasmonics—volume 7. Bioinspired systems and methods. | Summary: "This 21st Century Nanoscience Handbook will be the most comprehensive, up-to-date large reference work for the field of nanoscience. Handbook of Nanophysics, by the same editor, published in the fall of 2010, was embraced as the first comprehensive reference to consider both fundamental and applied aspects of nanophysics. This follow-up project has been conceived as a necessary expansion and full update that considers the significant advances made in the field since 2010. It goes well beyond the physics as warranted by recent developments in the field"—Provided by publisher.
Identifiers: LCCN 2019024160 (print) | LCCN 2019024161 (ebook) | ISBN 9780815384434 (v. 1 ; hardback) | ISBN 9780815392330 (v. 2 ; hardback) | ISBN 9780815384731 (v. 3 ; hardback) | ISBN 9780815355281 (v. 4 ; hardback) | ISBN 9780815356264 (v. 5 ; hardback) | ISBN 9780815356417 (v. 6 ; hardback) | ISBN 9780815357032 (v. 7 ; hardback) | ISBN 9780815357070 (v. 8 ; hardback) | ISBN 9780815357087 (v. 9 ; hardback) | ISBN 9780815357094 (v. 10 ; hardback) | ISBN 9780367333003 (v. 1 ; ebook) | ISBN 9780367341558 (v. 2 ; ebook) | ISBN 9780429340420 (v. 3 ; ebook) | ISBN 9780429347290 (v. 4 ; ebook) | ISBN 9780429347313 (v. 5 ; ebook) | ISBN 9780429351617 (v. 6 ; ebook) | ISBN 9780429351525 (v. 7 ; ebook) | ISBN 9780429351587 (v. 8 ; ebook) | ISBN 9780429351594 (v. 9 ; ebook) | ISBN 9780429351631 (v. 10 ; ebook)
Subjects: LCSH: Nanoscience—Handbooks, manuals, etc.
Classification: LCC QC176.8.N35 A22 2020 (print) | LCC QC176.8.N35 (ebook) | DDC 500—dc23
LC record available at https://lccn.loc.gov/2019024160
LC ebook record available at https://lccn.loc.gov/2019024161

Visit the Taylor & Francis Web site at
http://www.taylorandfrancis.com

and the CRC Press Web site at
http://www.crcpress.com

Contents

Editor

Klaus D. Sattler pursued his undergraduate and master's courses at the University of Karlsruhe in Germany. He earned his PhD under the guidance of Professors G. Busch and H.C. Siegmann at the Swiss Federal Institute of Technology (ETH) in Zurich. For three years he was a Heisenberg fellow at the University of California, Berkeley, where he initiated the first studies with a scanning tunneling microscope of atomic clusters on surfaces. Dr. Sattler accepted a position as professor of physics at the University of Hawaii, Honolulu, in 1988. In 1994, his group produced the first carbon nanocones. His current work focuses on novel nanomaterials and solar photocatalysis with nanoparticles for the purification of water. He is the editor of the sister references, *Carbon Nanomaterials Sourcebook* (2016) and *Silicon Nanomaterials Sourcebook* (2017), as well as *Fundamentals of Picoscience* (2014). Among his many other accomplishments, Dr. Sattler was awarded the prestigious Walter Schottky Prize from the German Physical Society in 1983. At the University of Hawaii, he teaches courses in general physics, solid state physics, and quantum mechanics.

Contributors

Jandro L. Abot
Department of Mechanical Engineering
The Catholic University of America
Washington, DC

Mashkoor Ahmad
Nanomaterials Research Group (NRG), Physics Division
PINSTECH
Islamabad, Pakistan

Jude C. Anike
Department of Mechanical Engineering
The Catholic University of America
Washington, DC

H. Bauer
Zentrum für Astronomie und Astrophysik
Technische Universität Berlin
Berlin, Germany

Somayeh Behzad
Department of Engineering Physics
Kermanshah University of Technology
Kermanshah, Iran

Luigi Bonacina
GAP-Biophotonics
University of Geneva
Geneva, Switzerland

Iván Cabria
Departamento de Física Teórica, Atómica y Óptica
Universidad de Valladolid
Valladolid, Spain

Luca Camilli
DTU Nanotech
Technical University of Denmark
Lyngby, Denmark

José M. Caridad
DTU Nanotech
Technical University of Denmark
Lyngby, Denmark

Ch. Chang
Zentrum für Astronomie und Astrophysik
Technische Universität Berlin
Berlin, Germany

Sanghyeon Choi
KU-KIST Graduate School of Converging Science and Technology
Korea University
Seongbuk-gu, Republic of Korea

Stuart Cornes
Department of Materials
University of Oxford
Oxford, United Kingdom

Panagiotis Dallas
Department of Materials
University of Oxford
Oxford, United Kingdom

Dongyang Deng
Department of Built Environment
College of Science and Technology
North Carolina Agricultural and Technical State University
Greensboro, North Carolina

Reuven Gordon
Department Electrical and Computer Engineering
University of Victoria
Victoria, Canada

Beth Guiton
Department of Chemistry
University of Kentucky
Lexington, Kentucky

Lin Guo
School of Chemistry, Beijing
Advanced Innovation Center for Biomedical Engineering
Beihang University
Beijing, China

Quanmin Guo
School of Physics and Astronomy
University of Birmingham
Birmingham, United Kingdom

Reuben Harding
Department of Materials
University of Oxford
Oxford, United Kingdom

Keea Stancato
Nanomaterials and Devices Laboratory
Department of Physics
University of Houston
Houston, Texas

Olof Hultin
Division of Solid State Physics
Lund University
Lund, Sweden

Muhammad Hussain
Nanomaterials Research Group (NRG), Physics Division
PINSTECH
Islamabad, Pakistan
and
Centre for High Energy Physics
University of the Punjab
Lahore, Pakistan

Changlong Jiang
Institute of Intelligent Machines
Chinese Academy of Sciences
Hefei, China

Ali Karatutlu
UNAM-National Nanotechnology Research Center
Institute of Materials Science and Nanotechnology
Bilkent University
Ankara, Turkey

Dogan Kaya
Department of Electronics and Automation
Vocational School of Adana, Cukurova University
Adana, Turkey

Ali Khademi
Department Electrical and Computer Engineering
University of Victoria
Victoria, Canada

Edward Laird
Department of Materials
University of Oxford
Oxford, United Kingdom
and
Department of Physics
Lancaster University
Lancaster, United Kingdom

Lidong Li
School of Chemistry, Beijing Advanced Innovation Center for Biomedical Engineering
Beihang University
Beijing, China

Thuc Hue Ly
Department of Chemistry and Center of Super-Diamond & Advanced Films (COSDAF)
City University of Hong Kong
Hong Kong SAR, China

Ahmed A. Maarouf
Department of Physics
Institute for Research and Medical Consultations
Imam Abdulrahman Bin Faisal University
Dammam, Saudi Arabia

Luis F. Mazadiego
Departamento de Energía y Combustibles, E.T.S. de Ingenieros de Minas y Energía
Universidad Politécnica de Madrid
Madrid, Spain

Sheng Meng
Beijing National Laboratory for Condensed Matter Physics, and Institute of Physics
Chinese Academy of Sciences
Beijing, China

Gerardo Morell
Institute for Functional Nanomaterials
University of Puerto Rico
San Juan, Puerto Rico
and
Department of Physics
University of Puerto Rico at Río Piedras
San Juan, Puerto Rico

Yannick Mugnier
SYMME
Université Savoie Mont Blanc
Annecy, France

Saira Naz
Nanomaterials Research Group (NRG), Physics Division
PINSTECH
Islamabad, Pakistan
and
Institute of Chemical Sciences
University of Peshawar
Peshawar, Pakistan

Ram Neupane
Nanomaterials and Devices Laboratory, Department of Physics
University of Houston
Houston, Texas

Marcin Opallo
Institute of Physical Chemistry
Polish Academy of Sciences
Warsaw, Poland

Bülend Ortaç
UNAM-National Nanotechnology Research Center
Institute of Materials Science and Nanotechnology
Bilkent University
Ankara, Turkey

Marcelo F. Ortega
Departamento de Energía y Combustibles, E.T.S. de Ingenieros de Minas y Energía
Universidad Politécnica de Madrid
Madrid, Spain

Javier Palomino
Department of Physics-Mathematics
Pontifical Catholic University of Puerto Rico
Ponce, Puerto Rico
and
Institute for Functional Nanomaterials
University of Puerto Rico
San Juan, Puerto Rico
and
Department of Physics
University of Puerto Rico at Río Piedras
San Juan, Puerto Rico

A. B. C. Patzer
Zentrum für Astronomie und Astrophysik
Technische Universität Berlin
Berlin, Germany

Maggie Paulose
Nanomaterials and Devices Laboratory, Department of Physics
University of Houston
Houston, Texas

Kyriakos Porfyrakis
Department of Materials
University of Oxford
Oxford, United Kingdom

Ilija Rašović
Department of Materials
University of Oxford
Oxford, United Kingdom

Alexandra J. Riddle
Department of Chemistry
University of Kentucky
Lexington, Kentucky

Lars Samuelson
Division of Solid State Physics
Lund University
Lund, Sweden

Sapna Sinha
Department of Materials
University of Oxford
Oxford, United Kingdom

Irena G. Stará
Institute of Organic Chemistry and Biochemistry
Czech Academy of Sciences
Prague, Czech Republic

Ivo Starý
Institute of Organic Chemistry and Biochemistry
Czech Academy of Sciences
Prague, Czech Republic

Kristian Storm
RISE Acreo AB
Lund, Sweden

Fabián Suárez-García
Department of Material Chemistry
Instituto Nacional del Carbón, INCAR-CSIC
Oviedo, Spain

D. Sülzle
Zentrum für Astronomie und Astrophysik
Technische Universität Berlin
Berlin, Germany

Oomman K. Varghese
Nanomaterials and Devices Laboratory, Department of Physics
University of Houston
Houston, Texas

Gunuk Wang
KU-KIST Graduate School of Converging Science and Technology
Korea University
Seongbuk-gu, Republic of Korea

Brad R. Weiner
Institute for Functional Nanomaterials
University of Puerto Rico
San Juan, Puerto Rico
and
Department of Chemistry
University of Puerto Rico at Río Piedras
San Juan, Puerto Rico

Katarzyna Winkler
Institute of Physical Chemistry
Polish Academy of Sciences
Warsaw, Poland

Lokwing Wong
Department of Applied Physics
The Hong Kong Polytechnic University
Hong Kong SAR, China

Dao Xiang
Department Electrical and Computer Engineering
University of Victoria
Victoria, Canada

Shihao Xu
Institute of Intelligent Machines
Chinese Academy of Sciences
Hefei, China

Elif Yapar Yildirim
UNAM-National Nanotechnology Research Center
Institute of Materials Science and Nanotechnology
Bilkent University
Ankara, Turkey

Peiwei You
Beijing National Laboratory for Condensed Matter Physics, and Institute of Physics
Chinese Academy of Sciences
Beijing, China

Jian Yu
School of Chemistry, Beijing Advanced Innovation Center for Biomedical Engineering
Beihang University
Beijing, China

Lifeng Zhang
Department of Nanoengineering
Joint School of Nanoscience and Nanoengineering, North Carolina Agricultural and Technical State University
Greensboro, North Carolina

Jiong Zhao
Department of Applied Physics
The Hong Kong Polytechnic University
Hong Kong SAR, China

Fangyuan Zheng
Department of Applied Physics
The Hong Kong Polytechnic University
Hong Kong SAR, China

Shen Zhou
Department of Materials
University of Oxford
Oxford, United Kingdom

1

Advances in 1D Materials

Javier Palomino
Pontifical Catholic University of Puerto Rico
University of Puerto Rico

Brad R. Weiner and
Gerardo Morell
University of Puerto Rico

1.1 Introduction

In the past decades, nanotechnology has attracted tremendous attention of researchers around the world, due to its wide range of innovative applications, creating new investigation fields such as nanomaterials, nanoelectronics, nanophotonics, nanosensors, and nanodevices, among others. In the field of nanomaterials, one-dimensional (1D) structures such as nanotubes (NTs), nanowires (NWs), and nanofibers (NFs) have been widely studied due to their unique properties that enable new-generation applications at nano-, micro-, and macroscale dimensions.

The carbon NT (CNT) is the most significant among NTs due to its high tensile strength, which is approximately 100 times greater than that of steel (Ruoff and Lorents, 1995). This extraordinary property is mainly attributed to carbon-to-carbon covalent bonds and also to the fact that each CNT is one large molecule. Furthermore, Young's modulus for CNTs is superior to that for the steel (Chandran and Gifty Honeyta A, 2017). Besides their strength and elasticity, CNTs are also lightweight with a density about one quarter that of steel (Chandran and Gifty Honeyta A, 2017). Moreover, CNTs possess very good thermal conductivity, which is five times that of silver (Thang, Khoi and Minh, 2015). CNTs have been used as electrically conducting components in polymer composites mostly for commercial applications in automotive and plastic industries (Xu et al., 2007). Because of the high surface area of porous NT arrays, combined with their high electronic conductivity and unique mechanical properties, CNTs are appealing in electrochemical devices such as supercapacitors, batteries, and actuators (Palomino et al., 2015). The industry and the academy have focused their research activity on using CNTs as field emission electron sources for panel displays (Choi et al., 1999). CNTs were also used in the fabrication of nanometer-sized electronic devices like NT field effect transistors (NT-FETs) and nonvolatile memory devices (Martel et al., 1998). Its nanolevel size makes CNT a highly sensitive element since it requires a small amount of material for a response. There are extensive studies conducted on high-resolution CNT scanning probe tips for atomic probe microscopes (Nguyen et al., 2004). The continuous increase in publications and patents on CNTs demonstrates the increase in the interest of industry and academy, and guarantees the development of new-generation applications.

The NWs are also prominent 1D nanostructures that possess unique electrical, thermoelectrical, optical, magnetic, mechanical, and chemical properties to name but a few. In 1D materials, the electron motion is constrained to 1D as a result of quantum effects, which depend on wire material, axis orientation, length, and diameter (Khare, Patil and Kodambaka, 2011). Because of quantum effects, NWs exhibit exclusive properties, which is different from their bulk form. Consequently, they have been exhaustively studied in the past decades by the academy and the industry for their promising applications in electronics, sensors, batteries, and light harvesting, among others (Palomino et al., 2018). It has been reported that 1D NWs show a ballistic-type electron transport mechanism, which depends mainly upon the wire length and diameter (Cho et al., 2008). The photoluminescence (PL) spectroscopic studies have shown that the energy peak and the bandgap of PL increase with a decrease in the wire diameter, validating the effect of quantum confinement in NWs when the wire diameter is reduced (Lee et al., 2016). Because of their enhanced surface-to-volume ratio, high aspect ratio, large curvature at the NW tips, and large number of surface atoms, NWs present high chemical reactivity, making them noteworthy for sensor device applications (Patolsky, Zheng and Lieber, 2006). NWs have been used in numerous electronic applications such as junction diodes, memory cells and switches, transistors, LEDs, inverters, logic gates, and field emission devices like flat panel displays (Shao, Ma and Lee, 2010). ZnO NWs were used for optical applications such as laser (Huang et al., 2001). Moreover, the large surface area and the high conductivity along the length of NWs were successfully used in high-efficiency solar cells (Kelzenberg et al., 2010). In addition, chemical and biological sensors made of NWs as sensing probe showed enhanced sensitivity and fast responsivity compared to the conventional sensors (Patolsky, Zheng and Lieber, 2006).

NFs also represent an important 1D material since they are less than 100 nm in diameter and possess unique properties such as large area-to-volume ratio, flexibility, high porosity, and enhanced mechanical properties, among others (Poveda and Gupta, 2016). Several methods have been used to synthesize NFs, including electrospinning and chemical vapor deposition (CVD). Carbon NFs (CNFs) are relatively more similar to multiwalled CNTs (MWCNTs), but instead of the graphitic concentric cylindrical tubes in CNTs, quasi-conically shaped stacked graphitic sheets are placed in CNFs. NFs have many technological and commercial applications such as tissue engineering, drug delivery, cancer diagnosis, and cartilage production and stimulation (Stout, Basu and Webster, 2011). Furthermore, CNFs have been used in a wide range of applications including lithium batteries, fuel cells, supercapacitors (Zhou et al., 2009), optical sensors, optoelectronics (Jang, Bae and Park, 2006), piezoelectric NFs, flame retardant, and filters.

This chapter gives a brief introduction to the main 1D materials including NTs, NWs, and NFs. Hence, this chapter is divided into three principal sections and is focused on CNTs, silicon NWs (SiNWs), and CNFs.

1.1.1 Density of States

The electronic density of states (DOS) gives us the number of possible energy states that can be occupied by electrons (Khare, Patil and Kodambaka, 2011). Thus, the DOS determines diverse physical and chemical properties of materials. Figure 1.1 demonstrates the DOS dependence on energy for materials, with different dimensions, by showing how the DOS changes from bulk (3D) to 2D, 1D, and 0D nanostructures. The DOS is continuous for 3D structures; meanwhile, for 1D, it has sharp spikes at different band edges, resulting in a strong signal in many optical and electronic measurements at these energies. Similarly, the chemical bonding and mechanical properties of NWs are affected due to the confinement of electrons in 1D. The dimensionality effect in nanomaterials is the reason for the increased interest in nanostructures, particularly in 1D materials for their enhanced mechanical, electric, and optical properties.

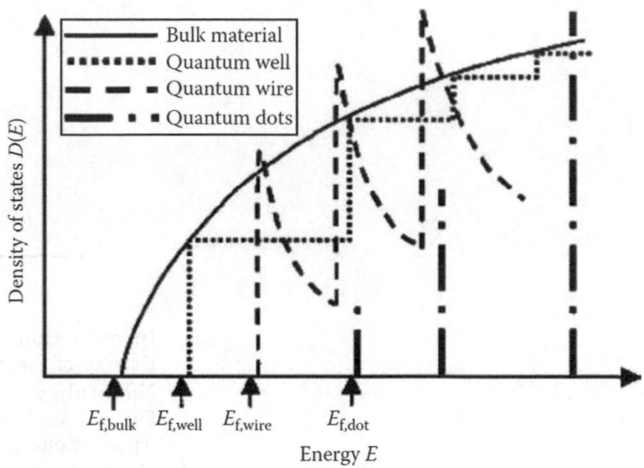

FIGURE 1.1 DOS for bulk and quantum nanostructures. (Reproduced from Khare et al., *Handbook of Nanophysics: Nanotubes and Nanowires* 16-1–16-14, 2011.)

1.2 Nanotubes

In general, an NT is a nanostructure with tubular shape and high aspect ratio, having diameter in nanometers but length in micrometers. This nanostructure has been synthesized, studied, and reported by several researchers since 1952, including Iijima who published a detailed analysis of CNTs in 1991 (Iijima, 1991). Researchers around the world reported various kinds of NTs such as boron and nitrogen NTs, boron nitride NT, gallium nitride NT, silicon NT (Pokropivny, 2001), and the exciting CNTs (Iijima, 1991). CNT exhibits extraordinary properties like it is many-fold stronger than steel and its electrical conductivity is superior to copper, thus capturing the attention of academic and industrial laboratories around the world to find its practical uses. Thus, researchers have produced thousands of publications and patents on innumerous potential applications in the fields of communication, transport, health, and environment, among others.

1.2.1 Fundamentals of Carbon Nanotubes

CNT is formed by sp^2 carbon atoms densely arranged in hexagonal configuration, in which carbon atom with atomic number 6 possesses six electrons that occupy the orbitals $1s^2$, $2s^2$, and $2p^2$. The orbitals can be hybridized into sp, sp^2, or sp^3 forms. The material composed of sp^3 carbon atoms is the recognized diamond. On the other hand, sp^2 carbon atoms are present in graphene, fullerenes, and CNTs. Theoretically, CNT is formed by a cylindrically rolled graphene sheet. Thus, most of the physical properties of CNTs are relatively similar to graphene. CNTs are organized in single-walled CNTs (SWCNTs) and MWCNTs (Iijima and Ichihashi, 1993), as shown in Figure 1.2a and b, respectively.

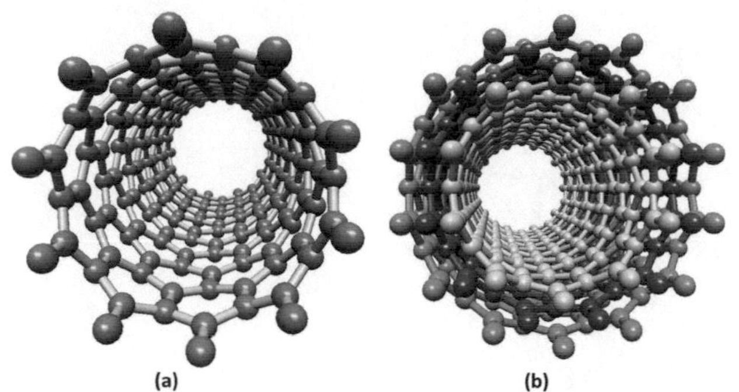

FIGURE 1.2 Schematic structure of (a) SWCNT and (b) MWCNT.

It has been observed in real space that the diameters of SWCNTs range from 0.4 to 3 nm, and their length is typically in the micrometer range (Eatemadi et al., 2014). On the other hand, MWCNTs show a range of interlayer spacing from 0.34 to 0.39 nm (Lehman et al., 2011). In addition, depending on the number of layers, the inner diameter is in the range from 0.4 nm to a few nanometers and outer diameter from 2 to 30 nm (Eatemadi et al., 2014), and axial size varies from 1 μm to a few centimeters. Moreover, MWCNTs usually exhibit closed tip by half-fullerene molecule.

There are three different forms of SWCNTs, namely, zigzag, armchair, and chiral (Figure 1.3), depending on how graphene is wrapped to form the NT. The SWCNT structure is characterized by a pair of indices (n, m) that describe the chiral vector. This chiral vector finds out the direction of rolling a graphene sheet (Figure 1.4). Therefore, the geometry of NT can be determined by the chiral vector of the original hexagonal lattice. The chiral vector is defined as $C_h = na_1 + ma_2$, where a_1 and a_2 are the base cell vectors of the hexagonal lattice and can be expressed in the Cartesian coordinate (x, y) as follows:

$$a_1 = \left(\frac{3}{2}a_{c-c}, +\frac{\sqrt{3}}{2}a_{c-c}\right)$$

$$a_2 = \left(\frac{3}{2}a_{c-c}, -\frac{\sqrt{3}}{2}a_{c-c}\right)$$

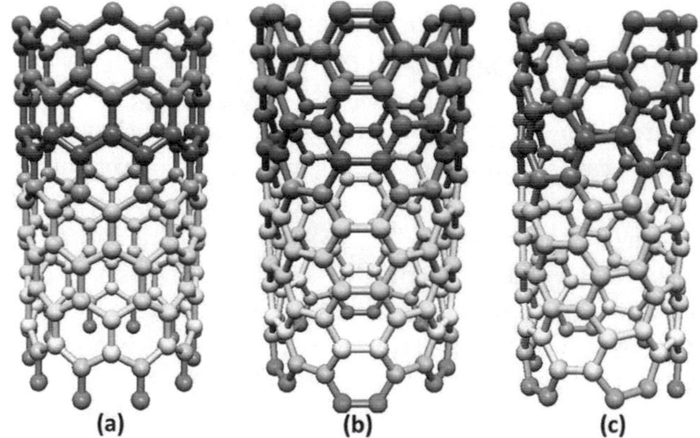

FIGURE 1.3 Models of atomically perfect SWCNT structures. (a) Zigzag, (b) armchair, and (c) chiral.

Here, $a_{c-c} = 1.421$Å is the bond length of carbon atoms in graphite. Hence, the length of the chiral vector C_h represents the circumference of the NT:

$$C_h = \sqrt{3}a_{c-c}\sqrt{n^2 + nm + m^2}$$

Consequently, the circumference for zigzag $(m = 0)$ is $C_h = \sqrt{3}na_{c-c}$ and for armchair $(n = m)$ is $C_h = 3na_{c-c}$. Thus, the CNT diameter can be calculated by $d = \frac{C_h}{\pi}$.

The chirality directly affects the electrical properties of NTs. Thus, when $(n-m)/3$ is an integer, the NT is described as "metallic" due to the occurrence of metallic conductance, and if not, then the NT is described as a semiconductor (Saito et al., 1992). The armchair configuration is intrinsically metallic; nevertheless, variations can make the NT a semiconductor.

On the other hand, when MWCNT is formed by coaxial NTs, it is called the Russian Doll model. On the contrary, when a single graphene sheet is wrapped around itself manifold times, it is called the Parchment model (Eatemadi et al., 2014). The MWCNTs allow incorporation of diverse defects, which significantly affect electrical and mechanical properties. In addition, the outer walls of MWCNT shield the inner CNTs from chemical interactions, and present high tensile strength properties, not observed in SWCNT (Yu et al., 2000).

1.2.2 Synthesis

Several techniques that have been developed for synthesizing CNT structures mainly involve gas-phase processes. The most common methods for producing CNTs are (i) the carbon arc discharge, (ii) the laser ablation, and (iii) the CVD (Landi et al., 2009). The first methods to produce CNTs were laser ablation and arc discharge, but currently, these high-temperature techniques have been substituted by low-temperature CVD methods (<800°C), since nanotube properties such as length, diameter, alignment, purity, and density can be accurately controlled at low temperatures (Kumar and Ando, 2010).

CNT Growth Mechanism

There are various CNT growth mechanisms; nevertheless, the most accepted mechanism based on catalyst particle is sketched as follows. When hydrocarbon vapor directly contacts the catalyst nanoparticle (hot metal like nickel), hydrocarbon decomposes into carbon and hydrogen species, as shown in Figure 1.5. Since carbon is highly soluble in catalyst metals, carbon atoms get dissolved into the metal until reaching the carbon solubility limit at low temperature (<800°C); then as-dissolved carbon atoms precipitate out and crystallize around the metal particle forming a ring structure without dangling bonds, hence becoming energetically stable (Kumar and Ando, 2010). Subsequently, fresh carbon atoms are dissolved into the metal and precipitate out forming a second ring structure, and so forth. This process results in the formation of a stable cylindrical

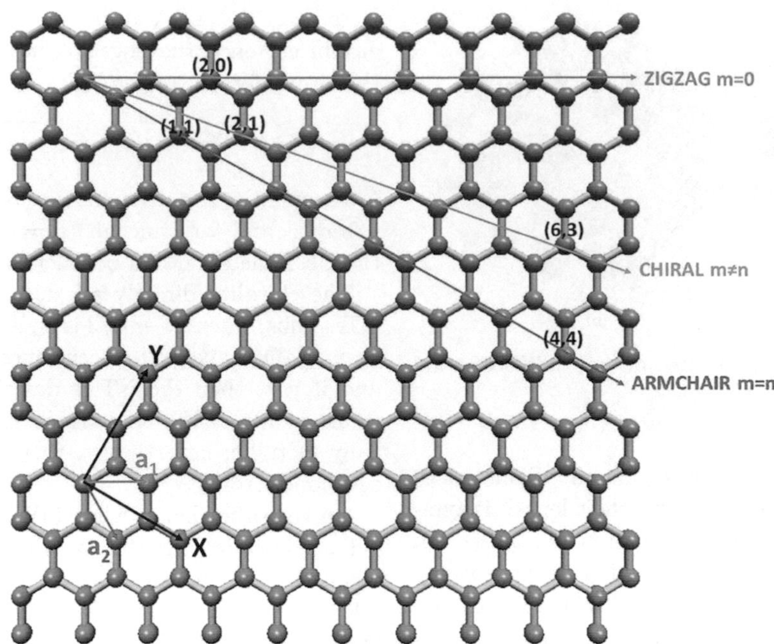

FIGURE 1.4 Representation of the chiral vector C on graphene layer.

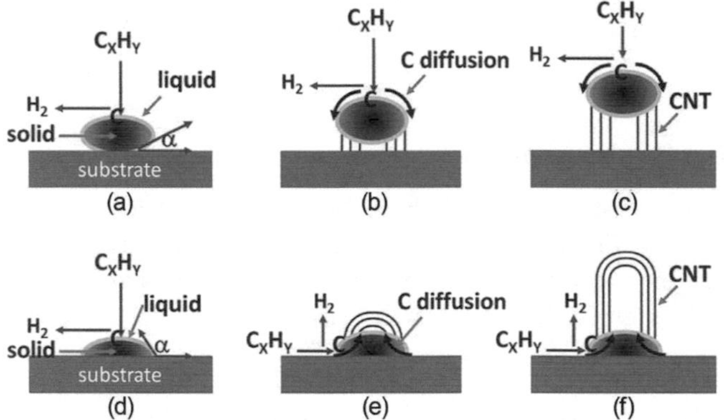

FIGURE 1.5 CNT growth mechanisms: (a–c) "tip-growth" and (d–f) "base-growth" models.

structure, which is well known as CNTs. Hydrocarbon decomposition is an exothermic process that releases heat to the metal and melts the exposed surface, whereas carbon crystallization is an endothermic process that absorbs heat from the precipitation zone of metal. This thermodynamic process inside the metal particle keeps the growth of CNTs active as long as the vapor components are supplied.

Depending on catalyst–substrate interaction, the mechanism presents two cases. If the interaction between catalyst and substrate is weak, the metal forms an acute contact angle with the substrate, as shown in Figure 1.5a. Hydrocarbon exothermically decomposes on the top surface of the metal, then carbon atoms diffuse down through the melted surface of the metal (Figure 1.5b), and carbon atoms precipitate out at the bottom, pushing the metal nanoparticle upward (Figure 1.5c). While the top surface of the metal is available for fresh hydrocarbon decomposition, carbon atoms diffuse and precipitate out forming the cylindrical structure of CNT. This is known as "tip-growth model" (Kumar and Ando, 2010).

If the interaction between catalyst and substrate is strong, the metal forms an obtuse contact angle with the substrate, as shown in Figure 1.5d. The hydrocarbon decomposes, and carbon atoms diffuse. Because of the strong catalyst–substrate interaction, carbon precipitation takes place at the metal apex. Initially, carbon precipitates out and crystallizes producing a spherical dome (Figure 1.5e). The subsequent hydrocarbon decomposition occurs at the lower surface of the metal, and as-dissolved carbon atoms diffuse upward and precipitate out forming the graphitic cylinder (Figure 1.5f). This mechanism is known as "base-growth model" because CNT grows up with the catalyst particle at the base (Kumar and Ando, 2010).

Electric Arc Discharge

The CNTs can be synthesized by arc-discharge method with fewer structural defects in comparison with other techniques. Arc-discharge method produces higher temperatures (above 1,700°C), resulting in expansion defects of CNTs. The arc-discharge system comprises a vacuum chamber, a graphite cathode and anode, as well as a power supply (Figure 1.6). In order to start the arcing process, the chamber is pressurized in He or Ar atmosphere and direct current is passed through the anode and cathode electrodes (arcing process), generating a plasma at temperatures over 4,000°C (Iijima, 1991). In the course of this process, the anode is consumed, and about half of the evaporated carbon from the anode is deposited on the cathode tip. This process is called cylindrical hard deposition, with a rate of around 1 mm/min, yielding either SWCNTs or MWCNTs and graphene particles (Eatemadi et al., 2014).

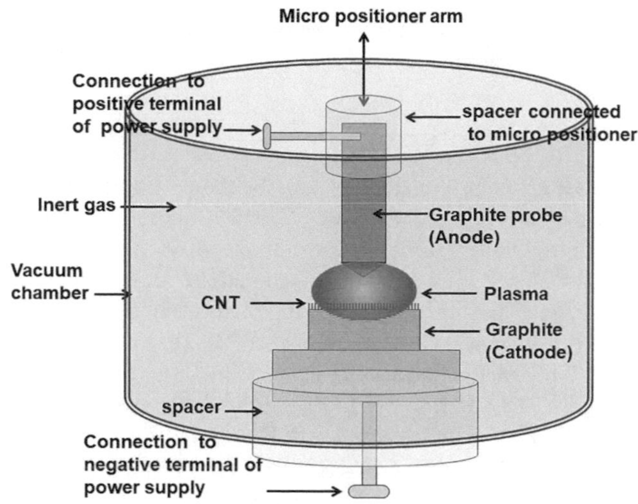

FIGURE 1.6 Electric arc-discharge system for CNT synthesis.

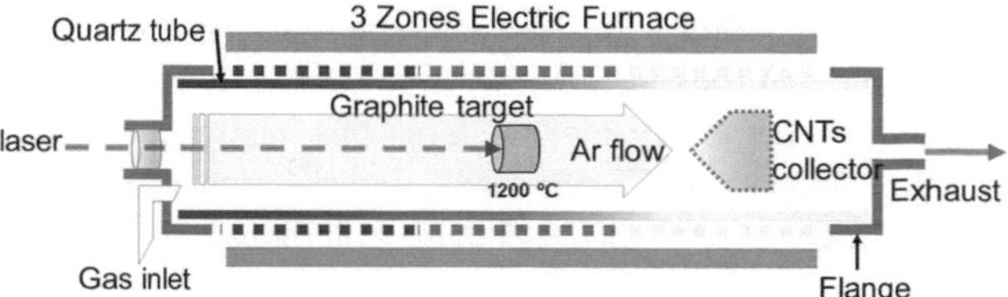

FIGURE 1.7 Schematic representation of laser ablation system for CNT synthesis.

The CNTs can be synthesized by the arc-discharge method through two main approaches. Generally, the synthesis of MWCNTs is performed without the use of catalyst precursors, but the synthesis of SWCNTs utilizes different catalyst precursors. Studies have demonstrated that Ni-Y-graphite mixtures can produce high yields (<90%) of SWCNTs with an average diameter of 1.4 nm (Saito, Dresselhaus and Dresselhaus, 1998). Nowadays, this mixture is used worldwide for the synthesis of SWCNTs. The main advantage of arc-discharge technique is its capability to produce a large quantity of NTs with fewer structural defects. On the other hand, the main disadvantage of this method is the difficulty to control the dimensions and morphology of NTs, which determine the physical and chemical properties of CNTs (Malekimoghadam, 2018).

Laser Ablation Method

The laser ablation technique uses a high-power YAG-type laser as energy source to evaporate the graphite target, which could contain metal particles as catalyst. The laser ablation system includes a tube furnace with a quartz tube containing a target of high-purity graphite which is heated at 1,200°C in an Ar atmosphere (Malekimoghadam, 2018), as shown in Figure 1.7. Then, the energetic laser pulses vaporize the graphite target surface, and the Ar carrier gas transports the carbon and metal atoms toward the collector, resulting in the deposition and the growth of SWCNTs and MWCNTs. The laser method leads to the management of various synthesis parameters such as laser properties (peak power, pulse frequency, energy fluence, oscillation wavelength), flow of the carrier gas, the chamber pressure and the chemical composition, distance between the target and the substrate, chemical composition of the target material, and temperature. Since these parameters affect the physical and chemical properties of CNTs, the laser ablation method has the potential for production of CNTs with high purity and high quality. The primary advantage of this technique is relatively low metallic impurities, due to the fact that the catalyst metallic atoms tend to evaporate from the end of the tube once it is closed. On the other hand, the principal disadvantage is that the fabricated NTs are not homogeneously straight. In addition, the laser ablation method is not economically advantageous because the procedure requires graphite rods, and the yield of NTs is lower compared to arc-discharge technique (Malekimoghadam, 2018).

Chemical Vapor Deposition

One of the most important techniques to generate CNTs is CVD. There are several types of CVD methods such as thermal (T-CVD) (Palomino et al., 2014), microwave plasma-enhanced (MPE-CVD) (Chen et al., 2007), radiofrequency (RF-CVD) (Wang et al., 2004), and hot filament (HF-CVD) (Palomino et al., 2014), among others. Our research group has demonstrated the performance of HF-CVD for the synthesis of carbon allotropes, including micro-, nano-, and ultrananocrystalline diamond (UNCD) (Varshney et al., 2014), CNTs, graphene, and hybrid materials (Palomino et al., 2015). In a CVD process, the reactant gases pass across energetic zones like hot filament, to be decomposed into reactive species, and then transported toward the substrate surface, either to be deposited directly or transform into condensable species via chemical reactions at the substrate surface to form nucleation centers (Palomino et al., 2015). At the surface of the catalyst particle, the hydrocarbon gas is broken apart and so the carbon after saturation precipitates at the edges of the nanoparticle, resulting in the formation of NTs. The process continues as long as the reactants are available. In CVD processing, the most used metal catalyst particles are nickel, cobalt, iron, or their combinations (Kumar and Ando, 2010). Depending on the interaction between the substrate and the catalyst particle, the growth of CNTs follows the base-growth or tip-growth model, as described above (Kumar and Ando, 2010). Similar to arc-discharge and laser ablation methods, the CVD technique is accompanied by the production of chemical residual products that are swept out of the chamber together with unreacted precursor gases. Compared to laser ablation, CVD is an economic method for large-scale production with the leading advantage of high-purity CNT production due to easy control of the reactions (Patole et al., 2008).

Our research group has studied the synthesis of carbon allotropes, including graphene and diamond; CNTs; and hybrids by the CVD method. In this chapter, we will describe a silicon–CNT (Si-CNT) hybrid structure, fabricated in a single step on Cu substrate by HF-CVD.

Experimental Setup for HF-CVD System

The Si-CNT hybrid nanostructure films were grown on Cu substrate, in a custom-made HF-CVD chamber, which incorporates a filament of Re wire of 0.5 mm diameter rolled as a helical spring and positioned at 8 mm above the substrate, as shown in Figure 1.8 (Morell, Canales and Weiner, 1999; Morell et al., 2006). We selected rhenium wire as the filament because it is not reactive with carbon and is not consumed during the growing process of Si-CNTs. Polycrystalline copper substrates (99.9% pure, 0.5 mm thick, 14 mm disk diameter) were hand-polished, then cleaned in an ultrasonic bath with 2-propanol for 15 min, and dried with argon gas. Nickel is a suitable catalyst to grow carbon nanostructures, because of its significant carbon solubility; thus, Cu substrates were coated with nickel by RF sputtering. The Ni film thickness after 30 min of deposition was ~50 nm. A mixture of 1 g of n-tetracosane (Alfa Aesar), 1 g of n-octacosane (Alfa Aesar), and silicon nanoparticles at the concentration of 20 wt% was melted in a glass beaker on a hot plate at a temperature of 100°C. A portion of this silicon–polymer melt was transferred onto the nickel-coated copper substrate. The substrate was then inserted in the HF-CVD chamber on a molybdenum holder, which is integrated with a graphite heater. Prior to each deposition, the CVD chamber was evacuated to 5×10^{-7} Torr and then filled with a mixture of 10.0% CH_4 and 90.0% H_2, preserving the combined flow of gases to 20 sccm and the total pressure at 20 Torr. A constant current of 20 A (~2,400°C) and 4 A (~500°C) was applied through the rhenium filament and graphite heater, respectively. The deposition time was optimized to 30 min in order to attain a film that covered the whole substrate.

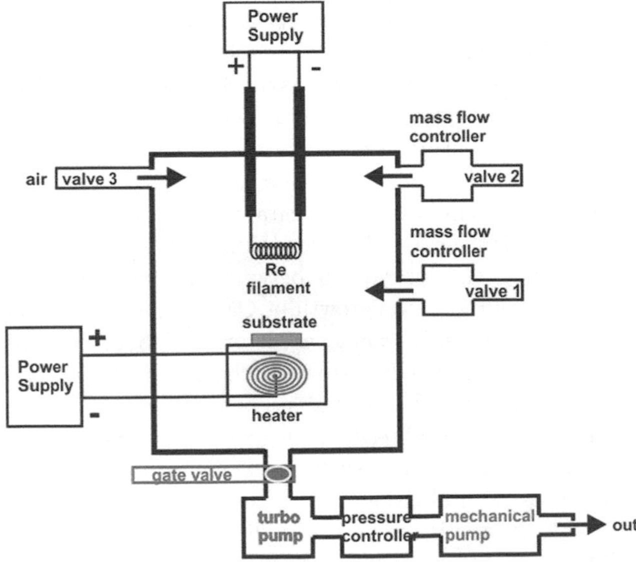

FIGURE 1.8 Schematic representation of HF-CVD system.

1.2.3 Characterization

The TEM images in Figure 1.9a,b show tubular nanostructures that consist of MWCNTs with diameters around ~100 nm, which are conformally coated with silicon nanoparticles. Figure 1.9c reveals the lattice fringes of the Si coating, with an interplanar spacing of about 0.31 nm, which matches with the (111) orientation of Si (Palomino et al., 2014). The carbon and silicon electron energy loss spectroscopy (EELS) spectra (Figure 1.9d,e) show the energy loss peaks of the 1s → π^* (~283 eV) and 1s → σ^* bonds (~291 eV), where the C K-edge shifted to lower energies, congruous with Si-C hybrid structures (Xie, Möbus and Zhang, 2010; Shakerzadeh et al., 2011), whereas the Si $L_{3,2}$ edge shifted to higher energies, attributed to a strong Si-C chemical bonding.

The Raman spectra of the Si-CNTs, described elsewhere (Palomino et al., 2015), show the G-band around 1,580 cm^{-1}, which is associated with the crystalline graphite-like materials (Dresselhaus et al., 2005; Lehman et al., 2011). The intense D-band at 1,351 cm^{-1} is attributed to defects in the crystalline graphite, consistent with the significant amount of mass disorder caused by the insertion of Si atoms into CNTs (Maurin, 2000). The broad band around 750 cm^{-1} is ascribed to the nanocrystalline SiC (Sohrabi, Nikniazi and Movl, 2013). The peaks at around 506 and 970 cm^{-1} represent, respectively, the first and second orders of the transverse optical (TO) phonon modes of Si nanoparticles of about 8 nm (Morell et al., 1996; Meier et al., 2006). The band around 195 cm^{-1} corresponds to the radial breathing mode (RBM) of Si-CNT (Lehman et al., 2011).

In addition, thermogravimetric analysis (TGA) was performed to determine the amount of silicon present in the Si-CNT material, as reported elsewhere (Palomino et al., 2015). The TGA profiles (weight loss and weight loss derivative) show the first weight loss at around 400°C, due to the sublimation of amorphous carbon and contaminants (Shanov, Yun and Schulz, 2006). The significant weight loss took place at ~685°C due to the oxidation of CNTs into CO_2 (Rinzler et al., 1998). After the sublimation of C was completed, 14.5 ± 0.3 wt% of the material remained, which corresponds to Si.

Moreover, X-ray photoelectron spectroscopy (XPS) spectra of Si-CNTs, as reported elsewhere (Palomino et al., 2015), revealed the chemical state of Si by an asymmetrical broad band Si 2p, which was deconvoluted into six peaks. The first two peaks at ~99.5 and ~100.2 eV are associated with Si^0 $2p_{3/2}$ and Si^0 $2p_{1/2}$, respectively. The third peak at ~100.7 eV is attributed to SiC (Watanabe and Hosoi, 2013). Additionally, the presence of silicon suboxides due to contamination was confirmed by three peaks at ~102.2 eV (SiO), ~103.7 eV (Si_2O_3), and ~104.7 eV (SiO_2) (Logofatu et al., 2011; Bashouti et al., 2012). Similarly, the C 1s band was deconvoluted into five peaks. The first peak

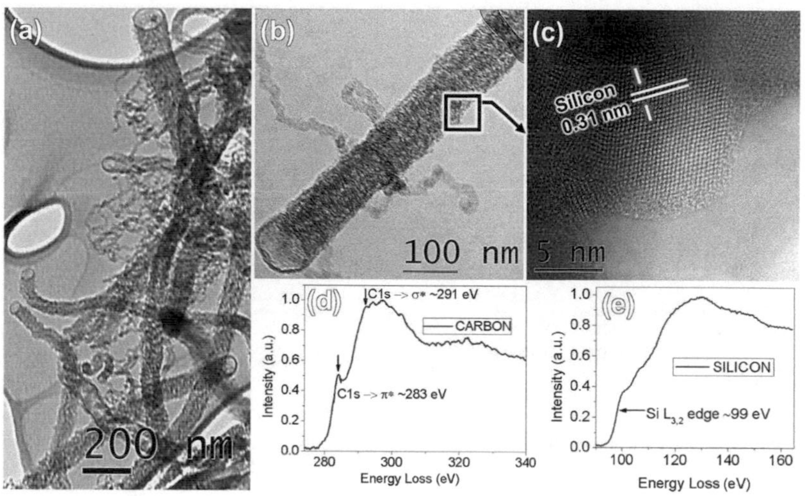

FIGURE 1.9 TEM images of (a) Si-coated CNT at low magnification, (b) tube cavity, and (c) Si coating lattice fringes. EELS spectra obtained on the Si-CNT for (d) carbon and (e) silicon. (Reproduced from Palomino et al., *J. Phys. Chem. C* **119**, 21125–21134, 2015.)

at ~283.4 eV is attributed to silicon carbide (Maruyama and Naritsuka, 2011; Watanabe and Hosoi, 2013), while the intense second peak at ~284.5 eV demonstrates the presence of graphitic carbon. The third peak centered at ~286.3 eV is associated with SiCO (Maruyama and Naritsuka, 2011), and the next two peaks C=O and COOH (Okpalugo et al., 2005; Li et al., 2008; Wepasnick et al., 2010; Maruyama and Naritsuka, 2011; Hawaldar et al., 2012) are related to contamination.

Growth Mechanism

The growth mechanism for the formation of Si-CNTs in the HF-CVD environment is described as follows. At the substrate temperature of ~500°C, the polymer $CH_3[CH_2]_nCH_3$ (Figure 1.10A-a) decomposes into hydrocarbon radicals (CH_x) leaving behind sp^2 fragments (Varshney et al., 2013) which act as nucleation centers for CNT growth, carbon gets dissolved into the Ni catalyst, and hydrogen escapes. The atomic hydrogen etches the oxide layer of the Si nanoparticles (Morell et al., 2000) that is in contact with the Ni catalyst. Consequently, both carbon and silicon atoms get dissolved into the Ni catalyst (Figure 1.10A-b). These processes generate the conditions for the growth of Si-CNTs via the standard base-growth model mechanism under a CH_4 atmosphere (Figure 1.10A-c) (Che et al., 1998; Teo et al., 2003; Kumar and Ando, 2010). The final result of the Si-CNT is shown in Figure 1.10B, which consists of MWCNTs with most of the Si atoms coating the NT surface and some Si atoms inserted in the NT walls.

1.2.4 Applications

The Si-CNT material was applied in lithium ion battery (LIB) anodes and studied by cyclic voltammetry (CV) and charge–discharge (C-D) measurements, as described elsewhere (Palomino et al., 2015). The cathodic region of the voltammogram (Figure 1.11a) shows a peak at 0.01, consistent with the insertion of Li^+ to form LiC_6 (Katar et al., 2010), a peak at around 0.21 V corresponding to the formation of Li_xSi (Green et al., 2003; Yang et al., 2005; Kang et al., 2010; Chen et al., 2011), and the two peaks at around 1.8 and 1.25 V demonstrating the formation of the solid–electrolyte interface (SEI) (Lin et al., 2006; Katar et al., 2008b, 2010). On the other hand, the anodic

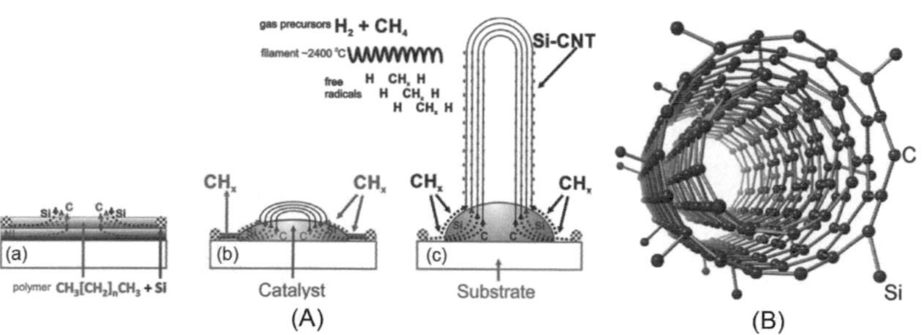

FIGURE 1.10 (A) Base-growth mechanism for Si-CNT: As the substrate temperature is incremented, (a) the aliphatic polymer decomposes into hydrocarbon radicals mixed with Si atoms, (b) Ni nano-islands catalyze the development of nucleation centers, and (c) the Si-CNT growth continues from the base with added H_2 and CH_4. (B) Schematic representation of Si-CNT. (Reproduced from Palomino et al., *J. Phys. Chem. C* **119**, 21125–21134, 2015.)

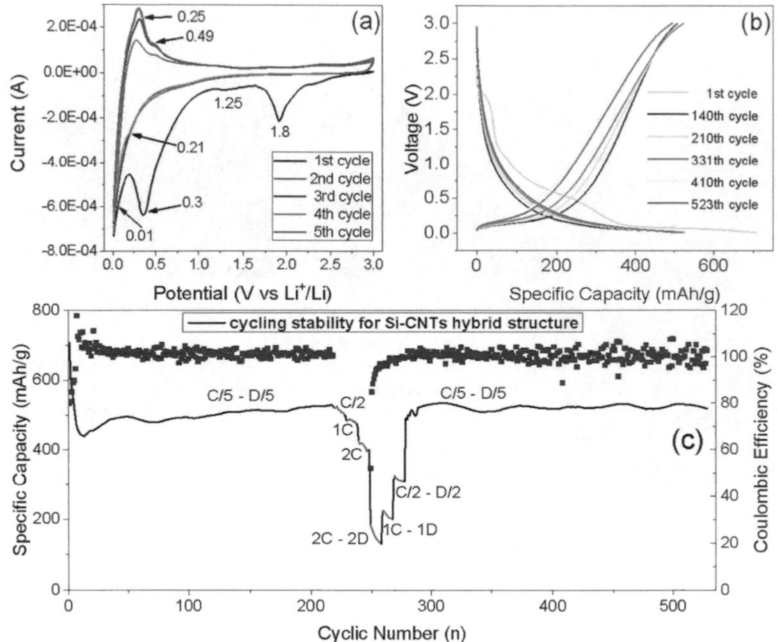

FIGURE 1.11 (a) Cyclic voltammograms of Si-CNT hybrid material. (b) C-D profiles of Si-CNT hybrid material anodes. (c) Cyclic performance and its columbic efficiency of the Si-CNT anode at different current rates. (Reproduced from Palomino et al., *J. Phys. Chem. C* **119**, 21125–21134, 2015.)

smooth band at 0.25 V represents the extraction of Li from LiC_6 sites (Guo et al., 2005), whereas the oxidation peak at around 490 mV confirms the extraction of Li from Li_xSi sites (Yoshio, Tsumura and Dimov, 2005; Kang et al., 2010). Moreover, the C-D profiles performed on coin cells suggest that Si-CNT nanostructures retain the reversible electrochemical performance of lithium even after 520 cycles, as shown in Figure 1.11b. The first discharge curve exhibits a high discharge capacity of ~707 mAh/g and a plateaus at 1.8 and 1.25 V, consistent with the first voltammogram and are related to the decomposition of the electrolyte and the formation of the SEI (Zuo et al., 2007; Varzi et al., 2011). The cycling stability profile of Si-CNT anodes over 520 cycles (Figure 1.11c) demonstrates a reversible and high specific capacity of 510 mAh/g, superior to the experimental and theoretical (372 mAh/g) graphite capacity (Yang, Song and Chen, 2006). By the systematic variation of the charge and discharge current densities, we determined that the resulting capacity depends on the current rates. Accordingly, the original stable capacity of 510 mAh/g was fully recovered after applying different current densities $C/5$, $C/2$, $1C$, $2C$, $1C$, $C/2$, $C/5$ ($C = 500$ mA/g), denoting good reversibility with Li upon cycling, and robustness and stability of the Si-CNT anode hybrid material.

1.3 Nanowires

NWs are also prominent 1D nanostructures and have been extensively studied because of their novel physical and chemical properties, and their prospective applications in device development (Morales and Lieber, 1998; Feng et al., 2000; Liu et al., 2000). Depending on their chemical composition, NWs are fabricated via several methods including CVD, laser ablation (Morales and Lieber, 1998), thermal evaporation (Feng et al., 2000), and chemical etching (Peng, Huang and Zhu, 2004). Fabrication of electron-emitting nanomaterials (Iijima, 1991; Rinzler et al., 1995; Chen et al., 2010) and their application to flat panel displays (Lee et al., 2001; Biaggi-Labiosa et al., 2008) have attracted great interest in the study of 1D materials having high aspect ratios, stable structures, and enhanced electron field emission (EFE) properties (Chen et al., 2002; Wu et al., 2002; Zhu et al., 2003; Xiang et al., 2005; Valentín et al., 2013). Since silicon plays a significant role in the microelectronics field, SiNW based emitters have been widely studied (Kulkarni et al., 2005; Huang et al., 2007). In order to improve the EFE properties, numerous modifications have been done on SiNWs, such as H_2 plasma surface treatment (Au et al., 1999), gold decoration (Zhao et al., 2011), and diamond coating (Liu et al., 1994; She et al., 1999; Tzeng et al., 2007; Thomas et al., 2012).

1.3.1 Fundamentals of Nanowires

Currently, there are a wide variety of NWs made of metals, semiconductors, conductors, insulators, and organic compounds. Since our research group has the experience of synthesizing silicon nanomaterials, this section is focused on SiNWs. The sp^3-bonded SiNW is a quasi 1D material, where the atoms in crystalline silicon are arranged in a diamond lattice structure with a lattice constant of 5.430 Å (Tang et al., 2012), as shown in Figure 1.12. Silicon is solid at room temperature, with a high melting point of 1,414°C.

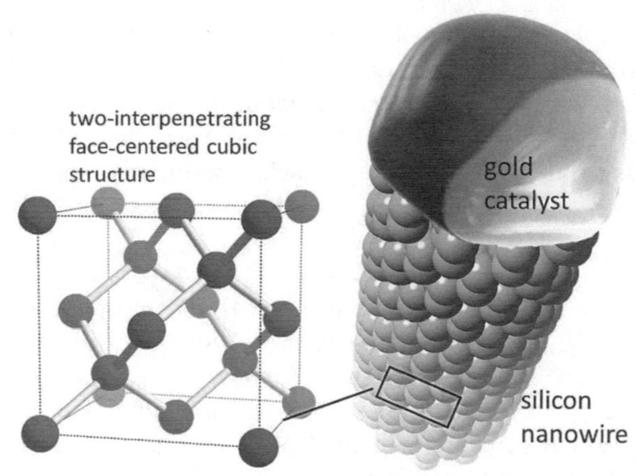

FIGURE 1.12 Representation of a SiNW with the gold catalyst at the tip.

Besides, silicon is a semiconductor material with a bandgap of ~1.12 eV and four valence electrons; the four bonding electrons provide the opportunity to combine with many other elements to form a wide range of compounds.

On the other hand, the reduction in dimensionality of SiNW results in dramatic quantum effects producing variations in the bandgap, which are related to the nanostructure dimensions, 0D (quantum dots), 1D (NWs), and 2D (quantum wells). The bandgap increases as the nanomaterial size decreases (Pedersen, 2005). Therefore, the quantum effects in NWs change the electrical, chemical, and mechanical properties. Thus, 1D SiNWs reveal properties that are extraordinarily different from their bulk form. In the past decade, they have been intensely studied by experimentalists and theorists because of their applicable properties and compatibility with conventional silicon microtechnology, already implemented in the industry (Shao, Ma and Lee, 2010; Won, 2010).

Due to their high surface-to-volume ratio and a quasi 1D structure, SiNWs possess properties that can outperform the current state of the art of silicon applications such as sensors with high sensitivity, efficient solar cells, and enhanced LIBs. In addition, the integration of diamond with SiNW leads to a composite material, having the right combination of functional properties of the constituent materials, that may have potential applications in enhanced LIBs, cold field emitters, and UHD flat panel displays, among others.

1.3.2 Synthesis

Silicon-based nanostructures can be synthesized by various methods including chemical etching by acid (Hochbaum et al., 2008), and CVD, which commonly uses the hazardous silane gas as silicon source (Palomino et al., 2014).

Chemical Etching

The processes of metal-assisted chemical etching (MACE) are represented in Figure 1.13, and they use noble metals (such as Au, Pt, and Ag) as a hole (h+) catalyst deposited

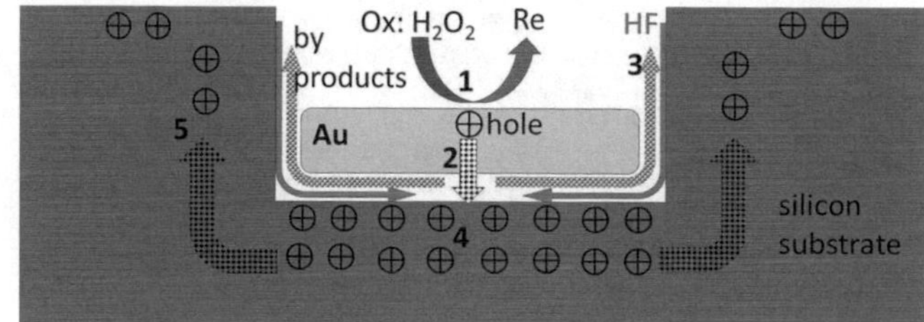

FIGURE 1.13 Scheme of processes involved in MACE. (Adapted from Huang et al., *Adv. Mater.* **23**, 285–308, 2011.)

on the surface of a semiconductor (Si). MACE is described briefly in the following: (i) In an acidic solution (such as HF), the metal accelerates the hole (h+) generation from an oxidant (such as H_2O_2) by providing electrons to the oxidant and producing holes at the metal (local oxidation and reduction reactions). (ii) The holes diffuse through the noble metal toward the metal–silicon interface, and the holes (h^+) are injected from metal catalyst to silicon by transferring electrons from silicon to metal. (iii) This process results in the oxidation and dissolution of silicon at the metal–silicon interface by hydrofluoric acid (HF), without net consumption of the metal. Subsequently, HF and the byproducts diffuse along the interface to be removed. (iv) Due to high concentration of holes under the noble metal, at the metal–silicon interface, the silicon is etched preferentially below the metal catalyst. As a result, metal descends into the semiconductor as the semiconductor is being etched right underneath. (v) Afterward, the excess of holes diffuses from the Si under the interface to off-metal areas. Ultimately, metal byproducts on the structures can be washed off with deionized water after the synthesis was completed. In addition, nitric acid bath for at least 1 h can be used to remove all residual metals from the surface of the fabricated nanostructures (Huang et al., 2011). In order to achieve tailored nanostructures with high aspect ratio like arrays of vertical NWs, the catalyst metal must be systematically patterned on the silicon substrate, to be etched into the semiconductor and produce the engineered nanostructure.

Thermal CVD

Our research group has demonstrated that using T-CVD method, silicon-based nanostructures can be synthesized from silicon nanoparticles to avoid the pyrophoric hazards associated with silane (LaDou, 1983). A schematic of a custom-built hot wall T-CVD reactor is shown in Figure 1.14. It includes a resistive heater at the wall of the electric tube furnace, a silicon source material consisting of a mixture of Si nanoparticles and graphite (1:1) placed at the center of a quartz tube, and a gold-coated copper substrate that is placed 7–10 cm away from the source material on the downstream side of the Ar flow, as described elsewhere (Palomino et al., 2014). Before each deposition, the T-CVD chamber is evacuated and then filled by a continuous flow (100 sccm) of Ar gas reaching a constant pressure of ~100 mTorr during the whole growth process, and

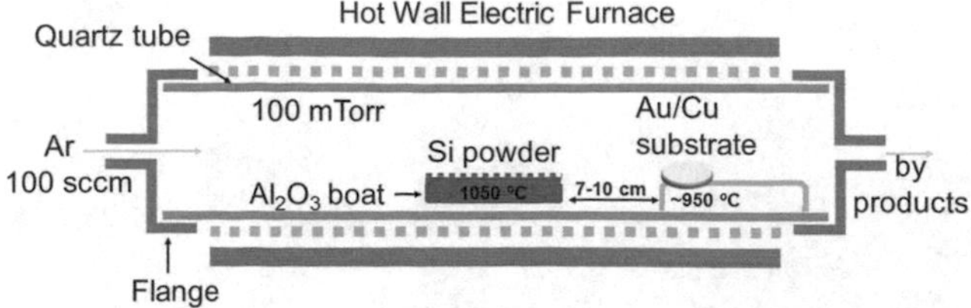

FIGURE 1.14 Schematic representation of the T-CVD system showing the configuration of raw materials and the corresponding temperatures. (Adapted from Palomino et al., *ACS Appl. Mater. Interfaces* **6**, 13815–13822, 2014.)

Ar is introduced from one closed end of the quartz tube and acts as a carrier gas. The temperature of the source material is kept constant at 1,050°C for ~3 h. In contact with graphite at high temperature, the native SiO_x contained in the source material is reduced to form Si and CO_x vapors, where Si vapor is transported by the carrier gas toward the substrate (950°C) to form nucleation centers and the subsequent growth of SiNWs. The byproducts, including CO_x, together with unreacted precursor gases are extracted from the chamber by the pump.

Growth Mechanism of Si Nanowires

The synthesis of SiNWs follows the vapour–liquid–solid (VLS) (Wagner and Ellis, 1964; Schmidt et al., 2009) growth mechanism, with Au as the catalyst. Figure 1.15 shows a schematic representation of the VLS mechanism. Au drops at high temperatures (950°C) and adsorbs vapors of the source material (Si) due to its high solubility, achieving a supersaturation stage, and the subsequent precipitation of the source material (Si) at the liquid–solid interface occurs, where the silicon atoms form sp^3 bonds resulting in a two-interpenetrating face-centered cubic, diamond structure. The 1D crystal growth begins and continues as long as the vapor components are supplied. Since vapor (carrying Si nanoparticles), liquid (gold catalyst), and solid (precipitated 1D structures) phases are involved, it is known as the VLS mechanism (Wagner and Ellis, 1964; Schmidt et al., 2009; Klimovskaya et al., 2011).

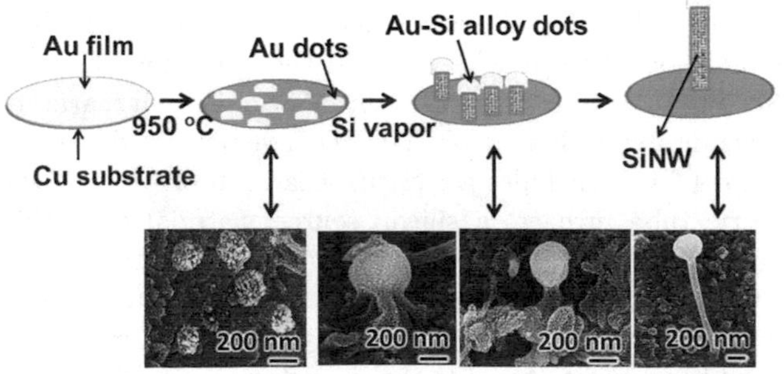

FIGURE 1.15 Schematic representation of the VLS growth mechanism of SiNWs, supported by FE-SEM images. (Reproduced from Palomino et al., *ACS Appl. Mater. Interfaces* **6**, 13815–13822, 2014.)

In order to improve the electron emission properties, the SiNWs were conformally decorated with UNCD by HF-CVD method, which is described in detail elsewhere (Palomino et al., 2014), where paraffin wax was used as a diamond seeding source, since this approach is more efficient in the creation of diamond nuclei than the traditional methods.

Characterization

FE-SEM images (Figure 1.16) reveal the morphology of the SiNWs, as described elsewhere (Palomino et al., 2014). Figure 1.16a shows a straight NW consisting of a core/shell structure of the SiNW/SiO_x. The low-magnification image (Figure 1.16b) exhibits high density of homogenous bare Si NWs, where the diameter of the NWs estimated from the image is in the range of 100–180 nm and the tip of each wire is bulbous with a diameter of about 200–250 nm due to the presence of (Au) catalyst, in agreement with the VLS growth mechanism. Figure 1.16c demonstrates that the UNCD coating is exceptionally uniform throughout the wire length, as compared to published reports (Liu et al., 1994; Kiselev et al., 2005; Tzeng et al., 2007).

Moreover, the room temperature Raman spectrum of the UNCD-coated SiNWs revealed a strong first-order Si TO phonon mode at 514 cm^{-1}, as described in detail elsewhere (Palomino et al., 2014). In addition, the spectrum shows a broad D-band around 1,347 cm^{-1} related to disordered sp^2 carbon (Wang et al., 2012) present at the UNCD grain boundaries. However, the highly optically absorbing sp^2-bonded carbon precludes observation of the sp^3 Raman signal, since the visible Raman signal is about 50–250 times more sensitive to sp^2-bonded carbon than to the sp^3-bonded carbon (Xiao et al., 2004).

Figure 1.17a shows TEM images of the SiNWs with a constant diameter and decorated with UNCD particles of sizes ranging from 3 to 10 nm dispersed uniformly along the whole wire. The image of the bulbous wire tip is also covered with UNCD nanoparticles (Figure 1.17b). Figure 1.17c clearly reveals the core/shell structure of the SiNW having a UNCD shell thickness of around 10–20 nm.

The HR-TEM micrograph of UNCD/SiNWs suggests that the UNCD is comprised of small grains, around 5 nm (Figure 1.18a,b). Figure 1.18c shows the lattice fringes of the NW crystalline core, revealing an interplanar spacing of about 0.31 nm, which matches with the (111) orientation of Si (Suzuki et al., 1991). Figure 1.18d depicts that the UNCD lattice spacing is about 0.205 nm, which is a typical lattice parameter for diamond (111) orientation (Tzeng et al., 2008).

The UNCD-coated SiNWs were further studied by EELS. Figure 1.19b shows a small peak at 107 eV attributed to the presence of SiO_x (Shakerzadeh et al., 2011). Figure 1.19c recorded for the UNCD coating shows the presence of a weak band feature at 285.23 eV, corresponding to the C 1s $\rightarrow \pi^*$ transition, demonstrating the presence of sp^2-C. Additionally, it shows the C 1s $\rightarrow \sigma^*$ transition around 292.5 eV consistent with sp^3-bonded carbon (Chen et al., 2013). The dip at ~302.4 eV reveals the second absolute

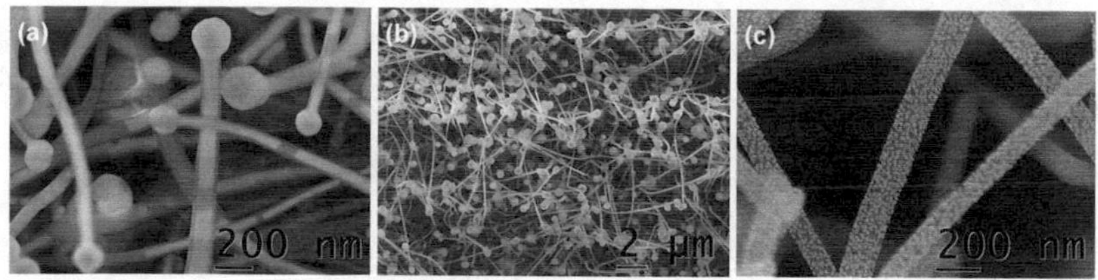

FIGURE 1.16 SEM images of the as-grown (a) core/shell structure of SiNW/SiO_x showing (Au) catalyst at its tip. (b) High density of SiNWs at low magnification and (c) UNCD conformally coated SiNWs. (Reproduced from Palomino et al., *ACS Appl. Mater. Interfaces* **6**, 13815–13822, 2014.)

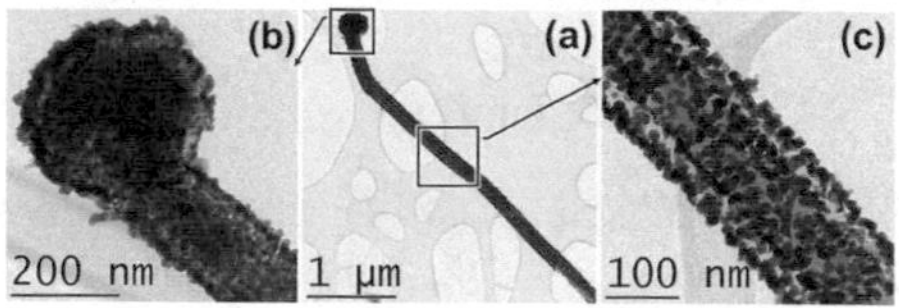

FIGURE 1.17 TEM images of (a) UNCD-coated SiNW, (b) bulbous tip of the wire, and (c) linear part of the wire. (Reproduced from Palomino et al., *ACS Appl. Mater. Interfaces* **6**, 13815–13822, 2014.)

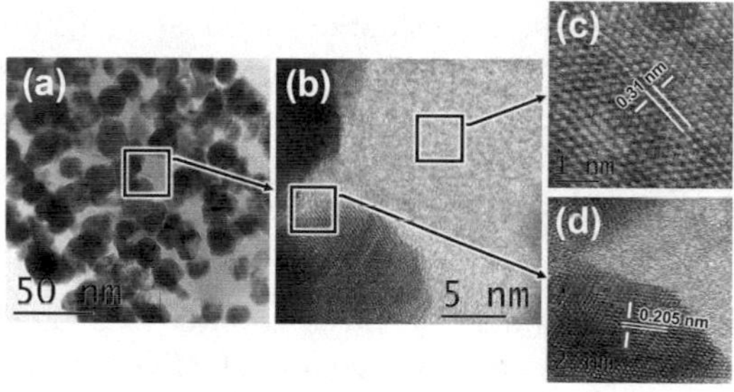

FIGURE 1.18 High-resolution transmission electron microscopic scanning mode (STEM) images of (a) UNCD-coated SiNW, (b) magnified image of the wire showing (c) interplanar spacing of the Si crystal and (d) diamond crystals. (Reproduced from Palomino et al., *ACS Appl. Mater. Interfaces* **6**, 13815–13822, 2014.)

bandgap of diamond, proving the diamond nature of the dense nanograins (Chen et al., 2008).

1.3.3 Applications

Fabrication of electron-emitting 1D materials having high aspect ratio and high EFE properties (Valentín et al., 2013) is attracting much attention for applications in ultra-high-resolution flat panel displays (Lee et al., 2001; Biaggi-Labiosa et al., 2008). Therefore, we explore the potential of 1D SiNWs and UNCD-coated NWs as electron field emitters, as described in detail elsewhere (Palomino et al., 2014). The measured current density as a function of the macroscopic electric field (Figure 1.20a) indicates that bare SiNWs can be turned on at $E_0 = 4.3 \pm 0.1$ V/µm, while the UNCD/SiNW heterostructures exhibit a lower threshold field of $E_0 = 3.7 \pm 0.1$ V/µm. Emission current density of bare SiNWs goes as high as ~0.1 mA/cm^2 at 25 V/µm, whereas the current density for UNCD/SiNW reaches ~2 mA/cm^2 at 25 V/µm. The superior field emission properties of UNCD/SiNW are attributed to the presence of sp^2-hybridized carbon around the diamond crystallites and to the geometrical enhancement factor (diameters ~5–10 nm), which play an important role in facilitating the electron emission process itself.

Fowler–Nordheim Plot Analysis for SiNW and UNCD/SiNW Field Emitters

From the slope of linear regions of Fowler–Nordheim (F-N) plots shown in Figure 1.20b, and assuming that the work function of SiNWs is close to 4.7 eV (Minami, Miyata and Yamamoto, 1998), the geometrical enhancement factor, β, can be calculated by $\beta = -\frac{B\phi^{\frac{3}{2}}}{\text{slope}}$, where $B = -6.83 \times 10^3 \text{V}(\text{eV})^{3/2}\ (\mu\text{m})^{-1}$ is a constant, and ϕ is the work function of the material. High values of β indicate a high local electric field due to the presence of 1D SiNWs and UNCD coating. For bare SiNWs, the F-N plot is fitted with a single slope with an enhancement factor β = 1,447. For UNCD/SiNWs, the F-N plot presents two slopes with enhancement factor β values of ~1,163 at low field and ~1,982 at high field. The first emission state of

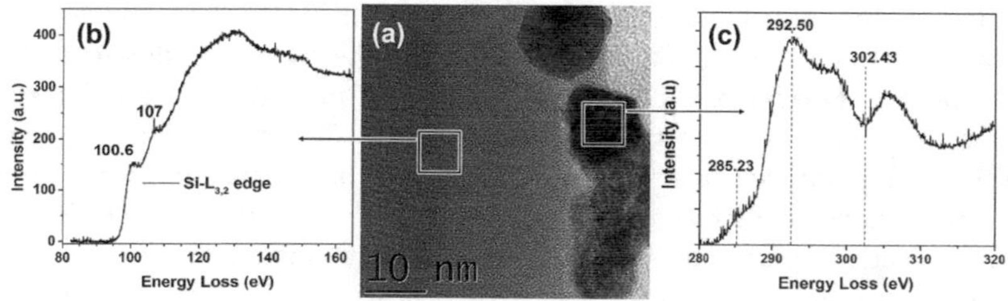

FIGURE 1.19 EELS of (a) High-resolution STEM image of UNCD-coated SiNW, (b) SiNWs, (c) UNCD crystals recorded at the region shown in Figure 1.19a. (Reproduced from Palomino et al., *ACS Appl. Mater. Interfaces* **6**, 13815–13822, 2014.)

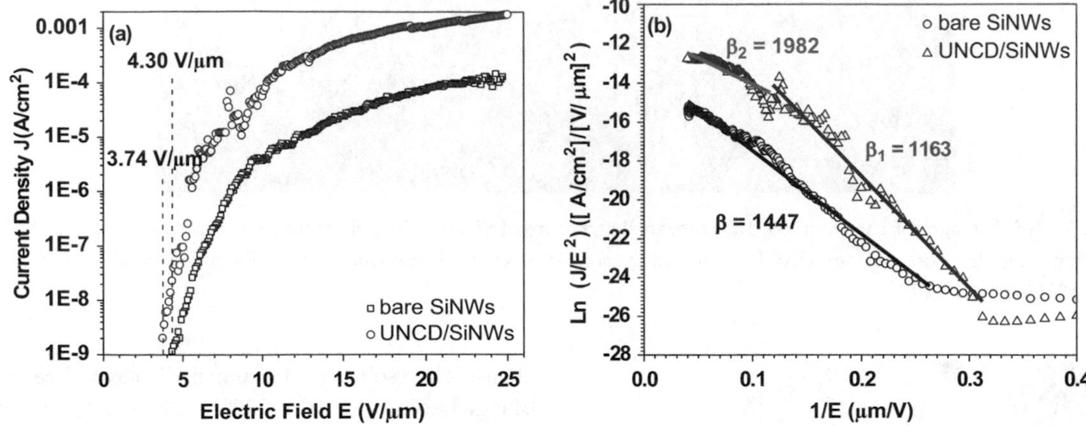

FIGURE 1.20 (a) Field emission J vs E plots for bare SiNW and UNCD/SiNW films. (b) Fowler–Nordheim Ln$[J/E^2]$ vs $[1/E]$ plots for bare SiNW and UNCD/SiNW films. (Adapted from Palomino et al., *ACS Appl. Mater. Interfaces* **6**, 13815–13822, 2014.)

UNCD/SiNWs arises at low field regime <7 V/m, where electrons are emitted mainly from the sp^2 carbon around the UNCD crystallites due to the direct contact between UNCD and Si, avoiding the oxide layer barrier and taking advantage of the negative electron affinity of diamond (Liao et al., 1998). The second regime at fields >7 V/m corresponds to higher local electric fields where the electrons are emitted directly from the SiNWs.

Silicon Nanowire-Based Sensors: The quasi 1D properties of SiNWs can be utilized in sensor devices as the active sensing part (translates the input into a temporary signal) and as transducers (decodes the temporary signal into an electrical response). Applications in chemical, biochemical, and biological sensors utilize the small geometry and the electrical and mechanical properties of NWs, as shown by Zheng et al. (2005), who demonstrated the capabilities of SiNWs to detect cancer markers electrically. Mechanical sensors take benefit from the high piezoresistance effect of silicon NWs (Kumar Bhaskar et al., 2013).

Silicon Nanowire-Based Solar Cells: Energy collecting from renewable sources and storage of electrical energy are among the most demanding challenges of our society (Armaroli and Balzani, 2007). Light harvesting can benefit from SiNWs with large surface area, because they have the capability to increase the optical absorption and collection efficiency in solar cells. Moreover, light trapping can be enhanced by forming high aspect ratio structures, with *pn* junction placed much closer to the carrier-generating regions, as reported by Tian et al. (2007).

Silicon Anodes for Li-Ion Batteries: LIBs are currently one of the best technologies to store electrical energy, with graphite as the standard anode material. However, theoretical calculations predict that silicon capacity is one order of magnitude superior to graphite, due to the accommodation of up to 4.4 Li atoms per silicon atom, producing a huge lattice expansion upon lithiation/delithiation that leads to the pulverization of the silicon anode. Nevertheless, researchers have reported that SiNWs can expand laterally and still maintain the current transport in the vertical direction. Furthermore, it has been reported that silicon below a critical size (150 nm) does not crack or fracture during lithiation/delithiation (Liu et al., 2012). Therefore, 1D SiNWs are potential candidates for new-generation LIBs.

1.4 Nanofibers

From the 1950s through the 1970s, several researchers have found involuntarily the occurrence of CNFs in metallic catalysts used for the conversion of carbon-containing gases into olefins (Melechko et al., 2005). Hence, the CNFs have attracted the attention of experimentalists and theoreticians, producing comprehensive studies to control its properties such as surface structure, diameter, length, chemical and mechanical properties. The studies revealed that CNFs possess high mechanical strengths and moduli, high stiffness, excellent electrical and thermal conductivities, as well as strong fatigue and corrosion resistance (Zhou et al., 2009). Presently, CNF is a material of great interest due to its modern potential applications including tips for scanning microscopy and field emission devices, biological probes, and nanoelectronics (Merkulov et al., 2001).

1.4.1 Fundamentals of Nanofibers

CNFs are cylindrical or conical structures that have diameters in the range of nanometers and lengths ranging in microns. Currently, there is no strict classification of NF structures. The main distinguishing characteristic of NFs from NTs is the stacking configuration of graphene sheets. The internal structure of CNFs varies and is comprised of different arrangements of modified graphene sheets (Melechko et al., 2005). In general, an NF consists of stacked curved graphite layers (Figure 1.21a) that form "cones" or "cups", as shown in Figure 1.21. The stacked "cone" structure is often referred to as *fishbone* (Figure 1.21c,d), while

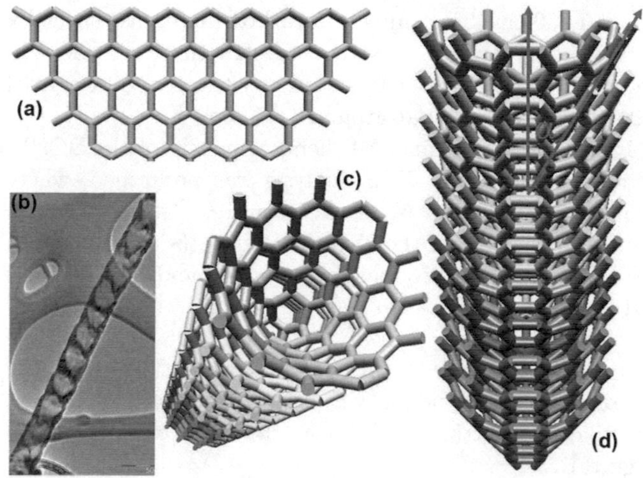

FIGURE 1.21 3D rendering of (a) cone graphene layer, (b) cup-stacked, and (c) cone-stacked CNF showing the (d) angle between graphene and fiber axis.

the stacked "cups" structure is often referred to as *bamboo-type fiber* (BCNF) (Figure 1.21b).

Figure 1.21 exhibits a monolayer graphene, which folded in conical shape resembling a CNF consisting of staked cones. The angle α between graphene sheet and the axis defines the properties of CNF (Figure 1.21d). The NF with $\alpha = 0$ is a special case where graphene layers form concentric cylinders with extraordinary properties, and this type of a NF is well known around the world as CNT. On the other hand, NFs with $\alpha \neq 0$ show mechanical, electrical, and chemical properties different from CNTs. CNF is assembled from graphene layers that are relatively short and poorly connected, similar to graphite. Thus, the charge transport along the NF is determined by both "in-plane" and "interplane" properties (Melechko et al., 2005). The stacking configuration also affects the mechanical properties since the van der Waals bonding between the graphene planes differs drastically from the in-plane covalent bonds. Moreover, NF offers exposed edges and unsaturated bonds of graphene planes; thus, the chemical properties of CNFs are more reactive than those of CNTs (Zhao et al., 2016).

Synthesis

Since CNFs and CNTs are related structures, both can be synthesized by the methods mentioned above in the section of CNTs, including laser vaporization, arc discharge, and CVD. Arc discharge and laser ablation are very efficient methods to produce high-quality NFs; however, they do not prove control over the spatial arrangement of the resulting NFs and require complex purification methods to remove amorphous carbon particles and entangled catalyst (Melechko et al., 2005). On the contrary, CVD method produces phase pure NFs, where the alignment, dimensions, shape, and location of each individual NF can be controlled during synthesis.

Chemical Vapor Deposition Method

CVD comprises chemical reactions, adsorption, diffusion, nucleation, growth, and desorption, which are described in detail elsewhere (Palomino et al., 2018). As we described above in the section of CNTs, in a catalytic growth, the deposition of carbon occurs on one side of the surface of a catalyst metal. Since two of the dimensions of the growing material are restricted by the size of the nanometal, the unlimited third dimension leads to quasi-1D growth. Thus, the diameter of the resultant carbon fiber is approximately equal to that of the catalyst metal nanoparticle. The growth mechanism of CNFs as well as CNTs has been carefully studied by many different groups. The mechanism to grow CNF is similar to that of CNT and includes the following steps: (i) adsorption of the reactant hydrocarbon molecule (CH_x) and decomposition into C and H atoms on the surface of catalyst metal nanoparticle, (ii) dissolution and diffusion of carbon atoms through the melted catalyst metal surface, and (iii) saturation and precipitation of carbon atoms on the opposite surface of the catalyst nanoparticle resulting in the formation of the NF structure (Melechko et al., 2005), as shown in Figure 1.22.

Since two different CNF growth modes have been observed, CNF growth mechanism can also be described by tip-growth and base-growth models. The growth mode depends on the interaction of the catalyst with the substrate. Such interaction is characterized by the contact angle of the catalyst with the substrate at a given growth temperature. An acute contact angle corresponds to weak interaction (Pavese et al., 2008), resulting in a tip-growth model, while an obtuse contact angle is indicative of strong interaction, leading to a base-growth model. The parameters that define the morphology and degree of crystallinity of CNFs include the chemical nature of the substrate and catalyst nanoparticle, the reaction temperature, the pressure, and the composition of the reactant gases.

Our research group has also developed BCNFs, which were grown directly on copper substrates by HF-CVD (Katar et al., 2010). The polycrystalline copper substrates (99.9% pure, 0.5 mm thick, 14 mm disk diameter) were polished to remove the oxides and smoothen the surface, then cleaned in an ultrasonic bath, and dried with argon gas.

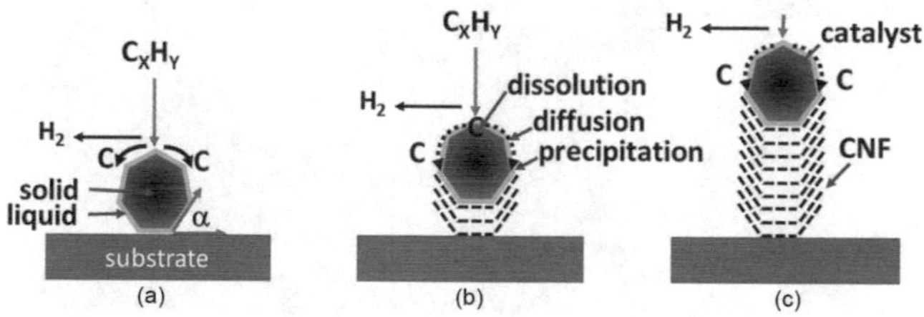

FIGURE 1.22 Mechanism of CNF formation. (a) Adsorption and decomposition of the hydrocarbon, (b) dissolution and diffusion of carbon atoms, and (c) precipitation of carbon atoms and incorporation into graphene layers. (Adapted from Melechko et al., *J. Appl. Phys.* **97**, 1–39 2005.)

Prior to each deposition, the CVD chamber was pumped down to 8×10^{-6} mbar. Afterward, the chamber was filled with gas mixture composed of 2.0% CH_4 and 98% H_2 and 500 ppm H_2S, with the combined flow of 100 sccm, and deposition pressure was kept constant at 27 mbar. The filament temperature was settled at ~2,500°C and copper substrate at 900°C; the filament–substrate distance was fixed at 8.0 mm. The deposition time was selected in order to achieve a film that covered the whole substrate. Subsequently, the fabricated BCNF was coated with SiN via RF sputtering. Before starting the sputtering, the residual pressure inside the chamber was $\sim 10^{-7}$ mbar. During the sputtering process, the argon pressure was kept at $\sim 10^{-2}$ mbar, the target was at a voltage of ~1,050 V with a forward power of ~50 W, and the substrate temperature was kept at ~80°C for the 30 min of deposition.

1.4.2 Characterization

The FE-SEM image (Figure 1.23a) exhibits entangled clusters of NF structures with diameters ranging from 50 to 200 nm and lengths in micrometers. The CNFs were analyzed in more detail by TEM, revealing a BCNFs with a diameter of around 150 nm, as shown in Figure 1.23b. It contains nanocavities that are stacked one over the other with closed graphene walls. Figure 1.23c shows the TEM image of Si/N/BCNFs confirming that the BCNFs were coated with nitrogen and silicon retaining the BCNF structure.

In addition, the Si/N-coated BCNFs have been studied by Raman spectroscopy, as described elsewhere (Katar et al., 2010). The Raman spectrum exhibited a broad band in the range of 800–1,200 cm^{-1} ascribed to a disordered or amorphous SiN film (Katar et al., 2010). Moreover, the presence of D- and G-bands of sp^2 carbon from the underlying BCNFs was observed. The D-band with a disordered peak at about 1,350 cm^{-1} is common in carbon materials especially in CNFs due to the existence of structural disorder. In contrast, the G-band corresponds to the graphite crystal with an ordered peak at around 1,580 cm^{-1}.

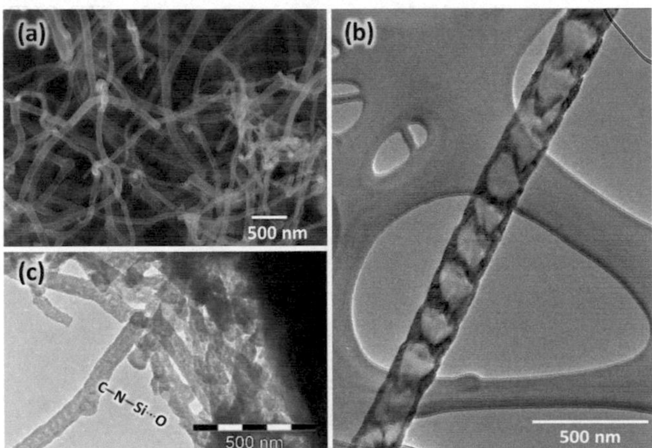

FIGURE 1.23 (a) FE-SEM image showing entangled BCNFs, (b) TEM image of the BCNF showing the graphitic sidewalls and nanocavities, and (c) SiN-coated BCNFs showing the bonding structure (C–N–Si–O). (Adapted from Katar et al., *Electrochim. Acta* **55**, 2269–2274, 2010.)

In order to determine the chemical nature of the BCNFs sputtered with Si/N, XPS analysis was performed, where the regions of Si 2p, N 1s, C 1s, and O 1s were studied via deconvolution. In the N 1s spectrum, the peak at 398.3 eV is related to silicon–nitrogen bond, and the peak at 401.5 eV corresponds to carbon–nitrogen bond. The Si 2p spectrum shows a peak related to SiN at 102.3 eV and a peak related to SiO_2 at 103.2 eV. Therefore, the silicon nitride is bonding with the carbon in the NFs, and the silicon is getting oxidized at the surface, as shown in Figure 1.23c.

1.4.3 Applications

Electron Field Emission Studies

1D materials like BCNFs are known to be excellent field emitters (Katar et al., 2008a, 2009), and SiN is expected to show poor field emission properties on its own. Therefore, the field emission properties of SiN/Cu, BCNFs/Cu, and SiN-coated BCNFs/Cu were studied by comparison. As expected, the bare BCNFs exhibited good field emission properties, with a turn-on field of 3.4 V/μm. On the other hand, the pure SiN shows poor field emission properties, with a turn-on field beyond the measurement range employed (>40 V/μm). On the other hand, the SiN/BCNF films revealed competitive field emission properties, very similar to those of bare BCNFs, with a turn-on field of 9.3 V/μm. These results are in good agreement with the XPS conclusions and prove that the thin SiN layer is in proper contact with the BCNFs, resulting in competitive electron field emission properties.

Electrochemical Performance Studies

The electrochemical performance of the SiN/BCNT as an anode material for LIBs was studied in detail, as described elsewhere (Katar et al., 2010). The first CV shows a peak at around 0.2 V consistent with the alloying of Li ions with SiN/BCNT. The peaks at 0.7, 1.68, and 2.2 V that appear in the first cycle and disappear in the subsequent cycles are associated with the formation of SEI film (Katar et al., 2010). In contrast, the anodic peaks at 1.2 and 0.5 V were related to the extraction of Li ions from SiN/BCNT. The CV curves from the second to the fifth cycles overlap, demonstrating a good reversibility upon lithiation/delithiation.

The C-D measurements of SiN/BCNT were taken on a coin cell at a constant current density of 0.02 mA/cm^2, where SiN/BCNT was used as a working electrode material and Li-metal foil as a counter and reference electrode. The measured open-circuit voltage of the fresh half-cell was about 2.7 V. The first discharge capacity of SiN/BCNT composite was found to be 2,000 mAh/g, which is superior to theoretical capacity of graphite (373 mAh/g). After ten cycles, 307 mAh/g was obtained, representing the actual reversible capacity of this anode material. The lithium alloying behavior mainly occurs under 1.5 V and delivers a reversible capacity above 300 mAh/g after the tenth cycle, representing a good reversibility for lithium alloying and dealloying.

The electrochemical performance of SiN/BCNF anodes was attributed principally to the good conductivity of BCNFs linked to SiN layer. Furthermore, BCNF acts as a support structure to alleviate the volume expansion caused by lithiation/delithiation, and the presence of nanocavities in BCNFs acts as additional lithium storage sites. Thus, the roles of BCNFs upon lithium storage result in an upgraded reversible capacity.

1.5 Summary

We have shown a brief theory that connects the physical dimensions (1D materials) to its properties (quantum effects). A brief overview of the fundamentals of CNTs, SiNWs, and CNFs has been provided. We have described the most important growth techniques for the synthesis of 1D materials. Moreover, we have presented the 1D materials synthesized by our research group, including SiNWs, CNTs, and CNFs. The characterization results (SEM, TEM, and Raman spectroscopy) have been used to propose the growth mechanisms for SiNWs, CNTs, and CNFs. We have demonstrated the applications of these materials in LIBs, as well as in electron field emitters. Therefore, the presented 1D materials are potential candidates for the next-generation applications in batteries, solar cells, microelectronics, photonics, and nanodevices. This chapter will serve as a reference for 1D materials, its growth methods and mechanisms, properties, and applications.

Acknowledgments

This research was made possible by funds from PR DOE EPSCoR (DOE Grant DEFG02-08ER46526, J.P.) and PR NASA EPSCoR (NASA Cooperative Agreement NNX15A-K43A). We would like to acknowledge the IFN Nanoscopy Facility (NSF Cooperative Agreement EPS-01002410) at UPR-RP for permitting us to use the electron microscopes. We also acknowledge the SpecLab Facilities for access to the Raman Spectrometer, and the Material Characterization Center in the Molecular Sciences Research Building for the TGA and XPS measurements. Special thanks to César E. Echevarría-Narváez and Leyinska Garay-Rodríguez for proofreading this manuscript. The authors are thankful to Project Director PINSTECH Phase-II project for the financial support.The authors are also thankful to Pakistan Science Foundation and TWAS for financial support through projects (PSF/Res/C-PINSTECH/Phys(172) and (13-319RG/MSN/ASC-UNESCO-FR:3240279202).

References

Armaroli, N. and Balzani, V. (2007) The future of energy supply: Challenges and opportunities, *Angewandte Chemie International Edition*, 46(1–2), pp. 52–66. doi: 10.1002/anie.200602373.

Au, F. C. K. et al. (1999) Electron field emission from silicon nanowires, *Applied Physics Letters*, 75(12), pp. 1700–1702. doi: 10.1063/1.124794.

Bashouti, Y. M. et al. (2012) Hybrid silicon nanowires: From basic research to applied nanotechnology. In: Peng, X. (ed.) *Nanowires: Recent Advances*. InTech, pp. 178–210. doi:10.5772/54383.

Biaggi-Labiosa, A. et al. (2008) Nanocrystalline silicon as the light emitting material of a field emission display device, *Nanotechnology*, 19, pp. 225202-1–6. doi: 10.1088/0957-4484/19/22/225202.

Chandran, R. and Gifty Honeyta A, M. (2017) Simplified equation for Young's modulus of CNT reinforced concrete, *AIP Advances*, 7(12). doi: 10.1063/1.5011319.

Che, G. et al. (1998) Chemical vapor deposition based synthesis of carbon nanotubes and nanofibers using a template method, *Chemistry of Materials*, 10(1), pp. 260–267. doi: 10.1021/cm970412f.

Chen, H. et al. (2011) Silicon nanowires coated with copper layer as anode materials for lithium-ion batteries, *Journal of Power Sources*, 196(16), pp. 6657–6662, Elsevier B.V. doi: 10.1016/j.jpowsour.2010.12.075.

Chen, H.-C. et al. (2013) The potential application of ultra-nanocrystalline diamond films for heavy ion irradiation detection, *AIP Advances*, 3(6), pp. 062113-1–20. doi: 10.1063/1.4811338.

Chen, J. et al. (2002) Field emission from crystalline copper sulphide nanowire arrays, *Applied Physics Letters*, 80(19), pp. 3620–3622. doi: 10.1063/1.1478149.

Chen, K. et al. (2010) Influence of Zn ion implantation on structures and field emission properties of multi-walled carbon nanotube arrays, *Science China Technological Sciences*, 53(3), pp. 776–781. doi: 10.1007/s11431-009-0384-x.

Chen, L.-J. et al. (2007) Effects of pretreatment processes on improving the formation of ultrananocrystalline diamond, *Journal of Applied Physics*, 101(6), pp. 064308-1–6. doi: 10.1063/1.2434008.

Chen, Y. C. et al. (2008) Synthesis and characterization of smooth ultrananocrystalline diamond films via low pressure bias-enhanced nucleation and growth, *Applied Physics Letters*, 92(13), pp. 133113-1–3. doi: 10.1063/1.2838303.

Cho, K. H. et al. (2008) Experimental evidence of ballistic transport in cylindrical gate-all-around twin silicon nanowire metal-oxide-semiconductor field-effect transistors, *Applied Physics Letters*, 92(5). doi: 10.1063/1.2840187.

Choi, W. B. et al. (1999) Fully sealed, high-brightness carbon-nanotube field-emission display, *Applied Physics Letters*, 75(20), pp. 3129–3131. doi: 10.1063/1.125253.

Dresselhaus, M. S. et al. (2005) Raman spectroscopy of carbon nanotubes, *Physics Reports*, 409(2), pp. 47–99. doi: 10.1016/j.physrep.2004.10.006.

Eatemadi, A. et al. (2014) Carbon nanotubes: Properties, synthesis, purification, and medical applications, *Nanoscale Research Letters*, 9(1), pp. 1–13. doi: 10.1186/1556-276X-9-393.

Feng, S. Q. et al. (2000) The growth mechanism of silicon nanowires and their quantum confinement effect, *Journal of Crystal Growth*, 209(2–3), pp. 513–517. doi: 10.1016/S0022-0248(99)00608-9.

Green, M. et al. (2003) Structured silicon anodes for lithium battery applications, *Electrochemical and Solid-State Letters*, 6(5), pp. A75–A79. doi: 10.1149/1.1563094.

Guo, Z. et al. (2005) Electrochemical lithiation and de-lithiation of MWNT–Sn/SnNi nanocomposites, *Carbon*, 43(7), pp. 1392–1399. doi: 10.1016/j.carbon.2005.01.008.

Hawaldar, R. et al. (2012) Large-area high-throughput synthesis of monolayer graphene sheet by hot filament thermal chemical vapor deposition, *Scientific Reports*, 2, pp. 682-1-682–9. doi: 10.1038/srep00682.

Hochbaum, A. I. et al. (2008) Enhanced thermoelectric performance of rough silicon nanowires, *Nature*, 451(7175), pp. 163–167. doi: 10.1038/nature06381.

Huang, C. T. et al. (2007) Er-doped silicon nanowires with 1.54 μm light-emitting and enhanced electrical and field emission properties, *Applied Physics Letters*, 91(9), pp. 093133-1–3. doi: 10.1063/1.2777181.

Huang, M. H. et al. (2001) Room-temperature ultraviolet nanowire nanolasers, *Science*, 292(5523), pp. 1897–1899. doi: 10.1126/science.1060367.

Huang, Z. et al. (2011) Metal-assisted chemical etching of silicon: A review, *Advanced Materials*, 23(2), pp. 285–308. doi: 10.1002/adma.201001784.

Iijima, S. (1991) Helical microtubules of graphitic carbon, *Nature*, 354(6348), pp. 56–58. doi: 10.1038/354056a0.

Iijima, S. and Ichihashi, T. (1993) Single-shell carbon nanotubes of 1-nm diameter, *Nature*, 363(6430), pp. 603–605, Nature Publishing Group. doi: 10.1038/363603a0.

Jang, J., Bae, J. and Park, E. (2006) Polyacrylonitrile nanofibers: Formation mechanism and applications as a photoluminescent material and carbon-nanofiber precursor, *Advanced Functional Materials*, 16(11), pp. 1400–1406. doi: 10.1002/adfm.200500598.

Kang, K. et al. (2010) Maximum Li storage in Si nanowires for the high capacity three-dimensional Li-ion battery, *Applied Physics Letters*, 96(5), pp. 053110-1–3. doi: 10.1063/1.3299006.

Katar, S. et al. (2009) Silicon encapsulated carbon nanotubes, *Nanoscale Research Letters*, 5(1), pp. 74–80. doi: 10.1007/s11671-009-9446-z.

Katar, S. L. et al. (2008a) Direct deposition of bamboo-like carbon nanotubes on copper substrates by sulfur-assisted HFCVD, *Journal of Nanomaterials*, 2008, pp. 1–7. doi: 10.1155/2008/515890.

Katar, S. L. et al. (2008b) Films of bamboo-like carbon nanotubes as electrode material for rechargeable lithium batteries, *Journal of the Electrochemical Society*, 155(2), pp. A125–A128. doi: 10.1149/1.2815675.

Katar, S. L. et al. (2010) SiN/bamboo like carbon nanotube composite electrodes for lithium ion rechargeable batteries, *Electrochimica Acta*, 55(7), pp. 2269–2274, Elsevier Ltd. doi: 10.1016/j.electacta.2009.11.070.

Kelzenberg, M. D. et al. (2010) Enhanced absorption and carrier collection in Si wire arrays for photovoltaic applications, *Nature Materials*, 9(3), pp. 239–244. doi: 10.1038/nmat2635.

Khare, S. V., Patil, S. K. R. and Kodambaka, S. (2011) Germanium nanowires 16.1. In: Sattler, K. D. (ed.) *Handbook of Nanophysics: Nanotubes and Nanowires*, Boca Raton: CRC Press, pp. 16-1–16-14.

Kiselev, N. A. et al. (2005) TEM and HREM of diamond crystals grown on Si tips: Structure and results of ion-beam-treatment, *Micron (Oxford, England: 1993)*, 36(1), pp. 81–88. doi: 10.1016/j.micron.2004.09.001.

Klimovskaya, A. et al. (2011) Study of the formation processes of gold droplet arrays on Si substrates by high temperature anneals, *Nanoscale Research Letters*, 6(1), pp. 151-1–8, Springer Open Ltd. doi: 10.1186/1556-276X-6-151.

Kulkarni, N. N. et al. (2005) Low-threshold field emission from cesiated silicon nanowires, *Applied Physics Letters*, 87(21), pp. 213115-1–3. doi: 10.1063/1.2136217.

Kumar Bhaskar, U. et al. (2013) Piezoresistance of nano-scale silicon up to 2 GPa in tension, *Applied Physics Letters*, 102(3), pp. 031911-1–4. doi: 10.1063/1.4788919.

Kumar, M. and Ando, Y. (2010) Chemical vapor deposition of carbon nanotubes: A review on growth mechanism and mass production, *Journal of Nanoscience and Nanotechnology*, 10(6), pp. 3739–3758. doi: 10.1166/jnn.2010.2939.

LaDou, J. (1983) Potential occupational health hazards in the microelectronics industry, *Scandinavian Journal of Work, Environment and Health*, 9(1), pp. 42–46. doi: 10.5271/sjweh.2444.

Landi, B. J. et al. (2009) Carbon nanotubes for lithium ion batteries, *Energy and Environmental Science*, 2(6), pp. 638–654. doi: 10.1039/b904116h.

Lee, N. S. et al. (2001) Application of carbon nanotubes to field emission displays, *Diamond and Related Materials*, 10(2), pp. 265–270. doi: 10.1016/S0925-9635(00)00478-7.

Lee, S. H. et al. (2016) Inorganic nano light-emitting transistor: p-type porous silicon nanowire/n-type ZnO nanofilm, *Small*, 12(31), pp. 4222–4228. doi: 10.1002/smll.201601205.

Lehman, J. H. et al. (2011) Evaluating the characteristics of multiwall carbon nanotubes, *Carbon*, 49(8), pp. 2581–2602, Elsevier Ltd. doi: 10.1016/j.carbon.2011.03.028.

Li, M. et al. (2008) Oxidation of single-walled carbon nanotubes in dilute aqueous solutions by ozone as affected by ultrasound, *Carbon*, 46(3), pp. 466–475. doi: 10.1016/j.carbon.2007.12.012.

Liao, M. et al. (1998) Field-emission current from diamond film deposited on molybdenum, *Journal of Applied Physics*, 84(2), pp. 1081–1084. doi: 10.1063/1.368096.

Lin, K. et al. (2006) The kinetic and thermodynamic analysis of Li ion in multi-walled carbon nanotubes, *Materials Chemistry and Physics*, 99(2–3), pp. 190–196. doi: 10.1016/j.matchemphys.2005.09.035.

Liu, J. et al. (1994) Electron emission from diamond coated silicon field emitters, *Applied Physics Letters*, 65(22), pp. 2842–2844. doi: 10.1063/1.112538.

Liu, X. H. et al. (2012) Size-dependent fracture of silicon nanoparticles during lithiation, *ACS Nano*, 6(2), pp. 1522–1531. doi: 10.1021/nn204476h.

Liu, Z. et al. (2000) Synthesis of silicon nanowires using AuPd nanoparticles catalyst on silicon substrate, *Journal of Physics and Chemistry of Solids*, 61(7), pp. 1171–1174. doi: 10.1016/S0022-3697(99) 00380-7.

Logofatu, C. et al. (2011) Study of SiO_2/Si interface by surface techniques. In: Basu, S. (ed.) *Crystalline Silicon: Properties and Uses*. InTech, 1(100), pp. 23–42. doi:10.5772/844.

Malekimoghadam, R. (2018) 3 – Carbon nanotubes processing. In: Rafiee, R. (ed.) *Carbon Nanotube-Reinforced Polymers*. Elsevier, Amsterdam, pp. 41–59. doi: 10.1016/B978-0-323-48221-9.00003-0.

Martel, R. et al. (1998) Single- and multi-wall carbon nanotube field-effect transistors, *Applied Physics Letters*, 73(17), pp. 2447–2449. doi: 10.1063/1.122477.

Maruyama, T. and Naritsuka, S. (2011) Initial growth process of carbon nanotubes in surface decomposition of SiC. In: Yellampalli, S. (ed.) *Carbon Nanotubes: Synthesis, Characterization, Applications*. InTech, 2, pp. 29–46. doi: 10.5772/17253.

Maurin, G. (2000) Electrochemical lithium intercalation into multiwall carbon nanotubes: A micro-Raman study, *Solid State Ionics*, 136–137(1–2), pp. 1295–1299. doi: 10.1016/S0167-2738(00)00599-3.

Meier, C. et al. (2006) Raman properties of silicon nanoparticles, *Physica E: Low-Dimensional Systems and Nanostructures*, 32(1–2), pp. 155–158. doi: 10.1016/j.physe.2005.12.030.

Melechko, A. V. et al. (2005) Vertically aligned carbon nanofibers and related structures: Controlled synthesis and directed assembly, *Journal of Applied Physics*, 97(4), pp. 1–39. doi: 10.1063/1.1857591.

Merkulov, V. I. et al. (2001) Alignment mechanism of carbon nanofibers produced by plasma-enhanced chemical-vapor deposition, *Applied Physics Letters*, 79(18), pp. 2970–2972. doi: 10.1063/1.1415411.

Minami, T., Miyata, T. and Yamamoto, T. (1998) Work function of transparent conducting multicomponent oxide thin films prepared by magnetron sputtering, *Surface and Coatings Technology*, 108–109, pp. 583–587. doi: 10.1016/S0257-8972(98)00592-1.

Morales, A. and Lieber, C. (1998) A laser ablation method for the synthesis of crystalline semiconductor nanowires, *Science (New York, N.Y.)*, 279(5348), pp. 208–211. Available at: www.ncbi.nlm.nih.gov/pubmed/9422689.

Morell, G. et al. (1996) Characterization of the silicon network disorder in hydrogenated amorphous silicon carbide alloys with low carbon concentrations, *Journal of Non-Crystalline Solids*, 194(1–2), pp. 78–84. doi: 10.1016/0022-3093(95)00459-9.

Morell, G. et al. (2000) In situ phase-modulated ellipsometry study of the surface damaging process of silicon under atomic hydrogen, *Solid State Communications*, 116(4), pp. 217–220. doi: 10.1016/S0038-1098(00)00289-1.

Morell, G. et al. (2006) Synthesis, structure, and field emission properties of sulfur-doped nanocrystalline diamond, *Journal of Materials Science: Materials in Electronics*, 17(6), pp. 443–451. doi: 10.1007/s10854-006-8090-y.

Morell, G., Canales, E. and Weiner, B. (1999) In situ measurements of methane and acetylene concentrations in a CVD reactor by infrared spectroscopy, *Diamond and Related Materials*, 8(2–5), pp. 166–170. doi: 10.1016/S0925-9635(98)00343-4.

Nguyen, C. V. et al. (2004) High lateral resolution imaging with sharpened tip of multi-walled carbon nanotube scanning probe, *The Journal of Physical Chemistry B*, 108(9), pp. 2816–2821. doi: 10.1021/jp0361529.

Okpalugo, T. I. T. et al. (2005) High resolution XPS characterization of chemical functionalised MWCNTs and SWCNTs, *Carbon*, 43(1), pp. 153–161. doi: 10.1016/j.carbon.2004.08.033.

Palomino, J. et al. (2014) Ultrananocrystalline diamond-decorated silicon nanowire field emitters, *ACS Applied Materials and Interfaces*, 6(16), pp. 13815–13822. doi: 10.1021/am503221t.

Palomino, J. et al. (2015) Study of the structural changes undergone by hybrid nanostructured Si-CNTs employed as an anode material in a rechargeable lithium-ion battery, *Journal of Physical Chemistry C*, 119(36), pp. 21125–21134. doi: 10.1021/acs.jpcc.5b01178.

Palomino, J. et al. (2018) Silicon nanowires as electron field emitters. In: Sattler, K. (ed.) *Silicon Nanomaterials Sourcebook*. CRC Press, Boca Raton, FL, pp. 435–454. Available at: www.crcpress.com/Silicon-Nanomaterials-Sourcebook-Low-Dimensional-Structures-Quantum-Dots/Sattler/p/book/9781498763776.

Patole, S. P. et al. (2008) Optimization of water assisted chemical vapor deposition parameters for super growth of carbon nanotubes, *Carbon*, 46(14), pp. 1987–1993. doi: 10.1016/j.carbon.2008.08.009.

Patolsky, F., Zheng, G. and Lieber, C. M. (2006) Fabrication of silicon nanowire devices for ultrasensitive, label-free, real-time detection of biological and chemical species, *Nature Protocols*, 1(4), pp. 1711–1724. doi: 10.1038/nprot.2006.227.

Pavese, M. et al. (2008) An analysis of carbon nanotube structure wettability before and after oxidation treatment, *Journal of Physics Condensed Matter*, 20(47). doi: 10.1088/0953-8984/20/47/474206.

Pedersen, T. G. (2005) Quantum size effects in ZnO nanowires, *Physica Status Solidi (C)*. 2(12), pp. 4026–4030, Wiley-Blackwell. doi: 10.1002/pssc.2005 62222.

Pokropivny, V. V. (2001) Non-carbon nanotubes (review). Part 2. Types and structure. In: *Powder Metallurgy and Metal Ceramics*. Kluwer Academic Publishers-Plenum Publishers, 40(11/12), pp. 582–594. doi: 10.1023/A:1015232003933, https://link.springer.com/journal/11106.

Poveda, R. L. and Gupta, N. (2016) Carbon nanofibers: Structure and fabrication, In: *Carbon Nanofiber Reinforced Polymer Composites*. SpringerBriefs in Materials. Springer, Cham, pp. 11–26. doi: 10.1007/978-3-319-23787-9_2.

Peng, K. Q, Huang, Z. P. and Zhu, J. (2004) Fabrication of large-area silicon nanowire p–n junction diode arrays, *Advanced Materials*, 16(1), pp. 73–76. doi: 10.1002/adma.200306185.

Rinzler, A. G. et al. (1995) Unraveling nanotubes: Field emission from an atomic wire, *Science (New York, N.Y.)*, 269(5230), pp. 1550–1553. doi: 10.1126/science.269.5230.1550.

Rinzler, A. G. et al. (1998) Large-scale purification of single-wall carbon nanotubes: Process, product, and characterization, *Applied Physics A: Materials Science and Processing*, 67(1), pp. 29–37. doi: 10.1007/s003390050734.

Ruoff, R. S. and Lorents, D. C. (1995) Mechanical and thermal properties of carbon nanotubes, *Carbon*, 33(7), pp. 925–930. doi: 10.1016/0008-6223(95)00021-5.

Saito, R. et al. (1992) Electronic structure of chiral graphene tubules, *Applied Physics Letters*, 60(18), pp. 2204–2206. doi: 10.1063/1.107080.

Saito, R., Dresselhaus, G. and Dresselhaus, M. S. (1998) *Physical Properties of Carbon Nanotubes, Carbon Nanotubes*. Imperial College Press, London, doi: 10.1063/1.56490.

Schmidt, V. et al. (2009) Silicon nanowires: A review on aspects of their growth and their electrical properties, *Advanced Materials*, 21(25–26), pp. 2681–2702. doi: 10.1002/adma.200803754.

Shakerzadeh, M. et al. (2011) Plasma density induced formation of nanocrystals in physical vapor deposited carbon films, *Carbon*, 49(5), pp. 1733–1744. doi: 10.1016/j.carbon.2010.12.059.

Shanov, V., Yun, Y. and Schulz, M. (2006) Synthesis and characterization of carbon nanotube materials, *Journal of the University of Chemical Technology and Metallurgy*, 41(4), pp. 377–390. Available at: http://pns.uctm.edu/joomla15/journal/j2006-4/01-Shanov-377-390.pdf (Accessed: 13 November 2013).

Shao, M., Ma, D. D. D. and Lee, S.-T. (2010) Silicon nanowires: Synthesis, properties, and applications, *European Journal of Inorganic Chemistry*, 2010(27), pp. 4264–4278. doi: 10.1002/ejic.201000634.

She, J. C. et al. (1999) Comparative study of electron emission characteristics of silicon tip arrays with and without amorphous diamond coating, *Journal of Vacuum Science and Technology B: Microelectronics and Nanometer Structures*, 17(2), pp. 592–595. doi: 10.1116/1.590600.

Sohrabi, F., Nikniazi, A. and Movl, H. (2013) Optimization of third generation nanostructured silicon-based solar cells, In: Morales-Acevedo, A. (ed.) *Solar Cells: Research and Application Perspectives*. InTech, pp. 1–26. doi: 10.5772/51616.

Stout, D. A., Basu, B. and Webster, T. J. (2011) Poly(lactic-co-glycolic acid): Carbon nanofiber composites for myocardial tissue engineering applications, *Acta Biomaterialia*, 7(8), pp. 3101–3112, Acta Materialia Inc., doi: 10.1016/j.actbio.2011.04.028.

Suzuki, M. et al. (1991) Monoatomic step observation on Si(111) surfaces by force microscopy in air, *Applied Physics Letters*, 58(20), pp. 2225–2227. doi: 10.1063/1.104934.

Tang, W. et al. (2012) Ultrashort channel silicon nanowire transistors with nickel silicide source/drain contacts, *Nano Letters*, 12(8), pp. 3979–3985. doi: 10.1021/nl3011676.

Teo, K. et al. (2003) Catalytic synthesis of carbon nanotubes and nanofibers. In: Nalwa, H. S. (ed.) *Encyclopedia of Nanoscience and Nanotechnology*, X, pp. 1–22. Available at: www-g.eng.cam.ac.uk/cnt/oldsite/papers/ken_enn.pdf (Accessed: 24 September 2013).

Thang, B. H., Khoi, P. H. and Minh, P. N. (2015) A modified model for thermal conductivity of carbon nanotube-nanofluids, *Physics of Fluids*, 27(3). doi: 10.1063/1.4914405.

Thomas, J. P. et al. (2012) Preferentially grown ultranano c-diamond and n-diamond grains on silicon nanoneedles from energetic species with enhanced field-emission properties, *ACS Applied Materials and Interfaces*, 4(10), pp. 5103–5108. doi: 10.1021/am3016203.

Tian, B. et al. (2007) Coaxial silicon nanowires as solar cells and nanoelectronic power sources, *Nature*, 449(7164), pp. 885–889. doi: 10.1038/nature06181.

Tzeng, Y.-F. et al. (2007) Fabrication of an ultrananocrystalline diamond-coated silicon wire array with enhanced field-emission performance, *Nanotechnology*, 18(43), pp. 435703-1–5. doi: 10.1088/0957-4484/18/43/435703.

Tzeng, Y.-F. et al. (2008) Electron field emission properties on ultra-nano-crystalline diamond coated silicon nanowires, *Diamond and Related Materials*, 17(7–10), pp. 1817–1820. doi: 10.1016/j.diamond.2008.03.023.

Valentín, L. A. et al. (2013) Field emission properties of single crystal chromium disilicide nanowires, *Journal of Applied Physics*, 113(1), pp. 014308-1–5, AIP Publishing. doi: 10.1063/1.4773105.

Varshney, D. et al. (2013) Single-step route to hierarchical flower-like carbon nanotube clusters decorated with ultrananocrystalline diamond, *Carbon*, 63, pp. 253–262. Elsevier Ltd. doi: 10.1016/j.carbon.2013.06.078.

Varshney, D. et al. (2014) New route to the fabrication of nanocrystalline diamond films, *Journal of Applied Physics*, 115(5). doi: 10.1063/1.4863822.

Varzi, A. et al. (2011) Study of multi-walled carbon nanotubes for lithium-ion battery electrodes, *Journal of Power Sources*, 196(6), pp. 3303–3309, Elsevier B.V. doi: 10.1016/j.jpowsour.2010.11.101.

Wagner, R. S. and Ellis, W. C. (1964) Vapor-liquid-solid mechanism of single crystal growth, *Applied Physics Letters*, 4(5), pp. 89–90. doi: 10.1063/1.1753975.

Wang, J. et al. (2004) Synthesis of carbon nanosheets by inductively coupled radio-frequency plasma enhanced chemical vapor deposition, *Carbon*, 42(14), pp. 2867–2872. doi: 10.1016/j.carbon.2004.06.035.

Wang, X. et al. (2012) Nanopatterning of ultrananocrystalline diamond nanowires, *Nanotechnology*, 23(7), pp. 075301-1–7. doi: 10.1088/0957-4484/23/7/075301.

Watanabe, H. and Hosoi, T. (2013) Physics and technology of silicon carbide devices. In: Hijikata, Y. (ed.) *Fundamental Aspects of Silicon Carbide Oxidation.* InTech, 9, pp. 235–250. doi:10.5772/51514.

Wepasnick, K. A. et al. (2010) Chemical and structural characterization of carbon nanotube surfaces, *Analytical and Bioanalytical Chemistry*, 396(3), pp. 1003–1014. doi: 10.1007/s00216-009-3332-5.

Won, R. (2010) Photovoltaics: Graphene–silicon solar cells, *Nature Photonics*, 4(7), pp. 411–411, Nature Publishing Group, a division of Macmillan Publishers Limited. All Rights Reserved. doi: 10.1038/nphoton.2010.140.

Wu, Z. S. et al. (2002) Needle-shaped silicon carbide nanowires: Synthesis and field electron emission properties, *Applied Physics Letters*, 80(20), pp. 3829–3831. doi: 10.1063/1.1476703.

Xiang, B. et al. (2005) Field-emission properties of TiO_2 nanowire arrays, *Journal of Physics D: Applied Physics*, 38(8), pp. 1152–1155. doi: 10.1088/0022-3727/38/8/009.

Xiao, X. et al. (2004) Low temperature growth of ultrananocrystalline diamond, *Journal of Applied Physics*, 96(4), pp. 2232–2239. doi: 10.1063/1.1769609.

Xie, W., Möbus, G. and Zhang, S. (2010) Carbon nanotube to SiC nanorod conversion in molten salt studied by EELS and aberration corrected HRTEM, *Journal of Physics: Conference Series*, 241, pp. 012093-1–4. doi: 10.1088/1742-6596/241/1/012093.

Xu, X. Bin et al. (2007) Ultralight conductive carbon-nanotube-polymer composite, *Small*, 3(3), pp. 408–411. doi: 10.1002/smll.200600348.

Yang, S., Song, H. and Chen, X. (2006) Electrochemical performance of expanded mesocarbon microbeads as anode material for lithium-ion batteries, *Electrochemistry Communications*, 8(1), pp. 137–142. doi: 10.1016/j.elecom.2005.10.035.

Yang, Z. H. et al. (2005) Lithium insertion into multi-walled raw carbon nanotubes pre-doped with lithium, *Materials Chemistry and Physics*, 89(2–3), pp. 295–299. doi: 10.1016/j.matchemphys.2004.08.021.

Yoshio, M., Tsumura, T. and Dimov, N. (2005) Electrochemical behaviors of silicon based anode material, *Journal of Power Sources*, 146(1–2), pp. 10–14. doi: 10.1016/j.jpowsour.2005.03.143.

Yu, M. F. et al. (2000) Strength and breaking mechanism of multiwalled carbon nanotubes under tensile load, *Science*, 287(5453), pp. 637–640. doi: 10.1126/science.287.5453.637.

Zhao, F. et al. (2011) Field emission enhancement of Au-Si nano-particle-decorated silicon nanowires, *Nanoscale Research Letters*, 6(1), pp. 176-1–5, Springer Open Ltd. doi: 10.1186/1556-276X-6-176.

Zhao, X. et al. (2016) Bionanofibers in drug delivery. In: Grumezescu, A.M. (ed.) *Nanobiomaterials in Drug Delivery: Applications of Nanobiomaterials*, William Andrew, pp. 403–445. doi: 10.1016/B978-0-323-42866-8.00012-5.

Zheng, G. et al. (2005) Multiplexed electrical detection of cancer markers with nanowire sensor arrays, *Nature Biotechnology*, 23(10), pp. 1294–1301, Nature Publishing Group. doi: 10.1038/nbt1138.

Zhou, Z. et al. (2009) Development of carbon nanofibers from aligned electrospun polyacrylonitrile nanofiber bundles and characterization of their microstructural, electrical, and mechanical properties, *Polymer*, 50(13), pp. 2999–3006. doi: 10.1016/j.polymer.2009.04.058.

Zhu, Y. W. et al. (2003) Efficient field emission from ZnO nanoneedle arrays, *Applied Physics Letters*, 83(1), pp. 144–146. doi: 10.1063/1.1589166.

Zuo, P. et al. (2007) Synthesis and electrochemical performance of Si/Cu and Si/Cu/graphite composite anode, *Materials Chemistry and Physics*, 104(2–3), pp. 444–447. doi: 10.1016/j.matchemphys.2007.04.001.

2

Survey of Low-Dimensional Nanomaterials

Shihao Xu and Changlong Jiang
Chinese Academy of Sciences

2.1 Low-Dimensional Materials

Nanomaterials continue to captivate and occupy the minds and labor of scientists and engineers across the globe. From a fundamental perspective, the ability to control catalytic or optoelectronic properties by tuning the composition, size, and morphology of low-dimensional materials is fascinating.

Nanostructured materials are of interest because they can bridge the gap between the bulk and molecular levels and lead to entirely new avenues for applications, especially in absorption, energy storage, electronics, optoelectronics, and biology.[1–4] When a solid exhibits a distinct variation in optical and electronic properties with a variation in particle size <100 nm, it can be called a nanostructure and is categorized as (i) two dimensional (2D), e.g., thin films or quantum wells, (ii) one dimensional (1D), e.g., quantum wires, or (iii) zero dimensional or dots. When material dimensions are reduced to the nanoscale, exceptional physical mechanics properties can be obtained that differ significantly from the corresponding bulk materials.

Here, we review the synthesis, characterization, theoretical calculation, and application of the low-dimensional nanomaterials, including zero-dimensional nanoparticles (quantum dots [QDs]), 1D nanoparticles (nanotubes and nanowires), and 2D nanoparticles (graphene and nanosheets).

2.2 Zero-Dimensional Nanoparticles

During the last two decades, a great deal of attention has been focused on the optoelectronic properties of nanostructured semiconductors or QDs as many fundamental properties are size dependent in the nanometer range. A QD is zero dimensional relative to the bulk, and the limited number of electrons results in discrete quantized energies in the density of states (DOS) for nonaggregated zero-dimensional structures[5,6] (Although it is zero dimensional to bulk, it is regarded as a box in quantum mechanics; size of the box is important and discussed later). Sometimes, the presence of one electronic charge in the QDs repels the addition of another charge and leads to a staircase-like I–V curve and DOS. The step size of the staircase is proportional to the reciprocal of the radius of the QDs. The boundaries, as to when a material has the properties of bulk, QDs or atoms, are dependent upon the composition and crystal structure of the compound or elemental solid. An enormous range of fundamental properties can be realized by changing the size at a constant composition, and some of these are discussed in this chapter. QDs can be broadly categorized into either elemental or compound systems. In this review, we emphasize inorganic, polymer, and composite nanostructured materials and their multimodal applications based on optoelectronic and optical properties.

2.2.1 Inorganic Nanoparticles

QDs are highly efficient multi-photon absorbers that can be potentially useful for three-dimensional (3D) multi-photon microscopy and imaging,[7–10] which is a rapidly developing area for both biological and medical applications. For the last few years, many researchers have also actively employed QDs for chemical sensing of small molecules or ions, demonstrating their broad applicability.[11–13]

Recently, CdTe QDs have received great attention due to their unique tunable optical properties for sensing and electronics applications (Figure 2.1). CdTe is one of the most important semiconductors in the II–VI group.[14] This well-studied semiconductor has a direct bandgap of 1.44 eV at room temperature, and it is currently used in solar cells,

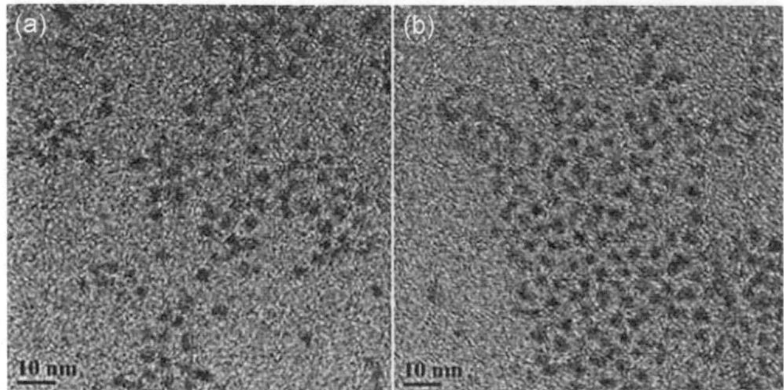

FIGURE 2.1 Transmission electron microscopic (TEM) images of monodispersed CdTe nanocrystals with sizes of (a) 3 nm and (b) 4 nm. The size of the nanocrystals can be tuned from 2 to 8 nm by further aging the reaction mixture.

in light-emitting diodes for flat-panel displays, and in other optical devices. For the past several years, several different synthesis methods have been developed to prepare CdTe semiconductor nanocrystals with the desired size and shape. CdTe QDs prepared directly in the aqueous phase generally have much higher quantum yield (QY) (>50%) compared to CdTe QDs (<30%) synthesized in organic solvents and later transferred to water.

CdTe QDs have shown significant advantages over traditional organic dyes and fluorescent proteins. QDs of multiple emission colors can be simultaneously excited with a single light source, with minimal spectral overlap, providing significant advantages for multiplex imaging of molecular targets.[15] The photoluminescence (PL) spectra of anti-thyroglobulin antibody (TGA)-capped CdTe QDs are tunable in the range of 525–625 nm. The QYs for these nanocrystals are usually 50%–70%, which is more than sufficient for bioimaging application (Figure 2.2). For example, Weng et al. synthesized CdTe QDs conjugated with either plant lectin or antibody anti-von Willebrand factors as biological labels for targeted cancer cell imaging.[16] Dong et al. reported the preparation of bioconjugated CdTe QDs for labeling prostate cancer cells.[17] Li et al. demonstrated the synthesis of PEGylated CdTe QDs for imaging human tongue cancer cells without observing any damage to the cells.[18]

2.2.2 Polymer Nanoparticles

Conjugated polymers have been developed with optical transitions that span the full range of the visible spectrum.[19] In particular, patterning of conjugated polymers has attracted considerable attention because the mechanical flexibility and luminescent characteristics of conjugated polymers make them attractive candidates for flexible organic displays.[20] Recently, conjugated polymer dots (Pdots) have been a subject of great interest because of their outstanding characteristics such as extraordinary fluorescence brightness, excellent stability, biocompatibility, and nontoxic features for biological imaging and biosensors.[21–25]

Invisible security inks are preferable to have minimal absorption in visible range, but provide fluorescent images readable only under special environments, such as excitation by ultraviolet (UV) light. Fluorene-based conjugated polymers exhibit great flexibility for the design of fluorescent materials as shown by the significant progress made so far in tuning their emission color from blue to deep red by the introduction of narrow-band-gap monomers into the polymer backbone.[26] Figure 2.3 provides the chemical structures and acronyms of the conjugated polymers employed in this study. Multicolor Pdots were prepared by the reprecipitation procedure described previously.[27] The aqueous Pdot solutions obtained from the three types of conjugated polymers are stable and clear (not turbid), with no sign of aggregation for months. All the Pdots exhibit dominant UV absorption spectra with peaks centered at 375 nm (Figure 2.3b), indicating that the resulting security patterns can be simultaneously excited by a single wavelength in the UV region.

2.2.3 Composited Nanoparticles

Metal–organic frameworks (MOFs), sometimes referred to as porous coordination polymers, are a class of crystalline microporous materials formed from transition metal ions or clusters and polytopic organic linkers that assemble into topologically diverse open network-type structures.[28–30] In this review, we focus on current progress in the

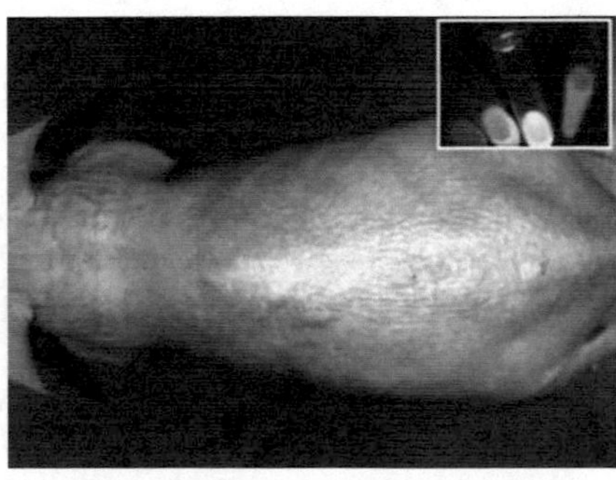

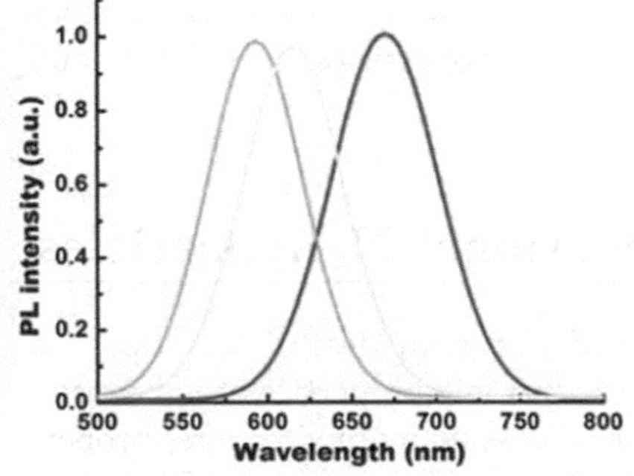

FIGURE 2.2 In vivo multiplex image of aqueous phase-prepared CdTe QDs subcutaneously injected at different spots in a live mouse. The back is injected with 570-, 605-, and 675 nm-emitting CdTe QDs. The image is taken under 488-nm excitation in the Maestro imaging system.

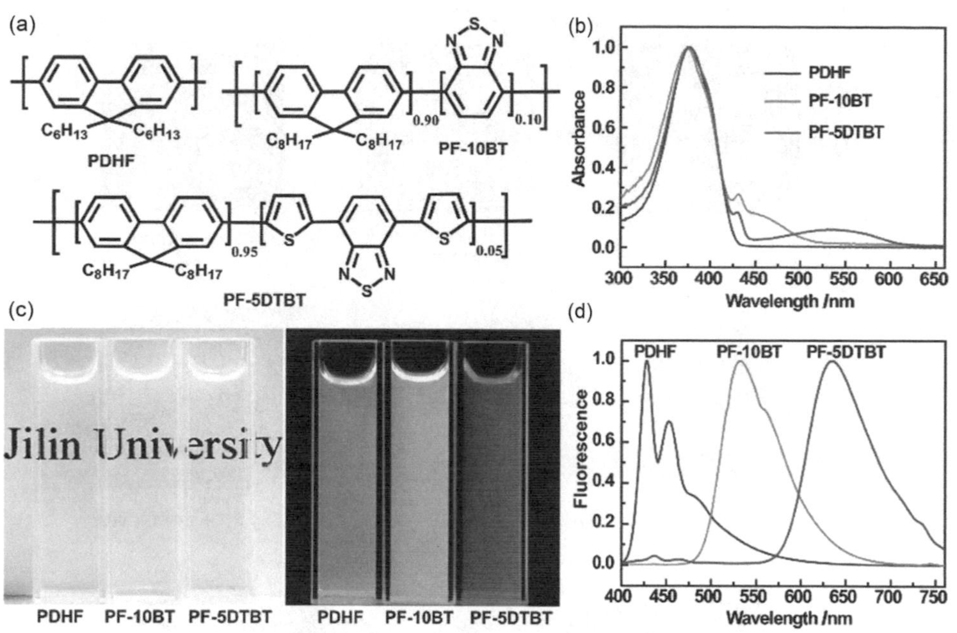

FIGURE 2.3 (a) Chemical structures of fluorescent Pdots. (b) Absorption spectra of the three types of Pdots. Their dominant absorption peaks are in the UV region. (c) Photographs of aqueous Pdot solutions (~2 ppm concentration) under room light (left) and UV lamp illumination (right). (d) Fluorescence spectra of Pdot solutions.

preparation of QD@MOF composites, where the size-dependent electronic and optical properties of the QDs can be stabilized when encapsulated in a MOF matrix and the resulting composites exploited for selective sensing, and as scaffolds to perform photocatalysis, including water splitting for hydrogen evolution.

There are two well-established routes that have been exploited to immobilize functional molecules or nanoscale objects within MOFs, known as "ship in the bottle" and "bottle around the ship" methods (Figure 2.4a,b), although they have been previously explored for other guests and matrices.[31–35] The "ship in the bottle" method (Figure 2.4a) consists of the immobilization of small molecules or nanoparticle precursors small enough to penetrate through the pore windows of the MOF, which by further treatment in situ results in their transformation to the desired functional structure. The "bottle around the ship" method (Figure 2.4b) is also known as the template synthesis method and usually occurs through an encapsulation process as previously reported for other materials.[33] Another approach to prepare QD@MOF composites that is not yet so well explored is the photochemical deposition of the semiconductor nanoparticles onto the framework surface (Figure 2.4c). The surface modification of MOFs with QDs can also be carried out by direct binding using a suitable linking group[36] (Figure 2.4d).

A priori, it might be expected that a composite material containing embedded photoactive particles such as QDs will be used mainly in light-dependent applications such as fluorescence sensing,[37,38] photocatalysis,[39,40] imaging,[41,42] drug release,[43] and other areas where light-induced mechanisms can be exploited. However, QDs have also proven effective as pseudocapacitors or supports for other molecules.[44,45] In this section, we will discuss the preparation and current state of the art in the applications of QD@MOF composites, which also provide an interesting platform to explore synergistic and/or symbiotic effects arising from the combination of the quantum confinement of semiconductor nanoparticles and the vast diversity of MOF compositions available, which can additionally confer higher stability to the nanoparticle.

2.3 One-Dimensional Nanomaterials

1D nanostructures, which include nanowires, nanofibers, nanoribbons, nanorods, and nanotubes among other morphologies, are recognized as one of the most promising material directions for energy-related applications.[46–49] Previous literature has introduced the various methods for synthesizing porous 1D nanomaterials, including electrospinning, liquid phase, template-assisted approaches, chemical deposition, and chemical etching methods. These preparation routes are highly effective and enable the controllable synthesis of porous 1D nanomaterials with different morphology, porosity, and inner structure. Since the groundbreaking discovery of carbon nanotubes in the 1990s, 1D nanostructures have attracted significant research interest because of their remarkable physical/chemical properties and their great potential in nanotechnology applications, such as adsorption, gas sensor, energy storage, and drug carrier.[50–55] In particular in electrochemical application, 1D nanostructures can provide direct current pathways, shorten the ion diffusion distance, lower the

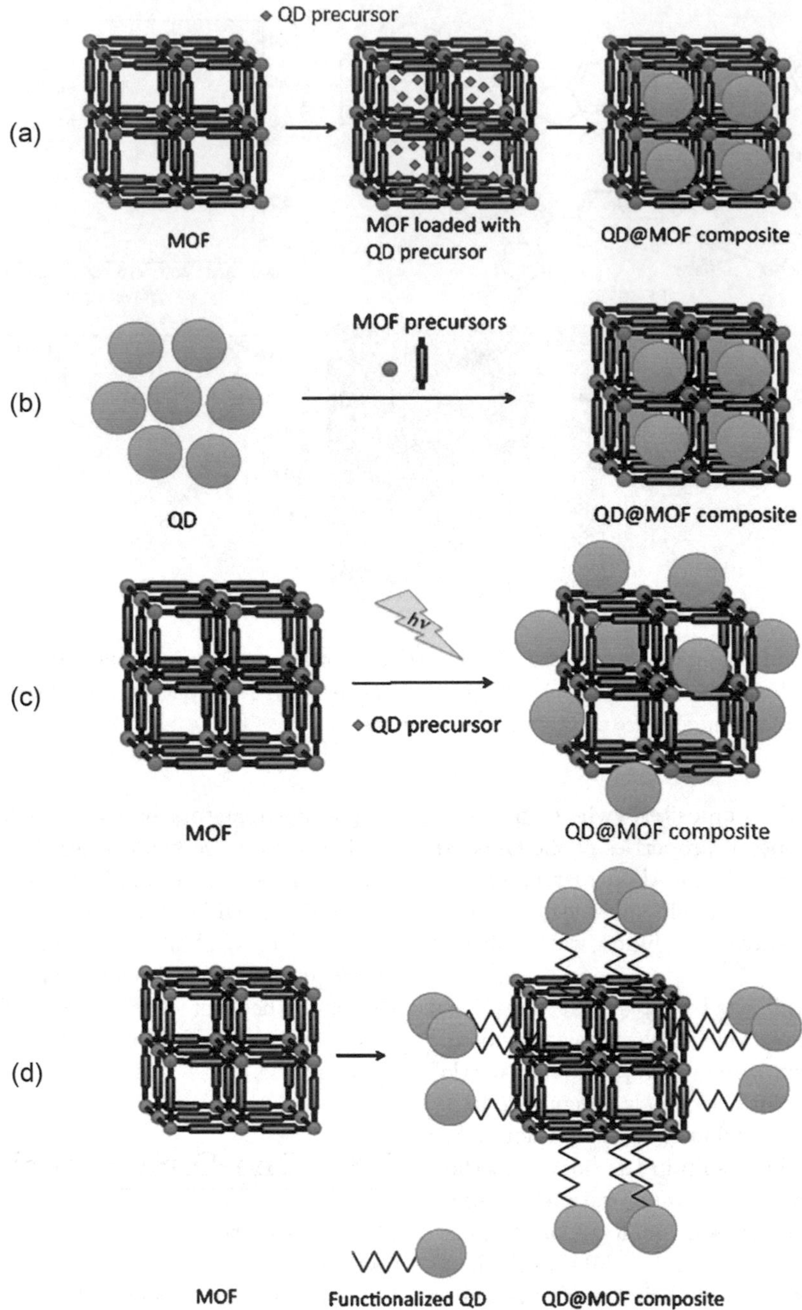

FIGURE 2.4 Main methodologies used to prepare QD@MOF composites. (a) Ship in the bottle, (b) bottle around the ship, (c) photochemical deposition, and (d) direct surface functionalization.

charge–discharge time, increase the electrolyte–electrode contact area, limit mechanical degradation, and accommodate volume.[56]

2.3.1 Inorganic Nanomaterials

Inorganic nanowires can also act as active components in devices as revealed by recent investigations. In the last 3–4 years, a variety of inorganic material nanowires have been synthesized and characterized. Thus, nanowires of elements, oxides, nitrides, carbides, and chalcogenides have been generated by employing various strategies. One of the crucial factors in the synthesis of nanowires is the control of composition, size, and crystallinity. Among the methods employed, some are based on vapor-phase techniques, while others are solution techniques. Compared to physical methods such as nanolithography and other patterning techniques, chemical methods have been more versatile and effective in the synthesis of these nanowires.

Lee et al.[57] propose that the growth of the Si nanowires is assisted by the Si oxide, where the Si_xO ($x > 1$) vapor generated by thermal evaporation or laser ablation plays the key role. Nucleation of the nanoparticles is assumed to occur on the substrate as shown in Eqs. (2.1) and (2.2).

$$Si_xO \rightarrow Si_{x-1} + SiO(x > 1), \quad (2.1)$$

and

$$2SiO \rightarrow Si + SiO_2 \quad (2.2)$$

These decompositions result in the precipitation of Si nanoparticles, which act as the nuclei of the silicon nanowires covered by shells of silicon oxide. As shown in Figure 2.5, the precipitation, nucleation, and growth of the nanowires occur in the area near the cold finger, suggesting that the temperature gradient provides the external driving force for the formation and growth of the nanowires.

Figure 2.5a–c shows the TEM images of the formation of nanowire nuclei at the initial stages. Figure 2.5a shows Si nanoparticles covered by an amorphous silicon oxide layer. The nanoparticles that are isolated, with the growth directions normal to the substrate surface, exhibit the fastest growth. The tip of the Si crystalline core contains a high concentration of defects, as marked by arrows in Figure 2.5c.

Solid materials such as polysulfurnitride, $(SN)_x$, grow into 1D nanostructures, the habit being determined by the anisotropic bonding in the structure. Other materials such as selenium,[58] tellurium,[59] and molybdenum chalcogenides[60,61] are easily obtained as nanowires due to anisotropic bonding, which dictates the crystallization to occur along the c-axis, favoring the stronger covalent bonds over the relatively weak van der Waals forces between the chains. Nanowires of a variety of oxides, nitrides, and carbides can be synthesized by carbothermal reactions. For example, carbon (activated carbon or carbon nanotubes) in mixture with an oxide produces sub-oxidic vapor species that reacts with C, O_2, N_2, or NH_3 to produce the desired nanowires. Thus, heating a mixture of Ga_2O_3 and carbon in N_2 or NH_3 produces GaN nanowires.[62]

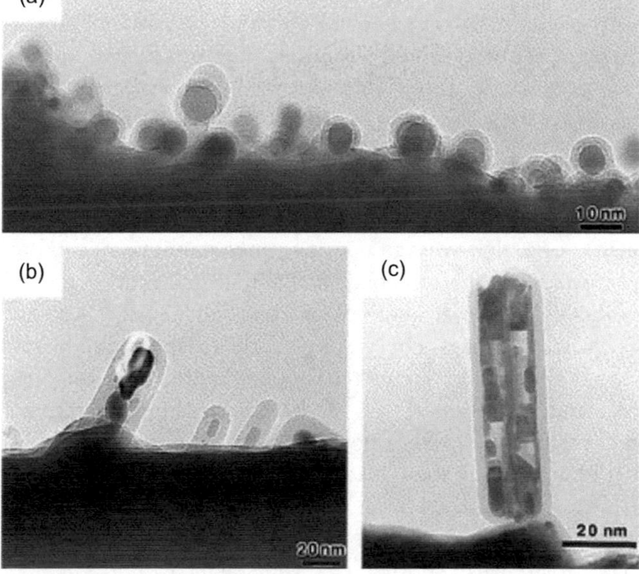

FIGURE 2.5 TEM micrographs of (a) Si nanowire nuclei formed on the Mo grid and (b), (c) initial growth stages of the nanowires.

2.3.2 Polymer Nanomaterials

The past decade has witnessed increasing attention in the synthesis, properties, and applications of 1D conducting polymer nanostructures, such as conjugated polypyrrole (PPy), polyaniline (PANI), polythiophene (PTh), poly (***p***-phenylenevinylene) (PPV), and derivatives thereof.

Regarding conducting polymers, the pioneering one is conducting polyacetylene (PA) doped with iodine discovered by MacDiarmid, Heeger, and Shirakawa and coworkers in 1977.[63,64] Since then, conducting polymers are rapidly gaining attraction in various applications due to the availability of more materials with good electrical, physical, and mechanical properties, as well as excellent solution processability. As shown in Figure 2.6, intrinsically conducting polymers include PPy, PANI, PTh, and its derivatives (i.e., poly(3-methylthiophene) [P3MT], poly(3-butylthiophene) [P3BT], poly(3-hexylthiophene) [P3HT], poly(3-octylthiophene) [P3OT], poly(3-pentylthiophene) [P3PT], poly(3,4-ethylenedioxythiophene) [PEDOT]), PPV and its derivatives, etc. In order to make PTh and PPV soluble and processable, soluble alkyl or alkoxy side chains are introduced (Figure 2.6). All these conjugated polymers have controllable conductivity ranging from insulator to semiconductors or conductors depending on the dopants and doping levels.[65]

Recently, several approaches, including vapor growth, liquid growth, and selfassembly, have been developed for the fabrication of 1D metallic, inorganic, and organometallic nanomaterials to achieve the desired structures.[66–70] As is well known, template-based synthesis represents one of the most popular strategies for fabricating different kinds of 1D nanomaterials such as metals, inorganic semiconductors, structural ceramics, and functional composites.[71–73] Herein, the template synthesis means a process employing

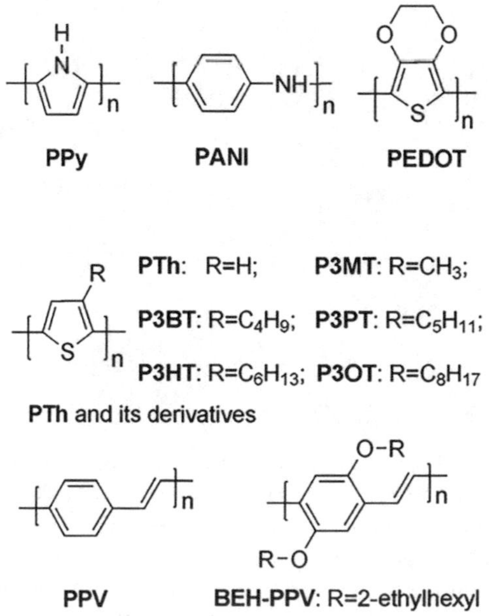

PPy PANI PEDOT

PTh: R=H; **P3MT**: R=CH_3;
P3BT: R=C_4H_9; **P3PT**: R=C_5H_{11};
P3HT: R=C_6H_{13}; **P3OT**: R=C_8H_{17}

PTh and its derivatives

PPV **BEH-PPV**: R=2-ethylhexyl

FIGURE 2.6 Molecular structures of some representative conducting polymers.

nanostructured matters as templates, which provide the spatial confinement function to promote 1D growth of conducting polymers on their inner pores and/or outer surfaces assisted by some physical, chemical, or multiple measures.[71]

In the case of MnO_2 nanowires, they served as both the wire template and the oxidant for the synthesis of PANI nanotubes.[74] The oxidation potential of MnO_2 was able to initiate the aniline polymerization, and as the polymerization proceeded, the PANI layers generated along the surfaces of the MnO_2 nanowires, while the MnO_2 core was consumed at the same time (Figure 2.7a). As a result, the morphology of the MnO_2 nanowires was finally transferred to that of PANI nanotubes, whose external size and shape were similar in dimensions to that of the MnO_2 nanowire template (Figure 2.7c–d). By simply varying the morphologies and sizes of the MnO_2 templates, the structural parameters of PANI products (i.e., diameter, length, and morphology) could be accurately controlled. When MnO_2 nanotubes were used as the alternative to MnO_2 nanowires, the double-shell PANI nanotubules (one round and closed, and the other opened end) were obtained by this self-degraded wire-template method (Figure 2.7b,e,f).

As discussed in previous sections, various 1D conducting polymer nanomaterials with desired morphology and size can be controllably synthesized by using an efficient template methodology[75,76] or a template-free methodology.[77–79] The unique electrical properties, charge transport properties, optical properties, and redox electrochemical properties of these functional materials are related to their anisotropic nanostructures. These advantages make them promising candidates for applications in high-performance energy devices. Therefore, this kind of advanced nanostructured materials can act as important active materials, or electrodes, or buffer materials, or catalysts, or electrode/catalyst supports, which play critical roles in energy conversion and storage applications such as various solar cells, fuel cells, rechargeable lithium batteries, and electrochemical supercapacitors.

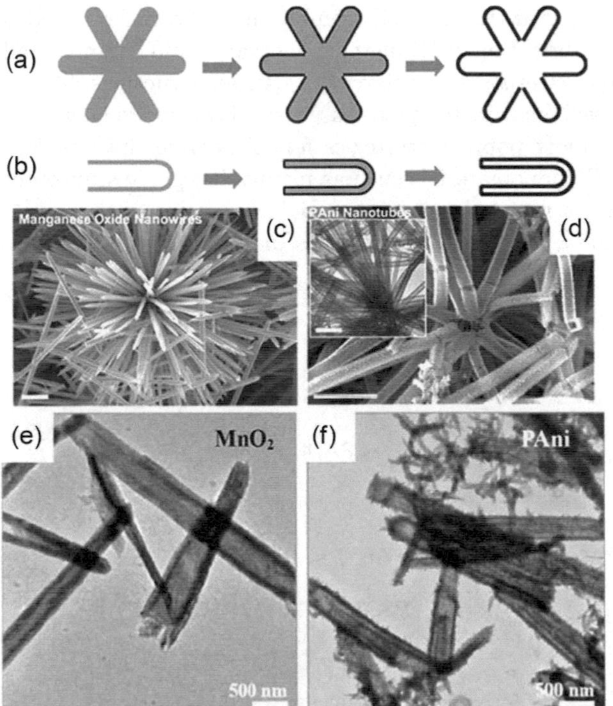

FIGURE 2.7 Schematic illustration of the formation mechanism of the PANI nanotubes fabricated from (a) MnO_2 nanowires and (b) MnO_2 nanotubes, respectively. Scanning electron microscopic (SEM) images of (c) the MnO_2 nanowire templates and (d) the resulting PANI nanotubes (inset: a TEM image of the PANI nanotubes). The scale bar is 1 μm. TEM images of (e) the MnO_2 nanotube templates and (f) the resultant double-shell PANI nanotubes.

2.3.3 Composited Nanomaterials

A large number of advanced techniques have been developed to fabricate 1D nanostructures with well-controlled morphology and chemical composition. Among these methods, electrospinning seems to be the simplest and most versatile technique capable of generating 1D nanostructures (mainly nanofibers; other types of 1D nanostructure, such as nanorods, can be cut from nanofibers by special methods[80,81]) from a variety of polymers.[72,82–86]

Electrospinning has been understood for almost a century and may be considered as a variation of the electrospray process. During the process of electrospraying, the liquid drop elongates with an increasing electric field. When the repulsive force induced by the charge distribution on the surface of the drop is balanced with the surface tension of the liquid, the liquid drop distorts into a conical shape. Once the repulsive force exceeds the surface tension, a jet of liquid ejects from the cone tip. Small droplets form as a result of the varicose breakup of the jet in the case of low-viscosity liquids. Compared to the commercial mechanical spinning process for generating microfibers, electrospinning mainly makes use of the electrostatic repulsions between surface charges to reduce the diameter of a viscoelastic jet or a glassy filament.

The basic setup for electrospinning is very simple and easily controlled, as illustrated in Figure 2.8a. Generally, it consists of a high-voltage power supply, a spinneret, and an electrically conductive collector (a piece of aluminum foil or silicon). However, not all of these parts are necessary for electrospinning. For example, electrospinning devices without the use of a spinneret have been described by Yarin et al. and Kameoka et al.[87,88] In a typical electrospinning experiment, the liquid (a polymer solution or melt) for electrospinning is pumped into a syringe with a thin nozzle with an inner diameter on the order of 100 mm to about 1 mm. The nozzle serves as one electrode and the collector is connected with another electrode, which affords a high electric field of 100–3,000 kV/m. The distance between the nozzle and the collector is usually 5–25 cm in most laboratory systems.

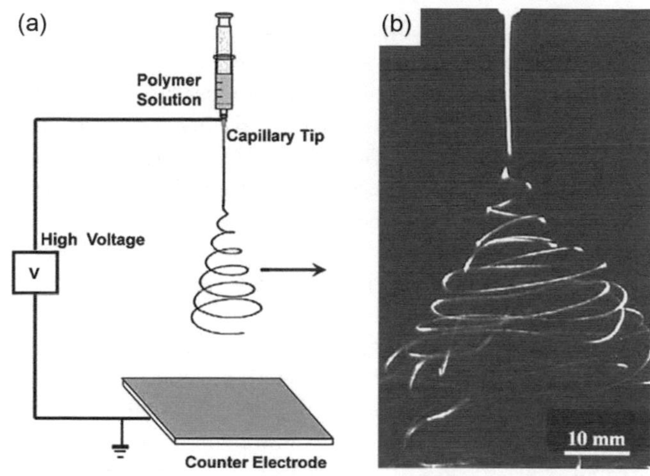

FIGURE 2.8 (a) Schematic diagram showing a laboratory setup for electrospinning. (b) Photographs of typical electrospinning jets captured by a high-speed video showing the bending instability of the jet.

Electrospinning has exhibited a strong ability to generate polymeric nanofibers in the past decade. Combined with calcination or carbonation, ceramic or other inorganic nanofibers could also be synthesized using the electrospinning technique. Haddon and coworkers prepared polystyrene/single-walled carbon nanotubes (PS/SWNT) and polyurethane/single-walled carbon nanotubes (PU/SWNT) composite nanofibers by electrospinning a mixture of polymer and ester (EST)-functionalized SWNTs, and studied their mechanical properties.[89] The results showed that the tensile strength of ESTSWNT-PU mats was enhanced by 104% as compared to electrospun pure PU mats, while an increase of only 46% was achieved by incorporating as-prepared SWNTs in the PU matrix. Reneker and coworkers demonstrated that polyacrylonitrile/multi-walled carbon nanotubes (PAN/MWNT) composite nanofibers could be prepared by electrospinning a mixture of the PAN and surface-oxidized MWNTs in N,N-Dimethylformamide (DMF).[90,91]

Traditional electrospinning technology included the direct dispersion method, gas–solid reaction, in situ photoreduction, sol–gel method, emulsion electrospinning method, solvent evaporation, and coaxial electrospinning. Wei's group first introduced a gas–solid reaction to the electrospinning technique to incorporate semiconductor nanostructures into polymer nanofibers; similar results were also reported by other groups.[92–96] The synthetic strategy involved three steps: (i) Co-dissolve metal salt and polymer into one solvent to make a homogeneous solution; (ii) electrospinning the above solution to obtain polymer/metal salt composite nanofibers. Compared to the direct dispersion method, gas–solid reaction affords a simple way to disperse inorganic nanoparticles into polymer nanofibers. When cadmium acetate was used to produce poly(vinylpyrrolidinone) (PVP)/CdS composite nanofibers instead of lead acetate, the obtained CdS nanostructures were nanorods, not dense spherical nanoparticles; we thought that this might be due to the different interactions between the polymer molecules and different metal ions[97] (Figure 2.9).

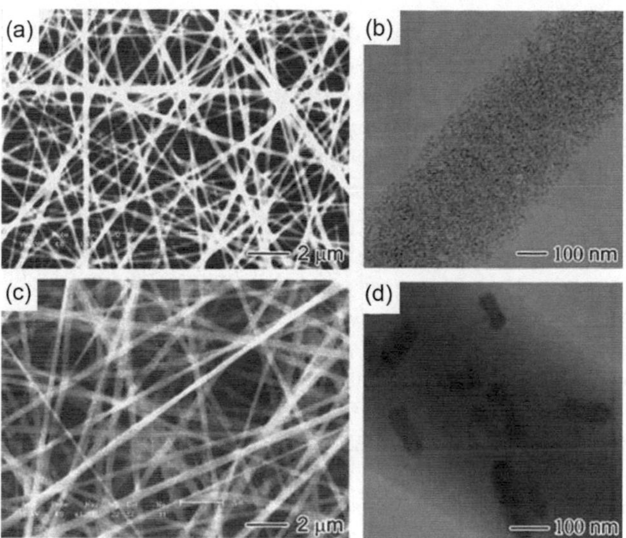

FIGURE 2.9 (a) SEM image of PVP/PbS composite fibers. (b) TEM image of spherical PbS nanoparticles incorporated into PVP fibers, showing that the PbS nanoparticles with a diameter of about 5 nm are well dispersed in PVP fibers. (c) SEM image of PVP/CdS composite fibers. (d) TEM image of CdS nanorods formed in PVP fibers after reaction with H_2S gas.

Electrospinning affords us a remarkably easy and versatile technique for the formation of very thin fibers with large surface areas and superior mechanical properties. Many desirable properties can be achieved by electrospinning multicomponent mixtures and post-modification with functional reagents to form functional 1D composite nanomaterials. Functional composite nanofibers achieved by these methods have been widely used in the applications of electronic and optical nanodevices, chemical and biological sensors, catalysis and electrocatalysis, superhydrophobic surfaces, environment, energy, and biomedical fields. Although many technical issues still need to be resolved or improved upon, there is no doubt that electrospinning has become one of the most powerful tools for fabricating functional 1D nanomaterials.

2.4 Two-Dimensional Nanomaterials

Since the exfoliation and identification of graphene in 2004,[98] layered ultrathin 2D nanomaterials have been the subject of intensive study over the last decade.[99–102] Ultrathin 2D nanomaterials are sheet-like structures with single- or few-layer thickness (typically less than 5 nm), but a lateral size larger than 100 nm or even up to tens of micrometers. Till now, besides graphene, a large number of graphene-like ultrathin 2D nanomaterials, such as transition metal dichalcogenides, layered metal oxides, transition metal carbides, and double-layered hydroxides, have been prepared via various methods.[103–106]

The advantages of ultrathin 2D nanostructures include the following aspects. First, the surface area of nanomaterials can be remarkably enlarged by reducing their thickness, leading to increasing accessible surfaces to facilitate the contact between foreign molecules and active sites on their surfaces. Second, nearly all interior surface atoms are exposed to the outside of such ultrathin nanostructures. These exposed sites (especially metal atoms) can serve as active centers in catalysis. Third, these ultrathin structures are prone to the formation of more defects to generate more active edges and coordination-unsaturated metal sites, which can reduce the reactive energy barrier by superior interfacial charge transfer and rapid mass transport. Fourth, when the thickness of nanosheets is reduced to atom-layered level, their conductivity can be promoted. Owing to the ultrahigh specific surface area and strong quantum confinement of electrons in two dimensions, these ultrathin 2D nanomaterials display many unconventional physical, optical, chemical, and electronic properties. They have also shown great potential in various applications such as electronic devices,[107,108] catalysis,[109–111] energy storage and conversion,[112,113] sensing,[114,115] and biomedicine.[116] Therefore, these ultrathin 2D materials can likewise serve as a potential and ideal platform for improved electrochemical performance in renewable energy fields.[117–119]

Layered compounds are those that possess strong lateral chemical bonding in planes but display weak van der Waals interaction between planes. One typical example is graphite that consists of weakly stacked graphene sheets forming 3D bulk crystals. However, many other materials form atomic bonding in three dimensions (e.g., metals), reflecting the non-layered nature of their bulk crystals. Inspired by the layered ultrathin 2D crystals, one can also anticipate that controlled synthesis of non-layer structured 2D materials may bring up some unique properties and advanced functions that cannot be achieved for their counterparts in other dimensionalities. A host of non-layer structured ultrathin 2D nanomaterials, such as noble metals (e.g., Au, Pd, and Rh), metal oxides (e.g., TiO_2, WO_3, CeO_2, In_2O_3, SnO_2, and Fe_2O_3), and metal chalcogenides (e.g., PbS, CuS, CuSe, SnSe, ZnSe, ZnS, and CdSe), have been prepared over the last few years; almost all of these non-layer structured ultrathin 2D nanomaterials are synthesized using wet-chemical synthesis approaches.

Similar to graphene, other 2D nanomaterials can also be used as templates to assist the growth of functional nanostructures, such as metals[120–122] and semiconductors.[123] Despite the great efforts made in graphene-based synthesis, the epitaxial growth of inorganic nanostructures was only realized on mechanical exfoliated graphene[124,125] or chemical vapor deposition (CVD) graphene.[126,127] The dispersible rGO nanosheets, on the other hand, lose the long range order of the graphitic lattice due to the pretreatment with strong oxidizing agents, and thus could not induce solution-phase epitaxial growth of nanocrystals. Importantly, previous article demonstrated that single-layer MoS_2 nanosheets can direct the growth of a series of noble metal nanostructures, such as Ag, Au, Pt, and Pd. Particularly, Pd nanoparticles with sizes of ca. 5 nm and Pt nanoparticles with sizes of 1–3 nm were epitaxially grown on the MoS_2 nanosheets (Figure 2.10a–d). Besides the spherical nanoparticles, anisotropic structures such as Ag nanoplates were also grown and epitaxially aligned on the MoS_2 surface (Figure 2.10e–h).[120] Depending on the kind of metals, different synthetic methods and conditions were applied. For example, Pd nanoparticles were synthesized in situ by reduction of K_2PdCl_4 with ascorbic acid in the presence of PVP and MoS_2, while Pt nanoparticles were

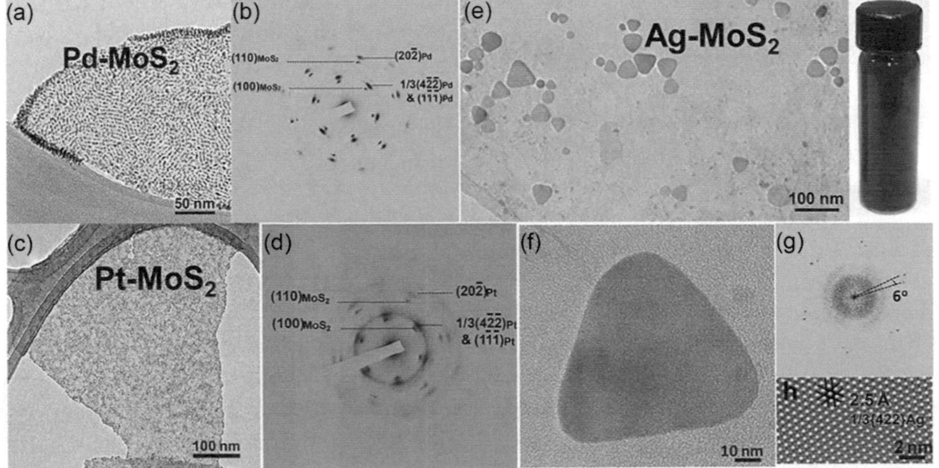

FIGURE 2.10 (a) TEM image of Pd nanoparticles synthesized on an MoS_2 nanosheet. (b) Selected area electron diffraction (SAED) pattern of a Pd-MoS_2 hybrid nanosheet with the electron beam perpendicular to the basal plane of the MoS_2 nanosheet. (c) TEM image of Pt nanoparticles synthesized on an MoS_2 nanosheet. (d) SAED pattern of a Pt-MoS_2 hybrid nanosheet with the electron beam perpendicular to the basal plane of the MoS_2 nanosheet. (e) TEM image of Ag nanoplates synthesized on MoS_2 nanosheet. (f) TEM image of a typical Ag nanoplate on MoS_2 nanosheet. (g) Fast fourier transform (FFT)-generated SAED pattern of (f). (h) Filtered high resolution transmission electron microscopy (HRTEM) image of the Ag nanoplate in (f). Inset in (e): photograph of the Ag-MoS_2 solution.

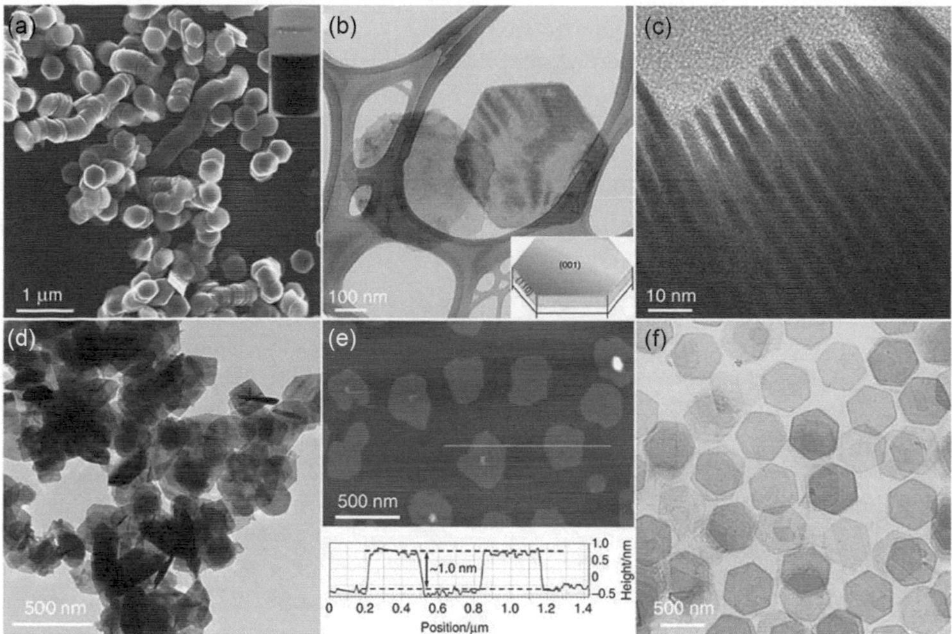

FIGURE 2.11 (a) SEM image of ultrathin CuS nanosheets. Inset: photograph of the colloid solution of CuS nanosheets. TEM images of ultrathin CuS nanosheets with (b) lying flat and (c) standing on the TEM grids. Inset in b: scheme of an ultrathin CuS nanosheet. (d) TEM and (e) AFM images of SnSe nanosheets. (f) TEM image of Pd nanosheets.

obtained by photochemical reduction of K_2PtCl_4 in the presence of sodium citrate. It is noteworthy that, compared to the conventional solid-state epitaxial growth of metals on bulk substrates, the wet-chemical epitaxial synthesis enables large-scale production at a relatively low cost, which is essential for many practical applications, such as fuel cells and photocatalysis.

Some other wet-chemical methods have also been developed for the synthesis of ultrathin 2D nanostructures, which cannot be categorized into the aforementioned methods.[128] Ultrathin CuS nanosheets with a thickness of 3.2 nm (two unit cells) were prepared in gram amount (Figure 2.11a–c). The resultant CuS nanosheets have a regular hexagonal shape with lateral size up to 453 nm (Figure 2.11a,b). As a typical example, Wang and coworkers presented a one-pot synthetic method for the synthesis of single-layer SnSe nanosheets in oil phase.[128] 1,10-Phenanthroline was used as the morphology control agent, which played a crucial role in controlling the morphology of SnSe nanocrystals. The obtained SnSe nanosheets have a lateral size of 300 and a thickness of 1.0 nm (Figure 2.11d,e). As an interesting example, Zheng and coworkers reported a CO-confined growth method to synthesize freestanding ultrathin hexagonal Pd nanosheets with thickness less than ten atomic layers (1.8 nm) and controllable edge length from 20 to 160 nm (Figure 2.11f).[129]

With the ability to prepare various nanosized materials of all dimensionalities, combined with the advanced nanofabrication techniques, the current research in the preparation of hybrid nanostructures lies on not only the choice of functional components, but also the spatial organization/assembly and geometric properties of the complex nanostructures. By using 2D nanomaterials with unique properties, the planar hybrids, porous hierarchical architectures, and vertically stacked heterostructures have been prepared which have showed much impressing properties, enhanced functions, and improved performance.

Specifically, for applications in catalysis, energy storage, and solar harvesting, an advanced hybrid nanostructure should generally meet the requirements of large specific surface area for reaction, ion exchange or light adsorption, a conductive network for charge transport, and an interface/heterojunction formed by two components for the effective channeling or separation of charge carriers. For electronic and optoelectronic devices, the vertically stacked heterostructures hold great promises.

Owing to their large lateral size and ultrathin thickness, 2D nanomaterials possess ultrahigh specific surface area[130] and thus are ideal candidates for surface-active applications. For instance, ultrathin 2D nanomaterials have been proved to be fascinating platforms for engineering high-efficient catalysts for various kinds of catalytic applications.[131,132] It was found that some of the synthesized ultrathin 2D nanosheets have excellent activities in a number of catalytic processes. In addition, the ultrahigh surface area of 2D nanomaterials also makes them very promising electrode materials for supercapacitors and photodetectors.

References

1. Henglein, A., Small-particle research: Physicochemical properties of extremely small colloidal metal and semiconductor particles. *Chem. Rev.* **1989**, *89* (8), 1861–1873.

2. Trindade, T.; O'Brien, P.; Pickett, N. L., Nanocrystalline semiconductors: Synthesis, properties, and perspectives. *Chem. Mater.* **2001**, *13* (11), 3843–3858.
3. Kuchibhatla, S. V.; Karakoti, A.; Bera, D.; et al. One dimensional nanostructured materials. *Prog. Mater. Sci.* **2007**, *52* (5), 699–913.
4. Bera, D.; Kuiry, S. C.; Seal, S., Synthesis of nanostructured materials using template-assisted electrodeposition. *JOM-US* **2004**, *56* (1), 49–53.
5. Alivisatos, A. P., Perspectives on the physical chemistry of semiconductor nanocrystals. *J. Phys. Chem. A* **1996**, *100* (31), 13226–13239.
6. Bera, D.; Qian, L.; Holloway, P. H., Phosphor quantum dots. *Lumin. Mater. Appl.* **2008**, *25*, 19–27.
7. Bierman, M. J.; Lau, Y. A.; Jin, S., Hyperbranched PbS and PbSe nanowires and the effect of hydrogen gas on their synthesis. *Nano Lett.* **2007**, *7* (9), 2907–2912.
8. Burda, C.; Chen, X.; Narayanan, R.; et al. Chemistry and properties of nanocrystals of different shapes. *Chem. Rev.* **2005**, *105* (4), 1025–1102.
9. Yong, K. T.; Sahoo, Y.; Choudhury, K. R.; et al. Control of the morphology and size of PbS nanowires using gold nanoparticles. *Chem. Mater.* **2006**, *18* (25), 5965–5972.
10. Yong, K. T.; Sahoo, Y.; Zeng, H.; et al. Formation of ZnTe nanowires by oriented attachment. *Chem. Mater.* **2007**, *19* (17), 4108–4110.
11. Moeno, S.; Nyokong, T., The photophysical studies of a mixture of CdTe quantum dots and negatively charged zinc phthalocyanines. *Polyhedron* **2008**, *27* (8), 1953–1958.
12. Wang, Y.; Zheng, J.; Zhang, Z.; et al. CdTe nanocrystals as luminescent probes for detecting ATP, folic acid and L-cysteine in aqueous solution. *Colloid. Surfaces A* **2009**, *342* (1), 102–106.
13. Farokhzad, O. C.; Langer, R., Nanomedicine: Developing smarter therapeutic and diagnostic modalities. *Adv. Drug Deliver. Rev.* **2006**, *58* (14), 1456–1459.
14. Bae, P. K.; Kim, K. N.; Lee, S. J.; et al. The modification of quantum dot probes used for the targeted imaging of his-tagged fusion proteins. *Biomatererias* **2009**, *30* (5), 836–842.
15. Kheng, K.; Besombes, L.; Mariette, H., Excited states and multi-exciton complexes in single CdTe quantum dots. *Physica E* **2005**, *26* (1), 262–266.
16. Weng, J.; Song, X.; Li, L.; et al. Highly luminescent CdTe quantum dots prepared in aqueous phase as an alternative fluorescent probe for cell imaging. *Talanta* **2006**, *70* (2), 397–402.
17. Dong, W.; Guo, L.; Wang, M.; et al. CdTe QDs-based prostate-specific antigen probe for human prostate cancer cell imaging. *J. Lumin.* **2009**, *129* (9), 926–930.
18. Li, Z.; Wang, K.; Tan, W.; et al. Immunofluorescent labeling of cancer cells with quantum dots synthesized in aqueous solution. *Anal. Biochem.* **2006**, *354* (2), 169–174.
19. Hide, F.; DÍaz-GarcÍa, M. A.; Schwartz, B. J.; et al. New developments in the photonic applications of conjugated polymers. *Acc. Chem. Res.* **1997**, *30* (10), 430–436.
20. Forrest, S. R., The path to ubiquitous and low-cost organic electronic appliances on plastic. *Nature* **2004**, *428* (6986), 911–918.
21. Wu, C.; Bull, B.; Szymanski, C.; et al. Multicolor conjugated polymer dots for biological fluorescence imaging. *ACS Nano* **2008**, *2* (11), 2415–2423.
22. Wu, C.; Chiu, D. T., Highly fluorescent semiconducting polymer dots for biology and medicine. *Angew. Chem., Int. Ed.* **2013**, *52* (11), 3086–3109.
23. Feng, L.; Zhu, C.; Yuan, H.; et al. Conjugated polymer nanoparticles: Preparation, properties, functionalization and biological applications. *Chem. Soc. Rev.* **2013**, *42* (16), 6620–6633.
24. Pecher, J.; Mecking, S., Nanoparticles of conjugated polymers. *Chem. Rev.* **2010**, *110* (10), 6260–6279.
25. Pu, K. Y.; Liu, B., Fluorescent conjugated polyelectrolytes for bioimaging. *Adv. Funct. Mater.* **2011**, *21* (18), 3408–3423.
26. Rong, Y.; Wu, C.; Yu, J.; et al. Multicolor fluorescent semiconducting polymer dots with narrow emissions and high brightness. *ACS Nano* **2013**, *7* (1), 376–384.
27. Wu, C.; Szymanski, C.; McNeill, J., Preparation and encapsulation of highly fluorescent conjugated polymer nanoparticles. *Langmuir* **2006**, *22* (7), 2956–2960.
28. Kitagawa, S.; Kitaura, R.; Noro, S. I., Functional porous coordination polymers. *Angew. Chem., Int. Ed.* **2004**, *43* (18), 2334–2375.
29. Li, H.; Eddaoudi, M.; O'Keeffe, M.; et al. Design and synthesis of an exceptionally stable and highly porous metal-organic framework. *Nature* **1999**, *402* (6759), 276–279.
30. Cheetham, A. K.; Rao, C.; Feller, R. K., Structural diversity and chemical trends in hybrid inorganic–organic framework materials. *Chem. Commun.* **2006**, *12* (46), 4780–4795.
31. Xu, C.; Mochizuki, D.; Hashimoto, Y.; et al. Luminescence of ortho-metalated iridium complexes encapsulated in zeolite supercages by the ship-in-a-bottle method. *Eur. J. Inorg. Chem.* **2012**, *2012* (19), 3113–3120.
32. Mukai, S. R.; Shimoda, M.; Lin, L.; et al. Improvement of the preparation method of "ship-in-the-bottle" type 12-molybdophosphoric acid encaged Y-type zeolite catalysts. *Appl. Catal. A-Gen* **2003**, *256* (1), 107–113.
33. Urrego, S.; Serra, E.; Alfredsson, V.; et al. Bottle-around-the-ship: A method to encapsulate enzymes

in ordered mesoporous materials. *Micropor. Mesopor. Mater.* **2010**, *129* (1), 173–178.

34. Zhan, B. Z.; Li, X. Y., A novel 'build-bottle-around-ship' method to encapsulate metalloporphyrins in zeolite-Y. An efficient biomimetic catalyst. *Chem. Commun.* **1998**, *136* (3), 349–350.
35. Herron, N., A cobalt oxygen carrier in zeolite Y. A molecular "ship in a bottle". *Inorg. Chem.* **1986**, *25* (26), 4714–4717.
36. Jin, S.; Son, H. J.; Farha, O. K.; et al. Energy transfer from quantum dots to metal-organic frameworks for enhanced light harvesting. *J. Am. Chem. Soc.* **2013**, *135* (3), 955–958.
37. Somers, R. C.; Bawendi, M. G.; Nocera, D. G., CdSe nanocrystal based chem-/bio-sensors. *Chem. Soc. Rev.* **2007**, *36* (4), 579–591.
38. Aguilera-Sigalat, J.; Casas-Solvas, J. M.; Morant-Miñana, M. C.; et al. Quantum dot/cyclodextrin supramolecular systems based on efficient molecular recognition and their use for sensing. *Chem. Commun.* **2012**, *48* (20), 2573–2575.
39. Wang, C. I.; Periasamy, A. P.; Chang, H. T., Photoluminescent C-dots@RGO probe for sensitive and selective detection of acetylcholine. *Anal. Chem.* **2013**, *85* (6), 3263–3270.
40. Das, A.; Han, Z.; Haghighi, M. G.; et al. Photogeneration of hydrogen from water using CdSe nanocrystals demonstrating the importance of surface exchange. *P. Natl. Acad. Sci.* **2013**, *110* (42), 16716–16723.
41. Michalet, X.; Pinaud, F.; Bentolila, L.; et al. Quantum dots for live cells, in vivo imaging, and diagnostics. *Science* **2005**, *307* (5709), 538–544.
42. Chen, F.; Gerion, D., Fluorescent CdSe/ZnS nanocrystal-peptide conjugates for long-term, nontoxic imaging and nuclear targeting in living cells. *Nano Lett.* **2004**, *4* (10), 1827–1832.
43. Yang, Y. J.; Tao, X.; Hou, Q.; et al. Fluorescent mesoporous silica nanotubes incorporating CdS quantum dots for controlled release of ibuprofen. *Acta Biomater.* **2009**, *5* (9), 3488–3496.
44. Cui, H.; Liu, Y.; Ren, W.; et al. Large scale synthesis of highly crystallized SnO_2quantum dots at room temperature and their high electrochemical performance. *Nanotechnology* **2013**, *24* (34), 345602–345605.
45. Aguilera-Sigalat, J.; Sanchez-SanMartín, J.; Agudelo-Morales, C. E.; et al. Further insight into the photostability of the pyrene fluorophore in halogenated solvents. *ChemPhysChem* **2012**, *13* (3), 835–844.
46. Dasgupta, N. P.; Sun, J.; Liu, C.; et al. 25th anniversary article: Semiconductor nanowires-synthesis, characterization, and applications. *Adv. Mater.* **2014**, *26* (14), 2137–2184.
47. Kempa, T. J.; Day, R. W.; Kim, S. K.; et al. Semiconductor nanowires: A platform for exploring limits and concepts for nano-enabled solar cells. *Energy Environ. Sci.* **2013**, *6* (3), 719–733.
48. Tian, B.; Kempa, T. J.; Lieber, C. M., Single nanowire photovoltaics. *Chem. Soc. Rev.* **2009**, *38* (1), 16–24.
49. Hochbaum, A. I.; Yang, P., Semiconductor nanowires for energy conversion. *Chem. Rev.* **2009**, *110* (1), 527–546.
50. Su, B.; Wu, Y.; Jiang, L., The art of aligning one-dimensional (1D) nanostructures. *Chem. Soc. Rev.* **2012**, *41* (23), 7832–7856.
51. Choi, J. W.; McDonough, J.; Jeong, S.; et al. Stepwise nanopore evolution in one-dimensional nanostructures. *Nano Lett.* **2010**, *10* (4), 1409–1413.
52. Li, Z.; Liu, Z.; Sun, H.; et al. Superstructured assembly of nanocarbons: Fullerenes, nanotubes, and graphene. *Chem. Rev.* **2015**, *115* (15), 7046–7117.
53. Yang, Z.; Ren, J.; Zhang, Z.; et al. Recent advancement of nanostructured carbon for energy applications. *Chem. Rev.* **2015**, *115* (11), 5159–5223.
54. Joshi, R. K.; Schneider, J. J., Assembly of one dimensional inorganic nanostructures into functional 2D and 3D architectures. Synthesis, arrangement and functionality. *Chem. Soc. Rev.* **2012**, *41* (15), 5285–5312.
55. Lei, D.; Benson, J.; Magasinski, A.; et al. Transformation of bulk alloys to oxide nanowires. *Science* **2017**, *355* (6322), 267–271.
56. Zhang, G.; Xiao, X.; Li, B.; et al. Transition metal oxides with one-dimensional/one-dimensional-analogue nanostructures for advanced supercapacitors. *J. Mater. Chem. A* **2017**, *5* (18), 8155–8186.
57. Teo, B. K.; Li, C.; Sun, X.; et al. Silicon-silica nanowires, nanotubes, and biaxial nanowires: Inside, outside, and side-by-side growth of silicon versus silica on zeolite. *Inorg. Chem.* **2003**, *42* (21), 6723–6728.
58. Gates, B.; Mayers, B.; Wu, Y.; et al. Synthesis and characterization of crystalline Ag_2Se nanowires through a template-engaged reaction at room temperature. *Adv. Funct. Mater.* **2002**, *12* (10), 679–686.
59. Jiang, X.; Wang, Y.; Herricks, T.; et al. Ethylene glycol-mediated synthesis of metal oxide nanowires. *J. Mater. Chem. A* **2004**, *14* (4), 695–703.
60. Song, J. H.; Messer, B.; Wu, Y.; et al. MMo_3Se_3 (M = Li^+, Na^+, Rb^+, Cs^+, NMe^{4+}) nanowire formation via cation exchange in organic solution. *J. Am. Chem. Soc.* **2001**, *123* (39), 9714–9715.
61. Mao, Y.; Wong, S. S., General, room-temperature method for the synthesis of isolated as well as arrays of single-crystalline ABO_4-type nanorods. *J. Am. Chem. Soc.* **2004**, *126* (46), 15245–15252.

62. Rao, C. N. R.; Deepak, F.; Gundiah, G.; et al. Inorganic nanowires. *Prog. Solid State Chem.* **2003**, *31* (1), 5–147.
63. Shirakawa, H.; Louis, E. J.; MacDiarmid, A. G.; et al. Synthesis of electrically conducting organic polymers: Halogen derivatives of polyacetylene,$(CH)_x$. *Chem. Commun.* **1977**, *26* (16), 578–580.
64. Chiang, C. K.; Fincher Jr, C.; Park, Y. W.; et al. Electrical conductivity in doped polyacetylene. *Phys. Rev. Lett.* **1977**, *39* (17), 1098–1101.
65. Yin, Z.; Zheng, Q., Controlled synthesis and energy applications of one-dimensional conducting polymer nanostructures: An overview. *Adv. Energy Mater.* **2012**, *2* (2), 179–218.
66. Murphy, C. J.; Sau, T. K.; Gole, A. M.; et al. Anisotropic metal nanoparticles: Synthesis, assembly, and optical applications. *J. Phys. Chem. B* **2005**, *109* (29), 13857–13870.
67. Murphy, C. J.; Gole, A. M.; Hunyadi, S. E.; et al. Chemical sensing and imaging with metallic nanorods. *Chem. Commun.* **2008**, *259* (5), 544–557.
68. Lee, J. Y.; Connor, S. T.; Cui, Y.; et al. Solution-processed metal nanowire mesh transparent electrodes. *Nano Lett.* **2008**, *8* (2), 689–692.
69. Liu, J.; Cao, G.; Yang, Z.; et al. Oriented nanostructures for energy conversion and storage. *ChemSusChem* **2008**, *1* (9), 676–697.
70. Yin, Z.; Wang, B.; Chen, G.; et al. One-dimensional 8-hydroxyquinoline metal complex nanomaterials: Synthesis, optoelectronic properties, and applications. *J. Mater. Sci.* **2011**, *46* (8), 2397–2409.
71. Xia, Y.; Yang, P.; Sun, Y.; et al. One-dimensional nanostructures: Synthesis, characterization, and applications. *Adv. Mater.* **2003**, *15* (5), 353–389.
72. Li, D.; Xia, Y., Electrospinning of nanofibers: Reinventing the wheel. *Adv. Mater.* **2004**, *16* (14), 1151–1170.
73. Liang, H. W.; Liu, S.; Yu, S. H., Controlled synthesis of one-dimensional inorganic nanostructures using pre-existing one-dimensional nanostructures as templates. *Adv. Mater.* **2010**, *22* (35), 3925–3937.
74. Pan, L.; Pu, L.; Shi, Y.; et al. Synthesis of polyaniline nanotubes with a reactive template of manganese oxide. *Adv. Mater.* **2007**, *19* (3), 461–464.
75. Han, M. G.; Foulger, S. H., 1-dimensional structures of poly (3, 4-ethylenedioxythiophene)(PEDOT): A chemical route to tubes, rods, thimbles, and belts. *Chem. Commun.* **2005**, *1* (24), 3092–3094.
76. Han, M. G.; Foulger, S. H., Facile synthesis of poly (3, 4-ethylenedioxythiophene) nanofibers from an aqueous surfactant solution. *Small* **2006**, *2* (10), 1164–1169.
77. Tran, H. D.; Li, D.; Kaner, R. B., One-dimensional conducting polymer nanostructures: Bulk synthesis and applications. *Adv. Mater.* **2009**, *21* (14), 1487–1499.
78. Kim, F. S.; Ren, G.; Jenekhe, S. A., One-dimensional nanostructures of π-conjugated molecular systems: Assembly, properties, and applications from photovoltaics, sensors, and nanophotonics to nanoelectronics. *Chem. Mater.* **2010**, *23* (3), 682–732.
79. Wan, M., A template-free method towards conducting polymer nanostructures. *Adv. Mater.* **2008**, *20* (15), 2926–2932.
80. Kriha, O.; Becker, M.; Lehmann, M.; et al. Connection of hippocampal neurons by magnetically controlled movement of short electrospun polymer fibers-a route to magnetic micromanipulators. *Adv. Mater.* **2007**, *19* (18), 2483–2485.
81. Stoiljkovic, A.; Agarwal, S., Short electrospun fibers by UV cutting method. *Macromol. Mater. Eng.* **2008**, *293* (11), 895–899.
82. Huang, Z. M.; Zhang, Y. Z.; Kotaki, M.; et al. A review on polymer nanofibers by electrospinning and their applications in nanocomposites. *Compos. Sci. Technol.* **2003**, *63* (15), 2223–2253.
83. Greiner, A.; Wendorff, J. H., Cover picture: Electrospinning: A fascinating method for the preparation of ultrathin fibers. *Angew. Chem., Int. Ed.* **2007**, *46* (30), 5633–5633.
84. Burger, C.; Hsiao, B. S.; Chu, B., Nanofibrous materials and their applications. *Annu. Rev. Mater. Res.* **2006**, *36* (1), 333–368.
85. Barnes, C. P.; Sell, S. A.; Boland, E. D.; et al. Nanofiber technology: Designing the next generation of tissue engineering scaffolds. *Adv. Drug Deliver. Rev.* **2007**, *59* (14), 1413–1433.
86. Xie, J.; Li, X.; Xia, Y., Putting electrospun nanofibers to work for biomedical research. *Macromol. Rapid Comm.* **2008**, *29* (22), 1775–1792.
87. Yarin, A.; Zussman, E., Upward needleless electrospinning of multiple nanofibers. *Polymer* **2004**, *45* (9), 2977–2980.
88. Kameoka, J.; Orth, R.; Yang, Y.; et al. A scanning tip electrospinning source for deposition of oriented nanofibres. *Nanotechnology* **2003**, *14* (10), 1124–1131.
89. Sen, R.; Zhao, B.; Perea, D.; et al. Preparation of single-walled carbon nanotube reinforced polystyrene and polyurethane nanofibers and membranes by electrospinning. *Nano Lett.* **2004**, *4* (3), 459–464.
90. Ge, J. J.; Hou, H.; Li, Q.; et al. Assembly of well-aligned multiwalled carbon nanotubes in confined polyacrylonitrile environments: Electrospun composite nanofiber sheets. *J. Am. Chem. Soc.* **2004**, *126* (48), 15754–15761.
91. Hou, H.; Ge, J. J.; Zeng, J.; et al. Electrospun polyacrylonitrile nanofibers containing a high concentration of well-aligned multiwall carbon nanotubes. *Chem. Mater.* **2005**, *17* (5), 967–973.
92. Lu, X.; Zhao, Y.; Wang, C., Fabrication of PbS nanoparticles in polymer-fiber matrices by

electrospinning. *Adv. Mater.* **2005**, *17* (20), 2485–2488.

93. Lu, X.; Mao, H.; Zhang, W.; et al. Synthesis and characterization of CdS nanoparticles in polystyrene microfibers. *Mater. Lett.* **2007**, *61* (11), 2288–2291.
94. Yang, Y.; Wang, H.; Lu, X.; et al. Electrospinning of carbon/CdS coaxial nanofibers with photoluminescence and conductive properties. *Mater. Sci. Eng. B-Adv.* **2007**, *140* (1), 48–52.
95. Bai, J.; Li, Y.; Yang, S.; et al. Synthesis of AgCl/PAN composite nanofibres using an electrospinning method. *Nanotechnology* **2007**, *18* (30), 305601–305607.
96. Dong, F.; Li, Z.; Huang, H.; et al. Fabrication of semiconductor nanostructures on the outer surfaces of polyacrylonitrile nanofibers by in-situ electrospinning. *Mater. Lett.* **2007**, *61* (11), 2556–2559.
97. Lu, X.; Zhao, Y.; Wang, C.; et al. Fabrication of CdS nanorods in PVP fiber matrices by electrospinning. *Macromol. Rapid Comm.* **2005**, *26* (16), 1325–1329.
98. Novoselov, K. S.; Geim, A. K.; Morozov, S. V.; et al. Electric field effect in atomically thin carbon films. *Science* **2004**, *306* (5696), 666–669.
99. Geim, A. K.; Novoselov, K. S., The rise of graphene. *Nat. Mater.* **2007**, *6* (3), 183–191.
100. Chhowalla, M.; Shin, H. S.; Eda, G.; et al. The chemistry of two-dimensional layered transition metal dichalcogenide nanosheets. *Nat. Chem.* **2013**, *5* (4), 263–275.
101. Xu, M.; Liang, T.; Shi, M.; et al. Graphene-like two-dimensional materials. *Chem. Rev.* **2013**, *113* (5), 3766–3798.
102. Nicolosi, V.; Chhowalla, M.; Kanatzidis, M. G.; et al. Liquid exfoliation of layered materials. *Science* **2013**, *340* (6139), 1420–1436.
103. Wang, Q. H.; Kalantar-Zadeh, K.; Kis, A.; et al. Electronics and optoelectronics of two-dimensional transition metal dichalcogenides. *Nat. Nanotechnol.* **2012**, *7* (11), 699–712.
104. Huang, X.; Zeng, Z.; Zhang, H., Metal dichalcogenide nanosheets: Preparation, properties and applications. *Chem. Soc. Rev.* **2013**, *42* (5), 1934–1946.
105. Sun, Y.; Gao, S.; Xie, Y., Atomically-thick two-dimensional crystals: Electronic structure regulation and energy device construction. *Chem. Soc. Rev.* **2014**, *43* (2), 530–546.
106. Naguib, M.; Mashtalir, O.; Carle, J.; et al. Two-dimensional transition metal carbides. *ACS Nano* **2012**, *6* (2), 1322–1331.
107. Liang, L.; Li, K.; Xiao, C.; et al. Vacancy associates-rich ultrathin nanosheets for high performance and flexible nonvolatile memory device. *J. Am. Chem. Soc.* **2015**, *137* (8), 3102–3108.
108. Hu, P.; Wang, L.; Yoon, M.; et al. Highly responsive ultrathin GaS nanosheet photodetectors on rigid and flexible substrates. *Nano Lett.* **2013**, *13* (4), 1649–1654.
109. Song, F.; Hu, X., Exfoliation of layered double hydroxides for enhanced oxygen evolution catalysis. *Nat. Commun.* **2014**, *5* (5), 4477–4486.
110. Voiry, D.; Yamaguchi, H.; Li, J.; et al. Enhanced catalytic activity in strained chemically exfoliated WS_2 nanosheets for hydrogen evolution. *Nat. Mater.* **2013**, *12* (9), 850–855.
111. Yang, S.; Gong, Y.; Zhang, J.; et al. Exfoliated graphitic carbon nitride nanosheets as efficient catalysts for hydrogen evolution under visible light. *Adv. Mater.* **2013**, *25* (17), 2452–2456.
112. Du, G.; Guo, Z.; Wang, S.; et al. Superior stability and high capacity of restacked molybdenum disulfide as anode material for lithium ion batteries. *Chem. Commun.* **2010**, *46* (7), 1106–1108.
113. Tsai, M. L.; Su, S. H.; Chang, J. K.; et al. Monolayer MoS_2 heterojunction solar cells. *ACS Nano* **2014**, *8* (8), 8317–8322.
114. Zhu, C.; Zeng, Z.; Li, H.; et al. Single-layer MoS_2-based nanoprobes for homogeneous detection of biomolecules. *J. Am. Chem. Soc.* **2013**, *135* (16), 5998–6001.
115. Perkins, F. K.; Friedman, A. L.; Cobas, E.; et al. Chemical vapor sensing with monolayer MoS_2. *Nano Lett.* **2013**, *13* (2), 668–673.
116. Cheng, L.; Liu, J.; Gu, X.; et al. PEGylated WS_2 nanosheets as a multifunctional theranostic agent for in vivo dual-modal CT/photoacoustic imaging guided photothermal therapy. *Adv. Mater.* **2014**, *26* (12), 1886–1893.
117. Sun, Y.; Sun, Z.; Gao, S.; et al. Fabrication of flexible and freestanding zinc chalcogenide single layers. *Nat. Commun.* **2012**, *3* (3), 1057–1063.
118. Zhou, M.; Lou, X. W. D.; Xie, Y., Two-dimensional nanosheets for photoelectrochemical water splitting: Possibilities and opportunities. *Nano Today* **2013**, *8* (6), 598–618.
119. Ten Elshof, J. E.; Yuan, H.; Gonzalez Rodriguez, P., Two-dimensional metal oxide and metal hydroxide nanosheets: Synthesis, controlled assembly and applications in energy conversion and storage. *Adv. Energy Mater.* **2016**, *6* (23) 1600355–1600389.
120. Huang, X.; Zeng, Z.; Bao, S.; et al. Solution-phase epitaxial growth of noble metal nanostructures on dispersible single-layer molybdenum disulfide nanosheets. *Nat. Commun.* **2013**, *4* (4), 1444–1452.
121. Zeng, Z.; Tan, C.; Huang, X.; et al. Growth of noble metal nanoparticles on single-layer TiS_2 and TaS_2 nanosheets for hydrogen evolution reaction. *Energy Environ. Sci.* **2014**, *7* (2), 797–803.
122. Shi, Y.; Huang, J. K.; Jin, L.; et al. Selective decoration of Au nanoparticles on monolayer MoS_2 single crystals. *Sci. Rep.* **2013**, *3* (5), 1839–1846.
123. Forticaux, A.; Hacialioglu, S.; DeGrave, J. P.; et al. Three-dimensional mesoscale heterostructures of

ZnO nanowire arrays epitaxially grown on $CuGaO_2$ nanoplates as individual diodes. *ACS Nano* **2013**, *7* (9), 8224–8232.
124. Kim, Y. J.; Yoo, H.; Lee, C. H.; et al. Position-and morphology-controlled ZnO nanostructures grown on graphene layers. *Adv. Mater.* **2012**, *24* (41), 5565–5569.
125. Chung, K.; Lee, C. H.; Yi, G. C., Transferable GaN layers grown on ZnO-coated graphene layers for optoelectronic devices. *Science* **2010**, *330* (6004), 655–657.
126. Shi, Y.; Zhou, W.; Lu, A. Y.; et al. Van der Waals epitaxy of MoS_2 layers using graphene as growth templates. *Nano Lett.* **2012**, *12* (6), 2784–2791.
127. Mohseni, P. K.; Behnam, A.; Wood, J. D.; et al. In_xGa_{1-x} as nanowire growth on graphene: Van der Waals epitaxy induced phase segregation. *Nano Lett.* **2013**, *13* (3), 1153–1161.
128. Li, L.; Chen, Z.; Hu, Y.; et al. Single-layer single-crystalline SnSe nanosheets. *J. Am. Chem. Soc.* **2013**, *135* (4), 1213–1216.
129. Huang, X.; Tang, S.; Mu, X.; et al. Freestanding palladium nanosheets with plasmonic and catalytic properties. *Nat. Nanotechnol.* **2011**, *6* (1), 28–32.
130. Stoller, M. D.; Park, S.; Zhu, Y.; et al. Graphene-based ultracapacitors. *Nano Lett.* **2008**, *8* (10), 3498–3502.
131. Morales-Guio, C. G.; Stern, L. A.; Hu, X., Nanostructured hydrotreating catalysts for electrochemical hydrogen evolution. *Chem. Soc. Rev.* **2014**, *43* (18), 6555–6569.
132. Sun, Y.; Gao, S.; Lei, F.; et al. Atomically-thin two-dimensional sheets for understanding active sites in catalysis. *Chem. Soc. Rev.* **2015**, *44* (3), 623–636.

3

Low-Dimensional Hybrid Nanomaterials

Mashkoor Ahmad
Nanomaterials Research Group (NRG)
Physics Division PINSTECH

Saira Naz
Nanomaterials Research Group (NRG)
University of Peshawar

Muhammad Hussain
Nanomaterials Research Group (NRG)
University of the Punjab

3.1 Low-Dimensional Materials

Low-dimensional materials refer to a new class of material with reduced dimensionality, i.e., with one or more physical dimensions constrained to the nanometer scale. Zero-dimensional (0D) fullerenes, one-dimensional (1D) nanotubes, and two-dimensional (2D) nanosheets represent typical examples of such materials. When compared to three-dimensional (3D) bulk substances, low-dimensional structures are anticipated to exhibit new properties due to quantum confinement and/or surface and interfacial effects. Therefore, in recent years, materials scientists have drawn particular attention to these nanosystems to make sense of their unusual physical and chemical properties that can promote novel applications in engineering (Figure 3.1).

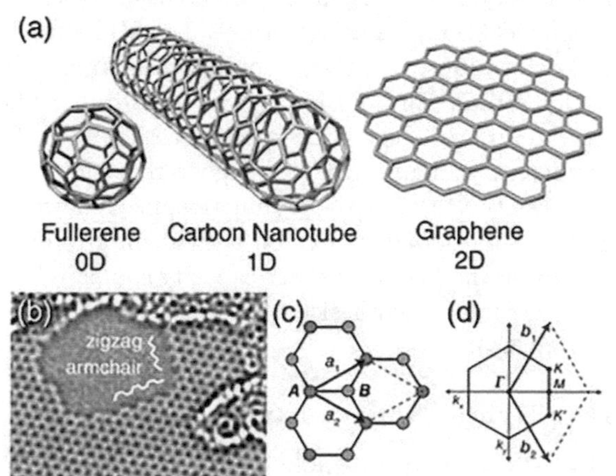

FIGURE 3.1 Honeycomb lattice. Allotropes of carbon based on (a) 0D fullerene, 1D armchair nanotube and 2D graphene sheet; (b) High resolution TEM image of suspended graphene; (c) Hexagonal crystal structure of graphene; and (d) Reciprocal lattice [1].

3.2 Hybrid Nanomaterials

Hybrid nanostructures have excellent properties in their pure or single form. The heterostructures consisting of different materials have attracted considerable attention with respect to the realization of multicomponent system of functional electronic devices.

3.2.1 Metal Oxide-Based Hybrid Nanomaterials

ZnO/ZnS Hybrid Nanomaterials

ZnO and ZnS, well-known direct bandgap II–VI semiconductors, are promising materials for fabricating photoelectrical, photochemical, photonic, optical, and electronic devices. ZnO/ZnS heterojunctions make possible to tailor the optical properties of nanostructures. These advantages make heterostructures one of the most promising candidates for the exploration of new applications in nanoscale heterostructured electronic devices, electromechanical systems, optoelectronics, field emitters, and light-emitting diodes [2,3]. Extensive effort has focused on the synthesis of ZnS and ZnO heterostructures such as ZnO/ZnS heterojunction nanoribbons, ZnO/ZnS nanobelt and nanowire heterostructures, and ZnO/ZnS coaxial cables [4] (Figure 3.2).

ZnO/ZnS hybrid nanostructures consist of a large number of 1D saw-like curved and straight nanostructures having one side flat and another side with teeth. Typically, saw-like nanostructures have been grown with a diameter of ≈100 nm, and the length extended to several hundred micrometers. ZnO/ZnS biaxial nanobelt heterostructures are grown in a horizontal furnace using chemical vapor deposition technique. These nanostructures have a diameter

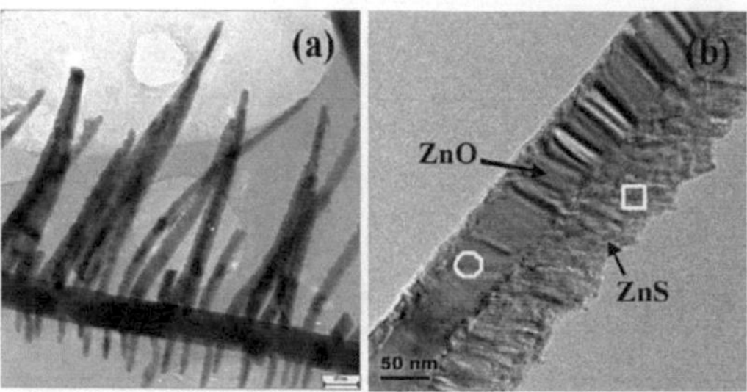

FIGURE 3.2 (a) Low-magnification transmission electron microscopic (TEM) images of brushlike heterostructures and (b) magnified image of single-branch nanowires [5].

varying from several tens on nanometers to 100 nm, and the length of these hybrid structures is up to tens of micrometers. The product is composed of two heterostructures, i.e., heterocrystalline ZnS/single-crystalline ZnO biaxial nanobelts. The documented visible spectrum in ZnO and ZnS lies between 400 and 650, and the band-edge emission is 368–390 nm for ZnO and 330–345 nm for ZnS [6,7]. Hence, the fabricated ZnO/ZnS composite nanomaterials are investigated to be of high promise for novel optoelectronic nanoscale devices [8]. The ZnO/ZnS nanowire heterostructure nanowire arrays have been synthesized, under controlled conditions, by thermal evaporation of ZnS powder in vacuum tube furnace. The vertically aligned nanowire arrays have been grown on the ZnS buffer layers. The structural characterization shows that the obtained hybrid nanowire arrays are of single-crystal structures of wurtzite ZnO and ZnS [9].

Indium-Doped ZnO Nanowires

ZnO is one of the most promising materials for the fabrication of optoelectronic devices due to its superior conducting properties based on oxygen vacancies; it has been investigated as transparent conducting and piezoelectric materials for fabricating solar cells, electrodes, and sensors. Doping in semiconductor offers an effective method to adjust the electrical, optical, and magnetic properties, crucial for their practical application [10,11]. Indium is recognized as one of the most efficient elements used to enhance the optoelectronic properties of ZnO. The presence of In dopant in ZnO wires induces a dramatic decrease in the electrical resistivity of nanowires, which makes it potentially applicable for optoelectronics devices [12]. Because In-doped ZnO has excellent optical transmission, high electrical conductivity, unique chemical stability, thermal stability than undoped ZnO and can be used in solar cells, transparent conducting electronics, etc [13] (Figure 3.3).

Au-Functionalized ZnO Nanoflowers

Oxide semiconductor/metal nanocomposites are functional materials that have potentials in optoelectronics, drug delivery, environmental monitoring, control of chemical processes, photocatalysis, energy storage, and biomedical diagnosis applications. ZnO nanostructures decorated with

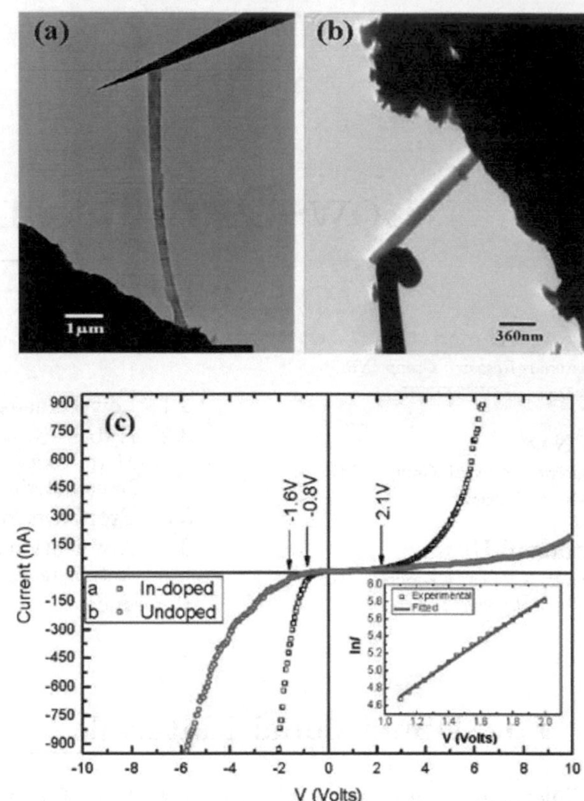

FIGURE 3.3 TEM images of (a) in-doped, (b) undoped ZnO NWs, and (c) corresponding I-V curves for the ZnO nanowires [12].

noble metal nanoparticles have been already found their widespread use in a range of applications, such as ZnO/Au nanocomposite for DNA and protein detection, ZnO/Ag for photocatalysis, ZnO/Pt for cholesterol detection [14,15]. Metal nanoparticles may alter the photocatalytic sites, and the noble metal itself can act as a catalyst to degrade the contaminants [16]. Various ZnO nanostructures have been synthesized, including 1D nanorods and nanowires, 2D nanobelts and nanosheets, 3D hierarchical nanospheres and hierarchical complex architectures. The 3D hierarchical flower-like structures with a large void space are more advantageous for sensors, lithium ion battery (LIB), and photocatalysis because of their large specific surface area, facile mass transport in materials, and higher reversible capacities and safety [12,17]. Au-NP-functionalized flower-like-ZnO hybrid nanostructures exhibit a great improvement in photocatalytic and lithium storage capacity compared to the pure ZnO structure synthesized with the same conditions [18] (Figure 3.4).

TiO_2/Ag Hybrid Nanostructures

The size, shape, and assembly of various anatase TiO_2 nanostructures have been intensively investigated to optimize the Li ion storage properties [19,20]. It was found that Li ion insertion was favored on the [001] surface of anatase TiO_2, which has a more open structure and short path for lithium ion diffusion along this direction [21,22]. It is

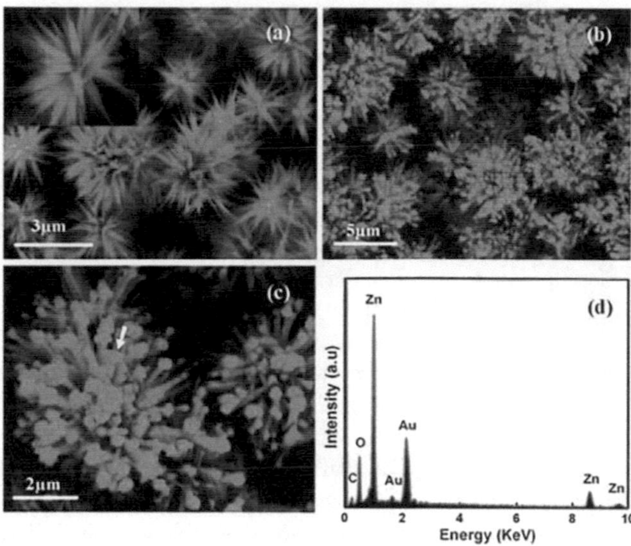

FIGURE 3.4 Field emission scanning electron microscopy (FESEM) images of (a–c) low and high magnification image of hierarchical flower-like Au/ZnO hybrid nanostructure, and (d) corresponding energy dispersive X-ray spectroscopy (EDS) spectrum [18].

thus believed that the synthesis of anatase TiO_2 exposed with [001] facets is beneficial for improving the performance of LIBs. Growth of nano/microanatase TiO_2 crystals with exposed high-energy [001] facets is highly challenging. A variety of TiO_2 nanocomposites such as Ag/TiO_2, Au/TiO_2, Pt/TiO_2, Pd/TiO_2, Co/TiO_2 have been reported for various applications, including CO oxidation, hydrosulfurization, degradation of organic toxic compounds, and energy harvesting [23,24]. Anatase Ag/TiO_2 nanostructures have been investigated for photocatalytic and electrochemical activities, and it was found that Ag/TiO_2 nanocomposite exhibits great prospect for the development of an efficient environmental remediator and non-enzymatic biosensor [25].

CuO/Au Hybrid Nanoparticles

Recently, core/shell nanoparticles have attracted considerable attention because they can give an improved tuning factor via the interface of the two nanoparticles, which leads to an enhancement of the catalytic activity and electrical properties [26]. Au has low toxicity, and particularly, the properties of Au nanoparticles are affected significantly by the presence of metal oxide nanoparticles. Au/CuO interfaces are assembled through heterostructure formations, which significantly change the catalytic properties. Au/CuO core/shell nanoparticles are crucial and hold great promise for a variety of physical and chemical processes, such as biomedical imaging, photothermal therapy, photocatalysis, optoelectronics, and plasmon-enhanced spectroscopies [26–28]. Au nanoparticles decorated with CuO nanowire arrays have been found to exhibit enhanced photocatalytic performance when compared to CuO nanowire arrays and commercial CuO powders [29]. The sensing activity of semiconductor metal oxides can be advantageously tailored by loading with noble metal nanoparticles (Pd, Pt, Au), resulting in an improved charge carrier separation with beneficial effects on sensitivity, response time, and operational temperatures. CuO/TiO_2 nanocomposites represent one of the most appealing systems due to the synergistic effect between the single-oxide properties, such as the low bandgap of p-type CuO_4 and the high reactivity of n-type TiO_2. The functional response was appreciably enhanced upon the introduction of Au nanoparticles, highlighting the present CuO/TiO_2/Au nanosystems as appealing candidates in nanotechnological applications [30] (Figure 3.5).

3.2.2 Metal: Metal-Based Hybrid Nanomaterials

Ni–Cu–Co Nanowires

1D nanowires of Cu, Co, Ni, Fe and their alloys like Ni–Cu, Ni–Co, Cu–Ni–Co, NiFe, CoFe exhibit unique and tunable magnetic properties either by controlling the morphology and crystalline structure or by tailoring the magnetic anisotropy [31]. Ni–Cu–Co composite magnetic nanowires have been successfully synthesized by electrochemical deposition with a strong diamagnetic contribution [32]. Magnetic nanowires have potential applications in the field of magnetic nanodevices, high-density patterned recording media, magneto-resistive nanosensors, and nonreciprocal microwave absorption devices, and offer a lot of applications in nanoelectronics as well [31,33]. The structure and growth direction of these nanowires need to be controlled because magnetic properties are strongly dependent on these parameters.

Au–Ni–Au Nanowires

1D magnetic nanowires have been extensively used for magnetic recording and spintronic applications. They are very promising candidates for a variety of applications in electronics, medical industries, and high-density magnetic storage devices [34,35]. Compared to single-segment nanowires, multi-segment nanowires offer more diverse

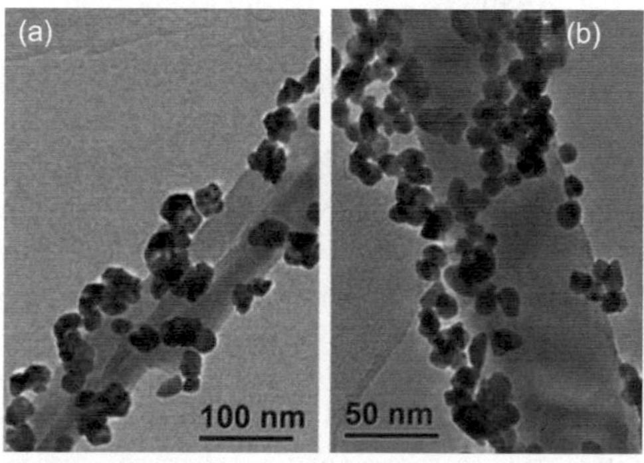

FIGURE 3.5 TEM images of the Au NPs decorated with CuO nanowires (a) low (b) high magnified [29].

functionality. Composition-modulated nanowires have been frequently reported, whereas diameter-modulated nanowires consisting of a bead-like structure have been rarely reported [36]. Diameter-modulated multilayer nanowires have interesting ferromagnetic ordering. The nanowires consisting of nanomagnets give rise to the interesting magnetic properties at room temperature and at lower temperatures that are different from their bulk counterparts. Arrays of segmented nanowires with a diameter of few hundred nanometers and length of few microns are suitable candidates for a variety of applications in recent technology [35]. Segmented Au–Ni–Au nanowires have been fabricated by electrochemical deposition in alumina templates with a diameter of 100 nm and Ni segment length of around 800 nm. The nanowires prepared using this technique have potential applications in ultra-high-density magnetic storage devices. Moreover, shape anisotropy also plays a dominant role in segmented nanowires [37]. Magnetic nanostructures can be produced using different techniques, among which electrochemical deposition is very common and easily achievable at room temperature. These wires can be electrodeposited by alternately changing the electrolytes during deposition, and the length of each segment can be tuned by controlling the time and charge deposited during the deposition process [38].

3.2.3 Carbon-Based Hybrid Nanomaterials

Carbon nanomaterials such as graphene, carbon nanotubes (CNTs), and fullerene have drawn great interest during the last decade due to their unique electrical, mechanical, thermal, and optical properties [39]. Nanostructured carbon materials have been well demonstrated for their superior electric conductivity, strong mechanical properties, unique optical properties, extremely high thermal conductivity, enriched nanocarbon surface chemistry, which make them being one of the most preferred substitute materials in the field of electronic devices, optical modulators, solar cells, thermal management devices, and biocompatible devices. By further combining these carbon nanomaterials with external nanocrystals of metal, semiconductor, ceramic, and quantum dot, it is possible to achieve the evolution of nanoscale hybrids and heterostructures with mutually improved performances and multifunctionalities [40–42].

The modification of nanomaterials has been exploited in recent decades, especially by nanochemists and nanotechnologists. The synthesis of carbon nanomaterials and alteration of their surfaces provide an opportunity to bolster scientific efforts in order to create a more resourceful world community capable of confronting its challenges. Functionalized carbon nanomaterials have unlocked an array of applications across a wide spectrum of fields. Among carbon nanomaterials, CNTs and graphene have many superior properties such as low weight, very high aspect ratio, high electrical conductivity, and extraordinary mechanical, optical, and thermal properties [43]. The application of carbon nanomaterials to various fields has been assisted by functionalization of their surfaces. The unparalleled physiochemical features of these functionalized nanomaterials have been exploited for energy, cancer treatment, antiviral drug development, drug transportation in biological systems, and biotechnological applications [44–47]. Nonfunctionalized carbon nanomaterials possess some drawbacks, including the tendency to form stable aggregates or bundles due to very strong intermolecular interactions such as van der Waals forces, dipole–dipole interactions. Strategies for the combination of carbon nanomaterials with foreign metallic, semiconducting, or ceramic nanostructures mainly include covalent methods (chemical bonding) and non-covalent methods (physical absorption) as well as the in situ synthesis approaches such as chemical reduction, electrostatic force-directed assembly, and electrodeposition. In situ synthesis strategy has been considered to be of importance for developing heterostructures with novel structural design, desirable compatibility, and multiple functionalities [48] (Figure 3.6).

Most of the functionalization approaches developed at present can be categorized into two categories: covalent and non-covalent functionalization. Covalent functionalization is to alter the state of bond connectivity. In this method, the translational symmetry of the CNTs and graphene is disrupted by changing sp^2 carbon atoms to sp^3 carbon atoms, and both the electronic and transport properties are influenced. In covalent functionalization, the functional units form a covalent linkage with a skeleton of the graphene or CNTs. H-bonding and π-π stacking play an important role in non-covalent functionalization and enhance the solubility and assembly without effecting π-π conjugation of the skeleton of the CNTs or graphene [49,50] (Figure 3.7).

The complex heterostructures based on the non-covalent reaction (π-π stacking) are attractive because both the nanocarbon-supporting materials are able to maintain their non-damaged intrinsic properties. Covalent strategies were widely proposed as reliable and promising approaches in the synthesis of complex heterostructures. The structural design and fabrication approaches of nanocarbon-based hybrid architectures or heterostructures have opened new sights for research and applications of novel carbon nanomaterials (e.g., graphene, CNTs, and fullerenes) in fields

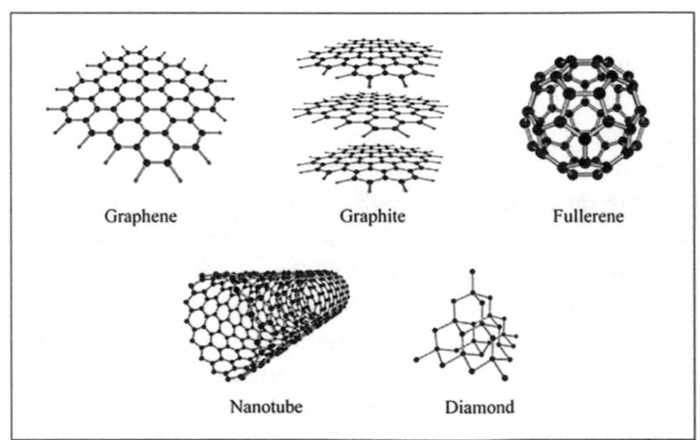

FIGURE 3.6 Different forms of carbon nanomaterials.

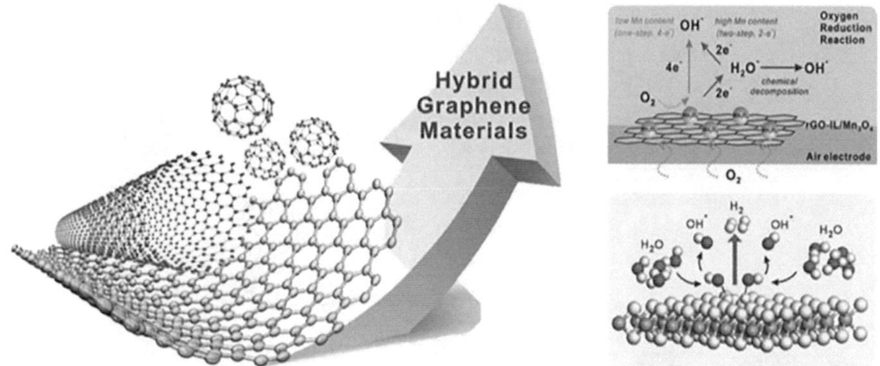

FIGURE 3.7 Synthesis and assembly of hybrid graphene materials for catalytic electrode.

of electronic/chemical sensors, catalysis, and energy conversion/storage [51] (Figure 3.8).

Owing to their electrical, magnetic, and catalytic properties, metal nanoparticles with well-defined shapes and sizes have received considerable attention. Metal catalysts supported by carbon nanomaterials have excellent characteristic properties, due to their high surface area, thermal stability, and porous surface, which act as a scaffold to prevent the agglomeration of the immobilized metal particles. Some of the important and efficient hybrid nanostructures of graphene and CNTs will be discussed in detail [52,53].

Fe_3O_4/Graphene

Graphene can host the nanostructured materials by providing a support for anchoring nanoparticles and work as a highly conductive matrix. Consequently, great efforts have been devoted to develop new synthetic methods of metal oxide/graphene composites. Graphene/magnetic nanoparticle composite was tailored by a number of effective methods. Iron oxide nanoparticles, mostly magnetite (Fe_3O_4) and maghemite (Fe_2O_3), are highly attractive for diverse applications in medical, biological, or even ecological fields. Besides, iron oxide is an attractive material because of its natural abundance, its low cost, and being environmentally friendly material. The association of the magnetic properties of iron oxide nanoparticles with the characteristics of carbon nanomaterials opens new possibilities in the development of multimodal imaging and therapy platforms. Hydrophilic graphene nanosheet-based magnetic nanocomposites were superparamagnetic. They responded quickly to an external magnetic field and exhibited an efficient adsorption toward methylene blue as a cationic dye without leaching [54]. Carbon-decorated single-crystalline Fe_3O_4 nanowires, Fe_3O_4-based Cu nanoarchitecture, magnetite/carbon core/shell nanorods, and iron oxide-based nanotube arrays have been used to improve the electrochemical performance of iron oxides [55]. The presence of the magnetic nanoparticles on the nanotube surface confers magnetic properties to the CNTs. The magnetic nanoparticle/CNT hybrids have the potential as a contrast agent for magnetic resonance imaging (MRI) and their interactions with cells. They are internalized into tumor cells without showing cytotoxicity [56] (Figures 3.9 and 3.10).

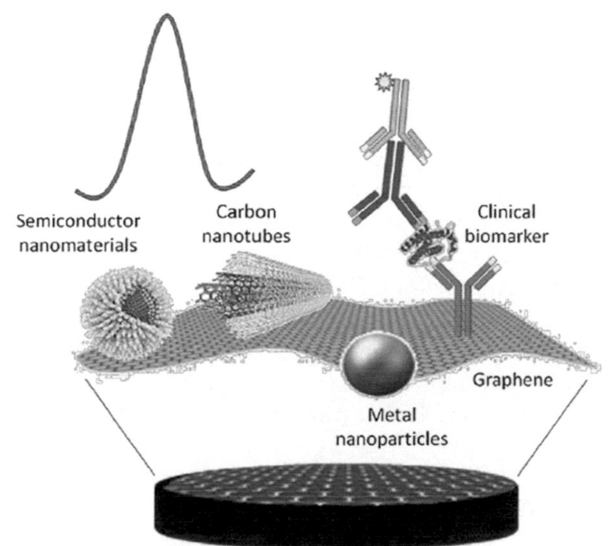

FIGURE 3.8 Carbon-based hybrid nanomaterials.

Superparamagnetic iron oxide nanoparticles with graphene can be used for numerous applications such as magnetic fluids, catalysis activity, magnetic storage media and MRI, tissue repair, hyperthermia, drug delivery, and cell separation [59,60]. All of the biomedical applications require the nanoparticles with high magnetization values, a size smaller than 100 nm, and a narrow particle size distribution. These applications also need a peculiar surface coating of the magnetic particles, which has to be nontoxic and biocompatible and must also allow for a targetable delivery with particle localization in a specific area. Such magnetic nanoparticles can bind to drugs, proteins, enzymes, antibodies, or nucleotides and can be directed to an organ, tissue, or tumor using an external magnetic field. A number of approaches have been described to produce magnetic nanoparticles [61]. Graphene with Fe_3O_4 nanoparticles can produce synergistic effects, resulting in an increased exposed surface area for the adsorptive degradation of organic pollutants [62].

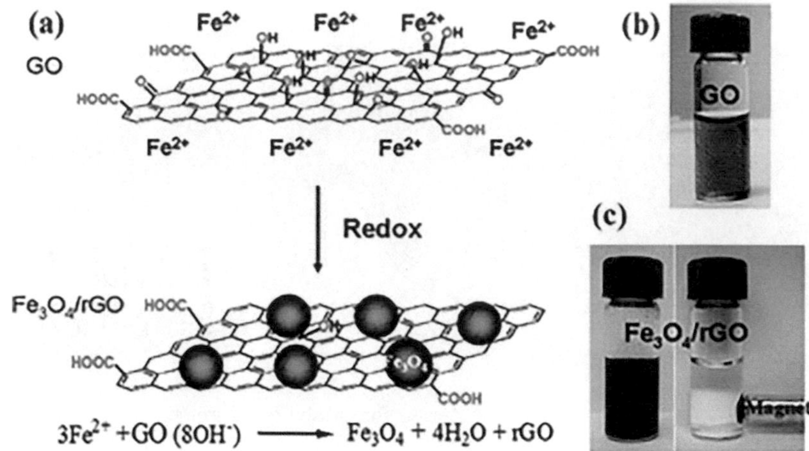

FIGURE 3.9 (a) A schematic representation of the preparation route to Fe_3O_4/rGO via redox reaction between GO and Fe^{2+}. Photos showing a water/NH4OH (pH = 9) solution of Fe_3O_4/rGO (b) before and (c, left panel) after the redox reaction with Fe^{2+}, and (c, right panel) with an applied magnet [57].

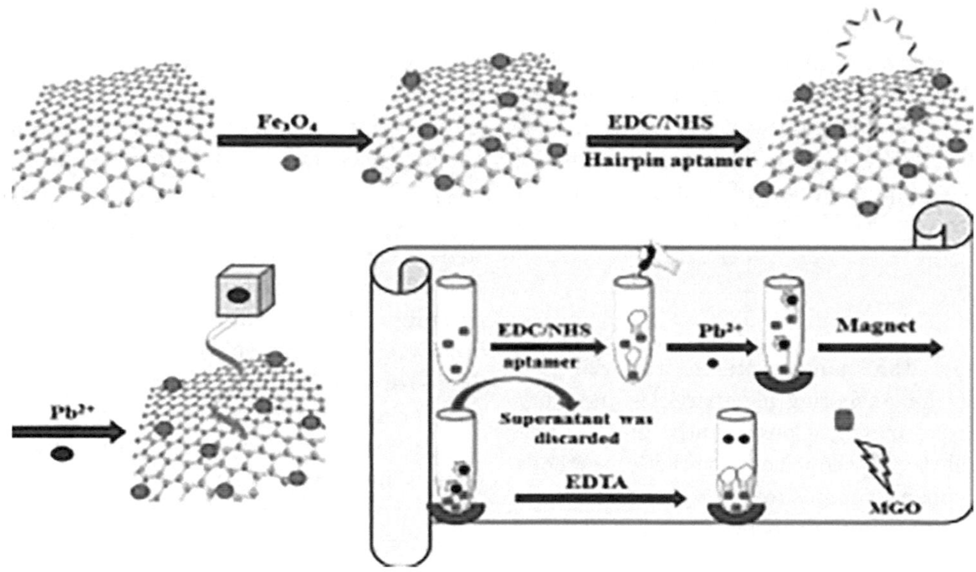

FIGURE 3.10 Functionalized Fe_3O_4/graphene oxide nanocomposites [58].

Pt/CNTs

Because of the 1D structure, CNTs can be easily dispersed or aligned in the composites by applying external forces through techniques such as mechanical stretching, electrical fields, magnetic fields, and spinning processes. Graphene and CNTs may be perfect substitutes as supporting materials for noble metal catalysts. The multiwalled CNTs (MWCNTs) with Pt nanoparticles using ethylene glycol as a reducing agent were developed with good analytical performance when used for the electrochemical detection of tartrazine [63]. An amperometric biosensor for sensitive and selective detection of glucose has been constructed by using highly dispersed Pt nanoparticles supported on CNTs (Pt-MWCNTs) as sensing interface is promising for the fabrication of nonenzymatic glucose sensors [64].

Platinum nanoparticles with a diameter of 2–3 nm in combination with single-walled CNTs (SWCNTs) were fabricated for electrochemical sensors with remarkably improved sensitivity toward hydrogen peroxide [65]. Platinum nanoparticles supported on acid-treated MWCNTs (Pt/MWCNT) hybrid catalyst were annealed at different temperatures to study the effects of heat treatment on the particle size, surface morphology, and the oxygen reduction reaction (ORR) activity; the specific activity for oxygen reduction on Pt/MWCNT catalysts was found higher than that of Pt/C [66]. The aptasensor developed by gold/platinum nanoparticles (Au/PtNPs) with acid-oxidized CNTs (aptamer-based electrochemical biosensor) exhibited good specificity, stability, and reproducibility [67] (Figure 3.11).

SnS_2/CNTs

Semiconductor nanomaterials have grabbed increasing attention in the electrochemical sensing fields due to

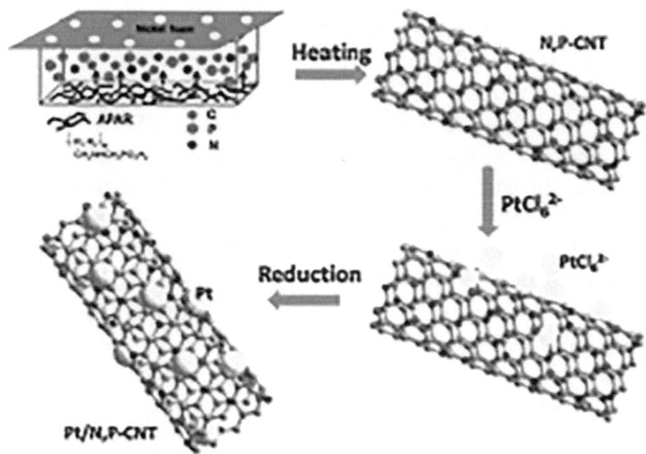

FIGURE 3.11 Pt/CNTs as a catalyst [68].

their excellent properties. SnS_2 is an n-type semiconductor with a bandgap of 2.18–2.44 eV having good stability in acid and neutral aqueous solutions, which makes it a promising visible light-sensitive photocatalyst [69]. In addition, because of fascinating electrical, optical, and gas-sensing properties, SnS_2 has been used in solar cells and optoelectronic devices, LIBs, and gas sensor [70]. CNTs and CNT-based composites have been used to modify electrode for electrochemical sensing of hydrogen peroxide, glucose, dopamine, uric acid, and ascorbic acid [71]. MWCNTs wrapped with nanoflake-like SnS_2 (MWCNTs/SnS_2) nanohybrid-modified electrode showed excellent electrocatalytic activity for hydrogen peroxide and oxygen [72].

SnO_2 anode can deliver a significant capacity at low potentials. Tin sulfide exists in various forms such as SnS, SnS_2, Sn_2S_3, Sn_3S_4, and Sn_4S_5, among which SnS_2 is the most common. However, pure SnS_2 anode presents inferior cyclic stability and poor rate capability because of its low conductivity, sluggish kinetics, and severe volume fluctuation. SnS_2/carbon composites inherit the advantage from both nanostructured SnS_2 and carbonaceous materials, with abundant active sites, short diffusion path, and enhanced electrical conductivity. The widely used carbon matrices include MWCNTs, graphene, carbon coating, and porous carbons. CNT and graphene, which have high electrical conductivity and good mechanical property, are also attractive carbon matrices to support SnS_2 [73] (Figure 3.12).

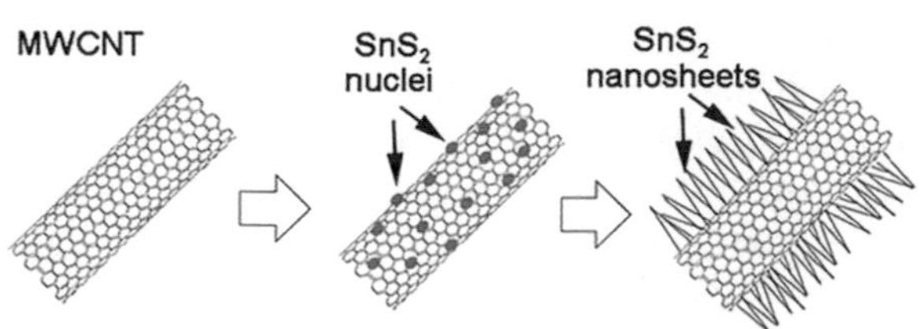

FIGURE 3.12 Carbon nanotubes anchored with SnS_2 as high-performance anode materials for LIBs [74].

$NiCo_2O_4$/Graphene

Among many metal oxides/hydroxides, $Co(OH)_2$ and $Ni(OH)_2$ are the best candidates as a pseudocapacitive electrode material with very high specific capacitances, owing to their layered structures with large interlayer spacing and characteristic redox reaction. Mixed metal hydroxide [$Co_xNi_{1-x}(OH)_2$] is advantageous in terms of environmental friendliness, low cost, and abundance in nature; therefore, it is considered a promising lucrative pseudocapacitive material in supercapacitors. It has been reported that the $Co_xNi_{1-x}(OH)_2$ demonstrates improved electrical conductivity along with electrochemical activity compared to solitary nickel hydroxide or cobalt hydroxide. $Co_xNi_{1-x}(OH)_2$ offers more affluent redox reactions, including contributions from both nickel and cobalt ions, than the solitary nickel hydroxide and cobalt hydroxide [75]. The ever-increasing energy demand has inspired current research to develop new high-performance electrode materials with various morphologies, from micro- to nanoscales, for LIBs. Binary metal oxides, such as $CuCo_2O_4$, $MnCo_2O_4$, $ZnCo_2O_4$, $NiCo_2O_4$, and $ZnMn_2O_4$, have also been reported as anode materials for LIBs [76]. Among different binary metal oxides, $NiCo_2O_4$ is a very promising electrode material because of its high theoretical capacity (890 mAh/g). More importantly, it has been reported that $NiCo_2O_4$ has much higher electrical conductivity and electrochemical performances than nickel oxides and cobalt oxides [77]. Because of the synergistic effect from higher electronic conductivity and porous structure, $NiCo_2O_4$ nanoflakes and nanobelts show much better electrochemical performances. Owing to its intriguing properties, it is a promising material for diverse applications, such as photodetectors, electrocatalytic water splitters, supercapacitors, and LIBs in order to develop high-performance LIBs [78–80]. Large volume changes and stresses commonly happen for $NiCo_2O_4$ during the Li-ion charge/discharge cycling, resulting in aggregation or pulverization of electrode materials. These imperfections partly induce a large increase in contact resistance and significant capacity fade, thereby limiting the commercial applications. Various methods have been developed to solve the problem of large volume changes in electrode materials. One effective way is to use carbonaceous materials, where the active materials adhered to or encapsulated in the carbonaceous materials [81,82] (Figure 3.13).

Co_3O_4/Graphene

Co_3O_4 attracts an extensive interest due to its high theoretical capacity (890 mAh/g), more than two times larger than that of graphite (372 mAh/g), which is expected to meet the requirements of future energy storage system. Graphene in hybrid with Co_3O_4 can be used to improve the electrochemical performance and can be used as an electrode material for LIBs. Graphene not only provides support for anchoring well-dispersed Co_3O_4 NPs and works as a highly conductive matrix for enabling good contact between them but also effectively prevents the volume

FIGURE 3.13 $NiCo_2O_4$ hexagonal nanoplates anchored on reduced graphene oxide sheets with enhanced electrocatalytic activity and stability for methanol and water oxidation [80].

expansion/contraction and aggregation of NPs during Li charge/discharge process [83]. Furthermore, the rGO-encapsulated Co_3O_4, referred to as rGO@Co_3O_4, showed higher capacity than the mixed Co_3O_4/rGO composite or pure Co_3O_4 and exhibited excellent cycling stability. Hybrid porous nanowire arrays composed of strongly interacting Co_3O_4 and carbon can be directly used as the working electrode for oxygen evolution reaction without employing extra substrates or binders [84]. Co_3O_4 1D nanostructure arrays have been grown firmly on insulating substrates, such as glass slides and ceramics, which is quite convenient for the construction of gas sensor devices without any extra electrode preparation process. The metal substrate-supported Co_3O_4 arrays could act as a promising electrode material and be straightforwardly integrated into electronic and electrochemical nanodevices [85]. The peapod-like Co_3O_4@carbon nanotube array (Co_3O_4@CNT) electrodes have high surface areas and large pore sizes, an excellent rate capacity, and cycling performance [86].

Polymer-Based Hybrid Nanomaterials

Polymer nanocomposites are advanced functional materials composed of nanomaterials as fillers dispersed inside the polymer matrix or coated by the polymer. The resulting material has combined suitable properties of both of its constituents. Carbon-based nanomaterials are known to be excellent candidates as filler materials because of their excellent mechanical, thermal, and optical properties. Most notably, the use of π-electron-rich polymers as matrices results in a more stable polymer–graphene dispersion due to its ability to form π-stacking. Functionalization plays a crucial role in developing high-performance polymer composites with CNTs and graphene. When introducing the individual nanomaterial into a polymer matrix, it is important to achieve thorough dispersion and alignment of the carbon nanomaterials (graphene and CNTs) as well as strong interfacial interactions between graphene/CNTs and the polymer in order to improve the load transfer across the nanofiller/polymer matrix interface. The functionalization has been considered one of the best approaches to prevent the aggregation and restacking of graphene and CNTs due to strong van der Waals forces. In fact, the functionalization of CNTs or graphene is a prerequisite in order to take advantage of most of the properties which enable facile fabrication of novel nanomaterials and nanodevices [87]. The functionalization of carbon nanomaterials with the requisite moieties is dependent on the chemistry of the base material. In turn, the mode of functionalization primarily depends on the nature of the problem and the intended use of the material. In recent times, efficient methods have been developed, most of which are classified as either covalent or non-covalent functionalization. Covalent interactions on the surfaces are favored due to the presence of hydroxyl, epoxy, and carboxylic groups, considered the best moieties for the functional group conversion [88]. Along with this, the presence of sp^2-hybridized π-network provides the opportunity for non-covalent interaction between the carbon nanomaterials and the host species. Graphenes and CNTs have been functionalized covalently by several approaches, including atom transfer radical polymerization (ATRP), reversible addition fragmentation chain transfer polymerization (RAFT), nitroxide-mediated radical polymerization (NMRP), anionic polymerization, and ring-opening polymerization (ROP) techniques. Graphene has been functionalized non-covalently with polymers via multiple π-π stacking, H-bonding, and hydrophobic interactions [87,89]. Similarly, CNTs also show surface modifications through non-covalent interactions such as polyaromatic adsorption π-π stacking, protein adoption, and lipid adsorption.

Over the past decades, quite a lot of studies have been conveyed on the use of polymer-based hybrids, allowing the achievement of desirable properties from the resulting materials [90]. Two main approaches, "grafting to" and "grafting from", have been reported for covalent grafting of polymers onto carbon nanomaterials [91]. The "grafting to" approach is based on the attachment of as-prepared or commercially available polymer molecules onto the carbon nanomaterial surface by a variety of chemical reactions, such as amidation, esterification, radical coupling (Figure 3.14).

The polymer must have suitable reactive functional groups for the preparation of nanocomposites in this approach. The "grafting to" approach allows grafting a polymer with functional end/side groups (e.g., $-OH$, $-NH_2$, $-COOH$, and $-COCl$) onto carbon nanomaterials. This technique is easy to carry out with both linear and dendritic polymers, but the grafting efficiency is always low owing to the steric hindrance of the pre-grafted macromolecular chains. In the case of "grafting from" approach, the polymer is bound to the carbon nanomaterial surface by in situ polymerization of monomers in the presence of reactive carbon nanomaterials [87,91,93]. Polymer-functionalized and polymer-based CNTs and graphene hybrid have been used as both catalyst supports and metal-free catalysts for the oxygen reduction reaction in fuel cell. Polymer composites with carbon nanomaterials have generated great interest in the last several decades due to the significant enhancement in the mechanical properties with a very low loading of carbon nanomaterials. The dispersion of carbon

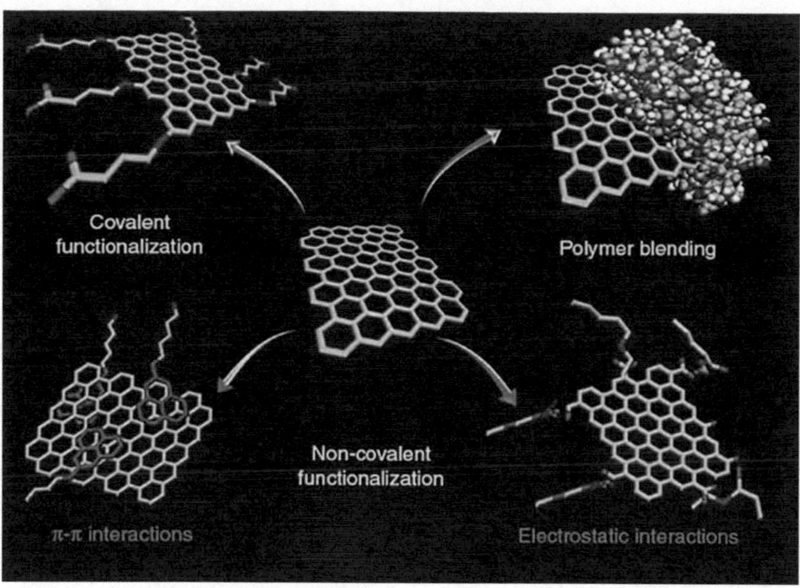

FIGURE 3.14 Graphene-based polymer composites [92].

nanomaterials and the interfacial interaction between those nanomaterials are the main factors that influence the mechanical properties of nanocomposites.

3.3 Properties of Hybrid Nanomaterials

By combining carbon nanostructures such as CNTs and graphene with metal, metal oxide, or other semiconducting nanocrystals, it is possible to achieve unique multifunctional hybrid nanoscale heterostructures with novel thermal, chemical, mechanical, electrical, and optical performances. The basic purposes and advantages of the hybridization or heterostructuring can be commented as follows [41]:

1. The obtained composite materials exhibit complementary and combined features or performance from the composing carbon and other external nanocrystals.
2. The incorporation of metal nanostructures with carbon materials significantly improves the chemical and thermal stability, making them feasible to be used in severe environments.
3. It provides possible approaches to achieve the large-scale yield of composite nanomaterials with low production cost [94].
4. The compositing extends the application of carbon nanomaterials in the areas associated with metals or metal compounds and improves their performance.

CNT displays unique electrocatalytic, electrical, and mechanical properties and was found to significantly increase the heat transport in polymer hybrids as a result of its 1D structure, high aspect ratio, and high thermal conductivity. However, there is a problem with a stable dispersion of CNT. For this purpose, graphene oxide could be a better dispersant to form a stable dispersion of CNT, which resulted in a novel hybrid dispersion named as GO/CNT. Graphene/CNT- and GO/CNT carbon-based hybrid nanomaterials show large specific area, higher electrical conductivities, and catalytic properties compared with either pristine GO/graphene or CNTs [95].

Since CNTs and graphene exhibit high aspect ratios and high electrical conductivity, they are the excellent fillers for the fabrication of electrically conducting composites. The carbon nanomaterial content can strongly affect the properties of the hybrid system. The change in properties of CNT-Al system shows the effect of CNT content in the hybrid because it can have pronounced implications on the physicomechanical properties. The high relative densities can be attributed to good dispersion of CNTs depending upon the mixing techniques. Overall, an increase in CNT content enhances the hardness of CNT-Al nanocomposites. Moreover, CNTs in Al-based systems act as load-bearing agents and remove the external load from the matrix. The CNT content can affect the mechanical properties of CNT-Al nanocomposites, and high tensile strength can be attributed to the incorporation of CNTs with a fine dispersion. Regardless of mixing techniques, adding CNT to Al matrix leads to a decrease in density and lowers the electrical conductivity due to the electron scattering by pores formed at CNT/Al interfaces. This is the case for the thermal conductivity. Herein, this decrement is attributed to (i) clustering, bending, and curving of nanotubes, (ii) the formation of pores originated from clustering, and (iii) the interfacial scattering. Thermal conductivity higher than that of pure Al through the incorporation of CNT is ascribed to the involvement of CNTs in the thermal conductivity throughout an agglomeration- or pore-free structure [96].

Incorporation of CNTs into polyurethane and polyimide matrices dramatically increases the tensile strength and modulus of polyurethane and polyimide [49].

3.4 Applications of Hybrid Nanomaterials

Hybrid nanostructures have received great research attention because they provide enhanced catalytic and tunable physical and chemical properties superior to a single system. Oxide semiconductor/metal nanocomposites are a new class of functional materials that have attracted tremendous interest in recent years owing to their potential in optoelectronics, drug delivery, environmental monitoring, control of chemical processes, photocatalysis, energy storage, and biomedical diagnosis applications. These advantages make them one of the most promising candidates for the exploration of new applications (Figure 3.15).

3.4.1 Antibacterial/Anticancer

In between carbon-based nanostructures, graphene is the best choice, which is introduced to detecting systems due to especial variety of properties such as having natural source, being biocompatible, and being cost-effective. One of the eye-catching advantages of this carbon-based nanostructure is the ability of surface treatment to be a good hostage for immobilizing ligands, nanoparticles, and single-stranded DNA in aptasensors. Besides unique electronic, mechanical, and thermal properties of graphene, some noticeable applications in nanosensors and nanomedicine make these materials in the center of attention. In recent years, there has been a growing interest among scientists in incorporating nanomaterial into nanocarbon matrices to make antibacterial and biocompatible carbon-based nanocomposites for a wide range of applications, such as biosensors and biomedical devices, wound dressing, water purification, and dispersions with antimicrobial properties [98,99]. To enhance the sensitivity of human immunodeficiency virus (HIV) detection, an electrochemical biosensor fabricated with graphene and CNTs were employed [100]. CNTs are playing an important role in preparing biosensors that can detect target molecules in trace amounts. This powerful aspect of CNTs is sourced from transduction of physical/chemical interactions and high surface area-to-volume ratio [101] (Figure 3.16).

Antibacterial reagents, including antibiotics, metal ions, enzymes, and quaternary ammonium compounds, have been extensively used in our daily life. However, the aforementioned materials have some drawbacks, such as antibiotic resistance, environmental damages, relatively high cost. Currently, a number of attempts have been made to develop novel, efficient, and environment-friendly antibacterial materials. For instance, graphene and graphene-related materials that exhibit strong antibacterial activity were extensively studied. Physical damages on cell membranes (Figure 3.15) are likely to occur in consequence of membrane stress induced by sharp edges of graphene nanosheets, thereby contributing to the loss of bacterial membrane integrity and the leakage of RNA [102,103]. Graphene and its composites have a wide range of potential applications in transistors, transparent conductors, polymer reinforcement, bioengineering, and biomaterials areas [104]. Adhering metal nanoparticles to a 2D graphene sheet by a chemical route through the reduction of metal precursors in graphene dispersion inhibits the aggregation of

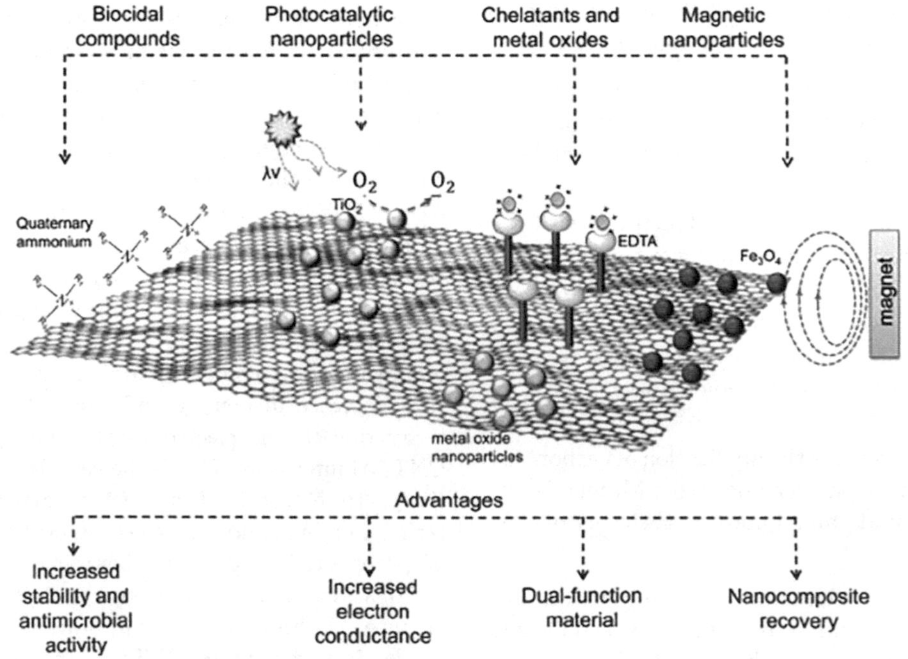

FIGURE 3.15 Different types of graphene-based antimicrobial nanocomposites [97].

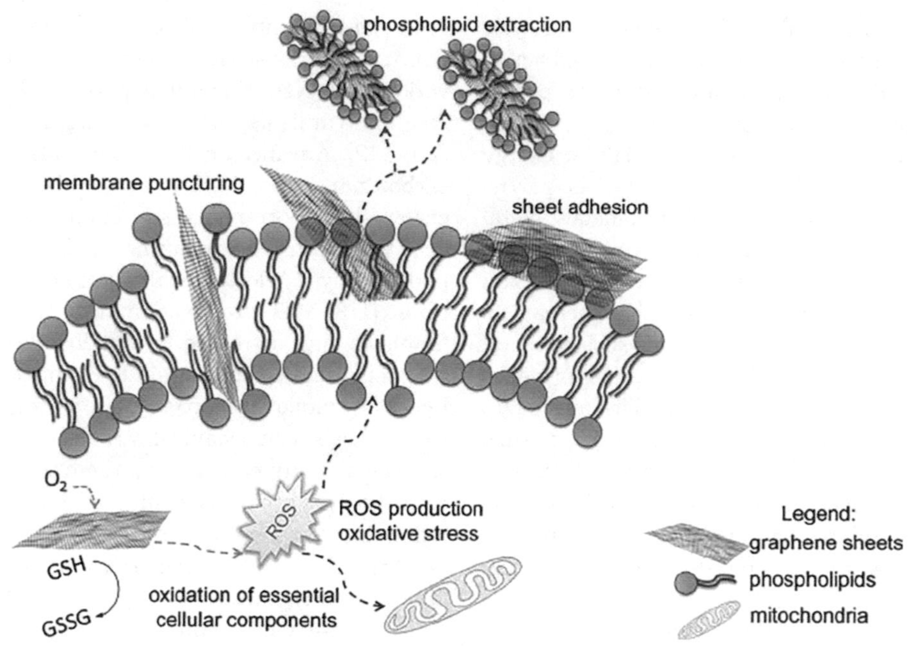

FIGURE 3.16 Mechanism of cellular interaction of graphene [97].

graphene sheets and results in an efficient antibacterial activity. Graphene nanohybrids displayed very low cytotoxicity and showed highly effective antibacterial activities against bacteria [105–107] (Figure 3.17).

3.4.2 Supercapacitors and Batteries

The increasing consumption and rapid exhaustion of fossil fuels has driven the major research attention to explore and utilize renewable energies such as wind energy, tidal energy, and solar energy for the past few decades. To provide prevalent usage of renewable energies, efficient energy storage and conversion technologies are required. Among these applications, portable electric vehicles and hybrid electric vehicles have been significantly developed. The electrochemical systems available for energy storage and conversion, including LIBs and electrochemical capacitors, are designed and developed for advanced energy conversion and storage devices. LIBs and supercapacitors are the devices with the most potential to address the energy storage need. Supercapacitors, also known as ultracapacitors, are energy storage devices that can often be safely

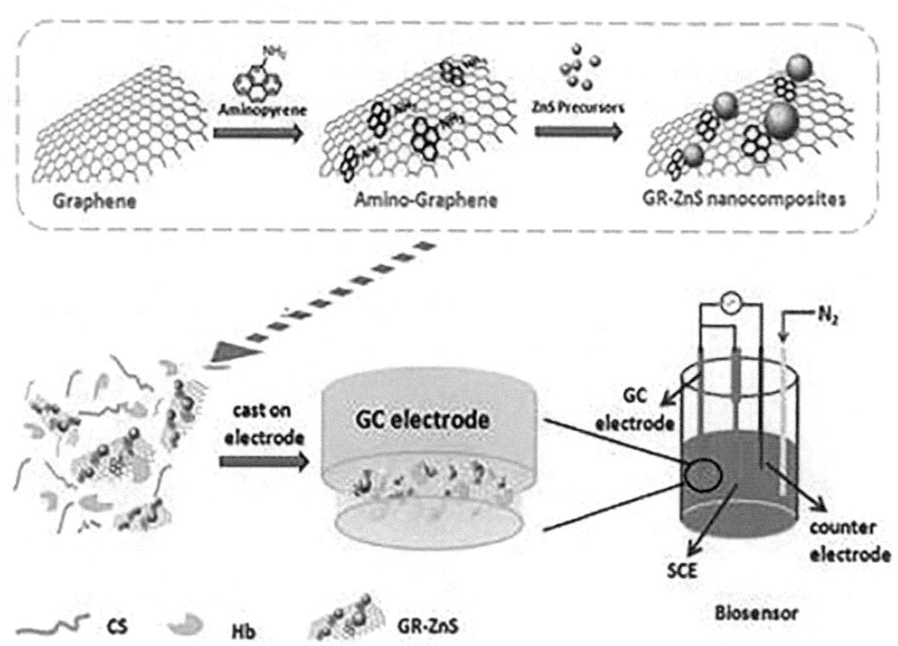

FIGURE 3.17 Graphene ZnS composite as biosensor [66].

charged or discharged in seconds with extremely long cycle life. Engineering carbon-based hybrid nanostructured active materials offers desirable functionality and great potential to achieve excellent energy storage, high rate capabilities, and long life span for electrode materials in energy devices and supercapacitors. Carbon nanomaterials, particularly graphene and CNts, in hybrid form with metals/metal oxides, polymers, and organic molecules are proven to be the promising electrode material for supercapacitors and batteries [41]. The hybrid structures that combine carbon nanomaterial with other functional materials such as metal oxides or organic molecules have shown better performance compared to the individual components. With regard to its unique structural features such as high surface area, flexibility, chemical stability, superior electric and thermal conductivity, graphene has been used as ideal building blocks for graphene-based materials with desirable functionality as alternative electrode materials in energy devices (Figure 3.18).

The extraordinary physicochemical properties of carbon nanomaterials make them ideal candidates to incorporate into conducting polymers for the development of high-performance supercapacitors. In addition to polyaniline (PANI)/CNTS, 3, 4-ethylenedioxythiophene (EDOT)/CNT and polypyrrole (PPy)/CNT nanocomposites have been used in supercapacitors. Graphene/conducting polymer nanocomposites have been investigated as electrode materials in supercapacitors. In short, the use of carbon nanohybrids in the development of the supercapacitors results in higher specific capacitance and stability. Their performance and stability depend not only on the materials used to design them, but also on the fabrication process, variation in the functionalization methods, comparative presence of various components, and other experimental variables.

3.4.3 Catalysis

More efficient catalysis and new catalytic technologies are critical for minimizing environmental pollution by reducing waste products, ensuring the development of clean energy applications, and from a broader perspective, enabling sustainable human development. Nowadays, various catalysts, including photo-, electro-, and organic-reaction catalysts, are becoming increasingly attractive [110–112]. Advanced carbon materials such as carbon dots, carbon nanowires, CNTs, carbon fibers, and graphene play extremely important roles in coming severe challenges and achieving dramatic breakthroughs in catalysis, including photocatalysis, electrocatalysis, and organic catalytic reactions [113]. 3D graphene materials assembled from 2D graphene have attracted tremendous attention, because 3D graphene can not only retain the properties of 2D graphene, but has unique properties such as adjustable pore structures, excellent mechanical strength, and outstanding electronic conductivity. Based on these properties, 3D graphene as a catalyst or catalyst support has fulfilled some of the requirements necessary for being considered as an advanced catalyst. Here are different roles offered to catalysis of 3D graphene in the field of catalysis. (i) 3D graphene modified by $-OH$, $-COOH$, and $-NH_2$ functional groups directly plays a catalytic role. (ii) Hybrid 3D graphene displays good performance in catalytic reactions. (iii) 3D graphene acts as a support for catalytic functionalities such as metals, metal oxides, chalcogenides, nitrides, and phosphides. (iv) 3D graphene can be treated as cocatalysts to facilitate photocatalytic reactions [114,115].

Well-dispersed copper nanoparticles on carbon nanomaterials (CRGO-CuI, CNT-CuI) are cyclable and reusable catalysts. CNTs exhibit electric, thermal, and chemical properties similar to those of graphene and thus have also been used for compositing with external nanostructures for catalysis and energy applications. Highly conductive rGO is an ideal material for the improvement of semiconductor photocatalytic materials, as it not only has the potential to serve as an excellent electron acceptor for the restriction of photo-generated electron/hole pair recombination, but also bears various inherent photocatalytic advantages such as its unique 2D carbon structure (sp^2 hybridization), excellent conductivity, and a large surface area [116,117] (Figure 3.19).

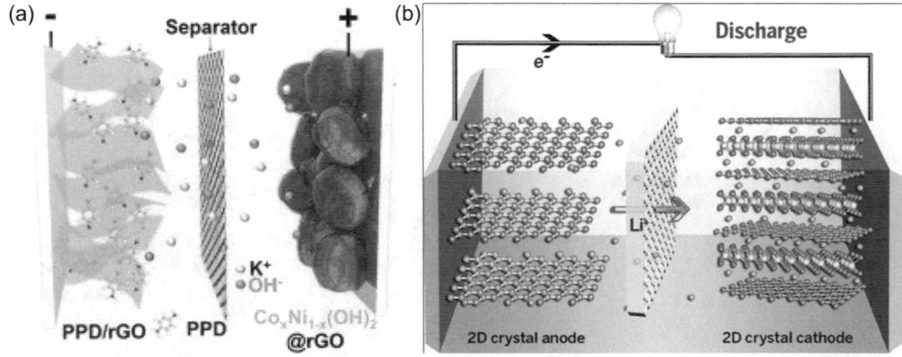

FIGURE 3.18 Hybrid supercapacitor based on graphene composite electrodes (a) [108] and graphene flakes anode and cathode is a hybrid graphene–lithium compound (b) [109].

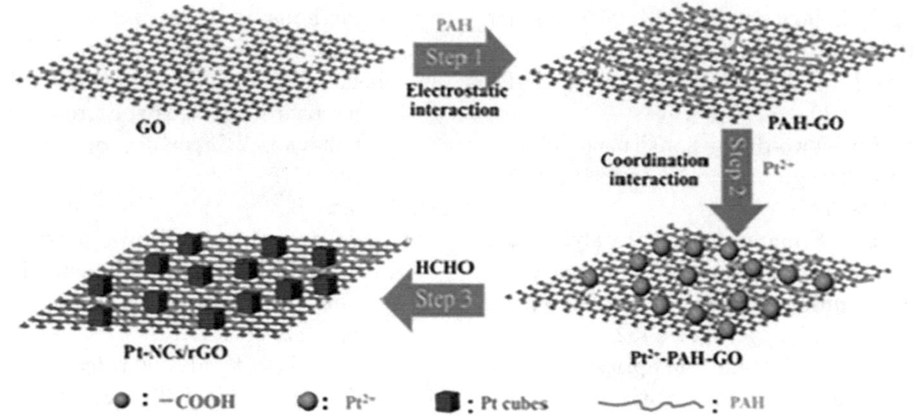

FIGURE 3.19 Pt/rGO for catalysis [118].

References

1. Weiss, N.O., et al., Graphene: An emerging electronic material. *Advanced Materials*, 2012. **24**(43): pp. 5782–5825.
2. Schrier, J., D.O. Demchenko, and A.P. Alivisatos, Optical properties of ZnO/ZnS and ZnO/ZnTe heterostructures for photovoltaic applications. *Nano Letters*, 2007. **7**(8): pp. 2377–2382.
3. Watt, J., et al., Synthesis and structural characterization of branched palladium nanostructures. *Advanced Materials*, 2009. **21**(22): pp. 2288–2293.
4. Fan, X., et al., ZnS/ZnO heterojunction nanoribbons. *Advanced Materials*, 2009. **21**(23): pp. 2393–2396.
5. Ahmad, M., X. Yan, and J. Zhu, Controlled synthesis, structural evolution, and photoluminescence properties of nanoscale one-dimensional hierarchical ZnO/ZnS heterostructures. *The Journal of Physical Chemistry C*, 2011. **115**(5): pp. 1831–1837.
6. Fang, X., et al., Multiangular branched ZnS nanostructures with needle-shaped tips: Potential luminescent and field-emitter nanomaterial. *The Journal of Physical Chemistry C*, 2008. **112**(12): pp. 4735–4742.
7. Jiang, Y., et al., Homoepitaxial growth and lasing properties of ZnS nanowire and nanoribbon arrays. *Advanced Materials*, 2006. **18**(12): pp. 1527–1532.
8. Yan, J., et al., Structure and cathodoluminescence of individual ZnS/ZnO biaxial nanobelt heterostructures. *Nano Letters*, 2008. **8**(9): pp. 2794–2799.
9. Lu, M.-Y., et al., ZnO− ZnS heterojunction and ZnS nanowire arrays for electricity generation. *ACS Nano*, 2009. **3**(2): pp. 357–362.
10. Saito, N., et al., Low-temperature fabrication of light-emitting zinc oxide micropatterns using self-assembled monolayers. *Advanced Materials*, 2002. **14**(6): pp. 418–421.
11. Rau, U. and M. Schmidt, Electronic properties of ZnO/CdS/Cu (In, Ga) Se_2 solar cells: Aspects of heterojunction formation. *Thin Solid Films*, 2001. **387**(1–2): pp. 141–146.
12. Ahmad, M., et al., Conductivity enhancement by slight indium doping in ZnO nanowires for optoelectronic applications. *Journal of Physics D: Applied Physics*, 2009. **42**(16): p. 165406.
13. Jung, Y.S., et al., Influence of DC magnetron sputtering parameters on the properties of amorphous indium zinc oxide thin film. *Thin Solid Films*, 2003. **445**(1): pp. 63–71.
14. Liu, Y., et al., Biocompatible ZnO/Au nanocomposites for ultrasensitive DNA detection using resonance Raman scattering. *The Journal of Physical Chemistry B*, 2008. **112**(20): pp. 6484–6489.
15. Ahmad, M., et al., Highly sensitive amperometric cholesterol biosensor based on Pt-incorporated fullerene-like ZnO nanospheres. *The Journal of Physical Chemistry C*, 2009. **114**(1): pp. 243–250.
16. Sano, T., et al., Photocatalytic degradation of gaseous acetaldehyde on TiO_2 with photodeposited metals and metal oxides. *Journal of Photochemistry and Photobiology A: Chemistry*, 2003. **160**(1–2): pp. 93–98.
17. Ahmad, M., C. Pan, and J. Zhu, Investigation of hydrogen storage capabilities of ZnO-based nanostructures. *The Journal of Physical Chemistry C*, 2010. **114**(6): pp. 2560–2565.
18. Ahmad, M., et al., Synthesis of hierarchical flower-like ZnO nanostructures and their functionalization by Au nanoparticles for improved photocatalytic and high performance Li-ion battery anodes. *Journal of Materials Chemistry*, 2011. **21**(21): pp. 7723–7729.
19. Wang, Y., X. Su, and S. Lu, Shape-controlled synthesis of TiO_2 hollow structures and their application in lithium batteries. *Journal of Materials Chemistry*, 2012. **22**(5): pp. 1969–1976.
20. Wu, H.B., et al., Nanostructured metal oxide-based materials as advanced anodes for lithium-ion batteries. *Nanoscale*, 2012. **4**(8): pp. 2526–2542.

21. Ding, S., et al., TiO_2 hollow spheres with large amount of exposed (001) facets for fast reversible lithium storage. *Journal of Materials Chemistry*, 2011. **21**(6): pp. 1677–1680.
22. Liu, J. and X.W. Liu, Two-dimensional nanoarchitectures for lithium storage. *Advanced Materials*, 2012. **24**(30): pp. 4097–4111.
23. Hussain, M., et al., Enhanced photocatalytic and electrochemical properties of Au nanoparticles supported TiO_2 microspheres. *New Journal of Chemistry*, 2014. **38**(4): pp. 1424–1432.
24. Kapilashrami, M., et al., Probing the optical property and electronic structure of TiO_2 nanomaterials for renewable energy applications. *Chemical Reviews*, 2014. **114**(19): pp. 9662–9707.
25. Hussain, M., et al., $AgTiO_2$ nanocomposite for environmental and sensing applications. *Materials Chemistry and Physics*, 2016. **181**: pp. 194–203.
26. Shaviv, E., et al., Absorption properties of metal–semiconductor hybrid nanoparticles. *ACS Nano*, 2011. **5**(6): pp. 4712–4719.
27. Zhang, L., D.A. Blom, and H. Wang, Au–Cu_2O core–shell nanoparticles: A hybrid metal-semiconductor heteronanostructure with geometrically tunable optical properties. *Chemistry of Materials*, 2011. **23**(20): pp. 4587–4598.
28. Kumar, D.R., et al., Au-CuO core-shell nanoparticles design and development for the selective determination of vitamin B6. *Electrochimica Acta*, 2015. **176**: pp. 514–522.
29. Yu, Y., et al., Au nanoparticles decorated CuO nanowire arrays with enhanced photocatalytic properties. *Materials Letters*, 2013. **108**: pp. 41–45.
30. Barreca, D., et al., Novel synthesis and gas sensing performances of CuO–TiO_2 nanocomposites functionalized with Au nanoparticles. *The Journal of Physical Chemistry C*, 2011. **115**(21): pp. 10510–10517.
31. Ahmad, N., et al., Magnetoelastic anisotropy induced effects on field and temperature dependent magnetization reversal of Ni nanowires and nanotubes. *Journal of Superconductivity and Novel Magnetism*, 2011. **24**(1–2): pp. 785–792.
32. Hussain, M., et al., Fabrication and temperature dependent magnetic properties of Ni–Cu–Co composite nanowires. *Physica B: Condensed Matter*, 2015. **475**: pp. 99–104.
33. Kuanr, B.K., et al., Nonreciprocal microwave devices based on magnetic nanowires. *Applied Physics Letters*, 2009. **94**(20): p. 202505.
34. Xu, C.-L., et al., Electrodeposition of ferromagnetic nanowire arrays on AAO/Ti/Si substrate for ultrahigh-density magnetic storage devices. *Materials Letters*, 2006. **60**(19): pp. 2335–2338.
35. Hurst, S.J., et al., Multisegmented one-dimensional nanorods prepared by hard-template synthetic methods. *Angewandte Chemie International Edition*, 2006. **45**(17): pp. 2672–2692.
36. Shen, G., et al., Pearl-like ZnS-decorated InP nanowire heterostructures and their electric behaviors. *Chemistry of Materials*, 2008. **20**(21): pp. 6779–6783.
37. Ishrat, S., et al., Fabrication and temperature-dependent magnetic properties of one-dimensional embedded nickel segment in gold nanowires. *Journal of Alloys and Compounds*, 2012. **541**: pp. 483–487.
38. Ishrat, S., et al., Fabrication and temperature-dependent magnetic properties of one-dimensional multilayer Au–Ni–Au–Ni–Au nanowires. *Journal of Solid State Chemistry*, 2014. **210**(1): pp. 116–120.
39. Jariwala, D., et al., Carbon nanomaterials for electronics, optoelectronics, photovoltaics, and sensing. *Chemical Society Reviews*, 2013. **42**(7): pp. 2824–2860.
40. Bonaccorso, F., et al., Production and processing of graphene and 2d crystals. *Materials Today*, 2012. **15**(12): pp. 564–589.
41. Li, Y., J. Wu, and N. Chopra, Nano-carbon-based hybrids and heterostructures: Progress in growth and application for lithium-ion batteries. *Journal of Materials Science*, 2015. **50**(24): pp.7843–7865.
42. Peng, X., et al., Carbon nanotube–nanocrystal heterostructures. *Chemical Society Reviews*, 2009. **38**(4): pp. 1076–1098.
43. Liu, W.-W., et al., Synthesis and characterization of graphene and carbon nanotubes: A review on the past and recent developments. *Journal of Industrial and Engineering Chemistry*, 2014. **20**(4): pp. 1171–1185.
44. Lim, D.-J., et al., Carbon-based drug delivery carriers for cancer therapy. *Archives of Pharmacal Research*, 2014. **37**(1): pp. 43–52.
45. Jang, H., et al., A new helicase assay based on graphene oxide for anti-viral drug development. *Molecules and Cells*, 2013. **35**(4): pp. 269–273.
46. Goenka, S., V. Sant, and S. Sant, Graphene-based nanomaterials for drug delivery and tissue engineering. *Journal of Controlled Release*, 2014. **173**: pp. 75–88.
47. Mao, H.Y., et al., Graphene: Promises, facts, opportunities, and challenges in nanomedicine. *Chemical Reviews*, 2013. **113**(5): pp. 3407–3424.
48. Jeong, S., et al., Efficient electron transfer in functional assemblies of pyridine-modified NQDs on SWNTs. *ACS Nano*, 2009. **4**(1): pp. 324–330.
49. Sahoo, N.G., et al., Polymer nanocomposites based on functionalized carbon nanotubes. *Progress in Polymer Science*, 2010. **35**(7): pp. 837–867.
50. Choi, E.-Y., et al., Noncovalent functionalization of graphene with end-functional polymers. *Journal of Materials Chemistry*, 2010. **20**(10): pp. 1907–1912.

51. Zhang, Q., et al., The road for nanomaterials industry: A review of carbon nanotube production, post-treatment, and bulk applications for composites and energy storage. *Small*, 2013. **9**(8): pp. 1237–1265.
52. Na, H.B., et al., Development of a T1 contrast agent for magnetic resonance imaging using MnO nanoparticles. *Angewandte Chemie*, 2007. **119**(28): pp. 5493–5497.
53. Besson, C., E.E. Finney, and R.G. Finke, A mechanism for transition-metal nanoparticle self-assembly. *Journal of the American Chemical Society*, 2005. **127**(22): pp. 8179–8184.
54. Namvari, M. and H. Namazi, Preparation of efficient magnetic biosorbents by clicking carbohydrates onto graphene oxide. *Journal of Materials Science*, 2015. **50**(15): pp. 5348–5361.
55. Zhou, G., et al., Graphene-wrapped Fe_3O_4 anode material with improved reversible capacity and cyclic stability for lithium ion batteries. *Chemistry of Materials*, 2010. **22**(18): pp. 5306–5313.
56. Lamanna, G., et al., Endowing carbon nanotubes with superparamagnetic properties: Applications for cell labeling, MRI cell tracking and magnetic manipulations. *Nanoscale*, 2013. **5**(10): pp. 4412–4421.
57. Xue, Y., et al., Oxidizing metal ions with graphene oxide: The in situ formation of magnetic nanoparticles on self-reduced graphene sheets for multifunctional applications. *Chemical Communications*, 2011. **47**(42): pp. 11689–11691.
58. Shamsipur, M., et al., Functionalized Fe_3O_4/graphene oxide nanocomposites with hairpin aptamers for the separation and preconcentration of trace Pb^{2+} from biological samples prior to determination by ICP MS. *Materials Science and Engineering: C*, 2017. **77**: pp. 459–469.
59. Hyeon, T., Y. Piao, and Y.I. Park, Method of preparing iron oxide nanoparticles coated with hydrophilic material, and magnetic resonance imaging contrast agent using the same. 2016, Google Patents. EP2673006A2.
60. Templier, V., et al., Ligands for label-free detection of whole bacteria on biosensors: A review. *TrAC Trends in Analytical Chemistry*, 2016. **79**: pp. 71–79.
61. Laurent, S., et al., Magnetic iron oxide nanoparticles: Synthesis, stabilization, vectorization, physicochemical characterizations, and biological applications. *Chemical Reviews*, 2008. **108**(6): pp. 2064–2110.
62. Wu, Q., et al., A facile one-pot solvothermal method to produce superparamagnetic graphene–Fe_3O_4 nanocomposite and its application in the removal of dye from aqueous solution. *Colloids and Surfaces B: Biointerfaces*, 2013. **101**: pp. 210–214.
63. Zhao, L., B. Zeng, and F. Zhao, Electrochemical determination of tartrazine using a molecularly imprinted polymer–multiwalled carbon nanotubes-ionic liquid supported Pt nanoparticles composite film coated electrode. *Electrochimica Acta*, 2014. **146**: pp. 611–617.
64. Rong, L.-Q., et al., Study of the nonenzymatic glucose sensor based on highly dispersed Pt nanoparticles supported on carbon nanotubes. *Talanta*, 2007. **72**(2): pp. 819–824.
65. Hrapovic, S., et al., Electrochemical biosensing platforms using platinum nanoparticles and carbon nanotubes. *Analytical Chemistry*, 2004. **76**(4): pp. 1083–1088.
66. Hussain, S., et al., Heat-treatment effects on the ORR activity of Pt nanoparticles deposited on multi-walled carbon nanotubes using magnetron sputtering technique. *International Journal of Hydrogen Energy*, 2017. **42**(9): pp. 5958–5970.
67. Beiranvand, Z.S., et al., Aptamer-based electrochemical biosensor by using Au-Pt nanoparticles, carbon nanotubes and acriflavine platform. *Analytical Biochemistry*, 2017. **518**: pp. 35–45.
68. Wang, J., et al., Rational design of three-dimensional nitrogen and phosphorus co-doped graphene nanoribbons/CNTs composite for the oxygen reduction. *Chinese Chemical Letters*, 2016. **27**(4): pp. 597–601.
69. Yang, C., et al., Preparation and photocatalytic activity of high-efficiency visible-light-responsive photocatalyst SnS_x/TiO_2. *Journal of Solid State Chemistry*, 2009. **182**(4): pp. 807–812.
70. Modarres, M.H., et al., Evolution of reduced graphene oxide–SnS_2 hybrid nanoparticle electrodes in Li-ion batteries. *The Journal of Physical Chemistry C*, 2017. **121**(24): pp. 13018–13024.
71. Sun, C.-L., et al., The simultaneous electrochemical detection of ascorbic acid, dopamine, and uric acid using graphene/size-selected Pt nanocomposites. *Biosensors and Bioelectronics*, 2011. **26**(8): pp. 3450–3455.
72. Li, J., X. Qin, and Z. Yang. Preparation, characterization and application of carbon nanotubes wrapped nanoflake-like SnS_2 composite. *In 2nd International Conference on Electronic and Mechanical Engineering and Information Technology*, Shenyang, 2012. Citeseer.
73. Zhao, Y., et al., Rational microstructure design of SnS_2-carbon composites for superior sodium storage performance. *Nanoscale*, 2018. **10**(17): pp. 7999–8008.
74. Zhai, C., et al., Multiwalled carbon nanotubes anchored with SnS_2 nanosheets as high-performance anode materials of lithium-ion batteries. *ACS Applied Materials and Interfaces*, 2011. **3**(10): pp. 4067–4074.
75. Patil, U.M., et al., Enhanced supercapacitive performance of chemically grown cobalt–nickel hydroxides on three-dimensional graphene foam electrodes.

ACS Applied Materials and Interfaces, 2014. **6**(4): pp. 2450–2458.

76. Zhou, L., et al., Facile preparation of $ZnMn_2O_4$ hollow microspheres as high-capacity anodes for lithium-ion batteries. *Journal of Materials Chemistry*, 2012. **22**(3): pp. 827–829.
77. Wang, J., Y. Xiong, and X. Zhang, Rational synthesis of $NiCo_2O_4$ meso-structures for high-rate supercapacitors. *Journal of Materials Science*, 2017. **52**(7): pp. 3678–3686.
78. Fu, F., et al., Hierarchical $NiCo_2O_4$ micro-and nanostructures with tunable morphologies as anode materials for lithium-and sodium-ion batteries. *ACS Applied materials and Interfaces*, 2017. **9**(19): pp. 16194–16201.
79. Mondal, A.K., et al., Highly porous $NiCo_2O_4$ nanoflakes and nanobelts as anode materials for lithium-ion batteries with excellent rate capability. *ACS Applied Materials and Interfaces*, 2014. **6**(17): pp. 14827–14835.
80. Umeshbabu, E. and G.R. Rao, $NiCo_2O_4$ hexagonal nanoplates anchored on reduced graphene oxide sheets with enhanced electrocatalytic activity and stability for methanol and water oxidation. *Electrochimica Acta*, 2016. **213**: pp. 717–729.
81. Peng, L., et al., Designed functional systems for high-performance lithium-ion batteries anode: From solid to hollow, and to core–shell $NiCo_2O_4$ nanoparticles encapsulated in ultrathin carbon nanosheets. *ACS Applied Materials and Interfaces*, 2016. **8**(7): pp. 4745–4753.
82. Liu, Y., et al., Hierarchical $CoNiO_2$ structures assembled from mesoporous nanosheets with tunable porosity and their application as lithium-ion battery electrodes. *New Journal of Chemistry*, 2014. **38**(7): pp. 3084–3091.
83. Wu, Z.-S., et al., Graphene anchored with Co_3O_4 nanoparticles as anode of lithium ion batteries with enhanced reversible capacity and cyclic performance. *ACS Nano*, 2010. **4**(6): pp. 3187–3194.
84. Ma, T.Y., et al., Metal–organic framework derived hybrid Co_3O_4-carbon porous nanowire arrays as reversible oxygen evolution electrodes. *Journal of the American Chemical Society*, 2014. **136**(39): pp. 13925–13931.
85. Jiang, J., et al., General synthesis of large-scale arrays of one-dimensional nanostructured Co_3O_4 directly on heterogeneous substrates. *Crystal Growth and Design*, 2009. **10**(1): pp. 70–75.
86. Gu, D., et al., Controllable synthesis of mesoporous peapod-like co3o4@ carbon nanotube arrays for high-performance lithium-ion batteries. *Angewandte Chemie International Edition*, 2015. **54**(24): pp. 7060–7064.
87. Punetha, V.D., et al., Functionalization of carbon nanomaterials for advanced polymer nanocomposites: A comparison study between CNT and graphene. *Progress in Polymer Science*, 2017. **67**: pp. 1–47.
88. Peng, H., et al., Sidewall carboxylic acid functionalization of single-walled carbon nanotubes. *Journal of the American Chemical Society*, 2003. **125**(49): pp. 15174–15182.
89. Gonçalves, G., et al., Graphene oxide modified with PMMA via ATRP as a reinforcement filler. *Journal of Materials Chemistry*, 2010. **20**(44): pp. 9927–9934.
90. Shen, J., et al., Synthesis of amphiphilic graphene nanoplatelets. *Small*, 2009. **5**(1): pp. 82–85.
91. Kumar, I., S. Rana, and J.W. Cho, Cycloaddition reactions: A controlled approach for carbon nanotube functionalization. *Chemistry-A European Journal*, 2011. **17**(40): pp. 11092–11101.
92. Idumah, C.I. and A. Hassan, Emerging trends in graphene carbon based polymer nanocomposites and applications. *Reviews in Chemical Engineering*, 2016. **32**(2): pp. 223–264.
93. Kim, B.H., et al., Surface energy modification by spin-cast, large-area graphene film for block copolymer lithography. *ACS Nano*, 2010. **4**(9): pp. 5464–5470.
94. Li, Y. and N. Chopra, Progress in large-scale production of graphene. Part 1: Chemical methods. *JOM*, 2015. **67**(1): pp. 34–43.
95. Mani, V., S.-M. Chen, and B.-S. Lou, Three dimensional graphene oxide-carbon nanotubes and graphene-carbon nanotubes hybrids. *International Journal of Electrochemical Science*, 2013. **8**(11641): p. e60.
96. Azarniya, A., et al., Physicomechanical properties of spark plasma sintered carbon nanotube-reinforced metal matrix nanocomposites. *Progress in Materials Science*, 2017. **90**: pp. 276–324.
97. Perreault, F., A.F. De Faria, and M. Elimelech, Environmental applications of graphene-based nanomaterials. *Chemical Society Reviews*, 2015. **44**(16): pp. 5861–5896.
98. Yousefi, M., et al., Anti-bacterial activity of graphene oxide as a new weapon nanomaterial to combat multidrug-resistance bacteria. *Materials Science and Engineering: C*, 2017. **74**: pp. 568–581.
99. Hasanzadeh, M., et al., Graphene quantum dot as an electrically conductive material toward low potential detection: A new platform for interface science. *Journal of Materials Science: Materials in Electronics*, 2016. **27**(6): pp. 6488–6495.
100. Fang, Y.-S., et al., An enhanced sensitive electrochemical immunosensor based on efficient encapsulation of enzyme in silica matrix for the detection of human immunodeficiency virus p24. *Biosensors and Bioelectronics*, 2015. **64**: pp. 324–332.
101. Yang, N., et al., Carbon nanotube based biosensors. *Sensors and Actuators B: Chemical*, 2015. **207**: pp. 690–715.

102. Yu, L., et al., Enhanced antibacterial activity of silver nanoparticles/halloysite nanotubes/graphene nanocomposites with sandwich-like structure. *Scientific Reports*, 2014. **4**: p. 4551.
103. Akhavan, O. and E. Ghaderi, Toxicity of graphene and graphene oxide nanowalls against bacteria. *ACS Nano*, 2010. **4**(10): pp. 5731–5736.
104. Kavitha, T., et al., Glucose sensing, photocatalytic and antibacterial properties of graphene–ZnO nanoparticle hybrids. *Carbon*, 2012. **50**(8): pp. 2994–3000.
105. Shao, W., et al., Preparation, characterization, and antibacterial activity of silver nanoparticle-decorated graphene oxide nanocomposite. *ACS Applied Materials and Interfaces*, 2015. **7**(12): pp. 6966–6973.
106. Pasricha, R., S. Gupta, and A.K. Srivastava, A facile and novel synthesis of Ag–graphene-based nanocomposites. *Small*, 2009. **5**(20): pp. 2253–2259.
107. Wang, Y.-W., et al., Superior antibacterial activity of zinc oxide/graphene oxide composites originating from high zinc concentration localized around bacteria. *ACS Applied Materials and Interfaces*, 2014. **6**(4): pp. 2791–2798.
108. Zhao, B., et al., A high-energy, long cycle-life hybrid supercapacitor based on graphene composite electrodes. *Energy Storage Materials*, 2017. **7**: pp. 32–39.
109. Bonaccorso, F., et al., Graphene, related two-dimensional crystals, and hybrid systems for energy conversion and storage. *Science*, 2015. **347**(6217): p. 1246501.
110. Schneider, J., et al., Understanding TiO_2 photocatalysis: Mechanisms and materials. *Chemical Reviews*, 2014. **114**(19): pp. 9919–9986.
111. Qiu, B., et al., Efficient solar light harvesting CdS/Co_9S_8 hollow cubes for Z-scheme photocatalytic water splitting. *Angewandte Chemie*, 2017. **129**(10): pp. 2728–2732.
112. Gao, M.-R., et al., An efficient molybdenum disulfide/cobalt diselenide hybrid catalyst for electrochemical hydrogen generation. *Nature Communications*, 2015. **6**: pp. 5982.
113. Wang, L., et al., Carbon dots modified mesoporous organosilica as an adsorbent for the removal of 2, 4-dichlorophenol and heavy metal ions. *Journal of Materials Chemistry A*, 2015. **3**(25): pp. 13357–13364.
114. Qiu, B., M. Xing, and J. Zhang, Mesoporous TiO_2 nanocrystals grown in situ on graphene aerogels for high photocatalysis and lithium-ion batteries. *Journal of the American Chemical Society*, 2014. **136**(16): pp. 5852–5855.
115. Han, A., et al., A robust hydrogen evolution catalyst based on crystalline nickel phosphide nanoflakes on three-dimensional graphene/nickel foam: High performance for electrocatalytic hydrogen production from pH 0 to 14. *Journal of Materials Chemistry A*, 2015. **3**(5): pp. 1941–1946.
116. Tsang, C.H.A., et al., The applications of graphene-based materials in pollutant control and disinfection. *Progress in Solid State Chemistry*, 2017. **45**: pp. 1–8.
117. Kumar, V., et al., Graphene and its nanocomposites as a platform for environmental applications. *Chemical Engineering Journal*, 2017. **315**: pp. 210–232.
118. Li, F., et al., Reduced graphene oxide supported platinum nanocubes composites: One-pot hydrothermal synthesis and enhanced catalytic activity. *Nanotechnology*, 2015. **26**(6): p. 065603.

4

Aromatic Helicenes

Irena G. Stará and Ivo Starý
Czech Academy of Sciences

4.1 Introduction

Parent helicenes are polyaromatic compounds that are composed of all-*ortho*-annulated benzene units. The name "helicene" is derived from its archetypal helical arrangement that is spontaneously adopted in order to release a steric repulsion between the termini of the backbone. The departure from planarity has fundamental consequences with respect to the structure and properties of these iconic three-dimensional aromatics. The first intentional synthesis of a helicene molecule was published by M. S. Newman and D. Lednicer in 1956 when they succeeded not only in the preparation of racemic hexahelicene **1** (Figure 4.1) but also in its resolution into enantiomers (Newman and Lednicer, 1956). From the current perspective, their seminal work gave birth to the modern helicene chemistry. Since then, however, helicenes were considered textbook stereochemical curiosities rather than subjects of a wide interest for more than five decades despite the seminal contributions by R. H. Martin et al., W. H. Laarhoven et al., H. Wynberg et al. (mostly in the seventies and eighties) and T. J. Katz et al. (mostly in the nineties). The situation has completely changed around the turn of the millennium when helicenes have been put under the spotlight, resulting in a rapid growth of the number of papers on this topic in particular after 2010. Chemistry of helicenes has recently been reviewed in a comprehensive way more than once (Chen and Shen, 2017; Gingras, 2013a,b; Gingras et al., 2013; Shen and Chen, 2011; Stará and Starý, 2009). From the historical perspective, it is worth noting that the first helicenes described in the literature were actually aza[5]helicenes **2** and **3** (twisted heteroaromatics, Figure 4.2) as reported by J. Meisenheimer and K. Witte (1903).

The numerical prefix (or a number in square brackets) before the helicene name expresses the number of fused cycles as exemplified by hexahelicene or simply [6]helicene **1**.

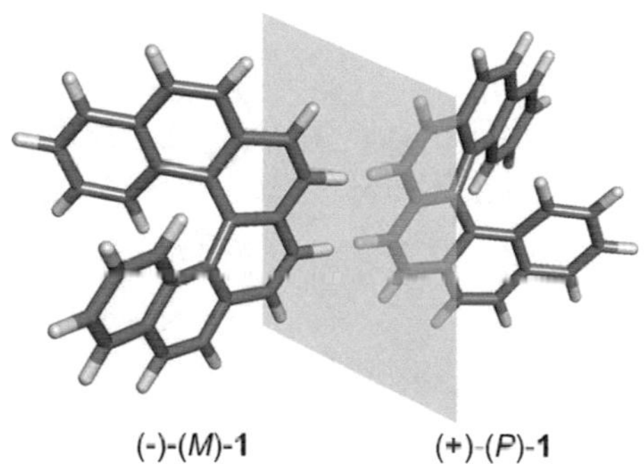

FIGURE 4.1 Molecular structure of [6]helicene enantiomers (−)-(*M*)- and (+)-(*P*)-**1**. (Reprinted from Šámal et al., 2015. With permission.)

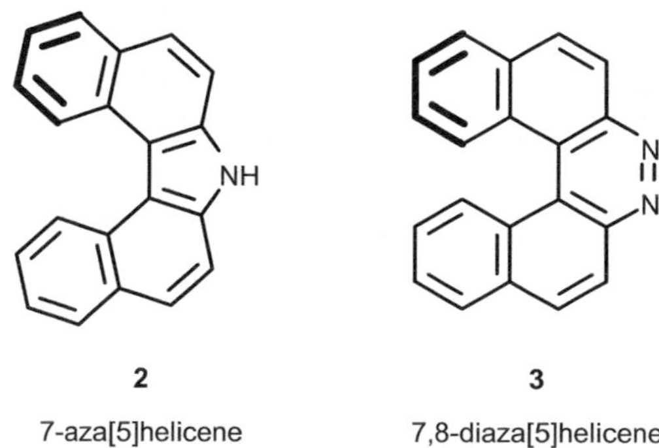

FIGURE 4.2 The first helicenes **2** and **3** reported in the literature by J. Meisenheimer and K. Witte in 1903 ((*P*)-enantiomers are shown) (Meisenheimer and Witte, 1903).

Provided all these rings are benzenes, such compounds are called carbohelicenes (**1**, Figure 4.3). If one (or more) benzene unit is formally displaced with a heterocycle, such a skeletal modification leads to heterohelicenes. Accordingly, sub-families of heterohelicenes are mentioned in the literature, such as azahelicenes or pyridohelicenes (**4**), azoniahelicenes (**5**) or thiahelicenes (**6**). Helicene-like compounds represent another type of abundant helicenes that preserve a typical helical shape of the molecule, but some (hetero)cycles forming the helical backbone are

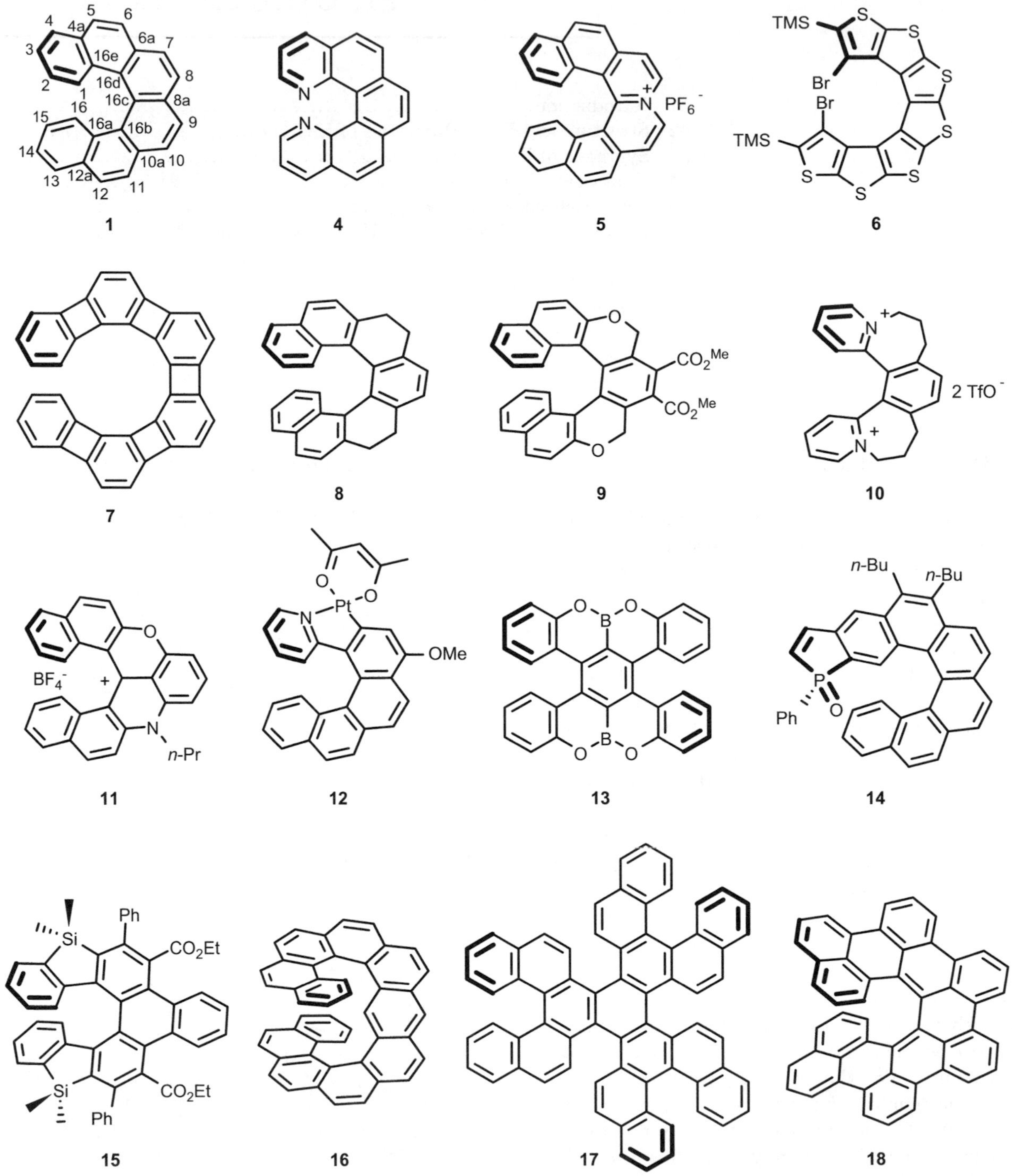

FIGURE 4.3 Examples of the structural diversity of helicenes ((*P*)-enantiomers are shown): carbohelicene **1** (Newman and Lednicer, 1956), aza- or pyridohelicene **4** (Míšek et al., 2008), azoniahelicene **5** (Nakai et al., 2013), thiahelicene **6** (Rajca et al., 2004), heliphene **7** (Han et al., 2002b), tetrahydrohelicene **8** (Stará et al., 1999), oxahelicene **9** (Tanaka et al., 2007), helquat **10** (Adriaenssens et al., 2009), cationic heterohelicene **11** (Torricelli et al., 2013), cyclometallated helicene **12** (Norel et al., 2010), borahelicene **13** (Katayama et al., 2016), phosphahelicene **14** (Yavari et al., 2012), silahelicene **15** (Shibata et al., 2012), *ortho*/*meta*-annulated helicene **16** (Sehnal et al., 2009), multipole helicene **17** (Berezhnaia et al., 2017), twisted nanographene **18** (Nakakuki et al., 2018).

contracted, expanded or non-aromatic (partially hydrogenated). The class of helicene-like compounds features a considerable structural diversity, and even specific terms were coined to some entities. It encompasses, e.g., heliphenes (**7**), tetrahydrohelicenes (**8**), oxahelicenes (**9**), helquats (**10**), cationic heterohelicenes (**11**), cyclometallated helicenes (**12**), borahelicenes (**13**), phosphahelicenes (**14**) and silahelicenes (**15**). A *meta*-annulated cycle might also be embedded into the otherwise all-*ortho*-annulated cycles (**16**, Figure 4.3). Finally, several helicene units might be combined within a single molecule to speak about multipole helicenes (**17**), or the helicene backbone can be laterally extended to form a twisted nanographene (**18**). Obviously, the family of helicenes has considerably expanded and diversified in the structure patterns as helicene chemistry has significantly developed over time. Nowadays, it is generally accepted that (i) helicene is a molecule composed of four and more carbo/heterocycles that are mostly aromatic and annulated normally in an *ortho* fashion, (ii) its curved and twisted backbone forms a helix owing to the steric repulsion between its termini and (iii) it resembles a molecular shape of the original Newman's hexahelicene **1** (at least in its part). Accordingly, in the following text, the generalized term "helicene" will be used in the broadest sense of its definition regardless of the specific structural variations.

Helicenes can exist in two enantiomeric forms regardless of their configurational stability; the handedness of the helix is specified by adding the (M) (minus) or (P) (plus) prefix (Figure 4.1). The atoms of the helicene backbone are routinely numbered in such a way that the innermost atom in the helicene "bay" holds number 1 and the numbering continues successively around the helicene backbone (**1**, Figure 4.3).

4.2 Structure and Basic Properties of Helicenes

Helicenes are in most cases thermally and chemically stable polyaromatic systems. As they feature a nonplanar three-dimensional architecture, the propensity for an extensive π-π stacking in solution is reduced, and therefore, they are significantly more soluble in respective organic solvents than their planar counterparts. They can be handled as normal organic compounds in air, in aqueous environment, in the presence of moderate oxidizing/reducing agents and acids/bases, at elevated temperature (usually well over 150°C but thermal racemization may occur, *vide infra*); they withstand an ultraviolet (UV) light irradiation for a limited period of time and can be purified by column chromatography on silica gel or sublimed under high vacuum. However, some exceptions from a generally good stability and solubility of helicenes may exist.

Helicenes belong to angular acenes that are more stable than their linear analogues. It can be documented by the energy difference between the simple structural isomers such as phenanthrene **19** and anthracene **20** as follows from various experimental and theoretical studies (Figure 4.4) (Poater et al., 2007). However, helicenes are forced to depart from planarity that installs a torsional strain into them. It does not lead to a significant destabilization of helicene molecules as such a deformation is almost evenly spread over the inner part of the backbone.

The helical architecture of helicenes renders them inherently chiral. They can be thermally racemized following the reversible first-order kinetics, and the energy barrier ($\Delta\Delta G_{\mathrm{rac}}$) to this process depends on the helicene structure. Illustratively, the racemization barrier within a homologous series of [4]- to [9]helicene ranges between 4.0 kcal/mol ([4]helicene **21**) and 43.5 ([9]helicene **25**), approaching a plateau at ca. 44 kcal/mol (Figure 4.5). This behavior is, however, counterintuitive, because molecular models indicate a pronounced steric congestion, which might be expected to prevent racemization. Although chemical pathways were also taken into account, a purely conformational process is now generally accepted to operate in helicene racemization (Grimme and Peyerimhoff, 1996; Janke et al., 1996; Barroso et al., 2017). The most recent calculations at the density functional theory (DFT) level of theory covering the dispersion effects show that racemization of lower [n]helicenes (n = 4–7) is a concerted process passing a single transition state, but racemization of higher [n]helicenes ($n \geq 8$) is a multistep process involving intermediates (Barroso et al., 2017). The barrier-determining transition states can be of various symmetries such as C_{2v} (planar, at [4]helicene **21**), C_s (achiral "*meso*", resembling a cut saddle-shaped structure, at [5]helicene **22** and its higher homologues such as [8]helicene **24**) and C_1 (chiral, in fact distorted transition state C_s, at some helicene analogues such as **16**) (Figure 4.6). It is worth noting that helicenes having C_1 symmetry in their ground state (for instance, monosubstituted carbohelicenes) can undergo a chiral pathway of racemization (Mislow, 1954; Mislow and Bolstad, 1955), where every intermediate or transition state is chiral and preserves a C_1 symmetry.

Nevertheless, starting from [6]helicene **1**, helicenes are fully configurationally stable at room temperature and can be heated at least to ca. 150°C for a few hours without a considerable racemization. Substituents installed in the innermost position 1 significantly increase configurational stability of helicenes such as the [4]helicene derivative **26** or 1-methyl[6]helicene **27** (Figure 4.7). The removal

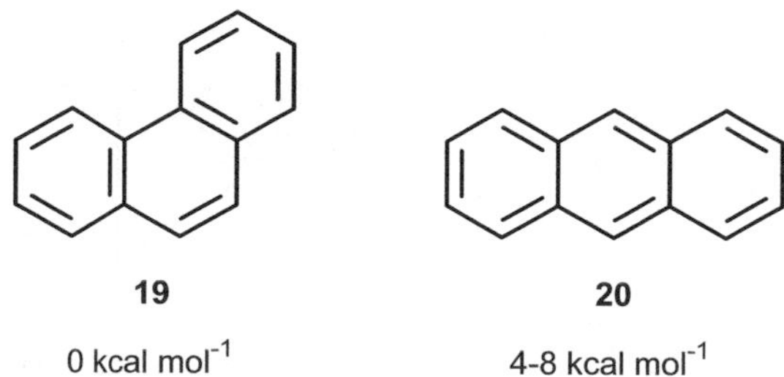

FIGURE 4.4 Relative energy of angular and linear acenes: phenanthrene **19** *versus* anthracene **20**.

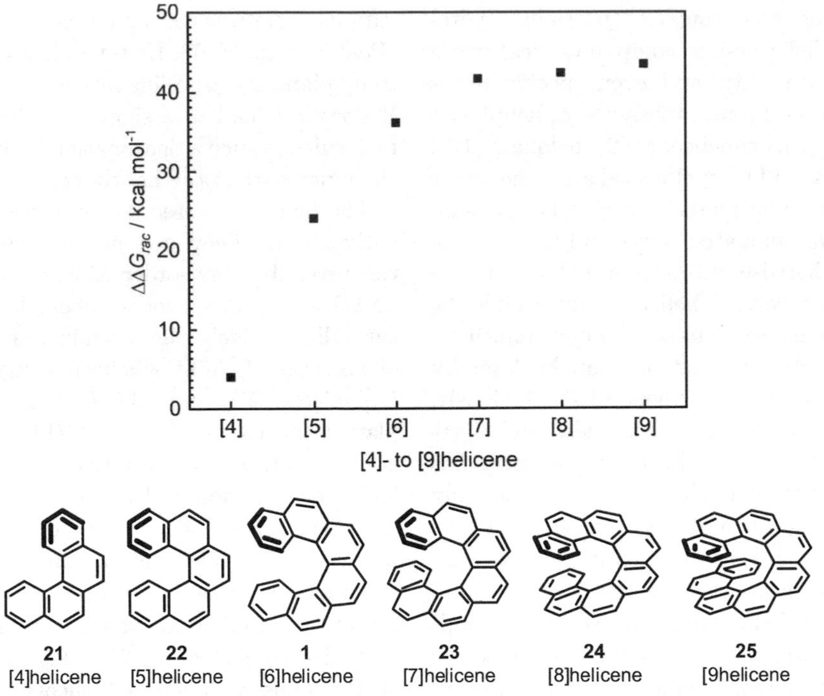

	$\Delta\Delta G_{\mathrm{rac}}$ (kcal/mol^{-1})	Temperature (°C)	Half-Life Time (min)	References
[4]helicene **21**	4.0	ΔE calcd by DFT-D		Barroso et al. (2017)
[5]helicene **22**	24.1	57	63	Goedicke and Stegemeyer (1970)
[6]helicene **1**	36.2	188	187	Martin and Marchant (1972)
[7]helicene **23**	41.7	239	743	Martin and Marchant (1974)
[8]helicene **24**	42.4	240	1761	Martin and Marchant (1974)
[9]helicene **25**	43.5	294	123	Martin and Marchant (1974)

FIGURE 4.5 Racemization barriers of [4]- to [9]helicene ((*P*)-enantiomers are shown).

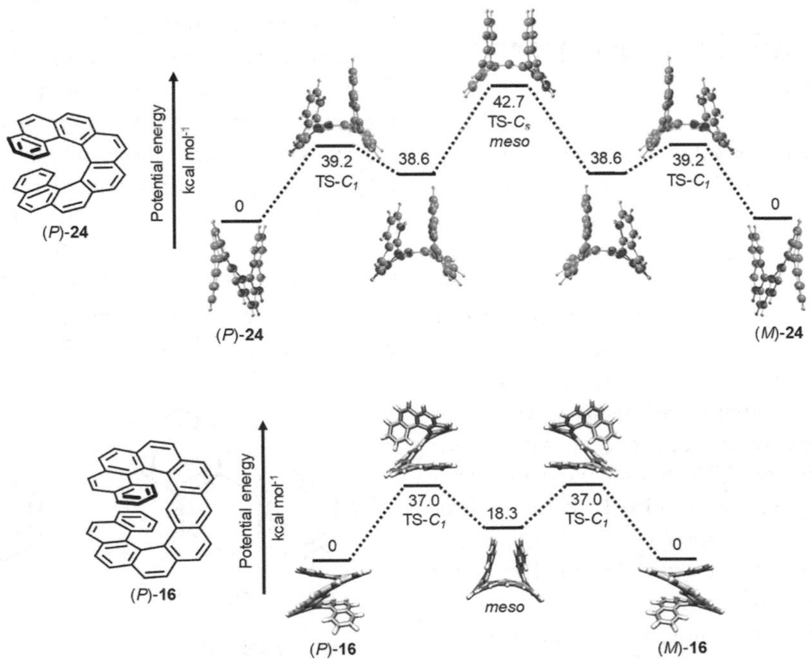

FIGURE 4.6 Complementary racemization pathways of [8]helicene **24** (Barroso et al., 2017) and [11]anthrahelicene **16** (Sehnal et al., 2009) calculated by DFT with an empirical dispersion correction.

of a hydrogen atom in this position as in 1-azahelicene **28** or displacement of the benzene unit(s) with a five-membered heterocycle as in dithia[6]helicene **29** (leading to a less-curved heterohelicene backbone due to geometric reasons) results in a faster racemization.

Helicenes exhibit remarkable chiroptical properties. They are known to possess giant values of specific optical rotation [α] that may reach several thousand degrees (Figure 4.8). The empirical rule of thumb says that laevorotatory helicenes have (*M*) helicity and dextrorotatory ones have

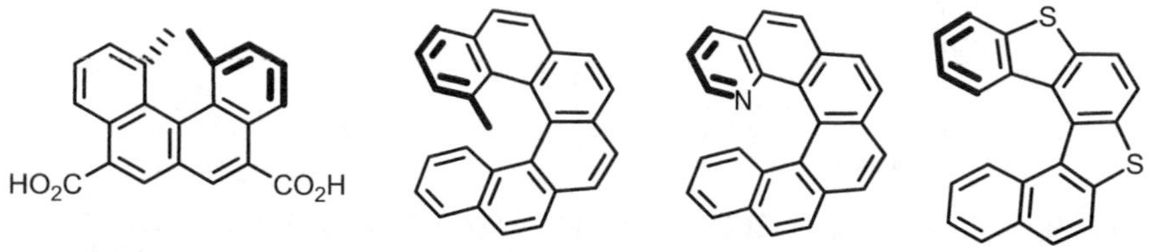

HO_2C CO_2H S S N

26 [4]helicene derivative | **27** 1-methyl[6]helicene | **28** 1-aza[6]helicene | **29** dithia[6]helicene

	$\Delta\Delta G_{rac}$ (kcal/mol^{-1})	Temperature (°C)	Half-Life Time (min)	References
[4]helicene deriv. **26**	stable under reflux in $CHCl_3$-MeOH			Okubo et al. (1998)
1-CH_3-[6]helicene **27**	43.8	269	231	Borkent and Laarhoven (1978)
1-aza[6]helicene **28**	32.2	140	72	Míšek et al. (2008)
dithia[6]helicene **29**	23.7	25	241	Wynberg and Groen (1969)

FIGURE 4.7 Effects of structural modifications on racemization barriers of [4]- and [6]helicene and its heteroanalogues ((*P*)-enantiomers are shown).

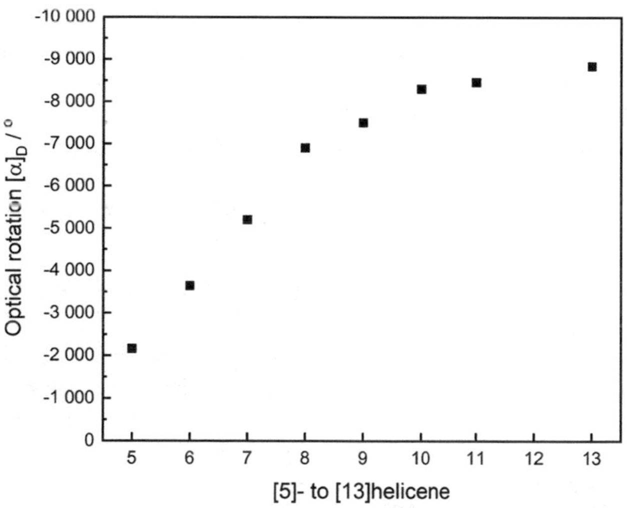

	$[\alpha]_D$ (°)[a]	Solvent	References
(-)-(*M*)-[5]helicene **22**	-2 160[b]	chloroform	Bestmann and Both (1974)
(-)-(*M*)-[6]helicene **1**	-3 640±10	chloroform	Newman and Lednicer (1956)
(-)-(*M*)-[7]helicene **23**	-5 200±200	chloroform	Martin et al. (1968)
(-)-(*M*)-[8]helicene **24**	-6 900±100	chloroform	Martin and Libert (1980)
(-)-(*M*)-[9]helicene **25**	-7 500±100	chloroform	Martin and Libert (1980)
(-)-(*M*)-[10]helicene **30**	-8 300±100	chloroform	Martin and Libert (1980)
(-)-(*M*)-[11]helicene **31**	-8 460±100	chloroform	Martin and Libert (1980)
(-)-(*M*)-[13]helicene **32**	-8 840±100	chloroform	Martin and Libert (1980)

[a] Measured at 589 nm, 20-26 °C, in chloroform (c = 0.004-0.29 g/100mL); [b] Based on $[\alpha]_D^2$ +2 160 of (+)-(*P*)-**22** (c = 2.90 mg/mL, chloroform).

FIGURE 4.8 Specific optical rotation $[\alpha]_D$ of (−)-(*M*)-[5]- to (−)-(*M*)-[13]helicene (in deg cm^3/g dm).

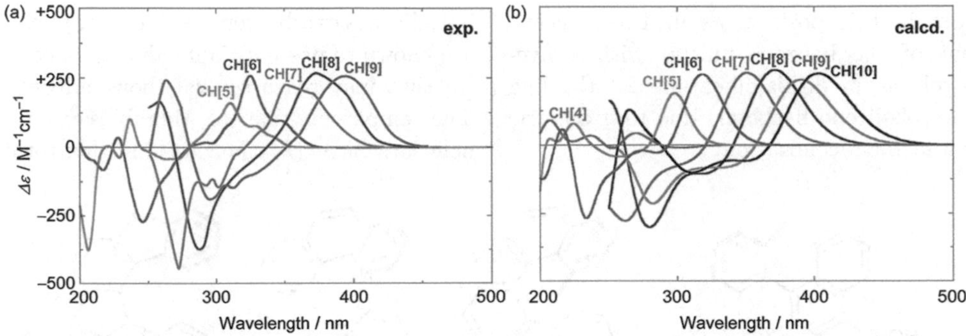

FIGURE 4.9 Experimental (a) and calculated (b) ECD spectra of [4]helicene **21** to [10]helicene **30** of P helicity (CH[4]-CH[10]). The ECD spectra were measured in chloroform ([7]helicene **23**, [8]helicene **24**, [9]helicene **25**), n-hexane/isopropanol (98:2, [5]helicene **22**) or acetonitrile ([6]helicene **1**); the theoretical spectra were calculated at the RI-CC2/TZVPP//DFT-D2-B97-D/TZVP level. (Reprinted from Nakai et al., 2012. With permission.)

(P) helicity but there are exceptions among helicene-like compounds such as helquats (Reyes-Gutiérrez et al., 2015). The value of the specific rotation [α] is possible to calculate by the time-dependent density functional theory (TD-DFT) method with B3-LYP or BH-LYP functional by using the Dunning's aug-cc-pVDZ (augD) basis set to reach a relatively good agreement between the theoretical and experimental data (Nakai et al., 2012): The measured $[\alpha]_D$ of (+)-(P)-[6]helicene **1** is +3,760 ± 20 (c 0.0025, acetonitrile) and calculated from +3,400 to +4,760 depending on the TD-DFT method employed.

Intense electronic circular dichroism (ECD) spectra are typical for parent fully aromatic enantiopure helicenes (Figure 4.9). They can be calculated at the TD-DFT (B97D/ccpVDZ) level to correlate experimental ECD spectra (Buchta et al., 2015) or by the RI-CC2/TZVPP// DFT-D2-B97-D/TZVP method to reproduce spectra in both excitation energy and rotational strength without any shift or scaling (Nakai et al., 2012). ECD spectra of helicenes are routinely used to determine their helicity (absolute configuration) by correlating the experimental ECD spectrum of an inspected molecule with the calculated one or with the known spectrum of a structurally related analogue of given helicity as in the case of (+)-(P)-1-aza[6]helicene **28** and (+)-(P)-2-aza[6]helicene **33** *versus* (+)-(P)-[6]helicene **1** (Míšek et al., 2008) (Figure 4.10). Helicity of a helicene derivative can be in most cases directly inferred from the respective experimental ECD spectrum according to the sign of the longest wavelength ECD band: Its positive sign indicates P helicity and *vice versa*. However, there are exceptions from this rule even though they are rare (Reyes-Gutiérrez et al., 2015).

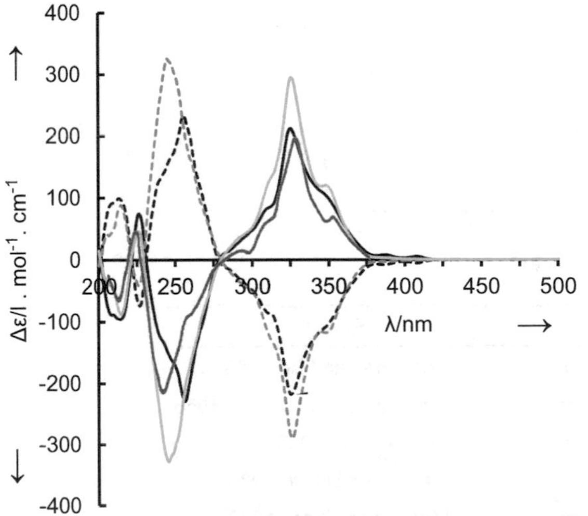

FIGURE 4.10 ECD spectra of (−)-(M)-1-aza[6]helicene **28** (black dashed line), (+)-(P)-1-aza[6]helicene **28** (black solid line), (−)-(M)-2-aza[6]helicene **33** (light gray dashed line), (+)-(P)-2-aza[6]helicene **33** (light gray solid line) in acetonitrile (4.70 × 10^{-4} M) as well as the spectrum of (+)-(P)-[6]helicene **1** (gray line) in methanol as a reference (1.65 × 10^{-5} M). (Reprinted from Míšek et al., 2008. With permission.)

4.3 Synthesis of Helicenes

Numerous synthetic methods for the preparation of helicenes, heterohelicenes and their analogues have been described in the literature since the pivotal contribution by M. S. Newman and D. Lednicer (1956). However, only the classical photocyclodehydrogenation of diaryl olefins and later [2 + 2 + 2] cycloisomerization of alkynes have so far been more widely exploited when building the helicene scaffolds.

A remarkable step forward in the synthesis of helicenes came in the late sixties when photodehydrocyclization of stilbene-type precursors was introduced as the first general method for preparing various helicenes (Martin, 1974; Mallory and Mallory, 1984; Laarhoven, 1989). Alternatively, terms such as photocyclodehydrogenation or photocyclization of stilbenes are also used. This general method for the synthesis of angular aromatics is based on UV light-induced *cis*/*trans* isomerization of 1,2-diarylethylenes **34** followed by conrotatory electrocyclization of the *cis* isomer to generate a primary dihydroaromatic product **35** with *trans* configuration (Figure 4.11). In the presence of an oxidant (iodine in a

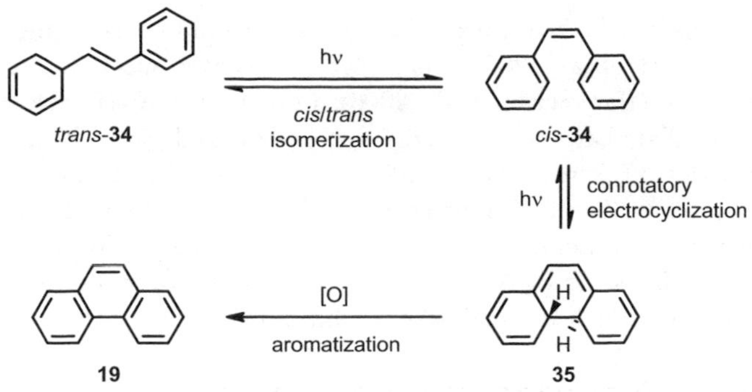

FIGURE 4.11 Mechanism of photodehydrocyclization of stilbene-type precursors to form angular aromatics.

stoichiometric amount is a superior one), the fully aromatic system **19** is furnished. This synthetic methodology has gained the widest popularity in the helicene community over decades. It benefits from an easy access to the key stilbene-type precursors (via the Wittig olefination reaction) without a need to control the *cis/trans* configuration of the alkene unit. Photodehydrocyclization was successfully applied to the synthesis of diverse carbohelicenes, pyridohelicenes, thiahelicenes and other related systems, including functionalized derivatives (Figure 4.12). For instance, the whole series of [5]- to [16]helicene (or some heterohelicene members) was prepared by the single/multiple photodehydrocyclization of stilbene-type precursors (Chen and Shen, 2017). Discouraging results might occasionally be obtained as the photodehydrocyclization reaction suffers from several drawbacks: (i) Regioselectivity of the ring closure can be low, (ii) high dilution condition is usually required to

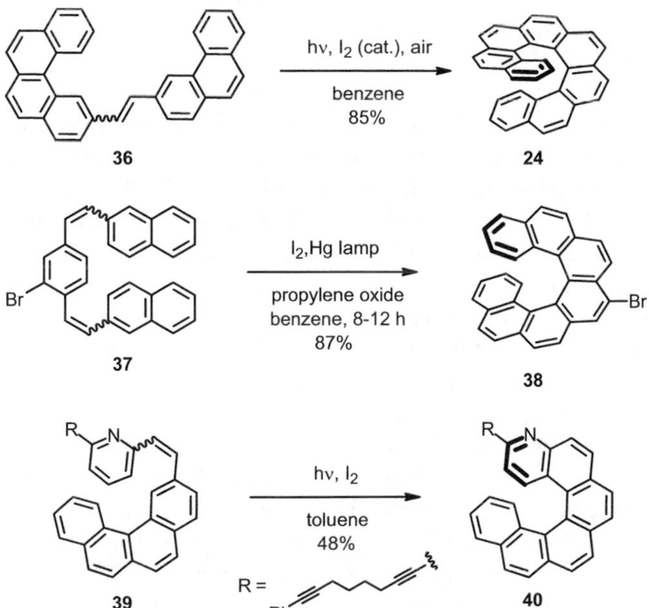

FIGURE 4.12 Photodehydrocyclization of stilbene-type precursors in the synthesis of [8]helicene **24** (Moradpour et al., 1975), [7]helicene **38** employing the directing effect of the bromine traceless auxiliary (Liu et al., 1991) and 4-aza[6]helicene derivative **40** (Graule et al., 2009).

prevent undesired photodimerization, (iii) some functional groups such as $N(CH_3)_2$, AcO or NO_2 can depopulate the required excited state, (iv) the high concentration of generated HI leads to photoreduction of double bonds, (v) the primary helical product can planarize via an intramolecular oxidative C–C coupling and (vi) asymmetric photocyclization to define helicity of the products is in the most cases difficult to control. Actually, some improvements of the methodology were developed: Propylene oxide or tetrahydrofuran is occasionally added as an instantaneous scavenger of HI (Liu et al., 1991), and the directing effect of the bromine traceless auxiliary can be used to control regioselectivity of the photocyclization (Liu and Katz, 1991). It is also possible to use a microwave reactor with an inner contactless UV/Vis lamp (Storch et al., 2013) or a flow reactor employing photoredox catalysis (Bédard et al., 2013).

The highly exergonic transition metal-catalyzed intramolecular [2 + 2 + 2] cycloisomerization of aromatic triynes (and related cyanodiynes or ynedinitriles) represents a new paradigm for the highly versatile nonphotochemical synthesis of helicenes, tetrahydrohelicenes, dibenzohelicenes, heterohelicenes, partially hydrogenated helicenes, helicene-like compounds and their functionalized derivatives (Stará et al., 1998; Teplý et al., 2002; Míšek et al., 2008) (Figure 4.13). The key cyclization can be catalyzed by various metal complexes (mostly by Co^I, Ni^0, Rh^I), allows for constructing three cycles of the helicene skeleton in a single operation (in contrast to other concurrent processes), is highly tolerant to a wide range of functional groups and can be done in an asymmetric fashion. The starting triynes (and their heteroanalogues) are usually easily accessible. The synthetic methodology originally developed by Stará et al. (1998) has recently attracted a considerable attention as other groups also recognized its synthetic potential such as Vollhardt et al. (Han et al., 2002a), Teplý et al. (Adriaenssens et al., 2011), Tanaka et al. (Sawada et al., 2012), Shibata et al. (Shibata et al., 2012), Carbery et al. (Crittall et al., 2011), Diederich et al. (Roose et al., 2013), Marinetti et al. (Aillard et al., 2014) and others. Discouraging results might occasionally be obtained if the key triyne intermediate tends to polymerize and if the starting di- and trisubstituted aromatic building blocks are difficult to obtain. It is worth noting that dibenzo[5]-, dibenzo[6]- and dibenzo[7]helicenes as well as their functionalized derivatives and heterocyclic analogues are easily accessible (optionally on a multigram scale) by employing a short sequence of reliable processes such as Sonogashira coupling, Suzuki–Miyaura coupling and [2 + 2 + 2] alkyne cycloisomerization as reported by Stará, Starý et al. (Jančařík et al., 2013) (Figure 4.14). Dibenzohelicenes have an advantage over the parent helicenes because of the simplicity of their non-photochemical preparation, and therefore, they have the potential to mimic or even substitute parent helicenes in envisaged applications.

Nevertheless, there are other non-photochemical approaches to helicenes that were developed mostly in the last two decades. Katz et al. published a robust

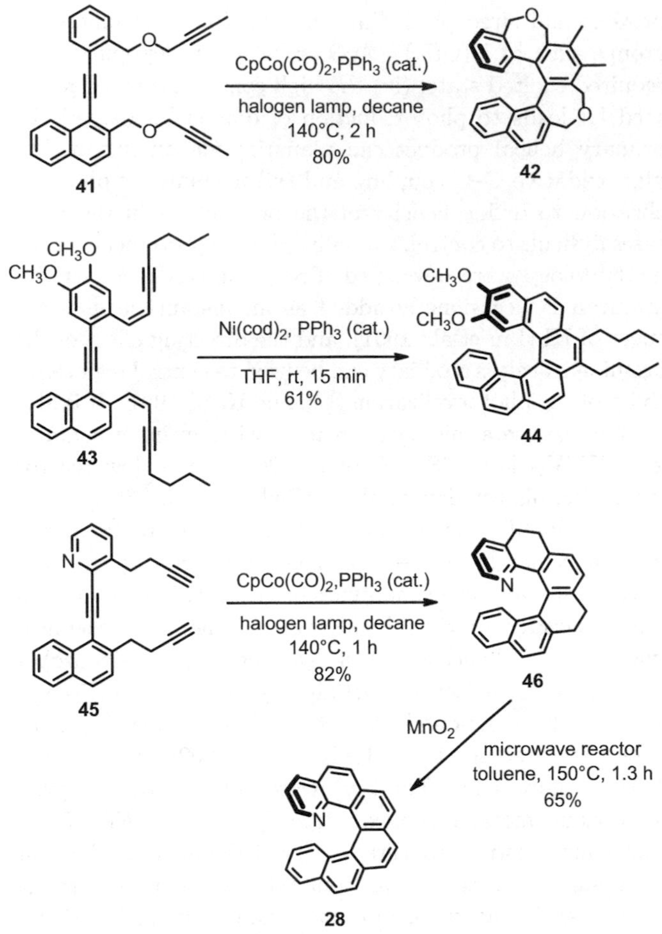

FIGURE 4.13 Transition metal-catalyzed intramolecular [2 + 2 + 2] cycloisomerization of triynes in the synthesis of the [6]helicene-like compound **42** (Stará et al., 1998), [6]helicene derivative **44** (Teplý et al., 2002) and 1-aza[6]helicene **28** (Míšek et al., 2008).

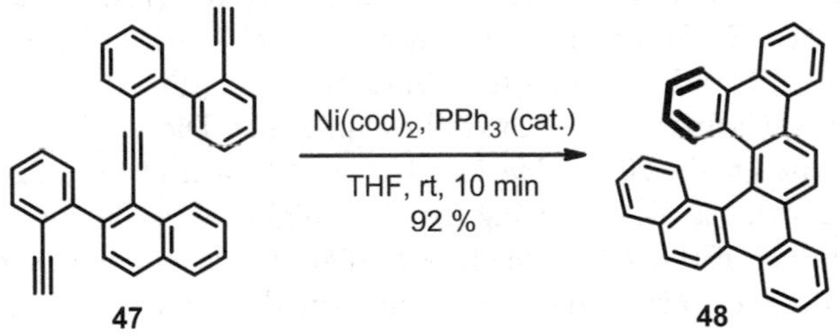

FIGURE 4.14 A versatile and short synthesis of dibenzohelicenes exemplified by the preparation of dibenzo[6]helicene **48** (Jančařík et al., 2013).

and versatile methodology based on thermal Diels–Alder reaction of aromatic bisvinylethers with *p*-benzoquinone in excess to afford helicenes with embedded terminal quinone moieties (Willmore et al., 1992). It allowed for the first time the preparation of various functionalized helicenes on a multigram scale (Phillips et al., 2001) (Figure 4.15). On top of that, there is a rich portfolio of other promising methodologies for the helicene synthesis, but they have not been widely employed so far. The construction of the helicene backbone is based on intramolecular processes such as Friedel–Crafts-type cyclization of 1,1-difluoro-1-alkenes (Ichikawa et al., 2008), homolytic aromatic substitution reaction (Harrowven et al., 2006), Pd-catalyzed C-H arylation (Kamikawa et al., 2007), Pd-catalyzed Stille–Kelly reaction (Takenaka et al., 2008), Ru-catalyzed ring-closing olefin metathesis (Collins et al., 2006), Pt-catalyzed alkyne arylation (Mamane et al., 2004; Storch et al., 2009; Weimar et al., 2013) or S$_N$Ar substitution reactions to synthesize cationic azahelicenes (Bosson et al., 2014).

4.4 Nonracemic Helicenes

Regardless of the synthetic methodology used, diverse helicenes were mostly prepared as racemates. Owing to the remarkable progress in the development and commercialization of chiral stationary phases for high-performance liquid chromatography (HPLC), racemic helicenes can easily be resolved into enantiomers by HPLC on chiral columns. It is a simple, general and straightforward approach to enantiopure (or highly enantioenriched) helicenes but semipreparative and namely preparative chiral columns are expensive. Nevertheless, if small amounts of nonracemic helicenes from a few milligrams to hundreds of milligrams are required, then resolution of racemate by HPLC on a chiral column is a method of choice (Figure 4.16). For analytical purposes, capillary electrophoresis (CE) with chiral selectors can also be used to resolve enantiomers of decently water-soluble derivatives of helicenes such as helquats (helical *N*-heteroaromatic dications) as demonstrated by Teplý, Kašička et al. (Koval et al., 2011).

If larger amounts of enantiopure or at least highly enantioenriched helicenes are needed (hundreds of milligrams or grams), racemates have to be resolved by using chiral resolving agents. Co-crystallization of a racemic helicene derivative (azahelicene or helquat) with (−)-*O,O′*-dibenzoyl-L-tartaric or (+)-*O,O′*-dibenzoyl-D-tartaric acid may work to separate the corresponding diastereomeric salts as reported by Stará, Starý et al. (Míšek et al., 2008), Teplý et al. (Vávra et al., 2013) and Takenaka et al. (Peng and Takenaka, 2013). Provided a small amount of an enantiopure helicene is available and its racemate crystallizes as conglomerate, it can be resolved into enantiomers by preferential crystallization as described by Teplý et al. in the case of [7]helquat (Vávra et al., 2013). Racemic helical cations (configurationally locked diaza[4]helicenium ions) combined with chiral hexacoordinated phosphorus-centered binphat (1,1′-binaphthalene-2,2′diolato)(bis(tetrachlor-1,2-benzenediolato)phosphat(V)) anions can be resolved into diastereomeric pair by liquid chromatography as demonstrated by Laursen, Lacour et al. (Herse et al., 2003).

Alternatively, racemic helicenes can be chemically transformed into diastereomeric pairs by employing chiral derivatizing agents. Then, after liquid chromatography resolution of diastereomers, the original now enantiopure helicenes are regenerated. This methodology was successfully manifested by Katz et al. (transforming helicen-1-ols into camphanates) (Thongpanchang et al., 2000) and

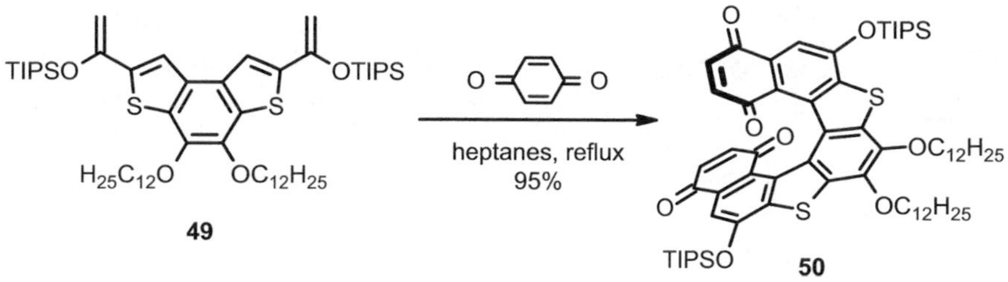

FIGURE 4.15 A general methodology for the preparation of helicene quinones by Diels–Alder reaction of aromatic bisvinylethers with *p*-benzoquinone (Phillips et al., 2001).

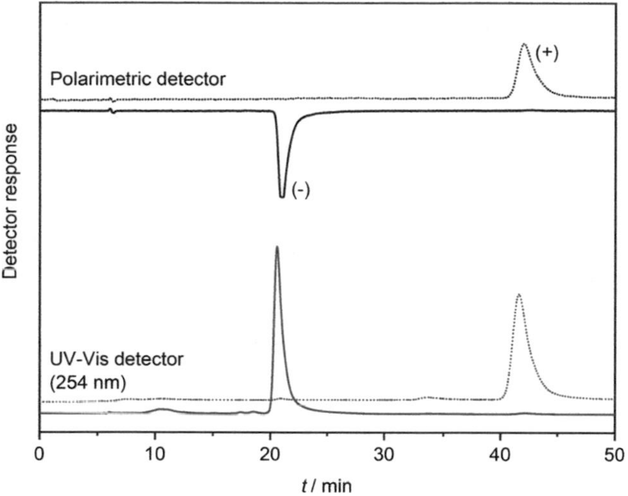

FIGURE 4.16 HPLC resolution of racemic [7]helicene-2-carboxylic acid **51** into the (−)-enantiomer (t_R = 20.5 min) and the (+)-enantiomer (t_R = 41.6 min) (Chirallica PST-4 column, 5 μm, 250 × 4.6 mm, heptane–isopropanol 95:5, flow rate 0.6 mL/min, repetitive 0.2 mg injections); the superposition of chromatograms of pure enantiomers is shown. (Adapted from Rybáček et al., 2011. With permission.)

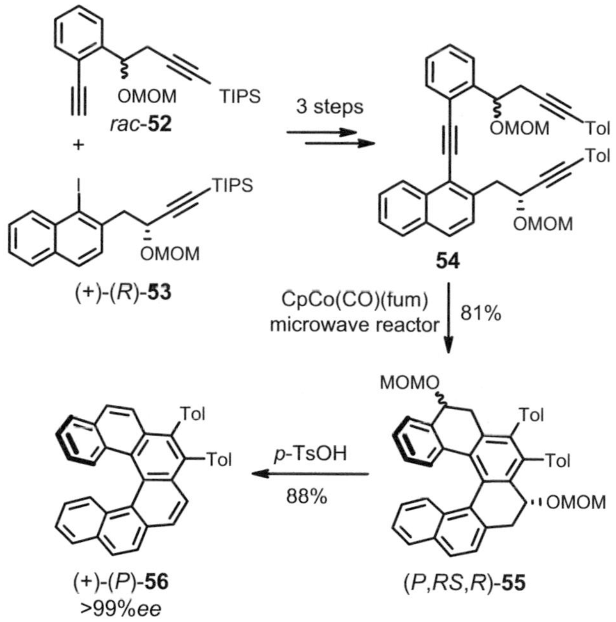

FIGURE 4.17 Asymmetric synthesis of enantiopure fully aromatic helicenes by alkyne [2 + 2 + 2] cycloisomerization employing chiral building blocks (obtained through enantioselective biocatalysis) is ultimately controlled by 1,3-allylic-type strain operating in the postcyclization thermodynamic equilibration of diastereomeric tetrahydrohelicene derivatives (Šámal et al., 2015).

Lacour et al. (transforming [4]heterohelicenium cations into adducts with a chiral sulfoxide) (Laleu et al., 2005).

As far as asymmetric synthesis (both catalytic and stoichiometric) of helicenes is concerned, various methodologies were explored to circumvent the necessity of racemate resolution that might fail or be laborious if enantiopure helicenes on a multigram scale are needed. Although there is still a way to go in order to develop a practical and general asymmetric synthesis (in particular catalytic) of enantiopure helicenes, the recent achievements indicate that a solution to this problem present since the birth of helicene chemistry in 1956 (Newman and Lednicer, 1956) is possible. Stará, Starý et al. developed a general methodology for the preparation of uniformly enantiopure fully aromatic [5]-, [6]-, and [7]helicenes through stoichiometric asymmetric synthesis (Šámal et al., 2015) (Figure 4.17). This approach is based on a tandem of [2 + 2 + 2] cycloisomerization of centrally chiral triynes and postcyclization thermodynamic equilibration of diastereomeric tetrahydrohelicene derivatives being ultimately controlled by the 1,3-allylic-type strain. The point-to-helical chirality transfer utilizing a traceless chiral auxiliary features a remarkable independence from the diverse structural perturbations allowing for the preparation of both parent and functionalized fully aromatic helicenes in enantiomer ratios of >99:<1. It is worth noting that the key building blocks such as (+)-(*R*)- or (−)-(*S*)-**53** can be received optically pure through enantioselective biocatalysis on a multigram scale. The same principle of stereocontrol can be applied to the versatile asymmetric synthesis of enantio- and diastereopure oxa[5]-, oxa[6]-, and oxa[7]helicenes (Žádný et al., 2012) (Figure 4.18). The diastereoselective [2 + 2 + 2] cycloisomerization of centrally chiral triynes in the presence of Co^I or Ni^0 complexes plays a key role in the formation of helical scaffolds with two 2*H*-pyran rings. The major advantages of this methodology are that (i) excellent diastereoselectivity is guaranteed (*d.r.* uniformly 100:0), (ii) the stereochemical outcome of the cyclization is highly tolerant to the structural diversity of the products, (iii) the synthesized 2*H*-pyran oxa[5]helicenes

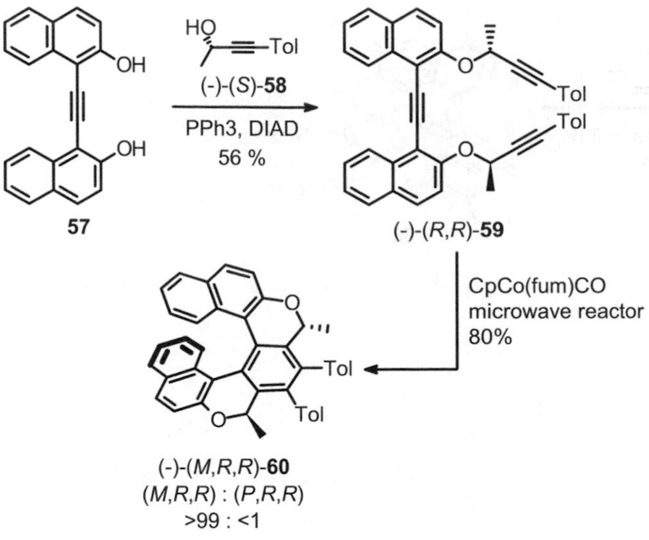

FIGURE 4.18 Asymmetric synthesis of enantio- and diastereopure 2*H*-pyran-modified helicenes by alkyne [2 + 2 + 2] cycloisomerization employing commercially available chiral building blocks (Žádný et al., 2012).

exist as single helices even at higher temperature (in contrast to the parent [5]helicene **22** that racemizes at room temperature), (iv) the helicity of the products can be easily predicted computationally and (v) both enantiomers of but-3-yn-2-ol (a key chiral building block) are commercially available. If 2*H*-pyran-modified helicenes can substitute the fully aromatic parent helicenes in a given application, the versatility of their asymmetric synthesis and its step economy makes them the most accessible enantiopure helicene surrogates in multigram amounts.

There are other examples of a stoichiometric asymmetric synthesis of nonracemic helicenes that met the success but a further development of some of them seemed to be discontinued. For instance, classical photodehydrocyclization of stilbene-type precursors can be carried out in an astonishingly stereoselective fashion. This was well demonstrated by the pioneering works by Vanest and Martin (1979), Katz et al. (1993) and later by Voituriez, Marinetti et al. (Yavari et al., 2014) who used stereocenter(s) external or internal to the helix to control stereoselectivity of helicene cyclizations. Carreño et al. developed an asymmetric version of the Diels–Alder approach to helicenes providing helical quinones with excellent optical purities (Urbano and Carreño, 2013).

Importantly, various synthetic methodologies were published to transform nonracemic biaryl precursors into nonracemic helicenes, thus demonstrating an excellent relay of stereochemical information when transforming axial chirality into helicity. An efficient synthetic route to highly enantioenriched 1-aza[6]helicenes **28** was developed by Fuchter et al. who succeeded in converting the separated atropisomers of the axially chiral biaryl (separated by semipreparative HPLC on a chiral column) through alkyne–arene cycloisomerization into azahelicene products (Weimar et al., 2013). Srebro-Hooper, Crassous, Guy et al. built an enantiopure backbone of the diazahelicene-like dibenzo[c]acridine derivative from the optically pure axially chiral bis-tetralone, which was obtained from racemate by preferential crystallization (it formed conglomerate) (Bensalah-Ledoux et al., 2016). Kamikawa et al. converted an enantiopure biaryl building block (obtained by resolution of the corresponding racemate by liquid chromatography on a chiral column) into enantiopure 6-aza[6]helicene in good yield by utilizing a palladium-catalyzed C-H annulation reaction (Kaneko et al., 2013). Nozaki et al. reported an efficient strategy for the synthesis of highly enantioenriched aza- and oxa[7]helicenes from the nonracemic biaryl precursor (4,4′-biphenanthryl-3,3′-diol) in good yields (Nakano et al., 2005).

Nevertheless, asymmetric synthesis of optically pure helicenes by an enantioselective catalytic process that forms the helical backbone in a stereocontrolled way is a challenging task of the highest priority. The first promising attempts at this were already reported, and therefore, there are prospects to develop a general, practical and short catalytic asymmetric synthesis of enantiopure helicenes. The synthesis of helicenes via alkyne [2 + 2 + 2] cycloisomerization mentioned above calls for an enantioselective fashion in order to control helicity of the resulting products by employing chiral metal catalysts. This approach to nonracemic helicenes introduced by Stará et al. (1999) developed into promising methodology providing highly enantioenriched helicenes in up to 85% *ee* under Ni^0 catalysis (Jančařík et al., 2013) or Rh^I catalysis (Sawada et al., 2012) (Figure 4.19).

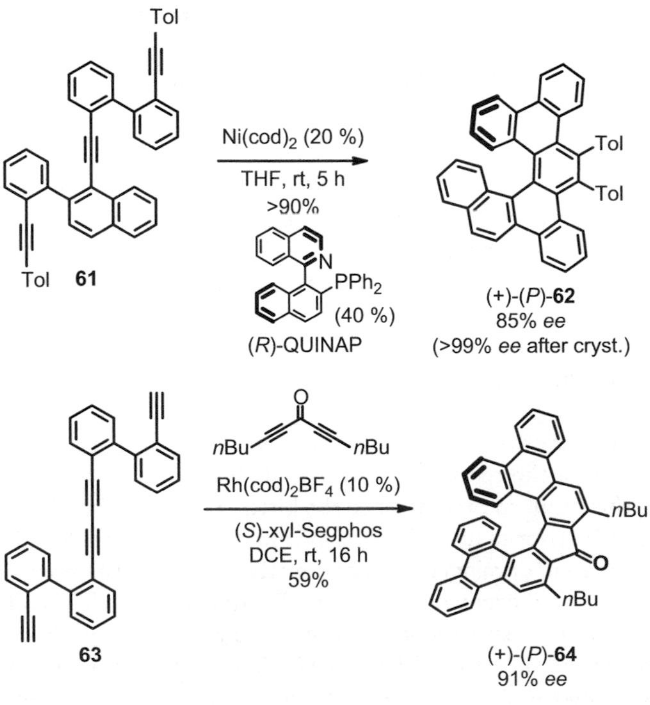

FIGURE 4.19 The synthesis of highly enantioenriched helicenes employing enantioselective alkyne [2 + 2 + 2] cycloisomerization (intramolecular or intermolecular) catalyzed by a chiral Ni complex (Jančařík et al., 2013) or Rh complex (Sawada et al., 2012).

A pioneering study published by List et al. reported on the asymmetric organocatalytic approach to indole/carbazole-derived azahelicenes starting from simple achiral materials (Kötzner et al., 2014). It employed enantioselective Fischer indolization reaction catalyzed by a chiral SPINOL (1,1′-spirobiindane-7,7′-diol)-derived phosphoric acid to form the helical backbone in good yield. The high level of stereocontrol in the synthesis of a series of azahelicene derivatives (receiving them in up to 92% *ee*) originated in a cleverly designed organocatalyst forming a deep chiral pocket to stabilize intermediates by π-π interactions. Furthermore, a highly enantioselective Au-catalyzed intramolecular hydroarylation of alkynes was employed by Alcarazo et al. in the synthesis of nonracemic substituted [6]helicenes in up to 99% *ee* (González-Fernández et al., 2017). Cationic TADDOL (2,2-dimethyl-α,α,α′,α′-tetraphenyldioxolane-4,5-dimethanol)-derived phosphonites served as tunable chiral ligands for gold.

4.5 Giant Helicenes and Their Congeners

Axially and laterally extended helicenes and their analogues are both challenging synthetic targets and potentially useful materials. Although numerous approaches to helicenes have been explored, only the multiple photocyclodehydrogenation of diaryl olefins and the [2 + 2 + 2] cycloisomerization of alkynes have so far passed the tough test for synthetic methods that are suitable for the preparation of molecular screws containing more than eleven all-*ortho* condensed rings in the molecular screw. The current record in the length of parent carbohelicenes is held by Mori et al., who developed the synthesis of the [16]helicene **67** and its derivative **66** by the use of a sextuple photocyclodehydrogenation to fold a cleverly designed single-stranded arylene–vinylene precursor **65** into a helix (Mori et al., 2015) (Figure 4.20). The final [16]helicene **67** was surprisingly insoluble in sharp contrast to other long but shorter helicenes. The low yield of the key multiple cyclization manifests, however, both the limits of the current synthetic methodologies and the structural complexity of the highest helicenes. Not surprisingly, the lower homologues, such as [13]thiahelicene **71** and [15]thiahelicene **72** reported by Yamada et al. (1981), [12]helicene **32** and [14]helicene **74** synthesized by Martin and Baes (1975) and [13]helicene **73** described by Martin et al. (1969), are also quite rare. The longest helicene analogue prepared to date as a monodispersed molecular material is oxa[19]helicene **70** designed by Stará, Starý et al. that comprises 19 *ortho*/*meta*-fused benzene/2*H*-pyran rings in its helical backbone (Nejedlý et al., 2017) (Figure 4.20). The use of a flow reactor in the key multiple cobalt(I)-mediated alkyne [2 + 2 + 2] cycloisomerization to form 12 C–C bonds and 12 rings in a single operation was found to be advantageous for the efficient folding of the branched aromatic oligoyne **69** into oxa[19]helicene **70** in reasonable preparative yield. Furthermore, the introduction of stereogenic centers of known absolute configuration into the oligoyne precursor resulted in complete stereoselectivity in the cyclization step to provide the inherently chiral product in enantio- and diastereomerically pure form. Similarly, the landmark synthesis of [17]heliphene **68** (angular [9]phenylene) reported by Vollhardt and coworkers relied also on the triple cobalt(I)-mediated [2 + 2 + 2] cycloisomerization of an aromatic nonayne, but the yield of the key multicyclization was in this case low (2%–3.5%) (Han et al., 2002a) (Figure 4.20). As far as polydispersed helicene materials are concerned, Morin et al. reported the synthesis of a well-defined, helical graphene nanoribbon (GNR)

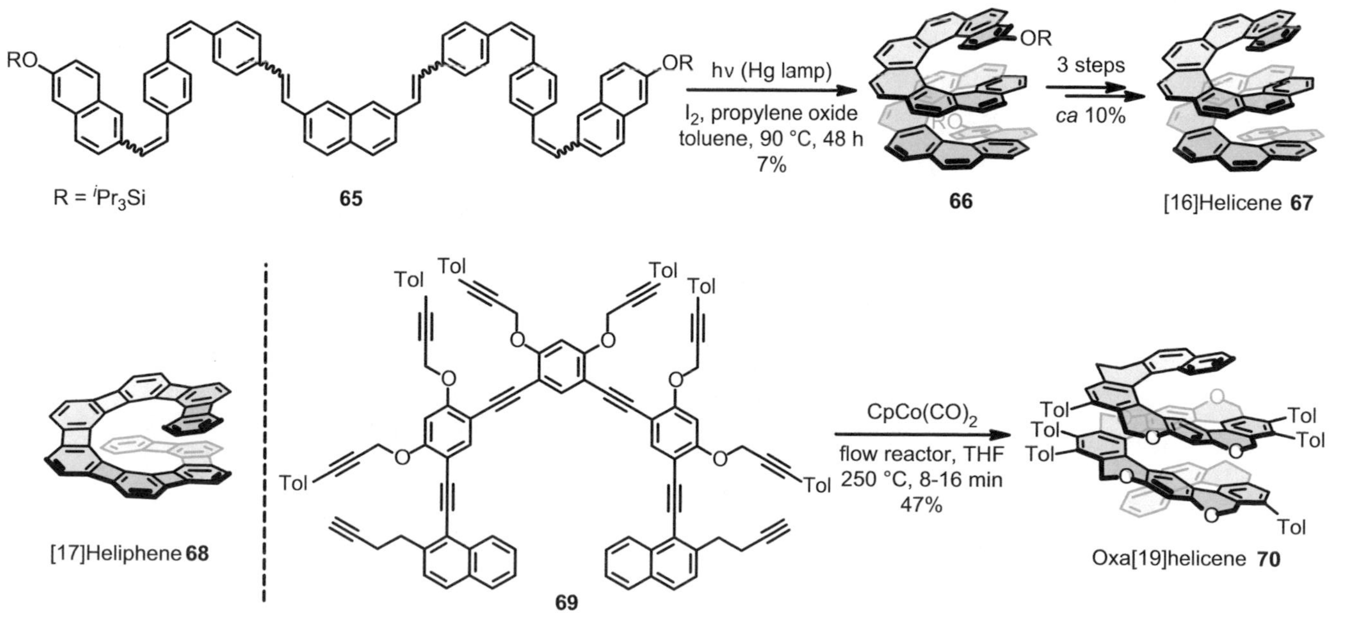

FIGURE 4.20 The longest helicenes and their analogues as monodispersed materials prepared to date.

76 from a polychlorinated poly(*m*-phenylene) **75** through a regioselective photochemical cyclodehydrochlorination (CDHC) reaction (Figure 4.21) (Daigle et al., 2017). The resulted orange polymer exhibits a degree of polymerization of 32 units, fluorescence and relatively good solubility in organic solvents (complete drying makes it difficult to dissolve again).

The helicene backbone can also be extended laterally to result in a widening of the helically coiled aromatic belt (Figure 4.22). Synthetic efforts in this respect are driven by envisaged screw dislocation in graphitic carbon materials (Xu et al., 2016) and their anticipated features such as nanometer-sized molecular inductors, spin filters and molecular spring materials responding to microscopic forces. Reliable synthetic methodologies such as [2 + 2 + 2] cycloisomerization of alkynes by Stará, Starý et al. (Buchta et al., 2015) (**78**) or Tilley et al. (Kiel et al., 2017) (**80**) and photocyclodehydrogenation of stilbene-type precursors by Hirose, Matsuda et al. (Nakakuki et al., 2018) (**83**), Nuckolls et al. (Schuster et al., 2018) (**85**) or Scott et al. (Fujikawa et al., 2016) (**86**) were employed to prepare diverse laterally extended helicenes. Instead of building the helicene backbone in a late stage of the synthesis, Martin, Crassous et al. extended the existing [6]helicene building block to receive helical bilayer nanographene (Evans et al., 2018) (**87**). Wide helicenes may exhibit a pronounced intramolecular π-π stacking and intriguing (chir)optical properties, e.g., one of the largest Cotton effects observed in the visible range that was found in **85** (due to a greatly amplified chirality).

There is another way of expanding the helicene backbone by fusing several helicene units into a single molecule. This concept of molecular design leads to multipole helicenes whose number in the literature is rapidly increasing. Hexabenzotriphenylene **89** synthesized from a corresponding *o*-trimethylsilyl (*o*-TMS) triflate as a benzyne precursor **88** by Pérez, Guitián et al. (Peña et al., 2000) represents a multipole helicene prototype in which a central aromatic unit (here benzene) is a joint component of the three surrounding helicene units (here [5]helicenes) (Figure 4.23). Similarly, Sygula et al. cyclotrimerized in situ generated corannulyne under palladium catalysis to provide a highly nonplanar hydrocarbon **90** with three both corannulene and [5]helicene subunits that prefers a twisted conformation of C_1 symmetry and exhibits dramatically different bowl-to-bowl inversion barriers (Yanney et al., 2011).

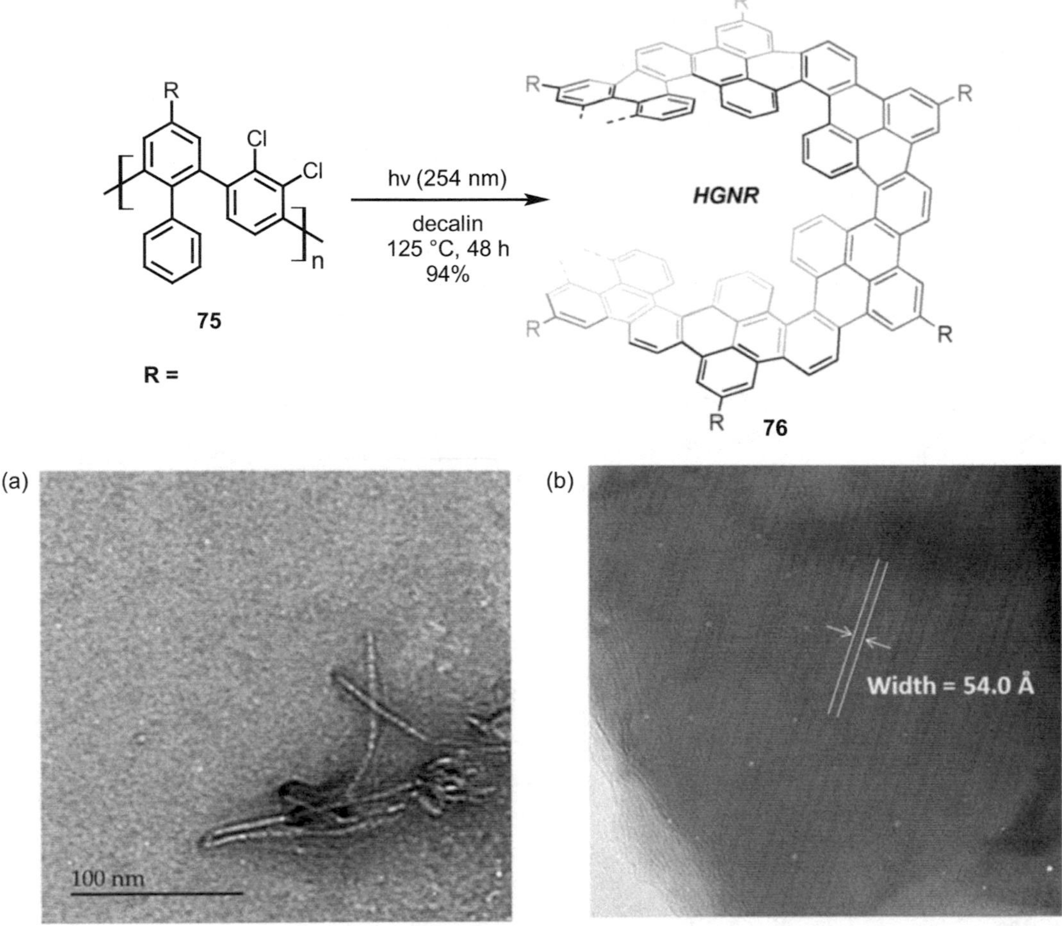

FIGURE 4.21 The polymeric helicenes: helically coiled GNRs **76** (HGNR). (a) Transmission electron microscopic (TEM) image of HGNRs prepared from suspensions in hexane/CH_2Cl_2. (b) TEM image of HGNRs prepared from suspensions in hexane/$CHCl_3$. Average measured width of HGNR = 54 Å. (Adapted from Daigle et al., 2017. With permission.)

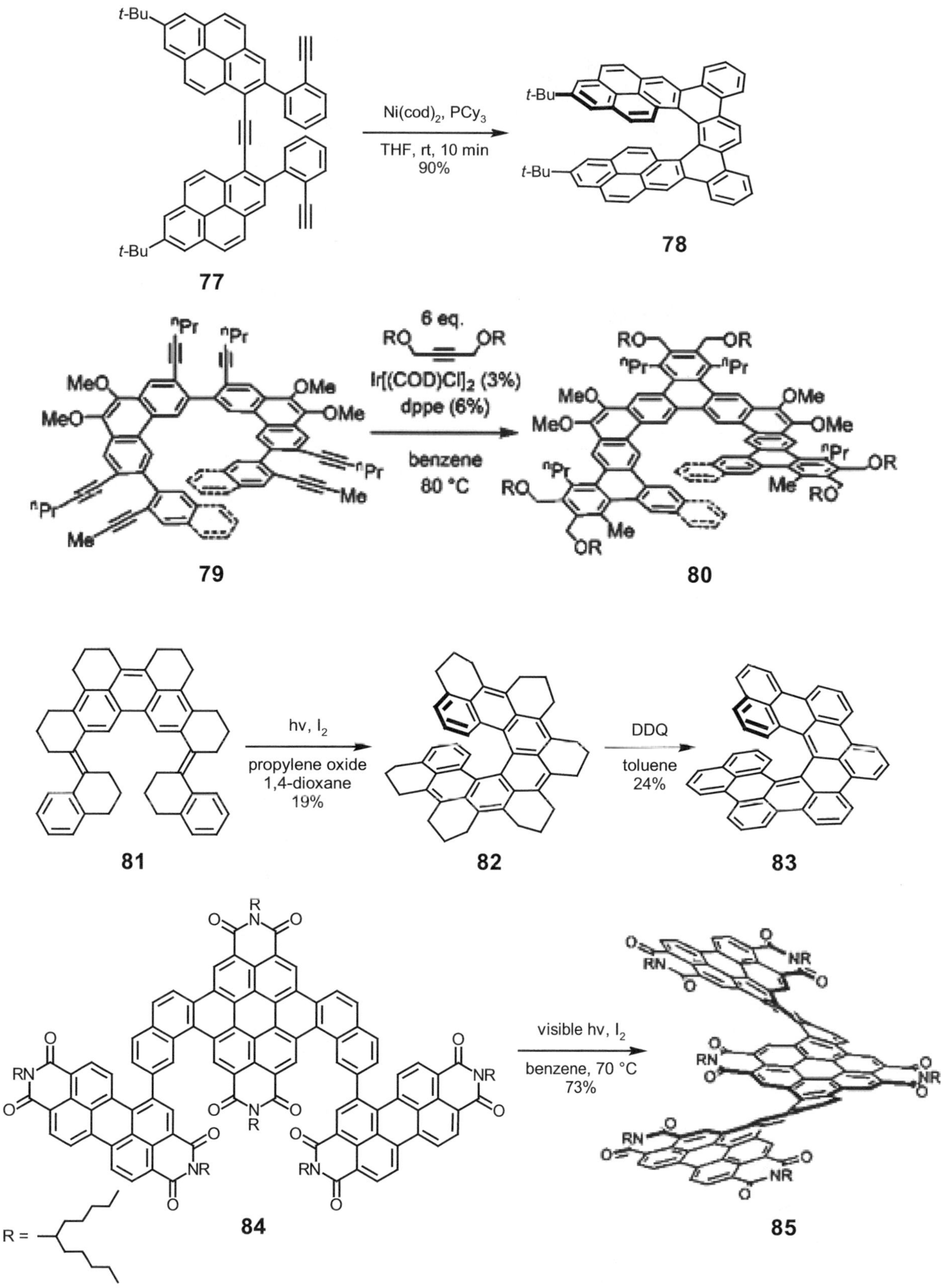

FIGURE 4.22 The laterally extended helicenes. (Parts adapted from Kiel et al., 2017 (**79**→**80**); Schuster et al., 2018 (**85**); Fujikawa et al., 2016 (**86**); Evans et al., 2018 (**87**). With permission.)

(Continued)

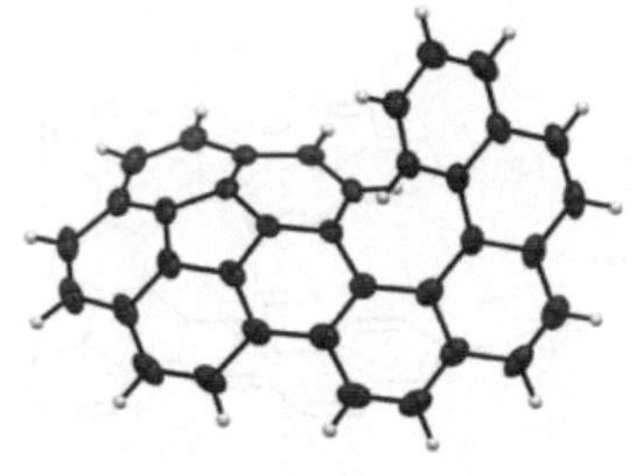

FIGURE 4.22 (CONTINUED) The laterally extended helicenes. (Parts adapted from Kiel et al., 2017 (**79**→**80**); Schuster et al., 2018 (**85**); Fujikawa et al., 2016 (**86**); Evans et al., 2018 (**87**). With permission.)

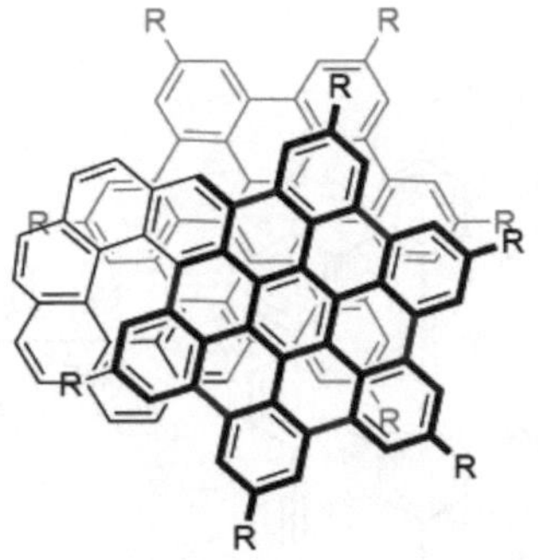

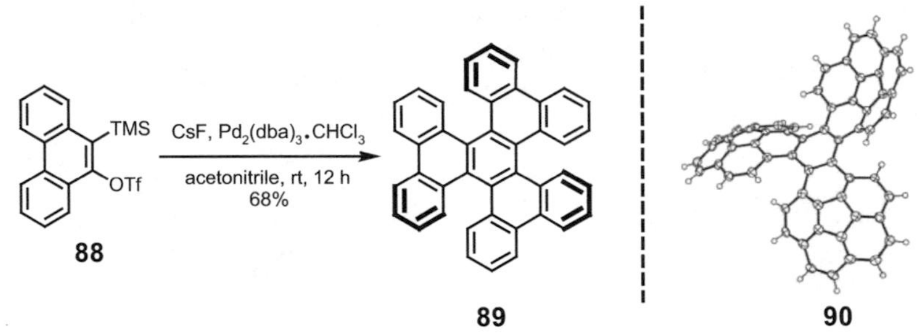

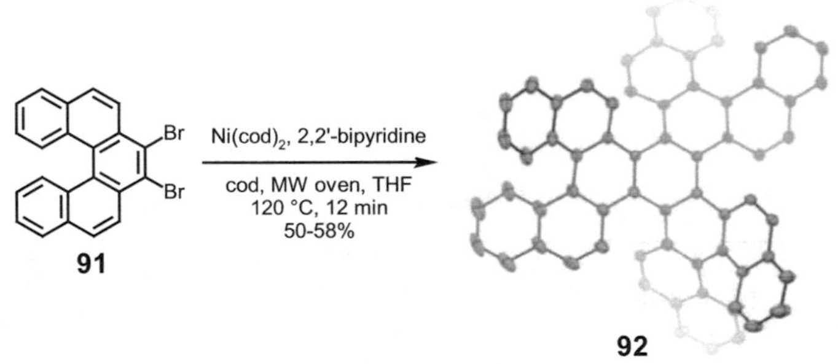

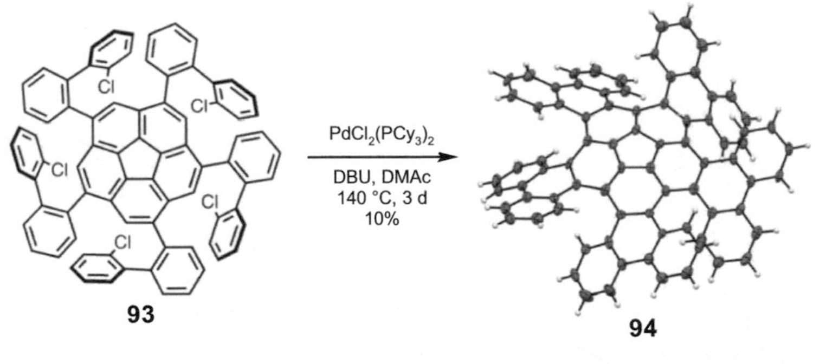

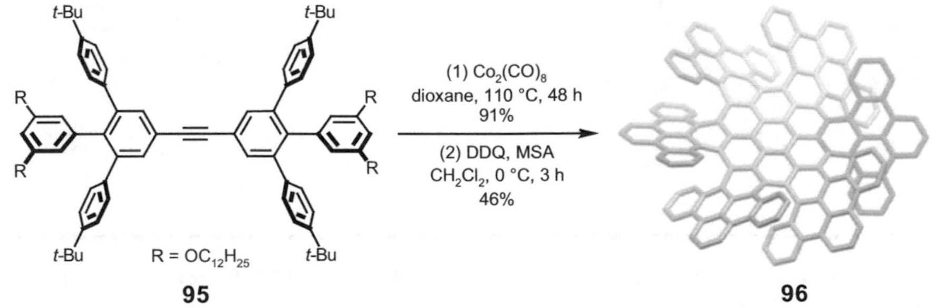

FIGURE 4.23 The multipole helicenes. (Parts adapted from Yanney et al., 2011 (**90**); Berezhnaia et al., 2017 (**92**); Kato et al., 2018 (**93**→**94**); Zhu et al., 2018 (**96**; substituents omitted for clarity). With permission.)

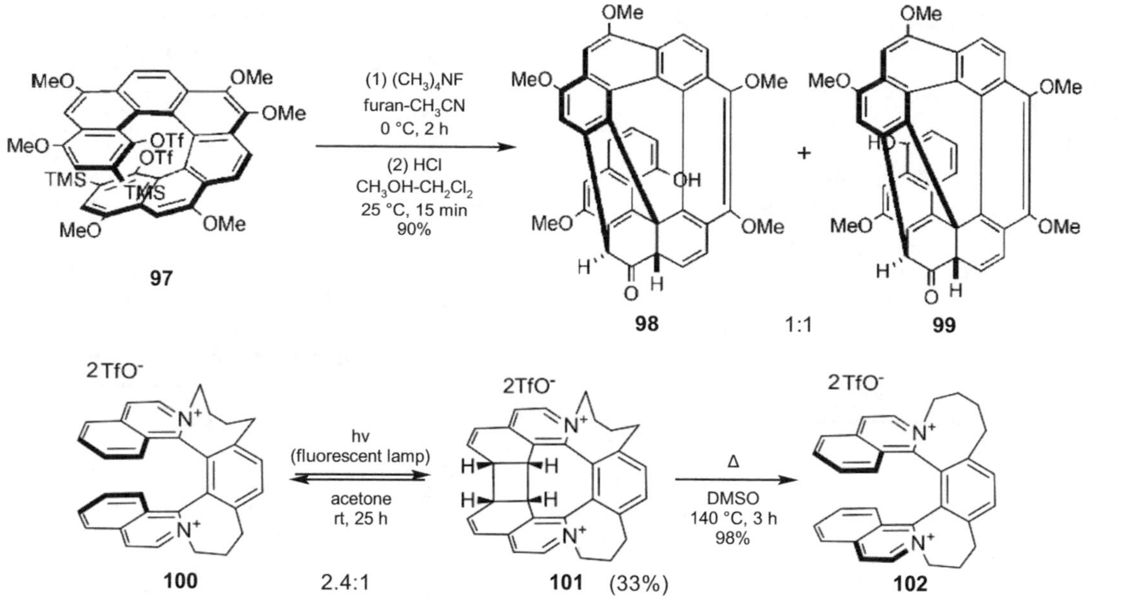

FIGURE 4.24 The helicene-derived circulene-like compounds **98**/**99** and **101**. (Adapted from Wang et al., 2004; Severa et al., 2012, respectively. With permission.)

Employing Yamamoto-type coupling, Coquerel, Gingras et al. succeeded in cyclotrimerizing [5]helicene dibromide **91** into hexapole [5]helicene **92** of D_3 symmetry that represents chiral nanographene propeller with six conformationally stable [5]helicene units (Berezhnaia et al., 2017). Independently, the same compound was prepared by Tsurusaki, Kamikawa et al. using the aforementioned Pd-catalyzed helicenyl aryne cyclotrimerization (Hosokawa et al., 2017). Not only benzene but also other polycyclic aromatic hydrocarbons can form a central core. To this end, Segawa, Itami et al. employed pentaborylated corannulene to synthesize, via multiple Suzuki–Miyaura coupling, pentaarylated corannulene **93** that underwent Pd-catalyzed fivefold intramolecular direct arylation to receive the quintuple [6]helicene **94** as a C_5-symmetric propeller-shaped structure (Kato et al., 2018). The most complex multiple helicene, here the hexapole [7]helicene **96**, was prepared by Wang et al. in two steps from arylated tolane **95** by employing Co-mediated alkyne [2 + 2 + 2] cycloisomerization followed by Scholl reaction to transform polyphenylene precursor into a circularly twisted chiral nanographene with a propeller structure (Zhu et al., 2018). Optically pure **96** displays high thermostability and a wealth of remarkable chiroptical and electronic properties.

Conceptually, helicenes are counterparts of circulenes (members of a larger cycloarene family) that can formally be designed by fusing the terminal rings of a helicene molecule (ideally in an *ortho* fashion) (C. Buttrick and T. King, 2017; Stępień et al., 2017; Rickhaus et al., 2016, 2017). Thus, circulenes are macrocyclic arenes in which a central polygon is completely surrounded and fused by mostly aromatic rings. There are two unique examples of helicene-derived circulene-like molecules that are formed by either intramolecular addition of an aryne intermediate to its own backbone providing a [6]circulene derivative **98**/**99** as observed by Katz et al. (Wang et al., 2004) or, after irradiation with a fluorescent lamp to undergo a reversible [6 + 6] photocycloaddition, [7]saddlequat **100** (an isolable conformer of the more stable corresponding helquat **102**), which forms a photostationary mixture with the chiral [8]circulenoid **101** as described by Teplý, Slavíček et al. (Severa et al., 2012) (Figure 4.24).

4.6 Computational Treatment of Helicenes and Their Properties

Various intriguing single-molecule properties of helicenes, including their spring-like mechanical behavior, charge transport, thermopower or piezoelectricity, were theoretically analyzed employing mostly the DFT calculations.

Rulíšek et al. studied low-lying vibrational states in a series of helicenes along with their behavior under the mechanical stress using DFT calculations (Rulíšek et al., 2007). They identified a low-lying vibrational mode as the symmetric stretching of the skeleton of [14]helicene **74** that behaves as a harmonic oscillator within the first-order approximation (Figure 4.25). The fact that helicenes exhibit an extraordinary deformation capacity when compressed/stretched along the helix axis owing to their elastic spring-like carbon scaffold was described also in other computational studies (Jalaie et al., 1997; Rempala and King, 2006; Guo et al., 2015; Šesták et al., 2015).

Electronic conductance of helicenes was theoretically studied by Treboux et al. by calculating the I/V characteristic for selected helicene classes and analyzing their ballistic electronic conductance (Treboux et al., 1999). It was found that helicenes can be made semiconducting or

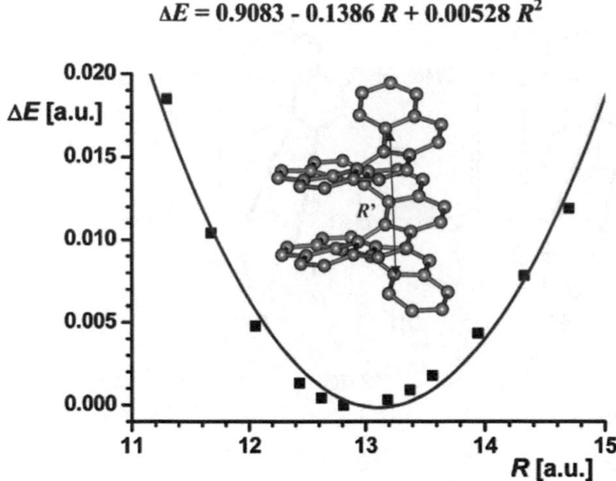

FIGURE 4.25 The calculated energy change corresponding to the stretching of [14]helicene **74** along the z-axis. (Adapted from Rulíšek et al., 2007. With permission.)

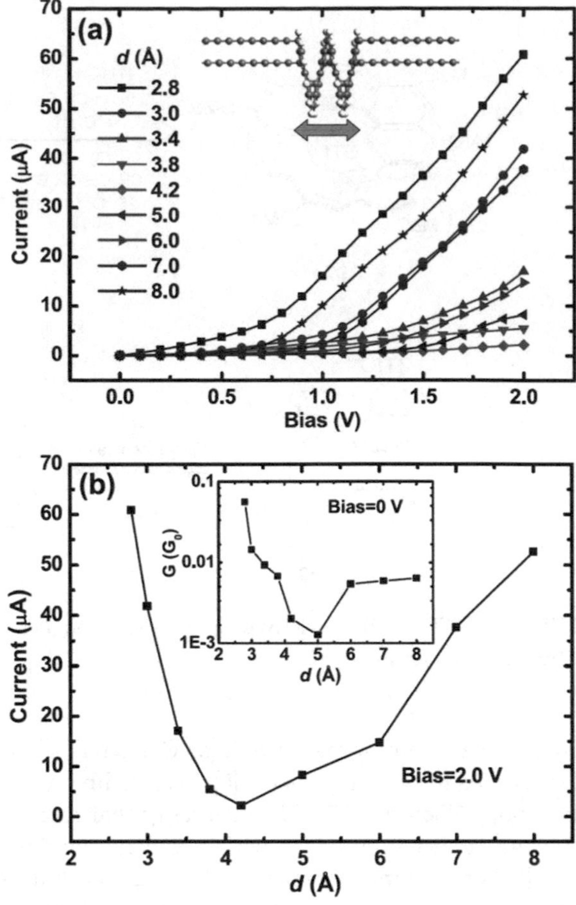

FIGURE 4.26 The I/V behavior of [12]helicene **32** contacted with carbon chain electrodes. The pitch of the gas-phase relaxed [12]helicene **32** is $d = 3.4$ Å. (a) I/V curve under different pitch d. (b) Current varies with d under the bias of 2.0 V. Inset: conductance under zero bias on a logarithmic scale (G_0 is the conductance quantum $G_0 = 2e^2/h$). (Adapted from Guo et al., 2015. With permission.)

metallic depending on the radius of the helix and the width of the helix ribbon. Intriguing results were obtained by Vacek et al. (2015) and Xiao et al. (Guo et al., 2015) who studied the electronic transport in helicenes under a mechanical stress. Using first-principles calculations (DFT, Landauer formalism, Green function), it was demonstrated that controlling the length of [12]helicene **32** by stretching or compressing the molecular junction can dramatically change the electronic properties of the helicene, leading to a tunable switching behavior of the conductance (and thermopower, *vide infra*) with on/off ratios of several orders of magnitude (Figure 4.26).

Furthermore, the spin-dependent electron transport through helicene molecules was studied by Guo et al. (Pan et al., 2016). The results indicated that the helicenes can present a significant spin-filtering effect even in the case of an extremely weak spin–orbit coupling. The underlying physics was attributed to the intrinsic chiral symmetry of helicenes.

Vacek, Dubi et al. studied thermopower in helicene molecular junctions using a combination of the DFT and tight-binding calculations (Vacek et al., 2015). By compressing or stretching the helicene molecular junction, its electronic properties can be controlled, leading to drastic changes in the charge transport and possible substantial increase in the thermopower and thermoelectric efficiency. Control over the helicene length and number of rings was shown to lead to a significant increase in the thermoelectric figure of merit of the helicene molecular junction (Figure 4.27).

Hutchison et al. identified helicenes as promising candidates for piezoelectrically active molecules (Quan et al., 2013): They lack inversion symmetry and possess a spring-like carbon scaffold, and a push–pull functionalization can install a properly oriented dipole moment. A set of asymmetrically substituted [6]helicenes and their extended congeners equipped with various electron-donating and electron-withdrawing groups having also varying polarizability of backbones as in **105** was theoretically analyzed. Importantly, highly responsive helicene molecules were identified that were foreseen to yield large piezoelectric coefficients in the range of tens to hundreds of pm/V (Figure 4.28). Thus, a successful development of the (single) molecule piezoelectrics would enable a construction of nanoelectro-mechanical systems being ultimately downsized.

Importantly, properties of not only single helicene molecules but also their supramolecular assemblies can be theoretically investigated. Jelfs et al. demonstrated an agreement between the experimentally observed and computationally simulated crystal packing of enantiopure [6]helicene **1** calculating also the respective electron and hole mobilities (Rice et al., 2018). It demonstrated the usefulness of the computational simulations to aid in the design of new molecules for organic electronics, through the a priori prediction of their likely solid-state form and properties.

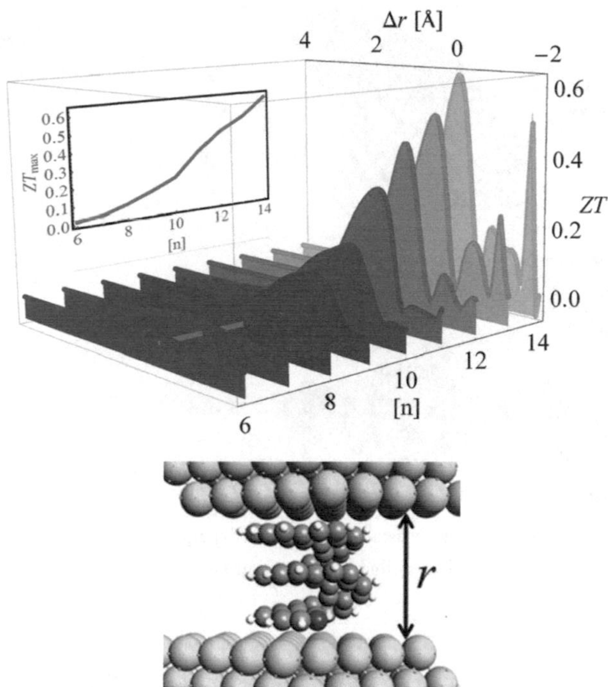

FIGURE 4.27 The thermoelectric figure of merit ZT as a function of Δr for a series of homologous diaza[n]helicenes (from 2,15-diaza[6]helicene **103** to 2,31-diaza[14]helicene **104**, r denoting the distance between the electrodes). Inset: the maximal thermoelectric figure of merit, ZT_{max}, as a function of helicene length [n] (number of fused (hetero)aromatic rings). (Adapted from Vacek et al., 2015. With permission.)

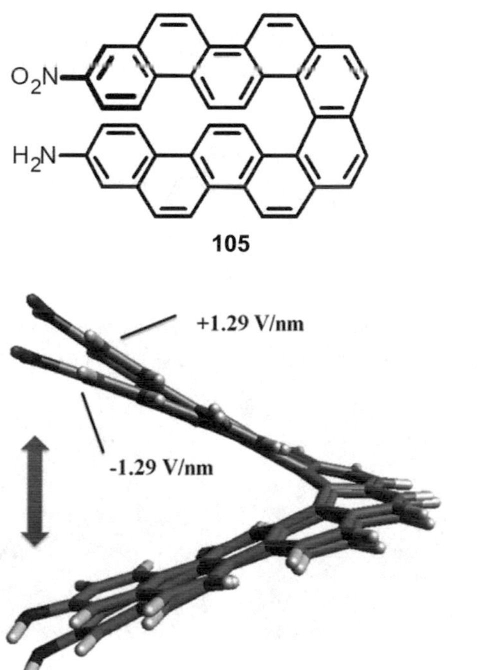

FIGURE 4.28 The computed geometrical deformation of the extended helicene ("clamphene") **105** under an applied electric field of ±1.29 V/nm parallel to the helix axis. (Adapted from Quan et al., 2013. With permission.)

4.7 Helicenes at Interfaces

An intriguing self-assembly of helicenes in Langmuir–Blodgett (LB) films was described by Katz et al. (Lovinger et al., 1998; Nuckolls et al., 1998) (Figure 4.29). Properly substituted nonracemic helicene quinones (−)-(M)-**106**, possessing both electron-rich inside and electron-deficient outside regions, can aggregate spontaneously to create columnar structures exhibiting enormous optical rotation values and remarkable nonlinear optical (NLO) properties (Verbiest et al., 1998). In LB films, an array of parallel columns can be observed directly by atomic force microscopy (AFM). Moreover, these columns are further organized into long micrometer-wide lamellar fibers visible under an optical microscope. The chiral supramolecular organization makes the second-order NLO susceptibility about 30 times larger for the nonracemic material than for the racemic one. The susceptibility of the nonracemic films is a respectable 50 pm/V, even though the helicene structure lacks features commonly associated with high nonlinearity. The chiroptical properties of such assemblies are remarkable that CD spectra could be measured for a monolayer.

Importantly, a significant attention has recently been paid to self-assembly of helicenes in two dimensions on solid surfaces as this 2D confined space might simplify the spatial arrangements of molecules and enable detailed studies on these systems even with submolecular resolution. Such a remarkable progress in surface science of molecular adsorbates was enabled by the recent development of high-resolution scanning tunneling microscopy (STM)/noncontact AFM (nc-AFM) techniques allowing for a detailed structural analysis of molecular arrays and investigation of manifold expressions of chirality in them (Ernst, 2012). Specifically, under ultra-high vacuum (UHV) conditions, an atomic (Sugimoto et al., 2007), bond (Gross et al., 2009) or orbital (Gross et al., 2011) resolution in the imaging of single-molecule adsorbates on monocrystal surfaces can now be achieved. On metals, helicenes tend to lie flat with their helix axis being approximately perpendicular to the surface plane (Ernst, 2016) or, more rarely, they are tilted (Ernst et al., 2001). On non-metallic surfaces (semiconducting or insulating), a dual adsorption geometry of helicenes was also found: They prefer to lie flat (with their helical axis approximately perpendicular to the surface plane) (Fuhr et al., 2017) or to stand upright/highly tilted (with their helical axis nearly parallel to the surface plane) (Hoff et al., 2014). At low or incomplete coverage, they form either ordered 2D islands (on the terraces) (Hoff et al., 2014), 1D chains at step edges (Sehnal et al., 2009) or terraces (Shchyrba et al., 2013; Rahe et al., 2010) or molecular clusters (quadruplets) (Seibel et al., 2014).

Fundamental achievements in this field were reported whose importance goes beyond the realm of helicenes. Thus, Fasel, Ernst et al. observed a strong amplification of chirality in 2D enantiomorphous lattices formed by [7]helicene *rac*-**23** on a Cu(111) surface (Parschau and Ernst, 2006)

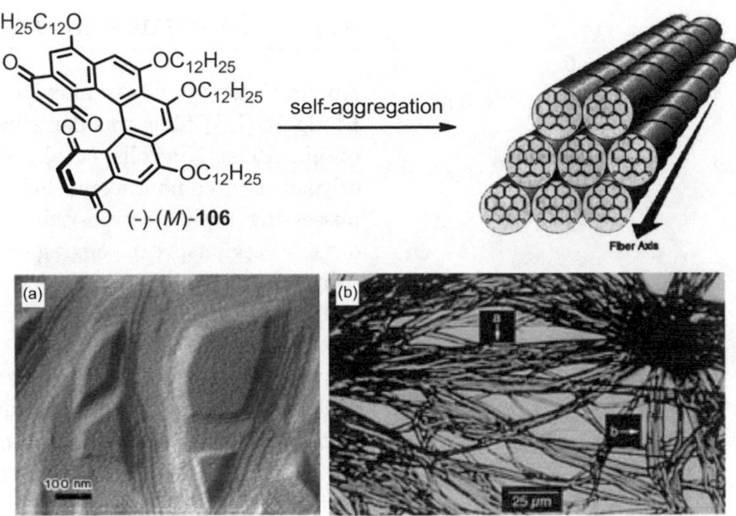

FIGURE 4.29 Self-assembly of helicene quinone (−)-(*M*,*R*,*R*)-**106** into columns; transmission electron micrograph (a) and optical microscopic view (b) of resulting lamellae/fibers (illustrating the continuity of individual lamellae over micrometer distance). (Adapted from Lovinger et al., 1998. With permission.)

(Figure 4.30). Although *rac*-**23** did not undergo spontaneous resolution into enantiomers, it self-assembled into large racemic domains with non-superimposable mirror-like lattice structures (chirality stemmed from an arrangement of the molecules). However, the addition of one enantiomer of **23** with a small excess (0.08% *ee*) resulted in a domination of only one non-superimposable, mirror-like lattice structure. Kühnle, Stará, Starý et al. described the formation of islands by enantiopure (−)-(*M*)-[7]helicene-2-carboxylic acid **107** on the calcite (1014) surface (Hauke et al., 2012) (Figure 4.31). In sharp contrast to that, racemic *rac*-**107** formed on the same surface unidirectional double rows and these wire-like structures were of well-defined width and lengths exceeding 100 nm. It demonstrated the influence of chirality on the on-surface self-assembly and indicated the preference of a heterochiral recognition.

FIGURE 4.30 Enantiomorphous domains of racemic [7]helicene **23** on Cu(111) and nonlinear amplification of chirality in nonracemic layers: detailed images (10×10 nm^2) of the two mirror domains featuring a zigzag row pattern (a); at 0% *ee*, the STM image (200×200 nm^2) shows the formation of extended close-packed (λ) (b) and (c) handed (ρ) domains (dark gray and light gray, respectively); at 0.08% *ee*, the ρ domains (light gray) have completely disappeared and only λ domains (gray) are observed. (Adapted from Parschau and Ernst, 2006. With permission.)

As far as the emerging on-surface chemistry at nanoscale is concerned, the first astonishing chemical transformations encompassing helicene molecules were described. Starý, Jelínek et al. reported recently on the chirality transfer from

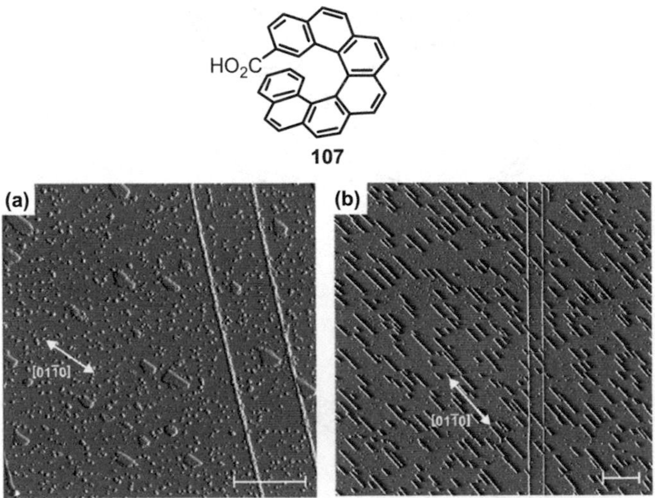

FIGURE 4.31 nc-AFM (Δf) images of enantiopure and racemic [7]helicene-2-carboxylic acid **107** on calcite (1014) (scale bars = 100 nm): (a) islands formed by enantiopure (−)-(*M*)-**107**; (b) unidirectional double rows formed by racemic *rac*-**107** (calcite step edges visible as straight lines running from top to bottom). (Adapted from Hauke et al., 2012. With permission.)

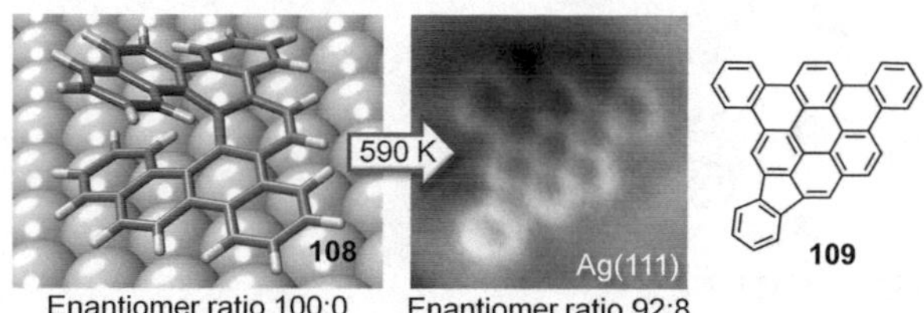

FIGURE 4.32 Starting with a single-handed dibenzo[7]helicene precursor **108** (left: molecular model), chiral adsorbates of planar coronene-derived hydrocarbon **109** (middle: high-resolution AFM image) with a prevailing single handedness are formed in on-surface reactions on the (111) surface of a silver single crystal. (Adapted from Stetsovych et al., 2017. With permission.)

a homochiral dibenzo[7]helicene precursor (+)-(*P*)-**108** to the enantiofacially adsorbed prochiral flattened coronene derivative **109** through a cascade of on-surface stereoconservative reactions (Stetsovych et al., 2017) (Figure 4.32). Notably, organic–inorganic chiral surfaces were formed that featured a high enantiomer ratio between chiral adsorbates (up to 92:8). Ernst et al. observed diastereoselective Ullmann coupling of the racemic 9-bromo[7]helicene **110** to a prevailing (*M,P*)-*meso* form of bishelicene **111** on the Au(111) surface (Mairena et al., 2018a). The stereochemical outcome of the reaction was controlled by the topochemical effect through the surface-induced alignment of the educts and the intermediates in combination with the different stereochemical constraints during the reaction. Gago et al. showed that the strength of the [5]helicene derivative–substrate interaction ruled the competitive reaction pathways (cyclodehydrogenation versus dehydrogenative polymerization) (Pinardi et al., 2013). Controlling the diffusion of the *N*-heteroaromatic helicene precursor by the nature of a metallic surface, the on-surface dehydrogenation can lead to monomolecular diazahexabenzocoronenes (*N*-doped nanographene) or *N*-doped oligomeric or polymeric networks. Ernst, Wäckerlin et al. described the autocatalytic combustion of sterically overcrowded polycyclic aromatic hydrocarbons exemplified by 9-bromo[7]helicene **110** on an oxygen-covered Cu(100) surface (Mairena et al., 2018b). It represents a new type of surface explosion chemistry having implications for better understanding of autocatalytic surface reactions as well as heterogeneous catalysis.

4.8 Applications of Helicenes

Functionalized derivatives of helicenes were already applied to different branches of science that were enabled by the remarkable development in their synthesis discussed above. The unique structure, inherent chirality, extended 3D aromatic system, intriguing physico-chemical properties as well as plasticity of their design make them not only challenging synthetic targets but also attractive materials for further exploration.

In physics, there is a bunch of seminal studies that could hardly be achieved with alternative systems. Focusing on molecular machinery, there is a fascinating application of a helicene structure by the Kelly et al. (2007). In a prototype of an artificial chemically driven molecular motor **112**, the [4]helicene ratchet controls a phosgene-powered unidirectional rotary motion about 120° to rotamer **113** (Figure 4.33). Furthermore, a converse piezoelectric effect in a single [7]helicene-2,17-dithiol-derived molecule *rac*-**114** on the Ag(111) surface using AFM and total energy DFT calculations was demonstrated by Starý, Jelínek et al. (Stetsovych et al., 2018) (Figure 4.34). The voltage bias-induced deformation of the spring-like scaffold of the helical polyaromatic molecule is attributed to the coupling of a soft vibrational mode of the molecular helix with a vertical electric dipole induced by molecule–substrate charge transfer. Naaman, Lacour et al. demonstrated the chirality-induced spin selectivity (CISS) effect in supramolecular assemblies of cationic [4]helicenes (−)-(*M*)- and (+)-(*P*)-**115** to form a new class of organic spin filter (Kiran et al., 2016)

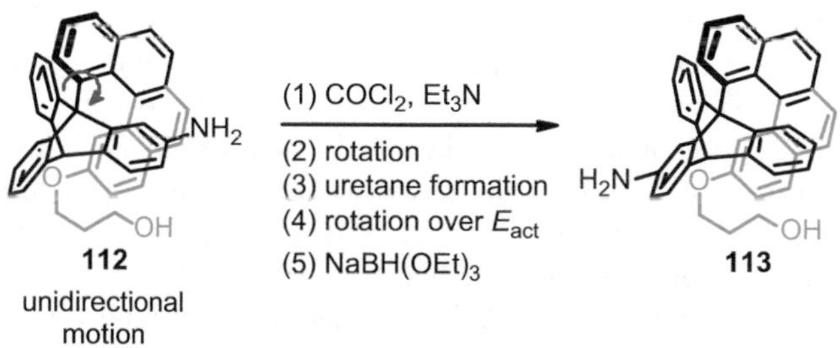

FIGURE 4.33 The phosgene-driven prototype of a molecular motor **112** containing a [4]helicene ratchet that controls the unidirectional rotary motion about 120° (Kelly et al., 2007.)

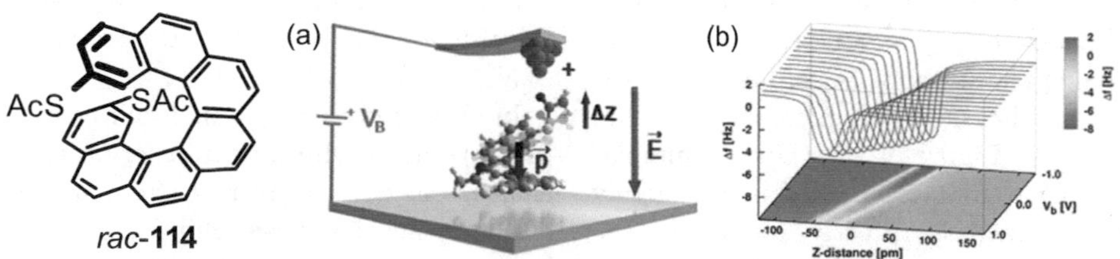

FIGURE 4.34 Measurements of the converse piezoelectric effect on the [7]helicene-2,17-dithiol-derived molecule **114** on the Ag(111) surface using AFM. (a) Ball-and-stick model of the experiment. (b) Measured Δf(z,VB) for the single molecule **114** with the same tip: the frequency shift Δf versus z-distance spectroscopies acquired for different bias voltages V_b. (Reprinted from Stetsovych et al., 2018. With permission.)

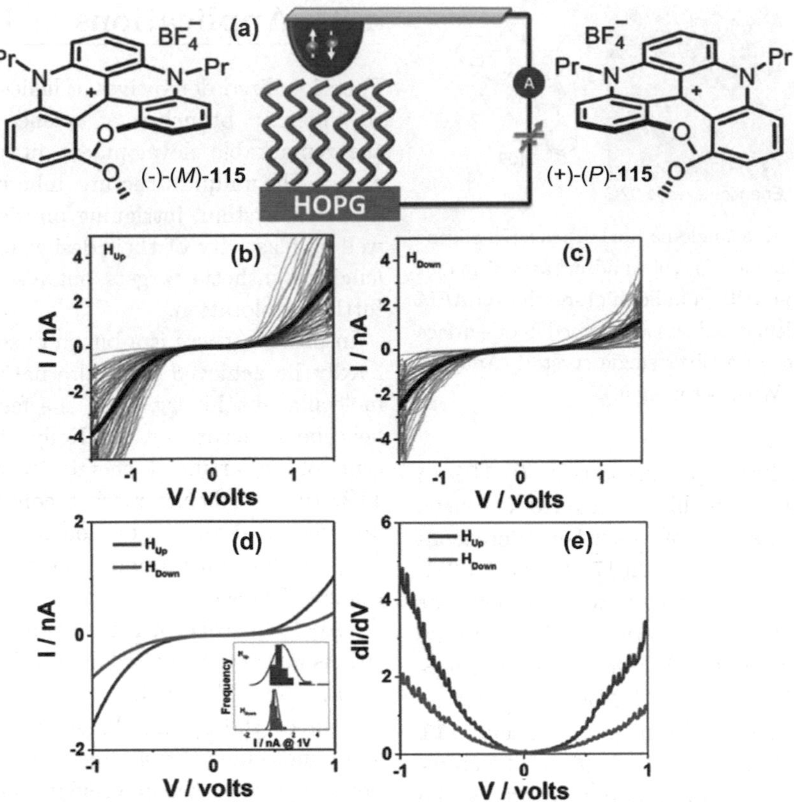

FIGURE 4.35 (a) Schematic representation of mCP-AFM measurements taken using Fe (magnetic) tip and the cationic helicene molecules (−)-(*M*)-**115** (left) and (+)-(*P*)-**115** (right). (b) and (c) represent I/V curves obtained from various helicene (−)-(*M*)-**115** junctions when tip is magnetized in UP and DOWN orientations, respectively. (d) and (e) depict averaged I/V data and corresponding dI/dV plot obtained from the data of (b) and (c). The inset of (d) gives a frequency histogram of the current for the two magnetic orientations. (Reprinted from Kiran et al., 2016. With permission.)

(Figure 4.35). Utilizing magnetic conductive probe AFM (mCP-AFM) measurements taken with a magnetic Fe tip, they demonstrated that the spin of electrons transferred through a monomolecular layer of enantiopure helicenes deposited on highly ordered pyrolytic graphite (HOPG) depends on their chirality and a degree of the molecule orientation in thin layers. The percentage of spin polarization measured at 1.0 V bias is significantly high about +49% or −45% for *P* and *M* enantiomers. Similarly, Ernst, Zacharias et al. described chirality-dependent electron spin filtering of photoelectrons transmitted through monolayers of enantiopure (−)-(*M*)- and (+)-(*P*)-[7]helicene **23** on Cu(332), Ag(110) and Au(111) surfaces (Kettner et al., 2018). Fuchter, Campbell et al. developed a new circularly polarized light-detecting organic field-effect transistor (OFET) containing enantiopure 1-aza[6]helicene (−)-(*M*)- or (+)-(*P*)-**28** in the active semiconducting thin film (Yang et al., 2013) (Figure 4.36). Upon illumination with circularly polarized light, they found a highly specific photoresponse, which is directly related to the handedness of the helicene molecule. Vacek, Stará, Starý et al. succeeded in measuring the single-molecule conductance of pyridooxa[9]helicene (−)-(*M*,*R*,*R*)-**116** repeatedly sandwiched between gold nanoelectrodes by mechanically controlled break-junction method (Nejedlý et al., 2017) (Figure 4.37). Its maximum was found at 8.8×10^{-4} G/G$_0$ in relation to the value of 4.5×10^{-3} G/G$_0$ calculated by DFT (PBE/SZP, DZP: Perdew, Burke and Ernzerhof pseudopotential with single-zeta or double-zeta polarized basis set) combined with non-equilibrium Green function (NEGF) and Landauer formalism. The diffuse character of the peak was attributed to the superposition of variable gold electrode geometry and the attachment configuration of the molecule.

Derivatives of nonracemic helicenes were found to respond reversibly to external stimuli constituting thus intriguing chiroptical switches. Various pH- or redox-responsive switches were described by Crassous et al. (Anger et al., 2012, 2014), Teplý et al. (Reyes-Gutiérrez et al., 2015; Pospíšil et al., 2014) and Diederich et al. (Schweinfurth et al., 2014).

In chemistry, helicene molecules can also play an important role in the chirality transfer processes. With respect to enantioselective catalysis, there is an increasing interest in designing helicene-based chiral ligands or catalysts. The field has not been so far thoroughly explored, but the first results have indicated the potential of helicenes in this perspective area of research (Narcis and Takenaka, 2014; Aillard et al., 2014; Demmer et al., 2017). To exemplify the use of helicene derivatives as chiral organocatalysts, Takenaka et al.

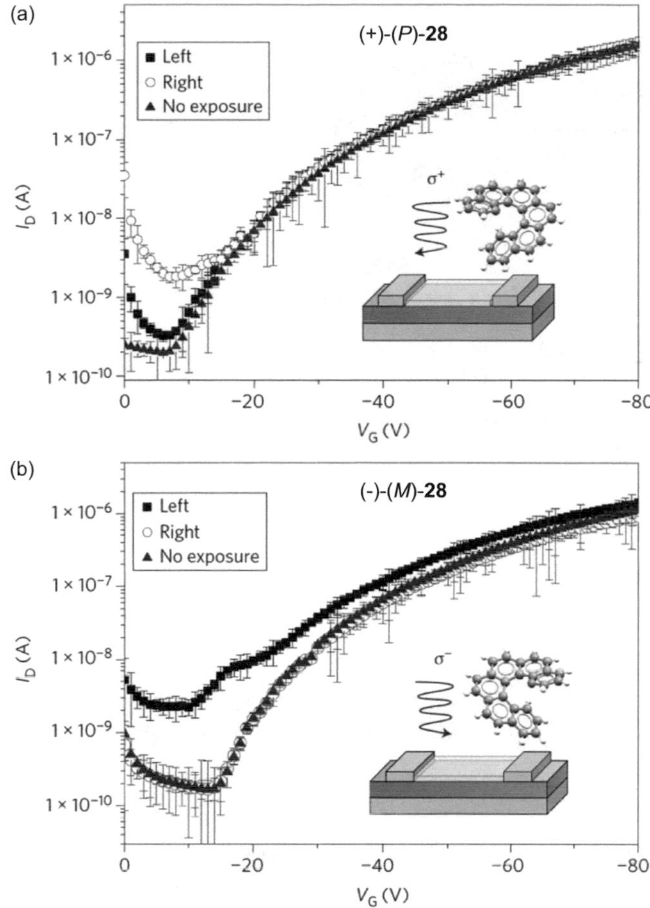

FIGURE 4.36 Response of helicene OFETs to circularly polarized light: variation in the transfer characteristics of 1-aza[6]helicene (−)-(M)-**28** (a) and (+)-(P)-**28** (b) OFETs upon exposure to left-handed (black squares) and right-handed (gray circles) circularly polarized illumination (the characteristics in the absence of circularly polarized illumination, dark gray triangles, are also shown as a reference). Insets: molecular structure of the enantiomer forming the active layer ((−)-(M)- or (+)-(P)-**28**) and the sign of circularly polarized light (right-handed (σ^+) or left-handed (σ^-)) to which the OFETs respond. (Reprinted from Yang et al., 2013. With permission.)

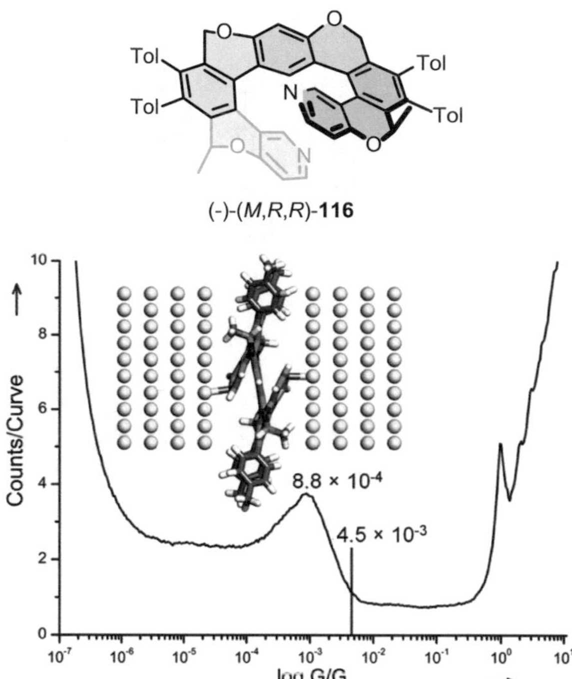

FIGURE 4.37 Single-molecule conductance of (−)-(M,R,R)-**116**, as measured by the mechanically controllable break-junction (MCBJ) method (histogram) and calculated by DFT (vertical mark). A theoretical model (DFT/PBE/SZP, DZP) of the Au(111)–single molecule–Au(111) junction used in the conductance calculations (NEGF, Landauer formalism, AtomistixToolKit software (ATK)) is shown in the inset. (Reprinted from Nejedlý et al., 2017. With permission.)

FIGURE 4.38 Asymmetric propargylation of aldehydes catalyzed by [6]helicene N-oxide derivative (+)-(P)-**118** (Chen et al., 2011).

developed helicene pyridine N-oxides such as (+)-(P)-**118** that served as excellent Lewis base catalysts in asymmetric ring opening of *meso*-epoxides (Takenaka et al., 2008) or asymmetric propargylation of aldehydes (Chen et al., 2011) (Figure 4.38). Attention has been paid also to chiral helicene-derived ligands for transition metals. Stará, Starý et al. developed a straightforward approach to enantiopure 2H-pyran-modified aminohelicenes that were converted to 1,3-disubstituted imidazolium salts such as (−)-(M,R,R),(M,R,R)-**121** and used as N-heterocyclic carbene ligand precursors in the Ni^0-catalyzed enantioselective [2 + 2 + 2] cycloisomerization of aromatic triynes such as **120** to obtain the model helicene derivatives such as (+)-(P)-**122** in up to 86% *ee* (Sánchez et al., 2017) (Figure 4.39). However, there were published other remarkable applications of helicenes and their congeners to enantioselective organo- or transition metal catalysis reporting on Pd-catalyzed asymmetric allylic alkylation (Reetz and Sostmann, 2000; Yamamoto et al., 2016), Ir-catalyzed asymmetric allylic amination (Krausová et al., 2011), Rh-catalyzed asymmetric hydrogenation (Nakano and Yamaguchi, 2003), asymmetric autocatalysis of pyrimidyl alkanol in addition

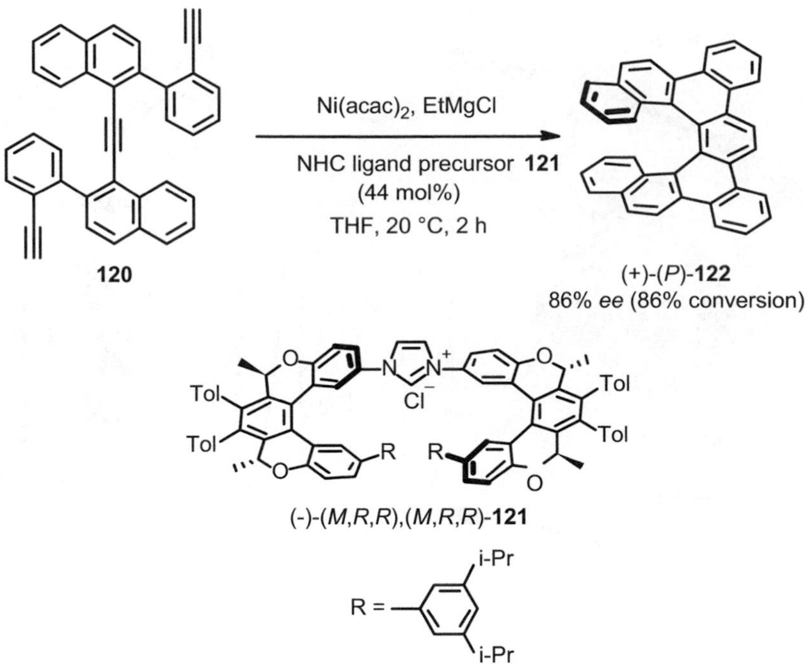

FIGURE 4.39 Enantioselective [2 + 2 + 2] cycloisomerization of the aromatic triyne **120** to receive the nonracemic [7]helicene derivative (+)-(*P*)-**122** by employing helicene-derived 1,3-disubstituted imidazolium salt (−)-(*M*,*R*,*R*),(*M*,*R*,*R*)-**121** as an *N*-heterocyclic carbene (NHC) ligand precursors (Sánchez et al., 2017).

of diisopropylzinc to carbaldehyde (Matsumoto et al., 2017), Ru-catalyzed asymmetric ring-closing metathesis (Karras et al., 2018), Pd-catalyzed enantioselective Suzuki–Miyaura cross-coupling reaction (Yamamoto et al., 2016), Au-catalyzed enantioselective cycloisomerization of an *N*-tethered 1,6-enynes (Yavari et al., 2014) or 1,6-allenenes (Aillard et al., 2015), asymmetric addition of dihydroindoles to nitroalkenes catalyzed by helicene-derived hydrogen-bond-donor catalysts (Takenaka et al., 2010), kinetic resolution of secondary alcohols catalyzed by helicene-derived dimethylaminopyridine (Crittall et al., 2012), organocatalytic enantioselective [3 + 2] cyclizations (Gicquel et al., 2015) and other processes (Demmer et al., 2017).

In biology, Sugiyama et al. demonstrated a selective interaction of a functionalized enantiopure thia[7]helicene derivative with *Z*-DNA (Xu et al., 2004) and telomerase inhibition by enantiopure bridged thia[7]helicenes (Shinohara et al., 2010).

4.9 Summary

As a reflection of fundamental discoveries of fullerenes, carbon nanotubes and graphene, a new field of the interdisciplinary research on advanced molecular nanocarbons has recently emerged. It definitely covers chemistry, physics and nanoscience of nonplanar polycyclic hydrocarbons, many of whom exhibit helical chirality such as iconic helicenes and their congeners. The combination of unique π-electron systems with the chirality issues makes them highly attractive across various branches of science. Indeed, the recent development in the synthesis and use of helically chiral aromatics is remarkable. The chemical bottom-up approach to them now benefits from a steadily widening portfolio of modern synthetic methods. They allow now for the preparation of helically chiral aromatics which we could previously only have dreamt about. Nevertheless, there is an urgent need for the development of new synthetic methodologies being suitable for the construction of nontrivial polyaromatic systems of the highest complexity: Such methods should be maximally efficient and selective, thus preventing a laborious and often impossible separation of target polyaromatics from a reaction mixture. In particular, optimized or even new reactions for a simultaneous construction of more carbon–carbon bonds in a single operation are requested. Ideally, such reactions should feature a favorable energetics to allow for imposing an inherent strain to the twisted product. Although the departure from planarity usually leads to an increased solubility of the polyaromatics, the control of solubility has remained an issue. There is room for a development of new easy-to-install/easy-to-remove traceless solubilizing groups as their presence could otherwise hamper a required tight packing of twisted polyaromatics. Alternatively, a significant progress in on-surface chemistry at nanoscale might be expected in the near future utilizing scanning probe microscopy techniques. A controlled synthesis of complex polyaromatic materials directly on a solid surface would circumvent a problem of a general insolubility of very large polyaromatics. An increasing attention is expected to be paid to chiral (helical) polyaromatics in the context of spintronics (CISS), circularly polarized light emission or enantioselective catalysis, recognition and sensing.

Acknowledgments

We acknowledge the financial support by the Czech Science Foundation (Reg. No. 16-08327S) and the Institute of Organic Chemistry and Biochemistry, Academy of Sciences of the Czech Republic (RVO: 61388963).

References

Adriaenssens, L., Severa, L., Koval, D., Císařová, I., Belmonte, M.M., Escudero-Adán, E.C., Novotná, P., Sázelová, P., Vávra, J., Pohl, R., Šaman, D., Urbanová, M., Kašička, V., Teplý, F. 2011. [6]Saddlequat: A [6]helquat captured on its racemization pathway. *Chem. Sci.* 2, 2314–2320. doi: 10.1039/C1SC00468A.

Adriaenssens, L., Severa, L., Šálová, T., Císařová, I., Pohl, R., Šaman, D., Rocha, S.V., Finney, N.S., Pospíšil, L., Slavíček, P., Teplý, F. 2009. Helquats: A facile, modular, scalable route to novel helical dications. *Chem. Eur. J.* 15, 1072–1076. doi: 10.1002/chem.200801904.

Aillard, P., Retailleau, P., Voituriez, A., Marinetti, A., 2014. A [2 + 2 + 2] cyclization strategy for the synthesis of phosphorus embedding [6]helicene-like structures. *Chem. Commun.* 50, 2199–2201. doi: 10.1039/C3CC48943D.

Aillard, P., Retailleau, P., Voituriez, A., Marinetti, A., 2015. Synthesis of new phosphahelicene scaffolds and development of gold(I)-catalyzed enantioselective allenene cyclizations. *Chem. Eur. J.* 21, 11989–11993. doi: 10.1002/chem.201501697.

Aillard, P., Voituriez, A., Marinetti, A., 2014. Helicene-like chiral auxiliaries in asymmetric catalysis. *Dalton Trans.* 43, 15263–15278. doi: 10.1039/C4DT01935K.

Anger, E., Srebro, M., Vanthuyne, N., Roussel, C., Toupet, L., Autschbach, J., Réau, R., Crassous, J., 2014. Helicene-grafted vinyl- and carbene-osmium complexes: An example of acid–base chiroptical switching. *Chem. Commun.* 50, 2854–2856. doi: 10.1039/C3CC47825D.

Anger, E., Srebro, M., Vanthuyne, N., Toupet, L., Rigaut, S., Roussel, C., Autschbach, J., Crassous, J., Réau, R., 2012. Ruthenium-vinylhelicenes: Remote metal-based enhancement and redox switching of the chiroptical properties of a helicene core. *J. Am. Chem. Soc.* 134, 15628–15631. doi: 10.1021/ja304424t.

Barroso, J., Cabellos, J.L., Pan, S., Murillo, F., Zarate, X., Fernandez-Herrera, M.A., Merino, G., 2017. Revisiting the racemization mechanism of helicenes. *Chem. Commun.* 54, 188–191. doi: 10.1039/C7CC08191J.

Bédard, A.-C., Vlassova, A., Hernandez-Perez, A.C., Bessette, A., Hanan, G.S., Heuft, M.A., Collins, S.K., 2013. Synthesis, crystal structure and photophysical properties of pyrene–helicene hybrids. *Chem. Eur. J.* 19, 16295–16302. doi: 10.1002/chem.201301431.

Bensalah-Ledoux, A., Pitrat, D., Reynaldo, T., Srebro-Hooper, M., Moore, B., Autschbach, J., Crassous, J., Guy, S., Guy, L., 2016. Large-scale synthesis of helicene-like molecules for the design of enantiopure thin films with strong chiroptical activity. *Chem. Eur. J.* 22, 3333–3346. doi: 10.1002/chem.201504174.

Berezhnaia, V., Roy, M., Vanthuyne, N., Villa, M., Naubron, J.-V., Rodriguez, J., Coquerel, Y., Gingras, M., 2017. Chiral nanographene propeller embedding six enantiomerically stable [5]helicene units. *J. Am. Chem. Soc.* 139, 18508–18511. doi: 10.1021/jacs.7b07622.

Bestmann, H.J., Both, W., 1974. Moleküle mit Helixstruktur, I. Synthese und absolute Konfiguration des (+)-Pentahelicens. *Chem. Ber.* 107, 2923–2925. doi: 10.1002/cber.19741070914.

Borkent, J.H., Laarhoven, W.H., 1978. The thermal racemization of methyl-substituted hexahelicenes. *Tetrahedron* 34, 2565–2567. doi: 10.1016/0040-4020(78)88387-2.

Bosson, J., Gouin, J., Lacour, J., 2014. Cationic triangulenes and helicenes: Synthesis, chemical stability, optical properties and extended applications of these unusual dyes. *Chem. Soc. Rev.* 43, 2824–2840. doi: 10.1039/C3CS60461F.

Buchta, M., Rybáček, J., Jančařík, A., Kudale, A.A., Buděšínský, M., Chocholoušová, J.V., Vacek, J., Bednárová, L., Císařová, I., Bodwell, G.J., Starý, I., Stará, I.G., 2015. Chimerical pyrene-based [7]helicenes as twisted polycondensed aromatics. *Chem. Eur. J.* 21, 8910–8917. doi: 10.1002/chem.201500826.

Buttrick, J.C., King, B.T., 2017. Kekulenes, cycloarenes, and heterocycloarenes: Addressing electronic structure and aromaticity through experiments and calculations. *Chem. Soc. Rev.* 46, 7–20. doi: 10.1039/C6CS00174B.

Chen, C.-F., Shen, Y., 2017. Structures and properties of helicenes. In: *Helicene Chemistry*. Springer, Berlin, Heidelberg, pp. 19–40.

Chen, J., Captain, B., Takenaka, N., 2011. Helical chiral 2,2′-bipyridine *N*-monoxides as catalysts in the enantioselective propargylation of aldehydes with allenyltrichlorosilane. *Org. Lett.* 13, 1654–1657. doi: 10.1021/ol200102c.

Collins, S.K., Grandbois, A., Vachon, M.P., Côté, J., 2006. Preparation of helicenes through olefin metathesis. *Angew. Chem. Int. Ed.* 45, 2923–2926. doi: 10.1002/anie.200504150.

Crittall, M.R., Fairhurst, N.W.G., Carbery, D.R., 2012. Point-to-helical chirality transfer for a scalable and resolution-free synthesis of a helicenoidal DMAP organocatalyst. *Chem. Commun.* 48, 11181–11183. doi: 10.1039/C2CC35583C.

Crittall, M.R., Rzepa, H.S., Carbery, D.R., 2011. Design, synthesis, and evaluation of a helicenoidal DMAP Lewis base catalyst. *Org. Lett.* 13, 1250–1253. doi: 10.1021/ol2001705.

Daigle, M., Miao, D., Lucotti, A., Tommasini, M., Morin, J.-F., 2017. Helically coiled graphene nanoribbons. *Angew. Chem. Int. Ed.* 56, 6213–6217. doi: 10.1002/anie.201611834.

Demmer, C.S., Voituriez, A., Marinetti, A., 2017. Catalytic uses of helicenes displaying phosphorus functions. *C. R. Chim.* 20, 860–879. doi: 10.1016/j.crci.2017.04.002.

Ernst, K.-H., 2012. Molecular chirality at surfaces. *Phys. Status Solidi B* 249, 2057–2088. doi: 10.1002/pssb.201248188.

Ernst, K.-H., 2016. Stereochemical recognition of helicenes on metal surfaces. *Acc. Chem. Res.* 49, 1182–1190. doi: 10.1021/acs.accounts.6b00110.

Ernst, K.-H., Neuber, M., Grunze, M., Ellerbeck, U., 2001. NEXAFS study on the orientation of chiral *P*-heptahelicene on Ni(100). *J. Am. Chem. Soc.* 123, 493–495. doi: 10.1021/ja003262+.

Evans, P.J., Ouyang, J., Favereau, L., Crassous, J., Fernández, I., Perles, J., Martín, N., 2018. Synthesis of a helical bilayer nanographene. *Angew. Chem. Int. Ed.* 57, 6774–6779. doi: 10.1002/anie.201800798.

Fuhr, J.D., van der Meijden, M.W., Cristina, L.J., Rodríguez, L.M., Kellogg, R.M., Gayone, J.E., Ascolani, H., Lingenfelder, M., 2017. Chiral expression of adsorbed (*MP*) 5-amino[6]helicenes: From random structures to dense racemic crystals by surface alloying. *Chem. Commun.* 53, 130–133. doi: 10.1039/C6CC0 6785A.

Fujikawa, T., Preda, D.V., Segawa, Y., Itami, K., Scott, L.T., 2016. Corannulene–helicene hybrids: Chiral π-systems comprising both bowl and helical motifs. *Org. Lett.* 18, 3992–3995. doi: 10.1021/acs.orglett. 6b01801.

Gicquel, M., Zhang, Y., Aillard, P., Retailleau, P., Voituriez, A., Marinetti, A., 2015. Phosphahelicenes in asymmetric organocatalysis: [3 + 2] cyclizations of γ-substituted allenes and electron-poor olefins. *Angew. Chem. Int. Ed.* 54, 5470–5473. doi: 10.1002/anie.201500299.

Gingras, M., 2013a. One hundred years of helicene chemistry. Part 1: Non-stereoselective syntheses of carbohelicenes. *Chem. Soc. Rev.* 42, 968–1006. doi: 10.1039/ C2CS35154D.

Gingras, M., 2013b. One hundred years of helicene chemistry. Part 3: Applications and properties of carbohelicenes. *Chem. Soc. Rev.* 42, 1051–1095. doi: 10.1039/ C2CS35134J.

Gingras, M., Félix, G., Peresutti, R., 2013. One hundred years of helicene chemistry. Part 2: Stereoselective syntheses and chiral separations of carbohelicenes. *Chem. Soc. Rev.* 42, 1007–1050. doi: 10.1039/C2CS35111K.

Goedicke, C., Stegemeyer, H., 1970. Resolution and racemization of pentahelicene. *Tetrahedron Lett.* 11, 937–940. doi: 10.1016/S0040-4039(01)97871-2.

González-Fernández, E., Nicholls, L.D.M., Schaaf, L.D., Farès, C., Lehmann, C.W., Alcarazo, M., 2017. Enantioselective synthesis of [6]carbohelicenes. *J. Am. Chem. Soc.* 139, 1428–1431. doi: 10.1021/jacs.6b12443.

Graule, S., Rudolph, M., Vanthuyne, N., Autschbach, J., Roussel, C., Crassous, J., Réau, R., 2009. Metal–bis(helicene) assemblies incorporating π-conjugated phosphole-azahelicene ligands: Impacting chiroptical properties by metal variation. *J. Am. Chem. Soc.* 131, 3183–3185. doi: 10.1021/ja809396f.

Grimme, S., Peyerimhoff, S., 1996. Theoretical study of the structures and racemization barriers of [n] helicenes (n = 3-6, 8). *Chem. Phys.* 204, 411–417.

Gross, L., Mohn, F., Moll, N., Liljeroth, P., Meyer, G., 2009. The chemical structure of a molecule resolved by atomic force microscopy. *Science* 325, 1110–1114. doi: 10.1126/science.1176210.

Gross, L., Moll, N., Mohn, F., Curioni, A., Meyer, G., Hanke, F., Persson, M., 2011. High-resolution molecular orbital imaging using a p-wave STM Tip. *Phys. Rev. Lett.* 107, 086101. doi: 10.1103/PhysRevLett.107.086101.

Guo, Y.-D., Yan, X.-H., Xiao, Y., Liu, C.-S., 2015. U-shaped relationship between current and pitch in helicene molecules. *Sci. Rep.* 5, 16731. doi: 10.1038/srep16731.

Han, S., Anderson, D.R., Bond, A.D., Chu, H.V., Disch, R.L., Holmes, D., Schulman, J.M., Teat, S.J., Vollhardt, K.P.C., Whitener, G.D., 2002a. Total syntheses of angular [7]-, [8]-, and [9]phenylene by triple cobalt-catalyzed cycloisomerization: Remarkably flexible heliphenes. *Angew. Chem. Int. Ed.* 41, 3227–3230. doi: 10.1002/1521-3773(20020902)41:17<3227::AID-ANIE3227>3.0.CO;2-T.

Han, S., Bond, A.D., Disch, R.L., Holmes, D., Schulman, J.M., Teat, S.J., Vollhardt, K.P.C., Whitener, G.D., 2002b. Total syntheses and structures of angular [6]- and [7]phenylene: The first helical phenylenes (heliphenes). *Angew. Chem. Int. Ed.* 41, 3223–3227. doi: 10.1002/1521-3773(20020902)41:17<3223::AID-ANIE3223>3.0.CO;2-G.

Harrowven, D.C., Guy, I.L., Nanson, L., 2006. Efficient phenanthrene, helicene, and azahelicene syntheses. *Angew. Chem. Int. Ed.* 45, 2242–2245. doi: 10.1002/anie.200504287.

Hauke, C.M., Rahe, P., Nimmrich, M., Schütte, J., Kittelmann, M., Stará, I.G., Starý, I., Rybáček, J., Kühnle, A., 2012. Molecular self-assembly of enantiopure heptahelicene-2-carboxylic acid on calcite (1014). *J. Phys. Chem. C* 116, 4637–4641. doi: 10.1021/jp2102258.

Herse, C., Bas, D., Krebs, F.C., Bürgi, T., Weber, J., Wesolowski, T., Laursen, B.W., Lacour, J., 2003. A highly configurationally stable [4]heterohelicenium cation. *Angew. Chem. Int. Ed.* 42, 3162–3166. doi: 10.1002/anie.200351443.

Hoff, B., Gingras, M., Peresutti, R., Henry, C.R., Foster, A.S., Barth, C., 2014. Mechanisms of the adsorption and self-assembly of molecules with polarized functional groups on insulating surfaces. *J. Phys. Chem. C* 118, 14569–14578. doi: 10.1021/jp501738c.

Hosokawa, T., Takahashi, Y., Matsushima, T., Watanabe, S., Kikkawa, S., Azumaya, I., Tsurusaki, A., Kamikawa, K., 2017. Synthesis, structures, and properties of hexapole helicenes: Assembling six [5]helicene substructures into highly twisted aromatic systems. *J. Am. Chem. Soc.* 139, 18512–18521. doi: 10.1021/jacs.7b07113.

Ichikawa, J., Yokota, M., Kudo, T., Umezaki, S., 2008. Efficient helicene synthesis: Friedel–crafts-type cyclization of 1,1-difluoro-1-alkenes. *Angew. Chem. Int. Ed.* 47, 4870–4873. doi: 10.1002/anie.200801396.

Jalaie, M., Weatherhead, S., Lipkowitz, K.B., Robertson, D., 1997. Modulating force constants in molecular springs. *Electron. J. Theor. Chem.* 2, 268–272. doi: 10.1002/ejtc.56.

Jančařík, A., Rybáček, J., Cocq, K., Vacek Chocholoušová, J., Vacek, J., Pohl, R., Bednárová, L., Fiedler, P., Císařová, I., Stará, I.G., Starý, I., 2013. Rapid access to dibenzohelicenes and their functionalized derivatives. *Angew. Chem. Int. Ed.* 52, 9970–9975. doi: 10.1002/anie.201301739.

Janke, R.H., Haufe, G., Würthwein, E.-U., Borkent, J.H., 1996. Racemization barriers of helicenes: A computational study. *J. Am. Chem. Soc.* 118, 6031–6035. doi: 10.1021/ja950774t.

Kamikawa, K., Takemoto, I., Takemoto, S., Matsuzaka, H., 2007. Synthesis of helicenes utilizing palladium-catalyzed double C–H arylation reaction. *J. Org. Chem.* 72, 7406–7408. doi: 10.1021/jo0711586.

Kaneko, E., Matsumoto, Y., Kamikawa, K., 2013. Synthesis of azahelicene N-oxide by palladium-catalyzed direct C–H annulation of a pendant (Z)-bromovinyl side chain. *Chem. Eur. J.* 19, 11837–11841. doi: 10.1002/chem.201302083.

Karras, M., Dąbrowski, M., Pohl, R., Rybáček, J., Vacek, J., Bednárová, L., Grela, K., Starý, I., Stará, I.G., Schmidt, B., 2018. Helicenes as chirality-inducing groups in transition-metal catalysis: The first helically chiral olefin metathesis catalyst. *Chem. Eur. J.* 24, 10994–10998. doi: 10.1002/chem.201802786.

Katayama, T., Nakatsuka, S., Hirai, H., Yasuda, N., Kumar, J., Kawai, T., Hatakeyama, T., 2016. Two-step synthesis of boron-fused double helicenes. *J. Am. Chem. Soc.* 138, 5210–5213. doi: 10.1021/jacs.6b01674.

Kato, K., Segawa, Y., Scott, L.T., Itami, K., 2018. A quintuple [6]helicene with a corannulene core as a C5-symmetric propeller-shaped π-system. *Angew. Chem. Int. Ed.* 57, 1337–1341. doi: 10.1002/anie.201711985.

Katz, T.J., Sudhakar, A., Teasley, M.F., Gilbert, A.M., Geiger, W.E., Robben, M.P., Wuensch, M., Ward, M.D., 1993. Synthesis and properties of optically active helical metallocene oligomers. *J. Am. Chem. Soc.* 115, 3182–3198. doi: 10.1021/ja00061a018.

Kelly, T.R., Cai, X., Damkaci, F., Panicker, S.B., Tu, B., Bushell, S.M., Cornella, I., Piggott, M.J., Salives, R., Cavero, M., Zhao, Y., Jasmin, S. 2007. Progress toward a rationally designed, chemically powered rotary molecular motor. *J. Am. Chemi.Soc.* 129, 376–386. doi: 10.1021/ja066044a.

Kettner, M., Maslyuk, V.V., Nürenberg, D., Seibel, J., Gutierrez, R., Cuniberti, G., Ernst, K.-H., Zacharias, H., 2018. Chirality-dependent electron spin filtering by molecular monolayers of helicenes. *J. Phys. Chem. Lett.* 9, 2025–2030. doi: 10.1021/acs.jpclett.8b00208.

Kiel, G.R., Patel, S.C., Smith, P.W., Levine, D.S., Tilley, T.D., 2017. Expanded helicenes: A general synthetic strategy and remarkable supramolecular and solid-state behavior. *J. Am. Chem. Soc.* 139, 18456–18459. doi: 10.1021/jacs.7b10902.

Kiran, V., Mathew, S.P., Cohen, S.R., Hernández Delgado, I., Lacour, J., Naaman, R., 2016. Helicenes: A new class of organic spin filter. *Adv. Mater.* 28, 1957–1962. doi: 10.1002/adma.201504725.

Kötzner, L., Webber, M.J., Martínez, A., De Fusco, C., List, B., 2014. Asymmetric catalysis on the nanoscale: The organocatalytic approach to helicenes. *Angew. Chem. Int. Ed.* 53, 5202–5205. doi: 10.1002/anie.201400474.

Koval, D., Severa, L., Adriaenssens, L., Vávra, J., Teplý, F., Kašička, V., 2011. Chiral analysis of helquats by capillary electrophoresis: Resolution of helical N-heteroaromatic dications using randomly sulfated cyclodextrins. *Electrophoresis* 32, 2683–2692. doi: 10.1002/elps.201100173.

Krausová, Z., Sehnal, P., Bondzic, B.P., Chercheja, S., Eilbracht, P., Stará, I.G., Šaman, D., Starý, I., 2011. Helicene-based phosphite ligands in asymmetric transition-metal catalysis: Exploring Rh-catalyzed hydroformylation and Ir-catalyzed allylic amination. *Eur. J. Org. Chem.* 2011, 3849–3857. doi: 10.1002/ejoc.201100259.

Laarhoven, W.H., 1989. *Organic Photochemistry.* Marcel Dekker, New York, p. 163.

Laleu, B., Mobian, P., Herse, C., Laursen, B.W., Hopfgartner, G., Bernardinelli, G., Lacour, J., 2005. Resolution of [4]heterohelicenium dyes with unprecedented Pummerer-like chemistry. *Angew. Chem. Int. Ed.* 117, 1913–1917. doi: 10.1002/anie.200462321.

Liu, L., Katz, T.J., 1991. Bromine auxiliaries in photosyntheses of [5]helicenes. *Tetrahedron Lett.* 32, 6831–6834. doi: 10.1016/0040-4039(91)80418-6.

Liu, L., Yang, B., Katz, T.J., Poindexter, M.K., 1991. Improved methodology for photocyclization reactions. *J. Org. Chem.* 56, 3769–3775. doi: 10.1021/jo00012a005.

Lovinger, A.J., Nuckolls, C., Katz, T.J., 1998. Structure and morphology of helicene fibers. *J. Am. Chem. Soc.* 120, 264–268. doi: 10.1021/ja973366t.

Mairena, A., Wäckerlin, C., Wienke, M., Grenader, K., Terfort, A., Ernst, K.-H., 2018a. Diastereoselective Ullmann coupling to bishelicenes by surface topochemistry. *J. Am. Chem. Soc.* 140, 15186–15189. doi: 10.1021/jacs.8b10059.

Mairena, A., Wienke, M., Martin, K., Avarvari, N., Terfort, A., Ernst, K.-H., Wäckerlin, C., 2018b. Stereospecific autocatalytic surface explosion chemistry of polycyclic aromatic hydrocarbons. *J. Am. Chem. Soc.* 140, 7705–7709. doi: 10.1021/jacs.8b04191.

Mallory, F.B., Mallory, C.W., 1984. *Organic Reactions.* Wiley, New York, p. 1.

Mamane, V., Hannen, P., Fürstner, A., 2004. Synthesis of phenanthrenes and polycyclic heteroarenes by transition-metal catalyzed cycloisomerization reactions. *Chem. Eur. J.* 10, 4556–4575. doi: 10.1002/chem.200400220.

Martin, R.H., 1974. The helicenes. *Angew. Chem. Int. Ed. Engl.* 13, 649–660. doi: 10.1002/anie.197406491.

Martin, R.H., Baes, M., 1975. Helicenes: Photosyntheses of [11], [12] and [14]helicene. *Tetrahedron* 31, 2135–2137. doi: 10.1016/0040-4020(75)80208-0.

Martin, R.H., Flammang-Barbieux, M., Cosyn, J.P., Gelbcke, M., 1968. 1-Synthesis of octa- and nonahelicenes. 2-New syntheses of hexa- and heptahelicenes. 3-Optical rotation and O.R.D. of heptahelicene (1). *Tetrahedron Lett.* 9, 3507–3510. doi: 10.1016/S0040-4039(01)99095-1.

Martin, R.H., Libert, V., 1980. Hélicènes: Utilisation de l'acide hexahélicène-2 carboxylique dédoublé en antipodes optiques comme précurseur commun à la synthèse photochimique des octa, nona, déca, undéca et tridécahélicènes optiquement purs. Racémisation thermique des déca et undécahélicènes. *J. Chem. Research (M)*, 1940–1950.

Martin, R.H., Marchant, M.J., 1972. Thermal racemisation of [6], [7], [8] and [9] helicene. *Tetrahedron Lett.* 13, 3707–3708. doi: 10.1016/S0040-4039(01)94141-3.

Martin, R.H., Marchant, M.J., 1974. Thermal racemisation of hepta-, octa-, and nonahelicene. Kinetic results, reaction path and experimental proofs that the racemisation of hexa- and heptahelicene does not involve an intramolecular double Diels-Alder reaction. *Tetrahedron* 30, 347–349. doi: 10.1016/S0040-4020(01) 91469-3.

Martin, R.H., Morren, G., Schurter, J.J., 1969. [13] Helicene and [13]helicene-10,21-d_2. *Tetrahedron Lett.* 10, 3683–3688. doi: 10.1016/S0040-4039(01)88487-2.

Matsumoto, A., Yonemitsu, K., Ozaki, H., Míšek, J., Starý, I., Stará, I.G., Soai, K., 2017. Reversal of the sense of enantioselectivity between 1- and 2-aza[6]helicenes used as chiral inducers of asymmetric autocatalysis. *Org. Biomol. Chem.* 15, 1321–1324. doi: 10.1039/C6OB02745H.

Meisenheimer, J., Witte, K., 1903. Reduction von 2-nitronaphtalin. *Ber. Dtsch. Chem. Ges.* 36, 4153–4164. doi: 10.1002/cber.19030360481.

Míšek, J., Teplý, F., Stará, I.G., Tichý, M., Šaman, D., Císařová, I., Vojtíšek, P., Starý, I., 2008. A straightforward route to helically chiral *N*-heteroaromatic compounds: Practical synthesis of racemic 1,14-diaza[5]helicene and optically pure 1- and 2-aza[6]helicenes. *Angew. Chem. Int. Ed.* 47, 3188–3191. doi: 10.1002/anie.200705463.

Mislow, K., 1954. Limitations of the symmetry criteria for optical inactivity and resolvability. *Science* 120, 232–233. doi: 10.1126/science.120.3110.232-a.

Mislow, K., Bolstad, R., 1955. Molecular dissymmetry and optical inactivity. *J. Am. Chem. Soc.* 77, 6712–6713. doi: 10.1021/ja01629a131.

Moradpour, A., Kagan, H., Baes, M., Morren, G., Martin, R.H., 1975. Photochemistry with circularly polarized light—III: Synthesis of helicenes using bis(arylvinyl) arenes as precursors. *Tetrahedron* 31, 2139–2143. doi: 10.1016/0040-4020(75)80209-2.

Mori, K., Murase, T., Fujita, M., 2015. One-step synthesis of [16]helicene. *Angew. Chem. Int. Ed.* 54, 6847–6851. doi: 10.1002/anie.201502436.

Nakai, Y., Mori, T., Inoue, Y., 2012. Theoretical and experimental studies on circular dichroism of carbo[n]helicenes. *J. Phys. Chem. A* 116, 7372–7385. doi: 10.1021/jp304576g.

Nakai, Y., Mori, T., Sato, K., Inoue, Y., 2013. Theoretical and experimental studies of circular dichroism of mono- and diazonia[6]helicenes. *J. Phys. Chem. A* 117, 5082–5092. doi: 10.1021/jp403426w.

Nakakuki, Y., Hirose, T., Sotome, H., Miyasaka, H., Matsuda, K., 2018. Hexa-*peri*-hexabenzo[7]helicene: Homogeneously π-extended helicene as a primary substructure of helically twisted chiral graphenes. *J. Am. Chem. Soc.* 140, 4317–4326. doi: 10.1021/jacs.7b13412.

Nakano, D., Yamaguchi, M., 2003. Enantioselective hydrogenation of itaconate using rhodium bihelicenol phosphite complex. Matched/mismatched phenomena between helical and axial chirality. *Tetrahedron Lett.* 44, 4969–4971. doi: 10.1016/S0040-4039(03)01183-3.

Nakano, K., Hidehira, Y., Takahashi, K., Hiyama, T., Nozaki, K., 2005. Stereospecific synthesis of hetero[7]helicenes by Pd-catalyzed double *N*-arylation and intramolecular O-arylation. *Angew. Chem. Int. Ed.* 44, 7136–7138. doi: 10.1002/anie.200502855.

Narcis, M.J., Takenaka, N., 2014. Helical-chiral small molecules in asymmetric catalysis: Helical-chiral small molecules in asymmetric catalysis. *Eur. J. Org. Chem.* 2014, 21–34. doi: 10.1002/ejoc.201301045.

Nejedlý, J., Šámal, M., Rybáček, J., Tobrmanová, M., Szydlo, F., Coudret, C., Neumeier, M., Vacek, J., Vacek Chocholoušová, J., Buděšínský, M., Šaman, D., Bednárová, L., Sieger, L., Stará, I.G., Starý, I., 2017. Synthesis of long oxahelicenes by polycyclization in a flow reactor. *Angew. Chem. Int. Ed.* 56, 5839–5843. doi: 10.1002/anie.201700341.

Newman, M.S., Lednicer, D., 1956. The synthesis and resolution of hexahelicene. *J. Am. Chem. Soc.* 78, 4765–4770. doi: 10.1021/ja01599a060.

Norel, L., Rudolph, M., Vanthuyne, N., Williams, J.A.G., Lescop, C., Roussel, C., Autschbach, J., Crassous, J., Réau, R., 2010. Metallahelicenes: Easily accessible helicene derivatives with large and tunable chiroptical properties. *Angew. Chem. Int. Ed.* 49, 99–102. doi: 10.1002/anie.200905099.

Nuckolls, C., Katz, T.J., Verbiest, T., Elshocht, S.V., Kuball, H.-G., Kiesewalter, S., Lovinger, A.J., Persoons, A., 1998. Circular dichroism and UV-visible absorption spectra of the Langmuir-Blodgett films of an aggregating helicene. *J. Am. Chem. Soc.* 120, 8656–8660. doi: 10.1021/ja981757h.

Okubo, H., Yamaguchi, M., Kabuto, C., 1998. Macrocyclic amides consisting of helical chiral 1,12- dimethylbenzo[*c*]phenanthrene-5,8-dicarboxylate. *J. Org. Chem.* 63, 9500–9509. doi: 10.1021/jo981720f.

Pan, T.-R., Guo, A.-M., Sun, Q.-F., 2016. Spin-polarized electron transport through helicene molecular junctions. *Phys. Rev. B* 94, 235448-1–235448-7. doi: 10.1103/PhysRevB.94.235448.

Parschau, M., Ernst, K.-H., 2006. Amplification of chirality in two-dimensional enantiomorphous lattices. *Nature* 439, 449–452. doi: 10.1038/nature04419.

Peña, D., Cobas, A., Pérez, D., Guitián, E., Castedo, L., 2000. Kinetic control in the palladium-catalyzed synthesis of C_2-symmetric hexabenzotriphenylene. A conformational study. *Org. Lett.* 2, 1629–1632. doi: 10.1021/ol005916p.

Peng, Z., Takenaka, N., 2013. Applications of helical-chiral pyridines as organocatalysts in asymmetric synthesis. *Chem. Rec.* 13, 28–42. doi: 10.1002/tcr.201200010.

Phillips, K.E.S., Katz, T.J., Jockusch, S., Lovinger, A.J., Turro, N.J., 2001. Synthesis and properties of an aggregating heterocyclic helicene. *J. Am. Chem. Soc.* 123, 11899–11907. doi: 10.1021/ja011706b.

Pinardi, A.L., Otero-Irurueta, G., Palacio, I., Martinez, J.I., Sanchez-Sanchez, C., Tello, M., Rogero, C., Cossaro, A., Preobrajenski, A., Gómez-Lor, B., Jancarik, A., Stará, I.G., Starý, I., Lopez, M.F., Méndez, J., Martin-Gago, J.Á., 2013. Tailored formation of *N*-doped nanoarchitectures by diffusion-controlled on-surface (cyclo)dehydrogenation of heteroaromatics. *ACS Nano* 7, 3676–3684. doi: 10.1021/nn400690c.

Poater, J., Visser, R., Solà, M., Bickelhaupt, F.M., 2007. Polycyclic benzenoids: Why kinked is more stable than straight. *J. Org. Chem.* 72, 1134–1142. doi: 10.1021/jo061637p.

Pospíšil, L., Bednárová, L., Štěpánek, P., Slavíček, P., Vávra, J., Hromadová, M., Dlouhá, H., Tarábek, J., Teplý, F., 2014. Intense chiroptical switching in a dicationic helicene-like derivative: Exploration of a viologen-type redox manifold of a non-racemic helquat. *J. Am. Chem. Soc.* 136, 10826–10829. doi: 10.1021/ja500220j.

Quan, X., Marvin, C.W., Seebald, L., Hutchison, G.R., 2013. Single-molecule piezoelectric deformation: Rational design from first-principles calculations. *J. Phys. Chem. C* 117, 16783–16790. doi: 10.1021/jp404252v.

Rahe, P., Nimmrich, M., Greuling, A., Schütte, J., Stará, I.G., Rybáček, J., Huerta-Angeles, G., Starý, I., Rohlfing, M., Kühnle, A., 2010. Toward molecular nanowires self-assembled on an insulating substrate: Heptahelicene-2-carboxylic acid on calcite (1014). *J. Phys. Chem. C* 114, 1547–1552. doi: 10.1021/jp911287p.

Rajca, A., Miyasaka, M., Pink, M., Wang, H., Rajca, S., 2004. Helically annelated and cross-conjugated oligothiophenes: Asymmetric synthesis, resolution, and characterization of a carbon–sulfur [7]helicene. *J. Am. Chem. Soc.* 126, 15211–15222. doi: 10.1021/ja0462530.

Reetz, M.T., Sostmann, S., 2000. Kinetic resolution in Pd-catalyzed allylic substitution using the helical PHelix ligand. *J. Organomet. Chem.* 603, 105–109. doi: 10.1016/S0022-328X(00)00173-X.

Rempala, P., King, B.T., 2006. Simulation of actuation by polymeric polyelectrolyte helicenes. *J. Chem. Theory Comput.* 2, 1112–1118. doi: 10.1021/ct600102r.

Reyes-Gutiérrez, P.E., Jirásek, M., Severa, L., Novotná, P., Koval, D., Sázelová, P., Vávra, J., Meyer, A., Císařová, I., Šaman, D., Pohl, R., Štěpánek, P., Slavíček, P., Coe, B.J., Hájek, M., Kašička, V., Urbanová, M., Teplý, F., 2015. Functional helquats: Helical cationic dyes with marked, switchable chiroptical properties in the visible region. *Chem. Commun.* 51, 1583–1586. doi: 10.1039/C4CC08967G.

Rice, B., LeBlanc, L.M., Otero-de-la-Roza, A., Fuchter, M.J., Johnson, E.R., Nelson, J., Jelfs, K.E., 2018. A computational exploration of the crystal energy and charge-carrier mobility landscapes of the chiral [6]helicene molecule. *Nanoscale* 10, 1865–1876. doi: 10.1039/C7NR08890F.

Rickhaus, M., Mayor, M., Juríček, M., 2016. Strain-induced helical chirality in polyaromatic systems. *Chem. Soc. Rev.* 45, 1542–1556. doi: 10.1039/C5CS00620A.

Rickhaus, M., Mayor, M., Juríček, M., 2017. Chirality in curved polyaromatic systems. *Chem. Soc. Rev.* 46, 1643–1660. doi: 10.1039/C6CS00623J.

Roose, J., Achermann, S., Dumele, O., Diederich, F., 2013. Electronically connected [n]helicenes: Synthesis and chiroptical properties of enantiomerically pure (*E*)-1,2-di([6]helicen-2-yl)ethenes. *Eur. J. Org. Chem.* 2013, 3223–3231. doi: 10.1002/ejoc.201300407.

Rulíšek, L., Exner, O., Cwiklik, L., Jungwirth, P., Starý, I., Pospíšil, L., Havlas, Z., 2007. On the convergence of the physicochemical properties of [n]helicenes. *J. Phys. Chem. C* 111, 14948–14955. doi: 10.1021/jp075129a.

Rybáček, J., Huerta-Angeles, G., Kollárovič, A., Stará, I.G., Starý, I., Rahe, P., Nimmrich, M., Kühnle, A., 2011. Racemic and optically pure heptahelicene-2-carboxylic acid: Its Synthesis and self-assembly into nanowire-like aggregates. *Eur. J. Org. Chem.* 2011, 853–860. doi: 10.1002/ejoc.201001110.

Šámal, M., Chercheja, S., Rybáček, J., Vacek Chocholoušová, J., Vacek, J., Bednárová, L., Šaman, D., Stará, I.G., Starý, I., 2015. An ultimate stereocontrol in asymmetric synthesis of optically pure fully aromatic helicenes. *J. Am. Chem. Soc.* 137, 8469–8474. doi: 10.1021/jacs.5b02794.

Sánchez, I.G., Šámal, M., Nejedlý, J., Karras, M., Klívar, J., Rybáček, J., Buděšínský, M., Bednárová, L., Seidlerová, B., Stará, I.G., Starý, I., 2017. Oxahelicene NHC ligands in the asymmetric synthesis of nonracemic helicenes. *Chem. Commun.* 53, 4370–4373. doi: 10.1039/C7CC00781G.

Sawada, Y., Furumi, S., Takai, A., Takeuchi, M., Noguchi, K., Tanaka, K., 2012. Rhodium-catalyzed enantioselective synthesis, crystal structures, and photophysical properties of helically chiral 1,1′-bitriphenylenes. *J. Am. Chem. Soc.* 134, 4080–4083. doi: 10.1021/ja300278e.

Schuster, N.J., Hernández Sánchez, R., Bukharina, D., Kotov, N.A., Berova, N., Ng, F., Steigerwald, M.L., Nuckolls, C., 2018. A helicene nanoribbon with greatly amplified chirality. *J. Am. Chem. Soc.* 140, 6235–6239. doi: 10.1021/jacs.8b03535.

Schweinfurth, D., Zalibera, M., Kathan, M., Shen, C., Mazzolini, M., Trapp, N., Crassous, J., Gescheidt, G., Diederich, F., 2014. Helicene quinones: Redox-triggered chiroptical switching and chiral recognition of the semiquinone radical anion lithium salt by electron nuclear double resonance spectroscopy. *J. Am. Chem. Soc.* 136, 13045–13052. doi: 10.1021/ja5069323.

Sehnal, P., Stará, I.G., Šaman, D., Tichý, M., Míšek, J., Cvačka, J., Rulíšek, L., Chocholoušová, J., Vacek, J., Goryl, G., Szymonski, M., Císařová, I., Starý, I., 2009. An organometallic route to long helicenes. *Proc. Natl. Acad. Sci. U. S. A.* 106, 13169–13174. doi: 10.1073/pnas.0902612106.

Seibel, J., Zoppi, L., Ernst, K.-H., 2014. 2D conglomerate crystallization of heptahelicene. *Chem. Commun.* 50, 8751–8753. doi: 10.1039/C4CC03574G.

Šesták, P., Wu, J., He, J., Pokluda, J., Zhang, Z., 2015. Extraordinary deformation capacity of smallest carbohelicene springs. *Phys. Chem. Chem. Phys.* 17, 18684–18690. doi: 10.1039/C5CP02043C.

Severa, L., Ončák, M., Koval, D., Pohl, R., Šaman, D., Císařová, I., Reyes-Gutiérrez, P.E., Sázelová, P., Kašička, V., Teplý, F., Slavíček, P., 2012. A chiral dicationic [8]circulenoid: Photochemical origin and facile thermal conversion into a helicene congener. *Angew. Chem. Int. Ed.* 51, 11972–11976. doi: 10.1002/anie.201203562.

Shchyrba, A., Nguyen, M.-T., Wäckerlin, C., Martens, S., Nowakowska, S., Ivas, T., Roose, J., Nijs, T., Boz, S., Schär, M., Stöhr, M., Pignedoli, C.A., Thilgen, C., Diederich, F., Passerone, D., Jung, T.A., 2013. Chirality transfer in 1D self-assemblies: Influence of H-bonding vs metal coordination between dicyano[7]helicene enantiomers. *J. Am. Chem. Soc.* 135, 15270–15273. doi: 10.1021/ja407315f.

Shen, Y., Chen, C.-F., 2011. Helicenes: Synthesis and applications. *Chem. Rev.* 112, 1463–1535. doi: 10.1021/cr200087r.

Shibata, T., Uchiyama, T., Yoshinami, Y., Takayasu, S., Tsuchikama, K., Endo, K., 2012. Highly enantioselective synthesis of silahelicenes using Ir-catalyzed [2 + 2 + 2] cycloaddition. *Chem. Commun.* 48, 1311–1313. doi: 10.1039/c1cc16762f.

Shinohara, K., Sannohe, Y., Kaieda, S., Tanaka, K., Osuga, H., Tahara, H., Xu, Y., Kawase, T., Bando, T., Sugiyama, H., 2010. A chiral wedge molecule inhibits telomerase activity. *J. Am. Chem. Soc.* 132, 3778–3782. doi: 10.1021/ja908897j.

Stará, I.G., Starý, I., 2009. Product class 21: Phenanthrenes, helicenes, and other angular acenes. In: Siegel, J.S. (Ed.) *Science of Synthesis, 45b: Category 6, Compounds with All-Carbon Functions.* Thieme Verlag, Leipzig. doi: 10.1055/sos-SD-045-00907.

Stará, I.G., Starý, I., Kollárovič, A., Teplý, F., Šaman, D., Tichý, M., 1998. A novel strategy for the synthesis of molecules with helical chirality. Intramolecular [2 + 2 + 2] cycloisomerization of triynes under cobalt catalysis. *J. Org. Chem.* 63, 4046–4050. doi: 10.1021/jo9801263.

Stará, I.G., Starý, I., Kollárovič, A., Teplý, F., Vyskočil, Š., Šaman, D., 1999. Transition metal catalysed synthesis of tetrahydro derivatives of [5]-, [6]- and [7]helicene. *Tetrahedron Lett.* 40, 1993–1996. doi: 10.1016/S0040-4039(99)00099-4.

Stępień, M., Gońka, E., Żyła, M., Sprutta, N., 2017. Heterocyclic nanographenes and other polycyclic heteroaromatic compounds: Synthetic routes, properties, and applications. *Chem. Rev.* 117, 3479–3716. doi: 10.1021/acs.chemrev.6b00076.

Stetsovych, O., Mutombo, P., Švec, M., Šámal, M., Nejedlý, J., Císařová, I., Vázquez, H., Moro-Lagares, M., Berger, J., Vacek, J., Stará, I.G., Starý, I., Jelínek, P., 2018. Large converse piezoelectric effect measured on a single molecule on a metallic surface. *J. Am. Chem. Soc.* 140, 940–946. doi: 10.1021/jacs.7b08729.

Stetsovych, O., Švec, M., Vacek, J., Chocholoušová, J.V., Jančařík, A., Rybáček, J., Kosmider, K., Stará, I.G., Jelínek, P., Starý, I., 2017. From helical to planar chirality by on-surface chemistry. *Nat. Chem.* 9, 213–218. doi: 10.1038/nchem.2662.

Storch, J., Církva, V., Bernard, M., Vokál, J., 2013. Způsob výroby [6]helicenů fotocyklizací. (Czech) Method and apparatus for production of [6]helicenes. Pat. No. 303997/PV 2012 – 245.

Storch, J., Sýkora, J., Čermák, J., Karban, J., Císařová, I., Růžička, A., 2009. Synthesis of hexahelicene and 1-methoxyhexahelicene via cycloisomerization of biphenylyl-naphthalene derivatives. *J. Org. Chem.* 74, 3090–3093. doi: 10.1021/jo900077j.

Sugimoto, Y., Pou, P., Abe, M., Jelinek, P., Pérez, R., Morita, S., Custance, Ó., 2007. Chemical identification of individual surface atoms by atomic force microscopy. *Nature* 446, 64–67. doi: 10.1038/nature05530.

Takenaka, N., Chen, J., Captain, B., Sarangthem, R.S., Chandrakumar, A., 2010. Helical chiral 2-aminopyridinium ions: A new class of hydrogen bond donor catalysts. *J. Am. Chem. Soc.* 132, 4536–4537. doi: 10.1021/ja100539c.

Takenaka, N., Sarangthem, R.S., Captain, B., 2008. Helical chiral pyridine *N*-oxides: A new family of asymmetric catalysts. *Angew. Chem. Int. Ed.* 47, 9708–9710. doi: 10.1002/anie.200803338.

Tanaka, K., Kamisawa, A., Suda, T., Noguchi, K., Hirano, M., 2007. Rh-catalyzed synthesis of helically chiral and ladder-type molecules via [2 + 2+ 2] and formal [2 + 1+ 2+ 1] cycloadditions involving CC triple bond cleavage. *J. Am. Chem. Soc.* 129, 12078–12079. doi: 10.1021/ja074914y.

Teplý, F., Stará, I.G., Starý, I., Kollárovič, A., Šaman, D., Rulíšek, L., Fiedler, P., 2002. Synthesis of [5]-, [6]-, and [7]helicene via Ni(0)- or Co(I)- catalyzed isomerization of aromatic *cis,cis*-dienetriynes. *J. Am. Chem. Soc.* 124, 9175–9180. doi: 10.1021/ja0259584.

Thongpanchang, T., Paruch, K., Katz, T.J., Rheingold, A.L., Lam, K.-C., Liable-Sands, L., 2000. Why (1*S*)-camphanates are excellent resolving agents for helicen-1-ols and why they can be used to analyze absolute configurations. *J. Org. Chem.* 65, 1850–1856. doi: 10.1021/jo9919411.

Torricelli, F., Bosson, J., Besnard, C., Chekini, M., Bürgi, T., Lacour, J., 2013. Modular synthesis, orthogonal post-functionalization, absorption, and chiroptical properties of cationic [6]helicenes. *Angew. Chem. Int. Ed.* 52, 1796–1800. doi: 10.1002/anie.201208926.

Treboux, G., Lapstun, P., Wu, Z., Silverbrook, K., 1999. Electronic conductance of helicenes. *Chem. Phys. Lett.* 301, 493–497. doi: 10.1016/S0009-2614(99)00085-8.

Urbano, A., Carreño, M.C., 2013. Enantioselective synthesis of helicenequinones and -bisquinones. *Org. Biomol. Chem.* 11, 699–708. doi: 10.1039/C2OB27108G.

Vacek, J., Chocholoušová, J.V., Stará, I.G., Starý, I., Dubi, Y., 2015. Mechanical tuning of conductance and thermopower in helicene molecular junctions. *Nanoscale* 7, 8793–8802. doi: 10.1039/C5NR01297J.

Vanest, J.M., Martin, R.H., 1979. Helicenes: A striking temperature dependence in a chemically induced asymmetric photosynthesis. *Rec. J. Roy. Neth. Chem. Soc.* 98, 113. doi: 10.1002/recl.19790980316.

Vávra, J., Severa, L., Císařová, I., Klepetářová, B., Šaman, D., Koval, D., Kašička, V., Teplý, F., 2013. Search for conglomerate in set of [7]helquat salts: Multigram resolution of helicene–viologen hybrid by preferential crystallization. *J. Org. Chem.* 78, 1329–1342. doi: 10.1021/jo301615k.

Verbiest, T., Elshocht, S.V., Kauranen, M., Hellemans, L., Snauwaert, J., Nuckolls, C., Katz, T.J., Persoons, A., 1998. Strong enhancement of nonlinear optical properties through supramolecular chirality. *Science* 282, 913–915. doi: 10.1126/science.282.5390.913.

Wang, D.Z., Katz, T.J., Golen, J., Rheingold, A.L., 2004. Diels–Alder additions of benzynes within helicene skeletons. *J. Org. Chem.* 69, 7769–7771. doi: 10.1021/jo048707h.

Weimar, M., Correa da Costa, R., Lee, F.-H., Fuchter, M.J., 2013. A scalable and expedient route to 1-aza[6]helicene derivatives and its subsequent application to a chiral-relay asymmetric strategy. *Org. Lett.* 15, 1706–1709. doi: 10.1021/ol400493x.

Willmore, N.D., Liu, L.B., Katz, T.J., 1992. A Diels-Alder route to [5]- and [6]-helicenes*Angew. Chem. Int. Ed.* 31, 1093–1095.doi: 10.1002/anie.199210931.

Wynberg, H., Groen, M.B., 1969. Racemization of two hexaheterohelicenes. *J. Chem. Soc. D*, 964–965. doi: 10.1039/C29690000964.

Xu, F., Yu, H., Sadrzadeh, A., Yakobson, B.I., 2016. Riemann surfaces of carbon as graphene nanosolenoids. *Nano Lett.* 16, 34–39. doi: 10.1021/acs.nanolett.5b02430.

Xu, Y., Zhang, Y.X., Sugiyama, H., Umano, T., Osuga, H., Tanaka, K., 2004. (*P*)-helicene displays chiral selection in binding to Z-DNA. *J. Am. Chem. Soc.* 126, 6566–6567. doi: 10.1021/ja0499748.

Yamada, K., Ogashiwa, S., Tanaka, H., Nakagawa, H., Kawazura, H., 1981. [7],[9],[11],[13], and [15]heterohelicenes annelated with alternant thiophene and benzene rings. Syntheses and NMR studies. *Chem. Lett.* 10, 343–346. doi: 10.1246/cl.1981.343.

Yamamoto, K., Shimizu, T., Igawa, K., Tomooka, K., Hirai, G., Suemune, H., Usui, K., 2016. Rational design and synthesis of [5]helicene-derived phosphine ligands and their application in Pd-catalyzed asymmetric reactions. *Sci. Rep.* 6, 36211. doi: 10.1038/srep36211.

Yang, Y., da Costa, R.C., Fuchter, M.J., Campbell, A.J., 2013. Circularly polarized light detection by a chiral organic semiconductor transistor. *Nat. Photonics* 7, 634–638. doi: 10.1038/nphoton.2013.176.

Yanney, M., Fronczek, F.R., Henry, W.P., Beard, D.J., Sygula, A., 2011. Cyclotrimerization of corannulyne: Steric hindrance tunes the inversion barriers of corannulene bowls. *Eur. J. Org. Chem.* 2011, 6636–6639. doi: 10.1002/ejoc.201101374.

Yavari, K., Aillard, P., Zhang, Y., Nuter, F., Retailleau, P., Voituriez, A., Marinetti, A., 2014. Helicenes with embedded phosphole units in enantioselective gold catalysis. *Angew. Chem. Int. Ed.* 53, 861–865. doi: 10.1002/anie.201308377.

Yavari, K., Moussa, S., Ben Hassine, B., Retailleau, P., Voituriez, A., Marinetti, A., 2012. 1*H*-phosphindoles as structural units in the synthesis of chiral helicenes. *Angew. Chem. Int. Ed.* 51, 6748–6752. doi: 10.1002/anie.201202024.

Žádný, J., Jančařík, A., Andronova, A., Šámal, M., Vacek Chocholoušová, J., Vacek, J., Pohl, R., Šaman, D., Císařová, I., Stará, I.G., Starý, I., 2012. A general approach to optically pure [5]-, [6]-, and [7]heterohelicenes. *Angew. Chem. Int. Ed.* 51, 5857–5861. doi: 10.1002/anie.201108307.

Zhu, Y., Xia, Z., Cai, Z., Yuan, Z., Jiang, N., Li, T., Wang, Y., Guo, X., Li, Z., Ma, S., Zhong, D., Li, Y., Wang, J., 2018. Synthesis and characterization of hexapole [7]helicene, a circularly twisted chiral nanographene. *J. Am. Chem. Soc.* 140, 4222–4226. doi: 10.1021/jacs.8b01447.

5

Supported Two-Dimensional Metal Clusters

Quanmin Guo
University of Birmingham

Dogan Kaya
Cukurova University

5.1 Introduction

On-surface synthesis of metal clusters can be performed by atomic deposition followed by nucleation and growth. This can be done on metallic, semiconducting or insulating substrates. During the initial phase of deposition, atomic dimer and trimer can be formed via either homogeneous or heterogeneous nucleation. Nucleation usually depends on the atomic flux, the substrate temperature and the density of surface defects [1]. Post-deposition ripening can take place by coarsening: growth of larger islands at the expense of the small islands via Ostwald ripening [2–4] or Smoluchowski ripening through island diffusion and coalescence [5–8]. The reason for producing two-dimensional (2D) or three-dimensional (3D) metallic clusters at the nanoscale is tailoring electronic [9], magnetic [10] and catalytic properties [11–13] of low-dimensional materials. Metal clusters supported on metal oxides, for example, are active components in catalytic, biological and microelectronic applications [14].

In this chapter, we will discuss the growth of 2D metal clusters produced on solid surfaces (metal, semiconductor or insulator) in ultra-high-vacuum conditions where the geometric shape of the cluster can be conveniently imaged with the scanning tunneling microscope (STM). The formation of 2D clusters depends on the details of atom-surface interaction, initial nucleation, atomic diffusion and cluster mobility. To understand the formation of 2D metallic clusters on surfaces, the interaction between the deposited metal atom and the substrate must be determined with a high degree of accuracy. The atom-surface interaction depends on the type of bonding (metallic, ionic, covalent) involved. Weak bonding usually leads to long diffusion lengths and hence favors heterogeneous nucleation and the growth of large clusters. Strong bonding, on the other hand, helps to immobilize atoms, promoting homogeneous nucleation. On crystalline substrates, an anisotropic potential surface usually causes the cluster to grow at different speeds along different crystallographic directions. Moreover, surfaces with a regular network of strain-relief pattern can be effectively used as templates for 2D cluster formation.

Metal atoms evaporated from Knudsen cells have thermal energy (k_BT) and will diffuse on the substrate surface until the excessive amount of thermal energy is dissipated [15]. An atom will stop diffusing once its thermal energy becomes less than the energy of the diffusion barrier. The diffusion of single atoms on crystalline substrate and the effect of temperature on diffusion have been extensively studied [16–19]. At sufficiently low temperatures, the diffusion of single atoms can be completely prohibited or requires external force [20–22]. Cluster formation, however, depends strongly on the ability of atoms to diffuse [23–25]. Homogeneous nucleation relies on the diffusing atoms to collide with each other to form dimers and trimers. Heterogeneous nucleation relies on the diffusing atoms to find the surface defects of steps. Growth of existing clusters is achieved by capturing diffusion atoms in the case of Ostwald ripening or by capturing diffusing clusters in the case of Smoluchowski ripening. Examples of Ostwald ripening are Au/Au(111) at 180 K [26], Ag/Ag(111) at 295 K [27], Cu/Cu(111) between 300 and 355 K [28] and Cu/Cu(100) at 343 K [29]. Typical examples of Smoluchowski ripening are given by Ag/Ag(100) and Cu/Cu(100) at 295 K [6,30]. In the following, we will discuss how the interactions between deposited atoms and the substrate affect the growth of 2D metal clusters. Here, 2D means cluster is just one atomic layer high.

Highly oriented pyrolytic graphite (HOPG) is a typical substrate used for the study of nucleation and growth of metal clusters. The basal plane of HOPG provides an atomic flat substrate, which is convenient for STM imaging. Due to the weak interaction between the metal atoms and the HOPG substrate, metal deposition usually leads to the formation of 3D clusters. Figure 5.1 shows Fe, Co and Ni

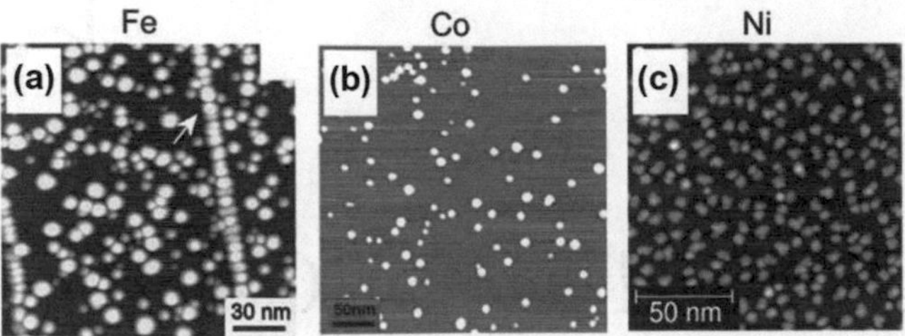

FIGURE 5.1 Deposition of transition metals on HOPG surface. (a–c) Fe atoms deposited at 300 K [31], Co atoms deposited at 300 K [32] and Ni atoms deposited below 200 K [33], respectively. (Image from Ref. [25].)

clusters grown on HOPG [31–33]. Similar 3D clusters are also found when metals are deposited onto metal oxide substrates such as TiO_2 [34,35]. The stability of the clusters depends on the cluster size. Clusters below a certain size are unstable and hence not observed at a specific temperature.

On metallic substrates, the interaction between the deposited atoms and the substrate is usually rather strong. This normally leads to the formation of 2D clusters. For homo-epitaxy, the formation of 2D clusters almost always takes place because of zero interfacial energy. For hetero-epitaxy, there are also many examples of 2D cluster formation. For example, deposition of Pd atoms on the Cu(111) surface at 98 K followed by annealing to 300 K gives rise to atomically flat Pd islands [36]. 2D clusters of cobalt have been produced on the Cu(100) surface for the investigation of CO oxidation [11]. 2D Ag clusters have been produced on the Fe(100) surface, and they are stable up to 800 K with a lateral lattice mismatch of 1% [37].

5.2 Three-Dimensional Metal Clusters on Non-metal Surfaces

The deposition of transition metal atoms is widely studied on semiconductor and semi-metal surfaces, for example, Ir [38], Fe [31], Co [32], Ni [33] clusters on HOPG (see Figure 5.1), Pb clusters on Si(111)-7 × 7 surfaces [39,40] and Ag clusters on GaAs(110) [41]. Metal atoms mostly bond weakly to the graphite surface; therefore, individual atoms and clusters may diffuse on the surface, coalescence and then form a compact shape. Due to weak interaction and long-range diffusion of atoms and clusters, clusters usually terminate at the step edges as well as defect sites on terraces. Transition metals usually form 3D clusters rather than 2D on HOPG [25]. During deposition, the sticking coefficient of many metal atoms on clean graphite can be as low as 0.1 [42].

5.3 Two-Dimensional Metal Clusters

For 2D metal clusters grown on metal substrates, the geometric shape of the cluster is usually controlled by the symmetry of the 2D lattice provided by the substrate. Taking the (100) surface of a face-centered cubic (fcc) metal as an example, the fourfold symmetry of the (100) surface favors the formation of square clusters. Figure 5.2a shows square-shaped Ag clusters after the deposition of 0.1 ML of Ag atoms onto Ag(100) at room temperature. Thermally induced ripening of the clusters is demonstrated with images in Figure 5.2b–d. Both Ostwald ripening and Smoluchowski ripening play important roles, with Smoluchowski ripening observed to be the dominant process at coverage less than 0.65 ML [43]. Other similar systems are Fe/Fe(100) [44], Cu/Cu(100) [45–47], Pd/Pd(100) [48], Ni/Ni(100) [49], Pt/Pt(100) [23], Ir/Ir(100) [23] and Al/Al(100) [50].

On the (111) surface of fcc metals, the near-equilibrium shape of 2D metal clusters is either hexagon or truncated triangle. At low temperatures, the cluster shape is affected by kinetics such as in diffusion-limited growth. Metal clusters grown on the (111) surface of fcc metallic substrates have been studied for many different materials, including Pt/Pt(111) [18,51], Ag/Pt(111) [52], Al/Al(111) [53,54], Ag/Ag(111) [55], Ir/Ir(111) [56], Rh/Rh(111) [56], Cu/Cu(111) [57], Au/Au(111) [26] and Au/Pt(111) [58]. In Figure 5.3, STM images taken at a range of sample temperatures show how the shape of the clusters changes from dendritic branches to regular hexagons. At the lowest temperature of 120 K, deposited Ag atoms form dendritic islands on Ag(111). Atoms landing on the surface are captured by the existing island via a hit-and-stick route. The island shape depends on edge and corner diffusion properties and may differ for different systems [58]. When the surface

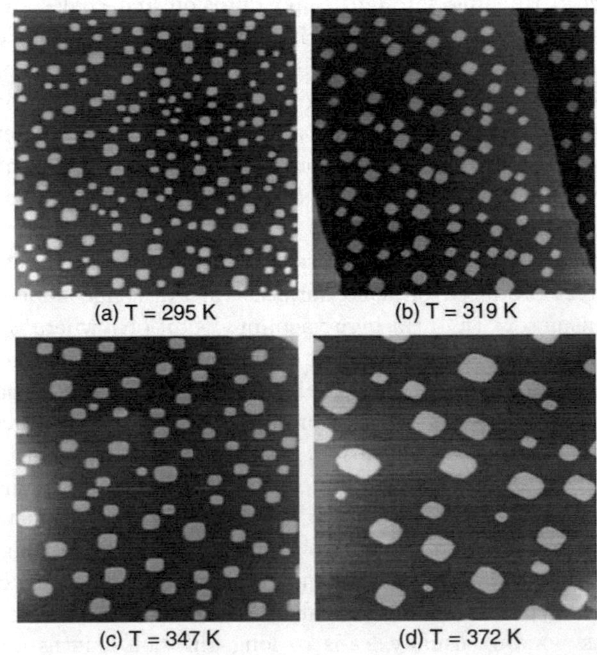

FIGURE 5.2 STM images of 0.1 ML atoms on Ag(100) surface presenting compact near-square islands. (a–d) STM images for elevated temperatures such as 295, 319, 347 and 372 K. (Islands are with a size of 176 × 176 nm^2 [43].)

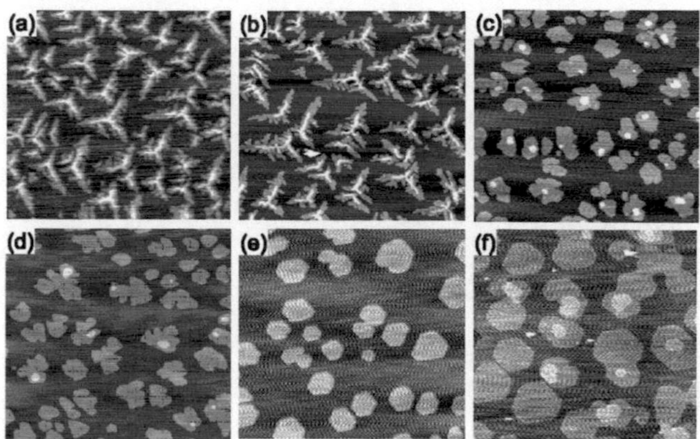

FIGURE 5.3 STM images of 0.3 ML Ag island distribution with a size of 30 × 30 nm^2 on Ag(111) surface at different temperatures. (a–f) 120, 135, 165, 180 and 200 K images, respectively [55].

temperature is increased gradually to 200 K, atoms at the edges of island are able to diffuse, leading to the formation of compact hexagonal islands [55]. Under extremely low sample temperatures, atoms landing on the surface are unable to diffuse and they exist as isolated atoms instead of forming clusters. The individual atoms can occupy either the fcc or the hexagonal close-packed (hcp) site as demonstrated with Pt at 20 K [59]. Migration of atoms from the hcp to the fcc site occurs when the sample is heated, as observed for Ir, Re, W and Pd adatoms on Ir(111) surface [60].

With high enough sample temperatures, the 2D clusters on the (111) fcc substrate appear circular in shape. This happens when the temperature is so high that the difference in the free energy of steps in different crystallographic directions becomes insignificant. For intermediate temperatures, the clusters are expected to have a polygonal shape. However, steps in different crystallographic directions do have different energies. It is natural for the cluster to have edges consisting of close-packed atomic rows because such rows have minimum free energy. On the (111) surface, there are two different close-packed rows with slightly different energies (Figure 5.4). The type B step has lower free energy than the type A step. Therefore, the thermodynamic equilibrium shape for the cluster is a triangle with type B steps only. In growth conditions, kinetic constraints can prevent the cluster reaching to its thermal equilibrium shape [19,53,61].

The (110) surface of fcc metals has a twofold symmetry. Homo-epitaxially grown metal clusters on the (110) surface tend to have an elongated shape due to the fast growth along the $[1\bar{1}0]$ direction. Figure 5.5 shows elongated Ag islands formed on the Ag(110) surface following the deposition of 0.3 ML of Ag atoms. Similar results are found for Cu/Pd(110) [19], Fe/Fe(110) [61], Ir/Ir(110) [24], Au/Au(110)-1 × 2 [16], Pt/Pt(110)-1 × 2 [16] and anisotropic surfaces such as Au/hex-Au(100) [62] and Pt/hex-Pt(100) [63]. The formation of elongated islands is facilitated by anisotropic diffusion with atoms diffusing much faster along the $[1\bar{1}0]$ direction than in the orthogonal [001] direction.

5.4 The Effect of Temperature

As seen in Figures 5.2, 5.3 and 5.5, surface temperature affects the island size distribution. Temperature is one of the most important parameters in nucleation growth, and it affects the critical size of the nucleus, diffusion of monomers on terraces, edge diffusion around an island and both the Ostwald and Smoluchowski ripening processes. Real-space imaging with the STM provides a detailed information on temperature dependence of cluster density and cluster size distribution. From high-resolution images acquired at a range of temperatures and deposition fluxes, useful methods have been developed to quantify the nucleation and growth process. Scaling relations have been derived for island size and island separation for a number of metals on metal

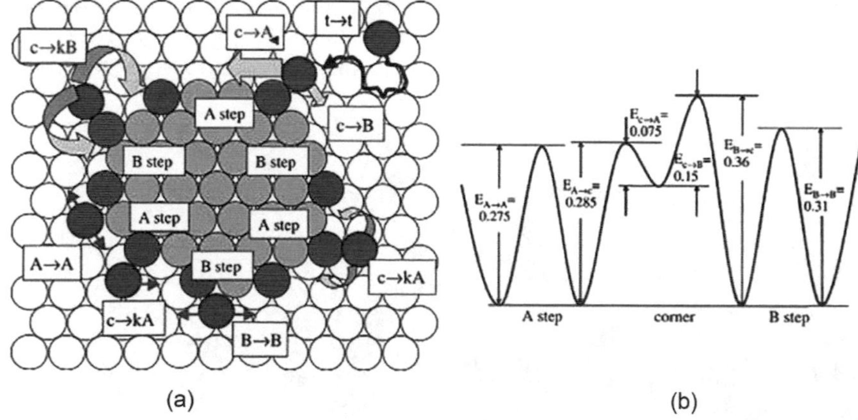

FIGURE 5.4 (a) Atomic ball model for Ag atomic diffusion and energy barriers involved in Ag island on Ag(111). Here, white circles represent surface atoms, and gray, dark gray and dark gray with arrows circles correspond to stable, corner and mobile Ag island, respectively. Two types of step edges are indicated as A and B as a function of surface direction. Possible diffusion processes are illustrated as A to A and B to B, and t, c and k denote terrace, corner and kink sites, respectively. Thus, c -> kA denotes a hopping process from a corner to A step. The large arrows indicate corner diffusion anisotropy, and bigger arrows indicate preferred processes. (b) Potential energy surface for corner rounding periphery diffusion [55].

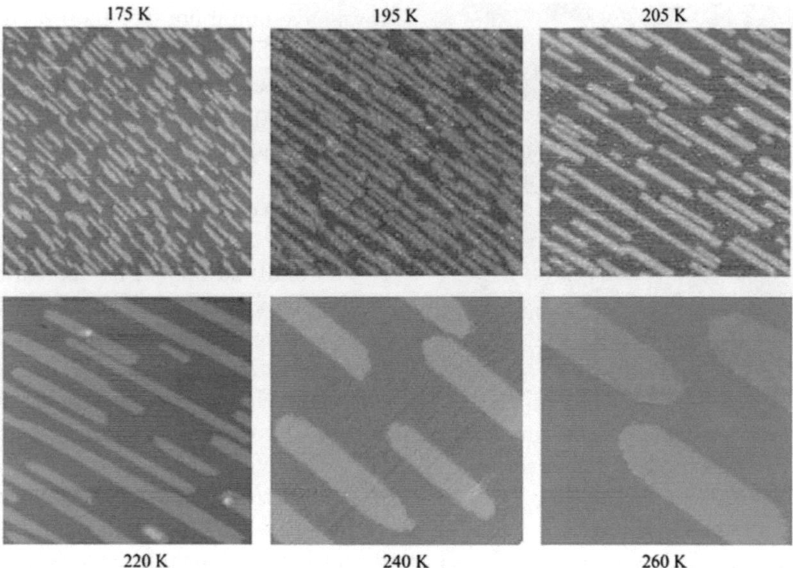

FIGURE 5.5 0.3 ML Ag deposited on Ag(110) surface. STM images show elongated Ag island formation with elevated temperatures ranging from 175 to 260 K [64].

systems. These include Ag/Ag(100) [43], Fe/Fe(100) [44], Cu/Cu(100) [45–47], Pd/Pd(100) [48], Ag/Ag(110) [65], Ni/Ni(100) [49] and Au/hex-Au(100) [62]. STM imaging has also been used to determine the diffusion parameters such as the activation energy and attempt frequency as demonstrated in Pt/Pt(111) [51].

5.5 Clusters Grown on a Template

In previous sections, we focused on the morphology of 2D metal clusters. In the following, we will demonstrate how to use the substrate as a template to control the location and orientation of clusters. Clean surfaces have a tendency to reconstruct, and one of the consequences of surface reconstruction is the formation of a strain-relief network within the top atomic layer. The strain-relief pattern can be used to control the nucleation of clusters. Among the many surfaces with strain-relief patterns, the Au(111) surface is the most extensively studied.

The Au (111) surface has a unique herringbone reconstruction as a result of alternating fcc and hcp stacking regions [66–69]. The reconstructed Au(111) substrate has been used for the growth of 2D atomic [70–72] and molecular [73–75] islands. High-resolution STM images [76] and the structural atomic model of the clean Au(111) surface [77] are presented in Figure 5.6. Figure 5.6a shows the clear herringbone discommensurate lines with the bulged and pinched elbow sites [76,78,79]. The area marked with a square in Figure 5.6a is magnified and shown with atomic resolution in Figure 5.6b, where A $(22 \times \sqrt{3})$ unit cell is highlighted. Within the unit cell, there is a 4.55% uniaxial compression along the $[1\bar{1}0]$ direction of the first layer. Atomic structural model of the (111) surface is given in Figure 5.6c, which shows 23 Au atoms in the topmost layer occupying a distance accommodating 22 atoms in the bulk, giving

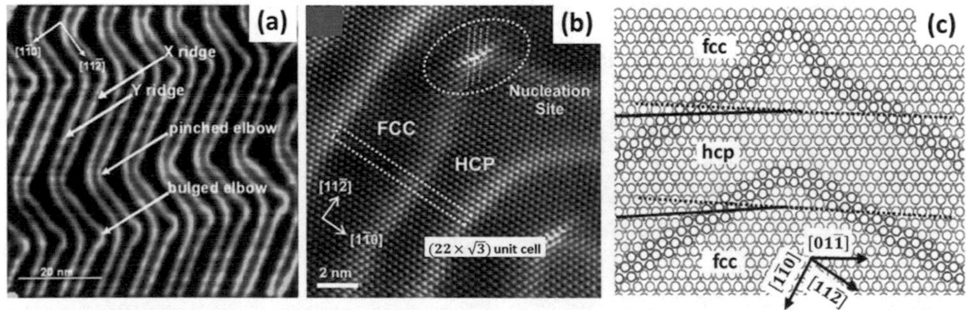

FIGURE 5.6 (a) High-resolution STM image defines herringbone reconstruction surface of Au(111) showing the fcc and hcp regions with pinched and bulged elbow sites with the surface directions of $[11\bar{2}]$ and $[1\bar{1}0]$ [76]. (b) Atomic resolution image of Au surface from (a) shows $(22 \times \sqrt{3})$ supercell white dashed rectangular nucleation site with fcc and hcp regions [76]. (c) Atomic structural model of the (111) surface showing herringbone ridges along the approximate $[01\bar{1}]$ azimuth differing a little from the exact $[01\bar{1}]$ direction, which are shown by dashed and solid lines [77].

rise to alternating fcc and hcp stacking regions with a 63 Å periodicity [66,68,80–82]. The fcc and hcp regions are ~3.5 and ~2.4 nm wide, respectively [68], with a surface potential difference on the order of 25 ± 5 meV [82]. The elbow site is the preferred nucleation site for atomic and molecular clusters. However, nucleation and growth can take place within the fcc or hcp domains if the sample temperature is low enough. An adsorbed layer of molecules can lift the surface reconstruction if sufficient charge transfer takes place between the molecules and the Au(111) substrate.

The formation of metal clusters on the Au(111) substrate has been investigated for a large number of metals, including Ni, Pt, Pd, Fe, Co, Rh, Ag, Cu and Al [70,79,83–92]. 2D metal clusters are mostly observed at the elbow site of the (111) surface, although nucleation and growth can also take place within the fcc or hcp region away from the elbow site. Figure 5.7 shows the formation of small metal clusters at the elbow sites on Au(111). Depending on the strength of interaction between the deposited metal atoms and the Au substrate, the deposited atoms can mix with Au atoms forming bimetallic clusters, for example, Ni/Au (570 K [83]), Pt/Au [85] and Pd/Au [84] bimetallic clusters.

There are two types of elbow sites, as shown in Figure 5.7: the bulged and the pinched elbow sites. Both types of elbow sites are effective nucleation sites for metal clusters. Due to the different locations of the surface defects at the elbows, clusters are formed in the fcc region next to the bulged elbows, but they are formed in the hcp region next to the pinched elbows. Figure 5.7 shows the locations of Au clusters formed on Au(111) at ~115 K [26]. The growth of 2D Au islands on the elbow sites of the Au(111) surface is possible at temperature less than 160 K. At higher temperatures, the elbow sites are unable to trap the clusters. Au atoms deposited at room temperature, for example, will all diffuse to step edges [26]. At 115 K, Au atoms landing on the substrate can diffuse. The diffusion is, however, confined within the discommensuration line. An atom landing into the fcc/hcp region cannot move into the neighboring hcp/fcc region by climbing over the walls of the discommensuration lines. The diffusing atoms are captured by the elbow sites. The average size of the metal clusters at the bulged elbow is larger than that at the pinched elbow. The size ratio is proportional to the fcc/hcp area ratio.

The Au clusters are single atomic layer high as shown by the height profile in Figure 5.7b. The size distribution of Au islands at the bulged elbow site is broader than that at the pinched elbow site, and the average size for islands is found to be 8.1 ± 3.5 and 4.5 ± 1.0 nm^2, respectively. A small number of Au islands with a size of 6.5 ± 4 nm^2 are also found at step edges. The binding of the Au cluster at the bulged elbow site is stronger than that at the pinched elbow site. By increasing the sample temperature, Ostwald ripening is observed. However, the ripening is highly selective with clusters at the bulged elbow site increasing in size at the expense of the clusters at the pinched elbows. By controlling the annealing time and temperature, it is hence possible to drive a complete migration of atomic clusters from the pinched elbow site to the bulged elbow site [26].

In Figure 5.8a–d, STM topographies at room temperature show self-organized Cu (~0.062 ML) [90], Pd

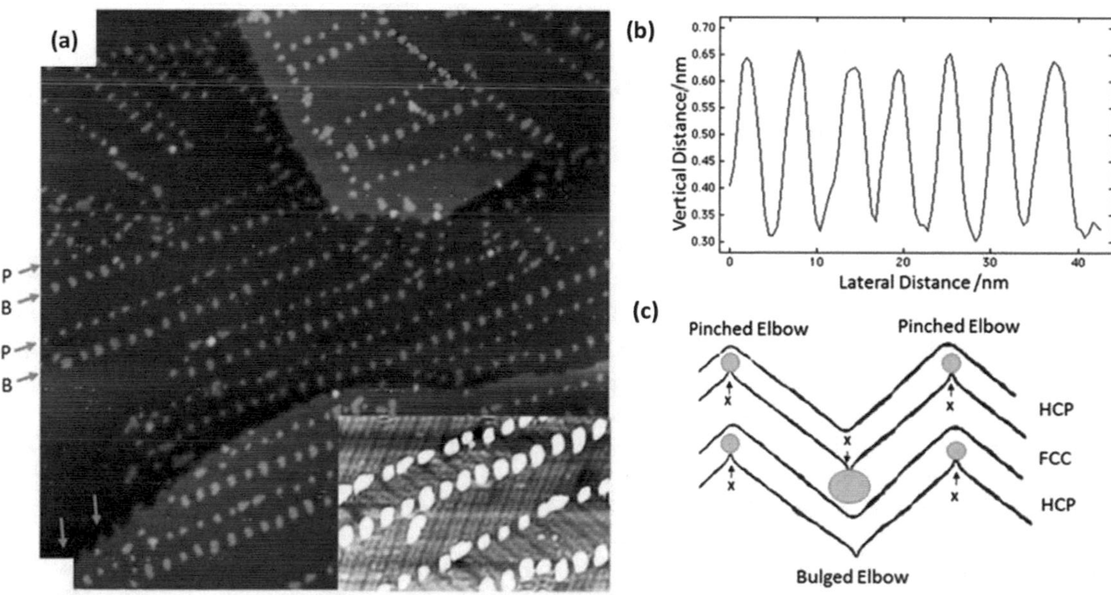

FIGURE 5.7 (a) Large area of STM image shows after the deposition of 0.041 ML Au on Au(111) surface at 115 K. At this coverage, Au island formed both pinched (P) and bulged (B) elbow sites as well as step edges. Inset image shows the herringbone reconstruction lines. (b) The gray dashed line shows the height profiles of the Au island along the elbow sites as an evidence of single atomic Au layer. (c) Schematic demonstration of individual Au island on the bulged (x) and pinched (y) elbow site. (Figure caption from Ref. [26].)

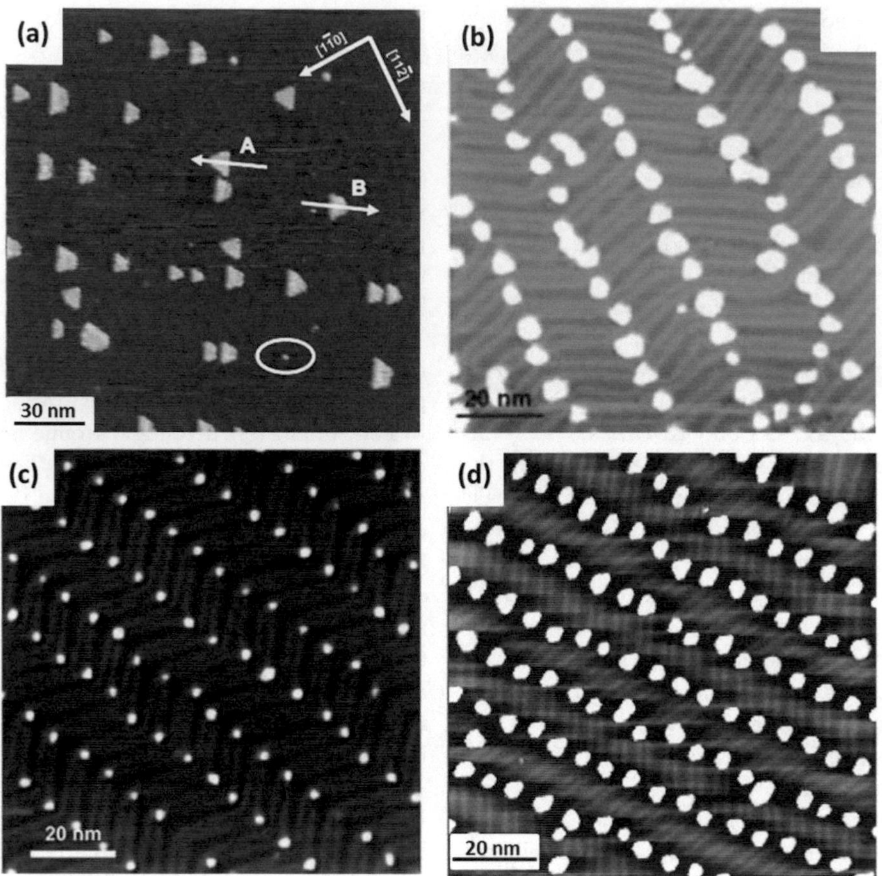

FIGURE 5.8 (a–d) STM topographies at room temperature showing self-organized Cu, Pd, Mo and Co clusters on the reconstruction of Au(111) surface, respectively. (a) Small Cu clusters (~0.062 ML) and triangular island shape formed on both pinched and bulged elbow sites of the surface given as A and B arrows, respectively [90]. (b) Pd clusters (~0.14 ML) formed on both pinched and bulged elbow sites of the surface [91]. Mo nanocluster (~0.1 ML) growth on both pinched and bulged elbow sites of the surface [76]. (d) Co clusters (~0.26 ML) with two atomic layers high [92].

(~0.14 ML) [76], Mo (~0.1 ML) [76] and Co (~0.26 ML) [92] clusters on the reconstruction of Au(111) surface, respectively. Mostly nucleation and growth of these atoms start at both pinched and bulged elbow sites of the surface as given in Figure 5.8 a A and B arrows, respectively.

Small 2D Au clusters are stable on Au(111) up to 240 K. Above 240 K, the clusters break apart with all Au atoms diffusing away to join existing step edges. The stability of Au clusters can be enhanced by capping them with molecules. For example, Xie et al. [93] produced magic number $(C_{60})_m$-$(Au)_n$ clusters by sequentially depositing Au atoms and C_{60} molecules on the Au(111) surface. In Figure 5.9a, different types of compact clusters can be seen in the STM image acquired at 293 K. Among these compact clusters, there are magic number $(C_{60})_m$-$(Au)_n$ clusters. One of the most abundant clusters, $(C_{60})_7$-$(Au)_{19}$, is formed with a 19 Au atom core and a 7 C_{60} molecule shell. The inset height profile confirms that the central C_{60} molecule of the $(C_{60})_7$-$(Au)_{19}$ magic number cluster sits on a disk of nineteen Au atoms arranged in a hexagon. The other height profile of a C_{60}-Au ring cluster reveals an Au island with island edges covered by C_{60} molecules. The close-packed magic number clusters have preferred azimuthal orientations. In Figure 5.9b, the structural model of the hexagonal Au island is given. There is one C_{60} molecule sitting next to each edge of the hexagonal Au cluster (Figure 5.9c), and one molecule sits directly above the hexagon. Other frequently observed magic number clusters are $(C_{60})_{10}$-$(Au)_{35}$ and $(C_{60})_{12}$-$(Au)_{49}$ (Figure 5.9d,e, respectively). The magic number clusters are stabilized by van der Waals interaction among the C_{60} molecules and the charge transfer reaction between the C_{60} molecule and the Au cluster [94–96]. As a result, even the smallest Au_{19} cluster is stable up to 400 K [97]. The interaction between atom-molecule and molecule-molecule can destabilize the cluster by removing Au atoms from the island.

The transformation from one magic number cluster to another has been studied using tip-triggered thermal cascade manipulation in the STM [98]. The manipulation is performed at room temperature by driving the STM tip 1.2 nm toward a selected C_{60} molecule in a cluster. This operation disrupts the cluster momentarily by displacing one

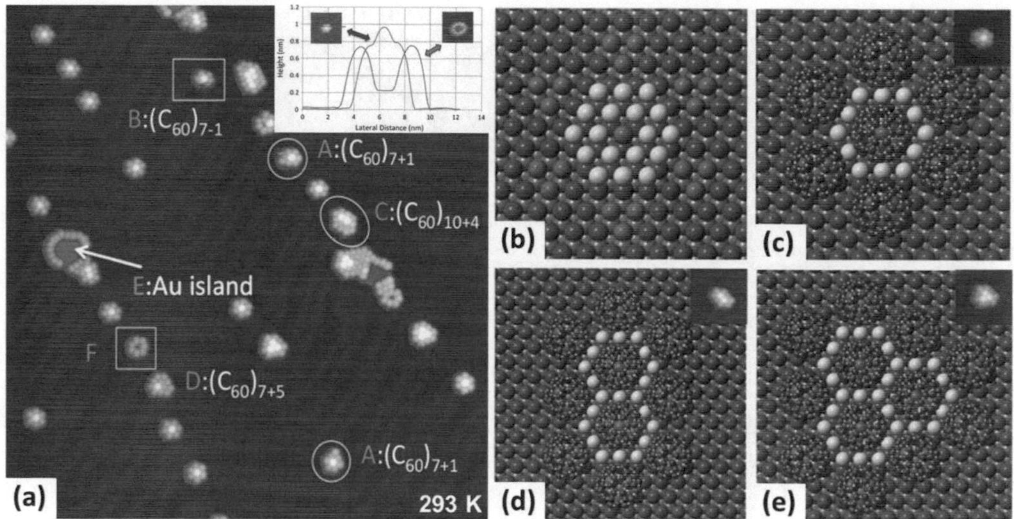

FIGURE 5.9 (a) STM image at 293 K for different types of clusters on the Au(111) elbow sites. Besides magic number $(C_{60})_m$-$(Au)_n$ clusters, irregular clusters observed with additional C_{60}, A, C, D or Au island decorated by additional C_{60} molecules, E, or missing molecule, B, F. Inset height profile of $(C_{60})_7$ magic number and C_{60}-Au island ring clusters. (b–e) Ball model of most observed magic number $(C_{60})_m$-$(Au)_n$ clusters. (b) 19 Au island, gray balls, on Au surface, dark gray balls. (c) 6 C_{60} molecules decorate Au island by bonding with three surface atoms and a C_{60} molecule decorates the top of Au island by bonding with a single Au atom. (d–e) $(C_{60})_{10}$-$(Au)_{35}$ and $(C_{60})_{12}$-$(Au)_{49}$ magic number clusters, respectively. (Figure caption adapted from Ref. [93].)

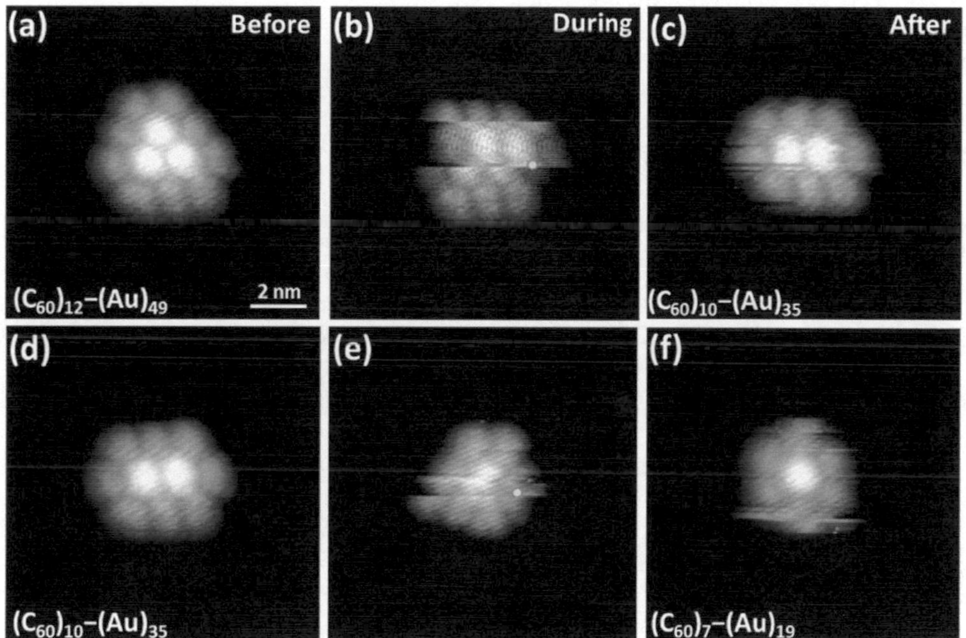

FIGURE 5.10 Tip-triggered cascade manipulation of hybrid clusters at room temperature. (a–c) Removing single C_{60} molecule (small gray dot) from $(C_{60})_{12}$-$(Au)_{49}$ and transforming to $(C_{60})_{10}$-$(Au)_{35}$ hybrid cluster. (d–f) Transformation from $(C_{60})_{10}$-$(Au)_{35}$ to $(C_{60})_7$-$(Au)_{19}$ hybrid cluster. (Figure caption adapted from Ref. [98].)

or more molecules away from the cluster. The remaining cluster is then given the opportunity to re-assemble. It is found that the stimulation of the STM tip is an effective method for downsizing the magic number cluster as shown in Figure 5.10. The STM images in Figure 5.10 demonstrate the successful transformation from a $(C_{60})_{12}$-$(Au)_{49}$ cluster to a $(C_{60})_{10}$-$(Au)_{35}$ cluster, and from a $(C_{60})_{10}$-$(Au)_{35}$ cluster to a $(C_{60})_7$-$(Au)_{19}$ cluster. It is clear that once a C_{60} molecule is removed from the edge of a magic number $(C_{60})_m$-Au_n cluster, the cluster spontaneously reorganizes to form another magic number cluster one size smaller. In order to confirm the existing 2D Au island surrounded by

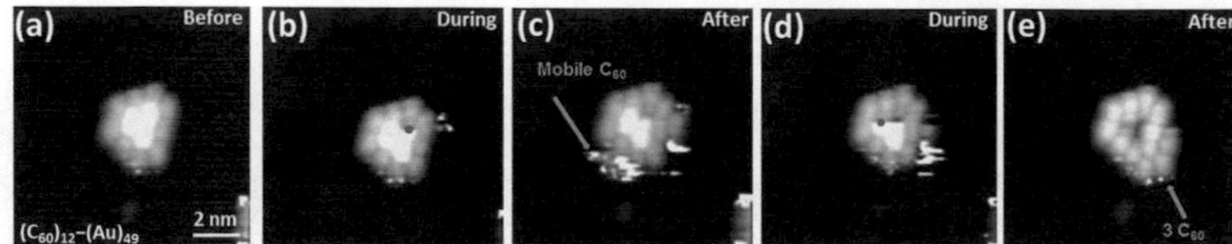

FIGURE 5.11 Manipulation of top three C_{60} molecules of $(C_{60})_{12}$-$(Au)_{49}$ cluster at 110 K. (a) $(C_{60})_{12}$-$(Au)_{49}$ cluster, (b) during manipulation, image on selected C_{60} molecule indicated as small dark gray dot, (c) mobile C_{60} around the cluster, (d) second C_{60} removed from the top, (e) all three removed without losing Au atom shape and cluster frame. (Figure caption adapted from Ref. [99].)

C_{60} molecules, the mechanical manipulation is performed on the top of $(C_{60})_{12}$-$(Au)_{49}$ cluster at 110 K by removing 3 C_{60} molecules which are seen as bright under STM topography in Figure 5.11. Manipulation performed on a selected C_{60} molecule is indicated as small dark gray dot in Figure 5.11b,d. After removing 3 C_{60} molecules without losing Au atoms, a hollow upper terrace confirms that cluster builds on 2D Au island (see Figure 5.11e).

5.6 Conclusion

The formation of the 2D clusters on surfaces starts with homo-nucleation or hetero-nucleation followed by the growth via kinetically or thermodynamically controlled process. Ostwald ripening or Smoluchowski ripening dominates 2D cluster growth on surfaces, and the size and the geometry of the cluster are affected by surface temperature and surface crystallography. Surface reconstruction such as the herringbone reconstruction on Au(111) can be used as a template for growing regular cluster arrays. Magic number 2D clusters can be produced by molecular capping. Such capped clusters have a much enhanced thermal stability.

References

1. J. Venables, Rate equation approaches to thin film nucleation kinetics, *Philosophical Magazine*, 27 (1973) 697–738.
2. P.W. Voorhees, The theory of Ostwald ripening, *Journal of Statistical Physics*, 38 (1985) 231–252.
3. V.P. Zhdanov, F.F. Schweinberger, U. Heiz, C. Langhammer, Ostwald ripening of supported Pt nanoclusters with initial size-selected distributions, *Chemical Physics Letters*, 631–632 (2015) 21–25.
4. R. Lorenz, Lehrbuch der allgemeinen Chemie, von W. OSTWALD. II. Band. 2. Teil: Verwandtschaftslehre. 1. Lieferung. (Leipzig, 1896.) 5 Mark, *Zeitschrift für anorganische Chemie*, 15 (1897) 239–239.
5. M. Von Smoluchowski, Drei vortrage uber diffusion. Brownsche bewegung und koagulation von kolloidteilchen, *Zeitschrift für Physik*, 17 (1916) 557–585.
6. J.M. Wen, J.W. Evans, M.C. Bartelt, J.W. Burnett, P.A. Thiel, Coarsening mechanisms in a metal film: From cluster diffusion to vacancy ripening, *Physical Review Letters*, 76 (1996) 652–655.
7. M. Vicanek, N.M. Ghoniem, The effects of mobility coalescence on the evolution of surface atomic clusters, *Thin Solid Films*, 207 (1992) 90–97.
8. M.J.J. Jak, C. Konstapel, A. van Kreuningen, J. Verhoeven, J.W.M. Frenken, Scanning tunnelling microscopy study of the growth of small palladium particles on $TiO_2(110)$, *Surface Science*, 457 (2000) 295–310.
9. D.V. Talapin, C.B. Murray, PbSe nanocrystal solids for n- and p-channel thin film field-effect transistors, *Science*, 310 (2005) 86–89.
10. N. Weiss, T. Cren, M. Epple, S. Rusponi, G. Baudot, S. Rohart, A. Tejeda, V. Repain, S. Rousset, P. Ohresser, F. Scheurer, P. Bencok, H. Brune, Uniform magnetic properties for an ultrahigh-density lattice of noninteracting Co nanostructures, *Physical Review Letters*, 95 (2005) 157204.
11. F. Falo, I. Cano, M. Salmerón, CO chemisorption on two-dimensional cobalt clusters: A surface science approach to cluster chemistry, *Surface Science*, 143 (1984) 303–313.
12. H. Poppa, Model studies in catalysis with uhv-deposited metal particles and clusters, *Vacuum*, 34 (1984) 1081–1095.
13. D.W. Goodman, Model studies in catalysis using surface science probes, *Chemical Reviews*, 95 (1995) 523–536.
14. B.V. Reddy, S.N. Khanna, Self-stimulated NO reduction and CO oxidation by iron oxide clusters, *Physical Review Letters*, 93 (2004) 068301.
15. S. Wang, G. Ehrlich, Atom condensation at lattice steps and clusters, *Physical Review Letters*, 71 (1993) 4174.
16. S. Günther, A. Hitzke, R. Behm, Low adatom mobility on the (1×2)-missing-row reconstructed Au(110) surface, *Surface Review and Letters*, 4 (1997) 1103–1108.
17. H. Brune, Microscopic view of epitaxial metal growth: Nucleation and aggregation, *Surface Science Reports*, 31 (1998) 125–229.
18. H. Brune, H. Roder, K. Bromann, K. Kern, J. Jacobsen, P. Stoltze, K. Jacobsen, J. Norskov, Anisotropic corner diffusion as origin for dendritic

growth on hexagonal substrates, *Surface Science*, 349 (1996) L115–L122.

19. H. Brune, C. Romainczyk, H. Röder, K. Kern, Mechanism of the transition from fractal to dendritic growth of surface aggregates, *Nature*, 369 (1994) 469.
20. D.M. Eigler, Positioning single atoms with a scanning tunnelling microscope, *Nature*, 344 (1990) 524.
21. G. Meyer, L. Bartels, S. Zöphel, E. Henze, K.-H. Rieder, Controlled atom by atom restructuring of a metal surface with the scanning tunneling microscope, *Physical Review Letters*, 78 (1997) 1512–1515.
22. G. Meyer, S. Zöphel, K.-H. Rieder, Scanning tunneling microscopy manipulation of native substrate atoms: A new way to obtain registry information on foreign adsorbates, *Physical Review Letters*, 77 (1996) 2113–2116.
23. P.J. Feibelman, Scaling of hopping self-diffusion barriers on fcc (100) surfaces with bulk bond energies, *Surface Science*, 423 (1999) 169–174.
24. C.-L. Chen, T.T. Tsong, Self-diffusion on the reconstructed and nonreconstructed Ir(110) surfaces, *Physical Review Letters*, 66 (1991) 1610.
25. D. Appy, H. Lei, C.-Z. Wang, M.C. Tringides, D.-J. Liu, J.W. Evans, P.A. Thiel, Transition metals on the (0001) surface of graphite: Fundamental aspects of adsorption, diffusion, and morphology, *Progress in Surface Science*, 89 (2014) 219–238.
26. M. Rokni-Fard, Q. Guo, Biased Ostwald ripening in site-selective growth of two-dimensional gold clusters, *The Journal of Physical Chemistry C*, 122 (2018) 7801–7805.
27. K. Morgenstern, G. Rosenfeld, G. Comsa, Decay of two-dimensional Ag islands on Ag(111), *Physical Review Letters*, 76 (1996) 2113.
28. G.S. Icking-Konert, G. Schulze Icking-Konert, M. Giesen, and H. Ibach, Decay of Cuad atom islands on Cu(111), *Surface Science*, 398 (1998) 37.
29. C.K. J. B. Hannon, M. Giesen, H. Ibach, N. C. Bartelt, J. C. Hamilton, Surface self-diffusion by vacancy motion: Island ripening on Cu(001), *Physical Review Letters*, 79 (1997) 2506.
30. W.W. Pai, A.K. Swan, Z. Zhang, J. Wendelken, Island diffusion and coarsening on metal (100) surfaces, *Physical Review Letters*, 79 (1997) 3210.
31. I.N. Kholmanov, L. Gavioli, M. Fanetti, M. Casella, C. Cepek, C. Mattevi, M. Sancrotti, Effect of substrate surface defects on the morphology of Fe film deposited on graphite, *Surface Science*, 601 (2007) 188–192.
32. S.W. Poon, A.T.S. Wee, E.S. Tok, Anomalous scaling behaviour of cobalt cluster size distributions on graphite, epitaxial graphene and carbon-rich $(6\sqrt{3} \times 6\sqrt{3})$R30°, *Surface Science*, 606 (2012) 1586–1593.
33. M. Marz, K. Sagisaka, D. Fujita, Ni nanocrystals on HOPG(0001): A scanning tunnelling microscope study, *Beilstein Journal of Nanotechnology*, 4 (2013) 406–417.
34. K. Luo, T. St. Clair, X. Lai, D. Goodman, Silver growth on $TiO_2(110)(1 \times 1)$ and (1×2), *The Journal of Physical Chemistry B*, 104 (2000) 3050–3057.
35. R.P. Galhenage, H. Yan, S.A. Tenney, N. Park, G. Henkelman, P. Albrecht, D.R. Mullins, D.A. Chen, Understanding the nucleation and growth of metals on TiO_2: Co compared to Au, Ni, and Pt, *The Journal of Physical Chemistry C*, 117 (2013) 7191–7201.
36. F. Calleja, J. Hinarejos, M. Passeggi Jr, A.V. de Parga, R. Miranda, Thermal stability of atomically flat metal nanofilms on metallic substrates, *Applied Surface Science*, 254 (2007) 12–15.
37. C.M. Wei, M.Y. Chou, Effects of the substrate on quantum well states: A first-principles study for Ag/Fe(100), *Physical Review B*, 68 (2003) 125406.
38. A.T. N'Diaye, S. Bleikamp, P.J. Feibelman, T. Michely, Two-dimensional Ir cluster lattice on a graphene moire on Ir(111), *Physical Review Letters*, 97 (2006) 215501.
39. E. Inami, I. Hamada, K. Ueda, M. Abe, S. Morita, Y. Sugimoto, Room-temperature-concerted switch made of a binary atom cluster, *Nature Communication*, 6 (2015) 6231.
40. M.M. Özer, C.-Z. Wang, Z. Zhang, H.H. Weitering, Quantum size effects in the growth, coarsening, and properties of ultra-thin metal films and related nanostructures, *Journal of Low Temperature Physics*, 157 (2009) 221–251.
41. A.R. Smith, K.-J. Chao, Q. Niu, C.-K. Shih, Formation of atomically flat silver films on GaAs with a "silver mean" quasi periodicity, *Science*, 273 (1996) 226–228.
42. E. Ganz, K. Sattler, J. Clarke, Scanning tunneling microscopy of Cu, Ag, Au and Al adatoms, small clusters, and islands on graphite, *Surface Science*, 219 (1989) 33–67.
43. C.-M. Zhang, M.C. Bartelt, J.-M. Wen, C.J. Jenks, J.W. Evans, P.A. Thiel, Submonolayer island formation and the onset of multilayer growth during Ag/Ag(100) homoepitaxy, *Surface Science*, 406 (1998) 178–193.
44. J.A. Stroscio, D.T. Pierce, Scaling of diffusion-mediated island growth in iron-on-iron homoepitaxy, *Physical Review B*, 49 (1994) 8522–8525.
45. A.K. Swan, Z.-P. Shi, J.F. Wendelken, Z. Zhang, Flux-dependent scaling behavior in Cu(100) submonolayer homoepitaxy, *Surface Science*, 391 (1997) L1205–L1211.
46. H. Dürr, J.F. Wendelken, J.K. Zuo, Island morphology and adatom energy barriers during homoepitaxy on Cu(001), *Surface Science*, 328 (1995) L527–L532.

47. H.-J. Ernst, F. Fabre, J. Lapujoulade, Nucleation and diffusion of Cu adatoms on Cu(100): A helium-atom-beam scattering study, *Physical Review B*, 46 (1992) 1929–1932.
48. J.W. Evans, D.K. Flynn-Sanders, P.A. Thiel, Surface self-diffusion barrier of Pd(100) from low-energy electron diffraction, *Surface Science*, 298 (1993) 378–383.
49. E. Kopatzki, S. Günther, W. Nichtl-Pecher, R.J. Behm, Homoepitaxial growth on Ni(100) and its modification by a preadsorbed oxygen adlayer, *Surface Science*, 284 (1993) 154–166.
50. P.J. Feibelman, Diffusion path for an Al adatom on Al (001), *Physical Review Letters*, 65 (1990) 729.
51. M. Bott, M. Hohage, M. Morgenstern, T. Michely, G. Comsa, New approach for determination of diffusion parameters of adatoms, *Physical Review Letters*, 76 (1996) 1304.
52. M. Hohage, M. Bott, M. Morgenstern, Z. Zhang, T. Michely, G. Comsa, Atomic processes in low temperature Pt-dendrite growth on Pt(111), *Physical Review Letters*, 76 (1996) 2366.
53. S. Ovesson, A. Bogicevic, B.I. Lundqvist, Origin of compact triangular islands in metal-on-metal growth, *Physical Review Letters*, 83 (1999) 2608.
54. T. Michely, J. Krug, *Islands, Mounds and Atoms*. Springer Science & Business Media, New York (2012).
55. E. Cox, M. Li, P.-W. Chung, C. Ghosh, T. Rahman, C.J. Jenks, J.W. Evans, P.A. Thiel, Temperature dependence of island growth shapes during submonolayer deposition of Ag on Ag (111), *Physical Review B*, 71 (2005) 115414.
56. F. Tsui, J. Wellman, C. Uher, R. Clarke, Morphology transition and layer-by-layer growth of Rh (111), *Physical Review Letters*, 76 (1996) 3164.
57. A. Bogicevic, S. Ovesson, P. Hyldgaard, B. Lundqvist, H. Brune, D. Jennison, Nature, strength, and consequences of indirect adsorbate interactions on metals, *Physical Review Letters*, 85 (2000) 1910.
58. S. Ogura, K. Fukutani, M. Matsumoto, T. Okano, M. Okada, T. Kawamura, Dendritic to non-dendritic transitions in Au islands investigated by scanning tunneling microscopy and Monte Carlo simulations, *Physical Review B*, 73 (2006) 125442.
59. A. Gölzhäuser, G. Ehrlich, Atom movement and binding on surface clusters: Pt on Pt(111) clusters, *Physical Review Letters*, 77 (1996) 1334–1337.
60. S. Wang, G. Ehrlich, Atom condensation on an atomically smooth surface: Ir, Re, W, and Pd on Ir (111), *The Journal of Chemical Physics*, 94 (1991) 4071–4074.
61. U. Köhler, C. Jensen, A. Schindler, L. Brendel, D. Wolf, Scanning tunnelling microscopy and Monte Carlo studies of homoepitaxy on Fe(110), *Philosophical Magazine B*, 80 (2000) 283–292.
62. S. Günther, E. Kopatzki, M. Bartelt, J.W. Evans, R. Behm, Anisotropy in nucleation and growth of two-dimensional islands during homoepitaxy on "hex" reconstructed Au(100), *Physical Review Letters*, 73 (1994) 553.
63. T.R. Linderoth, J.J. Mortensen, K.W. Jacobsen, E. Lægsgaard, I. Stensgaard, F. Besenbacher, Homoepitaxial growth of Pt on Pt(100)-hex: Effects of strongly anisotropic diffusion and finite island sizes, *Physical Review Letters*, 77 (1996) 87.
64. J.W. Evans, Y. Han, B. Ünal, M. Li, K. Caspersen, D. Jing, A.R. Layson, C. Stoldt, T. Duguet, P.A. Thiel, From initial to late stages of epitaxial thin film growth: STM analysis and atomistic or coarse-grained modeling, *AIP Conference Proceedings*, Dalian, China, AIP (2010) pp. 26–44.
65. C. De Giorgi, P. Aihemaiti, F.B. De Mongeot, C. Boragno, R. Ferrando, U. Valbusa, Submonolayer homoepitaxial growth on Ag (110), *Surface Science*, 487 (2001) 49–54.
66. C. Wöll, S. Chiang, R.J. Wilson, P.H. Lippel, Determination of atom positions at stacking-fault dislocations on Au(111) by scanning tunneling microscopy. In: H. Neddermeyer (Ed.) *Scanning Tunneling Microscopy*. Springer, Netherlands (1993), pp. 114–117.
67. S. Narasimhan, D. Vanderbilt, Elastic stress domains and the herringbone reconstruction on Au(111), *Physical Review Letters*, 69 (1992) 1564–1567.
68. L. Bürgi, H. Brune, K. Kern, Imaging of electron potential landscapes on Au(111), *Physical Review Letters*, 89 (2002) 176801.
69. C. Wöll, S. Chiang, R.J. Wilson, P.H. Lippel, Determination of atom positions at stacking-fault dislocations on Au(111) by scanning tunneling microscopy. In: H. Neddermeyer (Ed.) *Scanning Tunneling Microscopy*. Springer, Dordrecht (1993), pp. 114–117.
70. D.D. Chambliss, K.E. Johnson, R.J. Wilson, S. Chiang, Surface structure and metal epitaxy: STM studies of ultrathin metal films on Au(111) and Cu(100), *Journal of Magnetism and Magnetic Materials*, 121 (1993) 1–9.
71. J.A. Stroscio, D.T. Pierce, R.A. Dragoset, P.N. First, Microscopic aspects of the initial growth of metastable fcc iron on Au(111), *Journal of Vacuum Science and Technology A*, 10 (1992) 1981–1985.
72. S. Helveg, J.V. Lauritsen, E. Lægsgaard, I. Stensgaard, J.K. Nørskov, B.S. Clausen, H. Topsøe, F. Besenbacher, Atomic-scale structure of single-Layer MoS_2 nanoclusters, *Physical Review Letters*, 84 (2000) 951–954.
73. T. Yokoyama, S. Yokoyama, T. Kamikado, Y. Okuno, S. Mashiko, Selective assembly on a surface of supramolecular aggregates with controlled size and shape, *Nature*, 413 (2001) 619–621.

74. M. Böhringer, K. Morgenstern, W.-D. Schneider, M. Wühn, C. Wöll, R. Berndt, Self-assembly of 1-nitronaphthalene on Au(111), *Surface Science*, 444 (2000) 199–210.
75. Y. Wang, X. Ge, G. Schull, R. Berndt, C. Bornholdt, F. Koehler, R. Herges, Azo Supramolecules on Au(111) with controlled size and shape, *Journal of the American Chemical Society*, 130 (2008) 4218–4219.
76. F. Besenbacher, J.V. Lauritsen, T.R. Linderoth, E. Lægsgaard, R.T. Vang, S. Wendt, Atomic-scale surface science phenomena studied by scanning tunneling microscopy, *Surface Science*, 603 (2009) 1315–1327.
77. Z.-X. Xie, Z.-F. Huang, X. Xu, Influence of reconstruction on the structure of self-assembled normal-alkane monolayers on Au(111) surfaces, *Physical Chemistry Chemical Physics*, 4 (2002) 1486–1489.
78. J.V. Barth, H. Brune, G. Ertl, R.J. Behm, Scanning tunneling microscopy observations on the reconstructed Au(111) surface: Atomic structure, long-range superstructure, rotational domains, and surface defects, *Physical Review B*, 42 (1990) 9307–9318.
79. D.D. Chambliss, R.J. Wilson, S. Chiang, Nucleation of ordered Ni island arrays on Au(111) by surface-lattice dislocations, *Physical Review Letters*, 66 (1991) 1721–1724.
80. M.A. Van Hove, R.J. Koestner, P.C. Stair, J.P. Bibérian, L.L. Kesmodel, I. BartoŠ, G.A. Somorjai, The surface reconstructions of the (100) crystal faces of iridium, platinum and gold: I. Experimental observations and possible structural models, *Surface Science*, 103 (1981) 189–217.
81. P. Maksymovych, D.C. Sorescu, D. Dougherty, J.T. Yates, Surface bonding and dynamical behavior of the CH_3SH molecule on Au(111), *The Journal of Physical Chemistry B*, 109 (2005) 22463–22468.
82. F. Reinert, G. Nicolay, Influence of the herringbone reconstruction on the surface electronic structure of Au(111), *Applied Physics A*, 78 (2004) 817–821.
83. W.G. Cullen, P.N. First, Island shapes and intermixing for submonolayer nickel on Au(111), *Surface Science*, 420 (1999) 53–64.
84. L.A. Kibler, M. Kleinert, R. Randler, D.M. Kolb, Initial stages of Pd deposition on Au(hkl) part I: Pd on Au(111), *Surface Science*, 443 (1999) 19–30.
85. J.W.A. Sachtler, G.A. Somorjai, Influence of ensemble size on CO chemisorption and catalytic n-hexane conversion by Au-Pt(111) bimetallic single-crystal surfaces, *Journal of Catalysis*, 81 (1983) 77–94.
86. B. Voigtländer, G. Meyer, N.M. Amer, Epitaxial growth of thin magnetic cobalt films on Au(111) studied by scanning tunneling microscopy, *Physical Review B*, 44 (1991) 10354.
87. J.A. Stroscio, D.T. Pierce, R.A. Dragoset, P. First, Microscopic aspects of the initial growth of metastable fcc iron on Au (111), *Journal of Vacuum Science and Technology A: Vacuum, Surfaces, and Films*, 10 (1992) 1981–1985.
88. E.I. Altman, R.J. Colton, Growth of Rh on Au(111): Surface intermixing of immiscible metals, *Surface Science*, 304 (1994) L400–L406.
89. B. Fischer, H. Brune, J.V. Barth, A. Fricke, K. Kern, Nucleation kinetics on inhomogeneous substrates: Al/Au(111), *Physical Review Letters*, 82 (1999) 1732–1735.
90. F. Grillo, H. Früchtl, S.M. Francis, N.V. Richardson, Site selectivity in the growth of copper islands on Au (111), *New Journal of Physics*, 13 (2011) 013044.
91. C.S. Casari, S. Foglio, F. Siviero, A.L. Bassi, M. Passoni, C.E. Bottani, Direct observation of the basic mechanisms of Pd island nucleation on Au(111), *Physical Review B*, 79 (2009) 195402.
92. I. Chado, C. Goyhenex, H. Bulou, J. Bucher, Cluster critical size effect during growth on a heterogeneous surface, *Physical Review B*, 69 (2004) 085413.
93. Y.-C. Xie, L. Tang, Q. Guo, Cooperative assembly of magic number C_{60}-Au complexes, *Physical Review Letters*, 111 (2013) 186101.
94. E.I. Altman, R.J. Colton, Nucleation, growth, and structure of fullerene films on Au(111), *Surface Science*, 279 (1992) 49–67.
95. L. Tang, Y. Xie, Q. Guo, Probing the buried C_{60}/Au(111) interface with atoms, *The Journal of Chemical Physics*, 136 (2012) 214706.
96. X. Zhang, L. Tang, Q. Guo, Low-temperature growth of C_{60} monolayers on Au(111): Island orientation control with site-selective nucleation, *The Journal of Physical Chemistry C*, 114 (2010) 6433–6439.
97. R. Chakrabarty, P.S. Mukherjee, P.J. Stang, Supramolecular coordination: Self-assembly of finite two- and three-dimensional ensembles, *Chemical Reviews*, 111 (2011) 6810–6918.
98. D. Kaya, D.-L. Bao, R.E. Palmer, S. Du, Q. Guo, Tip-triggered thermal cascade manipulation of magic number gold-fullerene clusters in the scanning tunnelling microscope, *Nano Letters*, 17 (2017) 6171–6176.
99. D. Kaya, J. Gao, M.R. Fard, R.E. Palmer, Q. Guo, Controlled manipulation of magic number gold–fullerene clusters using scanning tunneling microscopy, *Langmuir*, 34 (2018) 8388–8392.

6

A Novel Class of Two-Dimensional Materials: Transition Metal Dichalcogenides*

Fangyuan Zheng,
Lokwing Wong, and
Jiong Zhao
The Hong Kong Polytechnic University

Thuc Hue Ly
City University of Hong Kong

6.1 Introduction

In the recent few years, two-dimensional (2D) material family is under explosive growth. Any materials that can be synthesized as or close to mono-atomic-layer scale can be counted as a 2D material in the broadest way of definition. Except for the few earliest 2D materials such as graphene [1–4] and hexagonal boron nitride [5–8], due to their peculiar atomic structures and physical properties, transition metal dichalcogenides (TMDs) serve as one group of compounds, which are promising in the context of functional 2D materials [9–13]. As we will introduce in the following, in terms of many electrical or optical properties, 2D TMDs are exceptional [14,15]. Long-distance spin transport and extraordinary mechanical properties make graphene popular among 2D materials [16,17]. However, it is not an ideal platform material for transistors because it lacks spin–orbit coupling and native bandgap [18]. On the contrary, 2D TMDs have been discovered owing to good thermal stability and superior photoluminescence (PL) properties, resulted from the direct bandgaps [1], and comparable with graphene; 2D TMDs preserve the high electron mobility and good mechanical flexibility [19–21].

The generalized chemical formula of TMD material is MX_2, where M stands for a transition metal of groups four to ten in the periodic table of elements and X presents a chalcogen (Figure 6.1). It includes three layers of atoms, which are X, M, and X, respectively. The interatomic bonding between M and X atoms is mainly constituted by van der Waals (vdW) forces, and the atoms are combined by covalent basal plane bonding [22]. In this chapter, we will start with their atomic structures, followed by the synthesis and processing, whereas some properties and the applications will be discussed in the end.

6.2 The Structures of 2D TMDs

2D TMDs are originated from their bulk semiconductor materials. The investigations on them were originated from the analogy with graphene at first [23–28]. Nevertheless, TMD monolayers own couples of novel and unique properties different from the bulk counterparts [22,29–32]. Similar to the other nanomaterials, their chemical, mechanical, electrical, and thermal properties highly depend on the atomic structures.

6.2.1 Polymorphic Nature in TMDs

Generally, the TMD monolayers can exist in either one of the following three types of structures: trigonal prismatic (space group P63/mmc), octahedral phases (space group $P\bar{3}m1$), and distorted octahedral structures (space group P21/m), which are denoted as 2H-, 1T-, and 1T'-MX_2. The structures or phases present in 2D TMDs are governed by many factors, including the relative stability at different temperatures as well as the synthesis/thermal treatment histories. The polymorphic structures in 2D TMDs provide sufficient freedom in the structure/property control for various applications.

*The first two authors contributed equally in this chapter.

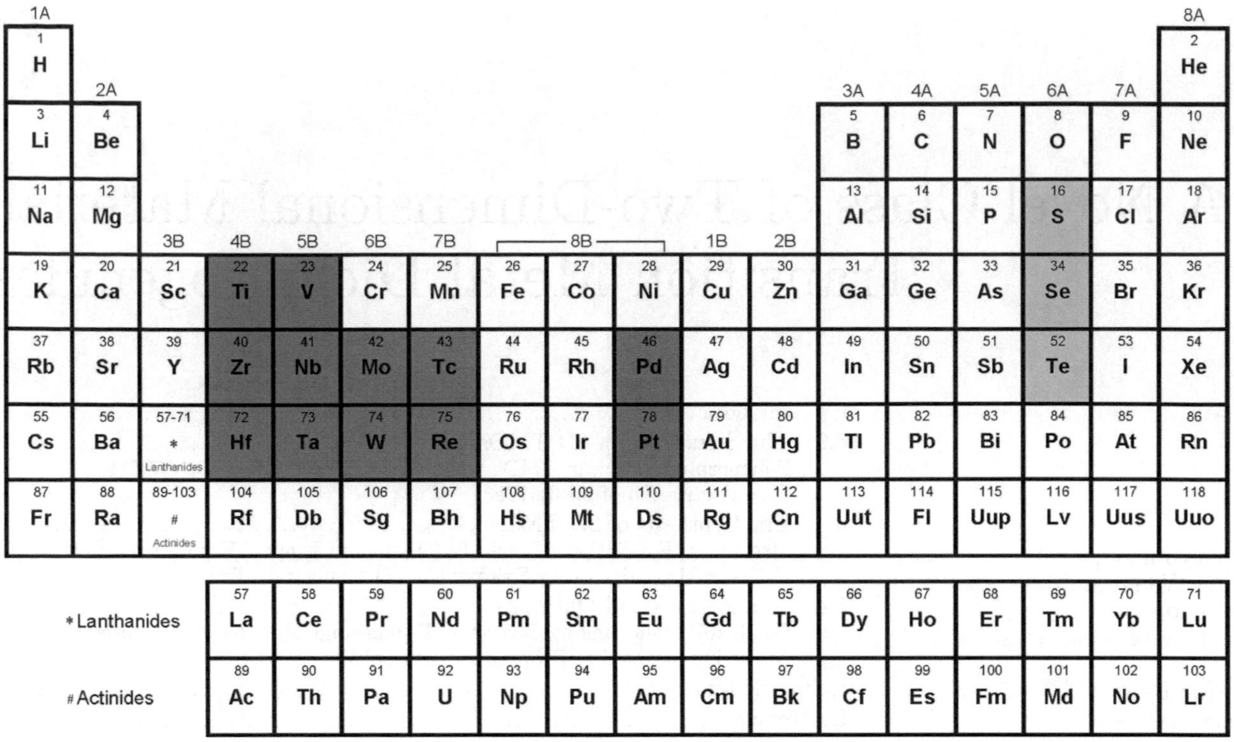

1A	2A	3B	4B	5B	6B	7B	8B			1B	2B	3A	4A	5A	6A	7A	8A
1 H																	2 He
3 Li	4 Be											5 B	6 C	7 N	8 O	9 F	10 Ne
11 Na	12 Mg											13 Al	14 Si	15 P	16 S	17 Cl	18 Ar
19 K	20 Ca	21 Sc	22 Ti	23 V	24 Cr	25 Mn	26 Fe	27 Co	28 Ni	29 Cu	30 Zn	31 Ga	32 Ge	33 As	34 Se	35 Br	36 Kr
37 Rb	38 Sr	39 Y	40 Zr	41 Nb	42 Mo	43 Tc	44 Ru	45 Rh	46 Pd	47 Ag	48 Cd	49 In	50 Sn	51 Sb	52 Te	53 I	54 Xe
55 Cs	56 Ba	57-71 * Lanthanides	72 Hf	73 Ta	74 W	75 Re	76 Os	77 Ir	78 Pt	79 Au	80 Hg	81 Tl	82 Pb	83 Bi	84 Po	85 At	86 Rn
87 Fr	88 Ra	89-103 # Actinides	104 Rf	105 Db	106 Sg	107 Bh	108 Hs	109 Mt	110 Ds	111 Rg	112 Cn	113 Uut	114 Fl	115 Uup	116 Lv	117 Uus	118 Uuo

	57	58	59	60	61	62	63	64	65	66	67	68	69	70	71
* Lanthanides	La	Ce	Pr	Nd	Pm	Sm	Eu	Gd	Tb	Dy	Ho	Er	Tm	Yb	Lu
	89	90	91	92	93	94	95	96	97	98	99	100	101	102	103
# Actinides	Ac	Th	Pa	U	Np	Pu	Am	Cm	Bk	Cf	Es	Fm	Md	No	Lr

FIGURE 6.1 The periodic table of elements. The highlighted boxes show the position of elements in TMDs.

The Crystal and Chemical Structures

A lot of 2D materials are originated from the vdW layered materials, which consist of the stacking of 2D layered building blocks. These building block layers are constituted by strong covalent/ionic bonding in basal planes and interact via vdW bonding between the layers. By mechanical exfoliation or liquid exfoliation processes [19,33–51], some layered TMDs can be easily isolated into 2D monolayers. The conventional 2D TMDs consist of three layers of atoms. The top and bottom atoms are chalcogen atoms, and the middle atoms belong to transition metals. For the hexagonal (2H) phases, the atoms in all the three atomic layers are arranged in a planar hexagonal manner, which is similar to graphene (Figure 6.2). Even though the vdW forces between two atoms are quite weak, sometimes the rippled structure in tetragonal (1T) or distorted tetragonal (1T') phases stabilizes the stacked structures [42]. In TMD monolayers, for example, in MoS_2, the S–S distance is about 3.12 Å, which is considerably larger than the bond length in the S2 dimer (1.89 Å). And the Mo–S bond distance is around 2.39 Å [43].

In transition metals, bonding of complexes is usually considered to exist between empty orbitals of the metal and long-pair electrons. However, in TMDs, the metal atoms provide four electrons to fill the bonding states. In this condition, the formal charge of transition metals and chalcogens is +4 and −2, respectively [28,44]. For the bulk TMDs, a prominent property is that they have a high degree of symmetry in the crystal structure. Different from the bulk, in TMD monolayers (as well as other thin TMD flakes with odd layer thickness), there is neither inversion symmetry nor an inversion center. The chalcogen atoms are mapped into empty positions if the metal atom is regarded as the inversion center.

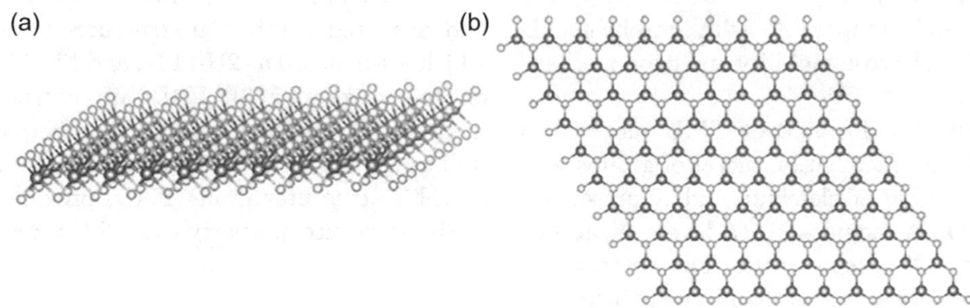

FIGURE 6.2 (a) A schematic diagram of a hexagonal TMD monolayer. M atoms are in black, and X atoms are in white. (b) A hexagonal TMD monolayer seen from (a).

The Electronic Structures

When TMDs are changed from bulk to monolayer, the electronic band structure correspondingly changes from non-direct to direct bandgap, which strongly influences the optical properties such as PL, electroluminescence, and absorption/transmissions [44–46]. Actually, the layered MX_2 possesses various kinds of electronic structures. If M belongs to groups of VIB transition metals, MX_2 are mostly semiconductors. On the other side, if M belongs to groups of VB transition metals, MX_2 are mostly metals [19,47]. The semiconductor 2D TMDs usually have a bandgap equivalent to that of visible light, and the forbidden band width varies with thickness of layers [44].

According to previous theoretical and experimental studies, the lattice parameters and band structures in MX_2 monolayer as well as bulk MX_2 are quite consistent. Kumar's group calculated the electronic band structure and states density of TMD compounds of Mo (Figure 6.3). The results showed that when the thickness of layers decreases from bulk to monolayer, the emerged bandgap monolayer evolved from the indirect bandgap in bulk after blue shifting [46, 51–53].

The electronic structures also depend on the class of transition metals and d-electron interactions. Quantum confinement effects on the electronic structure of MX_2 have been observed in its 2D forms. As the most famous example, MoS_2 has an indirect bandgap in its bulk form, and its indirect bandgap is transformed into a direct bandgap semiconductor when it is turned into the monolayer. For WS_2, it has band structures similar to those of MoS_2, but ReS_2 and NbS_2 behave differently, whereas their metallic properties are independent of the thickness of materials [54].

Phases in TMD Monolayers

As mentioned above, 2D TMDs have multiple stable phases. The semiconducting 2H phase has trigonal prismatic structure, while the metallic 1T and the semimetallic 1T' phases have octahedral and distorted octahedral structures, respectively [9,55–61] (Figure 6.4).

The trigonal prismatic phase is also regarded as the 2H phase, and it corresponds to hexagonal symmetry with trigonal prismatic coordination of the metal atoms. This geometry of the phase shows that in single layers, the X atoms are

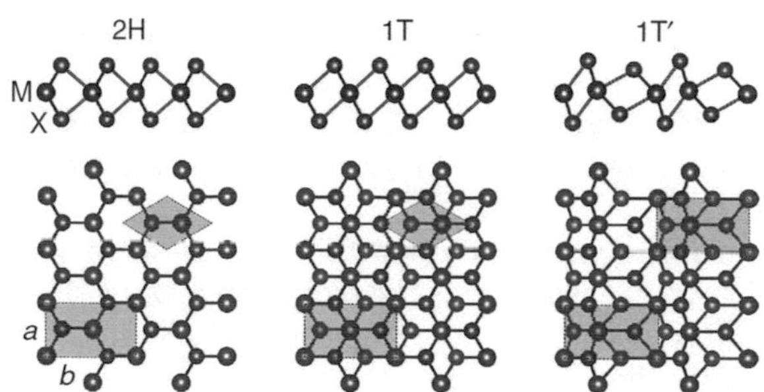

FIGURE 6.4 The top schematics show cross-sectional views, and the bottom schematics show basal plane views. The bigger atoms are transition metal atoms, and the smaller atoms are chalcogen atoms; in all three phases, a layer of transition metal atoms (M) is sandwiched between two chalcogenide layers (X). (Adapted from Ref. [56] with permission.)

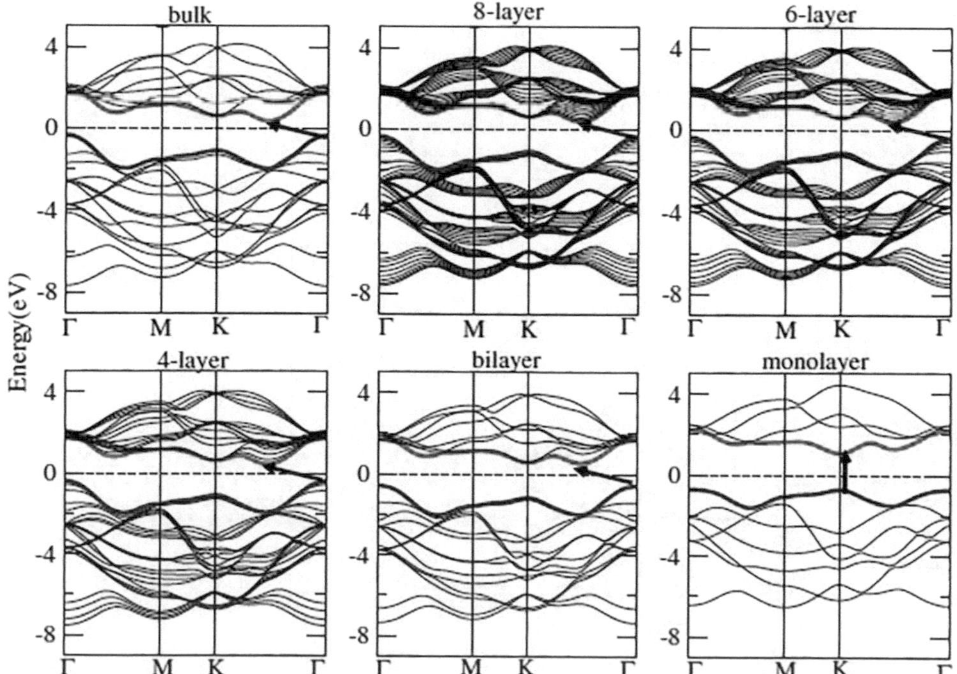

FIGURE 6.3 Electronic band structure of bulk MoS_2 monolayer as well as multilayers. The top of the valence band (curves beneath 0) and bottom of the conduction band (curves above 0) are highlighted. (Adapted from Ref. [53] with permission.)

vertically aligned along z-axis and the stacking sequence can be described as AbA. For the other type, octahedral phases, it is regarded as 1T phase and corresponds to the tetragonal symmetry. This phase is reflected in octahedral coordination of the metal atoms. In 1T phase, one of the sulfur layers is shifted compared with the others, which can be described as an AbC stacking sequence, where A is chalcogen atoms and b is metal atoms, respectively. Using Raman analysis, differences in symmetry between the monolayer 2H and 1T(1T') phases can be straightforwardly observed. In the 2H phases, the d orbital can be divided into three degenerate states, while in the 1T phase, the d orbitals of the metal degenerate in only two states.

Similar to graphene, the surface of TMD monolayer is not perfectly smooth. It usually contains many tiny ripples, which increase the monolayer height to around 6–10 Å. From Born–Oppenheimer molecular dynamics (MD) simulations, Miro's group found that the ripples resulting from the inherent dynamics of MoS_2 can even exist at low temperatures [57].

By density function theory (DFT), it has been verified that hexagonal structure is the most stable for most of the 2D TMDs except some Te compounds [37,38]. It has been found that the trigonal prismatic phase can transform into octahedral phase with the change of average lithium composition in the cathode during exfoliation synthesis [35,48–50]. And it can also change from octahedral phase to trigonal prismatic phase in the reverse direction [35,43]. Furthermore, the transition between the two phases can also be induced by the charge-transfer process between the inserted metal ions and the Mo layer itself (Figure 6.5) [59].

A similar study of phase transition in TMD monolayers works on $MoTe_2$. In the desired area, the phase transition from the 2H to 1T' phase in $MoTe_2$ is conducted by laser irradiation. It is clear to observe the structural phase transition from 2H to 1T', where the rectangular lattice symmetry as a feature of 1T' $MoTe_2$ appears (Figure 6.6). By theoretical calculations, it was verified that the energy per unit of Te vacancy drives the transition between 2H and 1T' phases [60].

In WTe_2 and $MoTe_2$, the imaginary frequencies and Fermi surface nesting might exist in the 1T phase, and the instability will lead to a stable 1T' phase [60]. In contrast, MoS_2 has a metastable metallic 1T phase and the traditional 2H phase is a stable phase [39,62]. Different Raman-active and infrared-active modes for these three phases have been clearly identified [61]. Thus, the various phase transitions in TMD monolayers offer the opportunity for phase engineering, including creating stable homojunction contacts.

6.2.2 Atomic Defects in 2D TMDs

In an ideal perfect crystal, the atoms are strictly arranged in a regular, periodic lattice in a certain order. However, in the real crystals, the arrangement of atoms cannot be perfect due to the contribution of entropy or other reasons during

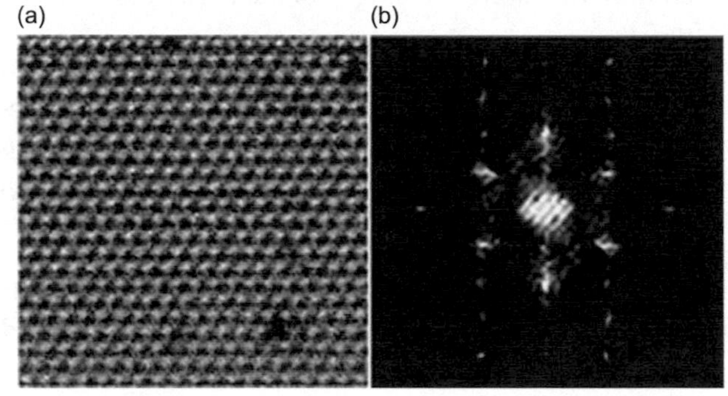

FIGURE 6.6 Atomic Te vacancies created artificially in $MoTe_2$ monolayer. Te single vacancy and divacancy are visible (a). Fast Fourier transform (FFT) of left image (b). (Adapted from Ref. [60] with permission.)

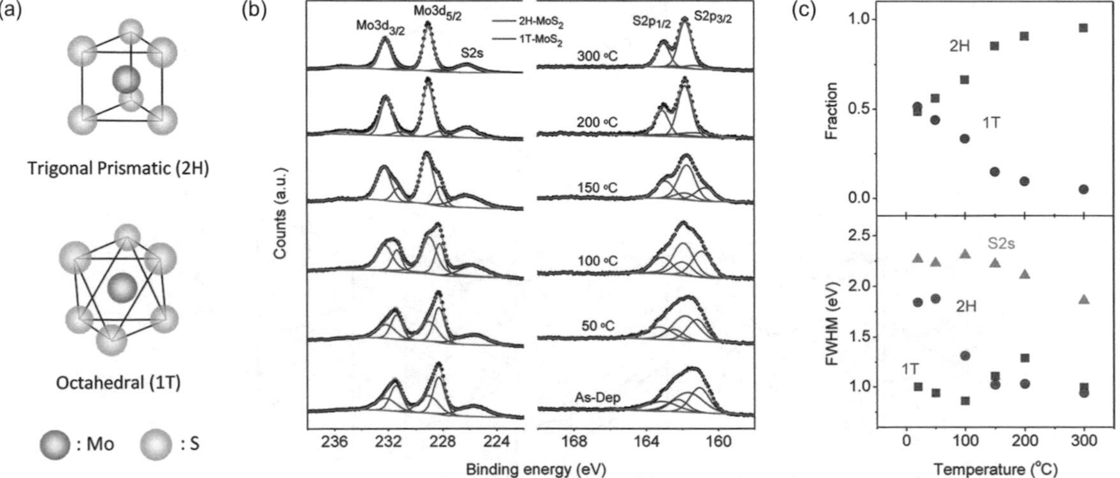

FIGURE 6.5 (a) Structure of 2H- and 1T-MoS_2. (b) X-ray photoelectron spectroscopy (XPS) spectra showing Mo 3d, S 2s, and S 2p core-level peak regions for samples annealed at various temperatures. (c) Extracted relative fraction of 2H and 1T components (top) and the linewidths of S 2s and Mo $3d_5/2$ peaks as a function of annealing temperature (bottom). (Adapted from Ref. [59] with permission.)

synthesis. These deviations from the perfectly periodic lattice structure undermine the symmetry of the crystal. In general, the crystal defects can be divided into many categories. Particularly, in 2D TMDs, there are mainly three kinds of defects [63–93], which are summarized in Figure 6.7 [63].

Point Defects in TMD Monolayers

Vacancies, adatoms, and dopants all belong to 0D point defect in 2D TMDs. Zhou et al reported a large amount of different point defects in monolayer MoS_2. V_S represents the mono-sulfur vacancy, which has the lowest formation energy; V_{S2} means di-sulfur vacancy; V_{MoS3} stands for vacancy complex of Mo and nearby three sulfur atoms, which has formation energy similar to S vacancies, and so on [64] (Figure 6.8). Using first-principles calculation to measure the formation energies and thermodynamic charge transfer in monolayer MoS_2, S vacancies are the most abundant defects in Mo-rich conditions. On the other hand, in S-rich conditions, formation energies are quite high with good-quality crystal growth [65].

Meanwhile, sometimes the lattice atoms may be replaced by other dopant atoms. In most cases, the atoms of M or X in MX_2 are replaced by substituting atoms dependent on the relative size of ions, electronegativity, valence, etc. From the previous studies, Mn, Nb, Fe, Re, Au, and Co are all potential dopants for the metallic atoms in 2D TMDs [65]. Especially, the chalcogen site doping may be favorable in sulfur-deficient crystals. From experiments and theoretical calculations, energy of different dopants in MoS_2 has been provided [67–69]. In brief, in n-type doping, the separation between Fermi level and the conduction band for Mn, Fe, and Co dopant atoms is 0.48, 0.77, and 0.78 eV, respectively [70,71]. In addition, Mn doping shows the highest potential for spintronics applications, which has the lowest separation value [72].

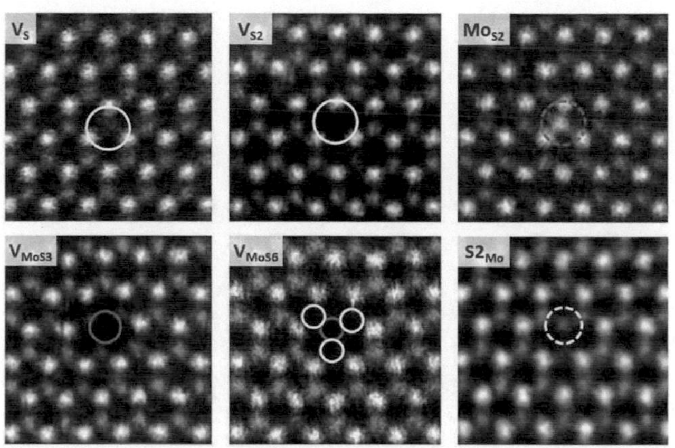

FIGURE 6.8 Atomic resolution scanning transmission electron microscope in Annular dark field (STEM-ADF) images of various intrinsic point defects present in monolayer chemical vapor deposition (CVD) MoS_2, including V_S, V_{S2}, MoS_2, V_{MoS3}, V_{MoS6}, and $S2_{MO}$. (Adapted from Ref. [64] with permission.)

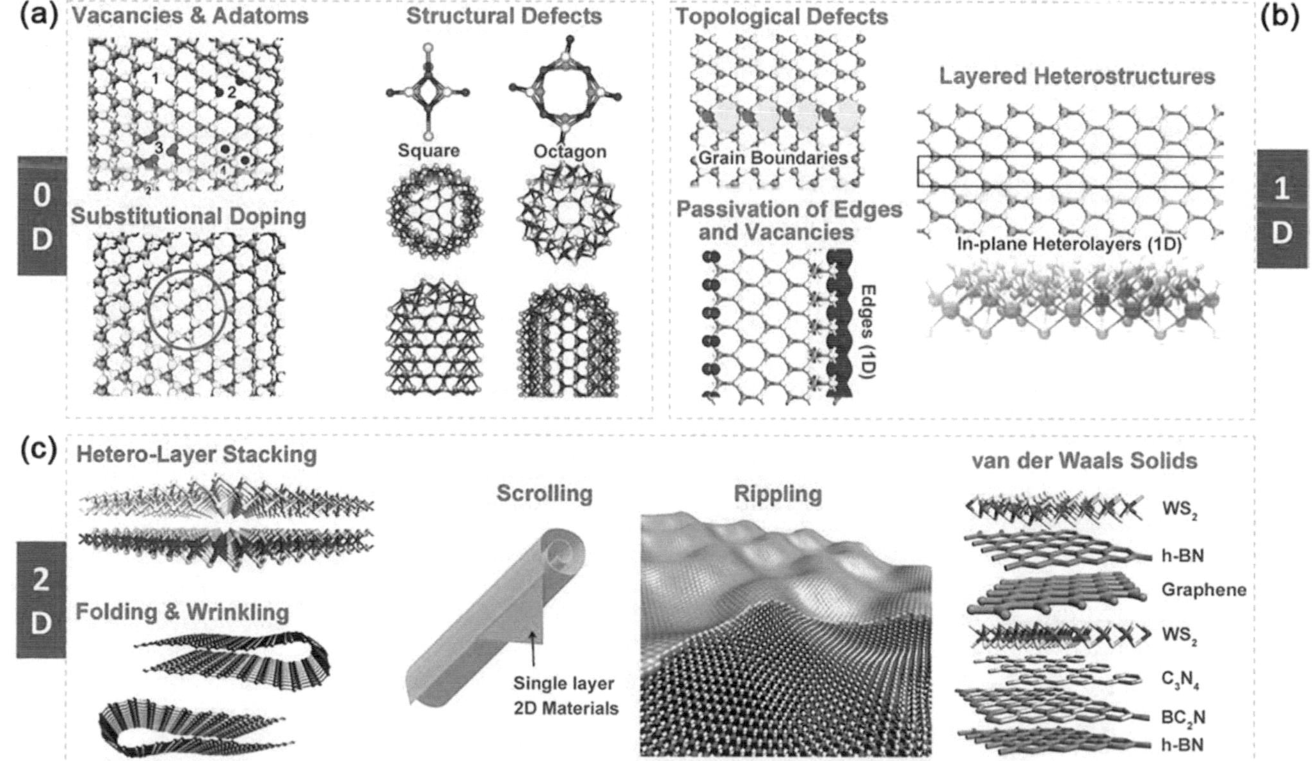

FIGURE 6.7 A schematic diagram of three kinds of defects in TMD monolayer. Typical zero-dimensional (0D) defects (a). Typical one-dimensional (1D) defects (b). Typical 2D defects (c). (Adapted from Ref. [63] with permission.)

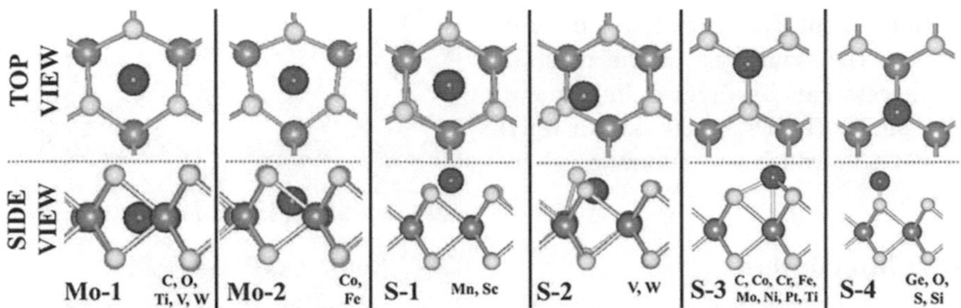

FIGURE 6.9 Top- and side-view schematic representations of possible adsorption geometries of adatoms obtained after structure optimization. Dark gray balls stand for adatoms; medium gray balls stand for host Mo atoms, and light gray balls stand for S atoms. (Adapted from Ref. [73] with permission.)

Except for substitutive doping, there are different kinds of atoms also playing the role as adatoms in TMD monolayers. In these cases, the foreign atoms can exist above the metal atoms as well as above the chalcogen atoms, by metal–chalcogen bonds, or above/within the center of hexagonal voids. It has been found that at diverse sites, adatoms can be adsorbed with significant binding energies in MoS_2. There are six distinct adsorption sites depicted, namely, Mo-1, Mo-2, S-1, S-2, S-3, and S-4. The schematic diagram shows the location of these adatoms (Figure 6.9) [73], referring to the squares, and octagons of defects result from the rotation of the lattice.

Apart from vacancies, adatoms, and dopants, dislocations in 2D materials can also belong to 0D point defects, which are different from the 1D dislocations in bulk. Experimental investigations of dislocations in 2D TMDs have figured out the factors of dislocation generation, which includes the growth/synthesis conditions, electron beam sputtering [75]. In particular, the deformation and cracking of the 2D TMDs have been observed to be closely related to the dynamics of these 0D dislocations [76].

Line Defects in TMD Monolayers

Line defects are also called 1D defects. Typical 1D defects include grain boundaries (GBs), edges, and phase interfaces. In many cases, the formation energies of line vacancies can be written as a function of the number of 0D defects [77,78]. It is well known that GBs are usually not perfectly straight and are linked by various GB kinks that can have important influences on the material properties. Previous studies working on defects and GBs in 2D TMDs use the transmission electron microscope (TEM) (Figure 6.10) [79], scanning electron microscope (SEM), and other optical microscopes to observe the morphology and atomic structures of them. In the experiment of GBs in TMD monolayers, ultraviolet (UV) irradiation is also used to generate generating oxygen and hydroxyl radicals and widen the size of GBs [79].

Besides the GBs, edges belong to line defects as well. It is another kind of prominent defect since the 2D flakes usually have limited sizes/dimensions. The theoretical results show that the edge bands can be deeply isolated in the middle of bandgap or totally metallic [81]. Figure 6.11 shows the zigzag edge of MX_2 for both the kinds of atoms. Since the 2D TMDs are mostly in triangle shapes by growth, the relative stability of different edge terminations can be deduced [82]. Nonetheless, there are some hexagonal flakes in CVD-grown WS_2, which consists of the ZZ edge types in both M-ZZ and X-ZZ types [83].

3D Defects in TMD Monolayers

Some special three-dimensional (3D) defects in 2D TMDs include folding, wrinkling, scrolling, rippling, and stacking. They are commonly related to the flexibility of the materials. As shown in Figure 6.12, these folded areas are formed after cracking of monolayers. Zhao et al used the 3D atomic mapping by TEM to analyze the folding behavior (Figure 6.12) [85].

Another AFM study on monolayer MoS_2 membrane measured the periodical rippling and wrinkles (Figure 6.13), in which the MoS_2 membrane was grown with ripples of different height and wavelength and the electrostatic properties were studied as well [86]. The result shows that the space-dependent surface potential and charge distributions are originated from the local tensile strain due to the mismatched lattice constant with that of substrate.

6.2.3 TEM Characterization of 2D TMDs

There are various structure characterization methods of 2D TMDs; however, the most popular and powerful among them is the TEM. With the development of modern TEM such as Cs correctors, it is becoming more and more important for atomic characterizations, especially for the 2D materials. Using TEM, the morphology of the samples and the crystal structures can be unambiguously resolved. Selected area electron diffraction (SAED) shows the crystal structures in reciprocal space, while in TEM, especially STEM images, the spatial and elemental information on a single atomic scale can be directly identified. Moreover, in diffraction contrast images, the defects, including dislocations, can be revealed. Figure 6.14 shows the morphology of MoS_2 with different layer thicknesses by HRTEM. The Moiré patterns are formed due to the rotation angle between the layers. Their SAED

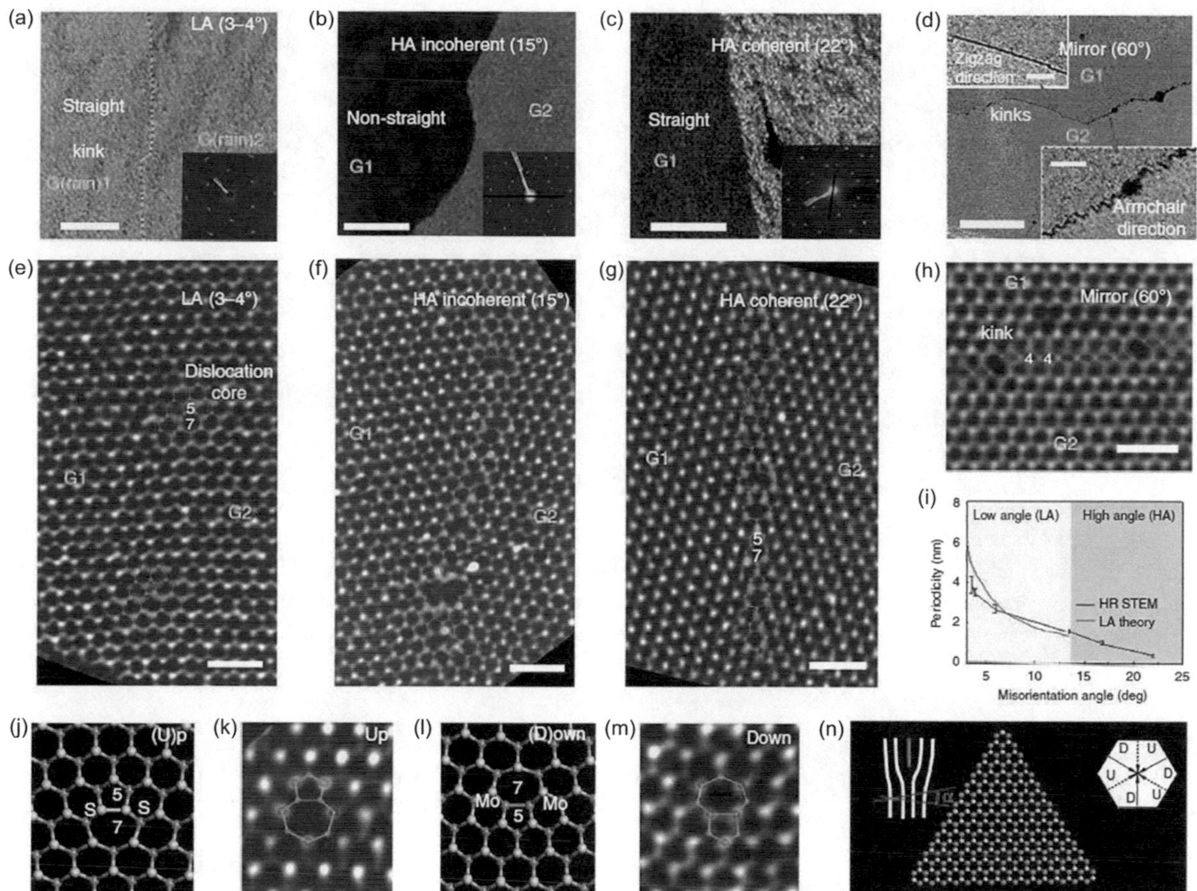

FIGURE 6.10 TEM characterization of the 2D MoS2 GBs. (a–d) Dark field images of different misorientation angle GBs. (e–h) High-resolution STEM-high-angle annular dark field (HAADF) images corresponding to (a–d). (i) The dislocation array structures as a function of misorientation angle. (j–m) The atomic image and structures for the dislocation cores. (n) The two different types of dislocations with respect to the flake edges. (Adapted from Ref. [70] with permission.)

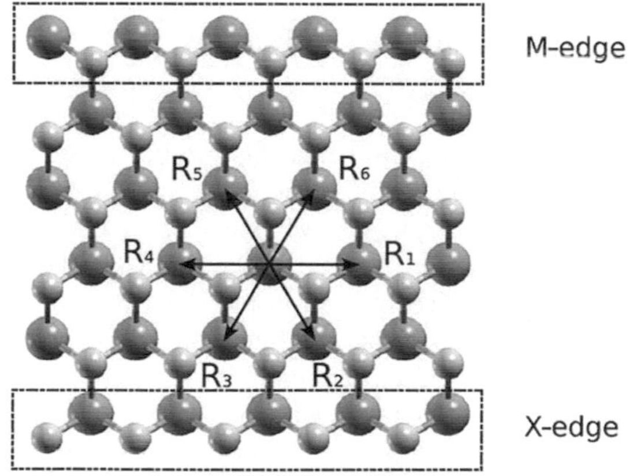

FIGURE 6.11 Top view of a monolayer MX2 zigzag ribbon. Ri are the vectors connecting the nearest M atoms. The ribbon is infinite in the R1 and R4 directions. The lower and upper edges of the ribbon are referred to as the X edge and M edge, respectively. (Adapted from Ref. [81] with permission.)

patterns are both six-folded symmetric with different diffraction spot sets corresponding to the different layers (Figure 6.14) [87].

Figure 6.15 provides the STEM-HAADF images of monolayer and bilayer MoS_2. The distance of lattice in two directions is highlighted in the figures. The monolayer and bilayer can be judged from the edge of the folding area. Since STEM-HAADF images are dependent on the number of the atoms, the positions of Mo and S atoms in the monolayer can be identified by the difference of contrast for the two kinds of atoms. Atoms with brighter contrast are Mo atoms, and the darker ones are S atoms [88].

6.3 The Synthesis of 2D TMDs

Actually, mono-atomic layer materials were first proposed in the 1850s [94]. However, there is a lack of synthesis method to realize them until 2004. The scotch tape mechanical exfoliation method used by Novoselov et al successfully prepared the first monolayer material, which was called graphene later [1]. Furthermore, they won the Nobel Prize in Physics in 2010 through their original researches on the physical properties of graphene [1,95,96]. Since then, the 2D vdW layered materials become hot topics [97–99]. About 700 materials were predicted that could be synthesized and stable in 2D

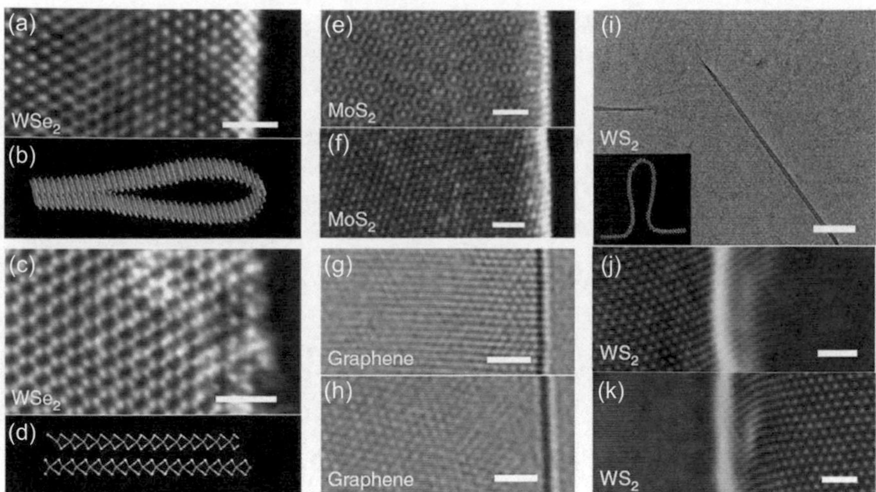

FIGURE 6.12 TEM images of different folded monolayers. (a,b) are the ADF image and atomic model of one armchair WSe_2 fold. (c,d) are the ADF image and atomic model of cracked WSe_2 fold. Original (e) and separated half ADF images (f) for one chiral MoS_2 fold. The high resolution transmission electron microscope (HRTEM) images for zigzag (g) and chiral (h) monolayer graphene folds. (i–k) are the images of vertical fold. (Adapted from Ref. [85] with permission.)

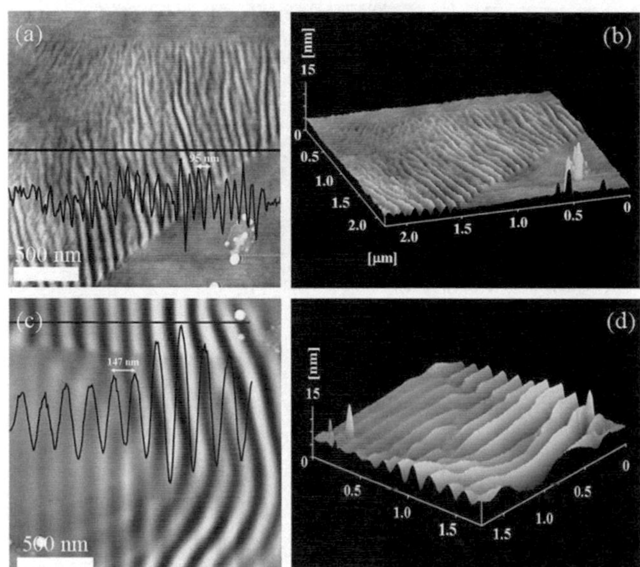

FIGURE 6.13 (a,c) AFM images of MoS2 ripples with different wavelengths and wave heights. (b,d) corresponding 3D modeling structure of (a,c), respectively. (Adapted from Ref. [86] with permission).

[100,101], among which the 2D TMDs have promising properties and grow rapidly. Until now, there are mainly four methods to synthesize the 2D TMDs. They can be categorized into two well-known groups, bottom-up approaches and top-down approaches.

In terms of bottom-up approaches, the materials are built from the fundamental building blocks, which are commonly down to atoms/molecules or atomic clusters. There are CVD [102–113] and molecular beam epitaxy (MBE) [102–104] methods. For top-down approaches, the monolayer is synthesized by breaking the bulk crystals. The weak vdW interactions between the atomic layers are gradually broken during this process. For instance, mechanical exfoliation [36,62,103–109,113–137] and liquid exfoliation [33,34,64,103–107,115,116] are widely used to scale down the dimension as well as the thickness of the 2D material. Basically, bottom-up approaches produce more defects, which require higher technique and longer time consumption than the top-down approaches. However, large-scale TMD monolayers can only be synthesized by bottom-up approaches. Different approaches can be selected based on our needs.

6.3.1 Mechanical Exfoliation

Scotch Tape Method

The first 2D material, graphene, was produced by mechanical exfoliation from a bulk graphite by using adhesive tapes [34]. Resembling this, mechanical exfoliation has also been applied to produce TMD monolayers [33–36,114]. This is the simplest and the fastest method to produce TMD monolayers with a good quality, but small flakes. As shown in Figure 6.16a, first of all, the scotch tape is attached to the TMD bulk crystals. Then, the exfoliation starts by removing the scotch tape, and repeated mechanical exfoliation eventually achieves monolayer TMDs.

The size and the production rate are very limited by using the original mechanical exfoliation method. Therefore, some modified mechanical exfoliation methods have been proposed. One modified method utilizes a power controllable motor instead of human hand to operate the adhesive tapes during mechanical exfoliation [116]. The 2D MoS_2 products are shown in Figure 6.16b. The flakes are relatively flat, and the quality of the samples has been significantly improved.

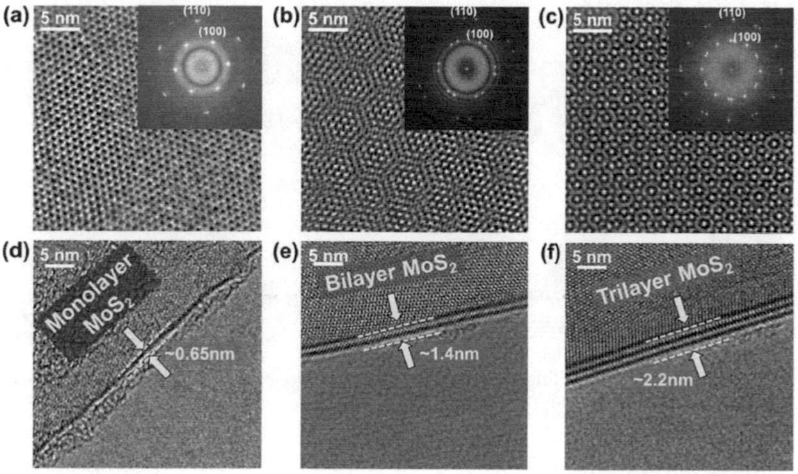

FIGURE 6.14 High resolution TEM images and the corresponding FFT selected area electron diffraction patterns (insets) for (a) monolayer MoS_2. (b) bilayer MoS_2 and (c) trilayer MoS_2. TEM images of folded layers for (d) monolayer MoS_2. (e) bilayer MoS_2 and (f) trilayer MoS_2. (Adapted from Ref. [87] with permission.)

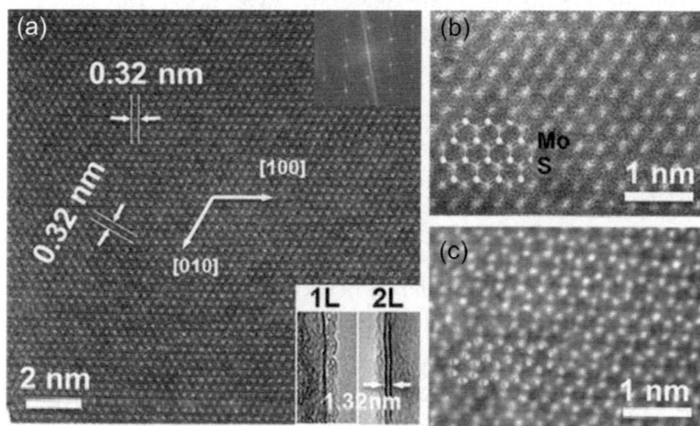

FIGURE 6.15 Structure of synthesized MoS2 monolayer and bilayer films. (a) HAADF image of typical MoS2 monolayer film. (b) HAADF image of MoS2 monolayer. (c) HAADF image of MoS2 bilayer. (Adapted from Ref. [88] with permission.)

Ball Milling

Based on the similar mechanism, there are several alternative methods for mechanical exfoliation. The interlayer vdW attractions can be easily overcome by the applied normal forces or shear forces. To increase the output of mechanical exfoliation, the "ball milling," which utilizes numerously different size steel balls in a rotating cylinder to continuously collide and grind the bulk layered materials, is shown in Figure 6.17. However, the quality of product is worse than the traditional methods because of the multi-collision-induced defects.

During this mechanical process, the continuous collision and grinding lead the crystal gradually to be peeled off as pictured in Figure 6.17. The collision and grinding break the weak vdW attractions between the atomic layers. In addition, some of the 2D materials are exfoliated by attaching on the surface of steel balls. The main factor to optimize the product quality and production rate of the process is to find an optimal rotational speed.

To further improve the quality of ball milling, some protective materials are added into the ball milling machine, such as liquid and gas. As Tan Xing et al. reported [117], the protective gas flattens the surface of the flakes. The protective gas can saturate the dangling bond to achieve recovery from the defects. It is a promising method for the ball milling to exfoliate TMD monolayer with high output.

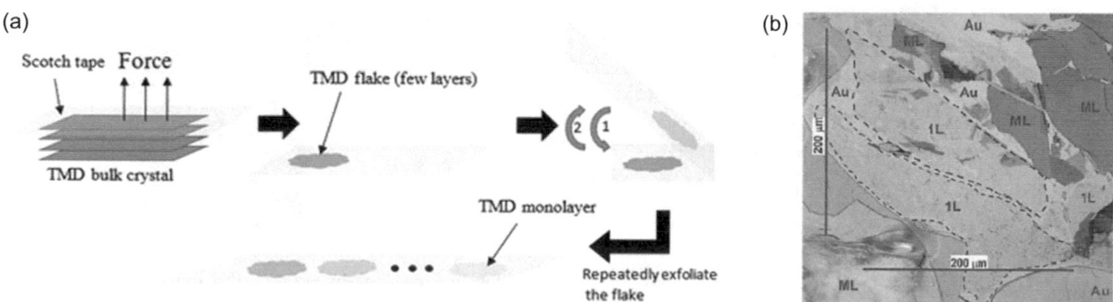

FIGURE 6.16 (a) The primitive mechanical exfoliation. (b) Optical image of MoS2 exfoliated by the modified scotch tape method. (Adapted from Ref. [116] with permission.)

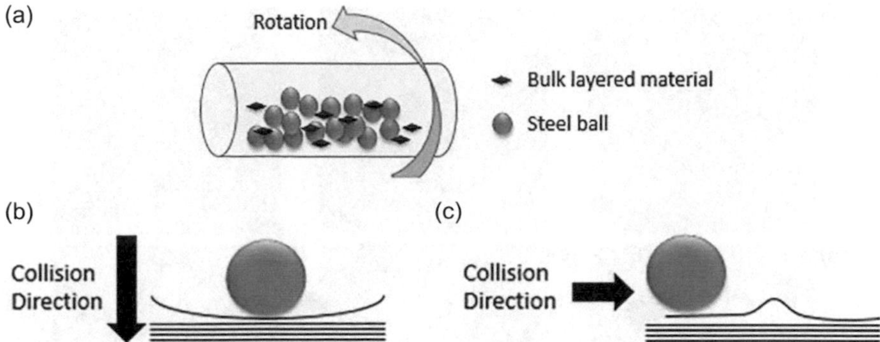

FIGURE 6.17 A schematic diagram of ball milling. (a) Rotate with steel ball. (b) Collision in vertical direction. (c) Collision in horizontal direction.

Manual Shaking

A recent research reported the manual shaking method for 2D TMD production. The original single crystals of TMDs which have to be larger than 1 mm are partially lithiated at first. Then, the crystal mixture is shaken manually and gently. Finally, the monolayer products were purified by removing the multilayer flakes and impurities using centripetal force. The morphology of the product is shown in Figure 6.18. The lateral size of TaS_2 nanosheets is up to 142 μm with an average size of around 21.9 μm. This method also provides a larger amount of output than the traditional adhesive tape approach. The SAED pattern clearly displays the hexagonal structure of the TaS_2, and the STEM-HAADF image presents the defect-free structures [118].

6.3.2 Liquid Exfoliation

General Liquid-Phase Exfoliation

Liquid exfoliation utilizes the same strategy with mechanical exfoliation, which breaks the vdW attraction between the atomic interlayers by using an appropriate solvent and sonication. The interlayer attraction can be easily compensated by isopropyl alcohol (IPA), dimethylformamide (DMF), or N-methyl-2-pyrrolidone (NMP) [120]. However, in this process, some unnecessary precipitates are generated as well, although some precipitates can be removed by centripetal force. At last, the dispersed layers are collected from the suspensions. Although liquid exfoliation can also achieve a high output of 2D materials, the quality of the products is still limited. Since 2D materials are suspended and allowed to freely move in the solvent during the process of liquid-phase exfoliation, they tend to sweep up, become uneven, or stack together again.

Ball Milling–Assisted Liquid Exfoliation

Ravindra et al. recently combined the method of ball milling and liquid exfoliation [104]. The crystal powder is first ball milled, followed by heat treatment in an oven. After drying, the mixture of diethyl ether and water is employed as a solvent. The samples are immersed into the solvent using sonication, and then the resulting solution is rested for a day to precipitate. Again, the centripetal force is applied to remove the impurity, and then, the products are collected by a micro-pipette at last. Although the sample is slightly wrinkled, the SAED pattern provides the evidence that the crystallinity is not damaged [104–122]. Hence, it is possible for improving the quality of liquid exfoliation by mixing with other mechanical approaches.

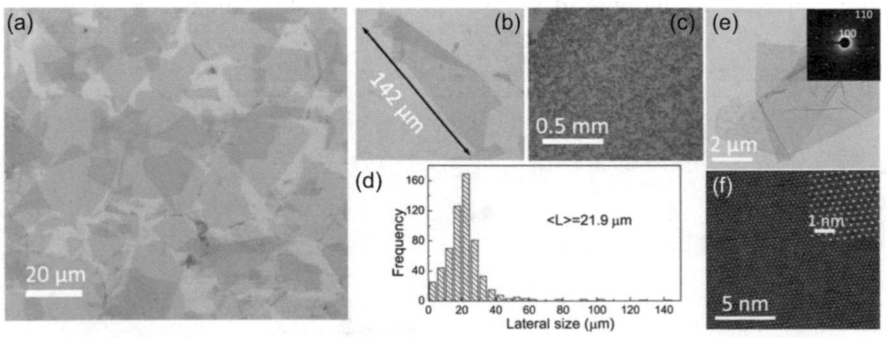

FIGURE 6.18 Morphology of the exfoliated TaS_2 monolayers. (a–c) Typical optical images of the exfoliated TaS_2 monolayers on SiO_2/Si substrate. (d) Statistical analysis of the lateral sizes of the TaS_2 monolayers. (e) TEM image of a TaS_2 nanosheet. The inset is the corresponding SAED pattern. (f) HAADF images of a TaS_2 nanosheet. The inset is the magnified image. (Adapted from Ref. [118] with permission.)

6.3.3 Chemical Vapor Deposition (CVD)

In terms of bottom-up approaches, CVD is widely used in industry to grow a large uniform thin film with high quality for device fabrication [123–128]. Typically, the reactants are first vaporized in the furnace under high temperature. Then, the gas-phase reactants are carried by one or more gas-phase precursors. In order to produce a desired deposition, the precursors will decompose or react on the substrate surfaces. Finally, physical and chemical adsorption will occur on the substrate surface to form a continuous thin film. Usually, the gas-phase byproducts will be produced in the reaction and then removed by the vacuum pump. The schematic diagram of CVD to synthesize 2D TMDs is shown in Figure 6.19. There are several control parameters that influence the quality of the product, including gas flow rate, reactants' ratio, substrate location, vacuum level, and temperature, which will be specifically addressed one by one as follows.

Gas Flow Rate

The unit of gas flow in CVD is standard cubic centimeter per minute (sccm) at 0°C and atmospheric pressure. The rate of precursor flow aims to transfer the reactants uniformly to the substrate surface. If the gas flow rate is fast, there is not enough time for the reactants to take part in the reactions. If the gas flow rate is slow, the substrate will not be uniform since the byproducts/containments may stick on the substrate and the deposition is done before the desired position. To optimize the quality and maximize the size of product, the gas flow rate should fit the deposition rate.

Dong et al. pointed out that the gas flow rate directly affects the quality and the size of MoS_2 [129]. As shown in Figure 6.20, the flakes cannot be synthesized when gas flux is too low such as 15 sccm. There are clean flakes when the gas flow rate reaches 25 and 30 sccm. However, when the gas flux is equal to or larger than 35 sccm, the impurities will emerge. The impurities are mainly MoS_{2-x} clusters that have not completed the reactions yet. The MoS_{2-x} tends to stay in the center of the flakes. Although the size and the coverage of flakes are the biggest and broadest, respectively, at 40 sccm, there are numerous flakes with incomplete growth.

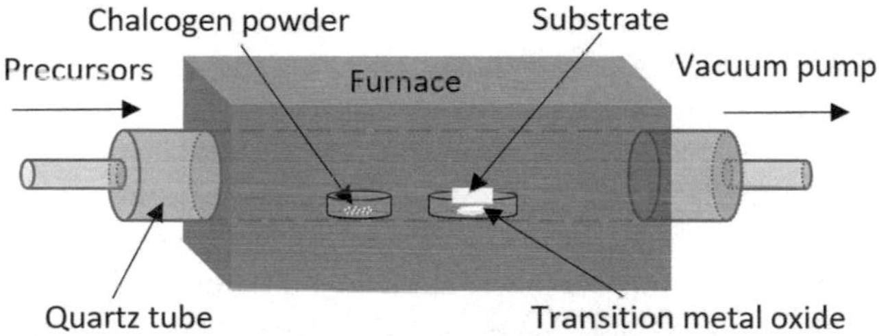

FIGURE 6.19 A schematic diagram of CVD growth of 2D TMDs.

Reactants' Ratio

In ideal case, reactants' ratio should follow the chemical reaction formula. However, Özden et al [131] indicated that the ratio of S to MoO_3 between 66 and 150 can achieve uniform and large size monolayers for MoS_2. Their work showed that the flakes are equally distributed on the substrate when the ratio of S to the MoO_3 is 20 and 66. There are very less overlapping zones and the flakes are largest about 80–90 μm when the ratio is 66. More specific results showed that when the ratio is 66, the thin film nearly covered the whole substrate. Therefore, a suitable ratio between reactants is critical to synthesize the homogeneous TMD monolayer.

Substrate Location

The substrates for CVD growth of MoS_2 monolayers can be located on the quartz tube or other containers, face-up, face-down, etc [91,128–134]. The distance between MoO_3 (reactants) and substrate also affects the scale of the TMD monolayer. Because the concentration of gaseous MoS_2 is inversely proportional to the distance to the substrate,

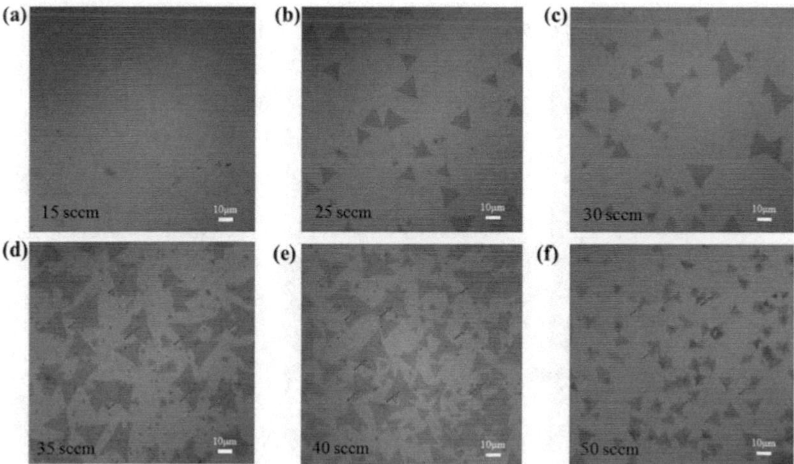

FIGURE 6.20 Optical images for CVD-grown MoS_2 flakes at (a) 15 sccm, (b) 25 sccm, (c) 30 sccm, (d) 35 sccm, (e) 40 sccm, and (f) 50 sccm. The arrows point to the impurities on the flakes. (Adapted from Ref. [129] with permission.)

the deposition is variable for different distances. A strong evidence has been provided that the distance between reactants and substrate not only influences the quality of the product, but also controls the size of the product [91,135].

Vacuum Level

The vacuum level is also an important parameter to control the quality of the samples because there are many containments in the surrounding. Some containments may stick on the substrate surface, which may induce defects. Therefore, it is recommended to conduct the CVD experiment under or $<10^{-5}$ Pa to have a relatively good result. Besides, the reactants can have a lower temperature to be vaporized/sublimed in lower pressure than atmospheric pressure, which can greatly shorten the time of the growth due to the acceleration of heating and cooling.

Temperature

Unsurprisingly, the quality of 2D materials is greatly influenced by the growth temperature. Figure 6.21a–d clearly demonstrates the influence of the 2D MoS_2 growth temperature. There is not any impurity on the substrate and MoS_2 flakes when the growth temperatures are 725°C and 750°C, while the impurities occur between 775°C and 800°C. Same with in Section 6.3.3.1, the impurities are MoS_{2-x}, which has not completed the reactions. However, the scale of flakes is larger when the growth temperature is bigger. Therefore, the size and the quality have to be compromised. The optimized temperature is 750°C for 2D MoS_2 synthesis [129].

6.3.4 Molecular Beam Epitaxy (MBE)

MBE requests a high-vacuum or ultra-high-vacuum environment for the control of product quality. During MBE, the pure elements are positioned in separated Knudsen cells or effusion evaporators. The elements are gradually heated until sublimation. The vaporized elements are concentrated on the substrate to employ epitaxy; in the meantime, the gaseous elements can have reactions with each other. Furthermore, the synthesis process can be monitored by reflection high-energy electron diffraction (RHEED), which can characterize the surface of crystalline materials so that the thickness of each single layer can be controlled precisely. Therefore, MBE can grow TMD monolayers with an extremely good quality [137–140]. The deposition rate of MBE has to be slow so that the 2D material can have a good epitaxial growth. However, the whole process costs relatively long time. Besides, the maximum deposition rate depends on the vacuum level without affecting the quality. By direct epitaxy without carrier gases and ultra-high vacuum, the greatest quality and best deposition rate can be achieved [141].

6.3.5 Substrate Transfer

Transfer of 2D materials on different substrates is unavoidable for processing and various future applications. Since poly(methyl methacrylate) (PMMA) has a high viscosity property, it can act as a binding material to assist the substrate transfer [142–144]. PMMA transfer is developed from etching-need to etching-free, which further reduces the possibility of sample damage [145,146].

Etching-Need PMMA Transfer

There are two types of transfer for etching-need PMMA transfer, which are called dry transfer and wet transfer (Figure 6.22) [147–151], targeting on different substrates. In terms of dry transfer, substrate with light cavity is preferred because the liquid is undesired to be trapped in the cavity. Besides, even substrate and porous substrate are suitable for wet transfer. Moreover, there is an advanced method, which is polyvinyl alcohol (PVA)-assisted transfer, which provides a high-quality transfer with containment-free. The details of methodology are discussed below.

FIGURE 6.21 Optical images of CVD-grown MoS_2 flakes with the growth temperature at (a) 725°C, (b) 750°C, (c) 775°C, and (d) 800°C. The arrows point to the impurities on the flakes. (Adapted from Ref. [129] with permission.)

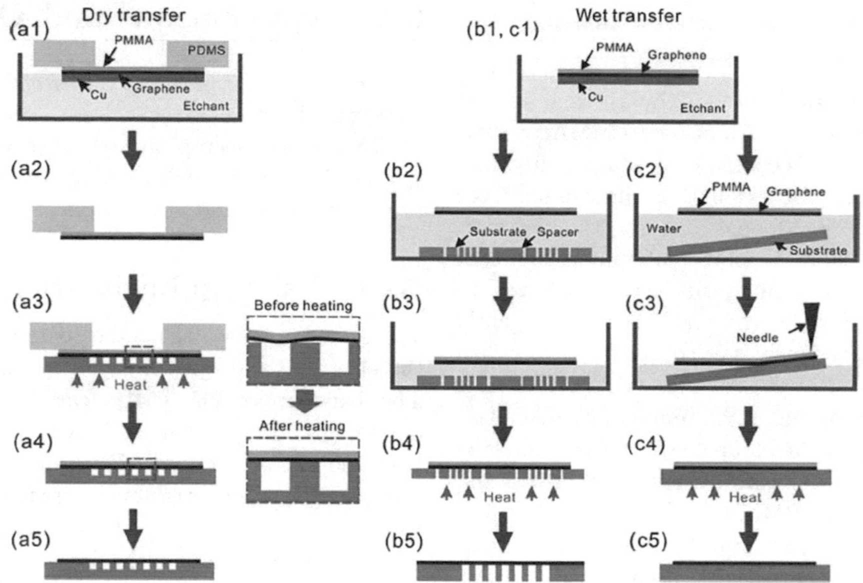

FIGURE 6.22 Schematic illustration of dry and wet transfer processes. (a) Dry transfer onto shallow depressions. Wet transfer onto (b) perforated substrates and (c) flat substrates. The boxes with dashed lines in (a3) and (a4) show magnified views. (Adapted from Ref. [151] with permission.)

For dry transfer, PMMA is first spin-coated onto a 2D TMD sample, followed by the attachment of two pieces of PDMS on two sides on the top of the PMMA layer in order to support the PMMA/sample/substrate system. To remove the original substrate, it is immersed into the etchant without other parts of the system. After the etching process, the system is transferred to a target substrate. The sample may be uneven after transfer due to previous steps. Considering the uniformity of the sample, the heat treatment is employed. The heat treatment also removes the PDMS. At the end, a furnace is used to eliminate the PMMA at 350°C with argon gas and hydrogen gas for 2 h to achieve dry transfer [151].

Wet transfer is mainly for two types of substrates: flat and porous. Similar to dry transfer, PMMA is coated onto a sample first. However, PDMS is unnecessary here. PMMA/sample/substrate system is immersed into etchant. After peeling off the substrate, the PMMA/sample flake is washed gently with deionized water. Then, the target substrate is placed under water, and the PMMA/sample flake is placed on the water above the target substrate in the same petri dish [151].

Except for the above two methods, there is another modified method to obtain a high quality with containment-free. First of all, the surface of 2D material is coated with PVA and then coated with PMMA. PVA is solvable by water, especially hot water (>130°C). PVA enhances the distance between PMMA and the etchant. Thus, it is better to control the wrinkling of the etching to prevent the PMMA layer from etching away when the original substrate is put into the etchant. Therefore, these two protective layers can increase the protection of the 2D material to obtain a high-quality substrate transfer. After etching the original substrate by the etchant, the PMMA/PVA/graphene is cleaned by deionized water several times and the remaining water is removed by bulb instead of pipette. Then, it is transferred to an objective substrate. PVA and PMMA layers are simultaneously peeled off by hot deionized water since the PMMA is stacked on the PVA layer. This method totally clears the polymer particles to achieve high-quality and containment-free substrate transfer without annealing or acetone treatment [145].

Etching-Free PMMA Transfer

Etching-free PMMA transfer generally relies on the bubbling effect to exfoliate a sample from an original substrate. Similar to other methods, PMMA is spin-coated onto a sample, followed by immersion of the PMMA/sample/substrate film into deionized water. However, the film is fully immersed into the water for the etching-free transfer. Tweezers are preferred to hold the film into water. To utilize the bubbling effect, ultrasonic energy is applied to the water. During the vibration of water, bubbles are steadily generated between PMMA/sample and the substrate. The bubbles then keep expansion and exfoliate between the PMMA/sample and the substrate from the edge to the centers. After peeling off the substrate, the exfoliated PMMA/sample will float on the water surface. Before peeling off the PMMA, heat treatment is preferred to remove the water residue and to make the film more uniform. Finally, the PMMA is removed by acetone or other methods. Most of the substrates are suitable for this method, which makes it widely applied [152].

A Novel Substrate Transfer Method: Thermal Release Tape Transfer

Lin et al. suggested a novel method by thermal release tape (TRT) transfer instead of traditional PMMA transfer [147]. In the beginning, Cu is coated onto sample/substrate by thermal evaporation. For exfoliating the sample from the substrate, the TRT is adhered onto the Cu film first, followed by slow exfoliation from the substrate. Now that there is a TRT/Cu film/sample which is transferred to a target substrate, and then, the stack is pressed gently for attachment. TRT is easily peeled off by heat when temperature is about 120°C. At last, the Cu film is dissolved by the mixture of 15% ammonium persulfate and deionized water, and the sample transfer is completed after cleaning. TRT transfer can achieve residual-free and damage-free rather than PMMA transfer. By comparisons in Figure 6.23, PMMA transfer not only remains residuals, but also folds the samples. Hence, the wrinkles and cracks are easily created by the folding. Conversely, TRT transfer provides damage-free environment. There is no difference between before and after the TRT transfer.

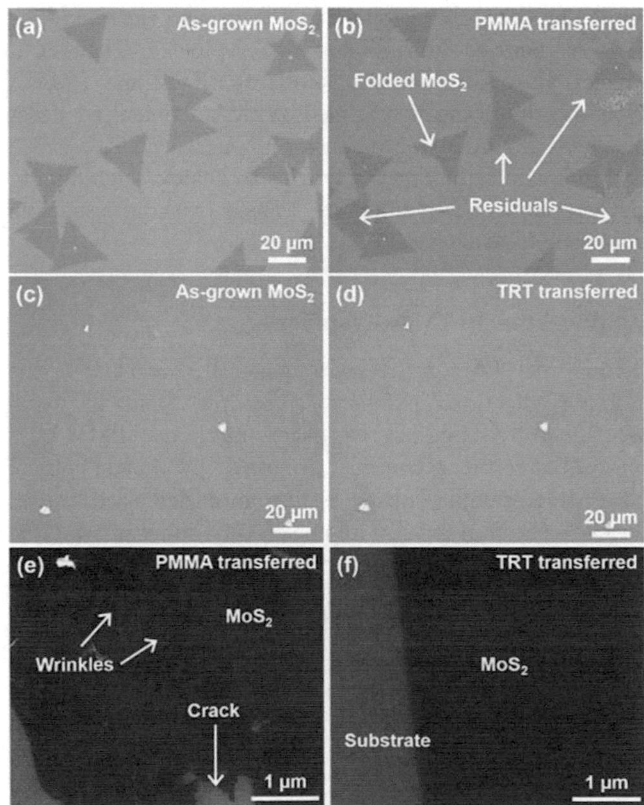

FIGURE 6.23 Optical images of the MoS_2 (a) before and (b) after the PMMA transfer. Optical images of the MoS_2 (c) before and (d) after the TRT transfer. SEM images of the MoS_2 transferred by (e) PMMA and (f) TRT. (Adapted from Ref. [147] with permission.)

6.4 Applications and Outlooks

As we have mentioned in the beginning of this chapter, currently the novel electrical and optical properties of the 2D TMDs have been proposed for various application possibilities. Herein, we will address several examples in their property control and novel applications.

6.4.1 Bandgap Engineering by Defects

We already know that the bandgap directly controls the electrical and optical performances of semiconductors. The bandgap of 2D TMDs can be easily controlled by doping [153–156], strain [157–161], and defect engineering [162–166]. As an example, Figure 6.24 shows the images of GBs, which own a brighter contrast in the pattern. Moire patterns are quite clear in these images resulting from the different orientations of the lattices in different layers. It also suggests that the appearance of bright line is influenced by the value of voltage and there is no pronounced state detected in bandgaps. The bandgaps and band energy are tunable related to the GBs. It has also been proved that in many TMD monolayers, the reconstructions at these GBs can form new states near the Fermi level [167–169]. During the mild oxidation process, it may induce a higher reactivity for oxidation or impurity adsorption. Apart from that, some optical properties such as PL can be enhanced by GBs [75,170–173].

The band structure is significantly influenced by GBs, and it is the reason that some optical and electronic properties change abruptly with occurrences of GBs. Apart from that,

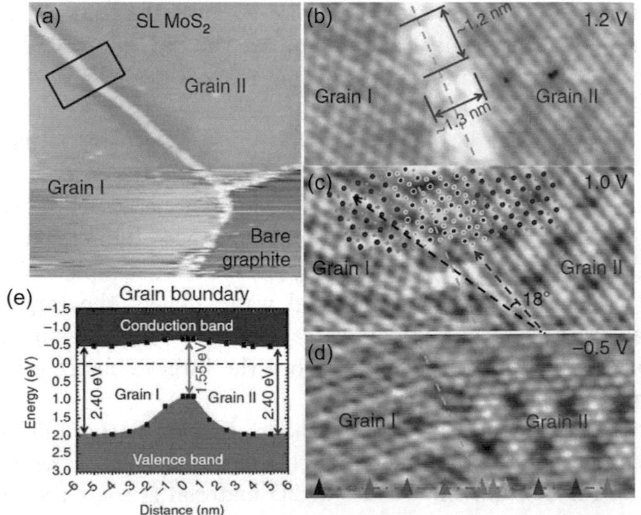

FIGURE 6.24 Bandgap tunability at the GB. (a) The STM image of a single-layer MoS2 island composed of two grains, where the GB appears as bright protrusions. (b–d) Images recorded at the boundary region are highlighted by a black rectangle in panel. The misorientations between the grains are in different angles. (e) A schematic diagram shows the change of bandgap at the GB. (Adapted from Ref. [170] with permission.)

2D TMDs own exotic properties, which include indirect-to-direct bandgap crossover as the number of atomic layers decreases. The previous study shows that the bandgaps are influenced by thickness and stress [82,84,172–176]. Strain effect has been proved of having great effect on the bandgap and transport properties; hence, it belongs to the main mechanism of 2D defects in TMD monolayers.

6.4.2 2D Semiconductor Heterostructures

The lateral and vertical heterostructures made of 2D TMDs have been realized, and their novel electrical/optical properties, especially at the sharp interfaces, have been experimentally verified. Lateral heterostructures can be synthesized mainly via CVD growth [177–180], while vertical heterostructures have been synthesized via both CVD and mechanical exfoliation [181,182].

Figure 6.25 shows the pattern of the transferred WS_2/MoS_2 heterostructure in a HRTEM image. A WS_2 monolayer can present higher image contrast than a MoS_2 monolayer. The bright and dark contrasts exhibit the 2H stacking of this kind of heterostructure, which proved that the orientation of the two kinds of materials can be well controlled by this method. It has also been observed that some Mo atoms are substituted into the WS_2 layer and W atoms are substituted into the MoS_2 layer [184].

For the vertical heterostructures, Sefaattin et al reported the interaction of interlayers tuned from no coupling to strong coupling. Different from the former one, the position of MoS_2 and WS_2 is not in the same plane. They prepared the top layer of MoS_2 and the bottom layer of WS_2 aligned in one area, while they prepared the bottom layer of MoS_2 and the top layer of WS_2 aligned in the other area. TEM images (Figure 6.26) show that there is neither phase change nor defects in the heterostructures, and no

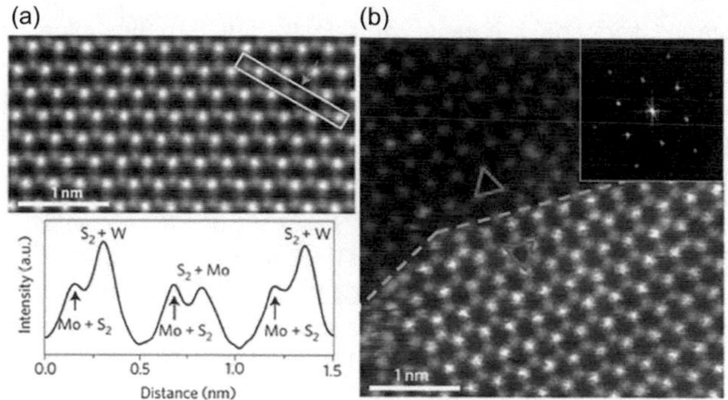

FIGURE 6.25 (a) Top: Z-contrast image of the bilayer region with a 2H stacking orientation. The marked position is the atomic in which a W atom is replaced by a Mo atom in the WS_2 layer. Bottom: Image intensity pattern of the white rectangle in the image above. (b) Z-contrast image of the step edge of the WS_2/MoS_2 bilayer. The gray dashed line indicates the step edge, and the two triangles indicate the orientation of the MoS_2 (top part of image) and WS_2 (bottom part) layers. Also, it contains its corresponding SAED pattern. (Adapted from Ref. [178] with permission.)

chemical reactions between the two materials as well. The split spots in diffraction patterns confirmed the presence of MoS_2 and WS_2 layers. This implies that the observed changes in the optical properties of heterostructures result from the changes of distance in interlayer during the process of annealing [183].

6.4.3 Appealing Mechanical Properties

For the emergent 2D materials [1,178,183,184–187], the morphologies are associated with elastic properties, while the duration and sustainability are affected by the plastic properties [188,189]. 2D materials are unique in nature,

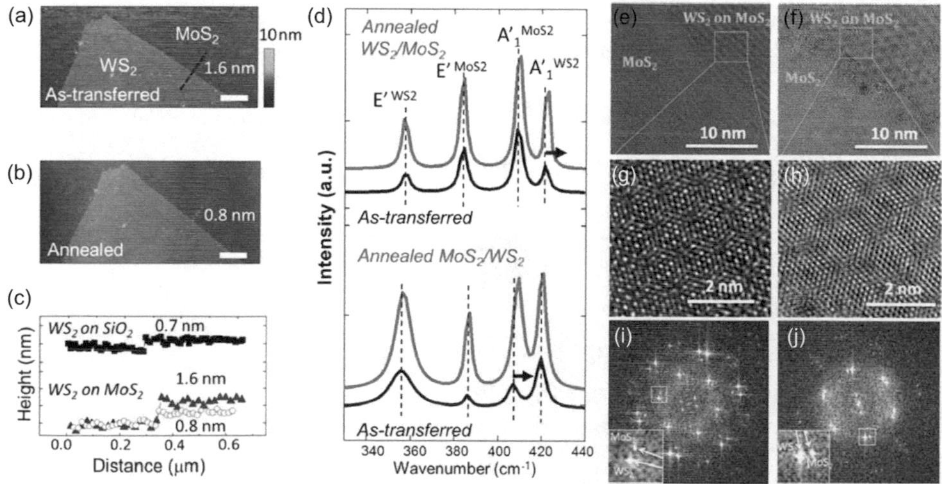

FIGURE 6.26 (a) AFM image of the heterostructure before the thermal annealing. (b) AFM image of the heterostructure after the thermal annealing (Scale bar is 2 μm). (c) Image of height on cross-section of a monolayer WS_2 on SiO_2 substrate. (d) Raman spectrum of WS_2/MoS_2 (top) structure, MoS_2/WS_2 (bottom), before and after annealing. (e,f) HRTEM images of MoS_2/WS_2 heterostructure, before and after annealing. (g,h) Zoomed-in images of the MoS_2/WS_2 heterostructure corresponding to (e) and (f), respectively. (i,j) FFT images of the MoS_2/WS_2 heterostructure before and after annealing. (Adapted from ref [183] with permission.)

because they only consist of one to few atomic layers in thickness. Owing to the anisotropic structure, graphene has an ultra-high in-plane modulus and strength [190–192] with ultra-low out-of-plane bending modulus [150, 193,194]. 2D TMDs like MoS_2 monolayer have similar properties and high flexibility as well. Their physical properties vary a lot depending on the bending/rippling structures [195,196], and the mechanical bending, rippling, and buckling of 2D materials are key factors in the emergent applications such as active materials for wearable devices [197,198], diffusive membranes for energy devices [199,200], nanoelectromechanical (NEMS) devices [201], adhesion/wearable layers or lubricants [202–204].

The ultra-high in-plane strength of monolayer TMDs has been verified by nano-indentation approaches (Figure 6.27) [205]. The ultimate strength is close to the theoretically predicted values owing to the absence of atomic defects. Figure 6.27b shows the elastic properties of some 2D membranes having MoS_2, graphene, and WS_2. The schematic diagram of the experiment process is shown in Figure 6.27a. For 2D materials, the strain energy is normalized by the sheet area, which gives rise to 2D stress and elastic modulus. And in this kind of monolayer membrane, clamps across a hole are touched at the central part by a tiny tip, so the bending modulus is small enough to ignore. The load is balanced by the pretension of the membrane and scales linearly with vertical deflection under small loads [205,206]. And when the load is large, it is dominated by the stiffness of the membrane with a cubic relationship. The elastic modulus of MoS_2 is similar to that of 2D material and is about half of that of graphene, which is the strongest 2D material. The pretension values of monolayers depend on the transfer process as well as their intrinsic mechanical properties, because this parameter relates to the elastic energy of pretension after the transferring [205].

For the plasticity and cracking of TMD monolayers, Ly et al. have also investigated the cracking processes in monolayer MoS_2 by in situ TEM (Figure 6.28) [76], and the emission of dislocations at the crack tip and the limited plasticity can be attributed to the unique 2D structures. The novel mechanical properties in these 2D TMDs are still under intensive research at present.

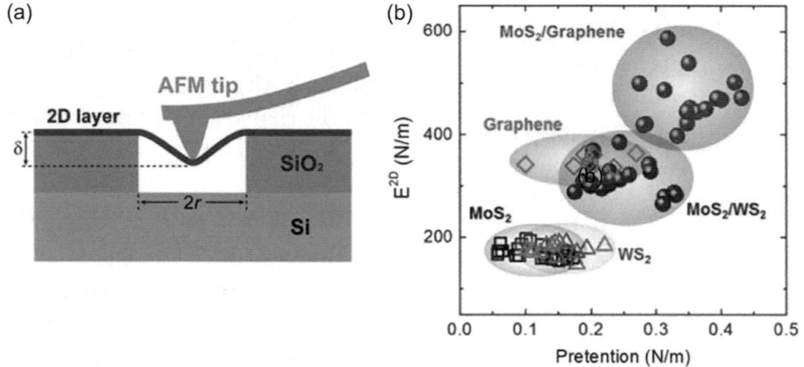

FIGURE 6.27 A schematic diagram of the indentation measurement and experimental data of 2D modulus and pretension for various 2D layers. (a) Illustration of the indentation measurement. (b) Experimental data of 2D modulus and pretension for various 2D layers and heterostructures. (Adapted from Ref. [205] with permission.)

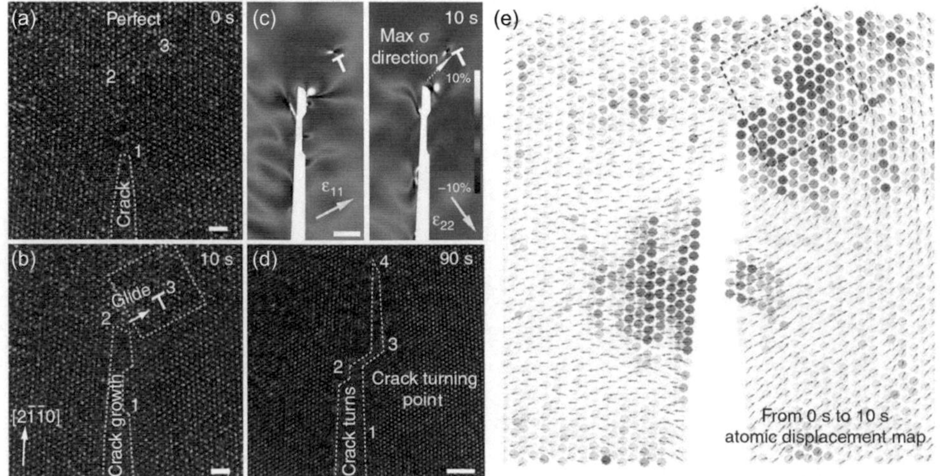

FIGURE 6.28 (a–d) The emission of dislocations at the crack tip in monolayer MoS_2 revealed by in situ HRTEM. Scale bars, 1 nm. (e) The strain evolution after a single dislocation emission at the crack tip. (Adapted from Ref. [76] with permission.)

6.4.4 Outlooks

So far, we have discussed some examples of 2D TMDs and their associated properties together with potential applications. This area is under fast development. Currently, the lack of reliable and reproducible synthesis and lack of deep understandings of some fundamental physical and chemical properties originated from their complex and nontrivial electronic structures in 2D TMDs are the most desirable and urgent targets for the whole research societies. Relying on the previous rapid development of 2D TMD area, the future promising directions of 2D TMDs can be comprised of, but not restricted to, the following: (i) Exploration of new 2D TMDs using appropriate synthesis methods. (ii) Defect control, phase control, doping, and bandgap engineering in 2D TMDs. (iii) Large-scale synthesis methods aiming for wafer-scale industrial productions. (iv) New physical properties in mesoscale or nanoscale owing to the reduced dimensions and broken symmetry and topology. (v) Catalysis-related applications in energy science and technologies.

Acknowledgments

This work was supported by Hong Kong Research Grant Council Early Career Scheme (Project no.25301018, 21303218, City University of Hong Kong (Project No. 9610387), The Hong Kong Polytechnic University (Project No. 1-ZE8C), National Science Foundation of China (51872248), the Shenzhen Science and Technology Innovation Commission (Project No. JCYJ20170818104717087) and Institute for Basic Science (IBS-R011-D1).

References

1. Novoselov, K. S., Geim, A. K., Morozov, S. V., Jiang, D., Zhang, Y., Dubonos, S. V., ... & Firsov, A. A. (2004). Electric field effect in atomically thin carbon films. *Science*, 306(5696), 666–669.
2. Geim, A. K. (2009). Graphene: Status and prospects. *Science*, 324(5934), 1530–1534.
3. Ferrari, A. C., Meyer, J. C., Scardaci, V., Casiraghi, C., Lazzeri, M., Mauri, F., ... & Geim, A. K. (2006). Raman spectrum of graphene and graphene layers. *Physical Review Letters*, 97(18), 187401.
4. Ramanathan, T., Abdala, A. A., Stankovich, S., Dikin, D. A., Herrera-Alonso, M., Piner, R. D., ... & Nguyen, S. T. (2008). Functionalized graphene sheets for polymer nanocomposites. *Nature Nanotechnology*, 3(6), 327.
5. Watanabe, K., Taniguchi, T., & Kanda, H. (2004). Direct-bandgap properties and evidence for ultraviolet lasing of hexagonal boron nitride single crystal. *Nature Materials*, 3(6), 404.
6. Geick, R., Perry, C. H., & Rupprecht, G. (1966). Normal modes in hexagonal boron nitride. *Physical Review*, 146(2), 543.
7. Xue, J., Sanchez-Yamagishi, J., Bulmash, D., Jacquod, P., Deshpande, A., Watanabe, K., ... & LeRoy, B. J. (2011). Scanning tunnelling microscopy and spectroscopy of ultra-flat graphene on hexagonal boron nitride. *Nature Materials*, 10(4), 282.
8. Yankowitz, M., Xue, J., Cormode, D., Sanchez-Yamagishi, J. D., Watanabe, K., Taniguchi, T., ... & LeRoy, B. J. (2012). Emergence of superlattice Dirac points in graphene on hexagonal boron nitride. *Nature Physics*, 8(5), 382.
9. Mak, K. F., Lee, C., Hone, J., Shan, J., & Heinz, T. F. (2010). Atomically thin MoS_2: A new direct-gap semiconductor. *Physical Review Letters*, 105(13), 136805.
10. Chernikov, A., Berkelbach, T. C., Hill, H. M., Rigosi, A., Li, Y., Aslan, O. B., ... & Heinz, T. F. (2014). Exciton binding energy and nonhydrogenic Rydberg series in monolayer WS_2. *Physical Review Letters*, 113(7), 076802.
11. Sun, Z., Martinez, A., & Wang, F. (2016). Optical modulators with 2D layered materials. *Nature Photonics*, 10(4), 227–238.
12. Ling, X., Lin, Y., Ma, Q., Wang, Z., Song, Y., Yu, L., ... & Bie, Y. (2016). Parallel stitching of 2D materials. *Advanced Materials*, 28(12), 2322–2329.
13. Ghorbani-Asl, M., Zibouche, N., Wahiduzzaman, M., Oliveira, A. F., Kuc, A., & Heine, T. (2013). Electromechanics in MoS_2 and WS_2: Nanotubes vs. monolayers. *Scientific Reports*, 3, 2961.
14. Koperski, M., Nogajewski, K., Arora, A., Cherkez, V., Mallet, P., Veuillen, J. Y., ... & Potemski, M. (2015). Single photon emitters in exfoliated WSe_2 structures. *Nature Nanotechnology*, 10(6), 503.
15. Arora, A., Koperski, M., Nogajewski, K., Marcus, J., Faugeras, C., & Potemski, M. (2015). Excitonic resonances in thin films of WSe_2: From monolayer to bulk material. *Nanoscale*, 7(23), 10421–10429.
16. Medhekar, N. V., Ramasubramaniam, A., Ruoff, R. S., & Shenoy, V. B. (2010). Hydrogen bond networks in graphene oxide composite paper: Structure and mechanical properties. *ACS Nano*, 4(4), 2300–2306.
17. Rozhkov, A. V., Giavaras, G., Bliokh, Y. P., Freilikher, V., & Nori, F. (2011). Electronic properties of mesoscopic graphene structures: Charge confinement and control of spin and charge transport. *Physics Reports*, 503(2–3), 77–114.
18. Zhang, Y., Tang, T. T., Girit, C., Hao, Z., Martin, M. C., Zettl, A., ... & Wang, F. (2009). Direct observation of a widely tunable bandgap in bilayer graphene. *Nature*, 459(7248), 820.
19. Johari, P., & Shenoy, V. B. (2012). Tuning the electronic properties of semiconducting transition metal dichalcogenides by applying mechanical strains. *ACS Nano*, 6(6), 5449–5456.
20. Akinwande, D., Petrone, N., & Hone, J. (2014). Two-dimensional flexible nanoelectronics. *Nature Communications*, 5, 5678.

21. Podzorov, V., Gershenson, M. E., Kloc, C., Zeis, R., & Bucher, E. (2004). High-mobility field-effect transistors based on transition metal dichalcogenides. *Applied Physics Letters*, 84(17), 3301–3303.
22. Knirsch, K. C., Berner, N. C., Nerl, H. C., Cucinotta, C. S., Gholamvand, Z., McEvoy, N., ... & Sanvito, S. (2015). Basal-plane functionalization of chemically exfoliated molybdenum disulfide by diazonium salts. *ACS Nano*, 9(6), 6018–6030.
23. Kuc, A. (2014). Low-dimensional transition-metal dichalcogenides. In: *Chemical Modelling*, Springborg, M., Joswig, J.-O. (Eds). Cambridge: Royal Society of Chemistry, 1–29.
24. Koppens, F. H. L., Mueller, T., Avouris, P., Ferrari, A. C., Vitiello, M. S., & Polini, M. (2014). Photodetectors based on graphene, other two-dimensional materials and hybrid systems. *Nature Nanotechnology*, 9(10), 780.
25. Bonaccorso, F., Sun, Z., Hasan, T., & Ferrari, A. C. (2010). Graphene photonics and optoelectronics. *Nature Photonics*, 4(9), 611.
26. Bonaccorso, F., Colombo, L., Yu, G., Stoller, M., Tozzini, V., Ferrari, A. C., ... & Pellegrini, V. (2015). Graphene, related two-dimensional crystals, and hybrid systems for energy conversion and storage. *Science*, 347(6217), 1246501.
27. Ju, L., Geng, B., Horng, J., Girit, C., Martin, M., Hao, Z., ... & Wang, F. (2011). Graphene plasmonics for tunable terahertz metamaterials. *Nature Nanotechnology*, 6(10), 630.
28. Wilson, J. A., & Yoffe, A. D. (1969). The transition metal dichalcogenides discussion and interpretation of the observed optical, electrical and structural properties. *Advances in Physics*, 18(73), 193–335.
29. Bernardi, M., Palummo, M., & Grossman, J. C. (2013). Extraordinary sunlight absorption and one nanometer thick photovoltaics using two-dimensional monolayer materials. *Nano Letters*, 13(8), 3664–3670.
30. Zhang, X., Qiao, X. F., Shi, W., Wu, J. B., Jiang, D. S., & Tan, P. H. (2015). Phonon and Raman scattering of two-dimensional transition metal dichalcogenides from monolayer, multilayer to bulk material. *Chemical Society Reviews*, 44(9), 2757–2785.
31. Berkelbach, T. C., Hybertsen, M. S., & Reichman, D. R. (2013). Theory of neutral and charged excitons in monolayer transition metal dichalcogenides. *Physical Review B*, 88(4), 045318.
32. Sahin, H., Tongay, S., Horzum, S., Fan, W., Zhou, J., Li, J., ... & Peeters, F. M. (2013). Anomalous Raman spectra and thickness-dependent electronic properties of WSe_2. *Physical Review B*, 87(16), 165409.
33. Coleman, J. N., Lotya, M., O'Neill, A., Bergin, S. D., King, P. J., Khan, U., ... & Shvets, I. V. (2011). Two-dimensional nanosheets produced by liquid exfoliation of layered materials. *Science*, 331(6017), 568–571.
34. Nicolosi, V., Chhowalla, M., Kanatzidis, M. G., Strano, M. S., & Coleman, J. N. (2013). Liquid exfoliation of layered materials. *Science*, 340(6139), 1226419.
35. Chhowalla, M., Shin, H. S., Eda, G., Li, L. J., Loh, K. P., & Zhang, H. (2013). The chemistry of two-dimensional layered transition metal dichalcogenide nanosheets. *Nature Chemistry*, 5(4), 263.
36. Li, H., Wu, J., Yin, Z., & Zhang, H. (2014). Preparation and applications of mechanically exfoliated single-layer and multilayer MoS_2 and WSe_2 nanosheets. *Accounts of Chemical Research*, 47(4), 1067–1075.
37. Komsa, H. P., Kotakoski, J., Kurasch, S., Lehtinen, O., Kaiser, U., & Krasheninnikov, A. V. (2012). Two-dimensional transition metal dichalcogenides under electron irradiation: Defect production and doping. *Physical Review Letters*, 109(3), 035503.
38. Chang, J., Register, L. F., & Banerjee, S. K. (2014). Ballistic performance comparison of monolayer transition metal dichalcogenide MX2 (M = Mo, W; X = S, Se, Te) metal-oxide-semiconductor field effect transistors. *Journal of Applied Physics*, 115(8), 084506.
39. Acerce, M., Voiry, D., & Chhowalla, M. (2015). Metallic 1T phase MoS_2 nanosheets as supercapacitor electrode materials. *Nature Nanotechnology*, 10(4), 313.
40. Ramakrishna Matte, H. S. S., Gomathi, A., Manna, A. K., Late, D. J., Datta, R., Pati, S. K., & Rao, C. N. R. (2010). MoS_2 and WS_2 analogues of graphene. *Angewandte Chemie International Edition*, 49(24), 4059–4062.
41. Meyer, J. C., Geim, A. K., Katsnelson, M. I., Novoselov, K. S., Booth, T. J., & Roth, S. (2007). The structure of suspended graphene sheets. *Nature*, 446(7131), 60–63.
42. Bertolazzi, S., Brivio, J., & Kis, A. (2011). Stretching and breaking of ultrathin MoS_2. *ACS Nano*, 5(12), 9703–9709.
43. Kadantsev, E. S., & Hawrylak, P. (2012). Electronic structure of a single MoS_2 monolayer. *Solid State Communications*, 152(10), 909–913.
44. Yun, W. S., Han, S. W., Hong, S. C., Kim, I. G., & Lee, J. D. (2012). Thickness and strain effects on electronic structures of transition metal dichalcogenides: 2H-M X_2 semiconductors (M = Mo, W; X = S, Se, Te). *Physical Review B*, 85(3), 033305.
45. Ramasubramaniam, A., Naveh, D., & Towe, E. (2011). Tunable band gaps in bilayer transition-metal dichalcogenides. *Physical Review B*, 84(20), 205325.
46. Radisavljevic, B., Radenovic, A., Brivio, J., Giacometti, I. V., & Kis, A. (2011). Single-layer MoS_2 transistors. *Nature Nanotechnology*, 6(3), 147–150.
47. Ugeda, M. M., Bradley, A. J., Shi, S. F., Felipe, H., Zhang, Y., Qiu, D. Y., ... & Wang, F. (2014). Giant

bandgap renormalization and excitonic effects in a monolayer transition metal dichalcogenide semiconductor. *Nature Materials*, 13(12), 1091.
48. Benavente, E., Santa Ana, M. A., Mendizábal, F., & González, G. (2002). Intercalation chemistry of molybdenum disulfide. *Coordination Chemistry Reviews*, 224(1–2), 87–109.
49. Fan, X., Xu, P., Zhou, D., Sun, Y., Li, Y. C., Nguyen, M. A. T., ... & Mallouk, T. E. (2015). Fast and efficient preparation of exfoliated 2H MoS_2 nanosheets by sonication-assisted lithium intercalation and infrared laser-induced 1T to 2H phase reversion. *Nano Letters*, 15(9), 5956–5960.
50. Chou, S. S., Huang, Y. K., Kim, J., Kaehr, B., Foley, B. M., Lu, P., ... & Dravid, V. P. (2015). Controlling the metal to semiconductor transition of MoS_2 and WS_2 in solution. *Journal of the American Chemical Society*, 137(5), 1742–1745.
51. Splendiani, A., Sun, L., Zhang, Y., Li, T., Kim, J., Chim, C. Y., ... & Wang, F. (2010). Emerging photoluminescence in monolayer MoS_2. *Nano Letters*, 10(4), 1271–1275.
52. Mattheiss, L. F. (1973). Band structures of transition-metal-dichalcogenide layer compounds. *Physical Review B*, 8(8), 3719.
53. Kumar, A., & Ahluwalia, P. K. (2012). Electronic structure of transition metal dichalcogenides monolayers 1H-MX_2 (M = Mo, W; X = S, Se, Te) from ab-initio theory: New direct band gap semiconductors. *The European Physical Journal B*, 85(6), 186.
54. Kuc, A., Zibouche, N., & Heine, T. (2011). Influence of quantum confinement on the electronic structure of the transition metal sulfide TS_2. *Physical Review B*, 83(24), 245213.
55. Kolobov, A. V., & Tominaga, J. (2016). *Two-Dimensional Transition-Metal Dichalcogenides* (Vol. 239). Springer, Berlin.
56. Li, Y., Duerloo, K. A. N., Wauson, K., & Reed, E. J. (2016). Structural semiconductor-to-semimetal phase transition in two-dimensional materials induced by electrostatic gating. *Nature Communications*, 7, 10671.
57. Miró, P., Ghorbani-Asl, M., & Heine, T. (2013). Spontaneous ripple formation in MoS_2 monolayers: Electronic structure and transport effects. *Advanced Materials*, 25(38), 5473–5475.
58. Ramana, C. V., Becker, U., Shutthanandan, V., & Julien, C. M. (2008). Oxidation and metal-insertion in molybdenite surfaces: Evaluation of charge-transfer mechanisms and dynamics. *Geochemical Transactions*, 9(1), 8.
59. Eda, G., Yamaguchi, H., Voiry, D., Fujita, T., Chen, M., & Chhowalla, M. (2011). Photoluminescence from chemically exfoliated MoS_2. *Nano Letters*, 11(12), 5111–5116.
60. Cho, S., Kim, S., Kim, J. H., Zhao, J., Seok, J., Keum, D. H., ... & Kim, S. W. (2015). Phase patterning for ohmic homojunction contact in $MoTe_2$. *Science*, 349(6248), 625–628.
61. Kan, M., Nam, H. G., Lee, Y. H., & Sun, Q. (2015). Phase stability and Raman vibration of the molybdenum ditelluride ($MoTe_2$) monolayer. *Physical Chemistry Chemical Physics*, 17(22), 14866–14871.
62. Kappera, R., Voiry, D., Yalcin, S. E., Branch, B., Gupta, G., Mohite, A. D., & Chhowalla, M. (2014). Phase-engineered low-resistance contacts for ultrathin MoS_2 transistors. *Nature Materials*, 13(12), 1128–1134.
63. Lin, Z., Carvalho, B. R., Kahn, E., Lv, R., Rao, R., Terrones, H., ... & Terrones, M. (2016). Defect engineering of two-dimensional transition metal dichalcogenides. *2D Materials*, 3(2), 022002.
64. Zhou, W., Zou, X., Najmaei, S., Liu, Z., Shi, Y., Kong, J., ... & Idrobo, J. C. (2013). Intrinsic structural defects in monolayer molybdenum disulfide. *Nano Letters*, 13(6), 2615–2622.
65. Komsa, H. P., & Krasheninnikov, A. V. (2015). Native defects in bulk and monolayer MoS_2 from first principles. *Physical Review B*, 91(12), 125304.
66. Hanada, N., Ichikawa, T., & Fujii, H. (2005). Catalytic effect of Ni nano-particle and Nb oxide on H-desorption properties in MgH_2 prepared by ball milling. *Journal of Alloys and Compounds*, 404, 716–719.
67. Zabinski, J. S., Donley, M. S., Walck, S. D., Schneider, T. R., & McDevitt, N. T. (1995). The effects of dopants on the chemistry and tribology of sputter-deposited MoS_2 films. *Tribology Transactions*, 38(4), 894–904.
68. Dolui, K., Rungger, I., Pemmaraju, C. D., & Sanvito, S. (2013). Possible doping strategies for MoS_2 monolayers: An ab initio study. *Physical Review B*, 88(7), 075420.
69. Lin, Y. C., Dumcenco, D. O., Komsa, H. P., Niimi, Y., Krasheninnikov, A. V., Huang, Y. S., & Suenaga, K. (2014). Properties of individual dopant atoms in single-layer MoS_2: Atomic structure, migration, and enhanced reactivity. *Advanced Materials*, 26(18), 2857–2861.
70. Lin, X., & Ni, J. (2014). Charge and magnetic states of Mn-, Fe-, and Co-doped monolayer MoS_2. *Journal of Applied Physics*, 116(4), 044311.
71. Alsaad, A. (2014). Structural, electronic and magnetic properties of Fe, Co, Mn-doped GaN and ZnO diluted magnetic semiconductors. *Physica B: Condensed Matter*, 440, 1–9.
72. Lu, S. C., & Leburton, J. P. (2014). Electronic structures of defects and magnetic impurities in MoS_2 monolayers. *Nanoscale Research Letters*, 9(1), 676.
73. Ataca, C., & Ciraci, S. (2011). Functionalization of single-layer MoS_2 honeycomb structures. *The Journal of Physical Chemistry C*, 115(27), 13303–13311.

74. Lin, Y. C., Björkman, T., Komsa, H. P., Teng, P. Y., Yeh, C. H., Huang, F. S., ... & Krasheninnikov, A. V. (2015). Three-fold rotational defects in two-dimensional transition metal dichalcogenides. *Nature Communications*, 6, 6736.
75. Warner, J. H., Margine, E. R., Mukai, M., Robertson, A. W., Giustino, F., & Kirkland, A. I. (2012). Dislocation-driven deformations in graphene. *Science*, 337(6091), 209–212.
76. Ly, T. H., Zhao, J., Cichocka, M. O., Li, L. J., & Lee, Y. H. (2017). Dynamical observations on the crack tip zone and stress corrosion of two-dimensional MoS_2. *Nature Communications*, 8, 14116.
77. Elkin, E. L., & Watkins, G. D. (1968). Defects in irradiated silicon: Electron paramagnetic resonance and electron-nuclear double resonance of the arsenic-and antimony-vacancy pairs. *Physical Review*, 174(3), 881.
78. Wan-Lun, Y., Xin-Min, Z., La-Xun, Y., & Bao-Qing, Z. (1994). Spectroscopic properties of Cr^{3+} ions at the defect sites in cubic fluoroperovskite crystals. *Physical Review B*, 50(10), 6756.
79. Ly, T. H., Perello, D. J., Zhao, J., Deng, Q., Kim, H., Han, G. H., ... & Lee, Y. H. (2016). Misorientation-angle-dependent electrical transport across molybdenum disulfide grain boundaries. *Nature Communications*, 7, 10426.
80. Zhang, Y., Zhang, Y., Ji, Q., Ju, J., Yuan, H., Shi, J., ... & Song, X. (2013). Controlled growth of high-quality monolayer WS_2 layers on sapphire and imaging its grain boundary. *ACS Nano*, 7(10), 8963–8971.
81. Chu, R. L., Liu, G. B., Yao, W., Xu, X., Xiao, D., & Zhang, C. (2014). Spin-orbit-coupled quantum wires and Majorana fermions on zigzag edges of monolayer transition-metal dichalcogenides. *Physical Review B*, 89(15), 155317.
82. Helveg, S., Lauritsen, J. V., Lægsgaard, E., Stensgaard, I., Nørskov, J. K., Clausen, B. S., ... & Besenbacher, F. (2000). Atomic-scale structure of single-layer MoS_2 nanoclusters. *Physical Review Letters*, 84(5), 951.
83. Deng, Q., Thi, Q. H., Zhao, J., Yun, S. J., Kim, H., Chen, G., & Ly, T. H. (2018). Impact of polar edge terminations of the transition metal dichalcogenide monolayers during vapor growth. *The Journal of Physical Chemistry C*, 122(6), 3575–3581.
84. Conley, H. J., Wang, B., Ziegler, J. I., Haglund Jr, R. F., Pantelides, S. T., & Bolotin, K. I. (2013). Bandgap engineering of strained monolayer and bilayer MoS_2. *Nano Letters*, 13(8), 3626–3630.
85. Zhao, J., Deng, Q., Ly, T. H., Han, G. H., Sandeep, G., & Rümmeli, M. H. (2015). Two-dimensional membrane as elastic shell with proof on the folds revealed by three-dimensional atomic mapping. *Nature Communications*, 6, 8935.
86. Luo, S., Hao, G., Fan, Y., Kou, L., He, C., Qi, X., ... & Zhong, J. (2015). Formation of ripples in atomically thin MoS_2 and local strain engineering of electrostatic properties. *Nanotechnology*, 26(10), 105705.
87. Jeon, J., Jang, S. K., Jeon, S. M., Yoo, G., Jang, Y. H., Park, J. H., & Lee, S. (2015). Layer-controlled CVD growth of large-area two-dimensional MoS_2 films. *Nanoscale*, 7(5), 1688–1695.
88. Yu, Y., Li, C., Liu, Y., Su, L., Zhang, Y., & Cao, L. (2013). Controlled scalable synthesis of uniform, high-quality monolayer and few-layer MoS_2 films. *Scientific Reports*, 3, 1866–1870.
89. Lee, Y. H., Zhang, X. Q., Zhang, W., Chang, M. T., Lin, C. T., Chang, K. D., ... & Lin, T. W. (2012). Synthesis of large-area MoS_2 atomic layers with chemical vapor deposition. *Advanced Materials*, 24(17), 2320–2325.
90. Berkdemir, A., Gutiérrez, H. R., Botello-Méndez, A. R., Perea-López, N., Elías, A. L., Chia, C. I., ... & Terrones, H. (2013). Identification of individual and few layers of WS_2 using Raman Spectroscopy. *Scientific Reports*, 3, 1755.
91. Shaw, J. C., Zhou, H., Chen, Y., Weiss, N. O., Liu, Y., Huang, Y., & Duan, X. (2014). Chemical vapor deposition growth of monolayer $MoSe_2$ nanosheets. *Nano Research*, 7(4), 511–517.
92. Huang, J. K., Pu, J., Hsu, C. L., Chiu, M. H., Juang, Z. Y., Chang, Y. H., ... & Li, L. J. (2013). Large-area synthesis of highly crystalline WSe_2 monolayers and device applications. *ACS Nano*, 8(1), 923–930.
93. Dong, N., Li, Y., Feng, Y., Zhang, S., Zhang, X., Chang, C., ... & Wang, J. (2015). Optical limiting and theoretical modelling of layered transition metal dichalcogenide nanosheets. *Scientific Reports*, 5, 14646.
94. Wallace, P. R. (1947). The band theory of graphite. *Physical Review*, 71(9), 622.
95. Geim, A. K., & Novoselov, K. S. (2007). The rise of graphene. *Nature Materials*, 6(3), 183.
96. Neto, A. C., Guinea, F., Peres, N. M., Novoselov, K. S., & Geim, A. K. (2009). The electronic properties of graphene. *Reviews of Modern Physics*, 81(1), 109.
97. Novoselov, K. S., Mishchenko, A., Carvalho, A., & Neto, A. C. (2016). 2D materials and van der Waals heterostructures. *Science*, 353(6298), aac9439.
98. Geim, A. K., & Grigorieva, I. V. (2013). Van der Waals heterostructures. *Nature*, 499(7459), 419.
99. Wang, X., & Xia, F. (2015). Van der Waals heterostructures: Stacked 2D materials shed light. *Nature Materials*, 14(3), 264.
100. Novoselov, K. S., Jiang, D., Schedin, F., Booth, T. J., Khotkevich, V. V., Morozov, S. V., & Geim, A. K. (2005). Two-dimensional atomic crystals. *Proceedings of the National Academy of Sciences of the United States of America*, 102(30), 10451–10453.
101. Ashton, M., Paul, J., Sinnott, S. B., & Hennig, R. G. (2017). Topology-scaling identification of layered

solids and stable exfoliated 2D materials. *Physical Review Letters*, 118(10), 106101.
102. Fasolino, A., Los, J. H., & Katsnelson, M. I. (2007). Intrinsic ripples in graphene. *Nature Materials*, 6(11), 858.
103. Jiao, L., Liu, H. J., Chen, J. L., Yi, Y., Chen, W. G., Cai, Y., ... & Xie, M. H. (2015). Molecular-beam epitaxy of monolayer $MoSe_2$: Growth characteristics and domain boundary formation. *New Journal of Physics*, 17(5), 053023.
104. Bhimanapati, G. R., Lin, Z., Meunier, V., Jung, Y., Cha, J., Das, S., ... & Liang, L. (2015). Recent advances in two-dimensional materials beyond graphene. *ACS Nano*, 9(12), 11509–11539.
105. Xu, L., McGraw, J. W., Gao, F., Grundy, M., Ye, Z., Gu, Z., & Shepherd, J. L. (2013). Production of high-concentration graphene dispersions in low-boiling-point organic solvents by liquid-phase noncovalent exfoliation of graphite with a hyperbranched polyethylene and formation of graphene/ethylene copolymer composites. *The Journal of Physical Chemistry C*, 117(20), 10730–10742.
106. Gupta, A., Sakthivel, T., & Seal, S. (2015). Recent development in 2D materials beyond graphene. *Progress in Materials Science*, 73, 44–126.
107. Wang, Q. H., Kalantar-Zadeh, K., Kis, A., Coleman, J. N., & Strano, M. S. (2012). Electronics and optoelectronics of two-dimensional transition metal dichalcogenides. *Nature Nanotechnology*, 7(11), 699.
108. Choi, W., Choudhary, N., Han, G. H., Park, J., Akinwande, D., & Lee, Y. H. (2017). Recent development of two-dimensional transition metal dichalcogenides and their applications. *Materials Today*, 20(3), 116–130.
109. Li, F., & Xue, M. (2016). Two-dimensional transition metal dichalcogenides for electrocatalytic energy conversion applications. In: *Two-Dimensional Materials-Synthesis*, Characterization and Potential Applications, Nayak, P. K. (Ed). London: InTech, 63–82.
110. Voiry, D., Mohite, A., & Chhowalla, M. (2015). Phase engineering of transition metal dichalcogenides. *Chemical Society Reviews*, 44(9), 2702–2712.
111. Naylor, C. H., Parkin, W. M., Ping, J., Gao, Z., Zhou, Y. R., Kim, Y., ... & Kikkawa, J. M. (2016). Monolayer single-crystal 1T′-$MoTe_2$ grown by chemical vapor deposition exhibits weak antilocalization effect. *Nano Letters*, 16(7), 4297–4304.
112. Park, J. C., Yun, S. J., Kim, H., Park, J. H., Chae, S. H., An, S. J., ... & Lee, Y. H. (2015). Phase-engineered synthesis of centimeter-scale 1T′-and 2H-molybdenum ditelluride thin films. *ACS Nano*, 9(6), 6548–6554.
113. Yang, Z., & Hao, J. (2016). Progress in pulsed laser deposited two-dimensional layered materials for device applications. *Journal of Materials Chemistry C*, 4(38), 8859–8878.
114. Gao, M., Zhang, M., Niu, W., Chen, Y., Gu, M., Wang, H., ... & Wang, X. (2017). Tuning the transport behavior of centimeter-scale WTe_2 ultrathin films fabricated by pulsed laser deposition. *Applied Physics Letters*, 111(3), 031906.
115. Li, H., Lu, G., Wang, Y., Yin, Z., Cong, C., He, Q., ... & Zhang, H. (2013). Mechanical exfoliation and characterization of single-and few-layer nanosheets of WSe_2, TaS_2, and $TaSe_2$. *Small*, 9(11), 1974–1981.
116. Magda, G. Z., Pető, J., Dobrik, G., Hwang, C., Biró, L. P., & Tapasztó, L. (2015). Exfoliation of large-area transition metal chalcogenide single layers. *Scientific Reports*, 5, 14714.
117. Xing, T., Mateti, S., Li, L. H., Ma, F., Du, A., Gogotsi, Y., & Chen, Y. (2016). Gas protection of two-dimensional nanomaterials from high-energy impacts. *Scientific reports*, 6, 35532. Peng, J., Wu, J., Li, X., Zhou, Y., Yu, Z., Guo, Y., ... & Wu, C. (2017).
118. Very large-sized transition metal dichalcogenides monolayers from fast exfoliation by manual shaking. *Journal of the American Chemical Society*, 139(26), 9019–9025.
119. Hernandez, Y., Nicolosi, V., Lotya, M., Blighe, F. M., Sun, Z., De, S., ... & Boland, J. J. (2008). High-yield production of graphene by liquid-phase exfoliation of graphite. *Nature Nanotechnology*, 3(9), 563.
120. Wu, J., Liu, M., Chatterjee, K., Hackenberg, K. P., Shen, J., Zou, X., ... & Ajayan, P. M. (2016). Exfoliated 2D transition metal disulfides for enhanced electrocatalysis of oxygen evolution reaction in acidic medium. *Advanced Materials Interfaces*, 3(9), 1500669.
121. Dong, Z., & Ye, Z. (2012). Hyperbranched polyethylenes by chain walking polymerization: Synthesis, properties, functionalization, and applications. *Polymer Chemistry*, 3(2), 286–301.
122. Vadukumpully, S., Paul, J., Mahanta, N., & Valiyaveettil, S. (2011). Flexible conductive graphene/poly (vinyl chloride) composite thin films with high mechanical strength and thermal stability. *Carbon*, 49(1), 198–205.
123. Davis, R. F., Kelner, G., Shur, M., Palmour, J. W., & Edmond, J. A. (1991). Thin film deposition and microelectronic and optoelectronic device fabrication and characterization in monocrystalline alpha and beta silicon carbide. *Proceedings of the IEEE*, 79(5), 677–701.
124. Meyerson, B. S. (1992). UHV/CVD growth of Si and Si: Ge alloys: Chemistry, physics, and device applications. *Proceedings of the IEEE*, 80(10), 1592–1608.
125. Li, Y., Mann, D., Rolandi, M., Kim, W., Ural, A., Hung, S., ... & Wang, Q. (2004). Preferential growth of semiconducting single-walled carbon nanotubes

by a plasma enhanced CVD method. *Nano Letters*, 4(2), 317–321.

126. Yu, J. Y., Chung, S. W., & Heath, J. R. (2000). Silicon nanowires: Preparation, device fabrication, and transport properties. *The Journal of Physical Chemistry B*, 104(50), 11864–11870.
127. Yoon, M. Y., Lee, S. I., & Lim, H. S. (2002). U.S. Patent No. 6,358,829. Washington, DC: U.S. Patent and Trademark Office.
128. Sun, Z., Liu, Z., Li, J., Tai, G. A., Lau, S. P., & Yan, F. (2012). Infrared photodetectors based on CVD-grown graphene and PbS quantum dots with ultrahigh responsivity. *Advanced Materials*, 24(43), 5878–5883.
129. Zhou, D., Shu, H., Hu, C., Jiang, L., Liang, P., & Chen, X. (2018). Unveiling the growth mechanism of MoS2 with chemical vapor deposition: from two-dimensional planar nucleation to self-seeding nucleation. *Crystal Growth & Design*, 18(2), 1012-1019.
130. Zhan, Y., Liu, Z., Najmaei, S., Ajayan, P. M., & Lou, J. (2012). Large-area vapor-phase growth and characterization of MoS_2 atomic layers on a SiO_2 substrate. *Small*, 8(7), 966–971.
131. Özden, A., Ay, F., Sevik, C., & Perkgöz, N. K. (2017). CVD growth of monolayer MoS_2: Role of growth zone configuration and precursors ratio. *Japanese Journal of Applied Physics*, 56(6S1), 06GG05.
132. Wang, X., Gong, Y., Shi, G., Chow, W. L., Keyshar, K., Ye, G., ... & Tay, B. K. (2014). Chemical vapor deposition growth of crystalline monolayer $MoSe_2$. *ACS Nano*, 8(5), 5125–5131.
133. Wang, S., Rong, Y., Fan, Y., Pacios, M., Bhaskaran, H., He, K., & Warner, J. H. (2014). Shape evolution of monolayer MoS_2 crystals grown by chemical vapor deposition. *Chemistry of Materials*, 26(22), 6371–6379.
134. Zhang, J., Yu, H., Chen, W., Tian, X., Liu, D., Cheng, M., ... & Shi, D. (2014). Scalable growth of high-quality polycrystalline MoS_2 monolayers on SiO_2 with tunable grain sizes. *ACS Nano*, 8(6), 6024–6030.
135. Sahatiya, P., Madhava, C., Shinde, A., & Badhulika, S. (2018). Flexible substrate based few layer MoS_2 electrode for passive electronic devices and interactive frequency modulation based on human motion. *IEEE Transactions on Nanotechnology*, 17(2), 338–344.
136. Wang, D. N., White, J. M., Law, K. S., Leung, C., Umotoy, S. P., Collins, K. S., ... & Maydan, D. (1991). U.S. Patent No. 5,000,113. Washington, DC: U.S. Patent and Trademark Office.
137. Liu, H., Jiao, L., Yang, F., Cai, Y., Wu, X., Ho, W., ... & Yao, W. (2014). Dense network of one-dimensional midgap metallic modes in monolayer $MoSe_2$ and their spatial undulations. *Physical Review Letters*, 113(6), 066105.
138. Liu, H. J., Jiao, L., Xie, L., Yang, F., Chen, J. L., Ho, W. K., ... & Xie, M. H. (2015). Molecular-beam epitaxy of monolayer and bilayer WSe_2: A scanning tunneling microscopy/spectroscopy study and deduction of exciton binding energy. *2D Materials*, 2(3), 034004.
139. Roy, A., Movva, H. C., Satpati, B., Kim, K., Dey, R., Rai, A., ... & Banerjee, S. K. (2016). Structural and electrical properties of $MoTe_2$ and $MoSe_2$ grown by molecular beam epitaxy. *ACS Applied Materials and Interfaces*, 8(11), 7396–7402.
140. Westover, R. D., Ditto, J., Falmbigl, M., Hay, Z. L., & Johnson, D. C. (2015). Synthesis and characterization of quaternary monolayer thick $MoSe_2$/SnSe/$NbSe_2$/SnSe heterojunction superlattices. *Chemistry of Materials*, 27(18), 6411–6417.
141. Miwa, J. A., Dendzik, M., Grønborg, S. S., Bianchi, M., Lauritsen, J. V., Hofmann, P., & Ulstrup, S. (2015). Van der Waals epitaxy of two-dimensional MoS_2–graphene heterostructures in ultrahigh vacuum. *ACS Nano*, 9(6), 6502–6510.
142. Li, X., Zhu, Y., Cai, W., Borysiak, M., Han, B., Chen, D., ... & Ruoff, R. S. (2009). Transfer of large-area graphene films for high-performance transparent conductive electrodes. *Nano Letters*, 9(12), 4359–4363.
143. Dean, C. R., Young, A. F., Meric, I., Lee, C., Wang, L., Sorgenfrei, S., ... & Hone, J. (2010). Boron nitride substrates for high-quality graphene electronics. *Nature Nanotechnology*, 5(10), 722.
144. Chen, X. D., Liu, Z. B., Zheng, C. Y., Xing, F., Yan, X. Q., Chen, Y., & Tian, J. G. (2013). High-quality and efficient transfer of large-area graphene films onto different substrates. *Carbon*, 56, 271–278.
145. Van Ngoc, H., Qian, Y., Han, S. K., & Kang, D. J. (2016). PMMA-etching-free transfer of wafer-scale chemical vapor deposition two-dimensional atomic crystal by a water soluble polyvinyl alcohol polymer method. *Scientific Reports*, 6, 33096.
146. Yang, S. Y., Oh, J. G., Jung, D. Y., Choi, H., Yu, C. H., Shin, J., ... & Choi, S. Y. (2015). Metal-etching-free direct delamination and transfer of single-layer graphene with a high degree of freedom. *Small*, 11(2), 175–181.
147. Lin, Z., Zhao, Y., Zhou, C., Zhong, R., Wang, X., Tsang, Y. H., & Chai, Y. (2015). Controllable growth of large–size crystalline MoS_2 and resist-free transfer assisted with a cu thin film. *Scientific Reports*, 5, 18596.
148. Chen, J., Duan, M., & Chen, G. (2012). Continuous mechanical exfoliation of graphene sheets via three-roll mill. *Journal of Materials Chemistry*, 22(37), 19625–19628.
149. Yi, M., & Shen, Z. (2015). A review on mechanical exfoliation for the scalable production of graphene. *Journal of Materials Chemistry A*, 3(22), 11700–11715.

150. Berman, D., Erdemir, A., & Sumant, A. V. (2014). Graphene: A new emerging lubricant. *Materials Today*, 17(1), 31–42.
151. Suk, J. W., Kitt, A., Magnuson, C. W., Hao, Y., Ahmed, S., An, J., ... & Ruoff, R. S. (2011). Transfer of CVD-grown monolayer graphene onto arbitrary substrates. *ACS Nano*, 5(9), 6916–6924.
152. Ma, D., Shi, J., Ji, Q., Chen, K., Yin, J., Lin, Y., ... & Guo, X. (2015). A universal etching-free transfer of MoS_2 films for applications in photodetectors. *Nano Research*, 8(11), 3662–3672.
153. Yang, L., Majumdar, K., Liu, H., Du, Y., Wu, H., Hatzistergos, M., ... & Ye, P. D. (2014). Chloride molecular doping technique on 2D materials: WS_2 and MoS_2. *Nano Letters*, 14(11), 6275–6280.
154. Gong, Y., Liu, Z., Lupini, A. R., Shi, G., Lin, J., Najmaei, S., ... & Terrones, H. (2013). Band gap engineering and layer-by-layer mapping of selenium-doped molybdenum disulfide. *Nano Letters*, 14(2), 442–449.
155. Ma, Y., Dai, Y., Guo, M., Niu, C., Lu, J., & Huang, B. (2011). Electronic and magnetic properties of perfect, vacancy-doped, and nonmetal adsorbed $MoSe_2$, $MoTe_2$ and WS_2 monolayers. *Physical Chemistry Chemical Physics*, 13(34), 15546–15553.
156. Chuang, H. J., Tan, X., Ghimire, N. J., Perera, M. M., Chamlagain, B., Cheng, M. M. C., ... & Zhou, Z. (2014). High mobility WSe_2 p-and n-type field-effect transistors contacted by highly doped graphene for low-resistance contacts. *Nano Letters*, 14(6), 3594–3601.
157. Lu, N., Guo, H., Li, L., Dai, J., Wang, L., Mei, W. N., ... & Zeng, X. C. (2014). MoS_2/MX_2 heterobilayers: Bandgap engineering via tensile strain or external electrical field. *Nanoscale*, 6(5), 2879–2886.
158. Wang, Y., Cong, C., Yang, W., Shang, J., Peimyoo, N., Chen, Y., ... & Yu, T. (2015). Strain-induced direct–indirect bandgap transition and phonon modulation in monolayer WS_2. *Nano Research*, 8(8), 2562–2572.
159. Ruppert, C., Aslan, O. B., & Heinz, T. F. (2014). Optical properties and band gap of single-and few-layer $MoTe_2$ crystals. *Nano Letters*, 14(11), 6231–6236.
160. Roldán, R., Castellanos-Gomez, A., Cappelluti, E., & Guinea, F. (2015). Strain engineering in semiconducting two-dimensional crystals. *Journal of Physics: Condensed Matter*, 27(31), 313201.
161. Gong, C., Zhang, H., Wang, W., Colombo, L., Wallace, R. M., & Cho, K. (2013). Band alignment of two-dimensional transition metal dichalcogenides: Application in tunnel field effect transistors. *Applied Physics Letters*, 103(5), 053513.
162. Tongay, S., Suh, J., Ataca, C., Fan, W., Luce, A., Kang, J. S., ... & Ogletree, F. (2013). Defects activated photoluminescence in two-dimensional semiconductors: Interplay between bound, charged, and free excitons. *Scientific Reports*, 3, 2657.
163. Santosh, K. C., Longo, R. C., Addou, R., Wallace, R. M., & Cho, K. (2014). Impact of intrinsic atomic defects on the electronic structure of MoS_2 monolayers. *Nanotechnology*, 25(37), 375703.
164. Nan, H., Wang, Z., Wang, W., Liang, Z., Lu, Y., Chen, Q., ... & Wang, J. (2014). Strong photoluminescence enhancement of MoS_2 through defect engineering and oxygen bonding. *ACS Nano*, 8(6), 5738–5745.
165. Zou, X., & Yakobson, B. I. (2014). An open canvas 2D materials with defects, disorder, and functionality. *Accounts of Chemical Research*, 48(1), 73–80.
166. McDonnell, S., Addou, R., Buie, C., Wallace, R. M., & Hinkle, C. L. (2014). Defect-dominated doping and contact resistance in MoS_2. *ACS Nano*, 8(3), 2880–2888.
167. Gong, C., Colombo, L., Wallace, R. M., & Cho, K. (2014). The unusual mechanism of partial Fermi level pinning at metal–MoS_2 interfaces. *Nano Letters*, 14(4), 1714–1720.
168. Huang, Y. L., Ding, Z., Zhang, W., Chang, Y. H., Shi, Y., Li, L. J., ... & Wee, A. T. (2016). Gap states at low-angle grain boundaries in monolayer tungsten diselenide. *Nano Letters*, 16(6), 3682–3688.
169. Yan, C., Dong, X., Li, C. H., & Li, L. (2018). Charging effect at grain boundaries of MoS_2. *Nanotechnology*, 29(19), 195704.
170. Huang, Y. L., Chen, Y., Zhang, W., Quek, S. Y., Chen, C. H., Li, L. J., ... & Wee, A. T. (2015). Bandgap tunability at single-layer molybdenum disulphide grain boundaries. *Nature Communications*, 6, 6298.
171. Schmidt, H., Wang, S., Chu, L., Toh, M., Kumar, R., Zhao, W., ... & Eda, G. (2014). Transport properties of monolayer MoS_2 grown by chemical vapor deposition. *Nano Letters*, 14(4), 1909–1913.
172. Rong, Y., He, K., Pacios, M., Robertson, A. W., Bhaskaran, H., & Warner, J. H. (2015). Controlled preferential oxidation of grain boundaries in monolayer tungsten disulfide for direct optical imaging. *ACS Nano*, 9(4), 3695–3703.
173. Tongay, S., Zhou, J., Ataca, C., Lo, K., Matthews, T. S., Li, J., ... & Wu, J. (2012). Thermally driven crossover from indirect toward direct bandgap in 2D semiconductors: $MoSe_2$ versus MoS_2. *Nano Letters*, 12(11), 5576–5580.
174. Dai, J., & Zeng, X. C. (2014). Bilayer phosphorene: Effect of stacking order on bandgap and its potential applications in thin-film solar cells. *The Journal of Physical Chemistry Letters*, 5(7), 1289–1293.
175. McDonnell, S., Brennan, B., Azcatl, A., Lu, N., Dong, H., Buie, C., ... & Wallace, R. M. (2013). HfO_2 on MoS_2 by atomic layer deposition: Adsorption mechanisms and thickness scalability. *ACS Nano*, 7(11), 10354–10361.
176. Ly, T. H., Chiu, M. H., Li, M. Y., Zhao, J., Perello, D. J., Cichocka, M. O., ... & Li, L. J. (2014).

Observing grain boundaries in CVD-grown monolayer transition metal dichalcogenides. *ACS Nano*, 8(11), 11401–11408.

177. Cong, C., Shang, J., Wu, X., Cao, B., Peimyoo, N., Qiu, C., ... & Yu, T. (2014). Synthesis and optical properties of large-area single-crystalline 2D semiconductor WS_2 monolayer from chemical vapor deposition. *Advanced Optical Materials*, 2(2), 131–136.
178. Gong, Y., Lin, J., Wang, X., Shi, G., Lei, S., Lin, Z., ... & Terrones, H. (2014). Vertical and in-plane heterostructures from WS_2/MoS_2 monolayers. *Nature Materials*, 13(12), 1135–1142.
179. Purica, M., Budianu, E., Rusu, E., Danila, M., & Gavrila, R. (2002). Optical and structural investigation of ZnO thin films prepared by chemical vapor deposition (CVD). *Thin Solid Films*, 403, 485–488.
180. Gong, Y., Lei, S., Ye, G., Li, B., He, Y., Keyshar, K., ... & Vajtai, R. (2015). Two-step growth of two-dimensional $WSe_2/MoSe_2$ heterostructures. *Nano Letters*, 15(9), 6135–6141.
181. Lee, G. H., Yu, Y. J., Cui, X., Petrone, N., Lee, C. H., Choi, M. S., ... & Taniguchi, T. (2013). Flexible and transparent MoS_2 field-effect transistors on hexagonal boron nitride-graphene heterostructures. *ACS Nano*, 7(9), 7931–7936.
182. Island, J. O., Steele, G. A., van der Zant, H. S., & Castellanos-Gomez, A. (2015). Environmental instability of few-layer black phosphorus. *2D Materials*, 2(1), 011002.
183. Tongay, S., Fan, W., Kang, J., Park, J., Koldemir, U., Suh, J., ... & Sinclair, R. (2014). Tuning interlayer coupling in large-area heterostructures with CVD-grown MoS_2 and WS_2 monolayers. *Nano Letters*, 14(6), 3185–3190.
184. Luo, B., Liu, G., & Wang, L. (2016). Recent advances in 2D materials for photocatalysis. *Nanoscale*, 8(13), 6904–6920.
185. Schmidt-Rohr, K., Clauss, J., & Spiess, H. W. (1992). Correlation of structure, mobility, and morphological information in heterogeneous polymer materials by two-dimensional wideline-separation NMR spectroscopy. *Macromolecules*, 25(12), 3273–3277.
186. Berhan, L., Yi, Y. B., Sastry, A. M., Munoz, E., Selvidge, M., & Baughman, R. (2004). Mechanical properties of nanotube sheets: Alterations in joint morphology and achievable moduli in manufacturable materials. *Journal of Applied Physics*, 95(8), 4335–4345.
187. Han, G. H., Gunes, F., Bae, J. J., Kim, E. S., Chae, S. J., Shin, H. J., ... & Lee, Y. H. (2011). Influence of copper morphology in forming nucleation seeds for graphene growth. *Nano Letters*, 11(10), 4144–4148.
188. Li, H., Oppenheimer, S. M., Stupp, S. I., Dunand, D. C., & Brinson, L. C. (2004). Effects of pore morphology and bone ingrowth on mechanical properties of microporous titanium as an orthopaedic implant material. *Materials Transactions*, 45(4), 1124–1131.
189. Evans, A. G., Gulden, M. E., & Rosenblatt, M. (1978). Impact damage in brittle materials in the elastic-plastic response regime. *Proceedings of the Royal Society of London: Series A*, 361(1706), 343–365.
190. Zandiatashbar, A., Lee, G. H., An, S. J., Lee, S., Mathew, N., Terrones, M., ... & Koratkar, N. (2014). Effect of defects on the intrinsic strength and stiffness of graphene. *Nature Communications*, 5, 3186.
191. Kim, Y., Lee, J., Yeom, M. S., Shin, J. W., Kim, H., Cui, Y., ... & Han, S. M. (2013). Strengthening effect of single-atomic-layer graphene in metal–graphene nanolayered composites. *Nature Communications*, 4, 2114.
192. Zhu, T., & Li, J. (2010). Ultra-strength materials. *Progress in Materials Science*, 55(7), 710–757.
193. Xu, P., Neek-Amal, M., Barber, S. D., Schoelz, J. K., Ackerman, M. L., Thibado, P. M., ... & Peeters, F. M. (2014). Unusual ultra-low-frequency fluctuations in freestanding graphene. *Nature Communications*, 5, 3720.
194. Liu, Y., Xie, B., Zhang, Z., Zheng, Q., & Xu, Z. (2012). Mechanical properties of graphene papers. *Journal of the Mechanics and Physics of Solids*, 60(4), 591–605.
195. Ghorbani-Asl, M., Borini, S., Kuc, A., & Heine, T. (2013). Strain-dependent modulation of conductivity in single-layer transition-metal dichalcogenides. *Physical Review B*, 87(23), 235434.
196. Sorkin, V., Pan, H., Shi, H., Quek, S. Y., & Zhang, Y. W. (2014). Nanoscale transition metal dichalcogenides: Structures, properties, and applications. *Critical Reviews in Solid State and Materials Sciences*, 39(5), 319–367.
197. Zheng, Z., Zhang, T., Yao, J., Zhang, Y., Xu, J., & Yang, G. (2016). Flexible, transparent and ultra-broadband photodetector based on large-area WSe_2 film for wearable devices. *Nanotechnology*, 27(22), 225501.
198. Baugher, B. W., Churchill, H. O., Yang, Y., & Jarillo-Herrero, P. (2014). Optoelectronic devices based on electrically tunable p–n diodes in a monolayer dichalcogenide. *Nature Nanotechnology*, 9(4), 262.
199. Chhowalla, M., Liu, Z., & Zhang, H. (2015). Two-dimensional transition metal dichalcogenide (TMD) nanosheets. *Chemical Society Reviews*, 44(9), 2584–2586.
200. Gao, M. R., Xu, Y. F., Jiang, J., & Yu, S. H. (2013). Nanostructured metal chalcogenides: Synthesis, modification, and applications in energy conversion and storage devices. *Chemical Society Reviews*, 42(7), 2986–3017.
201. Kirkpatirck, S. R., Siahmakoun, A., Adams, T. M., & Wang, Z. (2008). U.S. Patent No. 7,444,812. Washington, DC: U.S. Patent and Trademark Office.

202. Polcar, T., & Cavaleiro, A. (2011). Review on self-lubricant transition metal dichalcogenide nanocomposite coatings alloyed with carbon. *Surface and Coatings Technology*, 206(4), 686–695.
203. Polcar, T., Evaristo, M., & Cavaleiro, A. (2009). Comparative study of the tribological behavior of self-lubricating W–S–C and Mo–Se–C sputtered coatings. *Wear*, 266(3–4), 388–392.
204. Singer, I. L. (1988). Solid lubricating films for extreme environments. *MRS Online Proceedings Library Archive*, 140, 215–226.
205. Liu, K., Yan, Q., Chen, M., Fan, W., Sun, Y., Suh, J., ... & Ji, J. (2014). Elastic properties of chemical-vapor-deposited monolayer MoS_2, WS_2, and their bilayer heterostructures. *Nano Letters*, 14(9), 5097–5103.
206. Wan, K. T., Guo, S., & Dillard, D. A. (2003). A theoretical and numerical study of a thin clamped circular film under an external load in the presence of a tensile residual stress. *Thin Solid Films*, 425(1), 150–162.

7

Two-Dimensional Gallium Nitride

Somayeh Behzad
Kermanshah University of Technology

7.1 Synthesis of Two-Dimensional (2D) GaN

As a key semiconductor with a wide bandgap of 3.4 eV and technologically important applications in light-emitting diode (LED), laser, solar cells, and ultraviolet detector [1–5], GaN nanosheets with few layers attract specific attention for their peculiar optoelectronic properties predicted in theory, including the thickness-dependent energy bandgap and the p-type conductivity, and the diverse functionality in the construction of optoelectronic nanodevices [6–9].

In 2003, single-crystalline GaN nanoribbons were synthesized by nitriding Ga_2O_3 thin films deposited on sapphire (0001) substrates by radiofrequency magnetron sputtering [10]. Neither metal catalysts nor templates were used in this process. A detailed characterization of the synthesized ribbon-like one-dimensional nanostructures revealed that they are high-quality single-crystalline hexagonal wurtzite GaN. The thickness and width-to-thickness ratio of the grown GaN nanoribbons are in the range of 8–15 nm and ~5–10, respectively.

Crystalline gallium nitride nanoribbon (GaNNR) rings have been fabricated through a simple catalytic reaction using metallic Ga as a source material with ammonia using Ni as a catalyst on silicon substrate [11]. The X-ray powder diffraction and selected-area electron diffraction results revealed that the products are pure, crystalline GaN with hexagonal wurtzite structure. The widths of the nanoribbon rings were in the range of 150–350 nm, and the diameters were 3–8 mm. The formation of the unique structure is predominantly controlled by growth dynamics. The axial and radial growths exist simultaneously. The fast-growth direction is parallel to [1 0 0] direction, and the internal stress plays a key role in the formation process of rings.

Various traditional GaN crystal growth methods such as hydride vapor-phase epitaxy [12], metal-organic chemical vapor deposition [13], molecular beam epitaxy, and electrochemical deposition [14] have been used to synthesize GaN films. However, they are not applicable in the direct synthesis of GaN nanosheets, and cleavage of tetrahedral-coordinated GaN bulk crystal along the (0001) plane is also impossible owing to unsaturated dangling bonds on the surface [15]. If one could passivate such dangling bonds, then the synthesis of 2D forms of GaN could in fact be possible.

Recently, the challenge in preparing GaN layers has been overcome by the research groups led by Redwing and Robinson et al. [8], using a migration-enhanced encapsulated growth (MEEG) method, in which the GaN monolayer is created beneath the graphene layer, which is used as a capping or stabilizing layer during the synthesis [16].

The unique aspect of MEEG is that the GaN growth occurs in between the growth substrate (silicon carbide, SiC) and a graphene capping bilayer (Figure 7.1). This graphene capping sheet is formed by sublimation of Si from the SiC substrate followed by hydrogenation. Trimethylgallium is decomposed on the graphene/SiC surface, causing gallium atoms to penetrate through the graphene into the interstitial space between the SiC substrate and the graphene bilayer as illustrated schematically in Figure 7.1a. The researchers show that defects (such as carbon vacancies), wrinkles (tears), and gallium metal islands act as the favored sites for gallium to penetrate through the graphene. Once gallium atoms intercalate into the graphene/SiC interface, they appear to be stabilized as a bilayer of gallium, probably because of the highly passivated nature of the hydrogenated SiC/graphene interface. In the final growth steps, atomic nitrogen (created from the decomposition of ammonia) also penetrates through the defective graphene capping sheet

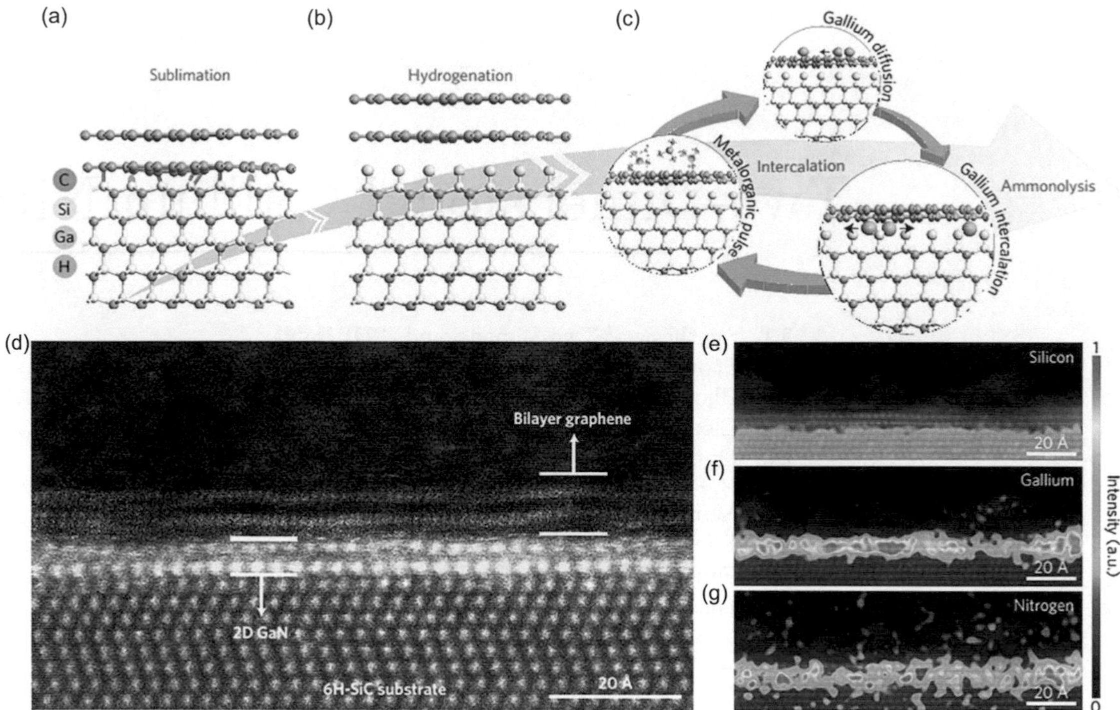

FIGURE 7.1 2D GaN formation via MEEG. (a–c) Schematic of the MEEG process that leads to the formation of 2D GaN (d). (a) The process of silicon sublimation from SiC(0001) to grow epitaxial graphene that consists of an initial partially bounded graphene-buffer layer (bottom) followed by a monolayer of graphene (top). The gray halos at the SiC/graphene interface represent Si dandling bonds. (b) Exposing the epitaxial graphene (a) to ultrahigh-purity hydrogen at elevated temperatures decouples the initial (bottom) graphene-buffer layer to form bilayer quasi-free standing epitaxial graphene. (c) The proposed MEEG process for the formation of 2D GaN: first, trimethylgallium precursor decomposition and gallium adatom surface diffusion; second, intercalation and lateral interface diffusion; finally, transformation of gallium to 2D GaN via ammonolysis. (d) High-angle annular dark-field scanning transmission electron microscopy (STEM) cross-section of 2D GaN, consisting of two sub-layers of gallium, between bilayer graphene and SiC(0001). (e–g) Elemental energy-dispersive X-ray mapping of silicon (e), gallium (f), and nitrogen (g) in 2D GaN [8].

and reacts with the intercalated gallium to form a patchwork of 2D GaN islands.

However, the complicated synthetic process and the inhomogeneity of GaN layers in size and layer thickness further hinder the property investigation and practical applications in optoelectronic nanodevices. Additionally, the removal of graphene layer on the top of GaN layer and the peeling off of GaN layers from SiC substrate still remain a key challenge.

An accessible template approach toward the rational synthesis of GaN nanosheets through the nitridation of metastable γ-Ga_2O_3 nanosheets synthesized from a hydrothermal reaction was reported [17]. The nitridation of γ-Ga_2O_3 nanosheets in NH3 gas leads to the direct structural transition from the cubic γ-Ga_2O_3 phase to the hexagonal GaN (h-GaN) phase with the sheet-like morphology partially collapsed, but the layer framework is well maintained. *Ultraviolet/visible* spectrum measurement reveals that the GaN nanosheets show a bandgap of 3.30 eV with strong visible absorption in the range of 370–500 nm.

More recently, a facile synthesis approach is reported to prepare GaN nanosheets using graphene oxide (GaO) as sacrificial template [18]. The approach constructs GaN nanosheets through only several facile solution-based steps. As shown schematically in Figure 7.2, ammonia ($NH_3 \cdot H_2O$) is used as the OH^- supplier; the released OH^- reacts with Ga^{3+}. GaOOH is anchored onto the GaO sheets via functional groups by a homogeneous precipitation method. GaN nanosheets are achieved by calcining in NH_3 and air atmosphere sequentially. Thermogravimetric analysis also reveals that GO disappeared during thermal annealing.

7.2 Structure, Energetics, and Mechanical Properties

In the GaN monolayer with a planar hexagonal honeycomb structure, Ga-sp^2 and N-sp^2 hybrid orbitals form σ-bonds along Ga–N bonds arranged as a hexagon. The structure has D_{3h} symmetry. Due to the planar sp^2 bonding, the bond angle between Ga–N bonds is 120°. In addition to three sp^2 hybrid orbitals of each constituent, Ga and N, their p_z orbitals are perpendicular to the plane of GaN monolayer. Due to the electronic charge transfer from Ga to N, the Ga–N bonds have covalent–ionic characteristic.

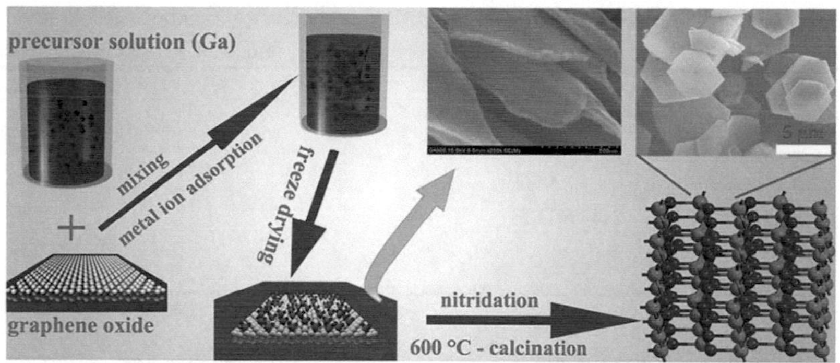

FIGURE 7.2 Schematic illustration for the synthesis and structure of GaN nanosheets [18].

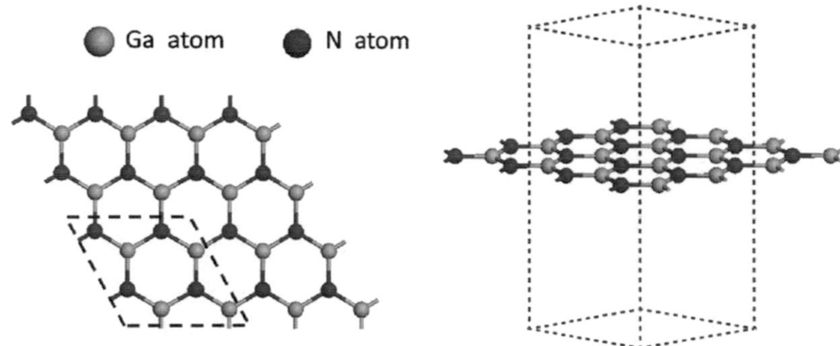

FIGURE 7.3 Top and side views of the optimized GaN monolayer. The rhombus plotted in dashed line represents the primitive cell [19].

The optimized atomic structure together with the primitive cell is shown in Figure 7.3.

Charge–density contour plot of Ga–N bond in the horizontal plane (in the atomic plane of GaN monolayer) is shown in Figure 7.4. Bader analysis yielding a charge transfer of 1.5 electrons from Ga to N indicates a significant ionic contribution in the binding [20].

Lattice constants $a = b$, bond length d, cohesive energy E_C, in-plane stiffness C, charge transfer, Born effective charge values, and fundamental bandgaps calculated by different methods are presented in Table 7.1.

It is noted that Ga–N bonds in 3D wz(zb)-GaN, which is constructed from tetrahedrally coordinated sp^3 hybrid orbitals, are 0.12 Å longer than the Ga–N bonds of GaN monolayer constructed from planar sp^2 hybrid orbitals $+p_z$ orbitals. This indicates that Ga–N bonds in GaN monolayer are stronger than those in wz(zb)-GaN [20].

The multilayer GaN bonds prefer a planar configuration rather than a buckled bulk-like configuration. The stacking sequence AA′AA′ ... (i.e., hexagons on the top of each other, with the Ga atom being above N) is energetically the most favorable configuration for multilayer GaN [20,24].

Recently, 2D GaN with Haeckelite structure containing square and octagonal rings has been reported (see Figure 7.5) [27].

The optimized structure of GaN Haeckelite monolayer is a planar structure with a 2D square lattice, and its binding energy is −3.376 eV per atom. Two different interatomic

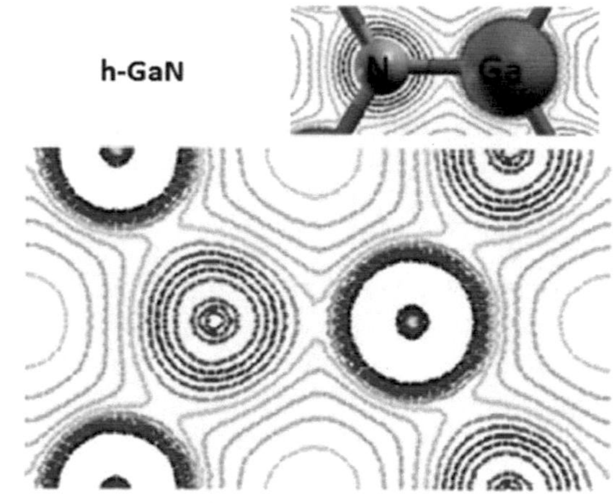

FIGURE 7.4 The charge–density plot of GaN monolayer [21].

distances were found: $d_{\mathrm{Ga-N}} = 1.85$ Å (bond belonging to octagons) and $d_{\mathrm{Ga-N}} = 1.95$ Å (bond belonging to squares). Beyond three layers, it was found that after structure relaxation, the Haeckelite structure is no longer a planar system.

The monolayer hexagonal GaNNRs consist of alternating N and Ga atoms with each inner N (Ga) atom having three Ga (N) atoms as its nearest neighbors. However, for each N (Ga) atom at two edges, there are only two Ga (N) atoms as its nearest neighbors, so there must be a dangling bond attached to each edged N (Ga) atom [28]. Li et al [29] demonstrated that the edge N atoms tend to relax closer

TABLE 7.1 Optimized Lattice Constant a; Ga–N Bond Length d, Cohesive Energy E_C per Ga–N Pair; In-plane Stiffness C, Poisson's Ratio ν, Charge Transfer Q_b^* from Ga to N, Born Effective Charge Z^*, and Indirect Bandgap E_{G-i} of GaN Monolayer [20]

	a (Å)	d (Å)	E_C (eV/GaN)	C (N/m)	ν (%)	Q_b^* (e)	Z^* (e)	E_{G-i} (eV)
GGA + d	3.21	1.85	8.04	109.8	0.43	1.50	3.08	2.16
HSE06($\alpha = 0.25/0.35$)								3.42/3.98
G_0W_0								4.55
LDA [22]	3.20	1.85	12.47	110	0.48	1.70		2.27 (GW_0)
LDA [23]	3.21	1.85		109.4	0.43			
LDA [24]		1.85	8.38					2.17
GGA [19]		1.87	8.06					1.87 (GW_0: 4.14)
G_0W_0 [25]	3.17							4.27 (LDA: 2.36)
PBE/HSE/G_0W_0 [26]	3.25					1.34	3.23	3.23 (HSE06)/4.00 (G_0W_0)

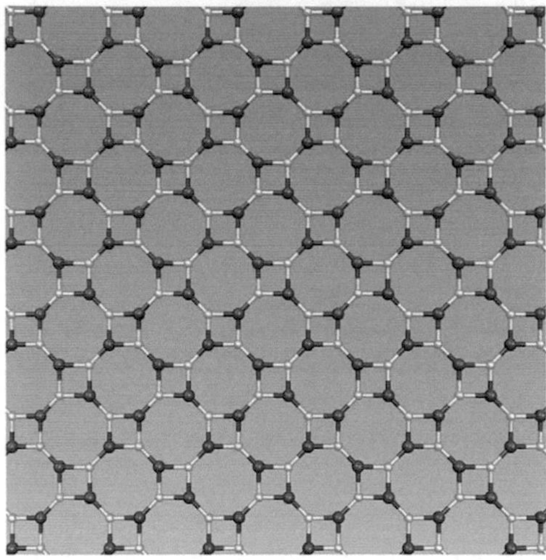

FIGURE 7.5 Two-dimensional Haeckelite eight to four structures containing square and octagonal rings [27].

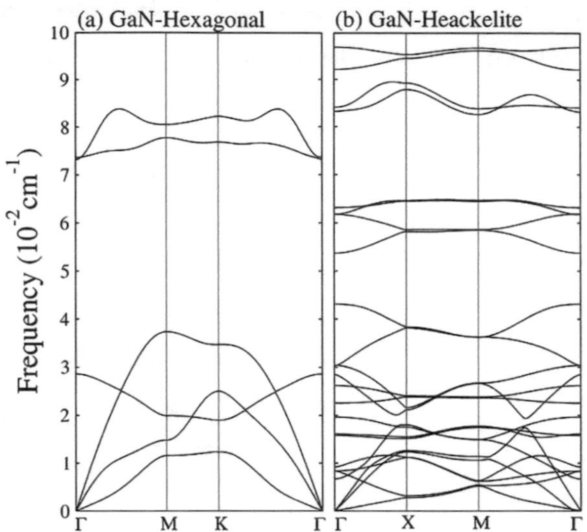

FIGURE 7.6 Vibration frequencies of phonon modes of GaN monolayers: (a) hexagonal and (b) Haeckelite. For both monolayers when K goes to zero, the acoustic modes LA and TA are linear.

to the ribbons at a Ga–N bond length of 1.76 Å. However, the edge Ga atoms relax significantly less than N atoms such that the Ga–N bond length changed to about 1.83 Å. Atoms in the inner part relax less as well, and the Ga–N bond length was retained at ~1.82 Å.

7.3 Phonon Spectra and Stability

It is well known that if the vibration frequency of specific modes, ($\boldsymbol{k}$), is imaginary, the corresponding displacements would result in an instability, since displacement cannot be restored. The calculated frequencies of phonon modes for the Haeckelite monolayer and honeycomb structures are shown in Figure 7.6. The hexagonal monolayer with two atoms in the unit cell contains six modes: one longitudinal acoustic (LA), one transversal acoustic (TA), one out-of-plane acoustic (ZA), and three optical modes. For the Haeckelite monolayer with 8 atoms in the primitive cell, 24 branches (one LA, one TA, one ZA, and 21 optical modes) were found; furthermore, when $k \to 0$, the behavior of LA and TA is linear, while that of ZA is quadratic. For both monolayers, it was found that all calculated frequencies in the Brillouin zone are positive, thus confirming the stability of the Haeckelite monolayer.

Although the calculated frequencies of the phonon modes are all positive, instabilities can be induced through thermal excitations. This situation occurs when the local minimum of a given phase is shallow and the structure dissociates at low temperatures. To show that GaN monolayer can survive at high temperatures and is suitable for technological applications above the room temperature, *ab initio* finite-temperature calculations have been made in the temperature range from 0 to 1,000 K [20]. The honeycomb structure did not dissociate even after 3 ps simulation at 1,000 K. This indicates that GaN monolayer is rather stable in a deep minimum on the Born-Oppenheimer (BO) surface, and hence, devices fabricated from GaN monolayer can sustain operations above room temperature.

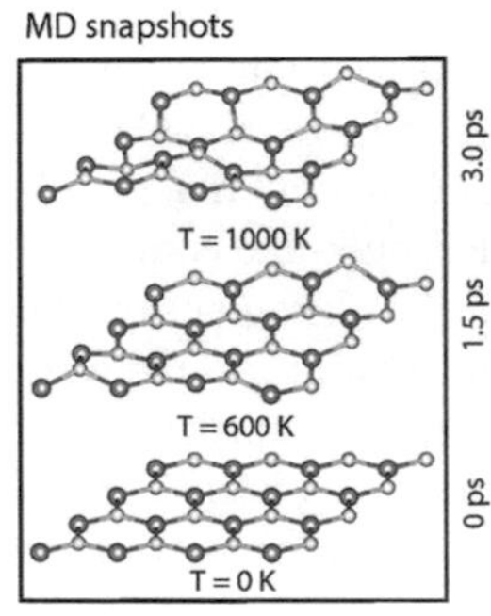

FIGURE 7.7 Snapshots of the atomic configurations in MD simulations at 0, 600, and 1,000 K, in which honeycomb-like structures are maintained [20].

In Figure 7.7, snapshots of the atomic configurations obtained from molecular-dynamics (MD) simulations at different temperatures are presented.

7.4 Electronic Structure and Magnetic Properties

Since an antibonding π^*-bond is separated from a π-bond by a significant energy, the π- and π^*-bands derived from these bonds open a significant bandgap. Accordingly, GaN monolayer is a nonmagnetic, wide-bandgap semiconductor. In Figure 7.8, the electronic energy band structure of GaN monolayer in the symmetry directions of the hexagonal Brillouin zone, as well as the corresponding total density of states (TDOS) and orbital-projected density of states (PDOS), is shown.

The conduction band minimum (CBM) is located at Γ-point, whereas the valence band maximum (VBM) is located at K-point [30]. Accordingly, the energy bands calculated by Perdew–Burke–Ernzerhof (PBE) mark an indirect bandgap of 2.16 eV from the K- to the Γ-point. This is a dramatic deviation from the bulk 3D wz(zb)-GaN, which has a PBE direct bandgap of 1.71 eV at Γ-point. Apparently, the fundamental bandgap increases by 0.45 eV as one goes from three dimensions to two dimensions of monolayer. The difference in the nature of the bandgap of GaN sheet and bulk wurtzite GaN can be attributed to a shift in the position of the VBM from Γ to K when the dimensionality is reduced, whereas the characteristics of the CBM remains unchanged. A higher value of the bandgap for the monolayer relative to that of the bulk wurtzite GaN can be attributed to the quantum confinement effect [31].

The band near the VBM is dominated by the N-p_z orbitals, while the lowest conduction band is dominated by the Ga-p_z orbitals. The overlap between the Ga's and N's p_z orbitals is weak, which originates from the larger difference in the atomic radii and in electronegativities. The corrected bandgaps by using HSE06 and G_0W_0 methods are shown in Figure 7.9. The indirect PBE bandgap increased to 3.42 eV after the HSE06 correction. This corresponds to a correction of 1.26 eV. On the other hand, the correction induced by the G_0W_0 method is larger than that of the HSE06 method by nearly 1 eV, revealing a bandgap of 4.55 eV. Spin–orbit coupling (SOC) at the top of the valence band at the Γ-point leads to the splitting of the degenerate bands by only 11 meV.

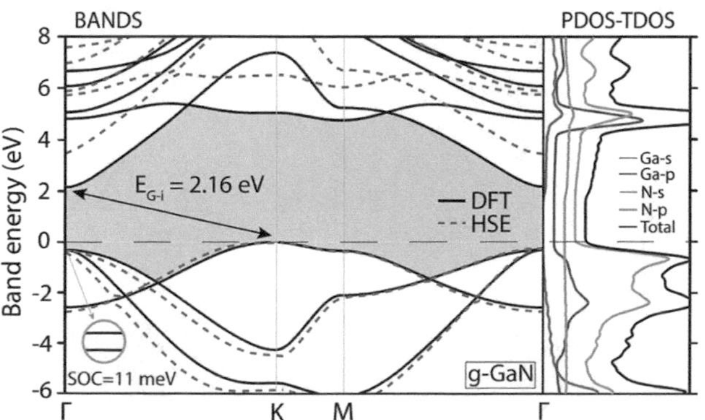

FIGURE 7.8 The electronic band structure of GaN monolayer is presented along the symmetry directions of the Brillouin zone. Zero of energy is set to the top of the valence band. The splitting of the degenerate bands at the top of the valence band at the Γ-point due to SOC is shown by the inset. PBE bands corrected by the HSE06 method are shown by the dashed lines [20].

The response of the conduction and valence bands to the applied strain ε, and the resulting changes in the fundamental gap, is of interest for the fabrication of devices operating under strain. Within PBE calculations, the bandgap of GaN monolayer was found to decrease monotonically from 2.16 to 0.21 eV, going from $\varepsilon = 0\%$ to 10%. Furthermore, the gap seemed to close and lead to a metallic band structure when biaxial tensile strain was further increased, up to 16%. The shifts in the conduction and valence bands under strain and the variation in the fundamental bandgap are shown in Figure 7.7ab, respectively. This is an important result predicting dramatic changes in the electronic structure with applied strain, once $\varepsilon_x = \varepsilon_y$ 10% is affordable in the GaN monolayer system.

As shown in Figure 7.10, applying the compressive strain by shortening the GaN bond length from its equilibrium value would strengthen the repulsive force between the neighboring Ga and N atoms and increase the overlap of neighboring Ga and N orbitals, causes an increase in the bandgap. Also, the monolayer GaN transitions from an indirect to a direct bandgap semiconductor when the bond length decreases to 1.78 A as VBM moves from K to Γ. The indirect-to-direct bandgap transition in monolayer GaN is an important result, which will guide future research in the field of LEDs.

The electronic structure of multilayer GaN is significantly different from that of the monolayer, and it is sensitive to the number of layers [32]. As illustrated in Figure 7.11, by increasing the number of layers, the bandgap of GaN layer decreases and the CBM at the Γ-point shifts downward, while the VBMs along the M-Γ and K-Γ directions move toward Γ. When the number of layers increased to ten, the

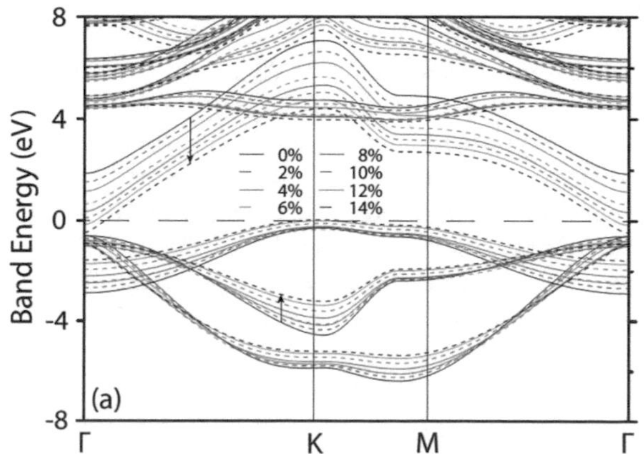

FIGURE 7.9 Variation in the energy bands of g-GaN near the fundamental bandgap under applied biaxial strain [20].

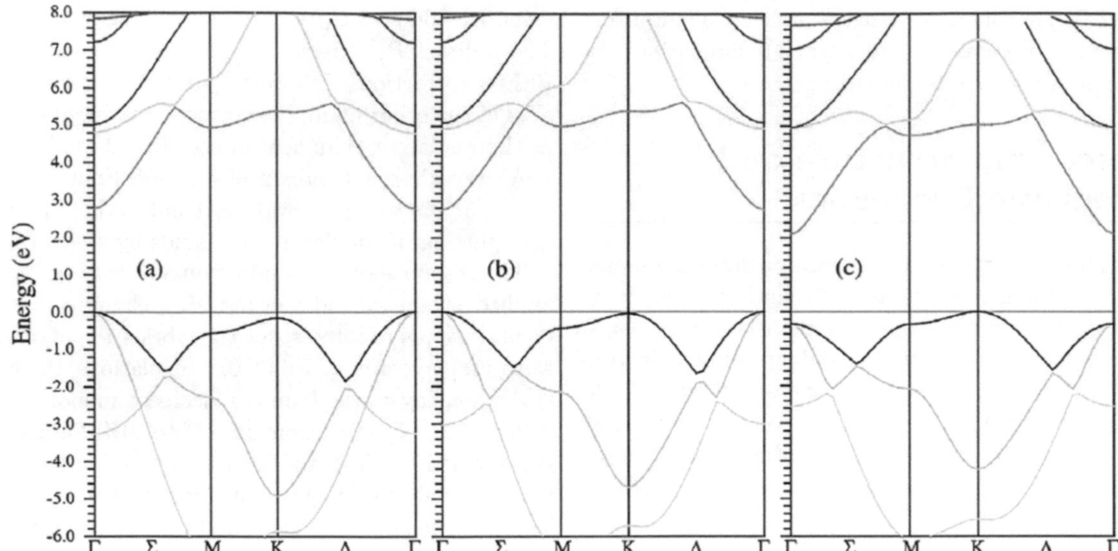

FIGURE 7.10 Calculated band structure of GaN monolayer under compressive strain for bond length (Å) of (a) 1.7, (b) 1.78, and (c) 1.86 (strain-free) [32].

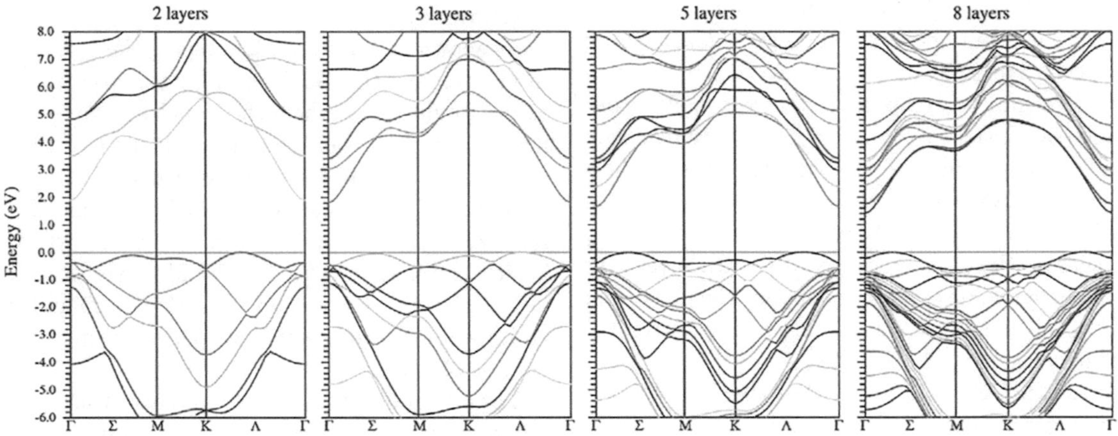

FIGURE 7.11 The electronic structures of GaN multilayers [32].

bandgap becomes direct. The band structures of the multilayer GaN sheets also show the characteristic of the splitting of bands due to the interlayer coupling.

The perpendicular strain is applied on bilayer GaN by tuning the interlayer distance from 5.00 to 2.20 Å [32]. For the interlayer distance of 5 Å, the interlayer interaction is negligible and the band structure of bilayer GaN is similar to that of the monolayer. By decreasing the interlayer distance, the interaction between the layers shifts the VBM toward. The position of the CBM in the Brillouin zone does not change with decreasing the interlayer distance and only moves toward higher energies due to the increase in interaction strength between the layers (see Figure 7.12). Thus, by applying the perpendicular strain (decreasing the interlayer distance from 5 to 2.2 Å), the indirect characteristic is conserved and the bandgap increases from 1.65 to 2.03 eV.

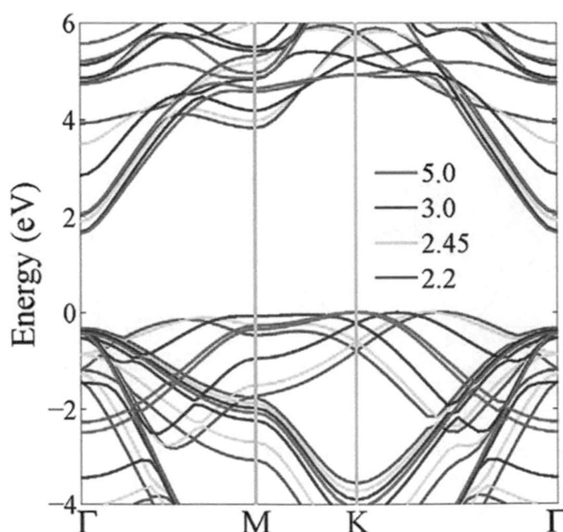

FIGURE 7.12 Calculated band structure of bilayer GaN under perpendicular strain [32].

The band-structure plot for the Haeckelite structure exhibits an indirect bandgap of 1.6 eV; here, the CBM is located at gamma point, whereas the VBM is set in two different points (see Figure 7.13) [27].

Pure GaNNR is a semiconductor. The zigzag GaNNR has an indirect bandgap, whereas the armchair GaNNR has a direct bandgap (see Figure 7.14) [33]. In addition, with increasing ribbon width and thus decreasing interaction between the two edges, both bandgaps decrease monotonically and become closer to each other approaching their asymptotic limit of GaN monolayer.

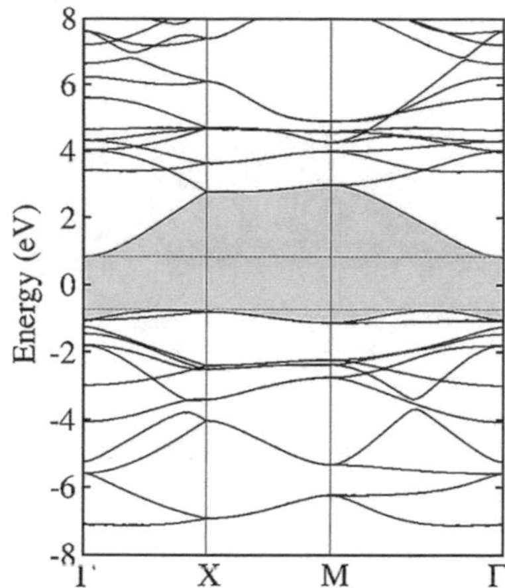

FIGURE 7.13 Band-structure plot of Haeckelite GaN monolayer [27].

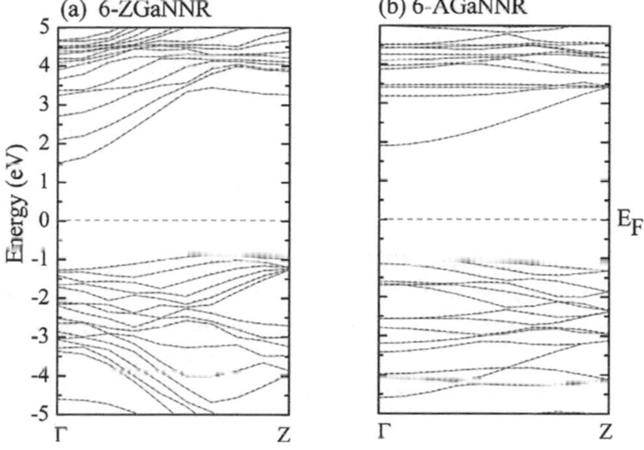

FIGURE 7.14 The band structures for (a) 6-ZGaNNR and (b) 6-AGaNNR. The solid and dashed lines represent the majority and minority spin bands, respectively [33].

First-principles calculations of magnetic properties of zigzag and armchair GaNNRs show that the armchair GaNNRs can be classified as nonmagnetic semiconductors. A ferromagnetic character occurs in bare zigzag GaNNRs with a width of about 1.7 nm. The spin density distribution shows that the unpaired spin mainly concentrates at Ga and N atoms at the edge, while inner Ga and N atoms contribute with a small amount of unpaired spins. When the width of GaNNRs is small, such as when $W = 7$, the magnetic moment of inner atoms allows a weak ferromagnetic coupling. By increasing the width of GaNNRs, it is expected that the interaction from magnetic moments at both edges becomes weaker, and the total magnetic moment is mainly determined by the edge effect [29].

7.5 Optical Properties

Figure 7.15 shows the real and imaginary parts of dielectric function for 2D GaN monolayer, calculated using PBE approximation [32]. Some of the features of spectrum can be summarized as follows: (i) The onset of the optical absorption in the imaginary part of dielectric function spectrum $\varepsilon_2(\omega)$ is in compliance with its band calculated within PBE. (ii) The sharp peak in $\varepsilon_2(\omega)$ of GaN monolayer occurs due to the optical transitions from flat occupied π-bands between K- and M-points to flat empty π^* bands ~5 eV above. (iii) The value of static dielectric function for parallel polarization is 2.37.

It is found that the tensile and compressive strains cause the red and blue shifts in the optical spectra, respectively, for both monolayer and bilayer GaN.

2D GaNNs made of about two to three monolayers using Na-4 mica nanochannels as template show pronounced surface optical (SO) phonon modes because of surface modulation or roughness [31]. A strong electron–SO phonon coupling for wave vectors smaller than the inverse of the

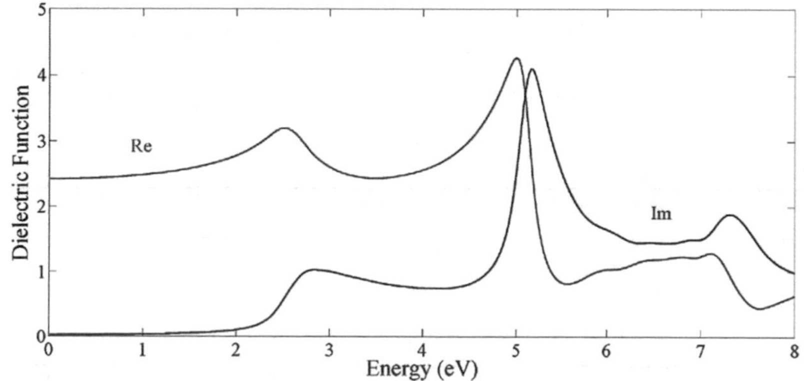

FIGURE 7.15 The real and imaginary parts of dielectric function for strain-free monolayer GaN [32].

nanoribbon cross-section is suggested to be responsible for the enhanced intensity of SO modes.

7.6 Heterostructures Consisting of GaN and Other Two-Dimensional Materials

van der Waals (vdW) heterostructures consisting of GaN sheet and other two-dimensional materials have been studied using density functional theory.

All the vdW heterostructures consisting of graphene and GaN sheet (graphene/GaN, graphene/graphene/GaN, and GaN/graphene/graphene/GaN) have small bandgaps at or near the Γ-point of the Brillouin zone [34]. The tunability of the bandgap is sensitive to the stacking sequence in bilayer graphene-based heterostructures. In particular, in the case of graphene/graphene/GaN, a bandgap of up to 334 meV is obtained under a perpendicular electric field. The bandgap of bilayer graphene between GaN sheets (GaN/graphene/graphene/GaN) shows a similar tunability and increases to 217 meV with the perpendicular electric field reaching 0.8 V/Å. Table 7.2 presents the stacking order, interlayer spacing, and binding energy for the considered structures.

The geometric heterostructures of h-GaN with graphene and hexagonal BN (h-BN) are shown in Figure 7.16. For h-GaN/graphene and h-GaN/h-BN heterostructures, the optimum interlayer spacing of both systems is 3.4 Å, irrespective of the substrate species [35]. Owing to the weak van der Waals interaction, the electronic structures of h-GaN/graphene and h-GaN/h-BN heterostructures may be obtained by superposing the electronic structure of each constituent monolayer. The h-GaN/graphene retains the characteristic electronic structure described by the electronic structure of each constituent unit near the Fermi level: The VBM and CBM states of graphene converge at the K-point with a linear dispersion, and h-GaN has an energy gap between the K- and Γ-points as in the case of an isolated sheet. The h-GaN/h-BN is a semiconductor with an indirect bandgap of 2.11 eV between the K- and Γ-points for the VBM and CBM states, respectively, which is slightly smaller than the value of 2.28 eV for monolayer h-GaN.

The blue phosphorene (BlueP)/GaN monolayer vdW heterostructure can be created by stacking BlueP and GaN monolayers together [36]. The most stable structure of the

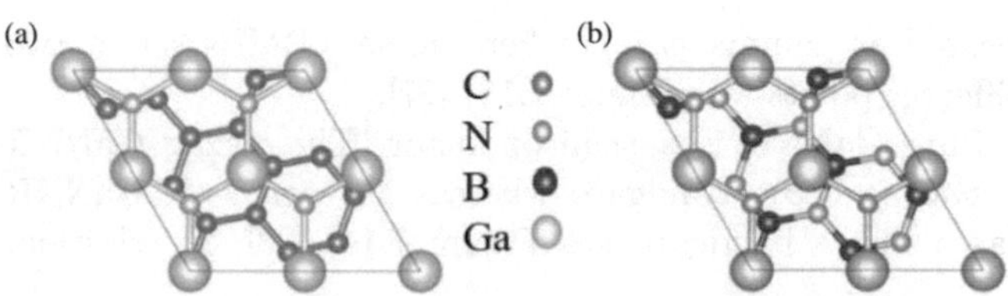

FIGURE 7.16 Geometric structures of (a) h-GaN/graphene and (b) h-GaN/h-BN [35].

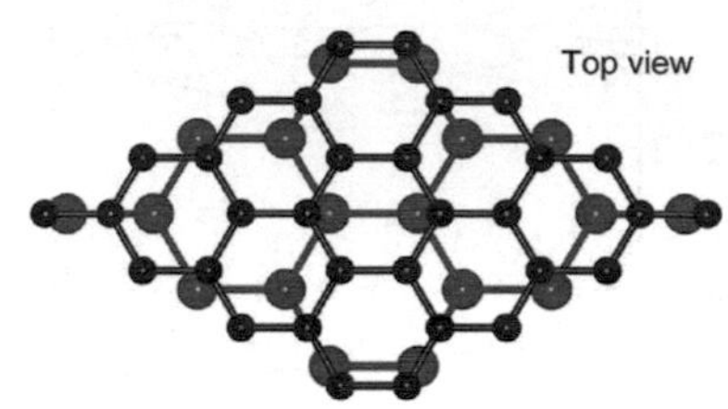

FIGURE 7.17 Schematic illustration of the crystal structure of BlueP/G vdW heterostructures.

BlueP/g-GaN bilayer (see Figure 7.17) has an A–A stacking pattern, in which the interlayer P–Ga distance is 3.03 Å.

The BlueP/GaN vdW heterostructure is an indirect semiconductor with the CBM located at the M-point, while the VBM is located at the K-point. Moreover, the CBM mainly originates from the BlueP layer, while the VBM mainly originates from the GaN layer. Consequently, the GaN layer significantly modifies the electronic properties of the freestanding BlueP. The bandgap is only 1.05 eV, which is much lower than that in both the BlueP and GaN monolayers (see Figure 7.18). When BlueP is stacked on a GaN monolayer, its bandgap is significantly reduced. When the BlueP monolayer is stacked with a GaN monolayer, they form a type-II band alignment heterostructure, which is promising for unipolar electronic device applications. Furthermore, a large redistribution of the electrostatic potential across the interface occurs, which may drastically affect the carrier dynamics and induce a low charge injection barrier in BlueP.

The pristine monolayer $Ni(OH)_2$ owns no macromagnetism with antiferromagnetic (AFM) coupling between two nearest Ni atoms [37]. The electronic structure can be modulated through the heterostructure. It is found that the $Ni(OH)_2$/GaN heterostructure retains the AFM coupling. More interestingly, the complete electron–hole separation is found in the $Ni(OH)_2$/GaN heterostructure.

TABLE 7.2 Structural Properties of vdW Heterostructures

System	Stacking Order	Interlayer Spacing d_1 d_2 (Å)	Binding Energy/ Supercell (eV)	Bandgap without Electric Field (meV)	Maximum Bandgap under Electric Field (meV (V/Å))
Graphene/GaN		3.28	−3.11	3.5	3.9(1.0)
G/GaN/G		3.27	−6.43	8.5	
G/G/GaN	AA	3.28 3.46	−6.22	15	
G/G/GaN	AB	3.29 3.20	−6.86	47	334(1.0) 125(0.8)
GaN/G/G/GaN	AA	3.28 3.44	−9.67	7.9	
GaN/G/G/CaN	AB	3.28 3.20	−10.31	43	217(0.7)

The bandgap without an electric field and the maximum bandgap under an external electric field are also given; the corresponding strength of external electric field for generating the maximum bandgap is written in brackets at the side in units of V/Å [34].

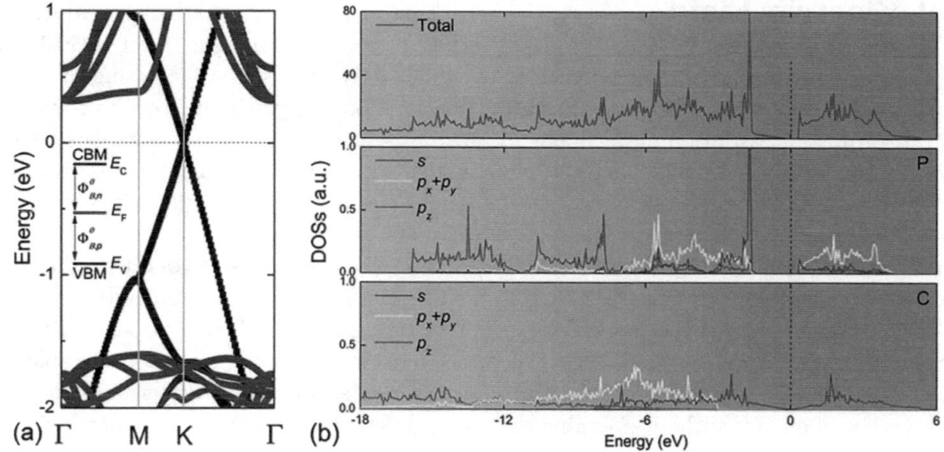

FIGURE 7.18 (a) Projected band structure of the BlueP/G vdW heterostructure. The blue and black symbols represent BlueP and G, respectively. (b) TDOS and PDOS of the BlueP/G vdW heterostructure. The Fermi level was set to zero and is indicated by the black dashed line [36].

The stable, in-plane composite materials (GaN)p/(AlN)q, or simply (p/q), can be constructed of periodically repeating stripes of GaN and AlN continuously (or commensurately) joined along their zigzag (Z) edges [38]. These composite materials exhibit a wide range of physical properties depending on the values of p and q: A structure having p = 1–2, but very large q, or vice versa, is identified as a δ doping, where very narrow ribbon of GaN is implemented commensurately in a large AlN stripe, or vice versa (see Figure 7.19). For p = 1–2 and q – 1–2, the composite structure behaves differently from its parent constituents as if a compound of lines of GaN and AlN. For small p and q, the fundamental bandgap of the composite can be tuned by varying p and q. On the other hand, for large p and q, the electronic states become confined to one of the constituent stripes, experiencing a transition from 2D to one-dimensional (1D). At the boundary region of the junction, charge is transferred from one region to another and normally sets the band lineup, constructing multiple quantum wells and quasi-one-dimensional (quasi-1D) quantum structures in a 2D atomically thin nanomaterial with diverse functionalities.

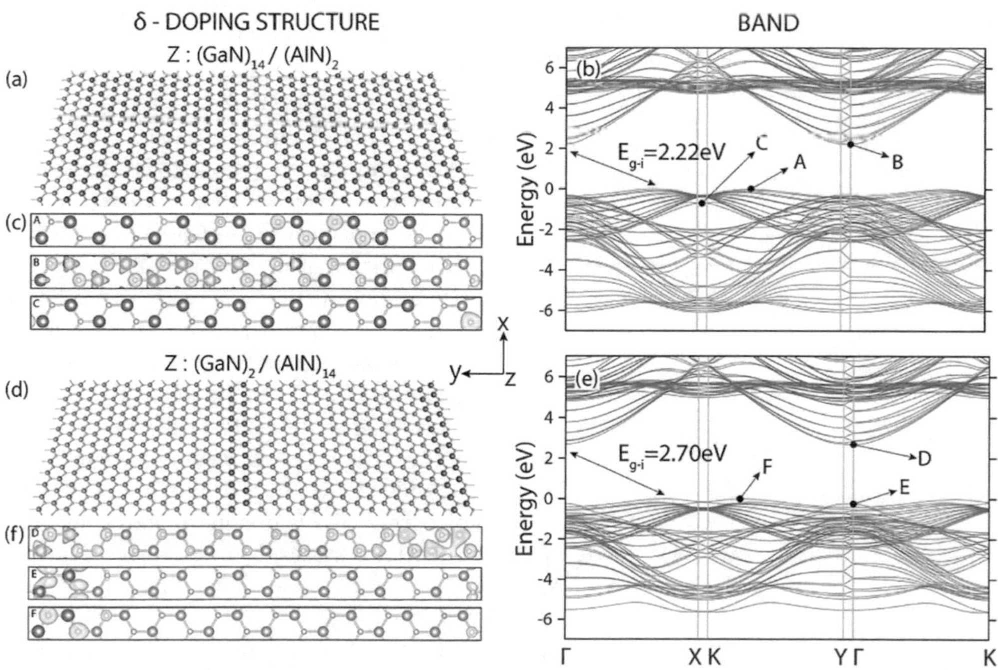

FIGURE 7.19 δ doping: (a) A perspective view of the composite structure Z:(14/2), where wide GaN stripes are δ doped by very narrow stripes of AlN. (b) Electronic energy bands with the indirect bandgap shown by an arrow. Zero of energy is set at the top of the valence band. (c) Isosurfaces of the charge density of the selected band states indicated in (b). (d)–(f) Same for Z:(2/14) [38].

7.7 Effect of Electric Field

The effects of an electric field applied perpendicular to the nitride nanomembranes on the electronic properties are shown in Figure 7.20. A prominent variation in the bandgap is predicted for AlN as compared to that for BN or GaN [21]. A higher degree of electronic polarization in the electric field is facilitated by the ionic bond, which in turn leads to a significant variation in the bandgap of AlN under the application of the external electric field. It is to be noted that the maximum value of 2 V/Å considered for the perpendicular electric field appears to be high.

Application of the perpendicular electric field only shifts the states relative to the Fermi energy without changing the nature of the bands (i.e., the Stark effect).

The application of an external perpendicular electric field induces distinct stacking-dependent features in the electronic properties of GaN multilayers: The bandgap of a monolayer does not change significantly for the small electric field, whereas that of a trilayer is significantly reduced [24]. A response to an external electric field applied perpendicular to the layer is the largest for the ABA (NGaN) trilayer. Furthermore, the calculated results predict tuning of the bandgap from 1.6 to 0.33 eV with an electric field varying from 0 to ±0.4 V/Å.

7.8 Functionalization of GaN

The functionalization of 2D GaN nanostructures has multi-effects on their electronic and magnetic properties.

When the GaN monolayer is modified by H and F atoms, the gap converts into a direct one and is enlarged by 0.81 eV. Furthermore, the gap of GaN monolayer modified with H and F atoms on both sides can be efficiently manipulated in a range of 1.8–3.5 eV by applying an external electric field [19].

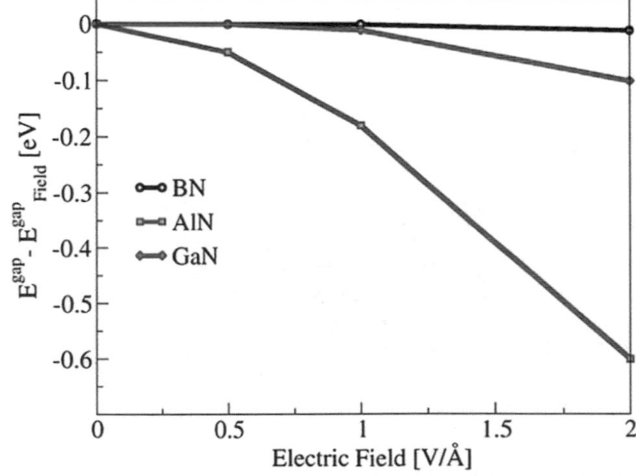

FIGURE 7.20 The variation in bandgap with the applied electric field for the BN, AlN, and GaN nanomembranes [21].

First-principles calculations reveal that the energy difference between ferromagnetic and AFM couplings increases significantly with strain increasing for half-fluorinated GaN sheet. More surprisingly, the half-fluorinated GaN sheet exhibits intriguing magnetic transitions between ferromagnetism and antiferromagnetism by applying strain, even giving rise to half-metal when the sheets are under compression of 6% [39].

Also, the metallic/semiconducting nature can be obtained in zigzag GaNNRs via controlled edge fluorination [40]. Interestingly, F-passivated zigzag GaNNRs are found to be the most stable, making them preferable for practical applications. The proposed two-terminal device exhibits a negative differential resistance behavior/switching characteristic with sufficiently large peak-to-valley current ratio/threshold voltage on the order of $10^2 - 10^{14}$/2.8–3.6 V, which is a function of selective edge passivation. Figures 7.21–7.23 show the device structure, I–V characteristics, and transmission spectrum of bare and F-passivated (F-GaN and GaN-F) zigzag GaNNR. In bare/F-GaN/GaN-F geometry-based devices, current initially increases linearly to its maximum value at a peak voltage of 0.3 V (−0.3 V) in positive (negative) bias.

In the 2D infinitely thick GaN nanoribbons, the surface state is excluded, and thus, only edge states and quantum size effect are important. Such nanoribbons are all nonmagnetic regardless of the edge shapes (armchair or zigzag), and their direct bandgaps increase with increasing ribbon width (or decreasing ratio of edge state). The edge hydrogenation partially or completely removes the edge states, and thus, the bandgap of the hydrogenated GaN nanoribbons decreases with increasing width. The armchair-edged GaNNRs, bare or hydrogenated, have wider bandgaps than their zigzagged analogues [41].

The buckled 2D GaN is accessed by hydrofluorination (FGaNH) and hydrogenation (HGaNH) [42]. It was predicted that the anisotropic carrier mobilities of buckled 2D GaN can exceed those of 2D MoS_2 and can be altered by an alterable surface chemical bond (converting from Ga-F–Ga bond of FGaNH to Ga–H bond of HGaNH).

Investigation of the half-metallic properties of semichlorinated gallium nitride (Cl—GaN) nanosheets under an electric field F shows that the electric field can modulate Cl–GaN nanosheets efficiently from ferromagnetic metals to half-metals [43]. More interestingly, under a broad range of electric field intensity (−0.10 to −1.30 V/Å), Cl–GaN nanosheets have the excellently half-metallic properties with the bandgaps (3.71 to 0.96 eV) and maximal half-metallic gaps with 0.30 eV in spin-up states and metallic behaviors in spin-down states. Moreover, the total magnetic moment decreases (increases) depending on the negative (positive) F, mainly induced by the unpaired N atoms.

As a result of the difference in the bond length between C–Ga (or C–N) and N–Ga bonds, local distortion of the atomic structure takes place when one N (or Ga) atom is substituted by one C atom in zigzag GaNNR or armchair GaNNR [33]. In detail, one C atom substituting for one

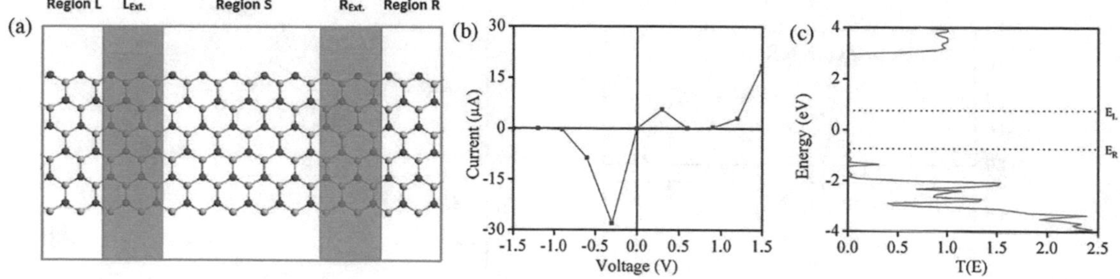

FIGURE 7.21 *I–V* characteristic for device based on bare zigzag GaNNR. (a) Device structure. (b) *I–V* characteristics. (c) Transmission spectrum [40].

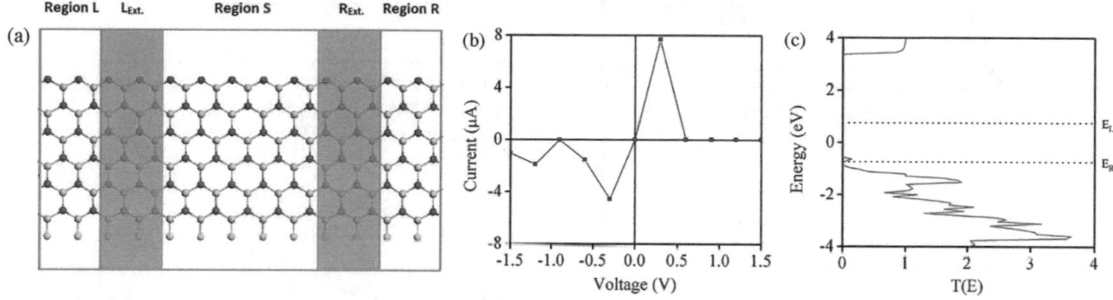

FIGURE 7.22 *I–V* characteristic for device based on F-GaN zigzag GaNNR. (a) Device structure. (b) *I–V* characteristics. (c) Transmission spectrum [40].

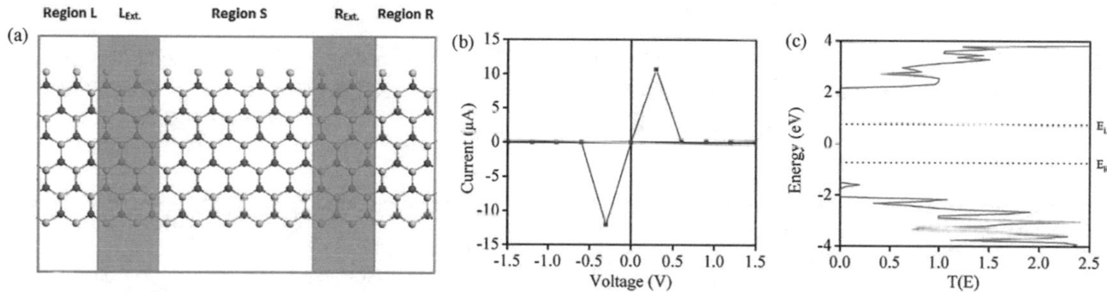

FIGURE 7.23 *I–V* characteristic for device based on GaN-F zigzag GaNNR. (a) Device structure. (b) *I–V* characteristics. (c) Transmission spectrum [40].

N atom causes a slight local expansion, while one C atom substituting for one Ga atom results in a large local contraction. This is because the covalent radius of C (0.77 Å) is slightly larger than that of N (0.75 Å), but much smaller than that of Ga (1.26 Å). Furthermore, the C atom is preferred to substitute for an edge N1 or Ga1 atom in either 6-ZGaNNR or 6-AGaNNR, especially edge Ga1 atoms in 6-AGaNNR. There exists about 0.65 mB magnetic moment, which arises mainly from doped C atom, while a single N atom is substituted by one C atom in either 6-ZGaNNR or 6-AGaNNR. This is because the interaction between the C atom and its nearest neighbor Ga atom is very weak, as indicated either by the larger bond lengths d^{A}_{C-Ga} and d^{Z}_{C-Ga}, or by the near-zero formation energies E_f (Figure 7.24).

The effect of adatom adsorption (i.e., F, O, and N) on the magnetism and band structure of GaN monolayer was studied by spin-polarized density functional theory [44]. Adsorption of fluorine adatoms at low coverage ($\Theta = 1/8$) caused the GaN monolayer to transform into a magnetic half-metal from a nonmagnetic indirect bandgap semiconductor, whereas the monolayer turned back to a semiconductor at high coverage ($\Theta = 1/2$). The Curie temperature for the low-coverage F-adsorbed GaN monolayer is estimated to be about 480 K, by Monte Carlo simulation based on Ising model. The GaN monolayer would become magnetic metal when adsorbing nitrogen adatoms at high coverage, while it would become magnetic half-metal at low coverage. Their Curie temperatures were estimated to be about 415 and 325 K. The induced magnetism mentioned above mainly came from the spin splitting of N 2p ($2p_z$ or $2p_x/2p_y$) orbitals. In contrast, the adsorption of oxygen adatoms did not induce magnetism but reduced the bandgap of GaN monolayer. Figures 7.25 and 7.26 show the band structures and p-projected density of states (pPDOS) of F-adsorbed and N-adsorbed GaN monolayer at the coverage of 1/8 and 1/2, respectively.

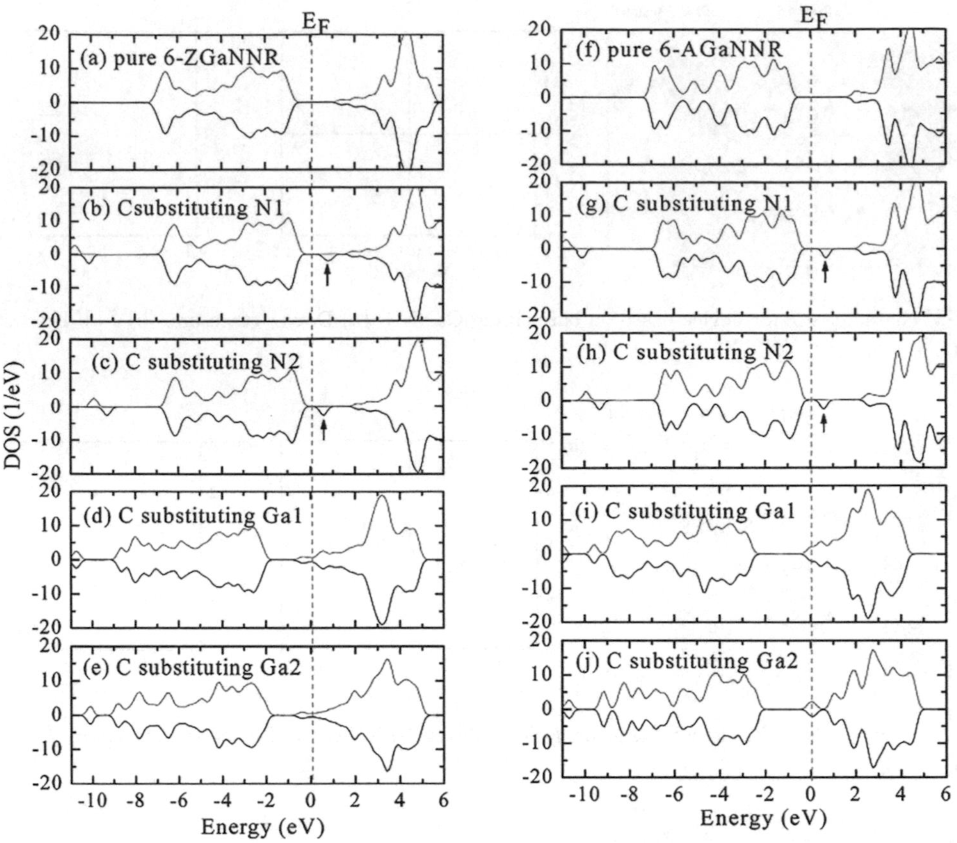

FIGURE 7.24 The DOS for (a) pure 6-ZGaNNR, (b) C substituting N1, (c) C substituting N2, (d) C substituting Ga1, and (e) C substituting Ga2 for 6-ZGaNNR (left panels); (f) pure 6-AGaNNR, (g) C substituting N1, (h) C substituting N2, (i) C substituting Ga1, and (j) C substituting Ga2 sites for 6-AGaNNR (right panels) [43].

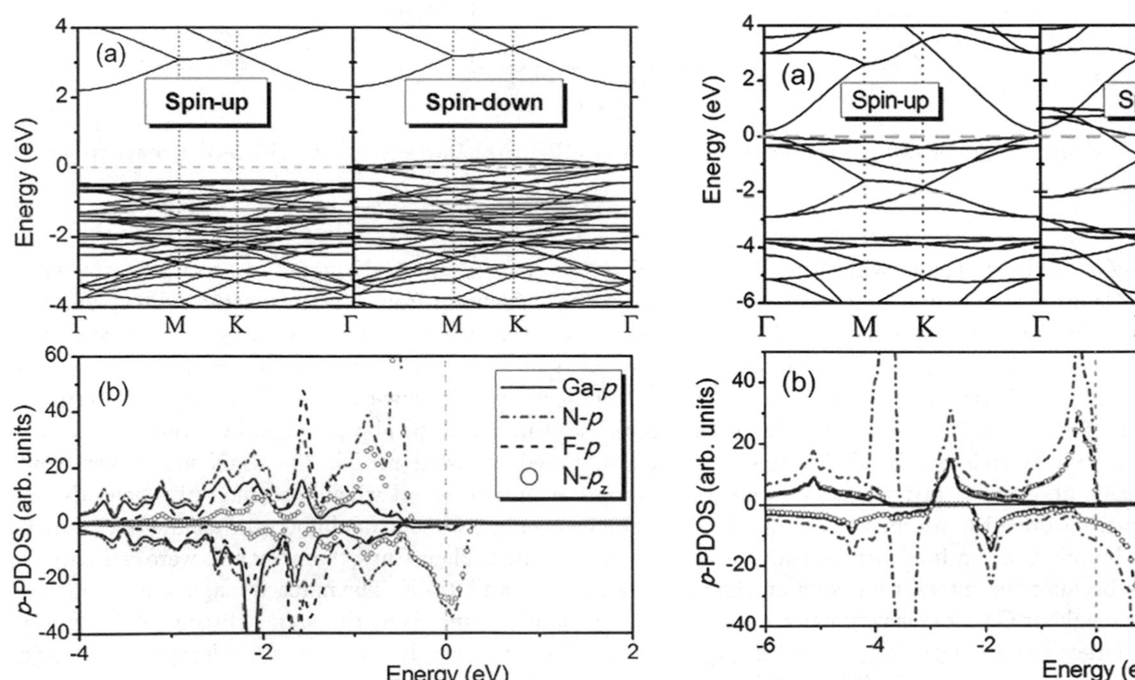

FIGURE 7.25 (a) Band structures and (b) pPDOS of F-adsorbed GaN monolayer at the coverage of 1/8 [44].

FIGURE 7.26 (a) Band structures and (b) pPDOS of N-adsorbed GaN monolayer at the coverage of 1/2 [44].

The structural, electronic, and magnetic properties of zigzag GaNNR (ZGaNNR) with period vacancy located at different sites across the ribbon width have been investigated by first-principles calculations [28]. The results show that the formation of the N-vacancy is easier than that of the Ga-vacancy at each equivalent geometrical site and both of them are endothermic. An inward relaxation of the three nearest Ga atoms around an N-vacancy occurs, while for the three nearest-neighbor N atoms around the Ga-vacancy, an outward relaxation occurs. Except for a typical nonedge N-vacancy, the N-, or Ga-vacancies at other sites induce magnetic moment and spin polarization, implying such vacancy-defective ZGaNNRs can be useful in spintronics and nanomagnets. The magnetic moment of the N-vacancy is dependent on defect sites, while for the Ga-vacancy, it is less dependent on the defect sites. The net magnetic moment of the vacancy-defective 8-ZGaNNR is mainly contributed by the atoms around a vacancy. In addition, the extended line defect is stable in GaN monolayer [45]. The electronic density of states calculations demonstrated that in gap states are emerged when extended line defect is incorporated into the honeycomb structure.

As a new anode material of Li-ion batteries, GaN nanosheet-based electrodes deliver a discharge capacity up to 702 mA h/g after 100 cycles at 0.1 A/g and above 600 mA h/g after 1,000 cycles at 1.0 A/g. More importantly, GaN nanosheet-based electrodes exhibit ultrahigh rate capability [18].

References

1. H. Amano, Growth of GaN layers on sapphire by low-temperature-deposited buffer layers and realization of p-type GaN by magnesium doping and electron beam irradiation (Nobel Lecture), *Angewandte Chemie International Edition*, 54 (2015) 7764–7769.
2. S. Nakamura, Nobel Lecture: Background story of the invention of efficient blue InGaN light emitting diodes, *Reviews of Modern Physics*, 87 (2015) 1139–1151.
3. I. Akasaki, Fascinating journeys into blue light (Nobel Lecture), *Annalen der Physik*, 527 (2015) 311–326.
4. Y. Huang, X. Duan, Y. Cui, C.M. Lieber, Gallium nitride nanowire nanodevices, *Nano Letters*, 2 (2002) 101–104.
5. F. Qian, Y. Li, S. Gradečak, D. Wang, C.J. Barrelet, C.M. Lieber, Gallium nitride-based nanowire radial heterostructures for nanophotonics, *Nano Letters*, 4 (2004) 1975–1979.
6. A.K. Singh, H.L. Zhuang, R.G. Hennig, *Ab initio* synthesis of single-layer III-V materials, *Physical Review B*, 89 (2014) 245431.
7. A.K. Singh, B.C. Revard, R. Ramanathan, M. Ashton, F. Tavazza, R.G. Hennig, Genetic algorithm prediction of two-dimensional group-IV dioxides for dielectrics, *Physical Review B*, 95 (2017) 155426.
8. Z.Y. Al Balushi, K. Wang, R.K. Ghosh, R.A. Vila, S.M. Eichfeld, J.D. Caldwell, X. Qin, Y.-C. Lin, P.A. DeSario, G. Stone, S. Subramanian, D.F. Paul, R.M. Wallace, S. Datta, J.M. Redwing, J.A. Robinson, Two-dimensional gallium nitride realized via graphene encapsulation, *Nature Materials*, 15 (2016) 1166–1171.
9. D. Wu, M.G. Lagally, F. Liu, Stabilizing graphitic thin films of wurtzite materials by epitaxial strain, *Physical Review Letters*, 107 (2011) 236101.
10. L. Yang, X. Zhang, R. Huang, G. Zhang, X. An, Synthesis of single crystalline GaN nanoribbons on sapphire (0001) substrates, *Solid State Communications*, 130 (2004) 769–772.
11. X. Xiang, C. Cao, F. Huang, R. Lv, H. Zhu, Synthesis and characterization of crystalline gallium nitride nanoribbon rings, *Journal of Crystal Growth*, 263 (2004) 25–29.
12. A.A. Yamaguchi, M. Takashi, S. Akira, S. Haruo, K. Akitaka, N. Masaaki, U. Akira, Single domain hexagonal GaN films on GaAs (100) vicinal substrates grown by hydride vapor phase epitaxy, *Japanese Journal of Applied Physics*, 35 (1996) L873.
13. G. Pozina, C. Hemmingsson, J.P. Bergman, T. Kawashima, H. Amano, I. Akasaki, A. Usui, B. Monemar, Optical properties of metastable shallow acceptors in Mg-doped GaN layers grown by metal-organic vapor phase epitaxy, *AIP Conference Proceedings*, 1199 (2010) 110–111.
14. A. Lahiri, N. Borisenko, A. Borodin, F. Endres, Electrodeposition of gallium in the presence of NH4Cl in an ionic liquid: Hints for GaN formation, *Chemical Communications*, 50 (2014) 10438–10440.
15. A.V. Andrianov, D.E. Lacklison, J.W. Orton, D.J. Dewsnip, S.E. Hooper, C.T. Foxon, Low-temperature luminescence study of GaN films grown by MBE, *Semiconductor Science and Technology*, 11 (1996) 366.
16. N.A. Koratkar, Materials synthesis: Two-dimensional gallium nitride, *Nature Materials*, 15 (2016) 1153–1154.
17. B. Liu, W. Yang, J. Li, X. Zhang, P. Niu, X. Jiang, Template approach to crystalline GaN nanosheets, *Nano Letters*, 17 (2017) 3195–3201.
18. C. Sun, M. Yang, T. Wang, Y. Shao, Y. Wu, X. Hao, Graphene-oxide-assisted synthesis of GaN nanosheets as a new anode material for lithium-ion battery, *ACS Applied Materials and Interfaces*, 9 (2017) 26631–26636.
19. Q. Chen, H. Hu, X. Chen, J. Wang, Tailoring band gap in GaN sheet by chemical modification and electric field: Ab initio calculations, *Applied Physics Letters*, 98 (2011) 053102.
20. A. Onen, D. Kecik, E. Durgun, S. Ciraci, GaN: From three- to two-dimensional single-layer crystal and its multilayer van der Waals solids, *Physical Review B*, 93 (2016) 085431.

21. G.A. Rodrigo, Z. Xiaoliang, M. Saikat, P. Ravindra, R.R. Alexandre, P.K. Shashi, Strain- and electric field-induced band gap modulation in nitride nanomembranes, *Journal of Physics: Condensed Matter*, 25 (2013) 195801.
22. H. Sahin, S. Cahangirov, M. Topsakal, E. Bekaroglu, A. Ethem, T. Senger, S. Ciraci, Monolayer honeycomb structures of group-IV elements and III-V binary compounds: First-principles calculations, *Physical Review B*, 80 (2009) 155453.
23. Q. Peng, C. Liang, W. Ji, S. De, Mechanical properties of g-GaN: A first principles study, *Applied Physics A*, 113 (2013) 483–490.
24. X. Dongwei, H. Haiying, P. Ravindra, P.K. Shashi, Stacking and electric field effects in atomically thin layers of GaN, *Journal of Physics: Condensed Matter*, 25 (2013) 345302.
25. C. Attaccalite, A. Nguer, E. Cannuccia, M. Gruning, Strong second harmonic generation in SiC, ZnO, GaN two-dimensional hexagonal crystals from first-principles many-body calculations, *Physical Chemistry Chemical Physics*, 17 (2015) 9533–9540.
26. H.L. Zhuang, A.K. Singh, R.G. Hennig, Computational discovery of single-layer III-V materials, *Physical Review B*, 87 (2013) 165415.
27. D.C. Camacho-Mojica, F. López-Urías, GaN Haeckelite single-layered nanostructures: Monolayer and nanotubes, *Scientific Reports*, 5 (2015) 17902.
28. G.-X. Chen, D.-D. Wang, J.-M. Zhang, K.-W. Xu, Structural, electronic, and magnetic properties of the period vacancy in zigzag GaN nanoribbons, *Physica Status Solidi B*, 250 (2013) 1510–1518.
29. H. Li, J. Dai, J. Li, S. Zhang, J. Zhou, L. Zhang, W. Chu, D. Chen, H. Zhao, J. Yang, Z. Wu, Electronic structures and magnetic properties of GaN sheets and nanoribbons, *The Journal of Physical Chemistry C*, 114 (2010) 11390–11394.
30. S. Behzad, Electronic structure, optical absorption and energy loss spectra of GaN graphitic sheet, *Journal of Materials Science: Materials in Electronics*, 26 (2015) 9898–9906.
31. S. Bhattacharya, A. Datta, S. Dhara, D. Chakravorty, Surface optical Raman modes in GaN nanoribbons, *Journal of Raman Spectroscopy*, 42 (2011) 429–433.
32. S. Behzad, Effects of strain and thickness on the electronic and optical behaviors of two-dimensional hexagonal gallium nitride, *Superlattices and Microstructures*, 106 (2017) 102–110.
33. F.-L. Zheng, Y. Zhang, J.-M. Zhang, K.-W. Xu, Structural, electronic, and magnetic properties of C-doped GaN nanoribbon, *Journal of Applied Physics*, 109 (2011) 104313.
34. H. Le, Y. Qu, K. Jun, L. Yan, L. Jingbo, Tunable band gaps in graphene/GaN van der Waals heterostructures, *Journal of Physics: Condensed Matter*, 26 (2014) 295304.
35. G. Yanlin, O. Susumu, Energetics and electronic structures of thin films and heterostructures of a hexagonal GaN sheet, *Japanese Journal of Applied Physics*, 56 (2017) 065201.
36. M. Sun, J.-P. Chou, J. Yu, W. Tang, Electronic properties of blue phosphorene/graphene and blue phosphorene/graphene-like gallium nitride heterostructures, *Physical Chemistry Chemical Physics*, 19 (2017) 17324–17330.
37. X.-L. Wei, Z.-K. Tang, G.-C. Guo, S. Ma, L.-M. Liu, Electronic and magnetism properties of two-dimensional stacked nickel hydroxides and nitrides, *Scientific Reports*, 5 (2015) 11656.
38. A. Onen, D. Kecik, E. Durgun, S. Ciraci, In-plane commensurate GaN/AlN junctions: Single-layer composite structures, single and multiple quantum wells and quantum dots, *Physical Review B*, 95 (2017) 155435.
39. Y. Ma, Y. Dai, M. Guo, C. Niu, L. Yu, B. Huang, Strain-induced magnetic transitions in half-fluorinated single layers of BN, GaN and graphene, *Nanoscale*, 3 (2011) 2301–2306.
40. S.V. Inge, N.K. Jaiswal, P.N. Kondekar, Realizing negative differential resistance/switching phenomena in Zigzag GaN nanoribbons by edge fluorination: A DFT investigation, *Advanced Materials Interfaces*, 4 (2017) 1700400.
41. Q. Tang, Y. Cui, Y. Li, Z. Zhou, Z. Chen, How do surface and edge effects alter the electronic properties of GaN nanoribbons? *The Journal of Physical Chemistry C*, 115 (2011) 1724–1731.
42. L. Tong, J. He, M. Yang, Z. Chen, J. Zhang, Y. Lu, Z. Zhao, Anisotropic carrier mobility in buckled two-dimensional GaN, *Physical Chemistry Chemical Physics*, 19 (2017) 23492–23496.
43. M.X. Xiao, H.Y. Song, Z.M. Ao, T.H. Xu, L.L. Wang, Electric field modulated half-metallicity of semichlorinated GaN nanosheets, *Solid State Communications*, 245 (2016) 5–10.
44. Y. Mu, Chemical functionalization of GaN monolayer by adatom adsorption, *The Journal of Physical Chemistry C*, 119 (2015) 20911–20916.
45. D.C. Camacho-Mojica, F. López-Urías, Extended line defects in BN, GaN, and AlN semiconductor materials: Graphene-like structures, *Chemical Physics Letters*, 652 (2016) 73–78.

8

Graphene Nanodot Arrays

Luca Camilli and
José M. Caridad
Technical University of Denmark

8.1 Introduction

8.1.1 Graphene

Carbon can hybridize in different forms, thus giving rise to materials, called allotropes, which exhibit distinct physical and chemical properties. For example, diamond is made of carbon atoms with sp^3 hybridization, while graphite is made of carbon atoms with sp^2 hybridization. Diamond is a wide bandgap insulator, while graphite is a semimetal.

Graphene is a one-atom-thick crystal made entirely of sp^2 hybridized carbon atoms arranged in a hexagonal lattice. It was isolated for the first time in the laboratories of the University of Manchester (Manchester, UK) in 2004 through a simple yet effective micromechanical cleavage of bulk graphite by means of scotch tape (Novoselov et al., 2004). Its most remarkable property – which gained notoriety – is perhaps its thermodynamical stability. This came rather as a surprise at that time, as a number of theoretical studies performed in the mid-1990s predicted that thermal fluctuations would destroy a long-range ordering of isotropic systems in less than three dimensions (Peierls, 1934; Landau and Lifsl, 1969; Mermin and Wagner, 1966). The experimentally proved stability of graphene, that is, the stability of a two-dimensional (2D) crystal, has later on triggered the interest of material scientists and physicists toward other 2D materials, such as hexagonal boron nitride and transition metal dichalcogenides (Novoselov et al., 2005a; Zou and Yakobson, 2015). The family of 2D materials is still growing, and today, they are not just studied as individual systems, but are often combined together in either planar or vertical heterostructures in order to give rise to novel, multifunctional devices (Zhang et al., 2015; Wang et al., 2014; Dean et al., 2010).

Yet, graphene is still the most largely studied 2D material due to the number of compelling properties it exhibits. For instance, being atomically thin, it is almost transparent, with only 2.3% absorption of visible light (Nair et al., 2008). However, when deposited on a silicon oxide substrate, graphene can be easily spotted, because of the refraction and interference of light between graphene and substrate. Moreover, the distinct color and contrast exhibited by graphene layers of different thickness is used to distinguish the number of layers (Blake et al., 2007). Despite its atomic thickness, graphene is completely impermeable to all molecules and gases, even to the small H_2 and He (Bunch et al., 2008). On the one hand, the aforementioned property makes graphene a promising barrier layer to be used, for instance, for anticorrosive applications (Chen et al., 2011). On the other hand, it could be used to engineer graphene into a nanosieve device for a selective transport of gas (Kim et al., 2013), liquids (Surwade et al., 2015), biomaterials (Schneider et al., 2010; Merchant et al., 2010) and even ions (O'Hern et al., 2014) or protons (Hu et al., 2014).

Graphene also exhibits outstanding mechanical properties, with a Young's modulus of 1.0 TPa and an intrinsic strength of 130 GPa, which make it the strongest material ever measured (Lee et al., 2008). The flexibility of the material is extremely important from an application point of view, because it enables graphene to deform and conform to any substrate.

Furthermore, with respect to thermal properties, graphene shows an unusually high thermal conductivity, even higher than carbon nanotubes, thus making it an attractive material for thermal management (Balandin et al., 2008).

Its electrical properties are also of great interest, which also make it a promising material for a number of electronic applications. For instance, graphene has mobility as high as 2×10^5 cm^2/Vs at room temperature (Morozov et al., 2008), which is more than two orders of magnitude higher than in silicon, and a sheet resistance of 30 Ω/sq (Bae et al., 2010), which, along with its optical transparency, makes it intriguing as a transparent electrode.

In addition, graphene exhibits ambipolar transport behavior (Neto et al., 2009), and thus, graphene devices are capable of conducting both electrons and holes. In fact, this type of conduction has also been observed in other novel materials, especially 2D ones (Bisri et al., 2014; Das et al., 2014; Steinberg et al., 2010; Zhang et al., 2012; Tao et al., 2015). Moreover, from a more fundamental point of view, graphene offers the possibility to explore and study not only spin (Tombros et al., 2007) but also valley degree (Rycerz et al., 2007) of freedom of charge carriers, owing to the weak spin-orbit coupling and the two inequivalent valleys occurring at the +K and −K points of the Brillouin zone (Figure 8.1). These topics are currently attracting a great deal of interest in the field of 2D materials (Han, 2016; Schaibley et al., 2016).

Graphene shows a unique band structure, with conical conduction and valence bands meeting at the K and K′ points of the Brillouin zone, so-called Dirac points (Neto et al., 2009). As a consequence, charge carriers in graphene behave like relativistic massless (Dirac) fermions (Novoselov et al., 2005b), thus giving rise to a range of unique and unconventional effects that can be exclusively observed in this material. Examples of such phenomena comprise the anomalous integer quantum Hall effect (Novoselov et al., 2005b; Zhang et al., 2005) observable even at room temperature (Novoselov et al., 2007); the occurrence of Klein tunneling (Katsnelson et al., 2006), which allows Dirac fermions to penetrate high and wide barriers without reflection; or the possibility of focusing electric currents by using graphene n-p-n junctions (Cheianov et al., 2007). More importantly, owing to the low density of states in this material around the Dirac point, the carrier transport in graphene can be easily tuned by electrostatic or chemical doping (Neto et al., 2009).

From an application point of view, the remarkable flexibility, electrical and thermal conductivities, transparency and carrier tunability pave the way for the use of graphene in fields such as high-frequency electronics, tunable photodetectors, flexible electronics, light-emitting devices and solar cells (Ferrari et al., 2015).

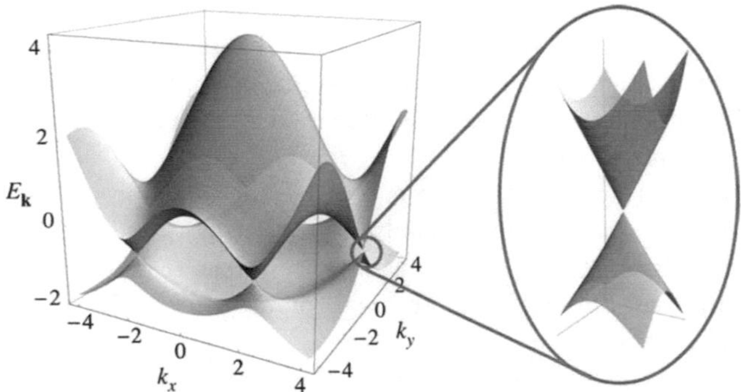

FIGURE 8.1 The electronic band structure of graphene. The inset shows the valence and conduction bands touching each other at one of the Dirac points. (Reprinted with permission from American Physical Society, Neto et al., 2009.)

8.1.2 Graphene Nanostructures

In its pristine form, graphene is already a fantastic playground for condensed matter physicists, as discussed earlier. Nevertheless, even more fascinating is the possibility of structuring this material at the nanoscale, thus reducing its dimensionality even further.

The interest here relies on the fact that generally nanostructures, owing to their characteristic size, exhibit physical and chemical properties that are different from those in their extended version. Graphene nanoribbons, for instance, have received a great deal of attention because they could have a bandgap in their electronic structure, a feature which missing in pristine graphene (Han et al., 2007; Son et al., 2006; Yang et al., 2007). Such feature is of vital importance for the realization, for instance, of conventional logic gates. If pristine graphene were to be used as a channel in field effect transistors (FETs), the on/off ratio – the ratio between current at on and at off state of the device – would be rather poor, thus leading to undesirable power consumption of the device.

Several strategies for the synthesis and fabrication of graphene nanoribbons have been developed over the years, such as unzipping of carbon nanotubes (Jiao et al., 2009, 2010), wet chemical processes (Li et al., 2008), on-surface reactions (Cai et al., 2010;Talirz et al., 2016) or electron beam lithography (Han et al., 2007). Scanning tunneling microscopy (STM) and scanning tunneling spectroscopy (STS) are often used for the characterization of graphene nanoribbons, as they allow to probe the structural and electronic properties of these nanostructures with atomic precision (Tao et al., 2011; Ruffieux et al., 2012; Chen et al., 2013). Ballistic graphene nanoribbons epitaxially grown on silicon carbide have also been note that the measurement or the measurement at room temperature is irrelevant in this context, dealing with fabrication (Baringhaus et al., 2014).

If graphene nanoribbons can be considered as one-dimensional (1D) forms of graphene, graphene nanodots (GNDs) are the zero-dimensional (0D) version. Here, we want to emphasize that in literature, there are two different systems that are either called GNDs or also referred sometimes as graphene quantum dots.

The first type is the one that is obtained when graphene is spatially confined in all three dimensions, in analogy to the case of classical semiconductor quantum dots (Koch, 1993) (Figure 8.2a). In this case, strong quantum confinement and edge effects are to be expected. For instance, because electrons in graphene behave like Dirac fermions, in such isolated GNDs one could observe the behavior of massless relativistic particles in confined space – the chaotic Dirac billiards (Berry and Mondragon, 1987). Besides, size confinement should lead to the opening of an energy gap for isolated GNDs that is predicted to scale down exponentially

FIGURE 8.2 (a) Two isolated GNDs (i.e., spatially confined GNDs); (b) two GNDs formed when local, nanometer-sized regions within a graphene sheet have a different doping compared to the surroundings (i.e., electrostatically confined GNDs); and (c) a graphene sheet with two nanometer-sized holes (nanoporous graphene).

as a function of the number of carbon atoms (Zhang et al., 2008; Güçlü et al., 2010).

The second type of GNDs reported in literature is formed when local, nanometer-sized regions within a continuous graphene sheet have a different doping level with respect to their surroundings. Here, both extended and nanometer-sized regions are separated by p-n (n-p, n-n' or p-p') junctions (Figure 8.2b). This second type of GNDs is particularly interesting because it can lead to novel phenomena such as focusing and/or guiding of Dirac electrons, as discussed further within the applications section.

For completeness, it is convenient to mention here that the presence of a bandgap in graphene does not only occur in nanodot structures but also occurs in nanoporous lattices (NPLs), that is, structures consisting of an electrically continuous graphene sheet periodically perforated with holes of nanometer sizes (Pedersen et al., 2008; Moreno et al., 2018) (Figure 8.2c). In particular, recent bottom-up fabrication approaches (Moreno et al., 2018) have demonstrated the synthesis of tunable periodic arrays of pores of nanometer size, separated by ribbons with atomic precision. These systems have anisotropic electrical properties, exhibiting 1D semiconducting channels with energy gaps of ~1 eV. In general, being a continuous 2D sheet with a tunable bandgap, NPL nanostructures may be crucial for the development of electronic devices made from graphene and based on conventional (i.e., complementary metal-oxide semiconductor, CMOS) design principles, such as logic gates or high-performance graphene transistors for high-frequency electronics (Wu et al., 2012).

8.2 Fabrication

Here, we will cover strategies used to prepare GNDs, as defined above, and that can be extended to deliver GND arrays with spatial periodicity. For this reason, we will not cover GNDs that are synthesized through wet chemical methods, as they do not easily allow for the fabrication of periodic structures. For these colloidal and solution-based GNDs, we refer the reader to reviews on the subject (Li et al., 2012, 2015; Bacon et al., 2014).

8.2.1 Lithography

Lithographic processes are largely used for the fabrication of micro- and nanostructures, as they are relatively easy to use and allow for the creation of architectures on large areas with control of the shape and with size down to tens of nanometers (Cabrini and Kawata, 2012). Indeed, GNDs have been fabricated through lithographic techniques.

In 2008, using electron beam lithography, Ponomarenko et al. produced GNDs with a characteristic size ranging from more than 100 nm down to only few nanometers. Specifically, a polymethyl methacrylate (PMMA)-protecting mask made using electron beam lithography was placed on the top of a graphene flake mechanically exfoliated on a Si/SiO_2 substrate. Next, oxygen plasma was used to remove the graphene areas that were left unprotected by the PMMA mask, which after removal with suitable solvents would leave several GNDs, with features as small as 10 nm. This method is very flexible, delivering samples with size and shape that can be controlled with great precision. Such systems display a large range of operational regimes from conventional single-electron detectors to Dirac billiards (Ponomarenko et al., 2008).

A similar strategy was used by scientists at ETH Zurich in Switzerland to make GNDs in which they observed Coulomb blockade effects (Stampfer et al., 2008). Here, after preparing the PMMA mask using electron beam lithography, they etched an exfoliated graphene flake via reactive ion etching (RIE) treatment to fabricate graphene nanostructures with a minimum size of approximately 50 nm. Other samples have been produced by this research group through the same process to study charge detection as well as localized resonant states in these carbon nanostructures (Schnez et al., 2010; Güttinger et al., 2008). It is worth mentioning, however, that these approaches based on the use of electron beam lithography are generally low throughput and suitable for delivering just few, high-quality samples for academic purposes.

An alternative method is to replace the PMMA mask defined by electron beam lithography with self-assembling block copolymers (BCP). By varying the employed polymers, their average molecular weight and their annealing conditions (Lee et al., 2012; Rasappa et al., 2015), arrays of nanostructures with different shapes and sizes can be defined on the top of graphene, such as nanoribbons (Rasappa et al., 2015), dots (Lee et al., 2012; Wang et al., 2016) and nanopores (Bai et al., 2010). In particular, shortly after nanoporous graphene were reported to be the first nanostructures made by BCP-assisted lithography (Bai et al., 2010), scientists from the Republic of Korea managed to fabricate also GNDs of uniform size (Lee et al., 2012). The general processing steps for these nanostructures comprise the following: First, a graphene sheet is spin-coated with Si-containing polystyrene-b-polydimethylsiloxane (PS-PDMS) BCP. Next, a regular pattern of PDMS cylinders develops within a PS matrix after a few chemical and annealing steps. At this point, an oxygen plasma treatment will etch the PS (and so graphene underneath), while the PDMS part will be hardened. Such hardened PDMS nano-cylinders will serve as a protecting mask for the synthesis of GNDs. Indeed, several GNDs of around 10 nm in diameter and with very narrow distribution size can be produced with this scalable method. Variants to this technique are reported in other studies in literature (Wang et al., 2016).

Another lithographic approach that takes advantage of self-assembly of polymeric structures was suggested by Oh et al. (2015) and Zhu et al. (2014). In this case, polystyrene nanospheres assemble within a sixfold symmetry array on a chemical vapor-deposited graphene film on copper to act as a protecting mask during the oxygen plasma etching. Here, the oxygen plasma plays a dual role: On the one hand, it reduces the diameter of the polymeric nanospheres; on the other hand, it etches away the unprotected graphene regions. The final result is the formation of an ordered array of uniform-sized GNDs that can be later peeled off from the copper slab with the help of a support polymer layer and placed on any target substrate. With this method, the size of the GNDs can be tuned by adjusting the etching time.

However, a general limitation of top-down approaches like lithography is the lack of control over the edge termination of the synthesized nanostructures. Regarding three-dimensional materials, the surface becomes increasingly more important with shrinking of the material size, owing to a higher surface-to-volume ratio. Similarly, in the case of 2D materials, edges become increasingly more important when the 2D material is further spatially confined, like in the case of nanoribbons (i.e., 1D structures) and nanodots (i.e., 0D structures). More importantly, this is particularly true for the case of carbon nanostructures. It is known, for instance, that carbon nanotubes show metallic or semiconducting character depending on the arrangement of graphitic rings in their walls, and the diameter (Mintmire et al., 1992; Hamada et al., 1992; Saito et al., 1992a,b; Odom et al., 1998). In analogy, graphene nanoribbons and GNDs also exhibit remarkably different properties depending on the crystallographic orientation of their edges. Notably, it has been verified experimentally that GNDs of 7–8 nm size and terminating mainly with zigzag edges are metallic, due to the presence of zigzag edge states, while graphene nanoribbons with armchair termination show semiconducting features (Ritter and Lyding, 2009).

Moreover, a recent study proved that not only the crystallographic orientation of the edge, but even its degree of order and smoothness is extremely important while considering the output of electronic devices based on graphene nanostructures (Caridad et al., 2018). In this context, by developing bottom-up approaches, it would be possible to synthesize graphene nanoribbons and GNDs with atomic precision through assembling of their atomic building blocks.

8.2.2 On-Surface Synthesis

In general, if fullerenes could be opened and made flat, then they would become GNDs. In 2011, a group of scientists from Singapore managed to fabricate geometrically well-defined GNDs by opening C_{60} molecules on the surface of a Ru catalyst (Lu et al., 2011). Once the C_{60} molecules are deposited on the metallic crystal, the strong interaction between the carbon nanostructures and the substrate leads to the formation of vacancies on the Ru surface, with a subsequent embedding of the C_{60} molecules. Then, annealing at high temperature fragments the molecular precursors and produces carbon clusters that diffuse on the Ru surface until they nucleate into well-defined GNDs. Additionally, the scientists showed that the final shape of the GNDs can be selected depending on the annealing temperature. Although extremely interesting, this approach cannot be easily scaled up, nor does it lend itself to the formation of periodic arrays of GNDs.

A similar concept, where molecular precursors are used as building blocks to build up GNDs, was recently proposed (Wang et al., 2017). Here, to begin with, 10,10′-dibromo-9,9′-bianthryl precursor monomers are deposited on the surface of an Au crystal in ultra-high-vacuum condition. A first series of heat treatments leads to dehydrogenation of the molecular precursor and the formation of seven-carbon-atom-wide armchair graphene nanoribbons. These subsequently merge after a further heat treatment to yield GNDs with a defined and atomically sharp termination (Figure 8.3). STS reveals that the fundamental gap of the GNDs can be tuned by the first order of magnitude within a length range of few nanometers.

The method just described has been very recently developed even further, and ordered arrays of nanosized pores have been formed (Moreno et al., 2018). First, following a series of on-surface synthesis reactions similar to what reported above, graphene nanoribbons are formed with atomic precision on an Au crystal starting from monomer precursors. Then, an additional thermally activated reaction step yields the lateral merging of the ribbons by means of a highly selective dehydrogenation cross-coupling. Remarkably, ordered nanopores whose size, density, morphology and chemical composition can be defined with atomic precision by the design of the molecular precursors are formed through this process within a graphene sheet on large area.

8.2.3 Self-Assembly of GNDs on Surfaces

An alternative to the bottom-up approach presented in the previous paragraph has been recently reported (Camilli et al., 2017). Here, the idea is to take advantage of the template action of a metal substrate and long-range

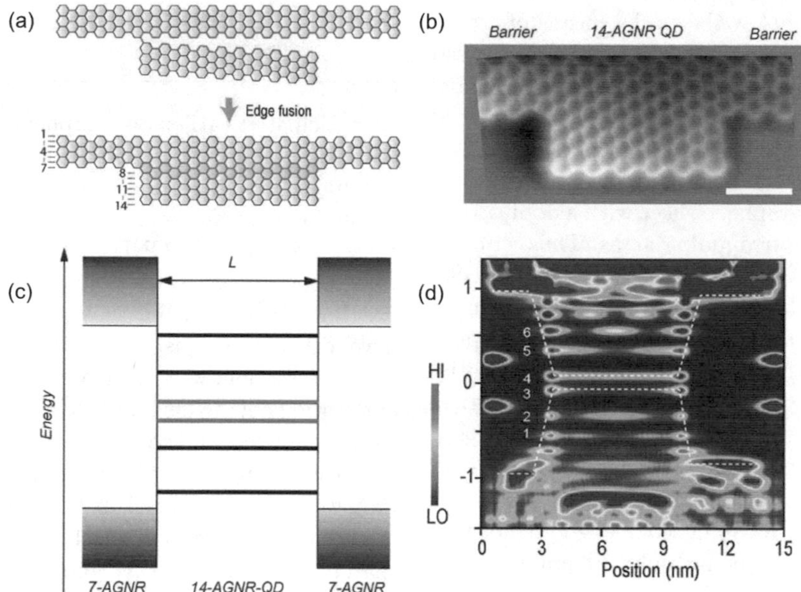

FIGURE 8.3 (a) Schematic illustration of the formation of a GND within a graphene nanoribbon by edge fusion of smaller graphene nanoribbons. (b) Non-contact atomic force microscopic image of such a GND. Scale bar: 2 nm. (c) Schematic illustration of the energy-level diagram of the GND; (d) Simulation based on density functional theory of the localized density of states in a GND embedded within graphene nanoribbons. (Picture adapted from Wang et al., 2017. Copyright (2017) The American Chemical Society.)

repulsion between dots in order to fabricate ordered arrays of GNDs epitaxially embedded in a 2D matrix. When the carbon content of a 2D B-C-N alloy reaches a critical value, GNDs of highly regular size, 1.6 ± 0.2 nm in diameter, segregate from the alloy and form periodic arrays over large areas, with the periodicity that can be tuned via the growth conditions. Moreover, the spontaneously assembled arrays behave like a 2D superlattice, showing characteristic defects such as vacancies and dislocations (Figure 8.4). The observed self-assembly behavior of the GNDs of uniform size is explained with a model that takes into account the dot-boundary energy and a moiré-modulated substrate interaction between dots. Notably, while the latter tends to give rise to dots with large diameter, the former tends to form domains that maximize the overlap between the carbon atoms and the atoms of the underlying iridium substrate. The competition between these two energetic contributions leads to a value of the dot size that minimizes its energy.

8.2.4 Doping/Electrostatics

The synthesis processes described above aim at producing graphene nanostructures by spatially confining graphene in nanometer-sized areas (i.e., spatial confinement). As discussed in paragraph 1.3, a conceptually different system of GNDs can be obtained by using electrostatics and creating nanometer-sized junctions within a continuous graphene sheet.

When a graphene sheet is deposited on a thin film of insulating hexagonal boron nitride, defects in this dielectric film can act as local gates that can locally dope the graphene overlayer if charged. Crommie and coworkers, for instance, used an STM tip to ionize defects in the region of

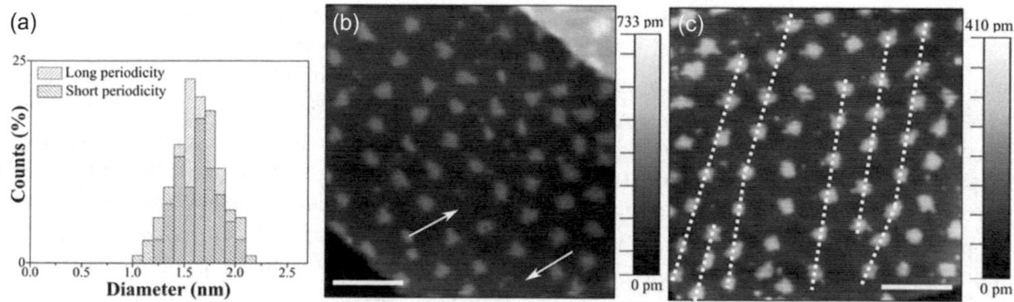

FIGURE 8.4 (a) Histogram of the size of the GNDs embedded in a 2D B-C-N matrix. The average diameter of the GNDs is approximately 1.6 nm. (b,c) STM image showing vacancy-like (c) and dislocation-like (d) defects within the GND arrays. Scale bars: 10 nm. (Picture adapted from Camilli et al., 2017, thanks to a Creative Commons Attribution 4.0 International License.)

the boron nitride directly below the tip by means of a relatively high-voltage pulse (Lee et al., 2016). The polarized defects can locally shield the graphene layer from a global back gate when the graphene/boron nitride stack is placed on a Si/SiO_2 substrate, with doped Si acting as the gate. This local shielding induces the formation of a stationary circular region within the graphene sheet with a doping level different from that of the surrounding areas. Thus, circular p-n-p (or n-p-n) junctions with a radius of ~100 nm can be formed with spatially high precision within a graphene sheet. As we will see later on in the section related to applications, this type of lateral junctions, occurring over spatial regions larger than the Fermi wavelength of charge carriers in graphene (Neto et al., 2009), enables the focusing and guiding of electron currents (Heinisch et al., 2013; Caridad et al., 2016).

Furthermore, n-p-n (or p-n-p) junctions in graphene, with the circular p (or n) region being only 10 nm in diameter or smaller, have been studied by Gutierrez et al. (2016). Here, authors take advantage of naturally occurring differences in a metallic substrate surface to create regions of diverse doping within a continuous graphene sheet. Graphene is grown on a Cu(111) single crystal. Due to the particularly high temperature used for growth, the copper surface exhibits a peculiar reconstruction, displaying a long-wavelength spiral pattern. However, in few, round nanometer-sized regions, the copper surface shows no reconstruction. As a consequence, the reconstructed and the non-reconstructed copper areas will have a slightly different value of the work function. Therefore, the overlayer graphene sheet will exhibit a different level of doping, even though the conical nature of the graphene bands close to the Dirac point is still preserved (Giovannetti et al., 2008). The size of these GNDs is comparable to or smaller than the Fermi wavelength of graphene charge carriers in common graphene devices (20–50 nm). Such conditions may lead to quasi-bound resonant states with highly suppressed Klein tunneling (Heinisch et al., 2013).

Even more interesting could be the possibility of creating arrays of such junctions (i.e., arrays of GNDs) that would then allow us to engineer the transport and optical properties of extended graphene over a large area. However, using an STM tip to create polarized defects or taking advantage of naturally occurring differences in a substrate's surface are not the methods that can be scaled up or reliably controlled on a large area.

An alternative and tunable method to create arrays of circular n-p-n (p-n-p) regions within a continuous graphene layer is represented by the deposition of metal dots of different sizes decorating the surface of graphene (Caridad et al., 2016; Echtermeyer et al., 2011; Han et al., 2014). Here, the deposition of metal nanoparticles directly on the graphene sheet induces local doping in the regions underneath the metal (Santos et al., 2011; Giovannetti et al., 2008). The creation of periodic arrays of these GNDs enables the design of novel architectures for appealing electronic and optoelectronic applications.

8.3 Properties and Applications of GNDs

As anticipated earlier on, graphene nanostructures such as graphene nanoribbons, GNDs and graphene NPL exhibit a number of remarkable properties that can be extremely useful for several applications. Notably, these nanostructures, and especially GNDs, not only may efficiently replace conventional semiconducting materials already used in existing technology, but may also enable the realization of novel concept designs.

In this section, we expand on both conventional and novel applications of GNDs and GND arrays, based on their electronic, optoelectronic and optical properties. To be coherent with the definition of GNDs that is found in literature and that we have introduced in paragraph 1.3, we will review, in the following section, both the applications of isolated, spatially confined GNDs and the applications of GNDs defined by nanometer-sized p-n junctions within a continuous graphene sheet.

8.3.1 Electronic Properties and Applications of GNDs

As a general rule, the transport of charge carriers in systems made of GNDs is highly dependent on both the structural parameters of GNDs, such as size, shape and edge termination, and the way GNDs are electrically connected. Specifically, to be electrically connected, GNDs must either have physical contact between them or be placed closely to allow the tunneling of charge carriers.

An interesting example is the case of a 1D chain of physically connected and spatially confined GNDs. Yuan and colleagues have studied such a system and considered different shapes (i.e., triangular, hexagonal or diamond-like), sizes and edge terminations (i.e., armchair or zigzag) of the GNDs (Figure 8.5) (Yuan et al., 2014). Notably, these calculations have shown that GNDs with armchair termination present semiconducting properties with a size-dependent bandgap, whereas GNDs with zigzag termination have a metallic character (Yuan et al., 2014). The precise controllability of the electronic properties of these systems shows that GNDs are promising elemental building blocks to develop devices with custom-tailored conducting and semiconducting nanochannels.

On the other hand, GNDs can be positioned (physically) separated from each other, but electrically connected via tunneling. Interestingly, when arranged in a 1D chain, these systems present a negative differential resistance due to a resonant tunneling occurring between the GNDs. The overall device constitutes a resonant tunneling diode, which can be used in terahertz frequency oscillators or memory devices (Al-Dirini et al., 2017).

Nevertheless, we emphasize that to reach a high level of controllability in terms of size, shape and edge termination of graphene nanostructures is extremely difficult

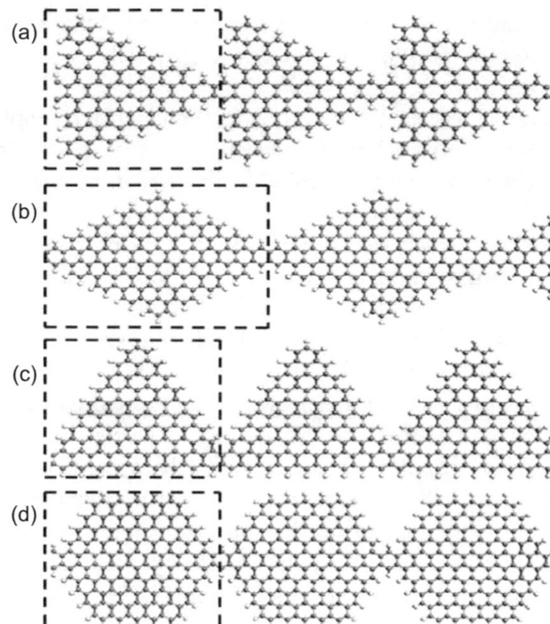

FIGURE 8.5 1D chain of GND arrays with different shape and edge termination: (a,b) armchair, (c,d) zigzag (c) edges. Dotted boxes indicate the individual GND (unit cell). (Picture adapted with permission from Yuan et al., 2014.)

from an experimental point of view. For example, it has been reported that quasi-1D structures (i.e., nanoribbons) produced by lithographically patterned graphene with large aspect ratios have a considerable degree of disorder associated with edge roughness and impurities in the conduction channel (Güttinger et al., 2012). In particular, disorder forms charged islands in these nanostructures, electrically behaving as consecutive chains of quantum dots of random sizes. Despite the absence of an actual bandgap in their electronic band structure, charge transport in these nanostructures exhibits diamond-shaped regions of suppressed conductance around the Dirac point signature of Coulomb-blockaded transport.

A special case of GNDs arranged in 1D chains can be constructed by the substitutional inclusion of pairs of boron atoms (BP) into graphene nanoribbons via bottom-up chemical reactions (Carbonell-Sanromà et al., 2016). As seen by STS, a region between two BPs shows discretization of the valence band into confined modes due to electron confinement, giving rise to quantum dots embedded in the nanoribbon. Interestingly, first-principles calculations have shown that the confinement is band selective; thus, only some electronic bands with specific symmetry are actually affected. A visionary application of such systems is the design of functional devices where only the transmission through some specific conduction channels is allowed (Carbonell-Sanromà et al., 2016).

Of great interest are also GND arrays formed when an electrically continuous graphene monolayer is locally doped in regions that are periodically spaced (Dubey et al., 2013; Drienovsky et al., 2017; Caridad et al., 2016). In these systems, pn junctions form tunnel barriers between the extended graphene sheet and the local regions diversely doped. The electrical conductivity of these systems having GNDs with characteristic sizes smaller than the wavelength of the electrons shows a large and tunable modulation (Fehske et al., 2015). Hence, these systems could be used as electrical switches (Fehske et al., 2015). In contrast, when the dot size is larger than the electron wavelength in graphene, it is possible to tune the spatial and energetic modulation of such GND arrays to guide the electron beam in particular directions (Figure 8.6a)

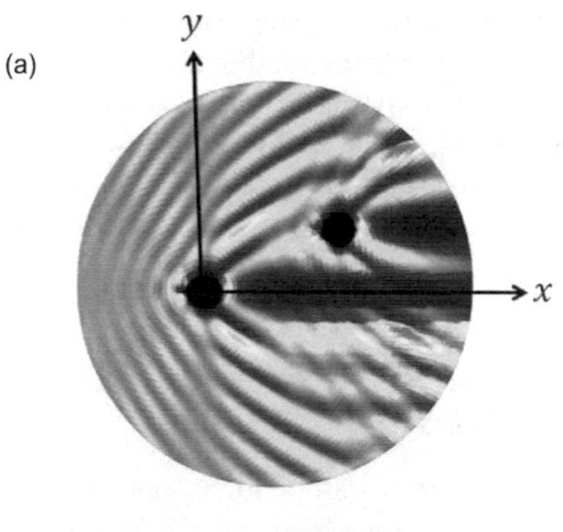

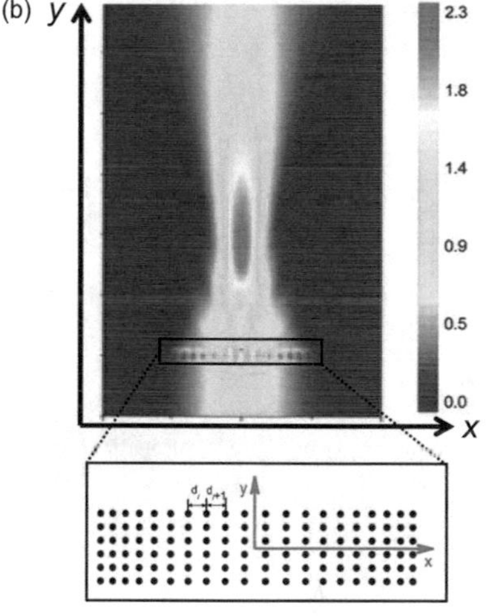

FIGURE 8.6 Focusing and guiding an electron beam in graphene using GNDs. (a) Electron density distribution for cascaded scattering on two canted GNDs (black, 100 nm in diameter) with respect to the direction of the incident plane wave. The electronic wave function is incident from the left along the x-axis. An asymmetric redistribution is found, implying the guiding of the electron beam toward positive values of the y-axis. (Picture adapted with permission from Caridad et al., 2016, thanks to a Creative Commons Attribution 4.0 International License.) (b) Electron density distribution for the scattering on an array of GNDs arranged gradually changing the lattice constant (black dots in the lower inset). The electronic wave function is incident from the bottom along the y-axis. An obvious focusing effect can be observed after the plane wave passes through the proposed array of GNDs. (Picture adapted with permission from Tang et al., 2016, thanks to a Creative Commons Attribution 4.0 International License.)

(Caridad et al., 2016), or to focus the electron beam (Tang et al., 2016) (Figure 8.6b). Switching, guiding and focusing electron beams can be seen as elemental building blocks of more complex device architectures and functionalities predicted to occur using the unique electrical properties of graphene (Cheianov et al., 2007) as, for instance, multifunction logic devices (Tanachutiwat et al., 2010).

Finally, it is worth mentioning that an additional control over the tunneling coupling between GNDs can be achieved in GNDs electrostatically defined within a gated bilayer graphene sheet (Eich et al., 2018; Banszerus et al., 2018). Indeed, since a remarkable bandgap in bilayer graphene can be opened by electrostatic gating, these systems can exhibit complex, rich and fully controllable charge density patterns. From a fundamental point of view, these structures will facilitate the search for Kondo and spin-blockade physics in bilayer graphene, while from a more application point of view, they offer promising perspectives in graphene-based quantum computation (Eich et al., 2018).

8.3.2 Optical Properties and Applications of GNDs

Graphene's wavelength-independent optical absorption of 2.3% in the visible range (Nair et al., 2008) is large when considering the monoatomic thickness of this material. However, it is still necessary to considerably enhance the light-graphene interaction if one wants to realize highly efficient optical and optoelectronic devices. Furthermore, selectivity to specific wavelengths is generally required in optical devices.

A possible and facile approach for enhancing the optical absorption at the selected wavelengths in graphene is to exploit the strong and controllable absorption properties of localized plasmon resonances (LSPRs) – light-driven collective oscillations of electrons – occurring intrinsically in graphene nanostructures (Ferrari et al., 2015). This can be understood by considering the following three facts (Fang et al., 2013a,b; Yan et al., 2012): First, the use of graphene nanostructures smaller than the wavelength of the incident radiation provides a straightforward and more direct optical excitation of LSPRs with respect to surface plasmons on an extended graphene sheet. Second, the system becomes resonant at specific wavelengths because the plasmon wavelengths strongly depend on the concentration of the charge carriers, which can be easily tuned by electrostatic or chemical doping, and the size of the GNDs. Third, the smaller damping of the plasmonic excitations in graphene with respect to common plasmonic materials, such as noble metals (Ferrari et al., 2015; Dean et al., 2010), results in narrow and highly enhanced resonant profiles.

As an example, tunable LSPRs at terahertz and mid-infrared wavelengths have been reported in several graphene nanostructures such as nanoribbons (Ju et al., 2011), rings (Fang et al., 2013b) and GNDs (Yan et al., 2012; Fang et al., 2013a). In the specific case of arrays of GNDs (Figure 8.7),

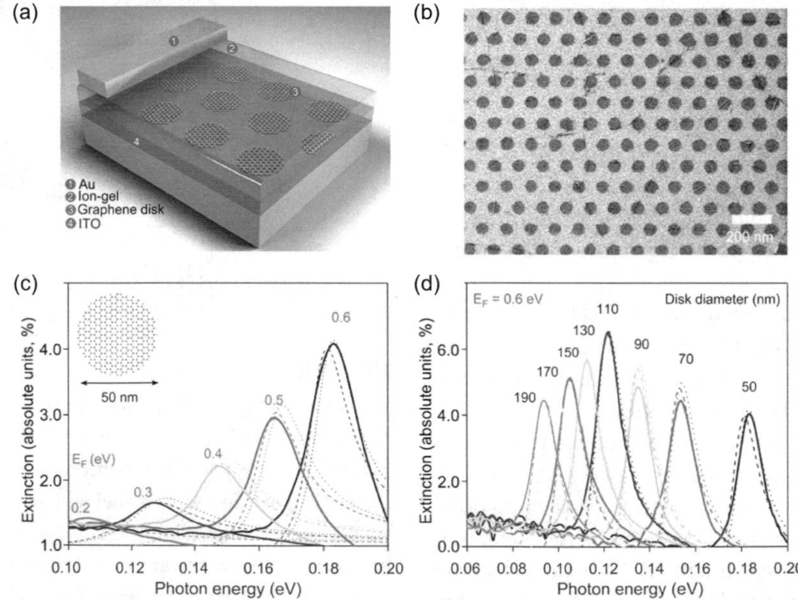

FIGURE 8.7 Electrical and geometrical tunability of GND's plasmons. (a) Schematic of the devices used in Fang et al, (2013a), consisting of a light transparent (indium tin oxide-silica) substrate (4) and an array of GNDs on top (3). The array is covered with ion gel (2), in order to dope the GNDs via a gate voltage, ΔV, applied between the ITO and a gold contact (1). (b) Scanning electron micrograph of a characteristic array of GNDs. (c) Measured (solid curves) and calculated (dashed curves and dotted curves) extinction spectra of a 50-nm disk array under different applied voltages ΔV. Here, the doping level of GNDs is indicated by the Fermi energy, EF. (d) Experimental and theoretical spectra for fixed doping when varying disk diameter (indicated by labels from 50 to 190 nm). The extinction is given in % of the difference between regions with and without graphene. (Picture adapted with permission from Fang et al., 2013a. Copyright (2013) The American Chemical Society.)

both the efficiency and the spectral position of the LSPR can be tuned by modifying structural parameters such their spacing and/or their doping level, in addition to the size of GND (Fang et al., 2013a,b; Yan et al., 2012). Moreover, the external environment affects LSPRs in GND arrays as well. For example, interactions with the substrate (Zhu et al., 2014) or the presence of stacked GNDs separated by a thin dielectric (Yan et al., 2012) have been demonstrated to make an impact in the LSPRs of GNDs, thus offering further possibilities to tune and control these localized plasmons. Specifically, GND's arrays are reported to shield up to 30% of the incoming electromagnetic radiation (Fang et al., 2013b), value that can be increased to about 50% when the individual dot structure in the array is composed by stacks of graphene and thin dielectric materials rather than a single graphene dot (Figure 8.7) (Yan et al., 2012). The wavelengths at which these LSPRs are excited approximately can vary from far-infrared (100 μm) down to mid-infrared (8 μm) (Yan et al., 2012; Fang et al., 2013b), with all of the nanostructures being fabricated via electron beam lithography. Additionally, recent studies have demonstrated LSPRs in GNDs working in the near-infrared (at 2 μm) by fabricating GND's arrays with dot diameters below 20 nm via BCP lithography (Wang et al., 2016).

These intrinsic localized plasmons occurring in arrays of GNDs show an excellent potential for device applications in optical sensing, modulators and other more sophisticated metamaterial systems (Ju et al., 2011; Yan et al., 2012; Fang et al., 2013a,b).

Similar to the previous section, also here, GND arrays formed within an electrically continuous graphene monolayer by local doping are of great interest. In particular, one can envisage metamaterials and transformation optical devices in two dimensions. For example, Vakil and Engheta investigated numerically the propagation of surface plasmon polaritons (SPP) along a graphene sheet with an array of GNDs. Each individual GND acts as a scatterer for the SPP surface wave, and the collection of these GNDs in a 2D periodic array under certain conditions indeed exhibits a backward wave propagation effect (Vakil and Engheta, 2011).

8.3.3 Optoelectronic Properties and Applications of GNDs

In general terms, the large room-temperature mobility and high Fermi velocity of carriers in graphene, together with the tunability of the light-matter interactions offered by nanopatterning this 2D material, make periodic graphene nanostructures highly attractive building blocks for efficient, compact and rapid optoelectronic devices (Grigorenko et al., 2012; Ferrari et al., 2015). Examples of such devices comprise photodetectors (Echtermeyer et al., 2011; Liu et al., 2011) and plasmonic nano-optoelectronic circuits (Vakil and Engheta, 2011; Fang et al., 2012). In this section, we focus on experimentally realized photodetection devices exclusively made from GNDs, leaving metal-graphene hybrid systems out of this chapter since their optical response is mainly dominated by the optical response of the metal nanoparticles (Echtermeyer et al., 2011; Liu et al., 2011).

In general, the field of periodic arrays of GNDs in photodetection and/or other optoelectronic applications is still at its infancy. Indeed, to the best of our knowledge, we have found only a very recent work by Tang et al. (2018). These authors study a system where physically etched arrays of GNDs are placed onto a semiconductor material, such as zinc oxide (ZnO). Owing to its wide bandgap (3.37 eV), ZnO is an excellent material for photodetection in the ultraviolet (UV) range by itself. However, the responsivity and efficiency of these ZnO-based devices in the UV region is twofold increased when combined with arrays of GNDs with dot diameters of about 20 nm. The authors ascribe this phenomenon to the formation of a favorable band alignment between ZnO and GNDs that prevents the recombination of the photogenerated carriers (Tang et al., 2018).

8.4 Conclusions

Graphene – an atomically thin sheet of carbon atoms arranged in a hexagonal lattice – exhibits a number of outstanding properties that make it attracting from both a fundamental and an application point of view. Even more interesting is the design of graphene nanostructures, and in particular, GNDs, that have been the subject of this chapter. Here, according to the definitions found in literature, two types of GNDs have been defined: The first type consists of graphene that has been physically confined in all three dimensions (i.e., spatial confinement), in analogy to the case of classical semiconductor quantum dots, such as Si or GaAs quantum dots. The second type, instead, is formed when nanometer-sized regions within a continuous graphene sheet have a doping level that is different from that of the surrounding graphene areas (i.e., electrostatic confinement).

In this chapter, we have made an overview of the synthesis processes that can be implemented to fabricate either type of GNDs. We have seen that both top-down and bottom-up approaches can indeed be used. While the former has the advantage of being easier to scale up, as, for instance, in the case of BCP lithography, the latter offers a higher degree of control, especially over edge termination, which has been shown to play an important role in defining the electronic and electric properties of the formed GNDs. Notably, in the last years, there has been a tremendous improvement in the synthesis of graphene nanostructures via bottom-up approaches such as on-surface synthesis. Here, by carefully choosing the molecular precursor(s) with a specific chemical composition, it is possible to create with atomic precision not only GND systems (also referred to as graphene quantum dots), but also graphene NPLs. We foresee that we will see more examples of advanced graphene nanostructures formed by this method.

In terms of applications, we have shown that arrays of GNDs constitute a promising building block for the design of novel and efficient electronic, optical and optoelectronic devices such as high-frequency and flexible transistors, non-volatile memories and photodetectors. This is possible due to the simple tunability of the optical and electronic properties of these graphene nanostructures via modification of their structural parameters (such as size and spatial arrangement) and/or electrostatic doping. We want to point out that these properties differ and complement those offered by non-structured (i.e., extended) graphene. In experimental terms, novel electronic functionalities such as the ability to guide electron beams or the control of the conductance in two-terminal devices via precise tunneling coupling have been already reported in arrays of doped GNDs. In contrast, the electronic properties observed in top-down (physically etched) graphene nanostructures suffer heavily from the effect (and, hence, the level) of edge disorder. It is worth noting here that, while the level of edge disorder achieved in the latter GNDs is critical for electronic applications, it is so less with respect to their optical properties. The optical properties of GNDs offer enhanced and selective absorption levels with respect to the extended graphene sheet, a behavior that can be predicted with precision by theoretical calculations. Finally, it is worth emphasizing that the optoelectronic applications of arrays of GNDs are still at its infancy. This is a promising field, which could lead to easily tunable, novel metamaterial concepts and/or nano-optoelectronic circuits. For this reason, we expect that, in the coming years, further experimental efforts will be dedicated to this topic.

References

Al-dirini, F., Mohammed, M. A., Jiang, L., et al. 2017. Negative differential resistance in planar graphene quantum dot resonant tunneling diodes. *IEEE-NANO 2017 (17th International Conference on Nanotechnology)*, Pittsburgh, PA, IEEE, 965–968.

Bacon, M., Bradley, S.J. & Nann, T. 2014. Graphene quantum dots. *Particle and Particle Systems Characterization*, 31, 415–428.

Bae, S., Kim, H., Lee, et al. 2010. Roll-to-roll production of 30-inch graphene films for transparent electrodes. *Nature Nanotechnology*, 5, 574.

Bai, J., Zhong, X., Jiang, S., Huang, Y. & Duan, X. 2010. Graphene nanomesh. *Nature Nanotechnology*, 5, 190.

Balandin, A. A., Ghosh, S., Bao, W., et al. 2008. Superior thermal conductivity of single-layer graphene. *Nano Letters*, 8, 902–907.

Banszerus, L., Frohn, B., Epping, A., et al. 2018. Gate-defined electron-hole double dots in bilayer graphene. *arXiv preprint arXiv:1803.10857*.

Baringhaus, J., Ruan, M., Edler, F., et al. 2014. Exceptional ballistic transport in epitaxial graphene nanoribbons. *Nature*, 506, 349.

Berry, M. V. & Mondragon, R. J. 1987. Neutrino billiards: time-reversal symmetry-breaking without magnetic fields. *Proceedings of the Royal Society London A*, 412, 53–74.

Bisri, S. Z., Piliego, C., Gao, J. & Loi, M. A. 2014. Outlook and emerging semiconducting materials for ambipolar transistors. *Advanced Materials*, 26, 1176–1199.

Blake, P., Hill, E., Castro Neto, A., et al. 2007. Making graphene visible. *Applied Physics Letters*, 91, 063124.

Bunch, J. S., Verbridge, S. S., Alden, J. S., et al. 2008. Impermeable atomic membranes from graphene sheets. *Nano Letters*, 8, 2458–2462.

Cabrini, S. & Kawata, S. 2012. *Nanofabrication Handbook*, CRC Press, Boca Raton, FL.

Cai, J., Ruffieux, P., Jaafar, R., et al. 2010. Atomically precise bottom-up fabrication of graphene nanoribbons. *Nature*, 466, 470–473.

Camilli, L., Jørgensen, J. H., Tersoff, J., et al. 2017. Self-assembly of ordered graphene nanodot arrays. *Nature Communications*, 8, 47.

Carbonell-Sanromà, E., Brandimarte, P., Balog, R., et al. 2016. Quantum dots embedded in graphene nanoribbons by chemical substitution. *Nano Letters*, 17, 50–56.

Caridad, J. M., Connaughton, S., Ott, C., Weber, H. B. & Krstic, V. 2016. An electrical analogy to Mie scattering. *Nature Communications*, 7, 12894.

Caridad, J. M., Power, S. R., Lotz, et al. 2018. Conductance quantization suppression in the quantum Hall regime. *Nature Communications*, 9, 659.

Cheianov, V. V., Falko, V. & Aatshuler, B. 2007. The focusing of electron flow and a Veselago lens in graphene pn junctions. *Science*, 315, 1252–1255.

Chen, S., Brown L., Levendorf, M., et al. 2011. Oxidation resistance of graphene-coated Cu and Cu/Ni alloy. *ACS Nano*, 5, 1321–1327.

Chen, Y.-C., De Oteyza, D. G., Pedramrazi, Z., Chen, C., Fischer, F. R. & Crommie, M. F. 2013. Tuning the band gap of graphene nanoribbons synthesized from molecular precursors. *ACS Nano*, 7, 6123–6128.

Cserti, J., Pályi, A. & Pèterfalvi, C. 2007. Caustics due to a negative refractive index in circular graphene p– n junctions. *Physical Review Letters*, 99, 246801.

Das, S., Demarteau, M. & Roelofs, A. 2014. Ambipolar phosphorene field effect transistor. *ACS Nano*, 8, 11730–11738.

Dean, C. R., Young, A. F., Merici, et al. 2010. Boron nitride substrates for high-quality graphene electronics. *Nat Nano*, 5, 722–726.

Drienovsky, M., Sandner, A., Baumgartner, et al. 2017. Few-layer graphene patterned bottom gates for van der Waals heterostructures. *arXiv preprint arXiv:1703.05631*.

Dubey, S., Singh, V., Bhat, A. K., et al. 2013. Tunable superlattice in graphene to control the number of Dirac points. *Nano Letters*, 13, 3990–3995.

Echtermeyer, T., Britnell, L., Jasnos, P., et al. 2011. Strong plasmonic enhancement of photovoltage in graphene. *Nature Communications*, 2, 458.

Eich, M., Pisoni, R., Pally, A., et al. 2018. Coupled quantum dots in bilayer graphene. *arXiv preprint arXiv:1805.02943*.

Fang, Z., Thongrattanasiri, S., Schlather, A., et al. 2013a. Gated tunability and hybridization of localized plasmons in nanostructured graphene. *ACS Nano*, 7, 2388–2395.

Fang, Z., Wang, Y., Liu, Z., et al. 2012. Plasmon-induced doping of graphene. *ACS Nano*, 6, 10222–10228.

Fang, Z., Wang, Y., Schlather, A. E., et al. 2013b. Active tunable absorption enhancement with graphene nanodisk arrays. *Nano Letters*, 14, 299–304.

Fehske, H., Hager, G. & Pieper, A. 2015. Electron confinement in graphene with gate-defined quantum dots. *Physica Status Solidi B*, 252, 1868–1871.

Ferrari, A. C., Bonaccorso, F., Falko, V., et al. 2015. Science and technology roadmap for graphene, related two-dimensional crystals, and hybrid systems. *Nanoscale*, 7, 4598–4810.

Giovannetti, G., Khomyakov, P., Brocks, G., Karpan, V. V., Van Den Brink, J. & Kelly, P. J. 2008. Doping graphene with metal contacts. *Physical Review Letters*, 101, 026803.

Grigorenko, A., Polini, M. & Novoselov, K. 2012. Graphene plasmonics. *Nature Photonics*, 6, 749.

Guinea, F. 2008. Models of electron transport in single layer graphene. *Journal of Low Temperature Physics*, 153, 359–373.

Gutierrez, C., Brown, L., Kim, C. J., Park, J. & Pasupathy, A. N. 2016. Klein tunnelling and electron trapping in nanometre-scale graphene quantum dots. *Nature Physics*, 12, 1069.

Güçlü, A. D., Potasz, P. & Hawrylak, P. 2010. Excitonic absorption in gate-controlled graphene quantum dots. *Physical Review B*, 82, 155445.

Güttinger, J., Molitor, F., Stampfer, C., et al. 2012. Transport through graphene quantum dots. *Reports on Progress in Physics*, 75, 126502.

Güttinger, J., Stampfer, C., Hellmuller, S., Molitor, F., Ihn, T. & Ensslin, K. 2008. Charge detection in graphene quantum dots. *Applied Physics Letters*, 93, 212102.

Hamada, N., Sawada, S.-I. & Oshiyama, A. 1992. New one-dimensional conductors: Graphitic microtubules. *Physical Review Letters*, 68, 1579.

Han, M. Y., Özyilmaz, B., Zhang, Y. & Kim, P. 2007. Energy band-gap engineering of graphene nanoribbons. *Physical Review Letters*, 98, 206805.

Han, W. 2016. Perspectives for spintronics in 2D materials. *APL Materials*, 4, 032401.

Han, Z., Allain, A., Arjmandi-Tash, et al. 2014. Collapse of superconductivity in a hybrid tin–graphene Josephson junction array. *Nature Physics*, 10, 380.

Heinisch, R., Bronold, F. & Fehske, H. 2013. Mie scattering analog in graphene: Lensing, particle confinement, and depletion of Klein tunneling. *Physical Review B*, 87, 155409.

Hu, S., Lozada-Hidalgo, M., Wang, F., et al. 2014. Proton transport through one-atom-thick crystals. *Nature*, 516, 227–230.

Jiao, L., Wang, X., Diankov, G., Wang, H. & Dai, H. 2010. Facile synthesis of high-quality graphene nanoribbons. *Nature Nanotechnology*, 5, 321.

Jiao, L., Zhang, L., Wang, X., Diankov, G. & Dai, H. 2009. Narrow graphene nanoribbons from carbon nanotubes. *Nature*, 458, 877.

Ju, L., Geng, B., Horng, J., et al. 2011. Graphene plasmonics for tunable terahertz metamaterials. *Nature Nanotechnology*, 6, 630.

Katsnelson, M., Novoselov, K. & Geim, A. 2006. Chiral tunnelling and the Klein paradox in graphene. *Nature Physics*, 2, 620.

Kim, H. W., Yoon, H. W., Yoon, S.-M., et al. 2013. Selective gas transport through few-layered graphene and graphene oxide membranes. *Science*, 342, 91–95.

Koch, S. W. 1993. *Semiconductor Quantum Dots*, World Scientific, Singapore.

Landau, L. D. & Lifsl, E. M. 1969. *Statistical Physics*. 2nd rev. Oxford: Pergamon Press and Reading: Addison-Wesley.

Lee, C., Wei, X., Kysar, J. W. & Hone, J. 2008. Measurement of the elastic properties and intrinsic strength of monolayer graphene. *Science*, 321, 385–388.

Lee, J., Kim, K., Park, W. I., et al. 2012. Uniform graphene quantum dots patterned from self-assembled silica nanodots. *Nano Letters*, 12, 6078–6083.

Lee, J., Wong, D., Velasco, J., et al. 2016. Imaging electrostatically confined Dirac fermions in graphene quantum dots. *Nature Physics*, 12, 1032–1036.

Li, H., Kang, Z., Liu, Y. & Lee, S.-T. 2012. Carbon nanodots: Synthesis, properties and applications. *Journal of Materials Chemistry*, 22, 24230–24253.

Li, X., Rui, M., Song, J., Shen, Z. & Zeng, H. 2015. Carbon and graphene quantum dots for optoelectronic and energy devices: A review. *Advanced Functional Materials*, 25, 4929–4947.

Li, X., Wang, X., Zhang, L., Lee, S. & Dai, H. 2008. Chemically derived, ultrasmooth graphene nanoribbon semiconductors. *Science*, 319, 1229–1232.

Liu, Y., Cheng, R., Liao, L., et al. 2011. Plasmon resonance enhanced multicolour photodetection by graphene. *Nature Communications*, 2, 579.

Lu, J., Yeo, P. S. E., Gan, C. K., Wu, P. & Loh, K. P. 2011. Transforming C60 molecules into graphene quantum dots. *Nature Nanotechnology*, 6, 247–252.

Merchant, C. A., Healy, K., Wanunu, M., et al. 2010. DNA translocation through graphene nanopores. *Nano Letters*, 10, 2915–2921.

Mermin, N. D. & Wagner, H. 1966. Absence of ferromagnetism or antiferromagnetism in one- or two-dimensional isotropic Heisenberg models. *Physical Review Letters*, 17, 1133–1136.

Mintmire, J., Dunlap, B. & White, C. 1992. Are fullerene tubules metallic? *Physical Review Letters*, 68, 631.

Molle, A., Goldberger, J., Houssa, M., Xu, Y., Zhang, S.-C. & Akinwande, D. 2017. Buckled two-dimensional Xene sheets. *Nature Materials*, 16, 163–169.

Moreno, C., Vilas-Varela, M., Kretz, B., et al. 2018. Bottom-up synthesis of multifunctional nanoporous graphene. *Science*, 360, 199–203.

Morozov, S., Novoselov, K., Katsnelson, M.,et al. 2008. Giant intrinsic carrier mobilities in graphene and its bilayer. *Physical Review Letters*, 100, 016602.

Nair, R. R., Blake, P., Grigorenko, A. N., et al. 2008. Fine structure constant defines visual transparency of graphene. *Science*, 320, 1308–1308.

Neto, A. C., Guinea, F., Peres, N. M., Novoselov, K. S. & Geim, A. K. 2009. The electronic properties of graphene. *Reviews of Modern Physics*, 81, 109.

Novoselov, K., Jiang, D., Schedin, F., et al. 2005a. Two-dimensional atomic crystals. *Proceedings of the National Academy of Sciences of the United States of America*, 102, 10451–10453.

Novoselov, K. S., Geim, A. K., Morozov, S., et al. 2005b. Two-dimensional gas of massless Dirac fermions in graphene. *Nature*, 438, 197.

Novoselov, K. S., Geim, A. K., Morozov, S. V., et al. 2004. Electric field effect in atomically thin carbon films. *Science*, 306, 666–669.

Novoselov, K. S., Jiang, Z., Zhang, Y., et al. 2007. Room-temperature quantum Hall effect in graphene. *Science*, 315, 1379–1379.

O'Hern, S. C., Boutilier, M. S., Idrobo, J.-C., et al. 2014. Selective ionic transport through tunable subnanometer pores in single-layer graphene membranes. *Nano Letters*, 14, 1234–1241.

Odom, T. W., Huang, J.-L., Kim, P. & Lieber, C. M. 1998. Atomic structure and electronic properties of single-walled carbon nanotubes. *Nature*, 391, 62.

Oh, S. D., Kim, J., Lee, D. H., et al. 2015. Structural and optical characteristics of graphene quantum dots size-controlled and well-aligned on a large scale by polystyrene-nanosphere lithography. *Journal of Physics D: Applied Physics*, 49, 025308.

Pedersen, T. G., Flindt, C., Pedersen, J., Mortensen, N. A., Jauho, A.-P. & Pedersen, K. 2008. Graphene antidot lattices: Designed defects and spin qubits. *Physical Review Letters*, 100, 136804.

Peierls, R. 1934. Bemerkungen über umwandlungstemperaturen. *Helvetica Physica Acta*, 7, 158.

Ponomarenko, L., Schedin, F., Katsnelson, M., et al. 2008. Chaotic Dirac billiard in graphene quantum dots. *Science*, 320, 356–358.

Rasappa, S., Caridad, J. M., Schulte, L., et al. 2015. High quality sub-10 nm graphene nanoribbons by on-chip PS-b-PDMS block copolymer lithography. *RSC Advances*, 5, 66711–66717.

Ritter, K. A. & Lyding, J. W. 2009. The influence of edge structure on the electronic properties of graphene quantum dots and nanoribbons. *Nature Materials*, 8, 235–242.

Ruffieux, P., Cai, J., Plumb, N. C., et al. 2012. Electronic structure of atomically precise graphene nanoribbons. *ACS Nano*, 6, 6930–6935.

Rycerz, A., Tworzydlo, J. & Beenakker, C. 2007. Valley filter and valley valve in graphene. *Nature Physics*, 3, 172.

Saito, R., Fujita, M., Dresselhaus, G. & Dresselhaus, M. S. 1992a. Electronic structure of graphene tubules based on C 60. *Physical Review B*, 46, 1804.

Saito, R., Fujita, M., Dresselhaus, G. & Dresselhaus, U. M. 1992b. Electronic structure of chiral graphene tubules. *Applied Physics Letters*, 60, 2204–2206.

Santos, J. E., Peres, N. M., Dos Santos, J. M. L. & Neto, A. H. C. 2011. Electronic doping of graphene by deposited transition metal atoms. *Physical Review B*, 84, 085430.

Schaibley, J. R., Yu, H., Clark, G., et al. 2016. Valleytronics in 2D materials. *Nature Reviews Materials*, 1, 16055.

Schneider, G. F., Kowalczyk, S. W., Calado, V. E., et al. 2010. DNA translocation through graphene nanopores. *Nano Letters*, 10, 3163–3167.

Schnez, S., Gurttinger, J., Huefner, M., Stampfer, C., Ensslin, K. & Ihn, T. 2010. Imaging localized states in graphene nanostructures. *Physical Review B*, 82, 165445.

Son, Y.-W., Cohen, M. L. & Louie, S. G. 2006. Energy gaps in graphene nanoribbons. *Physical Review Letters*, 97, 216803.

Stampfer, C., Guttinger, J., Molitor, F., Graf, D., Ihn, T. & Ensslin, K. 2008. Tunable Coulomb blockade in nanostructured graphene. *Applied Physics Letters*, 92, 012102.

Steinberg, H., Gardner, D. R., Lee, Y. S. & Jarillo-Herrero, P. 2010. Surface state transport and ambipolar electric field effect in Bi_2Se_3 nanodevices. *Nano Letters*, 10, 5032–5036.

Surwade, S. P., Smirnov, S. N., Vlassiouk, I. V., et al. 2015. Water desalination using nanoporous single-layer graphene. *Nat Nano*, 10, 459–464.

Talirz, L., Ruffieux, P. & Fasel, R. 2016. On-surface synthesis of atomically precise graphene nanoribbons. *Advanced Materials*, 28, 6222–6231.

Tanachutiwat, S., Lee, J. U., Wang, W. & Sung, C. Y. 2010. Reconfigurable multi-function logic based on graphene pn junctions. *Proceedings of the 47th Design Automation Conference*, Anaheim, California, ACM, 883–888.

Tang, R., Han, S., Teng, F., et al. 2018. Size-controlled graphene nanodot arrays/ZnO hybrids for high-performance UV photodetectors. *Advanced Science*, 5, 1700334.

Tang, Y., Cao, X., Guo, R., et al. 2016. Flat-lens focusing of electron beams in graphene. *Scientific Reports*, 6, 33522.

Tao, C., Jiao, L., Yazyev, O. V., et al. 2011. Spatially resolving edge states of chiral graphene nanoribbons. *Nature Physics*, 7, 616.

Tao, L., Cinquanta, E., Chiappe, D., et al. 2015. Silicene field-effect transistors operating at room temperature. *Nature Nanotechnology*, 10, 227.

Tombros, N., Jozsa, C., Popinciuc, M., Jonkman, H. T. & Van Wees, B. J. 2007. Electronic spin transport and spin precession in single graphene layers at room temperature. *Nature*, 448, 571.

Vakil, A. & Engheta, N. 2011. Transformation optics using graphene. *Science*, 332, 1291–1294.

Wang, H., Liu, F., Fu, W., Fang, Z., Zhou, W. & Liu, Z. 2014. Two-dimensional heterostructures: Fabrication, characterization, and application. *Nanoscale*, 6, 12250–12272.

Wang, S., Kharche, N., Costa Girao, E., et al. 2017. Quantum dots in graphene nanoribbons. *Nano Letters*, 17, 4277–4283.

Wang, Z., Li, T., Almdal, K., Mortensen, N. A., Xiao, S. & Ndoni, S. 2016. Experimental demonstration of graphene plasmons working close to the near-infrared window. *Optics Letters*, 41, 5345–5348.

Wu, Y., Jenkins, K. A., Valdes-Garcia, A., et al. 2012. State-of-the-art graphene high-frequency electronics. *Nano Letters*, 12, 3062–3067.

Yan, H., Li, X., Chandra, B., et al. 2012. Tunable infrared plasmonic devices using graphene/insulator stacks. *Nature Nanotechnology*, 7, 330.

Yang, L., Park, C.-H., Son, Y.-W., Cohen, M. L. & Louie, S. G. 2007. Quasiparticle energies and band gaps in graphene nanoribbons. *Physical Review Letters*, 99, 186801.

Yuan, P., Tian, W., Zeng, Y., Zhang, Z. & Zhang, J. 2014. Electronic properties of one-dimensional graphene quantum-dot arrays. *Organic Electronics*, 15, 3577–3583.

Zhang, X. Q., Lin, C. H., Tseng, Y. W., Huang, K. H. & Lee, Y. H. 2015. Synthesis of lateral heterostructures of semiconducting atomic layers. *Nano Letters*, 15, 410–415.

Zhang, Y., Tan, Y.-W., Stormer, H. L. & Kim, P. 2005. Experimental observation of the quantum Hall effect and Berry's phase in graphene. *Nature*, 438, 201.

Zhang, Y., Ye, J., Matsuhashi, Y. & Iwasa, Y. 2012. Ambipolar MoS_2 thin flake transistors. *Nano Letters*, 12, 1136–1140.

Zhang, Z. Z., Chang, K. & Peeters, F. M. 2008. Tuning of energy levels and optical properties of graphene quantum dots. *Physical Review B*, 77, 235411.

Zhu, X., Wang, W., Yan, W., et al. 2014. Plasmon–phonon coupling in large-area graphene dot and antidot arrays fabricated by nanosphere lithography. *Nano Letters*, 14, 2907–2913.

Zou, X. & Yakobson, B. I. 2015. An open canvas: 2D materials with defects, disorder, and functionality. *Accounts of Chemical Research*, 48, 73–80.

9

Graphene Nanomesh

Ahmed A. Maarouf
Imam Abdulrahman Bin Faisal University

Graphene nanomeshes (GNMs, also called graphene antidot lattices) are two-dimensional (2D) structures, created by forming a superlattice of pores in a graphene sheet. The sizes and geometries of the pores control the physical properties of the GNM.

9.1 Introduction

Since its experimental realization in 2004, graphene has been the focus of substantial theoretical and experimental research [8,21]. This monolayer of sp^2 carbon atoms possesses unique physical and chemical properties, including high electronic mobility, optical transparency, and supreme mechanical strength, which make it an ideal candidate for many applications in science and technology, such as nanoelectronics [32,33,38,53], chemical separation [27], catalysis [40], and radio frequency amplifiers [58].

Despite the remarkable success of graphene in many domains, it has two problems that can hinder its full utilization in technological applications. First, it has a zero electronic gap, which prevents its existence in the transistor world. Second, and due to its highly stable sp^2 structure, it is (relatively) chemically inert, which greatly limits its chemical applications.

Many attempts have been made to open a gap in the electronic band structure of graphene. A bilayer graphene fabricated with Bernal stacking will develop an electronic gap if placed in an electric field [59], but the gap is too small for room temperature devices. Other attempts include the exploitation of quantum confinement by forming graphene nanoribbons [51], which do have a reasonable gap, but they suffer from edge scattering, in addition to the absence of a stable and controlled doping mechanism. A gap can also be opened by subjecting graphene to strain [9,11]. All routes have some challenges, which make them not technologically feasible, hindering the fabrication of a graphene field effect transistor (FET).

The relative chemical inertness of graphene makes its doping, in general, unstable [28,34]. Dopants tend to physisorb on its surface, rendering the shift of the Fermi energy from the Dirac point only temporary. Without the development of a stable doping method for graphene, the use of many of its exotic physical properties will be severely limited.

GNMs (also called porous graphene) may come to the rescue by solving both of graphene's problems by providing the means to (i) overcome the relative chemical inertness, hence allowing for doping stability, and (ii) open a technologically lucrative bandgap, hence securing graphene's entry to the transistor world. GNMs are crown-like structures formed by creating a superlattice of pores in a graphene sheet. The pore edges enhance graphene's chemical activity, allowing it to bind with specific chemical dopants. Furthermore, and depending on the details of the superlattice and the geometry of the pores, these structures can either inherit graphene's semimetallicity or exhibit semiconducting behavior with a fractional eV gap [41]. If such semiconducting structures could be both controllably and stably doped, they could be used to fabricate graphene-based transparent electrodes [28], computer logic switches [59], and spintronic devices [17].

9.2 Electronic and Magnetic Properties of GNMs

In this section, we briefly discuss the electronic and magnetic properties of GNMs, with a focus on those properties that differentiate them from their ancestor, graphene.

Pores in the graphene skeleton lead to quantum confinement of the electrons in the surrounding regions. One can

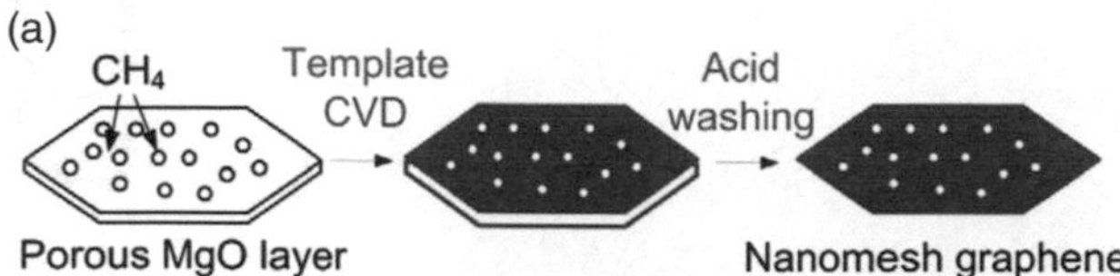

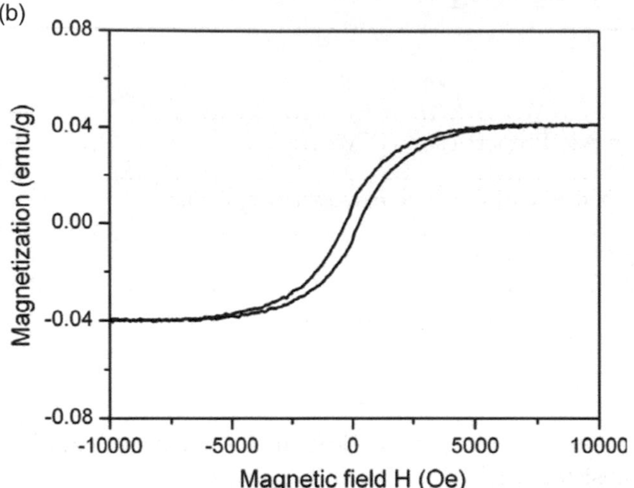

FIGURE 9.1 (a) CVD formation of GNM. (b) Magnetization hysteresis loop of the GNM at 300 K [42].

think of the pore lattice as a periodic potential. With the pore lattice, the unit cell of the system becomes bigger, which means that Brillouin zone becomes smaller. The pore lattice breaks the translation symmetry of the sheet. For certain pore lattices, the two distinct Dirac points of graphene are folded back on the Γ point, mixing the K and the $-K$ states, thus opening an electronic gap [20,37,41,44,47]. For other pore lattices, the two Dirac points do not mix, and the GNM is semimetallic. The size of the gap depends inversely on the width of the confinement region (sometimes called the *neck width*). Typically, a GNM with a nanometer-sized pore has a gap of about 0.5 eV.

The quest to create magnetic systems from graphene-based structures has been growing, as it would offer non-metallic materials with Curie temperature higher than

FIGURE 9.2 Optimal configuration for metal-doped N–GNM: (from left to right) trigonal (t) pore (top view), t-pore (side view), and hexagonal (h) pore (top view) [17].

room temperature, with other advantages, such as bio-computability, lightweight, and low cost.

Pristine graphene is non-magnetic. Nevertheless, some graphene structures can become magnetic upon the introduction of various defects, partial hydrogenation, and topological modifications [18,26,42,54,62,67].

A GNM produced using chemical vapor deposition (CVD) becomes ferromagnetic with a room temperature saturation magnetization of 0.04 emu/g (Figure 9.1) [42]. This is four times the value obtained for reduced graphene oxide. It is believed that the high density of defects at the edges of the pores contributes to the ferromagnetism of the GNM. First-principles calculations have shown that pore geometry strongly affects the magnetic properties of a GNM. For triangular-shaped hydrogen-passivated pores, the magnetic moment of the GNM increases with the difference in the numbers of removed A and B lattice sites (as expected from Lieb's theorem), and also with the size of the pore [61].

Another approach to producing magnetism in GNMs is through doping with 3d transition metal atoms [17]. In this context, GNMs with pores passivated with nitrogen are studied with spin-polarized first-principles calculations. High-symmetry trigonal (t) and hexagonal (h) pore shapes are considered, with various transition metal dopants (Figure 9.2). The t GNM has 6 C atoms removed and its pore edge is passivated with 3 N atoms, while the h GNM has 12 C atoms removed and its edge is passivated with 12 N atoms. The generated GNM structures are termed t–N-GNM and h–N-GNM, respectively.

Pristine t–N-GNM and h–N-GNM have zero magnetic moments. Figure 9.3 shows the density of states (DOS) for the t–N-GNM and h–N-GNM structures. For the t-pore, a broad band of states is created around the Fermi energy,

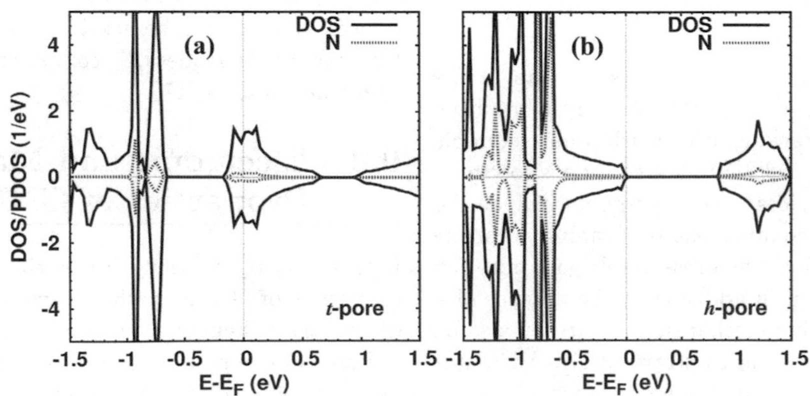

FIGURE 9.3 DOS and partial DOS (PDOS) of N atoms, for (a) t–N-GNM and (b) h–N-GNM [17].

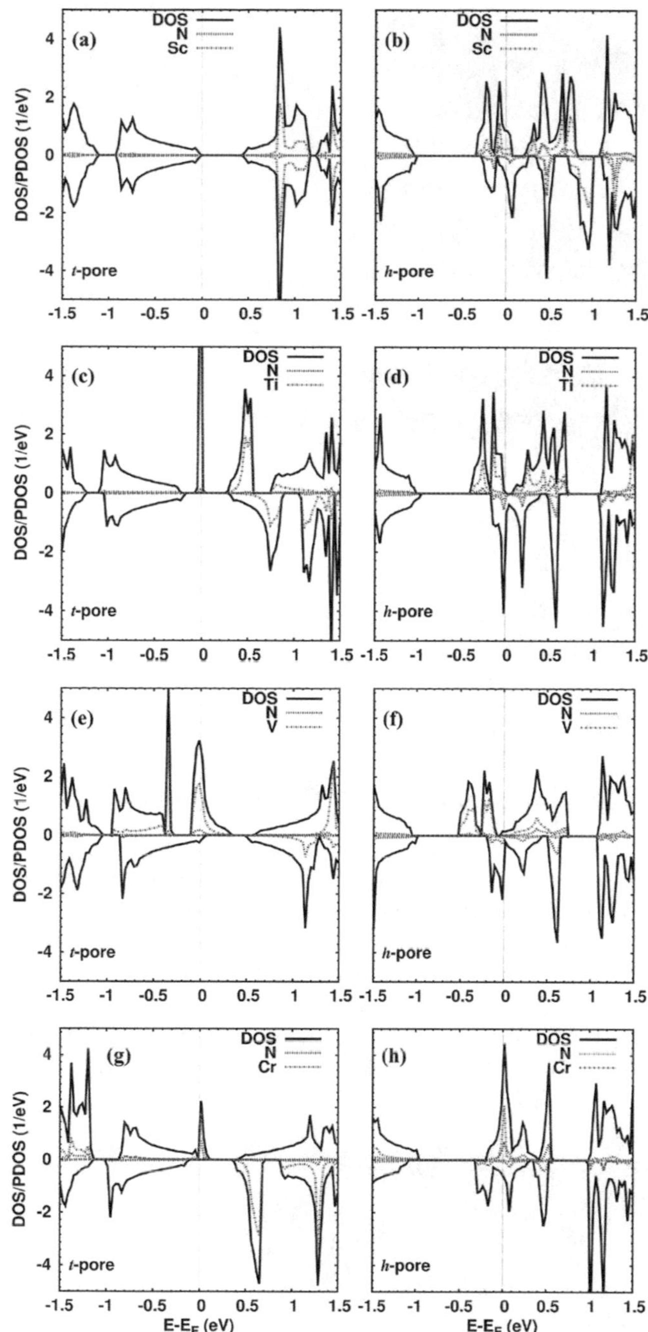

FIGURE 9.4 DOS and PDOS, for (a) and (b) Sc, (c) and (d) Ti, (e) and (f) V (e), and (g) and(f) Cr, doped N–GNM;t-pore (left) and h-pore (right) [17].

and the t–N-GNM system is metallic. On the other hand, the h–N-GNM exhibits a semiconducting behavior with a bandgap of about 0.8 eV. The N states have no significant contributions in the vicinity of the Fermi energy in both cases. The pore shape and geometry affect the electronic and magnetic properties, as seen with previously discussed H-passivated pores. The three closely interacting N atoms in the pore of the t–H–GNM structure p-dope it by acquiring three electrons from its carbon atoms, as is confirmed by integrating the DOS from the valence band edge to the Fermi energy. This is in contrast to the semiconducting behavior of the corresponding O-passivated GNM [1,39]

Upon doping with 3d transition metals (Sc through Zn), the metal atom binds to the closest N atoms at the pore edge (Figure 9.2), with an average bond length ranging from 1.8 to 1.98 for the t–N-GNM and from 1.96 to 2.22 for the h–N-GNM. The corresponding binding energy ranges are 4.7–10.0 eV and 1.4–7.4 eV, respectively. The covalent bonds formed between N and the 3d metal are the origin of the magnetic properties of metal-doped GNMs. The N–GNM systems are found to be magnetic for most 3d metals inside the trigonal pore, as shown in Figure 9.4 (exceptions are Sc, Fe, and Zn). For the hexagonal pore, and with the exception of Cu, all doped N–GNMs are magnetic except for Cu. A 100 % spin polarization at the Fermi level occurs with Ti for both pore types, as well as for Cr in the trigonal pore. In general, these structures are capable of sustaining a spin current and hence are suitable for use in spintronic devices. Some of them (V, Co, and Zn) can also be used to fabricate spin filters.

9.3 Fabrication of GNMs

There has been a growing interest in developing methods for fabricating GNM [3,4,6,7,36,55,63]. The main challenge is to develop systematic and reliable methods to form a highly periodic array of pores, such that disorder in the pore size/lattice does not quench the semiconducting gap, as well as the carrier mobility of the GNM.

9.3.1 Block Copolymer Method

Perhaps one of the earliest successful attempts to fabricate GNMs is reported by Bai et al [7]. The procedure (Figure 9.5) is summarized as follows: (a) Mechanically peeled graphene is used as the starting material (although the authors report that it can be extended to graphene films obtained through chemical exfoliation or CVD). (b) The graphene is then covered with a 10-nm-thick silicon oxide (SiO_x) film, which acts as a protecting layer and a grafting substrate for the subsequent block copolymer nanopatterning [23]. (c) The poly(styrene-block-methyl methacrylate) (P(S-b-MMA)) block copolymer film is annealed and developed, leaving the porous polystyrene (PS) matrix as the nanomesh template for further patterning [23]. (d) A CHF_3-based reactive-ion etching (RIE) process is followed by (e) an oxygen plasma etching process to create pores in the graphene layer. (f) Dipping in HF removes the oxide mask, and (g) GNM is obtained by etching away the silicon oxide substrate.

This method offers high control on the generated pore lattice, with its two main parameters, namely, the pore size and the pore lattice constant. For a fixed pore lattice constant, the pore size can be controlled by over-etching during the fabrication process. The pore lattice constant can be tuned by using smaller-molecular-weight block

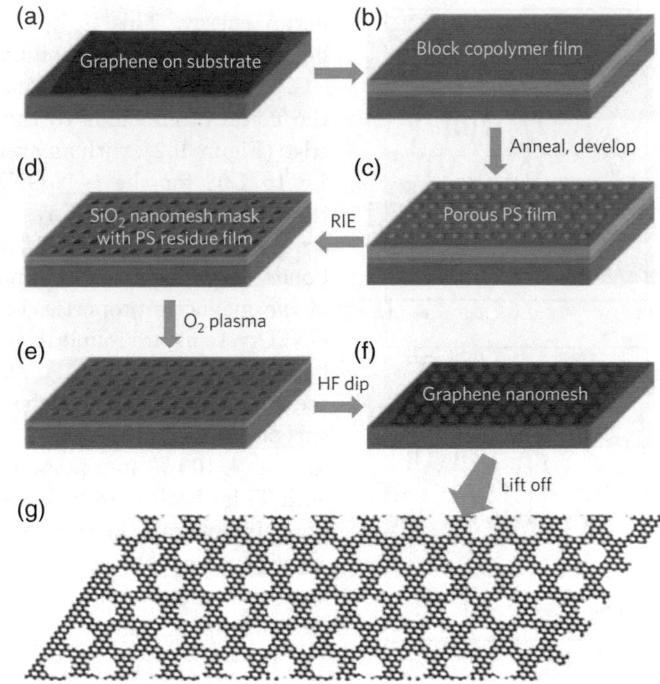

FIGURE 9.5 A schematic showing the fabrication process of a GNM using a block copolymer [7].

copolymer thin film as the mask template. The pore size can be alternatively described through the *neck width*, which measures the smallest carbon lattice width between two adjacent pores. Figure 9.6 shows transmission electron microscopic (TEM) images of a GNM with two lattice constants (39 and 27 nm), and multiple neck widths (14.6, 11.2, 7.1, and 9.3), obtained through controlled lateral over-etching of the PS mask. A statistical analysis of the neck widths is also shown [7].

9.3.2 Local Catalytic Hydrogenation Method

GNMs can also be fabricated with a catalytic approach, in which local hydrogenation of carbon atoms occurs by Cu nanoparticles (Figure 9.7). Graphene on SiO_2/Si is annealed at 400°C. A thin layer of Cu was then evaporated on the graphene sheet. Subsequent annealing at 500 °C leads to the formation of Cu nanoparticles on the graphene sheet. The size and density of the nanoparticles can be

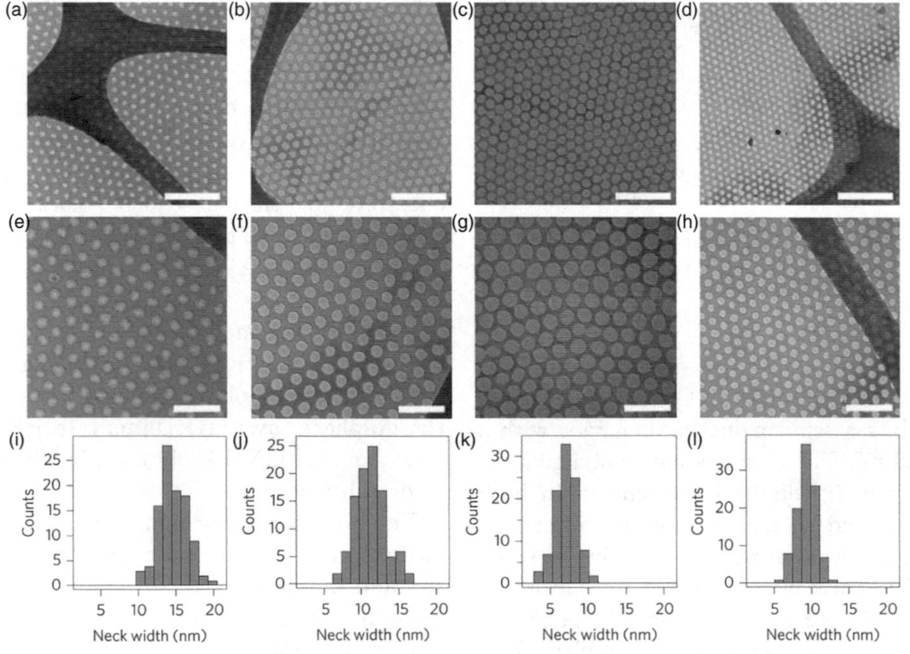

FIGURE 9.6 TEM studies of graphene and thin-layer graphite nanomesh [7].

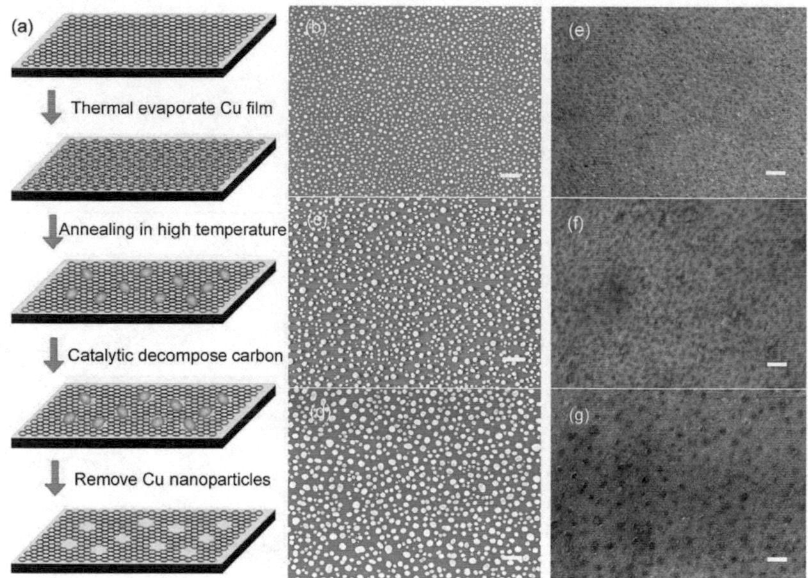

FIGURE 9.7 (a) A schematic diagram of the GNM fabrication through local catalytic hydrogenation of carbon at high temperature. (b-d) Scanning electron microscopic (SEM) images of Cu nanoparticles by deposited and annealed Cu film on graphene with 1, 2, and 4 nm thickness, respectively; (e-g) SEM images of the GNM corresponding to (b-d) after the Cu nanoparticles were removed. The scale bar is 200 nm [36].

controlled by the thickness of the evaporated Cu layer and the temperature of the annealing process, which occurs in forming gas (5% hydrogen, 95% ultrahigh-purity argon gas). The Cu nanoparticles catalyze the formation of methane gas, thus creating pores in the underlying graphene sheet. The Cu nanoparticles are later removed using Marble's reagent [36].

9.4 Doping of GNMs

In order to fabricate CMOS-type devices from GNMs, it is essential to both n- and p-dope the material most simply without major alteration of the band structure. Substitutional doping by ion implantation, which has been demonstrated in graphene [19,34,49,56], suffers from concentration fluctuations in nanoscale devices. An alternative strategy realized in graphene is surface doping, wherein non-volatile compounds are spread across a 2D material, but this approach is also subject to undesirable concentration fluctuations and the free moieties (dopants) can migrate across the device leading to stability problems [28,34]. It is therefore desirable to find an approach for doping GNMs that offers both high doping concentration control and thermal stability.

Here, a concept from chemistry is utilized, that is, ion chelation [24,31], to generate the desired doping physics in GNMs. The carbon atoms at the pore perimeter are chemically active, and unless the GNM is *in vacuo*, the pore perimeter will be passivated by a chemical moiety. Ions can be chelated in GNMs because an appropriate passivating species (such as H or O) forms a natural crown in which ions can selectively set or dock like a "jewel" (as in the well-known crown-ether materials [10,46].) Passivation with chosen chemical functional groups leads to charge polarization at the pore perimeters due to an electronegativity mismatch between the passivating entity and carbon, thus forming an alluring trap for ions. When an atom approaches a chelation site, the electrostatic energy gain becomes higher than the energy cost of charge transfer, allowing an electron or a hole to transfer to the GNM within a rigid band picture; the resulting ion is tightly bound in the pore. In Figure 9.8, atom X docks in the chelation site, which has been passivated by species Y. Charge transfer is stabilized by the electrostatic interaction between ionized species X and the local dipole moment (permanent plus induced) of the pore perimeter carbon-passivating moiety Y. We term these compounds "crown GNMs" and give them the symbol X@Y-GNM.

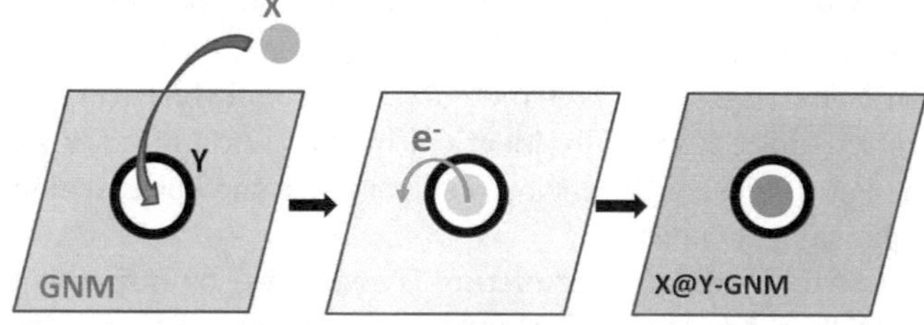

FIGURE 9.8 A three-state conceptual thermodynamic sketch of the ion chelation doping of a GNM; the chelant atom is brought close to the undoped passivated GNM, resulting in the docking of the chelant in the pore, the ionization of the chelant, and the doping of the GNM, followed by structural relaxation or solvation. Here, n-doping is shown; p-doping simply replaces the electron with a hole. The dynamical process involves partial charge transfer and raising/lowering of the HOMO/LUMO of the chelant [39].

In order to understand X@Y-GNM physics, two GNM passivations are considered: one with electronegativity lower than that of carbon ($\chi_C = 2.55$) and another with electronegativity greater than that of carbon. The two natural first choices are hydrogen ($\chi_H = 2.0$) and oxygen ($\chi_O = 3.44$). Pores passivated with hydrogen form pockets for electron-withdrawing p-dopants, such as fluorine and chlorine. On the other hand, pores passivated with oxygen form pockets for electron-donating n-dopants, such as sodium and potassium.

Figure 9.9a shows a hydrogen- and an oxygen-passivated GNM. The passivation saturates all atomic bonds, and therefore, a passivated GNM is chemically stable. Figure 9.9b compares the band structures of the two passivated systems and pristine graphene. Local density approximation (LDA) predicts a gap of 0.7 eV for H-GNM and of 0.4 eV for O-GNM. The bands of the passivated GNMs are linear away from the gap region, although with a group velocity that is half that of graphene. The DOS of H-GNM, O-GNM, and graphene is shown in Figure 9.9c. Projections of the DOS (PDOS) on the atomic orbitals show that neither passivating species have a significant contribution at the band edges or in the linear region, and therefore, one does not expect them to cause any resonant scattering [57]. Therefore, GNMs, if doped up to the linear portion of the band structure, may have mobilities comparable to those based on pristine graphene, which can be exploited in novel devices.

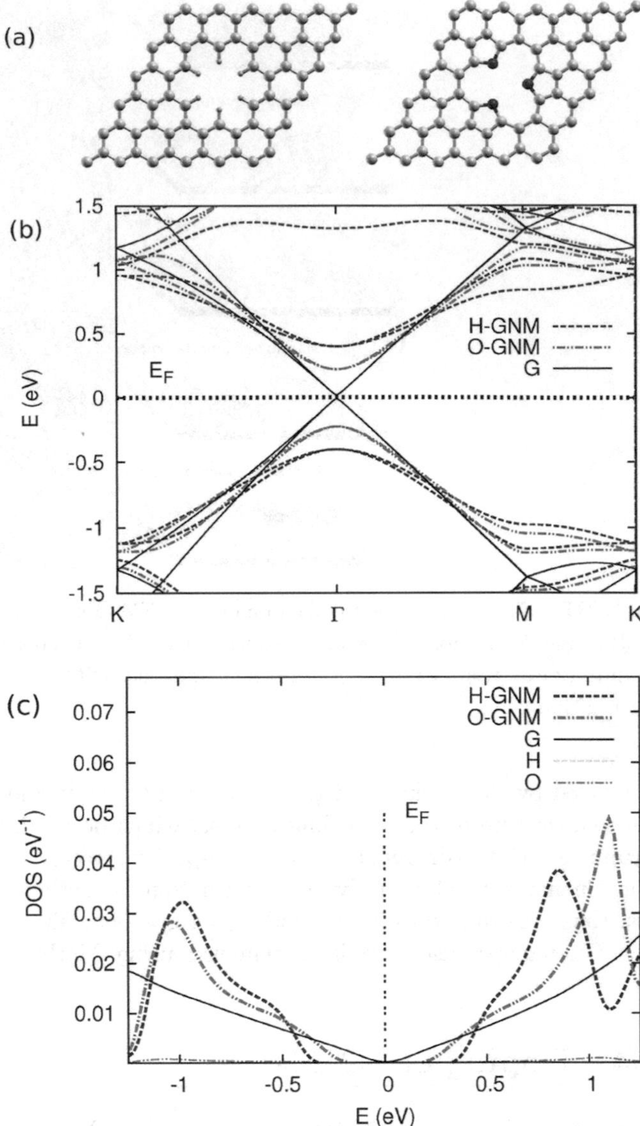

FIGURE 9.9 (a) Unit cells of a H-passivated GNM (H-GNM) and an O-passivated GNM (O-GNM), with a triangular lattice of pores and a pore size of about 0.6 nm. (b) Band structures of H-GNM (dark gray, dashed), O-GNM (light gray, dashed dotted), and pristine graphene (black, solid). (c) DOS of H-GNM (dark gray, dashed), O-GNM (light gray, dashed-dotted), and graphene (black, thin-solid). The hydrogen and oxygen contributions to the DOS of the H-GNM and O-GNM, respectively, are also plotted (both have negligible contributions in the energy range shown) [39].

9.4.1 p-Doping

Figure 9.10a shows the first p-doped system. If a fluorine atom is brought close to the pore of a H-GNM (Figure 9.10a), the system can lower its total energy if the fluorine docks in the center of the pore, forming a F@H-GNM, while an electron is transferred from the graphene sheet to the fluorine, ionizing it and doping the GNM. The resulting fluorine *ion* is now electrostatically bound to the pore (there is some small concomitant lattice relaxation). LDA gives a binding energy of 4.2 eV for the F@H-GNM, indicating very high thermal stability.

A comparison of the band structures of the doped and undoped systems (Figure 9.10c) shows that ion chelation doping occurs within the rigid band model. There is no significant change in the band curvatures, and the two band structures are essentially identical in the region of interest. The Fermi level of the chelated system is in the linear region of the valence band.

The rigid band doping picture is confirmed by inspecting the DOS of the F@H-GNM system. Figure 9.11a shows the total DOS and PDOS of that system, with contributions from carbon, hydrogen, and fluorine states. The total DOS of the H-GNM and that of pristine graphene are also shown for comparison. The Fermi level of the F@H-GNM system indicates that the GNM is now p-doped. The DOS of the chelated and unchelated systems is very similar, except for the two peaks on the left side of the gap, which correspond to the fluorine-*occupied* 2p states.

9.4.2 n-Doping

We now turn to the n-doping of crown GNMs. This is achieved by a reversal of the *chemical logic* underlying the p-doping approach. Oxygen, which has an electronegativity higher than carbon, is used as the pore perimeter passivator. This leads to carbon-oxygen bond polarization, with more negative charges at the oxygen side, making the pore suitable for hosting positive ions. Two electron-donating elements, potassium and sodium, are used to demonstrate n-doping of the O-GNM system. Due to the ionic sizes of

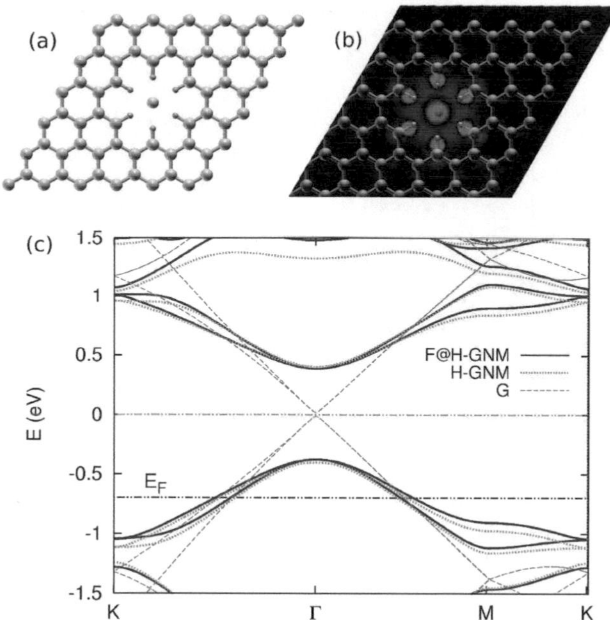

FIGURE 9.10 p-doped GNM system. (a) Unit cell of a hydrogen-passivated GNM with fluorine chelation (F@H-GNM), constructed from a 6×6 supercell of graphene, with a pore size of about 0.6 nm, and one fluorine atom per supercell. (b) A 2D map of the charge density *difference* between the doped passivated system and its neutral components (GNM, hydrogen, and fluorine). The atomic positions are overlayed on the map. Light gray (dark gray) color indicates an increase (decrease) in electronic charge density. The gray center shows that the fluorine is ionized. The gray color surrounding the center indicates that charge density is repelled away toward the pore perimeter (due to the ionization of the fluorine atom). (c) Band structures of the doped F@H-GNM system (black, solid), with its Fermi energy in the valence band (black, dashed-dotted line); the undoped O-GNM system (gray, dotted), with its Fermi energy at the origin (gray, dashed-dotted line); and pristine graphene (gray, dashed) [39].

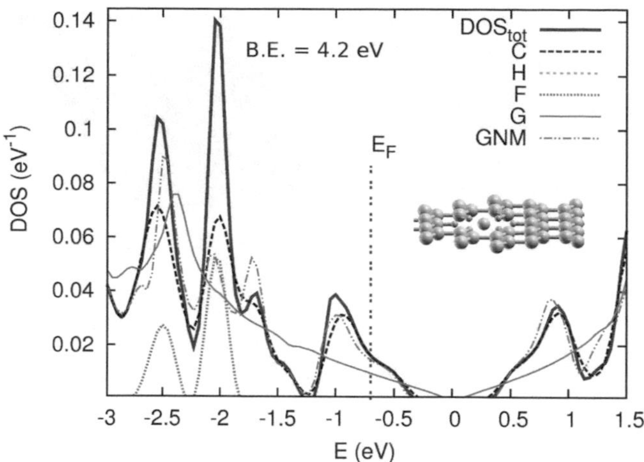

FIGURE 9.11 (a) Total and projected DOS of the p-doped F@H-GNM system. For the total DOS (black, solid), the Fermi level is below the gap (vertical line, black double dotted), indicating the p-doping of the GNM. Various contributions to the total DOS are shown: carbon 2 p_z states (black, dashed), hydrogen 1s states (gray, small-dashed), and fluorine 2p states (light gray, dotted). Hydrogen has a negligible contribution in the energy range shown. The first fluorine peak, located approximately 1.25 eV below the Fermi level, corresponds to $2p_x$ and $2p_y$ states, while the one lower in energy is mainly a $2p_z$ state. For comparison, undoped H-GNM total DOS is shown (gray, dashed-dotted), as well as pristine graphene (gray, solid). (b) Total and projected DOS of the p-doped Cl@H-GNM system. Same notation as in (a) with the F → Cl. The chlorine peak located 0.5 eV below the Fermi level corresponds to its 3p states [39].

these two elements, a GNM system with a bigger unit cell is used to accommodate a larger pore.

Figure 9.12a shows an oxygen-passivated GNM with a pore size of 0.7 nm. The relaxed configuration of the unchelated (undoped) O-GNM has a bandgap of 0.4 eV. By bringing the potassium atom to a distance of a few angstroms from the pore, the system lowers its energy by the docking and the ionization of the potassium atom in the center of the pore, with its 4s electron donated to the O-GNM π bands (see Figure 9.8). Figure 9.12b shows the 2D charge density difference map for the system.

Comparing the band structure of K@O-GNM and O-GNM shows that potassium chelation preserves the O-GNM band structure (Figure 9.12c) in the region of interest. The chelation merely shifts the Fermi level into the conduction band. This observation is further confirmed by inspection of the DOS and PDOS of the K@O-GNM system (Figure 9.13a), where we see that the DOSs of the two systems are very similar. The potassium 4s level does not contribute to the spectrum in the vicinity of the gap.

9.4.3 Larger Pores

Fabrication techniques are likely to yield GNMs with nanometer-sized pores, which can host more than one dopant atom. Experimental work on GNMs currently achieves pore diameters as small as 3 nm [66]. Therefore, one needs to consider multiple dopants per pore, studying the details of their packing and stability for various applications. Such studies also provide a framework for exploring the behavior of dopant ionic complexes for chemical separation applications. Dopant stability is of great importance, and hence, some questions arise regarding the doping of large pores. How would multiple dopants interact in a large pore? Would they uniformly distribute around the pore edge, or would they cluster together, with an obvious impact on the dopants' binding energy? In addition, would the dopants' structure in the pore alter the rigid band picture?

In GNMs with nanometer-sized pores, electrostatic dopant binding is expected to decrease due to four effects: (i) A weaker interaction between a dopant and the pore dipoles. (ii) The increase in the charge transfer to the sheet, and hence its energy cost. (iii) The repulsion between ionized dopants. (iv) The introduction of more dopants in large pores can lead to the clustering of dopant atoms. This implies that chemical binding, well known in metals, may

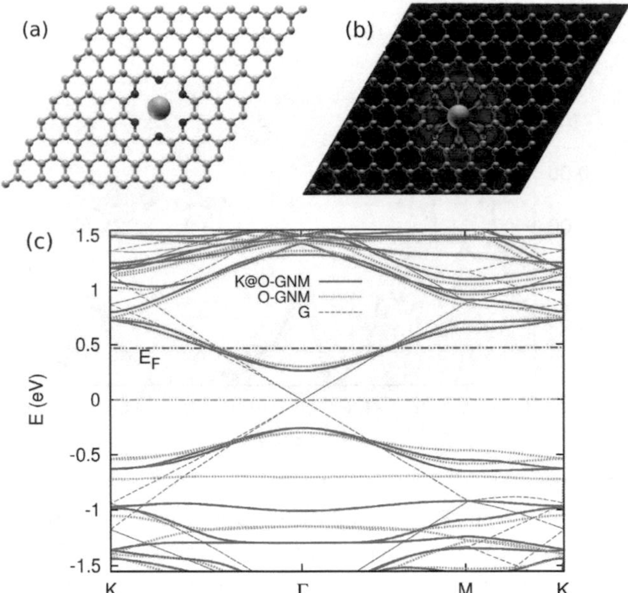

FIGURE 9.12 n-doped GNM system. (a). Unit cell of an oxygen-passivated GNM with potassium chelation (K@O-GNM), constructed from a 9×9 supercell of graphene. (b) A 2D map of the charge density *difference* between the doped passivated system and its neutral components (GNM, oxygen, and potassium). The atomic positions are overlayed on the map. The light gray center shows that the potassium atom is ionized. The dark gray color on the pore size of the oxygen atoms indicates that charge density is attracted toward the center of the pore (due to the ionization of the potassium atom). (c). Band structures of the doped K@O-GNM system (dark gray, solid), with its Fermi energy in the conduction band (dark gray, dashed-dotted line); the undoped O-GNM system (light gray, dotted), with its Fermi energy at the origin (light gray, dashed-dotted line); and pristine graphene (dark gray, dashed) [39].

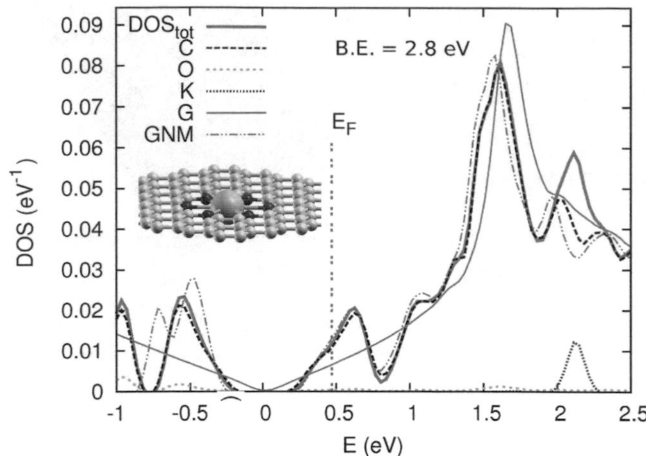

FIGURE 9.13 (a) Total and projected DOS of the n-doped K@O-GNM system. For the total DOS (dark gray, solid), the Fermi level is above the gap (vertical line, dark gray double dotted), indicating the n-doping of the GNM. Various contributions to the total DOS are shown: carbon $2p_z$ states (black, dashed), oxygen $2p$ states (light gray, small-dashed), and potassium $4s$ states (dark gray, dotted). Oxygen has a negligible contribution in the energy range of interest. The potassium peak located approximately 1.5 eV above the Fermi level corresponds to its $4s$ state. For comparison, undoped O-GNM total DOS is shown (medium gray, dashed-dotted), as well as pristine graphene (medium gray, solid). (b) Total and projected DOS of the n-doped Na@O-GNM system. Same notation as in (a) with the K → Na. The sodium $3s$ state is located 1.25 eV above the Fermi level [39].

occur between the dopant atoms, which may affect the doping physics observed with single dopants.

We first consider the cases of O–GNM9, with a pore diameter of 0.8 nm. A lithium atom brought close to the edge of the pore (inset of Figure 9.14a) loses its electron to the graphene skeleton, thereby ionizing and becoming electrostatically trapped in the field of the edge dipoles, forming a Li@O–GNM9. The DOS and PDOS of this system are shown in Figure 9.14a. The Fermi level location indicates that the O–GNM9 is n-doped. The DOS and PDOS of the 2Li@O–GNM9 system are shown in Figure 9.14b. The shift in the Fermi energy indicates the increase in the doping level of the 2Li@O–GNM9 over the Li@O-GNM9 system. Calculations confirm a charge transfer of $2e$ between the two lithium atoms and the graphene lattice.

Adding a third lithium atom to the pore introduces a qualitative change. Whereas the states of the single and double Li doping are *far* above the conduction band edge of the O–GNM9, one of the states of the triple Li doping now falls *in* the conduction band of the O–GNM9 (Figure 9.14c). The O–GNM9 perturbs the Li and 2Li line spectra in such a way so as to raise their atomic/molecular HOMO states, leading to the ionization of lithium atoms in both cases. In the 3Li case, the perturbation is such that the lowest lithium 2s-like state lies in the conduction band of the system *below* the Fermi level, and therefore, two electrons occupy that state, while the third electron is donated to the O–GNM9, thus doping it to a level similar to that of the Li@O–GNM9 system. The total binding energy in this case is 1.91 eV. The insets of Figures 9.14b,c compare the spectra of the two and three Li atoms in the pore to those of the Li_2 and Li_3 clusters.

To reach the maximum capacity of the pore, we add more lithium atoms. One interesting feature in the 4Li@O–GNM9 system (Figure 9.14d) is that the lowest Li_4 state is now below the *valence* band edge, with *two* electrons transferred from the cluster to the O–GNM. The binding energy of the 4Li@O–GNM9 is 1.59 eV. At higher doping (5Li@O–GNM9 and 6Li@O–GNM9), a Li state persists in the conduction band of the system.

We now turn to n-doping of the larger pore, where we consider six systems: Li@O–GNM12, 2Li@O–GNM12, 2Na@O–GNM12, 3Li@O–GNM12, 4Na@O–GNM12, and 6Na@O–GNM12. For 1 and 2 dopant atoms, the behavior is qualitatively similar to the small pore case. For 3 and 4 dopants, the situation is different. Figure 9.15a shows the structure and DOS of the 3Li@O–GNM12 system, indicating that all three Li atoms are now ionized, with three

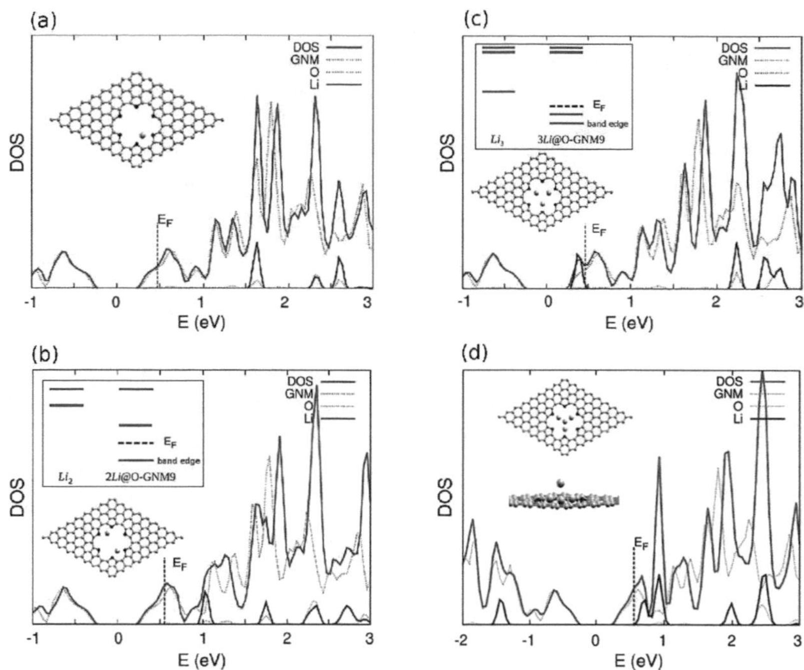

FIGURE 9.14 DOS of the (a) Li@O–GNM9, (b) 2Li@O–GNM9, (c) 3Li@O–GNM9, and (d) 4Li@O–GNM9 systems. The bottom insets refer to the relaxed structures. In (b,c) the top insets show the Li states within the corresponding O–GNM9 system, as compared to those of the isolated Li clusters. The DOS of the pristine O–GNM9 (GNM) is shown for comparison [15].

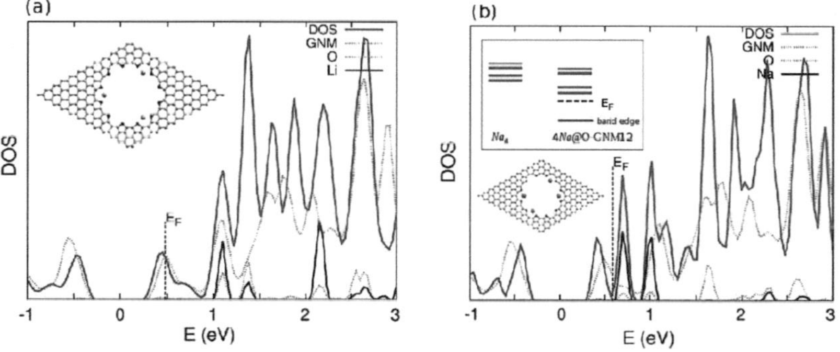

FIGURE 9.15 DOS of the (a) 3Li@O–GNM12 and (b) 4Na@O–GNM12 systems. The bottom insets refer to the relaxed structures. In (b), the top inset shows the Na states within the corresponding O–GNM12 system, as compared to those of the isolated Na_4 cluster. The DOS of the pristine O–GNM12 (GNM) is shown for comparison [15].

electrons transferred to the GNM. The system has a binding energy of 1.39 eV. All three Li-like states are now above the Fermi energy.

The higher load case of 4Na@O–GNM12 is shown in Figure 9.15b. All Na atoms are ionized, contrary to the small pore 4Na@O–GNM9 system, with 4 electrons now doping the GNM. The O–GNM12 pulls down the lowest two states of the Na_4 cluster, positioning them just above the Fermi level of the system, signaling that higher loads will result in pushing Na states down into the conduction band of the GNM. Indeed, with 6Na atoms, the 6Na@O–GNM12 system has its lowest Na state *below* the Fermi energy, now with only 4 electrons doping the GNM.

We now examine the p-doping of large pore GNM systems. We consider 6 doped H–GNM9 systems (pore size 0.8 nm) and 4 doped H–GNM12 systems (pore size 1.3 nm), using F. The reported binding energies are calculated from the energies of the most stable form of the dopants (F_2 molecule). Figure 9.16a shows the structure and DOS of the F@H–GNM9. The binding energy of F@H–GNM9 is found to be 1.07 eV. The H–GNM causes the F $2p_z$ state to be pinned about 0.3 eV below the valence band edge of the H–GNM, and 0.1 eV below the Fermi level.

The structure and DOS of the 2F@H–GNM9 system are shown in Figure 9.16b. The GNM is p-doped, with the Fermi energy located $\sim$ 0.3 eV below the valence band edge. The binding energy of the system is 1.72 eV. The two F atoms are now ionized, with their $2p_z$ states full and located $\sim$ 0.2 eV *below* the Fermi level. Doping with three fluorine atoms leads to a qualitatively similar scenario.

With four F atoms, two of the atoms form an *elongated* fluorine molecule, F_2^*, which protrudes out of the pore plane (Figure 9.16c), with a bond length of 1.68 Å (a F_2 molecule has a bond length of 1.43 Å). The F_2^* is 2.21 Å from the closest pore-edge hydrogen. The DOS of the 4F@H–GNM9 system shows that there is one fluorine state just *above* the Fermi energy. Inspection of this state indicates that it is a superposition of the $2p_z$ states of the two fluorine atoms forming the elongated molecule. These two fluorine atoms are *not* ionized. This is confirmed by the Löwdin charge analysis, which also shows that the other two fluorine atoms are each singly ionized.

Similar physics occurs for larger pore GNM, H–GNM12 (Figure 9.17a). The average F–F distance is 8.62 Å. The DOS of the system shows that the GNM12 is p-doped, with the Fermi energy placed $\sim$ 0.1 eV below the valence band edge. The fluorine $2p_z$ states are all full and are located at least 0.3 eV below the Fermi level. The integrated DOS shows that the GNM lost three electrons. The binding energy of the 3F@H–GNM12 is 2.72 eV.

Doping with 4 and 6 fluorine atoms removes 4 and 6 electrons from the GNM, respectively, confirming the full ionization of the fluorine atoms in both systems. We only show the 6F@H–GNM12 system here (Figure 9.17b). The highest filled fluorine $2p_z$ state is $\sim$ 0.5 eV away from the Fermi level, which is pinned 0.3 eV below the valence band edge. The binding energy of the 6F@H–GNM12 is 2.76 eV.

Fabrication of GNM pores is bound to produce pores with some level of disorder. As a result, a GNM will have a distribution of pores sizes and pore lattice constants, changing the symmetry requirement for the electronic gap formation. The effect of this is twofold. First, the gap will partially close, and second, the dopant load will vary for different pores. The latter effect will have a negligible impact on potential applications of the GNMs, as the resulting structures are still stably doped, whereas the former can seriously prevent the development of GNM-based transistors with technologically acceptable ON/OFF ratios.

Chelation-doped crown GNMs may be considered for many other applications, such as transparent electrodes. On the other hand, their use in microelectronic applications requires controlled doping of the sheets. Thus, the pores, post-fabrication, must be filled with the desired number of dopants, no more, no fewer. To achieve this result, it is convenient to select the binding energy dopant level with the lowest free energy such that sufficient thermal annealing, for instance, will allow the GNM/dopant system to relax to the correct number of dopants per pore. This approach avoids difficult stoichiometric control and other problems associated with achieving a doping state that is not the most thermodynamically stable. There are two caveats. First, choosing the right dopant whose lowest free energy state is a good fit for microelectronics. Second, the free energy of the next most stable dopant should be well separated from the lowest stable (large free energy gap) such that the thermal annealing will be effective. The results indicate that to achieve optimal control in transistor applications for large pores, dopants must be designed to fit in the pore of interest, allowing for carefully controlled device engineering.

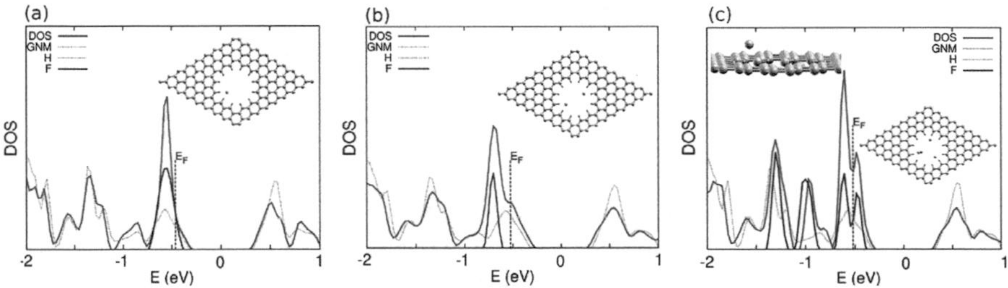

FIGURE 9.16 DOS of the (a) F@H–GNM9, (b) 2F@H–GNM9 and (c) 4F@H–GNM9 systems. The inset figures refer to the relaxed structures. In (c), the inset shows a side view of the 4F@H–GNM9 system, indicating the off-plane structure of the 4F dopants. The DOS of the pristine H–GNM9 (GNM) is shown for comparison [15].

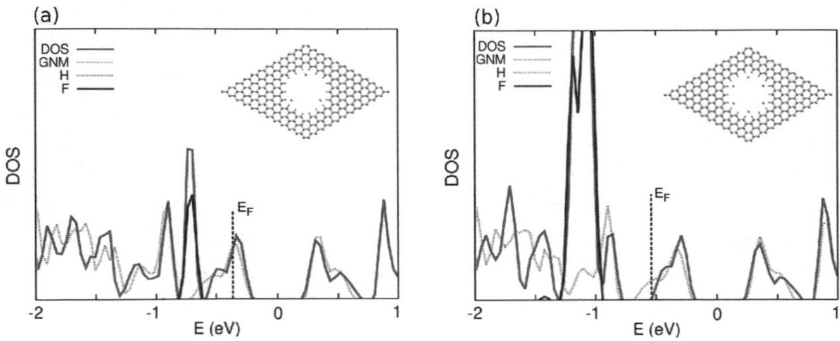

FIGURE 9.17 DOS of the (a) 3F@H–GNM12 and (b) 6F@H–GNM12 systems. The inset figures show the relaxed structures. The DOS of the pristine H–GNM12 (GNM) is shown for comparison [15].

9.5 Applications of GNMs

GNMs come with two main advantages, namely, an electronic gap and the possibility of chemical functionalization. Therefore, they provide graphene with a pass to many applications. In nanoelectronics, and with stable GNM doping [15,39], GNM-based FETs can be experimentally realized [43,50,64]. Utilizing the high optical transparency of GNMs, and their controlled doping, they can be used as transparent electrodes for various technologies [30,35,45,48]. Because of their pores, GNMs have also been considered for molecular separation [12,13,22,25,29,52] and energy storage [14,60].

GNMs can be utilized for chemical sensing applications. GNM pore edges can be functionalized with molecular receptors for matching molecular targets. If the target-receptor chemical encounter involves a charge transaction with the GNM skeleton, thereby moving its Fermi level and boosting its conductance, the GNM can be used as a sensor for the target molecule. An application of this concept to form a glucose sensor is shown in Figure 9.18, where the oxygen-passivated pore is functionalized with boronic acid. Upon reaction with a glucose molecule, charge is transferred to the GNM, doping it and increasing its electrical conductance [5].

GNM pores can also be used as docking stations for various nanoparticles. One example is to employ the pores to prevent the agglomeration of catalytic nanoparticles [65], by designing the pore edge chemistry that matches that of the catalytic nanoparticles [2]. Another example is to use GNMs in hydrogen storage, where the GNM pores are passivated with nitrogen to anchor Ni nanoparticles that can subsequently be loaded with H_2 molecules (Figure 9.19). It is found that nitrogen-passivated GNMs (N–GNMs) are capable of firmly anchoring Ni clusters through strong binding to the pore N atoms, unlike oxygen-passivated pores which clusters binding only weakly. DFT calculations show that Ni_6 and Ni_13 clusters supported on N–GNM are capable of adsorbing 6 and 8 H_2 molecules, respectively. With larger clusters, the H_2 storage capacity is expected to reach that of isolated Ni clusters, as the fraction of Ni atoms involved in the binding to the GNM pore decreases with increasing Ni cluster size. Furthermore, it is found that

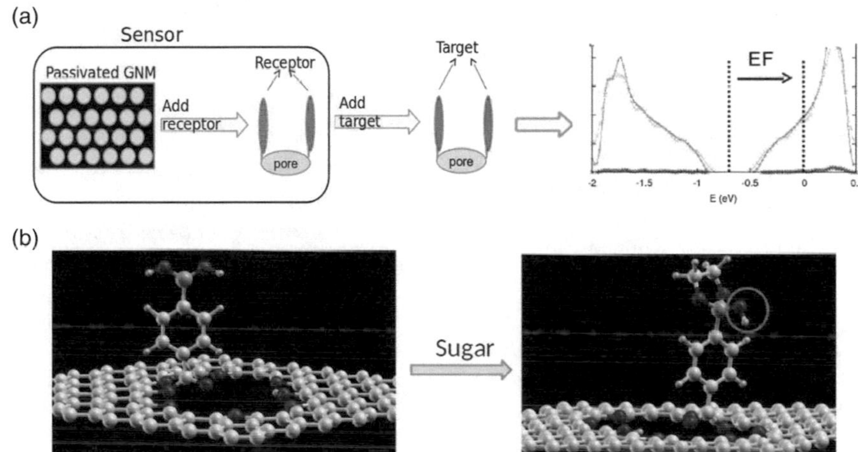

FIGURE 9.18 (a) A GNM-based charge sensor. (b) A realization for sensing glucose molecules [5].

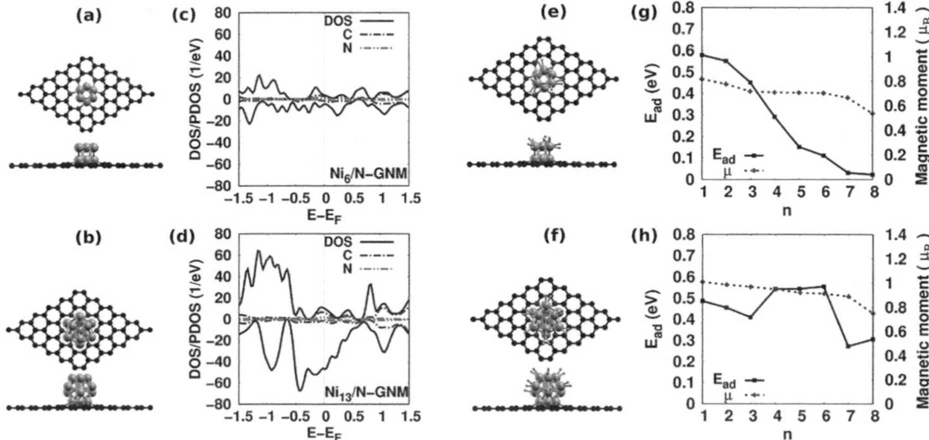

FIGURE 9.19 Optimized structures of (a) Ni_6/NGNM, (b) Ni_13 /N–GNM, and (c,d) corresponding DOS/PDOS. Optimized structures of (e) Ni_6 /NGNM, (f) Ni_13 /N–GNM with H_2 molecules adsorbed, corresponding (g) adsorption energy, and (h) magnetic moment per Ni atom as functions of the number of adsorbed H_2 molecules, n [16].

some Ni@N–GNM systems can be used for spin filtering, as one spin channel is dominated by delocalized graphene-like states, while the other has only localized Ni states [16].

Bibliography

1. A. Abou-Kandil, A. A. Maarouf, G.J. Martyna, H. Mohamed, and D.M. Newns. Doped, passivated graphene nanomesh, method of making the doped, passivated graphene nanomesh, and semiconductor device including the doped, passivated graphene nanomesh, 2015. US Patent 9,142,471.
2. A. Afzali-Ardakani, A. A. Maarouf, and G. J. Martyna. Controlled assembly of charged nanoparticles using functionalized graphene nanomesh, 2014. US Patent 8,835,686.
3. A. Afzali-Ardakani, A. A. Maarouf, and G. J. Martyna. Method of forming graphene nanomesh, 2014. US Patent 8,834,967.
4. A. Afzali-Ardakani, A. A. Maarouf, G. J. Martyna, and K. Saenger. Forming patterned graphene layers, 2015. US Patent 9,102,118.
5. A. Afzali-Ardakani, S.-J. Han, A. Kasry, A. A. Maarouf, G. J. Martyna, R. Nistor, and H. Tsai. Graphene nanomesh based charge sensor, 2015. US Patent 9,102,540.
6. O. Akhavan. Graphene nanomesh by ZnO nanorod photocatalysts. *ACS Nano*, 4(7):4174–4180, 2010.
7. J. Bai, X. Zhong, S. Jiang, Y. Huang, and X. Duan. Graphene nanomesh. *Nature Nanotechnology*, 5(3):190–194, 2010.
8. A. H. Castro Neto, F. Guinea, N. M. R. Peres, K. S. Novoselov, and A. K. Geim. The electronic properties of graphene. *Reviews of Modern Physics*, 81:109–162, 2009.
9. S. -M. Choi, S. -H. Jhi, and Y. -W. Son. Controlling energy gap of bilayer graphene by strain. *Nano Letters*, 10(9):3486–3489, 2010.
10. J. J. Christensen, J. O. Hill, and R. M. Izatt. Ion binding by synthetic macrocyclic compounds. *Science*, 174(4008):459–467, 1971.
11. G. Cocco, E. Cadelano, and L. Colombo. Gap opening in graphene by shear strain. *Phys. Rev. B*, 81:241412, 2010.
12. D. Cohen-Tanugi and J. C. Grossman. Water desalination across nanoporous graphene. *Nano Letters*, 12(7):3602–3608, 2012.
13. L. W Drahushuk and M. S. Strano. Mechanisms of gas permeation through single layer graphene membranes. *Langmuir: The ACS Journal of Surfaces and Colloids*, 28(48):16671–16678, 2012.
14. A. Du, Z. Zhu, and S. C. Smith. Multifunctional porous graphene for nanoelectronics and hydrogen storage: New properties revealed by first principle calculations. *Journal of the American Chemical Society*, 132(9):2876–2877, 2010.
15. M. S. Eldeeb, M. M. Fadlallah, G.J. Martyna, and A. A Maarouf. Doping of large-pore crown graphene nanomesh. *Carbon*, 133:369–378, 2018.
16. M. M. Fadlallah, A. Abdelrahman, U. Schwingenschlögl, and A. A. Maarouf. Graphene and graphene nanomesh supported nickel clusters: Electronic, magnetic, and hydrogen storage properties. *Nanotechnology*, 30(8): 085709, 2019.
17. M. M. Fadlallah, A. A. Maarouf,? U. Schwingenschlögl, and U. Eckern. Unravelling the interplay of geometrical, magnetic and electronic properties of metal-doped graphene nanomeshes. *Journal of Physics: Condensed Matter*, 29(5):055301, 2017.
18. J. Fernández-Rossier and J. J. Palacios. Magnetism in graphene nanoislands. *Physical Review Letters*, 99(17):177204, 2007.
19. L. Firlej, B. Kuchta, C. Wexler, and P. Pfeifer. Boron substituted graphene: Energy landscape for hydrogen adsorption. *Adsorption*, 15(3):312–317, 2009.
20. J. A. Fürst, J. G. Pedersen, C. Flindt, N. A. Mortensen, M. Brandbyge, T. G. Pedersen, and A. P. Jauho. Electronic properties of graphene antidot lattices. *New Journal of Physics*, 11(9):095020, 2009.
21. A. K. Geim and K. S. Novoselov. The rise of graphene. *Nat. Mater.*, 6:183–191, 2007.
22. J. Goldsmith and C. C. Martens. Molecular dynamics simulation of salt rejection in model surface-modified nanopores. *The Journal of Physical Chemistry Letters*, 1(2):528–535, 2010.
23. C. J. Hawker and T. P. Russell. Block copolymer lithography: Merging “bottom-up” with “top-down” processes. *MRS Bulletin*, 30(12):952–966, 2005.
24. Y. Hua and A. H. Flood. Click chemistry generates privileged CH hydrogen-bonding triazoles: The latest addition to anion supramolecular chemistry. *Chemical Society Reviews*, 39(4):1262–1271, 2010.
25. L. Huang, M. Zhang, C. Li, and G. Shi. Graphene-based membranes for molecular separation. *The Journal of Physical Chemistry Letters*, 6(14):2806–2815, 2015.
26. V. Hung Nguyen, M. Chung Nguyen, H. V. Nguyen, and P. Dollfus. Disorder effects on electronic bandgap and transport in graphene-nanomesh-based structures. *Journal of Applied Physics*, 113(1): 013702, 2013.
27. D. Jiang, V. R Cooper, and S. Dai. Porous graphene as the ultimate membrane for gas separation. *Nano Letters*, 9(12):4019–4024, 2009.
28. K. K. Kim, A. Reina, Y. Shi, H. Park, L.-J. Li, Y. H. Lee, and J. Kong. Enhancing the conductivity of transparent graphene films via doping. *Nanotechnology*, 21(28):285205, 2010.
29. S. P. Koenig, L. Wang, J. Pellegrino, and J. S. Bunch. Selective molecular sieving through porous

graphene. *Nature Nanotechnology*, 7(11):728–732, 2012.

30. D. Kuzum, H. Takano, E. Shim, J. C. Reed, H. Juul, A. G. Richardson, J. De Vries, H. Bink, M. A. Dichter, T. H. Lucas, D. A. Coulter, E. Cubukcu, and B. Litt. Transparent and flexible low noise graphene electrodes for simultaneous electrophysiology and neuroimaging. *Nature Communications*, 5(May):1–10, 2014.
31. Y. Li and A. H. Flood. Pure C-H hydrogen bonding to chloride ions: A preorganized and rigid macrocyclic receptor. *Angewandte Chemie International Edition*, 47(14):2649–2652, 2008.
32. Y. M. Lin, C. Dimitrakopoulos, K. A. Jenkins, D. B. Farmer, H. Y. Chiu, A. Grill, and P. Avouris. 100-GHz transistors from wafer-scale epitaxial graphene. *Science*, 327(5966):662, 2010.
33. Y.-M. Lin, K. A. Jenkins, A. Valdes-Garcia, J. P. Small, D. B. Farmer, and P. Avouris. Operation of graphene transistors at gigahertz frequencies. *Nano Letters*, 9(1):422–426, 2009.
34. H. Liu, Y. Liu, and D. Zhu. Chemical doping of graphene. *Journal of Materials Chemistry*, 21(10):3335, 2011.
35. J. Liu, G. Xu, C. Rochford, R. Lu, J. Wu, C. M. Edwards, C. L. Berrie, Z. Chen, and V. A. Maroni. Doped graphene nanohole arrays for flexible transparent conductors. *Applied Physics Letters*, 99(2):023111, 2011.
36. J. Liu, H. Cai, X. Yu, K. Zhang, X. Li, J. Li, N. Pan, Q. Shi, Y. Luo, and X. Wang. Fabrication of graphene nanomesh and improved chemical enhancement for Raman spectroscopy. *Journal of Physical Chemistry C*, 116(29):15741–15746, 2012.
37. W. Liu, Z. F. Wang, Q. W. Shi, J. Yang, and F. Liu. Band-gap scaling of graphene nanohole superlattices. *Physical Review B*, 80(23):2–5, 2009.
38. A. A. Maarouf, A. Kasry, B. Chandra, and G. J. Martyna. A graphene-carbon nanotube hybrid material for photovoltaic applications. *Carbon*, 102:74–80, 2016.
39. A. A. Maarouf, R. A. Nistor, A. Afzali-Ardakani, M. A. Kuroda, D. M. Newns, and G. J. Martyna. Crown graphene nanomeshes: Highly stable chelation-doped semiconducting materials. *Journal of Chemical Theory and Computation*, 9(5):2398–2403, 2013.
40. B. F. Machado and P. Serp. Graphene-based materials for catalysis. *Catalysis Science and Technology*, 2:54–75, 2012.
41. R. Martinazzo, S. Casolo, and G. Tantardini. Symmetry-induced band-gap opening in graphene superlattices. *Physical Review B*, 81(24):1–8, 2010.
42. G. Ning, C. Xu, L. Hao, O. Kazakova, Z. Fan, H. Wang, K. Wang, J. Gao, W. Qian, and F. Wei. Ferromagnetism in nanomesh graphene. *Carbon*, 51(1): 390–396, 2013.
43. R. C. Ordonez, C. K. Hayashi, C. M. Torres, J. L. Melcher, N. Kamin, G. Severa, and D. Garmire. Rapid fabrication of graphene field-effect transistors with liquid-metal interconnects and electrolytic gate dielectric made of honey. *Scientific Reports*, 7(1):1–9, 2017.
44. F. Ouyang, S. Peng, Z. Liu, and Z. Liu. Bandgap opening in graphene antidot lattices: The missing half. *ACS Nano*, 5(5):4023–4030, 2011.
45. I.-J. Park, T. I. Kim, T. Yoon, S. Kang, H. Cho, N. S. Cho, J.-I. Lee, T.-S. Kim, and S.-Y. Choi. Flexible and transparent graphene electrode architecture with selective defect decoration for organic light-emitting diodes. *Advanced Functional Materials*, 28(10):1704435, 2018.
46. C. J. Pedersen. Cyclic polyethers and their complexes with metal salts. *Journal of the American Chemical Society*, 89(157):7017–7036, 1967.
47. R. Petersen, T. G. Pedersen, and A.-P. Jauho. Clar sextet analysis of triangular, rectangular, and honeycomb graphene antidot lattices. *ACS nano*, 5(1):523–529, 2011.
48. Y. Qiang, P. Artoni, K. J. Seo, S. Culaclii, V. Hogan, X. Zhao, Y. Zhong, X. Han, P.-M. Wang, Y.-K. Lo, Y. Li, H. A. Patel, Y. Huang, A. Sambangi, J. S. V. Chu, W. Liu, M. Fagiolini, and H. Fang. Transparent arrays of bilayer-nanomesh microelectrodes for simultaneous electrophysiology and two-photon imaging in the brain. *Science Advances*, 4(9): eaat0626, 2018.
49. L. Qu, Y. Liu, J.-B. Baek, and L. Dai. Nitrogen-doped graphene as efficient metal-free electrocatalyst for oxygen reduction in fuel cells. *ACS Nano*, 4(3):1321–1326, 2010.
50. D. Reddy, L. F. Register, G. D. Carpenter, and S. K. Banerjee. Graphene field-effect transistors. *Journal of Physics D: Applied Physics*, 44(31):313001, 2011.
51. A.V. Rozhkov, G. Giavaras, Y. P. Bliokh, V. Freilikher, and F. Nori. Electronic properties of mesoscopic graphene structures: Charge confinement and control of spin and charge transport. *Physics Reports*, 503(2):77–114, 2011.
52. J. Ruparelia, S. Duttagupta, A. Chatterjee, and S. Mukherji. Potential of carbon nanomaterials for removal of heavy metals from water. *Desalination*, 232(1–3):145–156, 2008.
53. C. Sire, F. Ardiaca, S. Lepilliet, J.-W. T. Seo, M. C. Hersam, G. Dambrine, H. Happy, and V. Derycke. Flexible gigahertz transistors derived from solution-based single-layer graphene. *Nano Letters*, 12(3):1184–1188, 2012.
54. M. M. Ugeda, I. Brihuega, F. Guinea, and J. M. Gómez-Rodríguez. Missing atom as a source of carbon magnetism. *Physical Review Letters*, 104(9):096804, 2010.
55. M. Wang, L. Fu, L. Gan, C. Zhang, M. Rümmeli, A. Bachmatiuk, K. Huang, Y. Fang, and Z. Liu.

CVD growth of large area smooth-edged graphene nanomesh by nanosphere lithography. *Scientific Reports*, 3:1–6, 2013.

56. X. Wang, X. Li, L. Zhang, Y. Yoon, P. K. Weber, H. Wang, J. Guo, and H. Dai. N-doping of graphene through electrothermal reactions with ammonia. *Science (New York, N.Y.)*, 324(5928):768–771, 2009.
57. T. O. Wehling, S. Yuan, A. I. Lichtenstein, A. K. Geim, and M. I. Katsnelson. Resonant scattering by realistic impurities in graphene. *Physical Review Letters*, 105:056802, 2010.
58. Y. Wu, Y.-M. Lin, A. A. Bol, K. A. Jenkins, F. Xia, D. B. Farmer, Y. Zhu, and P. Avouris. High-frequency, scaled graphene transistors on diamond-like carbon. *Nature*, 472(7341):74–78, 2011.
59. F. Xia, D. B. Farmer, Y.-M. Lin, and P. Avouris. Graphene field-effect transistors with high on/off current ratio and large transport band gap at room temperature. *Nano Letters*, 10(2):715–718, 2010.
60. J. Xiao, D. Mei, X. Li, W. Xu, D. Wang, G. L. Graff, W. D. Bennett, Z. Nie, L. V. Saraf, I. A. Aksay, J. Liu, and J.-G. Zhang. Hierarchically porous graphene as a lithium-air battery electrode. *Nano Letters*, 11(11):5071–5078, 2011.
61. H.-X. Yang, M. Chshiev, D. W. Boukhvalov, X. Waintal, and S. Roche. Inducing and optimizing magnetism in graphene nanomeshes. *Physical Review B*, 84(21):214404, 2011.
62. O. V. Yazyev. Emergence of magnetism in graphene materials and nanostructures. *Reports on Progress in Physics*, 73(5):056501, 2010.
63. Z. Zeng, X. Huang, Z. Yin, H. Li, Y. Chen, H. Li, Q. Zhang, J. Ma, F. Boey, and H. Zhang. Fabrication of graphene nanomesh by using an anodic aluminum oxide membrane as a template. *Advanced Materials*, 24(30):4138–4142, 2012.
64. B. Zhan, C. Li, J. Yang, G. Jenkins, W. Huang, and X. Dong. Graphene field-effect transistor and its application for electronic sensing. *Small*, 10(20): 4042–4065, 2014.
65. C. Zhang, W. Lv, Q. Yang, and Y. Liu. Graphene supported nano particles of Pt-Ni for CO oxidation. *Applied Surface Science*, 258(20):7795–7800, 2012.
66. J. Zhang, H. Song, D. Zeng, H. Wang, Z. Qin, K. Xu, A. Pang, and C. Xie. Facile synthesis of diverse graphene nanomeshes based on simultaneous regulation of pore size and surface structure. *Scientific Reports*, 6, 2016. doi: 10.1038/srep32310.
67. J. Zhou, Q. Wang, Q. Sun, X. S. Chen, Y. Kawazoe, and P. Jena. Ferromagnetism in semihydrogenated graphene sheet. *Nano Letters*, 9(11):3867–3870, 2009.

10

Dielectric Harmonic Nanoparticles: Optical Properties, Synthesis, and Applications

Yannick Mugnier
Université Savoie Mont Blanc, Annecy

Luigi Bonacina
University of Geneva

10.1 Introduction

In the last decades, the quest for monitoring and controlling processes at the cellular and intracellular level has stimulated a plethora of nanophotonics approaches. To date, the most successful and widespread ones are those based on quantum dots and upconversion nanoparticles (NPs) for imaging, because of their bright luminescence emission, and those using metal particles for sensing, thanks to the sensitivity of surface plasmon scattering to the local environment (Smith & Gambhir, 2017). Recently, both Rayleigh scattering and Mie scattering from all-dielectric nanostructures have stimulated a lot of interest, primarily because of the reduced ohmic losses of these systems, their tailorable optical properties, and the compatibility of many of these materials with manufacturing techniques (Baranov et al., 2017; Kuznetsov, Miroshnichenko, Brongersma, Kivshar, & Luk'yanchuk, 2016). Along with nanotechnology, the photonics industry has steadily evolved offering new enabling tools, including compact and cost-effective ultrashort pulse sources in a wide range of frequencies, from visible to mid-infrared (mid-IR). The availability of these new instruments has greatly facilitated the investigation of the nonlinear response of nanomaterials. In particular, with respect to optical damage, which is often an issue under continuous excitation and with nanosecond pulses, the use of low-energy, high-peak power femtosecond pulses has opened the way to safely probe a variety of nanomaterials within their transparency range and even around their resonant frequencies. Nonlinear optical observables present various advantages, including the possibility to detect background-free parametric signals at excitation wavelengths which depose no or minimal energy on the sample under investigation. Second harmonic generation (SHG) is by far the most exploited among these processes, because it is the easiest to obtain and collect in terms of excitation and detection settings. Increasingly, the concurrent nonlinear emissions, which accompany SHG, are becoming accessible along with the development of new laser sources. The simultaneous collection and analysis of these signals allow developing refined detection approaches for increased sensitivity and selectivity in demanding imaging applications.

This chapter introduces and describes the use of a family of dielectric metal oxide NPs, harmonic NPs (HNPs), characterized by a noncentrosymmetric crystal structure and high nonlinear optical efficiencies, and displaying a very rich nonlinear response. We begin by providing a concise reminder of some fundamental nonlinear processes in optics needed to understand HNP responses to light excitation and their quantitative assessment from colloidal ensembles. Afterward, we outline specific properties of HNP materials along with their synthesis, and finally, we describe their applications for bioimaging.

10.2 Understanding the Nonlinear Optical Response of Harmonic Nanoparticles

This section provides a concise summary of nonlinear optical processes, which are observed when exciting HNPs by a pulsed femtosecond laser. Following the reasoning and symbol convention introduced by reference textbooks (Boyd, 2003; Gubler & Bosshard, 2002), we indicate with a tilde the quantities that are rapidly oscillating in time. We can physically understand the nonlinear response of a body invested by an external oscillating field $\tilde{E}(t)$ by modeling the position $\tilde{x}(t)$ of an electron in the medium as:

$$\ddot{\tilde{x}}(t) + 2\gamma\dot{\tilde{x}}(t) + \frac{F(\tilde{x})}{m} = \frac{e\tilde{E}(t)}{m} \tag{10.1}$$

This formula describes a harmonic oscillator with a dumping term proportional to γ, and a restoring force $F(\tilde{x}) = -\mathrm{m}\left(\omega_0^2\tilde{x}(t) + a\tilde{x}^2(t) - b\tilde{x}^3(t) + \cdots\right)$ constituted by the sum of one harmonic term, $\omega_0^2\tilde{x}(t)$ with ω_0, the single electron resonant frequency, and several anharmonic contributions, $a\tilde{x}^2(t) - b\tilde{x}^3(t) + \cdots$ where $a, b > 0$. The term on the right-hand side in Eq. 10.1 corresponds to the driving by the external field, with e and m representing the charge and the mass of an electron, respectively. The corresponding potential energy function defining the electron motion, $U(\tilde{x}) = -\int \tilde{F}(\tilde{x})\,d\tilde{x} = m\left(\frac{\omega_0^2\tilde{x}^2}{2} + \frac{a\tilde{x}^3}{3} - \frac{b\tilde{x}^4}{4} + \ldots\right)$, contains both even and odd powers of $\tilde{x}$. At low excitation intensities, the electron undergoes motions in the proximity of its equilibrium position, $\tilde{x} = 0$, and consequently, the harmonic term is sufficient for describing its oscillations. The induced dipole moment, $\tilde{p}(t) = e \cdot \tilde{x}(t)$, linearly depends on the instantaneous electric field strength $\tilde{E}(t)$. At higher intensities, however, the nonlinear components of the restoring force (*i.e.*, $a\tilde{x}^2(t) - b\tilde{x}^3(t) + \cdots$) must be included to account for the distorted motions exerted by the external field. Note that if even orders nonlinear terms are present (*e.g.*, $a \neq 0$), the potential energy function $U(\tilde{x})$ is not symmetric about 0 ($U(\tilde{x}) \neq U(-\tilde{x})$), whereas if only odd terms are present, the function is symmetric ($U(\tilde{x}) = U(-\tilde{x})$), although not harmonic in all cases. A perturbative expansion of Eq. 10.1 to the order n leads to a series of electron oscillation amplitudes $x^{(1)}, \ldots, x^{(n)}$ proportional to the nth power of the incident electric field amplitude. In the time domain, the nonlinear interaction between a single harmonic oscillator and $\tilde{E}(t)$ can be represented as an induced dipole moment $\tilde{p}_i(t)$ whose Cartesian components $i = 1, 2, 3$ are written as a power series in the strength of the instantaneous applied field (here and throughout the chapter, we use the Einstein summation convention):

$$\begin{aligned}\tilde{p}_i(t) = \epsilon_0 \Big(&\alpha_{ij}^{(1)}\tilde{E}_j(t) + \beta_{ijk}^{(2)}\tilde{E}_j(t)\tilde{E}_k(t) \\ &+\gamma_{ijkl}^{(3)}\widetilde{E_j}(t)\tilde{E}_k(t)\tilde{E}_l(t) + \ldots\Big)\end{aligned} \tag{10.2}$$

In Eq. 10.2, ϵ_0 is the vacuum permittivity in SI units, $\alpha_{ij}^{(1)}$ is the linear polarizability, $\beta_{ijk}^{(2)}$ is the second-order polarizability or first-order hyperpolarizability, and $\gamma_{ijkl}^{(3)}$ is the third-order polarizability or second-order hyperpolarizability. For a given medium, the manifestation of these different microscopic polarizabilities is directly related to the electron motion and symmetries of the potential energy function. Because accelerated charges behave as sources of electromagnetic fields, these motions are at the origin of the frequency components generated upon the interaction with the electromagnetic field. In Figure 10.1, different oscillating dipole traces $\tilde{p}_i(t)$ are plotted to illustrate the scattering of both linear and newly generated nonlinear frequencies at the two lowest nonlinear orders. In the first row, the electron oscillation amplitude is described by a sinusoidal function at the fundamental frequency ω: $\sin(\omega t)$. This situation simply results in the linear scattering of the incoming frequency. The case of an asymmetric oscillation about zero is reported in the second row; here, the distorted $\tilde{p}_i(t)$ trace is expressed as the sum of the previous sine function plus an even harmonic term, $\cos(2\omega t)$, of smaller amplitude. The presence of this new frequency component in the material response to the field, which is associated with the small peak at 2ω in the Fourier spectrum, corresponds to the SHG process. On the other hand, the addition of an odd harmonic term, $\sin(3\omega t)$ – third harmonic generation (THG), to the sine function leads to a symmetric distortion about zero as reported in the third row. Purely odd harmonics spectra are characteristic of the response of centrosymmetric media, where – as one intuitively expects – the oscillating electrons experience the same dielectric environment when moving in the positive and negative directions. In general, both even and odd harmonics are present for noncentrosymmetric crystals as displayed in the bottom row of Figure 10.1.

So far, our description assumed a single electron contribution as well as a lossless and dispersionless medium. This latter assumption is only valid when excitation and harmonic frequencies lie below the (multiple) electron resonances of the solid phase. Furthermore, additional parameters arising from the dense packing of dipoles within a lattice and from the symmetry properties of the relevant crystal class are to be considered. Instead of using electron-related variables such as $x^{(n)}$, these parameters are included in the macroscopic polarization $\tilde{P}_i(t) = N.\tilde{p}_i(t)$, where N stands for the number density of dipoles. In analogy to Eq. 10.2, the instantaneous polarization component along the i-axis is written as

$$\begin{aligned}\tilde{\boldsymbol{P}}_{\boldsymbol{i}}(\boldsymbol{t}) &= \epsilon_0\Big(\boldsymbol{\chi}_{\boldsymbol{ij}}^{(1)}\tilde{\boldsymbol{E}}_{\boldsymbol{j}}(\boldsymbol{t}) + \boldsymbol{\chi}_{\boldsymbol{ijk}}^{(2)}\tilde{\boldsymbol{E}}_{\boldsymbol{j}}(\boldsymbol{t})\tilde{\boldsymbol{E}}_{\boldsymbol{k}}(\boldsymbol{t}) \\ &\quad+\boldsymbol{\chi}_{\boldsymbol{ijkl}}^{(3)}\tilde{\boldsymbol{E}}_{\boldsymbol{j}}(\boldsymbol{t})\tilde{\boldsymbol{E}}_{\boldsymbol{k}}(\boldsymbol{t})\tilde{\boldsymbol{E}}_{\boldsymbol{l}}(\boldsymbol{t}) + \ldots\Big) \\ &= \tilde{\boldsymbol{P}}_{\boldsymbol{i}}^{(1)}(\boldsymbol{t}) + \tilde{\boldsymbol{P}}_{\boldsymbol{i}}^{(2)}(\boldsymbol{t}) + \tilde{\boldsymbol{P}}_{\boldsymbol{i}}^{(3)}(\boldsymbol{t}) + \ldots\end{aligned} \tag{10.3}$$

where we introduced the nonlinear susceptibilities tensors $\chi^{(n)}$ of rank $n+1$. In the linear regime, the proportionality between polarization and field is expressed through the first-order susceptibility $\chi_{ij}^{(1)}$, whereas second-order and third-order nonlinear effects are described by the third- and the fourth-rank tensors $\chi_{ijk}^{(2)}$ and $\chi_{ijkl}^{(3)}$, respectively. Importantly, the number of nonzero elements for each tensor is defined by the symmetry properties of each crystal class (Malgrange, Ricolleau, & Lefaucheux, 2012; Nye, 1985). Note also that the macroscopic electric fields used in Eq. 10.3 differ from the local fields defined in Eq. 10.2 and that appropriate local field factors accounting for the dense packing of atoms within a solid need to be introduced when replacing the microscopic polarizabilities with susceptibilities.

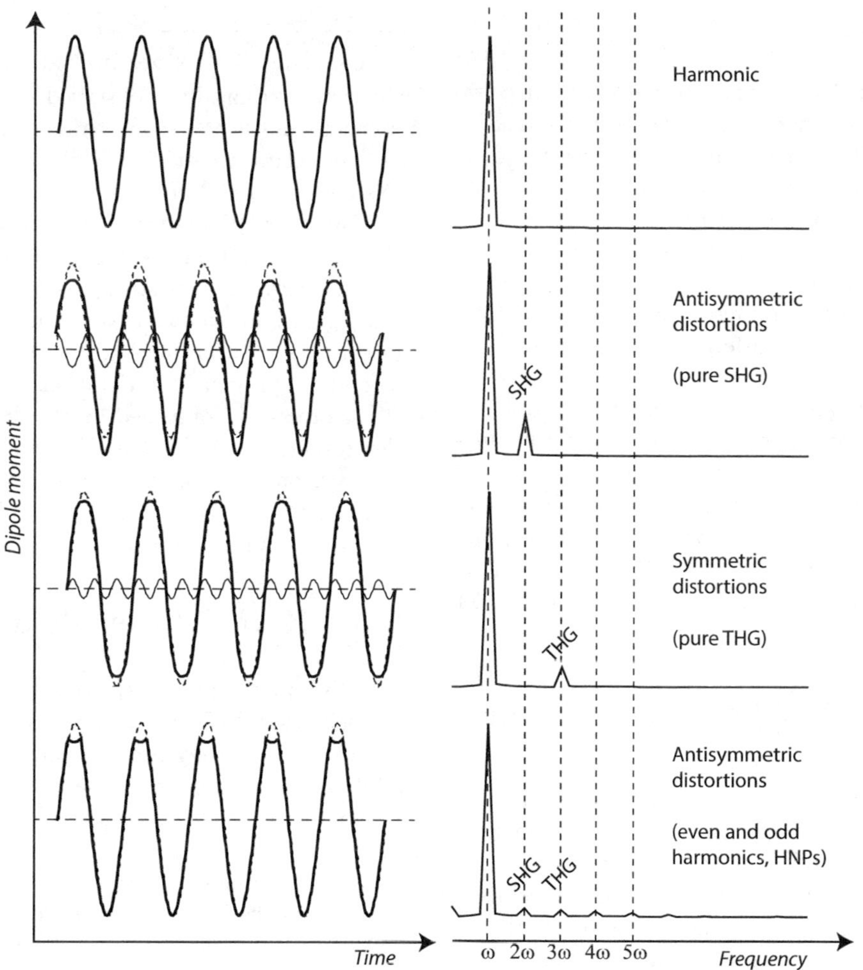

FIGURE 10.1 First row: the electric dipole response is linear with respect to the amplitude of the optical field. The dipole oscillations are harmonic, and only the fundamental frequency ω appears in the Fourier transform plot on the right. Second row: an antisymmetric distortion appears in the oscillation pattern (thick line) because of nonlinear dependence in the field strength. This distortion can be expressed as the sum of the harmonic component (dashed line, same as in the linear case) and a newly generated frequency at 2ω (thin line). Third row: a purely symmetric distortion can be expressed as the sum of the harmonic component (black dashed line) and a newly generated frequency at 3ω (thin line). Fourth row: the general case of a complex antisymmetric distortion featuring several components, similar to the case of HNPs simultaneously emitting multiple even and odd harmonics.

Equation 10.3 is the most comprehensive and informative expression for treating and understanding nonlinear phenomena in the perturbative regime. We point out that the presence of different conventions for defining the electric field and polarization amplitudes often complicates the comparison among experimental values from the literature. Here, the complex excitation field is described as a discrete sum of p monochromatic terms of frequency ω_p. The optical field strength is given by $\tilde{E}(r,t) = \frac{1}{2}\sum_p \left(E_p^{\omega_p} e^{-i\omega_p t} + c.c.\right)$, with $E_p^{\omega_p} = E_p^0 e^{ik_p.r}$ denoting the spatially varying amplitude and E_p^0 denoting the real part of the field amplitude oscillating at ω_p. Likewise, the nonlinear polarization is written as $\tilde{P}(r,t) = \frac{1}{2}\sum_p \left(P_p^{\omega_p'} e^{-i\omega' t} + c.c.\right)$ so that after expansion in Eq. 10.3, the Fourier components of the nonlinear polarization can be derived at each nonlinear order. For instance, in the simplest case of a single input frequency component ω, the amplitude of the second-order nonlinear polarization resulting from $\tilde{P}_{\mathrm{i}}^{(2)}(t) = \epsilon_0 \chi_{ijk}^{(2)} \tilde{E}_{\mathrm{j}}(t) \tilde{E}_{\mathrm{k}}(t)$ becomes

$$P_i^{2\omega} = \frac{1}{2}\epsilon_0 \chi_{ijk}^{(2)} E_j^{\omega} E_k^{\omega} = \epsilon_0 d_{ijk}^{(2)} E_j^{\omega} E_k^{\omega} \quad (10.4)$$

This expression accounts for SHG. $\chi_{ijk}^{(2)}$ is here real and frequency independent since optical frequencies are assumed in the material transparency range and well below electron resonance frequencies (Kleinman's assumption). $d_{ijk}^{(2)}$ is defined as the nonlinear optical coefficient for SHG.

If we now consider a two-frequency input field (ω_1 and ω_2), the second-order polarization contains the previous terms calculated for ω_1 and ω_2 independently, and new terms corresponding to their sum $\omega_1 + \omega_2$ (sum frequency generation, SFG) and difference $\omega_1 - \omega_2$ (DFG).

$$P_i^{\omega_1+\omega_2} = \epsilon_0 \chi_{ijk}^{(2)} E_j^{\omega_1} E_k^{\omega_2} \text{ and } P_i^{\omega_1-\omega_2} = \epsilon_0 \chi_{ijk}^{(2)} E_j^{\omega_1} (E_k^{\omega_2})^* \quad (10.5)$$

Note that the frequencies ω_1 and ω_2 are often just two distinct spectral components within the bandwidth of a femtosecond pulse. The derivation we just described to obtain the nonlinear processes at the second order can be readily repeated for the third order in the power expansion of $\tilde{P}(t)$. In this case, one obtains two families of processes: the so-called parametric ones where the total photon energy is conserved, and the nonparametric ones, where a fraction of the beam energy is transferred to the medium and which feature an imaginary component for the susceptibility $\chi^{(3)}$. The former family includes processes like THG ($\omega + \omega + \omega = 3\omega$, all possible sum and difference combinations of three frequencies (ω_1, ω_2, and ω_3), and the intensity modulation of the refractive index (Kerr effect). The second family includes phenomena such as two-photon absorption and stimulated Raman scattering. For THG, when a unique frequency is considered, the amplitude of the third-order polarization is given by

$$P_i^{3\omega} = \frac{1}{4} \epsilon_0 \chi_{ijkl}^{(3)} E_j^{\omega} E_k^{\omega} E_l^{\omega} \quad (10.6)$$

As oscillating polarization is a source of radiation, the relative strength of the nth order signal emitted by the nonlinear interaction in the medium scales as $(I_\omega)^n$, with I_ω being the intensity of the excitation field. This relationship is valid within the perturbative regime we used for all the derivations provided. Therefore, under moderate excitation conditions, the lowest order allowed by symmetry should prevail since for the anharmonic terms in the expression of the restoring force (see Eq. 10.1), it is expected that $\omega_0^2 \tilde{x}(t) \gg a\tilde{x}^2(t) \gg b\tilde{x}^3(t)$. This is the reason why HNPs are selected among materials possessing noncentrosymmetric crystal structure, so that they can exert second-order nonlinear response along with higher odd and even orders.

In this respect, dielectric HNPs are genuinely different from metal particles where the absence of inversion symmetry at the origin of second-order processes stems only from the particle surface and not from its inner structure (Nappa et al., 2005). Because the crystal volume of HNP materials is by itself devoid of an inversion center, this entails that the SHG emission intensity is essentially associated with the square of the HNP volume, V. Within the electric dipole approximation, this V^2 dependence comes from the coherent nature of signal summation over the individual nonlinear dipoles associated with each unit cell of the lattice (Sandeau et al., 2007; Staedler et al., 2012). Similar to their bulk counterparts, noncentrosymmetric nanocrystals are expected to exhibit large values for their second-order susceptibility tensor $\chi^{(2)}$ (Joulaud et al., 2013; Le Dantec et al., 2011). On the other hand, values of the different $\chi^{(n)}$ susceptibilities are not independent. There exist some heuristic relationships for estimating the average susceptibility value at order n from that at $n-1$, the most famous being Miller's rule, which states that $\frac{\chi^{(2)}(\omega_1+\omega_2,\omega_1,\omega_2)}{\chi^{(1)}(\omega_1+\omega_2)\chi^{(1)}(\omega_1)\chi^{(1)}(\omega_2)}$ is a constant for all noncentroysmmetric crystals if they are probed out of resonance. Similarly, some authors have observed that $|\chi^{(3)}| \propto |\chi^{(2)}|^2$ for frequencies in the transparent range of noncentrosymmetric materials characterized by $|\chi^{(2)}| \geq 10^{-12}$ pm/V (Morita & Yamashita, 1993). It is therefore a reasonable assumption that all HNP materials display both large $|\chi^{(2)}|$ and $|\chi^{(3)}|$ values simultaneously. This, in turn, means that HNPs simultaneously generate strong, volume-dependent SHG and THG signals. In quantitative terms, the SHG/THG ratio scales as $\frac{I_{3\omega}}{I_{2\omega}} \propto I_\omega$ so that high excitation intensities achievable under tight focusing conditions tend to favor the higher order, as recently confirmed by the detection of the fourth-order response of individual HNPs (Riporto et al., 2018). Moreover, one cannot fully neglect the existence of resonances in the material, which can selectively enhance the response at a specific range of frequencies.

10.3 HNP Materials and Synthesis

Most HNP materials belong to the family of metal oxides, and their crystal classes do not show inversion symmetry. Because of that, they are not only SHG active, but also exhibit a series of functional characteristics (piezoelectricity, ferroelectricity, electro-optical properties, etc) exploited in many sensor and transducer applications that typically require bulk crystals or ceramics. High-temperature and long-time processing methods have been developed for the preparation of these multifunctional materials. Unfortunately, crystal growth techniques and solid-state chemistry approaches leading to controlled size and shape are very limited at the nanoscale. Ideally, for HNPs, monodisperse and monocrystalline NPs of well-defined morphology and chemical composition corresponding to mixed-metal oxide phases are required such as those of the perovskite group (*e.g.*, barium titanate [$BaTiO_3$]) and the solid solutions belonging to the titanate, zirconate and PZT [$Pb(Zr_xTi_{1-x}O_3$] families. Lead-free niobate and tantalate materials derived from $KNbO_3$ and $KTaO_3$ are also potential candidates (Lombardi, Pearsall, Li, & O'Brien, 2016; Modeshia & Walton, 2010), and among other transition-metal perovskites, bismuth ferrite ($BiFeO_3$) possesses remarkable SH and TH properties (Clarke et al., 2018; Schmidt et al., 2016). Several non-perovskite structure materials such as $LiNBO_3$, ZnO, and $Fe(IO_3)_3$ have also been investigated.

For all these oxides and mixed-metal oxides, wet chemical routes are preferred since a homogenous, stoichiometric metal precursor containing two or more cations is more likely to be achieved from the liquid state. These approaches include chemical reactions in nanostructured media and sol-gel chemistry employing a large panel of organic additives and/or relying on the systematic variation of several parameters such as the composition of the reaction medium and its temperature (Danks, Hall, & Schnepp, 2016; Modeshia & Walton, 2010). Mild conditions associated with the hydro- and solvothermal processes combined with conventional furnaces or microwave heating are also increasingly applied for adjusting the nucleation and growth kinetics and for promoting crystallization without the need for a high-temperature calcination step (Danks et al., 2016). The production of HNPs of tailored size, shape, and morphology ultimately relies on the elucidation of the reaction (Niederberger & Garnweitner, 2006) and growth mechanisms, which may comprise a combination of several phenomena, including aggregation-induced crystallization, oriented attachment, dissolution-precipitation, and Ostwald ripening.

In practice, such a detailed understanding of the growth mechanisms requires extensive time-resolved measurements, and there is no general rule for the above-cited HNP materials in spite of the increasing available literature, which also includes several reviews (Rachid Ladj et al., 2013; Modeshia & Walton, 2010; Polking, Alivisatos, & Ramesh, 2015). A given material may indeed display a very different crystallization pathway depending on the details of the solution-mediated route applied. We outline below the most common ones present in the literature with some examples relevant for HNP synthesis.

1. Regarding coprecipitation reactions and the use of nanostructured reaction media like water-in-oil microemulsions, second harmonic scattering (SHS) (see next section) has proven to be very useful to probe in real time the crystallization dynamics of $Fe(IO_3)_3$. An aggregation-induced crystallization mechanism has been proposed to account for the particle growth and appearance of different hierarchically organized hybrid superstructures (Ladj et al., 2012; Mugnier et al., 2011). For $BiFeO_3$, solubility issues might explain the formation of stoichiometric amorphous hydroxide precursors after the room-temperature coprecipitation of Fe^{3+} and Bi^{3+} ions in homogeneous solutions, but the subsequent high-temperature crystallization step only results in poorly shape-defined NPs without the addition of organic additives (Tytus et al., 2018). Noteworthy, in the case of lithium niobate and lithium tantalate ($LiNbO_3$ and $LiTaO_3$), if the prerequisite of a homogeneous dispersion of the metal precursors at the atomic scale is not fulfilled, for instance when the Li:Nb (resp. Li:Ta) ratio is different from 1:1 with the Nb-rich hexaniobate (resp. Ta-rich hexatantalate) Lindqvist ions, the aqueous synthesis of phase-pure $LiNbO_3$ (resp. $LiTaO_3$) is prevented (Nyman, Anderson, & Provencio, 2009).
2. The traditional sol-gel chemistry route leading to monometallic oxide materials (Livage & Sanchez, 1992) from alkoxide precursors has been successfully extended to bimetallic oxides like $LiNbO_3$ films (Ono & Hirano, 1997). For this approach, the availability and cost of the metal elements that can be used as alkoxide precursors and their high degree of reactivity with water are the first to be considered. The initial hydrolysis and condensation reactions that are supposed to lead to the desired stoichiometric mixed-metal ABO_3 phase actually depend on several experimental parameters, including the element electronegativity, the amount of water and its pH, the solvent polarity, and the use of chelating agents. The detailed growth mechanisms are not always fully understood, but a series of very convincing results have recently been reported for $LiNbO_3$ (Mohanty et al., 2012), derivatives of $KNbO_3$ (Lombardi et al., 2016), and $BaTiO_3$ (Liu et al., 2015). Similarly, oleic acid-assisted hydrothermal treatments of precipitates resulting from the fast reaction between aqueous metal nitrates and butoxides result in well-defined cubic and spherical $BaTiO_3$ nanocrystals of varying sizes according to the nature of the solvent and cosolvent (Caruntu, Rostamzadeh, Costanzo, Parizi, & Caruntu, 2015). In the case of $LiNbO_3$, when benzyl alcohol is first applied to reduce the niobium ethoxide reactivity, the subsequent addition of triethylamine as a surfactant allows mediating the growth under solvothermal conditions after passivation of the nanocrystal surface. Transformation of the initial lithium and niobium hydroxide precursors to partially crystallized aggregates, and finally to monocrystalline NPs of different size, likely results in a combination of the aggregation-induced crystallization, oriented attachment, and Ostwald ripening processes (Ali & Gates, 2018).
3. Several chelating agents are also effective for stabilizing metal complexes from water-soluble metal salts to circumvent some of the inherent limitations of the alkoxide-based sol-gel route. In the case of $BiFeO_3$, this small-molecule approach (Danks et al., 2016) has been initiated by Ghosh, Dasgupta, Sen, & Sekhar Maiti (2005) and Selbach, Einarsrud, Tybell, & Grande (2007) with citric acid and a series of carboxylic acids with or without deliberate addition of extra hydroxyl groups. Both phase-pure and mixed-phase compounds can be obtained with an average crystallite size that increases with the annealing temperature as expected for an Ostwald ripening process for relatively aggregated products. This solvent evaporation route with metal complexes has then been further refined with mucic acid and NaCl in excess to promote the formation of phase-pure, monocrystalline, and SHG-efficient $BiFeO_3$ HNPs (Clarke et al., 2018). Note also that ethylenediaminetetraacetic acid (EDTA) is another example of commonly used chelating agent that has a key role in the solvothermal

synthesis of several perovskite-type nanomaterials (Modeshia & Walton, 2010).

4. Finally, the controlled formation of mixed-metal oxides from homogeneously dispersed metals in a polymeric precursor can be achieved from the Pechini method. Typically, citric acid is used as a chelating agent to form metal-citrate complexes, but a polymerizer like ethylene glycol is then added to initiate polyesterification upon heating. The as-obtained covalent organic network is expected to be more stable during annealing, thus giving more opportunities to control the crystallization and growth mechanisms. Note that many other available carboxylic acids and polyols can be used in place of citric acid and ethylene glycol, respectively. For $LiNbO_3$, calcination in air of the gel precursor requires high temperature (>450°C), and the resulting dried nanopowder then consists of aggregated nanocrystals (Yerlikaya, Ullah, Kamali, & Kumar, 2016). The lowest crystallization temperature achievable for $BiFeO_3$ has been found to strongly depend on the structure of the chelating agent, and when glycerol is added to promote polyesterification with the mucic acid, mixed-phase products and the absence of any size and shape control can be noticed after the gel combustion (Clarke et al., 2018; Selbach et al., 2007). Very likely, ill-defined local thermodynamic conditions and a non-uniform temperature during the combustion step account for the presence of impurities as already observed when the complexing agent is replaced by a fuel (Schwung et al., 2014).

10.4 Quantitative Assessments of $\chi^{(2)}$ and $\chi^{(3)}$ Values from Colloidal HNP Suspensions

Once HNP samples are obtained, the quantitative assessments of the orientation-averaged second- and third-order susceptibilities can be obtained by SH and TH scattering (SHS and THS) measurements provided that colloidal HNP suspensions of known particle size and mass concentration are available. Noteworthy, SH and TH efficiencies determined by this approach have been found in very good agreement with the literature values on bulk crystals, at least for nanocrystal size larger than 10 nm. We also want to emphasize that, contrary to this 'ensemble measurement' approach, imaging of individual HNPs based on SHG and THG microscopy techniques is less prone to provide reliable results on nonlinear efficiencies. In this case, in fact, a proper modeling of the focusing and collection properties at each nonlinear order needs to be introduced, especially when dealing with diffraction-limited objects. Moreover, when imaging individual NPs because of the tensorial nature of the different harmonic processes, the polarization-resolved traces one obtains should be fitted with mathematical expressions whose complexity rapidly increases with the nonlinear order and with the number of independent coefficients of the relevant crystal class (Bonacina et al., 2007; Le Floc'h, Brasselet, Roch, & Zyss, 2003; Schmidt et al., 2016).

Historically, harmonic light scattering from molecules in solutions (successively referred to hyper-Rayleigh scattering, HRS), first observed in 1965 (Terhune, Maker, & Savage, 1965), paved the way to SHS and THS experiments on colloidal suspensions of NPs. HRS was applied during the nineties to determine the second-order polarizability of nonpolar molecules (Clays & Persoons, 1991; Hendrickx, Clays, & Persoons, 1998). The HRS intensity $I_{2\omega}$ stems from the incoherent SH signal contributions scattered by the individual sources at concentrationN_m with a squared dependence on the incident intensity I_ω: $I_{2\omega} = G_{2\omega} \cdot \lambda_{2\omega}^{-4} \cdot N_m \cdot F_m^2 \cdot \left\langle \left(\beta^{(2)}\right)^2 \right\rangle \cdot I_\omega^2$. The local field factors included in $F_m = f_\omega^2 \cdot f_{2\omega}$ account for the microscopic optical field experienced by each molecule, $G_{2\omega}$ comprises the experimental collection efficiency and other constant factors, while $\left\langle \left(\beta^{(2)}\right)^2 \right\rangle$ is the squared hyperpolarizability isotropically averaged over all orientations (Cyvin, Rauch, & Decius, 1965). Keeping the original HRS experimental configuration, which implies the collection of $I_{2\omega}$ perpendicularly to the excitation beam, HRS was successively extended to larger objects (Deniset-Besseau et al., 2009; Jacobsohn & Banin, 2000; Russier-Antoine, Benichou, Bachelier, Jonin, & Brevet, 2007) and nanocrystal suspensions (Le Dantec et al., 2011; Rodriguez, de Araújo, Brito-Silva, Ivanenko, & Lipovskii, 2009). In the latter case, the introduction of an effective hyperpolarizability term proportional to the NP volume V conveniently upgrades the original HRS formalism (developed for molecular dipoles oscillating at 2ω) to the solid-state phase, filling the gap between microscopic and macroscopic entities. When HNPs are large enough to neglect surface contribution (roughly >20 nm) (Kim et al., 2013; Knabe, Buse, Assenmacher, & Mader, 2012), the scattered SH signal can be expressed as above, *viz.*, $I_{2\omega} = G_{2\omega} \cdot \lambda_{2\omega}^{-4} \cdot N_{\mathrm{NP}} \cdot T_{\mathrm{NP}}^2 \cdot \left\langle \left(\beta_{\mathrm{NP}}^{(2)}\right)^2 \right\rangle \cdot I_\omega^2$, where $\left\langle \beta_{\mathrm{NP}}^{(2)} \right\rangle$ is the volume-dependent hyperpolarizability given by $\left\langle \beta_{\mathrm{NP}}^{(2)} \right\rangle = \sqrt{\left\langle \left(\beta_{\mathrm{NP}}^{(2)}\right)^2 \right\rangle} = \left\langle d_{\mathrm{NP}}^{(2)} \right\rangle V$. Because of its size dependence, $\left\langle \beta_{\mathrm{NP}}^{(2)} \right\rangle$ is not an intrinsic material property as it represents – in the Rayleigh regime (Roke & Gonella, 2012) – the coherent contribution of the induced dipole moments within each unit cell of the NP lattice (Joulaud et al., 2013; Le Dantec et al., 2011). In line with the previous expression, $\left\langle d_{\mathrm{NP}}^{(2)} \right\rangle$ is the orientation-averaged SH coefficient linked to the second-order susceptibility by $\left\langle \chi_{\mathrm{NP}}^{(2)} \right\rangle = 2\left\langle d_{\mathrm{NP}}^{(2)} \right\rangle$. The SHS intensity $I_{2\omega}$ that linearly depends on the number density N_{NP} of HNPs can now be written as $I_{2\omega} = \frac{1}{4}G_{2\omega} \cdot \lambda_{2\omega}^{-4} \cdot N_{\mathrm{NP}} \cdot T_{\mathrm{NP}}^2 \left\langle \left(\chi_{\mathrm{NP}}^{(2)}\right)^2 \right\rangle \cdot V^2 I_\omega^2$, where the coherent contribution from the volume comes along with the V^2 term and where the local (molecular) field factors included in F_m are now replaced by reduction field factors included in T_{NP} to account for the refractive index change. The macroscopic optical excitation field E_j^ω within the NP volume is indeed given by $E_j^\omega = t_\omega \cdot E_{j,\mathrm{inc}}^\omega$, with $E_{j,\mathrm{inc}}^\omega$ denoting the incident optical field. Assuming quasi-spherical NPs, t_ω can be expressed as $t_\omega = \frac{3n_{\mathrm{sol}}^2(\omega)}{2n_{\mathrm{sol}}^2(\omega)+n_{\mathrm{NP}}^2(\omega)}$, where $n_{\mathrm{sol}}(\omega)$ and $n_{\mathrm{NP}(\omega)}$ stand for the solvent and NP average refractive index at ω, respectively. In addition, when dispersion of the refractive index is considered in the vicinity of electron resonances (*i.e.*, $n(\omega) \neq n(2\omega)$), the reduction field factor becomes $T_{\mathrm{NP}} = t_\omega^2 \cdot t_{2\omega}$.

Regarding THS measurements, a very similar formalism can be applied by using $I_{3\omega} = \frac{1}{16}G_{3\omega} \cdot \lambda_{3\omega}^{-4} \cdot N_{\mathrm{NP}} \cdot T_{\mathrm{NP}}^2 \left\langle \left(\chi_{\mathrm{NPs}}^{(3)}\right)^2 \right\rangle \cdot V^2 I_\omega^3$, where the 1/16 pre-factor is given within the convention used by Bosshard, Gubler, Kaatz, Mazerant, & Meier (2000) and with $T_{\mathrm{NP}} = t_\omega^3 \cdot t_{3\omega}$. The ratio between the two scattered harmonic signals $I_{2\omega}$ and $I_{3\omega}$ can be used as an approximate estimate of $\left\langle \chi_{\mathrm{NP}}^{(3)} \right\rangle = \sqrt{\left\langle \left(\chi_{\mathrm{NP}}^{(3)}\right)^2 \right\rangle}$, provided that the excitation intensity I_ω is precisely known (Multian et al., 2018; Schmidt et al., 2016) and after properly accounting for the collection and detection efficiencies at 2ω and 3ω. More generally, a quantitative assessment of the second-order susceptibility $\left\langle \chi_{\mathrm{NP}}^{(2)} \right\rangle$ can be obtained either by the internal reference method or by the external one, thanks to the availability of different reference molecules such as para-nitroaniline (pNa), which is suitable for Nd-YAG (Nd-doped yttrium aluminum garnet, Nd:$Y_3Al_5O_{12}$) laser excitation (1064 nm). Concerning the determination of third-order susceptibilities $\left\langle \chi_{\mathrm{NP}}^{(3)} \right\rangle$, the straightforward internal reference approach based on very common solvents has also been recently demonstrated (Van Steerteghem, Clays, Verbiest, & Van Cleuvenbergen, 2017).

From the experimental standpoint, it is essential to point out that exclusively orientation-averaged entities can be derived from SHS and THS measurements. In the most general case, the relations between $\left\langle \chi_{\mathrm{NP}}^{(2)} \right\rangle$ $\left(\text{resp. } \left\langle \chi_{\mathrm{NP}}^{(3)} \right\rangle\right)$ and the independent elements of the $\chi_{ijk}^{(2)}$ $\left(\text{resp. } \chi_{ijkl}^{(3)}\right)$ tensors depend on the scattering angle, the polarization of the excitation and harmonic signals, and the crystal symmetry. For instance, if we consider materials belonging to the orthorhombic lattice system and to the 222 crystal class, one of the simplest cases in terms of number of nonzero coefficients, any twofold rotation around one of the $i = 1, 2, 3$ crystalline axes keeps constant not only the macroscopic (bulk) crystal morphology but also all the physical properties after the application of the symmetry operator. It results that for each nonlinear order n, susceptibilities tensors $\chi^{(n)}$ of rank $n+1$ must be invariant in the new basis after the application of the tensor transformation. For the second-order susceptibility, this transformation operator assumes the form of $\chi_{ijk}^{(\mathrm{final})} = a_{ip}a_{jq}a_{kr}\chi_{pqr}^{(\mathrm{initial})}$. In the case of the 222 point group, this transformation rule can be applied successively with the a_{ip} coefficients of the twofold rotation matrix A_1 (around $i = 1$) and A_3 (rotation around $i = 3$). It results for a lossless material that the initial 27 (real) independent coefficients of the $\chi_{pqr}^{(\mathrm{initial})}$susceptibility tensor are reduced to only six nonvanishing tensor coefficients xyz, xzy, yzx, yxz, zxy, and zyx. If we now use the contracted Voigt notation, the initial 18 independent elements of the $3\times6\chi_{il}^{(2)}$ matrix can be reduced to three nonzero coefficients. In addition, $\chi_{14} = \chi_{25} = \chi_{36}$ when the excitation and SH frequencies are far from electron resonances (Boyd, 2003).

$$A_1 = \begin{pmatrix} 1 & 0 & 0 \\ 0 & -1 & 0 \\ 0 & 0 & -1 \end{pmatrix} \text{ and } A_3 = \begin{pmatrix} -1 & 0 & 0 \\ 0 & -1 & 0 \\ 0 & 0 & 1 \end{pmatrix}$$

$$\text{leads to } \chi_{il}^{(2)}(\text{class } 222) = \begin{pmatrix} 0 & 0 & 0 & \chi_{14} & 0 & 0 \\ 0 & 0 & 0 & 0 & \chi_{25} & 0 \\ 0 & 0 & 0 & 0 & 0 & \chi_{36} \end{pmatrix}$$

Among oxide materials belonging to high-symmetry point groups, the $\chi_{il}^{(2)}$ matrix of sillenite compounds like BSO ($Bi_{12}SiO_{20}$, ..., symmetry class 23) and the one of ZnO (symmetry class 6mm) are also given below in the nonresonant case as illustrative examples.

$$\chi_{il}^{(2)}(\text{BSO}) = \begin{pmatrix} 0 & 0 & 0 & \chi_{14} & 0 & 0 \\ 0 & 0 & 0 & 0 & \chi_{14} & 0 \\ 0 & 0 & 0 & 0 & 0 & \chi_{14} \end{pmatrix}$$

$$\chi_{il}^{(2)}(\text{ZnO}) = \begin{pmatrix} 0 & 0 & 0 & 0 & \chi_{31} & 0 \\ 0 & 0 & 0 & \chi_{31} & 0 & 0 \\ \chi_{31} & \chi_{31} & \chi_{33} & 0 & 0 & 0 \end{pmatrix}$$

Similarly, symmetry considerations and Kleinman's assumption greatly reduce the number of nonzero components for the third-order susceptibility tensors of each crystal class. After the application of the transformation formula $\chi_{ijkl}^{(\mathrm{final})} = a_{ip}a_{jq}a_{kr}a_{ls}\chi_{pqrs}^{(\mathrm{initial})}$ of a fourth-rank tensor, three independent coefficients $\left(\text{instead of 2 for } \chi_{il}^{(2)}\right)$ can be noticed in the case of ZnO in the following $3\text{x}10\chi_{im}^{(3)}$ (ZnO) third-order susceptibility matrix expressed using the convenient contracted notation introduced by Yang & Xie (1995).

$$\chi_{im}^{(3)}(\text{ZnO}) = \begin{pmatrix} \chi_{11} & 0 & 0 & 0 & 0 & \chi_{16} & 0 & 1/3\cdot\chi_{11} & 0 & 0 \\ 0 & \chi_{11} & 0 & \chi_{16} & 0 & 0 & 0 & 0 & 1/3\cdot\chi_{11} & 0 \\ 0 & 0 & \chi_{33} & 0 & \chi_{16} & 0 & \chi_{16} & 0 & 0 & 0 \end{pmatrix}$$

For SHS and THS measurements, the orientation-averaging procedure that allows deriving $\left\langle \chi^{(2)}_{\mathrm{NP}} \right\rangle$ and $\left\langle \chi^{(3)}_{\mathrm{NP}} \right\rangle$ is usually considered for an incident beam with a linear polarization angle γ defined in the laboratory frame. The incident (macroscopic) optical field in the laboratory frame $\{X,Y,Z\}$ can be expressed in the crystal frame $\{x,y,z\}$ of a single, randomly oriented HNP by applying the rotation matrix R_Ω, where $\Omega(\varphi,\theta,\psi)$ represent the Euler angles. Note that the excitation field is linked through f_ω to the local (microscopic) field e_j^ω experienced by each unit cell, as illustrated in Figure 10.2. The induced dipole moments oscillating at 2ω given by $p_i^{2\omega} = \frac{1}{2}\epsilon_0\beta^{(2)}_{ijk}e_j^\omega e_k^\omega$ can also be expressed as the product of the $3 \times 6\beta^{(2)}_{il}$ matrix and a six-component excitation-field vector defined as $\{e_x^2, e_y^2, e_z^2, 2e_ye_z, 2e_xe_z, 2e_xe_y\}$. The transformation of these nonlinear dipoles in the laboratory frame by the application of the operator S_Ω (*i.e.*, the transpose of R_Ω) allows calculating the macroscopic optical fields $E_I^{2\omega}$ through the dipole radiation formula where the unit vector (0,1,0) is used in Figure 10.3 (for a collection along the Y-axis direction).

The resulting vertically polarized SH intensity accounting for the orientation-averaging procedure is then simply calculated as $I_X^{2\omega} = \frac{1}{8\pi^2}\int_{\theta=0}^{\theta=\pi}\int_{\varphi=0}^{\varphi=2\pi}\int_{\psi=0}^{\psi=2\pi}\left(E_X^{2\omega}\right)^2 \cdot \sin\theta d\theta d\varphi d\psi$, whereas a similar expression applies for $I_z^{2\omega}$ with $E_Z^{2\omega}$. The above description is not only the most natural in terms of physical understanding, but also the most efficient to be numerically computed. It is of course consistent

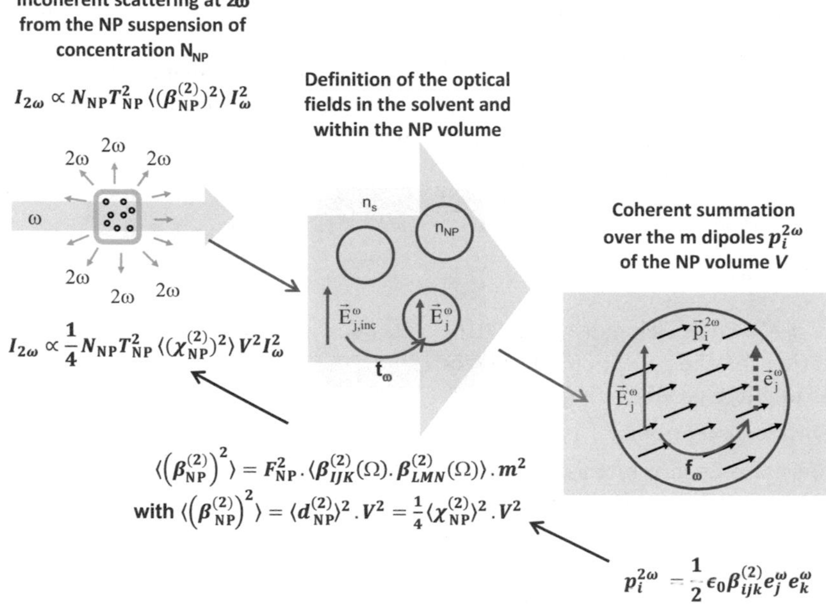

FIGURE 10.2 Principle of the SHS technique on colloidal suspensions of HNPs and definition of the different optical fields and physical entities.

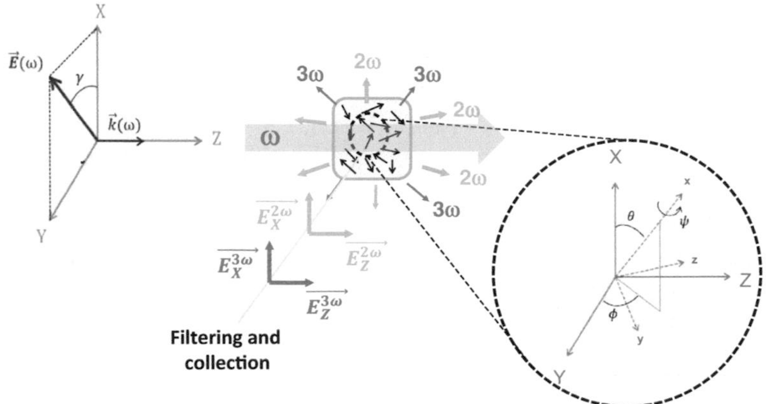

FIGURE 10.3 Experimental configuration for SHS and THS measurements with an incident linear polarization in the (X, Y) plane and a detection along the Y-axis direction. Definition of the Euler angles Ω (φ,θ,ψ) in the XYX convention leading to the orientation-averaged entities in the (macroscopic) laboratory frame $\{X,Y,Z\} = \{I,J,K\}$ from the susceptibility tensors expressed in the (microscopic) crystal frame $\{x,y,z\} = \{i,j,k\}$.

with the common alternative approach that consists in determining the macroscopic $\beta^{(2)}_{IJK}$ coefficients from the corresponding $\beta^{(2)}_{ijk}$ molecular components. This transformation defined by the usual tensor rotation formula $\beta^{(2)}_{IJK}(\Omega) = \sum_{ijk}(S_{Ii}S_{Jj}S_{Kk})\beta^{(2)}_{ijk}$ becomes

$$\left\langle \left(\beta^{(2)}\right)^2 \right\rangle = \left\langle \beta^{(2)}_{IJK}(\Omega) \cdot \beta^{(2)}_{LMN}(\Omega) \right\rangle$$
$$= \frac{1}{8\pi^2} \int_{\theta=0}^{\theta=\pi} \int_{\varphi=0}^{\varphi=2\pi} \int_{\psi=0}^{\psi=2\pi} \sum_{ijklmn} (S_{Ii}S_{Jj}S_{Kk}S_{Ll}S_{Mm}S_{Nn})$$
$$\times \beta^{(2)}_{ijk}\beta^{(2)}_{lmn} \cdot \sin\theta d\theta d\varphi d\psi$$

for the squared, orientation-averaged hyperpolarizability expressed in the laboratory frame (Cyvin et al., 1965). If we consider a colloidal suspension of NPs, the coherent contribution from the m dipoles of the HNP volume V is written as $\left\langle \left(\beta^{(2)}_{\mathrm{NPs}}\right)^2 \right\rangle = F^2_{\mathrm{NP}} \left\langle \beta^{(2)}_{IJK}(\Omega) \cdot \beta^{(2)}_{LMN}(\Omega) \right\rangle m^2 = \left\{ F^2_{\mathrm{NP}} \cdot \left\langle \beta^{(2)}_{IJK}(\Omega) \cdot \beta^{(2)}_{LMN}(\Omega) \right\rangle N^2 \right\} V^2$, where the material local field factors are included in $F_{\mathrm{NP}} = f^2_\omega \cdot f_{2\omega}$ and where the number density of formula units is $N = m/V$. The relation between the macroscopic susceptibility and the microscopic hyperpolarizability $d^{(2)}_{IJK} = \frac{1}{2}\chi^{(2)}_{IJK} = F_{\mathrm{NP}} \cdot \beta^{(2)}_{IJK}(\Omega) N$ finally results in a volume-dependent hyperpolarizability $\left\langle \left(\beta^{(2)}_{\mathrm{NPs}}\right)^2 \right\rangle = \left\langle d^{(2)}_{\mathrm{NPs}} \right\rangle^2 V^2 = \frac{1}{4} \left\langle \chi^{(2)}_{\mathrm{NPs}} \right\rangle^2 V^2$ as we previously discussed.

As an example, when the above formalism is applied to the ZnO case under non-resonant conditions and for a vertically polarized incident field ($\gamma = 0$) with no analyzer in the detection path, the square of the orientation-averaged second-order susceptibility becomes $\left\langle \left(\chi^{(2)}_{\mathrm{NP}}\right)^2 \right\rangle = \left\langle \left(\chi^{(2)}_{XXX}\right)^2 \right\rangle + \left\langle \left(\chi^{(2)}_{ZXX}\right)^2 \right\rangle$ with $\left\langle \left(\chi^{(2)}_{XXX}\right)^2 \right\rangle = \frac{1}{7}\chi^2_{33} + \frac{24}{35}\chi^2_{31} + \frac{12}{35}\chi_{33}\chi_{31}$ and $\left\langle \left(\chi^{(2)}_{ZXX}\right)^2 \right\rangle = \frac{1}{35}\chi^2_{33} + \frac{4}{21}\chi^2_{31} - \frac{4}{105}\chi_{33}\chi_{31}$.

Similarly, when $\beta^{(2)}_{ijk}\beta^{(2)}_{lmn}$ is now replaced by $\gamma^{(3)}_{ijkl} \cdot \gamma^{(3)}_{mnop}$ in the previously defined integration and the number of rotation matrix S_Ω consistently increased, the square of the orientation-averaged third-order susceptibility measured with the THS configuration is expressed as $\left\langle \left(\chi^{(3)}_{\mathrm{NP}}\right)^2 \right\rangle_{\mathrm{THS}} = \left\langle \left(\chi^{(3)}_{XXXX}\right)^2 \right\rangle_{\mathrm{THS}} + \left\langle \left(\chi^{(3)}_{ZXXX}\right)^2 \right\rangle_{\mathrm{THS}}$ with:

$$\left\langle \left(\chi^{(3)}_{XXXX}\right)^2 \right\rangle_{\mathrm{THS}} = \frac{1}{9}\chi^2_{33} + \frac{128}{315}\chi^2_{11} + \frac{32}{35}\chi^2_{16} + \frac{16}{315}\chi_{33}\chi_{11} + \frac{8}{21}\chi_{33}\chi_{16} + \frac{64}{105}\chi_{11}\chi_{16}$$

$$\left\langle \left(\chi^{(3)}_{ZXXX}\right)^2 \right\rangle_{\mathrm{THS}} = \frac{1}{63}\chi^2_{33} + \frac{8}{315}\chi^2_{11} + \frac{1}{7}\chi^2_{16} - \frac{8}{315}\chi_{33}\chi_{11} - \frac{2}{105}\chi_{33}\chi_{16} - \frac{8}{105}\chi_{11}\chi_{16}$$

where the independent third-order coefficients are here still expressed with the matrix notation (Alexiewicz, Ozgo, & Kielich, 1975; Multian et al., 2018).

10.5 Optical Bioimaging with HNPs

The choice of NPs in optical bioimaging applications over other methods (fluorescence immunostaining, fluorescent protein expression, etc) is usually motivated by one or several of the following criteria: (i) need of very bright and stable emission over time, (ii) access to multimodal detection, (iii) application of targeting strategy (antibodies, enhanced retention, etc), (iv) use of the NPs for *in situ* delivery of a molecular payload or as local triggers of a therapeutic effect. Clearly, not all these requirements are fully met by a single nanotechnological approach. The side effects of the use of NPs (in terms of toxicity or other perturbation of the sample under study) should also be carefully accounted for in the choice. Based on these general considerations, the following characteristics of HNPs, discussed in the previous sections, are relevant for their selection in specific bioimaging applications:

1. Typical size 50–200 nm, with in general a rather broad size distribution within this interval.
2. Absence of known toxic elements in their chemical composition.
3. Emission based on multiple simultaneous harmonic signals rather than luminescence.
   (a) The relatively large dimensions of HNPs follow from the chemical synthetic protocols and are very beneficial in terms of brightness since the signal intensity scales as V^2. This volume dependence poses limitations to the detection of HNPs with diameter $\ll$ 50 nm in a biological sample. Because of their typical size and not homogeneous shape distribution, HNPs are generally not the first choice for *intracellular* target delivery and, in general, for subcellular studies, with some noticeable exceptions (Liu, Cho, Cui, & Irudayaraj, 2014; Macias-Romero et al., 2014; Nakayama et al., 2007). The particles, in fact, tend to remain localized in specific compartments after internalization (endosomes) weakly interacting with the rest of the cell body (Staedler et al., 2012, 2015). Under this respect, they are better suited to label specific individual cells in a sample.
   (b) Differently from the most quantum dots, HNPs do not contain heavy elements (*e.g.*, Cd, Hg), known to be highly cytotoxic. However, their effects on cell's viability, proliferation, metabolism, etc., should be assessed for each sample type as the interaction depends not only on the particle

chemical composition, but is also affected by its structural stability at different pHs (to avoid the release of ions), its shape and size, surface charge, and concentration. Most importantly, HNP-cell interaction is critically sensitive to the cell type under study. Note that for most bioimaging applications, particles should be coated (e.g., with a polymer) and specific functionalization agents could be attached to the coating to selectively target cell membrane receptors (Débarre et al., 2005; Hsieh, Grange, Pu, & Psaltis, 2010; Nami, Sonay, Roxanne, E., & Periklis, 2018; Passemard et al., 2015). These additional procedures, common to other nanotechnology approaches, can further modify the toxicity profile of the particle. So far, most of the studies on HNP-cell interaction (including high-throughput multi-parameter analysis) have been performed on $BaTiO_3$, BFO, and ZnO NPs. The former two seem to affect weakly the cells, and notably, they have been proven compatible with stem cell differentiation (Li et al., 2016; Nami et al., 2018).

(c) HNP optical excitation and emission properties are the ones, which stand out making the difference with other approaches and could motivate their selection for specific imaging tasks. First, the fact that the detected signals rely on nonparametric optical generation (SHG and THG) implies that signal stability over time is not affected by blinking or bleaching (Le Xuan et al., 2008; Staedler et al., 2012). In fact, the light-particle interaction does not require sequential excitation and emission processes involving excited electronic states, which always possess a non-vanishing probability to lead the system into a dark state (e.g., surface trapping in quantum dots). Similarly, for HNPs excited within their spectral transparency range, minimal excess energy is deposed on the particle (contrary to the case of fluorescent probes), which helps in preventing sample photodegradation by cumulative heat deposition. A crucial role for motivating the selection of HNPs against other nanoprobes is played by the possibility to tune the excitation wavelength in a large spectral region spanning from near ultraviolet (UV) to mid-IR, according to the transparency region of each specific HNP material. The advantages opened by this possibility are multifold: the laser wavelength can be selected (i) to avoid autofluorescence excitation and overlap with the fluorescent staining agents present in the sample, and (ii) to best match the sample optical properties to increase imaging penetration depth (Grange, Lanvin, Hsieh, Pu, & Psaltis, 2011; Pantazis, Maloney, Wu, & Fraser, 2010; Rogov et al., 2015). In thick samples, the imaging depth critically depends on light scattering and absorption along the excitation path. As scattering is monotonously decreasing with wavelength, longer wavelengths are more favorable provided that the maxima of water absorption are avoided. With the advent of new light sources in the near infrared (NIR), new spectral regions amenable for imaging have been added to the traditional transparency tissue window (NIR I, 600–1,000 nm), namely, NIR II (1,100–1,350 nm) and NIR III (1,600–1,870 nm) (Sordillo, Pu, Pratavieira, Budansky, & Alfano, 2014). Very interestingly, HNPs can be excited within these NIR regions, and the harmonic signals produced conveniently lie in the visible, where water absorption is minimal. Note that the scattering properties of the sample do not constitute a limiting factor for the detection of the signal generated; on the contrary, scattering can be convenient for increasing the fraction of signal photons detected in the backward direction, which is the standard collection geometry for *in vivo* studies (Débarre, Olivier, & Beaurepaire, 2007).

The selectivity in the retrieval of HNPs in optically congested environments (e.g., tissue, body fluids), besides the aforementioned minimization of (auto)fluorescence hindrance by wavelength selection, can be further improved using a multiorder harmonic approach (Dubreil et al., 2017; Rogov et al., 2015). When excited in the NIR-II region, for example, at 1,300 nm, HNPs emit signals simultaneously at 650 nm (SHG) and 433 nm (THG). These two signals can be conveniently collected in parallel and correlated to identify NPs not only against any fluorescence background but also against endogenous sources of harmonic emission in tissues such as collagen (SHG) (Williams, Zipfel, & Webb, 2005) and lipids (THG) (Débarre et al., 2005). In fact, while these endogenous structures are known to emit efficiently only at one nonlinear order (either SHG or THG), HNPs generate both harmonic signals. As reported in Figure 10.4, this characteristic deems essential for identifying them in different tissue compartments. In panel A, an extended collagen network is visible by SHG. The eventual presence of structures or cells labeled by HNPs is not apparent in this channel; however, if observed in the THG channel, only HNPs appear thanks to their strong emission. Similarly, the lipid composition of myelin, a fatty substance surrounding nerve axons, appears in the THG channel hindering the presence of HNPs (panel E). In this case, the SHG channel provides background-free detection of

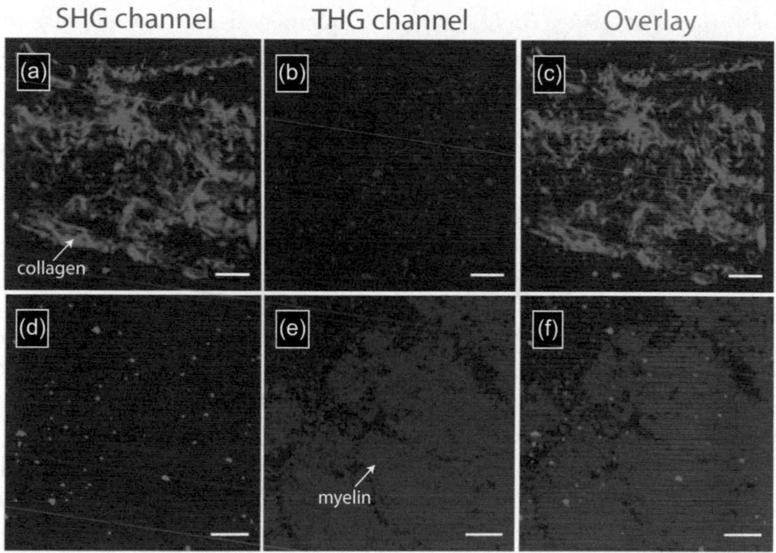

FIGURE 10.4 First row: HNPs on a collagen-rich tissue region. In the SHG channel, HNPs are not distinguishable from the background, while they appear with no hindrance in the THG channel. Second row: HNPs in a myelin-rich region. Here, HNPs are visible with high contrast exclusively the SHG channel. Laser excitation 1,300 nm. Scale bar 20 μm. (Adapted with permission from Dubreil et al., 2017. Copyright 2017 American Chemical Society.)

the particles. Note that the excitation and collection of SHG and THG channels are implemented simultaneously under standard conditions for multiphoton imaging. However, a THG signal can be detected only when using laser excitation above 1,100 nm, because of the transmission cutoff of microscopic regular optics <350 nm.

References

Alexiewicz, W., Ozgo, Z., & Kielich, S. (1975). Spectral theory of third-harmonic light scattering by molecular liquids. *Acta Physica Polonica A*, *48*, 243–252.

Ali, R. F., & Gates, B. D. (2018). Synthesis of lithium niobate nanocrystals with size focusing through an Ostwald ripening process. *Chemistry of Materials*, *30*(6), 2028–2035.

Baranov, D. G., Zuev, D. A., Lepeshov, S. I., Kotov, O. V., Krasnok, A. E., Evlyukhin, A. B., & Chichkov, B. N. (2017). All-dielectric nanophotonics: The quest for better materials and fabrication techniques. *Optica*, *4*(7), 814–825. doi:10.1364/OPTICA.4.000814

Bonacina, L., Mugnier, Y., Courvoisier, F., Le Dantec, R., Extermann, J., Lambert, Y., ... Wolf, J. P. (2007). Polar Fe(IO3)(3) nanocrystals as local probes for nonlinear microscopy. *Applied Physics B-Lasers and Optics*, *87*(3), 399–403. doi:10.1007/s00340-007-2612-z

Bosshard, C., Gubler, U., Kaatz, P., Mazerant, W., & Meier, U. (2000). Non-phase-matched optical third-harmonic generation in noncentrosymmetric media: Cascaded second-order contributions for the calibration of third-order nonlinearities. *Physical Review B*, *61*(16), 10688.

Boyd, R. W. (2003). *Nonlinear Optics*. Burlington, MA: Academic Press.

Caruntu, D., Rostamzadeh, T., Costanzo, T., Parizi, S. S., & Caruntu, G. (2015). Solvothermal synthesis and controlled self-assembly of monodisperse titanium-based perovskite colloidal nanocrystals. *Nanoscale*, *7*(30), 12955–12969.

Clarke, G., Rogov, A., McCarthy, S., Bonacina, L., Gun'ko, Y., Galez, C., ... Prina-Mello, A. (2018). Preparation from a revisited wet chemical route of phase-pure, monocrystalline and SHG-efficient $BiFeO_3$ nanoparticles for harmonic bio-imaging. *Scientific Reports*, *8*(1), 10473.

Clays, K., & Persoons, A. (1991). Hyper-Rayleigh scattering in solution. *Physical Review Letters*, *66*(23), 2980.

Cyvin, S., Rauch, J., & Decius, J. (1965). Theory of hyper-Raman effects (nonlinear inelastic light scattering): Selection rules and depolarization ratios for the second-order polarizability. *The Journal of Chemical Physics*, *43*(11), 4083–4095.

Danks, A., Hall, S., & Schnepp, Z. (2016). The evolution of 'sol–gel' chemistry as a technique for materials synthesis. *Materials Horizons*, *3*(2), 91–112.

Débarre, D., Olivier, N., & Beaurepaire, E. (2007). Signal epidetection in third-harmonic generation microscopy of turbid media. *Optics Express*, *15*(14), 8913–8924.

Débarre, D., Supatto, W., Pena, A.-M., Fabre, A., Tordjmann, T., Combettes, L., ... Beaurepaire, E. (2005). Imaging lipid bodies in cells and tissues using third-harmonic generation microscopy. *Nature Methods*, *3*(1), 47–53.

Deniset-Besseau, A., Duboisset, J., Benichou, E., Hache, F., Brevet, P.-F., & Schanne-Klein, M.-C. (2009). Measurement of the second-order hyperpolarizability of the collagen triple helix and determination of its physical origin. *The Journal of Physical Chemistry B*, *113*(40), 13437–13445.

Dubreil, L., Leroux, I., Ledevin, M., Schleder, C., Lagalice, L., Lovo, C., ... Rouger, K. (2017). Multi-harmonic imaging in the second near-infrared window of nanoparticle-labeled stem cells as monitoring tool in tissue depth. *ACS Nano*, *11*(7), 6672–6681.

Ghosh, S., Dasgupta, S., Sen, A., & Sekhar Maiti, H. (2005). Low-temperature synthesis of nanosized bismuth ferrite by soft chemical route. *Journal of the American Ceramic Society*, *88*(5), 1349–1352.

Grange, R., Lanvin, T., Hsieh, C. L., Pu, Y., & Psaltis, D. (2011). Imaging with second-harmonic radiation probes in living tissue. *Biomedical Optics Express*, *2*(9), 2532–2539.

Gubler, U., & Bosshard, C. (2002). Molecular design for third-order nonlinear optics. In: Lee, K.-S. (Ed.) *Polymers for Photonics Applications I. Advances in Polymer Science.* Berlin, Heidelberg: Springer, 123–191.

Hendrickx, E., Clays, K., & Persoons, A. (1998). Hyper-Rayleigh scattering in isotropic solution. *Accounts of Chemical Research*, *31*(10), 675–683.

Hsieh, C. L., Grange, R., Pu, Y., & Psaltis, D. (2010). Bioconjugation of barium titanate nanocrystals with immunoglobulin G antibody for second harmonic radiation imaging probes. *Biomaterials*, *31*(8), 2272–2277. doi:10.1016/j.biomaterials.2009.11.096

Jacobsohn, M., & Banin, U. (2000). Size dependence of second harmonic generation in CdSe nanocrystal quantum dots. *The Journal of Physical Chemistry B*, *104*(1), 1–5.

Joulaud, C., Mugnier, Y., Djanta, G., Dubled, M., Marty, J. C., Galez, C., ... Le Dantec, R. (2013). Characterization of the nonlinear optical properties of nanocrystals by Hyper Rayleigh scattering. *Journal of Nanobiotechnology*, *11*, S8. doi:10.1186/1477-3155-11-S1-S8

Kim, E., Steinbrück, A., Buscaglia, M. T., Buscaglia, V., Pertsch, T., & Grange, R. (2013). Second-harmonic generation of single $BaTiO_3$ Nanoparticles down to 22 nm diameter. *ACS Nano*, *7*(6), 5343–5349.

Knabe, B., Buse, K., Assenmacher, W., & Mader, W. (2012). Spontaneous polarization in ultrasmall lithium niobate nanocrystals revealed by second harmonic generation. *Physical Review B*, *86*(19), 195428.

Kuznetsov, A. I., Miroshnichenko, A. E., Brongersma, M. L., Kivshar, Y. S., & Luk'yanchuk, B. (2016). Optically resonant dielectric nanostructures. *Science*, *354*(6314), aag2472.

Ladj, R., Bitar, A., Eissa, M., Mugnier, Y., Le Dantec, R., Fessi, H., & Elaissari, A. (2013). Individual inorganic nanoparticles: Preparation, functionalization and in vitro biomedical diagnostic applications. *Journal of Materials Chemistry B*, *1*(10), 1381–1396.

Ladj, R., El Kass, M., Mugnier, Y., Le Dantec, R., Fessi, H., Galez, C., & Elaissari, A. (2012). SHG active Fe(IO3)3 particles: From spherical nanocrystals to urchin-like microstructures through the additive-mediated microemulsion route. *Crystal Growth and Design.* doi:10.1021/cg3009915

Le Dantec, R., Mugnier, Y., Djanta, G., Bonacina, L., Extermann, J., Badie, L., ... Galez, C. (2011). Ensemble and individual characterization of the nonlinear optical properties of ZnO and $BaTiO_3$ nanocrystals. *Journal of Physical Chemistry C*, *115*(31), 15140–15146. doi:10.1021/Jp200579x

Le Floc'h, V., Brasselet, S., Roch, J.-F., & Zyss, J. (2003). Monitoring of orientation in molecular ensembles by polarization sensitive nonlinear microscopy. *The Journal of Physical Chemistry B*, *107*(45), 12403–12410.

Le Xuan, L., Zhou, C., Slablab, A., Chauvat, D., Tard, C., Perruchas, S., ... Roch, J. F. (2008). Photostable second-harmonic generation from a single $KTiOPO_4$ nanocrystal for nonlinear microscopy. *Small*, *4*(9), 1332–1336. doi:10.1002/smll.200701093

Li, J. H., Qiu, J. C., Guo, W. B., Wang, S., Ma, B. J., Mou, X. N., ... Liu, H. (2016). Cellular internalization of $LiNbO_3$ nanocrystals for second harmonic imaging and the effects on stem cell differentiation. *Nanoscale*, *8*(14), 7416–7422. doi:10.1039/c6nr00785f

Liu, J., Cho, I.-H., Cui, Y., & Irudayaraj, J. (2014). Second harmonic super-resolution microscopy for quantification of mRNA at single copy sensitivity. *ACS Nano*, *8*(12), 12418–12427.

Liu, S., Huang, L., Li, W., Liu, X., Jing, S., Li, J., & O'Brien, S. (2015). Green and scalable production of colloidal perovskite nanocrystals and transparent sols by a controlled self-collection process. *Nanoscale*, *7*(27), 11766–11776.

Livage, J., & Sanchez, C. (1992). Sol-gel chemistry. *Journal of Non-Crystalline Solids*, *145*, 11–19.

Lombardi, J., Pearsall, F., Li, W., & O'Brien, S. (2016). Synthesis and dielectric properties of nanocrystalline oxide perovskites, $[KNbO_3]1-x$ [BaNi 0.5 Nb 0.5 O 3− δ] x, derived from potassium niobate $KNbO_3$ by gel collection. *Journal of Materials Chemistry C*, *4*(34), 7989–7998.

Macias-Romero, C., Didier, M. E. P., Zubkovs, V., Delannoy, L., Dutto, F., Radenovic, A., & Roke, S. (2014). Probing rotational and translational diffusion of nanodoublers in living cells on microsecond time scales. *Nano Letters*, *14*(5), 2552–2557.

Malgrange, C., Ricolleau, C., & Lefaucheux, F. (2012). *Symétrie et propriétés physiques des cristaux.* EDP Sciences, Les Ulis, France.

Modeshia, D. R., & Walton, R. I. (2010). Solvothermal synthesis of perovskites and pyrochlores: Crystallisation of functional oxides under mild conditions. *Chemical Society Reviews*, *39*(11), 4303–4325.

Mohanty, D., Chaubey, G. S., Yourdkhani, A., Adireddy, S., Caruntu, G., & Wiley, J. B. (2012). Synthesis and piezoelectric response of cubic and spherical $LiNbO_3$ nanocrystals. *RSC Advances*, *2*(5), 1913–1916. doi:10.1039/C2ra00628f

Morita, R., & Yamashita, M. (1993). Relationship between 2nd-order and 3rd-order nonlinear-optical susceptibilities due to electronic polarization. *Japanese Journal of Applied Physics Part 2-Letters, 32*(7A), L905–L907. doi:10.1143/Jjap.32.L905

Mugnier, Y., Houf, L., El-Kass, M., Le Dantec, R., Hadji, R., Vincent, B., ... Galez, C. (2011). In situ crystallization and growth dynamics of acentric iron iodate nanocrystals in w/o microemulsions probed by Hyper-Rayleigh scattering measurements. *Journal of Physical Chemistry C, 115*(1), 23–30. doi:10.1021/Jp105638s

Multian, V. V., Riporto, J., Urbain, M., Mugnier, Y., Djanta, G., Beauquis, S., ... Le Dantec, R. (2018). Averaged third-order susceptibility of ZnO nanocrystals from third harmonic generation and third harmonic scattering. *Optical Materials, 84*, 579–585. doi:10.1016/j.optmat.2018.07.032

Nakayama, Y., Pauzauskie, P. J., Radenovic, A., Onorato, R. M., Saykally, R. J., Liphardt, J., & Yang, P. D. (2007). Tunable nanowire nonlinear optical probe. *Nature, 447*(7148), U1098. doi:10.1038/Nature05921

Nami, S., Sonay, A. Y., Roxanne, T., Cohen, B. E., & Periklis, P. (2018). Effective labeling of primary somatic stem cells with $BaTiO_3$ nanocrystals for second harmonic generation imaging. *Small, 14*(8), 1703386. doi:10.1002/smll.201703386

Nappa, J., Revillod, G., Russier-Antoine, I., Benichou, E., Jonin, C., & Brevet, P. (2005). Electric dipole origin of the second harmonic generation of small metallic particles. *Physical Review B, 71*(16), 165407.

Niederberger, M., & Garnweitner, G. (2006). Organic reaction pathways in the nonaqueous synthesis of metal oxide nanoparticles. *Chemistry–A European Journal, 12*(28), 7282–7302.

Nye, J. F. (1985). *Physical Properties of Crystals: Their Representation by Tensors and Matrices*. Oxford: Oxford University Press.

Nyman, M., Anderson, T. M., & Provencio, P. P. (2009). Comparison of aqueous and non-aqueous soft-chemical syntheses of lithium niobate and lithium tantalate powders. *Crystal Growth and Design, 9*(2), 1036–1040.

Ono, S., & Hirano, S. I. (1997). Patterning of lithium niobate thin films derived from aqueous solution. *Journal of the American Ceramic Society, 80*(10), 2533–2540.

Pantazis, P., Maloney, J., Wu, D., & Fraser, S. E. (2010). Second harmonic generating (SHG) nanoprobes for in vivo imaging. *Proceedings of the National Academy of Sciences of the United States of America, 107*(33), 14535–14540. doi:10.1073/pnas.1004748107

Passemard, S. E., Staedler, D., Sonego, G., Magouroux, T., Schneiter, G. S. E., Juillerat-Jeanneret, L., ... Gerber-Lemaire, S. (2015). Functionalized bismuth ferrite harmonic nanoparticles for cancer cells labeling and imaging. *Journal of Nanoparticle Research, 17*(10), 1–13.

Polking, M. J., Alivisatos, A. P., & Ramesh, R. (2015). Synthesis, physics, and applications of ferroelectric nanomaterials. *MRS Communications, 5*(1), 27–44.

Riporto, J., Demierre, A., Kilin, V., Balciunas, T., Schmidt, C., Campargue, G., ... Wolf, J.-P. (2018). Bismuth ferrite dielectric nanoparticles excited at telecom wavelengths as multicolor sources by second, third, and fourth harmonic generation. *Nanoscale, 10*(17), 8146–8152.

Rodriguez, E. V., de Araújo, C. B., Brito-Silva, A. M., Ivanenko, V., & Lipovskii, A. (2009). Hyper-Rayleigh scattering from $BaTiO_3$ and $PbTiO_3$ nanocrystals. *Chemical Physics Letters, 467*(4–6), 335–338.

Rogov, A., Irondelle, M., Ramos Gomes, F., Bode, J., Staedler, D., Passemard, S., ... Wolf, J.-P. (2015). Simultaneous multiharmonic imaging of nanoparticles in tissues for increased selectivity. *ACS Photonics, 2*(10), 1416–1422. doi:10.1021/acsphotonics.5b00289

Roke, S., & Gonella, G. (2012). Nonlinear light scattering and spectroscopy of particles and droplets in liquids. *Annual Review of Physical chemistry, 63*, 353–378.

Russier-Antoine, I., Benichou, E., Bachelier, G., Jonin, C., & Brevet, P. (2007). Multipolar contributions of the second harmonic generation from silver and gold nanoparticles. *The Journal of Physical Chemistry C, 111*(26), 9044–9048.

Sandeau, N., Le Xuan, L., Chauvat, D., Zhou, C., Roch, J. F., & Brasselet, S. (2007). Defocused imaging of second harmonic generation from a single nanocrystal. *Optics Express, 15*(24), 16051–16060.

Schmidt, C., Riporto, J., Uldry, A., Rogov, A., Mugnier, Y., Le Dantec, R., ... Bonacina, L. (2016). Multi-order investigation of the nonlinear susceptibility tensors of individual nanoparticles. *Scientific Reports, 6*, 25415.

Schwung, S., Rogov, A., Clarke, G., Joulaud, C., Magouroux, T., Staedler, D., ... Galez, C. (2014). Nonlinear optical and magnetic properties of $BiFeO_3$ harmonic nanoparticles. *Journal of Applied Physics, 116*(11), 114306.

Selbach, S. M., Einarsrud, M. A., Tybell, T., & Grande, T. (2007). Synthesis of $BiFeO_3$ by wet chemical methods. *Journal of the American Ceramic Society, 90*(11), 3430–3434.

Smith, B. R., & Gambhir, S. S. (2017). Nanomaterials for in vivo imaging. *Chemical Reviews, 117*(3), 901–986.

Sordillo, L. A., Pu, Y., Pratavieira, S., Budansky, Y., & Alfano, R. R. (2014). Deep optical imaging of tissue using the second and third near-infrared spectral windows. *Journal of Biomedical Optics, 19*(5), 056004.

Staedler, D., Magouroux, T., Hadji, R., Joulaud, C., Extermann, J., Schwungi, S., ... Wolf, J. P. (2012). Harmonic nanocrystals for biolabeling: A survey of optical properties and biocompatibility. *ACS Nano, 6*(3), 2542–2549. doi:10.1021/Nn204990n

Staedler, D., Passemard, S., Magouroux, T., Rogov, A., Maguire, C. M., Mohamed, B. M., ... Wolf, J.-P. (2015). Cellular uptake and biocompatibility of bismuth ferrite harmonic advanced nanoparticles. *Nanomedicine: NBM, 11*(4), 815–828.

Terhune, R., Maker, P., & Savage, C. (1965). Measurements of nonlinear light scattering. *Physical Review Letters*, *14*(17), 681.

Tytus, T., Phelan, O., Urbain, M., Clarke, G., Riporto, J., Le Dantec, R., ... Mugnier, Y. (2018). Preparation and preliminary non-linear optical properties of $BiFeO_3$ nanocrystal suspensions from a simple, chelating agent-free precipitation route. *Journal of Nanomaterials*, 2018, 9.

Van Steerteghem, N., Clays, K., Verbiest, T., & Van Cleuvenbergen, S. (2017). Third-harmonic scattering for fast and sensitive screening of the second hyperpolarizability in solution. *Analytical Chemistry*, *89*(5), 2964–2971.

Williams, R. M., Zipfel, W. R., & Webb, W. W. (2005). Interpreting second-harmonic generation images of collagen I fibrils. *Biophysical Journal*, *88*(2), 1377–1386. doi:10.1529/biophysj.104.047308

Yang, X.-L., & Xie, S.-W. (1995). Expression of third-order effective nonlinear susceptibility for third-harmonic generation in crystals. *Applied Optics*, *34*(27), 6130–6135.

Yerlikaya, C., Ullah, N., Kamali, A. R., & Kumar, R. V. (2016). Size-controllable synthesis of lithium niobate nanocrystals using modified Pechini polymeric precursor method. *Journal of Thermal Analysis and Calorimetry*, *125*(1), 17–22.

11

Fundamentals of Laser-Generated Nanoparticles in Liquid-Phase

Ali Karatutlu,
Elif Yapar Yildirim, and
Bülend Ortaç
Bilkent University

In this chapter, fundamentals of the laser-generated nanoparticles (NPs) in liquid phase were highlighted. The pulsed laser ablation in liquid phase (PLA-LP) was attempted to be covered from its basics to the advanced topics. By discussing the key issues on the PLA-LP-synthesized NPs, we include theoretical background between the laser and material interaction, the PLA mechanism in liquid phase and examples of the applications. Variation in the laser parameters such as laser fluence is shown to produce a change in the properties of NPs. We also demonstrate the effect of environmental parameters such as temperature and pressure on the laser-generated NPs. The applications conducted using the ultrapure NPs formed by the PLA-LP method were demonstrated for solar cells, catalysts, sensors and biomedicine.

11.1 Introduction

Physical properties of materials can be enhanced when one has the opportunity to tailor the size of particles. After their size is reduced below 100 nm, such particles are classified as NPs according to *International Union of Pure and Applied Chemistry* (*IUPAC*) glossary. This is the scale where NPs are considered to initiate, demonstrating the unique properties compared to their bulk counterparts. Once it comes to choose a method to produce NPs for the criteria of a high quality and a good colloidal stability, one should consider how the purity, the surface termination and the size distribution are affected upon completing the chosen fabrication process.

There are two main approaches to produce NPs, namely, (i) *top-down* and (ii) *bottom-up*. A pulsed laser source can be operated for the purpose of processing and producing materials in nanoparticulate forms by fragmentation of their bulk counterparts or nucleation and growth of NPs and satisfying the aforementioned criteria as well. This process is called PLA after the pioneering studies (Ogale et al. 1987, Fojtik et al. 1993, Henglein 1993). Over the past decade, many developments in the formation of NPs using a pulsed laser source have come forth (Gökce et al. 2017, Zhang et al. 2017). The method can be conducted in a vacuum, air (solid), liquid or gas environment (phase) as an ambient atmosphere to surround the bulk target before the interaction between a pulsed laser source and the target material is held. Scalability of the continuous production of NPs obtained by the *PLA-LP* could be considered as an issue. Nevertheless, *gram*-scale synthesis of NPs by the *PLA-LP* was recently demonstrated to be available (Streubel et al. 2016a,b). This issue was overcome using a *femtosecond* (*fs*) pulsed laser system having high repetition rate in the order of megahertz (MHz) and a polygon scanner with the scanning speeds up to 500 m/s. Therefore, the *PLA-LP* method could be applied to any material for mass-scale production of NPs in liquid phase. For the purposes mentioned above, the easiest among those is the PLA-LP. Laser ablation in different environments is reviewed in the following studies (Tangwarodomnukun et al. 2015, Hamad et al. 2016). This chapter will include three subsections: (i) theoretical background for the laser and material interaction, (ii) ablation mechanism in liquid phase and (iii) applications.

Before moving on the subsections, some of the fundamental terms in the field of the laser ablation will be elucidated, including a pulsed laser, lasers and wavelengths, repetition rate, fluence, pulse duration, ablation plume, laser groove and colloid. For more detailed information about these terms, the readers are addressed to a more comprehensive study (Renk 2017).

11.1.1 Laser Parameters

When a laser delivering electromagnetic waves as a subsequent emission of pulses, such laser is generally called as a pulsed laser. Duration of a single pulse can be from 5 fs (1 fs = 10^{-15} fs) to ms (1 ms = 10^{-3} ms). In other words, an *fs* laser generates shortest pulses from laser oscillators. The laser light oscillates in a large number of longitudinal modes, and in the mode-locking regime, a fixed-phase relationship between the longitudinal modes of the laser cavity was obtained. An ultrashort (fs) laser is developed by creating an environment for steady-state solutions of stable pulses that exist due to the compensation of dispersion, nonlinearity, dissipation and gain (Ortaç et al. 2003, Schreiber et al. 2007). The repetition rate is described as a number of pulses per unit time, reported in number of pulses/s or simply in hertz (Hz). The repetition rate can be utilized in the order of kHz and MHz for the formation of NPs in liquid phase and in the order of Hz for the synthesis of the alloys of NPs (Yang 2007). Laser fluence is defined as the laser energy per the laser-exposed unit area on the target material. This parameter will affect the production efficiency the most. Thus, in order to produce NPs in mass scale in a relatively short time, in addition to total exposure time, the laser fluence should be set to the optimum value. Pulse duration in *ns*, *ps* and *fs* lasers is another important parameter affecting the production efficiency of NPs and their formation mechanism. Therefore, in most cases, it could be advantageous to have an *fs* laser for the further excitement of the ablation plume.

The visible spectra have a wavelength range from 400 to 700 nm. For example, in the visible region, an excimer laser can be operated for the PLA. Over 700 nm, in the near-infrared region (*NIR*), the pulsed lasers with a wavelength in the order of a *micrometer* (μm) can be utilized. Titanium-sapphire lasers ($\lambda = 650$–1,080 nm) or neodymium yttrium aluminum garnet (mostly referred as Nd:YAG at $\lambda = 1.06$ μm) lasers can be used for the PLA for the operation in the initial portion of the NIR. On the other hand, a fiber laser fabricated using a glass (*fused quartz*, SiO_2) fiber doped with rare-earth ions (e.g. Yb^{+3}, Er^{+3}) can be mode-locked. Therefore, fiber lasers can yield an opportunity for the production of NPs for wavelengths ranging from 0.7 to 3 μm by changing the dopants or the laser setup configuration.

11.1.2 Ablation Plume, Laser Groove, and Colloid

Ablation plume is formed after the exposure of a pulsed laser source on a target surface. Such ablation plume includes a plasma of ions from the target material, and later, it expands and leads an environment for the formation of NPs. In the text, the term "laser-induced plasma" is used interchangeably.

Laser groove is a form of pits fabricated after the ablation process started and passing the threshold of melting and solidification limit. This limit can be exceeded via increasing the laser fluence, resulting in spallation of NPs from the target materials and the formation of laser groove.

The term "colloid" can be explained when considering the size phenomenon of the solute particles in a liquid host (solvent) as put forward by Scottish Chemist Thomas Graham in 1861 (Graham 1861, Mokrushin 1962). The colloidal particles are 1–1,000 nm in size and lie between a true solution (size less 1 nm) and suspensions (larger than 1,000 nm). The color of a colloidal solution by naked eyes will be homogeneous along the solution due to uniform distribution of the colloids inside the liquid host. Nevertheless, depending on the size of the colloid, the reflection of light from these immiscible colloidal particles can be detected using an ultramicroscope (Zsigmond 1926). The pulsed laser-generated NPs are colloidally stable in a liquid host due to possible surface termination, target material type (metallic, semiconductor) and the size of the NPs formed after the ablation process.

11.2 Theoretical Background for the Laser and Material Interaction

Theoretical background of the interaction between a pulsed laser and a target material is highlighted in this subsection (Han and Yaguo 2011, Tangwarodomnukun et al. 2015, Hamad et al. 2016).

11.2.1 Laser–Material Interaction

It has been understood from the earliest work with a pulsed ruby laser that the laser and material interaction makes changes in material's properties permanently. This is done by the large amount of energy that is deposited on the surface of the material, over a short time. As soon as the pulsed laser beam interacts with a target material, it leads to the absorption of light depending on the material type and temperature changes as a response to the material. All these propagation and absorption properties of laser energy, plasma formation and all the material responses are beyond the scope of this single book chapter, so we will primarily focus on some thermodynamics, the effects of nonlinear interaction and ablation as a material response.

Thermodynamics

Thermal Effects

Incident laser energy will be absorbed by the subjected materials, resulting in an increase in temperature, which in turn leads to material expansion and material stress. Once this stress exceeds a certain value, the material deforms and/or fractures plastically. The expansion will cause changes in refractive index, the heat capacity of matter, etc.

When the laser beam is perpendicular to the flat surface of materials, the temperature change of material with respect to time t and depth x is given by:

$$\Delta T(x,t) = 2(1-R)\alpha I_0 \left(\frac{t}{\pi k \rho C}\right) \mathrm{ierfc}\left(\frac{x}{2\sqrt{\frac{kt}{\rho C}}}\right). \quad (11.1)$$

where t is the laser pulse irradiation time, R is the reflectivity, α is the absorptivity, I_0 is the spatial distribution of laser intensity, k is the thermal conductivity, and ρ is the density of irradiated materials.

When $x > 4\sqrt{\frac{kt}{\rho C}}$, the surface temperature will be simplified as:

$$\Delta T(t) = \frac{2\alpha I_0 \sqrt{t}}{\sqrt{\pi k \rho C}}. \quad (11.2)$$

Melting and Solidification. Increasing the laser pulse energy will increase the absorbed and deposited energy by the material, and this energy causes the material to increase temperature to the melting point. Melting, by laser-induced heating, is necessary – not all laser-induced heating, but some field of material modification because of the high atomic mobility and solubility of material when melted.

Solidification of the melted layer of material starts with the formation of the nuclei of the solid material. Solidification depends on the system and laser pulse, so the nucleation may take place according to these conditions. First, nuclei are formed in a liquid phase. Then, nuclei grow until the whole material is solidified. Solidification after melting will change the atomic structure of materials, and the mutual transformation between crystalline and amorphous state starts. The efforts on using the phenomenon of melting by a pulsed laser source were shown to separate "pulsed laser melting in liquid phase" from the *PLA-LP* method (Zhang et al. 2017).

Ionization and Gasification. Melted materials will be gasified and/or ionized as the temperature continues to increase to the vaporization point. Gasification by the laser energy can be divided into surface and bulk gasification, based on liquid-gas equilibrium. Gaseous particles with the Maxwell distribution will eject from the surface. These particles will be placed in the space that is known as Knudsen layer. Absorption and deposition of the laser energy will be improved by the ionization, and after ionization, the inverse Bremsstrahlung absorption will be the major absorption of plasma. After the ionization, recrystallization of the ionized materials starts and it causes changes in the structure of materials.

Phase Explosion. Another important thermal effect is phase explosion. As reported by Yoo (2000), the occurrence of this thermal effect involves three important requirements:

1. Fast formation of superheated liquid at least 0.8–0.9 critical temperature (Bleiner and Bogaerts 2006).
2. Order of tens of microns thickness to accommodate the nuclei which are grown in superheated liquid.
3. Enough time for nucleation to reach the critical size, generally several hundreds of picoseconds.

Figure 11.1 depicts the stages of phase explosion, which are given as follows:

I. A temperature increase in external layers of the target, when the target surface is irradiated by a high-power pulsed laser. Then, deposited energy diffuses into a certain depth of target material. The temperature of melted target materials increases to the temperature higher than the boiling point.
II. Formation of superheated liquid owing to laser energy deposition. Because of not generating the nucleation, boiling does not start.
III. The generation and growth of nucleation in superheated liquid.
IV. Explosion of nucleation.

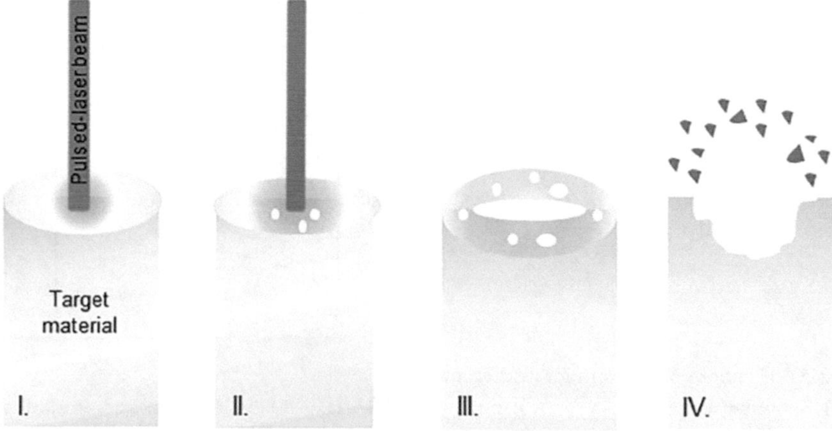

FIGURE 11.1 Stages of the phase explosion. (Reprinted from Han and Yaguo, 2011, under the CC-BY 4.0 license.)

Effects of Nonlinear Interaction

When material irradiated with a laser, it exhibits different kinds of nonlinear effects: self-focusing, multiphoton and avalanche ionization, etc.

Nonlinear Ionization. Nonlinear ionization occurs by absorbing high-energy lasers by electrons in the valence band and transition of these electrons from bound to free states. Photoionization and avalanche ionization are two different modes of nonlinear ionization. Multiphoton ionization and tunnel ionization are two constituents of the photoionization. The electric field is strong enough to make the electrons to overcome potential barrier and ionize in the tunnel ionization; electron absorbs more photons at a time to gain enough energy beyond potential trap and to be ionized in multiphoton ionization. Depending on the frequency and intensity of incident laser and material bandgap, the Keldysh parameter can be used to classify multiphoton ionization and tunnel ionization (Keldysh 1965) (see Figure 11.2).

$$\gamma = \frac{\omega}{e}\sqrt{\left[\frac{mcn\varepsilon_0 E_g}{I}\right]}. \qquad (11.3)$$

where ω is the laser frequency, I is the laser intensity at the focal point, m is the reduced mass, e is the electron charge, c is the speed of light, n is the refractive index, ε_0 is the material dielectric constant, and E_g is the material energy gap.

When the material is exposed to low-frequency and high-power laser, a tunnel ionization occurs; otherwise, multiphoton ionization takes place. Avalanche ionization is the process of producing a pair of conduction band electrons with lower kinetic energy by absorbing photons by the conduction band electrons and increasing their energy to the level, which in turn excite the electrons from the valence band to the conduction band through collisions.

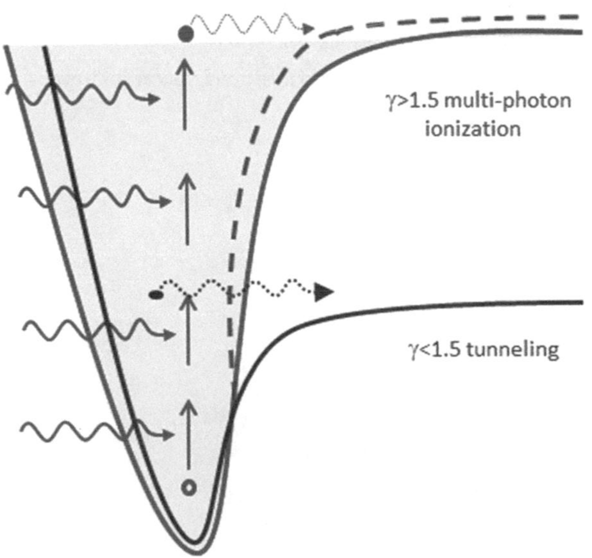

FIGURE 11.2 Schematic of photoionization for different Keldysh parameters. $\gamma < 1.5$: tunnel ionization, $\gamma > 1.5$: multiphoton ionization. (Reprinted from Han and Yaguo, 2011, under the CC-BY 4.0 license.)

Self-focusing. Self-focusing is a nonlinear optical process induced by the change in refractive index of materials exposed to intense electromagnetic radiation. The refractive index changes with the increasing laser density, which can be written as

$$n = n_0 + n_2 I. \qquad (11.4)$$

$$I = \frac{1}{2}\varepsilon_0 c n_0 \mathrm{IEI}^2. \qquad (11.5)$$

$$n_2 = \frac{3\chi(3)}{4\varepsilon_0 c n_0^2}. \qquad (11.6)$$

where I is the laser intensity at the focal point, c is the speed of light, n is the refractive index, and ε_0 is the dielectric constant of the material.

The refractive index of the medium is affected by the laser intensity, and the medium acts as a lens for an initially collimated beam. The focusing effect of the refractive index of the medium is named as a Kerr effect as demonstrated in Figure 11.3.

Kerr effect focuses $\Delta n = n_2 I$. Medium is considered as a lens, and when the beam travels through that medium, self-focusing occurs. This is the case when $n_2 > 0$.

Material Response

Material response to the laser process depends not only on the material system but also on the process conditions. For example, if the rates of excitation, induced by laser, are slower than the thermalization time, the process will be photothermal and all absorbed energy is transformed to the heat. In this chapter, we focus on the ablation as a material response to the pulsed laser irradiation. Further reading on the photothermal response of materials, including *energy absorption mechanism*, can be found herein (Brown and Arnold 2010).

Ablation

Ablation by laser is explained by the removal of material from target material by the absorption of laser energy. Target material system and some parameters related to a

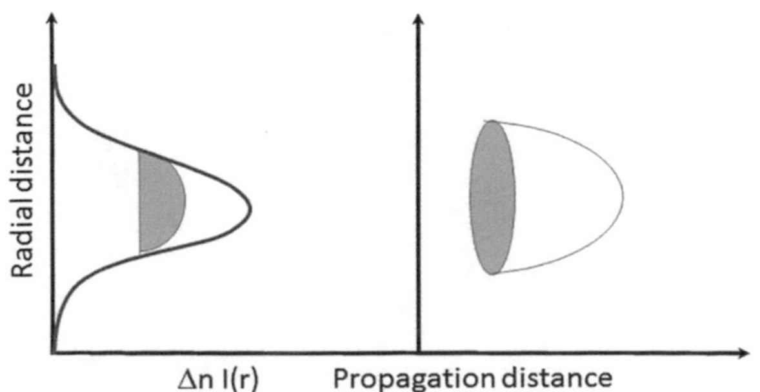

FIGURE 11.3 Kerr effect: Radial distance vs. propagation distance. (Adapted by permission from Couairon and Mysyrowicz, 2009, Copyright 2009.)

TABLE 11.1 The Typical Threshold Fluence Values for Different Material Types

Typical Threshold Fluence Values	J/cm^2
Metals	1–10
Inorganic insulators	0.5–2
Organic materials	0.1–1

Source: Bäuerle (2011).

TABLE 11.2 Ablation Mechanisms at Low and High Fluence Values

At Low Fluence Values	At High Fluence Values
Photothermal mechanisms of ablation:	1. heterogeneous nucleation of vapor bubbles
• material evaporation	2. normal boiling
• sublimation	1. if heating is sufficiently rapid for approaching its thermodynamic critical temperature
	2. homogenous nucleation and expansion of vapor bubbles
	3. phase explosion: explosive boiling carrying off solid and liquid material fragments

Source: Bulgakova and Bulgakov (2001).

laser such as its wavelength, the pulse duration and the fluence affect the mechanism of removal of material (Chrisey and Hubler 1994).

Table 11.1 shows the ablation threshold fluence values for different materials. Above the threshold fluence, ablation starts. The absorption mechanism (see Table 11.2) particularly affects material properties, microstructure, morphology, the presence of defects, laser wavelength and laser pulse duration. Other than the typical threshold fluence, multiple pulses from laser lead to accumulation defects, thus decreasing the ablation thresholds. By removing the thickness or volume of the target material after every pulse, the new removal rates increase with the fluence according to the Beer-Lambert law. Moreover, if the material is a multicomponent system, the more volatile species changes the composition of a remaining material by reducing material more rapidly (Mao et al. 1998).

11.3 Pulsed Laser Ablation Mechanisms in Liquid Phase

The basic mechanisms of the *PLA-LP* will be discussed henceforth. Some of the useful reviews can also be found in the following studies (Yang 2007, Amendola and Meneghetti 2013, Zhang et al. 2017). When a target material is immersed inside a liquid, the target material is surrounded by the liquid molecules. Then, the surface of the target material is exposed to the laser light. As soon as the material is exposed to the laser light in a liquid host such as water, the ablation process is initiated in the liquid phase. Figure 11.4 demonstrates the *PLA-LP* mechanisms in a liquid host for the generation of NPs. In step I, the laser-induced plasma or the ablation plume is formed after the laser source hits the target in the liquid host. Then, the laser-induced plasma will adiabatically expand to an extent due to the fact that the target material continuously absorbs the pulsed laser beam. It means that the energy of the plasma plume will be conserved and no heat or particles will diffuse inside the plume or leave from the plasma plume for a while (until step V). Since the ablation process is conducted in a liquid host, the laser-induced plasma will experience a confinement due to the surrounding liquid molecules. Therefore, in step II, the expanding laser-induced plasma will be ceased by the confinement effect applied by the surrounding liquid molecules, which is called *confinement of the liquid.* Here, it should be noted that from step I to II, the radius of plasma plume was changed from R_1 to R_2 and $R_2 > R_1$. As soon as the motion of the plasma plume is ceased by the confinement effect, the shock wave whose direction is opposite to the motion of expansion of the plasma plume is created inside the target material. This process is depicted in step III. Then, in step IV, the shock wave will create a pressure called plasma-induced pressure back to the plasma plume as formulated in Han and Yaguo (2011, p. 120,121). After this process, as shown in step V, the pressure P, the temperature T and the density ρ of the plasma plume are started to be increased from P_1, T_1 and ρ_1 to P_2, T_2 and ρ_2, respectively.

Later, there could be two different routes for the formation of NPs: In the first route, step VI(a) followed by step VII could directly yield NPs, and in the second route, step VI(b) after step V could directly yield NPs. In the first route, as schemed in step VI(a), the metastable phases are formed. These metastable phases stem from the chemical reactions between the plasma plume and the liquid host. Four kinds of the chemical reactions arising in this step are to be considered (Yang 2007, p. 656). Here, three of these reactions occur in the intermediate region between the ablation plume and the liquid. These three chemical reactions consist of the metastable phases formed by the ions and/or neutral atoms/molecules. The fourth reaction takes place only in the plasma plume that later deposits back to the target material when the ablation process is completed. Then, in step VII, the pulsed laser-generated NPs are formed by the metastable phases. On the other hand, NPs can directly form without the creation of metastable phases as represented in step VI(b). Here, as mentioned beforehand, exceeding the threshold limit of melting and solidification is essential so that the spallation of NPs from the target material could be achieved.

In step V, three key parameters, namely, the pressure P, the temperature T and the density ρ of the plasma plume, were described in terms of thermodynamics of the *PLA-LP*. These three key parameters are important to understand how the plasma plume is changed during the PLA process in a liquid host. For instance, from the expansion of the plasma plume, its density can be found. In other words, the temporal images of the expansion of the plasma plume can be recorded using an intensified charge coupled device (ICCD) camera, and the respective changes in the diameter (and thus in the volume) can be determined based on the assumption of the plasma plume to be a hemisphere. In addition, this approach can be used only when the NP

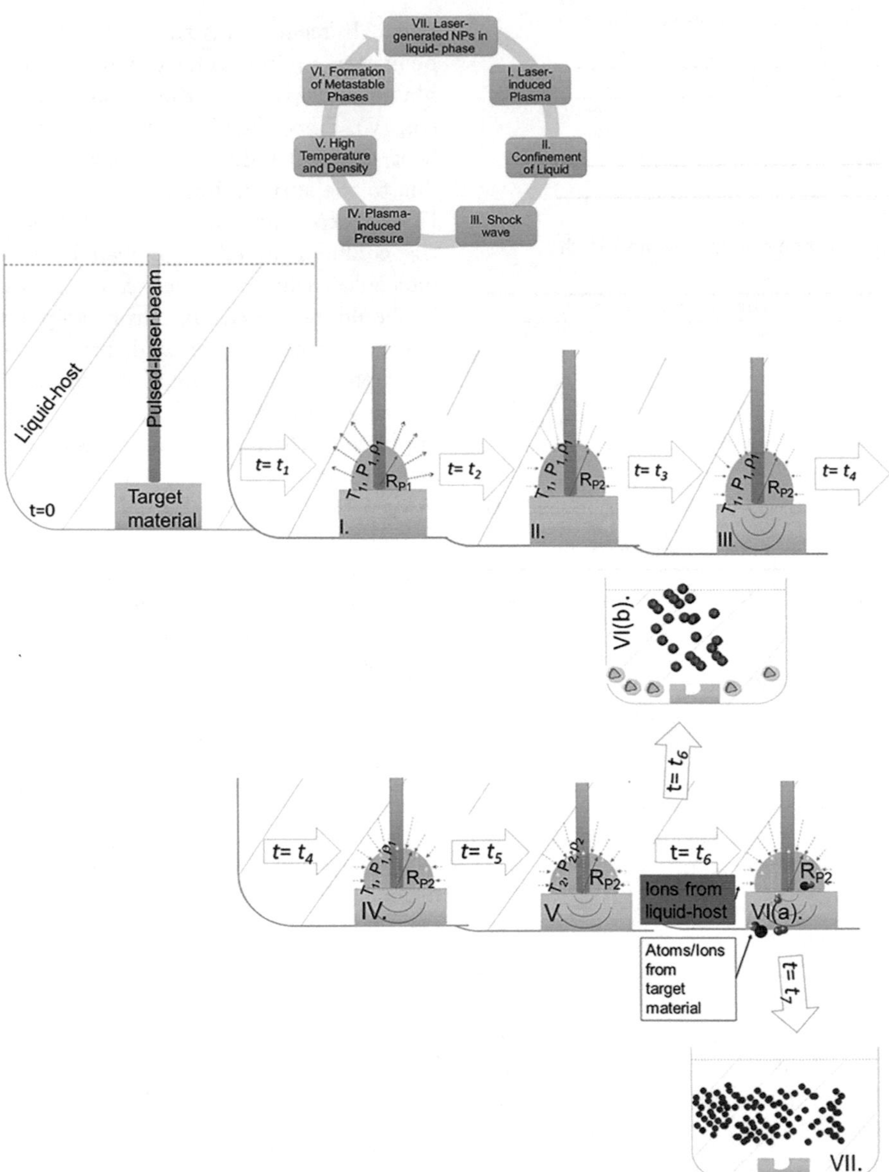

FIGURE 11.4 PLA mechanisms are demonstrated from the formation of laser-induced plasma to laser-generated NPs.

formation mechanism is valid as depicted in Figure 11.4. On the other hand, the expansion of the ablation plume was shown to be dependent on the medium where the ablation is held (Saito et al. 2002). For instance, Figure 11.5 shows the evolution of the plasma plume upon the ablation of a graphite target material both in the air and in the water. The effect called liquid confinement was considered to result in much (approximately ten times) shorter plasma duration than that conducted in air. This changes the return kinetics for the formation of NPs.

11.3.1 Nucleation and Growth of Nanoparticles in the PLA-LP

From henceforth, we will mention the nucleation and growth of NPs in the *PLA-LP*. For the nucleation and growth of NPs in liquid phase, there are various studies (DeYoreo and Vekilov 2003, Burda et al. 2005), including our recent study (Karatutlu et al. 2018). Some of the mechanisms will be given once more on the basis of understanding the concept of nucleation and growth of NPs in the PLA-LP. The reader should also keep in mind that there are several theories proposed until now for the mechanisms of nucleation and growth of NPs in liquid phase. These are as follows: *classical nucleation theory* (CNT), *LaMer nucleation and growth mechanism, the two-step nucleation and growth mechanism, pre-nucleation cluster mechanism.*

Nuclei known as the seed substance for the nucleation process are formed via consuming monomers. Nuclei are surrounded by the liquid layers, in which crystallization of nuclei takes place. This is the reason that nuclei can be counted to be the basis of crystals. In CNT, the clusters are aggregated in a way to form nuclei inside the bulk liquid. There will be a moment after reaching the nucleation

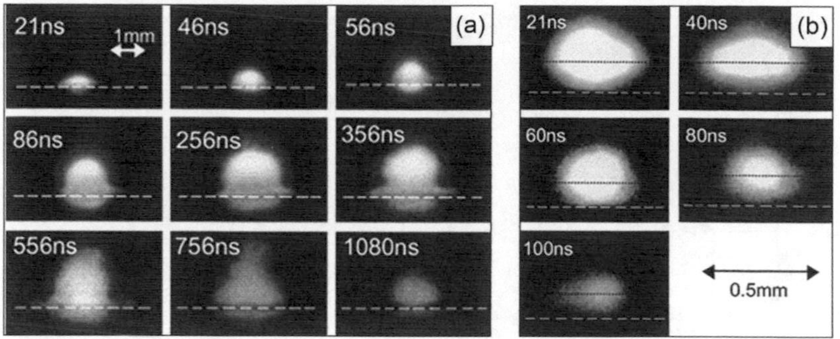

FIGURE 11.5 The images of the changes in the plasma plume (a) in air and (b) in water as a liquid host are demonstrated for the PLA of a graphite target material. (Adapted with permission from Saito et al., 2002. Copyright © 2001 Elsevier.)

work W, and thus, a new phase is formed spontaneously. This nucleation work can be considered as critical energy or energy barrier for the nucleation. Here, a new phase is called critical nucleus for which the surface and the volume can be defined within the parent bulk liquid. Gibbs free energy (after the study of J. W. Gibbs (1878)) for such new phase can be written as $\Delta G(r)$ based on the assumption of a new phase with a spherical shape having a radius r. The free energy $\Delta G(r)$ is equal to a positive surface contribution plus a negative volume contribution (see Figure 11.6a). In summary, overall free energy can be described as follows:

$$\Delta G(r) = \Delta G_S + \Delta G_V = 4\pi r^2 \sigma - \frac{4}{3V} \pi r^3 k_B T \ln(S). \quad (11.7)$$

where σ is the surface contribution in the free energy per unit area, V is the molecular volume of the nuclei in the crystal, k_B is the *Boltzmann constant*, T is the temperature (in *Kelvin*), and S is the saturation ratio.

The nucleation process starts when the solution reaches supersaturation limit. In the *PLA-LP*, this supersaturation limit is approached via the high pressure and the temperature provided by a pulsed laser source. When this condition is met, the nuclei will have a critical radius (r_{crit}). At this moment, maximum free energy (also known as thermodynamic energy barrier) will be obtained. Of course, this critical size will be limited either for the dissolution back to the solution or for the initiation to the growth of NPs. When an infinitesimal positive change occurs, a continuous growth process will be active and NPs in return will be formed. These processes are depicted by LaMer diagram in Figure 11.6b. This critical size is known as a metastable state $\left(\frac{\partial \Delta G^*}{\partial r} = 0 \text{ and } \frac{\partial^2 \Delta G^*}{\partial r^2} < 0\right)$. From here, r_{crit} will be denoted as:

$$r_{\text{crit}} = \frac{2V\sigma}{3k_B T \ln(S)}. \quad (11.8)$$

Nucleation thermodynamics of NPs formed by the *PLA-LP* method could give insights for finding the critical radius and the probability of the structural phases of the formed NPs (Liu 2005, Wang et al. 2005a,b). An example of nucleation thermodynamics is given in Figure 11.7a using the phase diagram for the cubic boron nitride (cBN) NPs (Liu et al. 2005). Following the supersaturation rate S from Shiba and Okawa (2005), Eq. (11.7) then becomes at high pressure, high density (gas) and high temperature (note that the nuclei are still isotropic, spherical, monodisperse particles):

$$\Delta G(r) = \Delta G_S + \Delta G_V = 4\pi r^2 \sigma + \frac{\frac{4}{3V}\pi r^3 \Delta g}{V_m}. \quad (11.9)$$

where Δg and V_m stand for the molar volume Gibbs free energy difference and the molar volume, respectively.

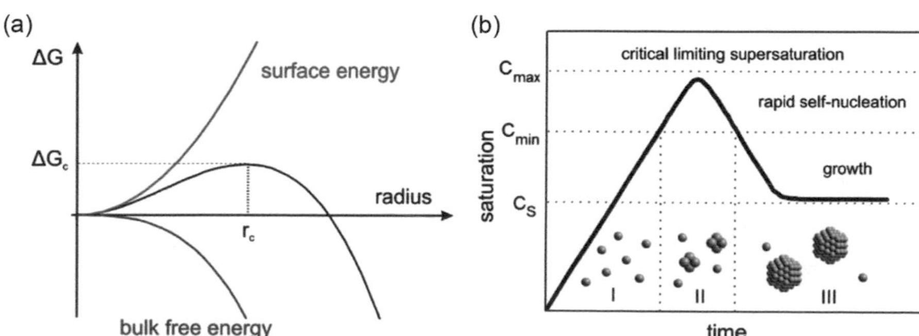

FIGURE 11.6 (a) The dependency of the cluster free energy, ΔG, on the cluster size, r, in accordance with the CNT. Reaching a maximum free energy ΔG at a critical cluster size, r_{crit}, defining the first stable particles – the nuclei. (b) The temporal separation of the nucleation and growth processes is schematically demonstrated by La'mer's diagram. The critical supersaturation level can be inferred from the peak position. The monomer concentration as a function of time is theoretically given by the qualitative curve. (Polte, 2015 – Published by The Royal Society of Chemistry.)

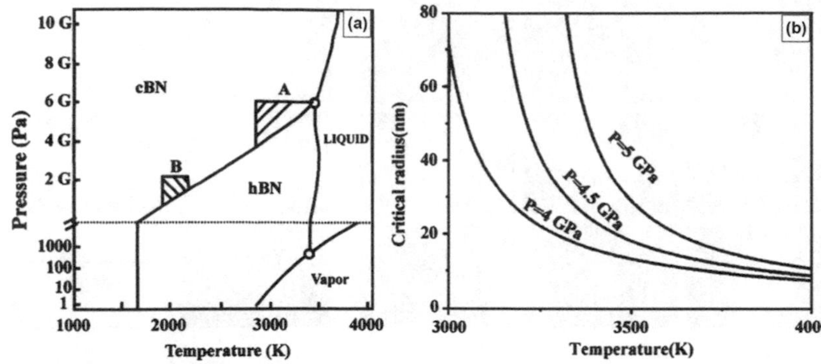

FIGURE 11.7 (a) The equilibrium-phase diagram of boron nitride. (Reprinted with permission from Solozhenko et al., 1999. Copyright 2018 American Chemical Society.) (b) Critical radius r_{crit} vs. temperature at different pressure values. (Reproduced from Liu et al. 2005, with permission from American Physical Society.)

Using $\left(\frac{\partial \Delta g_{T,P}}{\partial P}\right)_T = \Delta V$, one can have the following equation:

$$\Delta g_{T,P} = \Delta g_T^0 + \int_0^P \Delta V dP \tag{11.10}$$

where ΔV and Δg_T^0 represent the molar volume difference between two different phases and the molar Gibbs free energy difference at zero pressure, respectively. Here, as shown in Figure 11.4a, two different phases are referred to the example of cBN and hexagonal boron nitride (hBN) NPs. Considering $\Delta V \cong 3.79 \times 10^{-6}$ m^3/mol from Solozhenko et al. (1999), the size-induced additional pressure from the Laplace-Young equation $\Delta P = 2\sigma/r$ and the external pressure between hBN and cBN $P^e = 2.985 \times 10^6 T - 4.615 \times 10^9$ from Liu et al. (2005), the size-dependent equilibrium phase boundary becomes:

$$P^e = 2.985 \times 10^6 T - 4.615 \times 10^9 - \frac{2\sigma}{r}. \tag{11.11}$$

Equation (11.11) under the conditions of the equilibrium yields $\Delta g_{T,P} = 0$, and the molar Gibbs free energy difference for the transition from hBN to cBN becomes $\Delta g = \Delta V \times (P - 2.985 \times 10^6 T + 4.615 \times 10^9 + 2\sigma/r)$. From here, the Gibbs free energy becomes:

$$\Delta G(r) = 4\pi r^2 \sigma + \frac{4}{3V}\pi r^3 \Delta V \times \frac{P - 2.985 \times 10^6 T + 4.615 \times 10^9 + 2\frac{\sigma}{r}}{V_m}. \tag{11.12}$$

As mentioned before, the critical size is known as a metastable state where $\frac{\partial \Delta G^*}{\partial r} = 0$. From here, the critical size can be determined as follows:

$$r_c = 2\sigma \left(\frac{2\Delta V + 3V_m}{3\Delta V}\right) \times \left(\frac{1}{2.985 \times 10^6 T - 4.615 \times 10^9 - P}\right). \tag{11.13}$$

Finally, inserting r_{crit} into Eq. (11.6) results in the critical energy of the nuclei in cBN phase to be:

$$\Delta G(r_{\text{crit}}) = 4\pi r_{\text{crit}}^2 \sigma + \frac{4}{3V}\pi r_{\text{crit}}^3 \Delta V \times \frac{P - 2.985 \times 10^6 T + 4.615 \times 10^9 + 2\frac{\sigma}{r_{\text{crit}}}}{V_m}. \tag{11.14}$$

Dependence of r_{crit} of cBN on the temperature at different pressure values is given in Figure 11.7b when the *PLA-LP* method is conducted. It was shown that the critical size (given as radius in the graph) reduces by increasing the temperature, whereas it increases with the rise of the pressure.

11.4 Applications

Various applications and opportunities for the *PLA-LP* method have been demonstrated (Gökce et al. 2017, Zhang et al. 2017). Some of these applications are highlighted in this subsection. The exemplary applications are solar cells, catalysis, sensors and biomedicine.

11.4.1 Solar Cells

The NP colloids synthesized using the chemical synthesis methods are generally based on the ligand exchange mechanism. This precursor reaction affects the nanoproducts negatively because NPs coated with different ligands need extra purification for their use in solar cell applications.

Chemical vapor deposition (CVD) is often employed for material deposition in the preparation of solar cells, and it is an expensive method. NPs with high-purity (ligand-free) and film deposition devices (cheaper than CVD equipment) are desirable for solar cell preparation (Petridis et al. 2017).

Guo and Liu used liquid ablation in liquid (LAL) technique to fabricate Cu(In, Ga)Se_2 (CIGS) solar cells on Mo sheet substrates. They combined two methods: LAL and electrophoretic deposition (EPD) (a method for film deposition based on electrokinetic mobility of the charged colloids under an external electric field) techniques, and the resultant energy conversion efficiency is 7.37% for the solar cell (Guo and Liu 2012).

Au-NPs are used as a plasmonic metal, which is embedded inside bulk-heterojunction (BHJ) solar cells to trap incident light and increase the optical absorption (Gan et al. 2013). Therefore, they increase the performance of BHJ solar cells.

Kymakis et al. studied Au NPs with three different surface chemistry and their effect on the performance of BHJ solar cells. When comparing the power conversion efficiency of Au NPs, it was concluded that 20.3% increase in the power conversion efficiencies was observed for Au NPs (ligand-free) generated by LAL, which was much better than TOAB-terminated and P3HT-terminated Au NPs according to their performance as a plasmonic metal compared with the performance of poly(3-hexylthiophene-2,5-diyl): phenyl-C61-butyric acid methyl ester (P3HT:PCBM) BHJ solar cells. Atomic force microscopy (AFM) and time-resolved photoluminescence (PL) spectroscopy showed that solar cell performance can be enhanced by the direct contact of NPs with the active polymer layer and any ligand can cause a deterioration of the active layer (Kymakis et al. 2015).

These works show the importance and positive effect of ligand-free NPs on the performance of solar cells.

11.4.2 Catalysis

Photocatalysts have an important role in pollutant removal field due to their ability to mineralize organic pollutants and photocatalytic reduction of heavy metal ions (Herrmann 1999). The *PLA-LP* method gives an opportunity for the formation of different materials to be applied in photocatalysis as demonstrated in Table 11.3.

NPs formed by the *PLA-LP* method could also be applied in pollution control due to their surface charges and high catalytic activity (Zhang et al. 2013). Furthermore, the enhanced electrocatalytic activity of the *PLA-LP*-generated NPs was shown for various materials such as gold (Au), platinum (Pt) and Au-Pt alloys without any capping agents/surfactants (Oko et al. 2014, 2015).

TABLE 11.3 The PLA-LP Method Works for a Variety of Photocatalysts

Product	References
TiO_2	Herrmann (1999), Liqiang et al. (2006), Liu et al. (2009), Tian et al. (2009), Li et al. (2015), and Zimbone et al. (2015)
Au-$BiFeO_3$	Li et al. (2013)
CoO	Liao et al. (2014)
Pt/ZnO	Zeng et al. (2008b)
SnOx	Tian et al. (2011)
ZnO	Liqiang et al. (2006) and Zeng et al. (2008a)
α-Bi_2O_3	Lin et al. (2010)
TaxO@Ta_2O_5	Li et al. (2012)
Lanthanide-doped $NaYF_4$, TiO_2 and Au	Xu et al. (2015)
ZnS/Zn	Wang et al. (2016)
TiO_2 + graphite SiO_2	Ikeda et al. (2008)
ZnO/TiO_2	Gondal et al. (2016)
MWCNT/ZnO	Saleh et al. (2010)
$Sn_6O_4(OH)_4$	Xiao et al. (2014)
TiO_2-Graphine oxide	Li et al. (2016)
α-Ag_2WO_4	Lin et al. (2015)

Source: Extracted from Zhang et al. (2017).

11.4.3 Sensors

As mentioned before, the ligands tend to decrease the performance of sensors when compared with the chemically obtained Au-NPs. This is explained by the potential barrier width-dependent quantum-tunneling effect. This effect makes the Au-NP films use for highly sensitive strain gauges.

Burzhuev et al. showed that Au-NPs by the *PLA-LP* method were then used as films on polydimethylsiloxane (PDMS) substrate to obtain high gain sensitivity (Burzhuev et al. 2013). When the strain is applied to the sensor, the distance between Au-NPs begins to increase and the resistance of the film becomes ΔR, as shown in Figure 11.8b. The increased gain sensitivity is expressed by the more effective and stronger quantum-tunneling effect between purely obtained Au-NPs by the *PLA-LP* method.

11.4.4 Biomedicine

In the field of biomedicine, three main purposes are listed for the use of NPs formed by the *PLA-LP* method:

i. Bioimaging (both *in vitro* and *in vivo*)
ii. Targeted drug delivery and
iii. Other therapeutics such as hyperthermia.

For the abovementioned purposes, first, the *PLA-LP* yields NPs with high purity and the desired surface functionality. Due to the nature of NPs, its high emission efficiency and/or capability of acting as contrast agents, they are mostly used in bioimaging applications. Different types of NPs can be prepared as a bioimaging agent without any limitations. Some examples are semiconductor NPs, such as Si (Vaccaro et al. 2014, Momeni and Mahdieh 2016), Ge (Ghosh et al. 2013, Vadavalli et al. 2014) and CdS (Lalayan 2005); and metallic NPs, including silver (Ag) and gold (Au) (Restuccia and Torrisi 2018).

Second, NPs can be produced with high quality, for instance, to serve the targeting purpose via bioconjugation with biomolecules present in the liquid host. Therefore, the *PLA-LP* allows producing NPs whose surface can be terminated with nucleic acids (Petersen and Barcikowski 2009), biopolymers (Zamiri et al. 2011, Spano et al. 2012), peptides (Petersen et al. 2011) or proteins (Mutisya et al. 2013). On the other hand, some materials due to their nature (e.g. material type, size, surface termination and molecular structure) could be toxic and could not be utilized without surface or structural modification for biomedical applications. Ortac et al. demonstrated that ultrapure Ag-NPs formed by the *PLA-LP* could be toxic to the neural tissue (Kursungoz et al. 2017). Therefore, reducing the toxicity of materials (or increasing their biocompatibility) could be required before any biomedical applications. In other words, the *PLA-LP* method can provide an environment where NPs can be made less toxic or nontoxic regardless of the characteristics of the target material. For example, Ag-NPs can

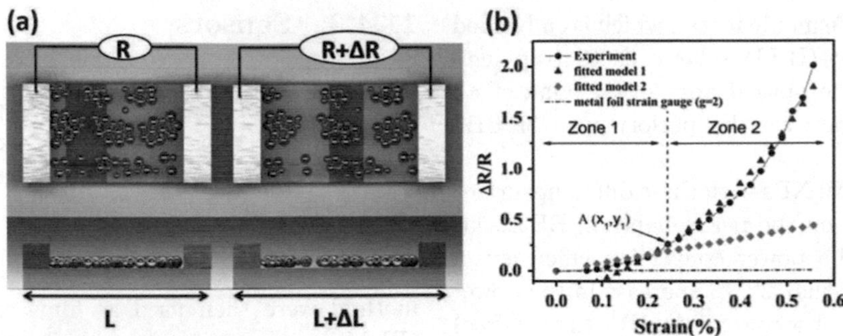

FIGURE 11.8 (a) Au-NP film-deposited strain sensors and (b) the resistive response of the Au-NP film. (Adapted with permission from Burzhuev et al., 2013. Copyright © 2013 Elsevier.)

be alloyed together with gold (Au) in order to reduce the toxicity of Ag (Taylor et al. 2015).

The third and final purpose of NPs generated by the *PLA-LP* is the formation of other therapeutics such as NPs for hyperthermia. In hyperthermia, conjugated NPs formed by the *PLA-LP* can transfer the heat locally to a specific organ. For example, Si/SiO_2 NPs generated by the *PLA-LP* were shown to be applied successfully for hyperthermia of Lewis lung carcinoma in vivo with a good biocompatibility, which was removed from the body in a week (Tamarov et al. 2015).

11.5 Conclusion

In summary, the fundamentals of NPs generated by the *PLA-LP* were demonstrated together with their promising applications. The laser and material interaction, and laser ablation mechanisms in liquid phase such as nucleation and growth of NPs are provided in the initial portion of the work. The *PLA-LP* method was shown to offer an opportunity (i) to synthesize NPs with high purity, (ii) to prepare NPs with the desired surface functionality and (iii) to use different materials in the same liquid host to permit alloys having enhanced physical and chemical properties. The NPs with these properties generated by the *PLA-LP* were shown to be applicable in the fields, but not limited to solar cells, catalysis, sensors and biomedicine. The design of innovative materials containing NPs formed by the *PLA-LP* together with the integration to the real-life applications was considered to result in novel contributions to many fields.

References

Amendola, V. and Meneghetti, M., 2013. What controls the composition and the structure of nanomaterials generated by laser ablation in liquid solution? *Physical Chemistry Chemical Physics: PCCP*, 15 (9), 3027–3046.

Bäuerle, D., 2011. Ultrashort-pulse laser ablation. In: *Laser Processing and Chemistry*. Berlin, Heidelberg: Springer, 196.

Bleiner, D. and Bogaerts, A., 2006. Multiplicity and contiguity of ablation mechanisms in laser-assisted analytical micro-sampling. *Spectrochimica Acta - Part B Atomic Spectroscopy*, 61 (4), 421–432.

Brown, M.S. and Arnold, C.B., 2010. Fundamentals of laser-material interaction and application to multiscale surface modification. In: K. Sugioka, M. Meunier, and L. Pique, eds. *Springer Series in Materials Science Volume 135: Laser Precision Microfabrication*. Berlin, Heidelberg: Springer, 91–120.

Bulgakova, N.M. and Bulgakov, A.V., 2001. Pulsed laser ablation of solids: Transition from normal vaporization to phase explosion. *Applied Physics A: Materials Science and Processing*, 73 (2), 199–208.

Burda, C., Chen, X., Narayanan, R., and El-Sayed, M.A., 2005. Chemistry and properties of nanocrystals of different shapes. *Chemical Reviews*, 105 (4), 1025–1102.

Burzhuev, S., Dâna, A., and Ortaç, B., 2013. Laser synthesized gold nanoparticles for high sensitive strain gauges. *Sensors and Actuators, A: Physical*, 203, 131–136.

Chrisey, D.B. and Hubler, G.K., 1994. *Pulsed Laser Deposition of Thin Films*. New York: John Wiley & Sons, Inc.

Couairon, A. and Mysyrowicz, A., 2009. Self-focusing and filamentation of femtosecond pulses in air and condensed matter: Simulations and experiments. In: R.W. Boyd, S.G. Lukishova, Y.R. Shen, eds. *Self-focusing: Past and Present*. New York: Springer, 297–322.

DeYoreo, J.J. and Vekilov, P.G., 2003. Principles of crystal nucleation and growth. *Reviews in Mineralogy and Geochemistry*, 54 (1), 57–93.

Fojtik, A., Giersig, M., and Henglein, A., 1993. Formation of nanometer-size silicon particles in a laser induced plasma in SiH_4. *Berichte der Bunsengesellschaft für physikalische Chemie*, 97 (11), 1493–1496.

Gan, Q., Bartoli, F.J., and Kafafi, Z.H., 2013. Plasmonic-enhanced organic photovoltaics: Breaking the 10% efficiency barrier. *Advanced Materials*, 25 (17), 2385–2396.

Ghosh, B., Sakka, Y., and Shirahata, N., 2013. Efficient green-luminescent germanium nanocrystals. *Journal of Materials Chemistry A*, 1 (11), 3747.

Gibbs, J.W., 1878. On the equilibrium of heterogeneous substances. *American Journal of Science*, S3–S16 (96), 441–458.

Gondal, M.A., Ilyas, A.M., and Baig, U., 2016. Pulsed laser ablation in liquid synthesis of ZnO/TiO_2 nanocomposite catalyst with enhanced photovoltaic and photocatalytic performance. *Ceramics International*, 42 (11), 13151–13160.

Gökce, B., Amendola, V., and Barcikowski, S., 2017. Opportunities and challenges for laser synthesis of colloids. *ChemPhysChem*, 18 (9), 983–985.

Graham, T., 1861. Liquid diffusion applied to analysis. *Philosophical Transactions of the Royal Society*, 151 (11), 183–224.

Guo, W. and Liu, B., 2012. Liquid-phase pulsed laser ablation and electrophoretic deposition for chalcopyrite thin-film solar cell application. *ACS Applied Materials and Interfaces*, 4 (12), 7036–7042.

Hamad, A.H., Khashan, K.S., and Hadi, A.A., 2016. Laser ablation in different environments and generation of nanoparticles. In: D. Yang, ed. *New Applications of Artificial Intelligence*. InTechOpen Limited, London, UK, 177–196.

Han, J. and Yaguo, L., 2011. Interaction between pulsed laser and materials. In: K. Jakubczak, ed. *Lasers - Applications in Science and Industry*. InTechOpen Limited, London, UK, 109.

Henglein, A., 1993. Physicochemical properties of small metal particles in solution: 'Microelectrode' reactions, chemisorption, composite metal particles, and the atom-to-metal transition. *Journal of Physical Chemistry*, 97 (21), 5457–5471.

Herrmann, J., 1999. Heterogeneous photocatalysis: Fundamentals and applications to the removal of various types of aqueous pollutants. *Catalysis Today*, 53 (1), 115–129.

Ikeda, M., Kusumoto, Y., Yang, H., Somekawa, S., Uenjyo, H., Abdulla-Al-Mamun, M., and Horie, Y., 2008. Photocatalytic hydrogen production enhanced by laser ablation in water–methanol mixture containing titanium(IV) oxide and graphite silica. *Catalysis Communications*, 9 (6), 1329–1333.

Karatutlu, A., Barhoum, A., and Sapelkin, A.V., 2018. Theories of nanoparticle and nanostructures formation in liquid-phase. In: A. Barhoum and A.S.H. Makhlouf, eds. *Fundamentals of Nanoparticles*. Elsevier Inc., Amsterdam, The Netherlands, 597–618.

Keldysh, L.V., 1965. Ionization in the field of a strong electromagnetic wave. *Soviet Physics JETP*, 20 (5), 1307–1314.

Kursungoz, C., Taş, S.T., Sargon, M.F., Sara, Y., and Ortaç, B., 2017. Toxicity of internalized laser generated pure silver nanoparticles to the isolated rat hippocampus cells. *Toxicology and Industrial Health*, 33 (7), 555–563.

Kymakis, E., Spyropoulos, G.D., Fernandes, R., Kakavelakis, G., Kanaras, A.G., and Stratakis, E., 2015. Plasmonic bulk heterojunction solar cells: The role of nanoparticle ligand coating. *ACS Photonics*, 2 (6), 714–723.

Lalayan, A.A., 2005. Formation of colloidal GaAs and CdS quantum dots by laser ablation in liquid media. *Applied Surface Science*, 248 (1–4), 209–212.

Li, L., Yu, L., Lin, Z., and Yang, G., 2016. Reduced TiO_2: Graphene oxide heterostructure as broad spectrum-driven efficient water-splitting photocatalysts. *ACS Applied Materials and Interfaces*, 8 (13), 8536–8545.

Li, L.H., Deng, Z.X., Xiao, J.X., and Yang, G.W., 2015. A metallic metal oxide (Ti_5O_9)-metal oxide (TiO_2) nanocomposite as the heterojunction to enhance visible-light photocatalytic activity. *Nanotechnology*, 26 (25), 255705.

Li, Q., Liang, C., Tian, Z., Zhang, J., Zhang, H., and Cai, W., 2012. Core–shell $Ta_xO@Ta_2O_5$ structured nanoparticles: Laser ablation synthesis in liquid, structure and photocatalytic property. *CrystEngComm*, 14 (9), 3236.

Li, S., Zhang, J., Kibria, M.G., Mi, Z., Chaker, M., Ma, D., Nechache, R., and Rosei, F., 2013. Remarkably enhanced photocatalytic activity of laser ablated Au nanoparticle decorated $BiFeO_3$ nanowires under visible-light. *Chemical Communications*, 49 (52), 5856.

Liao, L., Zhang, Q., Su, Z., Zhao, Z., Wang, Y., Li, Y., Lu, X., Wei, D., Feng, G., Yu, Q., Cai, X., Zhao, J., Ren, Z., Fang, H., Robles-Hernandez, F., Baldelli, S., and Bao, J., 2014. Efficient solar water-splitting using a nanocrystalline CoO photocatalyst. *Nature Nanotechnology*, 9 (1), 69–73.

Lin, G., Tan, D., Luo, F., Chen, D., Zhao, Q., Qiu, J., and Xu, Z., 2010. Fabrication and photocatalytic property of α-Bi_2O_3 nanoparticles by femtosecond laser ablation in liquid. *Journal of Alloys and Compounds*, 507 (2), L43–L46.

Lin, Z., Li, J., Zheng, Z., Yan, J., Liu, P., Wang, C., and Yang, G., 2015. Electronic reconstruction of α-Ag_2WO_4 nanorods for visible-light photocatalysis. *ACS Nano*, 9 (7), 7256–7265.

Liqiang, J., Yichun, Q., Baiqi, W., Shudan, L., Baojiang, J., Libin, Y., Wei, F., Honggang, F., and Jiazhong, S., 2006. Review of photoluminescence performance of nano-sized semiconductor materials and its relationships with photocatalytic activity. *Solar Energy Materials and Solar Cells*, 90 (12), 1773–1787.

Liu, P., Cai, W., Fang, M., Li, Z., Zeng, H., Hu, J., Luo, X., and Jing, W., 2009. Room temperature synthesized rutile TiO_2 nanoparticles induced by laser ablation in liquid and their photocatalytic activity. *Nanotechnology*, 20 (28), 285707.

Liu, Q.X., Wang, C.X., and Yang, G.W., 2005. Nucleation thermodynamics of cubic boron nitride in pulsed-laser ablation in liquid. *Physical Review B: Condensed Matter and Materials Physics*, 71 (15), 1–6.

Mao, L.X., Ciocan, C.A, and Russo, E.R., 1998. Preferential vaporization during laser ablation inductively coupled plasma atomic emission spectroscopy. *Applied Spectroscopy*, 52 (7), 913–918.

Mokrushin, S.G., 1962. Thomas graham and the definition of colloids. *Nature*, 195 (4844), 861.

Momeni, A. and Mahdieh, M.H., 2016. Photoluminescence analysis of colloidal silicon nanoparticles in ethanol produced by double-pulse ns laser ablation. *Journal of Luminescence*, 176, 136–143.

Mutisya, S., Franzel, L., Barnstein, B.O., Faber, T.W., Ryan, J.J., and Bertino, M.F., 2013. Comparison of in situ and ex situ bioconjugation of Au nanoparticles generated by laser ablation. *Applied Surface Science*, 264, 27–30.

Ogale, S.B., Patil, P.P., Roorda, S., and Saris, F.W., 1987. Nitridation of iron by pulsed excimer laser treatment under liquid ammonia: Mössbauer spectroscopic study. *Applied Physics Letters*, 50 (25), 1802.

Oko, D.N., Garbarino, S., Zhang, J., Xu, Z., Chaker, M., Ma, D., Guay, D., and Tavares, A.C., 2015. Dopamine and ascorbic acid electro-oxidation on Au, AuPt and Pt nanoparticles prepared by pulse laser ablation in water. *Electrochimica Acta*, 159, 174–183.

Oko, D.N., Zhang, J., Garbarino, S., Chaker, M., Ma, D., Tavares, A.C., and Guay, D., 2014. Formic acid electro-oxidation at PtAu alloyed nanoparticles synthesized by pulsed laser ablation in liquids. *Journal of Power Sources*, 248, 273–282.

Ortaç, B., Hideur, A., Chartier, T., Brunel, M., Özkul, C., and Sanchez, F., 2003. 90-fs stretched-pulse ytterbium-doped double-clad fiber laser. *Optics Letters*, 28 (15), 1305.

Petersen, S., Barchanski, A., Taylor, U., Klein, S., Rath, D., and Barcikowski, S., 2011. Penetratin-conjugated gold nanoparticles: Design of cell-penetrating nanomarkers by femtosecond laser ablation. *The Journal of Physical Chemistry C*, 115 (12), 5152–5159.

Petersen, S. and Barcikowski, S., 2009. In situ bioconjugation: Single step approach to tailored nanoparticle-bioconjugates by ultrashort pulsed laser ablation. *Advanced Functional Materials*, 19 (8), 1167–1172.

Petridis, C., Savva, K., Kymakis, E., and Stratakis, E., 2017. Laser generated nanoparticles based photovoltaics. *Journal of Colloid and Interface Science*, 489, 28–37.

Polte, J., 2015. Fundamental growth principles of colloidal metal nanoparticles: A new perspective. *CrystEngComm*, 17 (36), 6809–6830.

Renk, K.F., 2017. *Basics of Laser Physics: For Students of Science and Engineering*. Berlin Heidelberg: Springer-Verlag.

Restuccia, N. and Torrisi, L., 2018. Nanoparticles generated by laser in liquids as contrast medium and radiotherapy intensifiers. *EPJ Web of Conferences*, 167, 4007.

Saito, K., Takatani, K., Sakka, T., and Ogata, Y.H., 2002. Observation of the light emitting region produced by pulsed laser irradiation to a solid–liquid interface. *Applied Surface Science*, 197–198, 56–60.

Saleh, T.A., Gondal, M.A., and Drmosh, Q.A., 2010. Preparation of a MWCNT/ZnO nanocomposite and its photocatalytic activity for the removal of cyanide from water using a laser. *Nanotechnology*, 21 (49), 495705.

Shiba, F. and Okawa, Y., 2005. Relationship between supersaturation ratio and supply rate of solute in the growth process of monodisperse colloidal particles and application to AgBr systems. *The Journal of Physical Chemistry B*, 109 (46), 21664–21668.

Schreiber, T., Ortaç, B., Limpert, J., and Tünnermann, A., 2007. On the study of pulse evolution in ultra-short pulse mode-locked fiber lasers by numerical simulations. *Optics Express*, 15 (13), 8252.

Solozhenko, V.L., Turkevich, V.Z., and Holzapfel, W.B., 1999. Refined phase diagram of boron nitride. *The Journal of Physical Chemistry B*, 103 (15), 2903–2905.

Spano, F., Massaro, A., Blasi, L., Malerba, M., Cingolani, R., and Athanassiou, A., 2012. In situ formation and size control of gold nanoparticles into chitosan for nanocomposite surfaces with tailored wettability. *Langmuir*, 28 (8), 3911–3917.

Streubel, R., Barcikowski, S., and Gökce, B., 2016a. Continuous multigram nanoparticle synthesis by high-power, high-repetition-rate ultrafast laser ablation in liquids. *Optics Letters*, 41 (7), 1486.

Streubel, R., Bendt, G., and Gökce, B., 2016b. Pilot-scale synthesis of metal nanoparticles by high-speed pulsed laser ablation in liquids. *Nanotechnology*, 27 (20), 205602.

Tamarov, K.P., Osminkina, L.A., Zinovyev, S.V., Maximova, K.A., Kargina, J.V., Gongalsky, M.B., Ryabchikov, Y., Al-Kattan, A., Sviridov, A.P., Sentis, M., Ivanov, A.V., Nikiforov, V.N., Kabashin, A.V., and Timoshenko, V.Y., 2015. Radio frequency radiation-induced hyperthermia using Si nanoparticle-based sensitizers for mild cancer therapy. *Scientific Reports*, 4 (1), 7034.

Tangwarodomnukun, V., Chen, H., Tangwarodomnukun, V., and Chen, H., 2015. Laser ablation of PMMA in air, water, and ethanol environments. *LMMP*, 30 (5), 685–691.

Taylor, U., Tiedemann, D., Rehbock, C., Kues, W.A., Barcikowski, S., and Rath, D., 2015. Influence of gold, silver and gold–silver alloy nanoparticles on germ cell function and embryo development. *Beilstein Journal of Nanotechnology*, 6 (1), 651–664.

Tian, F., Sun, J., Yang, J., Wu, P., Wang, H.-L., and Du, X.-W., 2009. Preparation and photocatalytic properties of mixed-phase titania nanospheres by laser ablation. *Materials Letters*, 63 (27), 2384–2386.

Tian, Z., Liang, C., Liu, J., Zhang, H., and Zhang, L., 2011. Reactive and photocatalytic degradation of various water contaminants by laser ablation-derived SnO_x nanoparticles in liquid. *Journal of Materials Chemistry*, 21, 18242.

Vaccaro, L., Sciortino, L., Messina, F., Buscarino, G., Agnello, S., and Cannas, M., 2014. Luminescent silicon nanocrystals produced by near-infrared nanosecond pulsed laser ablation in water. *Applied Surface Science*, 302, 62–65.

Vadavalli, S., Valligatla, S., Neelamraju, B., Dar, M.H., Chiasera, A., Ferrari, M., and Desai, N.R., 2014. Optical properties of germanium nanoparticles synthesized by pulsed laser ablation in acetone. *Frontiers in Physics*, 2 (October), 1–9.

Wang, C.X., Liu, P., Cui, H., and Yang, G.W., 2005a. Nucleation and growth kinetics of nanocrystals formed upon pulsed-laser ablation in liquid. *Applied Physics Letters*, 87 (20), 1–3.

Wang, C.X., Yang, Y.H., and Yang, G.W., 2005b. Thermodynamical predictions of nanodiamonds synthesized by pulsed-laser ablation in liquid. *Journal of Applied Physics*, 97 (6), 1–4.

Wang, D., Zhang, H., Li, L., Chen, M., and Liu, X., 2016. Laser-ablation-induced synthesis of porous ZnS/Zn nano-cages and their visible-light-driven photocatalytic reduction of aqueous Cr(VI). *Optical Materials Express*, 6 (4), 1306.

Xiao, J., Wu, Q.L., Liu, P., Liang, Y., Li, H.B., Wu, M.M., and Yang, G.W., 2014. Highly stable sub-5 nm $Sn_6O_4(OH)_4$ nanocrystals with ultrahigh activity as advanced photocatalytic materials for photodegradation of methyl orange. *Nanotechnology*, 25 (13), 135702.

Xu, Z., Quintanilla, M., Vetrone, F., Govorov, A.O., Chaker, M., and Ma, D., 2015. Harvesting lost photons: Plasmon and upconversion enhanced broadband photocatalytic activity in core–shell microspheres based on lanthanide-doped $NaYF_4$, TiO_2, and Au. *Advanced Functional Materials*, 25 (20), 2950–2960.

Yang, G.W., 2007. Laser ablation in liquids: Applications in the synthesis of nanocrystals. *Progress in Materials Science*, 52 (4), 648–698.

Yoo, J.H., Jeong, S.H., Greif, R., and Russo, R.E., 2000. Explosive change in crater properties during high power nanosecond laser ablation of silicon. *Journal of Applied Physics*, 88 (3), 1638–1649.

Zamiri, R., Azmi, B.Z., Naseri, M.G., Ahangar, H.A., Darroudi, M., and Nazarpour, F.K., 2011. Laser based fabrication of chitosan mediated silver nanoparticles. *Applied Physics A: Materials Science and Processing*, 105 (1), 255–259.

Zeng, H., Cai, W., Liu, P., Xu, X., Zhou, H., Klingshirn, C., and Kalt, H., 2008a. ZnO-based hollow nanoparticles by selective etching: Elimination and reconstruction of metal−semiconductor interface, improvement of blue emission and photocatalysis. *ACS Nano*, 2 (8), 1661–1670.

Zeng, H., Liu, P., Cai, W., Yang, S., and Xu, X., 2008b. Controllable Pt/ZnO porous nanocages with improved photocatalytic activity. *The Journal of Physical Chemistry C*, 112 (49), 19620–19624.

Zhang, D., Gökce, B., and Barcikowski, S., 2017. Laser synthesis and processing of colloids: Fundamentals and applications. *Chemical Reviews*, 117 (5), 3990–4103.

Zhang, J., Chen, G., Chaker, M., Rosei, F., and Ma, D., 2013. Gold nanoparticle decorated ceria nanotubes with significantly high catalytic activity for the reduction of nitrophenol and mechanism study. *Applied Catalysis B: Environmental*, 132–133, 107–115.

Zimbone, M., Buccheri, M.A., Cacciato, G., Sanz, R., Rappazzo, G., Boninelli, S., Reitano, R., Romano, L., Privitera, V., and Grimaldi, M.G., 2015. Photocatalytical and antibacterial activity of TiO_2 nanoparticles obtained by laser ablation in water. *Applied Catalysis B: Environmental*, 165, 487–494.

Zsigmond, R.A., 1926. Properties of colloids [online]. Available from: www.nobelprize.org/nobel_prizes/chemistry/laureates/1925/zsigmondy-lecture.pdf [Accessed 18 Apr 2018].

12

Nanoparticles at the Polarized Liquid-Liquid Interface

Marcin Opallo and Katarzyna Winkler
Institute of Physical Chemistry, Polish Academy of Sciences

12.1 Introduction

Interfaces between two immiscible fluids are considered as flat, defect-free, and self-healing planes for immobilization, organization, and subsequent manipulation of nanoparticles (NPs) (Böker et al. 2007, Bresme and Oettel 2007, Kinge et al. 2008, Garbin et al. 2012, Booth et al. 2015, Edel et al. 2016, Giner-Casares and Reguera 2016, Toor et al. 2016). These fluid interfaces (FIs) include liquid/liquid interface (LLI) and liquid/gas interface (LGI). Interface between two immiscible electrolyte solutions (ITIES) belongs to the class LLI, and it is formed when both liquid phases contain ions. As a consequence of the presence of charged species in both phases, such LLI may be polarized.

From the beginning of 21st century, a number of researchers focused on the deposition of NPs at boundaries of fluid phases. They are convenient templates for two-dimensional (2D) ordering of various types of nanostructures, because of confinement in two dimensions and possibility of movement across interface. These self-assembled (SA) layers have attracted considerable attention because of their great application potential. For example, 2D structures consisting of metallic (Ag, Au) (Kiely et al. 2000, Lin et al. 2001) or semiconductor (CdSe, CdS) (Coe-Sullivan et al. 2005, Chen et al. 2009) NPs can be applied in a number of areas as solar cells (Gur et al. 2005, Talapin et al. 2010), sensors (Anker et al. 2008), light-emitting devices (Lin et al. 2001, Chen et al. 2009), storage devices (Paul et al. 2003, Chen et al. 2009), photodetectors (Chen et al. 2009), semiconductor nanorods' epitaxial growth (Kolasinski et al. 2006), electrode modification (Lesniewski et al. 2010), corrosion protection (Shen et al. 2005), antireflective films (Prevo et al. 2005), nanosphere lithography (Haynes and Van Duyne 2001), or heterogeneous catalysis (Lee et al. 2004). They can be also used for the stabilization of foams and emulsions (Lin et al. 2003b, Martinez et al. 2008) as well as nanoporous membranes (Lin et al. 2003a).

In general, SA is a type of organization of molecules or particles that does not require any external control. This process occurs due to entropic effects, intermolecular interactions (van der Waals, capillary, solvation, and fluctuation), and electrostatic and magnetic forces. In the case of organization on the FI, there are two levels of SA: (i) a spontaneous adsorption of the NPs at the interface and (ii) SA of the adsorbed NPs, leading to the fabrication of the ordered 2D structures. SA may be preceded by the formation of NPs at LLI by chemical reaction like in Brust-Schiffrin method (Brust et al. 1994). The formation of NPs at LLI is especially important at ITIES, where NPs are electrodeposited at interface with precursor and reducing agent present in different phases. Before the description of formation and properties of the NPs at ITIES, we will describe the fundamentals of NPs' SA at LLI.

12.2 Fundamentals of Nanoparticles' Self-Assembly at Liquid/Liquid Interfaces

12.2.1 Static and Dynamic Self-Assembly

There are two types of SA: static and dynamic (Whitesides and Grzybowski 2002). Static SA (SSA) occurs when structures organize autonomously into stable, ordered arrays. The driving force for this process is an energy decrease resulting in thermodynamic equilibrium. For this reason, structures formed by the SSA are quite stable and do not undergo further transformations. In turn, dynamic SA, also called dissipative SA, produces the organized structures far from the thermodynamic equilibrium. Their formation requires a continuous supply of energy, subsequently dissipated *via* the production of entropy (Fialkowski et al. 2006, Wang et al. 2012, Sashuk et al. 2013). In general, stable and applicable nanoparticulate structures are formed by SSA.

Although SSA at LLI is a spontaneous process, it requires a rational design of the building elements to obtain properly organized microscopic structures for technological applications (Grzelczak et al. 2010). This is achieved by proper choice of the size and shape of the NPs – efficient building blocks. Functionalization of these "bricks" is also important, because the ligands play a key role in providing stability at interfaces. Their type determines the steric, physical, and chemical interactions between NPs and their environment. For instance, by choosing charged ligands, one makes the NPs susceptible to an external directing field (e.g. electric). Another type of design of the SSA process is a modification of an environment where NPs' assembly occurs by templates. They induce steric factors resulting in the selective assembly of particles. Typical templates are small particles, acting like spacers between the NPs, as well as scaffolds in which assembling structures are immersed. Finally, temperature, pH, and the polarity of the environment affect the interparticle interactions and trigger the SSA of the NPs (Reincke et al. 2006, Grzelczak et al. 2010).

12.2.2 The Free Energy of Nanoparticles

Trapping of the NPs at the LLI occurs because of the decrease in surface free energy (ΔG), connected to the diminution of the contact surface between two fluids. This relationship can be described by the following equation (Garbin et al. 2012):

$$\Delta G = -\gamma \pi a^2 (|\cos\theta| - 1)^2, \qquad (12.1)$$

where γ is the surface tension between liquid phases, a is the radius of the NP, and θ is the equilibrium contact angle between three phases (two liquid phases, L1 and L2, and solid particle, S; see Figure 12.1), defined by the Young-Dupre equation:

$$\gamma_{\mathrm{SL1}} = \gamma_{\mathrm{SL2}} + \gamma_{\mathrm{L1L2}} \cos\theta. \qquad (12.2)$$

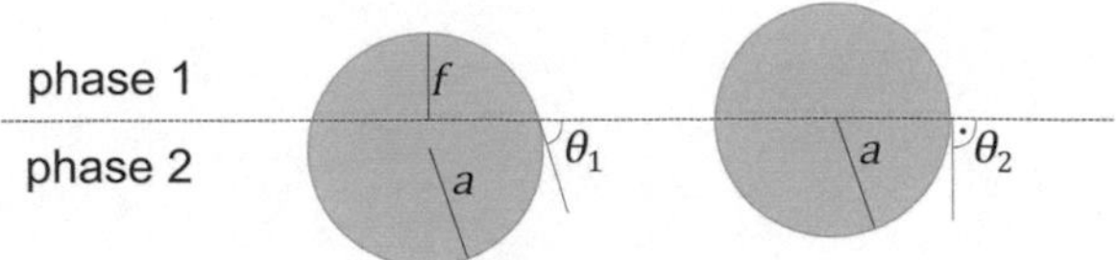

FIGURE 12.1 Two different spherical particles at their equilibrium positions, defined by the radii a and the different contact angles θ.

The phenomenon related to assembly on the interfaces is the energetically promoted size segregation of the NPs (Lin et al. 2003b, 2005). If there is a distribution of particle sizes and interfacial area is limited, trapping of particles with bigger radii is thermodynamically promoted, while smaller particles are pushed out into the bulk. This behavior results from Eq. (12.1) – reduction of the Helmholtz free energy is higher when the interface is decorated with the particles of bigger radius a. Moreover, smaller particles are more vulnerable to a thermally activated escape than bigger ones (Lin et al. 2003b) because thermal energy responsible for their spatial fluctuations is comparable to the interfacial energy. For example, for the spherical NPs of a diameter about 3 nm, deposited at the water/toluene interface, the activation energy for the detachment was calculated as $5k_BT$ (Lin et al. 2003b).

12.2.3 Interactions between Nanoparticles at Interfaces

SA of the NPs at FI, like air/water or oil/water, is a complex process that can lead to the formation of a number of types of 2D structures. As it was said above, there are many factors affecting this process. Unfortunately, a precise design of the arrangement of the NPs is still a challenge, among others, because of a relatively small number of detailed studies on this topic. Nevertheless, the fundamental knowledge about interactions between adjacent NPs as well as NPs and solvent molecules, or NPs and FI itself, is indispensable for predicting mechanisms of ordering of the NPs (Min et al. 2008). Among interactions affecting the formation of 2D structures at FI, van der Waals, electrostatic, and capillary interactions have been best studied.

van der Waals Interactions

Assembly of the NPs at the FI is regulated by the balance between intermolecular forces. Among them, the ubiquitous van der Waals interactions play an important role in all types of NP systems. These interactions at the interface are attractive or repulsive contrary to always attractive ones between NPs dispersed in a bulk (Williams and Berg 1992). van der Waals energy U_{vdW} of interaction between two spherical NPs depends on their radius a and separation distance r:

$$U_{\mathrm{vdW}}(r) = -\frac{A_{\mathrm{interf}}}{12}\frac{a}{r-2a}, \qquad (12.3)$$

where A_{interf} is an effective Hamaker constant (Williams and Berg 1992) determined as a function of a fractional height f of spherical NP immersed in the upper phase (see Figure 12.1) and the equilibrium three-phase contact angle θ measured in the bottom phase:

$$A_{\text{interf}} = A_b + f^2\,(3 - 2f)\,(A_u - A_b)\,, \tag{12.4}$$

where $f = (1 - \cos\theta)/2$, and A_b and A_u are the Hamaker constants across bottom fluid and upper fluid, respectively.

Attractive van der Waals interactions may be responsible for undesired aggregation and precipitation of the NPs from the system (Bishop et al. 2009). The omnipresence of these interactions may be therefore disadvantageous. However, employed properly, van der Waals interactions may allow for the fabrication of organized nanoparticulate systems in a bottom-up way (Figure 12.2). In 2D systems consisting of spherical monodispersed NPs, van der Waals interactions manifest themselves by the formation of NP arrays with hexagonal geometry. For polydispersed NPs, size-selective sorting is observed: larger particles are organized in the inner area of the assembly, while smaller (or differently shaped) ones are located at the peripheries (Ohara et al. 1995, Bishop et al. 2009).

Electrostatic Interactions

To avoid irreversible aggregation, the NPs at FIs can be functionalized with charged ligands. The electrostatic forces between similarly covered objects are repulsive and stabilize 2D dispersions in polar solvents. Since electrostatic properties of the nanomaterials deposited at LLI are easily tunable by changes in electrolyte composition, pH, or the ionic strength of the fluid phase, they can be applied as a convenient tool to control NP assembling. For example, change of one of these parameters allows for the precise change of a distance between adjacent NPs (see Figure 12.3).

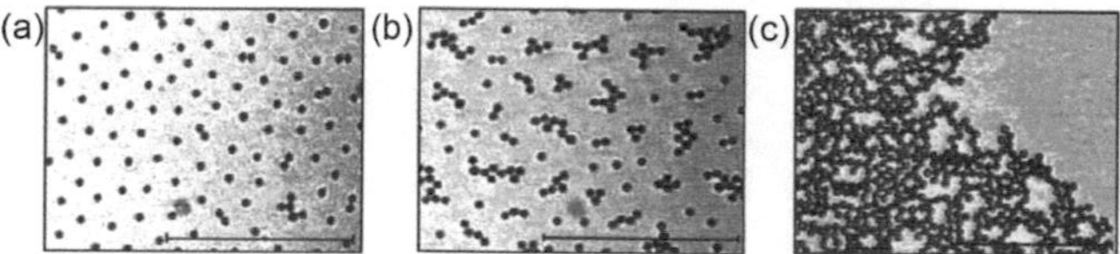

FIGURE 12.3 Monolayers of charged particles at air/aqueous NaCl interfaces of increasing ionic strength. NaCl concentrations are (a) 10 mM, (b) 100 mM, and (c) 1 M (Aveyard et al. 2000).

It should be noted that the electrostatic interactions around the particle immobilized between two phases of various dielectric constants is a complex phenomenon. The charge on the surface of the NP is asymmetric, which results in the appearance of an electrical dipole moment and dipole/dipole long-range interactions between nano-objects (Pieranski 1980). As it is shown in Figure 12.4, only ligands immersed in the polar phase, for example water, are able to release ions from their dissociable groups, and between as-achieved surface charges and counterions in water, the dipoles, normal to the interface, are formed. Interactions between them become dominant for $\kappa r \gg 10$, where κ is the inverse Debye screening length and r is the center-to-center separation distance (Hurd 1985). Assuming the existence of only diffuse double layer around NPs, the total interaction potential consists of both the Coulombic and dipolar contributions:

$$U_{\text{electrostatic}}\,(r) = \frac{b_1 k_B T}{3r} e^{-\kappa r} + \frac{b_2 k_B T}{r^3}, \tag{12.5}$$

where T is an absolute temperature, k_B is the Boltzmann constant, and b_1 and b_2 are the prefactors determining the order of magnitude of the screened Coulomb and the dipole/dipole interactions, respectively (Hurd 1985, Masschaele et al. 2010).

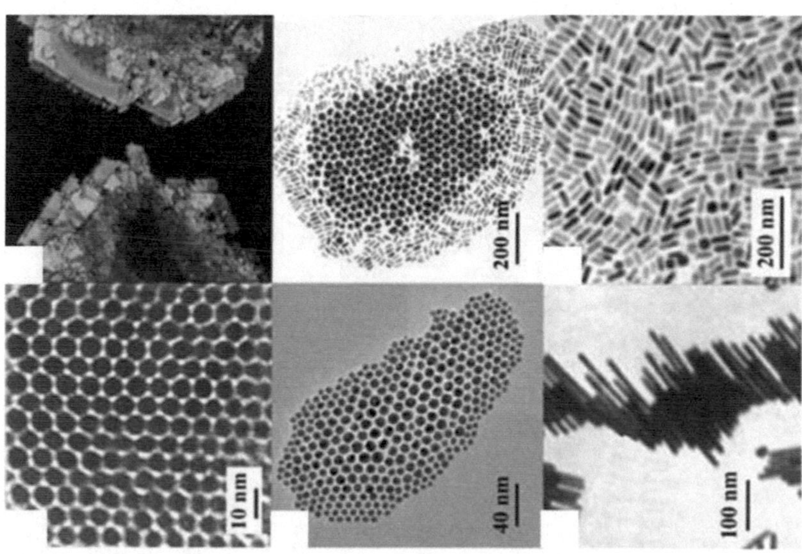

FIGURE 12.2 Various assemblies driven by van der Waals interactions. Not only close packing, but also shape segregation, size segregation of nanospheres, and side-by-side organization of nanorods are visible (Bishop et al. 2009).

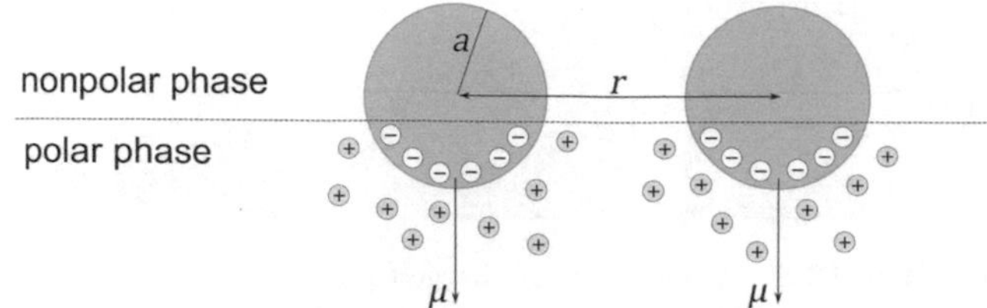

FIGURE 12.4 Charged, spherical NPs at the interface between nonpolar and polar liquid. The asymmetry of the ion cloud results in the appearance of an electrical dipole moment μ.

Capillary Interactions

The capillary forces between particles deposited on the FI are a consequence of the deformation of the interface. For macromolecules, this deformation is a result of gravity acting on them. NPs, however, are way too light to influence the interface due to the effects connected to their mass, but also for these small objects, a capillary interactions are observed. As it was previously explained (Stamou et al. 2000), NPs' impact on the interface is a result of the wetting by a fluid phase which contacts the surface of the particle at the given angle and is especially observed for the anisotropic or chemically heterogenic NPs. Energy of capillary interactions of such partially immersed NPs (so-called immersion forces) is much greater than kT (Paunov et al. 1993), and its force increases proportionally to the interfacial tension. The total interaction energy may be expressed as:

$$U_{\text{capillary}}(r) = -12\pi\gamma_{ff}H_p^2\omega\frac{a^4}{r^4}, \tag{12.6}$$

where γ_{ff} is the fluid/fluid surface tension, H_p is the amplitude of the deformation that can be estimated from typical values of contact angle hysteresis, and ω is the factor determined on the basis of the relative orientation of the interacting NPs (Stamou et al. 2000, Garbin et al. 2012).

Depending on homogeneity of the system, the forces could be attractive or repulsive: attraction is observed when menisci of both interacting NPs are concave or convex, while repulsion is typical for the interaction of two opposed types of meniscus (Danov et al. 2005, Bresme and Oettel 2007), as shown in Figure 12.5. Capillary attraction results in the aggregation of NPs and the formation of arrangements ranging from small clusters to ordered arrays of a macroscopic scale.

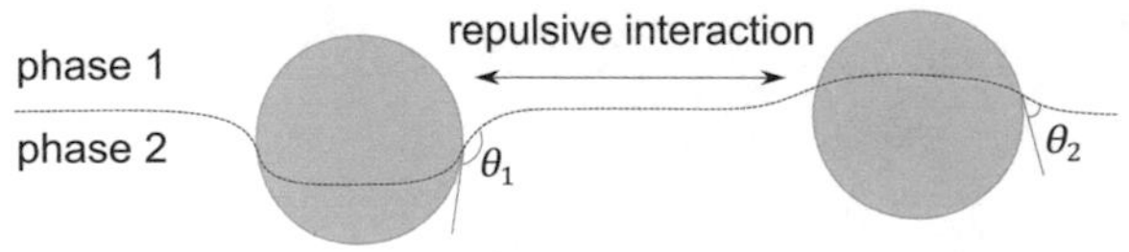

FIGURE 12.5 Two spherical particles of opposite wetting properties and different contact angles θ_1 and θ_2: the non-wettable particle gives rise to the formation of a concave meniscus, while the wettable one causes a convex meniscus. Between these two deformations of the interface, the repulsion occurs.

12.3 Interfaces between Two Immiscible Electrolyte Solutions

As it was said above, when both fluids in contact contain dissolved ions, polarized interface, often named by acronym ITIES, is formed. One of the solvents is always water, whereas the second one is an organic polar solvent with dielectric permittivity around ten. In the last two decades, 1,2-dichloroetane (DCE) is the most typical solvent used in this research area (Samec 2004). Hydrophobic ionic liquid is another example of organic phase (Samec et al. 2009). It is noteworthy that water and these solvents are not truly immiscible. For example, solubility of water in DCE reaches 0.1 M, and in equilibrium, both phases contain some amount of the second solvent (Trojanek et al. 2014). Therefore, before the experiment, one has to secure that both phases are mutually saturated.

The presence of ions in both phases allows for control of potential drop across LLI (Samec 1988, 2004, Samec et al. 2009, Girault 2010), adding new variable to the system. The potential drop across ITIES is in the range of a fraction of volt. However, local magnitude of electric field reaches 10^9 N/m. The control of the potential drop is realized either by external potential source (potentiostat) with two pairs of reference and counter electrodes immersed in both phases (Figure 12.6) or by an appropriate selection of electrolytes with common ion.

Typically, the electrolyte dissolved in organic phase is extremely hydrophobic, whereas the one dissolved in aqueous phase is extremely hydrophilic. This is to minimize the probability of an ion transfer within a certain range

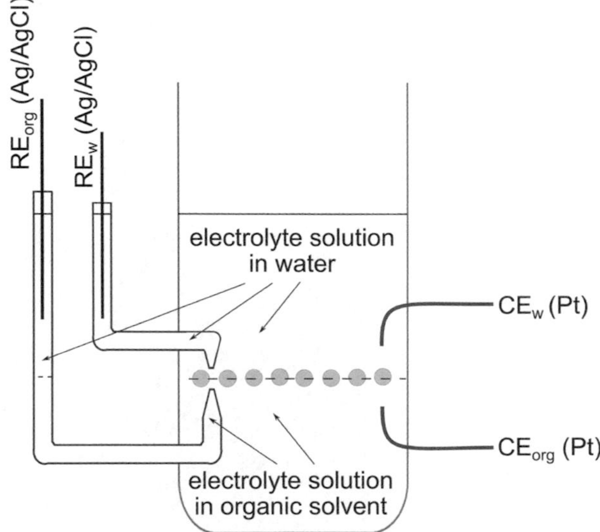

FIGURE 12.6 Scheme of electrochemical cell for studying ITIES. RE_{org}, CE_{org}, RE_w, and CE_w mark reference and counter electrodes immersed in organic and aqueous phases, respectively. Gray circles represent NPs at ITIES marked by a dashed line. The elements of the picture are not in scale.

of potential drop and keep these two electrolyte solutions truly immiscible. This potential window is called polarizable potential range and can reach up to *ca.* 1 V (Samec et al. 2004, 2009). If the externally applied potential difference is larger, ion transfer across ITIES occurs.

Research on physicochemical properties of ITIES was conducted since 1960s (Samec 1988, 2004, Samec et al. 2009, Girault 2010). The double-layer structure, adsorption at ITIES (Samec 1988), charge (ion or electron) transfer (Samec et al. 2004, 2009, Girault 2010), and (electro)catalytic reactions (Mendez et al. 2010, Scanlon et al. 2018) were studied.

As other LLIs, ITIES is macroscopically flat, fluidic, and self-healing and its geometry (meniscus) is affected by geometry of the cell. Although it exhibits microscopic fluctuating roughness, it is molecularly sharp (Michael and Benjamin 1998). The presence of ions allows to form a uniform film of NPs by electrodeposition – in situ growth driven by redox reaction (Dryfe et al. 2014). The other method is SA described above – spontaneous adsorption of preformed NPs driven by a decrease in surface energy (Booth et al. 2015). Furthermore, if conductive nanomaterials are present at ITIES, their Fermi level (Scanlon et al. 2016) and therefore their reactivity (Smirnov et al. 2015, Gschwend et al. 2017, Scanlon et al. 2018) and/or distribution at interface (Gschwend et al. 2017, Scanlon et al. 2018) can be manipulated.

12.4 Electrodeposition of Nanoparticles at ITIES

Electrodeposition is perhaps the most popular method to obtain conductive films on the conductive surface. Historically, this is also the case of ITIES. Although electrodeposition of metallic copper at LLI was demonstrated long time ago (Guanazzi et al. 1975), the first report on electrodeposition of relatively well-defined NPs (Au) at such interface was published 21 years later (Cheng and Schiffrin 1996). Since then, a wide range of nanomaterials was electrodeposited successfully at ITIES (Dryfe et al. 2014, Poltorak et al. 2017).

For this purpose, hydrophobic polar solvents such as DCE, 1,2-dichlorobenzene (DCB), or α,α,α-trifluorotoluene (TFT) were employed as organic phase. In the case of electrodeposition on the surface of metal, such material itself serves as a source of electrons to reduce metal cations present in a liquid phase, most frequently aqueous one. In the case of ITIES, reaction can be called as "electrodeposition without electrodes" (Johans et al. 2000). Electron donor is present in one phase (hydrophilic in an aqueous phase or hydrophobic in an organic phase) to reduce solvated metal cation (Ag^+ or Cu^{2+}) or complex anion ($AuCl_4^-$ or $PdCl_4^{2-}$) present in adjacent phase (Figure 12.7). In the case of ITIES, the number of electrons available is much smaller than that in the case of metal and the energy of electrons, and therefore, efficiency of electrodeposition depends on the redox potential of specific electron donor, not on the external potential adjusted by potentiostat.

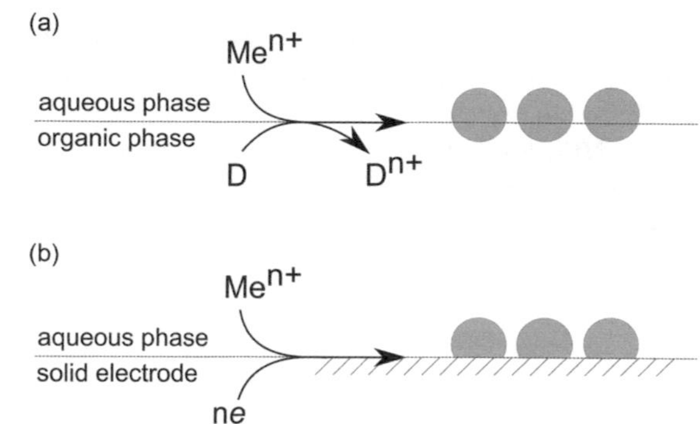

FIGURE 12.7 Scheme of electrodeposition of NPs at (a) ITIES and (b) solid electrode surface.

The oxidation potential of electron donor has to be low enough to reduce metal ion. The potential drop forces metal ion across ITIES to approach electron donor, and then, the reduction occurs. The mechanism of the electrodeposition without electrodes is more difficult to understand than that on the metal electrodes, because the transport of reactant occurs from both sides of interface (Dryfe et al. 2014). Obviously, metal nucleation and growth have to occur, but it is not clear how far from ITIES nucleation process starts.

Electrodeposition of Au was realized with hydrophilic ($Fe(CN)_6^{4-}$) or hydrophobic (decamethylferrocene [DMFc] or simple organic molecule easy to oxidize) donor present in the aqueous (Grunder et al. 2011) or the organic phase (Sefer et al. 2012, Zhu et al. 2014), respectively. Typically, $AuCl_4^-$ ions present in the aqueous phase are a source of gold due to their negative (−7.4 kJ/mol) Gibbs energy of transfer from DCE to water (Grunder et al. 2011). The mechanism of reduction is quite complex, because of the number of complexation and interfacial equilibria (Dryfe et al. 2014). To form Ag (Guo et al. 2003, Zhu et al. 2014), Pt (Platt et al. 2002, Trojanek et al. 2006, 2007, Nieminen et al. 2011, Zhu et al. 2014), Pd (Johans et al. 2000, Rahtinen et al. 2000, Platt et al. 2007, Nieminen et al. 2011, Izquierdo et al. 2012, Booth et al. 2017), $Pd_{80}Au_{20}$ (Grunder et al. 2013), and Cu (Aslan et al. 2015b) NPs, hydrophobic ferrocene derivatives dissolved in organic phase were applied as electron donors. With such procedure, spherical NPs may be obtained (Figure 12.8).

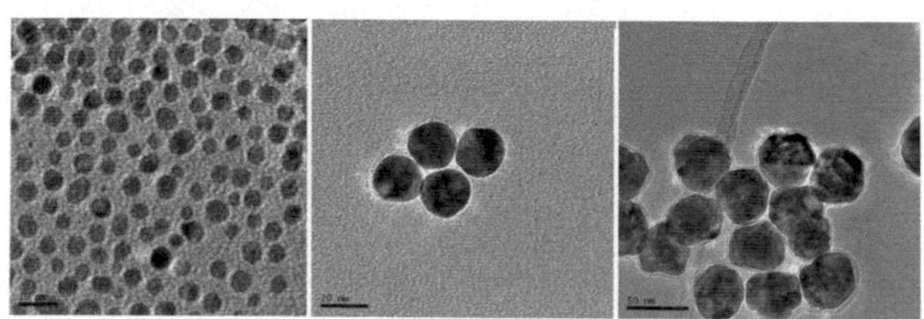

FIGURE 12.8 Transmission electron microscopic (TEM) image of Au (5 nm) particles, Au (19 nm) particles, and Pd80Au20 particles (Grunder et al. 2013).

Not surprisingly, the presence of impurities present at interface promotes electrodeposition (nucleation) of NPs at ITIES. This phenomenon was explored to prepare copper shells on AuNPs present at ITIES (Grunder et al. 2015).

The replacement of electrolyte solution in hydrophobic solvent with hydrophobic ionic liquid as an organic phase results in different geometry of Au nano-objects prepared at ITIES. In such conditions, large Au dendritic structures (Nishi et al. 2015) or needle mushrooms (Yao et al. 2016) are formed. The dendritic shape was attributed to a large viscosity of ionic liquid affecting the transport of Au complex ions toward ITIES (Kakinami et al. 2016, Nishi et al. 2016).

Metal nano-objects can be electrodeposited not only on fresh LLI, but also on species already assembled at ITIES. The electrodeposition of copper shell on AuNPs earlier assembled at ITIES is an example of preparation of core/shell NPs (Grunder et al. 2015). Polymer-nanomaterial composites were electrochemically prepared at ITIES by electrodeposition of NPs within polymer template assembled at ITIES or by interfacial electropolymerization followed by the assembly of nanomaterials at ITIES. PdNPs prepared in the pores of alumina membrane (Platt and Dryfe 2007) and polypyrrole-carbon nanotube films (Toth et al. 2015a) are examples of both approaches. Nanocomposites like Pt-silica (Janchenova et al. 2006, Lepkova et al. 2007) or AuNPs embedded in polymer matrix obtained from tyramine or resorcinol (Lepkova et al. 2008) can be obtained in one or two steps. In the latter case, the size of NPs can be controlled by the change of potential drop across ITIES by the addition of appropriate partitioning ions (Lepkova et al. 2008). Decoration of carbon nanomaterials like graphene with PdNPs (Toth et al. 2015b,c) (Figure 12.9) cobalt sulfide nanoparticles (Aslan et al. 2015a) or carbon nanotubes with CuNPs (Aslan et al. 2016) already assembled at ITIES was also demonstrated. In the former process, graphene embedded in polymer is assembled at ITIES, and then, polymer matrix is dissolved (Toth et al. 2015c).

ITIES can be also employed for electrodeposition of belts of metallic NPs on conductive surface (Kaminska et al. 2010, 2012). Here, NPs are deposited only at the three-phase junction: electrode/liquid/liquid. Such system consists of electrode (serving as an electron donor) immersed partially in organic and aqueous phases (Figure 12.10). The organic phase (electrolyte solution in toluene (Kaminska et al. 2012) or ionic liquid (Kaminska et al. 2010)) contains metal precursor in ionic form, whereas aqueous phase is also supported by electrolyte.

Similar setup based on three-phase junction electrode/liquid/liquid was employed to prepare carbon silicate sponge film on the vertically moving electrode. The electrode was always partially immersed in the hydrophilic carbon NPs in an aqueous solution and organic solvent containing sol-gel precursor (Szot et al. 2011). Here, electrogenerated protons catalyze sol-gel process, and NPs are trapped into the film deposited on moving vertically electrode.

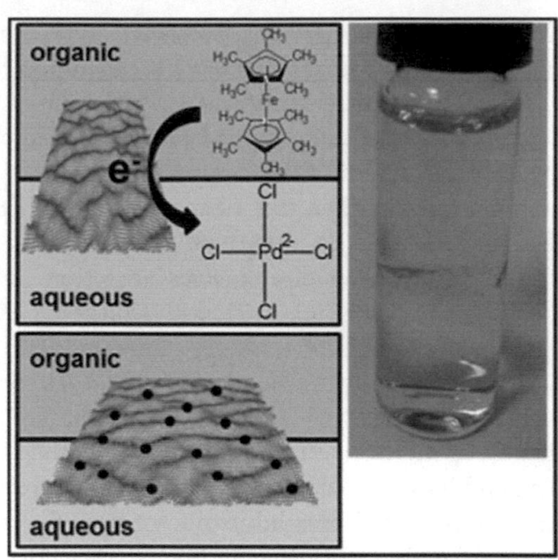

FIGURE 12.9 Schemes of the $PdCl_4^{2-}$ reduction by DMFc via assembled graphene at the interface (left upper image) by forming the PdNPs marked as black dots (left lower image) together with photograph of the cell (right image). The interfacial contact time of deposition was 1 min in all the cases to allow electrodeposition and dissolution of polymethylacrylate matrix. Organic phase consists of DCE and 5-nonanone with 1:4 ratio (Toth et al. 2015c).

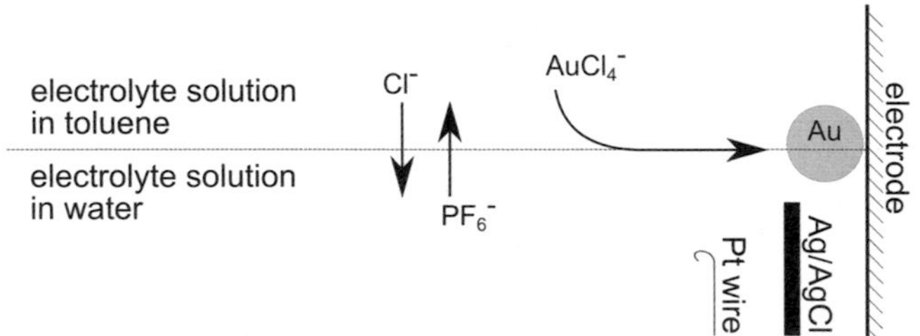

FIGURE 12.10 Scheme of electrochemical cell and mechanism of formation of AuNPs at three-phase junction electrode/liquid/liquid interface (Kaminska et al. 2012).

It is possible to place ITIES at the orifice of pipette as small as 10 nm in diameter, filled with organic electrolyte solution, and immersed in an aqueous electrolyte solution. When Au, Ag, and Pt are electrodeposited at such nanointerface, one can produce disk nanoelectrodes with such small diameter (Zhu et al. 2014).

Electrodeposition seems to be quite simple one-step method to prepare nanoparticulate film at ITIES. However, this approach is restricted to metals such as gold, platinum, palladium, and copper, where it is possible to find a suitable electron donor capable of reducing their precursors. Studies on the mechanism of electrodeposition of nanomaterials at ITIES and ion transfer voltammetry provide better understanding of NP formation in general. For example, it was concluded that reduction of $AuCl_4^-$ or $AuCl_2^-$ to metallic gold in Brust-Schiffrin method of formation of AuNPs occurs in the bulk organic phase (Uehara et al. 2015) and even more species formed during this reaction was identified (Uehara et al. 2016).

12.5 Assembly of Nanoparticles at ITIES

Earlier prepared nano-objects can be assembled at ITIES (Rodgers et al. 2014, Booth et al. 2015). This method of ITIES modification is more universal, because it allows to increase the range of assembled objects beyond the electrodeposited metals.

Most of attention was paid to SA of AuNPs (Su et al. 2004, Younan et al. 2010, Fang et al. 2013, Grunder et al. 2014, Gschwend et al. 2017, Smirnov et al. 2017a,b), because of the feasibility of their preparation and numerous potential applications (Scanlon et al. 2018).

Apart from spontaneous SA, AuNP films can be formed at ITIES by the precise injection of their methanol suspension (Smirnov et al. 2015) (Figure 12.11). The surface coverage is controlled by flow rate and time of the injection. Gold films obtained using the AuNPs of various sizes range from purple to metallic gold ones and exhibit different curvature (Smirnov et al. 2015). The assembly of core/shell Au-Ag (Abid et al. 2007) or Au-Pd (Grunder et al. 2014) NPs was also demonstrated. The process of SA can be reversibly driven by external voltage (Su et al. 2004, Abid et al. 2007), due to a change in the direction of electric field (Flatté et al. 2008).

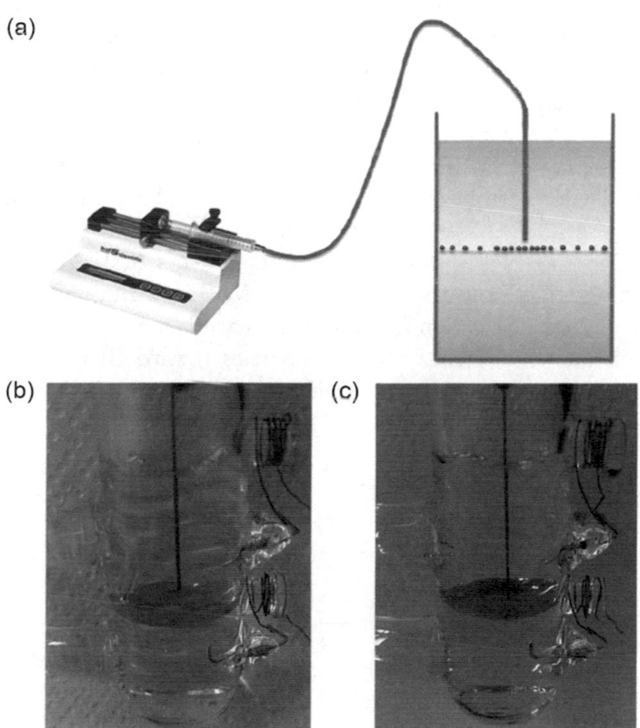

FIGURE 12.11 Functionalization of soft interfaces with Au nanofilms by the precise injection of AuNPs suspended in methanol at the interface. (a) Schematic of the capillary and syringe-pump setup used to settle the AuNPs directly at the interface between two immiscible liquids allowing a precise control over the AuNPs' surface coverage. Examples of AuNP films prepared at flat soft water/TNT interfaces in four-electrode electrochemical cells using AuNPs with mean diameters of (b) 12 nm and (c) 38 nm (Smirnov et al. 2015).

AuNP films at ITIES can be stabilized by redox functionalization in situ. When tetrathiofulvalene as an electron donor solution in DCE is emulsified with an aqueous suspension of citrate-stabilized AuNPs, the formation of golden (gray in print) color lustrous droplets containing organic solvents is seen (Figure 12.12) (Smirnov et al. 2014). This is because the AuNP acts as the electron acceptor and tetrathiofulvalene radical cations are formed. Both cations present in the aqueous phase and neutral molecules present in the organic phase form electrostatic "glue" for the negatively charged AuNPs, and multilayered film looking like bulk gold is formed. Such metallic droplets can be easily deformed and are stable for a longer period of time, in excess of a year. Interestingly, similar experiments can be done even with propylene carbonate as organic phase with solubility in water *ca.* 20% (Smirnov et al. 2017b).

Adsorption of PtNPs from diluted aqueous suspension at ITIES was also noted, and complexity of this process involving NPs' rolling and/or bouncing was emphasized (Stockmann et al. 2017).

Recently, it was shown that single-wall carbon nanotubes or graphene spontaneously assembles at ITIES and forms the film dense enough to block ion transfer (Toth et al. 2015c, Rabiu et al. 2017). More importantly, a number of inorganic nanomaterials of different geometry as Mo_2C nanowires (Bian et al. 2013), WC, W_2C, MoB, and Mo_2C particles (Scanlon et al. 2013), Ni_xS_yNPs (Akin et al. 2017), MoS_2 and WS_2 2D nanoplatelets (Hirunpinyopas et al. 2017), Co and Co_xB NPs (Camara et al. 2015), Cu_2CoSnS_4 nanofibers, Cu_2WS_4NPs (Ozel et al. 2016), and $BiVO_4$ nanocrystals (Rastgar et al. 2016) were successfully assembled at ITIES. Some of these nanomaterials such as MoS_2 (Ge et al. 2012), Mo_2C (Bian et al. 2013), C_xS_y (Aslan et al. 2016), or Ni_xS_y (Akin et al. 2017) were assembled at carbon nanotubes and other carbon nanomaterials already present at ITIES.

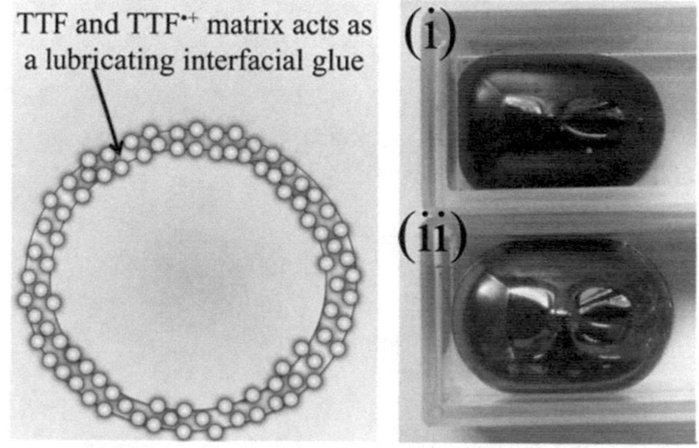

FIGURE 12.12 Scheme and optical images of metal liquid-like droplets formed by 14 nm (i) and 76 nm (ii) AuNPs (Smirnov et al. 2014).

12.6 Electrocatalysis at ITIES Modified with Nanoparticles

A number of effects related to the presence of NPs at ITIES were reported. Functionalized AuNPs increase interfacial capacitance as a result of an increase in electric charge or interfacial corrugation (Younan et al. 2010, Marinescu et al. 2012). The change in accessible potential window in the presence of NPs at ITIES was reported, and it was interpreted as a catalytic effect on ion transfer across ITIES (Camara et al. 2015).

More importantly, the conductivity of nanomaterials assembled at ITIES increases the efficiency of redox catalysis (Smirnov et al. 2014, 2015, Toth et al. 2015c, Gschwend et al. 2017). Catalytic processes can be divided into two classes: heterogeneous catalysis and homogeneous catalysis. In the heterogeneous catalysis, catalysts, for example NPs, are immobilized on a solid support. If the latter is conductive, its Fermi level may be shifted by applying potential difference between such electrode and reference electrode and such heterogeneous catalysis is called electrocatalysis. Consequently, if immobilized NPs are metallic, their Fermi level is also modified and electron(s) transfer occurs between them and reactant is in the close vicinity of the electrode. On the other hand, in homogeneous catalysis, electron transfer reaction occurs from donor to acceptor through catalyst molecule with fixed Fermi level. In the case of ITIES, electron is transferred from donor dissolved in organic phase to acceptor dissolved in aqueous phase (or opposite way) through conductive NPs or other objects assembled at interface (Figure 12.13a). The position of Fermi level in nano-objects assembled at ITIES can be modified (Figure 12.13b) by the application of external potential difference, like in conventional electrocatalysis or by setting of interfacial potential difference by an appropriate selection of electrolytes having common ion. Simply, NPs at ITIES act as conductive bipolar electrodes with equilibrated Fermi level.

Few examples of redox catalysis at ITIES involving conductive nano-objects like AuNPs (Smirnov et al. 2014, 2015, Gschwend et al. 2017) or carbon nanomaterials (Toth et al. 2015c) assembled at interface were reported. Such arrangement results not only in an increase in reaction efficiency, but also may induce phase transition in nanoparticulate film (Smirnov et al. 2017b). Moreover, it was suggested that it provides new methodology for studies of electrocatalytic properties of nano-objects in the absence of the electrode support (Peljo et al. 2017).

Not surprisingly, metallic and other NPs present at ITIES promote reactions involving electron transfer across LLI: hydrogen evolution reaction (HER) and oxygen reduction reaction (ORR).

They do not require any reactant except electron donor, typically hydrophobic derivative of ferrocene, and excess of protons, and these reactions occur at ITIES in the absence (HER) or presence of oxygen (ORR) in a system. Both HER and ORR at ITIES are driven by hydrophobic electron donor present in organic phase. HER was observed at ITIES with such nano-objects confined to ITIES as PtNPs (Nieminen et al. 2011), PdNPs (Nieminen et al. 2011), CuNPs (Aslan et al. 2014), and non-noble metal (molybdenum, cobalt, nickel or tungsten sulfides, borides, or carbides) nano-objects (Ge et al. 2012, Bian et al. 2013, Scanlon et al. 2013, Aslan et al. 2015a, 2015b, 2016, Ozel et al. 2015, Ozel et al. 2016, Akin et al. 2017) (Figure 12.14a). When reaction takes place at PtNPs confined to ITIES formed by aqueous and DCE electrolyte solutions, hydrogen bubble is observed (Figure 12.15) (Nieminen et al. 2011).

The role of 2D structure of nano-objects in HER was demonstrated on the example of MoS_2 (Hirunpinyopas et al. 2017). Selection of these nanomaterials was based on their known electrocatalytic properties toward HER when deposited on the electrode surface. ORR was studied with

FIGURE 12.13 Scheme of redox electrocatalysis (a) and scheme of Fermi level equilibration during redox electrocatalysis at ITIES with interface confined metallic NPs (b) following the concept proposed in literature (Peljo et al. 2017).

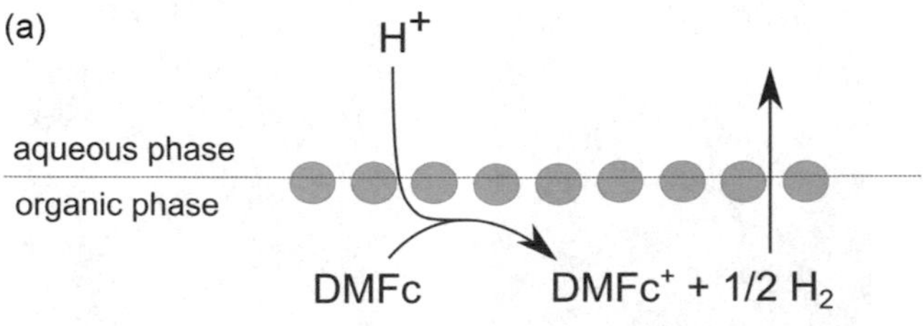

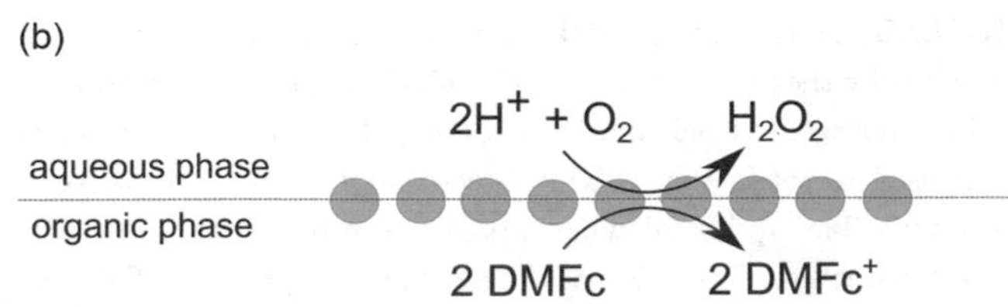

FIGURE 12.14 Scheme of (a) H_2 generation and (b) O_2 reduction at ITIES with DMFc as electron donor.

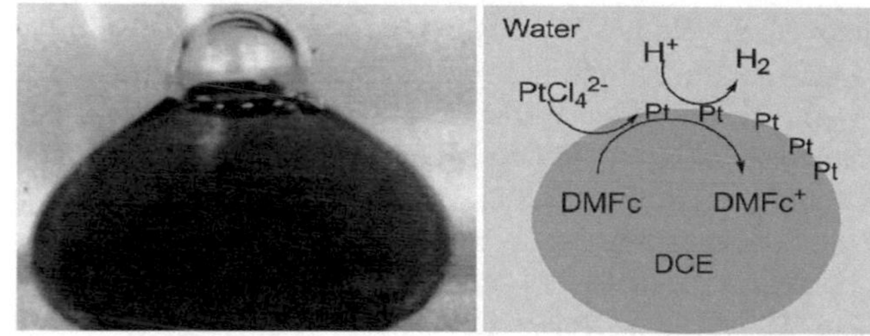

FIGURE 12.15 Photograph of hydrogen bubble formed at interface by acidic aqueous solution and concept of water splitting by electrolysis in a cell incorporating photodriven biphasic H_2 evolution in the cathodic compartment and photodriven biphasic oxidation in the anodic compartment (Nieminen et al. 2011).

the AuNPs (Trojanek et al. 2006, Grunder et al. 2013, Smirnov et al. 2016) confined to ITIES (Figure 12.14b) or the PtNPs suspended in an aqueous phase (Stockmann et al. 2017). The latter system was the first example of employment of micro-LLI for the detection of NPs. It allows for the detection of single impacts of NPs and demonstrates stochastic electrochemistry at ITIES. System based on ITIES is advantageous as compared to popular, for this purpose, carbon ultramicroelectrodes because it is stable and self-healing (Stockmann et al. 2017).

There were also few studies of reaction at ITIES driven by light. Acceleration of heterogeneous photoreduction of tetracyanoquinodimethane (TCNQ) by hydrophilic porphyrin by PdNPs assembled at ITIES was observed (Rahtinen et al. 2000). Increased efficiency of photocatalytic water oxidation at $BiVO_4$NPs assembled at ITIES with electron acceptor in organic phase was demonstrated (Rastgar et al. 2016, Rastgar and Wittstock 2017), and a mechanism of reaction was elucidated (Rastgar and Wittstock 2018). Such system is quite perspective from the point of view of practical applications; for example, the inclusion of such modified ITIES into water-splitting devices (Figure 12.16) was proposed (Scanlon et al. 2017).

12.7 Electrovariable Optical Phenomena

There are few examples of optical phenomena related to the presence of nanoparticulate film at ITIES, which can be driven by external electric field phase (Booth et al. 2014, Sikdar and Kornyshev 2016, Gschwend et al. 2017, Montelongo et al. 2017).

As assembled AuNPs of certain size (>15–50 nm) reflect the visible light (Hojeij et al. 2010, Smirnov et al. 2015), they form gold mirrors at ITIES (Figure 12.17). Their reflectivity increases with an increase in the nanoparticulate film conductivity (Fang et al. 2013). Application of external voltage in the range of hundreds millivolts causes assembly or disassembly of the film by pushing AuNPs in and out of aqueous phase (Montelongo et al. 2017) as predicted by theory (Sikdar and Kornyshev 2016). However, this process is extremely slow and takes thousands of seconds (Montelongo et al. 2017). The experiments demonstrating behavior of AuNP films at ITIES as electrically driven Marangoni shutter were reported (Gschwend et al. 2017).

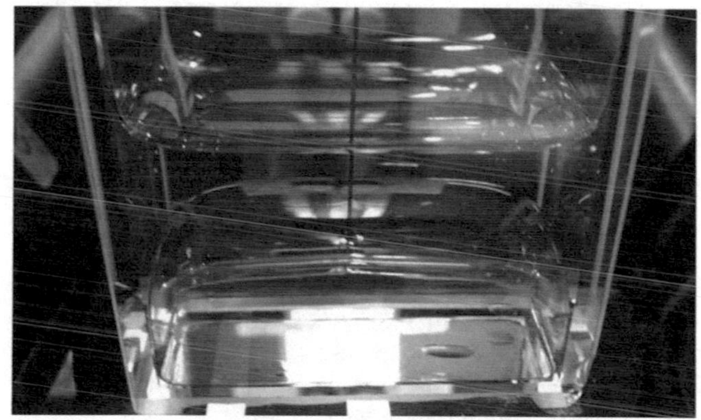

FIGURE 12.17 Film of AuNPs prepared on large curved ITIES (Smirnov et al. 2015).

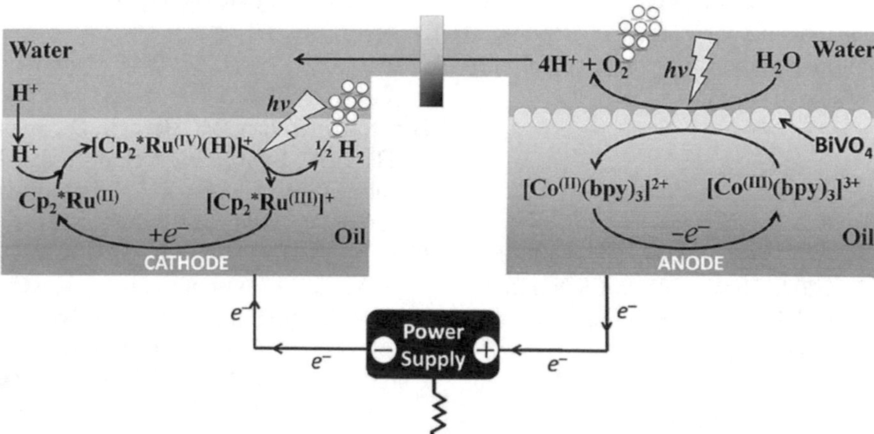

FIGURE 12.16 Concept of water splitting by electrolysis in a cell incorporating photodriven biphasic H_2 evolution in the cathodic compartment and photodriven biphasic oxidation in the anodic compartment (Scanlon et al. 2017).

Not surprisingly, the optical properties of metal NPs confined at LLI are different than in the bulk of both phases. It was calculated that absorption spectra of AuNPs exhibit two localized surface plasmon resonance (LSPR) peaks (Yang et al. 2014). Their position and magnitude depend on dielectric properties of both phases and NPs' position with respect to interface. The enhancement of surface-enhanced Raman spectroscopy (SERS) signal for Au nanospheres dimers was calculated in the range of 10^7–10^9. Adsorption of AgNPs at water/DCB interface was detected by SERS due to the enhancement of the signal related to a large organic cation soluble in organic phase (Booth et al. 2014). From the change of the magnitude of this signal upon the application of potential difference, the effect of potential on AgNPs' adsorption at ITIES was deduced. This is another proof of earlier discovered potential dependence of metallic NPs' adsorption at ITIES (Abid et al. 2007).

12.8 Conclusions

It was demonstrated that a wide range of NPs and other nano-objects can be confined to polarized interfaces. Moreover, the possibility of external control across ITIES allows to produce nanoparticulate metallic film directly at interface. Not only metallic NPs, but also the nano-objects prepared from inorganic materials containing more abundant non-precious metals such as Mo, Co, Ni, V, Cu, W and Bi were assembled at ITIES.

Apart from fundamental studies, research in this area was oriented toward the application of nanoparticulate film confined to ITIES in electrocatalysis. Metallic NPs (predominantly AuNPs) are employed as conductive paths for electron exchange between reagents present in different phases. On the other hand, the surface properties of nanomaterials not containing precious metals were explored, predominantly for hydrogen evolution. Particularly promising are studies of photodriven reactions at nanoparticulate film assembled at ITIES. Other important area is electrovariable optics at ITIES. Employment of ITIES-based systems for sensing seems to be promising, whereas slow response of optical properties of nanoparticulate film on external electric field is a strong obstacle for applications.

References

Abid, M., Abid, J.P., Bauer, C., Girault, H.H., Brevet, J.P., Krtil, P., Samec, Z. 2007. Reversible voltage-induced assembly of Au nanoparticles at liquid/liquid interfaces. *J. Phys. Chem. C.* 111: 8849–8855.

Akin, I., Aslan, E., Patir, I.H. 2017. Enhanced hydrogen evolution catalysis at the liquid/liquid interface by Ni_xS_y and Ni_xS_y/carbon nanotube catalysts. *Eur. J. Inorg. Chem.* 2017: 3961–3966.

Anker, J.N., Hall, W.P., Lyandres, O., Shah, N.C., Zhao, J., Van Duyne, R.P. 2008. Biosensing with plasmonic nanosensors. *Nat. Mater.* 7: 442–453.

Aslan, E., Akin, I., Patir, I.H. 2016. Enhanced hydrogen evolution catalysis based on Cu nanoparticles deposited on carbon nanotubes at the liquid/liquid interface. *ChemCatChem* 8: 719–723.

Aslan, E., Akin, I., Patir, I.H. 2015a. Highly active cobalt sulfide/carbon nanotube catalyst for hydrogen evolution at soft interfaces. *Chem. Eur. J.* 21: 5342–5349.

Aslan, E., Patir, I.H., Ersoz, M. 2014. Catalytic hydrogen evolution by tungsten disulfide at liquid-liquid interfaces. *ChemCatChem* 6: 2832–2835.

Aslan, E., Patir, I.H., Ersoz, M. 2015b. Cu nanoparticles electrodeposited at liquid–liquid interfaces: A highly efficient catalyst for the hydrogen evolution reaction. *Chem. Eur. J.* 21: 4585–4589.

Aveyard, R., Clint, J.H., Ness, D., Paunov, V.N. 2000. Compression and structure of monolayers of charged latex particles at air/water and octane/water interfaces. *Langmuir* 16: 1969–1979.

Bian, X., Scanlon, M.D., Wang, S., Liao, L., Tang, Y., Liu, B., Girault, H.H. 2013. Floating conductive catalytic nano-rafts at soft interfaces for hydrogen evolution. *Chem. Sci.* 4: 3432–3441.

Bishop, K.J.M., Wilmer, C.E., Soh, S., Grzybowski, B.A. 2009. Nanoscale forces and their uses in self-assembly. *Small* 5: 1600–1630.

Böker, A., He, J., Emrick, T., Russell, T.P. 2007. Self-assembly of nanoparticles at interfaces. *Soft Matter* 3: 1231–1248.

Booth, S.G., Cowcher, D.P., Goodacre, R., Dryfe, R.A.W. 2014. Electrochemical modulation of SERS at the liquid/liquid interface. *Chem. Comm.* 50: 4482–4484.

Booth, S.G., Dryfe, R.A.W. 2015. Assembly of nanoscale objects at the liquid/liquid interface. *J. Phys. Chem. C* 119: 23295–23309.

Booth, S.G., Chang, S.Y., Uehara, A., La Fontaine, C., Cibin, G., Schroeder, S.L.M., Mosselmans, F.W., Dryfe, R.A.W. 2017. In situ XAFS study of palladium electrodeposition at the liquid/liquid interface. *Electrochim. Acta* 235: 251–261.

Bresme, F., Oettel, M. 2007. Nanoparticles at fluid interfaces. *J. Phys. Condens. Matter* 19: 413101 (33 pp).

Brust, M., Walker, M., Bethell, D., Schiffrin, D.J., Whyman, R. 1994. Synthesis of thiol-derivatised gold nanoparticles in a two-phase liquid–liquid system. *J. Chem. Soc. Chem. Commun.* 801.

Camara, C.I., Monzon, L.M.A., Coey, J.M.D., Yudi, L.M. 2015. Assembly of magnetic nanoparticles at a liquid/liquid interface. Catalytic effect on ion transfer process. *J. Electroanal. Chem.* 756: 77–83.

Chen, J., Chan, Y.-H., Yang, T., Wark, S.E., Son, D.H., Batteas, J.D. 2009. Spatially selective optical tuning of quantum dot thin film luminescence. *J. Am. Chem. Soc.* 131: 18204–18205.

Cheng, Y., Schiffrin, D.J. 1996. Electrodeposition of metallic gold clusters at the water/1,2-dichloroethane interface. *J. Chem. Soc. Faraday Trans.* 92: 3865–3871.

Coe-Sullivan, S., Steckel, J.S., Woo, W.-K., Bawendi, M.G., Bulović, V. 2005. Large-area ordered quantum-dot monolayers via phase separation during spin-casting. *Adv. Funct. Mater.* 15: 1117–1124.

Danov, K.D., Kralchevsky, P.A., Naydenov, B.N., Brenn, G. 2005. Interactions between particles with an undulated contact line at a fluid interface: Capillary multipoles of arbitrary order. *J. Colloid Interface Sci.* 287: 121–134.

Dryfe, R.A.W., Uehara, A., Booth, S.G. 2014. Metal deposition at the liquid–liquid interface. *Chem. Rec.* 14: 1013–1023.

Edel, J.B., Kornyshev, A.A, Kucernak, A.R., Urbakh, M. 2016. Fundamentals and applications of self-assembled plasmonic nanoparticles at interfaces. *Chem. Soc. Rev.* 45, 1581–1596.

Fang, P.-P., Chen, S., Deng, H., Scanlon, M.D., Gumy, F., Lee, H.J., et al. 2013. Conductive gold nanoparticle mirrors at liquid/liquid interfaces. *ACS Nano* 7: 9241–9248.

Fialkowski, M., Bishop, K.J.M., Klajn, R., Smoukov, S.K., Campbell, C.J., Grzybowski, B.A. 2006. Principles and implementations of dissipative (dynamic) self-assembly. *J. Phys. Chem. B* 110: 2482–2496.

Flatté, M.E., Kornyshev, A.A., Urbakh, M. 2008. Understanding voltage-induced localization of nanoparticles at a liquid–liquid interface. *J. Phys. Condens. Matter* 20: 073102.

Garbin, V., Crocker, J.C., Stebe, K.J. 2012. Nanoparticles at fluid interfaces: Exploiting capping ligands to control adsorption, stability and dynamics. *J. Colloid Interface Sci.* 387: 1–11.

Ge, P., Scanlon, M.D., Peljo, P., Bian, X., Vrubel, H., O'Neill, A., et al. 2012. Hydrogen evolution across nano-Schottky junctions at carbon supported MoS_2 catalysts in biphasic liquid systems. *Chem. Commun.* 48: 6484–6486.

Giner-Casares, J.J., Reguera, J. 2016. Directed self-assembly of inorganic nanoparticles at air/liquid interfaces. *Nanoscale* 8: 16589–16595.

Girault, H.H. 2010. Electrochemistry at liquid-liquid interfaces. In: *Electroanalytical Chemistry*. Eds. A.J. Bard and C.G. Zoski, Vol. 23, 1–104. Boca Raton, FL: CRC University Press.

Grunder, Y., Fabian, M.D., Booth, S.G., Plana, D., Fermín, D.J., Hill, P.I., Dryfe, R.A.W. 2013. Solids at the liquid–liquid interface: Electrocatalysis with pre-formed nanoparticles. *Electrochim. Acta* 110: 809–815.

Grunder, Y., Ho, H.L.T., Mosselmans, F.W., Schroeder, S.L.M., Dryfe, R.A.W. 2011. Inhibited and enhanced nucleation of gold nanoparticles at the water|1,2-dichloroethane interface. *Phys. Chem. Chem. Phys.* 13: 15681–15689.

Grunder, Y., Ramasse, Q.M., Dryfe, R.A.W. 2015. A facile electrochemical route to the preparation of uniform and monoatomic copper shells for gold nanoparticles. *Phys. Chem. Chem. Phys.* 17: 5565–5568.

Grzelczak, M., Vermant, J., Furst, E.M., Liz-Marzan, L.M. 2010. Directed self-assembly of nanoparticles. *ACS Nano* 4: 3591–3605.

Gschwend, G., Peljo, P., Smirnov, E., Girault, H.H. 2017. Electrovariable gold nanoparticle films at liquid-liquid interfaces: From redox electrocatalysis to Marangoni-shutters. *Faraday Discuss.* 199: 565–583.

Guanazzi, M., Silvestri, G., Serravalle, G. 1975. Electrochemical metallization at the liquid-liquid interfaces of non-miscible electrolytic solutions. *J. Chem. Soc. Chem. Commun.* 200–202.

Guo, J.D., Tokimoto, T., Othman, R., Unwin, P.R. 2003. Formation of mesoscopic silver particles at micro and nano-liquid/liquid interfaces *Electrochem. Commun.* 5: 1005–1010.

Gur, I., Fromer, N.A., Geier, M.L., Alivisatos, A.P. 2005. Air-stable all-inorganic nanocrystal solar cells processed from solution. *Science* 310: 462–465.

Haynes, C.L., Van Duyne, R.P. 2001. Nanosphere lithography: A versatile nanofabrication tool for studies of size-dependent nanoparticle optics. *J. Phys. Chem. B* 105: 5599–5611.

Hirunpinyopas, W., Rodgers, A.N.J.S., Worrall, D., Bissett, M.A., Dryfe, R.A.W. 2017. Hydrogen evolution at liquid/liquid interfaces catalyzed by 2D materials. *ChemNanoMat* 3: 428–435.

Hojeij, M., Younan, N., Ribeaucourt, L., Girault, H.H. 2010. Surface plasmon resonance of gold nanoparticles assemblies at liquid vertical bar liquid interfaces. *Nanoscale* 2: 1665–1669.

Hurd, A.J. 1985. The electrostatic interaction between interfacial colloidal particles. *J. Phys. A: Math. Gen.* 18: L1055.

Izquierdo, D., Martinez, A., Heras, A., Palacio, J.L., Ruiz, V., Dryfe, R.A.W., Colina, A. 2012. Spatial scanning spectroelectrochemistry. Study of the electrodeposition of Pd nanoparticles at the liquid/liquid interface. *Anal. Chem.* 84: 5723–5730.

Janchenova, H., Stulik, K., Marecek, V. 2006. Preparation of a silicate membrane at a liquid/liquid interface and its doping with a platinum ion. *J. Electroanal. Chem.* 591: 41–45.

Johans, C., Lahtinen, R., Kontturi, K., Schiffrin, D.J. 2000. Nucleation at liquid | liquid interfaces: Electrodeposition without electrodes. *J. Electroanal. Chem.* 488: 99–109.

Kakinami, T., Nishi, N., Amano, K., Sakka, T. 2016. Preparation of dendritic gold nanofibers using a redox reaction at the interface between an ionic liquid and water: Correlation between viscosity and nanostructure. *Bunseki Kagaku* 65: 157–161.

Kaminska, I., Jonsson-Niedziolka, M., Kaminska, A., Pisarek, M., Holyst, R., Opallo, M., Niedziolka-Jonsson,

J. 2012. Electrodeposition of well-adhered multifarious Au particles at a SolidlToluenelAqueous electrolyte three-phase junction. *J. Phys. Chem. C* 116: 22476–22485.

Kaminska, I., Niedziolka-Jonsson, J., Roguska, A., Opallo, M. 2010. Electrodeposition of gold nanoparticles at a solid vertical bar ionic liquid vertical bar aqueous electrolyte three-phase junction. *Electrochem. Commun.* 12: 1742–1745.

Kiely, C.J., Fink, J., Zheng, J.G., Brust, M., Bethell, D., Schiffrin, D.J. 2000. Ordered colloidal nanoalloys. *Adv. Mater.* 12: 640–643.

Kinge, S., Crego-Calama, M., Reinhoudt, D.N. 2008. Self-assembling nanoparticles at surfaces and interfaces. *ChemPhysChem* 9: 20–42.

Kolasinski, K.W. 2006. Catalytic growth of nanowires: Vapor–liquid–solid, vapor–solid–solid, solution–liquid–solid and solid–liquid–solid growth. *Curr. Opin. Solid State Mater. Sci.* 10 : 182–191.

Lee, S., Drwiega, J., Wu, C.-Y., Mazyck, D., Sigmund, W.M. 2004. Anatase TiO_2 nanoparticle coating on barium ferrite using titanium bis-ammonium lactato dihydroxide and its use as a magnetic photocatalyst. *Chem. Mater.* 16: 1160–1164.

Lepkova, K., Clohessy, J., Cunnane, V.J. 2007. The pH-controlled synthesis of a gold nanoparticle/polymer matrix via electrodeposition at a liquid–liquid interface. *J. Phys. Condens. Matter* 19: 375106.

Lepkova, K., Clohessy, J., Cunnane, V.J. 2008. Electrodeposition of metal-based nanocomposites at a liquid–liquid interface controlled via the interfacial Galvani potential difference. *Electrochim. Acta* 19: 6273–6277.

Lesniewski, A., Paszewski, M., Opallo, M. 2010. Gold–carbon three dimensional film electrode prepared from oppositely charged conductive nanoparticles by layer-by-layer approach. *Electrochem. Commun.* 12: 435–437.

Lin, Y., Böker, A., Skaff, H., Cookson, D., Dinsmore, A.D., Emrick, T., Russel, T.P. 2005. Nanoparticle assembly at fluid interfaces: Structure and dynamics. *Langmuir* 21: 191–194.

Lin, X.M., Jaeger, H.M., Sorensen, C.M., Klabunde, K.J. 2001. Formation of long-range-ordered nanocrystal superlattices on silicon nitride substrates. *J. Phys. Chem. B* 105: 3353–3357.

Lin, Y., Skaff, H., Böker, A., Dinsmore, A.D., Emrick, T., Russel, T.P. 2003a. Ultrathin cross-linked nanoparticle membranes. *J. Am. Chem. Soc.* 125: 12690–12691.

Lin, Y., Skaff, H., Emrick, T., Dinsmore, A.D., Russel, T.P. 2003b. Nanoparticle assembly and transport at liquid-liquid interfaces. *Science* 229: 226–229.

Marinescu, M., Urbakh, M., Kornyshev, A.A. 2012. Voltage-dependent capacitance of metallic nanoparticles at a liquid/liquid interface. *Phys. Chem. Chem. Phys.* 14: 1471–1480.

Martinez, A.C., Rio, E., Delon, G., Saint-Jalmes, A., Langevin, D., Binks, B.P. 2008. On the origin of the remarkable stability of aqueous foams stabilised by nanoparticles: Link with microscopic surface properties. *Soft Matter* 4: 1531–1535.

Masschaele, K., Park, B.J., Furst, E.M., Fransaer, J., Vermant, J. 2010. Finite ion-size effects dominate the interaction between charged colloidal particles at an oil-water interface. *Phys. Rev. Lett.* 105: 048303.

Mendez, M., Partovi-Nia, R., Hatay, I., Su, B., Ge, P., Olaya, A., Younan, N., Hojeij, M., Girault, H.H. 2010. Molecular electrocatalysis at soft interfaces. *Phys. Chem. Chem. Phys.* 12: 15163–15171.

Michael, D., Benjamin, I. 1998. Molecular dynamics of the water|nitrobenzene interface. *J. Electroanal. Chem.* 450: 335–345.

Min, Y., Akbulut, M., Kristiansen, K., Golan, Y., Israelachvili, J. 2008. The role of interparticle and external forces in nanoparticle assembly. *Nat. Mater.* 7: 527–538.

Ohara, P.C., Leff, D.V., Heath, J.R., Gelbart, W.M. 1995. Crystallization of opals from polydisperse nanoparticles. *Phys. Rev. Lett.* 75: 3466–3469.

Montelongo, Y., Sikdar, D., Ma, Y., McIntosh, A.J.S., Velleman, L., Kucernak, A.R., Edel, J.B., Kornyshev, A.A. 2017. Electrotunable nanoplasmonic liquid mirror. *Nat. Mater.* 16: 1127–1135.

Nieminen, J.J., Hatay, I., Ge, P.Y., Mendez, M.A., Murtomaki, L., Girault, H.H. 2011. Hydrogen evolution catalyzed by electrodeposited nanoparticles at the liquid/liquid interface. *Chem. Commun.* 47: 5548–5550.

Nishi, N., Kakinami, T., Sakka, T. 2015. Dendritic nanofibers of gold formed by the electron transfer at the interface between water and a highly hydrophobic ionic liquid. *Chem. Commun.* 51: 13638–13641.

Ozel, F., Aslan, E., Sarlimaz, E., Patir, I.H. 2016. Hydrogen evolution catalyzed by Cu_2WS_4 at liquid-liquid interfaces. *ACS Appl. Mater. Interfaces* 8: 25881–25887.

Ozel, F., Yar, A., Aslan, E., Arkan, E., Abdalaziz, A., Can, M., Patir, I.H., Kus, M., Ersoz, M. 2015. Earth-abundant Cu_2CoSnS_4 nanofibers for highly efficient H_2 evolution at soft interfaces. *ChemNanoMat* 8: 477–481.

Paul, S., Pearson, C., Molloy, A., Cousins, M.A., Green, M., Kolliopoulou, S., Dimitrakis, P., Normand, P., Tsoukalas, D., Petty, M.C. 2003. Langmuir–Blodgett film deposition of metallic nanoparticles and their application to electronic memory structures. *Nano Lett.* 3: 533–536.

Paunov, V.N., Kralchevsky, P.A., Denkov N.D., Nagayama, K. 1993. Lateral capillary forces between floating submillimeter particles. *J. Colloid Interface Sci.* 157: 100–112.

Peljo, P., Scanlon, M.D, Olaya, A.J., Rivier, L., Smirnov, E., Girault, H.H. 2017. Redox electrocatalysis of floating nanoparticles: Determining electrocatalytic properties without the influence of solid supports. *J. Phys. Chem. Lett.* 8: 3564–3575.

Pieranski, P. 1980. Two-dimensional interfacial colloidal crystals. *Phys. Rev. Lett.* 45: 569.

Platt, M., Dryfe, R.A.W. 2007. Electrodeposition at the liquid/liquid interface: The chronoamperometric response as a function of applied potential difference. *J. Electroanal. Chem.* 599: 323–332.

Platt, M., Dryfe, R.A.W., Roberts, E.P.L. 2002. Controlled deposition of nanoparticles at the liquid–liquid interface. *Chem. Commun.* 2324–2325.

Platt, M., Dryfe, R.A.W., Roberts, E.P.L. 2007. Electrodeposition of palladium nanoparticles at the liquid–liquid interface using porous alumina templates. *Electrochim. Acta.* 48: 3037–3046.

Poltorak, L., Gamero-Quijano, A., Herzog, G., Walcarius, A. 2017. Decorating soft electrified interfaces: From molecular assemblies to nano-objects. *Appl. Mater. Today* 9: 533–550.

Prevo, B.G., Hwang, Y., Velev, O.D. 2005. Convective assembly of antireflective silica coatings with controlled thickness and refractive index. *Chem. Mater.* 17: 3642–3651.

Rabiu, A.K., Toth, P.S., Rodgers, A.N.J., Dryfe, R.A.W. 2017. Electrochemical Investigation of adsorption of single-wall carbon nanotubes at a liquid/liquid interface. *Chem. Open* 6: 57–63.

Rahtinen, R.M., Fermin, D.J., Jensen, H., Kontturi, K., Girault, H.H. 2000. Two-phase photocatalysis mediated by electrochemically generated Pd nanoparticles. *Electrochem. Commun.* 2: 230–234.

Rastgar, S., Pilarski, M., Wittstock, G. 2016. Polarized liquid-liquid interface meets visible light-driven catalytic water oxidation. *Chem. Commun.* 52: 11382–11385.

Rastgar, S., Wittstock, G. 2017. Polarized liquid-liquid interface meets visible light-driven catalytic water oxidation. *J. Phys. Chem. C* 121: 25941–25948.

Rastgar, S., Wittstock, G. 2018. In situ microtitration of intermediates of water oxidation reaction at nanoparticles assembled at water/oil interfaces. *J. Phys. Chem. C* 122: 12963–12969.

Reincke, F., Kegel, W.K., Zhang, H., Nolte, M., Wang, D., Vanmaekelbergh, D., Möhwald, H. 2006. Understanding the self-assembly of charged nanoparticles at the water/oil interface. *Phys. Chem. Chem. Phys* 8: 3828–3835.

Rodgers, A.N.J., Booth, S.G., Dryfe, R.A.W. 2014. Particle deposition and catalysis at the interface between two immiscible electrolyte solutions (ITIES): A mini-review. *Electrochem. Commun.* 47: 17–20.

Samec, Z. 1988. Electrical double layer at the interface between two immiscible electrolyte solutions. *Chem. Rev.* 88: 617–632.

Samec, Z. 2004. Electrochemistry at the interface between two immiscible electrolyte solutions. *Pure Appl. Chem.* 76: 2147–2180.

Samec, Z., Langmaier, J., Kakiuchi, T. 2004. Charge transfer processes at the interface between hydrophobic ionic liquid and water. *Pure Appl. Chem.* 81: 1473–1488.

Sashuk, V., Winkler, K., Żywociński, A., Wojciechowski, T., Górecka, E., Fiałkowski, M. 2013. Nanoparticles in a capillary trap: Dynamic self-assembly at fluid interfaces. *ACS Nano* 7: 8833–8839.

Scanlon, M.D., Bian, X., Vrubel, H., Amstutz, V., Schenk, K., Hu, X., Liu, B., Girault, H.H. 2013. Low-cost industrially available molybdenum boride and carbide as "platinum-like" catalysts for the hydrogen evolution reaction in biphasic liquid systems. *Phys. Chem. Chem. Phys.* 15: 2847–2857.

Scanlon, M.D., Peljo, P., Mendez, M.A., Smirnov, E.H., Girault, H.H. 2016. Charging and discharging at the nanoscale: Fermi level equilibration of metallic nanoparticles. *Chem. Sci.* 6: 2705–2720.

Scanlon, M.D., Peljo, P., Rivier, L., Vrubel, H., Girault, H.H. 2017. Mediated water electrolysis in biphasic systems. *Phys. Chem. Chem. Phys.* 19: 22700–22710.

Scanlon, M.D., Stockmann, T.J., Peljo, P. 2018. Gold nanofilms at liquid–liquid interfaces: An emerging platform for redox electrocatalysis, nanoplasmonic sensors, and electrovariable optics. *Chem. Rev.* 111: 3722–3751.

Sefer, B., Gulaboski, R., Mirceski, V. 2012. Electrochemical deposition of gold at liquid–liquid interfaces studied by thin organic film-modified electrodes. *J. Solid State Electrochem.* 16: 2373–2381.

Shen, G.X., Chen, Y.C., Lin, C.J. 2005. Corrosion protection of 316 L stainless steel by a TiO_2 nanoparticle coating prepared by sol–gel method. *Thin Solid Films* 489: 130–136.

Sikdar, D., Kornyshev, A.A. 2016. Theory of tailorable optical response of two-dimensional arrays of plasmonic nanoparticles at dielectric interfaces. *Sci. Rep.* 6: 33712.

Smirnov, E., Scanlon, M.D., Momotenko, D., Vrubel, H., Mendez, M.A., Brevet, P.-F., Girault, H.H. 2014. Gold metal liquid like droplets. *ACS Nano* 8: 9471–9481.

Smirnov, E., Peljo, P., Scanlon, M.D., Girault, H.H. 2015. Interfacial redox catalysis on gold nano films at interfaces. *ACS Nano* 9: 9565–9665.

Smirnov, E., Peljo, P., Scanlon, M.D., Gumy, F., Girault, H.H. 2016. Self-healing gold mirrors and filters at liquid-liquid interface. *Nanoscale* 8: 7723–7737.

Smirnov, E., Peljo, P., Girault, H.H. 2017a. Self-assembly and redox induced phase transfer of gold nanoparticles at a water–propylene carbonate interface. *Chem. Comm.* 53: 4108–4111.

Smirnov, E., Peljo, P., Scanlon, M.D., Girault, H.H. 2017b. Gold nanofilm redox catalysis for oxygen reduction at soft interfaces. *Electrochim. Acta* 197: 362–373.

Stamou, D., Duschl, C., Johannsmann, D. 2000. Long-range attraction between colloidal spheres at the air-water interface: The consequence of an irregular meniscus. *Phys. Rev. E* 62: 5263.

Stockmann, T.J., Angele, L., Brasiliense, V., Combellas, C., Kanoufi, F. 2017. Platinum nanoparticle impacts at a liquid j liquid interface. *Angew. Chem. Int. Ed.* 56: 13493–13497.

Su, B., Abid, J.P., Fermin, D.J., Girault, H.H., Hoffmanova, H., Krtil, P., Samec, Z. 2004. Reversible voltage-induced assembly of Au nanoparticles at liquid|liquid interfaces. *J. Am. Chem. Soc.* 126: 915–919.

Szot, K., Jonsson-Niedziolka, M., Palys, B., Niedziolka-Jonsson, J. 2011. One-step electrodeposition of carbon-silicate sponge assisted by a three phase junction for efficient bioelectrocatalysis. *Electrochem. Commun.* 13: 566–569.

Talapin, D.V., Lee, J.-S., Kovalenko, M.V., Shevchenko, E.V. 2010. Prospects of colloidal nanocrystals for electronic and optoelectronic applications. *Chem. Rev.* 110: 389–458.

Toor, A., Feng, T., Russel, T.P. 2016. Self-assembly of nanomaterials at fluid interfaces. *Eur. Phys. J. E* 39: 57.

Toth, P.S., Rabiu, A.K., Dryfe, R.A.W. 2015a. Controlled preparation of carbon nanotube-conducting polymer composites at the polarisable organic/water interface. *Electrochem. Commun.* 60: 153–157.

Toth, P.S., Ramasse, Q.M., Velický, M., Dryfe, R.A.W. 2015b. Functionalization of graphene at the organic/water interface. *Chem. Sci.* 6: 1316–1323.

Toth, P.S., Rodgers, A.N.J., Rabiu, A.K., Dryfe, R.A.W. 2015c. Electrochemical activity and metal deposition using few-layer graphene and carbon nanotubes assembled at the liquid–liquid interface. *Electrochem. Commun.* 50: 6–10.

Trojanek, A., Langmaier, J., Kvapilova, H., Zalis, Z., Samec, Z. 2014. Inhibitory effect of water on the oxygen reduction catalyzed by cobalt(I) porphyrin. *J. Phys. Chem. A* 118: 2018–2028.

Trojanek, A., Langmaier, J., Samec, Z. 2006. Electrocatalysis of the oxygen reduction at a polarised interface between two immiscible electrolyte solutions by electrochemically generated Pt particles. *Electrochem. Commun.* 8: 475–481.

Trojanek, A., Langmaier, J., Samec, Z. 2007. Random nucleation and growth of Pt nanoparticles at the polarised interface between two immiscible electrolyte solutions. *J. Electroanal. Chem.* 599: 160–166.

Uehara, A., Booth, S.G., Chang, S.Y., Schroeder, S.L.M., Imai, T., Hashimoto, T. et al. 2015. Electrochemical insight into the Brust–Schiffrin synthesis of Au nanoparticles. *J. Am. Chem. Soc.* 137: 15135–15144.

Uehara, A., Chang, S.Y., Booth, S.G., Schroeder, S.L.M., Mosselmans, F.W., Dryfe, R.A.W. 2016. Redox and ligand exchange during the reaction of tetrachloroaurate with hexacyanoferrate(II) at a liquid-liquid interface: Voltammetry and X-ray absorption fine-structure studies. *Electrochim. Acta* 137: 997–1006.

Wang, L., Xu, L., Kuang, K., Xu, C., Kotov, N.A. 2012. Dynamic nanoparticle assemblies. *Acc. Chem. Res.* 45: 1916–1926.

Whitesides, G.M., Grzybowski, B. 2002. Self-assembly at all scales. *Science* 295: 2418–2421.

Williams, D.F., Berg, J.C. 1992. The aggregation of colloidal particles at the air-water interface. *J. Colloid Interface. Sci.* 152: 218–229.

Yang, Z., Chen, S., Feng, P.-P., Ren, B., Girault, H.H., Tian, Z. 2014. LSPR properties of metal nanoparticles adsorbed at a liquid–liquid interface. *Phys. Chem. Chem. Phys.* 7: 5374–5378.

Yao, K., Huang, Q., Lu, W., Xu, A., Li, X., Zhang, H., Wang, J. 2016. A facile synthesis of gold micro/nanostructures at the interface of 1,3-dibutylimidazolium bis(trifluoromethylsulfonyl)imide and water. *J. Colloid Interface Sci.* 480: 30–38.

Younan, N., Hojeij, M., Ribeaucourt, L., Girault, H.H. 2010. Electrochemical properties of gold nanoparticles assembly at polarised liquid|liquid interfaces. *Electrochem. Commun.* 12: 912–915.

Zhu, X., Qiao, Y., Zhang, X., Zhang, S., Yin, X., Gu, J., Chen, Y., Zhu, Z., Li, M., Shao, Y. 2014. Fabrication of metal nanoelectrodes by interfacial reactions. *Anal. Chem.* 86: 7001–7008.

13

Terahertz Resonance of Nanoparticles in Water

Dao Xiang, Ali Khademi, and
Reuven Gordon
University of Victoria

13.1 Introduction

There is a long history of studying the third-order nonlinear optical response of nanoparticles. A gradient force acting on colloidal particles can produce a large optical nonlinear response (Smith et al. 1981, 1982). This response is a slow and needs wavelength-scale translation of the nanoparticles. Thus, it is not appropriate for frequency conversion or high-speed switching applications. Using the intrinsic nonlinearity of nanoparticles generates much faster responses. Silicon nanocrystals in silica (Vijayalakshmi et al. 2000, Ajgaonkar 1999) or metal inclusions (Vijayalakshmi et al. 1998, Meldrum et al. 2001) are examples of this phenomenon. However, the overall nonlinear response of these nanoparticles is still relatively small. Furthermore, using metals in this application causes absorption losses and Rayleigh scattering. Another method is the optomechanical response mediated by electrostriction, whose speed and strength of nonlinearity is between the other two regimes. It was observed that a single nanoparticles trapped in an aperture laser tweezer could exhibit a high-frequency mechanical response from electrostriction (Wheaton et al. 2015). Since this type of electrostriction does not use a plasmonic resonance to enhance the light-matter interaction, it is not as vulnerable to losses. It is analogous to what was observed for gold nanoparticles with the optical Kerr effect (Hartland 2006, Pelton et al. 2009) (even down to the single nanoparticle level (van Dijk et al. 2005, Yu et al. 2013)). Therefore, it is interesting to explore the use of electrostriction in four-wave mixing (FWM) using polystyrene and gold nanoparticles in solution.

This chapter focuses on applications of nanoparticles in enhanced optical nonlinearities. The role of acoustic resonance of nanoparticles in nonlinear optical response, where the electrostrictive force excites the acoustic modes of nanoparticles and produces a traveling periodic variation in refractive index of the sample, is investigated. The extraordinarily large optical nonlinearity is experimentally demonstrated in a continuous-wave (CW) FWM configuration to be a million times larger than typical electronic nonlinearities once a vacuum cavity is formed after the water is displaced by the oscillations. As a result, it works at low powers with low-cost lasers and is not damped out by the surrounding solution.

13.2 Background

13.2.1 Electrostriction in Light-Matter Interaction

Imposing an external electric field on a medium (a dielectric, which can be a liquid or a crystal lattice of solid) can change the pressure in the medium and deform it slightly. This phenomenon, which is happening due to interaction between external field and charges bounded in the medium, is called electrostriction (Landau and Lifshitz 1984, Delone 1993). In other words, electrostriction is a change of a material's shape under the application of an electric field due to a slight displacement of atoms. The energy in the electrostriction is proportional to the square of electric field. Thus, reversing the electric field will not reverse the direction of the deformation (Rennie and Law 2015, Cauz n.d.). The strong light field of laser radiation can be the source of external electric field required for electrostriction (Delone 1993).

It should be noted that the electrostriction and the converse piezoelectric effect are different. The converse piezoelectric effect is the property of only a particular class of nonconducting crystals and causes a much greater deformation for a given value of the electric field. Furthermore, reversal of the electric field reverses the direction of the deformation because the energy in the converse piezoelectricity is linearly proportional to the electric field (Rennie and Law 2015, Cauz n.d.).

13.2.2 Nonlinear Optics

Nonlinear optical effects (Armstrong et al. 1962) (for which Bloembergen was awarded the Nobel Prize in Physics in 1981) are typically observed only at very high intensities of light provided by lasers that are coherent light sources with a high degree of monochromaticity, high directionality, and high intensity. We are familiar with the linear optical response with a constant refractive index; that is, the induced polarization is linearly proportional to the amplitude of optical wave under weak illumination. However, if the illumination is made with high intensities, high-order terms in the power series of polarization cannot be ignored in mathematics. These nonlinear terms of polarization would work as sources in Maxwell's equations. As a result, the optical properties of materials, such as their refractive index $\boldsymbol{n}$, are varied as a function of the cycle-averaged intensity of intense laser $\boldsymbol{I}$. For example, the nonlinearity in refractive index $\Delta \boldsymbol{n} = \boldsymbol{n_2 I}$ ($\boldsymbol{n_2}$ is the nonlinear refraction coefficient) is harnessed to achieve the phase modulation and thus create the femtosecond laser pulses (Boyd 2008), which is the basis to temporally resolve the molecular dynamics (Zewail 2000).

Nonlinear optics may also play a role in high-speed digital information processing (Chraplyvy 1990, Cotter et al. 1999). Furthermore, it can determine the capacity of the digital photonic networks (Brackett et al. 1993). One example is the wavelength conversion techniques based on nonlinear effects, which can implement the wavelength translation and a high degree of wavelength reuse (Yoo 1996, D'ottavi et al. 1997), which facilitates networks to accommodate vast users. Another example is the optical switch based on nonlinear interferometer (Patel et al. 1996, Phillips et al. 1998, Robinson et al. 2002) or pulsed FWM (Andrekson et al. 1991, Koos et al. 2009).

These applications desire a large nonlinear coefficient and low optical attenuation (Stegeman and Miller et al. 1993). The optical fiber is the transparent material for the telecommunication band, but its optical nonlinearity is quite small ($n_2 \sim 10^{-20}$ m^2/W) (Agrawal 2007). The interband nonlinear coefficient of semiconductor optical amplifier is larger but with a slow carrier relaxation time in the order of hundreds picoseconds (Diez et al. 1997), which limits the fastest bit rate.

Metallic nanoparticles have long been considered as alternative nonlinear optical materials for their third-order nonlinear optical response enhanced by the surface plasmon resonance (Whelan et al. 2004, Liu et al. 2006, Danckwerts and Novotny 2007), but this application introduces absorption losses in addition to Rayleigh scattering. Large optical nonlinear responses can also be obtained by phase transitions (Khoo 2009) or gradient forces on nanoparticles (Smith et al. 1981, 1982). But they are all very slow processes, due to the slow response of phase transitions or transport of nanoparticles over macroscopic distances. As a result, they are not suitable for applications in high-frequency-shift wavelength conversion.

13.2.3 Four-Wave Mixing

FWM is a nonlinear effect, whereby interactions between three electromagnetic waves in a dielectric produce a new wave. The fields of first and second waves cause the polarization of the dielectric. Then, these two waves interfere with each other, which lead to harmonics in the polarization at the sum and difference frequencies. Finally, the field of third wave will also drive the polarization. As a result, third wave will beat with two other input waves as well as the sum and difference frequencies. This beating generates the fourth field in FWM. To model this nonlinear response, the induced polarization $\vec{P}$ can be expanded as a power series in the electric field strength $\vec{E}$, as follows:

$$\vec{P} = \chi^{(1)} \cdot \vec{E} + \chi^{(2)} \cdot \vec{E}\vec{E} + \chi^{(3)} \cdot \vec{E}\vec{E}\vec{E} + \dots \qquad (13.1)$$

where $\chi^{(1)}$, $\chi^{(2)}$, and $\chi^{(3)}$ are tensors of first-, second-, and third-order nonlinear susceptibility. $\chi^{(3)}$ is responsible for FWM processes (Bloembergen, 1979).

The third-order nonlinear polarization, which describes a coupling between four waves, can be written as:

$$P_i(\omega_4, \vec{r}) = \frac{1}{2}\chi^{(3)}_{ijkl}(-\omega_4, \omega_1, -\omega_2, \omega_3)\, E_j(\omega_1)\, E_k^*(\omega_2) \times E_l(\omega_3)\, e^{i(\vec{k_1} - \vec{k_2} + \vec{k_3})\vec{r} - i\omega_4 t} + c.c. \qquad (13.2)$$

It should be noted that energy and momentum conservation are necessary for efficient coupling between the four waves, which means $\omega_4 = \omega_1 - \omega_2 + \omega_3$ and $\vec{k_4} = \vec{k_1} - \vec{k_2} + \vec{k_3}$ (Bloembergen, 1980).[1]

13.2.4 Acousto-Optics

Acousto-optics is a branch of physics that deals with the interaction of sound and light. The diffraction of light by acoustic waves of short wavelengths was first postulated by Léon Brillouin in 1922 (Brillouin 1922). The first experimental demonstrations were performed in 1932, by Lucas and Biquard in France (Lucas and Biquard 1932) and Debye and Sears in the United States (Debye and Sears 1932). The physical basis of this diffraction phenomenon is the fact that the moving wave fronts of acoustic waves produce the density modulation that in turn writes a traveling index grating (the traveling periodic changes in the refractive index of medium), which scattered the light. The frequency of scattered light is actually shifted because of the Doppler shift associated with the acoustic velocity. The phenomenon of frequency shifting constitutes the basis of heterodyning

[1] Having a small mismatch is equivalent to the second condition $((\vec{k_1} - \vec{k_2} + \vec{k_3} - \vec{k_4}).\vec{L} = \vec{\Delta k}.\vec{L} \approx 0)$.

techniques in modern signal processing applications such as optical scanners (Dixon 1967), spatial light modulators (Abrams and Pinnow 1970), radio frequency (RF) pulse compressors (Kino and Matthews 1971), and programmable optical interconnectors (Cronin-Golomb 1989). In addition, the acousto-optic tunable filter based on the acousto-optic grating of electronically tunable frequency covers a wide range of applications, for example, the very rapid optical spectrum analysis (Chang 1981), the tunable laser (Taylor et al. 1971), and the wavelength-division multiplexing and demultiplexing in optical communication systems (Smith et al. 1990). In the time-domain signal processing, the acousto-optic modulator was developed to modulate the intensity of optical beams based on the dependence of diffraction efficiency on the strength of sound, paving the way to various applications, including the loss modulation to Q-switched or mode-locked lasers (Delgado-Pinar et al. 2006), and the time-domain convolution and correlation of wide-band RF signals in signal processing systems (Rhodes 1981). However, the electronic tunability between hundreds kilohertz and a few gigahertz and the complex modulation circuit limit the application of state-of-the-art commercial acousto-optic devices toward the future high-speed optical communication.

There has been much research at the boundary between nonlinear optics and acousto-optics to try to avoid those aforementioned bottlenecks. A typical faster approach, up to ~10 GHz, is the stimulated Brillouin scattering, which is extensively used in phononic band mapping (Maldovan 2013), stored light (Zhu et al. 2007), and microwave signal processing (Tomes and Carmon 2009). However, there are still challenges for higher frequency conversion in the range from 20 GHz to several terahertz frequencies, which is the target of all-optical data processing, faster than what is easily achievable with electronics alone.

13.2.5 Acoustic Vibrational Modes of Nanoparticles

Conduction electrons in the metal nanoparticles can absorb light from a pump laser pulse and impulsively transfer their excess energy to the lattice. This leads to a rapid rise of the lattice temperature and excites the phonon modes of metal nanoparticle. As a result, metal nanoparticle will expand, which causes a nearly spontaneous excitation of the acoustic vibrations (Pelton et al. 2009, Hartland 2006). Another mechanism for acoustic vibrations of metal nanoparticles is the electrostriction, which is the subject of this chapter and was described in Section 13.2.1.

For nanoparticles, two different modes of vibration can be considered: breathing and accordion. In the breathing mode ($l = 0$), nanoparticles reserve their sphere shape and their deformation occurs via the symmetric elongation of diameter in all axes. In the accordion mode ($l = 2$), nanoparticles deform from sphere to a spheroid and elongation of diameter occurs just in one axis (Xiang et al. 2016, Xiang and Gordon, 2016a).

13.2.6 Acoustic Cavitation

The formation and implosive collapse of bubbles under the influence of an intense sound is called acoustic cavitation (McNamara et al. 1999, Ashokkumar 2011). A cavity (i.e., a vacuum layer) can be formed around a vibrating nanoparticle in a liquid. When frequencies of oscillation are higher than the characteristic vibrational (Einstein) frequency of the fluid, even small-amplitude vibrations can form a cavity around nanoparticles. This cavity formation can intensify the quality factor of the oscillations significantly (Hsueh et al. 2018).

13.3 Nanoparticle Acoustic Resonance Enhanced Four-Wave Mixing

13.3.1 Introduction

Since light does not interact strongly with matter, the nonlinear optical response of materials is typically weak. For example, the nonlinear refractive index, n_2, is $\sim 2 \times 10^{-8}$ cm^2/MW in the hundreds of terahertz (for FWM applications) (Lin et al. 2007). This means that applications such as FWM wavelength conversion require high powers. The nonlinear responses for phase transitions (Khoo 2009), low-dimensional materials (Miller et al. 1983, Feuerbacher et al. 1991, Yang et al. 1994, Akiyama et al. 2000, Renger et al. 2009, 2010, Zhang et al. 2011, 2013, Kauranen and Zayats 2012), and gradient forces on nanoparticles (Smith et al. 1981, 1982), with responses of $\sim 3.6 \times 10^{-3}$ cm^2/MW, are much stronger. However, the slow response of phase transitions or transport of nanoparticles over macroscopic distances makes them very slow processes, which are not appropriate for applications like high-frequency-shift wavelength conversion. Stimulated Brillouin scattering (Ippen and Stolen 1972, Zhu et al. 2007) and nonlinear mixing in semiconductor amplifiers and waveguides (Geraghty et al. 1997, Fukuda et al. 2005, Dadap et al. 2008, Salem et al. 2008, Leuthold et al. 2010, Liu et al. 2010) are faster approaches, up to ~10 GHz. However, the higher range of above 20 GHz to several terahertz range is the target of all-optical data processing because it is faster than what is easily achievable with electronics alone. For frequency conversion in this range, processes that involve electrostriction can be considered.

While molecular vibrations typically produce a relatively weak response of stimulated Raman scattering, large-scale vibrations of nanoparticles can generate strong Raman response (Fujii et al. 1990, Nie and Emory 1997, Campion and Kambhampati 1998, Cao et al. 2002, Yadav et al. 2006). Past investigations about nanoparticles have been limited to the use of high peak intensity ultrafast lasers due to the fact that damping from the surrounding medium limits the nonlinear response (Bigot et al. 1995, Itoh et al. 2001, Muskens et al. 2006, Dhar et al. 1994, Ruhman et al. 1987, Zijlstra et al. 2008, Crut et al. 2015, Mongin et al. 2011,

Chakraborty et al. 2013, O'Brien et al. 2014). These oscillations are usually extremely damped in solid materials and provide only a small modification in the optical scattering (Bragas et al. 2004). High peak power femtosecond lasers are still needed for similar experiments in solution to have quality factors on the order of 10–100 and an appreciable damping (Pelton et al. 2009). The application of high-sensitivity sensors reaching the terahertz range, which demands the sufficiently high-quality factor, intrigued significant research interest in the nanoscale systems (Juvé et al. 2010). Likewise, study of the terahertz dynamics of proteins in solution (Turton et al. 2014), with observed quality factors below ten, has been performed using femtosecond optical Kerr effect spectroscopy (the ultrafast version of FWM).

Measurements of the FWM of polystyrene and gold nanoparticles in aqueous solution were encouraged by recent reports on large acoustic vibration responses of individual nanoparticles and proteins in an optical tweezer setup (Wheaton et al. 2015). In these measurements, even a weak laser can produce extraordinarily strong FWM signal above a critical threshold intensity, which means extremely high-quality factors for the nanoparticle vibrations, as if the water damping disappears entirely. This section explains these measurements, which were already published by authors in Refs (Xiang and Gordon, 2016a,b, Xiang et al. 2016).

13.3.2 Derivation of Four-Wave Mixing Induced by Electrostriction

The optomechanical responses are associated with the accordion mode ($l = 2$) and the breathing mode ($l = 0$) caused by the traveling electrostrictive forces exerting on individual nanoparticle. First, the nonlinear response for accordion mode will be explained assuming that the polarization direction of applied light is along y-axis and stretching of the particle does alter the volume of individual nanoparticle. Considering the elongation force, which is excited by the y-polarized optical wave, slightly stretches a nanosphere with a diameter of a via the elongation of major axis (y-axis) from $2a$ to $2(a + \Delta y)$, the time-average polarization energy change of individual bead can be calculated as (Jackson 1999)

$$\begin{aligned}\Delta U &= \mathrm{Re}\,(\alpha_s - \alpha_{e,a})\left\langle \tilde{E}^2\right\rangle_T \\ &= V_p\varepsilon_0 n_s^2 \mathrm{Re}\left(3\frac{m^2-1}{m^2+2} - \frac{m^2-1}{1+N_y(m^2-1)}\right)\left\langle \tilde{E}^2\right\rangle_T \end{aligned} \tag{13.3}$$

where V_p is the particle volume; α_s and $\alpha_{e,a}$ are the particle polarizability for nanosphere and that of nanospheroid, respectively; $m = n_p/n_s$ is the ratio of the refractive index of the particle (n_p) to that of the surrounding solvent (n_s); N_y is the depolarization factor of the ellipsoid in the polarization direction and is given by (Sihvola 1999)

$$N_y = \frac{1-e^2}{2e^3}\left(\ln\frac{1+e}{1-e} - 2e\right) \approx \frac{1}{3} - \frac{2}{15}e^2 \approx \frac{1}{3} - \frac{2}{5}\frac{\Delta y}{a} \tag{13.4}$$

where e is the eccentricity of spheroid. Therefore, the electrostrictive force that stretched the sphere is

$$\begin{aligned}F_{e,a} &= -\frac{\Delta U}{\Delta y} \approx \frac{24}{5}\pi a^2\varepsilon_0 n_s^2 \mathrm{Re}\left(\frac{m^2-1}{m^2+2}\right)^2\left\langle \tilde{E}^2\right\rangle_T \\ &= 9.6\pi a^2\varepsilon_0 n_s^2 \mathrm{Re}\left(\frac{m^2-1}{m^2+2}\right)^2 \cdot (S_{DC} + S_{12} + S_{23} + S_{31})\end{aligned} \tag{13.5}$$

with $S_{\mathrm{DC}} = |A_1|^2 + |A_2|^2 + |A_3|^2$

$S_{12} = A_1A_2^*e^{i((\vec{k}_1-\vec{k}_2)\vec{r}-\Omega t)} + c.c.$

$S_{23} = A_2^*A_3e^{i((\vec{k}_3-\vec{k}_2)\vec{r}-\Omega t)} + c.c.$

$S_{31} = A_3A_1^*e^{i(\vec{k}_3-\vec{k}_1)\vec{r}} + c.c.$

where $\Omega = \omega_3 - \omega_2$ is the frequency detuning. The acoustic vibration mode of nanoparticles will be resonantly excited when the beat frequency in S_{23} of Eq. (13.5) is tuned to the acoustic resonance of accordion mode. The resulting resonant oscillation of the spheroid eccentricity contributes to the refractive index perturbation and forms the moving grating.

The periodic force causes the local major axis elongation (i.e., a small departure y from the equilibrium a). Assuming that the acoustic vibration mode of nanoparticle can be explained as a simple harmonic oscillator with the acoustic resonance frequency ω_a and the damping rate γ_a, the equation of evolution of y can be written as

$$\frac{d^2y}{dt^2} + 2\gamma_a\frac{dy}{dt} + \omega_a^2 y = \frac{F_{e,a}}{m_a} \tag{13.6}$$

where m_a is the undetermined effective nanoparticle mass participating in such vibration. By solving Eq. (13.6) [See the Ref. (Xiang et al. 2016) for more details], the alternating current (AC) deformation of nanoparticle can be estimated as

$$y(\omega_a) = \frac{n_2\sqrt{I_2I_3}}{1.8v_i n_s \mathrm{Re}\left(\frac{m^2-1}{m^2+2}\right)^2}a \tag{13.7}$$

where n_2 is the nonlinear refractive index and v_i is the volume refraction of gold nanoparticles. If the medium would be sufficiently dilute, the total nonlinear polarization can be obtained by

$$\begin{aligned}\tilde{P} &= N\Delta\alpha\tilde{E} = N(\alpha_s - \alpha_e)\,\tilde{E} \\ &\approx \frac{24}{5}\pi a^2\varepsilon_0 n_s^2 \mathrm{Re}\left(\frac{m^2-1}{m^2+2}\right)^2 y\cdot\tilde{E}\end{aligned} \tag{13.8}$$

where N is the density of particles. Therefore, there are several different frequency components in the nonlinear polarization, and the component that can act as the phase-matched source for the signal wave is

$$\tilde{P}_4 = 6\varepsilon_0\chi_a^{(3)}A_1A_2A_3^*e^{i(\vec{k}_4\cdot\vec{r}-\omega_4 t)} + c.c. \tag{13.9}$$

where $\chi_a^{(3)} = \frac{7.68\varepsilon_0 N(\pi a^2 n_s^2)^2}{m_a(\omega_a^2-2i\Omega\gamma_a-\Omega^2)}\mathrm{Re}\left(\frac{m^2-1}{m^2+2}\right)^4$.

Assuming a constant permittivity of individual nanoparticle when the particle rhythmically beats for the breathing mode, the time-average polarization energy change of individual nanosphere with a diameter of a, which slightly deformed via the elongation of diameter from $2a$ to $2(a + \Delta r)$ by the electrostrictive force, is given by

$$\begin{aligned}\Delta U &= \mathrm{Re}\left(\alpha_s - \alpha_{e,b}\right)\left\langle \tilde{E}^2 \right\rangle_T \\ &= 4\pi\varepsilon_0 n_s^2 \mathrm{Re}\left(\frac{m^2-1}{m^2+2}\right)\left[a^3 - (a+\Delta r)^3\right]\left\langle \tilde{E}^2 \right\rangle_T \\ &\approx -12\pi a^2 \varepsilon_0 n_s^2 \mathrm{Re}\left(\frac{m^2-1}{m^2+2}\right)\Delta r \left\langle \tilde{E}^2 \right\rangle_T \qquad (13.10)\end{aligned}$$

where α_s and $\alpha_{e,b}$ are the particle polarizability for nanosphere with and without deformation, respectively; $m = n_p/n_s$ is the ratio of the refractive index of the particle (n_p) to that of the surrounding solvent (n_s). Thus, the electrostrictive force corresponding to the radial breathing mode is

$$\begin{aligned}F_{e,b} &= -\frac{\Delta U}{\Delta r} \approx 12\pi a^2 \varepsilon_0 n_s^2 \mathrm{Re}\left(\frac{m^2-1}{m^2+2}\right)\left\langle \tilde{E}^2 \right\rangle_T \\ &= 24\pi a^2 \varepsilon_0 n_s^2 \mathrm{Re}\left(\frac{m^2-1}{m^2+2}\right)\left(S_{\mathrm{DC}} + S_{12} + S_{23} + S_{31}\right) \qquad (13.11)\end{aligned}$$

The local diameter deformation can be considered as a small departure r from the equilibrium a influenced by the periodic force. Considering the acoustic vibration mode of nanoparticle as a simple harmonic oscillator with the acoustic resonance frequency ω_b and the damping rate γ_b, the equation of evolution of r is

$$\frac{d^2 r}{dt^2} + 2\gamma_b \frac{dr}{dt} + \omega_b^2 r = \frac{F_{e,b}}{m_b} \qquad (13.12)$$

where m_b is the undetermined effective nanoparticle mass participating in such vibration. Similar to accordion mode, the total nonlinear polarization for breathing mode is expressed by

$$\begin{aligned}\tilde{P} &= N\Delta\alpha\tilde{E} = N\left(\alpha_s - \alpha_{e,b}\right)\tilde{E} \\ &\approx 12\pi N a^2 \varepsilon_0 n_s^2 \mathrm{Re}\left(\frac{m^2-1}{m^2+2}\right) r \cdot \tilde{E} \qquad (13.13)\end{aligned}$$

The component can act as the phase-matched source for the signal wave, which is given by

$$\tilde{P}_4 = 6\varepsilon_0 \chi_b^{(3)} A_1 A_2 A_3^* e^{i(\vec{k_4}\cdot\vec{r} - \omega_4 t)} + c.c. \qquad (13.14)$$

where $\chi_b^{(3)} = \frac{48\varepsilon_0 N\left(\pi a^2 n_s^2\right)^2}{m_b\left(\omega_b^2 - 2i\Omega\gamma_b - \Omega^2\right)} \mathrm{Re}\left(\frac{m^2-1}{m^2+2}\right)^2$.

Using the coupling-wave equations in nonlinear optics (Boyd 2008), the nonlinear strength can be calculated as:

$$\frac{\partial A_4}{\partial z} = i\frac{3\omega}{n_s c}\chi^{(3)} A_1 A_2 A_3^* = i\kappa A_3^* \qquad (13.15)$$

Similarly,

$$\frac{\partial A_3}{\partial z} = -i\kappa A_4^* \text{ or } \frac{\partial A_3^*}{\partial z} = i\kappa^* A_4 \qquad (13.16)$$

Considering boundary condition $A_4(-L) = 0$, these output field amplitudes are given by

$$A_3^*(-L) = \frac{A_3^*(0)}{\cos\left(|\kappa|\, L\right)} \qquad (13.17)$$

$$A_4(0) = i\tan\left(|\kappa|\, L\right) A_3^*(0) \qquad (13.18)$$

The intensity of probe wave at $z = 0$ is given by

$$I_4(0) = \tan^2\left(|\kappa|\, L\right) I_3(0) \approx |\kappa|^2 L^2 I_3(0) \qquad (13.19)$$

where $|\kappa| = \frac{3\omega}{n_s c}\left|\chi^{(3)}\right| |A_1| |A_2| = \frac{3\pi}{\varepsilon_0 n_s^2 c\lambda}\left|\chi^{(3)}\right| \sqrt{I_1 I_2}$.

13.3.3 Experimental Results and Discussion

Experimental Setup

Figure 13.1 shows the FWM experimental setup, which is similar to degenerate FWM experiments that used gradient forces to create strong refractive index changes in degenerate FWM (Smith et al. 1981). However, an external cavity laser (ECL) and a distributed Bragg reflector (DBR) laser were used to tune to slightly different wavelengths for getting nondegenerate response. The sample, which was placed in a quartz cuvette, was illuminated by the counter-propagating optical beams. Co-polarized illumination was ensured by the polarization controller and polarizer. The angle between the laser beam with amplitude A_1 and the laser beam with amplitude A_3 was adjusted to 4° to allow over 1 mm light-matter interaction length (the length of the cuvette was 1 mm). To only permit the Rayleigh scattered light of that pump and the FWM beam from the counter-directional beam to be measured by an avalanche photodetector, the optical chopper modulated the intensity of one pump beam using a lock-in amplifier. The amplitude of Rayleigh scattered light was limited by fixing the power of the DBR laser to 25 mW. To excite two beams of the FWM process, the power of ECL was set to a relatively high value of 67 mW (total). The frequency difference between the two laser sources was scanned by temperature tuning the DBR laser. The power dependence of nonlinear response at acoustic resonance was obtained by using a variable optical attenuator after the ECL output (Xiang and Gordon, 2016a,b, Xiang et al. 2016).

The interference between the pump beam A_2 at frequency ω_2 and the beam A_3 at frequency ω_3 produces a traveling periodic variation in refractive index of the sample via the electrostrictive force that elongates individual nanoparticles along the polarization direction. Bragg diffraction of another pump wave A_1 at frequency ω_1 ($\omega_1 = \omega_3$) from the electrostriction induced moving grating, and its wavelength shifting due to the Doppler effect generates a signal wave A_4 with frequency $\omega_4 = \omega_2$. The acoustic vibration modes of the nanoparticles excited resonantly by tuning the beat frequency $\Omega = \omega_3 - \omega_2$ to the acoustic resonance. The moving grating, which is a strong refractive index perturbation, was created by the resulting resonant oscillation of the spheroid.

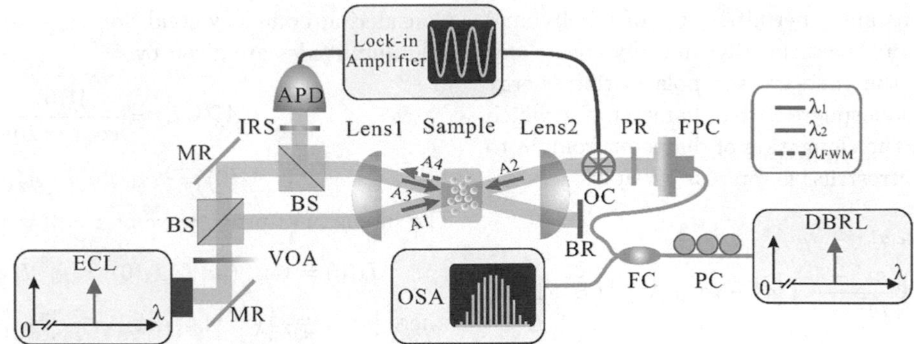

FIGURE 13.1 Experimental configuration of FWM. DBRL, distributed Bragg reflector laser; PC, polarization controller; FC, fiber coupler; OSA, optical spectrum analyzer; BR, blocker; FPC, fiber-port collimator; PR, polarizer; OC, optical chopper; IRS, iris; APD, avalanche photodetector; BS, beam splitter; MR, mirror; VOA, variable optical attenuator; ECL, external cavity laser. (Reprinted from Xiang and Gordon 2016b, Copyright 2016, SPIE.)

Considering the phase-matching condition in this scheme and assuming that the propagation directions of the pump wave A_1 and the wave A_3 are along $\mathbf{z}'$-axis and $\mathbf{z}$-axis, respectively, and the in-plane direction perpendicular to $\mathbf{z}$ is along the $\mathbf{y}$-axis, the wave vector mismatching term is

$$\overrightarrow{\Delta k} = \overrightarrow{k_1} + \overrightarrow{k_2} - \overrightarrow{k_3} + \overrightarrow{k_4} = n_s \frac{\omega_1 - \omega_2}{c} \hat{z} - n_s \frac{\omega_3 - \omega_4}{c} \widehat{z'}$$
$$= n_s \frac{\Omega}{c} (cos\theta - 1) \hat{z} + n_s \frac{\Omega}{c} \sin\theta \hat{y} \tag{13.20}$$

where n_s is the refractive index of solvent; c is the light velocity in vacuum; and $\hat{y}, \hat{z}, \widehat{z'}$ are the unit vectors. Since the angle between two laser beams θ is 4°, the phase mismatch $\vec{\Delta k} \cdot \vec{L}$ is small and ignorable (Xiang and Gordon, 2016a).

Results

Figure 13.2 depicts nonlinear response spectrum versus the beat frequency between the pump lasers for polystyrene nanoparticles and 2-nm-diameter gold nanoparticles. The acoustic resonant peaks are consistent with the theoretical prediction based on Lamb's theory (Lamb 1881).

The signal height of ~0.5 mV corresponded to the real generated power of ~4.5 nW. Based on this measured response of the APD, associated calculation was performed to estimate the nonlinearity strength at acoustic resonance. The third-order susceptibility $\chi^{(3)}$ of 3.8×10^{-17} m^2/V^2 for 22-nm polystyrene particles, and 1.6×10^{-16} m^2/V^2 and 0.9×10^{-16} m^2/V^2 corresponding to the $l = 2$ resonance and $l = 0$ resonance of 2-nm gold particles can be obtained by the following equation:

$$\chi^{(3)} \approx \frac{\varepsilon_0 n_s^2 c \lambda}{3\pi L} \sqrt{\frac{I_4}{I_1 I_2 I_3}} \tag{13.21}$$

where n_s is the refractive index of solvent; c is the light velocity in vacuum; I_i is the optical density for the beam $i(i = 1,2,3,4)$; ε_0 is the vacuum permittivity; λ is the wavelength of light; and L is the light-matter interaction length. This is approximately a million times larger than the typical electronic response of condensed matter (Boyd 2008). Interestingly, it can be seen in the measurement that a smaller bandwidth corresponds to smaller polystyrene nanoparticles due to the fact that larger particles have the resonances approaching to the collision rate for water of 20 GHz (Buchner et al. 1999), leading to the higher damping and a larger bandwidth.

The ECL power can be adjusted manually by a variable optical attenuator. Figure 13.3 shows its dependence of the nonlinear signal strength. A threshold behavior was observed. There is a negligible FWM signal below this

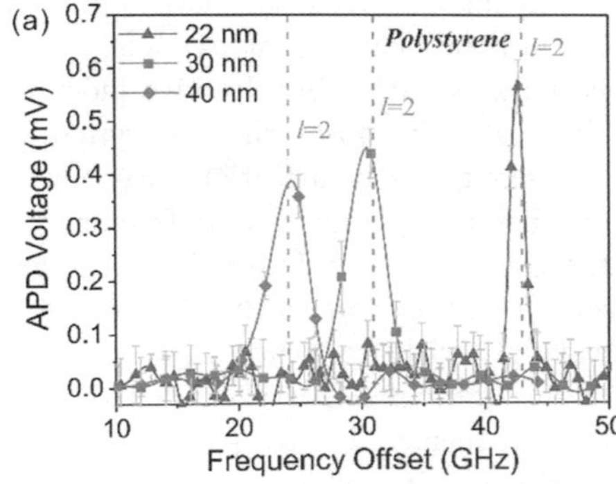

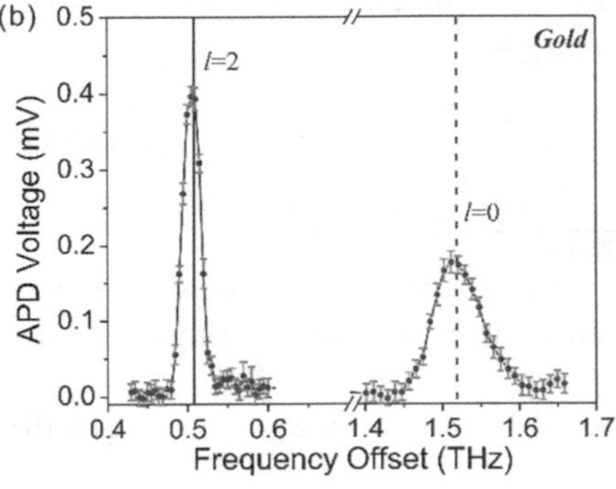

FIGURE 13.2 FWM response of an aqueous suspension as a function of beat frequency for (a) polystyrene and (b) gold nanoparticles. The nonresonant background from Rayleigh scattering is subtracted. (Reprinted from Xiang and Gordon 2016b, Copyright 2016, SPIE.)

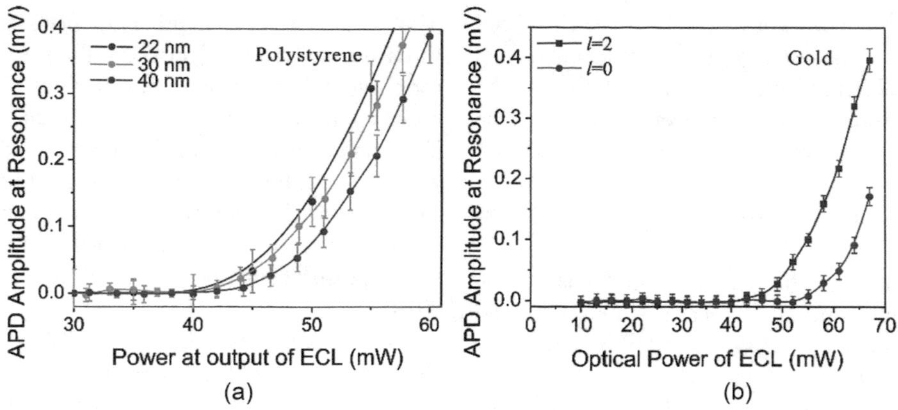

FIGURE 13.3 FWM signals measured by the avalanche photodetector versus ECL power for aqueous suspensions of (a) polystyrene and (b) gold nanoparticles. (Reprinted from Xiang and Gordon 2016b, Copyright 2016, SPIE.)

threshold. However, the signal increases with the ECL power above the threshold. While FWM would suggest a quadratic dependence on the power of ECL beam since it contributes two of the waves, it can be seen by the threshold dependence that the nonlinear effect abruptly turns on. Thus, a simple quadratic scaling does not apply. This will be discussed for gold in the next section.

Discussion

According to the viscoelastic theory's (Saviot et al. 2007) prediction for the water, the $l = 2$ resonance of gold nanoparticles occurs at 492 GHz with a quality factor of 7, and the $l = 0$ resonance occurs at 1.487 THz with a quality factor of 14. These values do not match with the experiment as good as conventional Lamb's theory for a sphere in vacuum. Furthermore, the damping suggested by low-quality factors makes it hard to achieve a strong nonlinear response.

Since the setup has a positive detuning of the tunable laser, which corresponds to stimulated Raman scattering, there is a possibility that the intrinsic threshold behavior of the nonlinear scattering can explain the observed threshold. However, the basic physical mechanism is electrostriction, which is common in acoustic Raman scattering, stimulated Brillouin scattering, and FWM. Using the generic formulation for $\chi^{(3)}$, the threshold gain parameter for $l = 2$ mode can be estimated as

$$G_{\text{th}} = |\kappa_{\text{th}}|\, L = \frac{3\omega}{2\varepsilon_0 n_s^2 c^2}\left|\chi^{(3)}\right| \sqrt{I_{1cr} I_2} L \approx 0.002 \quad (13.22)$$

where $\omega = 2.2 \times 10^{15}$ rad/s, $\varepsilon_0 = 8.85 \times 10^{12}$ F/m, $n_s = 1.33$, $c = 3 \times 10^8$ m/s, the critical intensity $I_{1cr} = 1.37 \times 10^6$ W/m^2, the DBR laser intensity $I_2 = 1.69 \times 10^7$ W/m^2, and $L = 1$ mm. Although an exceedingly large $\chi^{(3)}$ is used in this estimation of G_{th}, the value is four orders of magnitude smaller than the threshold required for stimulated Brillouin scattering (Boyd 2008) and stimulated Raman scattering (Agrawal 2007). Thus, those other mechanisms cannot alone explain the observations.

The formation of a cavity or bubble around the oscillating nanoparticles is another possible explanation of the observed threshold dependence and the strong nonlinear response. Strong electrostriction force can push away the water. Plugging the experimentally measured FWM signal into Eq. (13.10), the amplitudes of $l = 2$ and $l = 0$ vibrational modes can be found 3 pm and 1.4 pm at the maximum pump intensity. The time to collapse over this distance for a bubble of radius equal to the nanoparticle radius is called collapse time. The collapse times of 4.3 ps and 2 ps for the $l = 2$ and $l = 0$ vibrational modes were found by integration of the Rayleigh-Plesset equation (Brennen 2005). Since these times are larger than the oscillation periods, they may be viewed as the threshold condition to obtain oscillation with low damping (i.e., to maintain a stable bubble that does not collapse). Using Debye theory and permittivity measurements of water, the collision rate for water is around 20 GHz (Buchner et al. 1999). In oscillations faster than this rate, the nanoparticle pushes away the molecules before significant damping occurs. A displaced water molecule cannot return within one period of this oscillation. Thus, a stable cavity can form. Since the displacement is on the order of picometers, a tiny "bubble" is expected to form. Although this mechanism is analogous to the usual acoustic cavitation in low-pressure regions of water excited with acoustic waves (McNamara et al. 1999), molecular dynamics simulations may be more suitable due to having high-frequency and small-scale oscillations (Beckett and Hua 2001).

Table 13.1 shows that the measured acoustic resonant frequencies are more close to those in free space, indicating that the environment around the nanoparticle should be free space and so the bubbles have formed when the nonlinear response was measured. Therefore, the origin of

TABLE 13.1 Comparing Measured Values with Vibrational Resonances of 2-nm Gold Nanoparticle Estimated for Different Environments

Modes	Water	Free Space	Experiment
$l = 2$ (GHz)	492.1	507.8	504
$l = 0$ (THz)	1.487	1.518	1.511

Source: Reprinted with permission from Ref. (Xiang et al. 2016), Copyright 2016, American Chemical Society.

the threshold behavior should not be the intrinsic threshold of the nonlinear scattering, but the cavitation formation.

Considering how electrostriction theory relates to the observed nonlinear response and the expected damping, the quality factor of the resonance scales can be estimated as:

$$Q = \frac{y(\omega_a)}{y(0)} \tag{13.23}$$

where y is the amplitude of the displacement for the same driving field intensity. Using the experimentally measured FWM signal, $y(\omega_a)$ is calculated as 3×10^{-3} nm. In addition, $y(0)$ can be estimated as 6.09×10^{-9} nm by Eq. (13.9) from the electrostriction theory. The resulted quality factor of 5×10^5, which is four orders of magnitude larger than viscoelastic theory predicts, may lead us to the conclusion that bubbles are forming around the nanoparticles allowing them to oscillate with high-quality factors.

Damping factors of below 100 have been suggested by past works on nanoparticles, which considered acoustic coupling to a substrate (Ruijgrok et al. 2012). Many possible mechanisms such as capping layers and crystal defects were suggested for this internal damping. Quality factors exceeding 10,000 have been reported for nanomechanical resonators (Feng et al. 2007) that were still coupled mechanically to the environment via an anchor point. Thus, the intrinsic quality factor does still need further investigations. The mentioned measurement suggests that large quality factors can be obtained for nanoparticles as long as the damping from the surrounding environment is suppressed.

13.3.4 Numerical Calculations

The previous sections demonstrated the emergence of optically driven high-frequency acoustic vibrations of nanoparticles, which is a threshold behavior in the vibration amplitude. The fluid dewets the surface of the sphere above this threshold and the damping changes from continuous to intermittent with a dramatic increase in the quality factor of the vibrations. The fact that this threshold is achieved much earlier for larger frequencies has profound implications for the vibrations in this regime that are generally considered to be subject to substantial damping. In this section, this new phenomenon will be explained in a simple Lennard-Jones (LJ) fluid model with a deterministically vibrating sphere. This model confirms the hypothesis of formation of a cavity around the nanoparticles at a critical forcing amplitude and clearly shows the onset of a cavity in this nanoscale/gigahertz - terahertz regime (Hsueh et al. 2018).

Considering the LJ fluid composed of particles of mass m, the fundamental (Einstein) vibrational time scale is $\tau_0 = \sqrt{m\sigma/\epsilon}$, where ϵ and σ are the energy and length scales set by the LJ potential with a truncation of 2.5σ and vertically adjustment for continuity. To understand the interaction of the fluid with a nanoscale object vibrating on the same time scale, a sphere of radius $R_0 = 5\sigma$ composed of 423 LJ particles forming an face-centered cubic lattice (lattice constant 1.7σ) into the fluid is embedded. A purely repulsive LJ potential truncated at $2^{1/6}\sigma$ generates the interaction between the sphere atoms and the fluid atoms. To simulate a nanoparticle of 3 nm diameter immersed in water, the sphere's diameter is assumed ten times larger than that of the fluid particles. A static sphere is always in contact with the fluid molecules due to the positive hydrostatic pressure at the chosen fluid density (Hsueh et al. 2018).

Using a Langevin thermostat with a damping time of τ_0, the temperature of fluid is brought to a reduced temperature $T^* = 1$. Assuming the sphere particles are undergoing radial oscillatory motion with amplitude A about their equilibrium positions r_{i0}, integrating the equations of motion for the fluid particles only generates Eq. (13.24), which causes the radial expansion and contraction of the entire sphere with frequency ω and amplitude AR_0.

$$r_i(t) = r_{i0}[1 + A\sin(\omega t)] \tag{13.24}$$

A snapshot of the simulation setup with a sphere undergoing breathing mode oscillations immersed in a fluid of LJ particles is depicted in Figure 13.4a. Data is collected over 50 vibration periods after 1 τ_0 of relaxation for each value of A. Figure 13.4b demonstrates the temporally averaged fluid density as a function of radial distance to the center of the sphere. While the density profile is indistinguishable from that of a static sphere for small frequencies

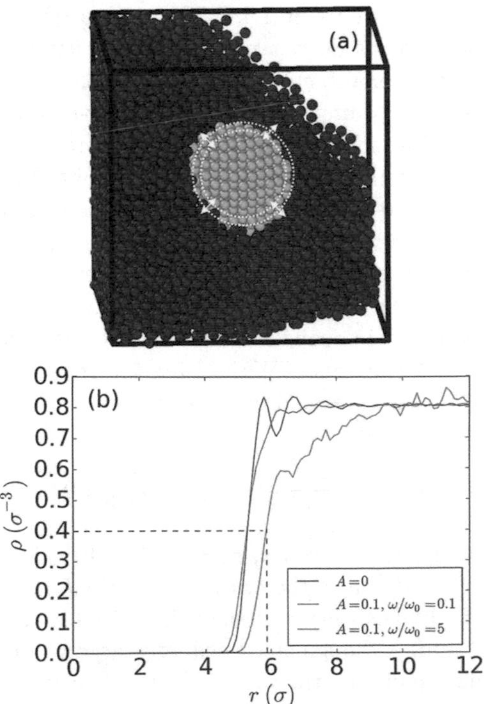

FIGURE 13.4 (a) A model nanosphere immersed in 13,500 LJ fluid particles in the simulation box of width 25.7 σ. (b) Profiles of radial fluid density for $\omega = 0.1\ \omega_0$ and $\omega = 5\omega_0$. (Reprinted with permission from Hsueh et al., 2018, Copyright 2016, American Chemical Society.)

$\omega \ll \omega_0 = 2\pi/\tau_0$, a visible gap opens up at larger frequencies $\omega > \omega_0$. The advent of the gap is an indication of a dewetting phenomenon, which is the main result of this simulation.

As can be seen in Figure 13.5, the thickness of the vacuum gap can be directly measured with $\Delta r_{1/2} = r_{1/2}(A, \omega) - r_{1/2}(0, 0)$, where $r_{1/2}(0, 0)$ is the static sphere of radius R_0 and $r_{1/2}(A, \omega)$ is the distance from the center of the sphere at which the fluid density reaches half of the bulk value, as denoted by the dashed lines in Figure 13.4b. In Figure 13.5, while the gap thickness enlarges after $A > 0.11$ in the low frequencies curve of $\omega/\omega_0 = 0.2$, the fluid particles completely wet the sphere for lower amplitudes followed by the sphere oscillations. On the other hand, the gap emerges at lower amplitudes for higher frequencies, to the extent that there is a maximum gap at $\omega/\omega_0 = 5$. Thus, there should be a critical velocity $v_c \sim R_0 A_c \omega$, at which the fluid loses the ability to follow the sphere. Using Figure 13.5, the critical velocities are estimated as $0.1\ \sigma/\tau < v_c < 0.2\ \sigma/\tau$ in the frequency range $0.2 < \omega/\omega_0 < 2$. The speed of the particles $v_p \simeq 1.3\ \sigma/\tau$ in the ballistic regime, which is estimated from the root mean square value of the distance traveled by the particles in the bulk fluid before colliding with other particles, is about first order of magnitude larger than these critical velocities. This is reasonable considering the fact that reflections decrease the speed of the particles at the surface.

After $\omega/\omega_0 > 5$, the gap thickness declines and saturates at a frequency-independent curve because the fluid in high frequencies $\omega \gg \omega_0$ faces essentially a static sphere of radius $R_0(1 + A_{\max})$, as indicated by the corresponding solid black line in Figure 13.5. However, the gap thickness surpasses that limit at finite frequencies, possibly due to inertial effects. On the other hand, the vanishing oscillation amplitudes always lead to the disappearance of the vacuum gap.

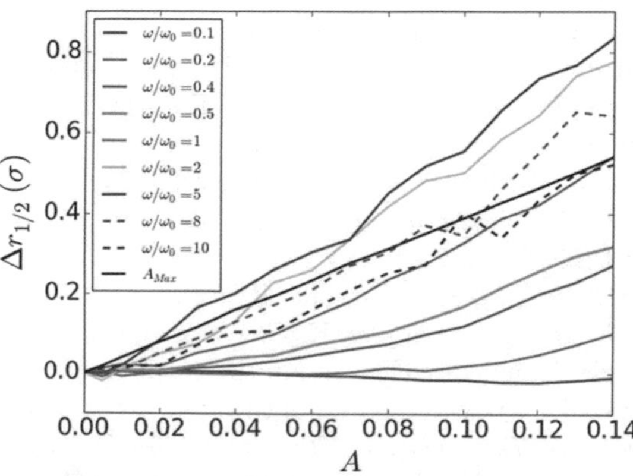

FIGURE 13.5 The vacuum gap thickness as a function of vibration amplitude A for different sphere vibration frequencies ω. $A_{\max}$ is coming from a static sphere with radius $R = R_0(1 + A_{\max})$. All data points' uncertainties are 0.02 σ. (Reprinted with permission from Hsueh et al., 2018, Copyright 2016, American Chemical Society.)

The bulk fluid pressure can be calculated from the usual virial expression

$$p_f = \frac{N k_B T}{V} + \frac{\sum r_{ij} \cdot f_{ij}}{3V} \tag{13.25}$$

where the summation runs over all fluid particles, V is the volume of the box subtracting the volume occupied by the sphere, and r_{ij} and f_{ij} indicate the separation and force vectors between particles i and j. Furthermore, a local pressure at the surface of the sphere is also computed from

$$p_{\text{loc}} = \sum \frac{f_i \cdot \hat{n}_i}{4\pi r_i^2} \tag{13.26}$$

where the summation runs over all sphere particles, f_i is the total force on sphere particle i by the fluid, and $\hat{n}_i$ denotes the radial unit vector.

The simulated profiles of these two pressures and the oscillating sphere radius $R(t)$ over several oscillation periods confirm the dewetting phenomenon observed in Figures 13.4 and 13.5. The sphere is always in direct contact with the fluid particles for $\omega \ll \omega_0$ because the fluid oscillates essentially in phase with the sphere and the local pressure always leads to the sphere oscillations. For $\omega > \omega_0$, the behavior is different. In this case, the sphere only comes in contact with the fluid when the radius is maximal in the cycle since the local pressure drops to almost zero during most of the period but spikes dramatically in maximum points in the cycle [See Ref. (Hsueh et al. 2018) for more details].

In order to quantify the dramatic decrease in the sphere vibrations' damping by the vacuum layer formation, the energy dissipated in one cycle is estimated as

$$W_{\text{diss}} = \int_0^T dt \sum f_i \cdot v_i \tag{13.27}$$

where the summation is carried out over all particles in the sphere and v_i indicate the particle velocities. According to the numerical results, W_{diss} reaches its maximum at $\omega/\omega_0 = 2$ for a given driving amplitude A and drops strongly for larger frequencies. This indicates that the rise and fall of W_{diss} roughly coincide with the growth and shrinkage of the cavity width as observed in Figure 13.5, because the frequency of maximum W_{diss} is only 2.5 times smaller than the frequency at which the largest vacuum gap is observed in Figure 13.5, $\omega/\omega_0 = 5$.

Assuming a harmonic system, which is not strictly the case here, the quality factor may defined as $Q = \tan^{-1}(\phi)$, where ϕ is the phase angle between the local pressure $P_{\text{loc}}(t)$ and the radius $R(t)$. This definition is justified by the hypothesis that the nanoparticle, which its vibrations are driven by light in this experiment, responds to the pressure at the surface. In the numerical results, this estimated Q demonstrates a stronger rise for frequencies $\omega > 2\omega_0$, which is an indication of the reduced dissipation when the sphere loses a contact with the fluid. Thus, this simulation confirms the speculated mechanism in Section 13.3.3, whereby the formation of a small cavity around the nanoparticle drastically decreases the damping of the surrounding fluid (Hsueh et al. 2018).

13.3.5 Summary

In summary, an unusually strong FWM scattering from nanoparticles in solution was observed in the gigahertz to terahertz range, as predicted by Lamb's theory. Thus, elastic vibrations explain the strong FWM peaks. The clear threshold behavior of FWM signal suggests the formation of a cavity around the nanoparticles that reduces the damping. This is promising for efficient nonlinear wavelength conversion and all-optical switching using nanoparticles in solution.

13.4 Conclusion

This chapter outlines a novel method of achieving the goals of high speed (up to terahertz range) and large strength of nonlinearity ($n_2 \sim 10^{-14}$ m^2/W, three orders of magnitude and six orders of magnitude larger than the intrinsic responses of silicon and silica, respectively) based on the electrostriction mechanism for nanoparticles in aqueous solution. Modulated electrostrictive force at resonant frequency, via beating two CW lasers with frequency difference, excites the acoustic vibration of nanoparticles, and thus, vibration-induced permittivity changes scattered incident photons. In this scene, there are actually four optical waves participating in the interaction, namely, two for writing the traveling grating along with the beam propagation, one for pump wave, and one for newly generated wave, so it is a FWM nonlinear process. Here, there are two physical origins of such fast and huge nonlinear optical process. First is the fact that sound travels in the dielectrics and metals at a kilometer per second, and thus, nanometer particles naturally have terahertz acoustic resonances. Another essential thing is the formation of stable acoustic cavitation as shown in the threshold behavior at which the nonlinearity suddenly "turns on". When the water molecules surrounding nanoparticles are unable to follow the oscillations of nanoparticles at such a rapid rate, much higher than the collision rate of water molecules of 20 GHz, the molecules will be pushed away and these displaced molecules cannot return during one period of oscillation. As a result, the stable cavity or bubble is formed, and thus, a huge enhancement in quality factor is realized (four orders of magnitude larger than usual hydrodynamic theory predicts), which breaks limits to the nonlinear response induced by the damping from the surrounding medium. This new physics of high-quality nanomechanical oscillations in solution may be extended to use in applications such as high bandwidth optical wavelength conversion with over terahertz wavelength shifting as desired optical wavelength-division multiplexing (Geraghty et al. 1997, Cotter et al. 1999), the spectroscopy of vibrational resonances of nanoparticles in solution (Wu et al. 2016a,b), or probing the vibrational resonances of proteins, for example, the large-scale elastic motions in the 100 GHz to terahertz range (Tirion 1996).

References

Abrams, R. L. and D. A. Pinnow. Acousto-optic properties of crystalline germanium. *Journal of Applied Physics 41(7)*, 1970: 2765–2768.

Agrawal, G. P. *Nonlinear Fiber Optics.* Oxford: Academic Press, 2007.

Ajgaonkar, M., Y. Zhang, H. Grebel and C. W. White. Nonlinear optical properties of a coherent array of sub-micron SiO_2 spheres (opal) embedded with Si nanoparticle. *Applied Physics Letters 75(11)*, 1999: 1532–1534.

Akiyama, T., O. Wada, H. Kuwatsuka, T. Simoyama, Y. Nakata, K. Mukai, M. Sugawara and H. Ishikawa. Nonlinear processes responsible for nondegenerate four-wave mixing in quantum-dot optical amplifiers. *Applied Physics Letters 77(12)*, 2000: 1753–1755.

Andrekson, P. A., N. A. Olsson, J. R. Simpson, T. Tanbun-Ek, R. A. Logan and M. Haner. 16 Gbit/s all-optical demultiplexing using four-wave mixing. *Electronics Letters 27(11)*, 1991: 922–924.

Armstrong, J. A., N. Bloembergen, J. Ducuing and P. S. Pershan. Interactions between light waves in a nonlinear dielectric. *Physical Review 127(6)*, 1962: 1918–1939.

Ashokkumar, M. The characterization of acoustic cavitation bubbles: An overview. *Ultrasonics Sonochemistry 18*, 2011: 864–872.

Beckett, M. A. and I. Hua. Impact of ultrasonic frequency on aqueous sonoluminescence and sonochemistry. *The Journal of Physical Chemistry A 105(15)*, 2001: 3796–3802.

Bigot, J. Y., J. Y. Merle, O. Cregut and A. Daunois. Electron dynamics in copper metallic nanoparticles probed with femtosecond optical pulses. *Physical Review Letters 75(25)*, 1995: 4702.

Bloembergen, N. Conservation laws in nonlinear optics. *Journal of the Optical Society of America 70(12)*, 1980: 1429–1436.

Bloembergen, N. Recent progress in four-wave mixing spectroscopy. In: *Laser Spectroscopy IV, Proceedings of the Fourth International Conference Rottach-Egern, Fed. Rep. of Germany*, H. WaltherKarl and W. Rothe (Eds), 340–348. Berlin, Heidelberg: Springer, 1979.

Boyd, R. W. *Nonlinear Optics*, 3rd ed. San Diego, CA: Academic Press, 2008.

Brackett, C. A., A. S. Acampora, J. Sweitzer, G. Tangonan, M. T. Smith, W. Lennon, K. C. Wang and R. H. Hobbs. A scalable multiwavelength multihop optical network: A proposal for research on all-optical networks. *Journal of Lightwave Technology 11(56)*, 1993: 736–753.

Bragas, A. V., C. Aku-Leh, S. Costantino, A. Ingale, J. Zhao and R. Merlin. Ultrafast optical generation of coherent phonons in $CdTe_1$-xSe_x quantum dots. *Physical Review B 69(20)*, 2004: 205306.

Brennen, C. E. *Fundamentals of Multiphase Flow.* Cambridge: Cambridge University Press, 2005.

Brillouin, L. Diffusion de la lumière et des rayons X par un corps transparent homogène. Influence de l'agitation thermique. *Annals of Physics (Paris) 17(88–122)*, 1922: 21.

Buchner, R., J. Barthel and J. Stauber. The dielectric relaxation of water between 0°C and 35°C. *Chemical Physics Letters 306(1)*, 1999: 57–63.

Campion, A. and P. Kambhampati. Surface-enhanced Raman scattering. *Chemical Society Reviews 27(4)*, 1998: 241–250.

Cao, Y. C., R. Jin and C. A. Mirkin. Nanoparticles with Raman spectroscopic fingerprints for DNA and RNA detection. *Science 297(5586)*, 2002: 1536–1540.

Cauz, J. (Ed). *Encyclopædia Britannica.* n.d. www.britannica.com/science/electrostriction (accessed January 28, 2018).

Chakraborty, D., E. van Leeuwen, M. Pelton and J. E. Sader. Vibration of nanoparticles in viscous fluids. *The Journal of Physical Chemistry C 117(16)*, 2013: 8536–8544.

Chang, I. C. Acousto-optic tunable filters. *Optical Engineering 20(6)*, 1981: 206824.

Chraplyvy, A. R. Limitations on lightwave communications imposed by optical-fiber nonlinearities. *Journal of Lightwave Technology 8(10)*, 1990: 1548–1557.

Cotter, D., R. J. Manning, K. J. Blow, A. D. Ellis, A. E. Kelly, D. Nesset, I. D. Phillips, A. J. Poustie and D. C. Rogers. Nonlinear optics for high-speed digital information processing. *Science 286(5444)*, 1999: 1523–1528.

Cronin-Golomb, M. Dynamically programmable self-aligning optical interconnect with fan-out and fan-in using self-pumped phase conjugation. *Applied Physics Letters 54(22)*, 1989: 2189–2191.

Crut, A., P. Maioli, N. Del Fatti and F. Vallée. Acoustic vibrations of metal nano-objects: Time-domain investigations. *Physics Reports 549*, 2015: 1–43.

Dadap, J. I., et al. Nonlinear-optical phase modification in dispersion-engineered Si photonic wires. *Optics Express 16(2)*, 2008: 1280–1299.

Danckwerts, M. and L. Novotny. Optical frequency mixing at coupled gold nanoparticles. *Physical Review Letters 98(2)*, 2007: 026104.

Debye, P. and F. W. Sears. On the scattering of light by supersonic waves. *Proceedings of the National Academy of Sciences 18(6)*, 1932: 409–414.

Delgado-Pinar, M., D. Zalvidea, A. Diez, P. Pérez-Millán and M. V. Andrés. Q-switching of an all-fiber laser by acousto-optic modulation of a fiber Bragg grating. *Optics Express 14(3)*, 2006: 1106–1112.

Delone, N. B. *Basics of Interaction of Laser Radiation with Matter.* France: Atlantica Séguier Frontières, 1993.

Dhar, L., J. A. Rogers and K. A. Nelson. Time-resolved vibrational spectroscopy in the impulsive limit. *Chemical Reviews 94(1)*, 1994: 157–193.

Diez, S., C. Schmidt, R. Ludwig, H. G. Weber, K. Obermann, S. Kindt, I. Koltchanov and K. Petermann. Four-wave mixing in semiconductor optical amplifiers for frequency conversion and fast optical switching. *Quantum Electronics 3(5)*, 1997: 1131–1145.

Dixon, R. W. Photoelastic properties of selected materials and their relevance for applications to acoustic light modulators and scanners. *Journal of Applied Physics 38(13)*, 1967: 5149–5153.

D'ottavi, A., et al. Four-wave mixing in semiconductor optical amplifiers: A practical tool for wavelength conversion. *IEEE Journal of Selected Topics in Quantum Electronics 3(2)*, 1997: 522–528.

Feng, X. L., R. He, P. Yang and M. L. Roukes. Very high frequency silicon nanowire electromechanical resonators. *Nano Letters 7(7)*, 2007: 1953–1959.

Feuerbacher, B. F., J. Kuhl and K. Ploog. Biexcitonic contribution to the degenerate-four-wave-mixing signal from a GaAs/Al_xGa1-xAs quantum well. *Physical Review B 43(3)*, 1991: 2439.

Fujii, M., S. Hayashi and K. Yamamoto. Raman scattering from quantum dots of Ge embedded in SiO_2 thin films. *Applied Physics Letters 57(25)*, 1990: 2692–2694.

Fukuda, H., K. Yamada, T. Shoji, M. Takahashi, T. Tsuchizawa, T. Watanabe, J. I. Takahashi and S. I. Itabashi. Four-wave mixing in silicon wire waveguides. *Optics Express 13(12)*, 2005: 4629–4637.

Geraghty, D. F., R. B. Lee, M. Verdiell, M. Ziari, A. Mathur, K. J. Vahala. Wavelength conversion for WDM communication systems using four-wave mixing in semiconductor optical amplifiers. *IEEE Journal of Selected Topics in Quantum Electronics 3(5)*, 1997: 1146–1155.

Hartland, G. V. Coherent excitation of vibrational modes in metallic nanoparticles. *Annual Review of Physical Chemistry 57*, 2006: 403–430.

Hsueh, C.-C., R. Gordon and J. Rottler. Dewetting during terahertz vibrations of nanoparticles. *Nano Letter 18(2)*, 2018: 773–777.

Ippen, E. P. and R. H. Stolen. Stimulated Brillouin scattering in optical fibers. *Applied Physics Letters 21(11)*, 1972: 539–541.

Itoh, T., T. Asahi and H. Masuhara. Femtosecond light scattering spectroscopy of single gold nanoparticles. *Applied Physics Letters 79(11)*, 2001: 1667–1669.

Jackson, J. D. *Classical Electrodynamics.* New York: Wiley, 1999.

Juvé, V., A. Crut, P. Maioli, M. Pellarin, M. Broyer, N. Del Fatti and F. Vallée. Probing elasticity at the nanoscale: Terahertz acoustic vibration of small metal nanoparticles. *Nano Letters 10(5)*, 2010: 1853–1858.

Kauranen, M. and A. V. Zayats. Nonlinear plasmonics. *Nature Photonics 6(11)*, 2012: 737–748.

Khoo, I. C. Nonlinear optics of liquid crystalline materials. *Physics Reports 471(5)*, 2009: 221–267.

Kino, G. S. and H. Matthews. Signal processing in acoustic surface-wave devices. *IEEE Spectrum 8(8)*, 1971: 22–35.

Koos, C., et al. All-optical high-speed signal processing with silicon–organic hybrid slot waveguides. *Nature Photonics 3(4)*, 2009: 216–219.

Lamb, H. On the vibrations of an elastic sphere. *Proceedings of the London Mathematical Society 1(1)*, 1881: 189–212.

Landau, L. D. and E. M. Lifshitz. *Electrodynamics of Continuous Media, Volume 8 of Course of Theoretical Physics*, 2nd revised ed. New York: Pergamon Press, 1984.

Leuthold, J., C. Koos and W. Freude. Nonlinear silicon photonics. *Nature Photonics 4(8)*, 2010: 535–544.

Lin, Q., J. Zhang, G. Piredda, R. W. Boyd, P. M. Fauchet and G. P. Agrawal. Dispersion of silicon nonlinearities in the near infrared region. *Applied Physics Letters 91(2)*, 2007: 021111.

Liu, T. M., et al. Measuring plasmon-resonance enhanced third-harmonic $\chi(3)$ of Ag nanoparticles. *Applied Physics Letters 89(4)*, 2006: 043122.

Liu, X., R. M. Osgood, Y. A. Vlasov and W. M. Green. Mid-infrared optical parametric amplifier using silicon nanophotonic waveguides. *Nature Photonics 4(8)*, 2010: 557–560.

Lucas, R. and P. Biquard. Optical properties of solids and liquids under ultrasonic vibrations. *Journal de Physique et Le Radium 3*, 1932: 464.

Maldovan, M. Sound and heat revolutions in phononics. *Nature 503(7475)*, 2013: 209–217.

McNamara, W. B., Y. T. Didenko and K. S. Suslick. Sonoluminescence temperatures during multi-bubble cavitation. *Nature 401(6755)*, 1999: 772–775.

Meldrum, A., L. A. Boatner and C. W. White. Nanocomposites formed by ion implantation: Recent developments and future opportunities. *Nuclear Instruments and Methods in Physics Research Section B: Beam Interactions with Materials and Atoms 178(1)*, 2001: 7–16.

Miller, D. A. B., D. S. Chemla, D. J. Eilenberger, P. W. Smith, A. C. Gossard and W. Wiegmann. Degenerate four-wave mixing in room-temperature GaAs/GaAlAs multiple quantum well structures. *Applied Physics Letters 42(11)*, 1983: 925–927.

Mongin, D., et al. Acoustic vibrations of metal-dielectric core–shell nanoparticles. *Nano Letters 11(7)*, 2011: 3016–3021.

Muskens, O. L., N. Del Fatti and F. Vallée. Femtosecond response of a single metal nanoparticle. *Nano Letters 6(3)*, 2006: 552–556.

Nie, S. and S. R. Emory. Probing single molecules and single nanoparticles by surface-enhanced Raman scattering. *Science 275(5303)*, 1997: 1102–1106.

O'Brien, K., N. D. Lanzillotti-Kimura, J. Rho, H. Suchowski, X. Yin and X. Zhang. Ultrafast acousto-plasmonic control and sensing in complex nanostructures. *Nature Communications 5*, 2014: 4042.

Patel, N. S., K. A. Rauschenbach and K. L. Hall. 40-Gb/s demultiplexing using an ultrafast nonlinear interferometer (UNI). *IEEE Photonics Technology Letters 8(12)*, 1996: 1695–1697.

Pelton, M., J. E. Sader, J. Burgin, M. Liu, P. Guyot-Sionnest and D. Gosztola. Damping of acoustic vibrations in gold nanoparticles. *Nature Nanotechnology 4(8)*, 2009: 492–495.

Phillips, I. D., D. Ellis, J. Thiele, R. J. Manning and A. E. Kelly. 40 Gbit/s all-optical data regeneration and demultiplexing with long pattern lengths using a semiconductor nonlinear interferometer. *Electronics Letters 34(24)*, 1998: 2340–2342.

Renger, J., R. Quidant, N. van Hulst, and L. Novotny. Surface-enhanced nonlinear four-wave mixing. *Physical Review Letters 104(4)*, 2010: 046803.

Renger, J., R. Quidant, N. van Hulst, S. Novotny, and L. Palomba. Free-space excitation of propagating surface plasmon polaritons by nonlinear four-wave mixing. *Physical Review Letters 103(26)*, 2009: 266802.

Rennie, R., and J. Law (Eds). *Oxford Dictionary of Physics*, 7th ed. Oxford: Oxford University Press, 2015.

Rhodes, W. T. Acousto-optic signal processing: Convolution and correlation. *Proceedings of the IEEE 69(1)*, 1981: 65–79.

Robinson, B. S., S. A. Hamilton and E. P. Ippen. Demultiplexing of 80-Gb/s pulse-position modulated data with an ultrafast nonlinear interferometer. *IEEE Photonics Technology Letters 14(2)*, 2002: 206–208.

Ruhman, S., A. G. Joly and K. A. Nelson. Time-resolved observations of coherent molecular vibrational motion and the general occurrence of impulsive stimulated scattering. *The Journal of Chemical Physics 86(11)*, 1987: 6563–6565.

Ruijgrok, P. V., P. Zijlstra, A. L. Tchebotareva and M. Orrit. Damping of acoustic vibrations of single gold nanoparticles optically trapped in water. *Nano Letters 12(2)*, 2012: 1063–1069.

Salem, R., M. A. Foster, A. C. Turner, D. F. Geraghty, M. Lipson and A. L. Gaeta. Signal regeneration using low-power four-wave mixing on silicon chip. *Nature Photonics 2(1)*, 2008: 35–38.

Saviot, L., C. H. Netting and D. B. Murray. Damping by bulk and shear viscosity of confined acoustic phonons for nanostructures in aqueous solution. *The Journal of Physical Chemistry B 111(25)*, 2007: 7457–7461.

Sihvola, A. H. *Electromagnetic Mixing Formulas and Applications*. London: IEE, 1999.

Smith, D. A., J. E. Baran, J. J. Johnson and K. W. Cheung. Integrated-optic acoustically-tunable filters for WDM networks. *IEEE Journal on Selected Areas in Communications 8(6)*, 1990: 1151–1159.

Smith, P. W., A. Ashkin and W. J. Tomlinson. Four-wave mixing in an artificial Kerr medium. *Optics Letters 6(6)*, 1981: 284–286.

Smith, P. W., P. J. Maloney and A. Ashkin. Use of a liquid suspension of dielectric spheres as an artificial Kerr medium. *Optics Letters 7(8)*, 1982: 347–349.

Stegeman, G. I. and A. Miller. Physics of all-optical switching devices. In: *Photonics in Switching, Vol 1:*

Background and Components, J. E. Midwinter (Ed), 81–145. Boston, FL: Academic Press, 1993.

Taylor, D. J., S. E. Harris, S. T. K. Nieh and T. W. Hansch. Electronic tuning of a dye laser using the acousto-optic filter. *Applied Physics Letters 19(8)*, 1971: 269–271.

Tirion, M. M. Large amplitude elastic motions in proteins from a single-parameter, atomic analysis. *Physical Review Letters 77*, 1996: 1905–1908.

Tomes, M. and T. Carmon. Photonic micro-electromechanical systems vibrating at X-band (11-GHz) rates. *Physical Review Letters 102(11)*, 2009: 113601.

Turton, D. A., H. M. Senn, T. Harwood, A. J. Lapthorn, E. M. Ellis and K. Wynne. Terahertz underdamped vibrational motion governs protein-ligand binding in solution. *Nature Communications 5*, 2014: 3999.

van Dijk, M. A., M. Lippitz and M. Orrit. Detection of acoustic oscillations of single gold nanospheres by time-resolved interferometry. *Physics Review Letters 95(26)*, 2005: 267406.

Vijayalakshmi, S., H. Grebel, Z. Iqbal and C. W. White. Artificial dielectrics: Nonlinear properties of Si nanoclusters formed by ion implantation in SiO_2 glassy matrix. *Journal of Applied Physics 84(12)*, 1998: 6502–6506.

Vijayalakshmi, S., H. Grebel, G. Yaglioglu, R. Pino, R. Dorsinville and C. W. White. Nonlinear optical response of Si nanostructures in a silica matrix. *Journal of Applied Physics 88(11)*, 2000: 6418–6422.

Wheaton, S., R. M. Gelfand and R. Gordon. Probing the Raman-active acoustic vibrations of nanoparticles with extraordinary spectral resolution. *Nature Photonics 9(1)*, 2015: 68–72.

Whelan, A. M., M. E. Brennan, W. J. Blau and J. M. Kellya. Enhanced third-order optical nonlinearity of silver nanoparticles with a tunable surface plasmon resonance. *Journal of Nanoscience and nanotechnology 4(1–2)*, 2004: 66–68.

Wu, J., D. Xiang, and R. Gordon. Characterizing gold nanorods in aqueous solution by acoustic vibrations probed with four-wave mixing. *Optics Express 24(12)*, 2016a: 12458–12465.

Wu, J., et al. Probing the acoustic vibrations of complex-shaped metal nanoparticles with four-wave mixing. *Optics Express 24(21)*, 2016b: 23747–23754.

Xiang, D. and R. Gordon. Nanoparticle acoustic resonance enhanced nearly degenerate four-wave mixing. *ACS Photonics 3(8)*, 2016a: 1421–1425.

Xiang, D. and R. Gordon. Nanoparticle electrostriction acoustic resonance enhanced nonlinearity. *SPIE Nanoscience + Engineering*. San Diego, CA: International Society for Optics and Photonics, September 2016b, 992223.

Xiang, D., J. Wu, J. Rottler, and R. Gordon. Threshold for terahertz resonance of nanoparticles in water. *Nano Letters 16(6)*, 2016: 3638–3641.

Yadav, H. K., V. Gupta, K. Sreenivas, S. P. Singh, B. Sundarakannan and R. S. Katiyar. Low frequency Raman scattering from acoustic phonons confined in ZnO nanoparticles. *Physical Review Letters 97(8)*, 2006: 085502.

Yang, L., et al. Size dependence of the third-order susceptibility of copper nanoclusters investigated by four-wave mixing. *JOSA B 11(3)*, 1994: 457–461.

Yoo, S. B. Wavelength conversion technologies for WDM network applications. *Journal of Lightwave Technology 14(6)*, 1996: 955–966.

Yu, K., P. Zijlstra, J. E. Sader, Q. H. Xu and M. Orrit. Damping of acoustic vibrations of immobilized single gold nanorods in different environments. *Nano Letters 13(6)*, 2013: 2710–2716.

Zewail, A. H. Femtochemistry: Atomic-scale dynamics of the chemical bond. *The Journal of Physical Chemistry A 104(24)*, 2000: 5660–5694.

Zhang, Y., F. Wen, Y. R. Zhen, P. Nordlander and N. J. Halas. Coherent Fano resonances in a plasmonic nanocluster enhance optical four-wave mixing. *Proceedings of the National Academy of Sciences 110(23)*, 2013: 9215–9219.

Zhang, Y., Z. Wang, Z. Nie, C. Li, H. Chen, K. Lu and M. Xiao. Four-wave mixing dipole soliton in laser-induced atomic gratings. *Physical Review Letters 106(9)*, 2011: 093904.

Zhu, Z., D. J. Gauthier and R. W. Boyd. Stored light in an optical fiber via stimulated Brillouin scattering. *Science 318(5857)*, 2007: 1748–1750.

Zijlstra, P., A. L. Tchebotareva, J. W. Chon, M. Gu and M. Orrit. Acoustic oscillations and elastic moduli of single gold nanorods. *Nano Letters 8(10)*, 2008: 3493–3497.

14

Water Photosplitting on Gold Nanoparticles: Quantum Selectivity and Dynamics

Sheng Meng and Peiwei You
Institute of Physics, Chinese Academy of Sciences

14.1 Introduction

It is generally believed that fossil fuels, the current primary but limited energy resources, will be replaced by cleaner and cheaper renewable energy sources for compelling environmental and economic challenges in the 21st century. Solar energy with its unlimited quantity is expected to be one of the most promising alternative energy sources in the future. It is expected that we are driven into a 100% renewable energy-based society within a few tens of years. Approaches and devices with low manufacturing cost and high efficiency are therefore a necessity for sunlight capture and light-to-energy conversion, since sunlight serves as one of the most promising and vast renewable energy sources.

Among various renewable energy technologies, direct splitting of water into H_2 fuel and O_2 gases by sunlight is the most promising and potentially low-cost sector. Solar water splitting is an essential step for artificial photosynthesis toward a sustainable energy future. Hydrogen gas production by solar water splitting could provide renewable fuels to potentially solve the world's ever-increasing energy demands. To generate H_2, photocatalysis utilizing plasmonic excitation in supported metal structures has gained increasing attentions, thanks to its dramatic light harnessing capability and easy tunability of plasmon excitations. Combining an oxide semiconductor with plasmonic metal nanoparticles (NPs) as co-catalyst for water splitting is prevalent in literature (Linic et al., 2011; Mukherjee et al., 2014). In this scenario, only hot electrons with sufficient energy to overcome the Schottky barrier can be collected by the conduction band of the semiconductor, and this bottleneck significantly limits the reaction efficiency.

Recently, Robatjazi et al. observed large photocurrents as a result of direct injection of hot electrons from plasmonic gold NP to molecules, driving solar water splitting in a Schottky-free junction (Robatjazi et al., 2015). Therefore, direct water splitting on plasmonic metal nanostructures upon photoexcitation can be achieved. So far, such experiments show a rather low photocatalytic activity improper for practical use. The key to achieve a significant efficiency improvement in this type of setup is strong light absorption and efficient carrier separation. Recent theoretical calculations start to attack the effect of size and shape of metal NPs on photocatalytic activity (Cottancin et al., 2006; Murray et al., 2007). However, the microscopic mechanism of photocatalytic water splitting, especially its dynamic processes at the atomic scale, has not been illustrated.

Gold NPs supported on titania exhibit an effective photocatalytic activity for water splitting under ultraviolet, visible, and near-infrared light (Awate et al., 2011; Liu et al., 2011; Boppella et al., 2017). Distinct from large particles, gold NPs show strong catalytic activities that depend on the size and shape of the supported clusters. A remarkable example is Au_{20} cluster with a large electronic energy gap of 1.77 eV. Its unique tetrahedral structure possesses a very high surface area and a large fraction of corner sites with low coordination, which provide ideal adsorption sites to bind molecules for catalysis.

In this chapter, we will present recent progresses in atomic-level understanding and engineering of water-photosplitting dynamics on plasmonic gold NPs. In Section 14.2, we present concepts and general considerations of water-photosplitting cells. In Section 14.3, we discuss the gold clusters supported on substrates. Then, we discuss water adsorption on supported gold NPs in Section 14.4. The elementary steps of water splitting on gold NPs at the microscopic level are discussed in Section 14.5. Quantum dynamics from nonadiabatic first-principles simulations of water photosplitting and hydrogen gas production are presented in Sections 14.6–14.8. At last, we present our conclusion and perspectives.

14.2 Water-Photosplitting Cells

A typical design of photochemical cells for producing solar hydrogen from water involves a set of different layers of components stacked in serial, including acrylic plate, spacer, photocatalytic sheet, liquid water film, and sealing acrylic plate on the top. A typical configuration (Goto et al., 2018), which uses $SrTiO_3$:Al as the photocatalyst, is shown in Figures 14.1 and 14.2. Materials and device processing to achieve splitting of water molecules upon sunlight irradiation proceeds as in the following steps. First, heat the mixture of $SrTiO_3$ (Wako), Al_2O_3 (Aldrich, nanopowder), and $SrCl_2$ (Kanto) in an alumina crucible at 1,423 K for 10 h in air, and separate $SrTiO_3$:Al from the flux by washing with deionized water. Second, impregnate $RhCrO_x$ into the $SrTiO_3$:Al from an aqueous solution of Na_3RhCl_6 and $Cr(NO_3)_3$ by calcination at 623 K for 1 h. Third, prepare a suspension of $SrTiO_3$:Al powder and nanometer-sized silica particles, drop-cast the suspension onto a frosted glass plate, and then dry on a hot plate at 323 K. Repeat the third process ten times, and ensemble the sheets as an array, which is shown in Figure 14.1b. However, it is still a long way to produce large-scale photosplitting cells owing to low efficiency of photocatalysts and many challenges such as the safe separation of H_2 and O_2, and the stability of the panels.

A variety of materials have been employed in experiments. These materials include some homogeneous photocatalysts (Si), many different metal oxides (TiO_2, $SrTiO_3$, etc.), metal sulfides (ZnS, MoS_2, etc.), metal nitrides (Ta_3N_5, Ge_3N_4, etc.), and some oxysulfides. Recently, a few groups focused on using metal-free photocatalysts such as C_3N_4 for an efficient water splitting (Wang et al., 2009; Liu et al., 2015). The first major issue during photosplitting of water on these substrates is the light absorption of the photocatalysts. In general, it requires that the conduction band is below the redox potential of $H+/H_2$ (0 V vs. normal hydrogen electrode, NHE), while the valence band is above the redox

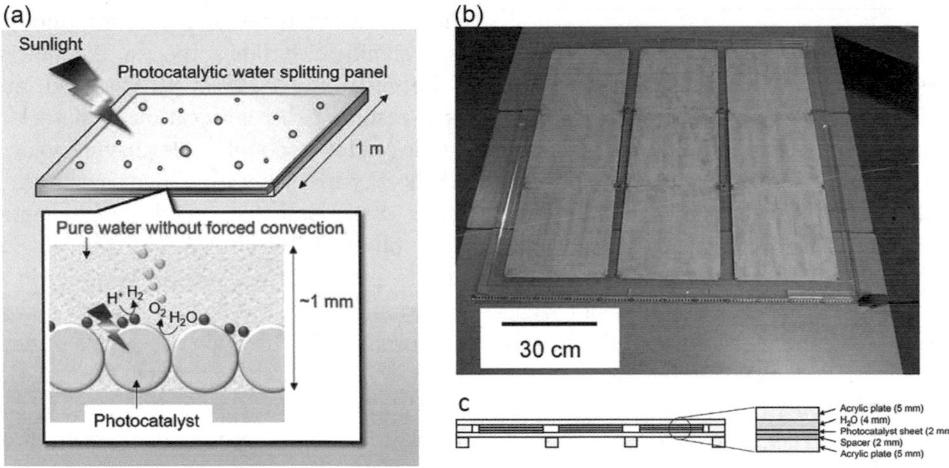

FIGURE 14.1 (a) A typical configuration of a solar-to-hydrogen (STH) cell. (b) The real-world photochemical plates. (c) Schematic of plate components and materials. (Adapted from Goto et al., 2018. Copyright: Cell Press.)

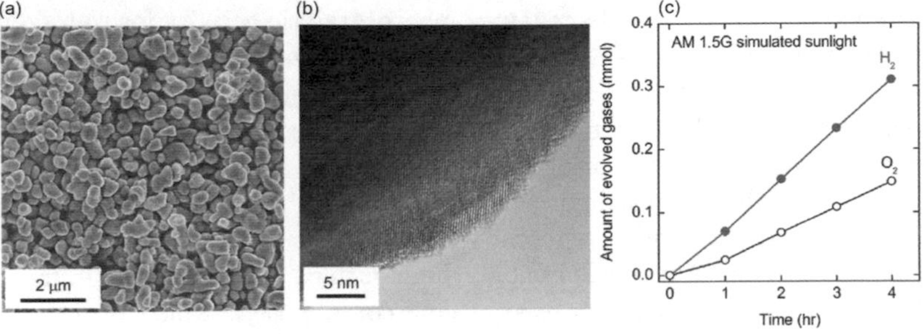

FIGURE 14.2 Transmission electron microscopic images (a,b) of photocatalytic materials used in water-photosplitting cells. (c) The production of H_2 and O_2 gases as a function of time under sunlight illumination. (Adapted from Goto et al., 2018. Copyright: Cell Press.)

potential of O_2/H_2O (1.23 V vs. NHE). The absorption region in visible light is better to nicely match the bandgap of semiconductors (Kudo and Miseki, 2009). Some materials (TiO_2, $SrTiO_3$, etc.) satisfy some of these requirements (Figure 14.3), but some substrates are only suitable for H_2 evolution (CdS, Si) or O_2 evolution (WO_3). The second important issue is the stability. The majority of photocatalyst developed for solar water splitting cannot sustain longer than 24 h. Here are some examples with exceptional performance: the $SrTiO_3$:Al photocatalyst (Goto et al., 2018) can maintain a good stability to produce stable H_2 flow for nearly 150 h, and after 1,000 h, the activity of $SrTiO_3$:Al decreases by 36%; the carbon dot–C_3N_4 composite (Liu et al., 2015) can exhibit long-term stability of 200 days after 200 times' reuse. Other challenges include charge carrier separation, high cost of water splitting, and toxicity to individuals and environments, which are in intensive laboratory exploration.

On the device level, a variety of multijunction photoelectrochemical STH cells have been developed. Typical STH conversion efficiencies range from $\ll 1\%$ to $\sim 20\%$; the recent reports on STH conversion efficiencies are summarized in Figure 14.4 (Ager et al., 2015). In fact, only eight values of STH in Figure 14.4 are $>10\%$, where the highest one

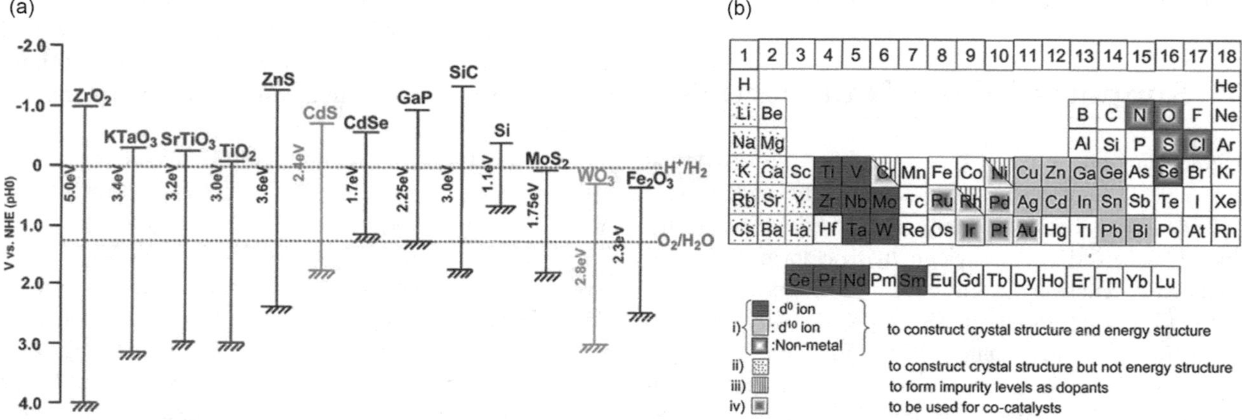

FIGURE 14.3 (a) Relationship between band structure of semiconductor and redox potentials of water splitting. (b) Elements for constructing heterogeneous photocatalysts. (Adapted from Kudo and Miseki, 2009. Copyright: Royal Society of Chemistry.)

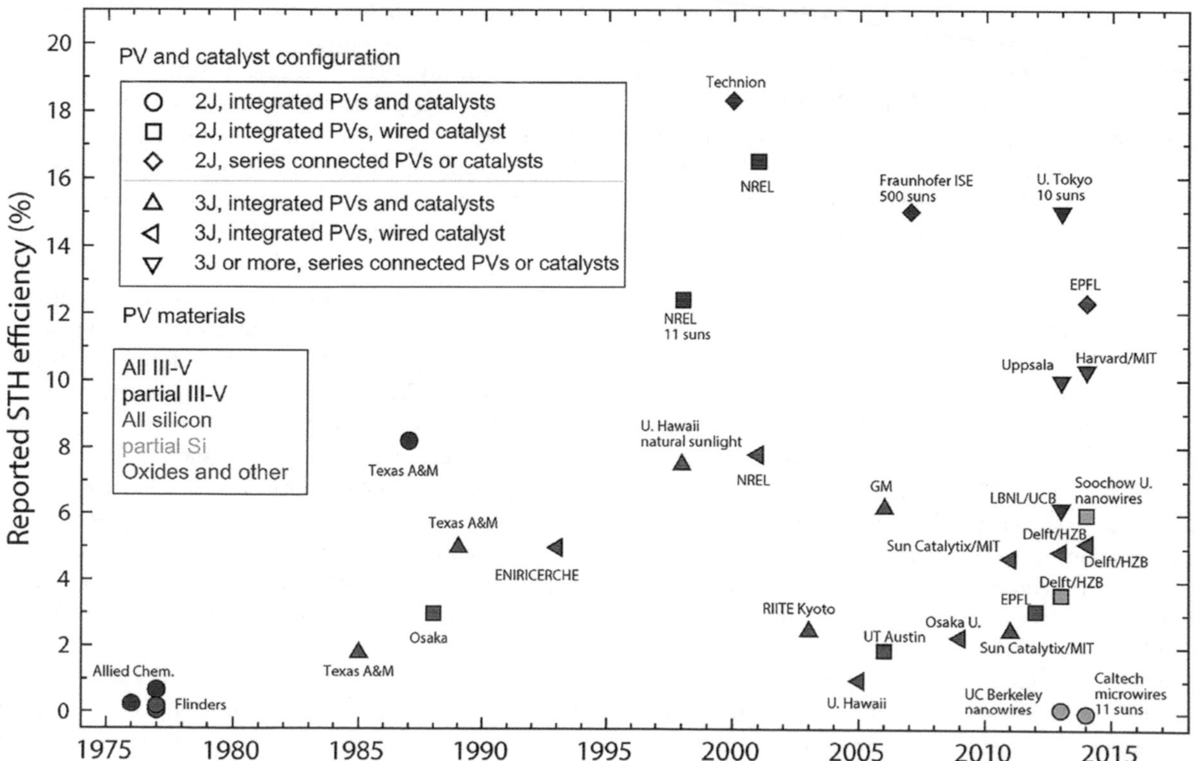

FIGURE 14.4 Reported STH efficiencies in multijunction photoelectrochemical cells. (Adopted from Ager et al., 2015. Copyright: Royal Society of Chemistry.)

is 18.3% by using $Al_{0.15}Ga_{0.85}As$(pn)//Si(pn) junctions. However, the efficiencies are still lower than the predicted theoretical limit for a single junction (>25%). Another drawback is the stability of these photovoltaic junctions in which the most stable one is a-Si(pin) that can last more than 31 days, while only a few cells can last more than 1 day.

It is obvious that both the exploration of materials and the design of device need more efforts to develop water-photosplitting cells of commercial interest. Some of these challenges attacked under the current research activities include band engineering, absorption control, carrier separation, selection of co-catalyst, low efficiency, long-term stability, large-scale production, toxicity to environment and individuals, and the economic cost of production.

14.3 Supported Gold Nanoparticles

Small gold clusters supported on surfaces are receiving intensive research interests since they can be widely used as efficient, active catalytic centers for various reactions, including CO oxidation, acetylene hydrochlorination, and water–gas shift reaction (Yoon et al., 2005; Herzing et al., 2008). Small gold NPs maintain high stability even in water solutions, making them suitable for catalytic applications. When landed on surface, different atomic configurations of gold NP have been found depending on the nature of the supporting substrate, for instance, atomic chains on NiAl(110), two-dimensional (2D) plates on alumina/NiAl(110), and three-dimensional (3D) clusters on iron oxide and titania (Herzing et al., 2008; Chen and Goodman, 2004; Wallis et al., 2002; Nilius et al., 2008). It is found that excess electrons are transferred from the substrate to supported metal NPs, due to electronic couplings between the substrate and NPs. Overall, the high catalytic activity of gold clusters on oxides has been attributed to structural effects (including particle size, shape), electronic states (charge transfer, status of metal oxidation), as well as to the influence of substrates.

Recently, it was shown that gold nanorods on TiO_2 produce stoichiometric O_2 evolution even under near-infrared illumination via plasmon-induced excitation and charge transfer (Nishijima et al., 2012). Dramatic enhancement of water splitting under visible light has been discovered upon Au thin film deposition on TiO_2, which was attributed to local field effects rather than the commonly assumed charge transfer mechanisms (Liu et al., 2011). Plasmon excitation of Ag nanocubes supported on alumina was also found to couple strongly with thermal energy facilitating rate-limiting O_2 dissociation at low temperatures (Linic et al., 2011). A full understanding of the mechanism of plasmon-enhanced photocatalytic reactions requires a detailed knowledge about adsorption geometry, charge states, electronic structure, and optical response of supported metal clusters.

In experiment, metal clusters supported by oxide thin film on metal substrate enable an in situ scanning tunneling microscopic (STM) investigation on their geometry, electronic structure, and even the oxidation processes directly. Examples include planar gold clusters with a magic number of atoms, Au_8, Au_{14}, and Au_{18}, imaged on a well-characterized MgO thin film of two monolayers (MLs) supported on Ag(001) (Lin et al., 2009). Quantum well states (QWSs) due to confinement effect in these artificial nanostructures were identified. Such studies provide atomistic information about NP/oxide interface and hint for potential ways for optimal control of reaction parameters. However, STM studies fail to provide precise identification and control of charge states of metal clusters, nor do they offer any information about optical excitation and photocatalysis mechanism in these clusters. Indeed, the excited states of supported metal cluster, especially about the influence of interface on photoabsorption and photocatalysis, are seldom investigated in either experiment or theory.

Small gold clusters exhibit unique size-dependent properties due to the presence of a large fraction of surface atoms and associated distinct electronic structures. We have investigated the supported gold clusters on MgO thin film based on first-principles density functional theory calculations (Ding and Meng, 2012). We first study isolated gold clusters Au_n ($n = 1$–10) in vacuum. We find that the average cohesive energy in Au_8 cluster is 1.96 eV per Au atom, which is higher than that in Au_7 (1.83 eV/Au) and Au_9 (1.92 eV/Au). It agrees with previous studies, where Au_8 is found more stable than A_7 (by 0.12 eV/Au) and Au_9 (by 0.03 eV/Au). In addition, it is also found that small 2D structures ($n \leq 9$) are more favorable than corresponding 3D structures in vacuum.

We then investigate the structure and properties of supported small gold clusters adsorbed on MgO/Ag(001). The optimized atomic configuration of Au_8 cluster on 2 ML MgO/Ag(001) is shown in Figure 14.5. On perfect MgO thin film, Au_8 cluster retains the planar geometry, with the most stable position being the center Au atom adsorbing at the top site of surface O atom. The adsorption energy is 0.28 eV/Au with reference to free Au_8. On the other hand, when Au_8 approaches an F-center defect (oxygen vacancy) of MgO, it adopts the 3D pyramid structure with five Au atoms close to MgO surface and three other Au staying above. On the F-center, the 3D structure is 0.82 eV more stable than the corresponding planar geometry, with the binding energy of 0.43 eV/Au. Furthermore, the O vacancy buried at the MgO and Ag(001) interface is stabilized by 0.72 eV compared to the O vacancy on the MgO surface (F-center). And we find that the Au_8 cluster above this interface defect also adopts the planar geometry, with the binding energy of 0.34 eV/Au.

To illustrate whether the surface defect or the excess charge is the driving force accounting for the 2D to 3D geometry change on MgO F-center, we compare energetics of Au_8 clusters in different conditions. The neutral Au_8 2D plate is 0.24 eV more stable than Au_8 3D cluster in vacuum; the energy difference is further enlarged when both

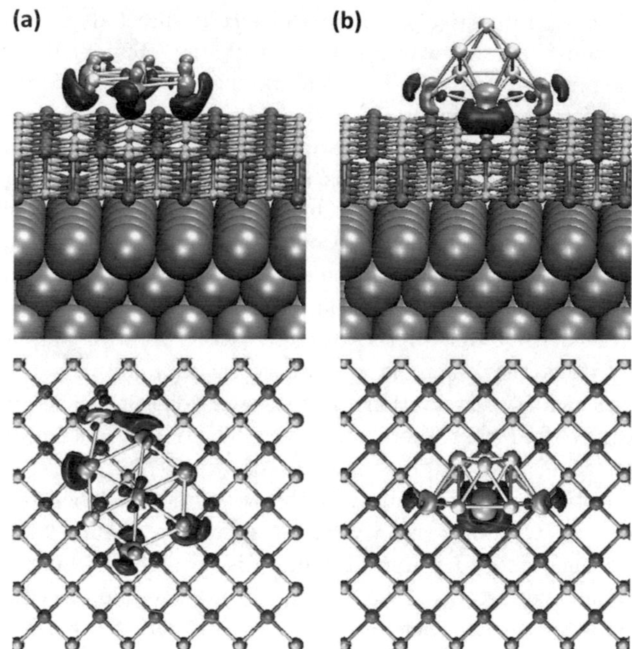

FIGURE 14.5 (a) Side and top views of the planar Au_8 NP on perfect MgO thin film. (b) Side and top views of the 3D Au_8 NP on F-center of MgO thin film. The MgO thin film (two MLs) is sitting on bulky silver (001) substrate.

structures are charged with two excess electrons (0.50 eV). Therefore, excess charge will favor 2D structures over 3D; that is, charge transfer is not the reason for Au_8 to adopt a 3D structure. The atomic structure of F-center defect is the main cause to change Au_8 plate into a 3D isomer. This fact manifests the importance of surface atomic structure for maintaining geometry, shape, and in turn electronic properties of metal clusters. We note that larger clusters Au_{14} and Au_{18} are 3D in vacuum, but all are planar on 2 ML MgO/Ag(001) (Lin et al., 2009).

The planar and 3D Au_8 NPs have different electronic characters when interacting with MgO/Ag(001) substrate. Based on Bader analysis, we find that there are excess electrons transferred from the substrate to Au_8, about $2.12e$ to Au_8 2D plate and $2.24e$ to Au_8 3D cluster. This electron transferring to Au particles is supported by projected density of states plotted in Figure 14.6c. For Au_8 plates on 2ML MgO/Ag(001), the energy level of 5d-orbital states (DS) of Au_8 overlaps with the valence band of MgO thin film. There are several QWSs presented within the MgO bandgap, originated from the quantum confinement effect in Au_8. They are labeled QWS1, QWS2, QWS3 ... running from lower to higher energies. Compared to isolated Au_8 and Au_8 on bulk MgO without Ag substrate, where only the first QWS1 is occupied, for Au_8 on 2ML MgO/Ag(001), both QWS1 and QWS2 are occupied because of $\sim 2e$ transferred to Au_8. And the energy-level spacings between QWS are smaller.

Interestingly, the distribution of excess charges is quite different for 2D and 3D Au_8, shown by the difference charge plotted in Figure 14.5. The excess electrons are

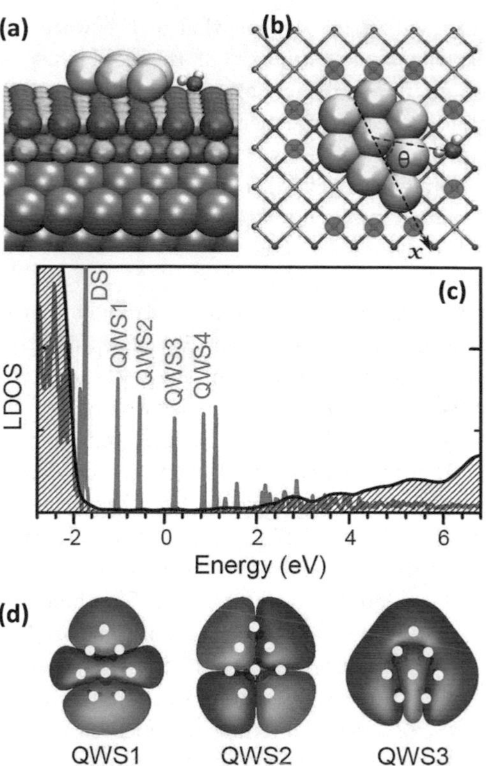

FIGURE 14.6 (a) Side and (b) top views of the planar Au_8 NP on MgO (2ML)/Ag(001), together with potential sites for water molecular adsorption (dark gray dots). (c) The electronic local density of states (LDOS) for the Au_8 on MgO (2ML)/Ag(001). (d) The wavefunctions for the QWSs (QWS1 to QWS3).

evenly distributed over every atom in Au_8 plate with s and d characters; however, in Au_8 3D cluster, all charges are strongly localized on the single Au atom in the closest contact of oxygen vacancy, due to stronger attraction between the Au atom and F-center. The different spatial distribution of excess charges may have profound implications in catalysis and photocatalysis applications.

14.4 Orbital-Dependent Water Adsorption

In this section, we investigate the interactions between water molecules and Au_8 plate supported on MgO thin film.

First of all, numerous site potentials for water adsorption are explored. We find that direct water adsorption on the top of Au_8 cluster is not favored, with binding energies E_{ads} being as small as 86 meV (above center Au atom) and 137 meV (on the Au atom at the edge of the cluster). This is consistent with weak interactions (110 meV) between water and Au(111) (Meng et al., 2004).

On the other hand, water adsorption is dramatically enhanced around MgO-supported Au NPs: E_{ads} increases from 360 meV on perfect MgO to 620 meV far away from the gold NP, and 780 meV for water on the periphery of Au_8, where water sits on the top of surface Mg atoms closest to Au_8. Depending on the location of the water molecule relative to the symmetry axis of the gold cluster, water adsorption energies fall in the ideal range of 0.6–0.8 eV, all larger than the cohesive energy of bulk water (0.58 eV), enabling prompt surface wetting to promote subsequent reactions (Ding et al., 2015). For simplicity, we characterize the orientation of water adsorption sites by the angle θ water makes to the symmetry axis (x-axis) of Au_8, as shown in Figure 14.6b.

Next, we find that water couples strongly to the QWS of Au_8, leading to orbital selectivity of water adsorption. The angular dependence of the binding energy of water is associated with the angular distribution of the electron density of QWS2 along the gray circle marked in the inset (Figure 14.7a). The density of QWS2 shows a four-lobe symmetry. Its wavefunction has four nodes at $\theta = 0°$, 90°, 180°, and 270°, respectively (Figure 14.6d). The bonding distance between the OH and the Au_8 NP, the H–Au bond length, is ≈2.2–2.4 Å. The binding energy shows angular dependence similar to that of the charge density. The maximum adsorption energy, 0.80 eV, appears at $\theta = 54°$, which corresponds to the wavefunction maximum (lobe) of QWS2. Other maxima are located at $\theta = 136°$, 228°, and 394°, respectively. The minimum of the binding energy is located at $\theta = 0°$, with $E_{ads} = 0.58$ eV, although geometrically the water molecule is closest to the apex gold atom in this configuration, whose coordination number is also the lowest (see Figure 14.6b). This indicates that the water–Au_8 interaction is governed by the global electronic structure of the QWS rather than by local atomic geometry. Other energy minima appear at $\theta = 78°$, 183°, and 253°, respectively, where the OH is directed toward the nodes of the wavefunction of the QWS2, resulting in a minimal coupling between water and QWS2. Water on a free gold cluster exhibits a similar angular dependence, showing that Mg–water bonding is not a cause of energy oscillation. This is also evidenced by the fact that Mg–O bond distances between the oxygen of the water molecule and the Mg atom are underneath constant, 2.11–2.14 Å, for all adsorption sites. Although the extrema of the adsorption energy deviate slightly from those of the charge density due to the discrete lattice sites of Mg, it is obvious that there is a close correlation between the charge density distribution and the water adsorption energy. We also find that there appears a linear and antilinear correlation between the energies of the QWS2 and the adsorption energy of water, at the wavefunction lobes and nodes of QWS2, respectively, as shown in Figure 14.8. This suggests that water adsorption is strongly modulated by the symmetry of the QWS. Strong coupling occurs when the overlap between the OH and QWS2 reaches maximum.

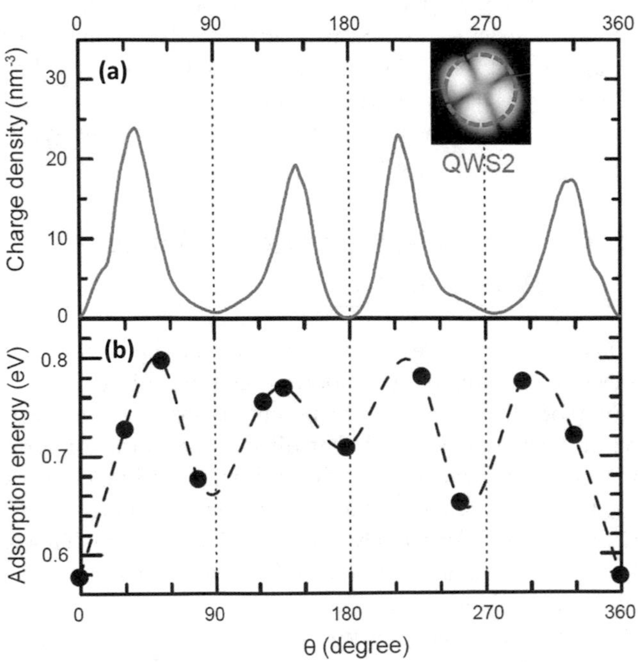

FIGURE 14.7 (a) The charge density distribution from the QWS2 at the periphery of Au_8 cluster. (b) Orientation dependence of water adsorption energies.

Now we will address why water adsorption at QWS2 lobe positions has larger adsorption energy than that at node configurations, despite their almost equivalent local atomic structure, which is the main feature of water–Au_8 interaction.

The interaction between water molecule and the Au_8 cluster involves mainly their frontier orbitals, namely, the highest occupied molecular orbital (HOMO) ($1b_1$) and the lowest unoccupied molecular orbital (LUMO) ($4a_1$) of water molecules and the QWS2 of the Au cluster. The LUMO of water has an s-like orbital symmetry on H, while the occupied HOMO states are p-like. When water adsorbs at the lobe positions of QWS2 (see geometry (a) of Figure 14.9), it forms bonding ($\psi^+ = \psi_{H_2O} + \psi_{QWS2}$) and antibonding ($\psi^- = \psi_{H_2O} - \psi_{QWS2}$) states. Only the LUMO state of water dominates in the interaction because of the wavefunction symmetry of water orbitals: there is

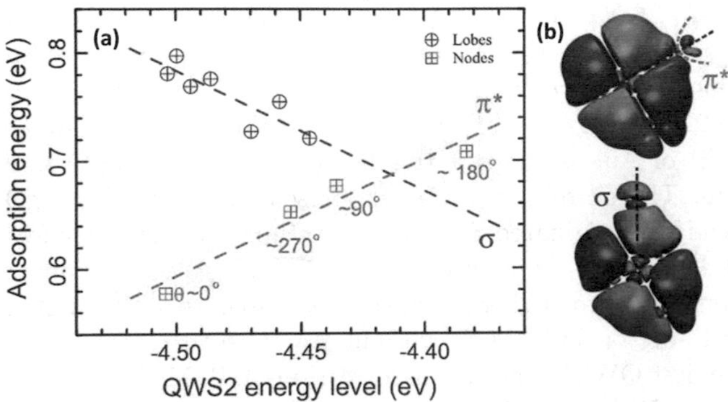

FIGURE 14.8 (a) Correlation between water adsorption energy and the QWS2 energy level for water binding at the node and lobe positions of QWS2. (b) The QWS2 wavefunction upon water adsorption shows σ and π* bonding at the lobe and node positions, respectively.

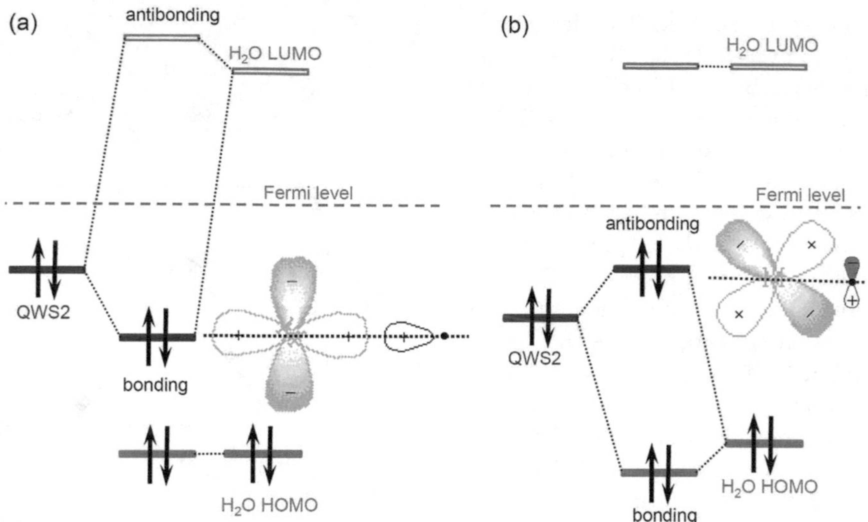

FIGURE 14.9 (a) Schematic energy diagram for the interaction between water and the QWS2 of Au_8 for water adsorption at the lobe positions. Inset shows schematic wavefunction of QWS2 upon interaction. (b) The same as (a) but for water adsorption at the node positions of QWS2.

substantial wavefunction overlap and hopping interaction $\langle\psi_{H_2O\_LUMO}|H|\psi_{QWS2}\rangle$ between s-like water LUMO and the QWS2, while there is the overlap for p-like water HOMO $\langle\psi_{H_2O\_HOMO}|H|\psi_{QWS2}\rangle \approx 0$. The large interaction between unoccupied water LUMO and the QWS2 leads to a low-energy bonding state, which in turn is a local energy minimum in the total energy of the combined system. However, when water adsorbs at node positions (geometry (b) in Figure 14.9), the hybridization between the LUMO of water and QWS2 of Au, $\langle\psi_{H_2O\_LUMO}|H|\psi_{QWS2}\rangle$, is almost vanishing due to the orbital symmetry (see Figure 14.8). The interaction is thus dominated by the HOMO ($1b_1$) states, leading to an antibonding states, and gives thus a higher total energy in the combined system. This is clearly shown from calculated wavefunctions of QWS2 upon water adsorption (Figure 14.8).

The linear and antilinear correlation between the energies of the QWS2 and the adsorption energy of water in Figure 14.8 results simply from the fact that the change of total energy of the system is simply dominated by the electronic energy contribution, reconfirming the electronic mechanism of water–QWS interaction. The sign of the slopes (+, −) is determined by the bonding and antibonding nature of this interaction, which leads to energy shifts in opposite directions, respectively. The energy shifts of QWS2 at different configurations in Figure 14.8 qualitatively support this picture. Interactions involving other QWSs can be analyzed in the same way.

At last, we have learned that water adsorption on Au_8 is dependent on molecular orbitals. The interaction between H_2O and QWS orients the OH group in water pointing to negatively charged Au atoms, elongating the OH bond length from 0.97 to 1.02 Å, and shortening H-Au length to 2.27 Å. This interaction would facilitate water splitting.

14.5 Elementary Atomic Steps in Water Photosplitting

14.5.1 Water Dissociation

We further investigate the ground-state kinetics of water dissociation on Au_8/MgO(2ML)/Ag(100), toward producing activated hydrogen atoms (denoted [H]) on the gold NP. The reaction energies of water dissociation on Au_8/MgO/Ag(100) are shown in Figure 14.10. Water dissociation on Au_8/MgO/Ag(100) with MgO thin film ranging from 1 to 3 ML, and that on different adsorption sites (S1: $\theta = 228°$ vs. S2: $\theta = 253°$) are compared. Water dissociation reaction starts with an intact water molecule adsorption (A). The attraction between H_2O and the QWS orients the OH bond of water molecule pointing to gold

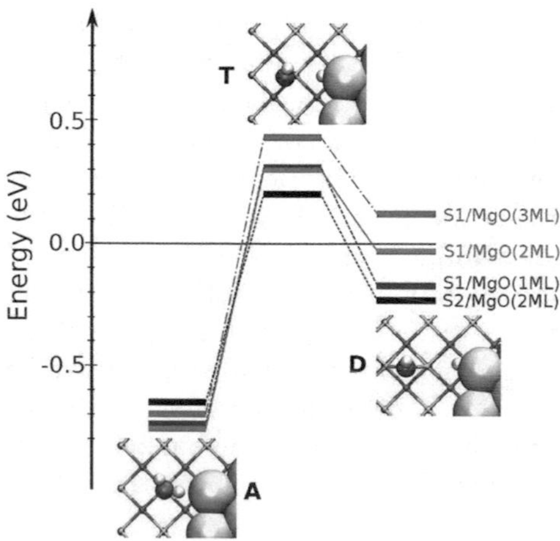

FIGURE 14.10 Reaction energy profiles and structures of water dissociation on Au_8/MgO/Ag(001) for the adsorption sites (S1: $\theta = 228°$ and S2: $\theta = 253°$) and 1–3 ML MgO thin film.

atoms with the H-Au length of ~2.3 Å. The dissociative (D) adsorption is the reaction product: the OH group binds onto the bridge site of Mg–Mg bond on MgO surface, and hydrogen atom binds on the gold cluster. The equilibrium distance between the [H] and nearest Au atom is ~1.6 Å. Bader analysis reveals that the charge on adsorbed OH is $-0.85e$ and that the charge on activated hydrogen atom is -0.04 to $-0.12e$. That is, the reaction product OH is anionic and the [H] is neutral (Ding et al., 2017). This fact is radically different from that for water dissociation on bare MgO, which produces an H charged by $+0.6e$. Unlike water dissociation catalyzed by gold cluster, the H atom is similar to H atoms in water molecule and is inactivated. The thickness of MgO thin film does not alter the water dissociation barriers significantly (ranging 1.05–1.13 eV), while adsorption on site S2 has a significantly lower water dissociation barrier (0.85 eV) than on site S1 (1.06 eV).

Next, we study how QWSs affect the kinetic pathway and energy barrier for water dissociation. Figure 14.11 shows the calculated activation energy (E_b) for water dissociation as a function of θ. The barrier varies dramatically between 0.85 and 1.33 eV and is highest at $\theta = 180°$ and lowest at $\theta = 250°$. This barrier is drastically reduced from that for the dissociation of gaseous water molecules (5 eV), and is close to the ideal value of 0.5–0.7 eV for efficient water splitting at near-ambient temperatures. Because the OH group of the dissociated state is relatively far away from the gold cluster in all configurations, the energy change caused by OH group displacement is roughly constant, at about 0.33–0.44 eV, for different adsorption sites. Therefore, the angular variation of energy barriers is mainly determined by the interaction between the dissociated H atom and the QWS in the transition state.

To illustrate this interaction in more detail, Figure 14.11c shows the binding energy of an H atom, relative to isolated H and supported Au_8, as a function of θ, which indeed follows angular dependence similar to that of the dissociation barrier, as shown in Figure 14.11b. The binding energy of the H atom is almost identical to half the H–H bond energy (4.5 eV), facilitating H_2 release. It also shows a larger variation, ~0.4–0.6 eV, than that for water. The angular dependence is in excellent agreement with the charge density distribution of the QWS3 state shown in Figure 14.11d, where the maxima of the barrier and binding energy correspond to the antinodes at $\theta = 90°$ and 270° (and an additional smaller antinode at $\theta = 0°$). From Figure 14.11, it is clear that the coupling between the H atom and QWS3 dominates the interaction between the gold cluster and the transition state and thus modulates the energy barrier for water dissociation. A detailed analysis shows that the energy barrier is linearly dependent on the binding energy of H on Au_8, a nice demonstration of the Bronsted–Evans–Polanyi principle.

It is interesting to note that the dissociation barrier is essentially modulated by QWS3 rather than QWS2 as for molecular water adsorption. The different trends in water adsorption energies and dissociation barriers result from

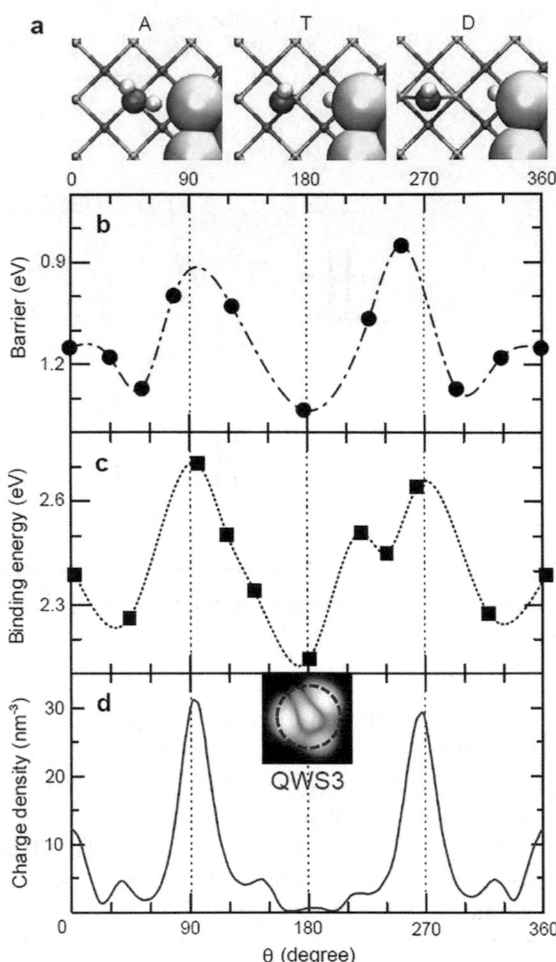

FIGURE 14.11 (a) Typical path for water dissociation starting from molecular adsorption of water (A) to the transition state of dissociation (T) to dissociated adsorption (D). (b) Water dissociation barriers and (c) H atom binding energies as a function of orientation angle. (d) Angular distribution of charge density of QW3 in the selected edge region. The inset shows the shape of QW3 and the edge zone.

additional charge transfer from silver substrate at the transition state and the dissociated state, which makes QWS3 occupied and responsible for changes in H binding energetics during water splitting. Our Bader charge analysis shows that an additional $0.6e$–$1.4e$, approximately, are transferred from silver substrate to adsorbates during water dissociation. The local density of state analysis suggests that QWS3 shows a large downward shift of energy by about 1–2 eV. We also find that this energy shift is the largest at $\theta = 90°$ and 270° (Figure 14.11c), which further confirms that the dissociation barrier is modulated by the interaction between the H atom and QWS3.

Together with insights gained into the water–QWS2 quantum interaction, we expect that water splitting can be optimized on adsorption sites where wavefunction nodes of QWS2 meet QWS3 lobes with large overlap, such that H binding is drastically enhanced and simultaneously water adsorption is slightly destabilized to further reduce their energy difference, the reaction barrier. We find $\theta = 90°$ and $270°$ to be such optimal sites.

14.5.2 Diffusion of Reaction Products

As a practical catalyst, the supported Au_8 on MgO should not suffer from poisoning and we perform the following simulations to establish this fact.

We verify that the activated H atom (denoted [H]) can be readily collected on the Au_8. Both the barrier (~1.06 eV) and the reaction path of producing the second [H] on the S1 site are very close to those of producing the first [H] on the same site (~1.05 eV). The effect of [H] on the gold cluster is highly localized and has little influence on water dissociation on other distinct sites. Collecting the [H]s is not correlated with the adsorbed [H]s on the other sites. All the reactions for [H]s' collection are similar to the production of first [H]. In the same mechanism of water dissociation, tuning the QWS's occupation by defect doping and excitation is helpful for producing the [H]s.

In addition, we confirm that a neighboring water molecule hydrogen-bonded to the water molecule on S1 site also enables an efficient water dissociation, whose product [H] also could transport to Au_8. The reaction barrier on MgO bulk is 1.07 eV (Shin et al., 2010; Jung et al., 2010), close to water dissociation on the same site of Au_8 on MgO(2ML)/Ag(001). Therefore, water molecule adsorption does not have a negative effect on water dissociation, and the hydroxyl spreads away from the gold cluster, facilitating the separation of H and OH products.

We further simulate the diffusion of OH radical between two water molecules on MgO(2ML)/Ag(001) without the gold cluster. The simulation reveals that a rapid proton transfer within the OH radical and water molecules could take place, similar to the proton transfer within water dimer and clusters (Hu et al., 2010). The barrier is only 0.24 eV, smaller than 0.60 eV, which is the case without hydrogen bonds. With hydrogen bond between water molecules, the OH radical spreads away from the gold cluster easily. We also find that the energy of OH removal is positively correlated with the MgO thickness (Jung et al., 2010).

Thanks to the significant overlap between 5d-orbital states (DS) of the gold cluster and the valence bands of MgO surface, the hole in the DS states of Au_8 generated by photoexcitation would transfer to MgO very efficiently. The holes gathered on MgO substrate would oxidize OH groups generated from water dissociation and produce H_2O_2 or O_2 gas on MgO. In reality, MgO is a good hole conductor and routinely used as an oxygen-storage material.

14.5.3 H_2 Formation

The activated [H] on the gold cluster can move easily between the different adsorption sites, with a barrier of 0.14 eV from S1 to S2 site. The [H] on the S2 site is the most stable. It is likely that H_2 gas is produced by combining two and more activated hydrogen atoms on Au_8.

In order to confirm the possibility of hydrogen production on Au_8/MgO/Ag(001), we design three reactant states (R) to synthesize hydrogen gas (H_2) as follows: (i) the dimer of OH belonging to hydroxylated MgO surface without the gold cluster, (ii) two [H]s adsorption on the neighboring sites of Au_8, and (iii) H_2O and one [H] adsorption on the same site of Au_8. The structures of the reactants are shown in Figure 14.12. Hydrogen production is performed to get the reaction product (P), which is H_2 molecule weakly adsorbed on MgO surface.

The transition states (T) and reaction paths are obtained by nudged elastic band (NEB) calculations:

When H_2 is synthesized from reactant state R_1: one hydrogen atom is split from surface hydroxyl, and combines with the other H with a barrier of 1.83 eV. Because of the highest barrier, this reaction path is difficult to occur.

i. From reactant state R_2: one [H] combines with another [H] with a barrier of 0.80 eV. Higher concentration of [H]s on Au_8 would result in an even lower barrier for H_2 desorption. Because ΔE is only 0.1 eV, the energy of each hydrogen atom in the R_2 state is close to that of H_2. This barrier is the lowest, which is the most probable path for H_2 production.

ii. From reactant state R_3: the hydrogen atom splits from the water molecule and combines with the [H] on Au_8. The energy of transition state (T) is close to that of final state (P), which is determined by the repulsion between H_2 molecule and the gold cluster. Because of high stability of the reactant state (R) ($\Delta E = 0.96$ eV), the barrier 1.58 eV is so high that the reaction can hardly take place.

Without the gold cluster, hydrogen gas production is unlikely to occur on bare MgO/Ag(001) surface owing to the large barriers. For the same reason, the dissociation of two water molecules on the same site is difficult for

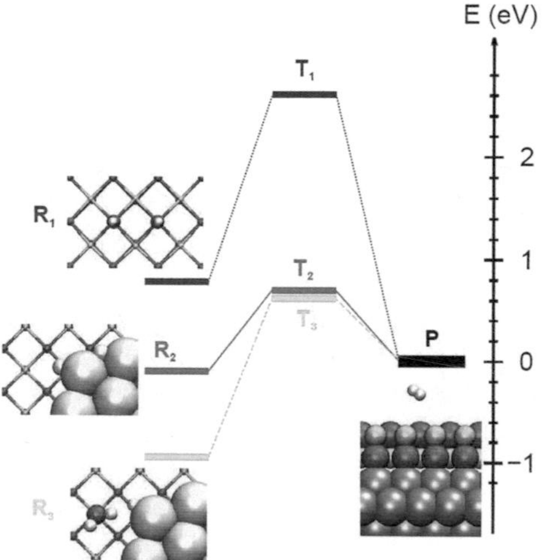

FIGURE 14.12 Reaction energy profiles for H_2 generation on MgO(2ML)/Ag(001) without/with the gold cluster.

H_2 generation. The hydrogen gas is produced most possibly by many activated hydrogen atoms on the same gold cluster. The gold NP serves as a reaction center to collect activated hydrogen atoms and produce H_2 gas.

Above all, these discussions are based on static calculations with density functional theory and NEB methods, to analyze the adsorption of water, the interaction between water and Au_8, the diffusion of OH radical, and the formation of H_2. Next, we will introduce the dynamic aspects of water photosplitting revealed by time-dependent density functional theory (TDDFT) simulations.

14.6 Quantum Dynamics of Water Photosplitting

In this section, we investigate the atomic-scale mechanism and real-time dynamics of water photosplitting on Au NPs, irradiated by femtosecond laser pulses, employing TDDFT. We explore the possibility of using supported Au magic clusters for water photosplitting. Decomposing water on oxide films has been a research focus currently under intensive investigation. It has been shown that MgO thin film can significantly decrease water dissociation barriers using thermal energies as input. However, the large bandgap of oxides (>6 eV) prevents any effective photon reactivity under solar irradiation. Since the substrate does not directly get involved in light absorption and photodriven reactions, we model only the photocatalysts without the substrates for simplicity and for the sake of affordable computational cost.

We first investigate the dynamic response of water on Au NP to laser field as shown in Figure 14.13, which displays time-dependent atomic distance d of water to the Au NP surface. Here, the Au NP is modeled by the jellium model for affordable electron dynamic simulations. All calculations are performed with adiabatic local density approximation and generalized gradient approximation of the exchange-correlation functional, and electron–nuclear interactions are described by the norm-conserving Troullier–Martins pseudopotentials (Troullier and Martins, 1991). Electron–ion dynamics is treated with the Ehrenfest scheme controlled at the electronic propagation timestep of typically 1–50 as (Alonso et al., 2008). We use a jellium sphere with electron density r_s = 4.5 a.u. to simulate the Au NP with a diameter D = 1.6–2.1 nm. This approximation is commonly used in literature (Zheng et al., 2004) since it captures the most important properties of metallic NPs.

External laser field polarized in the z-direction, which runs through the center of nanosphere (set as $z = 0$) and the oxygen (O) atom, is shaped by a Gaussian wave packet (Figure 14.13b)

$$E(\omega, t) = E_{\max}\exp\left[-\frac{(t-t_0)^2}{2\tau^2}\right]\cos(\omega t - \omega t_0 + \varphi)$$

where the phase φ is set to zero and the parameter τ is 1.6 fs. The maximum field strength of laser pulse is set to $E_{\max}$ = 1.6 V/Å and frequency $\hbar\omega$ = 2.62 eV (Figure 14.13b).

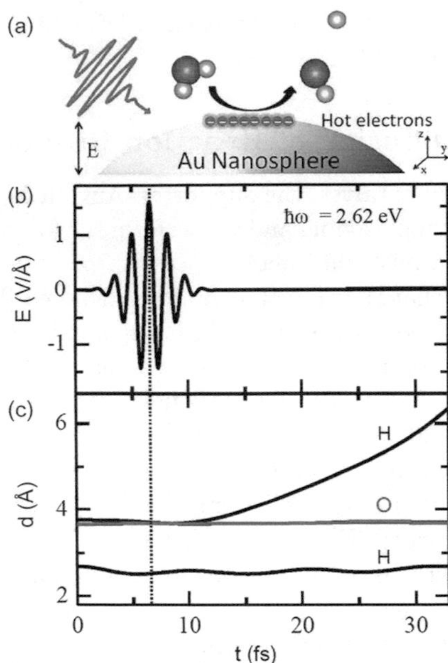

FIGURE 14.13 (a) Schematic showing plasmon-induced water splitting on Au nanosphere (D = 1.9 nm) under the laser field polarized in the z-direction. (b) Time evolution of the applied field and (c) atomic distance d of water along the z-direction to the Au surface. The vertical dotted line denotes the time t_0 = 6.6 fs at which laser field reaches its maximum strength.

This frequency matches the plasmonic absorption peak at 2.62 eV calculated for the Au NP (D = 1.9 nm). For comparison, the major absorption peak at 2.60 eV (Cottancin et al., 2006) is experimentally observed for Au NPs with the mean diameter of 1.9 nm embedded in alumina, in good agreement with our model. From Figure 14.13c, we find that the O atom is almost static and the H atom pointing to the NP oscillates in a 10-fs period during the simulation timespan. The height of the other H first oscillates, then keeps increasing from d = 3.7 Å at t = 10 fs to d = 6.4 Å at t = 33 fs. The corresponding OH bond length increases from 1.12 to 2.84 Å. That is, water molecule splits into hydroxyl group (OH) and hydrogen (H) within 30 fs (Yan et al., 2016). In contrast, for an isolated water molecule in the same laser field, two OH bonds oscillate continuously and never break. Therefore, we confirm that water splitting is mediated by plasmon excitation in the Au NP.

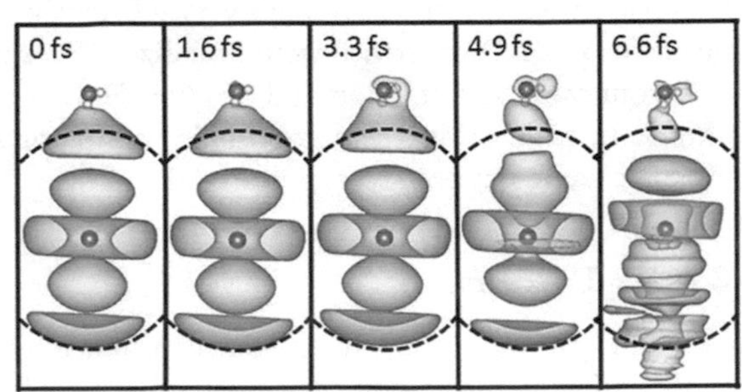

FIGURE 14.14 Snapshots of the simulated time evolution of charge density at the Fermi level. The gray dot denotes the center of the NP, and the dashed line indicates the NP surface.

We also examine the time-dependent charge density at the Fermi level E_F (Figure 14.14). Initially ($t = 0$ fs), there is almost no electron distributed on the water. At $t = 3$ and 7 fs, a small part of electrons indeed transfer from Au NP to water, implying state hybridization between Au NP and water. Thus, we provide direct evidences that water photosplitting on Au NP is induced by photoelectric conversion.

We find that the reaction rate strongly depends on the adsorption configuration of water on the NPs. In Figure 14.15, we show reaction rate and charge density variations during photoexcitation of plasmon modes at 2.62 and 2.36 eV. We see clearly that the H-up asymmetric configuration (second row in Figure 14.15) shows the highest reaction rate (62 ps^{-1} for 2.62 eV; 64 ps^{-1} for 2.36 eV) when one O–H bond points away from Au NP, while in

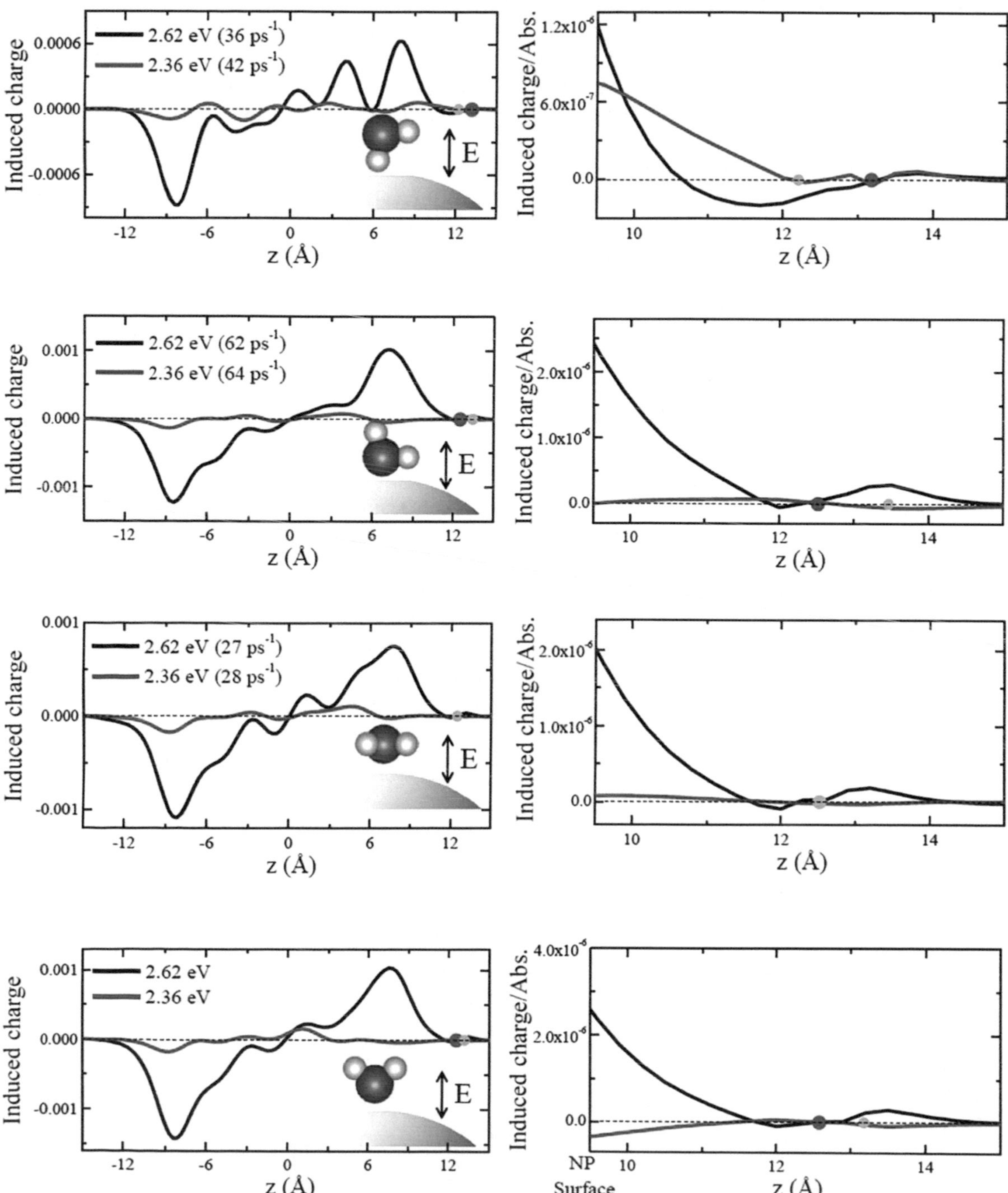

FIGURE 14.15 Average induced charge density for Au NP ($D = 1.9$ nm) at two plasmon resonances 2.36 and 2.62 eV, with the variation of water structure along the z-axis. The data in the bracket is water-splitting rates under the laser with $E_{\max} = 1.6$ V/Å and width $\tau = 1.6$ fs for the time span 50 fs. The water splitting from up to down is as follows: one top hydrogen atom splitting, one hydrogen atom away from NP splitting, two hydrogen atoms almost splitting at the same time, and no hydrogen atom splitting, respectively.

the H-up symmetric configuration (fourth row), there is no water splitting observed in the 50-fs simulation. We suppose that this phenomenon may be caused by the symmetry of water molecule. For symmetry protection, symmetrical configuration shows lower reaction rates than the asymmetric configurations (first and second rows), which have a lower symmetry. Therefore, the adsorption configurations may play an important role in water photosplitting.

14.7 Quantum Mode Selectivity

As we can see from above, our simulations strongly suggest that the rate of water splitting depends not only on optical absorption of Au NP, but also on its plasmonic excitation mode.

Further evidences implicating plasmon-induced reaction can be drawn from wavelength-dependent photocatalytic rates in Figure 14.16. For the Au NP with $D = 1.6$ nm, the absorption spectrum shows double peaks due to quantum size effect. Similar splitting of absorption spectrum has been observed experimentally for thiol-protected Au_{25} clusters (Zhu et al., 2008) and Ag NP with diameters ranging from 20 nm down to 1.7 nm (Scholl et al., 2012). No matter for Au NP with or without water adsorbed, low-energy absorption peak always exists. According to the previous analysis of electronic transition (Yan et al., 2011; Kale et al., 2014b), both major absorption peaks are plasmonic. Reaction rates also exhibit two major peaks, closely following the absorption spectrum. The reaction rate of 16 ps^{-1} for the plasmon excitation mode at $\hbar\omega = 2.34$ eV is higher than the reaction rate of 12 ps^{-1} for the mode at $\hbar\omega = 2.61$ eV, in coincidence with corresponding absorption intensity, which has a ratio of 1.25:1. Therefore, in the case of $D = 1.6$ nm, water-splitting rate compares well with absorption spectrum, in good agreement with previous experimental measurements (Linic et al., 2011; Mukherjee et al., 2013, 2014; Christopher et al., 2012; Shi et al., 2015; Lee et al., 2012; Ingram and Linic, 2011).

However, a surprising contrast in reaction rate and photoabsorption is found for Au NP of larger size. For Au NP with a diameter $D = 1.9$ nm, the intensity for absorption at resonance frequency $\hbar\omega = 2.62$ eV is almost doubled compared to that for $\hbar\omega = 2.36$ eV. The reaction rate with laser frequency $\hbar\omega = 2.62$ eV, however, is only 36 ps^{-1}, even smaller than the rate of 42 ps^{-1} at $\hbar\omega = 2.36$ eV (black dots in Figure 14.16b). Again, for Au NP with $D = 2.1$ nm, the reaction rate (38 ps^{-1}) at $\hbar\omega = 2.63$ eV, where the maximum absorption is reached, is almost same to that (38 ps^{-1}) at $\hbar\omega = 2.30$ eV with only weak absorption (Figure 14.16c). In the recent experiment using aluminum nanocrystals as a plasmonic photocatalyst for hydrogen dissociation, maximum HD (the molecule comprising a hydrogen atom and a deuterium atom) production is also observed at illumination wavelengths around 800 nm, even though its absorption peak is smaller than the plasmonic mode at around 460 nm (Zhou et al., 2016). However, the authors assigned the low-energy peak to the Al interband transition, not a plasmon. Also, the separation between the peaks is about 1.3 eV, much larger than ~0.3 eV in the present model.

To understand why plasmonic mode would affect the rate of water splitting, we display average induced charge density and Fourier transform of induced charge density for both modes in Figure 14.17. Take the Au NP with diameter $D = 1.9$ nm as an example. For the mode at $\hbar\omega = 2.36$ eV, there are two oscillating periods and one half-period from the NP center to both sides (Figure 14.17a); therefore, the oscillation period n is odd ($n = 5$). For the mode at $\hbar\omega = 2.62$ eV, there are only two oscillating periods from the center to both sides (Figure 14.17b), and this is even-period plasmon excitation ($n = 4$). Upon water adsorption, we find that the odd mode, indeed, induces more electrons around water than the even mode (Figure 14.17c), due to net electrons in the half-period of the odd mode strongly coupled to water. For different geometry of adsorption, the largest amount of net electrons appears in the H-up asymmetric configuration in Figure 14.16, and this also implies that the plasmon excitation mode can only couple with a certain geometry of water adsorption.

Additionally, the same phenomenon is observed for the Au NP with $D = 1.6$ nm. However, for the Au NP with $D = 1.6$ nm, there is a slightly larger induced charge at the odd mode, but the effect is much less pronounced than that for NP with $D = 1.9$ nm. We also examine the NP with $D = 2.1$ nm, and find that the mode at 2.63 eV

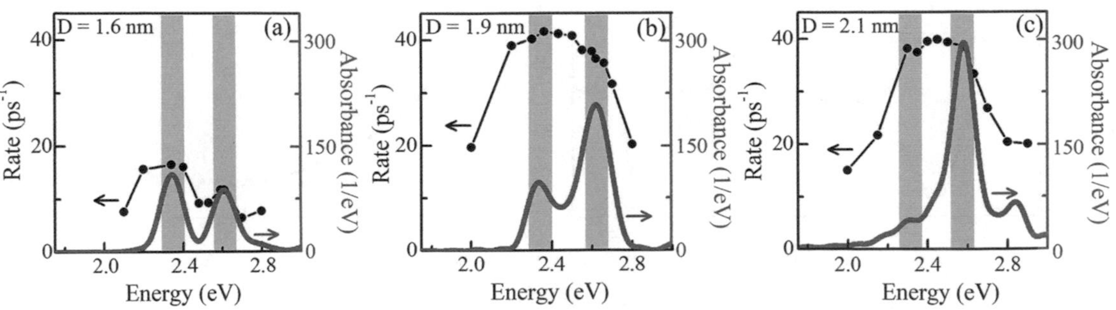

FIGURE 14.16 Water-splitting rates (black dots) on Au NP with the diameter D of 1.6 nm (a), 1.9 nm (b), and 2.1 nm (c), for the pulse shape shown in Figure 14.13b with varied laser frequency. The connected line is the guide to eyes. Corresponding absorption spectrum (gray lines) of Au NP to an impulse excitation in the z-direction is superimposed for comparison.

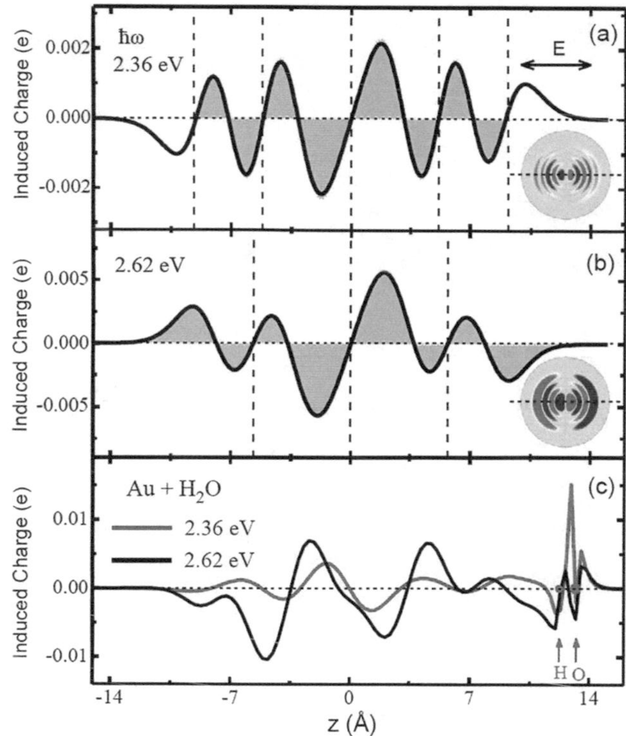

FIGURE 14.17 Fourier transform of induced charge density for Au NP (D = 1.9 nm) at two plasmon resonances 2.36 eV (a) and 2.62 eV (b), and for AuNP + H_2O (c) at the resonant frequencies 2.36 eV (gray line) and 2.62 eV (black line), to an impulse excitation along the z-axis. Inset shows corresponding induced density projected into the xz plane with the dotted line directing the z-axis. The dashed line in (a) and (b) is for partition of one oscillation period. The dark gray and light gray dots in (c) specify the position of oxygen and hydrogen atoms, respectively. The second hydrogen atom is at the same distance as the oxygen atom.

seems to induce more electrons around water than that at 2.30 eV. Yet, the reaction rates for both modes are almost the same. Therefore, we suppose that further research on mode-associated mechanisms is needed to explain all these behaviors.

14.8 "Chain Reactions" toward H_2 Generation

Now we simulate the direct production of H_2 molecules on gold NPs under solar irradiation. As Au_{20} nanocluster has a symmetric geometry and high stability (Zhao et al., 2006), we choose Au_{20} as a realistic model for gold NPs, which is enclosed with 52 water molecules to represent liquid water environment (see Figure 14.18). The stable geometry from the ground-state molecular dynamics (MD) simulations at room temperature (300 K) is selected as starting condition for TDDFT-MD simulations. In the initial configuration, three water molecules bind to the corner site of Au_{20} cluster by forming Au–O bonds (bond length: 2.3 Å). These bonds enhance the Lewis base character of Au atoms that are not

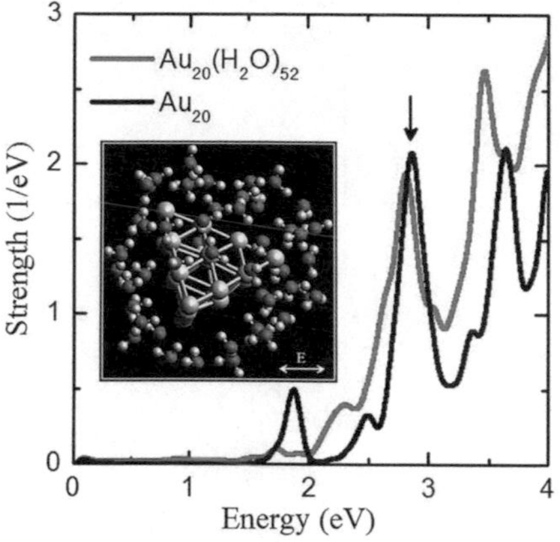

FIGURE 14.18 Optical absorption spectrum of gold cluster Au_{20} (black line) and Au_{20} in water (gray).

connected to water molecules, thereby preventing further bonding of water molecules to the Au_{20} cluster.

We first explore optical absorbance of the Au_{20}–water system (Figure 14.18). The adsorption spectrum shows a dominant peak at 2.81 eV (Figure 14.18), in good agreement with the value of 2.78 eV in literature (Idrobo et al., 2007). This peak is composed of multiple electronic excitations and is identified as plasmon resonance (Cushing and Wu, 2016; Wu et al., 2004). Other peaks at 1.73 and 2.30 eV are too weak in intensity to excite collective plasmon resonance. The peaks at 3.4–3.6 eV ascribe to the excitations from four vertex atoms of Au_{20} clusters (Idrobo et al., 2007; Wu et al., 2004). They are in the ultraviolet spectrum and therefore not our concern. The liquid water environment induces a redshift of 0.05 eV for the major absorption peak. The redshift in resonant energy can be understood by spontaneous charge transfer from Au_{20} to nearby water molecules (Sheldon et al., 2014).

To better understand plasmon-induced water-splitting reaction, we probe the dynamic response of the system using different laser fluence values shown in Figure 14.19a. In the following, we deem O–H bond split whenever its length reaches 2.0 Å. We find the number of O–H split is linearly dependent on the laser fluence, implying a single-photon process (Kale et al., 2014a). We note the linear line in Figure 14.19a does not go across the origin, which implies that a critical laser fluence is needed for water splitting (Mukherjee et al., 2013). In experiments, this linear relationship has been observed for hydrogen dissociation on Au NPs (Mukherjee et al., 2013) and ethylene epoxidation on Ag nanocubes (Christopher et al., 2011). Figure 14.19b shows wavelength dependence for the number of split O–H bonds, which compares well with the overall shape of optical absorption spectrum (Yan et al., 2018). This further confirms that water splitting is induced by plasmon excitation of Au_{20}, in good agreement with previous experiments (Christopher et al., 2012; Shi et al., 2015; Ingram and Linic, 2011). To illustrate the atomic processes during water splitting, Figure 14.19c shows the typical snapshots of time-dependent atomic configuration.

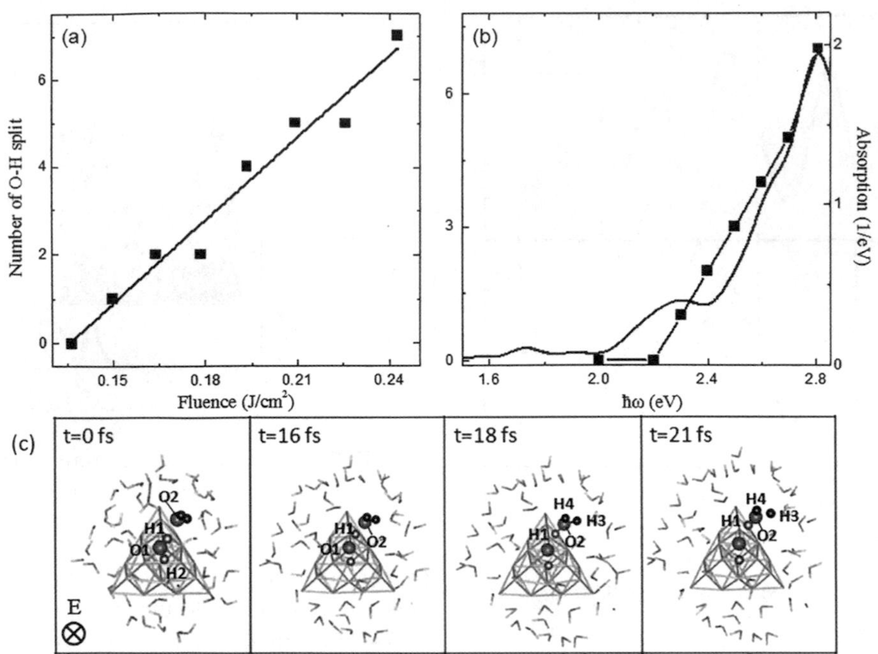

FIGURE 14.19 (a) Number of O–H bond split with varied laser fluence, corresponding to a field strength from $E_{\max} = 2.1$ to 2.8 V/Å. Linear fits to data points are shown as the black line. (b) Number of O–H bond split with varied laser frequency for the fixed field strength $E_{\max} = 2.8$ V/Å. Corresponding absorption spectrum (gray lines) is superimposed for comparison. (c) Atomic configurations at time $t = 0$, 16, 18, and 21 fs.

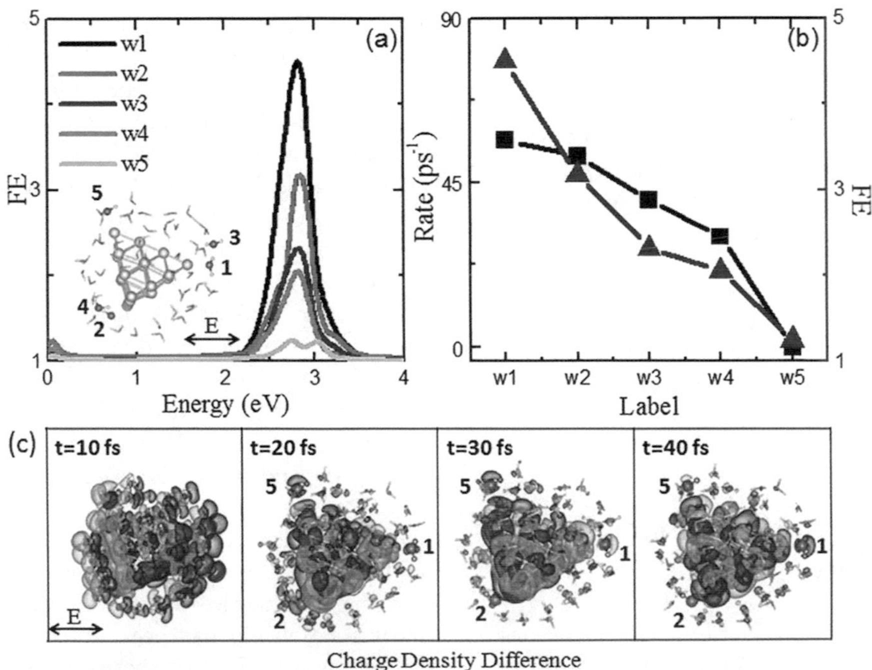

FIGURE 14.20 (a) Field enhancement (FE) along the laser polarization direction for Au_{20} cluster as a function of laser energy in different positions as defined in the inset. (b) Position dependence of the FE (gray line) and the corresponding rate for water splitting (black line) under laser field $E_{\max} = 2.7$ V/Å. (c) Charge density difference at time $t = 10$, 20, 30, and 40 fs.

To gain further insights into reaction mechanisms, we display FE spectrum at different positions around Au_{20} cluster in Figure 14.20a. These positions are adsorption sites for water molecules labeled as w1, w2, w3, w4, and w5, respectively. Here, the localized FE factor is defined as, $\text{FE(x)} = [v_{\text{eff}}(x + \delta\text{x}) - v_{\text{eff}}(\text{x})]/(\text{e}\delta\text{x}E_{\text{ext}})$ where x is the position, δx is the mesh size along the polarized direction, v_{eff} is the effective potential, and E_{ext} is the external field strength (Song et al., 2011). A peak appears in the FE spectrum around the laser frequency 2.81 eV for all cases. The peak value is 4.5, 3.2, 2.3, 2.0, and 1.2, decreasing in turn from w1 to w5, as shown in Figure 14.20b. Among them, w1 locates on the tip of Au_{20} cluster, showing the largest FE. We also show the FE in a transverse cut going through the tip atom, and find that the largest FE is ∼7.0 at the tip of Au_{20} (Figure 14.21). Similar tip-enhanced FE reaches a value of 7.1 on a larger cluster Ag_{489} (Christian et al., 2013) in the same magnitude as our result.

More importantly, we observe a rapid hydrogen production from water as displayed in Figure 14.22a. We define that a hydrogen molecule is formed when its H–H bond length is less than 1.0 Å. For laser field E_{max} = 2.90 V/Å, three hydrogen molecules are formed gradually after 20 fs. In other cases, two hydrogen molecules are first formed, and then one of them splits. This observation is universal for different MD trajectories, where we find that hydrogen molecule is generally produced within 35 fs as shown in Figure 14.22b. Considering the Au_{20} cluster with a size of ∼1.0 nm under the laser pulses with an intensity of ∼0.2 J/cm^2, the adsorbed photon number is ∼4,000. Thus, the quantum efficiency of H_2 production is about 0.08%, in the same magnitude as that for the 2-nm-thick Au NP for

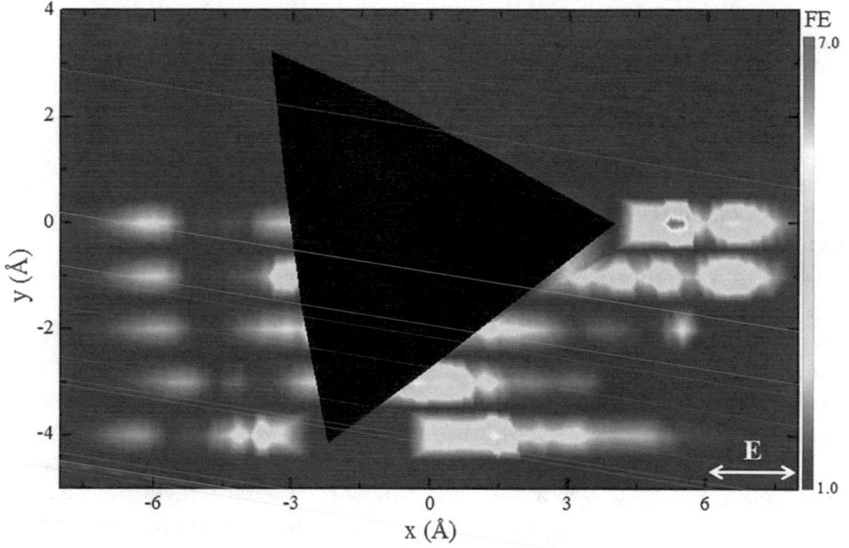

FIGURE 14.21 FE along the laser polarization direction in a transverse cut going through the tip atom of Au_{20} under the laser pulse with E_{max} = 0.05 V/Å. Electromagnetic field inside the gold cluster is set to zero in order to identify the NP boundary.

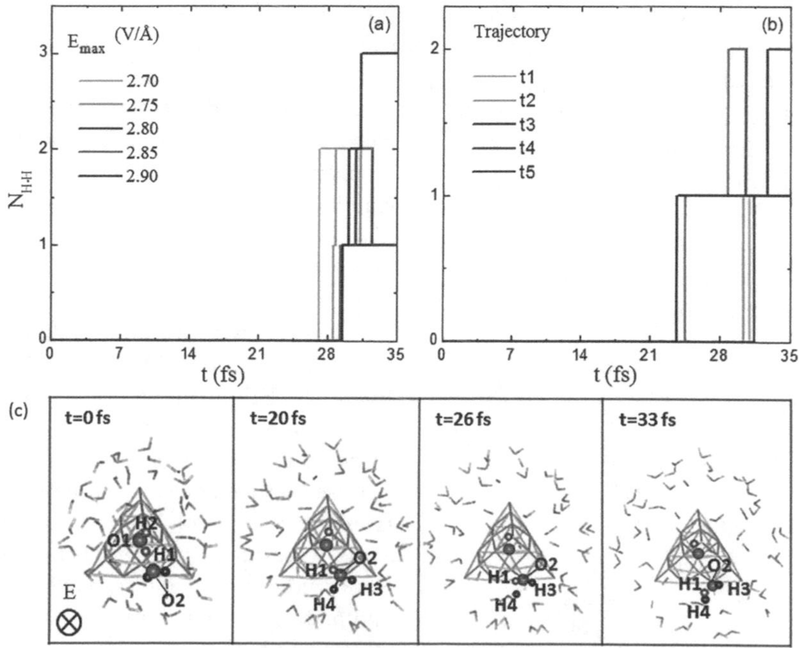

FIGURE 14.22 Time evolution of the number of hydrogen molecules (a) with varied laser intensity and (b) in separate MD trajectories with the field strength E_{max} = 2.8 V/Å. (c) Atomic configurations at time t = 0, 20, 26, and 33 fs.

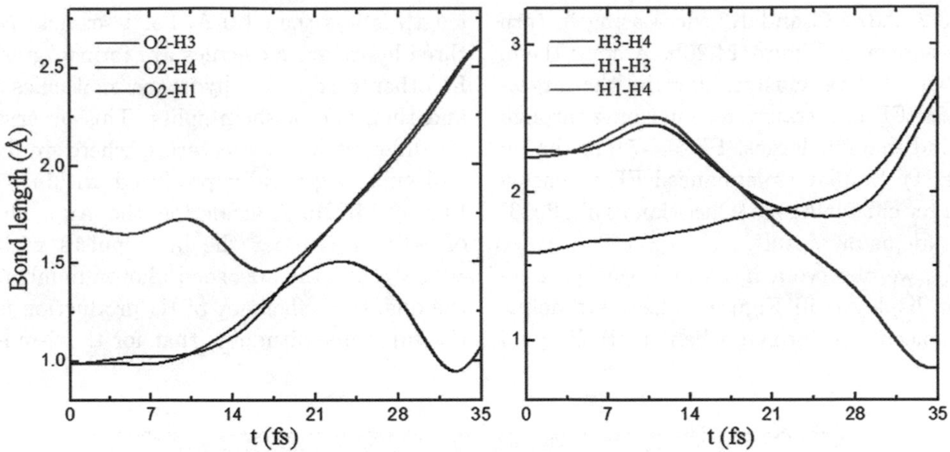

FIGURE 14.23 Time evolution of OH bond length associated with atoms labeled in Figure 14.22c, under the laser pulse with $E_{\max} = 2.8$ V/Å.

direct water splitting (Robatjazi et al., 2015). Without gold cluster, however, H_2 production is unlikely to occur owing to large energy barrier for water splitting and H recombination. Thus, gold cluster could serve as a photocatalytic reaction center to produce H_2 from liquid water.

The detailed process for hydrogen production is shown in Figure 14.22c. Initially, one water molecule with atoms labeled H_1, H_2, and O_1 locates near the tip of Au_{20} cluster. This molecule forms hydrogen bond with another water molecule consisting of H_3, H_4, and O_2 through O_2–H_1 hydrogen bond with a bond length $d_{O_2-H_1} = 1.72$ Å. At $t =$ 20 fs, the distance $d_{O_2-H_1}$ decreases to 1.55 Å, and O_1–H_1 bond length $d_{O_1-H_1}$ increases to 2.45 Å, indicating the split of O_1–H_1 bond. Simultaneously, H_1 atom comes closer to H_4 with $d_{H_1-H_4} = 1.89$ Å, and H_4 atom keeps going away from O_2 with $d_{O_2-H_4} = 1.51$ Å. At $t = 26$ fs, the distance $d_{H_1-H_4}$ decreases to 1.48 Å, and $d_{O_2-H_4}$ reaches 1.91 Å, implying O_2–H_4 bond breaks up. At $t = 33$ fs, $d_{H_1-H_4}$ reaches 0.86 Å, forming a H_2 molecule. The O_2–H_3 still keeps bonded with a bond length $d_{O_2-H_3} = 0.95$ Å. The details for time-evolved changes in bond length are shown in Figure 14.23. Therefore, hydrogen molecule is formed via a "chain reaction"-like mechanism (Figure 14.24): plasmon excitation produces intense hot electrons to activate OH bonds; an activated H then detaches from water and will be accelerated; the H from another water dissociates by the strong impact of the first H together with hot electrons, to produce a H_2 molecule.

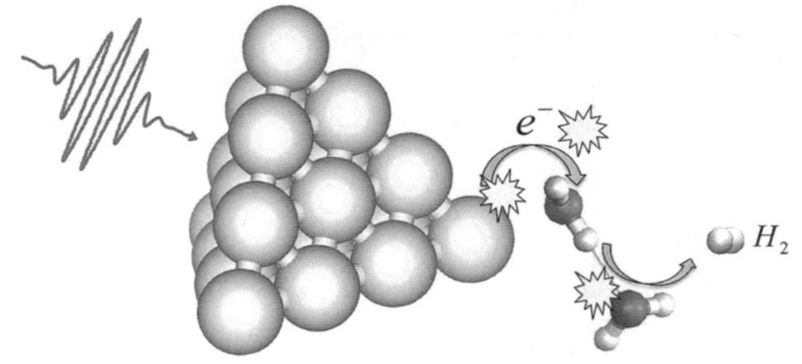

FIGURE 14.24 Schematics of "chain reactions" in plasmon-induced water photosplitting and H_2 generation.

14.9 Conclusion

In summary, after a brief introduction of the principles of STH photochemical cells, including their major components, the fabrication procedures, and recent developments, we try to focus on the atomic mechanisms of water adsorption and electron dynamics in STH cells, as obtained from accurate quantum mechanical simulations. We discover that water adsorption is strongly orbital dependent. Because of different orbital symmetry, water prefers to bond to the QWS upon adsorption on lobe positions, while antibonding interactions are formed upon water adsorption on node positions. Then, we study the dissociation of water and calculate the energy barrier of dissociation, the diffusion of OH radical, and the formation of H_2, where the transition states and reaction paths are obtained from NEB calculations. Next, we take a step forward to look at the quantum dynamics of water photosplitting by TDDFT-MD simulations. We identify the quantum oscillation modes of plasmonic excitation affecting the water-splitting rates. Or more specifically, the odd modes lead to more net electrons, which can facilitate the charge transfer whose energy matches well with the antibonding orbital of water. We also study the dynamics of H_2 generation in detail and also investigate the FE at different position of the magic gold particles. Finally, we propose a "chain reaction" mechanism of plasmon-induced H_2 generation from water splitting.

Based on the knowledge about the interface between electronic structure and dynamics at the molecular level, we also strive to design new materials for improved STH conversion. Such studies would open a way for a complete understanding and material design of photocatalysts to advance the performance of photochemical water-splitting cells. In some day, we hope we can design new materials and improve the efficiency of solar hydrogen production in a significant way.

Acknowledgment

We thank Dr. Zijing Ding and Dr. Lei Yan for discussion and collaboration. This work is financially supported by Ministry of Science and Technology (grant No. 2016YFA0300902) and National Natural Science Foundation of China (grant Nos. 11774396 and 91850120).

References

Ager, J. W.; Shaner, M. R.; Walczak, K. A.; et al. Experimental demonstrations of spontaneous, solar-driven photoelectrochemical water splitting. *Energy Environ. Sci.* 8, 2811 (2015).

Alonso, J. L.; Andrade, X.; Echenique, P.; et al. Efficient formalism for large-scale Ab initio molecular dynamics based on time-dependent density functional theory. *Phys. Rev. Lett.* 101, 096403 (2008).

Awate, S. V.; Deshpande, S. S.; Rakesh, K.; et al. Role of micro-structure and interfacial properties in the higher photocatalytic activity of TiO_2-supported nanogold for methanol-assisted visible-light-induced splitting of water. *Phys. Chem. Chem. Phys.* 13, 11329 (2011).

Boppella, R.; Kochuveedu, S. T.; Kim, H.; et al. Plasmon-sensitized graphene/TiO_2 inverse opal nanostructures with enhanced charge collection efficiency for water splitting. *ACS Appl. Mater. Interfaces* 9, 7075 (2017).

Chen, M. S.; Goodman, D. W. The structure of catalytically active gold on titania. *Science* 306, 252 (2004).

Christian, F. A. N.; Eduardo, M. P.; Eduardo, A. C.; et al. Quantum dynamical simulations of local field enhancement in metal nanoparticles. *J. Phys. Condens. Matter* 25, 125304 (2013).

Christopher, P.; Xin, H.; Linic, S. Visible-light-enhanced catalytic oxidation reactions on plasmonic silver nanostructures. *Nat. Chem.* 3, 467–472 (2011).

Christopher, P.; Xin, H.; Marimuthu, A.; Linic, S. Singular characteristics and unique chemical bond activation mechanisms of photocatalytic reactions on plasmonic nanostructures. *Nat. Mater.* 11, 1044–1050 (2012).

Cottancin, E.; Celep, G.; Lermé, J.; et al. Optical properties of noble metal clusters as a function of the size: Comparison between experiments and a semi-quantal theory. *Theor. Chem. Acc.* 116, 514–523 (2006).

Cushing, S. K.; Wu, N. Progress and perspectives of plasmon-enhanced solar energy conversion. *J. Phys. Chem. Lett.* 7, 666–675 (2016).

Ding, Z. J.; Gao, S. W.; Meng, S. Orbital dependent interaction of quantum well states for catalytic water splitting. *New J. Phys.* 17, 013023 (2015).

Ding, Z. J.; Meng, S. Promote water photosplitting via tuning quantum well states in supported metal clusters. *Phys. Rev. B* 86, 045455 (2012).

Ding, Z. J.; Yan, L.; Li, Z.; et al. Controlling catalytic activity of gold cluster on MgO thin film for water splitting. *Phys. Rev. Mater.* 1, 045404 (2017).

Goto, Y.; Hisatomi, T.; Wang, Q.; et al. A particulate photocatalyst water-splitting panel for large-scale solar hydrogen generation. *Joule* 2, 509 (2018).

Herzing, A. A.; Kiely, C. J.; Carley, A. F.; et al. Identification of active gold nanoclusters on iron oxide supports for CO oxidation. *Science* 321, 1331 (2008).

Hu, X. L.; Klimeš, J.; Michaelides, A. Proton transfer in adsorbed water dimers. *Phys. Chem. Chem. Phys.* 12, 3953 (2010).

Idrobo, J. C.; Walkosz, W.; Yip, S. F.; et al. Static polarizabilities and optical absorption spectra of gold clusters (Au_n, n = 2–14 and 20) from first principles. *Phys. Rev. B* 76, 205422 (2007).

Ingram, D. B.; Linic, S. Water splitting on composite plasmonic-metal/semiconductor photoelectrodes: Evidence for selective plasmon-induced formation of charge carriers near the semiconductor surface. *J. Am. Chem. Soc.* 133, 5202–5205 (2011).

Jung, J.; Shin, H. J.; Kim, Y.; et al. Controlling water dissociation on an ultrathin MgO film by tuning film thickness. *Phys. Rev. B* 82, 085413 (2010).

Kale, M. J.; Avanesian, T.; Christopher, P. Direct photocatalysis by plasmonic nanostructures. *ACS Catal.* 4, 116–128 (2014a).

Kale, M. J.; Avanesian, T.; Xin, H.; et al. Controlling catalytic selectivity on metal nanoparticles by direct photoexcitation of adsorbate–metal bonds. *Nano Lett.* 14, 5405–5412 (2014b).

Kudo, A.; Miseki, Y. Heterogeneous photocatalyst materials for water splitting. *Chem. Soc. Rev.* 38, 253 (2009).

Lee, J.; Mubeen, S.; Ji, X.; et al. Plasmonic photoanodes for solar water splitting with visible light. *Nano Lett.* 12, 5014–5019 (2012).

Lin, X.; Nilius, N.; Freund, H. J.; et al. Quantum well states in two-dimensional gold clusters on MgO thin films. *Phys. Rev. Lett.* 102, 206801 (2009).

Linic, S.; Christopher, P.; Ingram, D. B. Plasmonic-metal nanostructures for efficient conversion of solar to chemical. *Nat. Mater.* 10, 911–921 (2011).

Liu, J.; Liu, Y.; Lifshitz, Y.; et al. Metal-free efficient photocatalyst for stable visible water splitting via a two-electron pathway. *Science* 347, 970 (2015).

Liu, Z.; Hou, W.; Pavaskar, P.; et al. Plasmon resonant enhancement of photocatalytic water splitting under visible illumination. *Nano Lett.* 11(3), 1111 (2011).

Meng, S.; Wang, E. G.; Gao, S. W. Water adsorption on metal surfaces: A general picture from density functional theory studies. *Phys. Rev. B* 69, 195404 (2004).

Mukherjee, S.; Libisch, F.; Large, N.; et al. Hot electrons do the impossible: Plasmon-induced dissociation of H_2 on Au. *Nano Lett.* 13, 240–247 (2013).

Mukherjee, S.; Zhou, L.; Goodman, A. M.; et al. Hot-electron-induced dissociation of H_2 on gold nanoparticles supported on SiO_2. *J. Am. Chem. Soc.* 136, 64–67 (2014).

Murray, W. A.; Barnes, W. L. Plasmonic materials. *Adv. Mater.* 19, 3771–3782 (2007).

Nilius, N.; Ganduglia-Pirovano, M. V.; Brazdova, V.; et al. Counting electrons transferred through a thin alumina film into Au chains. *Phys. Rev. Lett.* 100, 096802 (2008).

Nishijima, Y.; Ueno, K.; Kotake, Y.; et al. Near-infrared plasmon-assisted water oxidation. *J. Phys. Chem. Lett.* 3, 1248 (2012).

Robatjazi, H.; Bahauddin, S. M.; Doiron, C.; et al. Direct plasmon-driven photoelectrocatalysis. *Nano Lett.* 15, 6155 (2015).

Scholl, J. A.; Koh, A. L.; Dionne, J. A. Quantum plasmon resonances of individual metallic nanoparticles. *Nature* 483, 421–427 (2012).

Sheldon, M. T.; van de Groep, J.; Brown, A. M.; et al. Plasmoelectric potentials in metal nanostructures. *Science* 346, 828–831 (2014).

Shi, Y.; Wang, J.; Wang, C.; et al. Hot electron of Au nanorods activates the electrocatalysis of hydrogen evolution on MoS_2 nanosheets. *J. Am. Chem. Soc.* 137, 7365–7370 (2015).

Shin, H. J.; Jung, J.; Motobayashi, K.; et al. State-selective dissociation of a single water molecule on an ultrathin MgO film. *Nat. Mater.* 9, 442 (2010).

Song, P.; Nordlander, P.; Gao, S. Quantum mechanical study of the coupling of plasmon excitations to atomic-scale electron transport. *J. Chem. Phys.* 134, 074701 (2011).

Troullier, N.; Martins, J. L. Efficient pseudopotentials for plane-wave calculation. *Phys. Rev. B* 43, 1993–2006 (1991).

Wallis, T. M.; Nilius, N.; Ho, W. Electronic density oscillations in gold atomic chains assembled atom by atom. *Phys. Rev. Lett.* 89, 236802 (2002).

Wang, X.; Maeda, K.; Thomas, A.; et al. A metal-free polymeric photocatalyst for hydrogen production from water under visible light. *Nat. Mater.* 8, 76–80 (2009).

Wu, K.; Li, J.; Lin, C. Remarkable second-order optical nonlinearity of nano-sized Au_{20} cluster: A TDDFT study. *Chem. Phys. Lett.* 388, 353–357 (2004).

Yan, J.; Jacobsen, K. W.; Thygesen, K. S. First-principles study of surface plasmons on Ag(111) and H/Ag(111). *Phys. Rev. B* 84, 235430 (2011).

Yan, L.; Wang, F. W.; Meng, S. Quantum mode selectivity of plasmon-induced water splitting on gold nanoparticles. *ACS Nano* 10, 5452 (2016).

Yan, L.; Xu, J. Y.; Wang, F. W.; Meng, S. Plasmon-induced ultrafast hydrogen production in liquid water. *J. Phys. Chem. Lett.* 9, 63 (2018).

Yoon, B.; Hakkinen, H.; Landman, U.; et al. Charging effects on bonding and catalyzed oxidation of CO on Au_8 clusters on MgO. *Science* 307, 403 (2005).

Zhao, L.; Jensen, L.; Schatz, G. C. Pyridine−Ag_{20} cluster: A model system for studying surface-enhanced Raman scattering. *J. Am. Chem. Soc.* 128, 2911–2919 (2006).

Zheng, J.; Zhang, C.; Dickson, R. M. Highly fluorescent, water-soluble, size-tunable gold quantum dots. *Phys. Rev. Lett.* 93, 077402 (2004).

Zhou, L.; Zhang, C.; McClain, M. J.; et al. Aluminum nanocrystals as a plasmonic photocatalyst for hydrogen dissociation. *Nano Lett.* 16, 1478–1484 (2016).

Zhu, M.; Aikens, C. M.; Hollander, F. J.; et al. Correlating the crystal structure of a thiol-protected Au_{25} cluster and optical properties. *J. Am. Chem. Soc.* 130, 5883–5885 (2008).

15

Onion-Like Inorganic Fullerenes from a Polyhedral Perspective

Ch. Chang, A. B. C. Patzer, D. Sülzle, and H. Bauer
Technische Universität Berlin

15.1 Introduction

The review will be mainly concerned with the geometric and structural properties of representative inorganic cage species even though not all of them could be strictly classified as fullerenes. It is their polyhedral shape usually accompanied with high symmetry and high sphericity that gives these molecular systems their beautiful aesthetic look. It is intriguing how polyhedral structures appear at different dimensional scales, from electron densities at the subatomic level up to everyday objects. Starting at the subatomic level, electron densities for different electron configurations in transition metal clusters may show cubic, octahedral, or tetrahedral shapes (Bouguerra et al. 2007; Jones, Eberhart, and Clougherty 2007). Also, representations of the density of d electrons around a metal atom in $[Cr(CO)_6]$ (Macchi and Sironi 2003), of its Laplacian in $[Fe_2(CO)_9]$ or in $[Mn(CO)_6]^+$ (Bo, Sarasa, and Poblet 1993; Bader, Matta, and Cortés-Guzmán 2004), and of the electron localization function in $[Re_2(CO)_{10}]$ reveal cubic shapes (Kohout, Wagner, and Grin 2002). At the polyatomic level, the coordination polyhedra around metal atoms have diameters of a few tenths of a nanometer, while typical metal clusters can approach 1 nm and large clusters can reach up to 2 nm in diameter. Very large multi-polyhedral systems assembled through bridging ligands can reach sizes of about 2–3 nm, as in Pd_{144} and Mo_{132} (Tran, Powell, and Dahl 2000; Müller et al. 1999). Icosahedral quasi-crystals, nanoclusters and nanoparticles are in the 10–20 nm size range, where we can also find a single-stranded DNA molecule folded into a hollow octahedron with a diameter of approximately 22 nm, as well as other DNA polyhedra including the tetrahedron, the cube, and the truncated octahedron reported by (Shih, Quispe, and Joyce 2004). The capsids of viruses may reach a size in an order of magnitude larger (between 10 and 100 nm), among which structures of the icosahedral symmetry have been widely reproduced (Twarock 2006).

Molecules which consist of atoms forming a closed cage have been known long since. However, it was not until the discovery of the celebrated C_{60} molecule (Osawa 1970; Kroto et al. 1985; Krätschmer, Fostiropoulus, and Huffman 1990; Krätschmer et al. 1990), which represented the hitherto unknown third allotropic and the first molecular form of carbon that initiated a tremendous growth of interest in similar species and opened up a whole new branch of molecular physics and material sciences. It was this breakthrough that made the explosive growth of a new area of molecular science possible. The C_{60} molecule was named buckminsterfullerene after R. Buckminster Fuller, a noted architect, developed the art of constructing buildings in the shape of geodesic dome-like polyhedra whose triangulated surface contains vertices where five and six faces meet. In the following time, a multitude of other smaller and larger carbon cage molecules C_n with n even, ranging from $n = 20$ to several hundreds which are now all generally called carbon fullerenes, were investigated both theoretically and experimentally. The great variety in structural architecture of these systems seems to be almost infinite. Until today, a diversity of carbon molecules with unusual shapes and geometric arrangements have been discovered, a wealth of morphologies, ranging from single or several nested polyhedral cages, graphitic sponges, tubes to even tori (Stephens 1993; Terrones et al. 2004), nanohorns, and nanocones (Yudasaka, Iijima, and Crespi 2008).

Since all these species are molecular aggregates formed by an assembly of atoms or well specified subunits of atoms (molecular units), the study of fullerenes can be considered as a sub-discipline of cluster physics. Eventually, a fullerene molecule itself could be regarded as such a subunit or a monomer of a larger aggregate. Depending on the cluster size (the number of monomers), these systems cover a range from the very small (microscopic) to the very large (macroscopic) with accordingly varying remarkable physical properties. They, therefore, provide a link between molecular, cluster, nanophysics, and eventually material science. The question of whether there might exist similar molecular systems composed of atoms other than carbon arises quite naturally. These so-called inorganic fullerenes, in particular, those consisting of more than a single cage, are the subject of this review article. The focus is thereby mainly on their geometric shape and symmetry aspects rather than on electronic and physical properties which are mentioned just in passing.

15.2 Onion-Like Inorganic Fullerenes

We begin with a loose definition of what is considered an inorganic fullerene. Inorganic implies that the molecule may consist of any atomic species apart from pure carbon. Fullerene-like implies that all contributing atoms can be considered as being located at the vertices of a not necessarily regular convex polyhedron giving the system a cage-like look. To be called strictly a fullerene, this polyhedron should obey certain geometric restrictions (cf. "A Special Family of Dual Polyhedra: Fullerenes and Frank–Kasper Cages" section). In the common literature, however, molecular systems that do not comply rigorously with these geometric restrictions, but merely exhibit a polyhedral cage structure, are nonetheless often called fullerenes as well. Therefore, the terms fullerene-like and cage-like are frequently used as synonyms.

There might be just one or several of such cages that might or might not interpenetrate each other. The latter will be termed onion-like as each cage could be regarded as a shell of an onion. We shall allow for the possibility of a single atom being placed inside at the center of the cage(s) and make use of the notation $X_i@Y_j@Z_k@...$ for a general onion-like inorganic fullerene where the atomic species X, Y, Z, etc., which might well be equal to each other, are located at the vertices of different polyhedral frameworks. The innermost cage X could, as mentioned, be just a single central atom. If unambiguous from the context, it is sometimes a convenient habit to just state the chemical sum formula.

There are different views of looking at these cage structures. The most common is to consider the actual physical bonds between neighboring atoms, for these interactions are responsible for the energetic stability of the molecular framework. Another view is a geometric idealization of molecular structures, which consists of associating the positions of sets of atoms to the vertices of one or several polyhedra. To describe a cage structure, we shall make use of the latter pure geometric concept. It should be noted that the interatomic distances between the vertices of the polyhedra, i.e. the edge lengths, do not necessarily represent physical bonds. The connection between geometric relationships and physical bonding situations in many kinds of cage molecules is not yet completely understood. Using the above definition, a well-known metal carbonyl such as $[Fe(CO)_5]$ could equally be written as $Fe@C_5@O_5$, where an iron atom lies at the center of two nested trigonal bipyramids, the inner one consisting of five carbon and the outer one of five oxygen atoms (see Figure 15.1). Clearly, this picture is entirely geometric and does not reflect any physical interactions. In fact this molecule is held together by five strong bonds between the central iron atom and the surrounding carbon monoxide molecules. Traditionally, this species would never be classified as a cage molecule and certainly not as an inorganic fullerene.

In the molecular world, the polyhedral shape is quite ubiquitous, in Particular, among inorganic species, e.g. coordination compounds, where the ligands around a central atom are generally lying at the vertices of some more or less regular convex polyhedron (González-Morage 1993). Equally, in the realm of atomic and molecular clusters, the most stable geometric arrangements disclose polyhedral patterns of high symmetry (Joyes 1990; Haberland 1994). Damasceno, Engel, and Glotzer (2012) studied the prospect to predict complex, but ordered, molecular structure like colloids and nanoparticles by exploring the propensity of self-assemblage of simple polyhedral building blocks. Monte Carlo simulations of self-assembling processes of numerous different polyhedra including the 'classical' ones viz. the Platonic, Archimedean, Catalan, and Johnson (cf. Appendix) solids showed that the polyhedral shape of the building blocks can be related to the formation of distinctly ordered categories of nanoscale structures. A very extensive treatise on multi-shell clusters using mainly a graph theoretical, topological approach has been given recently by Diudea (2017).

FIGURE 15.1 Classical view of $Fe(CO)_5$ and its representation as two nested trigonal bipyramids with a center Fe atom ($Fe@C_5@O_5$).

15.3 Basics of Polyhedra

Polyhedra have been known since antiquity. In about 520 B.C., Pythagoras already knew the existence of three of the five regular polyhedra viz. the cube, the tetrahedron, and the dodecahedron. Plato (about 350 B.C.) reported all five of them adding the octahedron and the icosahedron. He mystically related them as 'cosmic building stones' to the five so-called 'elements': Fire, air, water, earth, and 'heavenly bodies'. Therefore, these five regular bodies are today designated as Platonic solids. Archimedes (about 250 B.C.) studied the 13 uniform or semi-regular polyhedra, which today bear his name as Archimedean solids. Then, after a very long time period, Kepler (1619) arrived at an integrated description of the 5 Platonic and the 13 Archimedean solids. He also had the idea that it was possible to construct regular star-shaped so-called stellated polyhedra. This idea was completed two centuries later by Poinsot (1810) presenting the complete list of all fully regular stellated polyhedra. Subsequently, piece by piece, stellated and other general uniform polyhedra were being published. Today, the mathematical analysis of various kinds and families of polyhedra, as well as extensions to higher space dimensions greater than three (named polytopes) and an active area of mathematics are well established (cf. books by Coxeter (1973), Cromwell (1997), Ziegler (1995), Alexandrov (2005), and Grünbaum (2003)).

15.3.1 Geometric Properties

A general polyhedron in three-dimensional space is a solid bounded by a number of polygons (faces), which, two by two have a side in common (edges), while three or more polygonal faces join in common vertices. There is a solid angle Ω associated with each vertex. Broadly, all polyhedra fall into two categories: Convex or non-convex (concave). A convex polyhedron is a polyhedron with the property that for any two points inside of it, the line segment joining them is completely contained within the polyhedron. In general, any polyhedron subtending a solid angle $\Omega < 2\pi$ sr at each of its vertices is called convex while any polyhedron subtending a solid angle $\Omega > 2\pi$ sr at any of its vertices is concave. In other words, a convex polyhedron has each of its vertices protruding outward while a concave polyhedron has any of its vertices indented inwards the surface. Any polyhedron in three-dimensional space can be thoroughly characterized by the total number of its faces f, vertices v, and edges e, which are commonly related by Euler's theorem:

$$v - e + f = \chi = 2(1 - g).$$

where χ is called the Euler characteristic and g the genus (i.e. the number of holes in or tunnels through the polyhedral framework). The Euler characteristic is a topological invariant that describes the shape of a structure regardless of how it is bent. Positive/negative χ values indicate positive/negative curvature of the polyhedral structure. For every convex polyhedron, i.e. a polyhedron having no holes, tunnels, kinks etc. the genus g is equal to 0 and, therefore, χ is always equal to 2 and the Euler relation[1] becomes:

$$v - e + f = 2.$$

The polygonal faces are n-gons (n = 3, 4, 5, ...) which intersect at the edges and at the vertices. The vertices can be specified by their vertex configuration, which denotes the kind and number of faces that meet at a vertex. If all faces of a polyhedron are regular and congruent, i.e. they are all of the same kind and all vertex environments are equivalent, we have the well-known family of the five Platonic polyhedra (see also Table 15.2). Allowing for different regular n-gons but keeping the condition of equal vertex configurations, we arrive at the class of Archimedean solids (cf. Table 15.3), which are also called semi-regular. Other families of polyhedra are obtained by further relaxing the conditions and/or imposing special criteria on the nature of the faces and vertices. Another important geometric property of not necessarily convex polyhedra is called duality. Any polyhedron can be associated with a second, its so-called, dual polyhedron, where the vertices of one correspond to the faces of the other and the edges between pairs of vertices of one correspond to the edges between pairs of faces of the other. Thus, the number of faces and vertices is interchanged, while the number of edges remains constant.

The dual polyhedron can be constructed by connecting the midpoints of every face of the original polyhedron, which thereby becomes the vertices of the dual polyhedron. Starting with any given polyhedron, the dual of its dual is the original polyhedron. For example, if the midpoints of each of the six square faces of a cube having eight vertices are joined by a line, the resulting polyhedron is an octahedron having six vertices and eight faces. Thus, the octahedron is the dual solid of the cube and vice versa. If we do not directly connect the central points of each polygonal face but rather the points lying vertically above or below them, we arrive at the dual of the polyhedron that now lies completely outside or inside the original one. Going on that way, we can construct a shell structure polyhedron@dual@polyhedron@dual and so on. This is a possible symmetry preserving building principle to contrive onion-like inorganic fullerenes. In Section 15.4, this is outlined in greater detail. Duality preserves the symmetries of a polyhedron. Therefore, for many classes of polyhedra defined by their symmetry groups, the duals belong to the same symmetry group. Thus, e.g. the regular polyhedra – the (convex) Platonic solids – form dual pairs between themselves, where the regular tetrahedron is self-dual. The dual of an isogonal polyhedron, having equivalent vertices, is one which is isohedral, having equivalent faces. The dual of an isotoxal polyhedron (having equivalent edges) is also isotoxal. Duality is closely related to reciprocity or polarity,

[1]Euler's theorem does not hold for non-convex e.g. stellated polyhedra. They can have various different characteristics.

TABLE 15.1 Classification of Families of Polyhedra According to Their Framework Being Isohedral (i.e. Face-Transitive, Faces Are All of the Same Kind), Regular (i.e. All Faces Are Regular n–Gons), Isogonal (i.e. Vertex-Transitive, All Vertex Environments Are Equivalent), Isotoxal (i.e. Edge-Transitive, All Edge Environments Are Equivalent), and Equilateral (i.e. All Edges Have the Same Length). Even if the edge environments are not all equivalent, all edges are of the same length in polyhedra having only regular faces.

Family	Isohedral	Regular	Isogonal	Isotoxal	Equilateral
Platonic	+	+	+	+	+
Archimedean	−	+	+	+	+
n–prisms	−	+	+	+	+
n–antiprisms	−	+	+	+	+
Catalan (dual of Archimedean)	+	−	−	−	−
n–bipyramids (dual of n–prisms)	+	−	−	−	−
n–trapezohedra (dual of n–antiprisms)	+	−	−	−	−
Fullerenes	{5}, {6}	+[a]	(3)	−	+[a]
Frank Kasper (dual of fullerenes)	{3}	−[a]	(5), (6)	+	−[a]
Johnson	−	+	−	−	+

[a]Not necessarily required.

a geometric transformation that, when applied to a convex polyhedron, realizes the dual polyhedron as another convex polyhedron (cf. Alexandrov 2005). A summary of the most common classes of polyhedra that also appear in molecular arrangements is given in the following Table 15.1 (see also Appendix).

15.3.2 Symmetry Groups and Polyhedra

The symmetry present in a molecule is closely related to many of its fundamental physical properties. Molecular symmetry is predominately studied in the framework of group theory (cf. Hamermesh 1962; Herzberg 1966; Volatron and Chaquin 2017). Five different types of symmetry operations or elements can be used to describe the point group symmetry of a molecule:

- E: The identity[2] – every molecule or geometric object in general has this element regardless of its overall symmetry.
- σ: A mirror plane, which generates an identical copy of the initial structure by reflecting through a suitable plane passing through the framework.
- i: An inversion center[3] – it exists, if the points (x,y,z) and (−x,−y,−z) correspond to identical parts in the molecule.
- C_n: An n-fold rotation axis, which leaves the molecule unchanged after a $(2\pi/n)$ rotation.
- S_n: An n-fold improper rotation axis combines a C_n axis with a reflection in a plane σ perpendicular to it resulting in a structure identical to the original one.

Figure 15.2 shows a general flowchart of how to determine the point group of any molecule or geometric object by analyzing its different symmetry elements. As an illustration, we examine the symmetry of a tetrahedron, the self-dual Platonic body. Its symmetry elements are (Atkins and de Paula 2010): E, 6 σ, 3 C_2, 8 C_3, and 6 S_4, but there is no inversion center.

Following the scheme outlined in Figure 15.2, it is obvious that the tetrahedron is neither an atom nor a linear molecule. In addition, it has no C_5 or C_4 axis, so its point group cannot be I (I_h) or O (O_h). But it has at least four C_3 axis and more than one σ mirror plane, which indicates T_d or T_h as possible symmetry groups. Since the tetrahedron has no inversion center i, its molecular point group has to be finally T_d (cf. Atkins and de Paula 2010). In a similar way, the symmetry groups of the other polyhedra and their duals can be determined and are listed in Tables 15.2–15.4. Molecules with no other symmetry element but the identity belong to the point group C_1 as one can very easily infer from the flowchart in Figure 15.2. The relation between such low symmetry and chirality is quite noteworthy, because every C_1 molecule is chiral, but not every chiral molecule belongs to the point group C_1. The symmetry groups of single atoms and linear molecules represent continuous groups (or Lie groups) (see Hamermesh 1962), i.e. some of their elements depend on one or more continuous parameters and, therefore, possess an infinite number of symmetry operations, which is of course reflected by the respective point group designations, namely $K_{\infty h}$, $D_{\infty h}$, $C_{\infty v}$. All other molecular symmetry groups are discrete involving only a finite number of symmetry elements.

15.3.3 Sphericity and Convex Hull of Cages

A polyhedron P is convex if and only if, for any two points A and B inside the polyhedron, the line segment AB is inside P. One way to visualize this rather abstract concept of a convex hull is to put a 'rubber sheet' around all the vertices of the polyhedral point set and let it wrap as tight as it can. The resultant polyhedron is its convex hull (Goodman and O'Rourke 1997).

The convex hull of a finite set of points in k dimensions $S = \{\vec{x}_1, \vec{x}_2, ..., \vec{x}_n\} \subseteq \mathbb{R}^k$ is the smallest convex set containing S (Webster 1994; Boissonnat and Yvinec 1995), thus

[2]E is also required for the symmetry elements to form a mathematical group (see e.g. Hamermesh 1962).

[3]An atom can, but must not, be located at the inversion center.

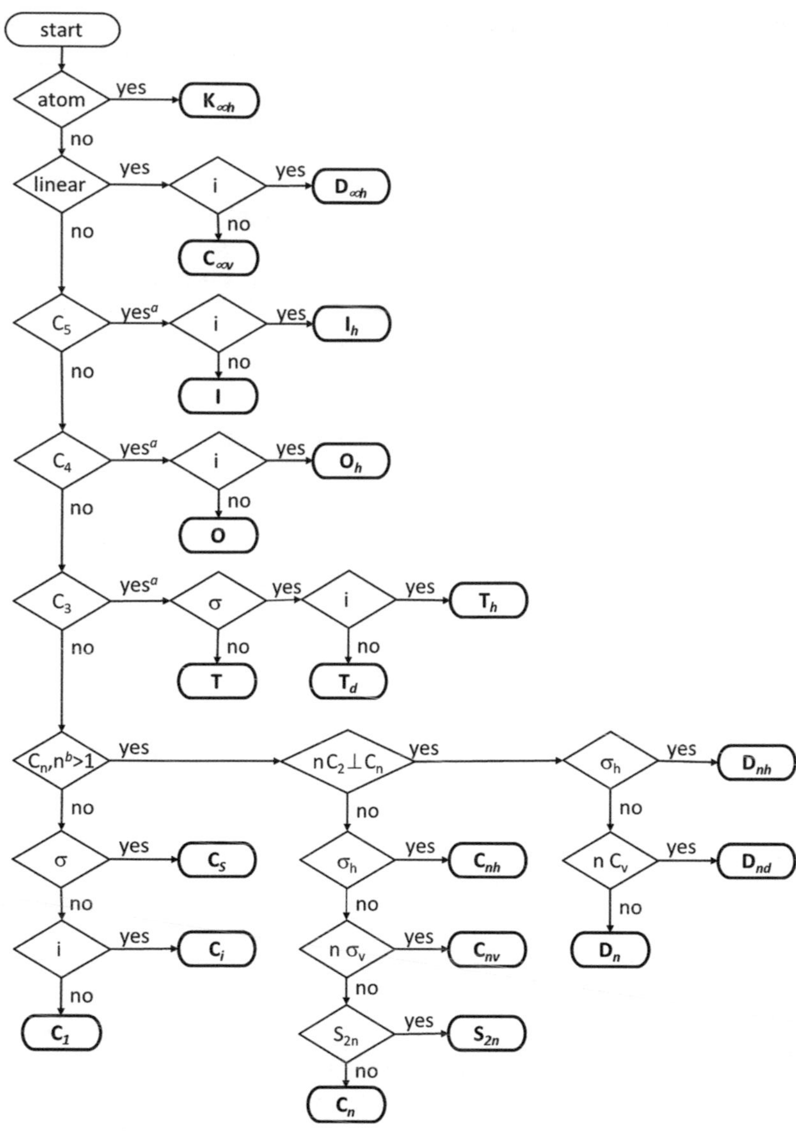

FIGURE 15.2 Flowchart for determining molecular point groups based on symmetry elements.

$$\mathrm{conv}(S) := \left\{ \sum_{i=1}^{n} \lambda_i \vec{x}_i \;\middle|\; \lambda_i \geq 0 \wedge \sum_{i=1}^{n} \lambda_i = 1 \right\}. \qquad (15.1)$$

In general, the convex hull of an arbitrary point set in three-dimensional space can only be determined numerically (Boissonnat and Yvinec 1995). The idea to define a sphericity measure for a convex polyhedron is simply to compare its surface area with that of a sphere of equal volume. Considering a general k-dimensional point set, a sphericity measure can be defined by referring to the hypersphere $\mathbb{S}^k$ which leads straightforward to the expression

$$\tau_k = \sqrt[k]{\frac{k^k \sqrt{\pi^k} V_{p,k}^{k-1}}{(\frac{1}{2}k)\Gamma(\frac{1}{2}k) A_{p,k}^k}} \qquad (0 \leq \tau_k \leq 1), \qquad (15.2)$$

where $V_{p,k}$ and $A_{p,k}$ denote, respectively, the generalised volume and surface area of a k-polytope. $\Gamma(\cdot)$ symbolizes the gamma function, as usual. For the special case of a three-dimensional point set, it then simply follows for τ_3 (Chang et al. 2002):

$$\tau_3 = \sqrt[3]{\frac{36\pi V_{p,3}^2}{A_{p,3}^3}} \qquad (0 \leq \tau_3 \leq 1), \qquad (15.3)$$

where $A_{p,3}$ and $V_{p,3}$ are the surface area and volume of the polyhedron, respectively. For a sphere, of course, having least surface area for any volume, $\tau_3 = 1$. In the following, the dimensional subscript on τ shall be dropped for the sake of simplicity, as only the case $k = 3$ will be of relevance. Generally, the sphericity τ is a real-valued function of the position vectors of the vertices (i.e. atoms) spanning the convex hull of the polyhedron. Therefore, it depends on a subset of the same independent variables as the potential energy hypersurface. There are at most $3n$ such independent variables, n being the number of vertices/atoms. However, because τ is invariant to all congruence (rotatory and translational motion) as well as similarity (scale) transformations, 6 + 1 degrees of freedom can be subtracted. Thus, $\tau : \mathbb{R}^{3n-7-3i} \longrightarrow [0,1]$, where i is the number of points not contributing to the convex hull. Sometimes the isoperimetric quotient (IQ) (Polya 1954), which is related to the sphericity τ simply by

$$\tau = \sqrt[3]{IQ}, \tag{15.4}$$

is used to analyze the spherical shape of polyhedra. Just like the sphericity τ, it is sufficiently sensitive to large shape changes but not too sensitive to small deformations. The two measures τ and IQ can be determined easily from the position vectors of the vertices of the polyhedron and are closely related to the shape factors appearing in the equation of state of hard convex molecular clusters (Nezbeda 1976). When there are, e.g., two polyhedral shells present, then we have two different cases for the contribution of their atoms to the convex hull of the system. First of all, if the two cages are well separated from each other, i.e one cage lies completely inside the other, then the convex hull is solely spanned by the atoms of the outer cage. In this instance, we have two really nested polyhedral cages. The second case arises when the two cages interpenetrate each other in such a way that the atoms of both equally span the convex hull. One could consider this situation as a new compounded polyhedron, the vertices of which emanate from both cages. At the same time, the polygonal faces break into smaller facets which generally increases the sphericity of the system.

There are several criteria that influence the stability of molecular frameworks. The most important and decisive of which is energetic stability, i.e. all molecules tend to reach the state of lowest possible energy. Besides a geometric configuration of high point group symmetry as well as high sphericity is also favorable but less determining. There exist quite a few molecular systems for which their lowest energy configuration does neither reflect the highest possible symmetry nor highest possible sphericity.

15.3.4 Convex Polyhedra Families

The Platonic Solids

The Platonic solids, also called the regular solids or regular polyhedra, are convex polyhedra with equivalent faces composed of congruent convex regular polygons. There are exactly five such solids: the cube, dodecahedron, icosahedron, octahedron, and tetrahedron (see Table 15.2). Platonic solids are further characterized by having three unique spheres associated. A circumsphere on which all vertices lie, an insphere which passes through all facial midpoints, and a midsphere which touches upon all edges. Furthermore, all dihedral angles (i.e. the angle between two adjacent faces) are equal, all solid angles as well as all vertex environments are equivalent because every vertex is surrounded by the same number and kind of faces. The dual of a Platonic solid is again a Platonic, i.e. they are dual between themselves, and in fact, the tetrahedron is self-dual. They are also characterized by high symmetry and high sphericity. The most spherical of them is the icosahedron (see Figure 15.3). A summary of their geometric features is given in Table 15.2.

The Archimedean Solids

An Archimedean solid is one of the 13 solids first enumerated by Archimedes. They are semi-regular convex polyhedra composed of regular polygons not all of the same kind meeting in identical vertices. Their geometric properties are summarized in Table 15.3. The family of Archimedean solids include as well the two infinite series of n–prisms and n–antiprisms, which will not be considered further here. The Archimedean polyhedra differ from the 92 so-called Johnson solids, whose regular polygonal faces do not meet in identical vertices (Johnson 1966). Among the Archimedean polyhedral, only the cuboctahedron (Figure 15.3) and the icosidodecahedron have all edge environments equivalent; all Archimedean solids have a unique circumsphere and midsphere but no unique insphere. There are several particular inspheres each of which passes through the midpoints of the same kind of faces. The solid with the highest sphericity is the snub dodecahedron. The snub cube (Figure 15.4) and the snub dodecahedron show chirality i.e. they possess two enantiomorphic forms. This is because their framework does not have any internal mirror symmetry.

Catalan Solids: The Duals of the Archimedean Polyhedra

The Catalan solids are all convex. They are face-transitive but not vertex-transitive, i.e. all faces are equivalent but the vertex environments are not. This is because the Archimedean solids are vertex-transitive and not face-transitive. Note that unlike Platonic solids and Archimedean solids, the faces of Catalan solids are not regular polygons. However, the vertex figures of Catalan solids are regular, and

TABLE 15.2 The Five Platonic Polyhedra. Number of vertices (v), number of n-gonal faces ($f\{n\}$), number of edges (e), vertex valency (q) (number of edges meeting at a vertex), point group symmetry (PG), sphericity (τ), and corresponding dual solid.

Polyhedron	v	$f\{n\}$	e	q	PG	τ	Dual
Tetrahedron	4	4 {3}	6	3	T_d	0.671139	Tetrahedron
Cube	8	6 {4}	12	4	O_h	0.805996	Octahedron
Octahedron	6	8 {3}	12	3	O_h	0.845583	Cube
Dodecahedron	20	12 {5}	30	5	I_h	0.910453	Icosahedron
Icosahedron	12	20 {3}	30	3	I_h	0.939326	Dodecahedron

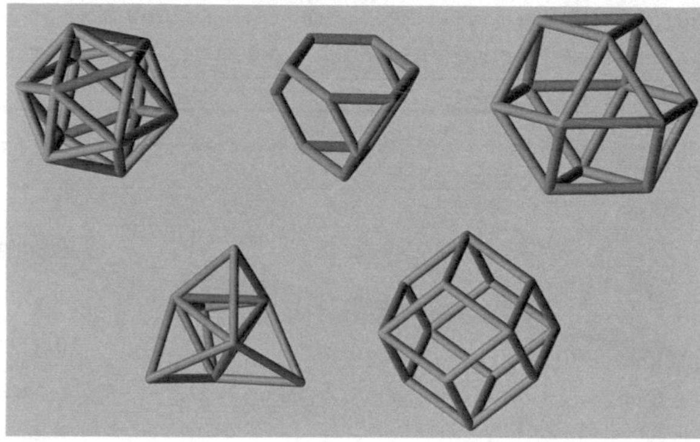

FIGURE 15.3 Convex polyhedra with 12 vertices or 12 faces taken from Tables 15.2–15.4. $v = 12$: icosahedron (Platonic), truncated tetrahedron (Archimedean), cuboctahedron (Archimedean) $f = 12$: dual of truncated tetrahedron (Catalan), dual of cuboctahedron (Catalan).

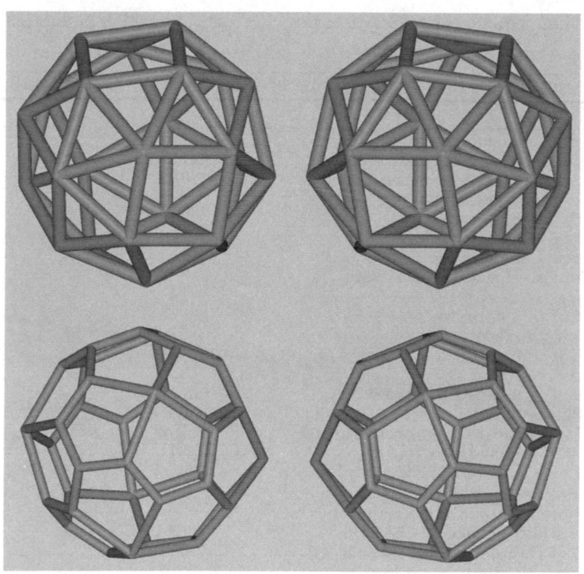

FIGURE 15.4 Enantiomorphs of the Archimedean snub cube (D, L) (upper row) and of its dual, the Catalan pentagonal icositetrahedron (D, L) (bottom row).

they have constant dihedral angles. Being face-transitive, Catalan solids are isohedral i.e. there is just one kind of face.

Among the Catalan polyhedra (Archimedean duals), only the rhombic dodecahedron (Figure 15.3) and the rhombic triacontahedron have all their edge environments equivalent. All Archimedian duals possess a unique insphere, and midsphere but no unique circumsphere. There are several particular circumspheres each of which passes through the vertices of the same kind.

The two infinite series of n–prisms and n–antiprisms give rise to two infinite dual series viz. the so-called bipyramids and the trapezohedra. The solid with the highest sphericity is the dual of the truncated rhombicosidodecahedron. It is interesting to note that, in general, all Catalan solids have a higher value of τ, i.e. have a higher sphericity than their corresponding Archimedeans. The geometric characteristics of these bodies are shown in Table 15.4.

Two of the Catalan solids like their corresponding Archimedeans are chiral:

- the pentagonal icositetrahedron (dual of the snub cube) (see also Figure 15.4) and
- the pentagonal hexacontahedron (dual of the snub dodecahedron),

i.e. each of these comes in two enantiomorphs as well.

TABLE 15.3 The 13 Archimedean Polyhedra. Number of vertices (v), number of n-gonal faces ($f\{n\}$), number of edges (e), vertex valency (q) (number of edges meeting at a vertex), point group symmetry (PG), and sphericity (τ). The corresponding dual solids are listed in Table 15.4.

Polyhedron	v	$f\{n\}$	e	q	PG	τ
Cuboctahedron	12	8 {3} 6 {4}	24	4	O_h	0.904997
Icosidodecahedron	30	20 {3} 12 {5}	60	4	I_h	0.951024
Truncated tetrahedron	12	4 {3} 4 {6}	18	3	T_d	0.775413
Truncated octahedron	24	6 {4} 8 {6}	36	3	O_h	0.909918
Truncated cube	24	8 {3} 6 {8}	36	3	O_h	0.849494
Truncated icosahedron	60	12 {5} 20 {6}	90	3	I_h	0.966622
Truncated dodecahedron	60	20 {3} 12 {10}	90	3	I_h	0.926013
Rhombicuboctahedron	24	8 {3} 18 {4}	48	4	O_h	0.954080
Rhombicosidodecahedron	60	20 {3} 30 {4} 12 {5}	120	4	I_h	0.979237
Truncated cuboctahedron	48	12 {4} 8 {6} 6 {8}	72	3	O_h	0.943166
Truncated icosidodecahedron	120	30 {4} 20 {6} 12 {10}	180	3	I_h	0.970313
Snub cube	24	32 {3} 6 {4}	60	5	O	0.965196
Snub dodecahedron	60	80 {3} 12 {5}	150	5	I	0.982011

TABLE 15.4 The 13 Archimedian Duals (Catalan Solids). Number of vertices (v), number of non-regular faces (f), number of vertices making up each face (v_f), point group symmetry of each face (PG_f), number of edges (e), vertex valency (q) (number of edges meeting at a vertex), point group symmetry (PG), and sphericity (τ).

Polyhedron	v	f	v_f	PG_f	e	q	PG	τ
Rhombic dodecahedron (dual of cuboctahedron)	14	12	4	D_{2h}	24	3, 4	O_h	0.904700
Rhombic triacontrahedron (dual of icosidodecahedron)	32	30	4	D_{2h}	60	3, 5	I_h	0.960890
Triakis tetrahedron (dual of truncated tetrahedron)	8	12	3	C_{2v}	18	3, 6	T_d	0.864385
Tetrakis hexahedron (dual of truncated octahedron)	14	24	3	C_{2v}	36	4, 6	O_h	0.944653
Triakis octahedron (dual of truncated cube)	14	24	3	C_{2v}	36	3, 8	O_h	0.924444
Pentakis dodecahedron (dual of truncated icosahedron)	32	60	3	C_{2v}	90	5, 6	I_h	0.979484
Triakis icosahedron (dual of truncated dodecahedron)	32	60	3	C_{2v}	90	3, 10	I_h	0.967339
Hexakis octahedron (dual of truncated cuboctahedron)	26	48	3	C_s	72	4, 6, 8	O_h	0.954558
Hexakis icosahedron (dual of truncated icosidodecahedron)	62	120	3	C_s	180	4, 6, 10	I_h	0.981615
Trapezoidal icositetrahedron (dual of rhombicuboctahedron)	26	24	4	C_{2v}	48	3, 4	O_h	0.969076
Trapezoidal hexacontrahedron (dual of rhombicosidodecahedron)	62	60	4	C_{2v}	120	3, 4, 5	I_h	0.985719
Pentagonal icositetrahedron (dual of snub cube)	38	24	5	C_{2v}	60	3, 4	O	0.955601
Pentagonal hexacontrahedron (dual of snub dodecahedron)	92	60	5	C_{2v}	150	3, 5	I	0.981630

A Special Family of Dual Polyhedra: Fullerenes and Frank–Kasper Cages

An important special family of polyhedra are designated as fullerenes, which involve only pentagonal and hexagonal faces. In a fullerene-like polyhedron, each vertex has three neighbors, so that the number v of vertices in the cage can be related to the number of edges e by:

$$2e = 3v.$$

Similarly, if f_n is the number of n-sided faces, it follows that

$$2e = \sum_n f_n.$$

Euler's theorem for convex polyhedra then gives a further condition

$$v + \sum_n f_n = e + 2.$$

If we now restrict the faces to be pentagons and hexagons only and eliminate v and e, we obtain

$$5f_5 + 6f_6 = 3((1/2)(5f_5 + 6f_6) + 2 - f_5 - f_6) \longrightarrow f_5 = 12,$$

i.e. we need 12 pentagons to produce a fullerene cage, with the number of hexagons given by

$$f_6 = (v - 20)/2.$$

The cages should be stable, if the curvature-related strain is symmetrically (geodesically) distributed and if the pentagons are isolated as much as possible by the hexagons to avoid the inherent instability of fused-pentagon configuration. This fact is often called the isolated pentagon rule (IPR). However, there exist quite a few exceptions to this rule (Beavers et al. 2006). Summarizing: In order for a species to belong strictly to the family of fullerenes, the main geometric characteristics are that

1. There are 12 pentagonal and any number of hexagonal faces ($f = 12 + f_6$)
2. All vertices are shared by three neighboring polygons.

The application of Euler's formula to a fullerene with a given number of hexagons, (f_6), fully determines its number of vertices (v=20+2f_6) and edges (e=30+3f_6). It follows that one can build fullerenes having an even number of vertices from 20 onward. Odd-numbered systems will always have at least one atom with a remaining dangling bond and will thus be very reactive and unstable.

The family of Frank–Kasper polyhedra is intimately related to that of the fullerenes. They form a subset of the deltahedra, which are characterized by having only triangular faces, not necessarily equilateral.[4] A particular family of deltahedra is formed by those with just five- or sixfold vertices (i.e. either five or six edges meeting at each vertex). The smaller members of this family (those with 14, 15, or 16 vertices, shown in Figure 15.5) are actually known as Frank–Kasper polyhedra (Frank and Kasper 1958,1959) and appear often as coordination polyhedra in intermetallic phases. To avoid confusion with other types of deltahedra, one could generally call Frank–Kasper polyhedra all those with

1. Only triangles as faces (not necessarily regular)
2. Only five- and sixfold vertices.

If v_5 and v_6 are the number of vertices at which five or six edges meet, respectively, it can be shown that v_5 = 12 in

[4]Sometimes deltahedra are also called simplicial polyhedra, because the triangle represents the simplex figure of two-dimensional space.

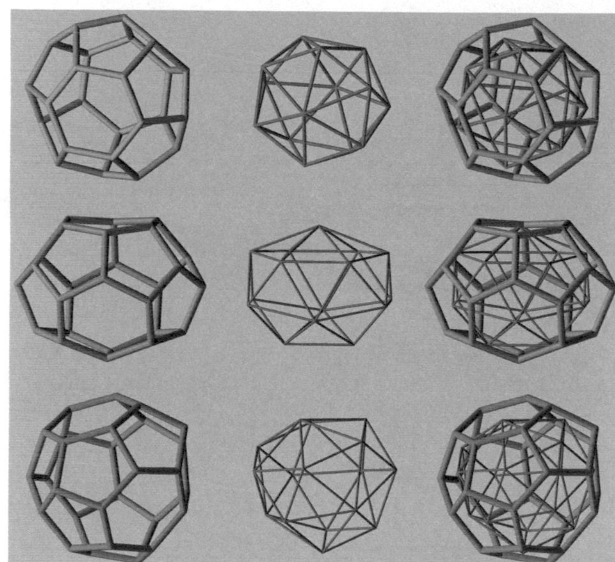

FIGURE 15.5 The original Frank–Kasper polyhedra ($v = 14$, 15, 16) (middle panels), the dual fullerenes (left panels), and their nested combinations (right panels).

TABLE 15.5 Relationship between the Number of Vertices v, Faces f, and Edges e, Where f_6 Is the Number of Hexagonal Faces ($f_6 = 1, 2, 3, \ldots$)

	$20 + 2f_6$	$12 + f_6$	$30 + 3f_6$
Fullerene	v	f	e
Frank–Kasper	f	v	e

all cases; hence, their total number of vertices is given by $v = 12+v_6$. From the definition of fullerenes given above and that for Frank–Kasper polyhedra, it follows that the properties of faces and vertices are interchanged among these two families of polyhedra. In other words, there is a duality relationship between them. Therefore, a given fullerene and its Frank–Kasper dual are identified by their common parameter $f_6 = v_6$ that corresponds to the number of hexagons and sixfold vertices, respectively. An obvious corollary of the duality relationship between Frank–Kasper polyhedra and fullerenes is that one can build up one of them by capping the faces of the other, thus forming a pair of intercalated polyhedra. The simplest fullerene is the dodecahedron $f_6 = 0$, and its Frank–Kasper dual is the icosahedron. In this special case, both of them equally belong to the more restricted family of Platonic solids (cf. Table 15.2). The relationship between the number of vertices v, faces f, and edges e of fullerene solids and their dual Frank–Kasper bodies is given in Table 15.5.

15.4 Building Principles of Nested Cage Molecules

15.4.1 Duality of Polyhedra as Building Principle

Dual polyhedra are likely to appear in molecular structures forming successive concentric shells, because occupation of the face centers of a polyhedron (as required for the formation of a dual) corresponds to situations that favor physical bonding. Three such possibilities can be mentioned:

1. Coordinative bonding between bridging ligands and the metal atoms in a polygonal face
2. Ionic bonding between ions of opposite sign in successive shells
3. Metallic bonding favored by the maximum coordination number.

Given the relationship between a polyhedral cage and its dual polyhedron, we can establish the compounding of concentric shells of such cages in a general way, which is shown diagrammatically in the flowchart scheme A (Figure 15.7).

If we start with a polyhedral framework having f_p faces and v_p vertices, which we shall symbolize by $X_{v_p}^{f_p}$, where the subscript indicates the number of vertices and the superscript the number of faces, the formation of its circumscribed dual polyhedron (step 1 in scheme A) will be characterized by $v_d = f_p$ vertices and $f_d = v_p$ faces, denoted by $Y_{v_d=f_p}^{f_d=v_p}$. The fact that in this notation the polyhedron and its dual solid have subscripts and superscripts interchanged follows from their dual nature. To this shell, we can now add a second polyhedral shell by capping all the faces of the dual polyhedron (step 2 in scheme A). This procedure could go on by adding shell after shell in the same way, i.e. repeating assiduously steps 1 and 2 (symbolically: Polyhedron@Dual@Polyhedron@Dual...). The convex hull of the solid composed in this way will always be spanned by the polyhedron last added, i.e. the outermost cage or onion layer. Furthermore, the symmetry of the original polyhedron will be conserved, i.e. remains unchanged by this procedure.

As an example, in the $[AsNi_{12}As_{20}]^{3-}$ anion (Moses, Fettinger, and Eichhorn 2003) and similar systems, a Ni_{12} icosahedron and a circumscribed As_{20} dodecahedron formed by face capping are of quite different sizes (see Figure 15.6). The convex hull of the system is therefore spanned by the atoms of the As_{20} dodecahedron only. We, therefore, have well-separated nested rather than interpenetrating cages, which correspond to a situation shown in step 1 of scheme A (Figure 15.7).

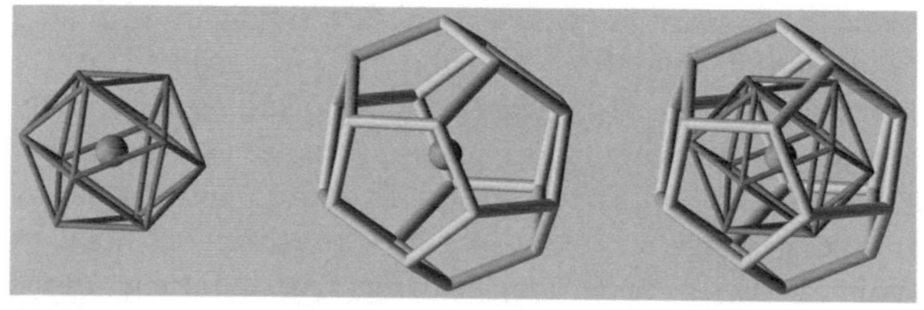

FIGURE 15.6 Illustration of step 1 of scheme A: Nested Keplerate cage $Z@X_{12}^{20}@Y_{20}^{12}$. A center atom located inside an icosahedron merged with its dual dodecahedron, which have both I_h symmetry.

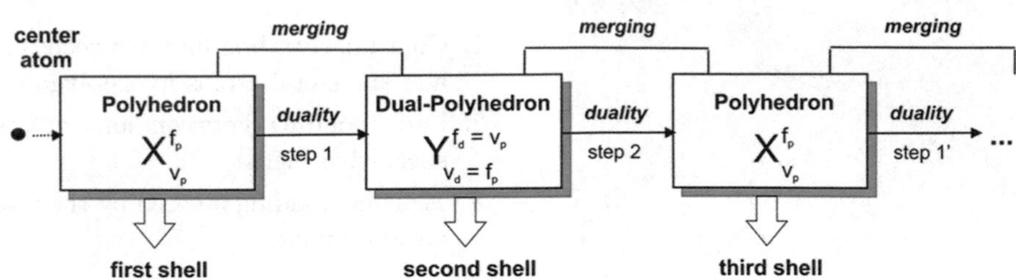

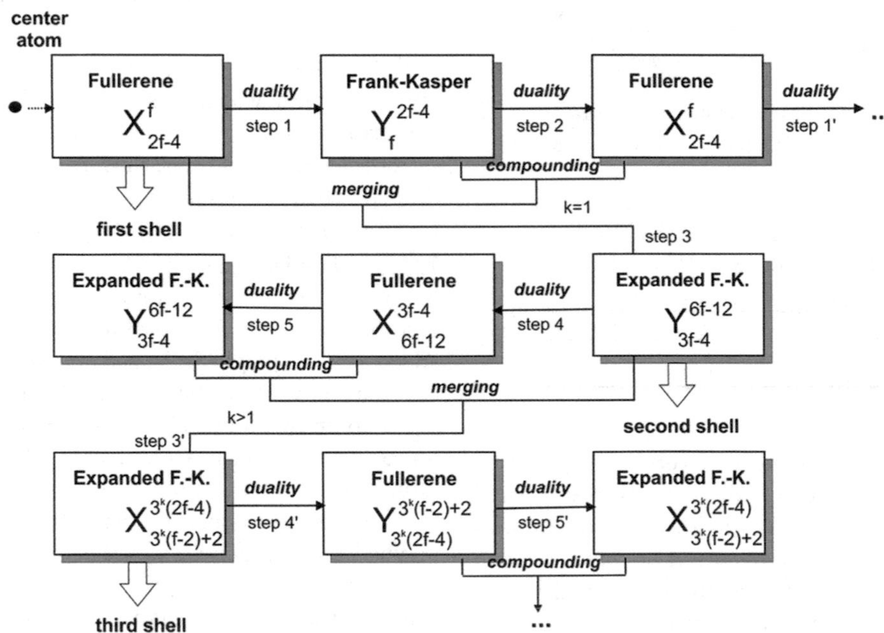

FIGURE 15.7 Building schemes of nested polyhedra: (A) General building scheme, (B) alternative building scheme for fullerene-like cages including extended Frank–Kasper polyhedra.

The building patterns just described imply that, whenever there is a polyhedral framework within a molecular structure, we may suspect the presence of an inscribed and/or circumscribed dual, and vice versa, not limited to just a dual pair, but larger numbers of shells can be arranged following the principles outlined in scheme A.

In extended crystal structures, the outermost onion skin layers of one set of intercalated polyhedra may overlap with corresponding ones from neighboring unit cells. Hence, a description of nested polyhedra does not necessarily imply that we are referring to independent molecular assemblies (see scheme B of Figure 15.7). However, another possibility is to merge a Frank–Kasper deltahedron obtained by duality with the underlying fullerene so that their atoms will equally span the convex hull of a combined polyhedron. The result is an expanded polyhedron of Frank–Kasper type with a larger number of vertices, $\tilde{Y}_{3f-4}^{6f-12)}$ (branching step 3 in scheme B, $k = 1$). Another shell can now be added by construction of the fullerene dual to the expanded Frank–Kasper (step 4 in scheme B, $k = 1$), which would be denoted as $\tilde{X}_{3f-4}^{6f-12}$. This procedure can equally continue shell after shell, i.e. running through successive values of k ($k = 1, 2, 3, ...$). Of course one could have started from a Frank–Kasper polyhedron at the beginning, then a similar scheme would apply. This means that the compositions of successive shells are accurately determined and correspond to a sequence of specific numbers. Thus, starting with a given fullerene having $f_6 = f - 12$ hexagonal faces, we can build a series of larger onion-like fullerenes by using the construction principle depicted in scheme B. Their general compositions $\tilde{X}_{\tilde{v}}^{\tilde{f}}$ correspond to the values of $\tilde{v}$ and $\tilde{f}$ that obey the following expressions $\tilde{v} = 3^k(2f - 4)$ and $\tilde{f} = 3^k(f - 2) + 2$, where k is any nonnegative integer (k = 0,1,2,...). The sequence of steps 3 and 4 that leads from a given fullerene to an expanded fullerene having three times as many vertices as the original is known as the *leapfrog relationship* (Fowler 1986). One can, thus, conclude that fullerenes related through a 'leapfrog' connection are likely to be found in a molecular structure separated by an intermediate layer shell of a dual Frank–Kasper polyhedron.

As an example of the first steps of these building principles, let us consider the simplest case that starts with a dodecahedron, the X_{20}^{12} (f_6=0) fullerene. Its dual is a special Frank–Kasper polyhedron, the Platonic icosahedron, and compounding of these two polyhedra gives a special expanded Frank–Kasper Y_{32}^{60} polyhedron: the 32-vertex rhombic triacontahedron, a Catalan solid with icosahedral symmetry. By face augmentation of this latter polyhedron, we obtain a X_{60}^{32} fullerene, nothing other than the Archimedean truncated icosahedron which is the well-known C_{60} structure (Figure 15.8). Such a process of compounding of a dodecahedron and an icosahedron to form a 32-vertex Frank–Kasper polyhedron is a $Li_{20}As_{12}$ arrangement found in the structure of stoichiometric composition $Li_{26}O[AsSiPrMe_2]_{12}$ (Driess et al. 1996) and a $Mo_{12}C_{20}$ unit in which the carbon atoms are capping the faces of a Mo_{12} icosahedron (Müller et al. 2002), for example. The relative size of the clusters participating in the compounding step 3 (scheme A) is not irrelevant. Thus, in the examples just discussed, the two polyhedra that form an expanded Frank–Kasper Y_{32} polyhedron have similar radii, and hence, their atoms all lie on the convex hull of the combined polyhedron.

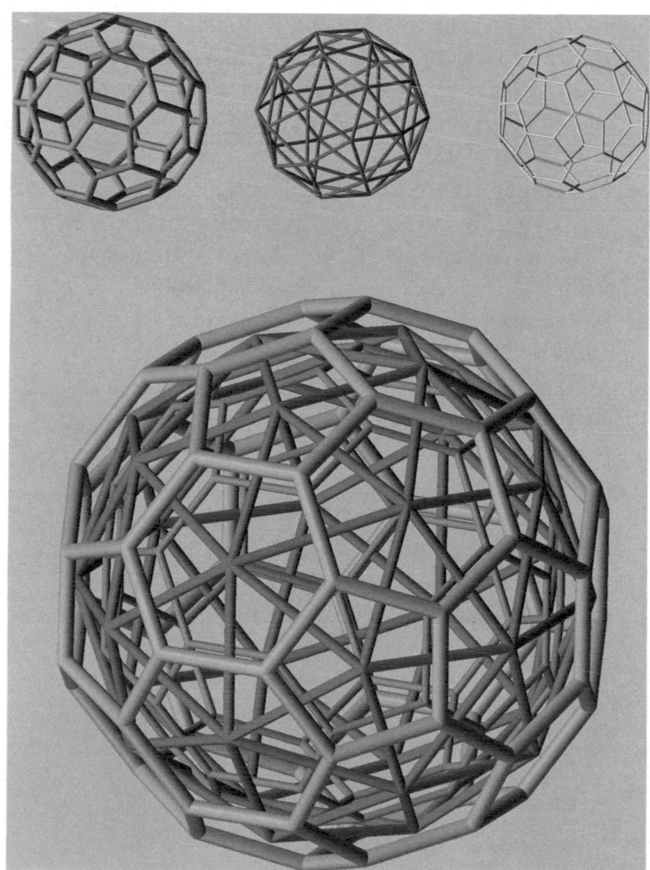

FIGURE 15.8 Illustration of the first line of scheme B: Fullerene X_{60}^{32} (left), the dual Frank–Kasper polyhedron Y_{32}^{60} (middle), fullerene X_{60}^{32} (right), and the combined three-shell polyhedron $X_{60}^{32}@Y_{32}^{60}@X_{60}^{32}$ (bottom).

15.4.2 Application: Polyhedra Families

Nested Keplerate Cages

Onion-like cages whose polyhedral shell structure consists of various combinations of Platonic, Archimedean, and/or their corresponding dual solids are known as Keplerates (Müller, Kögerler, and Dress 2001). They do not always strictly follow the building principle demonstrated in scheme A (Figure 15.7). In general, when embedding two or more polyhedra, it is important to note that the symmetry of a compound of two polyhedra is that of the common subgroup with maximum symmetry. Thus, the compound of a tetrahedron and an octahedron belongs to the T_d point group (a subgroup of the octahedral group O_h), whereas the composition of an octahedron (or a cube) and an icosahedron (or a dodecahedron) reduces the symmetry of the compound to that of a subgroup common to O_h and T_d, i.e. the T_h or D_{3d} point groups. In a similar way, compounded polyhedra with icosahedral and tetrahedral symmetries only retain the symmetry operations of a common subgroup T or D_{2d}.

In the following, a few selected examples of these Keplerate cages are presented:

A typical specimen for this kind of species is the series X_nY_n, where X = Al, Ga, or In, Y = N, As, or Sb, and n takes the values 12, 24, and 60 (Chang et al. 2001). However, for the most stable configuration of $(AlN)_n$, n = 12, 24, and 60 where the inner cage X and the outer cage Y are of comparable size, we have an exceptional case, observing an intercalation of two penetrating polyhedra, the vertices of which all lie on one triangular convex hull (Figure 15.9).

These triangles, though irregular, are distributed in such a way as to conserve a very high point group symmetry, viz. T_h, O, and I for n = 12, 24, and 60, respectively. On the other hand, for the heavier systems $(GaAs)_{24}$ and $(InSb)_{24}$ as well as for $(GaAs)_{60}$ and $(InSb)_{60}$, where the inner and outer cages are of quite different sizes, the most stable arrangement is a true onion-like system, where the convex hull consists uniquely of the atoms of the outer cage Y. The result is a structure of two nested snub cubes rotated by an angle of 45° against each other (n = 24) and two nested snub dodecahedra, where the rotation angle amounts to 36° (n = 60). For n = 12, e.g. $(GaAs)_{12}$ and $(InSb)_{12}$, we have a limiting case of two intercalated icosahedra rotated by an angle of 90°, where the vertices (X) of the inner cage lie almost in the polygonal faces of the outer one (Y), i.e. they almost equally span the convex hull of the species which is composed of triangles. More realizations of nested Keplerate cages are:

- The nesting of two Platonic polyhedra viz. an Si_{20} dodecahedron circumscribing an Na_{12} icosahedron ($Na_{12}@Si_{20}$), which occurs within the crystal structure of Na_8Si_{46} (Cros, Pouchard,

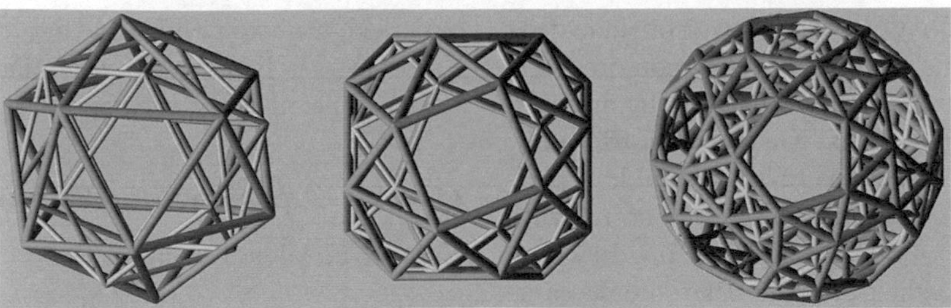

FIGURE 15.9 Series of penetrating polyhedra X_nY_n, $n = 12, 24, 60$.

and Hagenmuller 1971; Reny et al. 1998; Nolas 2014). This structural pattern icosahedron@dodecahedron is common and realized in many inorganic clusters.

- Another example of this type including a central atom is the intensely studied $As@Ni_{12}@As_{20}{}^{3-}$ anion (cf. Figure 15.6), which occurs in the solid state as a part of a complex salt (Moses, Fettinger, and Eichhorn 2003; Baruah et al. 2003). The heavier homologue $Sb@Pd_{12}@Sb_{20}{}^{3-}$ (Zhao and Xie 2004) as well as the neutral species $Kr@Ni_{12}@As_{20}$, $Ge@Zn_{12}@Ge_{20}$ and similar neutral and charged systems having the same structural properties were theoretically investigated by Chang et al. (2005,2006), Stegmaier and Fässler (2011), Wang et al. (2017), and Long et al. (2017).

  Furthermore, recently Hu and Kaltsoyannis (2018) investigated theoretically numerous systems having this kind of shell structure where the outer dodecahedral cage is formed by 20 actinide atoms (U, Pu, or Np). They found that the species $S@Mn_{12}@Np_{20}$ and $S@Mn_{12}@Pu_{20}$ have the highest ground state spin yet reported for a molecular cluster (viz. total spin quantum numbers $S = 80/2$ and $S = 100/2$, respectively).

- A species of different structure with more shells is the palladium complex system Pd_{145}, where the Pd_{145} core consists of a Pd atom surrounded by five concentric cage shells (Tran, Powell, and Dahl 2000). The composition and shape of these shells correspond to the following nested sequence of Platonic and Archimedean polyhedra: atom@icosahedron@icosidodecahedron@icosahedron@rhombicosidodecahedron@icosidodecahedron, the last shell is furthermore decorated by a phosphorus icosidodecahedron or alternatively $Pd@Pd_{12}@Pd_{30}@Pd_{12}@Pd_{60}@Pd_{30}@P_{30}$. Since in this case all the shells of the Pd_{145} core have icosahedral symmetry, it could be considered as an icosahedral nanoparticle with a diameter of about 2.2 nm. One might speculate about the possibility to grow crystalline materials without translational symmetry solely by a large number of atomic shells arranged concentrically having icosahedral symmetry.

- Transition metal icosidodecahedra encapsulating dodecahedra are a common motif in Keplerate cluster of *d*- and *f*-block elements. $La_{20}@Ni_{30}$ is realized in the cation of the salt $La_{20}Ni_{30}IDA_{30}(CO_3)_6(NO_3)_6(OH)_{30}(H_2O)_{12}$ (cf. Kong et al. 2007).

- A cation with overall cubic symmetry $Nd_8@Ni_{48}Nd_{24}@Ni_{24}$ is the realization of a Platonic cube nested in three Archimedean solids, a cuboctahedron, and two solids with 24 vertices, a truncated octahedron, and a rhombicuboctahedron (Peng et al. 2014).

- A cage molecule composed of four shells is the $[N_8P_{12}N_6S_{12}]^{6-}$ anion discovered by Fluck et al. (1976), see also Roth and Schnick (2001). The innermost cage is an N_8 cube surrounded by a P_{12} icosahedron. The outer shells are then an N_6 octahedron and finally another icosahedron S_{12}. The polyhedral structure of this $N_8@P_{12}@N_6@S_{12}$ system is depicted in Figure 15.10.

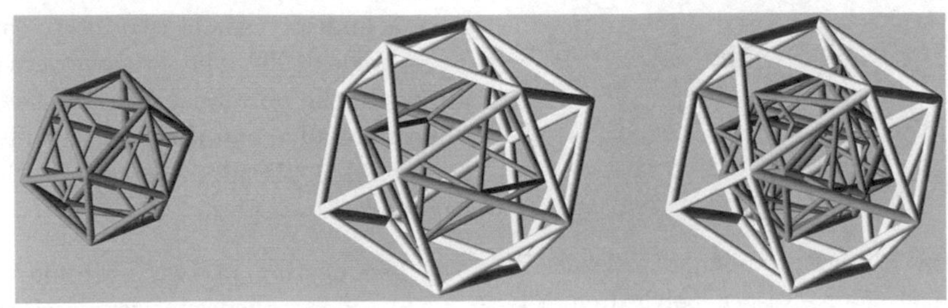

FIGURE 15.10 Nested Keplerate with four cages: $W_8^6@X_{12}^{20}@Y_6^8@Z_{12}^{20}$ built from different Platonic polyhedra, namely cube, octahedron, and two icosahedra.

- The $[O_8Ga_{12}O_6C_{12}]^{2+}$ cation (Swenson, Dagorne, and Jordan 2000) is shaped in a completely analogous way.
- Another cage system consisting of four polyhedral shells was described by Bai, Virovetz, and Scheer (2003). It is in fact the core of a larger rather complex molecule. This core with sum formula $Cl_{25}P_{60}N_{10}Cu_{25}Fe_{12}$ involves four embedded polyhedral cages ($Cl_{20}P_{60}Cu_{25}Fe_{12}$). However, although being well convex, the corresponding polyhedra are neither Platonic nor Archimedean solids but rather contain many non-regular faces. A particularity of this system is that there exist two planar polygonal rings, viz. a Cl_5 pentagon and an N_{10} decagon, which encircle the cages.
- The synthesis of high symmetric gadolinium polyhedra has been reported by Qin et al. (2017). Structural analysis reveals that the Gd_{20} cage resembles a dodecahedron, the Gd_{32} polyhedron can be described as a truncated octrahedron of 24 Gd atoms capped by a Gd_8 cube yielding $Gd_{24}@Gd_8$ with overall O_h symmetry, Gd_{50} displays an unprecedented combination of a Archimedean polyhedron inside a Platonic cage, in which an icosidodecahedron Gd_{30} core is encapsulated by an outer Gd_{20} dodecahedral shell, $Gd_{30}@Gd_{20}$, with approximate I_h symmetry, and the Gd_{60} shows a truncated octahedron geometry of O_h symmetry filled with a rhombicuboctahedron structure having a cuboctahedron inside, $Gd_{12}@Gd_{24}@Gd_{24}$, i.e. Gd_{60} is build up by three nested Archimedean polyhedra (see Figure 15.11).
- A single icosahedral cage anion Sn_{12}^{2-} was evidenced by Cui et al. (2007) using photoelectron spectroscopy and tentatively given the name 'stannaspherene'. The circumsphere of this icosahedron has a diameter which is only slightly smaller than that of the famous C_{60}. It can be further stabilized by placing a metal atom at its center resulting in a stable negatively charged $M^+@Sn_{12}^-$ and a neutral $M^{2+}@Sn_{12}^{2-}$ series of systems with M = Ti, V, Cr, Fe, Co, Ni, Cu, Y, Nb, Gd, Hf, Ta, Pt, and Au. A quite analogous series involving the heavier Pb_{12}^{2-} cage was studied and named 'plumbaspherene' (Cui et al. 2006).

  Another icosahedral shell molecule of the same type $M@Au_{12}$ (M = W or Mo) had also been predicted by Pyykkö and Runeberg (2002) and experimentally observed by Li et al. (2002). However, in this case, the central atom is critical for the stability of the icosahedral cage, because bare Au_{12} does not possess a stable cage structure.

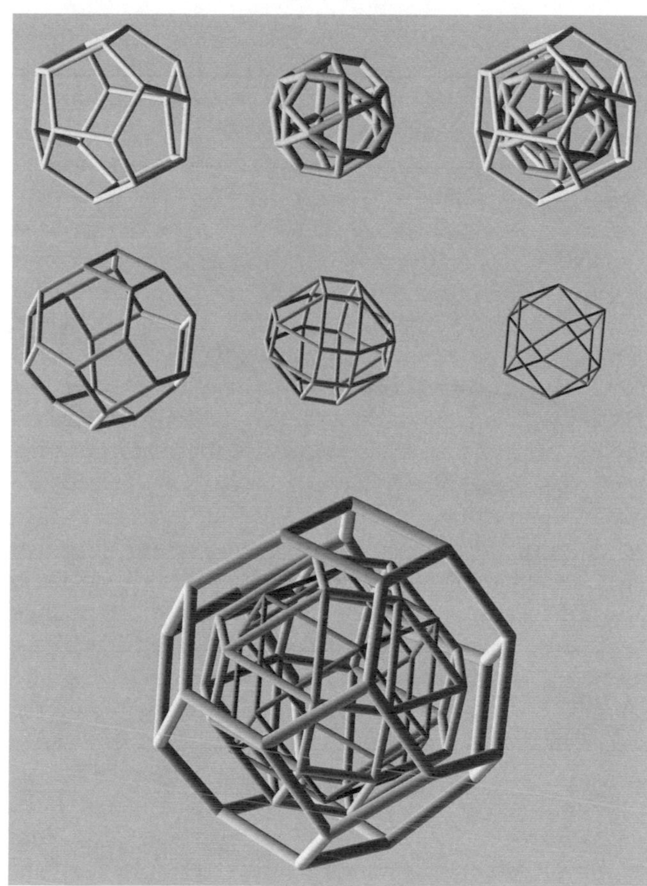

FIGURE 15.11 Gd_{50}: $Gd_{30}@Gd_{20}$ (upper panel) and Gd_{60}: $Gd_{12}@Gd_{24}@Gd_{24}$ (lower panel).

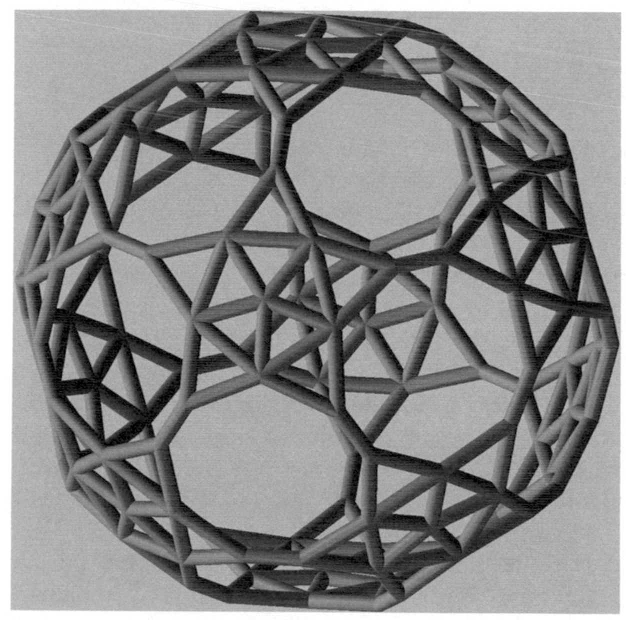

FIGURE 15.12 Mo_{132}: $[(Mo)Mo_5]_{12}[Mo_2]_{30}$.

A beautiful onion-like Keplerate with a complex shell structure is the nearly spherical giant polyoxomolybdate $Mo_{132}O_{372}(SO_4)_{30}(H_2O)_{72}$ of icosahedral symmetry (see Figure 15.12).

The pivotal molybdenum Mo_{132} cage can be written more specifically as $[(Mo)Mo_5]_{12}[Mo_2]_{30}$. It consists of 12 fundamental pentagonal $[(Mo)Mo_5]$ building blocks with a central Mo atom that are linked together by 30

$[Mo_2]$ units. A closer look by taking the oxygen atoms into account reveals that the central Mo atom of the pentagonal $[(Mo)Mo_5]$ unit in fact lies at the center of a pentagonal bipyramid (MoO_7), while the adjacent five Mo atoms are at the centers of edge-sharing octahedrons (MoO_6). Thus, the whole cage system represents a complex arrangement of intermingled shells spanned by molybdenum and several nonequivalent oxygen atoms as well as shells generated by the atoms of the ligand groups (SO_4, H_2O).

From a physical point of view, this spherical Mo/O cluster with an overall diameter of about 3 nm gives rise to promising perspectives in host–guest chemistry and size-selective catalysis because it possesses an open Mo/O framework architecture. The framework encloses an inner cavity with an approximate diameter of 1.7 nm and spans nine ring openings with an average ring aperture of 0.43 nm, which is comparable in size to the pores in zeolithic structures. Furthermore, the $[Mo_2]$ linking unit of the molybdenum cage represents an interesting key position of the cluster, for it might be replaced by a different atom or groups of atoms. In this way, the size of the inner cavity and the outer pores becomes tunable which offers the possibility to design specific nanopores that can act as substrate-specific receptors in order to form a composite of a unique kind. Besides, a replacement of the 30 diamagnetic $[Mo_2]$ linkages by 30 paramagnetic atoms such as Fe^{III} leads to the emergence of unusual molecular nanoferromagnets (Müller et al. 2001). An extensive description of the structural and physical properties of these fascinating objects was given by Müller, Kögerler, and Dress (2001) and Melgar Freire (2015).

Fullerenes@Frank–Kasper Cages

An example of such a dual relationship can be seen in one of the building blocks of the $Mg_{35}Cu_{24}Ga_{54}$ structure (Lin and Corbett 2005), formed by a Ga_{16} Friauf polyhedron (Friauf 1927) (Figure 15.5, bottom row, center panel) surrounded by its dual Mg_{28} fullerene, which is in turn circumscribed by another $Cu_{12}Ga_4$ Friauf polyhedron. This polyhedron is the specific case of the Frank–Kasper polyhedron with 28 triangular faces and 16 vertices, four of which are based on the midpoints of sustaining hexagons. A simpler realization involving just a single 16-vertex Frank–Kasper polyhedron is the species $M@Si_{16}$ (M = Ti or Ta) (Tsunoyama et al. 2017) and $Cu@Au_{16}^-$ (Wang et al. 2007), (see Figure 15.13).

It is to be noted that the composition of a fullerene and its dual Frank–Kasper, such that the vertices of both lie on the convex hull of the combined solid, results in a larger Frank–Kasper polyhedron. In the present example, an M_{44} Frank–Kasper is composed of the Mg_{28} fullerene and its $Cu_{12}Ga_4$ dual. Another seemly set of three different Frank–Kasper polyhedra inside their dual fullerenes is formed by $Na_{32}@In_{48}Na_{12}$, $Na_{37}@In_{70}$, and $Na_{39}In_2@In_{78}$ found in the crystal structure

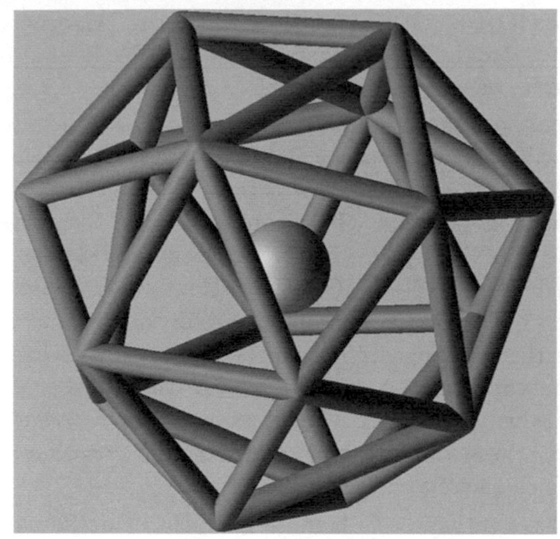

FIGURE 15.13 $Ti@Si_{16}$: a single 16-vertex Frank–Kasper polyhedron.

of $Na_{172}In_{197}Ni_2$ (Sevov and Corbett 1996). Table 15.6 summarizes other examples of onion-like inorganic fullerenes that appear concentric to their Frank–Kasper duals, including two characteristic structure types known to present typical Frank–Kasper polyhedra, viz. $MgCu_2$ and Cr_3Si.

We can go one step further by looking at the crystal structure of $Na_{13}(Cd_{1-x}Tl_x)_{27}$ (Li and Corbett 2004). There a Tl_{12} icosahedron (a Frank–Kasper polyhedron) is surrounded by a dual Na_{20} fullerene (a dodecahedron) and its Frank–Kasper dual, a Cd_{12} icosahedron. The compounding of the latter two polyhedra is in fact an M_{32} Frank–Kasper polyhedron, and capping its faces with Tl atoms result in a Tl_{60} fullerene. The game can go on by adding another Frank–Kasper set, Na_{32}, which is in fact a composition of a dodecahedron and an icosahedron (capping the hexagonal and pentagonal faces of Tl_{60}, respectively) similar to the inner $Na_{20}Cd_{12}$ cluster. It is easy to verify that these concentric polyhedra are built according to the rules shown in scheme B (Figure 15.7). One should notice that the symmetry is kept as more concentric shells are added by face augmentation, and both the inner Tl_{12} and the more outer Na_{12} clusters are concentric polyhedra. Although the building principles discussed so far do not require such a high symmetry, it might be interesting to sketch a systematic way of generating Frank–Kasper and fullerene polyhedra with high symmetry (Table 15.6).

It is worth noting here that the geodesic domes designed by Buckminster Fuller (Gorman 2005) are in fact formed by triangles and use a construction principle coincident in part with the one presented in scheme B. In contrast with the vast literature devoted to spherical carbon fullerenes, the hemispherical molecules that are closer to geodesic domes made by Buckminster Fuller have received relatively little attention.

TABLE 15.6 Some Examples of Onion-Like Fullerene Clusters Characterized by Their Number of Hexagons f_6 that Appear Intercalated with Their Dual Frank–Kasper Polyhedron in Inorganic Compounds

f_6	Fullerene	Frank–Kasper	Stoichiometric Composition	Ref.
0	Si_{20}	Na_{12}	Na_8Si_{46}	Cros, Pouchard, and Hagenmuller (1971), Reny et al. (1998), Nolas (2014)
0	Li_{20}	As_{12}	$Li_{26}O[AsSi(iPr)Me_2]_{12}$	Driess et al. (1996)
0	$Cr_{20}Si_4$	$Cr_{10}Si_4$	Cr_3Si	Jauch, Scultz, and Heger (1987)
0	Sn_{24}	Cs_{14}	Cs_8Sn_{46}	Grin et al. (1987)
4	$Mg_{12}Cu_{16}$	Mg_4Cu_{12}	$MgCu_2$	Ohba, Kitano, and Komura (1984)
4	Si_{28}	$Si_{28}Na_{16}$	$Cs_8Na_{16}Si_{136}$	Bobev and Sevov (1999)
4	Mg_{28}	Ga_{16}	$Mg_{35}Cu_{24}Ga_{53}$	Lin and Corbett (2005)
4	$Bi_{12}Sr_{16}$	Bi_{16}	$Sr_{33}Bi_{28}Al_{480}O_{147}$	Hervieu et al. (2004), Boudin et al. (2004)
4	U_{28}	K_{16}	$[K_{16}(H_2O)_4(O_2)_{44}(UO_2)_{29}]^{14-}$	Burns et al. (2005)
20	$In_{48}Na_{12}$	Na_{32}	$Na_{172}In_{197}Ni_2$	Sevov and Corbett (1996)
25	In_{70}	Na_{37}	$Na_{172}In_{197}Ni_2$	Sevov and Corbett (1996)
28	Al_{76}	M_{40}(M=Li,Mg)	$LiMgAl_2$	Nesper (1994)
29	In_{78}	$Na_{39}In_2$	$Na_{172}In_{197}Ni_2$	Sevov and Corbett (1996)
32	$Ga_{60}Cu_{24}$	$Mg_{28}Cu_{12}Ga_4$	$Cu_{24}Ga_{53.57}Mg_{35.12}$	Lin and Corbett (2005)
32	Al_{84}	$Sr_{32}Bi_{12}$	$Sr_{33}Bi_{28}Al_{480}O_{147}$	Hervieu et al. (2004), Boudin et al. (2004)

15.5 Other Types of Inorganic Fullerene-Like Cages

15.5.1 Borospherenes

Inorganic fullerenes consisting of boron with or without a possible central atom have been studied extensively. The results show that boron generally prefers planar (2D) arrangements rather than polyhedral cage-like structures (Yang et al. 2017). The first all-boron 3D cage computationally predicted and experimentally observed was B_{40} and its anion B_{40}^- named borospherene (Zhai et al. 2014; Martínez-Guajardo et al. 2015; Pan et al. 2018).

B_{40} has D_{2d} symmetry and could equally be written as B24@B16, because it is formed by two nested polyhedral cages with 24 and 16 vertices, respectively, but having non-regular faces (see Figure 15.14). When looking at the combined body, unusual hexagonal and heptagonal holes can be discerned. The possibility of placing a noble gas atom at the center of such a B_{40} cage (e.g. He@B_{40}) was recently shown by Pan et al. (2018).

A larger all-boron fullerene B_{80} was predicted by Szwacki, Sadrzadeh, and Yakobson (2007). It was supposed to represent a cage of perfect icosahedral (I_h) symmetry. A later study, however, revealed that the equilibrium structure of this molecular cluster has lower, namely, tetrahedral T_h symmetry (Baruah, Pederson, and Zope 2008). A further investigation (Zhao et al. 2010) found that the most stable B_{80} cluster has still lower symmetry. It consists of an inner B_{12} icosahedral cage linked to an outer shell mainly comprising triangular facets. A look at the outer surface reveals pentagonal, hexagonal, and heptagonal openings. However, an experimental evidence has still to come.

Another multi-shell cage system belonging to this family is B@Co_{12}@B_{80} reported by Wang et al. (2009). This species has perfect icosahedral (I_h) symmetry and is in fact a four- caged cluster, because the outermost shell (B_{80}) actually is composed of two boron shells viz. B_{20} and B_{60}. Therefore, this cage molecule could as well be written as B@Co_{12}@B_{20}@B_{60}.

15.5.2 Clathrates

Apart from the concentric onion-layer clusters, clathrates are another kind of nested or enclosed molecular compounds. The word clathrates is derived from the Latin word 'clathratus' meaning enclosed or protected by cross bars or gratings. In such compounds, one of the components has an open structure in the crystalline state containing cavities, holes, or channels in which atoms or molecules of another component with appropriate size are trapped. The trapped molecule is called the 'guest' while the other part is named the 'host'. Therefore, clathrates are host@guest complex molecular arrangements. Such complexation can occur both in a liquid solution and in the solid state. In the solid state, guests are retained by the host through crystal lattice forces. In each case, there is a complete enclosure of the guest molecules in a suitable cage structure formed by the molecules of the host. Such caged host–guest complexes are variously referred to as supramolecular assembly, extramolecular assembly, inclusion or occlusion compounds, or simply clathrates.

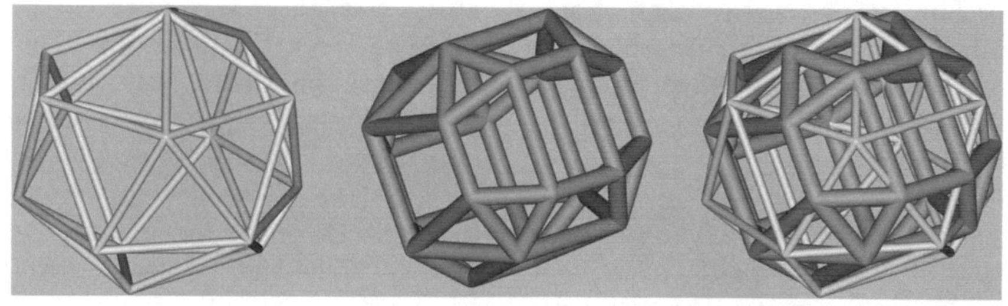

FIGURE 15.14 B_{40}: B_{24}@B_{16} being an irregular polyhedron inside a Frank–Kasper cage.

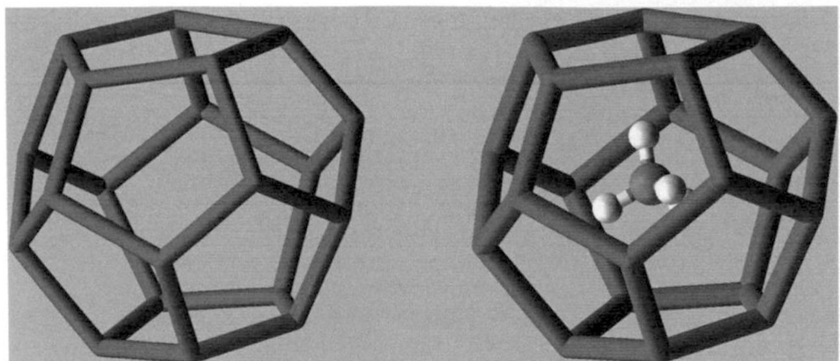

FIGURE 15.15 Clathrate $CH_4@(H_2O)_{20}$.

The guest atoms or molecules can be encapsulated into the host cages if their size is comparable to the size of the available empty space within the cage. The sole determining factor for the formation of clathrates is, therefore, proper molecular size. Molecules which are too large do not fit into the cage and those which are too small escape through the lattice framework. The cages can well be held together by very weak interactive forces such as hydrogen bonding, ion pairing, electric dipole–dipole interactions, or van der Waals attraction. Analysis indicates that the composition of clathrates is often slightly less than the expected stoichiometric ratio of the host to guest molecules. This suggests that all the available cavities may not be used. The reason for this can be understood from the enclosing process. A guest molecule must be properly oriented at the moment of enclosure, if not, it will be excluded and some unfilled holes are to be expected. As a simple example, a noble gas atom like He (the guest) encapsulated in a C_{60} fullerene (the host), $He@C_{60}$, could be considered as a clathrate of carbon (Weiske et al. 1991; Saunders et al. 1993). Other examples are $Ta@Si_{16}$ and $Cu@Au_{16}^-$ (Tsunoyama et al. 2017; Wang et al. 2007) where the cage is formed by a 16-vertex Frank–Kasper polyhedron. Well known is the methane hydrate clathrate, $CH_4@(H_2O)_{20}$ (Deible, Tuguldur, and Jordan 2014), in which a tetrahedral methane molecule is encapsulated by a dodecahedral cage of oxygen atoms of 20 water molecules (see Figure 15.15).

15.6 Concluding Remarks

Inorganic fullerene-like systems represent an active and challenging area of research with respect to both theory and experiment. However, apart from their beautiful shapes and their importance for fundamental scientific questions, only a few applications of the materials so far described have been reported in the literature. A major problem is that size and shape control in the synthesis is still in its infancy. One can summarize that the driving force of the product formation is the thermodynamic and kinetic stability of the highly symmetrical products. The synthesis of materials following the rules of construction described often starts with a serendipitous observation. The structures are assembled from smaller units, whereby the elementary steps of the building reactions are, in general, not known. Synthetic strategies involve the use of simple anions as weak templates in the presence of other multidentate ligands, preferably under solvothermal conditions. The synthesis of high symmetric gadolinium polyhedra, e.g., has been studied recently (Qin et al. 2017). Alternatively, high-power impulse magnetron sputtering coupled with a mass spectrometer and a soft landing apparatus can be used as an intensive, size-selected nanocluster source (Tsunoyama et al. 2017).

Appendix on Additional Geometric Properties of Polyhedron Families

In this appendix we give a succinct overview of geometric properties of polyhedron families which often appear in molecular structures adding a few other which were not explicitly discussed in the main text.

A polyhedron is called symmetric if it is stable under at least one non-trivial isometric transformation. Two commensurate components of a polyhedron (e.g. two vertices, two faces, or two edges) are said to be equivalent if there is an isometry (rotation or reflection) which transforms one into the other. When this results in only one equivalence class the following attributes are used to qualify the polyhedron:

1. isohedral (or face-transitive), if all faces are equivalent
2. isogonal (or vertex-transitive), if all vertex environments are equivalent
3. isotoxal (or edge-transitive), if all edge environments are equivalent
4. equilateral, when all edges have the same length.

A classification of some well-known families of polyhedra according to these properties is summarised in Table 15.1.

A polyhedron is uniform, if it is both isogonal and equilateral, which is true e.g. for all Archimedean solids, whereas the so called Johnson solids are not uniform polyhedra. There are exactly 92 polyhedra called Johnson solids (Johnson 1966). They generally bear quite exotic names and are, therefore, often just designated by the acronym Jnn (nn = 01 – 92). In geometry often the term 'near–miss Johnson solid' is used, which is a strictly convex polyhedron whose faces are close to being regular polygons, but some or all of which are not precisely regular. Thus, it fails to meet the definition of a Johnson solid, which is a polyhedron having only regular faces, though it 'can often be physically constructed without noticing the discrepancy' between its regular and irregular faces. The precise number of near misses depends on how closely the faces of such a polyhedron are required to approximate regular polygons. Some high symmetry near–misses are also known as symmetrohedra with some perfect regular polygon faces present.

A 'fullerene' is a polyhedron made from exactly 12 pentagons and any number of hexagons. Furthermore, they have rotational icosahedral symmetry and exactly three faces meet at each vertex. They are not necessarily achiral, therefore they might come in pairs of enantiomorphs. Icosahedral symmetry ensures that the pentagons are all always regular (all edges equal), although many of the hexagons may not be (semi-regular edges being alternatingly equal). Typically, but not infallibly, all of the vertices lie on a common sphere. The fullerene with only pentagonal faces is the Platonic dodecahedron and that with 20 hexagonal faces and 60 vertices is an Archimedean solid, the truncated icosahedron.

The isolated pentagon rule (IPR) established for the C_n fullerenes provides a criterion to predict, which system will have icosahedral symmetry, if not distorted by an electronic Jahn–Teller effect.[5] Those cluster that maintain I_h symmetry are special cases of polyhedra known as Goldberg polyhedra (Goldberg 1937). The number of vertices, faces, and edges of a Goldberg polyhedron can be calculated from $T = m^2 + mn + n^2 = (m + n)^2 - mn$, where m, n are non–negative integers. It follows that their number of vertices is $v = 20T$, their number of edges is $e = 30T$, and their number of faces is $f = 10T + 2$, which comprises 12 pentagons and $10(T - 1)$ hexagons. Goldberg polyhedra correspond to hexagonal close packing on the surface of an icosahedron. If $m = n$ and $mn = 0$ then the undistorted Goldberg polyhedra will have I_h symmetry otherwise the symmetry is reduced to its rotational subgroup I.

The duals of the fullerenes are called Frank–Kasper polyhedra or alternatively geodesic polyhedra or deltahedra (Frank and Kasper 1958, 1959). They are all convex with only triangular faces either all the same or different to each other. Five of the triangles meet at 12 vertices and six triangles meet at the remaining vertices. If the number of vertices is 12, 14, 15, and 16, there exists a close packing of 3-dimensional space by four Frank–Kasper polyhedra. The 16-vertex Frank–Kasper solid is also called Friauf polyhedron (Friauf 1927). A very special case is the Frank–Kasper polyhedron with $v = 12$ vertices whose triangular faces $f = 20$ are all regular which is in fact the Platonic icosahedron and its dual the simplest fullerene viz. the Platonic dodecahedron with $v = 20$ vertices and $f = 12$ pentagonal faces (the number of hexagons being zero). The relationship between the number of vertices v, faces f, and edges e of fullerenes and their Frank–Kasper duals is given in Table 15.5.

A more detailed account on polyhedron properties can be found in the books by Cromwell (1997), Coxeter (1973), Ziegler (1995), Alexandrov (2005), and Grünbaum (2003).

[5]The Jahn–Teller effect also called Jahn–Teller distortion is a mechanism of spontaneous symmetry breaking in molecular and solid state systems (Bersuker 2006)

Bibliography

Alexandrov, A. D. 2005. *Convex Polyhedra.* New York: Springer.

Atkins, P. W. and J. de Paula 2010. *Physical Chemistry.* Oxford: Oxford University Press.

Bader, R. F. W., C. F. Matta, and F. Cortés-Guzmán 2004. Where to draw the line in defining a molecular structure. *Organometallics* 23: 6253–6263.

Bai, J., A. V. Virovets, M. Scheer 2003. Synthesis of inorganic fullerene-like molecules. *Science* 300: 781–783.

Baruah, T., R. R. Zope, S. L. Richardson, and M. R. Pederson 2003. Electronic structure and rebonding in the onionlike $As@Ni_{12}@As_{20}{}^{3-}$ cluster. *Phys. Rev. B* 68: 241404-1–241404-4.

Baruah, T., M. R. Pederson, and R. R. Zope 2008. Vibrational stability and electronic structure of a B_{80} fullerene. *Phys. Rev. B* 78: 045408-1–045408-4.

Beavers, C. M., T. Zuo, J. C. Duchamp, K. Harich, H. C. Dorn, M. M. Olmstead, and A. L. Balch 2006. $Tb_3N@C_{84}$: An improbable, egg-shaped endohedral fullerene that violates the isolated pentagon rule. *J. Am. Chem. Soc.* 128: 11352–11353.

Bersuker, I. B. 2006. *The Jahn–Teller Effect.* Cambridge: Cambridge University Press.

Bo, C., J.-P. Sarasa, and J.-M. Poblet 1993. The Laplacian of charge density for binuclear complexes: Study of $V_2(\mu - \pi^2S_2)_2(S_2CH)_4$. *J. Phys. Chem.* 97: 6362–6366.

Bobev, S. and S. C. Sevov 1999. Synthesis and characterization of stable stoichiometric clathrates of silicon and germanium: $Cs_8Na_{16}Ge_{136}$. *J. Am. Chem. Soc.* 121: 3795–3796.

Boissonnat, J.-D. and M. Yvinec 1995. *Géométrie Algorithmique.* Paris: Ediscience.

Boudin, S., B. Mellenne, R. Retoux, M. Hervieu, and B. Raveau 2004. New aluminate with tetrahedral structure closely related to the C_{84} fullerene. *Inorg. Chem.* 43: 5954–5960.

Bouguerra, A., G. Fillion, E. K. Hlil, and P. Wolfes 2007. $Y_3Fe_5O_{12}$ yttrium iron garnet and lost magnetic moment (computing of spin density). *J. Alloys Compd.* 442: 231–234.

Burns, P. C., K.-A. Kubatko, G. Sigmon, B. J. Fryer, J. E. Gagnon, M. R. Antonio, and L. Soderholm 2005. Actinyl peroxide nanospheres. *Angew. Chem. Int. Ed.* 44: 2135–2139.

Chang, Ch., A. B. C. Patzer, E. Sedlmayr, T. Steinke, and D. Sülzle 2001. Computational evidence for stable inorganic fullerene-like structures of ceramic and semiconductor materials. *Chem. Phys. Lett.* 350: 399–404.

Chang, Ch., A. B. C. Patzer, E. Sedlmayr, and D. Sülzle 2002. Sphericity: a geometric approach to the internal rotation of C_2H_6, H_2O_2, and N_2H_4. *J. Molec. Struct. (Theochem)* 594: 71–77.

Chang, Ch., A. B. C. Patzer, E. Sedlmayr, and D. Sülzle 2005. Inorganic cage molecules encapsulating Kr: A computational study. *Phys. Rev. B* 72: 235402-1–235402-4.

Chang, Ch., A. B. C. Patzer, E. Sedlmayr, D. Sülzle, and T. Steinke 2006. Onion-like inorganic fullerenes of icosahedral symmetry. *Comp. Mater. Sci.* 35: 387–390.

Coxeter, H. S. M. 1973. *Regular Polytopes.* New York: Dover

Cromwell, P. R. 1997. *Polyhedra.* Cambridge: Cambridge University Press.

Cros, C., M. Pouchard, and P. Hagenmuller 1971. Sur deux nouvelles structures du silicium et du germanium de type clathrate. *Bull. Soc. Chim. Fr.* 2: 379–386.

Cui, L. F., X. Huang, L. M. Wang, J. Li, and L. S. Wang 2006. Pb_{12}^{2-}: Plumbaspherene. *J. Phys. Chem. A* 110: 10169–10172.

Cui, L. F., X. Huang, L. M. Wang, J. Li, and L. S. Wang 2007. Endohedral Stannaspherenes $M@Sn_{12}^{-}$: A Rich Class of Stable Molecular Cage Clusters. *Angew. Chem. Int. Ed.* 46: 742–745.

Damasceno, P. F., M. Engel, and S. C. Glotzer 2012. Predictive self–assembly of polyhedra into complex structures. *Science* 337: 453–457.

Deible, M. J., O. Tuguldur, and K. D. Jordan 2014. Theoretical study of the binding energy of a methane molecule in a $(H_2O)_{20}$ dodecahedral cage. *J. Phys. Chem. B* 118: 8257–8263.

Diudea, M. V. 2017. *Multi-shell Polyhedral Clusters*, Carbon Materials: Chemistry and Physics Series, Vol. 10, Eds. F. Cataldo and P. Milani, New York: Springer.

Driess, M., H. Pritzkow, S. Martin, S. Rell, D. Fenske, and G. Baum 1996. Molecular, shell-like dilithium (silyl) phosphanediide and dilithium (silyl) arsanediide aggregates with an $[Li_6O]^{4+}$ core. *Angew. Chem. Int. Ed.* 35: 986–1027.

Fluck, E., M. Lang, F. Horn, E. Hädicke, and G. M. Sheldrick 1976. Potassium closo-tetradecanitrogen dodecathio dodecaphosphate (6−), $K_6[P_{12}S_{12}N_{14}]$. *Z. Naturforsch.* 31B: 419–426.

Fowler, P. W. 1986. How unusual is C_{60}? Magic numbers for carbon clusters. *Chem. Phys. Lett.* 131: 444–450.

Frank, F. C., J. S. Kasper 1958. Complex alloy structures regarded as sphere packings. I. Definitions and basic principles. *Acta Cryst.* 11: 184–190.

Frank, F. C., J. S. Kasper 1959. Complex alloy structures regarded as sphere packings. II. Analysis and classification of representative structures. *Acta Cryst.* 12: 483–499.

Friauf, J. B. 1927. The crystal structure of magnesium dizincide. *Phys. Rev.* 29: 34–40.

Goldberg, M. 1937. A class of multi-symmetric polyhedra. *Tohoku Math. J.*: 43: 104–108.

González-Morage, G. 1993. *Cluster Chemistry.* New York: Springer.

Goodman, J. E. and J. O'Rourke, (Eds.) 1997. *Handbook of Discrete and Computational Geometry.* Boca Raton: CRC Press.

Gorman, M. J. 2005. *Buckminster Fuller-Designing for Mobility.* Torino: Skira.

Grin, Yu. N., L. Z. Melekhov, K. A. Chuntonov, and S. P. Yatsenko 1987. Crystal structure of cesium-tin (Cs_8Sn_{46}). *Kristallografiya* 32: 497–498.

Grünbaum, B. 2003. *Convex Polytopes.* New York: Springer.

Haberland, H., (Ed.) 1994. *Clusters of Atoms and Molecules.* New York: Springer.

Hamermesh, M. 1962. *Group Theory and its Application to Physical Problems.* Reading Massachusetts: Addison–Wesley.

Hervieu, M., B. Mellène, R. Retoux, S. Boudin, and B. Raveau 2004. The route to fullerenoid oxides. *Nat. Mater.* 3: 269–273.

Herzberg, G. 1966. *Molecular Spectra and Molecular Structure. III. Electronic Spectra and Electronic Structure of Polyatomic Molecules.* New York: Van Nostrand.

Hu, H. S. and N. Kaltsoyannis 2018. High spin ground states matryoshka actinide nanoclusters: A computational study. *Chem. Eur. J.* 24: 347–350.

Jauch, W., A. J. Schultz, and G. Heger 1987. Single-crystal time-of-flight neutron diffraction of Cr_3Si and MnF_2 comparison with monochromatic-beam techniques. *J. Appl. Crystallogr.* 20: 117–119.

Johnson, N. W. 1966. Convex polyhedra with regular faces. *Canad. J. Math.* 18: 169–200.

Jones, T. E., M. E. Eberhart, and D. P. Clougherty 2007. Topology of spin-polarized charge density in bcc and fcc iron. *Phys. Rev. Lett.* 100: 017208.

Joyes, P. 1990. *Les agrégats inorganiques élémentaires.* Les Ulis: Éditions de Physique.

Kepler, J. 1619. *Harmonices Mundi Libri V.* Linz: Lincii Austriae, Sumptibus G. Tampachii Bibl. Francof., Excudebat J. Plancus.

Kohout, M., F. R. Wagner, and Y. Grin 2002. Electron localization function for transition-metal compounds. *Theor. Chem. Acc.* 108: 150–156.

Kong, X. J., Y. P. Ren, L. S. Long, Z. Zheng, R. B. Huang, and L. S. Zheng 2007. A keplerate magnetic cluster featuring an icosidodecahedron of Ni(II) ions encapsulating a dodecahedron of La(III) ions. *J. Am. Chem. Soc.* 129 (22): 70167017.

Krätschmer, W., L. D. Lamb, K. Fostiropoulus, and D. R. Huffman 1990. Solid C_{60}: a new form of carbon. *Nature* 347: 354–358.

Krätschmer, W., K. Fostiropoulus, and D. R. Huffman 1990. The infrared and ultraviolet absorption spectra of laboratory-produced carbon dust: Evidence for the presence of the C_{60} molecule. *Chem. Phys. Lett.* 170: 170–176.

Kroto, H. W., J. R. Heath, S. C. O'Brien, R. F. Curl, and R. E. Smalley 1985. C_{60}: Buckminsterfullerene. *Nature* 318: 162–163.

Li, X., B. Kiran, J. Li, H. J. Zhai, and L. S. Wang 2002. Experimental observation and confirmation of icosahedral $W@Au_{12}$ and $Mo@Au_{12}$ molecules. *Angew. Chem. Int. Ed.* 41: 4786–4789.

Li, B. and J. D. Corbett 2004. Synthesis, structure and characterisation of a cubic thallium cluster phase of the Bergman Type, $Na_{13}(Cd_{\sim 0.70}Tl_{\sim 0.30})_{27}$. *Inorg. Chem.* 43: 3582–3587.

Lin, Q. and J. D. Corbett 2005. $Mg_{35}Cu_{24}Ga_{53}$: A three-dimensional cubic network composed of interconnected Cu_6Ga_6 icosahedra, Mg-centered Ga_{16} icosioctahedra, and a magnesium lattice. *Inorg. Chem.* 44: 512–518.

Long, F., H. Liu, D. Li, and J. Yan 2017. Spin-orbit coupling effects on ligand-free icosahedral matryoshka superatoms. *J. Phys. Chem. A* 121: 2420–2428.

Macchi, P. and A. Sironi 2003. Chemical bonding in transition metal carbonyl clusters: Complementary analysis of theoretical and experimental electron densities. *Coord. Chem. Rev.* 238–239: 383–412.

Martínez-Guajardo, G., J. L. Cabellos, A. Díaz-Celaya, S. Pan, R. Islas, P. K. Chattaraj, T. Heine, and G. Merino 2015. Dynamical behaviour of borospherene: A nanobubble. *Sci. Rep.* 5: 11287–11293.

Melgar Freire, M. D. 2015. Keplerates: From Electronic Structure to Dynamic Properties. Ph.D. Thesis, Universitat Rovira i Virgili, Tarragona.

Müller, A., P. Kögerler, and A. W. M. Dress 2001. Giant metal-oxide-based spheres and their topology: From pentagonal building blocks to keplerates and unusual spin systems. *Coord. Chem. Rev.* 222: 193–218.

Müller, A., E. Krickemyer, H. Bgge, M. Schmidtmann, S. Roy, and A. Berkle 2002. Changeable pore sizes allowing effective and specific recognition by a molybdenum-oxide based "nanosponge": En route to sphere-surface and nanoporous-cluster chemistry. *Angew. Chem. Int. Ed.* 41: 3604–3609.

Müller, A., M. Luban, C. Schröder, R. Modler, P. Kögerler, M. Axenovich, J. Schnack, P. Canfield, S. Bud'ko, and N. Harrison 2001. Classical and quantum magnetism in giant Keplerate magnetic molecules. *ChemPhysChem* 2: 517–521.

Müller, A., S. Sarkar, S. Q. N. Shah, H. Bgge, M. Schmidtmann, S. Sarkar, P. Kögerler, B. Hauptfleisch, A. X. Trautwein, and V. Schünemann 1999. Archimedian synthesis and magic numbers: "Sizing" giant molybdenum-oxide-based molecular spheres of the Keplerate type. *Angew. Chem. Int. Ed.* 38: 3238–3241.

Moses, M. J., J. C. Fettinger, and B. W. Eichhorn 2003. Interpenetrating As_{20} fullerene and Ni_{12} icosahedra in the onion-skin $[As@Ni_{12}@As_{20}]^{3-}$ ion. *Science* 300: 778–780.

Nesper, R. 1994. Fullercages without carbon - fulleranes, fullerenes, space-filler-enes? *Angew. Chem. Int. Ed.* 33: 843–846.

Nezbeda, I. 1976. Virial expansion and an improved equation of state for the hard convex molecule system. *Chem. Phys. Lett.* 41: 55–58.

Nolas, G. S. (Ed.) 2014. *The Physics and Chemistry of Inorganic Clathrates.* New York: Springer.

Ohba, T., Y. Kitano, and Y. Komura 1984. The charge-density study of the Lavesphases, $MgZn_2$ and $MgCu_2$. *Acta Crystallogr. Sect. C* 40: 1–5.

Osawa, E. 1970. Superaromaticity. *Kagaku* 25: 854–863 (in Japanese).

Pan, S., M. Ghara, S. Kar, X. Zarate, G. Merino, and P. K. Chattaraj 2018. Noble gas encapsulated B_{40} cage. *Phys. Chem. Chem. Phys.* 20: 1953–1963.

Peng, J. B., X. J. Kong, Q. C. Zhang, M. Orend, J. Prokleka, Y. P. Ren, L. S. long, Z. Zheng, and L. S. Zheng 2014. Beauty, symmetry, and magnetocaloric effectfour-shell keplerates with 104 lanthanide atoms. *J. Am. Chem. Soc.* 136 (52): 17938–17941.

Poinsot, L. 1810. Mémoire sur les polygones et les polyèdres. *J. de l'École Polytechnique* 10: 16–48.

Polya, G. 1954. *Induction and Analogy in Mathematics.* Princeton: Princeton University Press.

Pyykkö, P. and N. Runeberg 2002. Icosahedral WAu_{12}: A predicted closed-shell species, stabilised by aurophilic attraction and relativity and in accord with the 18-electron rule. *Angew. Chem. Int. Ed.* 41: 2174–2176.

Qin, L., G. J. Zhou, Y. Z. Yu, H. Nojiri, C. Schröder, R. E. P. Winpenny, and Y. Z. Zheng 2017. Topological self-assembly of highly symmetric lanthanide clusters: A magnetic study of exchange-coupling "Fingerprints" in Giant gadolinium(III) cages. *J. Am. Chem. Soc.* 138 (45): 16405–16411.

Reny, E., P.Gravereau, C. Cros, and M. Pouchard 1998. Structural characterisations of the Na_xSi_{136} and Na_8Si_{46} silicon clathrates using the Rietveld method. *J. Mater. Chem.* 8: 2839–2844.

Roth, S., and W. Schnick 2001. Synthese, Kristallstruktur und Eigenschaften der käfigartigen, sechsbasigen Säure $P_{12}S_{12}N_8NH_6{\cdot}14H_2O$ sowie ihrer Salze $Li_6P_{12}S_{12}N_{14}{\cdot}26H_2O$, $(NH_4)_6P_{12}S_{12}N_{14}{\cdot}8H_2O$ und $K_6P_{12}S_{12}N_{14}{\cdot}8H_2O$. *Z. Anorg. Allg. Chem.* 627: 1165–1172.

Saunders, M., H. A. Jiménez-Vásquez, R. J. Cross, and R. J. Poreda 1993. Stable compounds of helium and neon: $He@C_{60}$ and $Ne@C_{60}$. *Science* 259: 428–430.

Sevov, S. C. and J. D. Corbett 1996. A new indium phase with three stuffed and condensed fullerane-like cages: $Na_{172}In_{197}Z_2$ (Z=Ni, Pd, Pt). *J. Solid State Chem.* 123: 344–370.

Shih, W. M., J. D. Quispe, and G. F. Joyce 2004. A 1.7-kilobase single-stranded DNA that folds into a nanoscale octahedron. *Nature* 427: 618–621.

Stegmaier, S. and T. F. Fässler 2011. A bronze matryoshka: The discrete intermetalloid cluster $[Sn@Cu_{12}@Sn_{20}]^{12-}$ in the ternary phases $A_{12}Cu_{12}Sn_{21}$ (A = Na, K). *J. Am. Chem. Soc.* 133: 19758–19768.

Stephens, P. W., (Ed.) 1993. *Physics and Chemistry of Fullerenes.* A Reprint Collection. Singapore: World Scientific Publishing.

Swenson, D. C., S. Dagorne, and R. F. Jordan 2000. A dicationic gallium-oxo-hydroxide cape compound. *Acta Cryst. C* 56: 1213–1215.

Szwacki, N. G., A. Sadrzadeh, and B. I. Yakobson 2007. B_{80} Fullerene: An Ab initio prediction of geometry, stability, and electronic structure. *Phys. Rev. Lett.* 98: 166804-1–166804-4.

Terrones, H., M. Terrones, F. López-Urías, J. A. Rodgríguez-Manzo, and A. L. Mackay 2004. Shape and complexity at the atomic scale: The case of layered nanomaterials. *Phil. Trans. R. Soc. Lond. A* 362: 2039–2063.

Tran, N. T., D. R. Powell, and L. F. Dahl 2000. Nanosized $Pd_{145}(CO)_x(PEt_3)_{30}$ containing a capped three-shell 145-atom metal-core geometry of pseudo icosahedral symmetry. *Angew. Chem. Int. Ed.* 39: 4121–4125.

Tsunoyama, H., H. Akatsuka, M. Shibuta, T. Iwasa, Y. Mizuhata, N. Tokito, and A. Nakajima 2017. Development of integrated dry-wet synthesis method for metal encapsulating silicon cage superatoms of $M@_{16}$ (M = Ti and Ta). *J. Phys. Chem. C* 121: 20507–20516.

Twarock, R. 2006. Mathematical virology: A novel approach to the structure and assembly of viruses. *Phil. Trans. R. Soc. A* 364: 3357–3373.

Volatron F. and P. Chaquin 2017. *La Théorie des Groupes.* Paris: De Boeck Supérieur.

Wang, L. M., S. Bulusu, H. J. Zhai, X. C. Zeng, and L. S. Wang 2007. Doping golden buckyballs: $Cu@Au_{16}^-$ and $Cu@Au_{17}^-$ cluster anions. *Angew. Chem.* 119: 2973–2976.

Wang, J. T., C. Chen, E. G. Wang, D. S. Wang, H. Mizuseki, and Y. Kawazoe 2009. Highly stable and symmetric boron caged $B@Co_{12}@B_{80}$ core-shell cluster. *Appl. Phys. Lett.* 94: 133102-1–133102-3.

Wang, Y., M. Moses-DeBusk, L. Stevens, J. Hu, P. Zavalij, K. Bowen, B. I. Dunlap, E. R. Glaser, and B. Eichhorn 2017. $Sb@Ni_{12}@Sb_{20}^{n-/+}$ and $Sb@Pd_{12}@Sb_{20}^{n-/+}$ Cluster anions, where $n = +1, -1, -3, -4$: Multi-oxidation-state clusters of interpenetrating platonic solids. *J. Am. Chem. Soc.* 139: 619–622.

Webster, R. 1994. *Convexity.* Oxford: Oxford Univ. Press.

Weiske, T., D. K. Böhme, J. Hrusak, W. Krätschmer, H. Schwarz 1991. Endohedral cluster compounds: Inclusion of helium within C_{60}^+ and C_{70}^+ through collision experiments. *Angew. Chem. Int. Ed. Engl.* 30: 884–886.

Yang, Y., D. Jia, Y. J. Wang, H. J. Zhai, Y. Man, and S. D. Li 2017. A universal mechanism of the planar boron rotors B_{11}^-, B_{13}^+, B_{15}^+, and B_{19}^-: Inner wheels rotating in pseudo-rotating outer bearings. *Nanoscale* 9: 1443–1448.

Yudasaka, M., S. Iijima, and V. H. Crespi 2008. Single-wall carbon nanohorns and nanocones. In *Carbon Nanotubes: Advanced Topics in the Synthesis, Structure, Properties, and Applications*, Ed. A. Jorio, G. Dresselhaus, and M. S. Dresselhaus, 565–586. New York: Springer.

Zhai, H. J., Y. F. Zhao, W. L. Li, Q. Chen, H. Bai, H. S. Hu, Z. A, Piazza, W. J. Tian, H. G. Lu, Y. B. Wu, Y. W. Mu, G. F. Wei, Z. P. Liu, J. Li, S. D. Li, and L. S. Wang 2014. Observation of an all-boron fullerene. *Nature Chem.* 6: 727–731.

Zhao, J. and R.-H. Xie 2004. Density functional study of onion-skin-like $As@Ni_{12}@As_{20}^{3-}$ and $Sb@Pd_{12}@Sb_{20}^{3-}$ cluster ions. *Chem. Phys. Lett.* 396: 161–166.

Zhao, J., L. Wang, F. Li, and Z. Chen 2010. B_{80} and Other medium-sized boron clusters: Core-shell structures, not hollow cages. *J. Phys. Chem. A* 114: 9969–9972.

Ziegler, G. M. 1995. *Lectures on Polytopes.* New York: Springer.

16

Magnetic Properties of Endohedral Fullerenes: Applications and Perspectives

Panagiotis Dallas, Reuben Harding, Stuart Cornes, Sapna Sinha, Shen Zhou, and Ilija Rašović
University of Oxford

Edward Laird
University of Oxford
Lancaster University

Kyriakos Porfyrakis
University of Oxford

16.1 Introduction

Small children, when presented with hollow toys, will instinctively try to place different kinds of objects inside them. Chemists behave in much the same way. Soon after fullerene molecules were discovered (Kroto 1985), it was natural to ask whether the empty space inside them could host other atoms or molecules. Within a decade, the answer was known to be “yes”. This class of molecules are the endohedral fullerenes.

Although fullerene molecules are not magnetic, the incarcerated species may be. Endohedral fullerenes, thus, combine magnetic properties from the incarcerated species with chemistry that is often similar to that of empty fullerene molecules. This allows unique functionality, which cannot be attained in any other material. The magnetic properties of endohedral fullerenes offer advantages in two ways. Firstly, the fullerene protects the incarcerated species from its environment. This is beneficial when long spin lifetimes are desired, and possible applications are for quantum information and atomic clocks. Secondly, the fullerene protects the environment from the incarcerated species. This allows the otherwise toxic spin markers to be used inside the human body.

This critical review introduces the magnetic behavior of this class of materials with a view to possible applications. The remainder of this introduction summarizes the structure and synthesis of endohedral fullerenes. Section 16.2 describes basic magnetic behavior in nanomaterials and on substrates. Section 16.3 introduces peapods: chemically assembled spin arrays. Section 16.4 focuses on biomedical applications of endohedral fullerenes. Section 16.5 explains the uses of endohedral fullerenes as spin qubits. Section 16.6 introduces a new research topic, endohedral fullerenes for atomic clocks. We conclude by discussing challenges and future directions for the field.

Spin-bearing endohedral fullerenes come in two classes. The incarcerated species can be a metal atom or a metallic cluster (usually a transition metal or a rare earth), in which case the molecule is known as an endohedral metallofullerene (EMF). Alternatively, the incarcerated species can be a single atom from Group V (either N or P). The chemical formula is denoted $X@C_n$, where X is the incarcerated species and n is the number of atoms in the cage. Some of the most important endohedral fullerenes are listed in Table 16.1. These two classes behave in quite different ways. In the metallofullerenes, the incarcerated species generally interacts strongly with the cage, forming a charge transfer complex. The incarcerated species usually sits off-center in the cage and may have more than one stable position, leading to multiple isomers. By contrast, Group V endohedral fullerenes have almost no charge transfer, and the incarcerated atom sits in the center.

Different incarcerated species are compatible with different encapsulation structures. Both N and P are small enough to fit inside a C_{60} cage. This cage is nearly spherically symmetric (to be precise, it is icosahedral), and the spin, therefore, experiences a nearly isotropic environment. This symmetry contributes to the excellent spin coherence in these materials. While metals may be encapsulated in smaller cages, such as C_{60}, C_{70} or C_{76}, the solubility of these elusive fullerenes is very low and it is impossible to isolate them in respectable quantities. In contrast, EMFs with

TABLE 16.1 Selected Endohedral Fullerenes, Listing Electron and Nuclear Spin Quantum Numbers, Electronic g Factor and Hyperfine Coupling.

$^{14}N@C_{60}$	3/2	2.0021	1	5.70 G (Pietzak et al. 1998)
$^{15}N@C_{60}$	3/2	2.0020	1/2	7.98 G (Pietzak et al. 1998)
$^{31}P@C_{60}$	3/2	–	1/2	49.2 G (Knapp et al. 1998)
$^{45}Sc@C_{82}$ (I)	1/2	1.9999	7/2	3.82 G (Inakuma and Shinohara 2000)
$^{45}Sc@C_{82}$ (II)	1/2	2.0002	7/2	1.16 G (Inakuma and Shinohara 2000)
$^{45}Sc@C_{84}$	1/2	1.9993	7/2	3.78 G (Inakuma and Shinohara 2000)
$^{89}Y@C_{82}$ (I)	1/2	2.0006	1/2	0.49 G (Kikuchi et al. 1994)
$^{89}Y@C_{82}$ (II)	1/2	2.0001	1/2	0.32 G (Kikuchi et al. 1994)
$^{139}La@C_{82}$ (I)	1/2	–	7/2	1.15 G (Yamamoto et al. 1994)
$^{139}La@C_{82}$ (II)	1/2	–	7/2	0.83 G (Yamamoto et al. 1994)
$^{45}Sc_3C_2@C_{80}$	1/2	1.9985	[21/2]	6.51 G (Shinohara et al. 1994)

The nuclear spin value given for $^{45}Sc_3C_2@C_{80}$ is an effective value due to coupling between the identical Sc atoms, each with nuclear spin I=7/2. Molecular isomers are labeled by roman numerals, starting with the major isomer labeled (I).

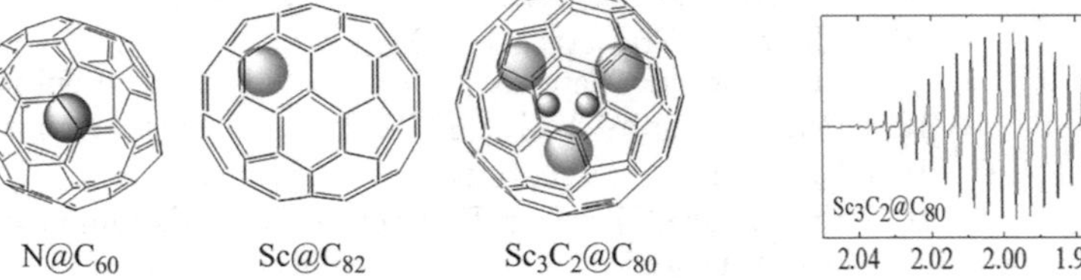

FIGURE 16.1 Structures of $N@C_{60}$, $Sc@C_{82}$ and $Sc_3C_2@C_{82}$. The EPR spectrum of $Sc_3C_2@C_{80}$. The number of resonances for each molecule is $N = 2I_\Sigma + 1$, where I_Σ is the sum of the nuclear spins. For example, in $Sc_3C_2@C_{80}$, three $I = 7/2$ atoms combine to give $I_\Sigma = 21/2$.

larger cages, for example C_{80}, C_{82} or C_{84}, have been isolated and thoroughly studied (Bartl et al., 2001). Some of these larger fullerenes, for example C_{82}, deviate substantially from spherical shape. The combination of charge transfer to the cage (allowing the electron spins to interact with the environment) and lower symmetry (meaning that the spins are perturbed by tumbling motion of the cage) makes the spin lifetimes of metallofullerenes lower than the Group V endohedral fullerenes. Figure 16.1 shows the structure and spin resonance spectrum of selected molecules.

Several ways to synthesize endohedral fullerenes are known, and three techniques have been particularly well developed. Metallofullerenes can be made by ablating graphite that has been doped with the desired species. This ablation is achieved either using powerful lasers or by arc discharge. This technique does not so far work for Group V endohedral fullerenes; instead, the molecules are synthesized using ion implantation to inject the N or P atom into an effusing beam of C_{60} molecules (Murphy et al., 1996; Pietzak et al., 1997; Cho et al., 2015). In all these techniques, the ratio of endohedral fullerene to fullerene in the product is low (typically less than 1%).

16.2 Magnetic Properties of EMFs

16.2.1 Nanomagnetic Materials

The paramagnetic properties of EMFs have attracted attention due to the presence of spin active transition metals and rare earth metals exhibiting different hyperfine couplings (HFCs) and the observation of ferromagnetic coupling with various functional molecules and substrates. Since the metal cation or the metal cluster can occupy different positions inside the cage, different g values and HFCs should be anticipated. Examples of transition metals with sharp electron paramagnetic resonance (EPR) signals and long spin lifetimes are scandium, yttrium and lanthanum. Rare earth elements such as gadolinium and erbium have applications in biomedicine as contrast agents (CAs) and in telecommunications, respectively. While erbium EMFs are appealing for optical communications because the characteristic 1,520 nm emission from the erbium ion falls within the telecommunications window, their EPR spectra are broad due to the location of the f-orbital electrons (Bondino et al. 2006). Previous works debate the location of the spin and the effect of the rotation of the metal cluster inside the cage, with both having a significant influence on the resulting EPR spectrum at lower temperatures. For example, the two isomers of $Sc@C_{82}$ and of $Y@C_{82}$ exhibit different HFCs and g values despite the fact that the cluster is inside the same, C_{82}, cage (Inakuma and Shinohara 2000). For a comprehensive analysis of the magnetic properties of EMFs, see the review by Zhao (2015). More complicated EPR spectra can be observed in scandium carbide $Sc_3C_2@C_{80}$, which exhibits a unique diamond-shaped EPR signal consisting of 22 lines (Roukala et al. 2017). This EMF has been used for the synthesis of molecular magnetic switchable dyads when connected with an organic molecule bearing a nitroxide radical (Figure 16.2, Wu et al. 2015). A similar EMF-nitroxide radical system based on a dysprosium EMF was used as a molecular compass with position-sensitive magnetoreception ability. EPR measurements show that the dipole–dipole interactions depend on the orientation of the Dy_3N cluster inside the cage (Li et al., 2017).

The rotation of ^{45}Sc inside the C_{80} cage and the dynamics of the fullerene and the endohedral cluster dictate

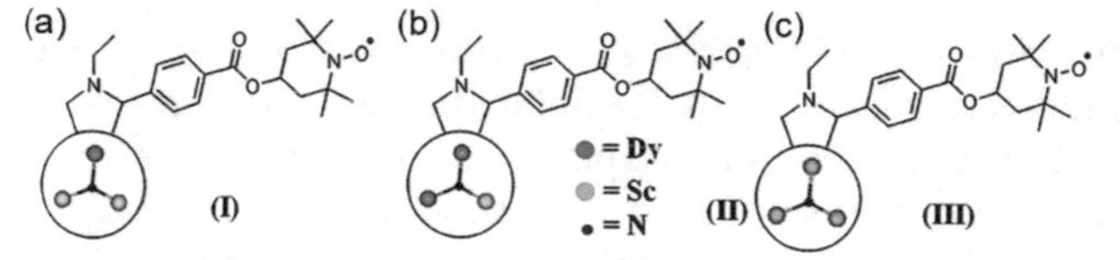

FIGURE 16.2 A dyad consisting of $DySc_2N@C_{80}$ and $Dy_2ScN@C_{80}$ and a nitroxide radical. (Reproduced with permission from Wu et al. 2016.)

the temperature dependence of the EPR spectrum. This rotation is found to be hindered by functionalization of the external surface. The observed temperature dependence is well explained by density functional theory (DFT), which predicts three-axis rotation of the carbide at high temperatures but only two-axis rotation at low temperatures. Detailed analysis of the EPR signals of scandium EMFs was performed and presented by the Shinohara group (Inakuma and Shinohara 2000). Three monometallic scandium incarcerating fullerenes, namely the two isomers of $Sc@C_{82}$ and $Sc@C_{84}$, were studied in a temperature range of 150–270 K. Unique spin dynamics are observed in EMF clusters. Furthermore, the creation of a radical on the surface of EPR silent EMFs can also give rise to an EPR spectrum with hyperfine features (Elliott et al., 2013).

Ferromagnetic interactions of endohedral fullerenes with organic molecules, such as copper porphyrin (Hajjaj et al. 2011), or a metallic substrate have been investigated—Svitova et al. (2014) reported the formation of an inclusion complex, where $La@C_{82}$ is incorporated between two Cu(II) porphyrin units. These studies demonstrated two different types of interactions: from a ferromagnetic coupling in the case of *cyclo*-$[P_{Cu}]_2$ acting as the inclusion host to a ferromagnetic coupling for a *cage*-$[P_{Cu}]_2$ cage host. Due to the long magnetic relaxation times, similar to those of single-molecule magnets, some dimetallic EMFs have been considered for spintronics' applications (Koltover, 2004; Koltover et al., 2004).

16.2.2 Magnetic Fullerenes on Substrates

With respect to their magnetic properties when EMFs form a layer on metallic substrate, interactions between the two components might take place (Theobald et al., 2003). As an example, Hermanns et al. (2013) demonstrated the coupling of a $Gd_3N@C_{80}$ EMF to a nickel substrate. The Gd magnetization is antiparallel to the Ni at low temperatures and parallel at high temperatures—the Gd atoms of the fullerenes are coupled with a ferromagnetic alignment to each other. By X-ray magnetic circular dichroism (XMCD) measurements, they revealed the magnetic coupling while the close self-assembly of the fullerenes was proven through STM images on a Cu(001) substrate. Dy_2S is a class of EMFs, synthesized using Dy_2S_3 as the inorganic metal component added in the graphite rods (Chen et al. 2017). These EMFs exhibit permanent magnetization as proven by hysteresis loops while DC and AC susceptibility measurements were employed for determining the magnetization relaxation times. In Figure 16.3, we present the

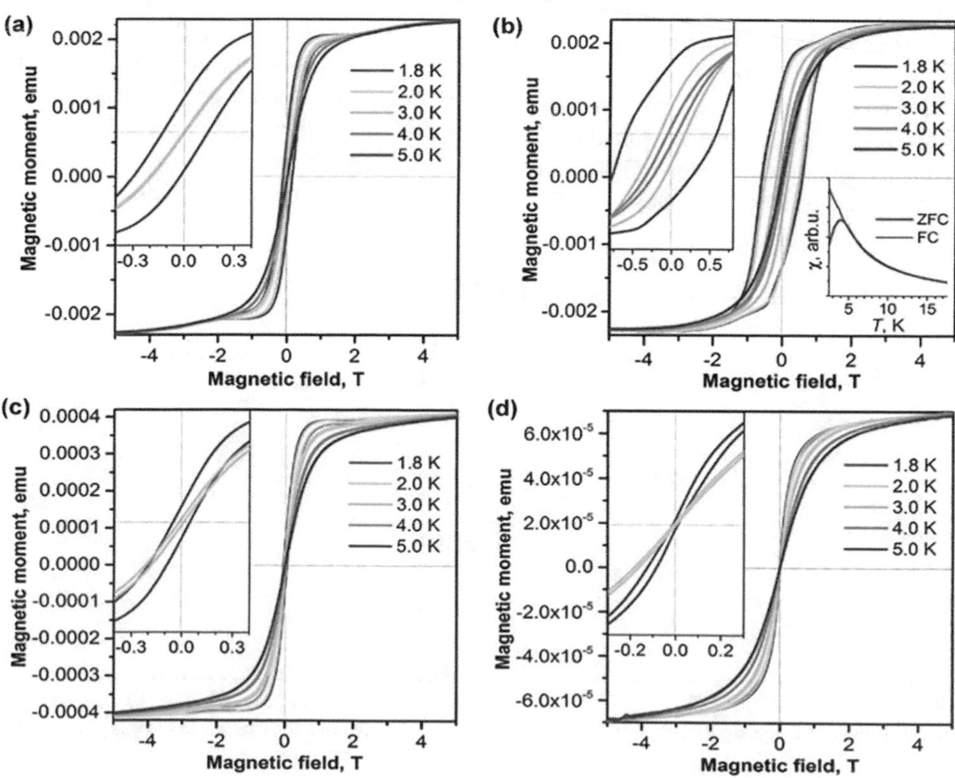

FIGURE 16.3 Magnetization curves for (a) $Dy_2S@C_{82}$-$C_s(6)$, (b) $Dy_2S@C_{82}$-$C_{3v}(8)$, (c) $Dy_2C_2@C_{82}$-$C_s(6)$ and (d) $Dy_2S@C_{72}$-$C_s(10528)$. The loops were recorded at $T = 1.8$–5 K. (Reproduced with permission from Chen et al. 2017.)

magnetization curves of the following EMFs: $Dy_2S@C_{82}$-C_{3v}(8), $Dy_2C_2@C_{82}$-C_s(6) and $Dy_2S@C_{72}$-C_s(10528). In very low temperatures, a clear hysteresis is observed in the curves of $Dy_2S@C_{82}$-C_s(6) and $Dy_2S@C_{82}$-C_{3v}(8).

16.3 Peapods: Spin–Active EMFs Inside Carbon Nanotubes

EPR is a powerful tool for studying and understanding the properties and behavior of electron spins. Through detailed EPR analysis, we can probe the structure, paramagnetic states and highest occupied molecular orbital (HOMO) and lowest unoccupied molecular orbital (LUMO) density distributions. Endohedral fullerenes give rise to some of the narrowest EPR lines to be detected under normal atmospheric conditions. For that reason, EMFs may be potential candidates for quantum information processing devices, with EPR as a tool for controlling qubits. The incarcerated electron spins offer both long coherence times and the potential for controlled spin–spin interactions.

In a carbon nanotube (CNT) peapod, fullerenes are arranged in a chain inside a nanotube. This offers a way to make a spin register—an array of metallofullerene spins with controlled interactions. Electrons delocalized on the carbon nanotube can themselves encode quantum bits (qubits) (Laird et al., 2013) and could ultimately act both as a spin bus and a readout line (Benjamin et al. 2006). The magnetic properties depend not only on the EMF species, (Kitaura et al., 2007), but also on their concentration inside the peapods (Figure 16.4, Círić et al., 2008). The EPR lines were deconvoluated in two components, a narrow and a broad component. In order to control the distance between the spin-active components, C_{60} have been used as a spacer between $La@C_{82}$ molecules inside a single-walled CNT. Below 70 K, the g_B factor of the broad EPR component rapidly decreases as the relative content of the EMFs increases and *vice versa* for measurements carried out at temperatures above 70 K. Spin loss can pose drawbacks while fabricating nano-peapods, and this should be taken into account when considering these hybrid materials as potential candidates for spintronics and quantum devices. An important point to note when considering electron spins of EMFs for quantum applications is that all the EPR measurements carried out so far on fullerene spins have been performed on ensembles of molecules.

16.4 Biomedical Applications

Derivatives of the gadolinium-containing endohedral metallofullerenes (Gd-EMFs) have been proposed as the next generation of T_1 CAs for ^1H magnetic resonance imaging (MRI) (Ghiassi et al., 2014; Mikawa et al., 2001). This is due to high relaxitivity, which is 10–40 times higher than commercial Gd-chelate CAs (Li and Dorn, 2017), and it allows sufficient image contrast at a low dose. In addition, the fullerene cage encapsulates the Gd(III) ions that would otherwise be toxic (Sosnovik and Caravan, 2013).

To achieve efficient water proton relaxation, we must engineer a strong dipolar interaction between the proton's nuclear spin and the encapsulated gadolinium's unpaired electrons. Since the water molecules cannot directly coordinate to the Gd(III) ion, as in Gd-chelate CAs, the interaction must occur *via* a second- or outer-sphere mechanism. This requires the fullerene surface to be functionalized with groups containing readily exchangeable protons. Hence, polyhydroxylated and polycarboxylated derivatives of $Gd@C_{60}$, $Gd@C_{82}$ and $Gd_3N@C_{80}$ have been investigated most widely (Zhang et al., 2014). The larger the number of functional groups attached to the surface, the higher the observed relaxivity (Zou et al., 2015). Furthermore, these derivatives allow the formation of aggregates by hydrogen bonding between molecules. Such aggregates,

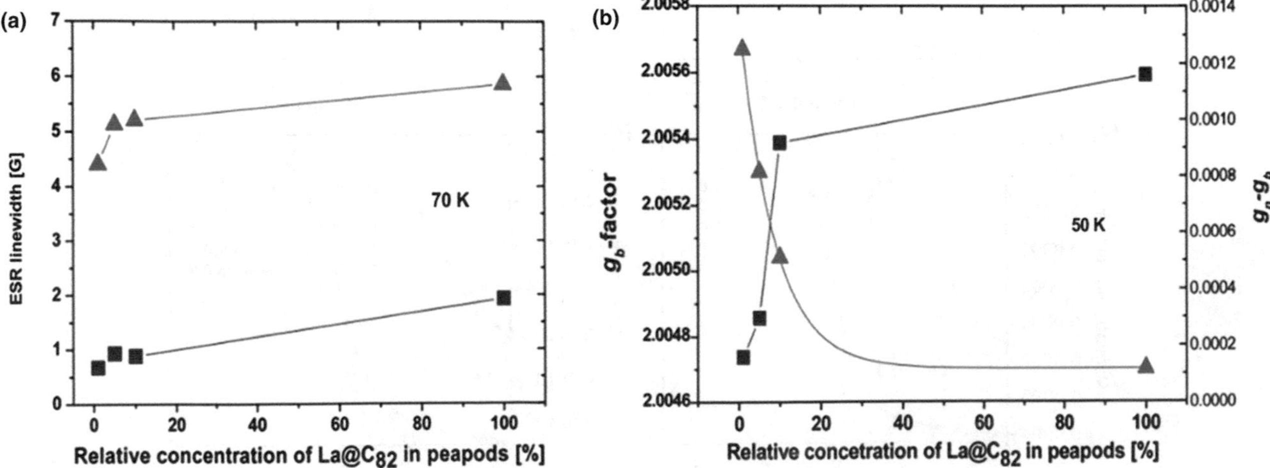

FIGURE 16.4 (a) The line width of the narrow (dark gray) and the broad (light gray) components of the EPR spectrum at different $La@C_{82}$ concentrations. (b) Relative content of the EMF in peapods, along with the evolution of the g_b factor of the broad ESR spectrum. (Reproduced with permission from Círić et al., 2008.)

which have typical diameters of 30–150 nm, have much longer rotational correlation time than isolated molecules, giving further enhanced relaxivity (Laus et al., 2007).

A number of reports have shown how the aggregation and, hence relaxivities, of Gd-EMFs can be altered either through the cage functionalization or through changes to the aqueous environment. Dorn and co-workers made a series of polyhydroxylated $Gd_3N@C_{80}$ derivatives, which also incorporated poly(ethylene glycol) (PEG) chains of varying lengths into the structure. Dynamic light scattering (DLS) measurements coupled with relaxivity studies revealed that the derivatives containing the shorter PEG chains (350/750 Daltons) formed larger aggregates with extremely high relaxivities, while those containing longer PEG chains (5,000 Da) formed smaller aggregates with lower relaxivities at clinical-range magnetic field strengths (Zhang et al., 2010).

In a similar fashion, Wang et al. prepared two polyhydroxylated $Gd@C_{82}$ derivatives containing differing numbers of hemiketal groups and demonstrated that the system with the larger number has significantly higher relaxivity (Zou et al., 2015). This was ascribed to the greater capacity for this system to large aggregates by hydrogen bonding, which was confirmed by DLS measurements. Wilson and co-workers investigated the effect of pH on $Gd@C_{60}(OH)_x$ and $Gd@C_{60}[C(COOH)]_{10}$ and found relaxivity to dramatically increase as pH decreased (Tóth et al., 2005). This was attributed to increased aggregate stability at low pH. The same group also demonstrated that aggregates of the same derivatives were destroyed under conditions of high salt concentration, with phosphate buffer causing the most noticeable effect (Laus et al., 2005).

Sun and co-workers have recently reported a graphene oxide –$Gd@C_{82}$ nanohybrid in which the unfunctionalized Gd-EMF is deposited on graphene oxide nanosheets through non-covalent π–π interactions (Cui et al., 2015). Interestingly, the structure was found to enhance proton relaxivity to an even greater extent than $Gd@C_{82}(OH)_x$, despite the lack of exchangeable protons directly attached to the fullerene surface. This is apparently due to "secondary spin–electron transfer" from the Gd(III) ion through the GO nanosheet onto the hydrophilic alcohol and carboxyl substituents, which in turn undergo exchange with the water protons. Maximum relaxivity was achieved by optimizing the equilibrium between the conductivity of the GO nanosheet and the number of proton exchange sites present (Li et al., 2016b).

Presently, efforts are being made to create multimodal imaging agents, which allow multiple, complementary imaging techniques to be used on a patient simultaneously, thereby improving diagnostic capability and accuracy. The ease with which EMFs can be functionalized makes them an ideal platform to be used for the preparation of such agents. Shultz and co-workers reported a bimodal positron emission tomography (PET)/MRI agent comprising a ^{124}I radiolabeled carboxylated and hydroxylated $Gd_3N@C_{80}$ derivative (Luo et al., 2012). Encouragingly the position and distribution of the agent within tumor-bearing rats was found to be comparable in both MRI and PET scans.

Gd-EMFs have also been conjugated to the exterior surface of nanoparticle (NP) systems to generate multimodal imaging agents. For example, Wang described a trimodal MRI/PET/photoacoustic imaging agent based on ^{64}Cu-radiolabeled polydopamine-Gd-EMF core–satellite NPs (Wang et al., 2017). Furthermore, the NPs were loaded with doxorubicin (DOX), a widely used chemotherapy drug, which imparted them with theranostic capabilities. This was demonstrated by using a near-infrared laser to induce DOX release inside mice and completely eliminate tumors. Li et al. have developed a bimodal MR/luminescence imaging agent through the conjugation of polyhydroxylated-$Gd@C_{82}$-PCBM to silica-coated $NaYF_4$ (Y = Yb, Er) NPs (Li et al., 2016a). The agent displayed good relaxivity in a 7 T magnetic field, while a cell viability study revealed it to have minimal cytotoxicity and good biocompatibility.

In addition to the preparation of multimodal imaging systems, functionalization of Gd-EMFs can be used to generate CAs with targeting capabilities. Dorn and co-workers recently reported a $Gd_3N@C_{80}$ derivative that targets glioblastoma multiforme (GBM) cells in mice (Li et al., 2015). This was achieved through conjugation of the amino-functionalized EMF with an interleukin-13 (IL-13) peptide chain, which binds to the IL-13Rα2 receptor on the surface of GBM cells. Furthermore, the positively charged amino groups of the fullerene were found to enhance the affinity of the agent for the cell surface relative to a negatively charged carboxylate analogue, which allowed endocytosis to occur more readily.

16.5 Endohedral Nitrogen Fullerenes: Towards Quantum Information Applications

Quantum computers that exploit the fundamental physical laws of superposition and entanglement would enormously accelerate important calculations that are intractable to existing classical computers (Mermin, 2007). To make such a computer, we need physical objects whose quantum states can be preserved and manipulated with high precision. Electron spins in Group V endohedral fullerenes may be one such object (Harneit, 2007, Benjamin, 2006), and $N@C_{60}$ has been particularly well studied for this purpose because of its excellent quantum coherence and the possibility of incorporating into nanoscale electronic devices.

A spin-based quantum computer encodes each quantum bit (qubit) in two spin energy levels with different M_S quantum numbers. By applying microwave bursts using the technique of EPR, we can create quantum superpositions of these two states; we can think of these bursts as rotating the spin axis relative to the static magnetic field. The first step to creating a molecular qubit is to find a pair of well-defined energy levels, which requires a detailed understanding of the electron and nuclear spin states.

In N@C_{60}, the spin states are identical to those of atomic nitrogen. Three unpaired p electrons combine to give electron spin quantum number $S = 3/2$. The resulting Hamiltonian is

$$\mathcal{H} = \mu_B \boldsymbol{B}\boldsymbol{g}\boldsymbol{S} + \mu_N \boldsymbol{B}\boldsymbol{g}_N \boldsymbol{I} + \boldsymbol{S}\boldsymbol{A}\boldsymbol{I} + \boldsymbol{S}\boldsymbol{D}\boldsymbol{S}. \quad (16.1)$$

The first two terms describe the Zeeman coupling to the magnetic field $\boldsymbol{B}$, with the first term arising from the electron spin (with spin operator $\boldsymbol{S}$ and gyromagnetic tensor $\boldsymbol{g}$) and the second term arising from the nuclear spin (with spin operator $\boldsymbol{I}$ and gyromagnetic tensor $\boldsymbol{g}_N$). The third term describes HFC between the electron and nuclear spins, parameterized by the hyperfine tensor $\boldsymbol{A}$, and the fourth term describes electron spin quadrupole coupling parameterized by the zero-field splitting (ZFS) tensor $\boldsymbol{D}$. Here, μ_B is the Bohr magneton and μ_B is the nuclear magneton.

For N@C_{60}, it is an excellent approximation to assume spherically symmetric confinement, in which case the tensors in Eq. (16.1) become scalars and the ZFS vanishes. Furthermore, the first term will always dominate over the second, and in most experiments, the magnetic field is set large enough that it also dominates over the third. For ^{14}N@C_{60}, this leads to the energy levels shown in Figure 16.5b. The resulting EPR spectrum (Figure 16.5a) shows three resonances, each corresponding to a different nuclear spin projection m_I. A fine scan over each resonance (Figure 16.5c) shows substructure arising partly from the hyperfine interaction in second order and partly from coupling (not included in Eq. (16.1)) to ^{13}C nuclear spins in the cage.

The scans in Figure 16.5c show how exceptionally sharp the EPR transitions can be in this material (Morton et al., 2006). This reflects the long electron spin lifetime, which is enabled by the structural symmetry of the molecule and the protection inside the cage (Knapp et al., 1997). For quantum computing, the most important lifetime is the decoherence time T_2 (also called the transverse relaxation time), which measures how long a quantum state can be preserved for. For carefully prepared solutions of ^{14}N@C_{60}, this time can reach as long as T_2 = 70μs at room temperature or T_2 = 250μs when cooled to 170K (Morton et al., 2006). This is among the longest coherence times for any molecular radical, surpassed only by (d_{20}-$Ph_4P)_2$ [$V(C_8S_8)_3$]. Remarkably, at 10 K, the value of T_2 is two orders of magnitude greater than for the other solvent systems, with T_2 = 675(7) μs (∼0.7 ms) in CS_2.(Zadrozny et al., 2015) The electron coherence time of N@C_{60} appears to be ultimately limited by Orbach relaxation, i.e. the interaction of the electron spin with phonons in the cage. (Morton et al., 2006). Even longer coherence times can be attained when the quantum state is transferred from the electron spins to the nuclear spins (Brown, 2010).

Quantum information processing requires more than isolated qubits. If fullerene qubits are to be coupled to one another, chemical reactions are needed to rationally modify and covalently link endohedral nitrogen fullerenes (ENFs). A set of chemical functionalization of ENFs has now been established (Zhou et al., 2015a). One price of functionalizing the cage is that one breaks spherical symmetry, thereby introducing significant ZFS. This introduces drawbacks, such as extra spin relaxation paths (Morton et al., 2006)

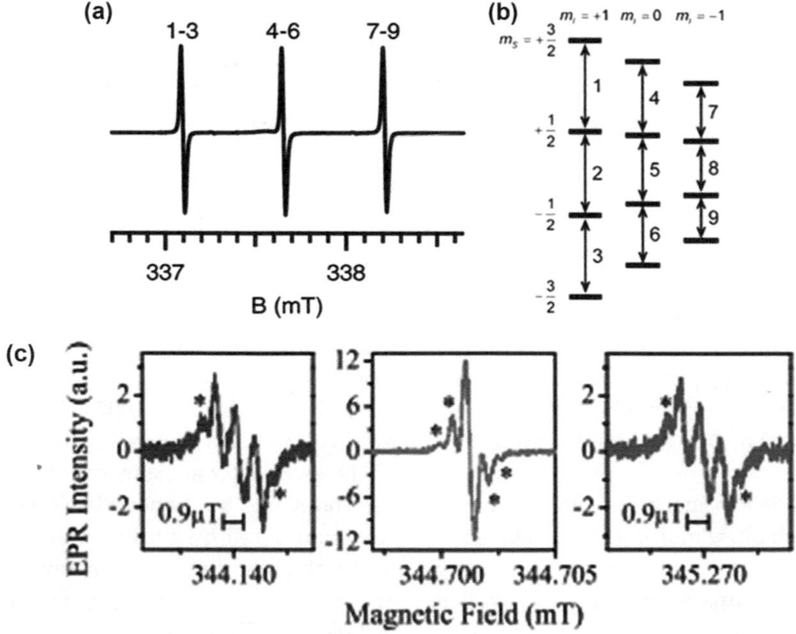

FIGURE 16.5 (a) EPR spectrum of N@C_{60} diluted in C_{60} powder. (b) Energy levels and allowed transitions of ^{14}N@C_{60} in the high field limit. For each of the three possible nuclear spin alignments m_I, there is a ladder of four electron spin levels. HFC gives each ladder a different energy spacing. (c) Zoom in for each of the three resonances (dark gray: $m_I = +1$, light gray: $m_I = 0$, medium gray: $m_I = -1$) showing the sharp linewidth and the second-order hyperfine pattern. (Asterisks denote resonances arising from coupling to cage ^{13}C spins.) Panels a and b from the study by Harneit (2007). Panel c from the study by Morton et al. (2006).

but also potential advantages, such as additional EPR transitions that allow the molecule to be used as a qubit, i.e. a system encoding more than one qubit of information (Gedik et al., 2015).

The strength of the ZFS is quantified by a traceless second-order tensor $\boldsymbol{D}$, which could be expressed as a diagonal matrix in its eigenframe with elements being $D_{xx} = -D/3 + E$, $D_{yy} = -D/3 - E$ and $D_{zz} = 2D/3$, where D and E are the ZFS parameters representing the axial and the non-axial components of the tensor. For typical cycloaddition products of ENFs, D is around 10 MHz and E is normally smaller than 1 MHz. Compared with other paramagnetic molecules, the ZFS effect is typically small. In addition to ZFS, dipolar coupling with a strength larger than 2.67 MHz between an N@C_{60} moiety and another covalently linked spin center has been reported (Farrington et al., 2012, Zhou et al., 2016). The implementation of controllable dipolar coupling with ENFs paves the way towards interacting qubits, which is required by quantum information applications.

16.6 Atomic Clocks: Fullerenes as Frequency Standards

Many modern technologies, such as communications and navigation, rely on precise and stable frequency standards (Vig, 1993). For example, high-frequency stability is necessary in communication systems to ensure that the transmitter and receiver remain synchronized. This is particularly important for jamming-resistant communications that work by coordinated hopping over different frequencies. In navigation applications, such as global navigation satellite system (GNSS) receivers, high-stability frequency standards could improve positional accuracy in signal-degraded environments (Misra, 1996). The most stable clocks work by locking an electronic oscillator to a reference frequency provided by an atomic transition (Riehle, 2004) as shown in Figure 16.6a. Since atomic transition frequencies are fixed by nature, this reduces the influence of manufacturing variation and drift and results in a highly stable and reproducible output frequency. Such a system is commonly called an "atomic clock". For portable atomic clocks, size, weight and power (SWaP) are important parameters in addition to stability (Vig, 1993).

The state-of-the-art miniaturized atomic frequency standard is the chip-scale atomic clock (CSAC) (Knappe, 2004). Such a clock operates by disciplining a local oscillator to the magnetic resonance signal of an optically probed alkali metal vapor, which is confined in a vacuum chamber. Microelectromechanical (MEM) fabrication allows construction of miniature vacuum chambers with integrated diode lasers and optical sensors, leading to commercially available atomic clocks with a total volume $<17\text{cm}^3$ (Microsemi Corporation, 2017). However, CSACs suffer long-term drifts caused by changing buffer gas pressure as the vapor cell ages. Moreover, further reduction of SWaP is necessary for broader adoption of portable atomic clocks (Lutwak et al., 2007). Another drawback is that some applications require robustness against acceleration and vibration (Vig, 1993), which is limited by the vacuum packaging that thermally isolates the vapor cell (Lutwak et al., 2007).

These problems may be solvable using frequency references based on condensed-matter systems, which obviate

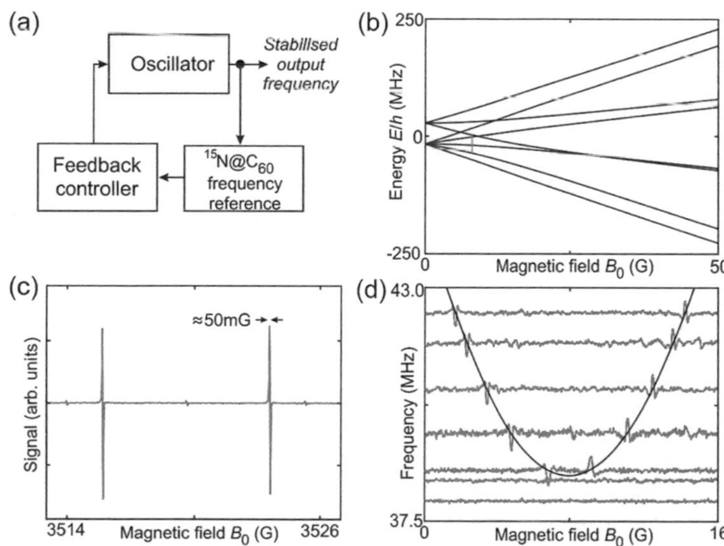

FIGURE 16.6 (a) Schematic of atomic clock using ^{15}N@C_{60} as the frequency reference. A radio signal generated by a local oscillator is used to probe a ^{15}N@C_{60} sample. The response of the sample is used for feedback to lock the frequency of the oscillator to the EPR resonance. (b) Low-field energy levels as a function of magnetic field. The clock transition (vertical light gray line) occurs in the region where the Zeeman energy is comparable to the HFC. (c) X-band spectrum of ^{15}N@C_{60}, showing clearly narrow resonance signals. (d) Field-frequency map of the clock transition frequency as a function of magnetic field, demonstrating $\mathrm{d}f/\mathrm{d}B = 0$ at the clock field $B_{\text{clock}} \approx 0.8\text{mT}$. (Based on the work by Harding et al. 2017.)

the need for the vapor cell. Proposed examples include V^{++} in MgO (White and Hajimir, 2005) and nitrogen vacancies in diamond (Hodges et al., 2014). However, in neither system has the necessary stability been demonstrated experimentally. For both these proposed frequency standards, the minimum resonance linewidth, and hence maximum stability, is limited by inhomogeneous broadening due to lattice strains.

Here we describe an alternative condensed-matter clock material: the Group V-containing endohedral fullerenes (Briggs and Ardavan, 2012). In contrast to vapor-based clocks, which use optical interrogation of the spin ensemble, the proposed fullerene-based clocks use radio-frequency (rf) measurement of the endohedral fullerene's EPR spectrum. As mentioned previously, resonances as narrow as 0.3μT have been observed at the X band in a carefully prepared sample (Morton et al., 2006). The importance of narrow resonances for a clock can be intuitively understood by considering them analogous to the markings on a ruler: the sharper the resonances, the more precisely the frequency can be measured.

Sharp resonances are a necessary but not sufficient condition for the frequency reference in an atomic clock. It is also necessary to select a transition whose frequency is insensitive to environmental noise (Audoin and Vanier, 1976). For example, all atomic clocks use "clock transitions", which are resonances with the property that their frequency f is first-order independent of magnetic field B: i.e. $df/dB = 0$. If B is set to the clock field B_{clock} at which such a clock transition exists, then small external magnetic fields barely perturb the clock frequency. At this clock field, the transition frequency is fixed by the isotropic HFC constant A.

Such a clock transition has long been predicted for $^{15}N@C_{60}$ (Briggs and Ardavan, 2012). As seen from Figure 16.6b, at a magnetic field of approximately 8 G, two of the spin energy levels run parallel, implying that the frequency of the transition between them satisfies the clock condition $df/dB = 0$. Experimental verification has been challenging due to its low frequency, which leads to a small spin polarization and weak signal. Recently, the clock transition has been measured using a custom built low-field EPR spectrometer (Harding et al., 2017), as shown in Figure 16.6d. Given the advances in spectrometer miniaturization developed for "1-chip" nuclear magnetic resonance (NMR) (Sun et al., 2011), a fullerene-based clock on an integrated circuit seems feasible.

One can predict the stability of an atomic clock by considering the frequency of the clock transition, the observed resonance linewidth and the signal-to-noise ratio (SNR) of the resonance (Riehle, 2004). Current measurements on fullerene-based systems imply a stability that is approximately seven orders of magnitude worse than existing CSACs (Knappe, 2004). However, there is scope for improvement by spectrometer and sample optimization, which could improve SNR and linewidth. The most significant improvements would likely come from sample purification to increase the spin density, and hence signal strength, or from reduction of the resonance linewidth. It is reasonable to expect an improvement to the linewidth, since the current value is approximately one order of magnitude worse than has been achieved in the literature. Furthermore, improved coherence times at the clock field have been observed in other systems, due to improved immunity to dipolar decoherence (Shiddiq et al., 2016), which indicates a potential for reduced linewidth at the clock field. Therefore, a fullerene-based clock could potentially be competitive with CSACs while achieving reduced SWaP. Furthermore, use of $^{31}P@C_{60}$may permit further improvements due to its larger HFC, which increases the clock transition frequency.

For long-term clock stability, it is also necessary to suppress the frequency drifts caused by temperature fluctuations. These arise because the hyperfine constant of $N@C_{60}$ depends on temperature, with a relative shift of $(1/A)[dA/dT] \approx 100$ ppm/K (Pietzak et al., 2002). It will be necessary either to stabilize the temperature of the sample cell or to compensate for this effect. One possibility is to offset the temperature dependence with a second control parameter such as pressure (Hodges et al., 2014). Another possibility is again to use $^{31}P@C_{60}$, which has a weaker temperature dependence (Pietzak et al., 2002).

16.7 Perspectives

While the well-known drawbacks of EMFs and nitrogen-containing endohedral fullerenes such as their low production yield and tedious purification processes remain unanswered, recent developments in the field, including the chemically activated reaction atmosphere, chemical doping of the graphite rods and the Lewis acid selective precipitation methods, provide hope for the scaled-up production of this class of exotic materials. The synthesis and surface functionalization of novel dimetallic EMFs that will exhibit stronger magnetization can present an important milestone regarding the applications of these materials as also the development of phosphorus-containing endohedral fullerenes for atomic clocks applications.

The conclusion is that the future of research in endohedral fullerenes and their derivatives is looking bright!

References

Audoin, C. and Vanier, J. 1976 Atomic frequency standards and clocks, *J. Phys. E: Sci. Instrum.* 9: 697–720.

Bartl, A. and Dunsch, L. 2001 Temperature dependent metal cluster position inside a C82 cage: $Sc_3@C_{82}$. *Synth. Met.* 121: 1147–1148.

Benjamin, S. C. et al. 2006 Towards a fullerene-based quantum computer. *J. Phys. Condens. Matter* 18: S867–S883.

Bondino, F., Cepek, C., Tagmatarchis, N., Prato, M., Shinohara, H., Goldoni, A. 2006 Element-specific probe of the magnetic and electronic properties of Dy incar-fullerenes. *J. Phys. Chem. B.* 110: 7289–7295.

Briggs, G.A.D. and Ardavan, A, 2012 US Patent No 8,217,724.

Brown, R.M. et al. 2010 Electron spin coherence in metallofullerenes: Y, Sc, and La@C_{28}. *Phys. Rev. B.* 82: 033410.

Chen, C.-H., Krylov, D.S., Avdoshenko, S.M., Liu, F., Spree, L., Yadav, R., Alvertis, A., Hozoi, L., Nenkov, K., Kostanyan, A., Greber, T., Wolter, A.U.B., Popov, A.A. 2017 Selective arc-discharge synthesis of Dy_2S-clusterfullerenes and their isomer-dependent single molecule magnetism. *Chem. Sci.* 8: 6451–6465.

Cho, S.C., Kaneko, T., Ishida, H. et al. 2015 Nitrogen-atom endohedral fullerene synthesis with high efficiency by controlling plasma-ion irradiation energy and C_{60} internal energy. *J. Appl. Phys.* 117: 123301.

Círić, L. et al. 2008 La@C_{82} as a spin-active filling of SWCNTs: ESR study of magnetic and photophysical properties. *Phys. Status Solidi Basic Res.* 245: 2042–2046.

Cui, R., Li, J., Huang, H., Zhang, M., Guo, X., Chang, Y., et al. 2015. Novel carbon nanohybrids as highly efficient magnetic resonance imaging contrast agents. *Nano Res.* 8: 1259–1268.

Elliott, B. et al. 2013 Spin density and cluster dynamics in $Sc_3N@C_{80}^-$ upon [5,6] exohedral functionalization: An ESR and DFT study. *J. Phys. Chem. C.* 117: 2344.

Farrington, B.J., Jevric, M., Rance, G.A., Ardavan, A., Khlobystov, A.N., Briggs, G.A. D., Porfyrakis, K. 2012 Chemistry at the nanoscale: Synthesis of an N@C_{60}–N@C_{60} endohedral fullerene dimer *Angew. Chem. Int. Ed.* 51 (15): 3587.

Gedik, Z., Silva, I.A., Çakmak, B., Karpat, G., Vidoto, E.L.G., Soares Pinto, D.O., DeAzevedo, E.R., Fanchini, F.F. 2015 *Sci. Rep.* 5: 1.

Ghiassi, K.B., Olmstead, M.M., and Balch, A.L. 2014 Gadolinium-containing endohedral fullerenes: Structures and function as magnetic resonance imaging (MRI) agents. *Dalton Trans.* 43: 7346–7358.

Hajjaj, F. et al. 2011 Ferromagnetic spin coupling between endohedral metallofullerene La@C_{82} and a cyclodimeric copper porphyrin upon inclusion *J. Am. Chem. Soc.* 133: 9290.

Harding, R.T., Zhou, S., Zhou, J. et al. 2017 Spin resonance clock transition of the endohedral fullerene $^{15}N@C_{60}$. *Phys. Rev. Lett.* 119: 140801.

Harneit, W., Huebener, K., Naydenov, B., Schaefer, S., Scheloske, M. 2007 N@C_{60} quantum bit engineering. *Phys. Status Solidi* 244: 3879–3884.

Hermanns, C.F. et al. 2013 Magnetic coupling of $Gd_3N@C_{80}$ endohedral fullerenes to a substrate. *Phys. Rev. Lett.* 111: 167203.

Hodges, J.S., Yao, N.Y., Maclaurin, D., Rastogi, C., Lukin, M.D., Englund, D. 2014 Timekeeping with electron spin states in diamon. *Phys. Rev. A.* 87: 032118.

Inakuma, M. and Shinohara, H. 2000a Structural and electronic properties of isomers of $Sc_2@C_{84}$(I, II, III): ^{13}C NMR and IR/Raman spectroscopic studies. *J. Phys. Chem. B.* 104: 7595.

Inakuma, M. and Shinohara, H. 2000b Temperature-dependent EPR studies on isolated scandium metallofullerenes: Sc@C_{82}(I, II) and Sc@C_{84}. *J. Phys. Chem. B* 104: 7595–7599.

Kikuchi, K., Nakao, Y., Suzuki, S., Achiba, Y., Suzuki, T., Maruyama, Y. 1994 Characterization of the isolated Y@C_{82}. *J. Am. Chem. Soc.* 116: 9367–9368.

Kitaura, R., Okimoto, H., Shinohara, H., Nakamura, T., Osawa, H. 2007 Magnetism of the endohedral metallofullerenes M@C_{82} (M=Gd,Dy) and the corresponding nanoscale peapods: Synchrotron soft x-ray magnetic circular dichroism and density-functional theory calculations. *Phys. Rev. B Condens. Matter Mater. Phys.* 76, 1–4.

Knapp, C., Dinse, K.P., Pietzak, B., Waiblinger, M., Weidinger, A. 1997 Fourier transform EPR study of N@C_{60} in solution. *Chem. Phys. Lett.* 272: 433–437.

Knapp, C., Weiden, N., Kass, H., Dinse, K.P., Pietzak, B., Waiblinger, M., Weidinger, A. 1998 Electron paramagnetic resonance stude of atomic phosphorus encapsulated in [60] fullerene. *Mol. Phys.* 95: 999–1004.

Knappe, S., Shah, V., Schwindt, P.D.D., Hollberg, L., Kitching, J. 2004 A microfabricated atomic clock. *Appl. Phys. Lett.* 85: 1460.

Koltover, V.K. 2004 Spin-leakage of the fullerene shell of endometallofullerenes: EPR, ENDOR and NMR evidences. *Carbon N. Y.* 42: 1179–1183.

Koltover, V.K. et al. 2004 Diamagnetic clusters of paramagnetic endometallofullerenes: A solid-state MAS NMR study. *J. Phys. Chem. B* 108: 12450–12455.

Laird, E.A., Pei, F., Kouwenhoven, L.P. 2013 A valley-spin qubit in a carbon nanotube. *Nat. Nanotechnol.* 8: 565.

Laus, S., Sitharaman, B., Tóth, É., Bolskar, R.D., Helm, L., Asokan, S., et al. 2005. Destroying gadofullerene aggregates by salt addition in aqueous solution of $Gd@C_{60}(OH)x$ and $Gd@C_{60}[C(COOH_2)]_{10}$. *J. Am. Chem. Soc.* 127: 9368–9369.

Laus, S., Sitharaman, B., Tóth, É., Bolskar, R.D., Helm, L., Wilson, L.J., et al. 2007 Understanding paramagnetic relaxation phenomena for water-soluble gadofullerenes. *J. Phys. Chem. C.* 111: 5633–5639.

Li, C., Cui, R., Feng, L., Li, J., Huang, H., Yao, H., et al. 2016a Synthesis of a UCNPs@SiO_2@gadofullerene nanocomposite and its application in UCL/MR bimodal imaging. *RSC Adv.* 6: 98968–98974.

Li, J., Cui, R., Chang, Y., Guo, X., Gu, W., Huang, H., et al. 2016b Adaption of the structure of carbon nanohybrids toward high-relaxivity for a new MRI contrast agent. *RSC Adv.* 6: 58028–58033.

Li, T. and Dorn, H.C. 2017 Biomedical applications of metal-encapsulated fullerene nanoparticles. *Small.* 13: 1603152.

Li, T., Murphy, S., Kiselev, B., Bakshi, K.S., Zhang, J., Eltahir, A., et al. 2015 A new Interleukin-13 amino-coated gadolinium metallofullerene nanoparticle for targeted MRI detection of glioblastoma tumor cells. *J. Am. Chem. Soc.* 137: 7881–7888.

Li, Y., Wang, T., Zhao, C., Qin, Y., Meng, H., Nie, M., Jiang, L., Wang, C. 2017 A magnetoreception system constructed by a dysprosium metallofullerene and nitroxide radical. *Dalton Trans.* 46: 8938.

Luo, J., Wilson, J.D., Zhang, J., Hirsch, J.I., Dorn, H.C., Fatouros, P.P., et al. 2012 A dual PET/MR imaging nanoprobe: 124I Labeled $Gd_3N@C_{80}$. *Appl. Sci.* 2: 465–478.

Lutwak, R., Rashed, A., Varghese, M., et al. 2007 The chip-scale atomic clock—prototype evaluation, in *Proceedings of the 36th Annuual Precise Time and Time Interval Systems and Applications Meeting*, Kona, Hawaii, USA.

Mermin, N.D. 2007 *Quantum Computer Science.* Cambridge: Cambridge University Press.

Microsemi Corporation, 2017 QuantumTM SA.45s CSAC chip scale atomic clock, Aliso Viejo, California, USA.

Mikawa, M., Kato, H., Okumura, M., Narazaki, M., Kanazawa, Y., Miwa, N., et al. 2001. Paramagnetic water-soluble metallofullerenes having the highest relaxivity for MRI contrast agents. *Bioconjugate Chem.* 12: 510–514.

Misra, P., 1996 The role of the clock in a GPS receiver. *GPS World*, 7: 60–66.

Morton, J.J.L., Tyryshkin, A.M., Ardavan, A., Porfyrakis, K., Lyon, S.A., Briggs, G.A.D, 2006 Electron spin relaxation of $N@C_{60}$ in CS_2. *J. Chem. Phys.* 124 (1): 014508.

Murphy, T.A., Pawlik, T., Weidinger, A. et al. 1996 Observation of atom like nitrogen in nitrogen-implanted solid C_{60}. *Phys. Rev. Lett.* 77: 1075–1078.

Pietzak, B., Waiblinger, M., Murphy, T.A. et al. 1997 Buckminsterfullerene C_{60}: A chemical Faraday cage for atomic nitrogen. *Chem. Phys. Lett.* 4: 259.

Pietzak, B., Waiblinger, M., Murphy, T.A., Weidinger, A., Hohne, M., Dietel, E., Hirsch, A., 1998 Properties of endohedral $N@C_{60}$ *Carbon* 36: 613–615.

Pietzak, B., Weidinger, A., Dinse, K.P., Hirsch, A. 2002 Group V Endohedral Fullerenes: $N@C_{60}$, $N@C_{70}$, and $P@C_{60}$. in T. Akasaka, S. Nagase (eds.) *Endofullerenes: A New Family of Carbon Clusters.* Dordrecht: Springer.

Riehle, F. 2004 *Frequency Standards: Basics and Applications.* Weinheim: Wiley.

Roukala, J. et al. 2017 Ratcheting rotation or speedy spinning: EPR and dynamics of $Sc_3C_2@C_{80}$. *Chem. Commun.* 53: 8992.

Shiddiq, M., Komijani, D., Duan, Y., Gaita-Arino, A., Coronado, E., Hill, S. 2016 Enhancing coherence in molecular spin qubits via atomic clock transitions. *Nature*, 531: 348–351.

Shinohara, H., Inakuma, M., Hayashi, N., Sato, H., Saito, Y., Kato, T., Bandow, S. 1994 Spectroscopic properties of isolated $Sc_3@C_{82}$ metallofullerene. *J. Phys. Chem.* 98(35): 8597.

Sosnovik, D.E. and Caravan, P. 2013 Molecular MRI of the cardiovascular system in the post-NSF Era. *Curr. Cardiovasc. Imaging Rep.* 6: 61–68.

Sun, N., Yoon, T.J., Lee, H., Andress, W., Weissleder, R., Ham, D. 2011 Palm NMR and 1-chip NMR. *IEEE J. Solid-State Circuits*, 46: 342–352.

Svitova, A.L., Krupskaya, Y., Samoylova, N., Kraus, R., Geck, J., Dunsch, L., Popov, A.A. 2014 Magnetic moments and exchange coupling in nitride cluster-fullerenes $Gd_xSc_{3-x}N@C_{80}$ (x =1–3). *Dalton Trans.* 43: 7387.

Theobald, J.A., Oxtoby, N.S., Phillips, M.A., Champness, N.R., Beton, P.H. 2003 Controlling molecular deposition and layer structure with supramolecular surface assemblies. *Nature* 424: 1029–1031.

Tóth, É, Bolskar, R.D., Borel, A., González, G., Helm, L., Merbach, A.E., et al. 2005 Water-soluble gadofullerenes: Toward high-relaxivity, PH-responsive MRI contrast agents. *J. Am. Chem. Soc.* 127: 799–805.

Vig, J.R., 1993 Military applications of high accuracy frequency standards and clocks. *IEEE Trans. Ultrason. Ferroelectr. Freq. Control* 40: 522.

Wang, S., Lin, J., Wang, Z., Zhou, Z., Bai, R., Lu, N., et al. 2017. Core–satellite polydopamine–gadolinium-metallofullerene nanotheranostics for multimodal imaging guided combination cancer therapy. *Adv. Mater.* 29: 1701013.

White, C.J. and Hajimir, A. 2005 A solid-state atomic frequency standard, in *Proceedings of the IEEE International Frequency Control Symposium and Exposition*, Vancouver, Canada.

Wu, B. et al. 2015 Molecular magnetic switch for a metallofullerene. *Nat. Commun.* 6: 6468.

Wu, B., Li, Y., Jiang, L., Wang, C., Wang T., 2016 Spin-paramagnet between nitroxide radical and metallofullerene. *J. Phys. Chem. C.* 120(11): 6252–6255.

Yamamoto, K., Funasaka, H., Takahashi, T., Akasaka, T. 1994 Isolation of an ESR-active metallofullerene of $La@C_{82}$. *J. Phys. Chem.* 98(8): 2008–2011.

Zadrozny, J.M., Niklas, J., Poluektov, O.G., 2015 Freedman D.E. *ACS Cent. Sci.* 1: 488.

Zhang, J., Fatouros, P.P., Shu, C., Reid, J., Owens, L.S., Cai, T., et al. 2010. High relaxivity trimetallic nitride (Gd_3N) metallofullerene MRI contrast agents with optimized functionality. *Bioconjugate Chem.* 21: 610–615.

Zhang, J., Ye, Y., Chen, Y., Pregot, C., Li, T., Balasubramaniam, S., et al. 2014. $Gd_3N@C_{84}(OH)_x$: A new egg-shaped metallofullerene magnetic resonance imaging contrast agent. *J. Am. Chem. Soc.* 136: 2630–2636.

Zhao, J. et al. 2015 Magnetic properties of atomic clusters and endohedral metallofullerenes. *Coord. Chem. Rev.* 289–290: 315.

Zhou, S., Rasovic, I., Briggs, G.A.D., Porfyrakis, K. 2015a Synthesis of the first completely spin-compatible $N@C_{60}$ cyclopropane derivatives by carefully tuning the DBU base catalyst. *Chem. Commun.* 51: 7096–7097.

Zhou, S., Yamamoto, M., Briggs, G.A.D., Imahori, H., Porfyrakis, K. 2016 Probing the dipolar coupling in a heterospin endohedral fullerene–phthalocyanine dyad. *J. Am. Chem. Soc.* 138: 1313–1319.

Zou, T, Zhen, M., Li, J., Chen, D., Feng, Y., Li, R., et al. 2015. The effect of hemiketals on the relaxivity of endohedral gadofullerenols. *RSC Adv.* 5: 96253–96257.

17

Titania Nanotubes

Oomman K. Varghese,
Keea Stancato, Maggie Paulose,
and Ram Neupane
University of Houston

17.1 Introduction

The new millennium has seen the advent of many new nanomaterials but only very few stimulated scientific curiosity and technological interest in a broad range of fields. Titanium dioxide (TiO_2; titania) nanotube (NT) is prominent among such materials. The following factors apparently helped titania NT to emerge as one of the most studied nanomaterials in recent years: (i) Titanium dioxide is an abundant material that has long been known for its practical applications ranging from paints and cosmetics to medicine and energy. Its relevance as an industrial material is evident from the fact that the global titanium dioxide market size was over 13 billion US dollars in 2015 and is growing fast [1]. It has been identified as a promising material for future technologies in different areas including photovoltaics, photocatalysis and environmental sensing. The NT geometry is expected to widen the application scope of the material and help solve some of the challenges associated with it. (ii) The NT architecture possesses highly useful attributes such as high geometric surface area per unit mass or the flat area it occupies (the geometric surface area of NTs can be higher than the flat area by a factor of hundred or thousand or even more) and nanoscale walls bound by inner and outer surfaces enabling low loss charge transfer to the surface for redox reactions or interfacial transport. The NT geometry offers possibilities of using it as an encapsulating material, high surface area support and low refractive index medium. The initial studies on the material conducted soon after its inception revealed its unique properties linked to its specific geometry [2–5]. (iii) Titania NTs can be prepared easily in a laboratory or industrial environment. Simple, scalable and low cost methods are available for fabrication. For example, anodic oxidation used for titania NT growth is more than a century old industrial technique known for the fabrication of low-cost protective and decorative coatings.

Based on the relevance of its exceptional properties for a number of applications and the extensive attention it has been receiving (over 1,000 research articles published per year), titania NT could be classified as a 21st century nanomaterial [5–8]. Therefore, in this chapter, the major developments in the fabrication of titania NT and knowledge about its properties in relation with applications in the fields of energy, environment and biology/medicine have been discussed.

17.2 Titania NT Fabrication

While a number of methods have been developed during the past couple of decades for growing NTs of different organic and inorganic materials, only a few template-based and two template-free methods have so far been successful in yielding titania NTs. The methods utilizing templates are discussed under "template-assisted fabrication". Hydrothermal and electrochemical anodization are the techniques that do not need templates for titania NT growth.

17.2.1 Template-Assisted Fabrication

The conventional wisdom earned from centuries and millennia of work on the fabrication of bulk materials of various shapes and sizes using removable molds/templates perhaps prompted the materials scientists to apply the same approach for the development of NTs and other nanoarchitectures. The method has been rigorously investigated for developing NTs of organic and inorganic materials since the discovery of carbon nanotubes (CNT) [9–11]. In general, the method involves coating a nanoporous membrane or a nanowire/rod with the material intended for NT formation and subsequently removing the template. NTs can be amorphous or crystalline depending upon the specific process employed. NTs in both array films and individually dispersed forms can be fabricated using this method. Another advantage of this method is that the length of the NT can be varied from a few tens of nanometers to hundreds of micrometers and the diameter from a few tens to a few hundreds of nanometers. On the downside, it involves multiple steps, consistency and integrity of the NTs are problems and it is difficult to scale up the process.

In 1996, Hoyer reported the fabrication of titania NT arrays using a template for the first time [12]. They used anodic porous aluminum oxide (alumina; Al_2O_3) as an initial template. Since then, a variety of recipes have been developed for making titania NTs using this method. The most commonly used template is still ordered porous alumina fabricated by anodic oxidation. Polymers, zeolite, glass and carbon-based porous and cylindrical structures are also used as templates. The templates are generally coated/filled with titania precursors using techniques such as sol–gel, chemical bath deposition and atomic layer deposition (ALD). A commonly used process for titania NT growth involves the deposition of titania sol in the pores of anodic alumina films/disks followed by heat treatment for solvent evaporation and stoichiometric oxide formation and the removal of alumina template by chemical dissolution.

In the original work, Hoyer coated a porous alumina film grown on an aluminum substrate using anodic oxidation with a 100 nm thick gold by evaporation [12]. Methyl methacrylate was polymerized inside the pores with the help of benzoyl peroxide under ultraviolet (UV) illumination. The aluminum substrate and Al_2O_3 template were then removed by treating them with 10% NaOH. The resulting poly(methyl methacrylate) (PMMA) negatype structure consisting of PMMA rod arrays on a PMMA film served as the template for titania NT formation. A gold film was first coated on the surface of the PMMA rods using electroless deposition. Amorphous titania was electrodeposited on this layer in a $TiCl_3/HCl/NaHCO_3$ electrolyte. The material was washed in dilute hydrochloric acid (HCl) and dried in diethylene glycol (DEG) at 80°C. Finally, the PMMA template was dissolved in acetone leaving the amorphous titania with an NT morphology. These samples were heat treated at 450°C to obtain anatase titania NT arrays.

Lakshmi et al. prepared titania NT arrays using an easier route [13]. They made a sol using titanium isopropoxide in ethanol with small amounts of HCl and water. An anodic porous alumina membrane was dipped in the sol for about 5 s and then removed and dried in air. The membrane was heat treated at 400°C for 6 h. The top surface of the membrane covered with a compact titania film was glued to a paper towel and immersed in aqueous NaOH (6 M) to dissolve the alumina template. The resulting titania NT array was found to have a highly crystalline anatase structure. Since this work, sol–gel preparation of titania NTs via porous alumina templates has become one of the commonly used methods [8,14]. Instead of porous alumina, Li et al. used CNT sponge as templates for the preparation of titania NTs through the hydrolysis–condensation reaction using a titanium alkoxide [15]. Ji and Shimizu used an iced peptidic lipid NT as the template for sol–gel titania NT fabrication [16]. Instead of alkoxides, titanium chlorides and fluorides were also used as precursors for NT growth. Liu et al. fabricated layered crystalline titania NTs (similar to carbon NTs) by hydrolyzing TiF_4 in an acidic solution (HCl and NH_4OH; pH 1.6) at 60°C in the pores of porous alumina templates [17]. Lee et al. used zinc oxide (ZnO) nanorod arrays as templates [18]. They dipped the ZnO nanorod array film in a solution consisting of ammonium hexafluorotitanate and boric acid in water so that the reaction proceeds through the hydrolysis reaction of the Ti-fluoro complex. After the formation of the ZnO/TiO_2 core–sheath structure, ZnO undergoes self-removal due to its reaction with H^+ in the depositing solution leaving titania NTs behind. Depending on the deposition time, a tubular or rod morphology was obtained.

Recently, techniques such as ALD and electrospinning have been used for template-assisted titania NT fabrication. ALD is best known for yielding conformal coatings on almost any morphologies, including large aspect ratio (length-to-diameter ratio) structures. The deposition can be carried out in vapor or liquid phase with coatings possible on any location of the template where the liquid or vapor precursors reach. A commonly used precursor is titanium chloride ($TiCl_4$). In a typical vapor phase process, the template is alternately exposed to $TiCl_4$ and water vapor. $TiCl_4$ gets adsorbed on the substrate surface during the first cycle. In the second cycle, this layer reacts with water vapor to form $Ti(OH)_x$, which condenses and becomes a TiO_2 layer. In this case also, anodic porous alumina substrates are generally used as templates.

Sander et al. fabricated titania NT arrays by ALD by exposing anodic porous alumina templates sequentially to $TiCl_4$ and H_2O at 105°C for a desired number of ALD cycles [19]. The duration of each cycle was less than 5 s. A titania overlayer formed on the top of the template was removed by mechanical polishing prior to the etching of the template. Shin et al. used nanoporous polycarbonate membranes and Kemell et al. used cellulose (ashless filter paper) as templates for the ALD growth of titania NTs (precursor: titanium methoxide, $Ti(OMe)_4$) [20,21]. Foong et al. developed a solution-based ALD, labeled as liquid ALD (LALD) [22]. A 40 nm anatase TiO_2 film was formed on a silicon (Si) or indium tin oxide (ITO) substrate through hydrolysis–condensation reaction of a spin-coated titanium isopropoxide/H_2O/ethanol/HNO_3 solution followed by heat treatment at 500°C. Aluminum deposited on this layer was anodized to get the porous template, which was then immersed in H_3PO_4 to remove the alumina barrier layer and widen the pores. The template was first dipped in a titanium isopropoxide–toluene solution, then in ethanol-10% water solution and finally heated at 90°C. The cycle was repeated a few times. Finally, the alumina template was removed using 1 M potassium hydroxide (KOH). Figure 17.1 shows the process and the resulting NT arrays. It is worth noting

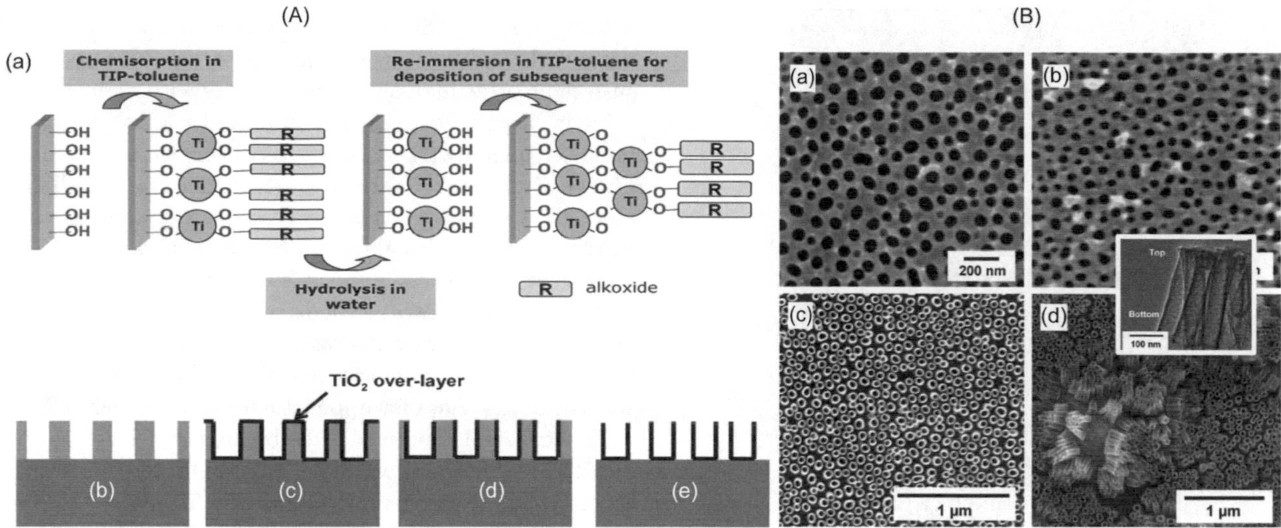

FIGURE 17.1 Alumina template-assisted LALD growth of titania NT arrays. (A) Schematics of (a) a typical ALD cycle showing the sequential and repeated chemisorption/hydrolysis reactions, (b and c) the growth of ALD titania layer inside alumina template pores, (d) the removal of the titania overlayer and (e) the NTs formed with the removal of the template. (B) Scanning Electron Microscope (SEM) images of the alumina template before (a) and after (b) ALD titania deposition and titania NT on Si substrate (c and d). Inset: The Transmission Electron Microscope (TEM) image of an NT cluster. The NTs are ~150 nm long with outer diameter 60–70 nm and wall thickness ~5 nm. (Reprinted with permission from Ref. [22]. Copyright (2010) John Wiley and Sons.)

that in ALD processes, vacuum or solution based, a titanium overlayer is formed at the top of the template and it is critical to remove this layer. Methods such as mechanical polishing and reactive ion etching can be used for the removal [19,22,23].

Huang and Kunitake used a sol–gel polymer–titania composite approach to fabricate flexible titania NTs [24,25]. A poly (ethyleneimine) (PEI)/poly (acrylic acid) (PAA) bilayer was first deposited in porous alumina. It was then subjected to treatments alternately using $Ti(O_nBu)_4$ solution and polyvinyl alcohol (PVA) n times ($n = 10$ for their experiments). After chemically etching the template, the resulting PEI/PAA/(titania/PVA)$_n$ NTs were subjected to heat or plasma treatment to remove the polymer and obtain titania NTs (Figure 17.2).

Various fibers, both natural and synthetic, were also used as templates. Caruso et al. fabricated titania NTs using a sol–gel coating on electrospun poly(L-lactide) fiber templates [26]. A titanium isopropoxide/isopropanol (1:19 v/v) solution was dripped onto the fiber mat and dried in a vacuum desiccator. The polymer template was removed by heat treating it at 450°C for 3 h in nitrogen and then for 10 h in oxygen. A mat like titania hollow fibers resulted. Ghadiri et al. made hollow titania fibers using cotton wool (natural cellulose) [27]. They coated the cotton fibers with supersaturated solutions of $(NH_4)_2TiF_6$ and H_3BO_3 at 50°C. The fiber was burned away by heating the sample at 500°C in air to obtain hollow fibers of TiO_2. While sol–gel precursors used are generally titanium alkoxides and $TiCl_4$, a variety of electrospun polymers are used as templates, and these include polyacrylonitrile (PAN), PVA and poly(vinyl pyrrolidone) (PVP) [28–32]. An interesting variation to the method was applied by Li and Xia [31]. They used continuous electrospinning of two immiscible liquids, a heavy mineral oil and an ethanol solution of PVP and titanium isopropoxide, through a coaxial two-capillary spinneret to make a mat of PVP/TiO_2 hollow fibers. Chen et al. used this technique to grow a titania nanowire-in-microtube structure [33]. Instead of sol gel, hot filament chemical vapor deposition (CVD) was used by Karaman et al. to make TiO_2 NTs with the aid of electrospun PMMA fibers [34]. In all cases, electrospun polymers were removed and titania NTs were recovered by heat treating the coated polymer in air/oxygen at high temperatures (typically, above 400°C).

Titania NTs could also be fabricated using supramolecular scaffolds. The method has been extensively used for the growth of nanoarchitectures of various materials [35,36]. These scaffolds can facilitate the self-assembly of target functional molecular units attached covalently or non-covalently to them to form a desired structure. Kobayashi et al. fabricated titania NTs (or hollow fibers) using the self-assembled organogelators, which served as templates for sol–gel polymerization of $Ti[OCH(CH_3)_2]_4$ (titanium isopropoxide) [37]. They utilized the supramolecular assembly of *trans*-(1*R*,2*R*)-1,2-Cyclohexanedi(11-aminocarbonylundecylpyridinium) hexafluorophosphate. Jung et al. used neutral dibenzo-30-crown-10-appended cholesterol gelator for the fabrication of double-walled titania NTs and helical ribbons [38]. Gundiah et al. used a tripodal cholamide having hydrophobic surfaces, which forms gel fibers in aqueous media, as hydrogelator to fabricate NTs of titania and various other metal oxides [39]. The advantage of this method is that no external template is required.

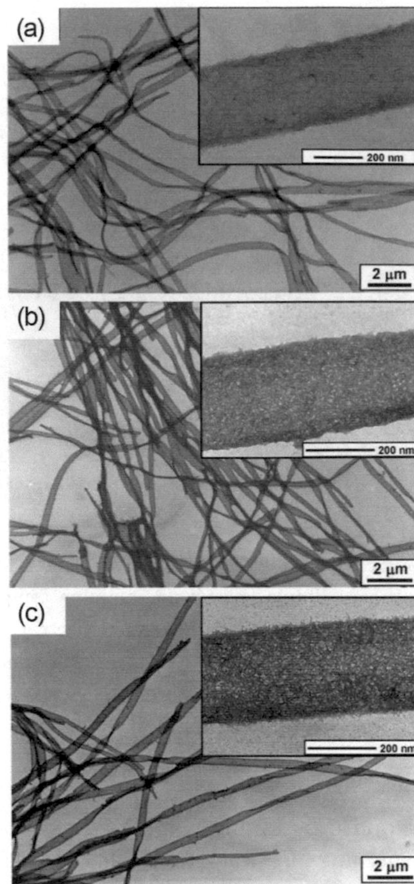

FIGURE 17.2 TEM images of flexible PEI/PAA/(titania/PVA) nanocomposite tubes: (a) as prepared, (b) calcined and (c) the low temperature O_2 plasma-treated tubes. (b and c) Titania NTs resulted after thermal or plasma treatment of the composite tubes. Insets: The high magnification images. (Reprinted with permission from Ref. [25]. Copyright (2006) Royal Society of Chemistry.)

17.2.2 Hydrothermal Method

Background

Hydrothermal method of material preparation involves carrying out a heterogeneous reaction in an aqueous medium at an elevated temperature, typically above the boiling point of the solution, and pressure in an autoclave. In case nonaqueous solutions are used, the method is termed solvothermal. Nature has used the hydrothermal process for the formation of minerals through various geological processes. The early reports of using the process for material fabrication date back to the middle of the 19th century [40–42]. By mid-20th century, it gained popularity as the major technique for zeolite synthesis. The first report on hydrothermal synthesis of titania NTs appeared in 1998 [43]. Since then, hydrothermal synthesis has become a highly explored technique for titania NT fabrication. The basic reason is that the process is simple, it generally involves only common chemicals like sodium hydroxide (NaOH) and hydrochloric acid (HCl), no expensive equipment is required for lab-scale experiments, the process is scalable (already used in the industry) and high aspect ratio (10–100s) high surface area (as high as 478 m^2/g) NTs with diameter typically in the range 3–15 nm and length a few nm to the hundred micrometer range (typical length is less than a micrometer) can be fabricated [44]. While such small diameter NTs are hydrothermal process specific, another unique feature of hydrothermal titania NTs is that walls are layered and the wall thickness is only a few layers of atoms (2–6 nm). Unlike the other two methods, it readily yields NTs with both ends open. Nevertheless, this method has not been proven effective for fabricating vertically aligned NTs on substrates like glass for applications such as solar cells.

Kasuga et al. obtained NTs by treating TiO_2 particles (crystallite size 6 nm) with 5–10 M NaOH aqueous solution at 110°C in an autoclave for 20 h [43]. The treated powder was washed with 0.1 N HCl aqueous solution and distilled water and then separated from the solution by centrifugation. The process of washing was repeated till the washing water showed a pH <7. The resulting NTs showed anatase phase of titania and had diameter 8 nm (inner diameter 5 nm) and length 100 nm. The NT-specific surface area was 400 m^2/g while the precursor powder had a surface area 150 m^2/g. Since then, different variations of this process have been investigated; however, the basic process that involves the hydrothermal treatment of a TiO_2 precursor dissolved in NaOH followed by washing in an acidic solution (typically HCl) or water largely remains the same. A hydrothermal temperature of 110°C–150°C is generally used for converting the precursor to crystalized tubular structures. The yield can be close to 100%.

The Structure and Morphology Evolution

There has been a significant debate over the crystal structure of the as-prepared NTs and their formation mechanism. Kasuga et al. reported HCl washing after hydrothermal process as critical to the NT crystalline phase formation [45]. After acid washing, NTs turned from amorphous to anatase phase regardless of the phase (anatase or rutile) of the precursor titania powder. Nevertheless, later studies showed that crystallization was indeed occuring along with morphology evolution during the hydrothermal process itself. Again, there were conflicting reports about the crystalline phase of the NTs. While some studies supported the claim that the NTs were crystallized in anatase phase, many observed a titanate phase rather than an anatase phase upon fabrication [46,47]. Figure 17.3 shows the images of NTs in anatase and titanate phases.

Among various titanate phases reported, monoclinic tritanic acid ($H_2Ti_3O_7$) phase was observed by serveral groups [44,47–50]. Chen et al. idenified the $H_2Ti_3O_7$ monoclinic structure in their NTs [47]. Huang et al. found x-ray diffraction (XRD) patterns of the NTs (8 M NaOH; 180°C; 12 h) matching with another protonic titanate, which is orthorhombic $H_2Ti_2O_5.H_2O$ [or $H_2Ti_2O_4(OH)_2$] [51]. The product was, however, predominantly monoclinic $NaTi_2O_4(OH)$ nanowires when the reaction time was

FIGURE 17.3 (a–c) Hydrothermally grown TiO_2 NTs. (a) SEM and (b) TEM images of a sample after washing only with water. Inset: selected area diffraction pattern showing anatase phase. (c) TEM image of a sample washed first with 0.1 M HCl aqueous solution and then with distilled water. Washing with acid did not change the phase (anatase) of the NTs. (Reprinted with permission from Ref. [46]. Copyright (2004) Cambridge University Press.) (d) High resolution TEM (HRTEM) image of a multi-walled $H_2Ti_3O_7$ NT with Fourier transform of the selected area (inset). The interlayer spacing is 0.78 nm. The structure has four layers on the upper side and five layers on the lower side indicating its formation by scrolling of layers. (Reprinted with permission from Ref. [47]. Copyright (2002) International Union of Crystallography.)

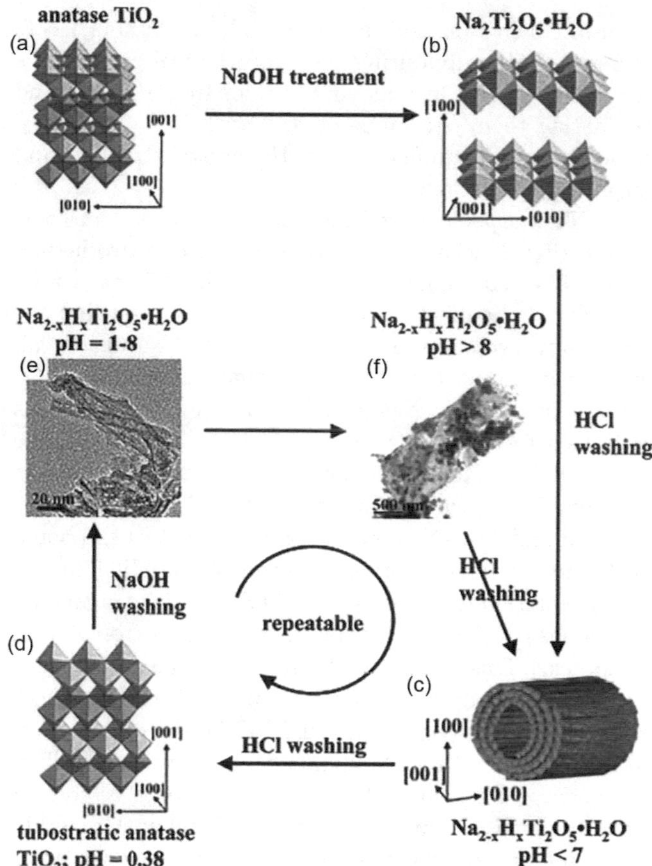

FIGURE 17.4 Scheme for the morphological and structural transformation of anatase TiO_2 nanoparticles with respect to sodium hydroxide treatment and post-treatment washing: (a–c) stages of transformation to titania NTs and (c–f) the cycle showing the conversion of titanate NTs to anatase TiO_2 particles and back to NTs by treating with solutions of appropriate pH. (Reprinted with permission from Ref. [53]. Copyright (2006) American Chemical Society.)

prolonged to 40 h. Yang et al. obtained $Na_2Ti_2O_4(OH)_2$ NTs after hydrothermal treatment (110°C, 20 h). The structure was converted to orthorhombic $H_2Ti_2O_4(OH)_2$ after HCl wash (pH 1) [52]. Tsai and Teng also found orthorhombic $Na_2Ti_2O_5.H_2O$ as the intermediate phase for NT formation, which would undergo an Na^+ exchange with H^+ in the post-treatment acid washing and form $Na_{2-x}H_xTi_2O_5.H_2O$ NTs [53]. The x value here would increase with decrease in pH and the phase would become anatase upon reducing pH to 1.6. Nonetheless, the NTs became defective due to dehydration-associated shrinkage during the process, and upon increasing the acidity, the NT structure was lost (anatase nanocrystals were formed). They could revert the anatase crystallites back to the titanate plates by washing with NaOH and finally to NTs by washing again with HCl. Figure 17.4 shows their scheme for NT formation and transformation. In another work, they observed NT formation when $Na_2Ti_2O_5.H_2O$-layered structure was washed with HNO_3, supporting the conclusion of Kasuga that the NTs are formed upon acid washing [54].

On the other hand, Wang et al. claimed that they could obtain anatase NTs upon hydrothermal treatment and the acid washing would not be required for NT formation [46]. The work of Poudel et al. showed that both synthesis and acid treatment conditions were important in deciding the morphology and phase [55]. They obtained both sodium titanate, hydrogen titanate and pure anatase depending upon the filling fraction in the hydrothermal cell. They concluded that acid washing is necessary for removing the impurities (sodium, for example) from the NTs and not necessary for the NT morphology. Deng et al. obtained brookite NTs using a double hydrothermal process [56]. They washed the layered $Na_xH_{2-x}Ti_3O_7$ NTs prepared using an initial hydrothermal reaction (120°C, 20 h) with dilute HCl to bring the pH value to the range from 5 to 12. This dispersion was used as a precursor for another hydrothermal treatment at 200°C for 1 day. The NT obtained from the precursors with pH < 7 was anatase. Nonetheless, brookite phase was observed for products from basic precursors. At pH $= 10.9$, pure brookte NTs were obtained. $H_2Ti_4O_9.H_2O$ and lepidocrocite $H_xTi_{2-x/4}\square_{x/4}\square O_4$ (where x is approximately 0.7 and the symbol $\square$ represents vacancy) are some other titananate phases reported in hydrothermal titania NTs [57–60]. In short, the fabrication and acid treatment conditions have strong influence on the NT morphology (discussed below) as well as crystal phase. Although NTs are formed primarily during the hydrothermal process, the acid washing helps in removing sodium from the NTs and protonating them (exchange of Na^+ with H^+). The pH of the acid solution is influensive in the crystal structure and morphology evolution.

The NT morphology evolution also came under considerable scrutiny. The layered growth is unique to hydrothermal titania NTs. The number of layers in the NT walls were found to be asymmetric, which gave clues regarding the morphology evolution. In an initial work, Kasuga et al. hyptothesized that during hydrothermal treatment of TiO_2 particles in aqueous NaOH, some of the Ti-O-Ti bonds are broken to form Ti-O-Na and Ti-OH bonds [45]. Some existing in the Na^+-O-Ti form could cause electrostatic repulsion. During dilute HCl wash, the charges gradually disappear and Ti-O-Na bond is converted into Ti-OH bond, which link with each other to form sheets. HCl in the washing solution dehydrates the Ti-OH bonds to create Ti-O-Ti bonds causing a reduction in the distance from one Ti to the next. This results in the folding of sheets. A slight electrostatic repulsion caused by some residual Ti-O-Na^+ bonds in the sheet can join the ends of the sheets to convert the folding into a tube. Nevertheless, as mentioned earlier, subsequent studies showed that the NTs could be formed during the hydrothermal treatment itself and washing could not notably influence the NT morphology unless the pH is too low [53].

Wang et al. who got anatase NTs after hydrothermal treatment (before washing) explained the formation of sheets as breaking of octahedra in the antase phase on reaction with NaOH and the linking of these free octahedras in a zigzag manner by sharing their edges with the formation of hydroxy bridges between the Ti ions [46]. These structures grow laterally into sheets through oxo bridges formed between the Ti centers. These sheets roll up to form NTs in order to reduce the high surface energy due to dangling bonds. Tsai et al. [53] noted that $Na_2Ti_2O_5.H_2O$ layered structures formed during hydrothermal treament would undergo conversion to NTs during acid wash (pH < 7). These sheets are formed by the intermediates Ti-O-Na and Ti-OH with the rearrangement of Na^+ and H^+ between the sheets. During acid washing, the ion exchange of Na^+ with H^+ causes surface charge fluctuations and scrolling of sheets to NTs. Most other reports show that Ti-O-Ti bond breaking, sodium titanate nanosheet formation, their exfoliation and curling up to NTs occur during the hydrothermal treatment [47,61–64]. Figure 17.5a depicts this mechanism for the NT formation [65]. The exfoliation is assumed to occur due to the hydrogen or Na^+ deficiency of the topmost tritanate layer [66].

While the general belief is that single sheets scroll up, some studies showed formation of onion structure indicating the scrolling up of multilayered nanosheets (see Figure 17.5b) [61,67]. Another model that is prevalent is the oriented crystal growth on a nanoloop seed [68]. Per this model, a very small amount of material is removed from the anatase crystallite forming terrraces on the surface. This material recrystallizes into a trititanate ($Na_2Ti_3O_7$) sheet, which curves up into a nanoloop of single-spiral, multiple spiral or onion-like cross section. This loop serves as a seed for oriented anatase crystal to grow and form NTs through the attachment of TiO_6 blocks released from the anatase raw material by the alkaline environment (see Figure 17.6). The authors believe that the sheet roll-up model can be applied only when local concentration fluctuations generate extreme conditions at the surface of the nanoparticles, and hence, their theory is more valid. Regardless, the rollup model is the most accepted one currently.

Influence of Hydrothermal Conditions

The hydrothermal parameters that were investigated primarily are the composition, morphology and crystalline nature of the titanium precursor, temperature and duration of the hydrothermal treatment, composition and concentration of alkaline medium, acid type and pH of the washing medium and post-fabrication annealing temperature [44,69]. Reports at the early stages of development of

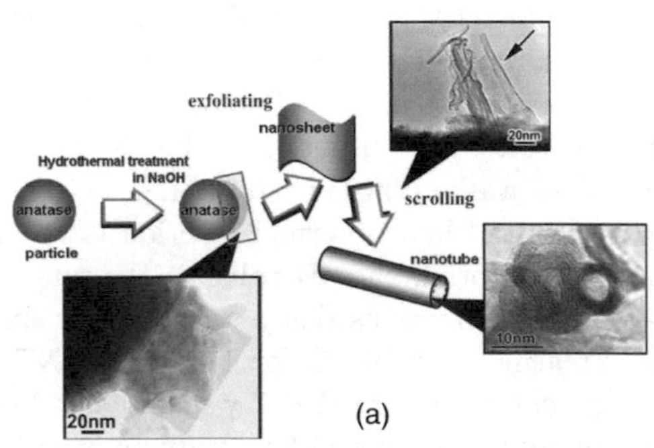

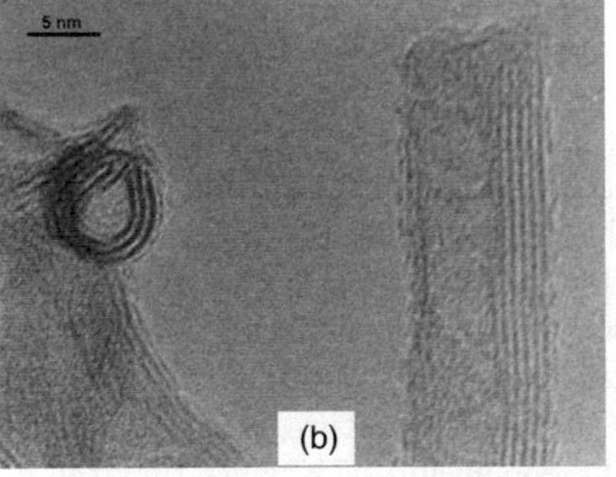

FIGURE 17.5 (a) A scheme showing the exfoliation of nanoparticles and scrolling of individual nanosheets to form NTs. (Reprinted with permission from Ref. [65]. Copyright (2010) American Chemical Society.) (b) TEM image showing the top (left side) and lateral (right side) views of NTs (left side) prepared by the hydrothermal treatment of anatase nanoparticles using NaOH at 120°C. The NT has higher number of walls on one side than the other. The top view shows a pronounced seam ("onion" type structure) indicating a multilayer nanosheet wrapping leading to NT formation. (Reprinted with permission from Ref. [61]. Copyright (2004) The Royal Society of Chemistry.)

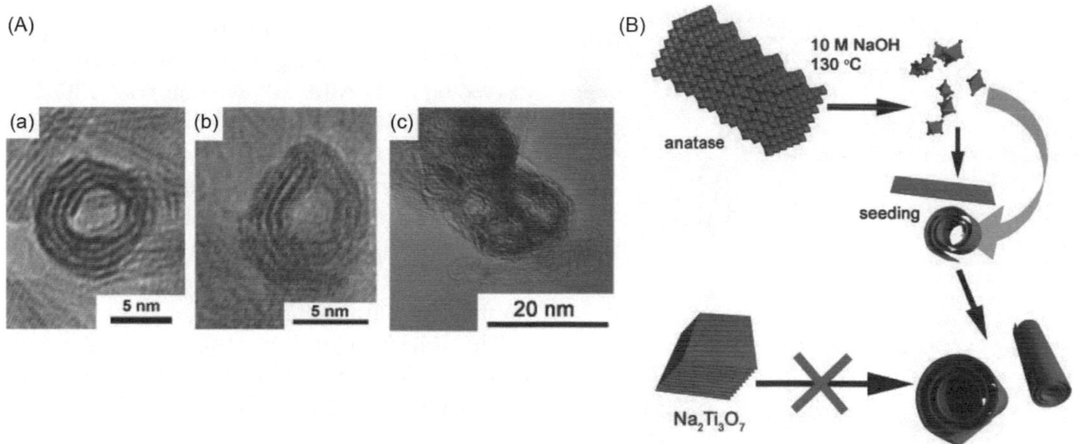

FIGURE 17.6 (A) TEM images of hydrothermally grown titanate NTs with (a) single spiral (b) onion-like and (c) entangled multiple spiral cross sections. The $Na_2Ti_3O_7$ NTs were prepared by treating anatase particles with 10 M NaOH at 130°C for 72 h and then washed with water. (B) A schematic representation of the NT formation mechanism. Nanoloops are formed from anatase starting material. These loops then serve as seeds for NT growth along length. (Reprinted with permission from Ref. [68]. Copyright (2005) The American Chemical Society.)

hydrothermal titania NTs emphasized the need of crystalline titania precursor powders for obtaining the NT structure [70]. Powders in anatase, rutile as well as mixed phases (Degussa P25) yielded NT morphology upon hydrothermal treatment in alkaline media [45,48,71–73]. Morgan et al. reported that the temperature window and NaOH concentration conditions required for obtaining the NTs would depend on the crystalline nature of the precursor powder. They developed "morphological phase diagrams" to show the temperature and NaOH concentration window for NT formation for anatase, rutile or Degussa P25 type starting powders (see Figure 17.7) [72]. Other studies show that the crystalline titania powder is not essential for obtaining NT morphology. Even an amorphous titanate [51] or molecular Ti^{IV} alkoxide [74] could yield NTs upon hydrothermal treatment in alkaline solution. Some reports indicate that the large size powder particles are necessary while others suggested that small particles are better. Liu et al. used a vapor phase growth using Ti foil as the starting material. They treated an NaOH-coated titanium foil using ammonia vapor at 150°C in an autoclave containing 28% ammonia solution for 1–3 days to get large diameter NTs (50–80 nm) [75]. Figure 17.8 shows the NT formed in 24 h. This work also demonstrated that nanosheet roll-up is the mechanimsm behind the formation of NTs (see Figure 17.8).

The temperature and medium used for hydrothermal treatment are critical for the successful growth of NTs. While the temperature window for getting NT morphology is generally 110°C–150°C, a combination of NaOH and KOH, however, yielded NTs at lower temperature than that used for pure NaOH [76]. Bavykin et al. studied the morphologies obtained by varying the fraction of NaOH in the mixed solution and mapped the temperature and solution fraction necessary for NT formation [77]. They

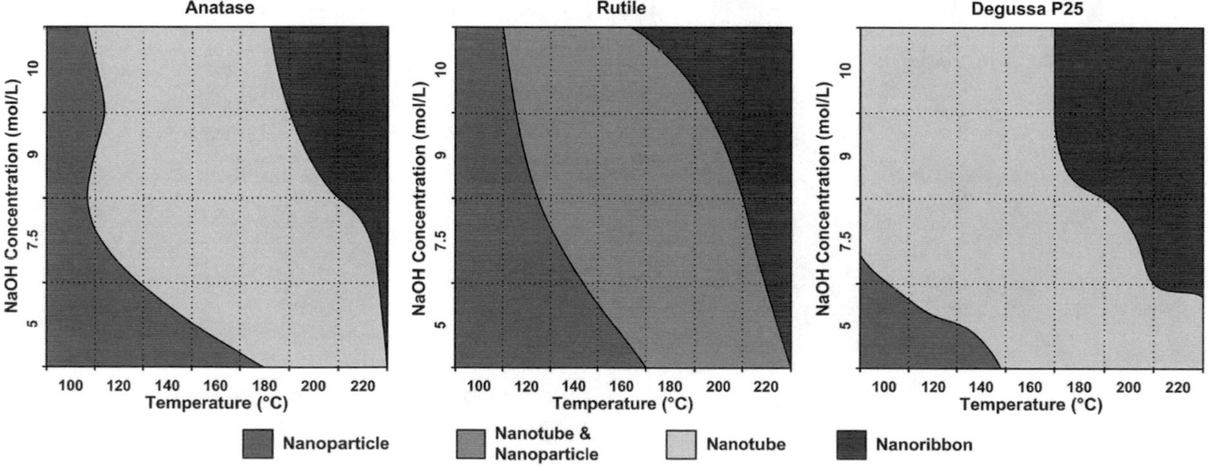

FIGURE 17.7 Morphological phase diagrams showing the shapes formed when anatase, rutile or P25 particles are treated hydrothermally with NaOH at different concentrations and temperature. The phase boundaries indicate the relative percentage of nanostructures formed within each condition. (Reprinted with permission from Ref. [72]. Copyright (2010) The American Chemical Society.)

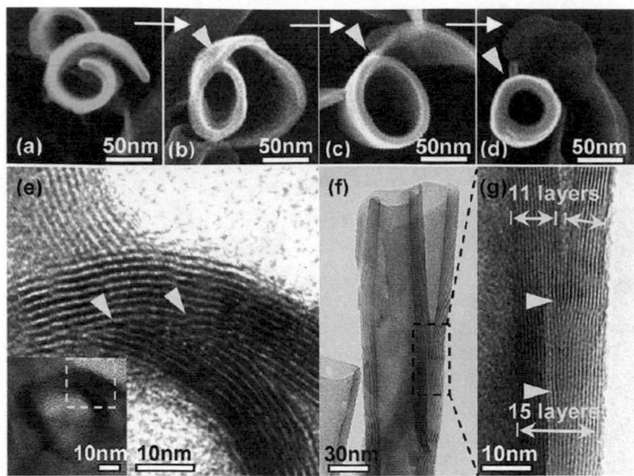

FIGURE 17.8 Titania NTs grown on titanium substrates using vapor phase hydrothermal method. (a–d) SEM images showing the morphological evolution of a nanosheet into an NT, (e) TEM image of a scrolled nanosheet observed along the scroll axis, (f) TEM image of an NT with a nanosheet attached to it and (g) a magnified image of the region marked in (f). (Reprinted with permission from Ref. [75]. Copyright (2011) The American Chemical Society.)

conducted reactions at atmospheric pressure. Figure 17.9a shows the TEM images of the nanostructures formed at various conditions when the reaction was conducted for 4 days. Figure 17.9b shows a map of the conditions used for getting the NT and other structures. At all NaOH fractional volumes >0.5, NTs were formed when the reaction was performed at 90°C. No NT is formed if the solution is pure KOH or any other solution. As indicated in the phase diagrams, the NTs become nanoribbons or fibers when the temperature exceeds about 180°C depending upon the NaOH concentration. The treatment duration also was varied from a few hours to several days (even up to 40 days). Other research groups demonstrated that pure KOH also could yield NTs. Sikhwivhilu tried LiOH, NaOH, KOH and NH_4OH as basic media and obtained NTs when NaOH or KOH was used as the hydrothermal treatment medium [78]. The NTs were grown when Degussa P25 particles were treated with either 5 M or 18 M KOH at 150°C. The titania NTs if treated hydrothermally again in an acidic medium, the tubes get converted into particles with or without a crystal structure change [79].

The time window for obtaining NTs depends on reaction conditions such as temperature, precursor amount and the alkaline solution concentration. When the time is too low, generally thin sheets or nanoparticles are the products and too long a reaction time results in structures such as nanofibers, nanoribbons and nanoparticles. For example, Sreekantan and Wei observed only nanosheets for a reaction temperature of 90°C for 24 h (NaOH toTiO_2 molar ratio ~80). NTs were found at temperatures 110°C–150°C; however, a minimum reaction time of 15 h was necessary for converting the precursor TiO_2 particles to NTs at 110°C. There are also studies that found NTs co-existing with nanoparticles and nanosheets after 4 h of reaction (10 M NaOH) in a sealed autoclave at 130°C [49]. The fraction of NTs increased and became complete after 36 h of reaction. Yu and Yu's experiment showed $H_2Ti_3O_7$ (titanate) NTs when the treatments were conducted at 140°C for 2–6 days (well-defined NTs after 6 days) [80]. The reaction time was reduced dramatically to about 1 h when microwave irradiation was used for the hydrothermal reaction [81]. Microwave power is an important parameter. Nevertheless, the determination of actual temperature is difficult.

The post-treatment washing is commonly done in dilute HCl as done by Kasuaga. HNO_3 or water also has been

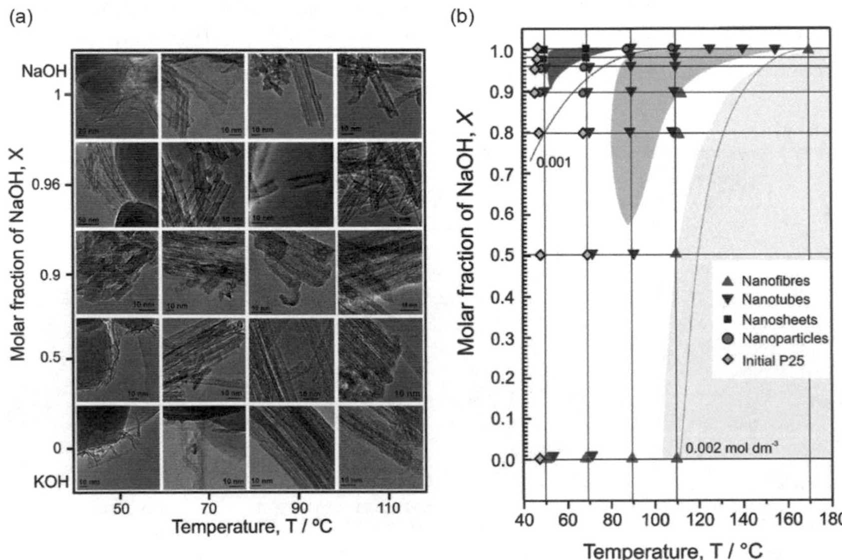

FIGURE 17.9 (a) TEM images of the nanostructures formed during the treatment of titania (P25) nanoparticles with NaOH–KOH mixtures for 4 days. (b) Morphologies of the products formed at different temperatures and concentrations of NaOH in the mixture. The three zones show the range of conditions favorable for the formation of titania nanosheets, NTs and nanofibers. (Reprinted with permission from Ref. [77]. Copyright (2010) The American Chemical Society.)

reported in place of HCl [80–83]. The washing is important as morphological evolution and/or elimination of sodium or structural water may happen during this process. The post-fabrication heat treatment is generally done at a temperature between 300°C and 500°C to remove the imperfections and transform the NTs to a highly crystalline state of the desired phase, typically anatase. Whether the initial phase is a titanate or anatase, anatase NTs can be obtained through heat treatment [84]. The anatase can be transformed to rutile at a temperature above this range. Deng et al. noticed that brookite phase in their NTs was stable even at 600°C except that the sample contained trace amount of rutile also [56]. Zavala et al. noticed that the anatase NTs were stable till 600°C but transformed to rutile particles at higher temperatures [85].

17.2.3 Anodic Oxidation

Background

Titanium dioxide is likely the first ever material discovered to form an NT array architecture, and electrochemical anodization of titanium (Ti) in a fluoride containing electrolyte is the method responsible for this. Assefpour-Dezfuly et al., in their article published in 1984, showed images of a tubular structure resulting from anodization of Ti in a chromic acid-ammonium fluoride (NH_4F) electrolyte; however, this work remained unnoticed till recently [86]. Anodization has been known since the middle of 19th century as a useful technique for the growth of protective oxide layers on metals such as aluminum, and fluoride containing electrolytes were studied widely for titanium anodization even before 1980s [87–92]. Perhaps, the NT architecture emerged in some of those earlier studies also, and the techniques used for imaging were not resolving the nanoscale features. With the development of highly ordered nanoporous aluminum oxide films by Masuda and Fukuda (1995), anodization started gaining popularity as a technique for growing ordered nanoarchitectures [93]. Four years later, Zwilling et al. studied the anodization behavior of Ti and its alloys in chromic acid–hydrofluoric acid (HF) electrolyte and reported nanoporous structures resembling NTs [94]. It was, however, with the demonstration by Gong et al. in 2001 that arrays of well-defined NTs could be fabricated using a simple electrolyte consisting of small amounts of HF in water and the subsequent publications revealing the NT properties and applications that the method received widespread attention [2,95]. This work proved that only the fluoride in the electrolyte and not the chromic acid was responsible for the growth of NT. Figure 17.10 shows the images of titania nanoarchitectures fabricated in the early stages of titania NT development [86,94,95].

Electrochemical anodization is a simple and scalable method useful for fabricating low-cost oxide films on surfaces of metals that form stable oxides. Environmentally friendly fabrication is possible with this method [96]. With the development of recipes to fabricate highly ordered nanoporous and NT architectures of different metals, it has become one of the most sought after techniques for nanomaterial synthesis. An account of NTs of various metals developed using anodic oxidation is given elsewhere [97]. The method yields NT arrays in the form of films on metal substrates as self-standing membranes as well as individual NTs in dispersed form. The NTs have the geometry of laboratory test tubes with an open top and a closed bottom (see Figure 17.10). Self-assembly of NTs into a vertical array is a unique feature of this method. The diameter of titania NTs grown by anodization can be typically varied from about 20 nm to 400 nm, length from a few tens of nanometers to a several hundred micrometers and wall thickness from 2 nm to about 100 nm. Furthermore, the spacing between the tubes can be manipulated. The array structure can be formed by close packed or well-spaced NTs. More importantly, large area NT films or membranes with the required quality and consistency for a commercial device can be fabricated using this method.

The process of anodization is performed in an electrochemical cell consisting of a positive (anode) and a negative electrode (cathode) dipped in an electrolyte (Figure 17.11). The metal on which the oxide layer is to be formed is used as the anode. Stable materials such as platinum, steel and graphite are generally used as the cathode in the cell. When a voltage is applied between the electrodes immersed in the electrolyte, the oxide starts forming on the metal anode. While chromic acid, sulfuric acid, etc. are traditionally used

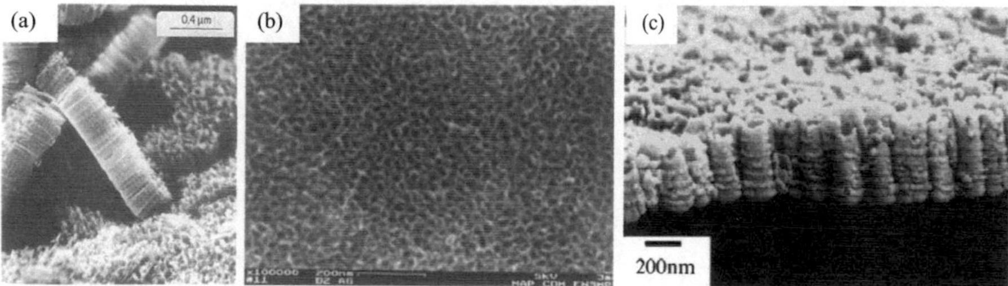

FIGURE 17.10 First-generation titania nanoarchitectures fabricated using anodic oxidation. The morphologies shown are obtained by anodizing titanium foil in aqueous (a) chromic acid–ammonium fluoride electrolyte at 10 V for 1.5 h (Reprinted with permission from Ref. [86]; copyright (1984) Springer Nature), (b) chromic acid–HF electrolyte at 5 V for 20 min (Reprinted with permission from Ref. [94]; copyright (1999) John Wiley and Sons) and (c) 0.5 wt% HF (only) at 20 V for 20 min. (Reprinted with permission from Ref. [95]; copyright (2001) Cambridge University Press.)

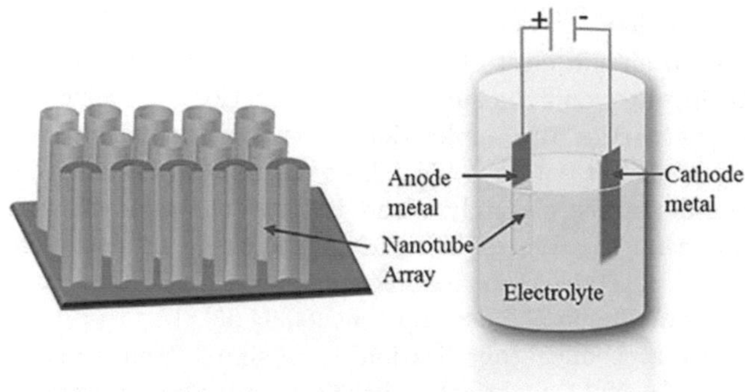

FIGURE 17.11 Sketch of a two-electrode anodization setup for fabricating oxide NTs. (Reprinted with permission from Ref. [97]. Copyright (2016) Cambridge University Press.)

as the electrolyte for compact or porous oxide formation in materials such as aluminum and titanium, highly ordered NT array formation typically requires a fluoride containing electrolyte (a chloride electrolyte was also reported to be useful in yielding NTs as discussed later). In all initial studies, aqueous electrolytes (either chromic acid containing fluoride or just dilute HF) were used [86,94,95]. While the anodization voltage in the first two cases was 10 V or below, which apparently prevented these structures from becoming fully developed NTs, Gong et al. used up to 40 V and obtained vertical NT arrays consisting of distinct NTs. They demonstrated that Ti foils anodized at constant anodizing voltages between 8 and 25 V in a 0.5 wt% to 1.5 wt% HF aqueous solution could yield NTs [4,95]. Low voltages (<10 V) used for anodization result in a porous film with a pore size a few nm to about 30 nm, similar to that of nanoporous alumina. At voltages exceeding 23 V, the NT arrays lose structure and resemble a porous spongy material. The duration of anodization for a fully grown NT structure is a few minutes to about 30 min depending upon the voltage. Later, Beranek et al. used H_2SO_4–HF electrolyte with a potential sweep from 0 to 20 V and Raja et al. used phosphoric acid–HF/NaF electrolyte (constant 20 V) to grow NTs [98,99]. All these titania NTs developed using aqueous fluoride electrolyte had their length limited to about 0.5 μm. These NTs were designated as first-generation titania NTs. Although the material showed promising properties for chemical sensing, the applications of this material were limited due to the low surface area of these short NTs [2].

Cai et al. pioneered the development of second-generation titania NTs with length in micron scale [100]. They developed a potassium fluoride (KF)-based aqueous electrolyte and varied the pH to grow NTs of different lengths. When the pH was increased from strongly acidic (pH < 1) to weakly acidic (pH 4.5), the NT length increased from 0.56 μm to 4.4 μm. This electrolyte composition yielded a maximum NT length of about 6 μm at pH 5 [100,101]. The images of NTs fabricated at different levels of electrolytes are shown in Figure 17.12 A pH >5 resulted in still

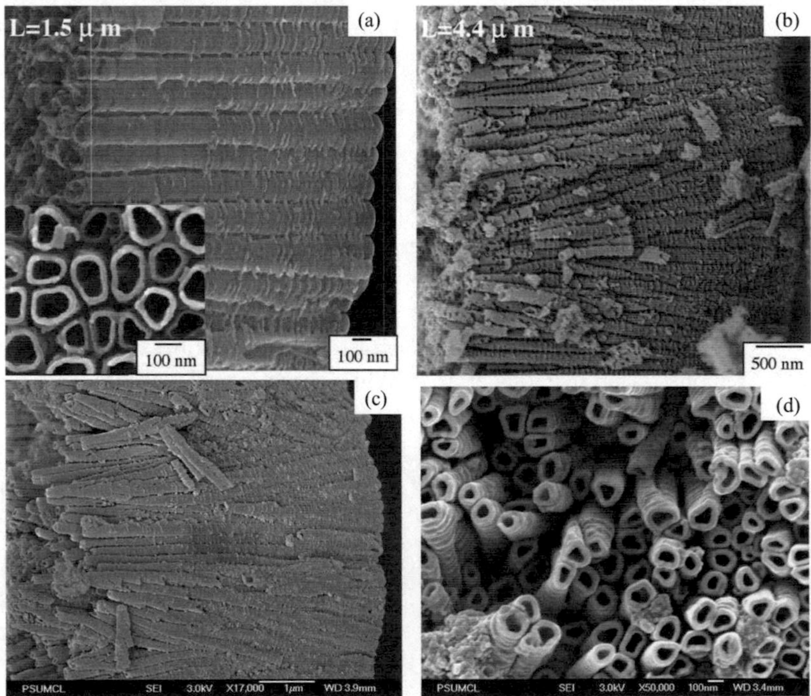

FIGURE 17.12 Titania NT arrays fabricated by anodic oxidation of titanium foil in aqueous electrolytes with different pH levels. Lateral views of NTs grown in a solution of (a) pH 2.8 at 25 V for 20 h, (b) pH 4.5 at 25 V for 20 h and (c) pH 5 at 25 V for 17 h (NT length 6 μm). (d) Top view of the NT of length 6 μm given in (c). (Images (a) and (b) are reprinted with permission from Ref. [100]. Copyright (2006) Cambridge University Press. Images (c) and (d) are reprinted with permission from Ref. [101]. Copyright (2005) Cambridge University Press.)

longer NTs but with undesirable hydrous titanic oxides as precipitates on top. The anodization voltage for NT growth was between 10 and 25 V. The pore size of the NT was determined by the applied potential and was independent of pH. A NaF-$(NH_4)_2SO_4$ aqueous electrolyte was introduced by Macak et al., and it could yield NTs of length up to 2.4 μm [102]. A similar NT length was attained in dimethyl sulfoxide (DMSO)–HF electrolyte used by Ruan et al., and this work became the first to successfully grow NTs using organic electrolytes [103]. HF–glycerol electrolyte was later introduced, but the NT length did not increase beyond 7 μm [104]. Finally, the major breakthrough came in 2006, when Paulose et al. invented that organic electrolytes consisting of NH_4F or HF in organic polar solvents such as ethylene glycol (EG), DMSO and formamide (FM) could yield NTs of lengths several tens to hundreds of microns [105–107]. Figure 17.13 shows the images of NTs of length up to 360 μm grown in EG-NH_4F electrolyte. The film prepared using EG electrolyte exhibited hexagonal close packing of NTs (Figure 17.13e). This electrolyte has since become one of the most extensively used electrolytes for the growth of these third-generation NTs. Later, Yoriya et al. introduced DEG and used in association with HF or NH_4F as the electrolyte for NT growth [108]. The interesting features of the resulting NTs were large tube-to-tube separation and crystallinity (anatase phase) in the as-fabricated condition. With all these innovations, NTs of almost any length can be fabricated using organic electrolytes, particularly using EG [107]. The development of third-generation NTs widened the application scope of titania NTs substantially.

Nakayama et al. succeeded in replacing fluoride electrolytes with a perchloric acid ($HClO_4$) solution [109–111]. The NTs were of diameter 18–23 nm, wall thickness 5–7 nm and length 20 μm. The diameter was significantly smaller than that of NTs fabricated using fluoride electrolytes. Richter et al. used NH_4Cl with oxalic acid, formic acid or sulfuric acid and NTs grew in all the electrolytes [112]. Nevertheless, the NTs grown using non-fluoride electrolytes exhibited disrupted NT array geometry and poor integrity compared to those from fluoride electrolytes.

Two major developments in the field are the fabrication of transparent films of titania on glass or other substrates and self-standing NT array membranes. Titania, due to the large bandgap (3.2 eV), is inherently transparent to visible light. Nonetheless, its visible light transmitting

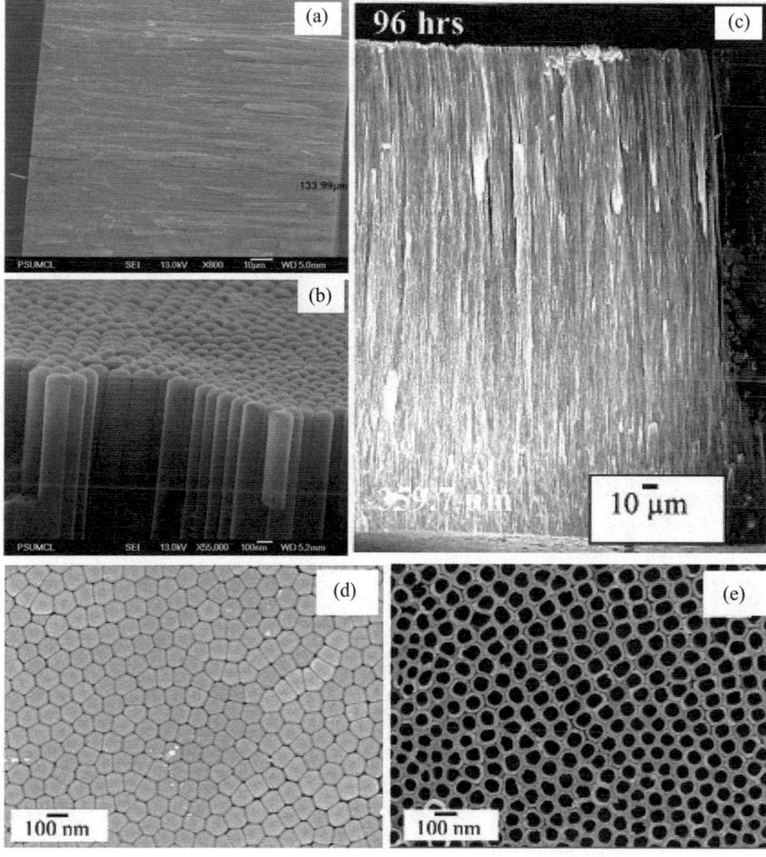

FIGURE 17.13 Field emission SEM images of titania NT arrays fabricated by anodic oxidation of titanium foil in EG–ammonium fluoride electrolyte at 60 V. (a and b) Lateral views of NTs obtained after 17 h of anodization (NT length ~134 μm) and (c) 96 h of anodization (NT length ~360 μm). (d and e) The bottom and top regions of the NTs, respectively. (Images (a) and (b) are reprinted with permission from Ref. [105]. Copyright (2006) American Chemical Society. Images (c), (d) and (e) are reprinted with permission from Ref. [106]. Copyright (2007) American Chemical Society.)

property could not be utilized for practical applications due to the presence of opaque titanium foil substrate. Growing titania NTs on transparent substrates like glass turned out to be a major challenge. This problem was solved by developing a process that involves depositing titanium films on a transparent substrates such as glass, anodizing it till the metal is almost entirely consumed and then heat treating it at high temperatures to oxidize the remaining metal and crystallizing the oxide NTs [113,114]. This opened doors for using the material in photovoltaics and similar applications [114]. Figure 17.14 shows the transmittance/reflectance spectra, an image of a 0.8 μm thick NT film on glass, the NT morphology and the selected area electron diffraction (SAED) pattern indicating the anatase crystal structure.

Fabrication of self-standing NT arrays and dispersed NTs by removing the substrate was another breakthrough work. Although NTs could be detached from the substrate using chemical etching or mechanical methods, fabrication of robust membrane became easy after EG-based electrolytes yielded thick closely packed NT arrays [105]. Prakasam et al. first reported the fabrication of self-standing NT array membrane [106]. Stress developed at the oxide metal interface during anodization enabled easy detachment of thick NT array films (a few tens of microns or higher) from titanium substrates. The membranes could be separated from substrates through mechanical, electrochemical or chemical means. For example, Paulose et al. dipped the as-fabricated NT array films bonded on Ti substrate in ethyl alcohol and subjected to ultrasonic agitation till the array film was separated from the substrate [107,115]. The closed bottom of the tube was opened by chemical etching. They demonstrated that the flow through membrane thus fabricated could be used for phenol red molecule separation. Figure 17.15 shows the digital images of the NT array film on Ti substrate and self-standing membranes detached from Ti substrates.

While hydrothermal method gives NTs with diameter and length in a narrow range, anodic oxidation offers tunability of the NT pore diameter, wall thickness, length, spacing and even morphology in a wide range. The size and morphology controlling factors primarily are voltage and electrolyte composition and pH, although factors like gas environment,

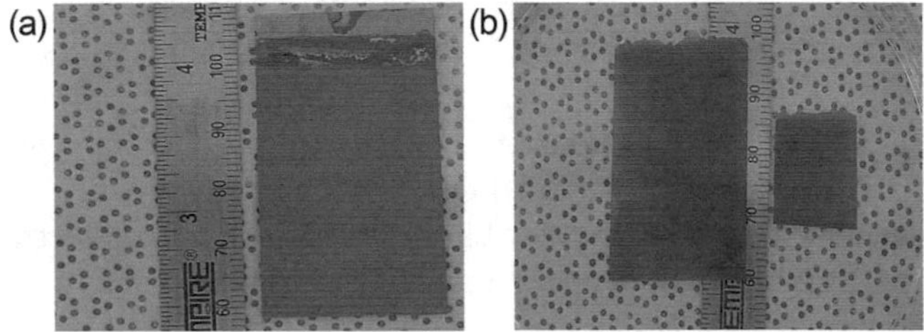

FIGURE 17.15 Photographs of (a) an as-prepared titania NT array film on titanium substrate and (b) flat membranes obtained after the removal of the titanium substrate. The membranes are kept in ethyl alcohol. (Reprinted with permission from Ref. [115]. Copyright (2008) Elsevier.)

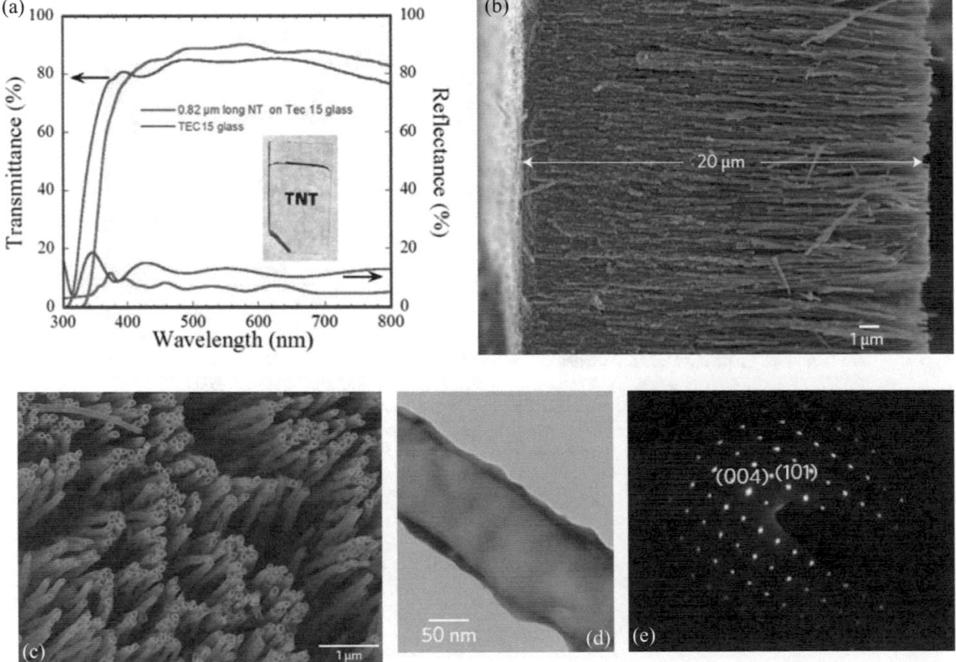

FIGURE 17.14 (a) Transmittance and reflectance spectra of a TiO_2 NT array film on fluorine-doped tin oxide (FTO) glass (TEC15) substrate with the corresponding spectra of the bare TEC15 substrate showing the antireflection properties of the NT film. Field emission SEM (FESEM) images showing the (b) lateral and (c) top views of a 20 μm thick NT array film. (d) TEM image and (e) selected area diffraction pattern of a part of an NT showing the anatase phase of titania. (Adapted/reprinted with permission from Ref. [114]. Copyright (2009) Springer Nature.)

humidity and temperature also could affect the process. By varying these parameters, NTs with various dimensions and morphologies were fabricated. NTs with conical shape, corrugated or small walls, close or loose packing and double walls are some examples.

Mechanism of Anodic Nanotube Formation

There are two major schools of thought regarding the mechanism of oxide layer formation during electrochemical anodization. One considers the porous film formation as a consequence of the interplay between three processes: the electric field-assisted oxidation of the metal, electric field-assisted dissolution of the oxide and chemical dissolution of the oxide [116–119]. This is generally called "field-assisted dissolution mechanism". Per this mechanism, the following processes can occur during titanium anodization in aqueous electrolytes: (i) Due to the interaction of the metal with O^{2-} ions from water in the electrolyte, oxide growth occurs on the surface of the metal. After the formation of the initial oxide layer, the anions migrate to the metal/oxide interface where they react with the metal. (ii) The metal ion (Ti^{4+}) at the metal/oxide interface is ejected from the metal/oxide interface under the applied electric field and migrates towards the oxide/electrolyte interface. (iii) Dissolution of the oxide at the oxide/electrolyte interface occurs due to the applied field. The Ti-O bond becomes polarized and is weakened promoting the dissolution of the metal cations. The Ti^{4+} cations are then dissolved into the electrolyte and the O^{2-} anions migrate to the metal/oxide interface to interact with the metal as seen in process (i). (iv) The chemical dissolution of metal or oxide by the acidic electrolyte.

Based on the electric field-assisted dissolution mechanism, Mor et al. proposed the first empirical model for the growth of titania NT architecture in aqueous HF electrolyte (Figure 17.16a) [120]. They studied the surface morphology change in the oxide layer with duration of anodization. According to the mechanism they proposed, pits are formed first in the initially grown compact oxide layer, which is followed by pore growth driven by the enhanced electric field at the pit bottom. Due to the relatively low ion mobility and high chemical solubility of the titanium oxide in the fluoride electrolyte, the pore walls could be thin, and this causes un-anodized metallic portions to remain in the inter-pore region. As the pores grow deeper, the electric field in the inter-pore metallic region increases enhancing the field-assisted oxide growth and dissolution in these regions. The inter-pore voids thus formed are grown along with the pores separating the pores and forming well-defined NTs. The authors later correlated the NT formation with the current–time characteristics and showed that the pore to NT conversion occurs during the second current decreasing phase (the region between P3 and P4 in Figure 17.16b) [113]. While the inward movement of the oxide layer at the bottom of the pore facilitates pore growth, the Ti^{4+} ions at the metal-oxide interface migrate outward to the oxide/electrolyte interface and dissolve in the HF electrolyte following either of the reactions below.

$$TiO_2 + 6F^- + 4H^+ \rightarrow [TiF_6]^{2-} + 2H_2O$$
$$Ti^{4+} + 6F^- \rightarrow [TiF_6]^{2-}$$

The process equilibrates when the rate of oxide growth and that of oxide dissolution are equal. The maximum thickness of the tube is reached when the chemical dissolution rate of the oxide at the top surface of the tube becomes equal to the rate of inward movement of the metal/oxide boundary at its base. A higher anodization voltage causes an increase in the rates of oxidation and field-assisted dissolution, this allows for a greater layer thickness of the NTs before the chemical dissolution reaches an equilibrium state. Small perturbations in the steady-state process associated with the slow material transport in the electrolyte region closer to the oxide surface could give rise to NTs with ring-shaped or corrugated walls [113,120].

The second school of thought is that a material flow mechanism is responsible for the porous oxide layer growth during anodization [121,122]. Per this model, widely known as "flow model", the flow of materials in the thin oxide layer under the pit towards the wall regions (the region between the pits) due to growth stresses and electric field-induced plasticity of the material is responsible for the pore formation. A theoretical work supporting this model was done by Hebert and Houser, and they considered the oxide as a Newtonian fluid with pores grown by the ionic migration in the oxide and stress-assisted diffusion of metal ions (see Figure 17.16c) [97,123]. In a later work, Hebert et al. ignored the plastic flow in thin oxide films and assigned the pore initiation to a morphological instability occurring in relation with oxide dissolution and nonlinear metal ion and oxygen ion transport [124]. Berger et al. argued that plastic flow could be responsible for the NT formation [125]. According to their hypothesis, a fluorine-rich layer is formed at the bottom of the pits/pores, which can then flow to the outer regions of pore walls and get dissolved in the electrolyte to create spacing between the pores and form NTs. In short, the flow model also has not been unambiguously verified. Su and Zhou believed that a titanium oxide hydroxide phase can grow at the oxide–metal interface and inter-pore regions and the dehydration of this phase leads to the separation of pores and formation of NTs [126]. Regonini et al. viewed the NT formation as due to the generation of cavities in the initially formed oxide layer due to the presence of fluoride ions [127]. The cavities align under the electric field and separate the pores to form NTs as depicted in Figure 17.16d. They attributed the ring formation in the NT wall to the disturbance caused by the evolution of oxygen bubbles at the anode and rib formation to the partial dissolution of the oxide rings. A number of models like these have been proposed for the anodic NT growth; nonetheless, the actual process of evolution of an initially formed porous oxide layer into NT array structure is still not clearly understood [128].

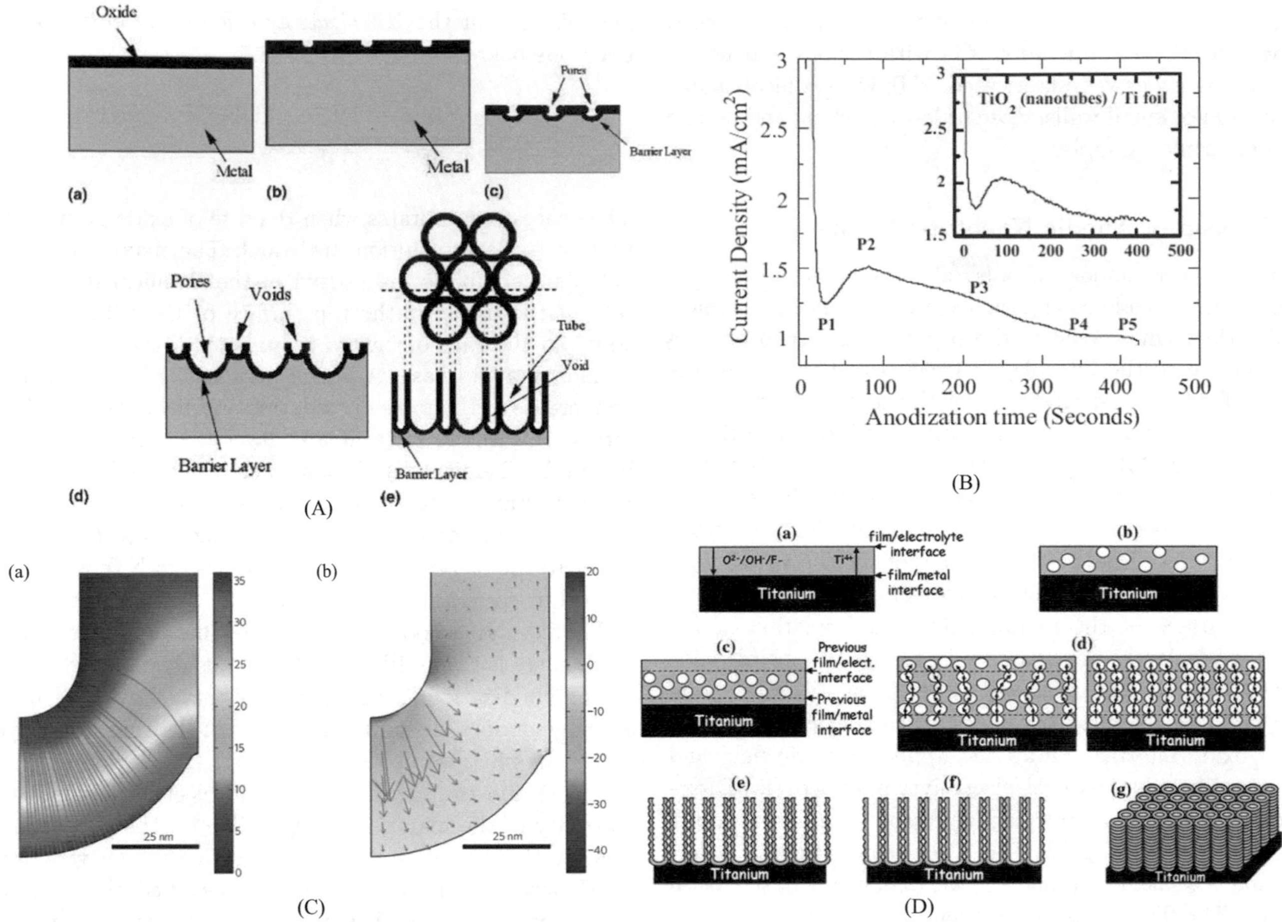

FIGURE 17.16 (A) Schematic diagram of NT formation process per field-assisted dissolution model: (a) oxide formation; (b) pit formation in the oxide layer; (c) growth of pits into scallop-shaped pores; (d) oxidation and field-assisted dissolution of the metallic region between the pores and (e) distinct NTs. (Reprinted with permission from Ref. [120]. Copyright (2003) Cambridge University Press.) (B) Variation of current during anodization of a 400 nm Ti thin film on glass and a Ti foil (inset) in an aqueous electrolyte containing HF and acetic acid at 10 V. Pits and cracks are formed in the initially formed oxide layer between points P1 and P2. A porous oxide structure is formed between P2 and P3. Transition of porous structure to NTs occurs between P3 and P4. Fully grown NTs can be obtained between P4 and P5. (Reprinted with permission from Ref. [113]. Copyright (2005) John Wiley and Sons.) (C) Simulated current lines and potential distribution for anodic film growth in oxalic acid electrolyte at 36 V (a) and corresponding velocity vectors and mean stress (b) supporting the "flow model". The color scale (gray in print) is potential in volts for (a) and dimensionless stress for (b). (Reprinted with permission from Ref. [123]. Copyright (2009) Springer Nature.) (D) Schematic diagram of the NT growth based on cavity formation. Anodic layer forms at the beginning and ions migrate through this layer under electric field (a); cavities are generated in the oxide layer by fluorine ions as the layer gets thickened (b and c); cavities orient under the action of electric field and link together (d) and development of NTs (e–g). (Reprinted with permission from Ref. [127]. Copyright (2007) Springer Nature.)

Crystallinity and Stability

The crystalline nature of the anodically formed nanoarchitectures and their thermal, mechanical and chemical stability are also interesting in a fundamental scientific standpoint as well as for device applications. Anodically formed oxides are generally amorphous. Thermal or chemical treatments make some materials crystalline. In the first comprehensive work on crystallization and high-temperature stability of titania NTs, Varghese et al. showed that titania NTs could crystallize in anatase phase at 280°C although the most stable phase of titania is rutile [129]. According to a model proposed by them, nucleation and growth of anatase crystallites in the walls of the NTs and that of rutile phase in the compact oxide layer (formed at the metal substrate–NT interface due to high temperature oxidation) take place quite independently above the crystallization temperature. At high temperatures (typically above 600°C), the anatase phase gets converted into rutile if NT walls fuse together and/or the rutile in the barrier layer grows and consumes the anatase phase. The NT morphology is lost in the case of short NTs generally at temperatures exceeding about 700°C. This model was proven correct later through experiments on transparent titania NT films and long NTs [113,130]. The NT films grown on glass substrates with almost no metal left after anodization showed only anatase phase after heat treatment at 500°C proving that the anatase phase was stabilized without an underlying metal layer. The oxidation of the metal is not substantial for NTs of length tens of microns or longer. X-ray diffractograms of such films provide only information from the top region of the NTs, and hence, the diffraction patterns need not show the rutile phase unless the annealing temperature is very high [130]. The high temperature stability is better for long NT films compared to short NT films.

Solvothermal/hydrothermal and other chemical treatments also can crystallize the anodically grown nanostructured oxides [131,132]. Titania NTs could be crystallized under hydrothermal conditions, but the morphology could be disturbed. Treating the crystallized NTs in ethanol vapor enhanced the crystallinity further without affecting the morphology [133]. It has also been shown that dipping the NTs inside water even at room temperature could induce crystallinity in NTs [134,135]. Nevertheless, the NT samples heat treated in conventional furnaces generally possess better mechanical, electrical and optical properties compared to those crystallized by other means.

17.3 Applications

17.3.1 Environmental Sensing

Sensors are an indispensable class of devices for today's technological world. They are used in common household to automobiles to space vehicles. These devices convert a physical or chemical information into a measurable signal and they are categorized accordingly. Titania, being a metal-oxide semiconductor, is well known for its chemiresistive properties. As in the case of metal oxides, such as tin oxide (SnO_2) and zinc oxide (ZnO), titania could reversibly change its resistance according to a chemical environment. Titania-based high-temperature chemiresistive oxygen sensors were extensively studied in the last few decades of the 20th century [136]. The devices are commercially used for various applications including automobile combustion control. The nanostructures of metal oxides brought unique functionalities to the devices [137]. Very high surface area provided by the nanostructures helped the development of chemical sensors with very high sensitivity to various gases and vapors. Titania NT hydrogen sensor with ultrahigh sensitivity is an example [138].

The pioneering work of Varghese et al. on the practical applications of titania NTs grown by anodic oxidation revealed that this material possessed remarkable characteristics as a chemiresistive hydrogen sensor [2,4]. The initial work performed using the first-generation anodic titania NTs grown using HF-H_2O electrolyte in the voltage range 12–20 V and annealed at 500°C showed that a 1,000 ppm hydrogen could cause a three-order reduction in the resistance of the NTs at 375°C [2]. Figure 17.17a and b shows the electrode configuration and sensor response. Upon removing the H_2 environment, the material could regain its original resistance. Platinum pressure contacts were used for electrical connection. The oxide layer formed at the interface of titania NT film and titanium substrate during the post-fabrication heat treatment served as a high resistive separation between the NTs and titanium (metal). The material was not responsive to CO or NH_3 gas. By reducing the anodization voltage to 10 V, NTs of length 200 nm, pore diameter 22 nm and wall thickness 13 nm could be fabricated, and these NTs changed the resistance

(a)

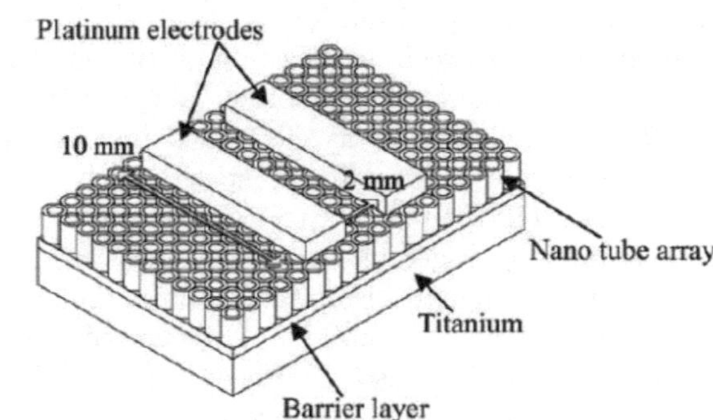

(b)

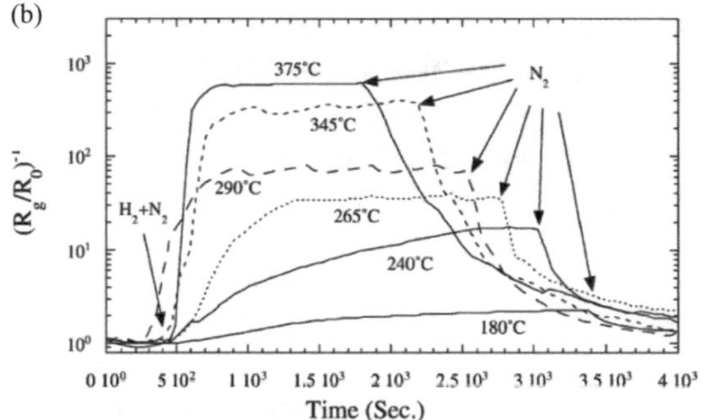

(c)

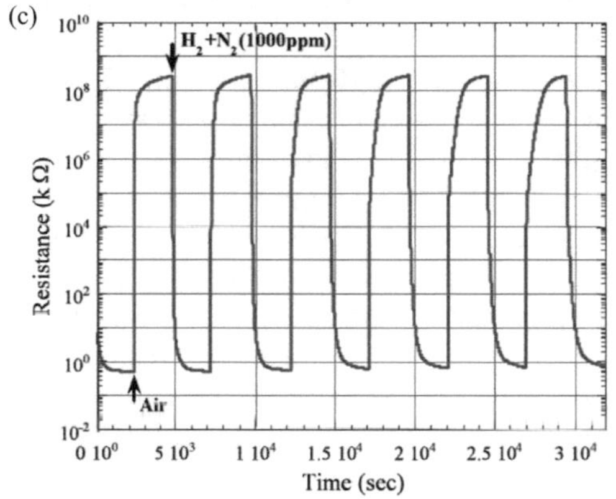

FIGURE 17.17 (a) An electrode configuration for NT-based hydrogen sensors. (b) Variation of normalized conductance of short NTs (~400 nm) grown in aqueous HF electrolyte with time when exposed to 1,000 ppm hydrogen at different temperatures. R_g and R_0 are the resistances in hydrogen–nitrogen mixture and pure nitrogen, respectively. (Reprinted with permission from Ref. [2]; copyright (2003) Elsevier.) (c) Room temperature resistance variation of an NT sample (length ~1 μm, pore diameter 30 nm and wall thickness 13 nm, grown in an electrolyte of pH 4 at 10 V an annealed at 480°C) upon cycling hydrogen (1,000 ppm) environment. (Reprinted with permission from Ref. [138]; copyright (2006) Institute of Physics Publishing.)

by four orders of magnitude (1,000 ppm H_2 in N_2) [4]. The interaction between hydrogen and titania NT was believed to be taking place through a spillover mechanism. Platinum films used as electrodes could dissociate hydrogen molecules and spill the species such as atoms (H^*) and ions (H^+) on the surface. The chemisorption of the spilled over hydrogen on titania NT surface, explained by a modified Langmuir isotherm, caused the dramatic decrease in resistance. Grimes et al. developed a wireless network using the NT-based hydrogen sensors [139]. Use of palladium as a catalyst (for hydrogen) on the surface made the NTs sense hydrogen at room temperature [140]. The sensitivity (change in resistance divided by baseline resistance) became even better and was 10^6–10^7 for 1,000 ppm hydrogen in an H_2-N_2 mixture. Titania NT hydrogen sensors showed exceptional self-cleaning properties also. The sensors that lost its hydrogen sensitivity completely due to a motor oil spill on the surface regained its original hydrogen sensitivity upon UV irradiation [141–143]. Titanium substrate beneath the NT films posed problems for devices by providing shunting paths. On the other hand, glass substrates provided a robust platform and the transparent NTs fabricated on glass were stable and responsive to hydrogen [144].

The advent of second- and third-generation anodic titania NTs made the development of hydrogen sensors with record sensitivity possible [100,105]. Paulose et al. demonstrated that a bare 1 μm thick NT array film (NT diameter 30 nm, wall thickness 13 nm) could offer about nine orders of magnitude change in resistance to 1,000 ppm H_2 in air [138]. Figure 17.17c shows the response. Remarkably, the material was selective to hydrogen (negligible sensitivity to reducing gases such as methane and CO) even without any hydrogen-specific catalyst such as Pd on the surface. The high surface area, low thickness of the walls (in the range of Debye length), tube interconnects and highly active surface states (see Figure 17.17c) were believed to be responsible for this unique sensing behavior. Increasing the NT length beyond a few microns was reported to have an adverse effect on the sensitivity [145]. Subsequently, devices for monitoring diseases were developed using this technology [146]. The devices were able to monitor hydrogen evolved out of skin. Studies on patients with lactose intolerance showed an excellent agreement between transcutaneous hydrogen levels measured using the sensors and breath hydrogen levels determined using gas chromatography.

Since these seminal efforts, a number of studies on gas sensing properties of titania NTs have been reported. Majority of the reports are on anodically grown titania NTs. Alev et al. studied the gas sensing characteristics of p-type Co_3O_4 loaded on anodic TiO_2 NTs using cathodic deposition [147]. They observed that after heterostructuring, the sensitivity to hydrogen increased while that to NO_2 (oxidizing gas) and volatile organic compounds (VOCs) such as acetone and toluene (reducing gas) decreased at 200°C. More recently, this group fabricated p-CuO/n-TiO_2 NT heterostructure by evaporating Cu on the NT film and thermally oxidizing it. The use of CuO loading improved the sensor specificity to hydrogen [148]. On the other hand, carbon doping was shown to reduce the hydrogen sensitivity [149]. SnO_2 loading on titania NT (length 3–4 μm, pore diameter 100 nm, wall thickness 10–20 nm) by soaking NTs in a $SnCl_2.H_2O$ solution improved the high-temperature hydrogen sensitivity [150]. Hydrogen sensing characteristics of titania NTs were studied by various other groups also [151–156].

Lu et al. demonstrated that amorphous titania NTs (2 μm) fabricated by anodic oxidation could sense low-concentration oxygen at 100°C [157]. Figure 17.18 shows the image and response. The anatase NTs showed much lower sensitivity to oxygen. Titania NTs were also shown to be sensitive to low concentrations of H_2S. Tong et al. fabricated titania NTs on Ti foils and transferred the NT film onto an alumina substrate with patterned gold electrodes (see Figure 17.19a–c) [158]. The resistance of a 3.8 μm long NTs (pore—110 nm and wall 16 nm) showed a 12-fold reduction in the presence of 10 ppm H_2S at 300°C. Response and recovery characteristics were also fast (22 s and 6 s, respectively). Earlier, Perillo and Rodriquez had fabricated anodic NT films on Kapton™ tape and found it sensitive to H_2S (6–38 ppm) at 70°C [159]. In separate articles, the team also reported the trimethylamine sensitivity of NTs on Kapton™ tape (at 70°C) and room temperature chloroform sensitivity of NTs (length ~2 μm and pore ~200 nm) on foil substrates. Jang et al. also used Kapton™ tape as the substrate for anodic titania NT growth and their work showed that the films were sensitive to CO and NH_3 at 350°C [160]. Galstyan et al. doped TiO_2 NTs with Nb [161]. At a temperature less than 200°C, the response to CO increased and that to acetone decreased as Nb concentration increased. At a higher temperature, ethanol sensitivity increased and CO and acetone sensitivities reduced. Comini et al. converted metallic Nb-Ti films into highly conducting Nb-TiO_2 NTs using anodic oxidation [162] and studied the gas sensing properties at 400°C. Compared to 50 nm and 75 nm pore size NTs, the 30 nm pore size NTs showed a higher response to ethanol. Zhang et al. used bare NTs to detect SO_2 as a dissociation product of SF_6. At temperatures above 200°C, the material showed a 76% sensitivity to 50 ppm SO_2 [163]. Sennik et al. demonstrated that anatase NTs detected alcohols better than rutile NTs [164]. High ethanol sensitivity (at 250°C) of short NT was reported by Kwon et al. [165]. The room temperature formaldehyde sensing characteristics of anodic titania NTs were reported by Lin et al. [166]. Titania NTs were also reported to have sensitivity to chloroform, trimethylamine and several organic compounds [167–169]. The titania NTs fabricated using HF-H_3PO_4 electrolyte and annealed at 600°C were reported to have high humidity sensitivity (Figure 17.19d) [170]. As the environment was changed between 11% and 95% relative humidity (RH), the response and recovery times were 100 s and 190 s, respectively. The rutile phase present in these samples was believed to have promoted the sensitivity to water molecules.

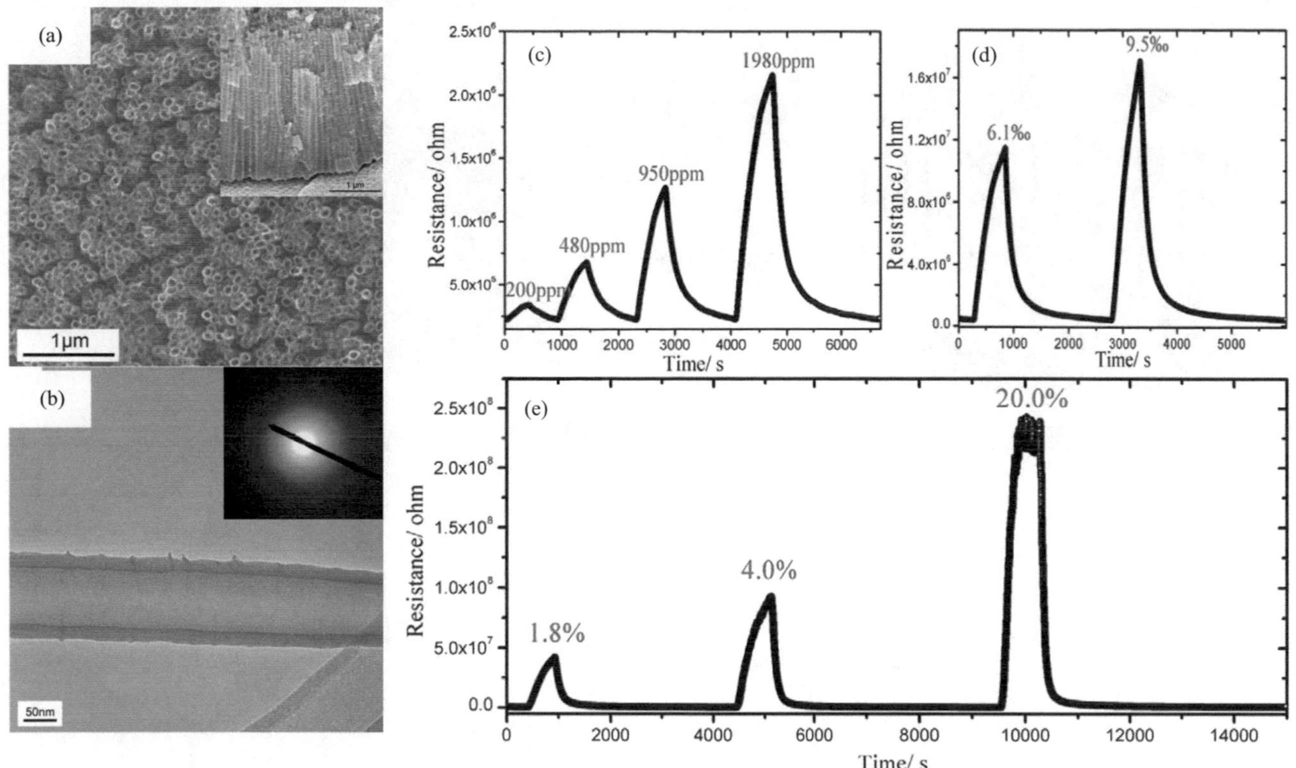

FIGURE 17.18 (a) SEM and (b) TEM images of amorphous titania NTs (as-prepared) grown by anodic oxidation in an electrolyte consisting of NH_4F and $(NH_4)_2SO_4$ in water at 20 V. Resistance variation of the NTs when exposed to (c) 200–1,980 ppm, (d) 6.1%–9.5% and (e) 1.8%–20.0% oxygen at 100°C. (Reprinted with permission from Ref. [157]; copyright (2008) Institute of Physics Publishing.)

A few efforts on gas sensing properties of NTs prepared by other two methods also have been reported. For example, Han et al. studied the effect of Pd and Pt catalysts on the hydrogen response of hydrothermally prepared titanate NTs [171]. They found that even dispersion of nanoscale Pd and Pt catalysts on titanate NTs could help improve the hydrogen adsorption and oxidation and, hence, the sensing characteristics. Seo et al. studied the toluene sensing properties of titania NTs prepared by hydrothermal method [172]. They demonstrated that upon reducing the length of the NTs by ball milling, the material could sense VOCs, particularly toluene, better than commercial TiO_2 nanoparticles at 500°C [173]. Figure 17.19e–h shows the images of NTs and the effect of ball milling time on the response of the sensor to toluene. In the NTs (250 nm long, pore diameter 70 nm) developed using ALD growth in porous alumina templates, Lee et al. observed a two-order change resistance to hydrogen at 100°C [174]. Lu et al. decorated the titania NTs obtained by templated method with Al_2O_3 particles and demonstrated that the material was responsive to NO_x in low ppm range [175].

Use of titania NTs for the detection of toxic heavy metal ions and organic contaminants in water and soil also have been reported [176]. For example, DNA-modified titania NT sensor was used for the detection of Pb^{2+} in water [177]. The process involves applying Pb^{2+} to the senor via submersion, then rinsing the sensor and transferring it to an electrolytic cell without Pb^{2+}. A differential pulse anodic stripping voltammetry method was then utilized to determine the Pb^{2+} concentration. Another metal that could be detected by titania NTs is arsenic. This metal requires both rapid and sensitive detection. Yang et al. developed an Au-modified titania NT as a sensor for arsenic [178]. The composite nanomaterial yielded a high surface area and was sensitive to As in μg scale. The Au/NT showed a sensitivity of almost 2.5 times bare NTs. A polymer-modified TNT sensor was reported to be useful for the detection of perfluorooctane sulfonate in water samples [179]. Electrocatalytic reduction process was employed for the detection. The detection limit was 86 ng/mL. Additionally, the development of an octachlorostyrene (OCS) photoelectrochemical immunosensor (detection limit – 2.58 pM) by cross-linking anti-OCS antibodies onto CdTe/CdS-sensitized TNTs was also reported [180].

In short, as a sensor material, titania NT has exhibited tremendous potential. The application scope of the material in the sensor field is still expanding.

17.3.2 Energy Storage Devices

Battery

Titania NTs have been extensively studied for use as a battery electrode material. Most studies have been aimed at improving the characteristics of lithium-ion batteries (LIBs). LIBs have become the front runner for energy storage; these

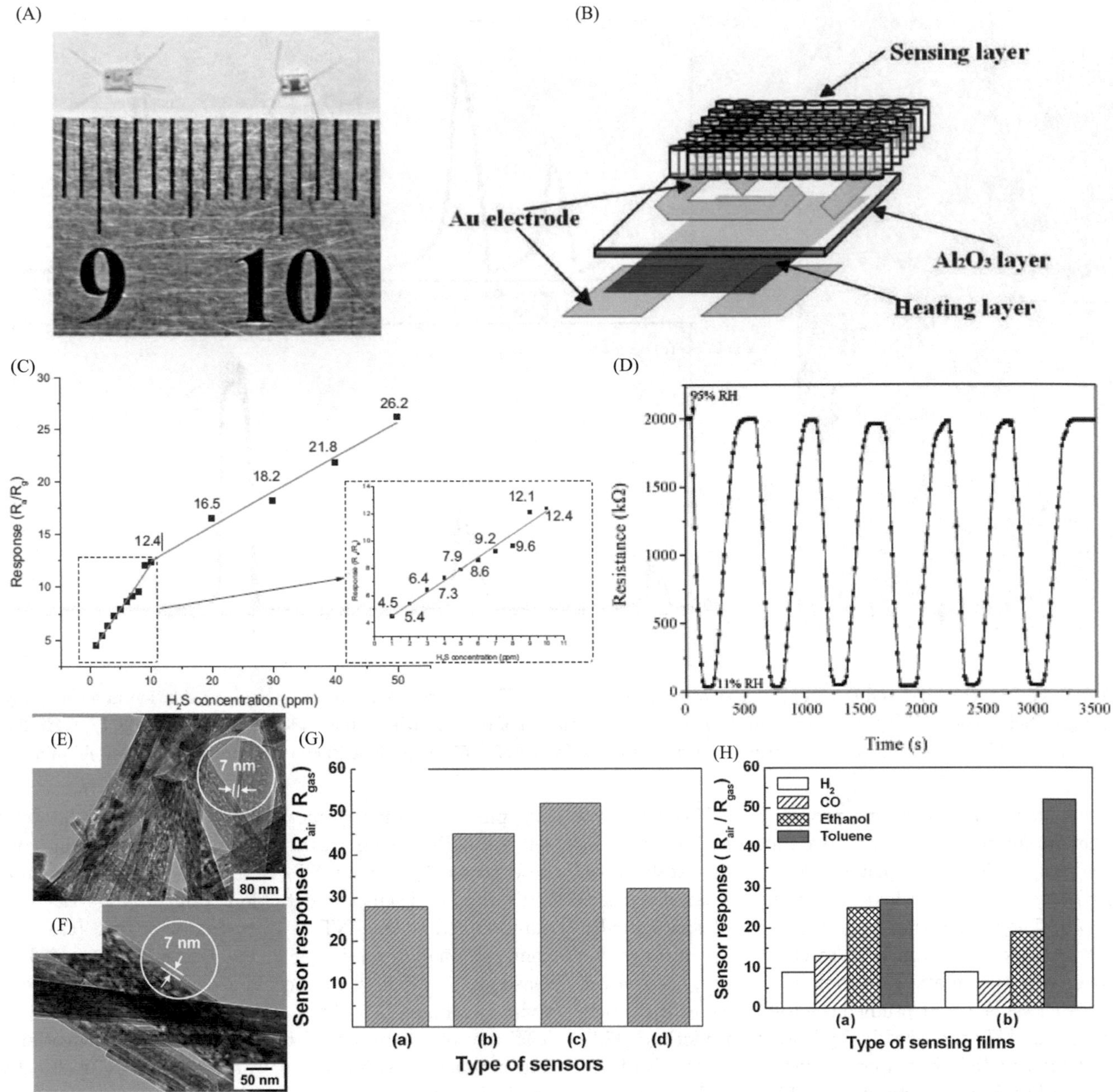

FIGURE 17.19 (A) Photograph and (B) schematic diagram of the configuration of H_2S sensors fabricated using membranes of anodic titania NT arrays. (C) Response of the sensor to H_2S at 300°C. (Reprinted with permission from Ref. [158]; copyright (2017) Elsevier.) Humidity response of an anodic titania NT sample grown in H_3PO_4-HF electrolyte and annealed at 600°C. (Reprinted with permission from Ref. [170]; copyright (2008) Elsevier.) TEM image of titania NTs grown by hydrothermal treatment: (E) as-prepared and (F) those subjected to ball milling for 3 h. (G) Response of (a) commercial TiO_2 nanoparticles and (b) 1 h, (c) 3 h and (d) 5 h ball-milled NTs to 50 ppm toluene at 500°C. (H) Response of (a) commercial TiO_2 nanoparticles and (b) 3 h ball-milled NTs to H_2 (500 ppm), CO (500 ppm), ethanol (47 ppm) and toluene (50 ppm). (Reprinted with permission from Ref. [173]; copyright (2011) Elsevier.)

are being used as the primary power sources for electric vehicles and portable devices. The properties that make them so valuable to the modern world are their long life cycle, high energy density, low manufacturing cost and design flexibility. Batteries convert chemical energy to electrical energy, through electrochemical reactions between an anode and a cathode separated by an electrolyte creating a potential difference between the electrodes. The main research focus is on developing a low-cost, low-mass, high-performance anodes. Graphite has been primarily used as the anode material; however, it suffers from problems such as structural collapse and exfoliation over cycling resulting in rapid capacity decay as well as temperature sensitivity. Furthermore, the Li intercalation in graphite happens near the Li^+/Li redox potential, and this could lead to Li plating during charging, triggering safety hazards. Metal oxides like titania are suitable alternative anode materials to graphite [181].

Titania is stable and has a safe Li intercalation voltage, and hence, it has been considered as a promising material for LIBs. Compared to other oxides, titania shows low reversible capacity; however, it offers better cycling stability and Coulombic efficiency. Storage of Li ions into the TiO_2 framework is reduced due to low Li^+ diffusivity in TiO_2 resulting from the strong repulsive forces among the ions. Li ion insertion and removal in TiO_2 is a diffusion-controlled process that depends on the diffusion coefficient and diffusion length of TiO_2. While the diffusion coefficient is determined by the nature of the materials, the diffusion length is dependent on particle size. TiO_2 NT arrays show promise for decreasing the diffusion paths of Li ions in these batteries. The ability to tune the TiO_2 NT dimensions in producing well-ordered high surface area NTs and array films have opened up various strategies for reducing weight and improving the performance of LIBs.

Using hydrothermally prepared titania NTs, Zhou et al. apparently performed the first tests of titania NTs in LIBs [182]. Their study showed that Li ion intercalation/deintercalation could occur reversibly in the titania NT electrode. Their electrochemical cell consisted of a titania NT–acetylene black (15%)-teflon binder (5%) composite anode and a Li metal cathode dipped in a 1.0 M $LiPF_6$ in ethylene carbonate/dimethyl carbonate electrolyte. The cell separator was Celgard 2,400 membrane. The Li ion insertion occurred between 2.2 and 1.4 V. The cell yielded a maximum rate capacity of 195 mAh/h. Armstrong et al. found that in TiO_2(B) NTs synthesized using hydrothermal treatment of anatase particles in 15M NaOH at 150°C for 72 h and heat treated at 400°C could intercalate lithium up to a composition of $Li_{0.9}TiO_2$ [183]. Tang et al. prepared NTs of length tens of microns (max ~30 μm) by stirring the solution during hydrothermal treatment, and these long NT-based LIBs yielded a capacity of 368 mAh/g initially, which reduced to 227 mAh/g after 100 cycles. Even at a current rate of 25°C, the capacity of these cells could reach 147 mAh/g [184]. The process schematic and battery performance are given in Figure 17.20. Wang et al. tested titania NTs with mesoporous walls for LIB application (see Figure 17.21) [185]. They prepared NTs using a porous alumina template-assisted method. A sol consisting of titanium isopropoxide and structure directing triblock copolymer $EO_{20}PO_{70}EO_{20}$ was used as the titanium source.

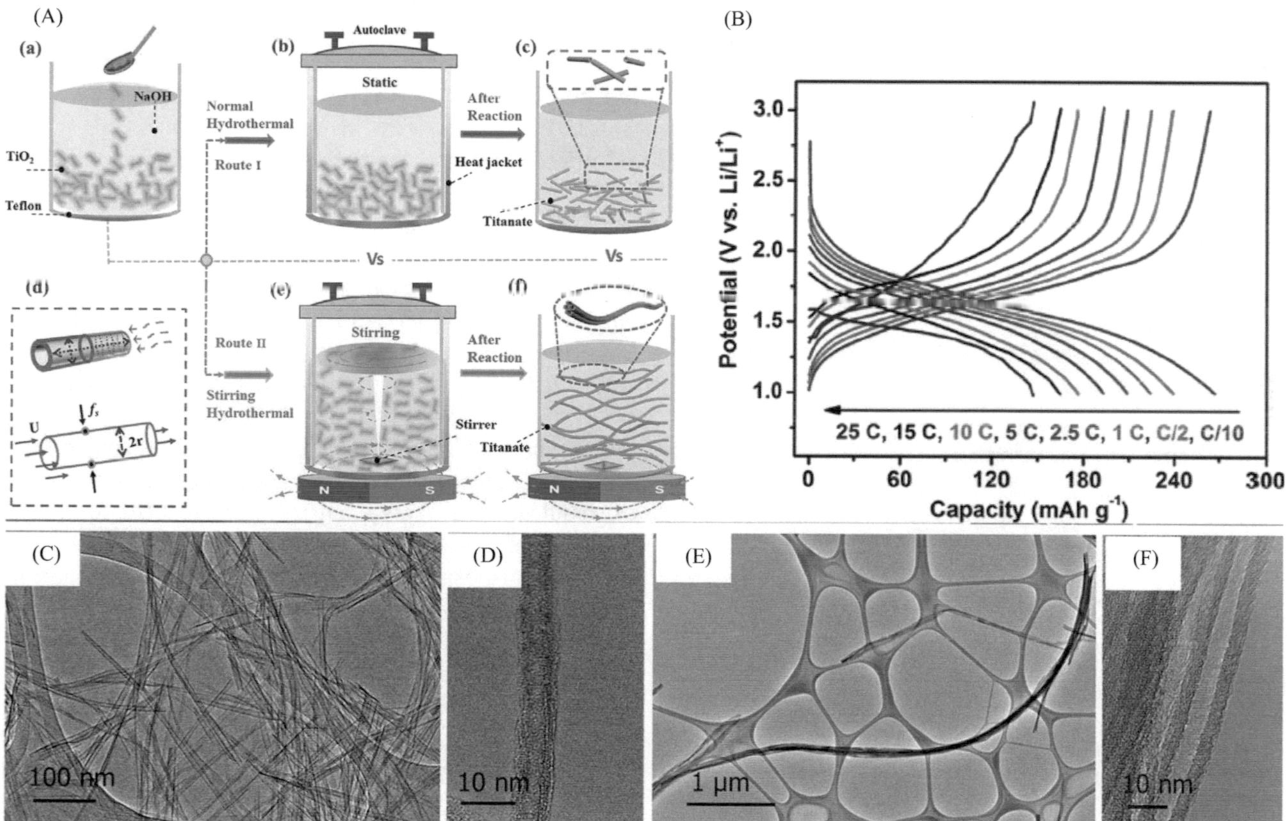

FIGURE 17.20 (A) Schematic representation of the formation of short and long NTs under normal (static) and stirring hydrothermal process (at 130°C for 24 h): (a) TiO_2 nanoparticles dispersed in aqueous NaOH, (b and c) formation of short NTs under static conditions, (e and f) formation of elongated NTs under stirring and (d) a model for the growth of NTs along axial and radial directions and the force analysis of an individual NT formed in (e). U is solution velocity, f_s the side force and r is the diameter of the tube. (B) Electrochemical performance of the elongated NT electrodes: discharge curves at different current rates of C/10 – 25 C. TEM images of (C and D) NTs formed under static conditions and (E and F) those formed at a stirring rate of 500 rpm. (Reprinted with permission from Ref. [184]; copyright (2014) John Wiley and Sons.)

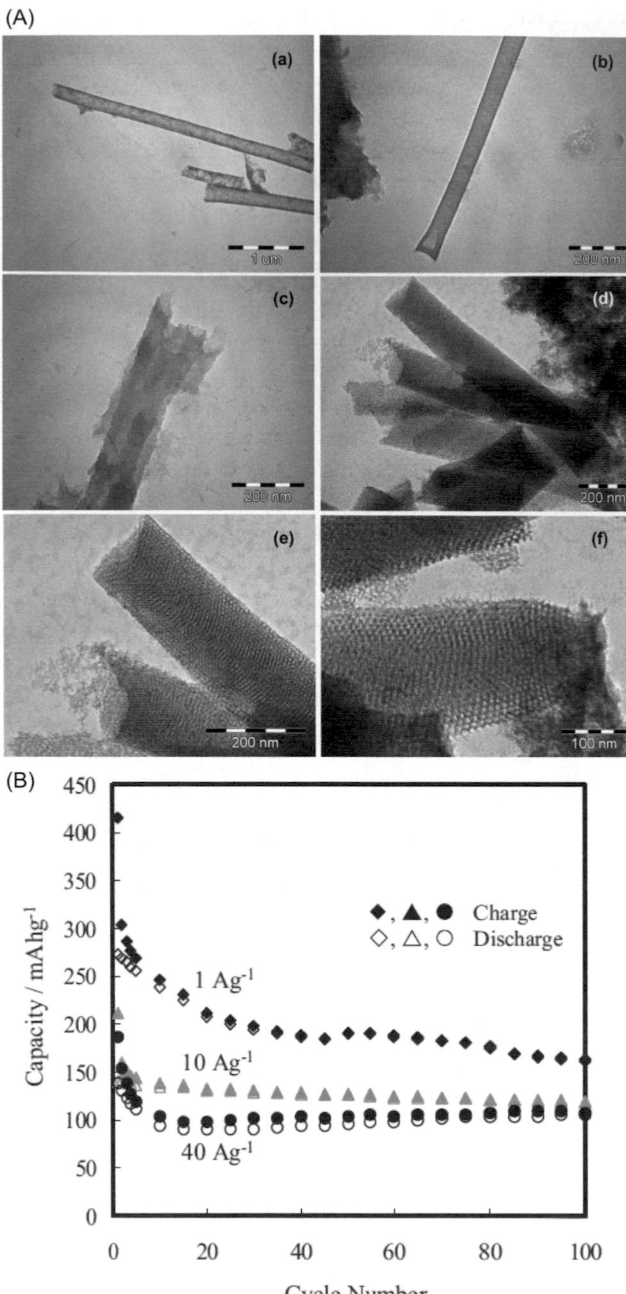

FIGURE 17.21 (A) TEM images of 150°C annealed TiO_2 NTs fabricated using template-assisted process. The images of (a) 200, (b) 100 and (c) 50 nm diameter NTs and the NTs calcined at 450°C revealing hexagonal mesoporous walls (d–f) are shown. (B) The performance cycle of charge and discharge capacities of mesoporous titania NTs. (Reprinted with permission from Ref. [185]; copyright (2007) John Wiley and Sons.)

The electrode made of mesoporous titania NTs showed high rate of charge discharge, which was attributed to the 3D network structure of mesoporous titania. A stable capacity of 162 mAh/g was obtained at a current density 1 A/g after about 100 discharges (Figure 17.21).

Liu et al. reported use of anodic titania NTs in LIBs [186]. They used NTs of length ~1 μm, pore size ~90 nm and wall thickness ~15–20 nm prepared by anodizing titanium foil in 0.1 M KF–1.0 M $NaHSO_4$ electrolyte at 20 V for the study. The cells constructed using 300°C annealed NTs as anodes showed a discharge capacity of 240 mAh/g initially but reduced to 160 mAh/g after 50 cycles (current density 320 mA/g). On the other hand, those with NTs annealed at 400°C delivered almost a constant capacity of 160 mAh/g during cycling. Higher annealing temperature (500°C) reduced the capacity. The work of Wei et al. showed that the capacity of titania NT-based LIBs would increase with NT length or wall thickness [187]. The LIB consisting of NTs of length up to 9 μm was stable for 500 cycles (capacity 0.46 mAh/cm^2 or 184 mAh/cm^2 at current density 0.05 mA/cm^2). They demonstrated that the freestanding membrane NTs also could be as the anode. These membrane-based LIBs exhibited cycle stability over 500 cycles maintaining 94% of the original capacity. Ivanov et al. preformed experiments on the phases of TiO_2 NTs and determined that an amorphous structure promotes a higher diffusion rate of the lithium ion [188]. The disorders and defects found in the amorphous state offer larger channels and more diffusion paths for the Li ions to travel; additionally, the lower mechanical stress encourages intercalation in larger amounts [189]. Although several studies have been reported on bare NTs prepared by all three methods (hydrothermal, template assisted and anodic oxidation), the stable capacity was generally less than 250 mAh/g [186,190–193].

Various strategies have been tried to improve the characteristics of titania NT as an LIB anode material. For example, Ag loading on anodic NTs was found to be improving the charge–discharge capacity [194]. The strategy of coating Li_3PO_4 on anodic titania NT was also reported, but the capacity of these LIBs was less than bare NT-based LIBs [195,196]. Annealing atmosphere of the TiO_2 NTs plays a role in the electrochemical storage properties. Annealing in different atmospheres affects the morphology and crystalline phases of the TiO_2 NT arrays. Annealing in argon versus air resulted in different morphologies; however, electrochemical tests showed better performances for the Ar-annealed samples [197]. Lu et al. used H_2/95% Ar as an annealing atmosphere for TiO_2 NTs [198]. The improved electrochemical rate performance was caused by the large number of resulting oxygen vacancies in the TiO_2 crystal lattice, which allowed for active sites for the Li ions to diffuse and intercalate. Figure 17.22 shows the NT images and the battery performance. Various similar studies have been reported using titania NT-based LIBs [199–202]. The studies to date in LIBs have aimed at increasing the storage performance of lithium ions through optimization of TiO_2 preparation conditions to reduce the internal resistivity of the material. Nonetheless, the poor electrical conductivity of TiO_2 NTs must be addressed through further investigations on doping and coating the NTs with other materials such as carbon to overcome this obstacle.

Titania NT has been tried in other battery types also. Effect of NT length on the performance of LIBs and

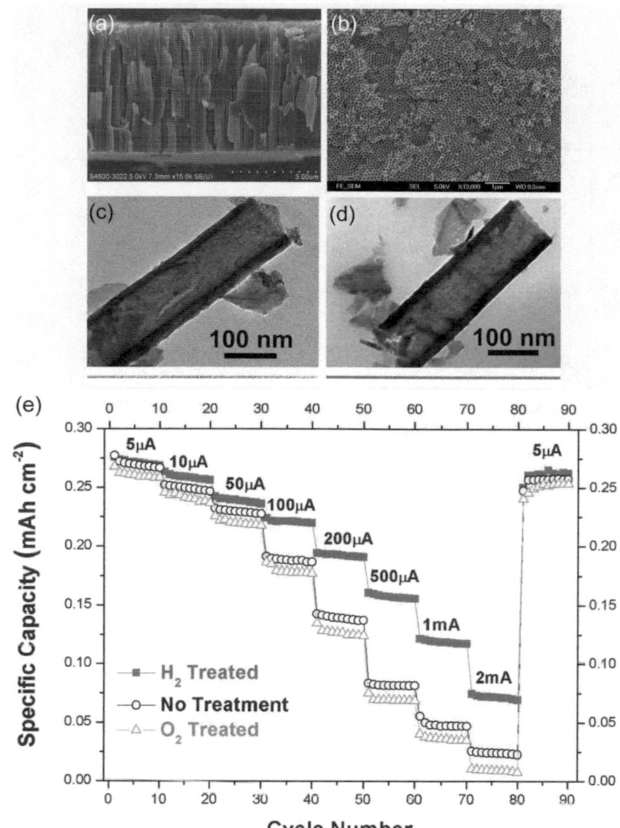

FIGURE 17.22 (a and b) SEM images of hydrogenated titania NTs. TEM images of NTs (c) before and (d) after H_2 treatment are also shown. (e) The specific capacities of untreated, H_2-treated and O_2-treated (after H_2 treatment) titania NT array electrodes. (Reprinted with permission from Ref. [198]; copyright (2012) John Wiley and Sons.)

sodium-ion batteries was studied by Gonzalez et al. [203]. Amorphous NTs prepared by anodic oxidation in EG-NH_4F electrolyte were used for the study. The capacity increased nonlinearly with length up to 200 μm. The areal capacity, in general, was lower for the sodium-ion battery (2–4 mAh/cm^2 vs <1 mAh/cm^2), which was attributed to the slower kinetics for sodium intercalation and irreversible surface reactions in long NTs. Recently, Mohanty et al. made composites of titania NT and sulfur and investigated their properties as a cathode in lithium sulfur battery [204]. The rationale was to address the problem of dissolution of lithium polysulfides formed at the cathode during discharge. The hydrothermally prepared NTs were dispersed in a solution of sulfur-dissolved CS_2. The mixture was sonicated, dried at 60°C for 12 h and finally crushed into a fine powder in a mortar. The cell fabricated using this composite with an NT:S mass ratio of 1:2 as cathode yielded a discharge capacity of 1,365 mAh/g, which reduced only by 4% after 50 cycles. The cycling performance of these cells was better than those with nanoparticle–sulfur composite or sulfur-based cathodes, and this was believed to be due to the suppression of polysulfide dissolution in the NT-S composite cathode-based cells.

Supercapacitor

Supercapacitors are high-capacity electrochemical capacitors with capacitance values significantly higher than that of conventional capacitors. These devices have a high specific power (>10 kWkg^{-1}), fast discharge rates, high stability and long cycle life. They are widely studied due to their promise as a bridge in the gap between conventional capacitors and long-term energy storage devices. The primary drawback of supercapacitors is their low-energy density; thus, the primary area of research is to enhance the energy density without reducing the high-power density. TiO_2 NT as an electrode in supercapacitors has seen vast progress in storage capacity in recent years. Titania NT offers a variety of advantages as an electrode material in supercapacitors [205]. The large surface area and highly ordered architecture, as in the case of anodic NTs, offer superior electron transport pathways and enhanced ion absorption. While the improved charge transport and interfacial ion movement are generally attributed to the tubular structure, the porosity of the structure enhances the ionic mass transfer through the NT channels. Additionally, the electrodes can act as substrates for the deposition of other materials to improve the conductivity or electrochemical activity. Other factors favoring the selection of TiO_2 NTs as an electrode material for supercapacitor are that they are mechanically robust, are chemically stable and can eliminate the need of a binder material.

Fabregat-Santiago et al. performed the capacitance studies on TiO_2 NT grown by anodic oxidation that revealed its potential as a supercapacitor material [206]. The material showed different characteristics in acidic and basic media (see Figure 17.23). The chemical capacitance and electronic conductivity increased exponentially with bias potential that indicated a Fermi-level displacement. In an acidic or a basic medium with a strong negative bias, a large increase in capacitance and conductivity was observed indicating a Fermi-level pinning. The authors believed that the difference in behavior was due to the proton intercalation in titania. The measured capacitance was in the order of 100 mF/cm^2 at bias of −0.5 V. As prepared, TiO_2 NTs exhibit storage behavior in aqueous electrolytes due to the fact that the Ti^{4+} to Ti^{3+} redox contribution is negligible [205]. The architecture provides an electron transport pathway, and the underlying substrate is utilized as a current collector. The capacitance and discharge rate are dependent on the crystallinity, tube length and acidity of the electrolyte. Overall, electrodes comprised of NTs have an enhanced capacitance compared to nanoparticle films. This increased capacitance is caused by the improved charge transport characteristics, increased surface area for ion absorption, electrical conductivity and carrier density, inherently found in TiO_2 NTs.

Since the low electrochemical capacitance of TiO_2 NT arrays as electrodes was found to be insufficient for supercapacitor applications, different strategies were used to increase the capacitance. Heat, electrochemical and

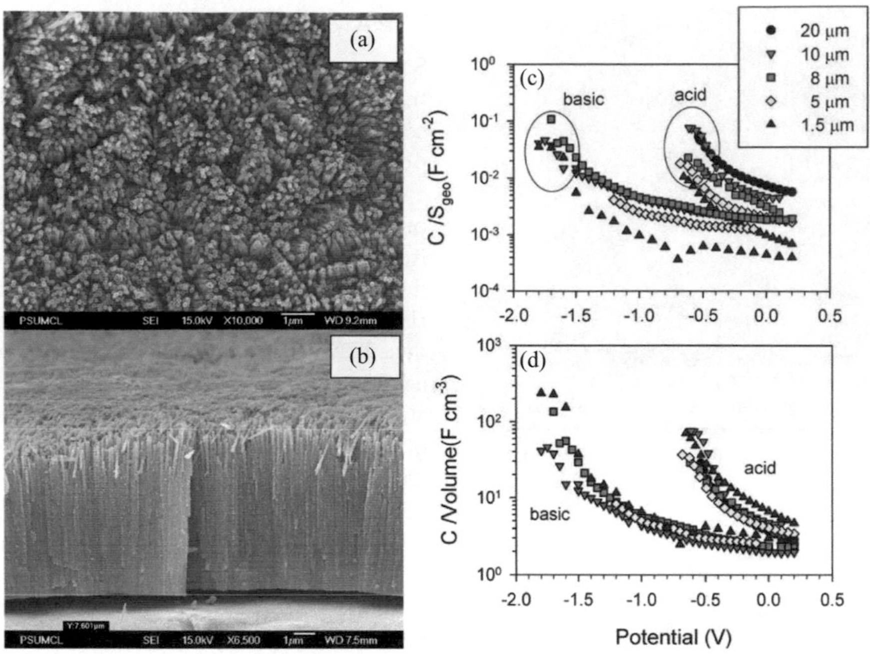

FIGURE 17.23 (a and b) SEM images of anodic titania NTs of length ~8 μm fabricated in an electrolyte consisting of tetrabutylammonium fluoride and FM and water. (c) Capacitance of TiO_2 NTs in basic and acidic media. (d) The capacitance data in (c) rescaled to volume density by dividing with NT length. (Reprinted with permission from Ref. [206]; copyright (2008) American Chemical Society.)

plasma treatments were demonstrated to increase the electrochemical capacitance. Heat treatments were aimed to convert the amorphous nature of as prepared NTs into a crystalline form, anatase, rutile, or brookite [207]. By optimizing the annealing parameters in a reductive atmosphere, for example H_2, which leads to the generation of oxygen vacancies and the partial reduction of ions from Ti^{4+} to Ti^{3+}, enhanced capacitance can be achieved. The oxygen vacancies act as electron donors and impact the conductivity of the TiO_2 NTs. Lu et al. proposed a hydrogen annealing method on NT arrays that produced an increase in the specific capacitance and discharge rate to 3.24 $mFcm^{-1}$ and 100 mVs^{-1}, respectively (see Figure 17.24) [208]. Heat treating the titania NTs and converting them to TiN NTs through nitridation was also attempted [209]. The chemical capacitance became 3.1 mF/cm^2 at 20 mV/s, due to the improved conductivity of TiN.

Low-temperature hydrogen plasma treatment performed in a CVD system resulting in enhanced capacitance in TiO_2 NT arrays was reported by Wu et al. [210]. The rough surface and incorporation of the defect states (Ti^{3+}) by the plasma resulted an enhanced capacitance of 7.22 $mFcm^{-2}$ at 0.05 $mAcm^{-2}$. An additional benefit found in using this method is that the plasma-treated NT arrays exhibit no degradation after 10,000 cycles. Electrochemical reduction is relatively a simple and fast method for enhancing electrode conductivity. By reducing the Ti^{4+} ions in the barrier layer at the bottom of titania NTs to Ti^{3+}, the conductivity at the bottom of the NTs could be improved. The reduction of Ti^{4+} to Ti^{3+} requires 2.0 eV–2.4 eV meaning that a potential of more than 0.9 V is required to make the reaction energetically favorable [205]. Using electrochemical doping, Wu et al. achieved a capacitance of 20.08 $mFcm^{-2}$ at 0.05 $mAcm^{-2}$ in TiO_2 NTs [211]. MnO_2-deposited titania NT was reported to show an areal capacitance 1.5 times that of bare MnO_2 [212]. The capacitance demonstrated was 24.07 $mFcm^{-2}$ at 10 mVs^{-1}. Further studies on the treatment of as-prepared NTs led to the development of black TiO_2 with a capacitance of 15.6 $mFcm^{-1}$ at 100 mVs^{-1} due to enhanced charge carrier density and lower transport resistance [213].

The random orientation and polycrystalline nature of anatase TiO_2 NT arrays were reported to cause electron scattering at crystal interfaces [214]. Thus, most of the charges are lost before reaching the current collector. C-axis-oriented NT arrays allowed for longer electron diffusion lengths. Anodization with low water concentration electrolytes and annealing in oxygen poor environments allowed for oriented growth along the [001] axis. Growth along other axes are prohibited in oxygen-deficient environments. Using the c-axis orientation method, the capacitance was increased to 8.21 $mFcm^{-2}$ at 100 mVs^{-1}. In NTs with randomly oriented crystallites, the capacitance was 0.35 $mFcm^{-2}$. For increasing the conductivity of NTs, Zhong et al. developed a method for fabricating Al-doped TiO_2 NTs [215]. Al^{3+} ions were introduced to the TiO_2 lattice via a bath containing $AlCl_3$. The resulting Al-doped NTs exhibited a reduction in resistance from 10 kΩ to 20 Ω; thus, the capacitance increased to 3.51 $mFcm^{-2}$ at 10 mVs^{-1}. Among the various strategies tried so far for increasing the capacitance of titania NTs, treatments like hydrogenation appears most promising.

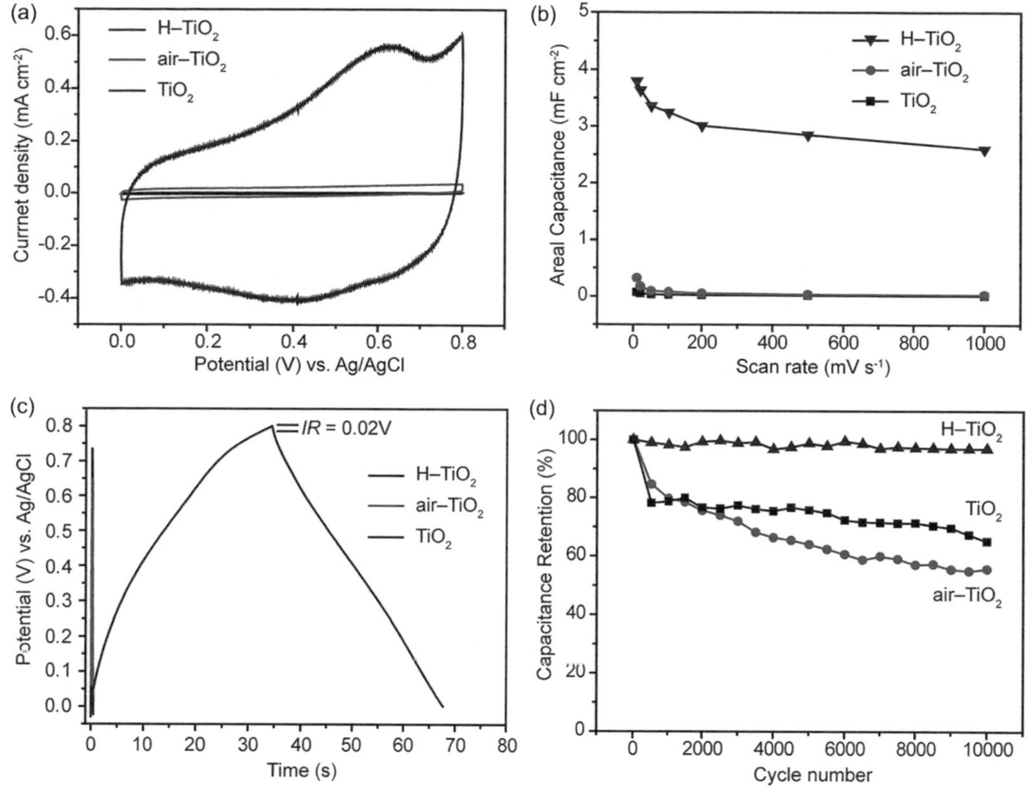

FIGURE 17.24 (a) Cyclic voltammetry curves, (b) areal capacitance, (c) galvanostatic charge/discharge curves (current density – 100 $\mu A/cm^{-2}$) and (d) cycle performance of TiO_2 NTs. (Reprinted with permission from Ref. [208]; copyright (2012) American Chemical Society.)

17.3.3 Biomedical

TiO_2 NTs have gained attention for use in biological systems due to their high biocompatibility and corrosion resistance. It has been explored extensively as an implantable material. Biocompatibility is key to "embedding" a host organism when used in regenerative medicine and tissue engineering. Due to excellent biocompatibility, titania NTs are considered an ideal instrument for growth and differentiation of osteoblasts. The passivation layer of TiO_2 allows for titanium and titanium alloy materials to be corrosion resistant as TiO_2 is not readily soluble even when exposed to an acidic medium. Bioactive biomaterials can be considered osteoconductive and osteoproductive. Combining complex hydroxyapatite and biological macromolecules with titania NTs could increase the bioactivity [216].

Due to the increasing instances of antimicrobial resistance to antibiotics, the bacterial infection of medical implants must be avoided. Some of the earlier studies on the biomedical applications of anodic titania NTs were focused on addressing this issue. Popat et al. demonstrated that titania NTs grown by anodic oxidation could be used for local delivery of antibiotics at the site of implantation to prevent bacterial infection after an orthopedic implant surgery [217,218]. The NTs were of length 400 nm and diameter 80 nm and were filled with gentacin. The drug-eluting NTs reduced the bacterial adhesion on the implanted surface. It was demonstrated that nanoscale Ti surfaces are more useful in reducing adhesion and bacteria activity than regular Ti. Zhang et al. showed that the physicochemical properties of titania NTs could be adjusted to limit the bacterial adhesion [219]. While NT length was shown to be not influencing the antibacterial activity, superhydrophobic modification doubled the effects to inhibit surface antibacterial ability and bacterial adhesion. Li et al. used surface plasmon resonance (SPR) effect in gold nanoparticle-coated titania NT to destroy bacteria [220]. The antibacterial functioning of the material was believed to be arising from the loss of electrons from bacteria to gold nanoparticles and to titania causing their death.

Objects implanted into the body experience adsorption of large amount of protein onto the surface immediately. This adsorbed protein acts as an intermediate layer between the implanted surface and the cell membrane [216,221]. In the case of titania NTs, proteins can be adsorbed more readily due to the high surface area of the NT. TiO_2 NTs are also favored by cell adhesion attributed to the selective protein adsorption. Protein adsorption occurs via adhesion of the interacting molecules at the surface and then their slow rearrangement. Surface charge density, topography and protein charge influence protein adsorption of implanted biomaterials [216]. A study showed that protein adsorption is maximum in NTs of length 10 μm and diameter 100 nm [222]. The charge of the protein also plays a role in the adsorption on TiO_2 NTs. Since the normal body pH level is greater than the isoelectric point of titania, the material has a negative surface charge. Positively charged proteins are limited by NT length, while negatively charged proteins are limited by charge. Implanted titanium and its alloys could lead to an immunological effect in the body resulting in implantation failure and this problem could be eliminated by coating TiO_2 NTs. Studies showed that titania NT has excellent blood compatibility, and hence, the blood contacting implants can have titania NT coatings to minimize the failure [223]. On the other hand, Roy et al. found that blood clotting can be made faster by dispersing titania NTs in the blood [224]. They suggested use of NT functionalized bandage to stop hemorrhage.

Several studies have been reported on ascertaining the interaction between titania NTs and different kinds of cells, including osteoblast, fibroblasts, chondrocytes, endothelial cells, muscle cells, epidermal, keratinocytes and mesenchymal stems cells [225]. Factors such as NT diameter, crystalline phases, surface wettability, surface chemical components and surface charge can affect the nature of interaction of NTs with cells [216]. TiO_2 NTs with diameters of approximately 30 nm promote cell adhesion while those with diameter of 70 nm–100 nm induce cell elongation, which leads to the transformation of the cytoskeletal stress and the differentiation towards osteoblasts [226]. A study on the adhesion and differentiation of Mesenchymal stem cells showed that NT diameters larger than 50 nm were not suitable for proliferation, migration and cell adhesion, while a 100 nm diameter would cause cell apoptosis [227]. Osteoblasts were highest on 15 nm diameter NTs. Mesenchymal stem cell adhesion was reported to be impaired when the tube diameter was greater than 50 nm [228]. Brammer et al. studied the behavior of bovine cartilage chondrocyte (BCC) on titania NTs ranging in diameters from 30 nm to 100 nm annealed at 500°C [229]. Most spherical BCCs were observed on titania NTs of 70 nm diameter. The amount of adhered cells was not affected by NT diameter, but the secretion of extracellular matrix in BCCs was enhanced on NTs of diameters ranging from 50 nm to 100 nm. The crystalline phase of NTs also reported to have an impact on their bioperfomance. It was shown that anatase could promote the growth of osteoblast [230]. Yu et al. observed that preosteoblast had maximum proliferation, spreading and mineralization with titania NT comprising a mixture of anatase and rutile phases [231]. Further studies showed that the rutile and anatase phases are more advantageous for the formation of hydroxyapatite in body fluids [216].

The NT surface wettability and chemical composition affect the bioactivity of titania NTs. Hydrophilic surfaces enhance cell adhesion and proliferation when compared with hydrophobic surfaces. A study by Lai et al. showed that the fibroblast cell adhesion could be selectively controlled by patterning superhydrophobic and superhdrophilic regions appropriately on a substrate [232]. Cell adhesion, extension and differentiation are affected by the hydrophilicity and free energy on the TNTs surface [233,234]. Therefore, these processes are also affected by surface treatment on Ti with TNTs. Park et al. showed that a high fluorine concentration on the surface of titania NTs could improve the cell proliferation [228]. The effect of surface chemical components of TNTs on various cells has not yet been fully studied.

Research has shown that titania NTs can promote the proliferation of osteoblasts both *in vitro* and *in vivo* [235]. NTs of diameter 100 nm were reported to have significantly higher bone implant contact value than those with diameter 15 nm [226]. Silicon-doped titania NT was shown to have promising cytocompatibility to both osteoblasts and endothelial cells [236]. Use of titania NTs in dental implants, particularly the screw shaped ones, has also been widely explored [237–240]. Titania NTs were grown on the titanium implants by anodic oxidation and tested on animals (see Figure 17.25). For example, from the studies done on rabbits, Jang et al. found that the miniscrews with titania NT coating had a greater bone to implant contact ratio than the control (52.8% vs 29.3%) [240]. These *in vivo* studies proved that the screw implants could substantially enhance osseointegration and improve the stability of the miniscrew.

Titania NTs have been found useful for cardiovascular stents and cancer therapy also. Studies showed that titania NT coatings on stents could improve endothelialization [241]. Titania NTs grown vertically on substrates showed no adverse effects on folding and expanding, which is essential for use in stents as these are commonly deployed using balloon catheters (Figure 17.26) [242]. Destruction of cancer cells using the photocatalytic as well as photothermal properties of titania NTs was reported [243–245]. Anticancer drug loading on NTs and its controlled release *in vivo* for localized destruction of cancer cells was also found feasible [246]. In a comprehensive *in vitro* study performed to assess the toxicity of anodic titania NTs, Mohamed et al. found that low concentrations of NTs could cause toxicity to the cells. The study revealed normal and functioning mitochondria; however, there was significant production of reactive oxygen species (ROS) in the nuclear compartment (Figure 17.27) [247]. The NTs were believed to have gained access to the nucleus and increased the stress level and caused genotoxicity. This property could be used for destroying cancer cells without cancer-specific drugs.

Titania NTs were investigated for use in instruments and devices in the biomedical field. For example, Wijeratne et al. used titania NTs grown radially on titanium wires for the enrichment of phosphorylated peptides [248]. Their preliminary studies on phosphopeptides generated from mouse liver complex tissue extracts showed that NTs provide efficacy comparable to the commonly used expensive titania beads from G. L. Sciences Inc., Japan (Figure 17.28). The development of titania NTs as biosensors for various diseases such as diabetes has been reported. Lee et al.

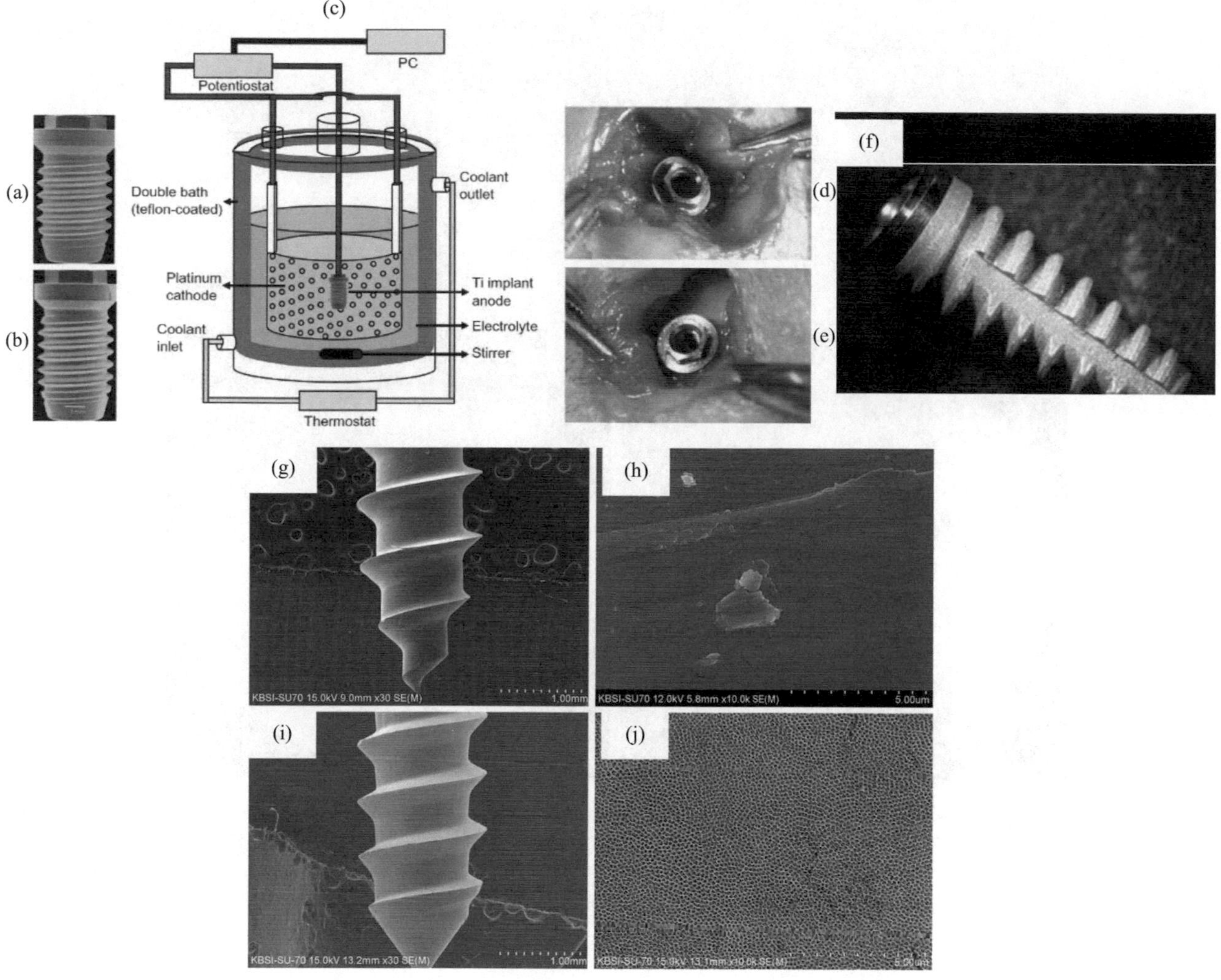

FIGURE 17.25 Images of (a) a test implant coated with TiO_2 NT and (b) a blasted control implant with thread pitch 600 µm. (c) Schematic of the experimental setup used for titania NT coating. (d and e) Images showing NT implant and blasted implant inserted in the femur condyle close to the knee joint of rabbits. (Reprinted from Ref. [237].) (f) A dental implant covered by titania NT. (Reprinted from Ref. [238].) SEM images of a miniscrew used for *in vivo* bone implant (g and h) before and (i and j) after anodization. (Reprinted with permission from Ref. [240]; copyright (2015) Springer Nature.)

developed amperometric glucose and uric acid sensors using titania NTs [249]. The sensor response was linear in the range between 2 and 14 mg/dL of uric acid concentrations and 50 and 125 mg/dL glucose concentrations. Glucose sensing properties of NTs were studied by other groups also [250,251]. Studies showed that NTs could detect H_2O_2 also [251].

17.3.4 Photocatalytic

Since Fujishima and Honda reported the photocatalytic water splitting and carbon dioxide conversion using titanium dioxide, rigorous investigations have been carried out for efficiently generating fuels using sun light and semiconductor photocatalysts [252,253]. Among a large number of semiconductor photocatalysts investigated so far, TiO_2 is still one of the most promising semiconductor photocatalysts due to its high oxidation potential, nontoxicity, high photostability, chemical inertness and the ability to form nanostructures possessing unique electronic properties. In a study on the self-cleaning cleaning properties of anodic titania NT-based hydrogen sensor, Mor and co-workers found that the material had strong photocatalytic properties [141]. As discussed in Section 17.3.1, they contaminated a hydrogen sensor with motor oil and destroyed the hydrogen sensitivity. Upon irradiation with UV light (wavelengths 254 nm and 365 nm), the NTs removed the motor oil completely and regained the original sensitivity. Titania is a wide bandgap material (3.2 eV for anatase titania), and hence, it absorbs photons in the UV region (wavelengths less than about 400 nm) only. This study revealed that the small thickness of the NT wall facilitated holes generated in the material to reach the surface without undergoing recombination and oxidize the organics. Subsequent studies showed that NTs could efficiently generate hydrogen and oxygen from water in a

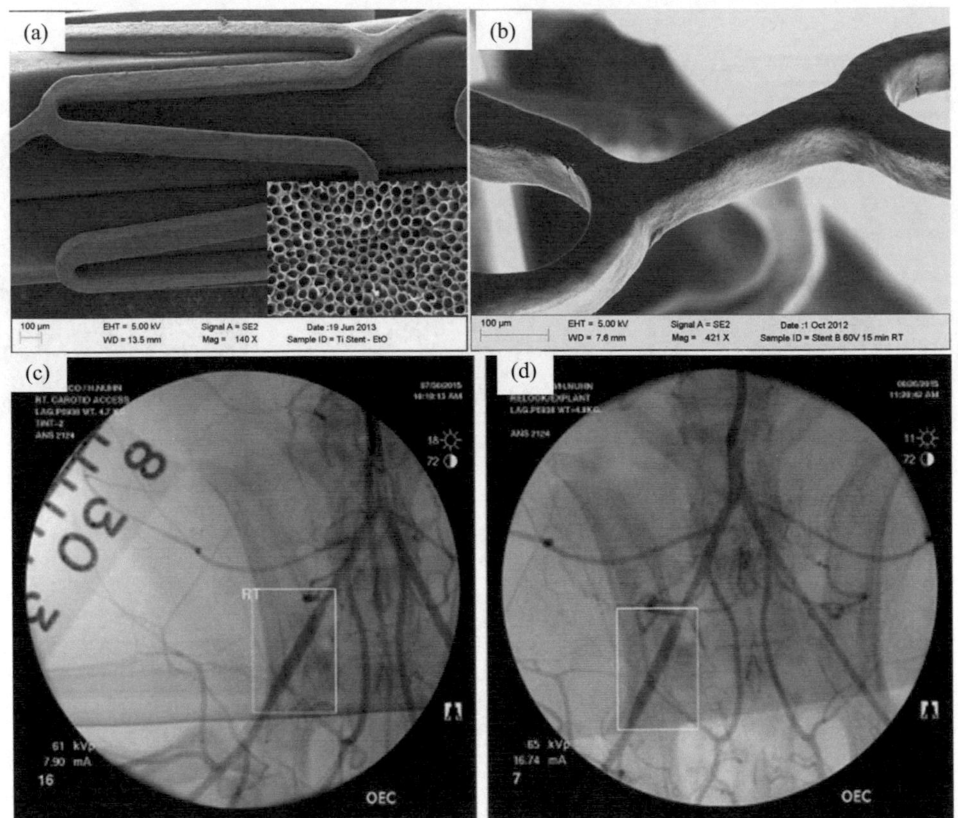

FIGURE 17.26 (a) A titania NT-coated crimped stent. Inset: image of an NT. (b) A TiO_2-covered stent with right half inflated to 3.5 mm ID and left kept at an ID of 1.8 mm with no critical damage to NTs at bridges, struts or turns. Fluoroscopy image taken (c) immediately and (d) 27 days after the deployment of stent in animals. White frame shows the stent location. (Reprinted with permission from Ref. [242]; copyright (2017) American Chemical Society.)

photoelectrochemical cell under UV irradiation [142,254]. The water photoelectrolysis experiments performed using first- and second-generation NTs grown by anodic oxidation showed an increase in light to hydrogen conversion efficiency (ratio of energy from hydrogen generated to energy from sunlight) with NT length up to about 6 μm [103,255,256]. The annealing temperature also strongly influenced the efficiency. The 600°C annealed NTs of length 6 μm yielded the maximum photocurrent and sunlight to hydrogen efficiency (∼0.6 %). The photocurrent and efficiency obtained from NTs of different lengths are shown in Figure 17.29. Shankar et al. performed studies using third-generation NTs of length several tens of microns grown in different electrolytes and found that the efficiency (under UV illumination) was maximum for NTs of length 30 μm and outer diameter 205 nm grown in a ammonium fluoride–FM electrolyte [130].

The photocatalytic properties of titania NTs grown by all three methods, anodic oxidation, hydrothermal and template-assisted growth, were determined to be at par or better than titania nanoparticles and other nanostructures as verified through a number of studies conducted for solar fuel generation, degradation of organic pollutants from liquid and gaseous media and removal of metal ions [15,257–267]. Unlike the nanoparticles, the surface area of NTs can be manipulated in a wide range and the NTs can be used in the dispersed or immobilized (grown on substrates) form as required.

A remarkable feature of anodic titania NT is that as a photoanode in an electrochemical cell, it could yield high incident photon to current conversion efficiency (IPCE). IPCE indicates the number of electrons driven to the external circuit per incident photon. In the case of titania NTs, IPCE was shown to be high (up to ∼80%) in the UV region of the solar spectrum indicating that the material could convert photons very efficiently to useful electrons [267]. Nevertheless, the sunlight to hydrogen conversion efficiency that can be obtained using pure titania NTs is less than 1% as these reactions were driven by UV light, which covers only less than 5% of the solar spectrum [268].

In order to enhance the efficiency of the photocatalytic fuel generation processes employing TiO_2 NTs, several strategies have been investigated. These include doping, sensitization and fabrication of heterostructures with other materials. Doping of various anions including nitrogen, carbon, boron and phosphorous were done in TiO_2 NT arrays with the aim of improving visible light response [269–272]. Varghese et al. demonstrated that moist carbon dioxide could be converted to hydrocarbons such as methane

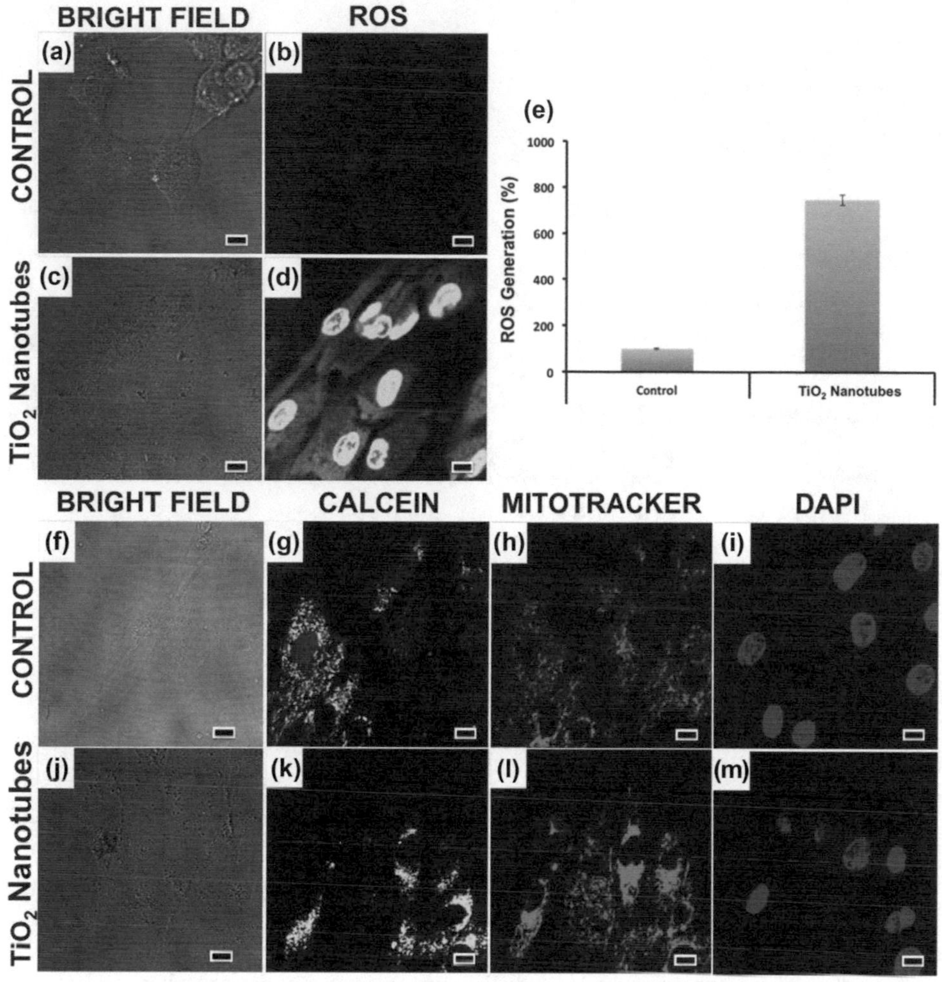

FIGURE 17.27 Effect of titania NT treatment on cells. (a–e) ROS generation in titania NT-treated cells. The nuclear ROS in the NT treated cells was increased sevenfold with respect to the control (e). The mitochondrial pore induction studies showed absence of any distorted signals from the mitochondrial region (l) with discrete calcein signaling (k) indicating a properly functioning mitochondrial system. This means, mitochondrial apparatus appears healthy while nuclear ROS is enhanced on treatment with TiO_2 NTs. (Reprinted from Ref. [247].)

at high rates using nitrogen-doped titania NTs illuminated with sunlight [269]. The nitrogen doping performed in situ with anodization shifted the light absorption edge of titania NTs from UV to the (gray in print) green wavelength region (~500 nm) of the solar spectrum (Figure 17.30). Nevertheless, nitrogen doping also introduced carrier recombination centers that prevented the NTs from making the solar carbon dioxide to hydrocarbon conversion process highly efficient. Shankar et al. tried propane flame annealing of NTs to incorporate carbon in the titania NT lattice [270]. Though the photocurrent increased in NTs of length 0.2 μm and 4.4 μm after flame annealing, the improvement was not substantial. This is because the carbon doping made only a slight improvement in the IPCE (up to 5%) in the region of the visible spectrum. Hydrogen treatment (normal or high pressure treatment) of titania NTs was also found useful in enhancing the photocatalytic hydrogen evolution rates. For example, Liu et al. implanted protons in titania NTs and demonstrated that the material, called black titania, could improve photocatalytic hydrogen evolution from methanol/water solution [273]. The Ti^{3+} defect states created by the process were believed to serve as catalytic sites for the hydrogen evolution reaction.

Doping by several metals was also tried. For example, Dong et al. fabricated Ni-doped TiO_2 NTs photoanodes and improved photoelectrochemical water splitting compared to bare TiO_2 NTs [274]. Isimjan et al. co-doped anodic titania NTs with iron, carbon and nitrogen and enhanced visible light photocatalytic response [275]. They anodized Ti first in NH_4F/EG electrolyte and grew a titania NT film. This NT film was removed with adhesion tape, and a second anodization was done on the textured titanium in a solution of EG/NH_4F/H_2O/$K_2Fe(CN)_6$. The NTs thus grown contained Fe, C and N as verified by x-ray photoelectron spectroscopy (XPS) and Energy dispersive x-ray (EDX) spectroscopy. Momeni and Ghayeb used potassium chromate in the anodization electrolyte to grow chromium (Cr)-doped titania NTs with visible light photoelectrochemical

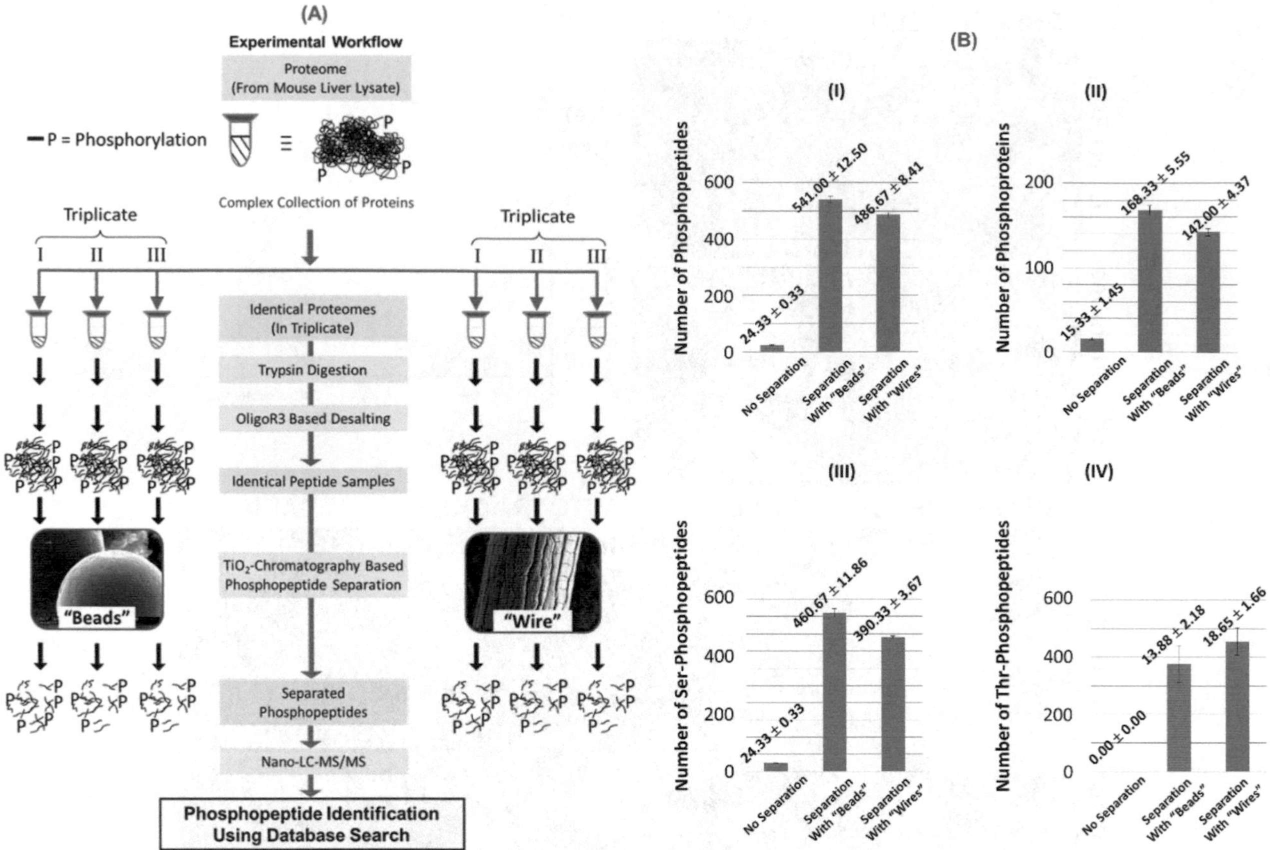

FIGURE 17.28 Liquid chromatography–mass spectrometry/tandem mass spectrometry analysis and identification of mouse liver proteome tryptic peptides to compare phosphopeptide separation capacity of TiO_2 beads (commercial) and titanium wire coated with titania NTs. (a) Experimental workflow. (b) Bar graphs representing unique and high confidence (>95%): (I) average number of phosphopeptides, (II) respective average number of phosphoproteins and (III) average number of Ser-phosphopeptides and (IV) average number of Thr-phosphopeptides. (Reprinted with permission from Ref. [242]; copyright (2017) American Chemical Society.)

water splitting [276]. Niobium is another metal that has been proven useful for enhancing the visible light responsiveness of titania NT photocatalysts [277].

Sensitization of NT surface with nanoparticles of metals, particularly Ag, Au and Pt, was widely studied. Metals can serve as co-catalysts for photocatalytic reactions. For example, sensitization using Pt or Cu or both was necessary for nitrogen-doped titania NTs to generate methane from CO_2 and H_2O using sunlight [269]. These metals were believed to trap the photogenerated electrons in titania and facilitate carbon dioxide or water reduction. Loading of nanoparticles of Ag or Au on NT was reported to enhance the photocatalytic degradation of organic materials as well as catalytic/photocatalytic destruction of bioorganisms such as bacteria [278–281] (see Figure 17.31). Metal nanoparticles, particularly gold nanoparticles, were also used to enhance the visible light photo-response of titania NTs through SPR effects. Zhang et al. showed that Au nanoparticle loading on titania NT could increase the IPCE up to 8% in a narrow wavelength region between 500 and 650 nm and cause an improvement in the overall photoconversion efficiency [282]. The observed IPCE region was consistent with the peak positions in the absorbance spectrum of small (peak at 556 nm) and large (peak at 590 nm) Au nanoparticle-loaded NTs. A porous titania layer was applied atop the NTs to serve as a plasmonic crystal and improve the SPR.

Construction of heterostructures is another approach attempted to enhance the visible light phototcatalytic activity of titania NTs. When a large bandgap semiconductor is coupled with a small bandgap semiconductor with a more negative CB level, conduction band electrons can be injected from the small to the large bandgap semiconductor. Thus, the heterostructure can facilitate broad spectrum light absorption and effective electron-hole separation. There have been large number of studies on the construction of heterojunctions of TiO_2 NT array photocatalysts with narrow bandgap semiconductors, including CdS, CdSe, WO_3, g-C_3N_4 and so on. Chen et al. [283] loaded CdS on anodic titania NT and showed a dramatic improvement in the photoelectrochemical performance. The visible light photocurrent increased by a factor of 16 when CdS/TiO_2 NT annealed at 400°C in nitrogen was used as the photoanode compared to bare TiO_2 NT photoanodes in a photoelectrochemical cell. Xie et al. sensitized CdS quantum dots (QDs) into TiO_2 NT arrays and extended

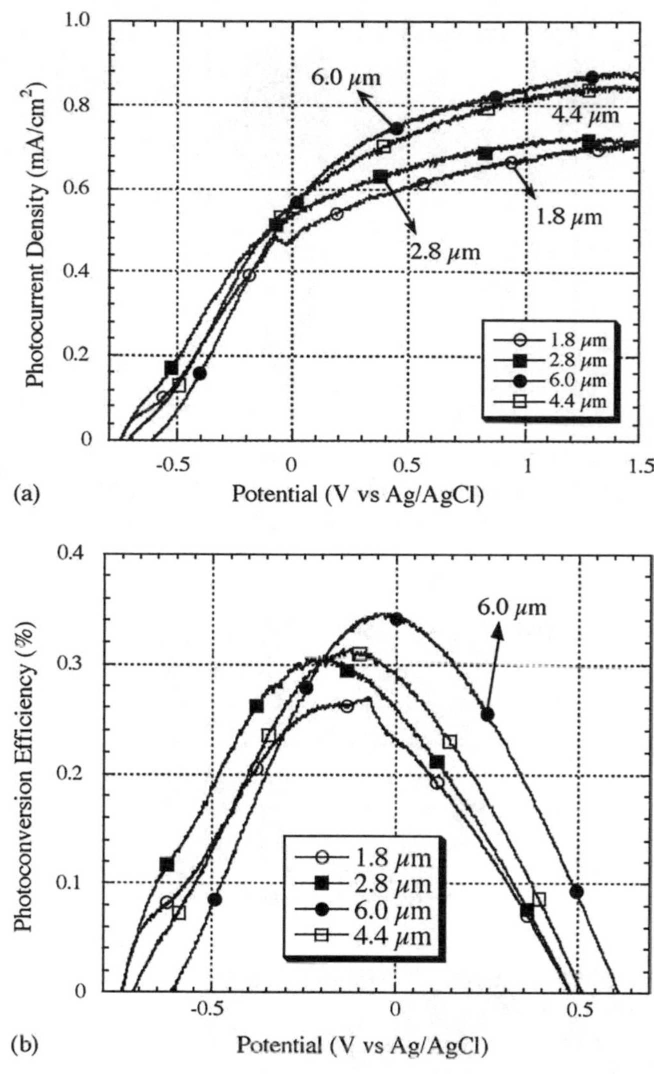

FIGURE 17.29 (a) Photocurrent generated under AM1.5 conditions from titania NTs of different lengths annealed at 530°C and (b) corresponding photoconversion efficiencies. (Reprinted with permission from Ref. [256]; copyright (2006) Elsevier.)

the absorption spectrum of the TiO_2 NTs into visible region up to 550 nm [284]. The CdS QDs promoted the effective transfer of photoexcited electrons from the conduction band of CdS to that of TiO_2 and favored electron-hole separation and exhibited much enhanced photocurrent generation and photocatalytic efficiency. Similarly, Li et al. incorporated CdS nanoparticle inside the TiO_2 NTs to develop a highly active photocatalyst for hydrogen production in the presence of sulfide and sulfite ions as hole scavengers under visible light irradiation [285]. Wang and co-workers developed a $CdSe/TiO_2$ heterojunction photocatalyst for PEC hydrogen generation to improve the photoelectrochemical activities [286]. Reyes-Gil and Robinson coupled WO_3 with TiO_2 NT arrays for solar water splitting and waste water treatment [287]. They showed that the IPCE rose from 0% to ~25% in the visible light region without reducing the conversion efficiency in the UV region. Liu et al. fabricated a $g\text{-}C_3N_4/TiO_2$ NT heterostructure that exhibited a photoelectrochemical water splitting activity twice that of pristine NTs [288]. It was also demonstrated that $BiVO_4$-loaded nitrogen-doped titania NTs grown by template method could exhibit a substantially higher photocatalytic activity than $BiVO_4$-loaded undoped NTs or bare $BiViO_4$ films [289].

Very recent studies showed that 1D/2D and 1D/0D heterostructures formed by titania NTs with a few new narrow bandgap materials possess unique characteristics for visible light photoelectrochemical water splitting. Balan et al. exfoliated hematite ore and obtained 2D non-van der Waals sheets of Fe_2O_3 called hematene. The bandgap was about 2.2 eV [290]. Hematene was loaded on titania NTs to form a 1D/2D heterostructure. The material showed an IPCE of up to about 80% in the UV and 12% in the visible light region. Interestingly, the conduction band of hematene was more positive (vs NHE) than TiO_2 NT, still IPCE was high, which indicated that there was transfer of hot electrons from hematene to titania (see Figure 17.32). Similar studies were reported on exfoliated 2D structure of $FeTiO_3$ (ilmenite) called ilmenene and manganese telluride [291,292]. Nonetheless, highest IPCE was reported in the case of fluorinated boron nitride (BN) QD-sensitized titania NT (the 1D/0D heterostructure) [293]. In this case, IPCE increased up to 24% in the visible region of the solar spectrum. Although BN is inherently a wide bandgap material, fluorination created gap states that mediated light absorption and charge transfer to NTs up to a wavelength of 600 nm (see Figure 17.33). All these heterostructures yielded a higher photocurrent than bare titania NTs. Thus, combining TiO_2 with other semiconductor materials with suitable band structures to form TiO_2-semicondcutor heterojunctions not only promotes the separation of photogenerated charges but also enhances the visible light absorption of TiO_2.

17.3.5 Solar Cell

Titania has become a key solar cell material since the development of first efficient dye-sensitized solar cells (DSSCs) by Graetzel et al. in 1991 [294–297]. In DSSCs, titania nanoparticle films with high surface area is used in the negative electrode (or photoanode) to support dye molecules and transport photogenerated electrons to the external circuit. Uchida et al. first reported the use of hydrothermally grown NTs in DSSCs replacing the titania nanoparticle film as high surface area electron transport medium [3]. This was immediately followed by the work of Adachi et al. [298,299]. Since then, titania NTs have been used in different categories of solar cells with DSSC, with either liquid or solid electrolyte, as the major type. Among the three general methods used for NT fabrication (discussed in Section 17.2), anodization received substantial attention as it facilitated the formation of NT arrays. Charge recombination at the interparticle regions was identified as a problem in titania nanoparticle-based DSSCs, and vertically aligned titania NT structure was believed to solve the problem by enabling vectorial electron transport along the length. Majority of the published

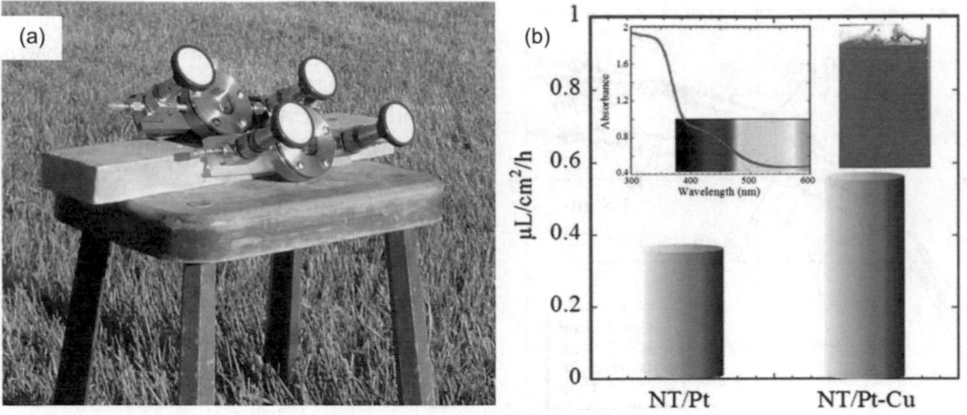

FIGURE 17.30 (a) Reaction chambers kept under sunlight for photocatalytic CO_2 generation. (b) The concentrations of methane generated from CO_2 and water vapor when a nitrogen-doped NT film (inset) was irradiated with sunlight. NT/Pt and NT/Pt-Cu represent, respectively, NT loaded with platinum and copper–platinum co-catalysts. (Reprinted with permission from Ref. [269]; copyright (2009) American Chemical Society.)

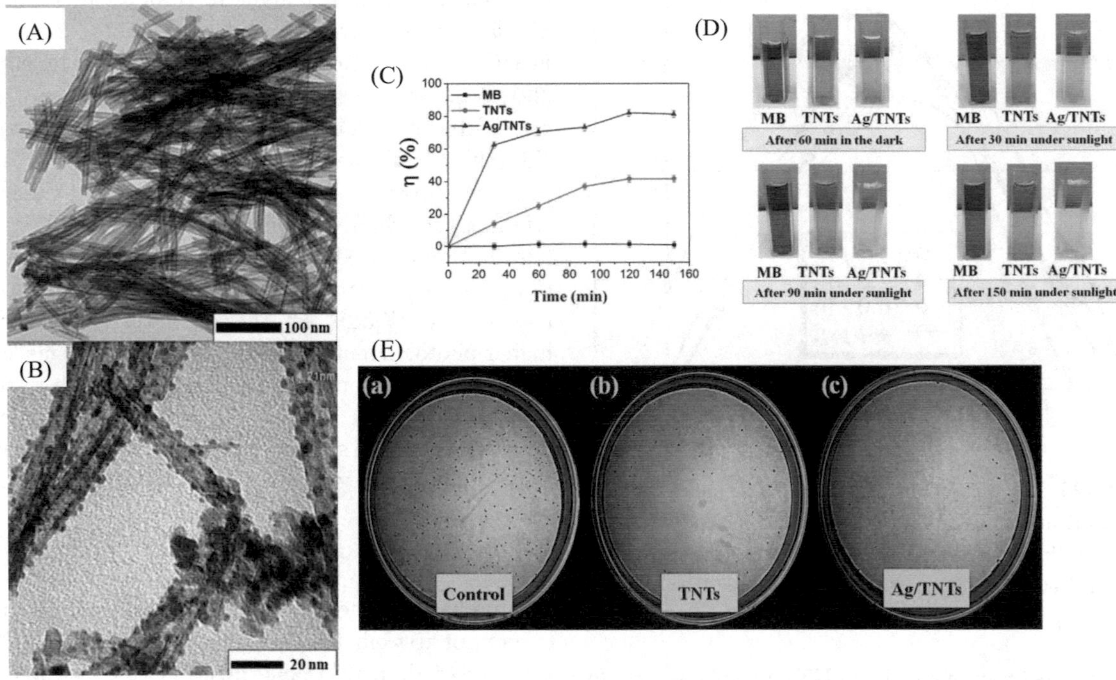

FIGURE 17.31 TEM images of (A) bare and (B) silver nanoparticle-loaded NTs. (C) Methylene blue photodegradation efficiency of the materials under sunlight and (D) the change in color during dye degradation process. (E) Antibacterial activities of samples against *Staphylococcus aureus* upon exposure to sunlight. (Reprinted with permission from Ref. [280]; copyright (2018) Elsevier.)

works with anodized titania NTs are on backside-illuminated geometry light entering the cell through the positive electrode because of the ease in fabrication of NTs on titanium foil. With the development of transparent films of titania NTs, front illumination geometry and flexible solar cells also became subjects of investigation [113,114]. Later, other types of NTs including polymer, QD and hybrid perovskite solar cells also employed titania NTs as the anode. The published works on titania NT-based solar cells are compiled in Table 17.1.

In order to alleviate the charge transport problems in films consisting of random network of titania nanoparticles, Adachi et al. prepared single crystalline anatase NTs by molecular assembly method by mixing laurylamine hydrochloride and titanium alkoxide with tetraisopropyl orthotitanate modified with acetylacetone (template) [298,299]. The DSSC made of a 4 μm thick film showed 4.9 % efficiency, indicating promising charge transport properties of the NT film. To obtain highly efficient DSSCs employing low extinction coefficient dyes such as N719, the thickness of titania NT electrode must be in the range of 10–20 μm. Blending titania NT gel with P25 titania nanoparticles was proved to be an effective way for increasing the thickness [300]. A mixed system

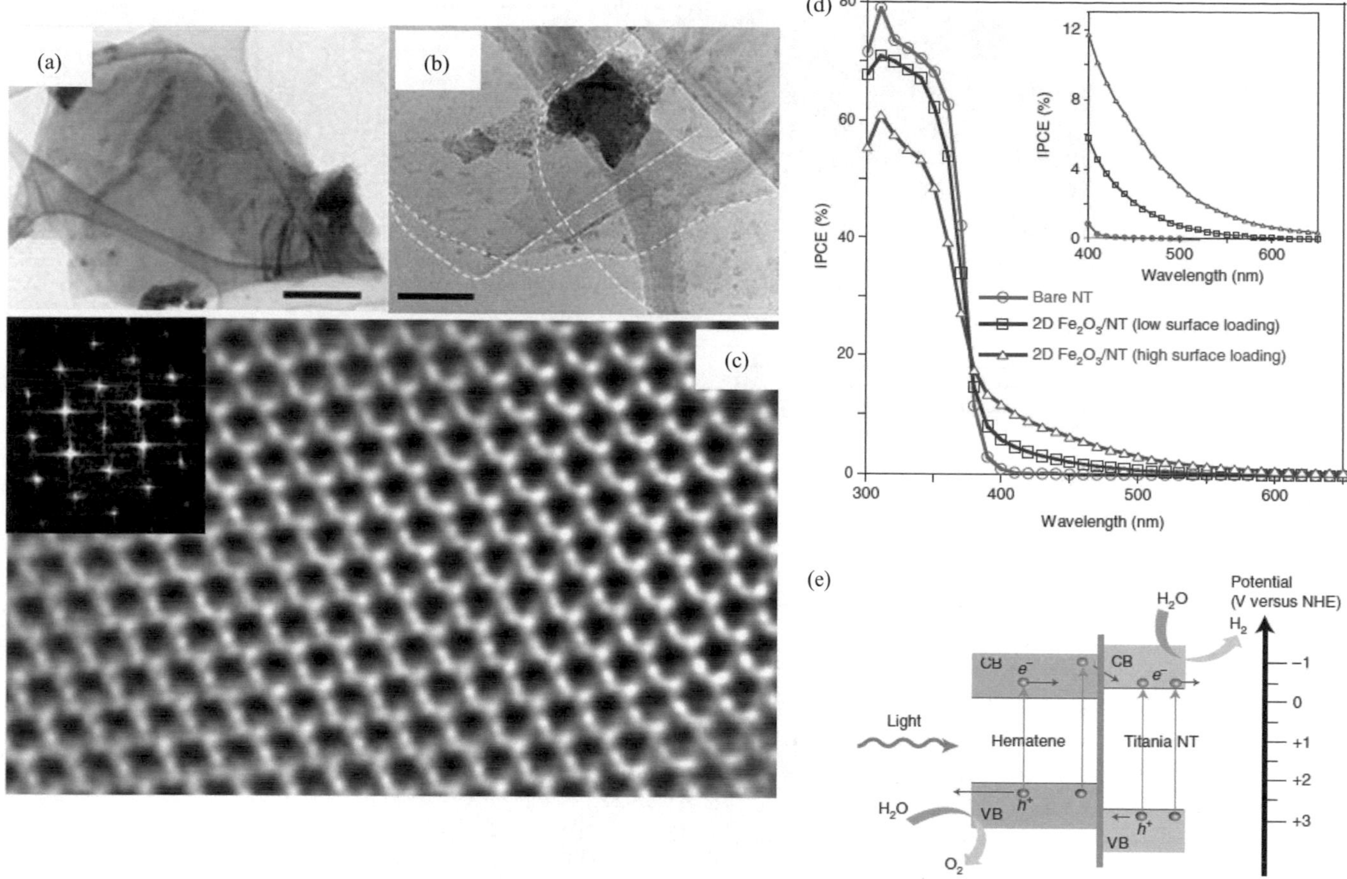

FIGURE 17.32 (a and b) TEM images of single and bilayer hematene (a single sheet of hematite; scale bar 0.5 μm for A and 50 nm for (b). (c) Scanning TEM image of hematene in the [001] orientation (scale bar 2 nm) with its Fourier transform in the inset. (d) Incident photon to current conversion efficiency spectra of bare NT film and those loaded with hematene. (e) Schematic diagram of interfacial charge transfer between hematene and TiO_2 NTs. (Reprinted with permission from Ref. [290]; copyright (2018) Springer Nature.)

of NT and 2% P25 yielded 8.4% efficiency for solar cells (j_{sc}—18.1 mAcm^{-2}, V_{oc}—720 mV and ff—0.64). Increasing the amount of P25 by more than 5% decreased the short circuit current density of solar cells. Macak et al. conducted photoelectrochemical studies in a three-electrode configuration on short and long titania NTs grown on Ti foil using anodization and confirmed higher dye adsorption in longer NTs [301]. Mor et al. reported the performance of the first device fabricated using anodic titania NTs [302]. Their study using transparent short NT array films prepared by anodizing titanium-coated FTO glass in water–HF electrolyte demonstrated promising characteristics of the material as an alternative to nanoparticle films. The photovoltage decay measurements indicated superior charge recombination characteristics (longer electron lifetime) in NT-based DSSCs compared to nanoparticle-based cells. At the same time, Paulose et al. used NTs of length up to 6 μm (second-generation anodic NTs) on Ti foil to fabricate backside-illuminated DSSCs [303,304]. They obtained a remarkable open circuit voltage (V_{oc}) of 841 mV and a J_{sc} of 8.8 mA/cm^2 with N719 as the absorber dye on NTs. The backside illumination restricted the efficiency to 4.2% because a fraction of light was absorbed by the electrolyte causing less light to reach the dye (see Figure 17.34). At the initial stages of its development, NT-based cells suffered from low fill factor (FF) compared to nanoparticulate films.

Using intensity-modulated photocurrent and photovoltage spectroscopies (IMPS and IMVS, respectively), Zhu et al. studied the dynamics of electron transport and recombination in DSSCs fabricated with anodic NT array as the electron transport medium [305]. They found that the titania nanoparticulate and NT films had comparable transport times, but NT-based DSSCs had higher recombination time supporting the observation of Mor et al. [302] (see Figure 17.35). That means, use of NT array films suppressed the recombination rate in DSSCs by offering fewer recombination centers. They also found that the charge collection and light harvesting efficiencies were, respectively, 25% and 20% higher in NT-based DSSCs. In another work, they showed that the electron transport time in the anodic titania NTs could be shortened by 50% by removing the disorders in the NTs such as those caused by NT bundling and film microcracks [306]. From the relations of electron diffusion coefficient and lifetime with electron quasi-Fermi level, Jennings et al. estimated the electron diffusion length in titania NTs as 100 μm, which is 3 times

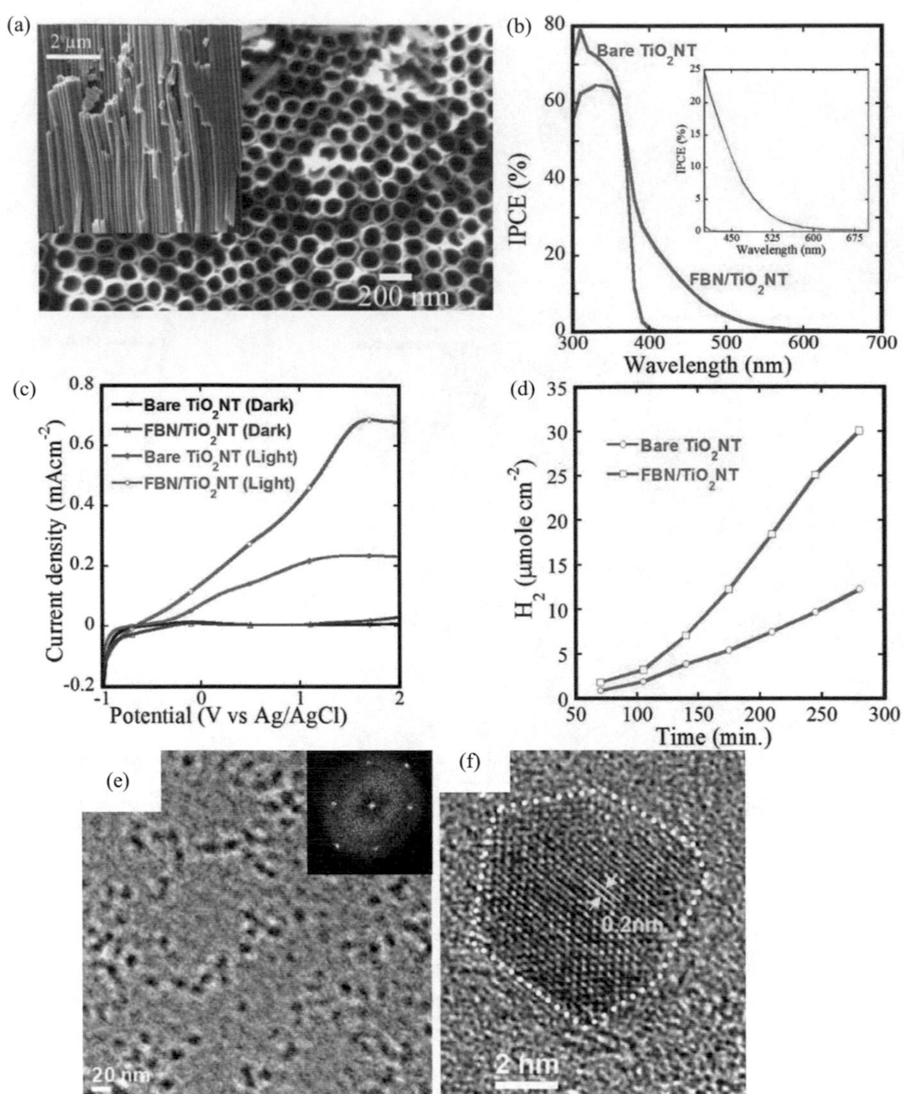

FIGURE 17.33 (a) SEM of TiO_2 NTs. (b) IPCE spectra of bare and BN QD-loaded titania NT. (c) Current–voltage characteristics of the bare and BN QD-loaded NTs. (d) Time-dependent hydrogen evolution from photoelectrochemical cells with these materials as photoanodes. (e) Low and (f) high resolution images of BN QDs. (Reprinted with permission from Ref. [293]; copyright (2019) John Wiley and Sons.)

better than that offered by nanoparticle films [307]. Shankar et al. reported the performance of DSSCs employing third-generation long NTs in backside illumination configuration [130]. They obtained an efficiency of 6.9% in 20 μm long NTs. Self-standing titania membranes of long anodic NTs transferred to FTO glass were also used for fabricating DSSCs. Park et al. used titanium isopropoxide while Chen and Xu used doctor-bladed titania nanoparticle layer (3 μm thick) to bond NT membranes to the FTO substrates and fabricated DSSCs with efficiencies, respectively, 7.6% (35 μm long NTs) and 5.5% [308,309]. NTs with rough walls were reported to have better dye adsorption than those with smooth walls [310].

The problem of integration of titania NTs with conducting glass substrates was solved by Varghese et al. by developing an anodization method for fabricating transparent films of short and long NTs on FTO glass using anodization [114]. These NTs when used in DSSC could scatter light and cause a substantial enhancement in the quantum efficiency in the long wavelength region (red wavelength region of the solar spectrum). In DSSCs employing 20 μm long NTs and N719 dye, the quantum efficiency reached 80% in a broad wavelength range in the visible light region up to the absorption edge (750 nm) of the N719 dye (see Figure 17.36). In standard DSSCs, a layer consisting of large (size 400–800 nm) titania particles is used atop the titania nanoparticle films to scatter light and enhance the quantum efficiency in the long wavelength region. No such layer is required if NTs are used. Nevertheless, enhancing the length beyond the range of 20 μm was reported to lower the efficiency [114,130].

Zhang et al. studied the efficiency of DSSCs with titania NT photoanodes filled with nanoparticles [311]. Maximum

TABLE 17.1 Solar Cells Developed Using Titania Nanotubes

Solar Cell/Solid State (S) or Liquid Electrolyte (L)	Absorber	NT Growth Method/Substrate	NT Film Thickness (μm)	$TiCl_4$ Treated (Y/N)	I_{sc} (mA/cm^2)/V_{oc} (V)/ FF/Efficiency (%)	References	Comments
DSSC/L	N719	Hydrothermal/FTO	60		1.26/0.704/0.66/2.9	Uchida et al. [3]	–
DSSC/L	N3	Template (mol. assembly)/FTO	4		15.3/0.58/0.54/4.88	Adachi et al. [299]	–
DSSC/L	N3	NT (template) + 2% NP/ITO	8.2		18.1/0.72/0.642/8.43	Yoshikawa [300]	–
PEC (3-electrode)/L	N3	Anodization/Ti foil	2.5		3.3 (IPCE)	Macak et al. [301]	–
DSSC/L	N719	Anodization/FTO	0.36	Y	7.87/0.75/0.49/2.9	Mor et al. [302]	–
DSSC/L	N719	Anodization/foil	6	Y	8.79/0.84/0.57/4.24	Paulose et al. [303]	–
DSSC/L	N719	Anodization/foil	6.2	Y	10.6/0.82/0 51/4.4	Paulose et al. [304]	–
			0.36		2.4/0.786/0.69/1.3		–
		Anodization/FTO	3.6	Y	10.3/0.84/0.54/4.7		–
DSSC/L	N719	Anodization/foil	1.9	N	4.4/0.64/0.6/1.7	Zhu et al. [305]	–
			5.7	N	5.7/0.61/0.55/3		–
DSSC/L	N3	Anodization/foil	2	N	3.28/0.64/0.38/0.8	Wang et al. [348]	–
DSSC/L	N719	Anodization/foil	20	Y	12.72/0.817/0.66/6.89	Shankar et al. [130]	–
DSSC/S	N3	Hydrothermal/FTO	6 (NP) + 6 (NT)	N	9.63/0.846/0.49/4.03	Flores et al. [316]	–
		Hydrothermal/FTO	6 (NP) + 6 (NT)		12.47/0.829/0.55/5.64		–
Hybrid/S	MEH-PPV	Sol–gel/FTO	0.3	N	0.88/0.62/0.37/0.2	Na et al. [327]	–
DSSC/L	Ru-TPA-NCS	Anodization/foil	14.4	N	13.44/0.723/0.63/6.1	Shankar et al. [315]	–
		Anodization/FTO	1	Y	10.1/0.743/0.55/4.1		–
Hybrid/S	P3HT/PCBM	Anodization/foil	0.27	N	3.91/0.324/0.43/0.54	Shankar et al. [325]	–
		Anodization/FTO	0.27	Y	12.4/0.641/0.51/4.1		–
Hybrid/S	P3HT/PCBM	Anodization/ITO	–	N	9.33/0.59/0.48/2.65	Yu et al. [326]	–
DSSC/L	D205	Anodization/foil	14	Y	6.11/0.743/0.725/3.29	Kuang et al. [320]	–
					7.8/0.705/0.58/3.19		Flexible/ITO-PEN
DSSC/L	N719	Anodization/foil	20	N	9.0/-/-/2.61	Jennings et al. [307]	–
DSSC/L	Ru based	Anodization/foil	8	N	5.9/0.59/0.51/1.9	Kim et al. [310]	–
					8.76/0.62/0.52/2.96		Bamboo-type NT
DSSC/L	N3	Anodization/foil	6	N	6.4/0.76/0.62/3.0	Chen et al. [349]	–
			18		9.7/0.781/0.61/4.3		–
			30		12.5/0.72/0.58/5.2		–
			19	Y	14.25/0.771/0.64/7		–
DSSC/S	C203	Anodization/FTO	2	Y	2.3/0.703/0.716/1.67	Chen et al. [317]	Spiro-MeOTAD
DSSC/L			2.8		7.9/0.677/0.693/3.8		Ionic electrolyte
DSSC/S	–	Anodization/FTO	0.6–0.7	Y	1.8/0.41/0.46/0.34	Mor et al. [328]	P3HT/PEDOT:PSS
	SQ-1		0.6–0.7		10.75/0.55/0.55/3.2		P3HT/PEDOT:PSS
DSSC/L	N719	Anodization/foil	10	N	-/-/0.52/1.64	Ghicov et al. [350]	–
			8		-/-/0.52/2		90 nm OD (outer diameter)
			8		-/-/0.51/1.7		140 nm OD
			16		-/-/0.44/2.4		120 nm OD
			16		-/-/0.47/1.7		220 nm OD
DSSC/L	N719	Anodization/foil	20	N	9.36/0.65/0.48/2.93	Roy et al. [351]	–
				Y	11.2/0.68/0.49/3.8		–
DSSC/L	N719	Anodization/FTO	17.6	Y	15.8/0.73/0.59/6.86	Varghese et al. [114]	–
			1.2		4.9/0.84/0.61/2.57		–
DSSC/S	SQ-1	Anodization/FTO	0.5–0.6	Y	4.05/0.68/0.67/1.64	Mor et al. [318]	Spiro- OMeTAD + DCM pyran (4-(dicyanomethylene)-2-methyl-6-(p-dimethylaminostyryl)-4H-pyran)
DSSC/L	N719	Anodization/FTO	35	N	16.8/0.73/0.62/7.6	Park et al. [308]	Membrane/TIP/FTO
DSSC/L	N719	Anodization/foil	25	N	9.22/0.614/0.495/2.8	Chen and Xu [309]	Membrane/3 um NaN crystals/FTO
				Y	12.4/0.701/0.633/5.5		

(Continued)

TABLE 17.1 (*Continued*) Solar Cells Developed Using Titania Nanotubes

Solar Cell/Solid State (S) or Liquid Electrolyte (L)	Absorber	NT Growth Method/Substrate	NT Film Thickness (μm)	$TiCl_4$ Treated (Y/N)	I_{sc} (mA/cm^2)/V_{oc} (V)/ FF/Efficiency (%)	References	Comments
DSSC/L	N719	Anodization/Ti mesh	40	N	4.95/0.52/0.56/1.47	Liu et al. [321]	Pt/FTO
					4.66/0.512/0.5/1.23		Pt/ITO/PET
DSSC/L	N719	Anodization/foil	14	N	12.16/0.7/0.51/4.34	Wang and Lin [352]	–
				Y	12.48/0.75/0.68/6.36		–
				Y	15.44/0.77/0.62/7.37		Plasma treated
DSSC/L	N719	Anodization-RBA/ (foil RBA – Rapid breakdown anodization)	10	N	2–6/0.75/2–3	Nakayama et al. [110]	Perchloric acid
DSSC/L	PEC-D03	Anodization-RBA/FTO	10	N	10.2/0.78/0.62/4.93	Ishibashi et al. [353]	NT paste, measured at 650 mW/cm^2 intensity
DSSC/L	N719	Hydrothermal/ITO/PEN (polyethylene naphthalate)	10	N	6.2/0.79/0.7/3.4	Xiao et al. [322]	without titania NT
					6.8/0.82/0.71/4		With titania NT
DSSC/L	N719	Anodization/ membrane/FTO	20.8	N	15.46/0.814/0.641/8.07	Lei et al. [354]	Front illumination
					13.18/0.82/0.675/7.29		Back illumination
DSSC/L	N719	Anodization/foil	30	Y	14.63/0.741/0.7/7.6	Li et al. [355]	–
				N	12.89/0.737/0.67/6.4		–
Hybrid/S	P3HT	LALD/FTO	0.15	N	0.82/0.65/0.5/0.3	Foong et al. [22]	–
DSSC/L	N719	Anodization/foil	20	Y	9.56/0.76/0.64/4.61	Li et al. [356]	–
		NT upright/NP/ITO	2(NP) + 20	Y	10.21/0.74/0.64/4.84		–
		NT-inverted/NP/ITO		Y	12.78/0.75/0.65/6.24		–
DSSC/Quasi-S	N719	Degussa P-25/FTO	12–15	Y	10.3/0.761/0.57/4.42	Stergiopoulos et al. [329]	Polymer electrolyte only
					10.8/0.843/0.64/5.42		Electrolyte with NTs
DSSC/L	N719	Anodization + ZN/Foil	15	N	7.28/0.65/0.6/2.8	Xie et al. [314]	
DSSC/L	N719	NP + Anodized Ti	17	Y	18.89/0.75/0.621/8.8	Baik et al. [357]	Two-layer NP+TNT
			8		14.05/0.76/o.588/6.28		Two-layer NP
			9		9/5.05/0.75/0.612/2.33		NT only
DSSC/L	N3	Anodization/foil	14	N	3.9/0.755/0.567/1.71	Liu et al. [358]	NT only
					5.96/0.895/0.704/3.755		NT/ZN
DSSC/L	D719	Anodization/foil	15	N	0.035/0.571/0.465/0.0093	Liu et al. [359]	NT-amorphous
					0.077/0.5/0.0935/0.0036		Nt-water anneal
					13.03/0.705/0.5095/4.68		Nt-thermal anneal
					7.99/0.787/0.536/3.37		NT (thermal + hydrothermal annealing)
DSSC/L	N719	Anodization/foil	1 + 15	N	9.96/0.73/0.63/4.57	Lin et al. [360]	NP(1um)+NT
		Anodization/FTO			12.67/0.73/0.65/6.06		NP + NT (watercrystal-lized) + 400°C thermal anneal (2 h)/F
DSSC/L	N719	Anodization/foil	–	N	4.09/0.71/0.65/1.89	Lee et al. [361]	–
		Electrodeposition(ZN)/ anodization(tio2)/foil			6.77/0.72/0.65/3.17		
DSSC/L	N719	ALD-template (15 nm TiO_2)/NT/FTO	7	N	9.1/0.69/0.616/3.9	Kurien et al. [362]	Untreated
					11.7/0.72/0.572/4.9		Water treated
DSSC/L	N719	Anodization (membrane)/FTO	33 ± 5	N	12.27/0.62/0.63/4.76	Zhang et al. [311]	–
				Y	21.5/0.66/0.62/8.96		–
				N	24.78/0.71/0.56/9.86		Sol–gel particle filled
DSSC/L	N719	Anodization (membrane)/FTO	33	N	11.47/0.63/0.668/4.86	Zhang et al. [312]	–
				Y	22.76/0.65/0.614/9.02		
DSSC/L	N719	Anodization (membrane)/FTO	34	N	12.04/0.713/0.66/5.67	Zeng et al. [313]	–
					16.45/0.725/0.68/8.11		Hydrothermal anneal
DSSC/L	N719 (at 70°C in autoclave)	Anodization/foil	9	N	7.7/0.7/0.594/3.2	Suhadolnik et al. [363]	–
				Y	10/0.69/0.571/4		–
				N	8.4/0.71/0.619/3.7		P25 treatment

(*Continued*)

TABLE 17.1 (*Continued*) Solar Cells Developed Using Titania Nanotubes

Solar Cell/Solid State (S) or Liquid Electrolyte (L)	Absorber	NT Growth Method/Substrate	NT Film Thickness (µm)	$TiCl_4$ Treated (Y/N)	I_{sc} (mA/cm^2)/V_{oc} (V)/ FF/Efficiency (%)	References	Comments
DSSC/L	N719	Electrospinning/ hydrothermal/ ITO/PET	9 µm	N	11.5/0.722/0.566/4.7	Song et al. [323]	CNT/TiO_2 NaN rod + branched NT
DSSC/S	N3	Anodization/foil	15	N	6/0.536/0.4/1.3	Peedikakkandy et al. [319]	Cs_2In_{I6} as HTM (hole transport material)
DSSC/L					9.1/0.68/0.54/3.3		–
DSSC/L	N719	Anodization/foil	16.5	N	8.82/0.7/0.57/3.54	Wang et al. [364]	Hydrothermal-water
Perovskite/L	$CH_3NH_3PbI_3$	Anodized membrane/FTO	2.3	N	17.9/0.63/0.578/6.52	Gao et al. [331]	–
Perovskite/S	$CH_3NH_3PbI_3$	Anodization/Foil	0.3	N	13.89/0.8/0.63/5.68	Wang et al. [332]	Flexible-25 µm Ti foil, CNT/spiro as HTM
				Y	14.36/0.99/0.68/8.31		
Perovskite/S	$CH_3NH_3PbI_3$	Anodization/FTO	0.4	N	19.6/0.67/0.37/5	Salazar et al. [333]	$CH_3NH_3PbI_3$/Au
Perovskite/S	$CH_3NH_3PbI_3$	Anodization/FTO	0.4–0.45	Y	22.6/1.007/0.64/14.8	Qin et al. [334]	Spiro-MeOTAD/Au
QDSSC/L(QD co-sensitized solar cells)/sulfide+KCl electrolyte	TiO_2	Anodization/foil	22	N	0.28/0.27/0.19/0.014	Lai et al. [335]	Thermal annealing
	TiO_2 + CdS(QD)				0.62/0.78/0.28/0.14		
	TiO_2 + CdSe				1.48/0.5/0.27/0.093		
	TiO_2 + CdS-CdSe core shell				10.82/0.53/0.38/2.2		
				Y	11.6/0.59/0.35/2.4		
					14.3/0.58/0.33/2.74		Sealed annealing
QDSSC/L (aqueous polysulfide)	CdS (SILAR)/ZnS	Anodic membrane/ QD TiO_2/FTO	16.7	Y	10.38/0.662/0.47/3.22	Lan et al. [336]	CuS/Pt counter electrode
QD PEC solar cells	CdSe	Anodization/foil	8	–	-/0.65/-/<1	Kongkanand et al. [337]	Aq. Na_2S/calomel
QD PEC solar cells	NT	Anodization/foil	19.2	N	0.22/-0.94/-/-	Sun et al. [338]	1 M Na_2S/Ag/AgCl
	NT + CdS				7.82/-1.27/0.578/4.15		
QD PEC solar cells	NT	Anodization/foil	11.3	N	0.17/-0.94/-/-	Gao et al. [339]	0.1 M Na_2S/Ag/AgCl
	NT + CdTe				6/-1.21/-/-		
QD PEC solar cells	NT only	Anodization/foil	25.5	N	0.16/-/-/-	Gao et al. [340]	0.1 M Na_2S/Ag/AgCl
	NT + CdS				4.8/-/-/-		
	NT + CdSe				6/-/-/-		
	NT + CdS + CdSe				13/-/-/-		
	NT + CdSe + CdS				1.8/-/-/-		
QD PEC solar cells	NT + CdTe + ZnS	Anodization/foil	–	N	11.15/-0.95/-/-	Wang et al. [341]	0.1 M Na_2S and 0.1 M Na_2SO_3/calomel
QD PEC solar cells	NT + CdS0.5Se0.46	Anodization/foil	–	N	15.58/-/-/-	Gakhar et al. [342]	0.35 M Na_2SO_3 and 0.25 M Na_2S/calomel
QD PEC solar cells	NT + CdS – CdTe	Anodization/foil	–	N	14.7/-/-/-	Wang et al. [343]	0.1 M Na_2S +1.1 M Na_2SO_3
	NT + CdS – CdSe				22.6/-/-/-		

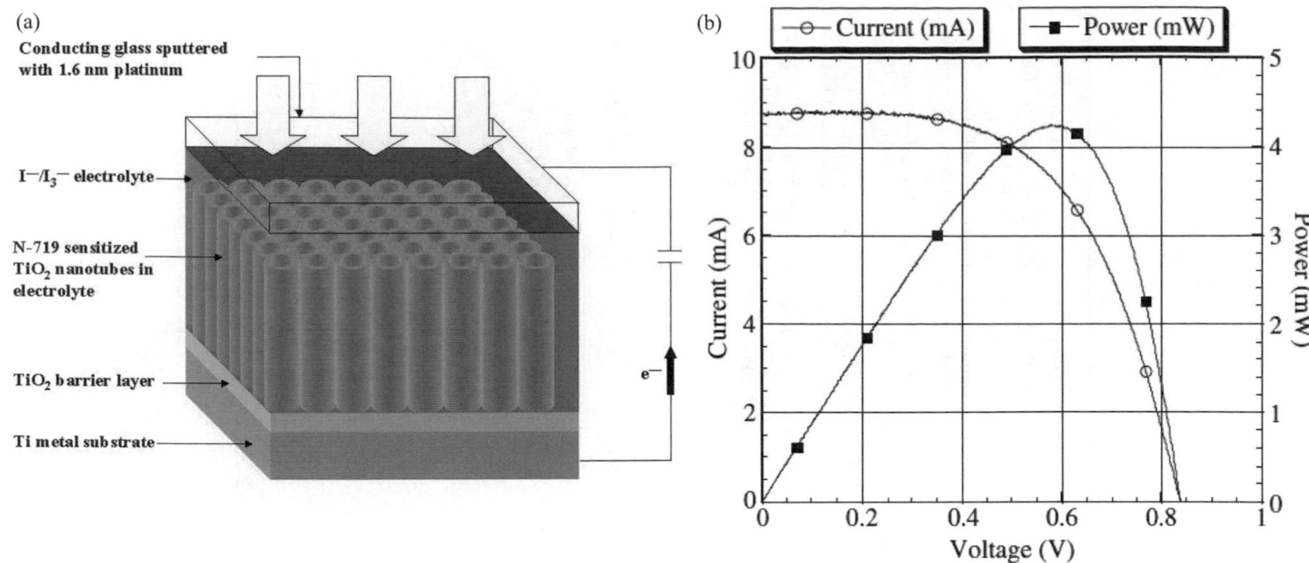

FIGURE 17.34 (a) Schematic diagram of backside-illuminated titania NT array DSSCs. (b) Current voltage characteristics of the solar cells under 1 sun illumination. (Reprinted with permission from Ref. [303]. Copyright (2006) Institute of Physics Publishing.)

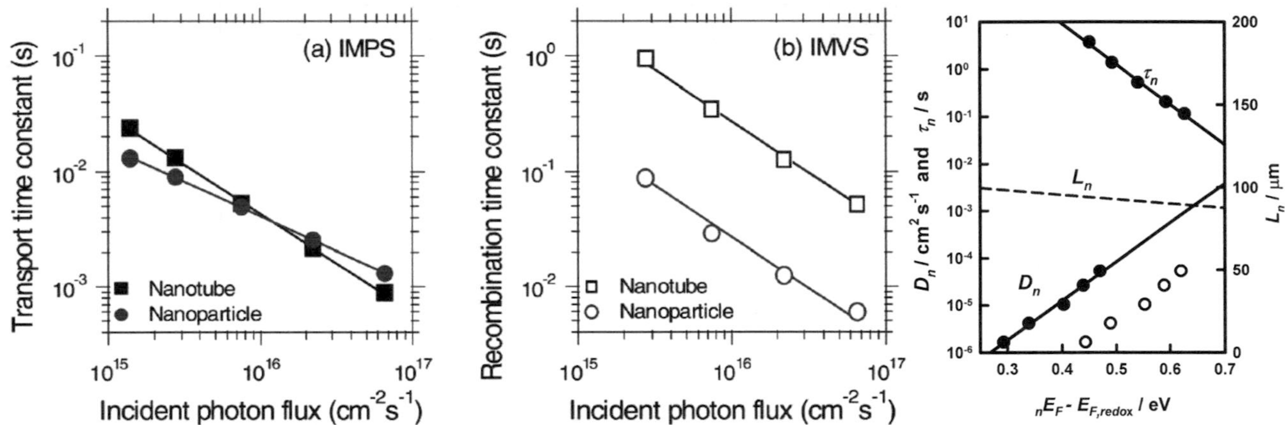

FIGURE 17.35 (a) Transport and (b) recombination time constants for NT and nanoparticle-based DSSCs. The incident photon flux is for 680 nm laser illumination. (Reprinted with permission from Ref. [303]; copyright (2007) American Chemical Society.) (c) Diffusion length for a 20 μm NT cell based on experimental diffusion coefficient and time constant. (Reprinted with permission from Ref. [307]; copyright (2008) American Chemical Society.)

efficiency was 9.86% when NT arrays were filled with particles of size 20 nm through hydroloysis and condensation of titanium tetrachloride in an aqeous solution containing alcohol and ammonia. In another article, they discussed the transfer and assembly of large-area NT arrays and the effect of $TiCl_4$ treatment [312]. Jiang et al. prepared front illuminated DSSC from hydrothermally crystallized anodic NT [313]. The DSSCs consisting of NTs of length 34 um treated hydrothermally at 180°C for 6 h gave an efficiency of 8.11%. Xie et al. deposited ZnO electrochemically on the titania NTs prepared by anodization [314]. These TiO_2/ZnO NT arrays were used as photoanodes in DSSCs. The heterojunction at the TiO_2/ZnO interface favors charge separation and reduces the probability of charge recombination. With the coaxial TiO_2/ZnO NTs as the photoanode, DSSC with an overall 2.8% energy-conversion efficiency was obtained. Shankar et al. used donor antenna dyes with high molar extinction coefficients in association with titania NTs and reduced recombination of photogenerated charges [315]. Better performances were observed in both front and back illuminated geometries. Donor antenna dyes appeared to provide a useful route to improving the efficiency of DSSCs owing to their high molar extinction coefficients and the effective spatial separation of charges in the charge-separated state that reduces the recombination of photogenerated charges. These dyes enable use of short NTs in high efficiency liquid and solid electrolyte DSSCs. The advantage is that short anodic NTs can be fabricated easily using aqueous electrolytes.

Flores et al. introduced a plasticized electrolyte consisting of poly(ethylene oxide-co-epychlorohydrin) and NaI–I_2 and used it in conjunction with hydrothermally grown NTs

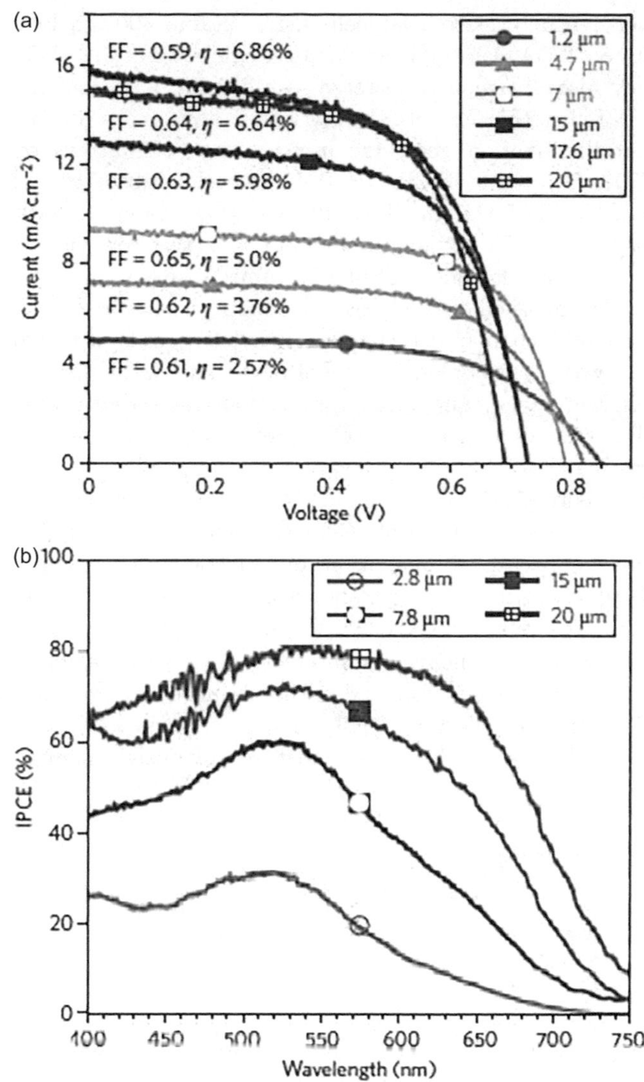

FIGURE 17.36 (a) Current–voltage characteristics and (b) IPCE spectra of DSSCs fabricated using transparent NT array films of various tube lengths. (Reprinted with permission from Ref. [114]; copyright (2009) Springer Nature.)

(Kasuga's method) with tehraethoxysilane and titanium isopropoxide as precursors [316]. Gratzel et al. fabricated for the first time solid-state DSSC with transparent NTs on FTO glass using metal-free C203 dye and spiro-MeOTAD [317]. Power conversion efficiency (1.67%) was low compared to nanoparticle electrodes. They also made a solvent-free DSSC cell with C203 dye with an efficiency of 3.8%. It was demonstrated by Mor et al. that Forster resonance energy transfer (FRET) process could be used for broad-spectrum light absorption in NT-based solid-state DSSCs [318]. They sensitized TiO_2 NTs of length 500–600 nm with near infrared (NIR) absorbing SQ1 (squaraine) acceptor dye and then intercalated with Spiro-OMeTAD blended with a visible light absorbing DCM-pyran donor dye. Due to FRET, the IPCE in the visible light region increased to about 25%. The excitation transfer efficiency was 67.5%. Peedikakkandy et al. used a perovskite variant Cs_2SnI_6 as the hole transport material in solid-state DSSCs based on NTs [319]. The cells gave an efficiency of 1.3% in back illuminated geometry. The performance was compared with a nanoparticle-based solid-state DSSC and an NT-based liquid electrolyte DSSC.

Kuang et al. developed a new binary ionic liquid (1-propyl-3-methylimidazolium iodide and 1-ethyl-3-methylimidazolium tetracyanoborate) electrolyte and used in DSSCs employing titania NT on flexible Ti foil. Polyethylene naphthalate (ITO/PEN) was used as the counter electrode. This flexible DSSC gave an efficiency of 3.6% in back illuminated geometry [320]. Impedance studies showed that low FF of flexible DSSC based on Pt/ITO-PEN substrate was due to higher R_s (series resistance) and lower activity of Pt catalyst at the Pt/ITO-PEN. Liu et al. made flexible solar cells using NTs on a titanium mesh and obtained an efficiency of 1.23% [321]. They used flexible ITO/PEN as the counter electrode. Xiao et al. developed 4% efficient flexible DSSCs using a mixture of hydrothermally synthesized titania NT and P25 nanoparticles on ITO/PEN substrates [322]. Song et al. fabricated branched anatase NTs using a combined electrospinning-hydrothermal-calcination process [323]. They used this material as an overlayer for the carbon NT/TiO_2 nanorod electrode in flexible DSSCs and obtained an efficiency of 4.7%. The efficiency was only 2.3% without this overlayer.

Conventional polymer solar cells use a thin compact titania film as electron transport layer (ETL). It was expected that the charge separation would be enhanced if these polymer light absorbers could be inserted into titania NTs of diameters comparable to or lower than the exciton diffusion length. Grimes et al. developed double heterojunction polymer solar cells with anodic NTs (length ~350 nm) grown on FTO glass as well as on titanium foil [324,325]. Double heterojunction indicates TiO_2/P_3HT and P_3HT/PCBM interfaces in the cell (see Table 17.1 for results). While backside illumination geometry did not work well with these type of solar cells, front illuminated cells gave an efficiency of 4%. A polymer cell with device structure ITO/TiO_2/P3HT:PCBM/V_2O_5/Al in which ITO and Al functioned as cathode and anode, respectively, was developed by Yu et al. [326]. Cells yielded a power conversion efficiency of 2.71%. V_2O_5 served as a buffer layer preventing the electrode metal diffusion to the polymer. Na et al. employed titania NTs fabricated using electrodeposited ZnO templates to develop hybrid solar cells [327]. Poly[2-methoxy,5-(2 -ethyl-hexyloxy)1,4-phenylenevinylene] (MEH-PPV) was used as the light absorber. They obtained an improved photocurrent compared to a compact titania film generally used in such cells. Mor et al. tried to increase the absorption in the NIR range by the use of unsymmetrical squaraine dye (SQ-1) that absorbs photos in the red and NIR portion of solar spectrum [328]. NTs of lengths 500 nm to 1.2 μm on FTO glass were first sensitized with SQ-1 and then uniformly infiltrated with p-type regioregular poly(3-hexylthiophene-2,5-diyl) (P3HT) to achieve typical device photoconversion efficiencies of 3.2%, which was significantly higher than cells with polymer alone. Foong et al. demonstrated the use of LALD process to form NTs of varied dimensions in an anodic alumina template, which was later etched off [22]. NTs of 150 nm length were used to make solar cells with infiltrated P3HT, but the efficiency was too low (0.3%). Falaras et al. studied the use of NTs prepared by anodization as fillers for the polymer electrolytes plasticized with polyethylene oxide [329,330]. NTs were ground with particle aggregates. The efficiency of the solar cells increased from 4.4% to 5.4% with NT fillers on a nanoparticle substrate.

Titania NTs have been used in perovskite hybrid solar cells also. Gao et al. fabricated $CH_3NH_3PbI_3$-sensitized solar cells with freestanding TiO_2 NT arrays of different lengths with iodine containg liquid electrolyte [331]. The cell utilizing 2.3 μm long NTs was 6.5% efficient and the efficiency decreased with tube length reaching 3.3% for cells with 9.4 μm long NTs. Hybrid solid-state flexible solar cells were fabricated with titania NTs of length 300 nm on 25 μm thick titanium foil by Wang et al. [332]. They used $CH_3NH_3PbI_3$ as the absorber, spiro-OMeTAD as hole conductor and transparent carbon NT as positive electrode. An efficiency of 8.31% was obtained for these solar cells. Salazar et al. fabricated hole transporting layer-free perovskite solar cells using titania 400 nm long NTs on FTO glass and obtained an efficiency of 5% [333]. They also observed a decrease in efficiency with increase in NT length. Qin et al. fabricated solar cells in the front illumination geometry using titania NT arrays on FTO infiltrated with $CH_3NH_3PbI_3$ perovskite absorber and spiro-MeOTAD as the hole transporting material [334]. They used NTs of length 400 nm and the cell yielded a maximum efficiency of 14.8%. Impedance spectroscopy revealed more efficient charge collection/extraction in NT-based cells compared to nanoparticle-based cells although there was relatively low loading of perovskite in the pores of NTs. More importantly, this work demonstrated that titania NT-based perovskite solar cells were highly stable as NTs served as scaffolds protecting the organometallic absorber (see Figure 17.37).

Lai et al. fabricated CdS/CdSe QD-sensitized titania NT DSSCs in backside illumination configuration [335]. The NT length was 22 μm, the electrolyte was a mixture of methanol and water containing 0.6 M Na_2S, 0.2 M sulfur powder and 0.2 M KCl and the counter electrode was platinized FTO. The efficiency obtained in the optimized cell in which NTs were subjected to $TiCl_4$ treatment was 2.7%. The CdS/CdSe co-sensitized NT structure obviously enhanced

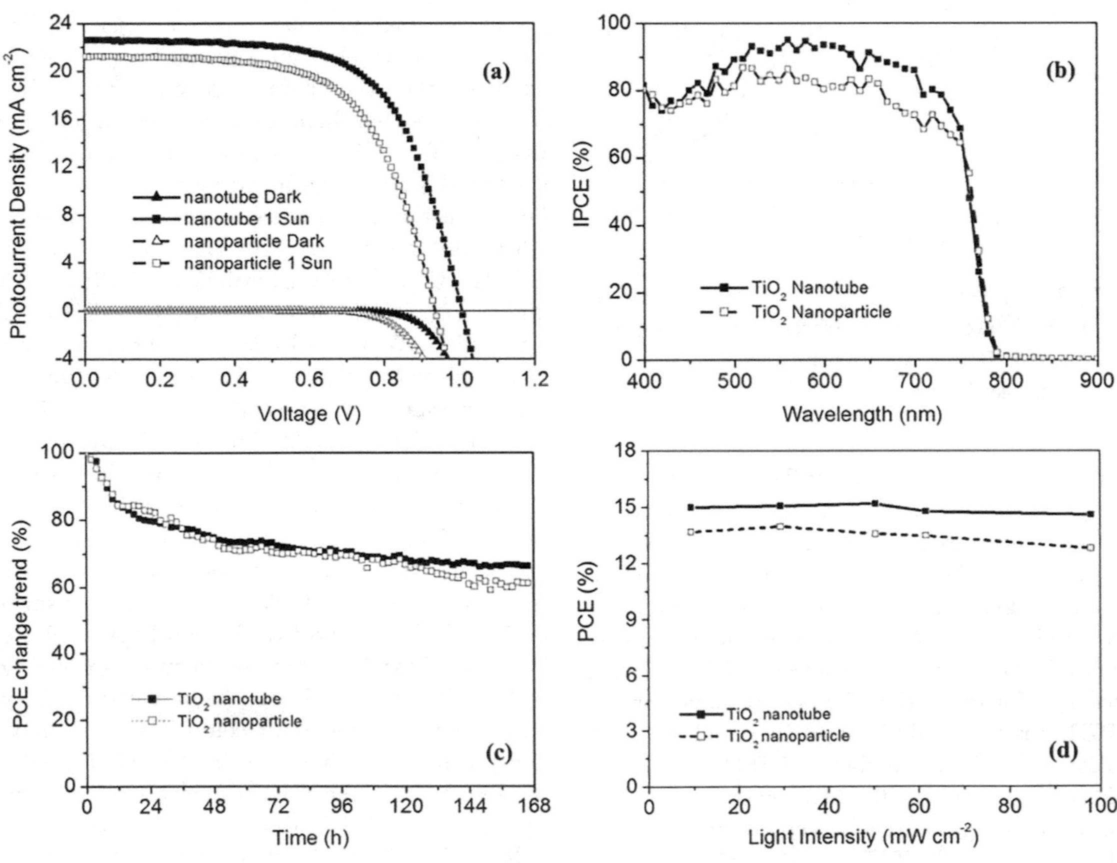

FIGURE 17.37 (a) Current–voltage characteristics and (b) IPCE spectra of $CH_3NH_3PbI_3$ perovskite solar cells based on anodic titania NTs and nanoparticles in the dark and under AM1.5G illumination. (c) Evolution of efficiencies of these cells under constant illumination at 100 mW cm^{-2} at 45°C. (d) The effect of different light intensities on efficiencies. (Reprinted with permission from Ref. [334]; copyright (2015) Johan Wiley and Sons.)

the optical absorption in the visible region and offered a suitable band-edge structure for improved charge separation compared to cells sensitized with CdS or CdSe QDs only. Lan et al. fabricated CdS-sensitized solar cells in the front illumination configuration by successive ionic layer adsorption and reaction (SILAR) method [336]. With optimized electrolyte (poly sulfide), electrode and number of coatings, an efficiency of 3.28% was achieved. Kongkanand et al. performed a photoelectrochemical study of CdSe-TiO_2 system with platinum gauze as counter electrode and aqueous solution of Na_2S as the redox electrolyte [337]. Larger CdSe particles with better absorption in the visible light region could not inject electrons into TiO_2 as effective as smaller particles. Maximum power conversion efficiency obtained was <1% with 3 nm diameter QDs on titania NTs. Based on these results, they proposed a rainbow solar cell with different sized QDs along the length of the titania architecture. Sun et al. studied QD-sensitized TiO_2 NT array photoelectrochemical solar cells [338]. There was a significant improvement in the photocurrent and photovoltage generated in QD sensitized compared to bare titania NTs. The maximum cell efficiency was 4.15%. The improvement in the efficiency was attributed to the multiple exciton generation in QDs. Also, the crystallinity and the geometry of the NTs allow fast and efficient transfer of the photogenerated electrons from CdS QDs to the Ti substrate. Gao et al. obtained a better photocurrent of 6 mA/cm^2 on annealing the CdTe QD-sensitized NT electrodes due to the improvements in the interfacial electron transport [339]. Here also they observed the size effect of CdSe QDs with maximum photocurrent for QDs with absorption peak at 530 nm. The same group studied composite systems of NT/CdS/CdSe and NT/CdSe/CdS also and found that the former had the best photocurrent [340]. They attributed this to the improved charge collection due to favorable Fermi-level alignment in the three semiconductors. Wang et al. fabricated NT/CdTe cells by SILAR method [341]. They used an additional ZnS QD coating for protecting the CdTe QDs from photocorrosion. Maximum current obtained in NT/CdTe system was 11.15 $mAcm^{-2}$. The ZnS QD coating improved the stability of the CdTe/TiO_2 NT photoelectrode in 0.1 M Na_2SO_4 electrolyte containing no redox system. Gakhar et al. sensitized NTs with CdSSe QD with SILAR technique [342]. The phase of the film after annealing at 400°C was determined by XRD as CdS0.54Se0.46 which gave a maximum current of 15.58 mA/cm^2. They found that the photocurrent increased with film thickness and annealing temperature. Increased crystallinity at higher temperature improved current transport and light harvesting ability. CdS–CdTe(CdSe) core–shell QD-sensitized titania NTs were prepared by Gao et al. using anion exchange method with TNT/CdS as sacrificial electrode [343]. Photocurrents measured were 22.6 and 14.7 mA/cm^2 for theTiO_2 NT/CdS–CdSe and TiO_2 NT/CdS–CdTe, respectively, in a mixed electrolyte of 0.1MNa_2S and 0.1MNa_2SO_3. They also found that the ZnS layer coated by SILAR atop the core–shell structure protected the QDs from photocorrosion in the redox couple free 0.1MNa_2SO_4 electrolyte.

17.4 Conclusion

The application scope of titania NTs is widening, even decades after its discovery. This chapter covered only a few major applications of the material. The material is being explored for various other applications. For example, its surface wetting properties have been investigated for microfluidics, heat transfer, etc. [344,345]. Its unique properties are getting revealed even now. For example, recent thermal, electrical and structural studies on individual titania NTs reveale that upon reducing the lattice, the material gets converted from n to p type at temperatures less than about 175 K and the change is reversible [346,347]. Theoretical studies also indicate titania NT as an interesting material. The 21st century is expected to witness the use of this material in various commercial products.

References

1. Grand View Research, Titanium Dioxide (TiO2) Market Size, Share & Trends Analysis Report By Application (Paints & Coatings, Plastics, Pulp & Paper, Cosmetics), By Region (North America, Europe, APAC, MEA, CSA), And Segment Forecasts, 2019 2025, Market Research Report (2018).
2. O.K. Varghese, D. Gong, M. Paulose, K.G. Ong, and C.A. Grimes, Hydrogen sensing using titania nanotubes, *Sensors and Actuators B*, **93** (2003) 338–344.
3. S. Uchida, R. Chiba, M. Tomiha, N. Masaki, and M. Shirai, Applications of titania nanotubes to a dye sensitized solar cell, *Electrochemistry,* **70** (2002) 418–420.
4. O.K. Varghese, D. Gong, M. Paulose, K.G. Ong, E.C. Dickey, and C.A. Grimes, Extreme changes in the electrical resistance of titania nanotubes with hydrogen exposure, *Advanced Materials*, **15** (2003) 624–627.
5. G.K. Mor, O.K. Varghese, M. Paulose, K. Shankar, and C.A. Grimes, A review on highly-ordered TiO_2 nanotube-arrays: Fabrication, material properties, and solar energy applications, *Solar Energy Materials and Solar Cells*, **90** (2006) 2011–2075.
6. Y. Fu and A. Mo, A review on the electrochemically self-organized titania nanotube arrays: Synthesis, modifications and biomedical applications, *Nanoscale Research Letters*, **13** (2018) 187.
7. X. Chen and S. Mao, Titanium dioxide nanomaterials: Synthesis, properties, modifications, and applications, *Chemical Reviews*, **107** (2007) 2891–2949.

8. K. Lee, A. Mazare, and P. Schmuki, One-dimensional titanium dioxide nanomaterials: Nanotubes, *Chemical Reviews*, **114** (2014) 9385–9454.
9. S. Iijima, Helical mircotubules of graphitic carbon, *Nature,* **354** (1991) 56–58.
10. J.C. Hulteen and C.R. Martin, A general template-based method for the preparation of nanomaterials, *Journal of Materials Chemistry,* **7** (1997) 1075–1087.
11. C.R. Martin, Nanomaterials: A membrane-based synthetic approach, *Science,* **266** (1994) 1961–1966.
12. P. Hoyer, Formation of a titanium dioxide nanotube array, *Langmuir,* **12** (1996) 1411–1413.
13. B.B. Lakshmi, P.K. Dorhout, and C.R. Martin, Sol-gel template synthesis of semiconductor nanostructures, *Chemistry of Materials,* **9** (1997) 857–862.
14. C.C. Chen, C. Cheng, and C. Lin, Template assisted fabrication of TiO_2 and WO_3 nanotubes, *Ceramics International,* **39** (2013) 6631–6636.
15. H. Li, Q. Zhou, Y. Gao, X. Gui, L. Yang, M. Du, E. Shi, J. Shi, A. Cao and Y. Fang, Templated synthesis of TiO_2 nanotube macrostructures and their photocatalytic properties, *Nano Research,* **8** (2015) 900–906.
16. Q. Ji and T. Shimizu, Chemical synthesis of transition metal oxide nanotubes in water using an iced lipid nanotube as a template, *Chemistry of Materials,* **2005** (2005) 4411–4413.
17. S.M. Liu et al., Synthesis of single-crystalline TiO_2 nanotubes, *Chemistry of Materials,* **14** (2002) 1391–1397.
18. J. Lee et al., Fabrication of aligned TiO_2 one-dimensional nanostructured arrays using a one-step templating solution approach, *The Journal of Physical Chemistry B,* **109** (2005) 13056–13059.
19. M.S. Sander et al., Template-assisted fabrication of dense, aligned arrays of titania nanotubes with well-controlled dimensions on substrates, *Advanced Materials,* **16** (2004) 2052–2057.
20. H. Shin et al., Formation of TiO_2 and ZrO_2 nanotubes using atomic layer deposition with ultraprecise control of the wall thickness, *Advanced Materials,* **16** (2004) 1197–1200.
21. M. Kemell et al., Atomic layer deposition in nanometer-level replication of cellulosic substances and preparation of photocatalytic TiO_2/cellulose composites, *Journal of the American Chemical Society,* **127** (2005) 14178–14179.
22. T.R.B. Foong, Y. Shen, X. Hu, and A. Sellinger, Template-directed liquid ALD growth of TiO_2 nanotube arrays: Properties and potential in photovoltaic devices, *Advanced Functional Materials,* **20** (2010) 1390–1396.
23. M. Knez, K. Nielsch, and L. Niinistö, Synthesis and surface engineering of complex nanostructures by atomic layer deposition, *Advanced Materials,* **19** (2007) 3425–3438.
24. J. Huang and T. Kunitake, Nanotubings of titania/polymer composite: Template synthesis and nanoparticle inclusion, *Journal of Materials Chemistry,* **16** (2006) 4257–4264.
25. J. Huang and T. Kunitake, Latex particle-encapsulated titania/polymer composite nanotubings: Free-standing, one-dimensional package of colloidal particles, *Chemical Communications,* **2005** (2005) 2680–2682.
26. R.A. Caruso, J.H. Schattka, and A. Greiner, Titanium dioxide tube from sol-gel coating of electrospun polymer fibers, *Advanced Materials,* **13** (2001) 1577–1579.
27. E. Ghadiri et al., Enhanced electron collection efficiency in dye-sensitized solar cells based on nanostructured TiO_2 hollow fibers, *Nano Letters,* **10** (2010) 1632–1638.
28. Y. Qui and J. Yu, Synthesis of titanium dioxide nanotubes from electrospun fiber templates, *Solid State Communications,* **148** (2008) 556–558.
29. K. Nakane et al., Formation of TiO_2 nanotubes by thermal decomposition of poly(vinyl alcohol)-titanium alkoxide hybrid nanofibers, *Journal of Materials Science,* **42** (2007) 4031–4035.
30. K. Nakane and N. Ogata, Photocatalyst nanofibers obtained by calcination of organic-inorganic hybrids, in *Nanofibers,* Ashok Kumar (Ed.), ISBN: 978-953-7619-86-2, InTech (2010).
31. D. Li and Y. Xia, Direct fabrication of composite and ceramic hollow nanofibers by electrospinning, *Nano Letters,* **4** (2004) 933–938.
32. S. Homaeigohar et al., The electrospun ceramic hollow nanofibers, *Nanomaterials,* **7** (2017) 1–32.
33. H. Chen et al., Nanowire-in-microtube structured core/shell fibers via multifluidic coaxial electrospinning, *Langmuir,* **26** (2010) 11291–11296.
34. M. Karaman et al., Template assisted synthesis of photocatalytic titanium dioxide nanotubes by hot filament chemical vapor deposition method, *Applied Surface Science,* **283** (2013) 993–998.
35. X. Du et al., Supramolecular hydrogelators and hydrogels: From soft matter to molecular biomaterials, *Chemical Reviews,* **115** (2015) 13165–13307.
36. F. Ishiwari, Y. Shoji, and T. Fukushima, Supramolecular scaffolds enabling the controlled assembly of functional molecular units, *Chemical Science,* **9** (2018) 2028–2386.
37. S. Kobayashi et al., Preparation of TiO_2 hollow-fibers using supramolecular assemblies, *Chemistry of Materials,* **12** (2000) 1523–1525.
38. J.H. Jung et al., Creation of novel helical ribbon and double-layered nanotube TiO_2 structures using an organogel template, *Chemistry of Materials,* **14** (2002) 1445–1447.
39. G. Gundiah et al., Hydrogel route to nanotubes of metal oxides and sulfates, *Journal of Materials Chemistry,* **13** (2003) 2118–2122.

40. A. Rabenau, The role of hydrothermal synthesis in preparative chemistry, *Angewandte Chemie International Edition*, **24** (1985) 1026–1040.
41. M.H. Sainte-Claire-Deville, Chimie Mineralogique–Reproduction de la Levyne, *Comptes Rendus Hebdomadaires Des Seances De L'Academie Des Sciences,* **54** (1862) 324–327.
42. C.S. Cundy and P.A. Cox, The hydrothermal synthesis of zeolites: History and development from the earliest days to the present time, *Chemical Reviews*, **103** (2003) 663–701.
43. T. Kasuga et al., Formation of titanium oxide nanotube, *Langmuir*, **14** (1998) 3160–3163.
44. N. Liu et al., A review on TiO_2-based nanotubes synthesized via hydrothermal method: Formation mechanism, structure modification, and photocatalytic applications, *Catalysis Today*, **225** (2014) 34–51.
45. T. Kasuga et al., Titania nanotubes prepared by chemical processing, *Advanced Materials*, **11** (1999) 1307–1311.
46. W. Wang et al., A study on the growth and structure of titania nanotubes, *Journal of Materials Research*, **19** (2004) 417–422.
47. Q. Chen et al., The structure of trititanate nanotubes, *Acta Crytallgraphica Section B*, **B58** (2002) 587–593.
48. G.H. Du et al., Preparation and structure analysis of titanium oxide nanotubes, *Applied Physics Letters*, **79** (2001) 3702–3704.
49. P. Dong et al., A study on the $H_2Ti_3O_7$ sheet-like products during the formation process of titanate nanotubes, *Journal of The Electrochemical Society*, **158** (2011) K183–K186.
50. D. Bavykin, J. Friedrich, and F. Walsh, Protonated titanates and TiO_2 nanostructured materials: Synthesis, properties, and applications, *Advanced Materials*, **18** (2006) 2807–2824.
51. J. Huang et al., Tailoring of low-dimensional titanate nanostructures, *Journal of Physical Chemistry C*, **114** (2010) 14748–14754.
52. J. Yang et al., Study on composition, structure and formation process of nanotube $Na_2Ti_2O_4(OH)_2$, *Dalton Transactions*, **20** (2003) 3898–3901.
53. C. Tsai and H. Teng, Structural features of nanotubes synthesized from NaOH treatment on TiO_2 with different post-treatments, *Chemistry of Materials*, **18** (2006) 367–373.
54. C. Tsai and H. Teng, Nanotube formation from a sodium titanate powder via low-temperature acid treatment, *Langmuir*, **24** (2008) 3434–3438.
55. B. Poudel et al., Formation of crystallized titania nanotubes and their transformation into nanowires, *Nanotechnology*, **16** (2005) 1935–1940.
56. Q. Deng et al., Brookite-type TiO_2 nanotubes, *Chemical Communications*, **31** (2008) 3657–3659.
57. A. Nakahira, W. Kato, M. Tamai, T. Isshiki, K. Nishio, and H. Aritani, Synthesis of nanotube from a layered $H_2Ti_4O_9.H_2O$ in a hydrothermal treatment using various titania sources, *Journal of Materials Science,* **39** (2004) 4239–4245.
58. R. Ma, Y. Bando and T. Sasaki, Nanotubes of lepidocrocite titanates, *Chemical Physics Letters,* **380** (2003) 577–582.
59. T. Gao, H. Fjelvag, and P. Norby, Crystal structures of titanate nanotubes: A Raman scattering study, *Inorganic Chemistry,* **48** (2009) 1423–1432.
60. H.-H. Ou and S.-L. Lo, Review of titania nanotubes synthesized via the hydrothermal treatment: fabrication, modification, and application, *Separation and Purification Technology,* **58** (2007) 179–191.
61. D. Bavykin et al., The effect of hydrothermal conditions on the mesoporous structure of TiO_2 nanotubes, *Journal of Materials Chemistry*, **14** (2004) 3370–3377.
62. M. Wei, Y. Konishi, H. Zhou, H. Sugihara, and H. Arakawa, Formation of nanotubes TiO_2 from layered titanate particles by a soft chemical process, *Solid State Communications,* **133** (2005) 493–497.
63. D.L. Morgan, G. Triani, M.G. Blackford, N.A. Raftery, R.L. Frost, and E.R. Waclawik, Alkalaine hydrothermal kinetics in titanate nanostructure formation, *Journal of Materials Science,* **46** (2011) 548–557.
64. R. Ma, Y. Bando, and T. Sasaki, Directly rolling nanosheets into nanotubes, *Journal of Physical Chemistry B,* **104** (2004) 2115–2119.
65. A. Nakahira, T. Kubo, and C. Numako, Formation mechanism of TiO_2-derived titanate nanotubes prepared by the hydrothermal process, *Inorganic Chemistry*, **49** (2010) 5845–5852.
66. Y.Q. Wang, G.Q. Hu, X.F. Duan, H.L. Sun, and Q.K. Xue, Microstructure and formation mechanism of titanium dioxide nanotubes, *Chemical Physics Letters* **365** (2002) 427–431.
67. Y. Guo et al, Preparation of titanate nanotube thin film using hydrothermal method, *Thin Solid Films*, **516** (2008) 8363–8371.
68. Á. Kukovecz et al, Oriented crystal growth model explains the formation of titania nanotubes, *The Journal of Physical Chemistry B*, **109** (2005) 17781–17783.
69. C.L. Wong, Y.N. Tan, and A.R. Mohamed, A review on the formation of titania nanotube photocatalysts by hydrothermal treatment, *Journal of Environmental Management*, **92** (2011) 1669–1680.
70. B.D. Yao et al., Formation mechanism of TiO_2 nanotubes, *Applied Physics Letters*, **82** (2003) 281–283.
71. Y. Lan et al., Titanate nanotubes and nanorods prepared from rutile powder, *Advanced Functional Materials*, **15** (2005) 1310–1318.

72. D. Morgan et al., Implications of precursor chemistry on the alkaline hydrothermal synthesis of titania/titanate nanostructures, *Journal of Physical Chemistry*, **114** (2010) 101–110.
73. Q. Zhang, J. Sun, and S. Zheng, Preparation of long TiO_2 nanotubes from ultrafine rutile nanocrystals, *Chemistry Letters*, **31** (2002) 226–227.
74. Z.V. Saponjic et al., Shaping nanometer-scale architecture through surface chemistry, *Advanced Materials*, **17** (2005) 965–971.
75. P. Liu et al., A facile vapor-phase hydrothermal method for direct growth of titanate nanotubes on a titanium substrate via a distinctive nanosheet roll-up mechanism, *Journal of the American Chemical Society*, **133** (2011) 19032–19035.
76. D. Bavykin et al., An aqueous, alkaline route to titanate nanotubes under atmospheric pressure conditions, *Nanotechnology*, **19** (2008) 1–5.
77. D. Bavykin, A. Kulak, and F. Walsh, Metastable nature of titanate nanotubes in an alkaline environment, *Crystal Growth & Design*, **10** (2010) 4421–4427.
78. L.M. Sikhwivhilu, S.S. Ray, and N.J. Coville, Influence of bases on hydrothermal synthesis of titanate nanostructures, *Applied Physics A*, **94** (2009) 963–973.
79. N. Murakami et al., Control of the crystal structure of titanium(IV) oxide by hydrothermal treatment of a titanate nanotube under acidic conditions, *CrystEngComm*, **12** (2010) 532–537.
80. J. Yu and H. Yu, Facile synthesis and characterization of novel nanocomposites of titanate nanotubes and rutile nanocrystals, *Materials Chemistry and Physics*, **100** (2006) 507–512.
81. X. Wu et al., Synthesis of titania nanotubes by microwave irradiation, *Solid State Communications*, **136** (2005) 513–517.
82. M. Li, Z. Chi, and Y. Wu, Morphology, Chemical composition and phase transformation of hydrothermal derived sodium titanate, *Journal of The American Ceramic Society*, **95** (2012) 3297–3304.
83. X. Zhang, Y. Gui, and X. Dong, Preparation and application of TiO_2 nanotube array gas sensor for SF_6-insulated equipment detection: A review, *Nanoscale Research Letters*, **11** (2016) 1–13.
84. D. Bavykin et al., Application of magic-angle spinning NMR to examine the nature of protons in titanate nanotubes, *Chemistry of Materials*, **22** (2010) 2458–2465.
85. M. Zavala et al., Synthesis of stable TiO_2 nanotubes: Effect of hydrothermal treatment, acid washing and annealing temperature, *Heliyon*, **3** (2017) 1–18.
86. M. Assefpour-Dezfuly, C. Vlachos, and E.H. Andrews: Oxide morphology and adhesive bonding on titanium surfaces. *Journal of Materials Science* **19** (1984) 3626.
87. H. Buff, Ueber das electrische verhalten des aluminiums, *Justus Liebigs Annalen der Chemie*, **102** (1857) 265.
88. T. Kujirai and S. Ueki, Process of coating metallic aluminum or aluminum alloys with aluminum oxide skin. United States Patent # 1735286 (1923).
89. L. Lerner, History of aluminum hard coating, *Aluminium International Today*, (2004) 33–34.
90. A. Caprani and I. Epelboin, Comportement electrochimique du titane en milieu sulfurique fluore contenant de L'Oxygene, *Journal of Electroanalytical Chemistry*, **29** (1971) 335–342.
91. M.J. Mandry and G. Rosenblatt, The effect of fluoride ion on the anodic behavior of titanium in sulfuric acid, *Journal of the Electrochemical Society*, **119** (1972) 29–33.
92. J.J. Kelly, The influence of fluoride ions on the passive dissolution of titanium, *Electrochimica Acta*, **24** (1979) 1273–1282.
93. H. Masuda and K. Fukuda, Ordered metal nanohole arrays made by a 2-step replication of honeycomb structures of anodic alumina, *Science*, **268** (1995) 1466.
94. V. Zwilling, E. Darque-Ceretti, A. Boutry-Forveille, D. David, M.Y. Perrin, and M. Aucouturier, Structure and physicochemistry of anodic oxide films on titanium and TA6V alloy. *Surface and Interface Analysis*, **27** (1999) 629.
95. D. Gong, C.A. Grimes, O.K. Varghese, W. Hu, R.S. Singh, Z. Chen, and E.C. Dickey, Titanium oxide nanotube arrays prepared by anodic oxidation. *Journal of Materials Research*, **16** (2001) 3331.
96. G. Katwal, M. Paulose, I.A. Rusakova, J.E. Martinez, and O.K. Varghese, Rapid growth of zinc oxide nanotube-nanowire hybrid architectures and their use in breast cancer-related volatile organics detection. *Nano Letters*, **11** (2016) 3014.
97. M.R. Banki, A. Torabi, and O.K. Varghese, Anodically grown functional oxide nanotubes and applications, *MRS Communications*, **6** (2016) 375–396.
98. R. Beranek, H. Hildebrand, and P. Schmuki, Self-organized porous titanium oxide prepared in H_2SO_4/HF electrolytes, *Electrochemical and Solid-State Letters*, **6** (2003) B12.
99. K.S. Raja, M. Misra, and K. Paramguru, Formation of self-ordered nano-tubular structure of anodic oxide layer on titanium, *Electochimica Acta*, **51** (2005) 154–165.
100. Q.Y. Cai, M. Paulose, O.K. Varghese, and C.A. Grimes, The effect of electrolyte composition on the fabrication of self-organized titanium oxide nanotube arrays by anodic oxidation, *Journal of Materials Research*, **20** (2005) 230.
101. M. Paulose, O.K. Varghese, K. Shankar, G.K. Mor, and C.A. Grimes, Photoelectrochemical properties of highly-ordered titania nanotube-arrays,

Materials Research Society Symposium Proceedings, **837** (2005) N3.13.1–N3.13.6.

102. J.M. Macak, H. Tsuchiya, and P. Schmuki, High-aspect-ratio TiO_2 nanotubes by anodization of titanium, *Angewandte Chemie International Edition*, **44** (2005) 2100–2102.
103. C.M. Ruan, M. Paulose, O.K. Varghese, G.K. Mor, and C.A. Grimes, Fabrication of highly ordered TiO_2 nanotube arrays using an organic electrolyte. *Journal of Physical Chemistry B*, **109** (2005) 15754.
104. J.M. Macak, H. Tsuchiya, L. Taveira, S. Aldabergerova, and P. Schmuki, Smooth anodic TiO_2 nanotubes, *Angewandte Chemie International Edition*, **19** (2005) 946–948.
105. M. Paulose, K. Shankar, S. Yoriya, H.E. Prakasam, O.K. Varghese, G.K. Mor, T.A. Latempa, A. Fitzgerald, and C.A. Grimes, Anodic growth of highly ordered TiO_2 nanotube arrays to 134 μm in length, *Journal of Physical Chemistry,* **110** (2006) 16179.
106. H.E. Prakasam, K. Shankar, M. Paulose, O.K. Varghese, and C.A. Grimes, A new benchmark for TiO_2 nanotube array growth by anodization. *The Journal of Physical Chemistry C*, **111** (2007) 7235.
107. M. Paulose, H.E. Prakasam, O.K. Varghese, L. Peng, K.C. Popat, G.K. Mor, T.A. Desai, and C.A. Grimes, TiO_2 nanotube arrays of 1,000 μm length by anodization of titanium foil: Phenol red diffusion. *The Journal of Physical Chemistry C*, **111** (2007) 14992.
108. S. Yoriya, G.K. Mor, S. Sharma, and C.A. Grimes, Synthesis of ordered arrays of discrete, partially crystalline titania nanotubes by Ti anodization using diethylene glycol electrolytes, *Journal of Materials Chemistry*, **18** (2008) 3332–3336.
109. K. Nakayama, T. Kubo, A. Tsubokura, Y. Nishikitani, and H. Masuda, Abstract 819, The Electrochemical Society Meeting Abstracts, Vol. 2005-2, Los Angeles, CA, Oct 16–21 (2005).
110. K. Nakayama, T. Kubo, T. Asano, A. Tsubokura, and Y. Nishikitani, Preparation of Normally-Aligned Titania NanotubeLayer Formed on Ti Substrate and Its Application to Dye-Sensitized Solar Cells, Abstract 843, The Electrochemical Society Meeting Abstracts, Vol. 2005-2, Los Angeles, CA, Oct 16–21 (2005).
111. K. Nakayama, T. Kubo, and Y. Nishikitani, Anodic formation of titania nanotubes with ultrahigh apsect ratio, *Electrochemical and Solid State Letters*, **11** (2008) C23–C26.
112. C. Richter, Z. Wu, E. Panaitescu, R.J. Willey, and L. Menon, Ultrahigh aspect ratio titania nanotubes, *Advanced Materials*, **19** (2007) 946–948.
113. G.K. Mor, O.K. Varghese, M. Paulose, and C.A. Grimes, Transparent highly-ordered TiO_2 nanotube-arrays via anodization of titanium thin films, *Advanced Functional Materials*, **15** (2005) 1291–1296.
114. O.K. Varghese, M. Paulose, and C.A. Grimes, Long vertically aligned titania nanotubes on transparent conducting oxide for highly efficient solar cells, *Nature Nanotechnology*, **4** (2009) 592–597.
115. M. Paulose, L. Peng, K.C. Popat, O.K. Varghese, T.J. LaTempa, N. Bao, T. Desai, and C.A. Grimes, Fabrication of mechanically robust, large area, polycrystalline nanotubular/porous TiO_2 membranes, *Journal of Membrane Science*, **319** (2008) 199–205.
116. T.P. Hoar and N.F. Mott, A mechanism for the formation of porous anodic oxide films on aluminum, *Journal of Physics and Chemistry of Solids*, **9** (1959) 97.
117. J.P. O'Sullivan and G.C. Wood, The morphology and mechanism of formation of porous anodic films on aluminum, *Proceedings of the Royal Society of London A*, **317** (1970) 511.
118. W. Lee and S.J. Park, Porous anodic aluminum oxide: Anodization and templated synthesis of functional nanostructures, *Chemical Review*, **25** (2000) 1258.
119. V.P. Parkhutik and V.I. Shershulsky, Theoretical modelling of porous oxide growth on aluminum, *Journal of Physics D: Applied Physics*, **25** (2000) 1258.
120. G.K. Mor, O.K. Varghese, M. Paulose, N. Mukherjee, and C.A. Grimes, Fabrication of tapered, conical-shaped titania nanotubes, *Journal of Materials Research*, **18** (2003) 2588.
121. P. Skeldon, G.E. Thompson, S.J. Garcia-Vargara, L. Iglesias-Rubianes, and C.E. Blanco-Pinzon, A tracer study of porous anodic alumina, *Electrochemical Solid-State Letters*, **9** (2006) B47.
122. S.J. Garcia-Vergara, P. Skeldon, G.E. Thompson, and H. Habazaki, A flow model of porous anodic film growth on aluminum, *Electrohimica Acta*, **52** (2006) 681.
123. J.E. Houser and K.R. Hebert, The role of viscous flow of oxide in the growth of self-ordered porous anodic alumina films, *Nature Materials*, **8** (2009) 415–420.
124. K.R. Hebert, S.P. Albu, I. Paramasivam, and P. Schmuki, Morphological instability leading to formation of porous anodic oxide films, *Nature Materials*, **11** (2012) 162–166.
125. S. Berger, S.P. Albu, F. Schmidt-Stein, H. Hildebrand, P. Schmuki, J.S. Hammond, D.F. Paul, and S. Reichlmaier, The origin of tubular growth of TiO_2 nanotubes: A fluoride rich layer between tube-walls, *Surface Science*, **605** (2011) L57.
126. Z. Su and W. Zhou, Formation mechanism of porous anodic aluminium and titanium oxides, *Advanced Materials*, **20** (2008) 3663.
127. A. Jaroenworaluck, D. Regonini, C.R. Bowen, R. Stevens, and D. Allsopp, Macro, micro and nanostructure of TiO_2 anodised films prepared in a

fluorine-containing electrolyte, *Journal of Materials Science*, **42** (2007) 6729–6734.

128. D. Regonini, C.R. Bowen, A. Jaroenworaluck, and R. Stevens, A review of growth mechanism, structure and crystallinity of anodized TiO_2 nanotubes, *Materials Science and Engineering R*, **74** (2013) 377.
129. O.K. Varghese, D. Gong, M. Paulose, C.A. Grimes, and E.C. Dickey, Crystallization and high-temperature structural stability of titanium oxide nanotube arrays, *Journal of Materials Research*, **18** (2003) 156–165.
130. K. Shankar, G.K. Mor, H.E Prakasam, S. Yoriya, M. Paulose, O.K. Varghese, and C.A. Grimes, Highly-ordered TiO_2 nanotube arrays up to 220 μm in length: Use in water photoelectrolysis and dye-sensitized solar cells, *Nanotechnology*, **18** (2007) 065707.
131. A. Jagminas, G. Niaura, J. Kuzmarskyte-Jagminiene, and V. Pakstas, Crystallization peculiriaties of titania nanotube films under hydrothermal and solvothermal conditions, *Solid State Sciences*, **26** (2013) 97–104.
132. J. Ding, Z. Huang, S. Kou, X. Zhang, and H. Yang, Low-temperature synthesis of high-ordered anatase TiO_2 nanotube array films coated with exposed {001} facets, *Scientific Reports*, **5** (2015) 17773–17780.
133. S. Sharma, O.K. Varghese, G.K. Mor, T.J. LaTempa, N.K. Allam, and C.A. Grimes, Ethanol vapor processing of titania nanotube array films: Enhanced crystallization and photoelectrochemical performance, *Journal of Materials Chemistry*, **19** (2009) 3895–3898.
134. D. Wang, L. Liu, F. Zhang, K. Tao, E. Pippel, and K. Domen, Spontaneous phase and morphology transformations of anodized titania nanotubes induced by water at room temperature, *Nano Letters*, **11** (2011) 3649–3655.
135. B.M. Rao and S.C. Roy, Water assisted crystallization, gas sensing and photo-chemical properties of electrochemically synthesized TiO_2 nanotube arrays, *RSC Advances*, **4** (2014) 49108–49114.
136. A. Takami, Development of titania heated exhaust-gas oxygen sensor, *Ceramic Bulletin*, **67** (1988) 1956–1960.
137. O.K. Varghese and C.A. Grimes, Metal oxide nanoarchitectures for environmental sensing, *Journal of Nanoscience and Nanotechnology*, **3** (2003) 277–293.
138. M. Paulose, O.K. Varghese, G.K. Mor, C.A. Grimes, and K.G. Ong: Unprecedented ultra-high hydrogen gas sensitivity in undoped titania nanotubes, *Nanotechnology*, **17** (2006) 398.
139. C.A. Grimes, K.G. Ong, O.K. Varghese, X. Yang, G. Mor, M. Paulose, E.C. Dickey, C. Ruan, M.V. Pishko, J.W. Kendig, and A.J. Mason, A sentinel sensor network for hydrogen sensing, *Sensors*, **3** (2003) 69–82.
140. O.K. Varghese, G.K. Mor, C.A. Grimes, M. Paulose, and N. Mukherjee, A titania nanotube-array room-temperature sensor for selective detection of hydrogen at low concentrations, *Journal of Nanoscience and Nanotechnology*, **4** (2004) 733–737.
141. G.K. Mor, O.K. Varghese, M. Paulose, and C.A. Grimes, A self-cleaning, room temperature titania-nanotube hydrogen gas sensor, *Sensors Letters*, **1** (2003) 42–46.
142. G.K. Mor, M.A. Carvalho, O.K. Varghese, M.V. Pishko, and C.A. Grimes, A room temperature TiO_2-nanotube hydrogen sensor able to self-clean photoactively from environmental contamination, *Journal of Materials Research*, **19** (2004) 628–634.
143. O.K. Varghese, G.K. Mor, M. Paulose, and C.A. Grimes, A titania nanotube-array room-temperature sensor for selective detection of low hydrogen concentrations, *Materials Research Society Symposium Proceedings*, **835** (2005) K4.1.1/A3.1.1–K4.1.9/A3.1.9.
144. G.K. Mor, O.K. Varghese, M. Paulose, K.G. Ong, and C.A. Grimes, Fabrication of hydrogen sensors with transparent titanium oxide nanotube-array thin films as sensing elements, *Thin Sold Films*, **496** (2006) 42–48.
145. S. Yoriya, H.E. Prakasam, O.K. Varghese, K. Shankar, M. Paulose, G.K. Mor, T.A. Latempa, and C.A. Grimes, Initial studies on the hydrogen gas sensing properties of highly-ordered high aspect ratio TiO_2 nanotube-arrays 20 μm to 222 μm in length, *Sensor Letters*, **4** (2006) 334–339.
146. O.K. Varghese, X. Yang, J. Kendig, M. Paulose, K. Zeng, C.Palmer, K.G. Ong, and C.A. Grimes, A transcutaneous hydrogen sensor: From design to application, *Sensor Letters*, **4** (2006) 120–128.
147. O. Alev, A. Kilic, C. Cakirlar, S. Buyukkose, and Z.Z. Ozturk, Gas sensing properties of p-Co_3O_4/n-TiO_2 nanotube heterostructures, *Sensors*, **18** (2018) 956.
148. O. Alev, E. Sennik, and Z.Z. Ozturk, Improved gas sensing performance of p-copper oxide thin film/n-TiO_2 nanotubes heterostructure, *Journal of Alloys and Compounds*, **749** (2018) 221–228.
149. N. Kilinc, E. Sennik, M. Isik, A.S. Ahsen, O. Ozturk, and Z.Z. Ozturk, Fabrication and gas sensing properties of C-doped and un-doped TiO_2 nanotubes, *Ceramics International*, **40** (2014) 109–115.
150. H. Xun, Z. Zhang, A. Yu, and J. Yi, Remarkably enhanced hydrogen sensing of highly ordered SnO_2 decorated TiO_2 nanotubes, *Sensors and Actuators B*, **273** (2018) 983–990.
151. Y. Kimura, S. Kimura, R. Kojima, M. Bitoh, M. Abe, and M. Niwano, Micro-scaled hydrogen gas sensors with patterned anodic titanium oxide

nanotube film, *Sensors and Actuators B*, **1771** (2013) 1156–1160.

152. K. Chen, K. Xie, X. Feng, S. Wang, R. Hu, H. Gu, and Y. Li, An excellent room temperature hydrogen sensor based on titania nanotube arrays, *International Journal of Hydrogen Energy*, **37** (2012) 13602–13609.
153. Y. Li, X. Yu, and Q. Yang, Fabrication of TiO_2 nanotube thin films and their gas sensing properties, *Journal of Sensors*, **2009** (2009) 402174.
154. H. Gu, Z. Wang, and Y. Hu, Hydrogen gas sensors based on semiconductor oxide nanostructures, *Sensors*, **12** (2012) 5517–5550.
155. J. Moon, H.-P. Hedman, M. Kemell, A. Tuominen, and R. Punkkinen, Hydrogen sensor of Pd decorated tubular TiO_2 layer prepared by anodization with patterned electrodes on SiO_2/Si substrate, *Sensors and Actuators B*, **222** (2016) 190–197.
156. V. Galstyan, E. Comini, G. Faglia, and G. Sberveglieri, TiO_2 nanotubes: Recent advances in synthesis and gas sensing properties, *Sensors*, **13** (2013) 14813 14838.
157. H.F. Lu, F. Li, G. Liu, Z.-G. Chen, D.-W. Wang, H.-Tao Fang, G.Q. Lu, Z.H. Jiang, and H.-M. Cheng. Amorphous TiO_2 nanotube arrays for low-temperature oxygen sensors, *Nanotechnology*, **19** (2008) 405504.
158. X. Tong, W. Shen, X. Chen, and J.-P. Corriou, A fast response and recovery H_2S gas sensor based on free-standing TiO_2 nanotube array films prepared by one-step anodization method, *Ceramics International*, **43** (2017) 14200–14209.
159. P.M. Perillo and D.F. Rodriguez, TiO_2 nanotubes membrane flexible sensor for low-temperature H_2S detection, *Chemosensors*, **4** (2016) 4030015.
160. N.-S. Jang, M.S. Kim, S.-H. Kim, S.-K. Lee, and J.-M. Kim, Direct growth of titania nanotubes on plastic substrates and their application to flexible gas sensors, *Sensors and Actuators B*, **199** (2014) 361–368.
161. V. Galstyan, E. Comini, G. Faglia, A. Vomiero, L. Borgese, E. Bontempi, and G. Sberveglieri, Fabrication and investigation of gas sensing properties of Nb doped TiO_2 nanotubular arrays, *Nanotechnology*, **23** (2012) 235706.
162. E. Comini, V. Galstyan, G. Faglia, E. Bontempi, and G. Sberveglieri, Highly conductive titanium oxide nanotubes chemical sensors, *Microporous and Mesoporous Materials*, **208** (2015) 165–170.
163. X. Zhang, J. Zhang, Y. Jia, P. Xiao, and J. Tang, TiO_2 nanotube array sensor for detecting the SF_6 decomposition product SO_2, *Sensors*, **12** (2012) 3302–3313.
164. E. Sennik, N. Kilinc, and Z.Z. Ozturk, Electrical and VOC sensing properties of anatase and rutile TiO_2 nanotubes, *Journal of Alloys and Compounds*, **616** (2014) 89–96.
165. Y. Kwon et al., Enhanced ethanol sensing properties of TiO_2 nanotube sensors, *Sensors and Actuators B*, **173** (2012) 441–446.
166. S. Lin, J. Wu, and S.A. Akbar, A selective room temperature formaldehyde gas sensor using TiO_2 nanotube arrays, *Sensors and Actuators B*, **156** (2011) 505–509.
167. P.M. Perillo and D.F. Rodríguez, A room temperature chloroform sensor using TiO_2 nanotubes, *Sensors and Actuators B*, **193** (2014) 263–266.
168. P.M. Perillo and D.F. Rodríguez, Low temperature trimethylamine flexible gas sensor based on TiO_2 membrane nanotubes, *Journal of Alloys and Compounds*, **657** (2016) 765–769.
169. F. Fedorov et al., Toward new gas-analytical multisensor chips based on titanium oxide nanotube array, *Scientific Reports*, **7** (2017) 1–9.
170. Y. Zhang et al., Synthesis and characterization of TiO_2 nanotubes for humidity sensing, *Applied Surface Science*, **254** (2008) 5545–5547.
171. C. Han et al., Synthesis of Pd or Pt/titanate nanotube and its application to catalytic type hydrogen gas sensor, *Sensors and Actuators B*, **128** (2007) 320–325.
172. M. Seo et al., Gas sensing characteristics and porosity control of nanostructured films composed of TiO_2 nanotubes, *Sensors and Actuators B*, **137** (2009) 513–520.
173. M. Seo et al., Microstructure control of TiO_2 nanotubular films for improved VOC sensing, *Sensors and Actuators B*, **154** (2011) 251–256.
174. J. Lee et al., A hydrogen gas sensor employing vertically aligned TiO_2 nanotube arrays prepared by template-assisted method, *Sensors and Actuators B*, **160** (2011) 1494–1498.
175. R. Lü et al., Alumina decorated TiO_2 nanotubes with ordered mesoporous walls as high sensitivity NO_x gas sensors at room temperature, *Nanoscale*, **5** (2013) 8569–8576.
176. Q. Zhou et al., Applications of TiO_2 nanotube arrays in environmental and energy fields: A review, *Microporous and Mesoporous Materials*, **202** (2015) 22–35.
177. M. Liu et al., A simple, stable and picomole level lead sensor fabricated on DNA-based carbon hybridized TiO_2 nanotube arrays, *Environmental Science Technology*, **44** (2010) 4241–4246.
178. L. Yang et al., Carbon-nanotube-guiding oriented growth of gold shrubs on TiO_2 nanotube arrays, *Journal of Physical Chemistry C*, **114** (2010) 7694–7699.
179. T. Tran et al., Molecularly imprinted polymer modified TiO_2 nanotube arrays for photoelectrochemical determination of perfluorooctane sulfonate (PFOS), *Sensors and Actuators B*, **190** (2014) 745–751.
180. J. Cai et al., Label-free photoelectrochemical immunosensor based on CdTe/CdS co-sensitized

TiO_2 nanotube array structure for octachlorostyrene detection, *Biosensors and Bioelectronics*, **50** (2013) 66–71.

181. M. Madian et al., Current advances in TiO_2 based nanostructure electrodes for high performance lithium ion batteries, *Batteries*, **4** (2018) 1–36.
182. Y. Zhou et al., Lithium insertion into TiO_2 nanotube prepared by the hydrothermal process, *Journal of the Electrochemical Society*, **150** (2003) A1246–A1249.
183. G. Armstrong et al., Nanotubes with the TiO_2-B structure, *Chemistry Communications*, **19** (2005) 2454–2456.
184. Y. Tang et al., Mechanical force-driven growth of elongated bending TiO_2-based nanotubular materials for ultrafast rechargeable lithium ion batteries, *Advanced Materials*, **26** (2014) 6111–6118.
185. K. Wang et al., Mesoporous titania nanotubes: Their preparation and application as electrode materials for rechargeable lithium batteries, *Advanced Materials*, **19** (2007) 3016–3020.
186. D. Liu et al, TiO_2 Nanotube arrays in N_2 for efficient lithium-ion intercalation, *Journal of Physical Chemistry C*, **112** (2008) 11175–11180.
187. W. Wei et al., High energy and power density TiO_2 nanotube electrodes for 3D Li-ion microbatteries, *Journal of Materials Chemistry A*, **1**(28) (2013) 8160–8169.
188. S. Ivanov et al., Electrochemical behavior of anodically obtained titania nanotubes in organic carbonate and ionic liquid based Li ion containing electrolytes, *Electrochimica Acta*, **104** (2013) 228–235.
189. H. Li et al., High cyclability of ionic liquid produced TiO_2 nanotube arrays as an anode material for lithium ion batteries, *Journal of Power Sources*, **218** (2012) 88–92.
190. G. Ortiz et al., Alternative Li-ion battery electrode based on self-organized titania nanotubes, *Chemistry of Materials*, **21** (2009) 63–67.
191. S. Kim et al., Fabrication and electrochemical characterization of TiO_2 three-dimensional nanonetwork based on peptide assembly, *ACS Nano*, **3** (2009) 1085–1090.
192. A. Tighineanu et al., Conductivity of TiO_2 nanotubes: Influence of annealing time and temperature, *Chemical Physics Letters*, **494** (2010) 260–263.
193. D. Guan et al., Amorphous and chystalline TiO_2 nanotube arrays for enhanced Li-ion intercalation properties, *Journal of Nanoscience and Nanotechnology*, **11** (2011) 3641–3650.
194. D. Fang et al., Electrochemical properties of ordered TiO_2 nanotube loaded with Ag nano-particles for lithium anode material, *Journal of Alloys and Compounds*, **464** (2008) L5–L9.
195. G. Ortiz et al., Exploring a Li-ion battery using surface modified titania nanotubes versus high voltage cathode nanowires, *Journal of Power Sources*, **303** (2016) 194–202.
196. M. López et al., Improving the performance of titania nanotube battery materials by surface modification with lithium phosphate, *ACS Applied Materials & Interfaces*, **6** (2014) 5669–5678.
197. Y. Wang et al., Electrochemical properties of free-standing TiO_2 nanotube membranes annealed in Ar for lithium anode material, *Journal of Solid State Electrochemistry*, **16** (2012) 723–729.
198. Z. Lu, C.-T. Yip, L. Wang, H. Huang, L. Zhou, Hydrogenated $TiO_{2=}$ nanotube arrays as high-rate anodes for lithium ion microbatteries, *ChemPlusChem*, **77** (2012) 991–1000.
199. M. Zhang, c. Wang, H. Li, J. Wang, M. Li and X. Chen, Enhanced performance of lithium ion batteries from self-doped TiO_2 nanotube anodes via an ajdustable electrochemical process, *Electrochimica Acta*, **326** (2019) 134972.
200. H. Han et al., Dominant factors governing the rate capability of a TiO_2 nanotube anode for high power lithium ion batteries, *ACS Nano*, **6** (2012) 8308–8315.
201. Z. Wei et al., TiO_2 nanotube array film prepared by anodization as anode material for lithium ion batteries, *Journal of Solid State Electrochemistry*, **14** (2010) 1045–1050.
202. Y. Liu and Y. Yang, Recent progress of TiO_2-based anodes for Li ion batteries, *Journal of Nanomaterials*, **2015** (2015) 1–15.
203. J. González et al., Controlled growth and application in lithium and sodium batteries of high-aspect-ratio, self-organized titania nanotubes, *Journal of the Electrochemical Society*, **160** (2013) A1390–A1398.
204. S.P. Mohanty, B. Kishore, and M. Nookala, Composites of sulfur-titania nanotubes prepared by a facile solution infiltration route as cathode material in lithium-sulfur battery, *Journal of Nanoscience and Nanotechnology*, **18** (2018) 6830–6837.
205. C. Raj and R. Prasanth, Review-advent of TiO_2 nanotube as supercapacitor electrode, *Journal of the Electrochemical Society*, **165** (2018) E345–E358.
206. F. Fabregat-Santiago et al., High carrier density and capacitance in TiO_2 nanotube arrays induced by electrochemical doping, *JACS Articles*, **130** (2008) 11312–11316.
207. P. Xiao et al., Electrochemical and photoelectrical properties of titania nanotube arrays annealed in different gases, *Sensors and Actuators B: Chemical*, **134** (2008) 367–372.
208. X. Lu et al., Hydrogenated TiO_2 nanotube arrays for supercapacitors, *Nano Letters*, **12** (2012) 1690–1696.
209. Y. Xie et al., Electrochemical capacitance performance of titanium nitride nanoarray, *Material Science and Engineering: B*, **178** (2013) 1443–1451.

210. H. Wu et al., Enhanced supercapacitance in anodic TiO_2 nanotube films by hydrogen plasma treatment, *Nanotechnology*, **24** (2013) 1–7.
211. H. Wu, D. Li, X. Zhu, C. Yang, D. Liu, X. Chen, Y. Song, and L. Lu, High performance and renewable supercapacitors based in TiO_2 nanotube array electrodes treated by an electrochemical doping approach, *Electrochimica Acta*, **116** (2014) 129.
212. H. Zhou and Y. Zhang, Enhanced electrochemical performance of manganese dioxide spheres deposited on a titanium dioxide nanotube array substrate, *Journal of Power Sources*, **116** (2014) 5626–5636.
213. C. Kim et al., Capacitive and oxidant generating properties of black-colored TiO_2 nanotube array fabricated by electrochemical self-doping, *ACS Applied Material & Interfaces*, **7** (2015) 7486–7491.
214. D. Pan et al., C-axiz preferentially oriented and fully activated TiO_2 nanotube arrays for lithium ion batteries and supercapacitors, *Journal of Materials Chemistry A*, **29** (2014) 11454–11465.
215. W. Zhong et al., Electrochemically conductive treatment of TiO_2 nanotube arrays in $AlCl_3$ aqueous solution for supercapacitors, *Journal of Power Sources*, **294** (2015) 216–222.
216. Y. Cheng et al., Progress in TiO_2 nanotube coatings for biomedical applications: A review, *Journal of Materials Chemistry B*, **6** (2018) 1862–1886.
217. K.C. Popat, M. Eltgroth, T.J. La Tempa, C.A. Grimes, and T.A. Desai, Decreased Staphylococus epidermis adhesion and increased osteoblast functionality on antibiotic loaded titania nanotubes, *Biomaterials*, **28** (2007) 4880–4888.
218. K.C. Popat, M. Eltgroth, T.J. La Tempa, C.A. Grimes, and T.A. Desai, Titania nanotubes: A novel platform for drug-eluting coatings for medical implants, *Small*, **11** (2007) 1878–1881.
219. L. Zhang et al., Inhibitory effect of siperhydrophobicity on silver release and antibacterial properties of super-hydrophobic Ag/TiO_2 nanotubes, *Journal of Biomedical Materials Research Part B*, **105** (2015) 1004–1012.
220. J. Li, H. Zhou, S. Qian, Z. Liu, J. Feng, P. Jin, and X. Liu, Plasmonic gold nanoparticles modified titania nanotubes for antibacterial application, *Applied Physics Letters*, **104** (2014) 261110.
221. M.M. Gentleman et al., The role of surface free energy in osteoblast-biomaterial interactions, *International Materials Review*, **59** (2014) 417–429.
222. M. Kulkarni et al., Protein interactions with layers of TiO_2 nanotube and nanopore arrays: Morphology and surface charge influence, *Acta Biomaterialia*, **45** (2016) 357–366.
223. B.S. Smith et al., Hemocompatibility of titania nanotube arrays, *Journal of Biomedical Materials Research, Part A*, **95** (2010) 350–360.
224. S.C. Roy, M. Paulose, and C.A. Grimes, The effect of TiO_2 nanotubes in the enhancement of blood clotting for the control of hemorrhage, *Biomaterials* **28** (2007) 4667–4672.
225. A.W. Tan et al., Review of titania nanotubes: Fabrication and cellular response, *Ceramics International*, **38** (2012) 4421–4435.
226. C.V. Wilmowsky et al., The diameter of anodic TiO_2 nanotubes affects bone formation and correlates with the bone morphogenetic protein-2 expression in vivo, *Clinical Oral Implants Research*, **23** (2011) 359–366.
227. J. Park et al., Synergistic control of mesenchymal stem cell differentiation by nanoscale surface geometry and immobilized growth factors on TiO_2 nanotubes, *Small*, **5** (2009) 666–671.
228. J. Park et al., Nanosize and vitality: TiO_2 nanotube diameter directs cell fate, *Nano Letters*, **7** (2007) 1686–1691.
229. K. Brammer et al., Nanotube surface triggers increased chondrocyte extracellular matrix production, *Materials Science and Engineering C*, **30** (2010) 518–525.
230. S. Oh, C. Daraio, L.H. Chen, T.R. Pisanic, R.R. Finones, and S. Jin, Significantly accelerated osteoblast cell growth on aligned TiO_2 nanotubes, *Journal of Biomedical Materials Research*, **78** (2006) 97–103.
231. W.Q. Yu et al., In vitro behavior of MC 3T3-EI preosteoblast with different annealing temperature titania nanotubes, *Oral Diseases*, **16** (2010) 624–630.
232. Y. Lai et al., Nanotube arrays: Bioinspired patterning with extreme wettability contrast on TiO_2 nanotube array surface: A versatile platform for biomedical applications, *Small*, **9** (2013) 2945–2953.
233. L. Zhao et al., The influence of hierarchical hybrid micro/nano-textured titanium surface with titania nanotubes on osteoblast functions, *Biomaterials*, **31** (2010) 2055–2063.
234. H.Y. Seo et al., Cellular attachment and differentiation on titania nanotubes exposed to air- or nitrogen-based non-thermal atmospheric pressure plasma, *PLoS One*, **9** (2014) e113477.
235. L. Xia et al., In vitro and in vivo studies of surface-structured implants for bone formation, *International Journal of Nanomedicine*, **4** (2012) 1423–1436.
236. L. Bai, R. Wu, Y. Wang, X. Wang, X. Zhang, X. Huang, L. Qin, R. Hang, L. Zhao, and B. Tang, Osteogenic and angiogenic activities of silicon-incorporated TiO_2 nanotube arrays, *Journal of Materials Chemistry B*, **4** (2016) 5548–5559.
237. Y.T. Sul, Electrochemical growth behavior, surface properties, and enhanced in vivo bone response of TiO_2 nanotubes on microstructured surfaces of

blasted, screw-shaped titanium implants, *International Journal of Nanomedicine*, **5** (2010) 87–100.

238. T. Monetta et al, TiO_2 Nanotubes on Ti dental implant. Part 1: Formation and aging in hank's solution, *Metals*, **7** (2017) 167.
239. S.H. Nemati and A. Hadjizadeh, Gentamicin-eluting titanium dioxide nanotubes grown on the ultrafine-grained titanium, *AAPS PharmSciTech*, **18** (2017) 2180–2187.
240. I. Jang, S.C. Shim, D.S. Choi, B.K. Cha, J.K. Lee, B.H. Choe, and W.Y. Choi, Effect of TiO_2 nanotubes arrays on osseointegration of orthodontic miniscrew, *Biomedical Microdevices*, **17** (2015) 76.
241. L. Peng et al., The effect of TiO_2 nanotubes on endothelial function and smooth muscle proliferation, *Biomaterials*, **30** (2009) 1268–1272.
242. H. Nuhn, C.E. Blanco, and T.A. Desai, Nanoengineered stent surface to reduce in-stent restenosis in vivo, *Applied Materials & Interfaces*, **19** (2017) 19677–19686.
243. M. Kalbacova et al., TiO_2 nanotubes: Photocatalyst for cancer cell killing, *Physica Status Solidi RRL*, **2** (2008) 194–196.
244. K. Tamura, Y. Ohko, H. Kawamura, H. Yoshikawa, T. Tatsuma, A. Fujishima, and J.I. Mizuki, X-ray induced photoelectrochemistry on TiO_2. *Electrochimica Acta*, **52** (2007) 6938–6942.
245. C. Hong, J. Kang, J. Lee, H. Zheng, S. Hong, D. Lee, and C. Lee, Photothermal therapy using TiO_2 nanotubes in combination with near-infrared laser, *Journal of Cancer Therapy*, **1** (2010) 52–58.
246. K. Gulati, M.S. Aw, and D. Losic, Nanoengineered drug-releasing Ti wires as an alternative for local delivery of chemotherapeutics in the brain, *International Journal of Nanomedicine*, **7** (2012) 2069–2076.
247. M.S. Mohamed, A. Torabi, M. Paulose, D.S. Kumar, and O.K. Varghese, Anodically grown titania nanotube induced cytotoxicity has genotoxic origins, *Scientific Reports*, **7** (2017) 41844.
248. A.B. Wijeratne, D.N. Wijesundera, M. Paulose, I.B. Ahiabu, W.-K. Chu, O.K. Varghese, K.D. Greis, Phosphopeptide separation using radially aligned titania nanotubes on titanium wire, *ACS Applied Materials & Interfaces*, **7** (2015) 11155–11164.
249. H.-C. Lee, L.-F. Zhang, J.-L. Lin, Y.-L Lin, and T.-P. Sun, Development of anodic titania nanotubes for application in high sensitivity amperometric glucose and uric acid biosensors, *Sensors*, **13** (2013) 14161–14174.
250. Y.B. Xie, L.M. Zhou, and H.T. Huang, Biosensor application of enzyme-functionalized titania/titanium composite, *Biosensors and Bioelectronics*, **22** (2007) 2812–2818.
251. J. Zhu et al, Preparation of polyaniline-TiO_2 nanotube composite for the development of electrochemical biosensors, *Sensors and Actuators B*, **221** (2015) 450–457.
252. A. Fujishima, K. Honda, Electrochemical photolysis of water at a semiconductor electrode. *Nature*, **238** (1972) 37–38.
253. T. Inoue, A. Fujishima, S. Konishi, and K. Honda, Photoelectrochemical reaction of carbon dioxide in aqueous suspensions of semiconductor powders, *Nature*, **277** (1979) 637–638.
254. G.K. Mor, K. Shankar, M. Paulose, O.K. Varghese, and C.A. Grimes, Enhanced photocleavage of water using titania nanotube arrays, *Nanoletters*, **5** (2005) 191–195.
255. O.K. Varghese, M. Paulose, K. Shankar, G.K. Mor, and C.A. Grimes, Water-photoelectrolysis properties of highly-ordered titania nanotube-arrays, *Journal of Nanoscience and Nanotechnology*, **5** (2005) 1158–1165.
256. M. Paulose, G.K. Mor, O.K. Varghese, K. Shankar, and C.A. Grimes, Visible light photoelectrochemical and water-photoelectrolysis properties of titania nanotube arrays, *Journal of Photochemistry and Photobiology A: Chemistry*, **178** (2006) 8–15.
257. C. Ruan, M. Paulose, O.K. Varghese, and C.A. Grimes, Enhanced photoelectrochemical-response in highly ordered TiO_2 nanotube arrays anodized in boric acid containing electrolyte, *Solar Energy Materials & Solar Cells*, **90** (2006) 1283–1295.
258. N.H. Nguyen and B. Bai, Effect of washing pH on the properties of titanate nanotubes and its activity for photocatalytic oxidation of NO and NO_2, *Applied Surface Science*, **355** (2015) 672–680.
259. W.-K. Jo, Y.Y. Lee, and H.-H Chun, Titania nanotubes grown on carbon fibers for photocatalytic decomposition of gas phase aromatic pollutants, *Materials*, **7** (2014) 1801–1813.
260. S. Sreekantan and L.C. Wei, Study on the formation and photocatalytic activity of titanate nanotubes synthesized via hydrothermal method, *Journal of Alloys and Compounds*, **490** (2010) 436–442.
261. H. Kmentova et al., Photoelectrochemical and structural properties of TiO_2 nanotubes and nanorods grown on FTO substrate: Comparatie study between electrochemical anodization and hydrothermal method used for the nanostructures fabrication, *Catalysis Today*, **287** (2017) 130–136.
262. T. Tachikawa, S. Tojo, M. Fujitsuka, T. Sekino and T. Majima, Photoinduced charge separation in titania nanotubes, *Journal of Physical Chemistry B*, **110** (2006) 14055–14059.
263. M. Ge, Q. Li, C. Cao, J. Huang, S. Li, S. Zhang, Z. Chen, K. Zhang, S.S. Al-Deyab, and Y. Lai, One dimensional TiO_2 nanotube photocatalysts for solar water splitting, *Advanced Science*, **4** (2017) 1600152.

264. Y. Xie, Photoelectrochemical application of nanotubular titania photoanode, *Electrochimica Acta*, **51** (2006) 3399–3406.
265. Z. Zhang, M.F. Hossain, and T. Takashashi, Photoelectrochemical water splitting on highly smooth and ordered TiO_2 nanotube arrays for hydrogen generation, *International Journal of Hydrogen Energy*, **35** (2010) 8528–8535.
266. Y.R. Smith, R.S. Ray, K. Carlson, B. Sarma, and M. Misra, Self-ordered titanium dioxide nanotube arrays: Anodic synthesis and their photo/electrocatalytic applications, *Materials*, **6** (2013) 2892–2957.
267. K. Shankar, J.I. Basham, N.K. Allam, O.K. Varghese, G.K. Mor, X. Feng, M. Paulose, J.A. Seabold, K.-S. Choi, and C.A. Grimes, Recent advances in the use of TiO_2 nanotube and nanowire arrays for oxidative photoelectrochemistry, *Journal of Physical Chemistry C*, **113** (2009) 6327–6359.
268. O.K. Varghese and C.A. Grimes, Appropriate strategies for determining the photoconversion efficiency of water photoelectrolysis cells: A review with examples using titania nanotube array photoanodes, *Solar Energy Materials and Solar Cells*, **92** (2008) 374–384.
269. O.K. Varghese, M. Paulose, T.J. LaTempa, and C.A. Grimes, High-rate solar photocatalytic conversion of CO_2 and water vapor to hydrocarbon fuels, *Nano Letters*, **9** (2009) 731–737.
270. K. Shankar, M. Paulose, G.K. Mor, O.K. Varghese, and C.A. Grimes, A study on the spectral response and photoelectrochemical properties of flame- annealed titania nanotube-arrays, *Journal of Physics D: Applied Physics*, **38** (2005) 3543–3549.
271. M. Szkoda, A. Lisowska-Oleksiak, and K. Siuzdak, Optimization of boron-doping process of titania nanotubes via electrochemical method toward enhanced photoactivity, *Journal of Solid State Electrochemistry*, **20** (2016) 1765–1774.
272. R. Asapu, V.M. Palla, B. Wang, Z. Guo, R. Sadu, and D.H. Chen, Phosphorous-doped titania nanotubes with enhanced photocatalytic activity, *Journal of Photochemistry and Photobiology A*, **225** (2011) 81–87.
273. N. Liu, V. Haublein, X. Zhou, U. Venkatesan, M. Hartmann, M. Mackovic, T. Nakajima, E. Spiecker, A. Osvet, L. Frey, and P. Schmuki, “Black” TiO_2 nanotubes formed by high energy proton implantation show noble-metal-co-catalyst free photocatalytic H_2 evolution, *Nano Letters*, **15** (2015) 6815–6820.
274. Z. Dong, D. Ding, T. Li, and C. Ning, Ni-doped TiO_2 nanotubes photoanode for enhanced photoelectrochemical water splitting. *Applied Surface Science*, **443** (2018) 321–328.
275. T.T. Isimjan, A. El Ruby, S. Rohani, and A.K. Ray, The fabrication of highly ordered and visible light responsive Fe-C-N codoped TiO_2 nanotubes, *Nanotechnolgy*, **21** (2010) 055706.
276. M.M. Momeni and Y. Ghayeb, Photoelectrochemical water splitting on chromium-doped titanium doxide nanotube photoanodes prepared by single-step anodizing, *Journal of Alloys and Compounds*, **637** (2015) 393–400.
277. T. Cottineau, N. Bealu, P.-A. Gross, S.N. Pronkin, N. Keller, E.R. Savinova, and V. Keller, One step synthesis of niobium doped titania nanotube arrays to form (N, Nb) co-doped TiO_2 with high visible light photoelectrochemical activity, *Journal of Materials Chemistry A*, **1** (2013) 2151–2160.
278. L. Sun, J. Li, C. Wang, S. Li, Y. Lai, H. Chen, and C. Lin, Ultrasound aided photochemical synthesis of Ag loaded TiO_2 nanotube arrays to enhance photocatalytic activity, *Journal of Hazardous Materials*, **171** (2009) 1045–1050.
279. M.-Y. Lan, C.-P. Liu, H.-H. Huang, and S.-W. Lee, Both enhanced biocompatibility and antibacterial activity in Ag decorated TiO_2 nanotubes, *PLoS One*, **8** (2013) e75364.
280. P.V. Viet, B.T. Phan, D. Mott, S. Maenosono, T.T. Sang, C.M. Thi, and L.V. Hieu, Silver nanoparticle loaded TiO_2 nanotubes with high photocatalytic and antibacterial activity synthesized by photoreduction method, *Journal of Photochemistry and Photobiology A*, **352** (2018) 106–112.
281. F. Xiao, Layer-by-layer self assembly construction of highly ordered metal TiO_2 nanotube arrays heterostructures (M/TNTs, M = Au, Ag, Pt) with tunable catalytic activities, *Journal of Physical Chemistry C*, **116** (2012) 16487–16498.
282. Z. Zhang, L. Zhang, M.N. Hedhili, H. Zhang, and P. Wang, Plasmonic gold nanocrystals coupled with photonic crystal seamlessly on TiO_2 nanotube photoelectrodes for efficient visible light photoelectrochemical water splitting, *Nano Letters*, **13** (2013) 14–20.
283. S. Chen, M. Paulose, C. Ruan, G.K. Mor, O.K. Varghese, D. Kouzoudis, and C.A. Grimes, Electrochemically synthesized CdS Nanoparticle-modified TiO_2 nanotube-array photoelectrodes: preparation, characterization and application to photoelectrochemical cells, *Journal of Photochemistry and Photobiology A: Chemistry*, **177** (2006) 177–184.
284. Y. Xie, G. Ali, S.H. Yoo, and S.O. Cho, Sonication-assisted synthesis of CdS quantum-dot-sensitized TiO_2 nanotube arrays with enhanced photoelectrochemical and photocatalytic activity, *ACS Applied Materials Interfaces*, **2** (2010) 2910–2914.
285. C. Li, J. Yuan, B. Han, L. Jiang, and W. Shangguan, TiO_2 nanotubes incorporated with CdS for photocatalytic hydrogen production from splitting water under visible light irradiation, *International Journal of Hydrogen Energy*, **35** (2010) 7073–7079.

286. W. Wang, F. Li, D. Zhang, D.Y.C. Leung, and G. Li, Photoelectrocatalytic hydrogen generation and simultaneous degradation of organic pollutant via CdSe/TiO_2 nanotube arrays, *Applied Surface Science*, **362** (2016) 490–497.

287. K.R. Reyes-Gil and D.B. Robinson, WO_3-enhanced TiO_2 nanotube photoanodes for solar water splitting with simultaneous wastewater treatment, *ACS Applied Materials and Interfaces*, **5** (2013) 12400–12410.

288. C. Liu, F. Wang, J. Zhang, K. Wang, Y. Qiu, Q. Liang, and Z. Chen, Efficient photoelectrochemical water splitting by g-C_3N_4/TiO_2 nanotube array heterostructures, *Nano-Micro Letters*, **10** (2018) 37.

289. R. Wang, J. Bai, Y. Li, Q. Zeng, J. Li, and B. Zhou, $BiVO_4$/TiO_2 (N_2) nanotubes heterojunction photoanode for highly efficient photoelectrocatalytic applications, *Nano-Micro Letters*, **9** (2017) 14.

290. A.P. Balan, S. Radhakrishnan, C.F. Woellner, S.K. Sinha, L. Deng, C.A. de los Reyes, B.M. Rao, M. Paulose, R. Neupane, A. Apte, V. Kochat, R. Vajtai, A.R. Harutyunyan, C.-W. Chu, G. Costin, D.S. Galvao, A.A. Marti, P.A. Aken, O.K. Varghese, C.S. Tiwary, M.R. Anantharaman, and P.M. Ajayan, Exfoliation of a non-van der Waals material from iron ore hematite, *Nature Nanotechnology*, **13** (2018) 602–609.

291. A.P. Balan, S. Radhakrishnan, R. Kumar, R. Neupane, S.K. Sinha, L. Deng, C.A. de los Reyes, A. Apte, B.M. Rao, M. Paulose, R. Vajtai, C.-W. Chu, G. Costin, A.A. Marti, O.K. Varghese, A.K. Sing, C.S. Tiwary, M.R. Anantharaman, and P.M. Ajayan, A non-van der Waals two-dimensional material from natural titanium mineral ore ilmenite, *Chemistry of Materials*, **30** (2018) 5923–5931.

292. A.P. Balan, S. Radhakrishnan, R. Neupane, S. Yazdi, L. Deng g, C.A. de los Reyes, A. Apte, A.B. Puthirath, B.B.M. Rao, M. Paulose, R.Vajtai, C.-W. Chu, A.A. Marti, O.K. Varghese, C.S. Tiwary, M.R. Anantharaman, and P.M. Ajayan, Magnetic properties and photocatalytic applications of 2D sheets of nonlayered manganese telluride by liquid exfoliation, *ACS Applied Nano Materials*, **1** (2018) 6427–6434.

293. S. Radhakrishnan, J.H. Park, R. Neupane, C.A. de los Reyes, P.M. Sudeep, M. Paulose, A.A. Marti, C.S. Tiwary, V.N. Khabashesku, O.K. Varghese, B.A. Kaipparettu, and P.M. Ajayan, Fluorinated boron nitride quantum Dots: A new 0D material for energy conversion and detection of cellular metabolism, *Particle & Particle Systems Characterization*, **36** (2019) 1800346.

294. B. O'Regan and M. Gratzel, A low-cost, high efficiency solar cell based on dye sensitized colloidal TiO_2 films, *Nature*, **353** (1991) 737–740.

295. M. Grätzel, Dye-sensitized solar cells, *Journal of Photochemistry and Photobiology C: Photochemistry Reviews*, **4** (2003) 145–153.

296. Q. Zhang and G. Cao, Nanostructured photoelectrodes for dye-sensitized solar cells, *Nano Today*, **6** (2011) 91–109.

297. J. Weickert, R.B. Dunbar, H.C. Hesse, W. Wiedemann, and L.S. Mende, Nanostructured organic and hybrid solar cells, *Advanced Materials*, **23** (2011) 1810–1828.

298. M. Adachi, I. Okada, S. Ngamsinlapasathian, Y. Murata, and S. Yoshikawa, Dye sensitized solar cells using semiconductor thin film composed of titania nanotubes, *Electrochemistry*, **70** (2002), 449–452.

299. M. Adachi, Y. Murata, I. Okada, and S. Yoshikawa, Formation of titania nanotubes and applications for dye-sensitized solar cells, *Journal of The Electrochemical Society*, **150** (2003) G488–G493.

300. S. Ngamsinlapasathian, S. Sakulkhaemaruethai, S. Pavasupree, A. Kitiyanan, T. Sreethawong, Y. Suzuki, and S. Yoshikawa, Highly efficient dye-sensitized solar cell using nanocrystalline titania containing nanotube structure, *Journal of Photochemistry and Photobiology A: Chemistry,* **164** (2004) 145–151.

301. M. Macak, H. Tsuchiya, A. Ghicov, P. Schmuki, Dye-sensitized anodic TiO_2 nanotubes, *Electrochemistry Communications*, **7** (2005) 1133–1137.

302. G.K. Mor, K. Shankar, M. Paulose, O.K. Varghese, and C.A. Grimes, Use of highly-ordered TiO_2 nanotube arrays in dye-sensitized solar cells, *Nano Letters*, **6** (2006) 215–218.

303. M. Paulose, K. Shankar, O.K. Varghese, G.K. Mor, B. Hardin, and C.A. Grimes, Backside illuminated dye-sensitized solar cells based on titania nanotube array electrodes, *Nanotechnology*, **17** (2006) 1446–1448.

304. M. Paulose, K. Shankar, O.K. Varghese, G.K. Mor, and C.A. Grimes, Application of highly-ordered TiO_2 nanotube-arrays in heterojunction dye-sensitized solar cells, *Journal of Physics D: Applied Physics*, **39** (2006) 2498–2503.

305. K. Zhu, N.R. Neale, A. Miedaner, and A.J. Frank, Enhanced charge-collection efficiencies and light scattering in dye-sensitized solar cells using oriented TiO_2 nanotubes arrays, *Nano Letters*, **7** (2007) 69–74.

306. K. Zhu, T.B. Vinzant, N.R. Neale, and A.J. Frank, Removing structural disorder from oriented TiO nanotube arrays: Reducing the dimensionality of transport and recombination in dye sensitized solar cells, *Nano Letters,* **7** (2007) 3739–3746.

307. J.R. Jennings, A. Ghicov, L.M. Peter, P. Schmuki, and A.B. Walker, Dye-sensitized solar cells based on oriented TiO_2 nanotube arrays: transport, trapping, and transfer of electrons, *Journal of the American Ceramic Society*, **130** (2008) 13364–13372.

308. J.H. Park, T.-W. Lee, and M.G. Kang, Growth, detachment and transfer of highly-ordered TiO_2 nanotube arrays: use in dye-sensitized solar cells, *Chemical Communications*, (2008) 2867–2869.
309. Q. Chen and D. Xu, Large-Scale, Noncurling, and free-standing crystallized TiO_2 nanotube arrays for dye-sensitized solar cells, *The Journal of Physical Chemistry C*, **113** (2009) 6310–6314.
310. D. Kim, A. Ghicov, S.P. Albu, and P. Schmuki, Bamboo-type TiO_2 nanotubes: Improved conversion efficiency in dye-sensitized solar cells, *Journal of the American Chemical Society*, **130** (2008) 16454–16455.
311. J. Zhang, Q. Li, S. Li, Yi Wang, C. Ye, P. Ruterana, and H. Wang, An efficient photoanode consisting of TiO_2 nanoparticle-filled TiO_2 nanotube arrays for dye sensitized solar cells, *Journal of Power Sources*, **268** (2014) 941–949.
312. J. Zhang, S. Li, H. Ding, Q. Li, B. Wang, X. Wang, and H. Wang, Transfer and assembly of large area TiO_2 nanotube arrays onto conductive glass for dye sensitized solar cells, *Journal of Power Sources*, **247** (2014) 807–812.
313. T. Zeng, H. Ni, X. Su, Y.Chen, and Yi Jiang, Highly crystalline Titania nanotube arrays realized by hydrothermal vapor route and used as front-illuminated photoanode in dye sensitized solar cells, *Journal of Power Sources,* **283** (2015) 443–451.
314. Y.-L. Xie, Z.-X. Li, Z.-G. Xu, and H.-L. Zhang, Preparation of coaxial TiO_2/ZnO nanotube arrays for high-efficiency photo-energy conversion applications, *Electrochemistry Communications*, **13** (2011) 788–791.
315. K. Shankar, J. Bandara, M. Paulose, H. Wietasch, O.K. Varghese, G.K. Mor, T.J. LaTempa, M. Thelakkat, and C.A. Grimes, Highly efficient solar cells using TiO_2 nanotube arrays sensitized with a donor-antenna dye, *Nano Letters*, **8** (2008) 1654–1659.
316. I.C. Flores, J.N.de Freitas, C.Longoa, M.-A. De Paoli, H.Winnischofer, and A.F. Nogueira, Dye-sensitized solar cells based on TiO_2 nanotubes and a solid-state electrolyte, *Journal of Photochemistry and Photobiology A: Chemistry*, **189** (2007) 153–160.
317. P.Chen, J. Brillet, H. Bala, P. Wang, S.M. Zakeeruddin, and M. Gratzel, Solid-state dye-sensitized solar cells using TiO_2 nanotube arrays on FTO glass, *Journal of Materials Chemistry*, **19** (2009) 5325–5328.
318. G.K. Mor, J. Basham, M. Paulose, S. Kim, O.K. Varghese, A. Vaish, S. Yoriya, and C.A. Grimes, High-efficiency Forster resonance energy transfer in solid state dye sensitized solar cells, *Nano Letters*, **10** (2010) 2387–2394.
319. L. Peedikakkandy, J. Naduvath, S. Mallick, and P. Bhargava, Lead free, air stable perovskite derivative Cs_2SnI_6 as HTM in DSSCs employing TiO_2 nanotubes as photoanode, *Materials Research Bulletin*, **108** (2018) 113–119.
320. D. Kuang, J. Brillet, P. Chen, M. Takata, S. Uchida, H. Miura, K. Sumioka, S.M. Zakeeruddin, and M. Gratzel, Application of highly ordered TiO_2 nanotube arrays in flexible dye-sensitized solar cells, *ACS Nano*, **2** (2008) 1113–1116.
321. Z. Liu, V.R. Subramania, and M. Misra, Vertically oriented TiO_2 nanotube arrays grown on Ti meshes for flexible dye-sensitized solar cells, *The Journal of Physical Chemistry C*, **113** (2009) 14028–14033.
322. Y. Xiao, J. Wu, G. Yue, G. Xie, J. Lin, and M. Huang, The preparation of titania nanotubes and its application in flexible dye-sensitized solar cells, *Electrochimica Acta*, **55** (2010) 4573–4578.
323. L. Song, Y. Zhou, Y. Guan, P. Du, J. Xiong, and F. Ko, Branched open-ended TiO_2 nanotubes for improved efficiency of flexible dye-sensitized solar cells, *Journal of Alloys and Compounds,* **724** (2017) 1124–1133.
324. G.K. Mor, K. Shankar, M. Paulose, O.K. Varghese, and C.A. Grimes, High efficiency double heterojunction polymer photovoltaic cells using highly ordered TiO_2 nanotube arrays, *Applied Physics Letters*, **91** (2007) 152111.
325. K. Shankar, G.K. Mor, M. Paulose, O.K. Varghese, and C.A. Grimes, Effect of device geometry on the performance of TiO_2 nanotube array-organic semiconductor double heterojunction solar cells, *Journal of Non-Crystalline Solids*, **354** (2008) 2767–2771.
326. B.-Y. Yu, A. Tsai, S.-P. Tsai, K-T. Wong, Y. Yang, C.-W. Chu, and J.-J. Shyue, Efficient inverted solar cells using TiO_2 nanotube arrays, *Nanotechnology*, **19** (2008) 255202.
327. S.-I. Na, S.-S. Kim, W.-K. Hong, J.-W. Park, J. Jo, Y.-C Nah, T. Lee, and D.-Y. Kim, Fabrication of TiO_2 nanotubes by using electrodeposited ZnO nanorod template and their application to hybrid solar cells, *Electrochimica Acta*, **53** (2008) 2560–2566.
328. G.K. Mor, S. Kim, M. Paulose, O.K. Varghese, K. Shankar, J. Basham, and C.A. Grimes, Visible to near-infrared light harvesting in TiO_2 nanotube array-P3HT based heterojunction solar cells, *Nano Letters*, **9** (2009) 4250–4257.
329. T. Stergiopoulos, E. Rozi, R. Hahn, P. Schmuki, and P. Falaras, Enhanced open-circuit photopotential in quasi-solid-state dye-sensitized solar cells based on polymer redox electrolytes filled with anodic titania nanotubes, *Advanced Energy Materials*, **1** (2011) 569–572.
330. P. Roy, D. Kim, K. Lee, E. Spiecker, and P. Schmuki, TiO_2 nanotubes and their application in dye-sensitized solar cells, *Nanoscale*, **2** (2010) 45–59.
331. X. Gao, J. Li, J. Baker, Y. Hou, D. Guan, J. Chen, and C. Yuan, Enhanced photovoltaic performance of perovskite $CH_3NH_3PbI_3$ solar cells with

freestanding TiO_2 nanotube array films, *Chemical Communications*, **50** (2014) 6368.

332. X. Wang, Z. Li, W. Xu, S.A. Kulkarni, S.K. Batabyal, S. Zhang, A. Cao, and L.H. Wong TiO_2 nanotube arrays based flexible perovskite solar cells with transparent carbon nanotube electrode, *Nano Energy*, **11** (2015) 728–735.
333. R. Salazar, M. Altomare, K. Lee, J. Tripathy, R. Kirchgeorg, N.T. Nguyen, M. Mokhtar, A. Alshehri, S.A. Al-Thabaiti, and P. Schmuki, Use of anodic TiO_2 nanotube layers as mesoporous scaffolds for fabricating $CH_3NH_3PbI_3$ perovskite-based solid-state solar cells, *ChemElectroChem*, **2** (2015) 824–828.
334. P. Qin, M. Paulose, M.I. Dar, T. Moehl, N. Arora, P. Gao, O.K. Varghese, M. Grätzel, and M.K. Nazeeruddin, stable and efficient perovskite solar cells based on titania nanotube arrays, *Small*, **11** (2015) 5533–5539.
335. Y. Lai, Z. Lin, D. Zheng, L. Chi, R. Du, and C. Lin, CdSe/CdS quantum dots co-sensitized TiO_2 nanotube array photoelectrode for highly efficient solar cells, *Electrochimica Acta*, **79** (2012) 175–181.
336. Z. Lan, W. Wu, S. Zhang, L. Que, and J. Wu, Preparation of high-efficiency CdS quantum-dot-sensitized solar cells based on ordered TiO_2 nanotube arrays, *Ceramics International*, **42** (2016) 8058–8065.
337. A. Kongkanand, K. Tvrdy, K. Takechi, M. Kuno, and P.V. Kamat, Quantum dot solar cells. Tuning photoresponse through size and shape control of CdSe-TiO_2 architecture, *Journal of the American Chemical Society*, **130** (2008) 4007–4015.
338. W.-T. Sun, Y. Yu, H.-Y. Pan, X.-F. Gao, Q. Chen, and L.-M. Peng, CdS quantum dots sensitized TiO_2 nanotube-array photoelectrodes, *Journal of the American Chemical Society*, **130** (2008) 1124–1125.
339. X.-F. Gao, H.-B. Li, W.-T. Sun, Q. Chen, F.-Q. Tang, and L.-M. Peng, CdTe quantum dots-sensitized TiO_2 nanotube array photoelectrodes, *Journal of Physical Chemistry C*, **113** (2009) 7531–7535.
340. X.-F. Gao, W.-T. Sun, G. Ai, and L.-M. Penga, Photoelectric performance of TiO_2 nanotube array photoelectrodes cosensitized with CdS/CdSe quantum dots, *Applied Physics Letters*, **96** (2010) 153104.
341. Q. Wang, X. Yang, L. Chi, and M. Cui, Photoelectrochemical performance of CdTe sensitized TiO_2 nanotube array photoelectrodes, *Electrochimica Acta*, **91** (2013) 330–336.
342. R. Gakhar, Y.R. Smith, M. Misra, and D. Chidambaram, Photoelectric performance of TiO_2 nanotube array photoelectrodes sensitized with CdS0.54Se0.46 quantum dots, *Applied Surface Science*, **355** (2015) 1279–1288.
343. Q. Wang, S. Li, J. Qiao, R. Jin, Y. Yu, and S. Gao, CdS–CdSe (CdTe)core–shell quantum dots sensitized TiO2 nanotube array solar cells, *Solar Energy Materials & Solar Cells*, **132** (2015) 650–654.
344. C. Xiang, L. Sun, Y. Wang, G. Wang, X. Zhao, and S. Zhang, Large-scale, uniform and superhydrophobic titania nanotubes at the inner surface of 1,000 mm long titanium tubes, *Journal of Physical Chemistry C*, **121** (2017) 15448–15455.
345. H. Liu and K. Tan, Adsorption of water on single-walled TiO_2 nanotubes: A DFT investigation, *Computation and Theoretical Chemistry*, **991** (2012) 98–101.
346. H. Brahmi, R. Neupane, L. Xie, S. Singh, M. Yarali, G. Katwal, S. Chen, M. Paulose, O.K. Varghese, and A. Mavrokefalos, Observation of a low temperature n-p transition in individual titania nanotubes, *Nanoscale*, **10** (2018) 3863–3870.
347. H. Brahmi, G. Katwal, M. Khodadadi, S. Chen, M. Paulose, O.K. Varghese, and A. Mavrokefalos, Thermal-structural relationship of individual titania nanotubes, *Nanoscale*, **7** (2015) 19004–19011.
348. H. Wang, C.T. Yip, K.Y. Cheung, A.B. Djurišić, and M.H. Xie, Titania-nanotube-array-based photovoltaic cells, *Applied Physics Letters*, **89** (2006) 023508.
349. C.-C. Chen, H.-W. Chung, C.-H. Chen, H.-P. Lu, C.-M. Lan, S.-F. Chen, L. Luo, C.-S. Hung, and E.W.-G. Diau, Fabrication and characterization of anodic titanium oxide nanotube arrays of controlled length for highly efficient dye-sensitized solar cells, *The Journal of Physical Chemistry C*, **112** (2008) 19151–19157.
350. A. Ghicov, S.P. Albu, R. Hahn, D. Kim, T. Stergiopoulos, J. Kunze, C.-A. Schiller, P. Falaras, and P. Schmuki, TiO_2 nanotubes in dye-sensitized solar cells: Critical factors for the conversion efficiency, *Chemistry an Asian Journal*, **4** (2009) 520–525.
351. P. Roy, D. Kim, I. Paramasivam, and P. Schmuki, Improved efficiency of TiO_2 nanotubes in dye sensitized solar cellsby decoration with TiO_2 nanoparticles, *Electrochemistry Communications*, **11** (2009) 1001–1004.
352. J. Wang and Z. Lin, Dye-sensitized TiO_2 nanotube solar cells with markedly enhanced performance via rational surface engineering, *Chemistry of Materials*, **22** (2010) 579–584.
353. K.-I. Ishibashi, R.-T Yamaguchi, Y. Kimura, and M. Niwano, Fabrication of titanium oxide nanotubes by rapid and homogeneous anodization in perchloric acid/ethanol mixture, *Journal of The Electrochemical Society*, **155** (2008) K10–K14.
354. B.-X. Lei, J.-Y. Liao, R. Zhang, J. Wang, C.-Y. Su, and D.-B. Kuang, Ordered crystalline TiO_2 nanotube arrays on transparent FTO glass for efficient dye-sensitized solar cells, *The Journal of Physical Chemistry C*, **114** (2010) 15228–15233.
355. L.-L. Li, C.-Y. Tsai, H.-P. Wu, C.-C. Chen, and E.W.-G. Diau, Fabrication of long TiO_2 nanotube arrays in a short time using a hybrid anodic method

for highly efficient dye-sensitized solar cells, *Journal of Materials Chemistry*, **20** (2010) 2753–2758.

356. L.-L. Li, Y.-J. Chen, H.-P. Wu, N.S. Wang, and E.W.-G. Diau, Detachment and transfer of ordered TiO_2 nanotube arrays for front-illuminated dye-sensitized solar cells, *Energy & Environmental Science*, **4** (2011) 3420.
357. Q. Zheng, H. Kang, J. Yun, J. Lee, J.H. Park, and S. Baik, Hierarchical construction of self-standing anodized titania nanotube arrays and nanoparticles for efficient and cost-effective front-illuminated dye-sensitized solar cells, *ACS Nano*, **5** (2011) 5088–5093.
358. R. Liu, W.-D. Yang, L.-S. Qiang, and H.-Y. Liu, Conveniently fabricated heterojunction ZnO/TiO_2 electrodes using TiO_2 nanotube arrays for dye-sensitized solar cells, *Journal of Power Sources*, **220** (2012) 153–159.
359. N. Liu, S.P. Albu, K. Lee, S. So, and P. Schmuki, Water annealing and other low temperature treatments of anodic TiO_2 nanotubes: A comparison of properties and efficiencies in dye sensitized solar cells and for water splitting, *Electrochimica Acta*, **82** (2012) 98–102.
360. J. Lin, X. Liu, M. Guo, W. Lu, G. Zhang, L. Zhou, X. Chen, and H. Huang, A facile route to fabricate an anodic TiO_2 nanotube–nanoparticle hybrid structure for high efficiency dye-sensitized solar cells, *Nanoscale*, **4** (2012) 5148.
361. K.M. Lee, E.S. Lee, B. Yoo, and D.H. Shin, Synthesis of ZnO-decorated TiO_2 nanotubes for dye-sensitized solar cells, *Electrochimica Acta*, **109** (2013) 181–186.
362. S. Kurian, P. Sudhagar, J. Lee, D. Song, W. Cho, S. Lee, Y.S. Kang, and H. Jeon, Formation of a crystalline nanotube–nanoparticle hybrid by post water-treatment of a thin amorphous TiO_2 layer on a TiO_2 nanotube array as an efficient photoanode in dye-sensitized solar cells, *Journal of Materials Chemistry A*, **1** (2013) 4370.
363. L. Suhadolnik, I. Jerman, A. Kujan, K. Zagar, M. Krivec, and M. Ceh, Morphological, optical and electrical characterization of titania-nanotubes-based dye-sensitized solar cells, *Solar Energy*, **127** (2016) 232–238.
364. X. Wang, L. Sun, S. Zhang, X.Wang, K. Huo, J. Fu, H. Wang, and D. Zhao, A composite electrode of TiO_2 nanotubes and nanoparticles synthesised by hydrothermal treatment for use in dye-sensitized solar cells, *RSC Advances*, **3** (2013) 11001–11006.

18

Nanotubes, Nanowires, and Nanofibers: Carbon Nanotubes and Carbon Nanotube Fibers

Jude C. Anike and
Jandro L. Abot
The Catholic University of America

18.1 Carbon Nanotubes

18.1.1 Introduction to Carbon Nanotubes (CNTs)

Carbon plays an important role in living organisms. By mass, it is the second most abundant element after oxygen in the human body. It is the 15th most abundant element on the Earth's crust and the fourth most abundant element in the universe. It is this abundance alongside the outstanding properties of carbon allotropes that makes the study of carbon nanotubes (CNTs) an intriguing subject in scientific research.

CNTs are one of the allotropes of carbon. The others are graphite, diamond, graphene, and Buckminster fullerene (Figure 18.1). Fullerenes are molecules where carbon atoms are arranged spherically, and hence, from the physical point of view, are zero-dimensional (0D) objects with discrete energy states. Graphite is the allotrope of sp^2 carbon shown as a stack of graphene sheets, hence the three-dimensional (3D) structure, while graphene is a one atomic layer of graphite with a two-dimensional (2D) structure. CNTs are quasi one-dimensional (1D) structure of graphene, formed by rolling a graphene layer into a seamless cylinder. The simplest form consists of one folded sheet of graphene and is simply known as a single-walled carbon nanotube (SWCNT). A SWCNT has a diameter of around 0.8–2 nanometers (nm) [1–3]. When two graphene sheets are rolled in concentric cylinders or an SWCNT nested inside another graphene sheet, the formed nanotubes are called double-walled carbon nanotube (DWCNT). The term "wall" signifies how many concentric layers of nanotubes are in the structure. Additional walls or greater than one or two folds of graphene sheets are generally classified as multi-walled carbon nanotube (MWCNT). The properties of carbon allotropes are compared in Table 18.1. CNTs have outstanding mechanical, electrical, thermal, and optical properties [1–6]. CNTs could withstand axial strains of up to 15%, which exceeds the capability of conventional strain gauges. They also possess a unique sensitivity to mechanical strain that alters their electrical resistance. This piezoresistive behavior could be tailored towards strain sensing. CNTs have a unique combination of properties that can be utilized for simultaneous use unlike any other smart materials. Such properties include ballistic electron

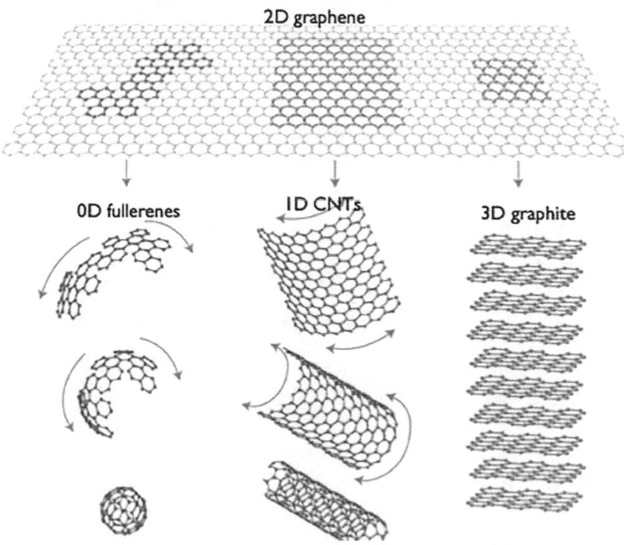

FIGURE 18.1 Schematic structure of graphene and structural transformations to graphite, CNT, and fullerene [3].

TABLE 18.1 Physical Properties of Carbon Allotropes [4]

Properties	Carbon Allotropes			
	Graphite	Diamond	Fullerene	CNT
Specific gravity (g/cm^3)	1.9–2.3	3.5	1.7	0.8–1.8
Electrical conductivity (S/cm)	$4{,}000^p$, 3.3^c	10^{-2}–10^{-15}	10^{-15}	10^2–10^6
Electron mobility ($cm^2/(V\ s)$)	$\sim 10^4$	1800	0.5–6	10^4–10^6
Thermal conductivity (W/(m K))	298^p, 2.2^c	900–2,320	0.4	2,000–6,000
Coefficient of thermal expansion (K^{-1})	-1×10^{-6p}, 2.9×10^{-5c}	$(1\sim3) \times 10^{-6}$	6.2×10^{-5}	Negligible
Thermal stability in air (°C)	450∼650	<600	<600	>700

Note: p: in-plane; c: c-axis.

transport, load-carrying capability, high specific strength and stiffness (100 times stronger than stainless steel yet at least, six times lighter), thermally more conductive than pure diamond yet as hard, thermal stability, corrosion resistance, current-carrying capacity that is 1,000 times higher than that of copper and exceeds that of superconductors, and customizable electronic properties. CNT can be metallic or semiconducting, depending on their diameter and chirality. Due to these properties, and their size (average diameter of about 0.7–10 nm), the discovery of CNT was revolutionary for nanotechnology both in timeline and advancement of nanomaterial applications in today's technology. It is unsurprising they have found use in various applications such as in composites [5,7–10], paints and coatings [11,12], microelectronic devices [13–21], wearable electronics [22–24], water purification [25–28], energy storage [29–37], biotechnology and biosensing [38–51], drug delivery [52], chemical and gas sensing [53,54], strain sensing [55,56], and electrical cables [57], among others.

History of CNTs

The history of CNTs is neither totally nor generally conclusive. The initial story was that in the 1950s, while trying to measure the triple point of carbon in a carbon arc furnace, Roger Bacon, a physicist at the Parma Technical Center of National Carbon Company, Cleveland, Ohio, could have possibly seen the first CNTs [58]. Unfortunately, there were no high-powered electron microscope that would have allowed the observation of the structure. In 1958 however, Roger Bacon invented the graphite fiber, which was patented in 1960 [59]. The graphite fiber is a graphite sheet rolled into cylindrical forms with circular or elliptical cross section. It had a tensile strength of about 19.7 GPa and an elastic modulus of 700 GPa close to single crystal values [59]. There was early skepticism about the potential of the discovery and its impact on future scientific research. Two decades later, Morinobu Endo, a PhD student at Nagoya University, Japan, could have seen CNTs while working on researching carbon filaments prepared by flowing hydrogen gas and benzene into an electric furnace containing a substrate in the 1970s [60,61]. After the experiment, it took him almost 3 days to wash and prepare the substrate blackened by carbon deposits for reuse. While cleaning the substrate with sand paper to avoid the long time used in washing it, he surprisingly found a large amount of carbon fibers in the electric furnace. The sand paper was made of iron oxide. In 1974, while doing research at University of Orleans in France, he discovered fine tubes in the middle of aggregates of carbon fibers with iron particles at the tips of tubes. However, the carbon filaments did not match CNTs' width dimensions, being later classified as barrelenes. Several other groups reported similar findings like the discovery of carbon clusters by Kroto and Smalley in 1985, which led to the discovery of fullerenes [62]. It wasn't until 1991 when an IBM scientist, Sumio Iijima, accidentally observed a material with hollow structures while studying the surface of graphite electrodes in an electric arc [7]. That observation was found to be MWCNTs. SWCNTs were to follow in 1993, when Sumio Iijima and Donald Bethune reported their discovery. Morinobu Endo later confirmed that the catalyst in his observation in 1974 was iron oxide while the tubes were CNTs [60]. This, however, was after Iijima's 1991 discovery was reported. The discovery of SWCNT was very important in the advancement of CNT research as its basic unit shed light on the structure and properties of CNTs.

In 1993, Ebbesen et al. synthesized aligned CNTs. It would be seminal to the production of macroscopic CNT assemblages. Earlier forms of macroscopic CNT assemblies were powders suspended in a solvent [63,64] or Bucky papers obtained by drying CNT suspensions [65]. Their assembly into continuous fibers at this point had been achieved only through post-processing methods. The breakthrough for CNT fibers came from the pioneering work of Vigolo et al. [66]. It was the first reported article on the spinning process that can be modified to produce fibers comprising mostly of SWCNTs. Unlike CNTs, the history of CNT yarn is well documented. The drawing of a continuous length of interconnected CNT web from vertically aligned CNT array was accidentally discovered in 2002 [67]. While trying to pull out a bundle of CNTs from a 100-μm-high CNT array, Jiang et al. obtained a continuous ribbon or web of pure CNTs, in a much similar way as drawing thread from a silk cocoon. In 2004, Li et al. reported about direct spinning of fibers and ribbons of CNTs directly from the chemical vapor deposition (CVD) synthesis zone of a furnace using a liquid source of carbon and an iron catalyst [68]. Application of CNT in strain measurement was first experimentally reported by Paulson et al. [69].

18.1.2 Synthesis of CNTs

The production methods of CNTs have improved over the years for optimal properties or requirements such as high yield of specified length, diameter, number of walls, and chirality. The synthesis of CNTs involves growing CNTs by excitation of carbon atoms in the presence of metallic catalyst particles. The methods include electric arc discharge, laser ablation, and CVD. This section gives an overview of them.

Arc discharge is the oldest technique and the first method used to synthesize CNTs [5]. The arc discharge method involves the evaporation of graphite electrodes in electric arcs at a very high temperature (~4,000°C) and pressure. This method yields a large quantity of nanotubes with good crystallinity. However, the as-synthesized nanotubes are highly impure with up to 70% of the arc-grown product containing amorphous carbon and metal particles due to the metallic catalyst included in the reaction. This means that the final products need to be purified afterwards. Purification methods such as oxidation, centrifugation, filtration, and acid treatment have been employed. Laser ablation method uses a high-power laser to vaporize a graphite rod in a tubular reactor in a high-temperature furnace. An argon gas is typically used to drive the reaction. The gas carries away the vaporized carbon, and the condensed particles are deposited on the walls of the reactor tube [63]. The condensed particles contain nanotubes. This method produces CNTs with a relatively high purity and low metallic impurities, since the metallic atoms involved tend to evaporate from the end of the tube once it is closed. One disadvantage is that their production yield is very low and, the nanotubes produced from this method are not necessarily uniformly straight.

The CVD process involves the chemical breakdown of a hydrocarbon on a substrate with the aid of a metal catalyst at a high temperature (600–1,200°C) [70–72]. This method's approach allows CNTs to grow on a variety of materials, which makes it more viable to integrate into already existent processes for synthesizing electronics. This method is low cost, scalable, and can produce long and well-aligned CNTs.

18.1.3 Structure of CNTs

Geometric Structure

The geometry of the nanotubes can be defined by the chiral vector, C, of the hexagonal lattice of the graphene sheet. The expression for C is given as:

$$C = na_1 + ma_2 \tag{18.1}$$

where a_1 and a_2 are the unit vectors of the graphene; and the integers (n,m) represent the chiral indices that define the roll orientation of the nanotubes.

In simple terms, this chiral vector is the vector that describes the circumference of the tubes generated by connecting two ends of graphene sheets. This arrangement of the carbon atoms at the circumference is referred to as chirality. The chirality is marked by the number of steps along the unit vectors a_1 and a_2 of the graphene lattice. The unit vectors a_1 and a_2 are often represented with 120° separation (the formation angles of the orbital in graphene)

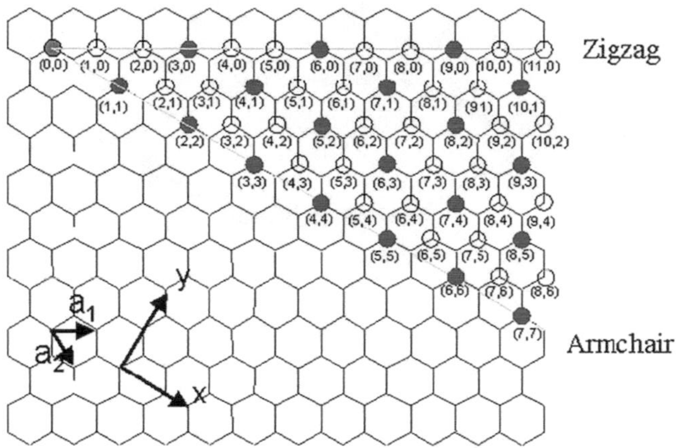

FIGURE 18.2 Chiral vector of CNTs [73].

as defined in Figure 18.2. There are three types of chirality in CNTs: (a) armchair when $m = n$; (b) zigzag when $m = 0$; and (c) chiral for other combinations ($m \neq n$ or $m \neq 0$).

The angle, θ, between the zigzag direction and C is called the chiral angle and it is given by the equation:

$$\tan\theta = \frac{\sqrt{3}m}{(m+2n)}. \tag{18.2}$$

Substituting the values of the indices n and m into the above equation gives the chiral angles of the nanotubes.

For armchair, $n = m$:

$$\theta = \tan^{-1}\left\{\frac{\sqrt{3}m}{(m+2m)}\right\} \tag{18.3}$$

by substitution,

$$\tan^{-1}\left\{\frac{1}{\sqrt{3}}\right\} = 30° \tag{18.4}$$

Zigzag, when $m = 0$:

$$\theta = \tan^{-1}\left\{\frac{0}{(0+2n)}\right\} \tag{18.5}$$

which reduces to

$$\tan^{-1}\{0\} = 0° \tag{18.6}$$

$$\text{Chiral, when } 0 < \theta < 30° \tag{18.7}$$

Electronic Structure of CNTs

The mechanical properties of the CNTs are derived from the covalent chemical bond between carbon atoms. Carbon atoms have six electrons: two $1s$ orbital close to the nucleus, another two in the $2s$ orbital, and the rest in $2p$ orbitals ($1s^2 2s^2 2p^2$). Graphene is formed by the combination of one s orbital and two p orbitals to form three hybrid sp^2 orbitals all at 120° angles. The bond within the plane is sigma (σ) (Figure 18.3). σ-bonds are the strongest types of covalent chemical bond with a bond energy of 346 KJ/mol. A weaker type of bond is formed perpendicularly to the

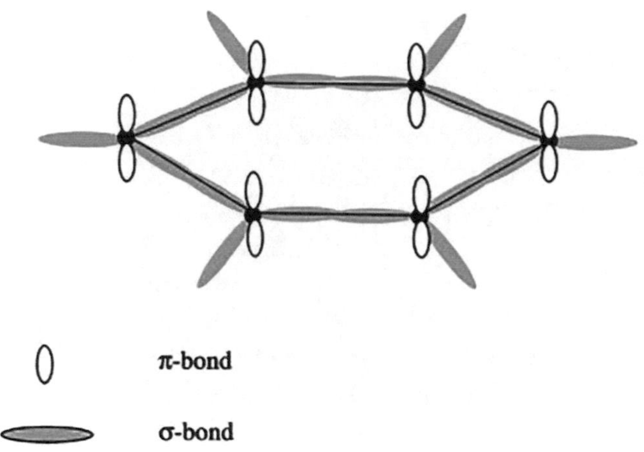

FIGURE 18.3 Basic hexagonal bonding structure for a graphene sheet. Carbon nuclei shown as filled circle, out-of-plane π-bonds, and σ-bonds connect the C nuclei in plane [74].

plane of the σ bond by the unattached/remaining p orbital. This out-of-plane bond known as pi (π-bond) contains π electrons that are unconfined to the atom and are free to roam the nanotube structure. Thus, the π bonds are mostly responsible for the electronic properties while the σ bonds mostly account for the mechanical properties. The zero gap in a semiconductor material like graphene makes them behave like a metal. The electronic structure of an SWCNT is derived from graphene, with the planar 2D honeycomb lattice providing a periodic boundary condition that allows the existence of only certain electronic states of graphene when it is folded into one-dimensional (1D) nanotube. Therefore, the transformation from 2D graphene to 1D CNT confines the electrons in the carbon atoms in one plane of the graphene sheet along the radial direction.

18.1.4 Properties of CNTs

Mechanical Properties of CNTs

Individual nanotubes have a very small diameter (0.8–10 nm), which makes it extremely challenging to subject them to direct tensile testing. The mechanical properties of SWCNTs are even more difficult to measure than those of MWCNTs, partly because as-produced SWCNTs are held together in bundles by van der Waals forces and to separate them into individual SWCNTs is not simple. A measurement technique would involve an in situ tensile displacement stage within a high-resolution electron microscope (HREM). Common techniques utilize the high-resolution and contrast mechanisms of a transmission electron microscope (TEM) with the probes of an atomic force microscopy (AFM).

The average maximum tensile strength of SWCNTs is about 30 GPa with a Young's modulus (elastic modulus) close to 1 TPa [75–78] while the Young's modulus of MWCNTs is 1.28 TPa [76] with tensile strengths of 0.01–0.15 TPa and ultimate tensile strains of approximately 6–12% [76–81].

Table 18.2 displays the reported measured values of the mechanical properties of individual CNTs. There has been discrepancy in the reported values of the elastic modulus of CNTs. This discrepancy is related to the measurement method and the estimation of the cross-sectional area of the nanotube. A higher Young's modulus is obtained for CNTs of smaller diameter. If the tube is considered as a hollow cylinder, a higher elastic modulus will be obtained than that of a solid cylinder.

Electrical Properties of CNTs

The electronic states vary according to the chirality of the nanotubes. Thus, the chirality determines the transport properties of the nanotubes. Therefore, in terms of electronic properties, CNTs are categorized into metallic (or semimetallic) and semiconducting. The theoretical electronic conductivity of SWNTs with respect to the chirality is shown in Figure 18.4.

The condition for the CNT to be metallic is when $\frac{n-m}{3}$ is an integer.

In armchair nanotubes, the valence and conduction bands overlap making all armchair CNTs to behave metallic.

For other nanotubes with a gap at the Fermi energy level, their behavior is semiconducting.

Semiconducting nanotubes are grouped into further two: small bandgap and wide bandgap. In wide bandgap nanotubes, $\frac{n-m}{3}$ is not an integer.

Small bandgap nanotubes, however, have a very small bandgap at room temperature and thus, can be considered metallic. Considering this, a common conclusion is that 1/3 of all nanotubes behave metallic while the remaining 2/3 are semiconducting. The dark gray-filled circles in Figure 18.2 represent metallic nanotubes while the black ring represents semiconducting nanotubes.

TABLE 18.2 Estimated Young's Modulus of CNTs and Measurement Methods

	Young's Modulus (TPa)	Measurement Method	References
SWCNTs	0.32–1.47	Direct measurement with AFM tip	[75]
SWCNTs	1.0	Direct measurement with AFM tip	[78]
SWCNTs	1.25	TEM-thermally vibrating cantilever	[6]
MWCNTs	1.28	AFM-cantilever	[76]
MWCNTs	0.9–1.23	AFM-cantilever	[77]
MWCNTs	0.81	AFM-end clamped	[78]
MWCNTs	1.26	TEM-thermally vibrating beam	[79]
MWCNTs	0.1–1	TEM-electrostatic deflection in cantilever	[80]
MWCNTs	0.27–0.95	SEM-AFM cantilevers	[81]
MWCNTs	0.91	TEM-direct tension	[82]
MWCNTs	1.8	TEM-thermally vibrating cantilever	[83]

(n, m)	Form of SWNTs	Electrical Conductivity
(n, 0)	 Zigzag	Metallic when n is the multiple of 3, otherwise, semiconducting
(n, n)	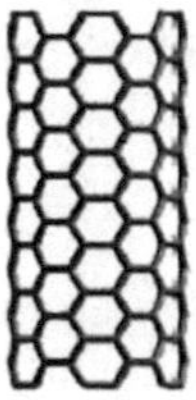 Armchair	Metallic
(n, m) when m ≠ 0, and n	 Chiral	Metallic when (2n+m)/3 is an integer, otherwise, semiconducting

FIGURE 18.4 SWCNT chirality and their electronic conductivity [84].

Most MWCNTs can be described as complex structures of many parallel weakly interacting SWNTs with varying chirality that are randomly distributed. Their electrical properties differ from the SWCNTs in the sense that electrical measurements on the tubes involve connection to the outermost shell. Thus, the outermost shell properties will mostly determine the electrical properties of the MWCNTs. It has been shown that when multiple electrodes are connected to different shells of MWCNT, the hopping of charge carriers of adjacent shells would play a key role in carrier transport and quantum conductance determined by the number of metallic shells that are directly in contact with the electrodes (Figure 18.5) [85].

However, it should be noted that due to the inverse proportionality between the bandgap and the diameter of the semiconducting tubes, large-diameter tubes will tend to behave metallic at room temperature. This is especially important about large-diameter MWCNTs. Since probing the electrical properties typically involves electrodes contacting the outermost shell, this shell will be dominating the transport properties.

Thermal Properties of CNTs

CNTs have axial thermal conductivity that exceeds the best-known bulk heat conductors and twice the heat capacity of diamond. They are also thermally stable up to 4,000 K. Experimentally measured thermal conductivities of isolated SWCNT and MWCNT, in the axial direction, were 3,500 and 3,000 $Wm^{-1}K^{-1}$, respectively [57]. These values exceed those of highly conductive metals including aluminum, copper, and gold with their highest conductivity all below 430 $Wm^{-1}K^{-1}$.

Optical Properties of CNTs

The common techniques to determine the optical properties of CNTs include fluorescence or photoluminescence, light absorption and Raman scattering. The optical properties of SWCNTs are linearly polarized along the tube axis permitting the determination of their orientation without the aid of electron microscope. Photoluminescence is useful in quantifying the semiconducting nanotubes in a CNT batch. It can also identify the symmetry and chirality of the CNT. This is possible because photoexcited semiconducting SWCNTs emit and absorb near infrared light. Photoexcitation occurs when an electron in the CNT absorbs light and goes to an excited state, creating an electron-hole pair, which can recombine to emit light. Metallic CNTs do not exhibit fluorescent excitation. Due to their overlapping bandgap, when electrons in metallic CNTs are excited, the holes are immediately occupied by free electrons in the metal.

The optical absorption in CNTs is due to transitions of electrons from valence to conduction bands. The transitions are relatively sharp depending on the energy level. The absorption spectrum can then be used to identify various CNT types. A strong light absorbance of 0.98–0.99 from the far-ultraviolet (200 nm) wavelength to far-infrared (200 μm) wavelength has been observed for vertically aligned SWCNTs [4]. The strong light absorption was

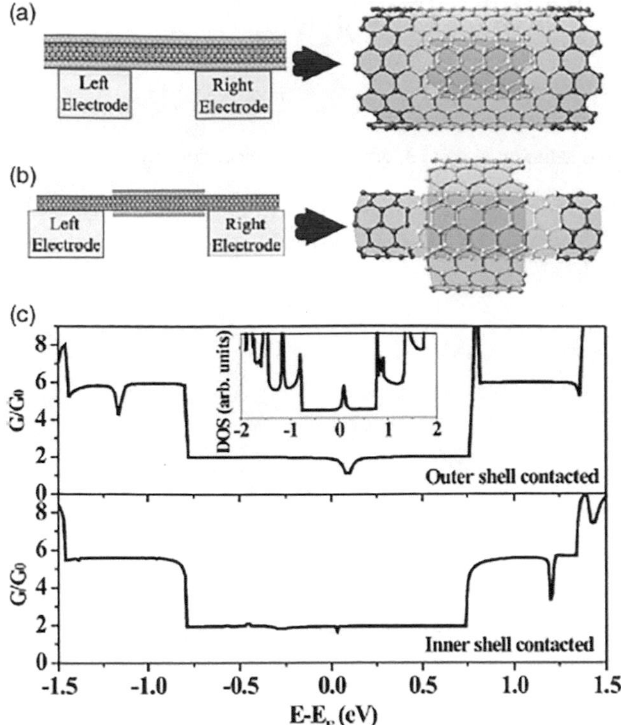

FIGURE 18.5 (a) Schematic of the contact condition in realistic side-contact four-point measurement of a (5,5) @ (10,10) DWCNT contacted with two (5,5) electrodes and the outer and/or inner layer number of DWCNT in the central region is twelve and eight, respectively. (b) Schematic of a hypothetical shuttle nanotube system, where a (5,5) @ (10,10) DWCNT is contacted with two (5,5) electrodes and thus only the inner shell directly contacts with electrodes. (c) Conductance of the side contact system with the outer (top panel) or inner shell contacted to the electrodes, and the corresponding density of states (DOS) spectrum is shown in the inset of top panel [85].

attributed to the various bandgaps of different CNTs due to variation in chirality and light that are trapped within CNT arrays due to multiple reflections.

Raman spectroscopy has provided an exceedingly powerful tool for characterization of CNTs with respect to their diameters and quality of the sample properties. SWCNT and MWCNT can be discriminated via this method while semiconducting and metallic CNTs can be detected by Raman mapping [1,4]. In optical absorption and Raman scattering, the sample preparation is easier compared to the fluorescence method. This is because agglomeration of CNTs leads to a broad optical spectrum especially in fluorescence measurements.

18.2 CNT Fibers

18.2.1 Macroscopic CNT Assemblies

CNT's high aspect ratio provides multiple opportunities for use in engineering applications. However, their nanoscale size and discrete length limit their utilization in industrial applications. Thus, it is important to have a macroscopic assembly of CNTs that could be handled with available technology. Currently CNT assemblies have been produced in different forms including ropes, arrays, ribbons, scrolls, fibers or yarns, braids, films or sheets, and tapes. In terms of structure, there are 3D (aligned CNT arrays/forests, CNT foams) [57,86], 2D (CNT sheets/films, CNT tapes) [87], and 1D assemblies (CNT fibers, CNT braids) [75,86,88].

Due to the weak van der Waals interactions between CNTs, they adhere to each other and self-assemble into groups or bundles otherwise called ropes to minimize mechanical strain energy. CNT bundles contain hundreds to thousands of individual nanotubes. Other forms of CNTs are mostly formed from the manipulation of the basic forms or bundles. CNT bundles with the weak inter-tube bonding can show properties different than individual CNTs. It has been challenging to produce macroscale pristine CNT fiber to fully utilize the outstanding mechanical and electrical properties of individual CNTs. The decrease in mechanical strength and stiffness, for example, in CNT bundles is primarily due to the adherence properties of the CNTs. However, the shear interaction of CNTs in a fiber or CNT bundles due to inter-tube slippage is also a contributing factor. Figure 18.6 displays the micrographs of a dry-spun CNT yarn, a CNT sheet drawn from an array and the schematic of an array.

18.2.2 Production of CNT Yarns

Methods of Production of CNT Yarns

Most CNT fibers are produced by spinning [86–105]. CNT fibers can be spun in both liquid (wet spinning) and solid state (dry spinning). The common methods for CNT fiber fabrication include direct spinning from an aerogel formed from CVD process [86,87], fiber extrusion from a CNT dispersed in a polymer solution [88,91], from foldable CNT films or sheets [89,90], and spinning from vertically aligned CNT array grown on a substrate [86,87]. Wet spinning process involves the dispersion of CNTs in a solution and injecting the well-dispersed CNTs into a coagulation bath to obtain continuous fibers [88]. The fibers are obtained through precipitation as the coagulant does not disperse them. The limitation for fibers produced in liquid state, especially extrusion from CNT–polymer solution, is that they contain high amounts of polymer remains [92]. To eliminate the residual polymers, the as-produced CNT fibers are annealed at very high temperatures to obtain neat CNT fibers. Currently, all spinnable CNT arrays are grown via CVD system.

Dry spinning of CNT fibers is achieved by directly drawing the CNT fibers from a CVD reactor (direct spinning) or by spinning of fibers from other CNT macro-assemblies produced from the CVD reactors. Of such assemblies, CNT yarns that were spun from vertically aligned arrays have shown to possess the best mechanical properties [92–94]. In this method, a CVD reactor is used to grow the CNT arrays. The process involves

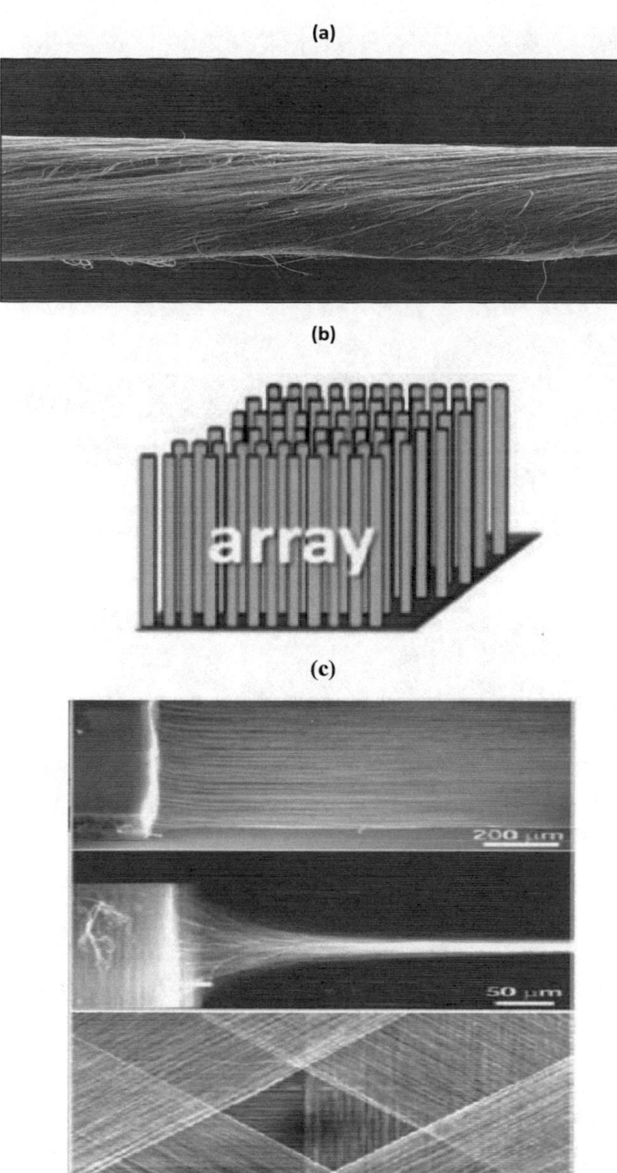

FIGURE 18.6 CNT macroscopic assemblies: (a) SEM image of CNT yarn. (b) Schematic of CNT array [57]. (c) SEM image CNT sheet from an array {top: top view; middle: side view; bottom: layers of sheet} [87].

the decomposition of a hydrocarbon on a catalyst-coated substrate by the means of a reactor at sufficiently high temperature (600–1,200°C). [70–72]. A carbon source in the presence of a carrier gas is passed through the reactor and a substrate containing catalyst particles placed in the reactor allows for a decomposition process of the carbon compound on the catalyst sites/locations. CNTs grow on the catalyst site in the reactor and are collected upon cooling the system to room temperature. The hydrocarbon can be in either liquid (benzene, alcohol, etc.) or solid form (ferrocene, camphor, etc.) [70]. Liquid hydrocarbons are vaporized in a container and then expunged into the reaction zone of the CVD by an inert gas. Solid hydrocarbon can be directly placed in the low-temperature zone of the reaction tube. Since most of the solid hydrocarbons are volatile and are readily vaporizable at relatively low temperature, they can be placed directly in the low-temperature zone of the reaction tube while a catalyst is passed from the high-temperature zone of the CVD reactor. This method is low cost, scalable, and can produce long and well-aligned CNTs. A vertically aligned array of CNTs known as forests is deposited on the substrate. The diameter of the nanotubes depends on the size of the catalyst particle used. From the array as seen in Figure 18.7a and b, a connecting network of CNT bundles known as CNT web is drawn. The drawing of a CNT web from an array could be simply done by using a tweezer and pulling from the top of the side wall of the array. When a nanotube is drawn from the side of an array, the nanotube drags other nanotubes along in an end-to-end connection due to van der Waals interactions. The CNT web can be used to form CNT sheet or films by aligning their axial dimensions horizontally in the plane of the pull. To produce a yarn, the CNT web is pulled to a spinner or a spinning machine (Figure 18.7c). Alternatively, the web from two arrays is passed through a hook to apply some tension to condense the fiber into a one-dimensional thread (Figure 18.7d). The CNT thread is passed through a spinner to produce a fiber or yarn, which is then collected on a spool. This process is faster and the tension from the hook provides additional tension and condensation that improves the mechanical and electrical properties of the CNT yarn. CNT yarns spun in solid state usually contain high amounts of catalyst impurities. Nonetheless, the CVD method is the most promising for mass production since the process is easily controllable, can operate in moderate conditions, and has lower cost of operation. Single CNT yarns could be twisted to produce plied yarns. Depending on applications, they could also be knitted, knotted, or woven to produce variants of CNT yarns (Figure 18.8).

A special technique from the CVD method is the pyrolysis or the floating catalyst method. This involves burning of hydrocarbon at a very high temperature (above 1,000°C) on a catalyst-coated substrate placed in the hot zone of the furnace in a gas flow to catalyze the CNT growth. This process involves the thermal decomposition of the catalyst vapor, which liberates the metal nanoparticles in situ. Hydrogen is used to control the size of the catalyst. Carbon produced by the pyrolysis of the hydrocarbons is reduced in the presence of the hydrogen supporting nanotubes to grow on the catalyst sites. The CNTs produced in this method form an aerogel. Aerogels as the name suggests are materials derived from gel in which the liquid component of the gel has been replaced with gas. The result is an extremely low-density solid that looks like a sock. Due to the highly porous nature of the aerogel, they will need to be condensed to be passed through a spinner. The yarns produced using the floating catalyst CVD method contain shorter CNTs. Unlike the CNT yarn drawn from a forest; where long CNTs can be produced by prolonging the growth time and by addition of H_2O, there is no report so far on controlling the growth of

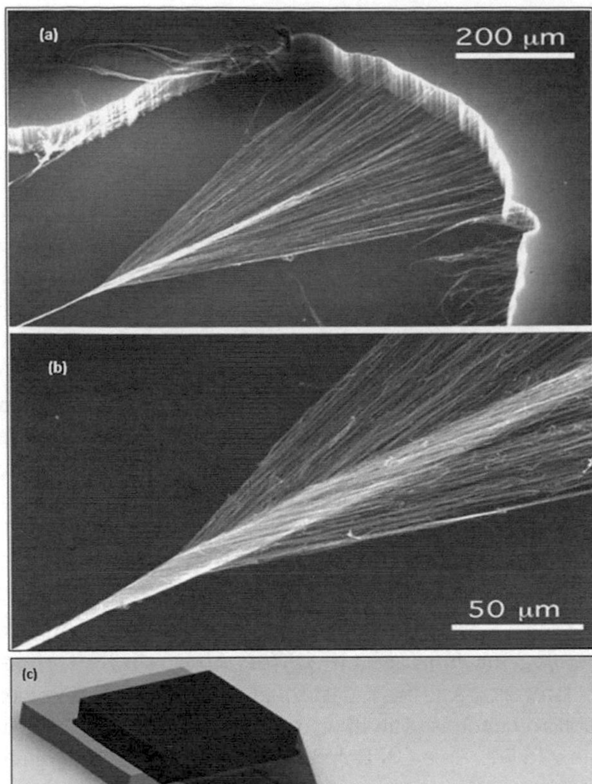

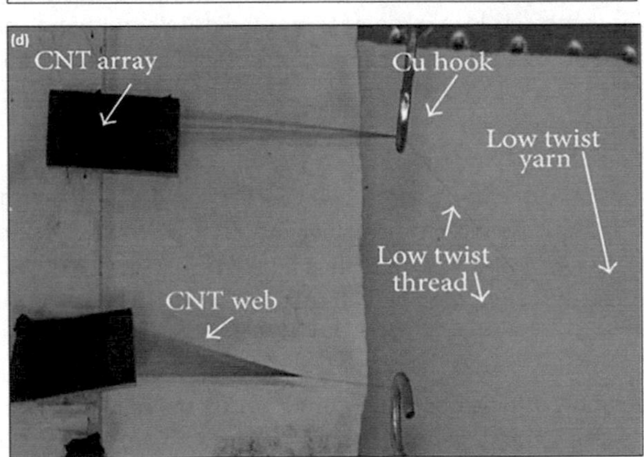

FIGURE 18.7 (a and b) SEM images, at two different magnifications, of a CNT yarn in the process of being simultaneously drawn and twisted during spinning from a nanotube array [86]. (c) Schematic of the dry spinning process of CNT yarns from vertical arrays to collection bobbin (spool). (d) Two CNT arrays spun into a yarn using the spinning method (HTS—heating and tension spinning technique) [92].

nanotubes using this method [105,106]. This method being a one-step process is very fast and easily scalable.

CNT yarns could also be spun from other CNT assemblies. The most convenient of them being the CNT sheet.

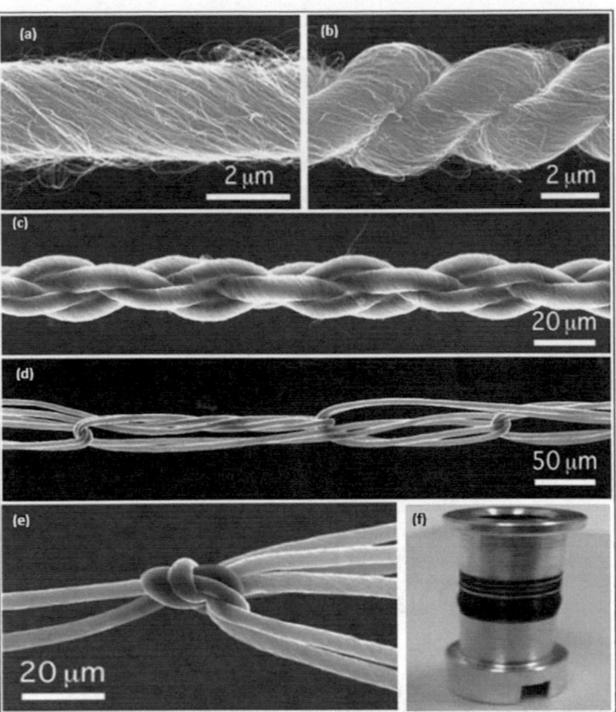

FIGURE 18.8 (a–e) SEM images of MWCNT yarns: (a) Single. (b) Two-ply. (c) Four-ply. (d) Knitted. (e) Knotted [86]. (f) Optical image of single-ply MWCNT yarn wound on a spool.

The CNT sheet is first prepared via the CVD system. The CNTs are directly drawn from the CVD reactor and stacked into layers to form of CNT sheet. The CNT sheet can be spun directly like the CNT array [57] or passed through wire-drawing dies to condense them before spinning [87]. The die also serves to improve the alignment of the CNT sheet. One observed disadvantage of this method is that owing to the discontinuous nature of CNT sheets, it is difficult to spin a continuous CNT fiber from them.

CNT composite fibers comprise of a matrix with CNT fibers serving as a reinforcement and sensing material. CNT composite fibers are used in creation of hybrid and functional materials with high strength and stiffness. Epoxy resins are the commonly used polymer matrices in making CNT composite fibers. Composite fibers can be produced mainly through: (i) melt-spinning of CNTs dispersed in polymer melts, which are then sonicated, blended, and extruded [95–97]; (ii) solution spinning of dispersed nanotubes in a polymer matrix (CNT solution) [98–100]; and (iii) electrospinning solution of nanotubes dispersed in polymer [101–104]. These types of fibers, however, have increased weight differential compared to neat CNT fibers and could be less conductive due to the presence of the polymer.

CNT Spinning

To successfully produce dry-spun CNT yarns, the CNT arrays must be spinnable. Spinnability of CNT arrays depends on the vertical alignment of the CNT, the uniformity of the CNTs (the morphology of the CNTs), and the height of the array [105–108]. This is because the production of continuous CNT yarns can only be successful when they are drawn from super-aligned CNT arrays in which the CNTs are aligned parallel to one another and are held together by van der Waals interactions to form aggregates or bundles [67]. In an unaligned CNT array, the top and bottom parts of the array are highly disordered and entangled unlike in vertically aligned array, where the bottom part is highly ordered without entanglement. CNTs with high degree of misalignment are unspinnable. Misalignment is often a consequence of the entanglement of the nanotubes due to high strain energy in the tubes. CNTs try to normalize the high strain energy by bonding together in bundles. This further explains why very long CNTs tend to have disorderly structure and over-waviness that limit their spinnability. Shorter length CNTs are preferred for dry spinning because they are less wavy and have lower curvature. Typical lengths of CNT arrays for spinning CNT fibers are few hundred micrometers [92]. From Figure 18.9, it is observed that CNTs from unaligned arrays are isolated from each other while CNTs from vertically aligned arrays form tight bundles, indicating strong van der Waals interactions between the tubes.

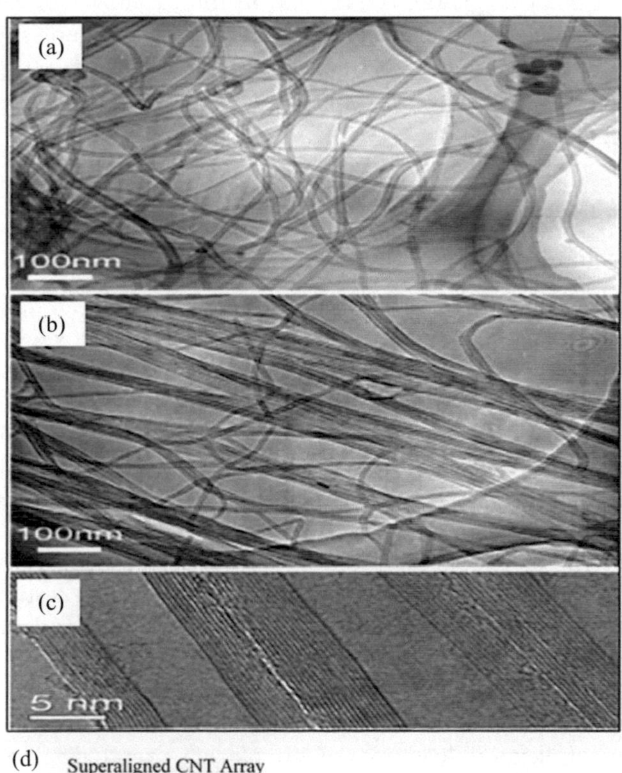

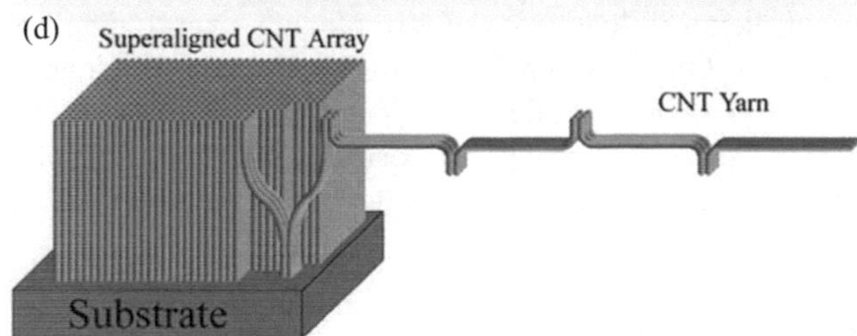

FIGURE 18.9 TEM images of CNTs in: (a) Normal arrays. (b) Super-aligned arrays. (c) HRTEM image of as-grown CNTs from super-aligned arrays. (d) Schematic of the model depicting the mechanism of pulling CNT yarn from an array [109].

Vertically aligned CNT arrays also produce CNTs that have very clean surfaces as seen in Figure 18.9c. The clean surfaces of yarns produced from the vertical arrays make them sticky and difficult to handle [109]. This is a challenge considering that CNT yarns are manually drawn from the arrays with a picking tool like tweezers or drawn directly by an electric motor. However, this enables the formation of a yarn. When pulling the CNTs from a vertically aligned array, van der Waals force connect the CNTs in an end-to-end fashion, forming a continuous yarn, as illustrated in Figure 18.9d.

It is estimated that an array area of about 1 cm^2 can produce approximately 10 m of yarn [67]. To grow a very long CNT array, the growth time can be increased [70,105]. Water also acts as a weak oxidizer to remove amorphous carbon, thereby aiding the growth of longer CNTs [106]. This process has led to an incredibly high length of spinnable CNT arrays currently reported up to 6 mm [106]. These control parameters are however for CVD-grown CNT arrays. Direct spinning of CNT yarns from CVD furnace has only reported up to 1 mm-long CNTs [107]. Spinning longer CNT yarns without high degree of misalignment using other methods has not been extensively reported so far.

Factors Affecting the Production of CNT Yarns

The morphology of the CNT array can be managed by controlling the pretreatment time of the catalyst or the addition of hydrogen or oxygen during growth. The length of the CNT array can be controlled by varying the growth time. Typically, water vapor can be used to prolong CNT growth time to produce longer CNT arrays [105–108].

The distribution of the CNT diameters and, to an extent, controlling the number of walls in the arrays can be achieved by adjusting the thickness of the catalyst layer [72,108]. The diameter of the yarn depends on the number of CNT filaments or bundles in the yarn. As a rule of thumb, the size of the CNT yarn can be controlled by the tip size of the tool that is used in picking up the CNT array to spin the yarn; smaller tip will produce thinner CNT yarns [67]. The twist speed of the spinner is used to control the twist angle on the yarn. High twisting speed produces high twist angle, while low twisting speed produces low twist angle.

SWCNTs have a higher energy of formation than MWCNTs, which is why they are grown at much lower temperature (600°C–900°C) than SWCNTs (900°C–1,200°C) [70]. This is assumed to be due to small diameters. Small-diameter CNTs have less thermodynamic stability and high curvature which lead to high strain energy. This might also be the reason why SWCNTs can only be grown from certain hydrocarbons that are stable in the temperature range of 900°C–1,200°C while most of the hydrocarbons used in growing MWCNTs (viz. acetylene, benzene, etc.) are unstable at elevated temperature and produce large amounts of residual metals at such temperatures [70–72].

18.2.3 Structure of CNT Yarns

A CNT yarn is comprised of bundles of CNTs. The structure of a CNT yarn is derived from the structure of the individual CNTs and CNT bundles. The basic unit of the yarn used in this study is the MWCNT. The representation of CNT aggregates in length scales and structure evolves from: MWCNT ($\times10^{-9}$ m), MWCNT bundles ($\times10^{-8}$ m), and MWCNT single yarn ($\times10^{-6}$ m). The hollow structure of individual CNTs accounts for the low mass exhibited by CNT yarns. As seen in Figure 18.10a, CNT yarns also have low packing density (volume fraction or specific weight) emanating from the CNTs' separation or contact length. This separation in contact length brings about low friction under tension. Low inter-tube friction causes a reduction in the tensile strength of CNT fibers [109–112]. To minimize this effect, the spun fibers are densified or condensed to increase the actual contact length. Condensation can be achieved via twist, solvent densification or both. Twists induce cohesion between CNTs and CNT bundles to prevent disintegration under lateral forces. With improved cohesion, the elasticity of the fiber is enhanced by initiating uniform load bearing capability across the fiber's length. Twisted and often densified fiber bundles are called CNT yarns. In the context of this work, CNT yarns refer to fibers spun from a CNT array and condensed. The use of organic solvent alongside twist not only shrinks the porous structure of the yarn and improves their mechanical and electrical properties, the densification reduces the sticking problem discussed as well. Zhang et al. reported that dry-spun CNT yarn passed through droplets of ethanol shrank several centimeters wide yarn into a tight yarn of about 20–30 μm in diameter [109]. Due to its reduced surface area, the authors observed that post-treated yarn was no longer as sticky as the fresh yarn and could be easily wound on and off a spool using a motor or by hand. Due to the rough nature of the yarn, even after densification, uniform radial properties cannot be guaranteed (Figure 18.10b). However, it has been studied that crosslinking or bridging CNT bundles in the yarn contributes to the robustness of the yarn [109]. The associated junction could also improve the mobility of charges carriers.

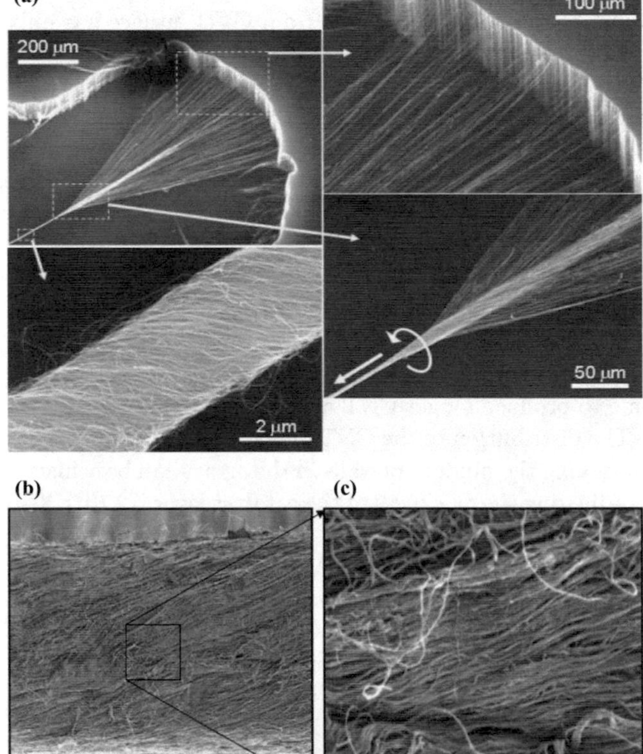

FIGURE 18.10 (a) SEM micrographs showing the structures formed during the draw-twist process. Top left: Bundled MWCNTs being pulled from the side of the array. Top right: Magnification of spinnable CNT array being drawn into web. Bottom right: Initial part of the draw-twist process. Bottom left: A spun CNT yarn [86]. (b and c) SEM images of CNT yarn: (b) after densification with acetone {×1,500}. (c) Insight into the internal structure of the spun yarn {×8,000}. (Images taken with JEOL JSM-7100FA FE SEM [113].)

18.2.4 Characterization

Mechanical Characterization

Mechanical characterization of CNT fibers is not as difficult as that of individual nanotubes. They can be viewed with the naked eyes and easily manipulated with laboratory handling tools like tweezers for characterization. Also, diameters of the CNT fibers can be viewed with an optical microscope unlike individual nanotubes with diameters of orders of magnitude smaller than the wavelengths of visible light (200–600 nm). However, electron microscopes provide a better resolution to characterize the morphology and surface features of both CNTs and CNT fibers. Universal testing machines (UTMs) are typically used to perform the mechanical tests. A force transducer attached to the UTM can be used to apply the load to the sample via a control software which is programmed to apply uniaxial tension, compression or flexural load to the sample at several displacement rates. Once the maximum load is reached, the loading machine stops and gradually unloads the sample until the crosshead extension reaches its original position. A typical setup for a uniaxial tension test and a test sample is shown in Figure 18.11.

Electrical Characterization

The electrical measurements are normally recorded concurrently with the tensile tests through the data acquisition systems. In the setup shown in Figure 18.12, an inductance–capacitance–resistance (LCR) reader unit mounted in a

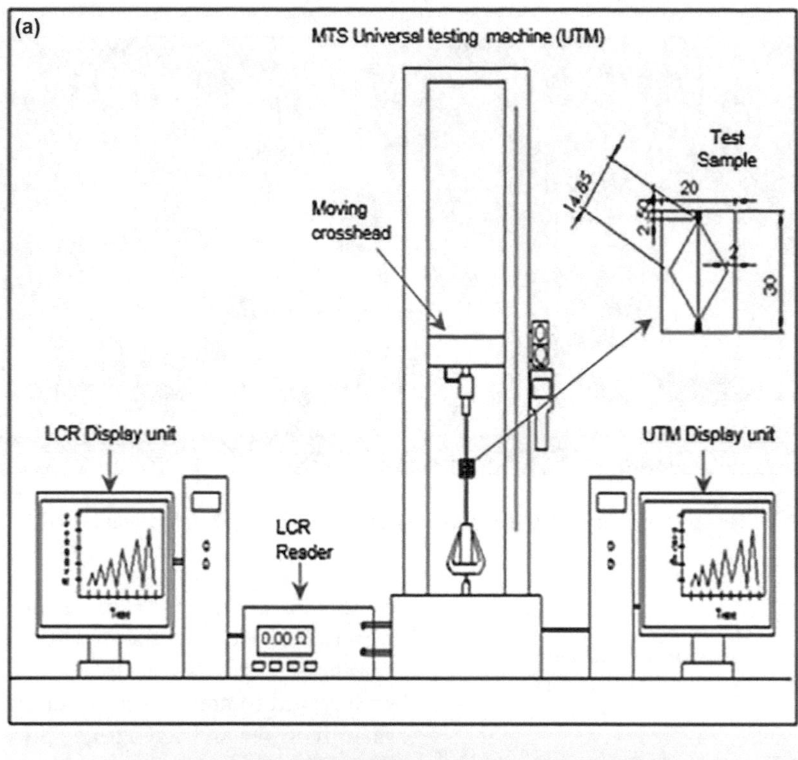

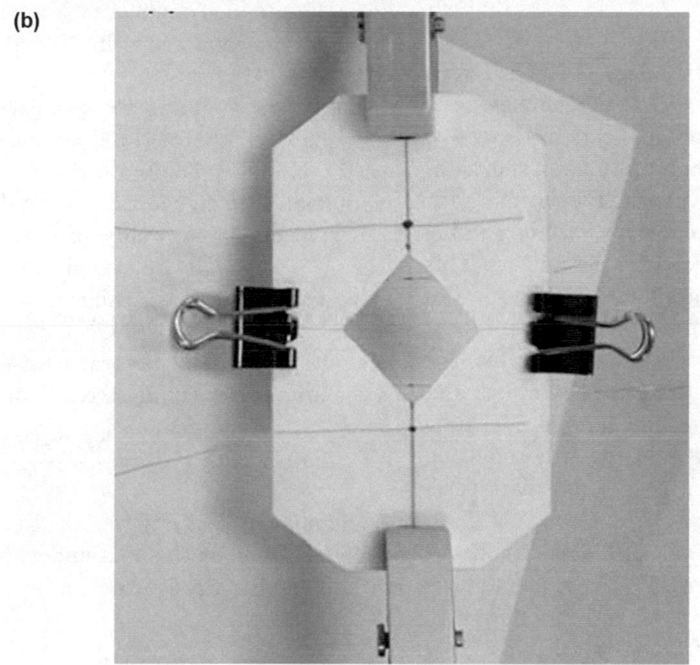

FIGURE 18.11 (a) Schematic of the experimental setup. (b) Optical image of CNT yarn sample.

National Instruments PXI-1033 chassis was used for the electrical measurements. The LCR was connected to the CNT yarn sample through wire clips attached at both ends (Figure 18.12). The connection of the clips was attached to the test frame to avoid dangling. The wires moved along with the testing frame because the length was enough to ensure they measured up to the displacement stage during testing. To ensure that the load was properly transferred to the sample, the adapter grips from the MTS machine were clipped at the edges of the gauge length to eliminate any gap. This way, any distance between the gauge length and the load connection could be neglected. The LCR has a capacity to measure resistances up to 10 MΩ. The electrical measurements from the LCR reader were outputted through a LabVIEW^{TM} virtual instrument. The electrical wires attached to the samples were connected to Bayonet Neill–Concelman (BNC) cables with mini-hook electrical leads. For accuracy, four-point probe measurements are used to eliminate contact resistance from the connecting wires.

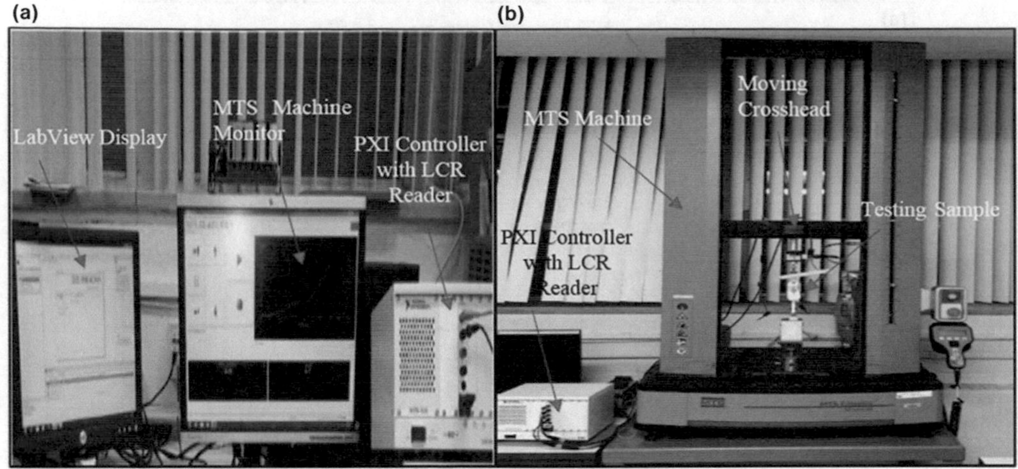

FIGURE 18.12 Experimental setup showing: (a) LCR device and displays. (b) LCR meter connected to the MTS machine.

Methodology

The CNT yarn's aspect ratio, s, is obtained using the relation between the length of fiber, l_y, to its radius, r_y:

$$s = \frac{l_y}{2r_y}. \tag{18.8}$$

The determination of the cross-sectional area of a CNT yarn to calculate the properties of CNT yarn is not simple due to the structure of the CNT yarn. The application of twist to the yarn introduces variability or nonuniformity in the diameter of the yarn across the radial length. Porosity or yarn volume density will also differ from yarn to yarn depending on the production method and processing parameters. This brings about variations in calculated material properties including yarn mechanical strength. The implication is that the elastic modulus of a yarn is a volume-average value. For this reason, using the cross-sectional area of a solid cylinder for the yarns is an unreliable value. The adoption of linear density generally used in the textile industry by normalizing the dimension of the yarns makes it easy to compare CNT yarns of varying production methods.

The International System of Units (SI unit) for linear density is gram per kilometer (g/km). However, it is tedious to obtain a lab measurement of the mass of 1 km-long yarn. An adequate method is to weigh a few meters, then multiply and normalize the values up to the unit weight.

The linear density, *LD*, of the CNT yarn is expressed in tex,

$$1\,\text{tex} = 1\text{g}/1,000\,\text{m} \tag{18.9}$$

The linear density can also be expressed in denier as mass in gram per 9 km,

$$1\,\text{den} = 1\,\text{g}/9,000\,\text{m} \tag{18.10}$$

The specific strength otherwise known as tenacity, T, is the tensile strength divided by the volumetric specific density with units of GPa/SG. The specific gravity, SG, is a dimensionless parameter for normalizing the density of the yarn by that of water which is 1 g/cm^3, i.e., the density of the material divided by the density of water. The use of the specific strength is due to the complex geometry of yarns with regard to nonuniformity in twist and porosity affecting their diameter and strength as previously mentioned for the area. The tenacity can also be expressed as the breaking force per linear density in N/tex. N/tex is equivalent to GPa/SG.

The elastic modulus was calculated from the stress–strain curve as the slope of the linear elastic region. The specific elastic modulus is obtained by the ratio of the elastic modulus (GPa) to the volumetric density (g/cm^3).

Resilience is the ability of a material to absorb energy when it is elastically deformed. This energy can be recovered by relaxing the stress as shown in Figure 18.13.

The resilience of the CNT yarn was calculated as the energy absorbed by the yarn up to the yielding point. It is the area under the hysteresis curve, given by:

$$\text{Resilience } (\%) = \frac{\text{ABC} - \text{ABD}}{\text{ABC}} \times 100. \tag{18.11}$$

For nonlinear elastic materials, the modulus of resilience is easily obtained as the area under the stress–strain curve up to yielding. For CNT yarns that are nonlinear elastic and the

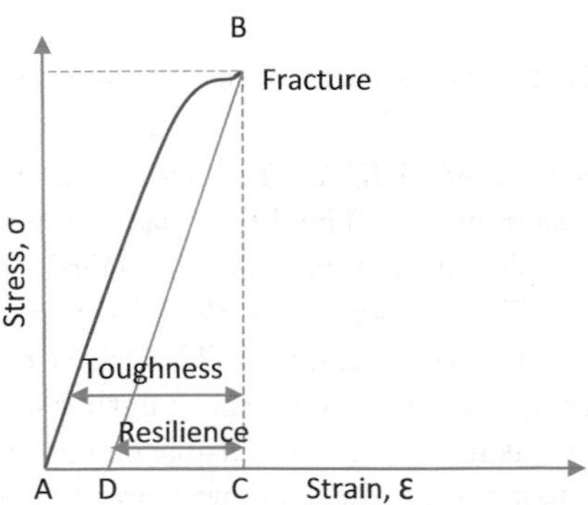

FIGURE 18.13 Schematic of the stress–strain curve showing the toughness and resilience method.

yield point is not truly defined, it is better to calculate the resilience value by integrating the stress–strain curve from zero up to the elastic limit.

Toughness is the ability of the CNT yarn to absorb mechanical energy up to the point of failure or the energy of mechanical deformation per unit volume prior to fracture. Since energy is defined as the ability of a body to do work and work itself is the product of force and the distance the force acts through, the energy absorbed by the CNT yarn can be determined by integrating the stress–strain curve or the area under the stress–strain curve (ABC).

$$\text{Toughness: } \frac{\text{energy}}{\text{volume}} = \int_0^{\varepsilon_f} \sigma d\varepsilon \qquad (18.12)$$

where ε_f is the failure strain. The area under the curve could easily be obtained in MATLAB™. The gravimetric yarn toughness is calculated by dividing the volumetric yarn toughness by the volumetric density in gm^{-3}.

It could be difficult to differentiate between the resilience and toughness of the CNT yarn due to their low elongation-to-break and indistinct yielding features. Another direct approach for comparing the rudimentary tensile strengths of yarns is the tensile factor. The tensile factor is an approximate measure of resilience, given by the product of the yarn's tenacity and the square root of their percent elongation, i.e.:

$$\text{tensile factor} = \left[(T)\,(E)^{1/2}\right] \qquad (18.13)$$

A yarn with higher tensile factor is considered better for applications in high strength and lightweight textiles as it provides improved wearing properties in wearable materials especially woven fabrics.

The yarn volume fraction or the concentration of CNT bundles in the yarn represents the compactness of the yarn and is expressed as the ratio of the volumetric density of the yarn structure to that of the constituent individual CNTs [110].

$$V = \frac{V_{\text{yarn}}}{V_{\text{CNT}}}. \qquad (18.14)$$

The volumetric density of the single CNT is calculated by evaluating the linear density of the individual CNTs using a hollow tube model [111].

$$V_{\text{CNT}} = \pi r^2 h \qquad (18.15)$$

The yarn's porosity, ϕ, is a measure of the compactness of the yarn. It is defined as the volume fraction of voids between the individual CNTs, their bundles, and the structure of the yarn.

$$\phi = 1–V \qquad (18.16)$$

where V is yarn volume fraction.

Electrical conductivity in Siemens per meter can be calculated from the electrical resistance by:

$$\sigma_e = \frac{l}{RA}. \qquad (18.17)$$

Specific conductivity (in Siemens per meter over gram per cubic centimeter),

$$\sigma_{sp} = \frac{l}{RxL_D}10^9 \qquad (18.18)$$

where l is the length of the yarn (in meter), R is the electrical resistance, and L_D is linear density (in tex).

18.2.5 Properties of CNT Yarns

Mechanical Properties of CNT Yarns

CNT fibers with tensile strength of up to 1.9 GPa and stiffness of up to 300 GPa have been reported [110], which exceed those of conventional conductors like copper. However, the mechanical properties of CNT fibers do not come close to the exceptional properties of the individual nanotubes. The difference in properties is due to the weak van der Waals interactions between CNTs in a yarn structure and the nature of association of the CNT bundles [111–113]. While the properties of the CNTs are primarily derived from the covalent carbon bonds, the CNT yarn relies on the nature of the van der Waals forces and the interfacial contact between the CNT bundles. Therefore, the strength of a CNT yarn is derived from the strength of its constituent CNTs and the frictional forces between them. By comparing the mechanical properties of CNT fibers produced via different methods as reported by various authors, Lekawa-Raus et al. reported that CNT fibers with the best mechanical properties have so far been obtained from fibers dry-spun from CNT arrays [57]. This underlines the role of mechanical densification or twist in improving fiber effectiveness or the bond between CNTs or their bundles through friction. In doing so, coherency (degree of packing) is imparted into the yarn to withstand significant strain under tension. The frictional forces from the lateral pressure exerted on the yarns under axial tensile stress depend on this degree of packing. Therefore, the strength of a yarn depends on the coherency and the degree of axial order of aligned CNTs or parallel alignment of the bundles in the yarn known as obliquity. In simple terms, twist produces the frictional force required to bind the fibers together to induce coherency in the yarn structure. The twist-induced pressure provides a lateral constraint that depends on tensile loading. By so doing, the yarn structure exerts some strength to withstand disintegration under pressure. In fact, a yarn has almost zero or no measurable strength without this pressure [112].

CNT Yarn Mechanics

The twist in a yarn is characterized using three main parameters: (i) twist factor or multiplier (otherwise called twist coefficient), (ii) twist or helix angle made by the fiber bundle with respect to yarn axial direction or the twist level in turns/unit length; and (iii) twist direction denoted by the letter "S" or "Z", which represents the orientation of the fibers on the surface of the yarn with respect to the

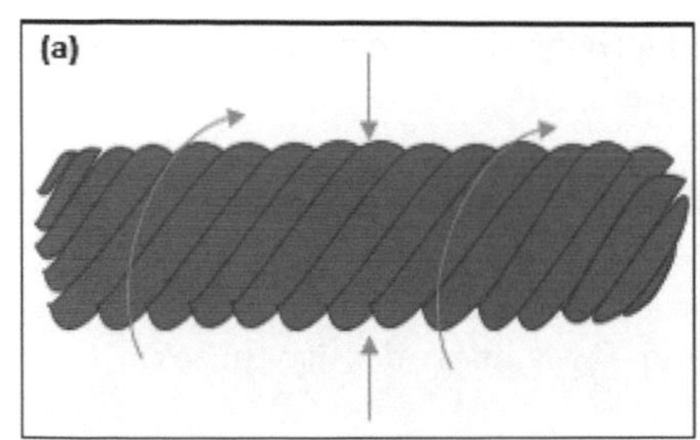

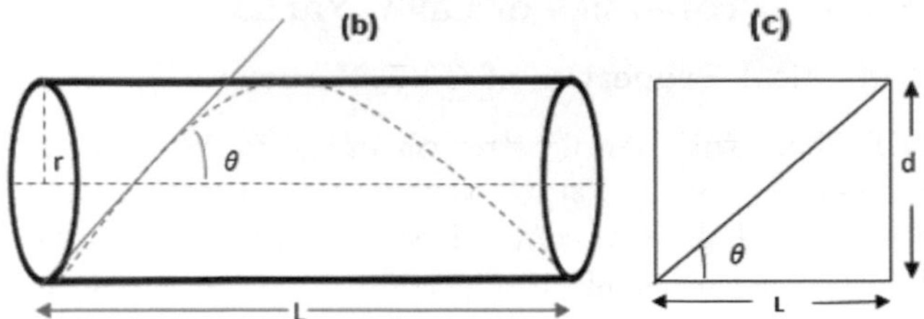

FIGURE 18.14 Schematics of: (a) twist on CNT yarn. (b) An arbitrary ideal yarn with length, L, diameter, d, and helix angle, θ. (c) 2D illustration of angle and axes.

yarn when placed in a vertical position. Twist direction is important in textiles as it affects fabric properties such as lustrousness, firmness, and softness. Twist direction of a yarn does not necessarily affect its strength but certainly its performance. In practice, it is best explained when the yarns are plied. A ply yarn of two single yarns with opposite twist directions will expand (loosen) while that of similar twist direction will contract (tighten) when woven into a fabric. In the same vein, a sewing thread with an S-twist will unravel through a single needle unlike Z-twist because the sewing process increases the tension in the thread. An S-twisted yarn was used in this study.

It is often important to correlate the properties of the yarn with twist. Using Hearle's model of classical idealized helical geometry of a circular yarn seen in Figure 18.14 [114], an expression can be derived for the twist with respect to the yarn parameters.

For an ideal yarn with twist angle, θ, as shown in Figure 18.14b,

$$\tan\theta = \frac{\pi d}{L} \tag{18.19}$$

where d is the diameter and L is the length of one turn of a twist in the yarn.

Twist is defined as the number of turns per unit length L. If L is in cm, then from Figure 18.14c, the turns per unit length L (let's say 1 cm) of the yarn would be:

$$L = \frac{1}{\tau} \tag{18.20}$$

where τ represents the number of twists per unit length or the twist level.

Substituting Eq. (18.20) into Eq. (18.19) gives:

$$\tan\theta = \pi d.\tau \tag{18.21}$$

$$\tau = \frac{\tan\theta}{\pi d} \tag{18.22}$$

Since mass,

$$m = \rho V, \tag{18.23}$$

linear density, L_D can be expressed as:

$$L_D = \frac{m}{l} \tag{18.24}$$

or

$$L_D = \frac{\rho V}{l} \tag{18.25}$$

where the volumetric density, ρ, of the yarn is given as the mass in grams divided by the volume of a solid cylinder, V, in cm^3.

$$\rho = \frac{L_D l}{V} \tag{18.26}$$

$$= \frac{L_D}{A} \tag{18.27}$$

where A is the cross-sectional area.

Assuming a yarn with cylindrical structure and a circular crosssection,

$$\rho = \frac{L_D.10^{-5}}{(\pi\frac{d^2}{4})}. \tag{18.28}$$

Equation (18.28) can be expressed in terms of the yarn's diameter:

$$d = \sqrt{\left(4\frac{L_D.10^{-5}}{\pi\rho}\right)}. \tag{18.29}$$

Substituting for d in Eq. (18.29) into Eq. (.22) gives:

$$\tau = \frac{\tan\theta}{2\pi\sqrt{\frac{L_D.10^{-5}}{\pi\rho}}}. \tag{18.30}$$

If

$$T_F \equiv \frac{1}{2}\tan\theta\sqrt{(10^5\pi\rho)} \tag{18.31}$$

then,

$$T_F = L_D^{1/2}\tau \tag{18.32}$$

where T_F is the twist factor.

When ρ is constant, T_F becomes proportional to θ. The unit of τ is tpm$\sqrt{\text{tex}}$ where tpm is turns per meter.

Twist–Strength Relationship of CNT Yarns

Figure 18.15 shows the strength–twist relationship of CNT yarn and conventional yarns. Figure 18.15a and b shows a similarity in pattern from low tenacity at low twist to high tenacity at the medium twist range before showing a decline in tenacity at further twist ranges. This trend could be explained by the numerous parameters in the structure of spun yarns that affect their strength, from coherency to obliquity. Such enormous variability in the structure of spun yarns makes it very difficult to make the different parameters affecting yarn's strength uniform.

This is partly the reason why the relationship between twist and strength of staple and continuous filament yarns, which have been studied over the years [115–122], has no single universal model to predict their failure behavior.

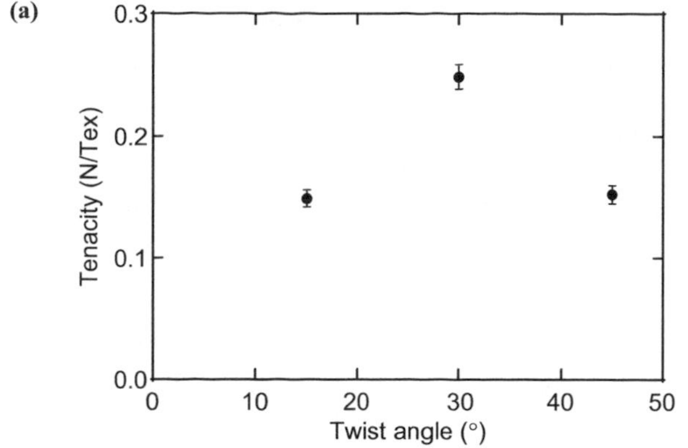

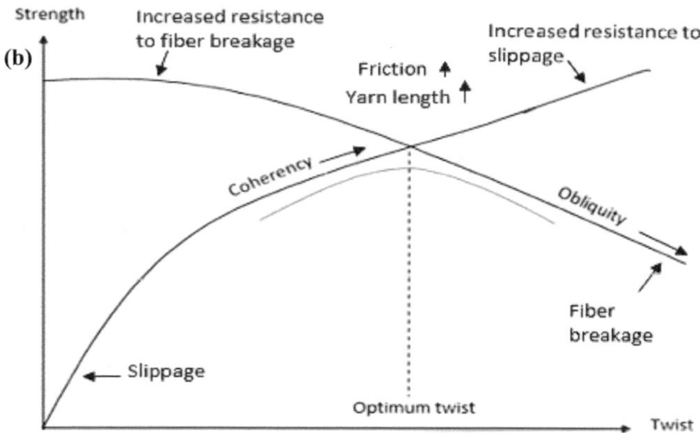

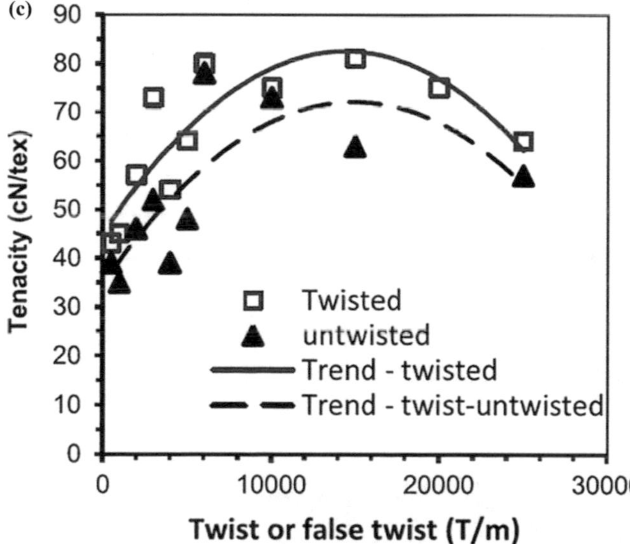

FIGURE 18.15 (a) Strength–twist relationship of CNT yarn. (b) Schematic of the strength–twist relationship for textile yarn [114], depicting the mechanisms of the association. (c) Flyer-spun CNT yarn tenacity as a function of twist [111].

To understand the role of both coherency and obliquity on the tenacity of the CNT yarn, it is important to discuss the role they play in twist formation.

CNT ribbons drawn from an array have very low mechanical strength and, thus, are twisted to form stronger yarns. The lack of strength observed in the (untwisted) ribbons is associated with the parallel orientation of the fiber bundles to the yarn axis. Untwisted yarns have low interfacial contact between the fiber bundles and, hence, low-to-zero binding force. Such yarns will exhibit similar characteristics to parallel fibers that fail under a sheer amount of slippage. Under the action of twist, the induced lateral force increases the interfacial contact leading to increased inter-tube friction. As the twist level increases, the inter-tube friction increases, so does the yarn's packing density. The increased packing density means that more fiber bundles are connected in bonds along the length of the yarn, increasing their resistance to slippage through friction. Further increase in twist will continue to bind the fiber bundles until they start to interlock.

Fiber locking occurs by the transfer of tensile stress to transverse stress during tensile deformation due to high twist action. As stress is continuously built up in the yarn under tension, the fiber locking effect becomes dominant limiting any inter-fiber bundle shear motion or slippage, consequently, improving the strength of the yarn. During tensile loading of the yarn, the apparent stress–strain behavior will generally have a short linear region characterized by slippage of the fiber bundles followed by nonlinearity with considerable extension while the fiber bundles continue to slip, and finally failure occurs.

Hearle's equation can be used to quantify the twist–strength effect in yarn structures [114]:

$$\frac{\sigma_y}{\sigma_f} = \cos^2\alpha\,(1 - k\csc\alpha) \tag{18.33}$$

where $\frac{\sigma_y}{\sigma_f}$ is the ratio of the yarn strength to the fiber bundle strength and α is the helix angle that the constituent fiber/bundles make with the yarn axis. The term $(1-k\csc\alpha)$ represent the fiber locking, where k is a constant given by:

$$k = \left(\frac{dQ}{\mu}\right)^{\frac{1}{2}} /3L \tag{18.34}$$

where d is the diameter of the fiber, Q is the fiber migration length, μ is the coefficient of friction, and L is the length of the fiber.

From Eqs. 18.33 and 18.34, increasing the coefficient of friction and length of the fiber and decreasing the diameter and the migration length will result in an increase in the strength of the yarn. However, there is a level of twist associated with the maximum strength of the yarn. The tensile strength of CNT yarns increases as twist increases to an optimum twist level, beyond which the strength of the yarn starts to decrease. Beyond this optimum, fiber bundles are oriented far away from the yarn's axis that the contribution of their strength to the yarn's strength becomes ineffective. This will reduce the overall strength of the yarn. The two effects are contrasting. The increase in strength with twist is associated with increase in fiber cohesion under twist while the decrease in strength with twist is associated with fiber obliquity under high twist reducing the contribution of the fiber strength to that of the yarn.

Below this optimum, the yarn will fail through fiber slippage by the failure of the bundles while above the optimum, the yarn will fail by yarn fracture (Figure 18.15a).

The CNT yarn of varying twist showed an optimum strength at 30° twist as seen in Figure 18.15a. Miao et al. [111] observed that the tenacity of a flyer-spun twisted yarn reached its peak at about 15,000 T/m and then decreased slowly as twist level was further increased, as shown by the trend curve in Figure 18.15c. The optimum strength close to intermediate twist angle seems to agree with that of conventional yarn as seen in Figure 18.15b. The key difference, however, is that the optimum strength of the CNT yarn is achieved at about 30° twist angle while that of textile yarns is closer to 50° twist angle. Stronger fibers can produce yarns with optimum strength at lower twist angles [120]. For example, polyester fibers (breaking strength of 35–60 cN/tex) have optimum yarn strength at lower twist level compared to cotton fibers (breaking strength of 15–40 cN/tex). The difference could also be related to the difference in mechanisms of the interfiber bundle association, degree of alignment, and the interfacial surface energy of CNTs. Whereas a cotton yarn has between 30 and 100 individual fibers in a yarn cross section and a synthetic fiber yarn up to 100 fibers, a 10 μm-CNT yarn is estimated to comprise between 51,000 and 115,000 bundles in its cross section [123]. This is over three orders of magnitude difference in the fibrils in the cross section of CNT yarns to that of conventional textile yarns [93]. The van der Waals interconnections between CNTs play a key role in the load transfer mechanism in a CNT yarn. Typically, in conventional yarns, the interfiber shear forces dictate the strength. CNT yarns, mostly dry-spun yarns, depend on this weak van der Waals connections between the CNTs to form and maintain a yarn structure. The CNT yarns used in this study were produced by highly aligned CNT arrays. The high alignment of CNT arrays is ensured by controlling the length or height of the array through growth time, for example. The longer CNTs tend to be entangled due to high strain energy of CNTs. It is well established that very long CNT arrays become very misaligned that they cannot be spun into yarns [105–107]. This is because the high surface energy of CNTs also leads to their existence in bundles to minimize surface energy. Bundling will result in minimal twist to achieve cohesion.

The tenacity of the yarn would be lowest at the lowest twist level as observed in Figure 18.16. The tenacity of the CNT yarn with medium twist was significantly higher than that of the high twist. This is in agreement with conventional understanding of twist–strength relationship in textile yarns. It is also important in the selection of twist level based on applications. The failure mode of low-twist CNT yarn is dominated by fiber pullout due to slippage while the high-twist CNT yarn fails as a unit in what resembles a brittle-like failure observed in metals as seen in Figure 18.17.

Toughness

Unlike in strength measurements, the highest value of volumetric toughness (4.7 MJ/m^3) was observed in the high-twist CNT yarns (Figure 18.18). It was then followed by the medium-twist (3.6 MJ/m^3) and lastly the low-twist CNT yarns (2.3 MJ/m^3). Toughness of the CNT yarns increases with twist level [124]. Although medium-twist CNT yarns showed the highest strength, their low strain values bring

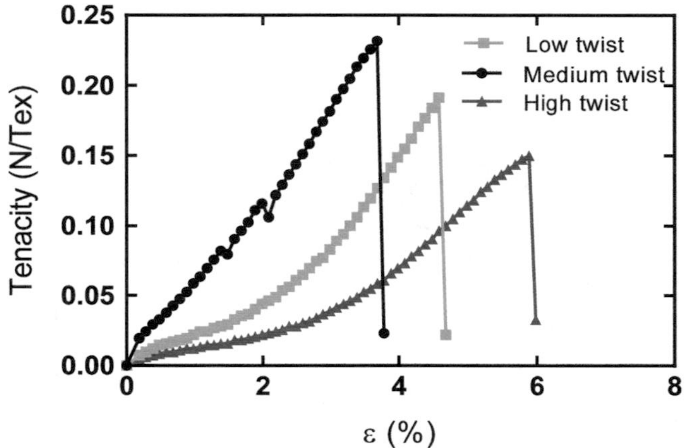

FIGURE 18.16 Tenacity–strain curves for different twist levels [114].

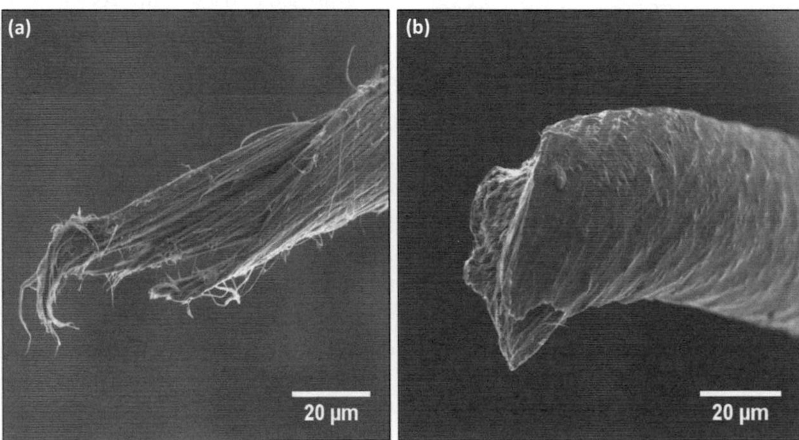

FIGURE 18.17 Fractography of the bare and composite CNT yarn: (a) Bare CNT yarn. (b) Composite CNT yarn (×1,200).

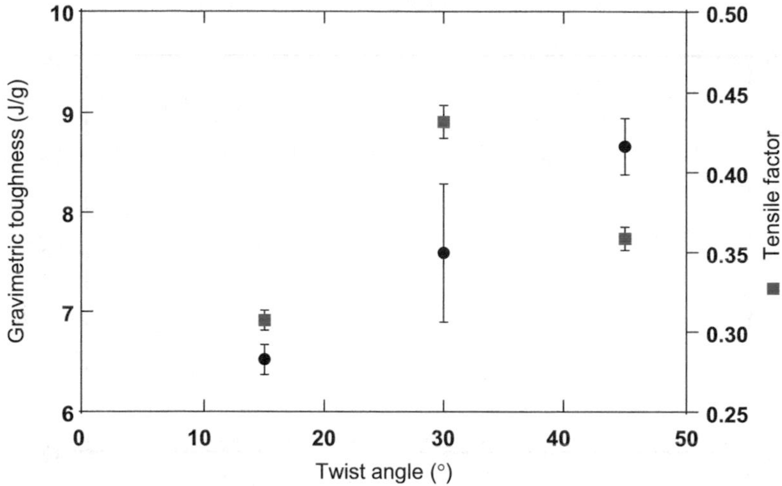

FIGURE 18.18 Gravimetric toughness and tensile factor versus twist for CNT yarn.

about a relatively lower toughness to the high-twist CNT yarn. The gravimetric toughness follows the same pattern as the volumetric toughness, with the least value of gravimetric toughness (6.5 J/g) obtained at low twist and the highest value at high twist (8.7 J/g).

Toughness of CNT yarn is strongly dependent upon the alignment of CNTs in the axial direction of the yarn [124–127]. Since coherency of fiber bundles in the yarn increases with twist level, it is expected that increased degree of packing leads to increased toughness. However, improving toughness in CNT yarns can be challenging. This is because it is difficult to produce a very strong yarn that can withstand a significant amount of strain. While increasing the twist level can help with elongation, large stress at high twist (above the optimum) decreases the tensile strength.

However, the medium-twist CNT yarn showed the highest tensile factor of 0.43 compared to the 0.31 and 0.36 obtained for the low- and high-twist CNT yarns, respectively. The tensile factor provides a better insight into the resilience of the yarn. This makes a medium twist better in absorbing elastic energy. Since tougher yarns provide better resistance to elastic and plastic deformations, they are mechanically suitable for extreme environmental conditions including impact, shock, and vibrations. They can be incorporated into flexible and stretchable smart textiles.

Electrical Properties of CNT Yarn

Porosity and diameter of the CNT yarn are functions of the twist as seen in Figure 18.19a. CNT yarns spun directly from an array are generally porous [128–131]. Twist insertion strengthens dry-spun CNT yarns by binding the fiber bundles together. The decrease in diameter with increasing twist is due to the increased tension from the outside to the interior of the yarn as twist increases. The higher the twist, the higher the tensions that are exerted on the yarn. This axial tension creates a counteracting lateral compressive force on the yarn, which increases the packing density. As the fiber bundles condense in the yarn, the diameter of the yarn decreases. Solvent densification, by shrinking the yarn, achieves the same purpose. This condensation process reduces the pore size in between the fiber bundles and, more importantly, brings the CNTs in the bundles closer.

Bearing in mind that tunnel junctions are significant to electron transport in CNT yarns, this increased packing density obtained at higher twist explains the increase in conductivity of CNT yarns. An inverse relationship of the porosity of the CNT yarn to the twist angle (Figure 18.19a), and of the conductivity of the CNT yarn to the porosity (Figure 18.19b), means that when CNTs are closely packed in the yarn, electron mobility is more efficient, increasing the conductivity of higher diameter and having less variation of mass or linear density. This observation is accurately explained by the porosity of the CNT yarn. The electrical conductivity of the flyer-spun twisted and twist-untwisted yarns as shown in Figure 18.19c increased rapidly as the level of twist increased [112].

Piezoresistivity of CNT Yarns

CNTs are known to exhibit ballistic conductivity due to minimal electron scattering in their 1D structure with mean free paths of the order of tens of microns [132]. Under deformation, the charge carriers are separated leading to an increase in resistance. For very small strains, the deformation has shown to be elastic and the conductive network is fully recovered when the strain is removed, leading to a decrease in resistance. Plastic deformation, however, has proven to be different. Although the resistance goes to zero when the strain is removed, hysteresis has been observed [130].

The piezoresistivity of CNT yarns comes from two types of resistance changes: (i) The intrinsic resistance of the CNTs and (ii) the inter-tube resistance of nanotubes in proximity or contact [133]. The intrinsic resistance, R_i, is the resistance of the CNT yarns due to the stretching of

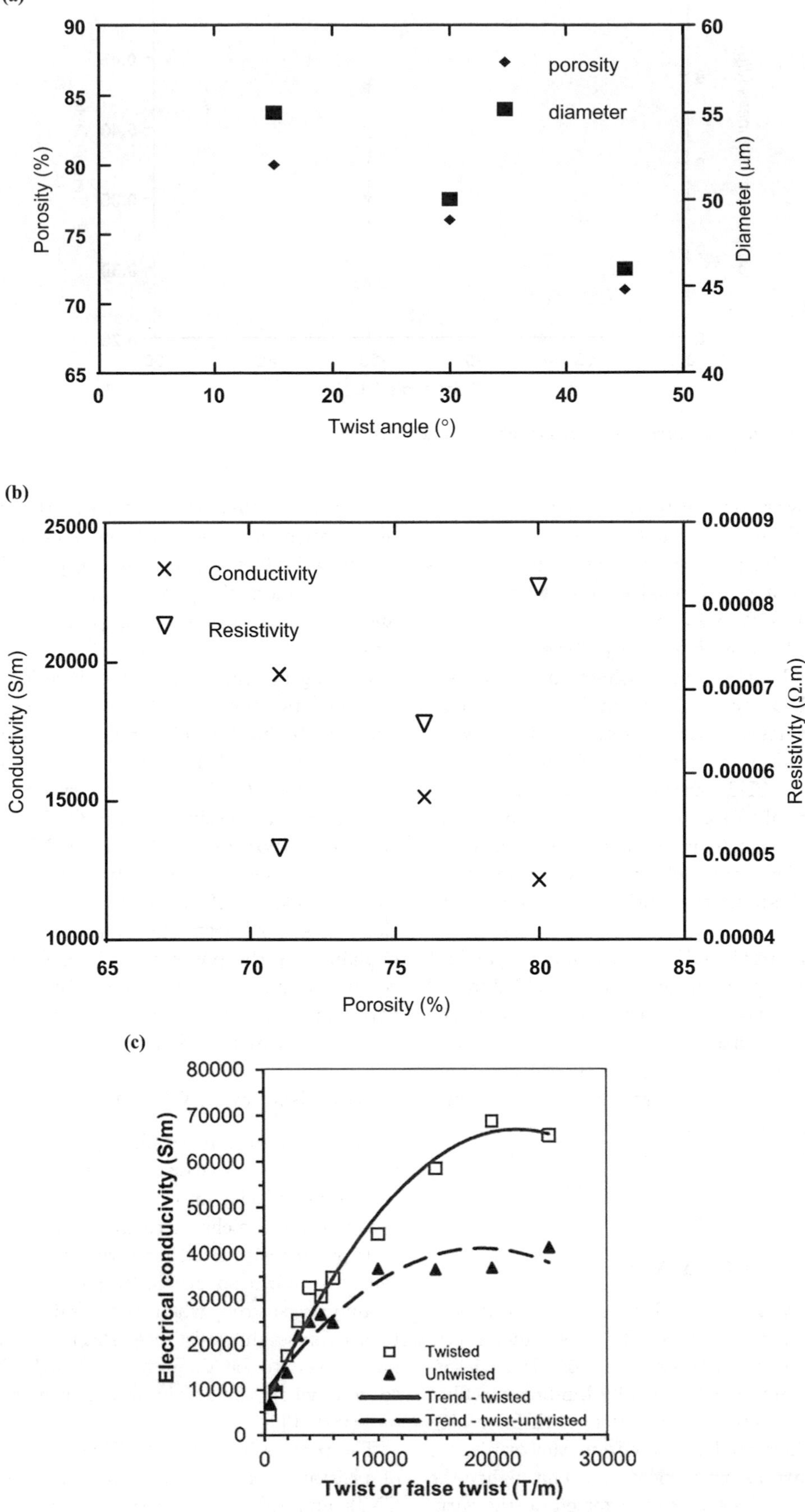

FIGURE 18.19 (a) Porosity and diameter of the CNT yarn versus twist angle. (b) Conductivity and resistivity of the CNT yarn versus porosity. (c) Electrical conductivity of flyer-spun twisted and twist-untwisted CNT yarns as a function of twist [112].

their carbon to carbon (C–C) bonds or separation of their charge carriers. The inter-tube resistance is broken down into contact resistance, R_C, of nanotubes in physical contact or tunneling resistance, R_T, when nanotubes are separated by small gap. According to J. G. Simmons [134], the conditions for tunneling to occur through the insulating region between the two electrodes are that: "(i) the electrons in the electrodes have enough thermal energy to surmount the potential barrier and flow in the conduction band. (ii) The barrier is thin enough to permit its penetration by the electric tunnel effect".

The tunneling resistance is expressed as:

$$R_T = \frac{dh^2}{Ae^2\varphi} e^{\frac{4\pi d}{h}\varphi} \tag{18.35}$$

where d is the tunneling distance between CNT, h is Planck's constant, A is the effective cross-sectional area, e is the quantum of electricity, and φ is given by:

$$\varphi = \sqrt{2m\delta} \tag{18.36}$$

where m is the electron mass and δ is the height of the potential barrier between adjacent CNTs.

From Eq. (18.35), R_T increases nonlinearly, resulting in a nonlinear piezoresistivity.

Inter-tube resistance is higher in CNT yarns due to the number of CNTs in contact. The short or discrete length of the CNTs means that junction resistance will play a part in their piezoresistivity under axial strain. Considering that CNTs do not span the entire length of the fiber, intrinsic resistance is expected to play a minimal role in the piezoresistive response, and it can be concluded that the piezoresistivity of CNT yarns is inter-tube resistance-driven. Increase in CNT length will increase the contribution of intrinsic resistance in the yarn and reduce the effect of inter-tube resistance due to contact. When a free or neat CNT fiber is stretched, we expect the deformation mechanism to be dominated by: (i) breaking of contact due to fiber unraveling and bond breaking and (ii) slippage [113]. The first phenomenon leads to an increase in contact resistance. In the presence of a matrix, tunneling seems to drive the piezoresistivity due to matrix infiltration of the porous fibers creating barriers for electron tunneling to occur.

One important factor to consider when characterizing a sensor is the sensitivity, an output change for a given change in an input parameter. For a strain gauge, this is represented by the ratio of relative change in electrical resistance, $\Delta R/Ro$, to the mechanical strain.

The sensitivity of the CNT yarn represented by the gauge factor is given as:

$$\text{GF} = \frac{\Delta R/Ro}{\varepsilon} = \frac{\Delta R/Ro}{\Delta L/Lo} \tag{18.37}$$

where Ro is the initial resistance, ΔR is the change in resistance, ε is strain, which is defined as the ratio of the change in length ΔL over the original length L.

Although the GF for SWCNT-based piezoresistive strain sensors has been shown to be >2,900 [47], CNT fiber-based sensors have shown values lower than that. The reported gauge factor of neat CNT fibers was around 0.5 [50,113]. The dominance of contact resistance between CNT bundles in the yarn means that the contribution of intrinsic resistance in the individual CNTs is minimal in a CNT fiber.

18.3 Applications

18.3.1 Electrical Wire

There is current interest in utilizing CNT's ballistic conductive profile as electrical cables.

Individual CNTs are capable of ballistic electron transport and at room temperature have a micrometer range of free mean path, i.e., they can travel up to a micrometer without scattering. The conductivity of CNT fibers has been reported up to 67,000 S/cm [57]. Recent studies have shown that the specific conductivity of CNT yarns may reach a maximum of 19.6×10^6 S/m/g cm^3 when doped exceeding 14.15×10^6 S/m/g cm^3 and 6.52×10^6 S/m/g cm^3, which are the specific conductivity of aluminum and copper, respectively [135,136]. Due to the abundance of carbon, CNTs have a cost advantage over conventional conductors like copper as shown in Table 18.3.

However, there are still challenges to the optimization of the properties of CNTs as electrical wires, which include improvements in control of chirality, morphology, and dimensions of the as-produced CNTs. Due to their

TABLE 18.3 Brief Comparison between CNT Wires and Traditional Copper Wires [57]

		CNT Wires	Copper Wires
Cost	Production	Low	High
	Maintenance	Low	Low
Fabrication/recycling	Availability of resources	Abundant	Limited
	Complexity of processing	Low	High
	Environmental impact	Minimal	High
	Recycling	Good	Good
Maintenance	Working life in normal operating conditions	Long	Long
	Corrosion resistance	High	Low
	Temperature stability in oxygen atmosphere	Low	Medium
	Temperature stability in inert atmosphere	High	Medium
Properties	Density (g/cm^3)	0.28–2	8.9
	Electrical conductivity (S/cm)	10–67,000	580,000
	Specific electrical conductivity (S/m/g cm^3)	0.07×10^4–19.6×10^6	6.5×10^6
	Current-carrying capacity (S/cm^2)	10^4–10^7	3×10^4
	Tensile strength (GPa)	0.013–1.91	0.220
	Specific tensile strength (GPa/SG)	0.027–1.94	0.025
	Axial thermal conductivity (W/mK)	5–1,230	400

metallic property, armchair CNT fibers should provide the best electrical conductivity. There are still challenges in producing chiral-controlled macroscopic CNTs. Also, the conductivity of armchair CNTs will not be affected by the diameter of the CNTs [57]. This is not true for other types of CNTs as the bandgap of semiconducting CNTs decreases with an increase in the diameter.

18.3.2 CNT Yarn Torque Sensors

Most applications that require rotational positioning and high torque generation for mechanical performance are often non-compact with a complex design that is not ideal for nanotechnological applications. Twist-spun CNT yarn can serve as an actuator for high-performance motion systems like artificial muscles that require torsional rotation in addition to bending and contraction and micromechanical devices. In addition to their relatively high strength [137], their nanoscale dimensions and aspect ratio are attractive for torsional sensing. Torsional acceleration in CNT yarns can be driven in both directions for conversion of mechanical energy to electrical energy [138–140]. This can find application in sensors that generate electrical signals through applied torsional rotation. The change in electrical resistance upon the twist loading of the CNT yarn is shown in Figure 18.20 [124,141]. The decrease in electrical resistance upon application of torsional loading to the CNT yarns demonstrates that applied twist increases fiber compaction, resulting in increased electrical contact between nanotubes and a negative piezoresistance. The changes in electrical resistance due to twist are mostly reversible, but irreversible resistance changes at higher shear strains exceeding 12.9% have been reported for composite CNT fibers due to matrix failure (Figure 18.20b) [141].

Untwisted CNT yarn shows a very large resistance increase in torsional displacement compared to CNT yarns with twist. Figure 18.21 shows the relative resistance change–time curve and the corresponding angular displacement–time curves for untwisted CNT yarn. This

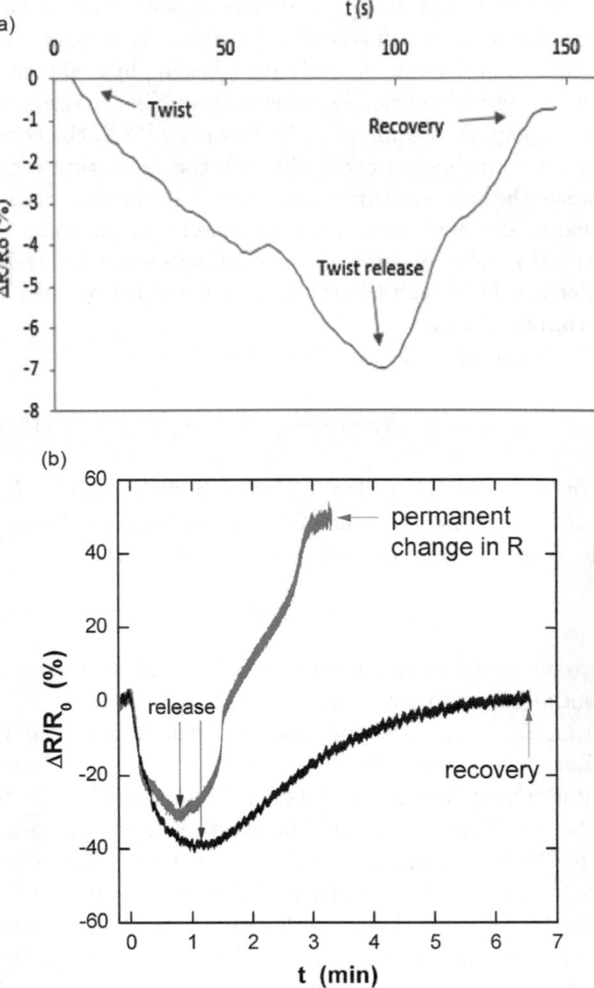

FIGURE 18.20 (a) Changes in electrical resistance (normalized) due to the CNT yarn twist. Twisting initiates at $t = 0$, then R changes due to applied shear strain in the yarn. Strain in the yarn is released by counterrotation of the disk to equivalent angle [124]. (b) Relative resistance change versus time of CNT yarn showing that the electrical resistance decreases due to applied shear strain in composite CNT yarns [141].

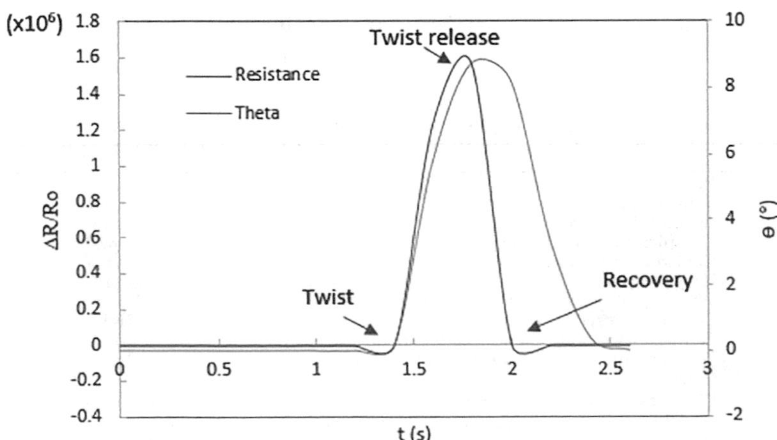

FIGURE 18.21 Relative resistance change–time curve and corresponding angular displacement–time curves for untwisted CNT yarn in torsional rotation [124].

is because twisted CNT yarns provide more resistance to torsion since some degree of twist has already been imparted to it.

18.3.3 Wearable Sensors

The development of CNT stretchable sensors has increased the research effort on using them in flexible functional electronic devices including strain gauges. CNT stretchable strain gauges can detect very large strains (typically >10%) that exceed the limit of conventional metallic foil strain gauges. The unique mechanical compliance of CNT fiber-based sensors renders them suitable for the applications involving interfacing with biological tissue such as health monitoring and rehabilitation, soft robotics, and human motion detection.

Many forms of CNTs have been utilized in human motion detection [142–158]. For example, Suzuki et al. [142] used dry-spun MWCNT sheet from spinnable array bonded to an elastomeric resin: a polycarbonate-urethane resin (PCU)/polytetramethylene ether glycol-urethane (PTMGU) substrate. The sensor achieved good stability up to 180,000 cycles. Yamada et al. [143] demonstrated a resistive strain gauge that explores the lateral fracture of as-grown SWCNT arrays, where the controlled opening and closing of cracks lead to reproducible resistive response upon repetitive stretching and releasing. The device can detect strains as high as 280% with very small amount of overshoot and relaxation. Shin et al. [144] explored MWCNT array infiltrated with urethane. A polyurethane (PU) substrate solution was directly filtered into the array to create a sensor that demonstrated 1,400% maximum strain on axial stretch. Foroughi et al. [145] knitted dry-spun MWCNT sheet on a spandex achieving a 900% stretchability and a stability of up to 1,000 cycles. However, the sensitivity of the sensor was less than that of metallic foil strain gauges like the results of Yamada et al. [143]. Since both sensors were fabricated using CNT sheet, it is safe to assume that aspect ratio of the fiber or an array is more suitable for high-sensitivity strain sensors. In another study, Cai et al. [146] reported a capacitive strain gauge, which is assembled into a parallel-plate capacitor using two layers of CVD-grown CNT thin films as stretchable electrodes and a piece of silicone elastomer as the dielectric layer. When stretched, the transverse deformation of the elastomer in the device results in an increase in capacitance, which is found to be proportional to the applied strain. Due to the excellent stretchability of the CNT substrate/electrodes, the strain gauge exhibits a stable and reliable piezoresistive response throughout the course of repetitive stretching with a maximum strain of up to 300%. Takahashi et al. [147] reported a flexible pressure sensor array using a stretchable SWCNT active-matrix backplane. However, the sensor cannot reach the strain measurement capability of stretchable sensors that use flexible substrates. Ryu et al. [148] fabricated a highly stretchable and wearable device from dry-spun CNT fiber for human motion detection. The device was fabricated on a flexible Ecoflex substrate that could measure strains greater than 900% with high sensitivity and exhibited a fast response and good durability. The authors incorporated a biaxial strain gauge rosette configuration into the sensor to measure complex human motion. The sensor was portable and demonstrated a good sensitivity and repeatability up to 10,000 cycles.

By embedding CNT yarn into Ecoflex^{TM} substrate, it has been reported that the change in resistance can substantially reach higher ranges [124,148]. This configuration will also prevent the CNT yarn from any form of deterioration due to harsh use or weather. Figure 18.22 shows that the CNT yarn-Ecoflex^{TM} sensor could measure strains greater than 1,000%. A stable interface between different components also plays a critical role as minute interfacial sliding or debonding may lead to failure of the entire device.

The CNT yarn strain sensor was used to demonstrate this motion monitoring capability by attaching them to body parts to measure movements or motion. The sensor was used to measure touch and wrist motions as shown in Figure 18.23. When the sensor was touched, the resistance decreased and started to recover once the touch ceased. Figure 18.23g shows that the sensor could capture the movement of the wrist. The resistance increased linearly with arching and decreased upon release. It can be seen from Table 18.4 that the strain sensors from a 1D CNT assembly demonstrated one of the highest reported GFs for a stretchable wearable sensor.

18.3.4 CNT Yarn Foil Strain Gauges

Recently, the fabrication of foil strain gauge comprising of CNT yarns was reported [159,160]. Metallic foil strain gauge sensors comprise of a piezoresistive membrane layer attached to a flexible substrate. The commonly used metallic component is constantan, an alloy of Cu (55%–60%) and Ni (45%–40%). Constantan has a low temperature coefficient suitable for resistance coils and a constant resistivity over a wide range of temperature. They are highly piezoresistive. The flexible substrate acts as a compliant structure that translates an input force into localized strain in the piezoresistive layer so that changes in electrical resistivity can be monitored and correlated to strain using the piezoresistivity effect. The strain in the piezoresistive layer can be electrically transduced by connecting to a Wheatstone bridge to improve the sensors' sensitivity and compensate undesirable temperature effects. Metallic foil strain gauges can capture very low fluctuations of strain with a maximum range of about 5% [15,18] while semiconductor strain gauges have higher gauge factors than metallic strain gauges. However, semiconductor strain gauges are sensitive to temperature limiting their efficiency. These piezoresistive strain gauges cannot detect initiating damage in composite materials with high compaction or multifaceted construction. There is also a form factor where these strain gauges can only be applied either on the surface or lack the aspect ratio to be integrated into complex structural components. More critically,

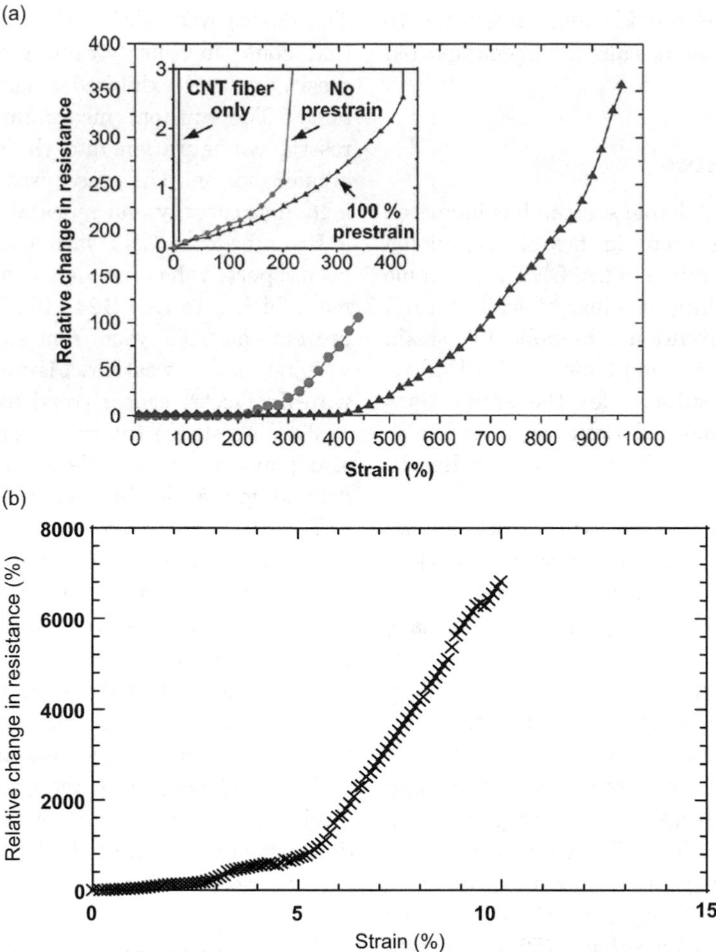

FIGURE 18.22 Relative change in resistance versus strain for: (a) unsupported CNT fibers (CNT fiber only), CNT fibers on an unstrained Ecoflex substrate (no pre-strain), and CNT fibers on an Ecoflex substrate pre-strained by 100%; the strain ranged from 0% to 450% strain (inset) [148]. (b) CNT yarn-Ecoflex^{TM} strip sensor [124].

they fail to achieve damage detection without altering the microstructure of the composite material.

The fabrication of a prototype CNT yarn strain gauge sensor that exhibits a very high sensitivity was done based on a parametric study by Abot et al. [159]. The model used in the parametric study considers a single layer where the piezoresistive layer comprising of the CNT fibers is immersed in a polymer substrate (Figure 18.24). The iso-strain or Voigt model is used to obtain the mechanical properties in the piezoresistive layer and considers the fact that the CNT yarns do not form a continuous phase. The effective properties of the strain gauge sensor are then calculated by homogenizing the properties of the CNT yarns and those of the polymer. The polymer is an isotropic material with elastic modulus, E_m, Poisson's ratio, ν_m, and electrical resistivity, ρ_m (m and f subscripts denote the polymer matrix and the CNT fiber, respectively). The CNT fiber was considered a transversely isotropic material with elastic moduli, E_{1f} and E_{2f}, Poisson's ratio, ν_f, shear modulus, G_{12f}, and electrical resistivities, ρ_{11f} and ρ_{22f}. The sensitivity of the strain gauge sensors was calculated by changing several geometrical and material parameters including the shape and dimensions of the strain gauge sensors and the exerted load. Abot et al. concluded that the highest sensitivity could be achieved in the case of a square sensor with CNT yarns oriented at 70° and spaced as close to each other as possible [159]. The strain gauge sensor is sensitive to all tractions although the highest sensitivity is achieved when a normal traction is relatively aligned with the CNT yarn direction.

It has been reported that size, geometry, and the relative arrangement of the piezoresistive membrane within the sensor significantly affect the performance of strain gauge sensors [160]. The higher the Poisson's ratio and the lower the spacing factor, the higher the sensitivity of the strain gauge. The spacing factor is represented as the normalized separation between the CNT yarns, i.e., the ratio of distance between CNT yarns and the diameter of the CNT yarn.

The strain gauge design objective is to obtain the location and configuration of the material that maximizes its sensitivity to external loading. From the relative resistance change and strain history curves (Figure 18.25a) and the GF obtained (Figure 18.25b), it is shown that these foil strain gauge sensors comprising of CNT yarn are sensitive enough to capture strain and can replicate the loading and unloading cycles.

Based on these findings, there is an ongoing effort to design flexible and stretchable foil strain gauge using CNT yarn. The goal is to have a foil strain gauge that can measure strain has a gauge factor that is at least ten times larger than that of conventional foil strain gauges using constantan wires. To achieve that, the first consideration is the substrate of the sensor. The substrate used in the design of the prototype CNT yarn foil strain gauge is Kapton HN^{TM} polyimide films. These films are very thin and are easily machinable, i.e., they can sustain micro-channels (grooves) that are created by laser prototype machine drills. Most importantly, their coefficient of thermal expansion (CTE) is close to that of CNT yarns. The use of elastomers as a substrate could also make it possible to achieve all the requirements needed from the substrate. The study is still underway and could be very seminal in the production of large-strain surface foil strain gauges.

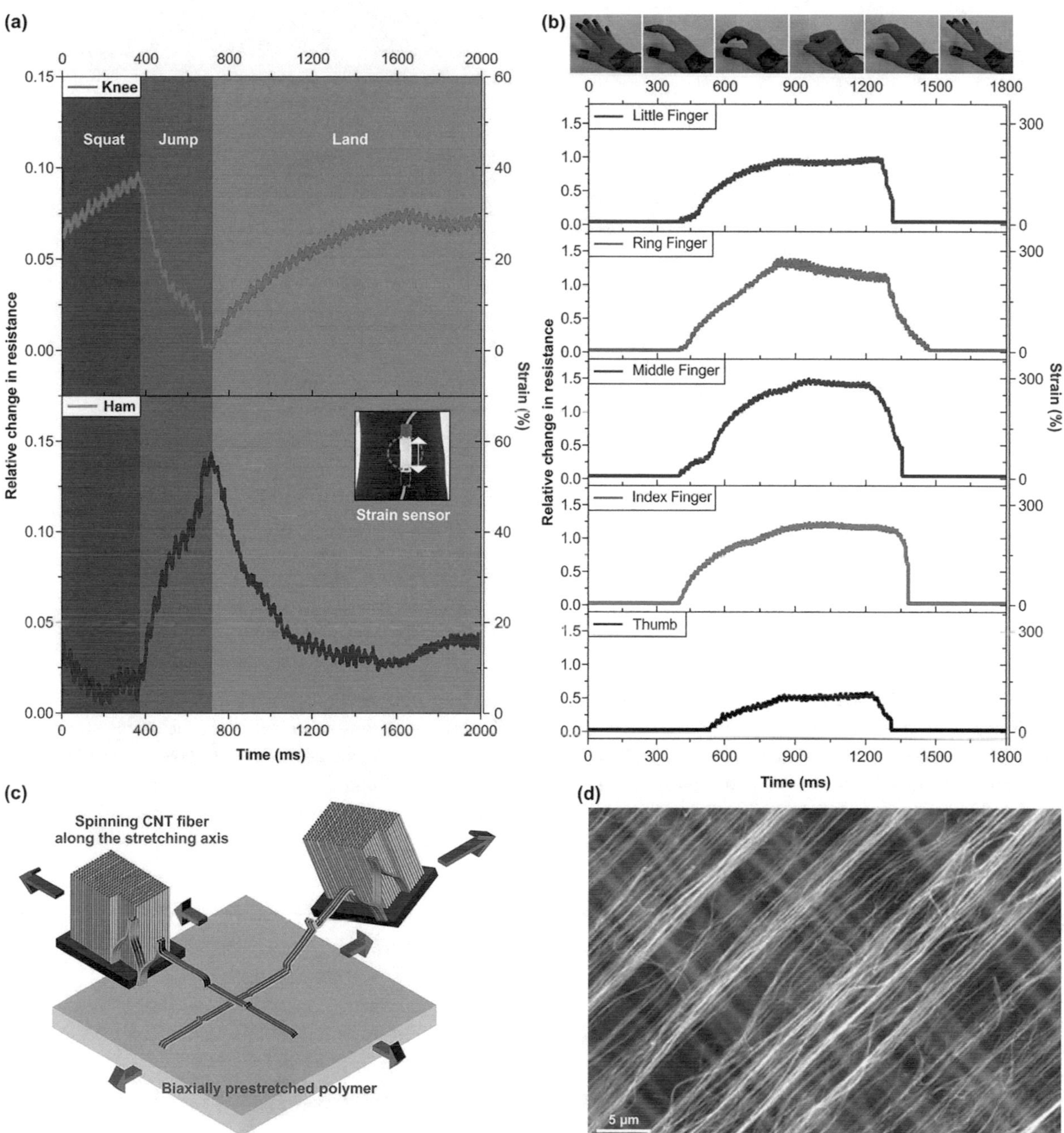

FIGURE 18.23 Wearable, extremely elastic CNT fiber-based strain sensor (substrate pre-strained by 100%). (a) Relative change in resistance (and estimated strain) versus time while jumping for strain sensors placed over a knee joint and hamstring muscles. (b) Photograph of a wearable strain sensor glove and relative changes in resistance (and estimated strain) versus time for grasping motion. (c) Illustration of the process for fabricating a biaxial strain sensor. (d) SEM image of biaxial strain sensor.

(*Continued*)

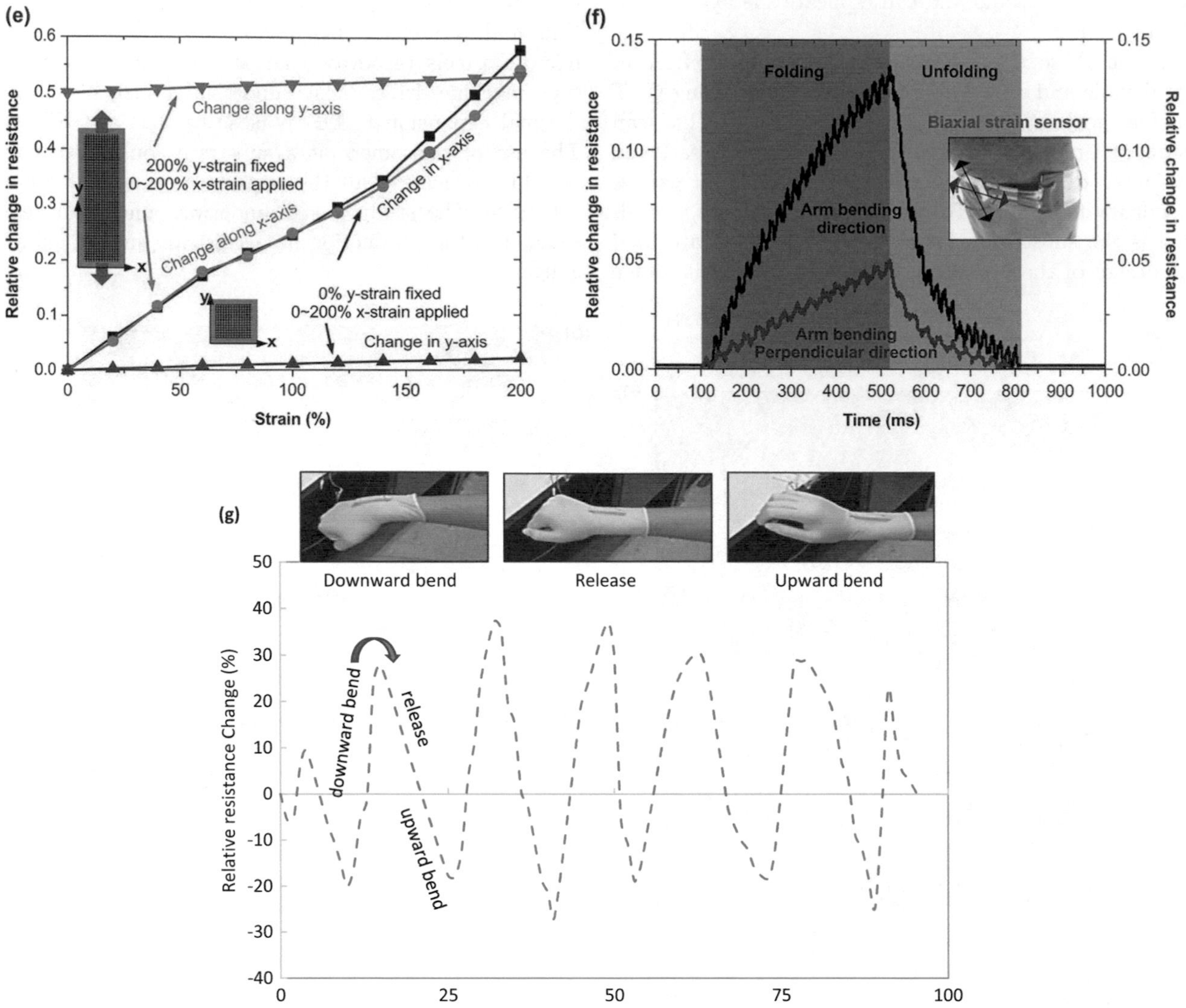

FIGURE 18.23 (CONTINUED) (e) Relative changes in resistance versus strain for a biaxial CNT-fiber strain sensor: x-axis resistance with variable x-axis strain and y-axis at constant 0% strain (black), y-axis resistance with variable x-axis strain and y-axis at a constant 0% strain (dark gray), y-axis resistance with variable x-axis strain and y-axis at a constant 200% strain (light gray), and y-axis resistance with variable x-axis strain and y-axis at a constant 200% strain (dark gray). (f) Relative changes in resistance versus time for a biaxial strain sensor placed over an elbow, with the x-axis parallel to the arm (black) and the y-axis parallel to joint's axis of rotation (gray) [148]. (g) Relative resistance change–time curves for wrist motion [124].

TABLE 18.4 Comparison of Piezoresistive Strain Sensors from Macroscopic CNT Assemblies

Reference	[148]	[142]	[143]	[144]	[145]	[124]
Production	Dry-spun MWCNT fiber from array	Dry-spun MWCNT sheet from spinnable array	Dry-spun SWCNT film	MWCNT array infiltrated with urethane	Dry-spun MWCNT sheet	Dry-spun MWCNT yarn from array
Substrate	Elastic Ecoflex	Elastomeric resin (PCU resin/ PTMGU)	Polydimethylsiloxane (PDMS)	PU	Spandex	Silicone rubber
Bonding method	van der Waals/ coating	Resin coating	van der Waals interaction	Direct infiltration of PU solution	Knitting	Embedded in substrate
Maximum strain (%)	440	200	280	1,400	900	50
Maximum GF	47	10	0.82 (0–40% strain)	1.07 (0–300% strain)	0.4	>1,000
Stability (cyclic loading)	Up to 10,000 cycles	180,000 cycles	10,000	100	1,000	–
Response time	10–12 ms	<15 ms	14 ms	–	–	10–20 ms

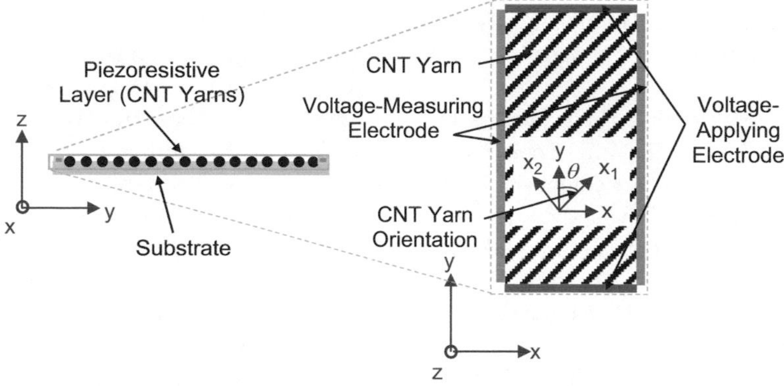

FIGURE 18.24 Schematic of cross section of a foil strain gauge sensor comprising CNT yarns. Inset: top schematic view of the arrangement of the CNT yarns in a unidirectional configuration [160].

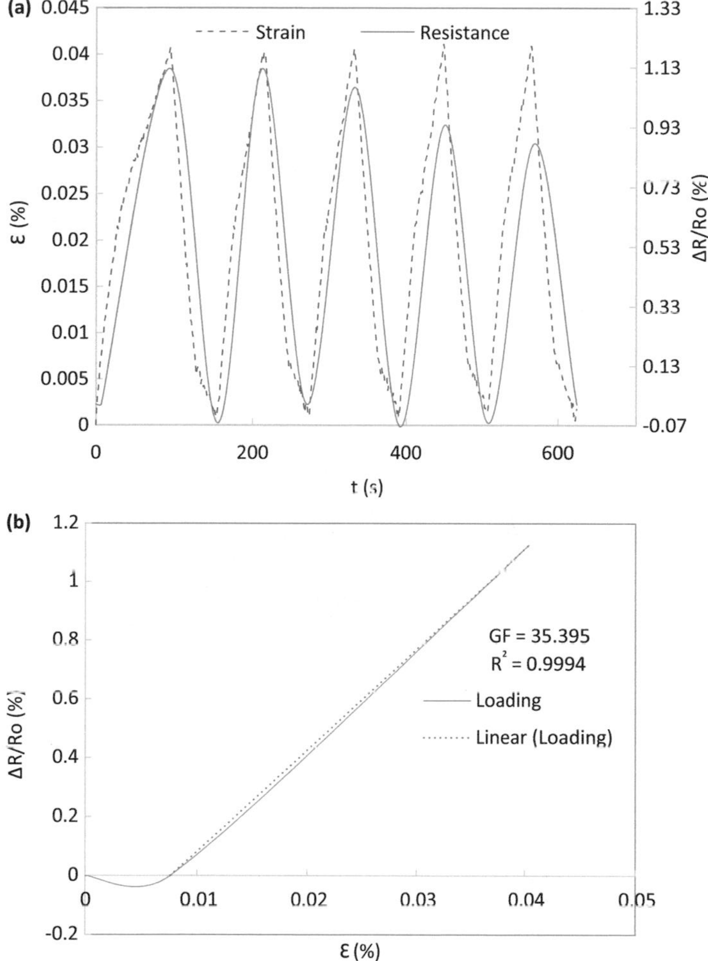

FIGURE 18.25 (a) Electromechanical response of a foil strain gauge prototype under cyclic loading at a displacement rate of 300 μm/min: (a) Strain and relative resistance change histories during five loading–unloading cycles and (b) Relative resistance change versus strain curve of first loading cycle and corresponding gauge factor [160].

18.3.5 Damage Detection with CNT Yarn Sensors

One of the proposed applications of CNTs is in sensing for real-time structural health monitoring (SHM) of composites. SHM methods provide constant and immediate feedback of the state of health of a structure including potential damage [37]. SHM methods may include vibration analysis, strain gauges, fiber-optic sensors, stress wave propagation techniques as well as several other methods [37]. SHM methods that utilize micro-strain sensors can capture strain variation due to piezoresistive effects, resonance monitoring, piezoelectric effects, capacitance variation, or changes in optical properties [15].

The hollow structure of CNTs makes them possess low weight, which when combined with the high aspect ratio and multifunctional properties is ideal for composite structures [161–164]. In SHM, a grid of CNT yarn-based sensors can be created over a large area of a structure and use it to monitor the strain field or damage in the structure. When a load is applied to the structure and a strain reaches the CNT sensor, it will trigger a change in CNT's resistance value depending on the magnitude of the stress. The resistance value of the CNT yarn would be restored when the load is removed. The concept is that CNT yarns are integrated in a laminated composite material forming a continuous sensor circuit, and their inherent piezoresistive sensitivity would capture small amounts of strain within the host material [164].

In laminated composites, delamination occurs due to the separation of their layers mostly due to interlaminar stress and matrix failure. Delamination represents a significant damage risk to their integrity [130] and can occur almost any place in the laminate: on the edge, near the surface, or at the center of the laminated composite. While the range in size of delamination and damage can vary drastically, it may elude the detection capabilities of most techniques that monitor change in material geometry such as metallic foil strain gauges, which are more suitable for surface strain detection or optical fiber monitoring that requires complex equipment and data analysis. The ability of the CNT yarn sensors to detect mode II-dominated delamination in laminated composite materials had been previously shown [163]. The determination of the exact location of delamination and its progression can be achieved with a configuration consisting of a combination of different CNT yarn sensors like the one shown in Figure 18.26, which includes stitched

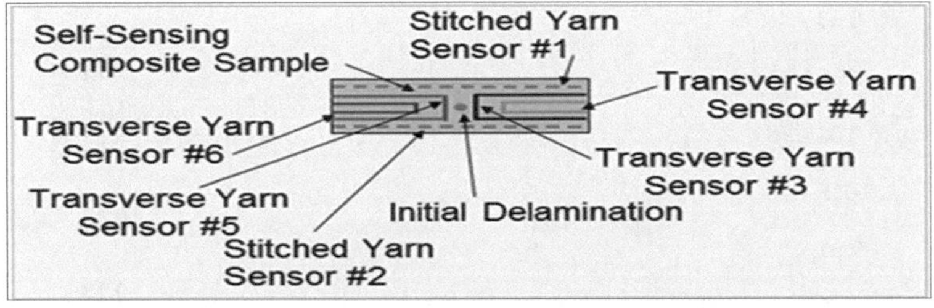

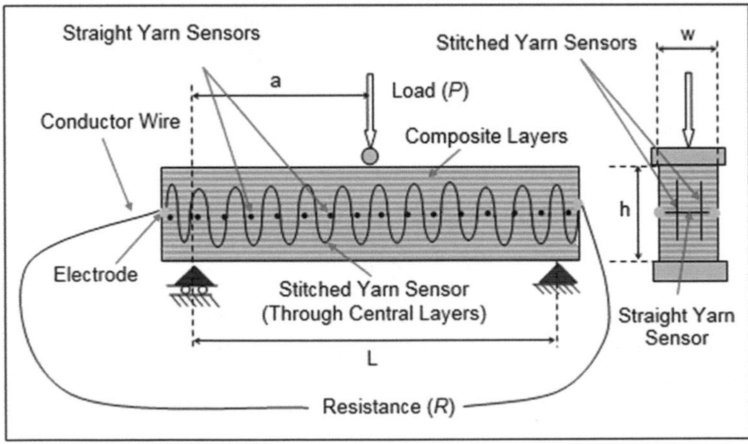

FIGURE 18.26 (a) Schematic of a 32-layer glass/epoxy laminated composite of the integrated yarn sensors including stitched ones (through layers 12–21) and straight ones (between layers 16 and 18), with a 25 × 15 mm central delamination. Wires are later connected to the CNT yarns for resistance measurements. (b) Optical image of the laminated composite samples instrumented with CNT yarn sensors and multimode fiber-optic sensor. (c) Schematic of experimental setup of self-sensing composite sample subjected to three-point bending: side and end cross-sectional views of laminated composite beam sample instrumented with stitched and straight yarn sensors [165].

and transverse, or longitudinal, yarn sensors. The yarn sensors stitched through the thickness of the laminates allow for the determination of delamination only; additional transverse yarn sensors parallel to the composite laminate layers and along the beam's width direction are required to establish the precise location of the delamination or the damage. It is worth mentioning that damage detection based on a significant resistance change increase does not require highly precise resistance measurements, and thus, two-point probe measurements are deemed appropriate and sufficient.

Figure 18.27 shows the results obtained with CNT yarn sensors validated using two experimental techniques: integrated optical fibers monitored through optical time-domain reflectometry (OTDR), and x-ray tomography of the entire laminated composite samples post-testing. High-performance plastic multimode optical fibers were integrated in the laminated composite plates. The localization of damage was achieved through OTDR, and the time dependence was monitored through a photodiode system. Once a growing crack or a delamination occurs within the composite sample, it damages or cracks the optical fiber as well, allowing OTDR to calculate the location of the damage along the length of the optical fiber.

CNT yarn sensors integrated in laminated composite materials have been shown capable of delamination detection [163 167]. The mechanical/electrical response of self-sensing composite samples with a glass plain weave architecture and combined sensor configurations is presented in Figure 18.27. The load, P, in terms of time, t, represents the load history, and the resistance change, ΔR, or the difference between the actual resistance and the initial resistance in terms of time, represents the resistance change history. Delamination in the sample is identified by the sudden decrease of the maximum load in the load history curve (event A). Delamination is detected by the stitched yarn sensors as evidenced by the increase of the resistance to infinity (events B1/B2). Event A and B1 occurred almost simultaneously, while the time difference between events B2 and A was 30 s. This difference implies that the yarn sensor can capture the delamination instantaneously. This response of the yarn sensor demonstrates its ability to not only detect the delamination but also to anticipate it (events B1/B2) even before the load history response of the laminated composite sample indicates it (event A). In addition, the yarn sensor could withstand loading well more than the maximum load as well as capture the delamination without its circuit failing.

18.3.6 Other Applications

Artificial muscles: CNT yarns could be used as an electrochemically powered, all-solid-state torsional and tensile artificial yarn muscles that provide attractive performance. Recently, it has been shown that highly twisted MWCNT yarns produce a unique mechanical actuation involving coupled rotation and axial contraction [168]. The torsional and tensile actuation is achieved by reversible yarn volume changes driven by electrolyte ion influx/release during electrochemical charge/discharge or thermal expansion/contraction of a guest material like paraffin wax. Figure 18.28 shows the structures of MWCNT yarn for torsional and tensile actuation. Single yarns are plied and twisted to form coiled yarns. From Figure 18.28e, the surface of a single-ply non-coiled yarn used for the

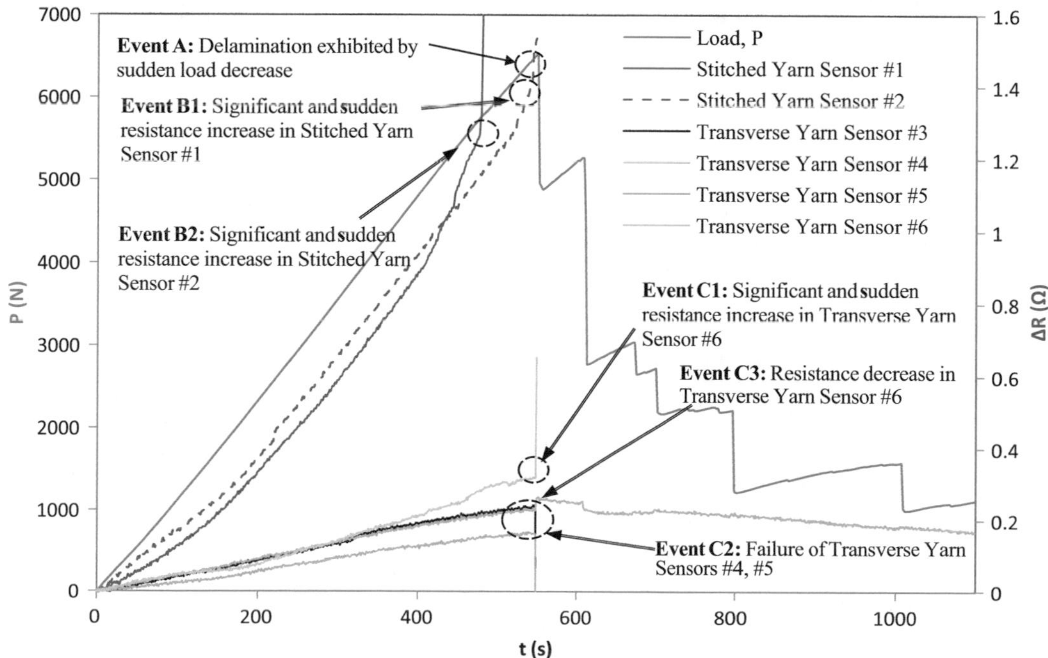

FIGURE 18.27 Localized detection of major delamination in 32-layer glass/epoxy composite sample using a combined stitched and transverse/longitudinal yarn sensors configuration: load and resistance change versus time curves [167].

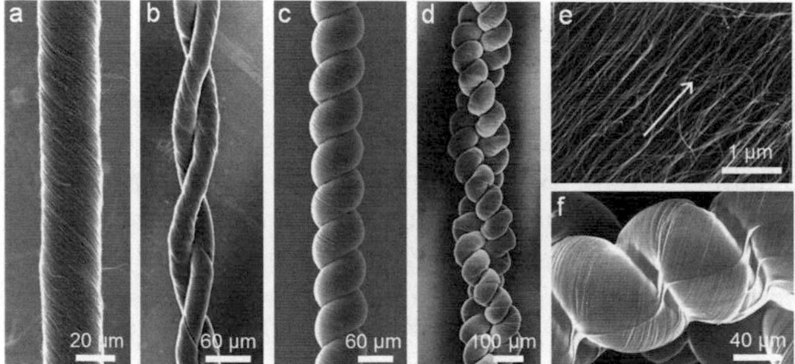

FIGURE 18.28 (a and b) SEM images of neat, single, and two-ply yarns, respectively, which are used in two-ply form for torsional actuation when electrolyte filled. (c and d) SEM images of neat, single, and plied, coiled yarns, respectively, which are used in two-ply form for tensile actuation when electrolyte filled. (e) A magnified SEM image of the lateral surface of a neat, twist-spun yarn, where the arrow indicates the fiber direction. (f) SEM image of a plied, coiled yarn that is fully infiltrated with PVA/H_2SO_4 solid gel electrolyte [168].

torsional muscle exhibits high porosity that allows the electrolyte to infiltrate the yarn, as well as a highly oriented fibrous structure, which contributes to yarn direction strength and electrical conductivity.

Energy harvesting: Energy could be harvested by twisting CNT yarns from both torsional and tensile motion. By stretching and relaxing tightly twisted yarns made of CNT yarns, Kim et al. [169] devised a way to harvest and store the motion as electrical energy. When coated or submerged in an electrolyte, CNT yarns are charged by the electrolyte to form supercapacitors. As electrical charges are moved by the twisting motion of the fibers within the supercapacitor, they attain and store potential. Figure 18.29 shows a Twistron harvester configuration, structure, and performance for tensile energy harvesting in 0.1 M HCl. The production of electrical energy can be optimized by controlling the twist intensity and a combination of homochiral and heterochiral coiled yarns.

Biosensors: CNT fibers have good electrocatalytic properties that are ideal for use in electrochemical devices. The porosity and high surface area of CNT fibers have been exploited to drive molecular-scale interactions with enzymes and other chemicals to efficiently capture and promote electron transfer reactions [170]. CNT fiber-based electrodes are being considered as alternatives to carbon fiber microelectrodes for the detection of neurotransmitters [171,172]. CNT fibers are more sensitive, and the microelectrodes could exhibit fast electron transfer kinetics.

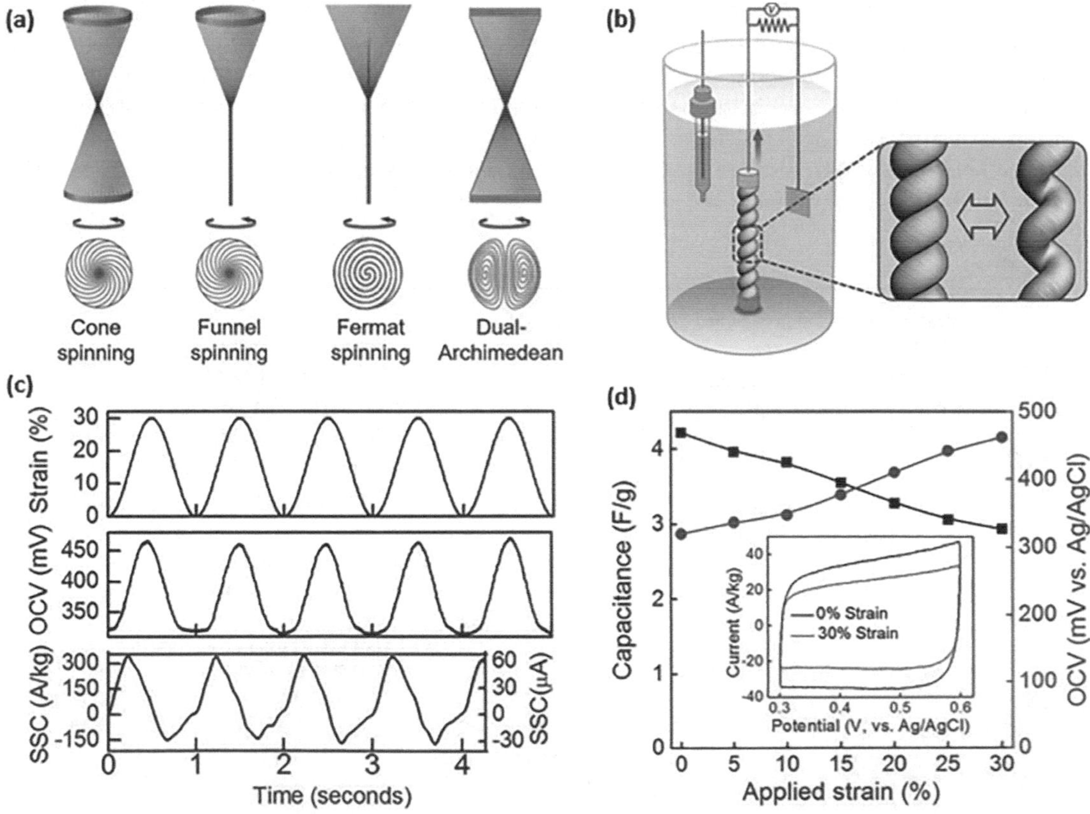

FIGURE 18.29 (a) Illustrations of cone, funnel, Fermat, and dual-Archimedean spinning (top) and resulting yarn cross sections (bottom). (b) Illustration of a torsionally tethered coiled harvester electrode and counter and reference electrodes in an electrochemical bath, showing the coiled yarn before and after stretch. (c) Sinusoidal applied tensile strain and resulting change in open-circuit voltage (OCV) and short-circuit current (SCC) before (right) and after (left) normalization for a cone-spun coiled harvester. (d) Capacitance and OCV versus applied strain for the harvester of (c). (Inset) Cyclic voltammetry curves for 0% and 30% strain.

(*Continued*)

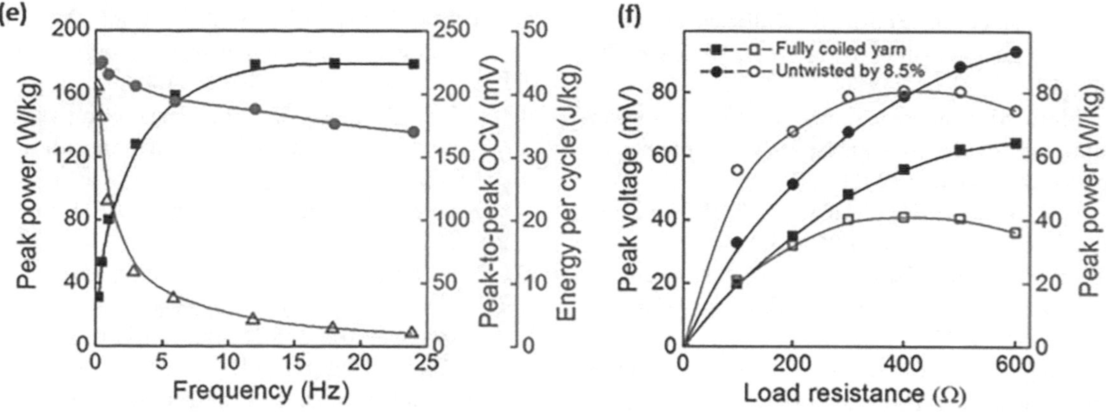

FIGURE 18.29 (CONTINUED) (e) Frequency dependence of peak power (solid squares), peak-to-peak OCV (solid circles), and energy per cycle (unfilled triangles) for 50% stretch of an 8.5% untwisted coiled harvester. (f) Generated peak power (solid black symbols) and peak voltage (unfilled black symbols) versus load resistance for a coiled yarn (squares) and a partially untwisted coiled yarn (circles) when stretched at 1 Hz to the maximum reversible elongation [169].

Zhu et al. demonstrated that microelectrodes made from CNT fibers could accelerate the redox process of these biomolecules allowing for high sensitivity of low-potential detection [173,174]. The glucose-sensing electrodes were built by adsorption of a mediator on the surface of a CNT fiber.

Other potential applications of CNT fibers include as electrodes for batteries, radio antennas, high strength composites, bullet vest proof, Electromagnetic interference shields, and much more [93].

References

1. R. Saito, G. Dresselhaus, M. S. Dresselhaus, *Physical Properties of Carbon Nanotubes*, Imperial College Press, London 1(998).
2. M. Dresselhaus, G. Dresselhaus, P. Eklund, *Science of Fullerenes and Carbon Nanotubes*, Academic Press, San Diego, CA (1996).
3. R. Saito, M. Fujita, G. Dresselhaus, M. S. Dresselhaus, Electronic structure of chiral graphene tubules. *Appl. Phys. Lett.* 60 (1992) 18.
4. P. C. Ma, J. K. Kim, *Carbon Nanotubes for Polymer Reinforcement*, CRC Press and Taylor & Francis, Boca Raton, FL (2011).
5. S. Iijima, Helical microtubules of graphitic carbon. *Nature* 354 (1991) 56–58.
6. A. Krishnan, E. Dujardin, T.W. Ebbesen, P.N. Yianilos, M.M.J. Treacy, Young's modulus of single-walled nanotubes. *Phys. Rev. B* 58 (20) (1998) 14013–14019.
7. F. H. Gojny, M. H. G. Wichmann, U. Kopke, B. Fiedler, K. Schulte, Carbon nanotube-reinforced epoxy-composites: Enhanced stiffness and fracture toughness at low nanotube content. *Compos. Sci. Technol.* 64 (2004) 2363.
8. W. Bauhofer, J. Z. Kovacs, A review and analysis of electrical percolation in carbon nanotube polymer composites. *Compos. Sci. Technol.* 69 (2009) 1486.
9. J. N. Coleman, U. Khan, W. J. Blau, Y. K. Gun'ko, Small but strong: A review of the mechanical properties of carbon nanotube-polymer composites. *Carbon* 44 (2006) 1624
10. E. J. Garcia, B. L. Wardle, A. J. Hart, N. Yamamoto, Fabrication and multifunctional properties of a hybrid laminate with aligned carbon nanotubes grown In Situ. *Compos. Sci. Technol.* 68 (2008) 2034.
11. A. Beigbeder et al., Preparation and characterisation of silicone-based coatings filled with carbon nanotubes and natural sepiolite and their application as marine fouling-release coatings. *Biofouling* 24 (2008) 291.
12. S. De, J. N. Coleman, The effects of percolation in nanostructured transparent conductors. *MRS Bull.* 36 (2011) 774.
13. A. M. Ionescu, H. Riel, Tunnel field-effect transistors as energy-efficient electronic switches. *Nature* 479 (2011) 329.
14. T. Rueckes et al., Carbon nanotube-based nonvolatile random-access memory for molecular computing. *Science* 289 (2000) 94.
15. P. Chen et al., Fully printed separated carbon nanotube thin film transistor circuits and its application in organic light emitting diode control. *Nano Lett.* 11 (2011) 5301.
16. M. Jung et al., All-printed and roll-to-roll-printable 13.56-MHz-operated 1-bit RF tag on plastic foils. *IEEE Trans. Electron. Dev.* 57 (2010) 571.
17. S. J. Tans, A. R. M. Verschueren, C. Dekker, Room-temperature transistor based on a single carbon nanotube. *Nature* 393 (1998) 49.

18. M. C. LeMieux et al., Self-sorted, aligned nanotube networks for thin-film transistors. *Science* 321 (2008) 101.
19. M. A. McCarthy et al., Low-voltage, low-power, organic light-emitting transistors for active matrix displays. *Science* 332 (2011) 570.
20. K. Jensen, J. Weldon, H. Garcia, A. Zettl, Nanotube radio. *Nano Lett.* 7 (2007) 3508.
21. A. D. Franklin et al., Variability in carbon nanotube transistors: Improving device-to-device consistency. *ACS Nano* 6 (2012) 1109.
22. N. K. Chang, C. C Su, S. H. Chang, Fabrication of single-walled carbon nanotube flexible strain sensors with high sensitivity. *Appl. Phys. Lett.* 92 (2008) 063501.
23. Q. Cao et al., Medium-scale carbon nanotube thin-film integrated circuits on flexible plastic substrates. *Nature* 454 (2008) 495.
24. D. M. Sun et al., Flexible high-performance carbon nanotube integrated circuits. *Nat. Nanotechnol.* 6 (2011) 156.
25. M. S. Rahaman, C. D. Vecitis, M. Elimelech, Electrochemical carbon-nanotube filter performance toward virus removal and inactivation in the presence of natural organic matter. *Environ. Sci. Technol.* 46 (2012) 1556.
26. B. Corry, Designing carbon nanotube membranes for efficient water desalination. *J. Phys. Chem. B* 112 (2008) 1427.
27. J. K. Holt et al., Fast mass transport through sub-2-nanometer carbon nanotubes. *Science* 312 (2006) 1034.
28. G. Gao, C. D. Vecitis, Electrochemical carbon nanotube filter oxidative performance as a function of surface chemistry. *Environ. Sci. Technol.* 45 (2011) 9726.
29. A. R. Köhler, C. Som, A. Helland, F. Gottschalk, Studying the potential release of carbon nanotubes throughout the application life cycle. *J. Clean. Prod.* 16 (2008) 927.
30. T. Matsumoto et al., Reduction of Pt usage in fuel cell electrocatalysts with carbon nanotube electrodes. *Chem. Commun.* 2004 (2004) 840.
31. N. M. Gabor, Z. Zhong, K. Bosnick, J. Park, P. L. McEuen, Extremely efficient multiple electron-hole pair generation in carbon nanotube photodiodes. *Science* 325 (2009) 1367.
32. Y. Gogotsi, P. Simon, Materials science. True performance metrics in electrochemical energy storage. *Science* 334 (2011) 917.
33. K. Evanoff et al., Towards ultrathick battery electrodes: Aligned carbon nanotube-enabled architecture. *Adv. Mater.* 24 (2012) 533.
34. A. Le Goff et al., From hydrogenases to noble metal-free catalytic nanomaterials for H_2 production and uptake. *Science* 326 (2009) 1384.
35. L. Dai, D. W. Chang, J.-B. Baek, W. Lu, Carbon nanomaterials for advanced energy conversion and storage. *Small* 8 (2012) 1130.
36. A. Izadi-Najafabadi et al., Extracting the full potential of single-walled carbon nanotubes as durable supercapacitor electrodes operable at 4 V with high power and energy density. *Adv. Mater.* 22 (2010) 235.
37. C. Sotowa et al., The reinforcing effect of combined carbon nanotubes and acetylene blacks on the positive electrode of lithium-ion batteries. *Chem. Sus. Chem.* 1 (2008) 911.
38. D. A. Heller, S. Baik, T. E. Eurell, M. S. Strano, Single-walled carbon nanotube spectroscopy in live cells: Towards long-term labels and optical sensors. *Adv. Mater.* 17 (2005) 2793.
39. E. W. Keefer, B. R. Botterman, M. I. Romero, A. F. Rossi, G. W. Gross, Carbon nanotube coating improves neuronal recordings. *Nat. Nanotechnol.* 3 (2008) 434.
40. A. De La Zerda et al., Carbon nanotubes as photoacoustic molecular imaging agents in living mice. *Nat. Nanotechnol.* 3 (2008) 557.
41. N. W. S. Kam, M. O'Connell, J. A. Wisdom, H. J. Dai, Carbon nanotubes as multifunctional biological transporters and near-infrared agents for selective cancer cell destruction. *Proc. Natl. Acad. Sci. U.S.A.* 102 (2005) 11600.
42. M. Endo, S. Koyama, Y. Matsuda, T. Hayashi, Y. A. Kim, Thrombogenicity and blood coagulation of a microcatheter prepared from carbon nanotube-nylon-based composite. *Nano Lett.* 5 (2005) 101.
43. G. M. Mutlu et al., Biocompatible nanoscale dispersion of single-walled carbon nanotubes minimizes in vivo pulmonary toxicity. *Nano Lett.* 10 (2010) 1664.
44. D. A. X. Nayagam et al., Biocompatibility of immobilized aligned carbon nanotubes. *Small* 7 (2011) 1035.
45. A. Bianco, K. Kostarelos, M. Prato, Making carbon nanotubes biocompatible and biodegradable. *Chem. Commun.* 47 (2011) 10182.
46. C. A. Poland et al., Carbon nanotubes introduced into the abdominal cavity of mice show asbestos-like pathogenicity in a pilot study. *Nat. Nanotechnol.* 3 (2008) 423.
47. S. Y. Hong et al., Filled and glycosylated carbon nanotubes for in vivo radioemitter localization and imaging. *Nat. Mater.* 9 (2010) 485.
48. T. Kurkina, A. Vlandas, A. Ahmad, K. Kern, K. Balasubramanian, Label-free detection of few copies of DNA with carbon nanotube impedance biosensors. *Angew. Chem. Int. Ed.* 50 (2011) 3710.
49. X. Shi, A. von dem Bussche, R. H. Hurt, A. B. Kane, H. Gao, Cell entry of one-dimensional nanomaterials occurs by tip recognition and rotation. *Nat. Nanotechnol.* 6 (2011) 714.

50. D. A. Heller et al., Multimodal optical sensing and analyte specificity using single-walled carbon nanotubes. *Nat. Nanotechnol.* 4 (2009) 114.
51. Z. Chen et al., Protein microarrays with carbon nanotubes as multicolor Raman labels. *Nat. Biotechnol.* 26 (2008) 1285.
52. Z. Liu, X. Sun, N. Nakayama-Ratchford, H. Dai, Supramolecular chemistry on water-soluble carbon nanotubes for drug loading and delivery. *ACS Nano* 1 (2007) 50.
53. E. S. Snow, F. K. Perkins, E. J. Houser, S. C. Badescu, T. L. Reinecke, Chemical detection with a single-walled carbon nanotube capacitor. *Science* 307 (2005) 1942.
54. B. Esser, J. M. Schnorr, T. M. Swager, Selective detection of ethylene gas using carbon nanotube-based devices: utility in determination of fruit ripeness. *Angew. Chem. Int. Ed.* 51 (2012) 5752.
55. C. Stampfer, A. Jungen, R. Linderman, D. Obergfell, S. Roth, C. Hierold, Nano-electromechanical displacement sensing based on single-walled carbon nanotubes. *Nano Lett.* 7 (2006) 1449–1453.
56. J. R. Wood, Q. Zhao, M. D. Frogley, E. R. Meurs, A. D. Prins, T. Peijs, et al., Carbon nanotubes: From molecular to macroscopic sensors. *Phys. Rev. B* 62 (2000) 7571–7575.
57. A. Lekawa-Raus, L. Kurzepa, X. Peng, K. Koziol, Towards the development of carbon nanotube based wires. *Carbon* 68 (2014) 597–609.
58. R. Bacon, Growth, structure and properties of graphite whiskers. *J. Appl. Phys.* 2 (1960) 283–290.
59. R. Bacon, Filamentary graphite and method for producing the same, US Patent # 2 957 756 (1960).
60. M. Endo, Opening an era of carbon nanotubes through large-scale production: "Nanonet Interview" Nanotechnology Researchers Network Center of Japan (2005).
61. M. A. Oberlin, M. Endo, T. Koyama, Filamentous growth of carbon through benzene decomposition. *J. Cryst. Growth* 3 (1976) 335–349.
62. H. W. Kroto, J. R. Heath, S. C. O'Brien, R. F. Curl, R. E. Smalley, C60: Buckminsterfullerene. *Nature* 6042 (1985) 162–163.
63. A. Thess, R. Lee, P. Nikolaev, H. Dai, P. Petit, J. Robert, et al., Crystalline ropes of metallic carbon nanotubes. *Science* 273 (1996) 483–487.
64. C. Journet, W. K. Maser, P. Bernier, A. Loiseau, M. Lamy de la Chapelle, S. Lefrant, et al., Large-scale production of single-walled carbon nanotubes by the electric-arc technique. *Nature* 388 (1997) 756.
65. J. Chen, M. A. Hamon, H. Hu, Y. Chen, A. M. Rao, P. C. Eklund, et al., Solution properties of single-walled carbon nanotubes. *Science* 282 (1998) 95.
66. B. Vigolo, A. Pénicaud, C. Coulon, C. Sauder, R. Pailler, C. Journet, et al., Macroscopic fibers and ribbons of oriented carbon nanotubes. *Science* 290 (2000) 1331–1334.
67. K. L. Jiang, Q. Q. Li, S. S. Fan, Nanotechnology: Spinning continuous carbon nanotube yarns. *Nature* 419 (2002) 801.
68. Y. L. Li, I. A. Kinloch, A. H. Windle, Direct spinning of carbon nanotube fibers from chemical vapor deposition synthesis. *Science* 304 (2004) 276–278.
69. S. Paulson, M. R. Falvo, N. Snider, A. Helser, T. Hudson, A. Seeger, et al., In situ resistance measurements of strained carbon nanotubes. *Appl. Phys. Lett.* 75 (1999) 2936–2938.
70. M. Kumar, Carbon nanotube synthesis and growth mechanism. *Carbon Nanotubes - Synthesis, Characterization, Applications*, Dr. Siva Yellampalli (Ed.), InTech: Vienna (2011).
71. K. Hata, D.N. Futaba, K. Mizuno, T. Namai, M. Yumura, S. Iijima, Water-assisted highly efficient synthesis of impurity-free single-walled carbon nanotubes. *Science* 306 (2004) 1362–1364.
72. F. Ding, P. Larsson, J.A. Larsson, R. Ahuja, H. Duan, A. Rosen, et al., The importance of strong carbon-metal adhesion for catalytic nucleation of single-walled carbon nanotubes. *Nano Lett.* 8 (2008) 463–468.
73. A. K. Geim, K. S. Novoselov, The rise of graphene. *Nature Mater.* 6 (2007) 183–191.
74. R. S. Ruoff, D. Qian, W. K. Liu, Mechanical properties of carbon nanotubes are discussed based on recent advances in both modeling and experiment. *C. R. Phys.* 4 (2003) 993–1008.
75. M. F. Yu, B. S. Files, S. Arepalli, R. S. Ruoff, Tensile loading of ropes of single wall carbon nanotubes and their mechanical properties. *Phys. Rev. Lett.* 84 (2000) 5552. (doi:10.1103/PhysRevLett.84.5552)
76. E. Wong, P. Sheehan, C. Lieber, Nanobeam mechanics: Elasticity, strength, and toughness of nanorods and nanotubes. *Science* 277 (1997) 1971.
77. H. J. Qi, K. B. K. Teo, K. K. S. Lau, M. C. Boyce, W. I. Milne, J. Robertson, et al., Determination of mechanical properties of carbon nanotubes and vertically aligned carbon nanotube forests using nanoindentation. *J. Mech. Phys. Solids* 51 (2003) 2213–2237.
78. J. P. Salvetat, J. M. Bonard, N. H. Thomson, A. J. Kulik, L. Forro, W. Benoit, et al., Mechanical properties of carbon nanotubes. *Appl. Phys. A.* 69 (1999) 255–260.
79. N. G. Chopra, PhD, Dissertation, Department of Physics, University of California, Berkeley, CA (1996).
80. P. Poncharal, Z.L. Wang, D. Ugarte, W.A. DeHeer, Electrostatic deflections and electromechanical resonances of carbon nanotubes. *Science* 283 (1999) 1513–1516. (doi:10.1126/science.283.5407.1513)
81. M. F. Yu, O. Lourie, M. J. Dyer, K. Moloni, T. F. Kelly, R. S. Ruoff, Strength and breaking mechanism of

multiwalled carbon nanotubes under tensile load. *Science* 287 (2000) 637–640.

82. B. G. Demczyk, Y. M. Wang, J. Cumings, M. Hetman, W. Han, and A. Zettl, et al., Direct mechanical measurement of the tensile strength and elastic modulus of multiwalled carbon nanotubes. *Mater. Sci. Eng. A* 334 (2002) 173–17.
83. M. M. J. Treacy, T. W. Ebbesen, J. M. Gibson, Exceptionally high Young's modulus observed for individual carbon nanotubes. *Nature* 6584 (1996) 678–680.
84. S. Ogata, Y. Shibutani, Ideal tensile strength and band gap of single-walled carbon nanotubes. *Phys. Rev. B* 68 (2003) 165409.
85. Q. Yan, J. Wu, G. Zhou, W. Duan, B. L. Gu, Ab initio study of transport properties of multiwalled carbon nanotubes. *Phys. Rev. B* 72 (2005) 155425–5.
86. M. Zhang, K. R. Atkinson, R. H. Baughman, Multifunctional carbon nanotube yarns by downsizing an ancient technology. *Science* 306 (2004) 1358–1361.
87. L. Liu, W. Ma, Z. Zhang, Macroscopic carbon nanotube assemblies: Preparation, properties, and potential applications. *Small* 7 (2011) 1504–1520.
88. S. Zhang, K. K. K. Koziol, I. A. Kinloch, A. H. Windle, Macroscopic fibers of well-aligned carbon nanotubes by wet spinning. *Small* 4 (2008) 1217–1222.
89. W. Ma, L. Song, R. Yang, T. Zhang, Y. Zhao, L. Sun, et al., Directly synthesized strong, highly conducting, transparent single-walled carbon nanotube films. *Nano Lett.* 7 (2007) 2307–2311.
90. W. Ma, L. Liu, R. Yang, T. Zhang, Z. Zhang, L. Song, et al., Monitoring a micromechanical process in macroscale carbon nanotube films and fibers. *Adv. Mater.* 21 (2009) 603–608.
91. W. Ma, L. Liu, Z. Zhang, R. Yang, G. Liu, T. Zhang, et al., High-strength composite fibers: Realizing true potential of carbon nanotubes in polymer matrix through continuous reticulate architecture and molecular level couplings. *Nano Lett.* 9 (2009) 2855–2861.
92. C. Jayasinghe, T. Amstutz, M.J. Schulz, V. Shanov, Improved processing of carbon nanotube yarn. *J. Nanomater.* (2013) 309617. (doi:10.1155/2013/309617)
93. M. Schulz, V. Shanov, Z. Yin, *Nanotube Superfiber Materials: Changing Engineering Design*, Elsevier: Oxford (2014).
94. M. Miao, Yarn spun from carbon nanotube forests: Production, structure, properties and applications. *Particuology* 11 (2013) 378–393.
95. W. Chen, X. M. Tao, and Y. Y. Liu, Carbon nanotube-reinforced polyurethane composite fibers. *Compos. Sci. Technol.* 15 (2006) 3029–3034.
96. T. D. Fornes, J. W. Baur, Y. Sabba, E. L. Thomas, Morphology and properties of melt-spun polycarbonate fibers containing single- and multi-wall carbon nanotubes. *Polymer* 5 (2006) 1704–1714.
97. R. Andrews, D. Jacques, A. M. Rao, T. Rantell, F. Derbyshire, Y. Chen, et al., Nanotube composite carbon fibers. *Appl. Phys. Lett.* 9 (1999) 1329–1331.
98. P. Xue, K. H. Park, X. M. Tao, W. Chen, X. Y. Cheng, Electrically conductive yarns based on PVA/carbon nanotubes. *Compos. Struct.* 78 (2007) 271–277.
99. J. Steinmetz, M. Glerup, M. Paillet, P. Bernier, M. Holzinger, Production of pure nanotube fibers using a modified wet-spinning method. *Carbon* 43 (2005) 2397–2400.
100. J. M. Razal, K. J. Gilmore, G. G. Wallace, Carbon nanotube biofiber formation in a polymer-free coagulation bath. *Adv. Funct. Mater.* 18 (2008) 61–66.
101. F. Ko, Y. Gogotsi, A. Ali, N. Naguib, H. Ye, G. Yang, et al., Electrospinning of continuons carbon nanotube-filled nanofiber yarns. *Adv. Mater.* 15 (2003) 1161–1165.
102. G. M. Kim, G. H. Michler, P. Potschke, Deformation processes of ultrahigh porous multiwalled carbon nanotubes/polycarbonate composite fibers prepared by electrospinning. *Polymer* 46 (2005) 7346–7351.
103. M. B. Bazbouz, G. K. Stylios, Novel mechanism for spinning continuous twisted composite nanofiber yarns. *Eur. Polym. J.* 44 (2008) 1–12.
104. R. Sen, B. Zhao, D. Perea, M. E. Itkis, H. Hu, J. Love, et al., Preparation of single-walled carbon nanotube reinforced polystyrene and polyurethane nanofibers and membranes by electrospinning. *Nano Lett.* 4 (2004) 459–464.
105. Y. Nakayama, Synthesis, nanoprocessing, and yarn application of carbon nanotubes. *Jpn. J. Appl. Phys.* 47 (2008) 8149.
106. C. Jayasinghe, S. Chakrabarti, M. J. Schulz, V. Shanov, Spinning yarn from long carbon nanotube arrays. *J. Mater. Res.* 26 (2011) 645.
107. K. Koziol, J. Vilatela, A. Moisala, M. Motta, P. Cunniff, M. Sennett, et al., High-performance carbon nanotube fiber. *Science* 318 (2007) 1892.
108. N. T. Alvarez, P. Miller, M. Haase, N. Kienzle, L. Zhang, M. J. Schulz, et al., Carbon nanotube assembly at near-industrial natural-fiber spinning rates. *Carbon* 86 (2015) 350–357.
109. X. Zhang, K. Jiang, C. Feng, P. Liu, L. Zhang, J. Kong, et al., Spinning and processing continuous yarns from 4-inch wafer scale super-aligned carbon nanotube arrays. *Adv. Mater.* 18 (2006) 1505–1510.
110. X. Zhang, Q. Li, Y. Tu, Y. Li, J. Y. Coulter, L. Zheng, et al., Strong carbon-nanotube fibers spun from long carbon-nanotube arrays. *Small* 3 (2007) 244.
111. M. Miao, The role of twist in dry spun carbon nanotube yarns. *Carbon* 96 (2016) 819–826.
112. M. Miao, Electrical conductivity of pure carbon nanotube yarns. *Carbon* 49 (2011) 3755–3761.

113. J. C. Anike, K. Belay, J. L. Abot, Piezoresistive response of carbon nanotube yarns under tension: Rate effects and phenomenology. *New Carbon Mater.* 33 (2) (2018) 140–154.
114. J. W. S. Hearle, P. Grosberg, S. Backer, *Structural Mechanics of Fibers, Yarns, and Fabrics*, Wiley-Interscience, New York (1969).
115. P. R. Lord, M. E. Perez, The behavior of twistless and low-twist staple yarns in a plain-weave fabric. *Textile Res. J.* 51 (1981) 45–51.
116. N. Pan, Development of a constitutive theory for short fiber yarns: Mechanics of staple yarn without slippage effect. *Textile Res. J.* 62 (1992) 749–765.
117. N. Pan, Development of a constitutive theory for short fiber yarns. Part II: Mechanics of staple yarn with slippage effect. *Textile Res. J.* 63 (1993) 504–514.
118. F. T. Peirce, Tensile tests for cotton yarns. Part 5: "Weakest link" theorems on the strength of long and of composite specimens. *J. Text. I.* 17 (1926) 355–368.
119. M. L. Realff, N. Pan, M. Seo, M. C. Boyce, S. Backer, A stochastic simulation of the failure process and ultimate strength of blended continuous yarns. *Textile Res. J.* 70 (2000) 415–430.
120. Y. Elmogahzy, Yarn engineering. *Indian J. Fiber Text. Res.*, Special Issue on Emerging Trends in Polymers & Textiles, 31 (2006) 150–160.
121. A. Kelly, N. H. MacMillan, *Strong Solids*, Clarendon Press: Oxford (1986).
122. N. Pan, T. Hua, Relationship between fiber and yarn strength, *Textile Res. J.* 71 (2001) 960–964.
123. J. Terrones, A. H Windle, J. A. Elliott, The electro-structural behaviour of yarn-like carbon nanotube fibres immersed in organic liquids. *Adv. Mater.* 14 (2014) 055008.
124. J. C. Anike, PhD, Dissertation, Department of Mechanical Engineering, The Catholic University of America, Washington, DC, USA (2018).
125. P. Miaudet, S. Badaire, M. Maugey, A. Derré, V. Pichot, P. Launois, et al., Hot-drawing of single and multiwall carbon nanotube fibers for high toughness and alignment. *Nano Lett.* 5 (2005) 2212–2215.
126. M. K. Shin, B. Lee, S. H. Kim, J. A. Lee, G. M. Spinks, S. Gambhir, G. G. Wallace, et al., Synergistic toughening of composite fibres by self-alignment of reduced graphene oxide and carbon nanotubes. *Nat. Commun.* 3 (2012) 650. (doi:10.1038/ncomms1661)
127. N. Pan, Prediction of statistical strengths of twisted fibre strucutres. *J. Mater. Sci.* 28 (1993) 6107–6114.
128. J. C. Anike, H. H. Le, G. E. Brodeur, J. L. Abot, Piezoresistive response of integrated cnt yarns under compression and tension: The effect of lateral constraint. *C* 3 (2017) 14. (doi:10.3390/c3020014)
129. J. L. Abot, T. Alosh, K. Belay, Strain dependence of electrical resistance in carbon nanotube yarns. *Carbon* 70 (2014) 95–102.
130. J. C. Anike, A. Bajar, J. L. Abot, Time-dependent effects on the coupled mechanical-electrical response of carbon nanotube yarns under tensile loading. *C* 2 (2016) 3.
131. J. L. Abot, C. P. Rajan, Carbon nanotube fibers. *Carbon Nanomaterials Sourcebook: Graphene, Fullerenes, Nanotubes, and Nanodiamonds,* K. D. Sattler (Ed.), Taylor and Francis: London (2016).
132. C. Berger, P. Poncharal, Y. Yi, W. de Heer, Ballistic conduction in multiwalled carbon nanotubes. *J. Nanosci. Nanotechnol.* 3 (2003) 171–177.
133. W. Obitayo, T. Liu, A review: Carbon nanotube-based piezoresistive strain sensors. *J. Sensors* (2012) 652438. (doi:10.1155/2012/652438)
134. J. G. Simmons, Generalized formula for the electric tunnel effect between similar electrodes separated by a thin insulating film. *J. Appl. Phys.* 34 (1963) 1793–1803.
135. J. Zhao, X. Zhang, J. Di, G. Xu, X. Yang, X. Liu, et al., Double-peak mechanical properties of carbon-nanotube fibers. *Small* 6 (2010) 2612–2617.
136. Y. Zhao, J. Wei, R. Vajtai, P. M. Ajayan, E. V. Barrera, Iodine doped carbon nanotube cables exceeding specific electrical conductivity of metals. *Sci. Rep.* 1 (2011) 83.
137. K. Liu, Y. Sun, R. Zhou, H. Zhu, J. Wang, L. Liu, et al., Carbon nanotube yarns with high tensile strength made by a twisting and shrinking method. *Nanotechnology* 2 (2010) 045708.
138. C. L. Tsai, I. M. Daniel, Determination of shear modulus of single fibers. *Exp. Mech.* 39 (1999) 284.
139. E. Kreyszig, *Advanced Engineering Mathematics*, 4th ed., John Wiley: New York (1979).
140. A. Hanks, *Torsional Pendulum EX-5521*, Pasco Scientific: Roseville, CA(2011).
141. A. S. Wu, X. Nie, M. C. Hudspeth, W. W. Chen, T.W. Chou, D. S. Lashmore, et al., Carbon nanotube fibers as torsion Sensors. *Appl. Phys. Lett.* 100 (2012) 201908. (doi:10.1063/1.4719058)
142. K. Suzuki, K. Yataka, Y. Okumiya, S. Sakakibara, K. Sako, H. Mimura, et al., Rapid-response, widely stretchable sensor of aligned mwcnt/elastomer composites for human motion detection. *ACS Sens.* 1 (2016) 817–825. (doi:10.1021/acssensors.6b00145)
143. T. Yamada, Y. Hayamizu, Y. Yamamoto, Y. Yomogida, A. Izadi-Najafabadi, D.N. Futaba, et al., A stretchable carbon nanotube strain sensor for human-motion detection. *Nat. Nanotechnol.* (2011) 6. (doi:10.1038/NNANO.2011.36)
144. M. K. Shin, J. Oh, M. Lima, M. E. Kozlov, S. J. Kim, R. H. Baughman, Elastomeric conductive composites based on carbon nanotube forests. *Adv. Mater.* 22 (2010) 2663–2667. (doi:10.1002/adma.200904270)
145. J. Foroughi, G. M. Spinks, S. Aziz, A. Mirabedini, A. Jeiranikhameneh, G. G. Wallace, et al.,

Knitted carbon-nanotube-sheath/spandex-core elastomeric yarns for artificial muscles and strain sensing. *ACS Nano* 10 (2016) 9129–9135. (doi:10.1021/acsnano.6b04125)

146. L. Cai, L. Song, P. Luan, Q. Zhang, N. Zhang, Q. Gao, et al., Super-stretchable, transparent carbon nanotube-based capacitive strain sensors for human motion detection. *Sci. Rep.* 3 (2013) 3048.
147. T. Takahashi, K. Takei, A. G. Gillies, R. S. Fearing, A. Javey. Carbon nanotube active-matrix backplanes for conformal electronics and sensors. *Nano Lett.* 11 (2011) 5408–5413.
148. S. Ryu, P. Lee, J. B. Chou, R. Xu, R. Zhao, J. H. Anastasios, et al., Extremely elastic wearable carbon nanotube fiber strain sensor for monitoring of human motion. *ACS Nano* 9 (2015) 5929–5936. (doi:10.1021/acsnano.5b00599)
149. S. Park, M. Vosguerichian, Z. Bao, A review of fabrication and applications of carbon nanotube film-based flexible electronics. *Nanoscale* 5 (2013) 1727–1752.
150. Q. Cao, J. A. Rogers, Ultrathin films of single-walled carbon nanotubes for electronics and sensors: A review of fundamental and applied aspects. *Adv Mater.* 21 (2009) 29–53. (doi:10.1002/adma.200801995)
151. L. Cai, P. Luan, Q. Zhang, N. Zhang, Q. Gao, D. Zhao, et al., Highly transparent and conductive stretchable conductors based on hierarchical reticulate single-walled carbon nanotube architecture. *Adv. Funct. Mater.* 22 (2012) 5238–5244.
152. K. L. Jiang, J. P. Wang, Q. Q. Li, L. A. Liu, C. H. Liu, S. S. Fan, Superaligned carbon nanotube arrays, films, and yarns: A road to applications. *Adv Mater.* 23 (2011) 1154–1161. (doi:10.1002/adma.201003989)
153. C. L. Wang, R. Cheng, L. Liao, X. F. Duan, High performance thin film electronics based on inorganic nanostructures and composites. *Nano Today* 8 (2013) 514–530. (doi:10.1016/j.nantod.2013.08.001)
154. M. F. L. De Volder, S. H. Tawfick, R. H. Baughman, A. J. Hart, Carbon nanotubes: Present and future commercial applications. *Science* 339 (2013) 535–539. (doi:10.1126/science.1222453)
155. D. J. Lipomi, M. Vosgueritchian, B. C. Tee, S.L. Hellstrom, J. A. Lee, C. H. Fox, et al., Skin-like pressure and strain sensors based on transparent elastic films of carbon nanotubes. *Nat Nanotechnol.* 12 (2011) 788–792.
156. B. L Liu, C. Wang, J. Liu, Y. C. Che, C. W. Zhou, Aligned carbon nanotubes: From controlled synthesis to electronic applications. *Nanoscale* 20 (2013) 9483–9502. (doi:10.1039/c3nr02595k)
157. C. Wang, K. Takei, T. Takahashi, A. Javey, Carbon nanotube electronics-moving forward. *Chem. Soc. Rev.* 7 (2013) 2592–2609.
158. L. B. Hu, D.S. Hecht, G. Gruner, Carbon nanotube thin films: Fabrication, properties, and applications. *Chem. Rev.* 10 (2010) 5790–5844. (doi:10.1021/cr9002962).
159. J. L. Abot, C. Y. Kiyono, G. P. Thomas, E. C. N. Silva, Strain gauge sensors comprised of carbon nanotube yarn: Parametric numerical analysis of their piezoresistive response. *Smart Mater. Struct.* 24 (2015) 075018.
160. J. L. Abot, C. Y. Kiyono, J. C. Anike, M. R. Góngora-Rubio, L. A. M. Mello, V. F. Cardoso, et al., Foil strain gauges using piezoresistive carbon nanotube yarn: Fabrication and calibration. *Sensors* 2 (2018) 464. (doi:10.3390/s18020464)
161. C. Hellier, *Handbook of Nondestructive Evaluation*, 2nd ed., McGraw-Hill: Westminster (2012).
162. J. Park, K. H. Lee, Carbon nanotube yarns. *Korean J. Chem. Eng.* 29 (2012) 277–287.
163. J. L. Abot, Y. Song, M. Sri Vatsavaya, S. Medikonda, Z. Kier, C. Jayasinghe, et al., Delamination detection with carbon nanotube thread in self-sensing composite materials. *Compos. Sci. Technol.* 70 (2010) 1113–1119.
164. J. L. Abot, K. Wynter, S. P. Mortin, H. Borges de Quadros, H. H. Le, D. C. Renner, et al., Localized detection of damage in laminated composite materials using carbon nanotube yarn sensors. *J. Multifunc. Compos.* Special Issue: Novel Sensing Techniques and Approaches in Composite Materials 2 (2014) 217–226.
165. J. C. Anike, J. L. Abot, J. Bills, D. L. Gonteski, T. Kvelashvili, M. S. Alsubhani, et al., Integrated structural health monitoring of composite laminates using carbon nanotube fibers: static/dynamic loading and validation. *Proceedings of the 21st International Conference on Composite Materials*, Xian, China (2017).
166. J. L. Abot, J. C. Anike, From carbon nanotube yarns to fibers into sensors: Recent findings and challenges. *8th International Conference on Nanotechnology*, University of Cambridge, Cambridge, United Kingdom (2017).
167. J. L. Abot, J. C. Anike, J. H. Bills, Z. Onorato, D. L. Gonteski, T. Kvelashvili, et al., Carbon nanotube yarn sensors for precise monitoring of damage evolution in laminated composite materials: Latest experimental results and in-situ and post-testing validation. *Proceedings of the 32nd American Society for Composites Conference*, West Lafayette, IN (2017).
168. J. A. Lee, Y. T. Kim, G. M. Spinks, D. Suh, X. Lepró, M. D. Lima, et al., All-solid-state carbon nanotube torsional and tensile artificial muscles. *Nano Lett.* 14 (2014) 2664–2669.
169. H. H. Kim, C. S. Haines, N. Li, K. J. Kim, T. J. Mun, C. Choi, Harvesting electrical energy from carbon nanotube yarn twist. *Science* 357 (2017) 773–778.

170. Z. Zhu, L. Garcia-Gancedo, A. J. Flewitt, F. Moussy, Y. L. Li, W. I. Milne, Design of carbon nanotube fiber microelectrode for glucose biosensing. *J. Chem. Technol. Biotechnol.* 87 (2012) 256–262.
171. C. Jiang, L. Li, H. Hao, Carbon nanotube yarns for deep brain stimulation electrode. *IEEE Trans. Neural Syst. Rehabil. Eng.* 19 (2011) 612–616.
172. A. C. Schmidt, X. Wang, Y. Zhu, L. A. Sombers, Carbon nanotube yarn electrodes for enhanced detection of neurotransmitter dynamics in live brain tissue. *ACS Nano* 7 (2013) 7864–7873.
173. Z. Zhu, W. Song, K. Burugapalli, F. Moussy, Y. L. Li, X. H. Zhong, Nano-yarn carbon nanotube fiber based enzymatic glucose biosensor. *Nanotechnology* 21 (2010) 165501.
174. Z. Zhu, L. Garcia-Gancedo, A. J. Flewitt, H. Xie, F. Moussy, W. I. Milne, A critical review of glucose biosensors based on carbon nanomaterials: Carbon nanotubes and graphene. *Sensors* 12 (2012) 5996–6022.

19

Hall Effect Characterization of Nanowires

Olof Hultin
RISE Research Institutes of Sweden AB

Kristian Storm
Hexagem AB

Lars Samuelson
Lund University

19.1 Introduction

Nanowire-based technology is maturing and is being commercialized in a number of applications, primarily in electronics and electro-optics. An important part of understanding and optimizing semiconductor devices is to control and have methods to accurately measure doping and electrical properties. The doping concentration not only controls the conductivity of a semiconductor but also affects several other key parameters such as the band structure, depletion widths, carrier lifetime and current distribution in a device. Significant progress has been made in doping characterization in recent years. This chapter reviews the progress in nanowire Hall effect characterization and gives a tutorial style guide to the subject.

Measurements of doping in nanowires introduce a number of new challenges compared to bulk materials. Electrical characterization at the nanoscale necessitates advanced fabrication methods and careful methodology to account for effects that may be negligible at the mm or µm scale but significant at the nanoscale. Many different methods to measure doping in nanowires have been proposed and demonstrated. The first measurements of mobility in nanowires were made in the year 2000, when the Lieber group fabricated back-gated nanowire field effect transistors (FETs) and calculated the mobility from the measured transconductance (dI_{SD}/dV_G) (Cui et al. 2000). This approach has been widely used since then and is probably the most used characterization method to date. However, while FET measurements can be carried out on devices that are relatively simple to fabricate, it is not always straightforward to interpret the results and extract a doping concentration. Perhaps the largest issue is the estimation of the gate capacitance for which the geometry, electronic structure of the semiconductor (Khanal and Wu 2007) and surface states (Dayeh, Soci, et al. 2007) should be accounted for. The contact resistance of the source and drain can also affect the transconductance (Chou and Antoniadis 1987, Dayeh, Aplin, et al. 2007, Gül et al. 2015, Hultin et al. 2016). All these effects lead to an overestimation of the carrier concentration if not accounted for. In addition, it is only possible to fully deplete relatively thin and low-doped nanowires. Thicker and more heavily doped nanowires cannot be fully depleted and, thus, only measure the properties close to the surface.

In contrast, Hall effect characterization measures the carrier concentration over the entire electrically active cross section of the nanowire, allows measurements with spatial resolution along the nanowire and is less sensitive to critical input parameters. However, Hall effect measurements on nanowires require very precise fabrication methods, and it was not until 2012 they were first demonstrated (Blömers et al. 2012, Storm et al. 2012). This chapter reviews previous research in the field, describes sample fabrication in detail, provides basic theory for interpreting the results, and discusses the limitations of the methods.

Other electrical characterization methods that have been used on nanowires include C–V measurements (Tu et al. 2007, Roddaro et al. 2008, Garnett et al. 2009), thermoelectric measurements (Tchoulfian et al. 2013, Schmidt et al. 2014) and photocurrent (Richter et al. 2008, Allen et al. 2009, Mansfield et al. 2009). The reviews by Wallentin and Borgström (2011) and Dayeh et al. (2017) cover even more methods.

19.2 Hall Effect

From the early days of classical electromagnetism, it has been known that a conductor that carries a current across magnetic field lines is subjected to a force perpendicular to both the magnetic field and the direction of the current. In 1879, Edwin H. Hall set out to test whether the force acts on the conductor or the current itself. He devised an experiment where a current was conducted through a thin strip of gold and the voltage between two equipotential points on each side of the strip was measured using a galvanometer as a magnetic field was applied (Hall 1879). Hall observed a voltage across the sample when a magnetic field was present and concluded that the charge carriers in the metal are deflected, giving rise to a counteracting electric field. A few years later, J. J. Thomson studied the behavior of the moving charges in cathode rays, and the force from magnetic and electric fields on a point charge (Lorentz force) was derived by O. Heaviside and later by H. Lorentz.

The Hall effect is maybe best understood by deriving it from the Lorentz force,

$$\boldsymbol{F} = q\boldsymbol{E} + q\boldsymbol{v} \times \boldsymbol{B} \tag{19.1}$$

where F is the force, q is the charge, E is the electric field, v is the velocity of the charge carrier and B is the magnetic field. The cross product means that a moving charge subjected to a magnetic field is deflected from its path in a direction perpendicular to both its velocity and the magnetic field. This is illustrated in Figure 19.1. Due to various scattering processes, moving charge carriers in a semiconductor or metal will be slowed down and we can write

$$\boldsymbol{F} = m^*\boldsymbol{a} = m^* \left(\frac{d\boldsymbol{v}}{dt} + \frac{\boldsymbol{v}}{\tau}\right) = q\boldsymbol{E} + q\boldsymbol{v} \times \boldsymbol{B} \tag{19.2}$$

where τ is the scattering time and m^* is the effective mass. If the current is applied in the x-direction and the magnetic field is applied in the z-direction, the equation can be separated into the spatial components

$$m^* \left(\frac{dv_x}{dt} + \frac{v_x}{\tau}\right) = q\left(E_x + v_y B\right) \tag{19.3a}$$

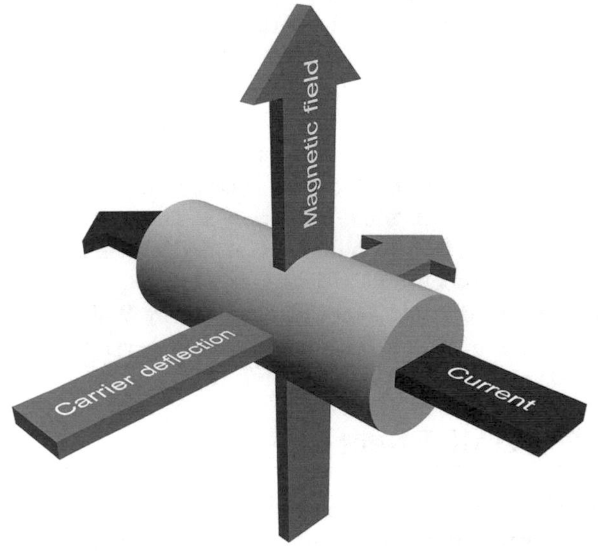

FIGURE 19.1 Hall effect in a cylindrical sample.

$$m^* \left(\frac{dv_y}{dt} + \frac{v_y}{\tau}\right) = q(E_y - v_x B) \tag{19.3b}$$

$$m^* \left(\frac{dv_z}{dt} + \frac{v_z}{\tau}\right) = qE_z. \tag{19.3c}$$

Since we are primarily interested in the effect of the magnetic field on the applied current, i.e. charge carriers moving along the x-direction, we focus on Eq. 19.3b. At steady state, there is no net acceleration of the charge carriers, $\frac{dv_y}{dt} = 0$. Furthermore, no current can float in the y-direction, so $v_y = 0$. We get

$$0 = q(E_y - v_x B) \tag{19.4}$$

$$qE_y = qv_x B. \tag{19.5}$$

By using the definition of current density (J),

$$J = qnv \tag{19.6}$$

where n is the charge carrier density. We arrive at

$$E_y = \frac{J_x B}{qn}. \tag{19.7}$$

In a cylindrical body, $I = J\pi r^2$ and $V_H = E_y 2r$. Where V_H is the voltage across the sample (Hall voltage), I is the applied current and r is the radius of the cylinder. We get

$$V_H = \frac{2IB}{\pi qrn}. \tag{19.8}$$

This is the basic relation between Hall voltage and carrier concentration in a wire.

19.3 Modeling the Hall Effect in Nanowires

Nanostructures often have complex shapes and geometries that are not possible to represent accurately using simple analytic models. In addition, the electric contacts often cover a relatively large part of the device, making the traditional assumption of point-like contacts and models such as Eq. 19.8 inaccurate. For a more accurate representation, finite element method (FEM) simulations can be used to simulate the potential and currents. FEM is a powerful tool by which semiconductor transport equations can be solved in an arbitrary geometry. We assume here that the nanowires are large enough not to exhibit any significant quantum confinement effects for the charge carriers and that their properties can be described by semiclassical 3D models.

To model the carrier transport, we start from the current continuity equation, stating that electric charge is a conserved quantity.

$$\nabla \cdot \boldsymbol{J} = -\frac{\partial \rho}{\partial t} \tag{19.9}$$

where ρ is the charge density. In other words, the current continuity equation states that the charge in a point does not change unless there is a difference in the current going in and out of the point.

The current in a semiconductor can be described by drift and diffusion. Drift current is due to the electrons' response to an electric field, whereas diffusion current is caused by diffusion of carriers between areas with different carrier densities. The drift and diffusion currents for electrons can be written as

$$\boldsymbol{J_n} = qn\mu \boldsymbol{E} + \mu kT\frac{dn}{dr} \tag{19.10}$$

where q is the charge of the charge carrier, n is the carrier density, μ is the mobility, E is the electric field, k is the Boltzmann constant, T is the temperature and dn/dr describes the carrier density gradient. The hole current is defined in a similar way. In some cases, a semiconductor device can be considered to have an isotropic carrier density and be modeled by just the current continuity equation and drift current (the first term in the equation above). In some cases, it is, however, necessary to consider differences in charge carrier concentration and potential. This can be calculated using another divergence theorem, Poisson's equation. This equation states that the electric flux density through a closed surface is proportional to the electric charge density (ρ) inside the surface

$$\nabla^2\psi = -\frac{\rho}{\varepsilon} \tag{19.11}$$

where ψ is the electric potential and ε is the permittivity. The electric charge density in a semiconductor is in turn described by the Fermi distribution

$$n = \int g(E)\, f_{FD} dE \tag{19.12}$$

where $g(E)$ is the density of states and f_{FD} is the Fermi distribution.

These equations give us the means to calculate charge distribution, currents and potentials in our devices. The effect of a magnetic field can be accounted for by introducing an anisotropic conductivity tensor (Storm et al. 2012). This model is valid for the case of an extrinsic semiconductor where the carrier scattering time (τ) is independent of energy (i.e. the Hall factor = 1). To derive the conductivity tensor, we rearrange Eq. 19.3 to find the velocities in different directions. $\frac{dv_y}{dt} = 0$ since we are looking at a steady-state case.

$$v_x = \frac{\frac{q\tau}{m^*}\left(E_x + \frac{q\tau}{m^*}E_yB\right)}{1+\left(\frac{q\tau}{m^*}B\right)^2} \tag{19.13a}$$

$$v_y = \frac{\frac{q\tau}{m^*}\left(E_y - \frac{q\tau}{m^*}E_xB\right)}{1+\left(\frac{q\tau}{m^*}B\right)^2} \tag{19.13b}$$

$$v_z = \frac{q\tau}{m^*}E_z. \tag{19.13c}$$

Since $\boldsymbol{J_{n,\mathrm{drift}}} = \boldsymbol{v}en = \boldsymbol{\sigma E}$, the above equations can be expressed in the conductivity tensor

$$\boldsymbol{\sigma} = \sigma_0 \begin{pmatrix} \frac{1}{1+\left(\frac{q\tau}{m^*}B\right)^2} & \frac{\frac{q\tau B}{m^*}}{1+\left(\frac{q\tau}{m^*}B\right)^2} & 0 \\ \frac{-\frac{q\tau B}{m^*}}{1+\left(\frac{q\tau}{m^*}B\right)^2} & \frac{1}{1+\left(\frac{q\tau}{m^*}B\right)^2} & 0 \\ 0 & 0 & 1 \end{pmatrix} \tag{19.14}$$

where $\sigma_0 = (q^2\tau n)/m^*$. By using the conductivity tensor in the relation $\boldsymbol{J_{n,\mathrm{drift}}} = \boldsymbol{\sigma E}$, we can simulate the Hall effect in virtually any semiconductor structure using an FEM software, such as COMSOL Multiphysics, and thus determine the charge carrier concentration from a Hall effect measurement.

Other authors have taken slightly different approaches to modeling the Hall effect in nanowires. Blömers et al. (2012) derived an analytic model for their surface 2D electron gas (2DEG). Barbut et al. (2014) simulated various nanowire Hall devices and pointed out that the contact may introduce a shunt path for the current along the nanowire and influence the measurement. This effect was also discussed by Hultin et al. (2017). Fernandes et al. (2014) showed that the diffusion current could reduce the Hall voltage in nanowires if the nanowire radius approaches the Debye screening length, since diffusion currents may prevent the formation of charge buildup at the edges of the sample. If not accounted for, this effect can cause an overestimation of the carrier concentration in thin nanowires with low doping level. Murata et al. (2017) carried out more detailed modeling of their bismuth nanowire system, including the band structure and different scattering mechanisms.

Using the model described above, we have simulated the Hall voltage for a typical case: An n-type InP nanowire with a diameter of 200 nm. The parameter of a real device that causes the largest deviation from the analytic expression is most often the contact overlap, i.e. how far the Hall contact extends onto the nanowire. The key question is where on the Hall contact the potential is actually measured. We, therefore, simulate devices with different contact gap and contact resistance and suggest using these results as correction factors when FEM simulations are not available. The contact gap is defined as the distance between the Hall contacts. For a 200 nm diameter nanowire, a contact gap of 200 nm corresponds to the contact just touching the nanowire on opposite sides of the outer perimeter. A contact gap of 0 nm would mean the Hall contacts are short-circuited. The contact is defined to be 100 nm thick, 80 nm wide, ohmic and having a conductivity of 4×10^7 S/m (similar to gold). The simulated voltages presented in Figure 19.2a are normalized with Eq. 19.8. Also plotted in Figure 19.2 are analytic functions describing the potential at the inner edge and the center point of the contact. The potential in the cross section of a device is shown in Figure 19.2b.

Two effects are apparent in the simulated data. First, a metal contact deposited on a nanowire creates a parallel shunt path for the current. If the contact resistivity is very low, some of the current passing through the nanowire will be shunted through the Hall contact, thereby reducing the current density in the nanowire and, consequently, the Hall

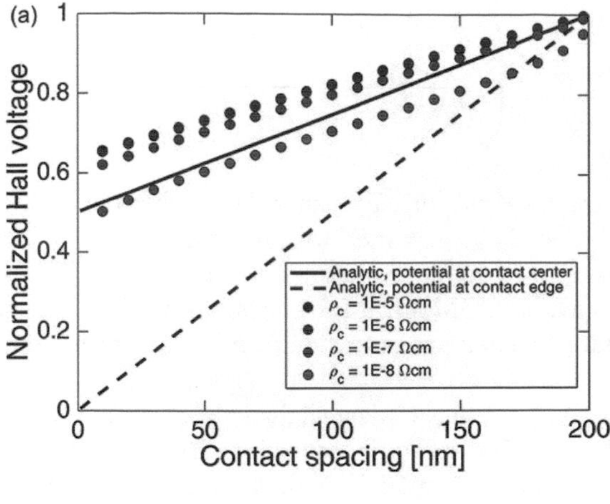

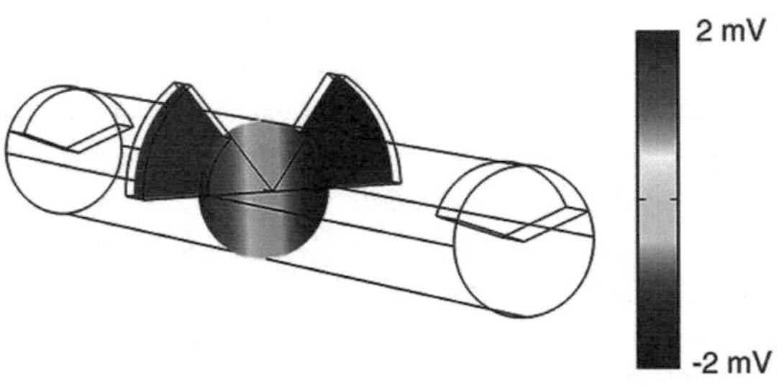

FIGURE 19.2 Simulation of Hall voltage in a nanowire with a diameter of 200 nm. (a) Simulated Hall voltages normalized with the analytic expression for Hall voltage in a cylinder. (b) Simulated potential in nanowire Hall device.

voltage. This is visible as a general reduction of the Hall voltage in the device with the lowest contact resistance. Second, the smaller the contact gap, the closer to the center of the nanowire the potential is probed resulting in a reduced Hall voltage. Compared to the analytic solution for the potential in the center of the contact, we see that the simulated voltages are larger (except for the lowest contact resistance, where much of the current is shunted through the contact). The potential must be effectively probed further out in the nanowire. Although it is not absolutely accurate, these simulations show that it can be a reasonable approximation to assume that the potential is probed at the center of the contact and use an analytic expression accordingly if FEM simulations are not available.

$$V_H = \left(\frac{2r + d_c}{4r}\right)\frac{2IB}{\pi q r n} \tag{19.15}$$

where d_c is the contact gap.

19.4 Fabrication of Nanowire Hall Devices

The challenge of placing contacts on the side facets of nanowires with sufficient precision is what held back the application of Hall effect measurements in nanowire characterization initially. However, with high-resolution electron beam lithography (EBL) systems and good alignment methods, it is now possible. Several research groups have proposed different ways of fabricating Hall devices. First presented here is the general EBL-based method used by the Samuelson group at Lund University (Storm et al. 2012, Heurlin et al. 2014, Lindgren et al. 2015, Hultin et al. 2016, Lindelöw et al. 2016). After that, a review of other methods is presented.

The first step of the fabrication process is to transfer nanowires from the growth substrate to a substrate more suitable for measurements. The measurement substrate should have pre-patterned alignment markers with features that can be used for EBL alignment and metallic bond pads. The surface of the measurement substrate should be insulating. A doped Si chip covered with 100 nm SiO_2 and 10 nm of HfO_2 is a suitable choice. Both oxides provide electric insulation, and the latter is added due to its etch resistance. The contact pads and alignment markers can be defined by UV lithography and EBL, respectively, followed by evaporation of metal (Figure 19.4a and b).

A simple yet effective way to transfer nanowires to the measurement substrate is to rub a small piece of cleanroom tissue on the growth substrate to pick up nanowires and then on the measurement substrate to deposit them. This results in a large amount of nanowires dispersed in random orientations over a relatively large area on the measurement substrate (Figures 19.3a and 19.4b).

To form a stable contact with a nanowire, the metal should form a stable bridge over the step between the nanowire and the substrate, meaning that the thickness of the metal has to be similar to the diameter of the nanowire. Due to the shadowing effect of the resist, it is hard to make thick contacts narrow (in the direction of the nanowire axis). To circumvent this and also make narrow contacts with thicker nanowires, a thin polymer spacer layer is spin coated on the chip in order to lift the contacts. The spacer layer is designed to just barely cover the nanowires. A mix resulting in a viscosity suitable for nanowires with a diameter around 200 nm is e.g. Shipley S1805:PGMA (1:1). The sample is then gently etched in oxygen plasma (Figures 19.3b and c and 19.4c), leaving the top part of the nanowire exposed.

The next step is to define the contacts to the nanowire. Selected areas of the measurement substrate are recorded with an SEM to determine the position of the nanowires with respect to the alignment markers. The electrical contacts can be designed from the images and then be exposed using EBL (Figure 19.3d). After contact definition, the sample is etched to remove native oxide from the nanowire surface and the metal is deposited to form the contacts. Excess metal is removed in a liftoff process. Finished devices are shown in Figures 19.3e, 19.4d and 19.5.

This method is versatile and can be used to make devices for many different applications, not Hall effect measurements alone. The constraints to the contact design mainly lie in the resolution and overlay accuracy of the lithography

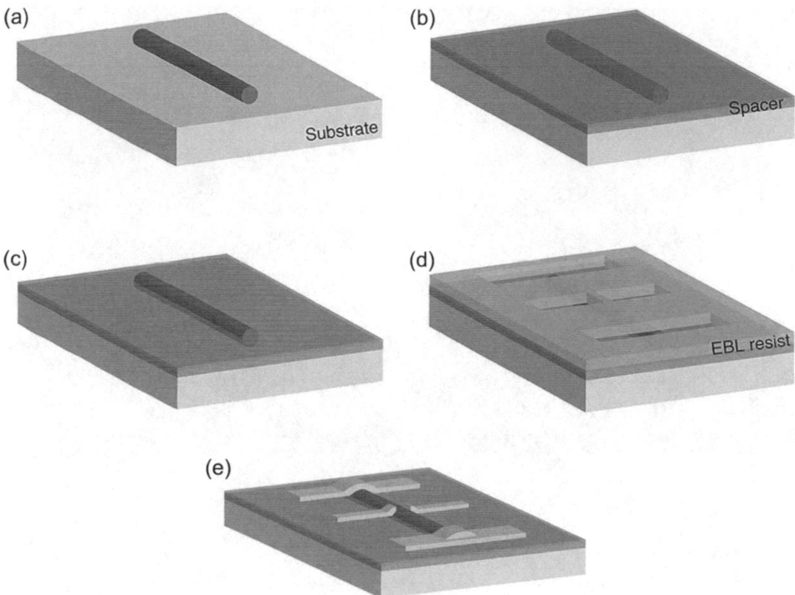

FIGURE 19.3 Nanowire Hall device fabrication. (a) Transfer of nanowires to a measurement substrate. (b) Spin coating a spacer layer on the nanowires. (c) Etching of spacer layer to allow access for contacts. (d) Definition of contacts with EBL. (e) Finished device.

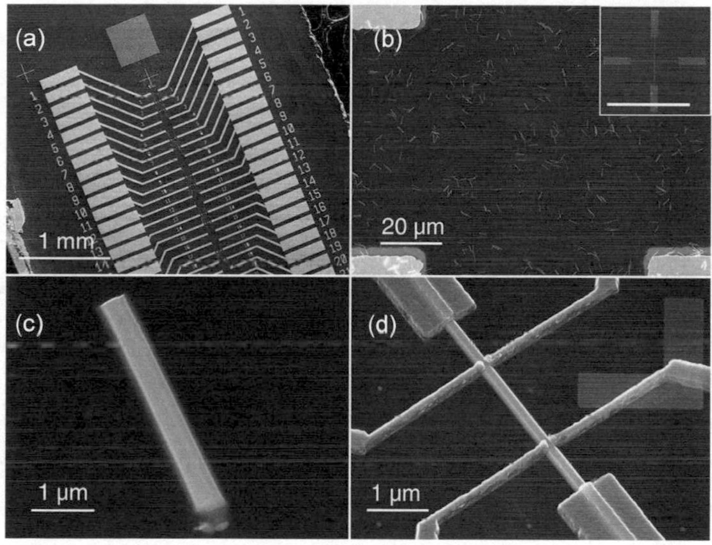

FIGURE 19.4 Lateral nanowire contacting. (a) Overview image of measurement substrate. (b) Overview of one EBL writefield area. The bond pads can be seen in the corners of the image. Inset: Close-up on alignment marker, scale bar 5 μm. (c) Nanowire surrounded by lifting layer. (d) Finished single nanowire Hall device.

process. With a good EBL system, it is possible to fabricate devices with contact spacing down to a few tens of nanometers. The alignment precision of the contacts is also on the order of tens of nanometers.

Blömers et al. used a similar approach but fabricated their InAs nanowire Hall devices without a spacer layer (Blömers et al. 2012). DeGrave et al. (2013) took a very different approach and fabricated Hall devices from FeS_2 strips by evaporating the contacts at an angle so that the nanowire itself is a shadow mask, thereby relaxing the alignment requirements in the lithography process. This fabrication is done by first depositing insulating alumina with an incidence angle perpendicular to the substrate (Figure 19.6a). This layer insulates the top of the nanowire so that only the side walls are contacted. Next, the contacts are deposited in two separate EBL and evaporation steps (Figure 19.6b and c). The evaporation is done at an angle, so that the nanowire acts as a shadow mask and only one of the side walls is exposed to metal. To avoid shorting the contacts, they must have an offset along the nanowire. A comparison to a reference structure (Figure 19.6f) was done to verify that the contact offset does not affect the results of the measurement. In a later paper by Liang et al., a process combining high-resolution EBL and tilted evaporation was used to fabricate

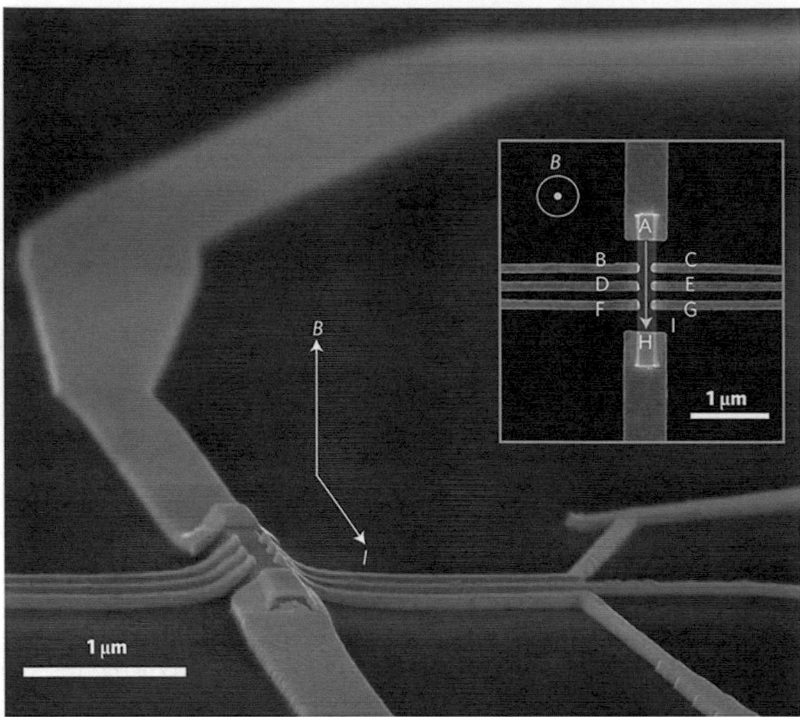

FIGURE 19.5 Nanowire hall device. (Reprinted with permission from Storm et al. 2012.)

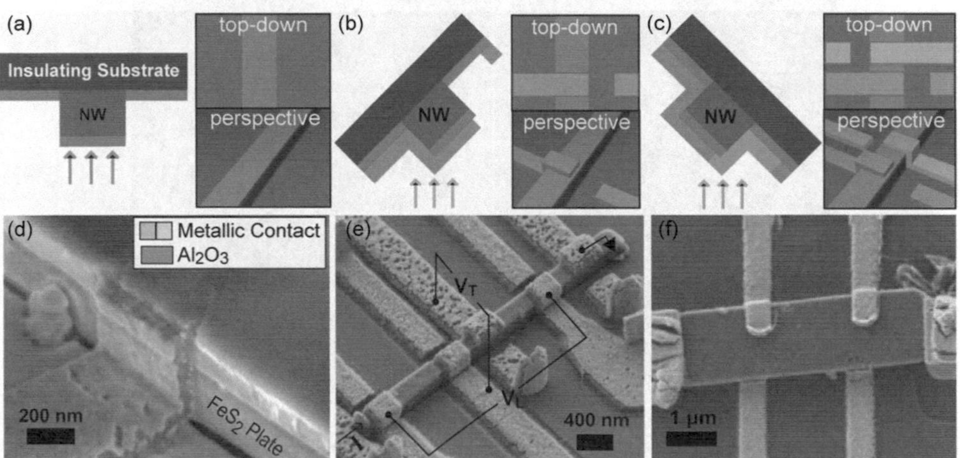

FIGURE 19.6 Nanowire Hall device fabrication by DeGrave et al. (a) Evaporation of alumina. (b) Evaporation of the first contact at an angle of 45°. (c) Device flipped for evaporation of the second contact. (d) Close-up of side wall contact. (e) Completed device. (f) Conventional Hall device based on a larger FeS_2 plate for comparison. (Reprinted with permission from DeGrave et al. 2013. Copyright 2013 American Chemical Society.)

devices without an offset and contact to the side wall (Liang et al. 2015).

Other approaches include focused ion beam (FIB) processing (Murata and Hasegawa 2013, Murata et al. 2017) and selectively grown lateral nanowires with post-processed contacts (Lindelöw et al. 2017).

19.5 Measurements

Measurement of Hall effect in nanowires requires a magnet and a basic set of electric test equipment to source current and measure voltage. Since nanowire devices tend to be sensitive to electrostatic discharge (ESD) and discharges from the test equipment, precautions should be taken to protect the sample.

A basic set of test equipment can consist of a source measure unit (SMU) capable of sourcing current in the microampere range through the nanowire and a good digital multimeter capable of measuring the Hall voltage. Care should be taken to ensure that the voltmeter has much higher input impedance than the nanowire device. Alternatively, another SMU can be used to measure the Hall voltage.

However, even with this type of specialized test equipment there is a fundamental limit to how small signals can be measured. This limit is most often set by noise generated from movement of charge carriers over the resistances in the circuit due to thermal energy, Johnson noise. Metallic conductors approach this theoretical limit, but semiconductors produce somewhat higher noise (Keithley 2004), especially if the contacts are not ohmic. The theoretical peak-to-peak voltage noise produced in a resistance, R, is

$$V_{\text{noise},p-p} = 5\sqrt{4kTRB} \quad (19.16)$$

where k is Boltzmann's constant, T is the absolute temperature and B is the noise bandwidth. The noise limit at room temperature for a typical measurement bandwidth (10 Hz) is plotted in Figure 19.7.

Nanowire devices are sensitive to electrical discharges due to their small dimensions. Figure 19.8 shows a discharge damaged device. A simple way to protect the nanowire devices is to always ground all contacts to the nanowire through a 10 kΩ resistor when no measurements are ongoing. Other precautions include avoiding switching on or off test equipment when the nanowire device is connected, making sure all contacts to the nanowire are shorted when stored if the chip is bonded to a chip holder and using a grounding bracelet when handling samples to avoid ESD.

Nanowire devices are sensitive to many phenomena that may affect the measured voltage: charge trapping, adsorption of foreign molecules on the surface, damages to the contacts or crystal, etc. It is, therefore, necessary to design the electrical measurements so that such a signal cannot be misinterpreted. Two basic considerations are as follows:

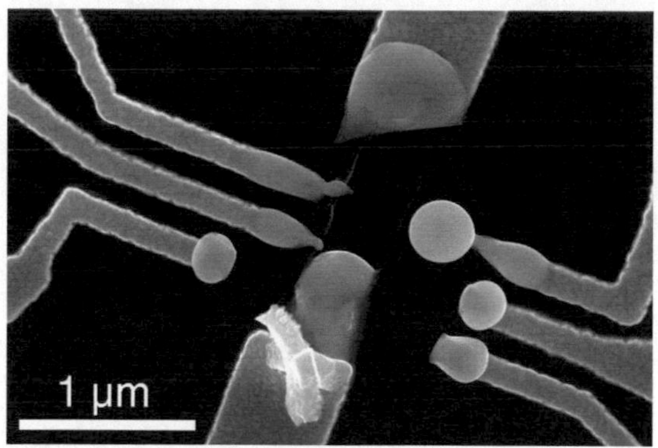

FIGURE 19.8 ESD damaged nanowire Hall device.

Multiple voltage measurements should be taken at each magnetic field strength to show how large the voltage fluctuations are, and measurements should be carried out at several different magnetic field strengths in a random order. This minimizes the risk of drift in the measured voltage due to changes to the nanowire, burn-in effects of the contacts, Joule heating, etc. Measurements on three InP nanowire Hall devices with different dopant molar flows (χ_{H_2S}) are shown in Figure 19.9. Here, 400 voltage measurements were taken at each magnetic field strength, and the standard deviation of the measured voltages is indicated by the error bars. The magnetic field was applied in a random order. It can be seen that the data fits well with a linear regression (the dashed lines). The measurement error can be estimated by calculating a confidence interval of the linear fit. Note that this error only reflects the noise in the measurement; it does not account for systematic errors such as surface depletion or contact effects.

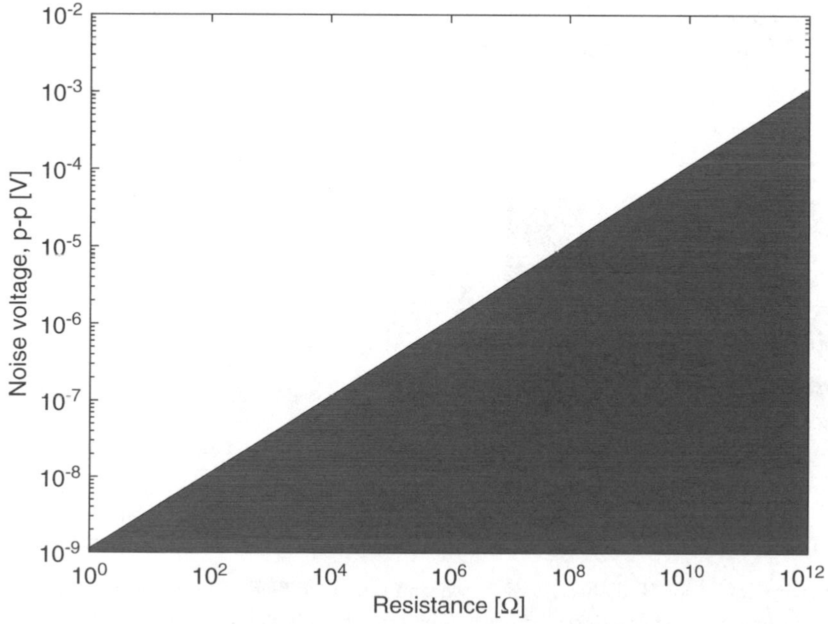

FIGURE 19.7 Peak-to-peak voltage noise at various resistances. The gray area is below the theoretical noise limit.

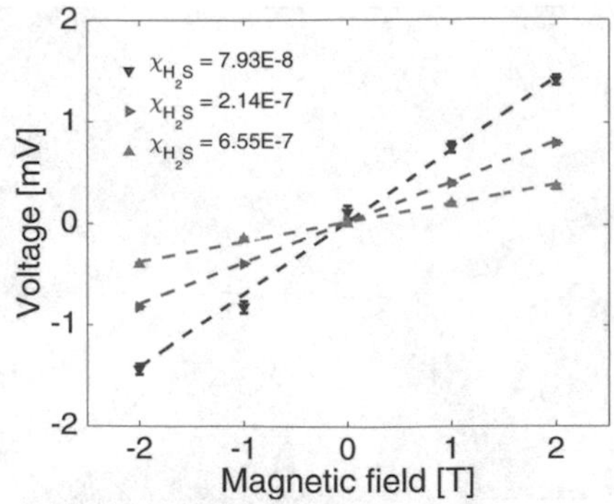

FIGURE 19.9 Hall effect measurements on three nanowires with different dopant molar fraction at growth. The voltage was measured 400 times at each magnetic field strength. The error bars indicate the standard deviation of the voltage measurements. The dashed lines are linear fits to the data.

19.6 Comparison to Other Characterization Techniques, Challenges and Limits

The most important thing about a characterization technique is that it is accurate and precise. Several studies have investigated the validity and limitations of the Hall effect measurements on nanowires. Here we give a short summary of comparisons to other measurement techniques, challenges, limits and unknowns of nanowire Hall measurements.

The proof-of-concept papers by Storm et al. (2012) and Blömers et al. (2012) both featured a comparison to another measurement technique. Storm et al. found an excellent correlation to the carrier concentration extracted from the peak shift in the cathodoluminescence (CL) spectrum. Blömers et al. found that the values from field effect measurements exceeded the values from the Hall measurement by about a factor of four. They attribute the high values from the field effect measurements to surface states. It should be noted that Blömers et al. only considered the carriers in the surface 2DEG of their InAs nanowires, whereas Storm et al. measured the 3D charge carrier concentration.

A correlation study between nanowire Hall measurements, cathodoluminescе and μ-photoluminescence (μPL) was presented by Lindgren et al. (2015). A series of Sn-doped InP nanowires grown under different conditions were investigated, first with Hall measurements and then with luminescence measurements. The nanowires had carrier concentrations (extracted by Hall measurements) ranging from high 10^{17} to mid 10^{19} cm^{-3}. Good correlation between the measurement techniques when full-width-half-max (FWHM) fitting was done for lower doping levels (below 4×10^{18} cm^{-3}) and Fermi-tail fitting was done for higher doping levels. These findings were further corroborated by a comparison to a set of planar samples.

Hultin et al. (2016) conducted a study in which Hall effect measurements were compared to back-gated and top-gated field effect measurements on the same single nanowires (Figure 19.10a). A series of InP nanowires with different dopant (H_2S) molar fraction were characterized. After accounting for contact and series resistance in the ungated parts of the nanowire as well as calculating the gate capacitance using an FEM model, the field effect measurements and Hall effect measurements correlate well (Figure 19.10b). The top-gated devices with a gate oxide thickness of only 11 nm, however, show slightly higher values than the other techniques. The authors attribute this to difficulties in controlling the native oxide on the nanowire surface and in determining the gate capacitance when the semiconductor capacitance becomes comparable to the oxide capacitance.

In another paper, Hultin et al. (2017) proposed a three-probe device to measure Hall effect in nanowires. Instead of measuring the Hall voltage as a difference between the potentials on the sides of the nanowire, the Hall voltage is measured as the change in potential in a single contact

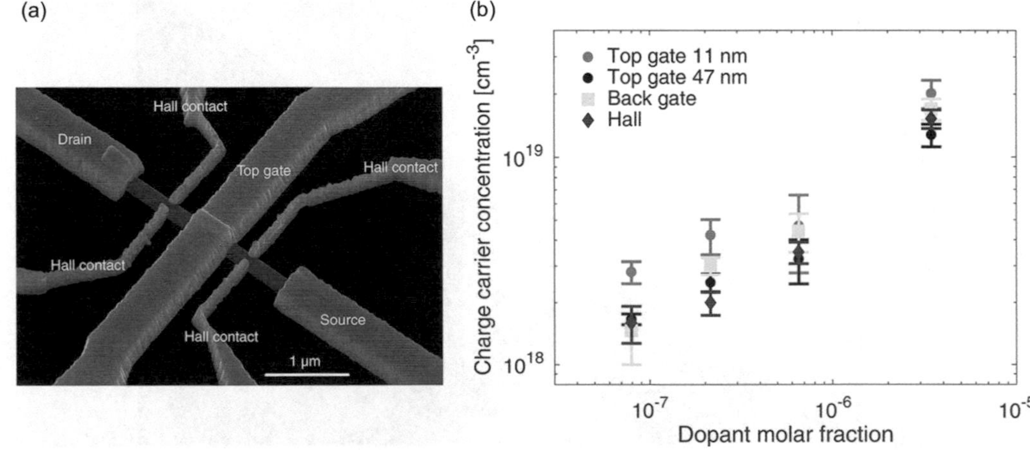

FIGURE 19.10 (a) Nanowire device for Hall effect, back-gated and top-gated field effect measurements. (b) Measured charge carrier concentration as a function of dopant molar fraction at growth. The error bars indicate the standard deviation of the measured charge carrier concentrations. (Modified with permission from Hultin et al. (2016).)

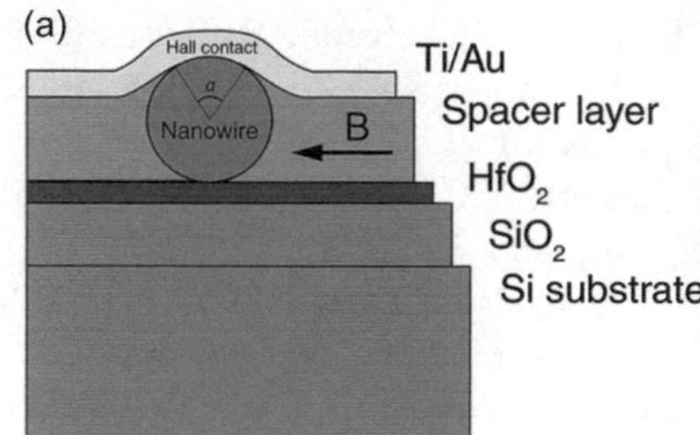

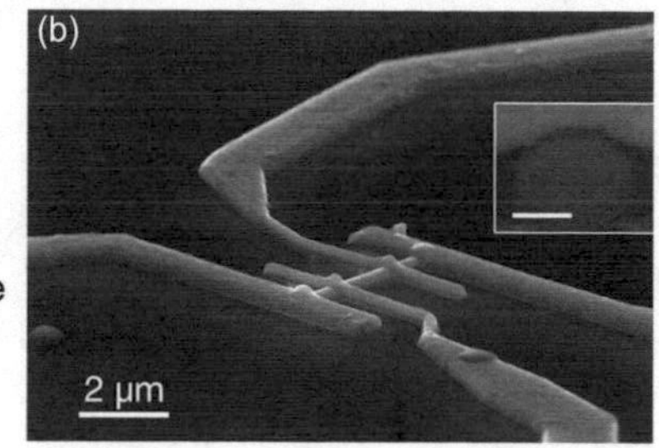

FIGURE 19.11 Three-probe Hall device. (a) Sketch of device cross section. (b) Overview SEM of three-probe Hall device with two Hall contacts, enabling measurement of carrier concentration at two different points along the nanowire. Inset: FIB cross section. (Reprinted with permission from Hultin et al. 2017.)

when the magnetic field strength is varied. By aligning the magnetic field to the plane of the substrate, perpendicular to the direction of the current, the Hall voltage appears in the vertical direction and a single contact can be placed on top of the nanowire to measure the change in potential (Figure 19.11). The contact overlap can be controlled by the thickness of a spacer layer instead of by a high-resolution EBL process, simplifying the fabrication process significantly. Since the requirements on the lithography are relaxed, it is also possible to conduct Hall measurements on much thinner nanowires with this method.

The downside with the three-probe method is that it is more sensitive to fluctuations in device resistance and magnetoresistance compared to a four-probe device in which these effects are mostly canceled out in the Hall voltage measurement. A method to conduct measurements to rule out these effects is suggested in the paper. Also an experimental comparison to four-probe Hall measurements is done. The two methods show excellent agreement on a series of InP nanowires.

Temperature-dependent Hall measurements have been carried out by Liang et al. (2014, 2015), Murata et al. (2017) and Haas et al. (2017), showing that characterization of carrier concentration, mobility and scattering mechanisms is possible over a wide range of temperatures.

Gated Hall effect measurements, in which a gate is used to modulate the carrier concentration while a Hall effect measurement is taken, have also been demonstrated by several authors. Blömers et al. (2012) used this to determine the surface defect density in InAs nanowires. Liang et al. (2014) used an electrolyte gate to characterize the bulk and surface properties of FeS_2 nanowires. Lindelöw et al. (2017) used the technique to study the gate–voltage dependence of the mobility in InGaAs nanowire transistors. Also, Haas et al. (2017) use a gate to modulate the carrier concentration in their devices.

19.6.1 Limits of Nanowire Hall Effect Measurements

The experimental comparisons on n-type InP show good correlation between different characterization techniques, verifying the validity of the Hall effect method. However, n-type InP is a material with good surface properties where formation of ohmic contacts is relatively simple compared to many other III–V materials. Hall effect measurements require good contacts to measure the Hall voltage without too much noise. This is not trivial to achieve in a nanowire Hall device where the contact area often is no larger than 50 nm × 50 nm. Thermal annealing processes that are often employed to improve contact properties should be used very cautiously in nanowire Hall devices, since diffusion of material in or out of the nanowire may completely change the properties of the nanowire that is to be characterized. Unfortunately, this limits the practical applicability of nanowire Hall measurements in many p-type III–V semiconductors.

Paradoxically, the Hall contacts should not have too low resistance either. When Hall contacts are placed on a nanowire, it introduces a parallel path for the current (Barbut et al. 2014). Some of the current will enter the contacts and flow through the low-resistive metal, leading to a reduction of the current density in the nanowire and thus a reduction of the Hall voltage. In most practical cases, this effect will, however, be negligible because the contact resistance is so large (Hultin et al. 2017). The application of a metal contact can also be intrusive in the sense that it may induce band bending in the semiconductor, thereby altering the charge carrier concentration. This will again be most prominent in thin nanowires with low doping level. A study fully evaluating this effect is yet to be undertaken.

Nanowires with very small diameter are very challenging to characterize with Hall measurements, not only because of the high alignment precision that would be necessary in the device fabrication but also because the Hall voltage is fundamentally limited by diffusion currents when the nanowire radius approaches the Debye screening length. The Hall voltage appears because nonequilibrium charge carriers displaced by the Lorentz force form charged layers at the sample edges. The thickness of the charged layer is approximately equal to the Debye screening length. If the nanowire is not thick enough for the charged layer to be formed, then consequently the Hall voltage will be suppressed (Fernandes et al. 2014). This puts a fundamental limit on the applicability of Hall effect measurements.

19.6.2 When Should Nanowire Hall Effect Measurements Be Considered?

The great advantage of Hall effect measurements compared to other electrical measurement techniques is that it can be used to characterize the entire electrically active cross section of the nanowire and that it is less sensitive to the surface properties of the nanowire, given that it is possible to make a good contact. The downside is that it may affect the semiconductor to place an electric contact right where the carrier concentration is measured. Thin nanowires with low doping level are most sensitive to this type of intrusion. Nanowire Hall effect measurements excel for relatively thick (>∼100 nm) nanowires with fairly high doping concentration (>∼5 × 10^{17} cm^{-3}). Such structures are attractive for use in optoelectronic devices such as solar cells and LEDs.

In summary, nanowire Hall measurements is a powerful and reliable characterization method with many uses and advantages. The applicability is, however, somewhat limited for very thin and low-doped nanowires. Further studies are needed to fully understand these limitations.

References

Allen, J.E., Perea, D.E., Hemesath, E.R., and Lauhon, L.J., 2009. Nonuniform nanowire doping profiles revealed by quantitative scanning photocurrent microscopy. *Advanced Materials*, 21 (30), 3067–3072.

Barbut, L., Jazaeri, F., Bouvet, D., and Sallese, J., 2014. Mobility measurement in nanowires based on magnetic field-induced current splitting method in H-shape devices. *IEEE Transactions on Electron Devices*, 61 (7), 2486–2494.

Blömers, C., Grap, T., Lepsa, M.I., Moers, J., Trellenkamp, S., Grützmacher, D., Lüth, H., and Schäpers, T., 2012. Hall effect measurements on InAs nanowires. *Applied Physics Letters*, 101 (15), 152106.

Chou, S.Y. and Antoniadis, D.A., 1987. Relationship between measured and intrinsic transconductances of FET's. *IEEE Transactions on Electron Devices*, 34 (2), 448–450.

Cui, Y., Duan, X., Hu, J., and Lieber, C.M., 2000. Doping and electrical transport in silicon nanowires. *The Journal of Physical Chemistry B*, 104 (22), 5213–5216.

Dayeh, S.A., Aplin, D.P.R., Zhou, X., Yu, P.K.L., Yu, E.T., and Wang, D., 2007. High electron mobility InAs nanowire field-effect transistors. *Small*, 3 (2), 326–332.

Dayeh, S.A., Chen, R., Ro, Y.G., and Sim, J., 2017. Progress in doping semiconductor nanowires during growth. *Materials Science in Semiconductor Processing*, 62, 135–155.

Dayeh, S.A., Soci, C., Yu, P.K.L., Yu, E.T., and Wang, D., 2007. Influence of surface states on the extraction of transport parameters from InAs nanowire field effect transistors. *Applied Physics Letters*, 90 (16), 162112.

DeGrave, J.P., Liang, D., and Jin, S., 2013. A general method to measure the hall effect in nanowires: Examples of FeS_2 and MnSi. *Nano Letters*, 13 (6), 2704–2709.

Fernandes, C., Ruda, H.E., and Shik, A., 2014. Hall effect in nanowires. *Journal of Applied Physics*, 115 (23), 234304.

Garnett, E.C., Tseng, Y.-C., Khanal, D.R., Wu, J., Bokor, J., and Yang, P., 2009. Dopant profiling and surface analysis of silicon nanowires using capacitance-voltage measurements. *Nature nanotechnology*, 4 (5), 311–3114.

Gül, Ö., Woerkom, D.J. Van, Weperen, I. Van, Car, D., Plissard, S.R., Bakkers, E.P.A.M., and Kouwenhoven, L.P., 2015. Towards high mobility InSb nanowire devices. *Nanotechnology*, 26 (21), 215202.

Haas, F., Zellekens, P., Lepsa, M., Rieger, T., Grutzmacher, D., Lüth, H., and Schäpers, T., 2017. Electron interference in Hall effect measurements on GaAs/InAs core/shell nanowires. *Nano Letters*, 17 (1), 128–135.

Hall, E.H., 1879. On a new action of the magnet on electric currents. *American Journal of Mathematics*, 2 (3), 287.

Heurlin, M., Hultin, O., Storm, K., Lindgren, D., Borgström, M.T., and Samuelson, L., 2014. Synthesis of doped InP core-shell nanowires evaluated using hall effect measurements. *Nano Letters*, 14 (2), 749–753.

Hultin, O., Otnes, G., Borgström, M.T., Björk, M., Samuelson, L., and Storm, K., 2016. Comparing hall effect and field effect measurements on the same single nanowire. *Nano Letters*, 16 (1), 205–211.

Hultin, O., Otnes, G., Samuelson, L., and Storm, K., 2017. Simplifying nanowire hall effect characterization by using a three-probe device design. *Nano Letters*, 17 (2), 1121–1126.

Keithley, 2004. *Low Level Measurements Handbook*. Sixth. e-book.

Khanal, D.R. and Wu, J., 2007. Gate coupling and charge distribution in nanowire field effect transistors. *Nano Letters*, 7 (9), 2778–2783.

Liang, D., Cabán-Acevedo, M., Kaiser, N.S., and Jin, S., 2014. Gated hall effect of nanoplate devices reveals surface-state-induced surface inversion in iron pyrite semiconductor. *Nano Letters*, 14 (12), 6754–6760.

Liang, D., DeGrave, J.P., Stolt, M.J., Tokura, Y., and Jin, S., 2015. Current-driven dynamics of skyrmions stabilized in MnSi nanowires revealed by topological Hall effect. *Nature Communications*, 6, 8217.

Lindelöw, F., Heurlin, M., Otnes, G., Dagytė, V., Lindgren, D., Hultin, O., Storm, K., Samuelson, L., and Borgström, M., 2016. Doping evaluation of InP nanowires for tandem junction solar cells. *Nanotechnology*, 27 (6), 65706.

Lindelöw, F., Zota, C.B., and Lind, E., 2017. Gated Hall effect measurements on selectively grown InGaAs nanowires. *Nanotechnology*, 28 (20).

Lindgren, D., Hultin, O., Heurlin, M., Storm, K., Borgström, M.T., Samuelson, L., and Gustafsson, A., 2015. Study of carrier concentration in single InP nanowires by luminescence and Hall measurements. *Nanotechnology*, 26 (4), 45705.

Mansfield, L.M., Bertness, K.A., Blanchard, P.T., Harvey, T.E., Sanders, A.W., and Sanford, N.A., 2009. GaN nanowire carrier concentration calculated from light and dark resistance measurements. *Journal of Electronic Materials*, 38 (4), 495–504.

Murata, M. and Hasegawa, Y., 2013. Focused ion beam processing to fabricate ohmic contact electrodes on a bismuth nanowire for hall measurements. *Nanoscale Research Letters*, 8 (1), 1–10.

Murata, M., Yamamoto, A., Hasegawa, Y., and Komine, T., 2017. Experimental and theoretical evaluations of the galvanomagnetic effect in an individual bismuth nanowire. *Nano Letters*, 17 (1), 110–119.

Richter, T., Meijers, H.L.R., Calarco, R., and Marso, M., 2008. Doping concentration of GaN nanowires determined by opto-electrical measurements. *Nano Letters*, 8 (9), 3056–3059.

Roddaro, S., Nilsson, K., Astromskas, G., Samuelson, L., Wernersson, L.E., Karlström, O., and Wacker, A., 2008. InAs nanowire metal-oxide-semiconductor capacitors. *Applied Physics Letters*, 92 (25), 253509.

Schmidt, V., Mensch, P.F.J., Karg, S.F., Gotsmann, B., Das Kanungo, P., Schmid, H., and Riel, H., 2014. Using the Seebeck coefficient to determine charge carrier concentration, mobility, and relaxation time in InAs nanowires. *Applied Physics Letters*, 104 (1), 12113.

Storm, K., Halvardsson, F., Heurlin, M., Lindgren, D., Gustafsson, A., Wu, P.M., Monemar, B., and Samuelson, L., 2012. Spatially resolved Hall effect measurement in a single semiconductor nanowire. *Nature Nanotechnology*, 7 (11), 718–722.

Tchoulfian, P., Donatini, F., Levy, F., Amstatt, B., Dussaigne, A., Ferret, P., Bustarret, E., and Pernot, J., 2013. Thermoelectric and micro-Raman measurements of carrier density and mobility in heavily Si-doped GaN wires. *Applied Physics Letters*, 103 (20), 202101.

Tu, R., Zhang, L., Nishi, Y., and Dai, H., 2007. Measuring the capacitance of individual semiconductor nanowires for carrier mobility assessment. *Nano Letters*, 7 (6), 1561–1565.

Wallentin, J. and Borgström, M.T., 2011. Doping of semiconductor nanowires. *Journal of Materials Research*, 26 (17), 2142–2156.

20

Metal Oxide Nanowire Arrays

Alexandra J. Riddle and
Beth Guiton
University of Kentucky

20.1 Introduction

Nanomaterial synthesis and fabrication methods may be classified into two general categories: bottom-up and top-down. Structures formed using bottom-up methods are prepared using atomic or molecular components to ultimately build the full structure (think, piecing together atomic/molecular Legos). Top-down methods begin with a bulk material and typically use lithography to modify the structure (think, cutting out shapes to make a paper snowflake). In the semiconductor industry, top-down approaches dominate due to the low associated costs and capability to fabricate large areas quickly. Using these approaches, the industry has established an expectation for ever-decreasing feature size, described by "Moore's Law," a prediction from Gordon Moore in 1965 that the number of transistors in an integrated circuit will roughly double every 2 years (Moore 1998, Waldrop 2016). Resolution can be defined as the ability to distinguish two different points and for fabrication methods depending on electromagnetic radiation such as lithography; therefore, the Abbe (diffraction) limit determines the minimum distance between two lithographically defined features (Zheludev 2008, Beyond the diffraction limit 2009). The diffraction limit is dependent on the wavelength of incident light, and decreasing the wavelength of light is one way to improve the resolution, though this is typically accompanied by an increase in the cost associated with manufacturing. As the resolution of lithographic methods has increasingly been pushed to approach its theoretical (and economical) limit, interest in bottom-up techniques has increased. Additionally, bottom-up methods allow for an endless possibility of complex structures. In this chapter, we will review top-down approaches using lithography (Section 20.3), bottom-up approaches such as those using templates (Section 20.2), and those to assemble post-synthesized nanowires (NWs) (Section 20.4).

NWs have promising attributes for future incorporation into devices due to their crystalline structure, large surface areas, and high aspect ratios. More specifically, metal oxide NWs have a wide variety of unique properties (depending on the material) with high potential in a wide range of applications such as electronics, photonics, energy storage, transistors, and gas sensors. NWs can be fabricated by a range of methods, which we generally categorize into three groups: templated growth, vapor-phase growth, and solution-based methods. Vapor-phase growth and solution-based methods often lead to NWs with high crystallinity, monodispersity, and controlled morphology. Typically, these techniques are not able to control the location of the NWs, which is critical in order to utilize them in devices. Both lithography and templating can aid in the alignment of NW arrays grown in vapor phase, or from solution, if they are utilized to define catalyst or seed particles from which the NW array is subsequently grown. Template-based methods used to grow NWs directly can control the location of each NW, but this approach can have drawbacks. In addition to the inefficiency of designing templates with the correct dimensions, the use of amorphous templates typically produces polycrystalline NWs, which can render them less effective for the desired application, and removing the template often requires harsh solvents that can damage the NWs (Cao and Liu 2008). Further, dissolution of the template means that it is only effective for single use. The production of properly aligned, closely spaced, and defect-free single-crystalline NWs is necessary for various devices. Thus, there is a crucial need to understand and improve upon methods that precisely align NWs.

As an example of a vapor-phase method, the vapor–liquid–solid (VLS) mechanism is an extensively used technique, which allows for precise control of NW size and offers versatility in material choice. The VLS method was developed by Wagner and Ellis in 1964 for their work with silicon (Si) whisker growth (Wagner and Ellis 1964). In this process, a metal catalyst particle adsorbs reactant vapor forming a liquid metal alloy until it becomes supersaturated. Precipitation of the NW occurs at the liquid–solid interface (which ceases if the source of reactant vapor is removed), and the diameter of the NW is determined by the diameter of the original catalyst. One challenge to VLS is the difficulty in achieving the desired nanoparticle (NP) catalyst size and placement due to aggregation and Ostwald ripening. NPs aggregate due to their strong van der Waals attractive forces to form large masses of NPs. And, Ostwald ripening occurs spontaneously at temperatures that are high enough to induce diffusion of metal NPs because larger NPs are more energetically favorable. Lithography is one method that can be used to precisely place the NPs for subsequent oriented NW growth via VLS. In order to produce *well-aligned arrays* of NWs without external support, the materials selected (for the substrate and the NW) should be homoepitaxial or heteroepitaxial, meaning the same crystalline material or different crystals with similar lattice parameters, respectively. There are some reported cases where highly ordered NWs are grown independent of substrate choice, such as the p–n junction containing zinc oxide (ZnO) NWs by Li et al. (2013) as shown in Figure 20.1a and b. In this case, radial growth is inhibited by the addition of polyethyleneimine (PEI), which selectively adheres to the {1010} ZnO facets allowing axial growth to dominate.

Solution-based methods can be performed at constant pressure, such as hot injection methods, or at constant volume, such as hydrothermal or solvothermal syntheses. Surfactants are commonly employed in all types of solution methods to stabilize specific crystal facets. Hydrothermal synthesis involves heating reactants in water inside a Teflon-lined autoclave. The crystal formation is possible due to the increase in pressure when the sealed container is heated, increasing the solubility of an originally insoluble precursor; and in some cases, a salt such as sodium hydroxide is added to further increase the solubility (Demazeau and Largeteau 2015). Some advantages to hydrothermal methods are low processing temperatures (compared to vapor methods), low cost, and the ability for large-scale production.

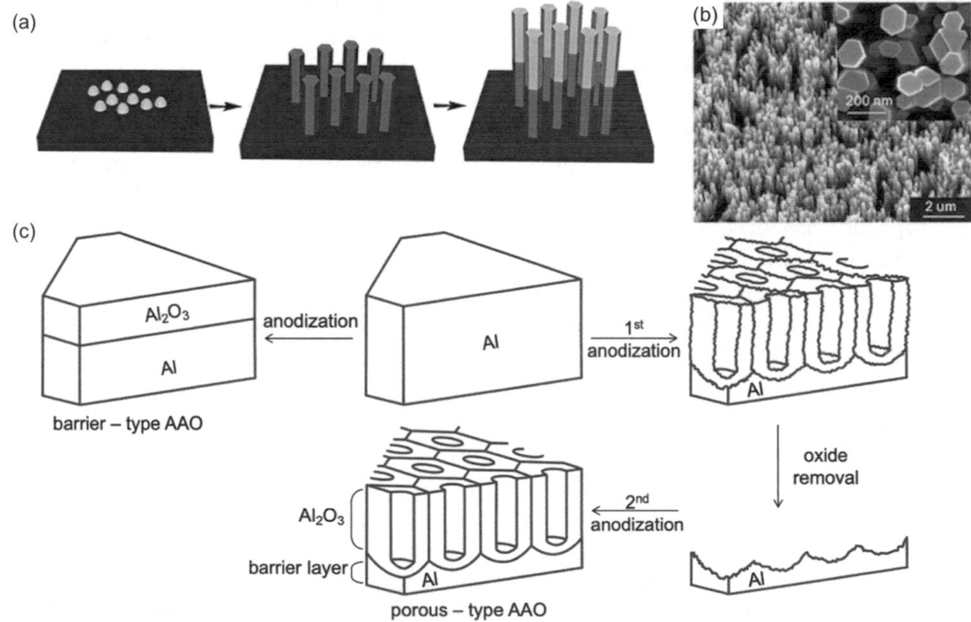

FIGURE 20.1 (a) Schematic diagram of the synthesis of ZnO p–n junction NW arrays. (b) SEM images showing the as-synthesized p–n junction containing ZnO NWs on indium tin oxide (ITO)-coated glass; inset shows the hexagonal cross section of the NWs. Very similar results were also observed on a c-sapphire and (001) Si substrates. (Reproduced with permission from Li et al. 2013. Copyright 2013 The Royal Society of Chemistry; permission conveyed through Copyright Clearance Center, Inc.) (c) Schematic diagram showing the two types of anodized Al: barrier-type and porous-type with the detailed two-step anodization process to form highly ordered pores for use as templates. The first step consists of a long anodization process, which forms disordered pores on the Al base. The removal of the disordered anodic aluminum oxide (AAO) leaves hemispherical shapes in the Al. In the second anodization, these concave features aid in the formation of highly ordered pores with the depth tunable by altering the anodization time. The side view formation of AAO showing the well-aligned hexagonally closed-packed structure (similar to a honeycomb) with the concave barrier layer before the Al support.

20.2 Template-Based Approaches

A simple but effective approach to the formation of NW arrays is the use of templates with a specific diameter and depth, to control the placement of NWs during growth. There are two main approaches to form NW arrays via templating: (i) the template can be used to only pattern catalytic metal NPs for the subsequent catalyzed growth of NWs or (ii) the NW arrays may be grown directly within the template. Furthermore, there are several techniques that are typically associated with nanomaterial fabrication by utilizing templates: electrochemical deposition (ECD), electrophoretic deposition (EPD), template filling, or reactive chemical conversion (Cao and Liu 2008). Template-based NW growth has the advantage of controlling the diameter, spacing, and length of the NWs based on the parameters of the template and enabling the facile production of NW heterostructures. The major challenge to these approaches is typically the removal of the template without disruption of the synthesized nanostructure. Furthermore, construction of templates with specific characteristics can prove difficult since template fabrication may be sensitive to delicate changes in the synthetic process, although an increasing variety of differently sized templates is now commercially available. Also, templates used for NW growth are typically amorphous, and NWs grown directly within these guides are mostly polycrystalline, though single-crystalline NWs formed by templating have been reported (Cao and Liu 2008).

For the manufacturing of materials, there are two categories of templates: hard and soft. Soft templates are typically a form of surfactant molecules in the shape of a single layer, multilayers, or spheres. In the materials community, hard templates are more commonly used. Some examples of hard templates are AAO, zeolites, polycarbonate (PC) membranes, and carbon nanotubes. Our focus here will be on the most commonly used template, porous AAO. In the late 1850s, Buff et al. realized that it was possible to induce oxidation via electrochemistry to expand the thickness of the native oxide layer (2–3 nm) on aluminum (Al) for further protection (Lee and Park 2014). This process was termed anodization and was commercialized by the 1920s, most notably for corrosion protection for airplanes (Poinern et al. 2011). In 1953, it was discovered that AAO contained circular pores in a hexagonal arrangement, and in the years following, the anodization procedure was industrialized for applying a colored coating on a variety of metals for protection and aesthetic reasons, termed electroplating (Keller et al. 1953). It wasn't until the 1990s that a two-step process was able to form a self-ordered hexagonal structure as determined by Masuda and Fukuda (1995). The two-step process improves the ordering of the self-assembled pores as seen in Figure 20.1c. The pores in AAO are formed by anodization of Al metal in acidic conditions with pore diameters of 5–250 nm and densities of 10^8–10^{11} pores/cm^2 (Xu and Guo 2003). The diameter of the pores is tunable by changing the conditions during formation such as electrolyte solution, voltage, and temperature (Na et al. 2009). Two morphologies of AAO exist, determined by the pH of the electrolyte used: nonporous barrier-type (neutral electrolyte) and porous-type (acidic electrolyte) (Mankotia et al. 2014). For templating purposes, the porous AAO is vital and the electrolyte solution is typically oxalic acid, but sulfuric and phosphoric acids have also been used (Yang et al. 2018). The specific electrochemical process of spontaneous pore formation is complex and driven by the reduction of strain associated with the even distribution of pores with a hexagonally close-packed (HPC) arrangement – a pattern which is commonly observed in nature (e.g. honeycomb) (Jessensky et al. 1998, Zhang et al. 2015). After pore formation, the template consists of an Al support with a barrier layer on top of which sits the pore structures, and the template is therefore only open on one end. A through-hole membrane is often needed in which case the pore side may be coated with a protecting polymer and the Al substrate dissolved by soaking in a hydrochloric acid/copper chloride solution (Cao and Liu 2008). Further etching by phosphoric acid removes the barrier layer and polymer support and can be used to widen the pores if necessary.

20.2.1 Electrochemical Deposition (ECD)

ECD or electrodeposition occurs when an external electric field is applied to an electrode, driving charged species to diffuse through a solution to be reduced on the surface of an electrode, thereby forming a film of the desired material. The material to be deposited *must be conductive* since the initially deposited monolayer forms a barrier, preventing contact between the electrode and the electrolyte solution. First, a conductive metal film is deposited on one side of the AAO template and serves as the cathode. The cathode is submerged in an ionic solution containing cations of the reactant metal for deposition. When an external electric field is applied, the ions flow between the electrodes converting electrical energy to chemical potential energy, in a process called electrolysis. Current density is measured to monitor the amount of growth that has occurred. The complete filling of pores can be detected as a sudden rise in current due to the formation of hemispherical caps, which merge to form a thin film deposited on top of the AAO template. As an example, Sander and Tan (2003) used an alumina template to electrodeposit nano-arrays of gold to serve as the catalyst for the growth of NWs. Figure 20.2 shows scanning electron microscopy (SEM) images of the gold (Au) nanorod array, where the diameter and height are easily controlled by the template dimensions and synthesis conditions. Thus, this technique is easily used to form small NPs for catalytic growth of NWs. In an unusual example of highly crystalline NW growth using a template-based approach, Gómez et al. used templated-assisted ECD method to grow ZnO NW arrays for dye-sensitive solar cells (Gómez et al. 2014). The AAO layer was prepared by electron beam evaporation of an Al layer on the ITO substrate, anodization of the Al layer, and chemical etching to remove the barrier

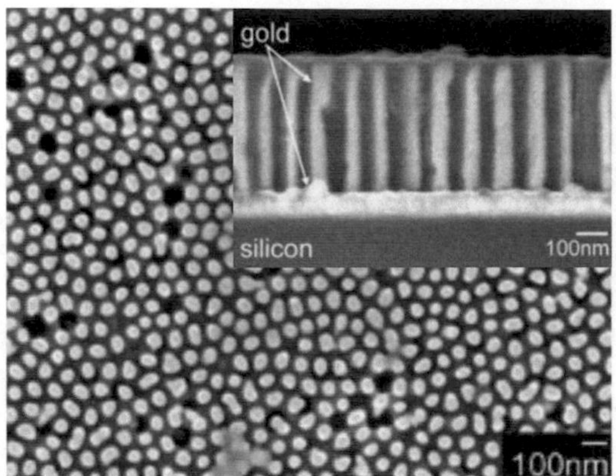

FIGURE 20.2 SEM images showing the results of ECD in an AAO template with metal deposition to form a highly aligned gold nanorod array; inset shows side view in alumina template. (Reproduced with permission from Sander and Tan 2003. Copyright 2003 John Wiley and Sons.)

layer and open the pores. ZnO was electrodeposited inside the AAO from a solution of zinc chloride ($ZnCl_2$) and oxygen (O_2) in dimethyl sulfoxide (DMSO). The NWs grew along the c-axis perpendicular to the substrate, evidenced by the observation of only the (0002) peak in the (X-ray diffraction) XRD indicative of a unique growth direction.

20.2.2 Electrophoretic Deposition (EPD)

An alternative approach to ECD is that of EPD. This technique has been utilized for the deposition of ceramic materials from colloidal dispersions (Boccaccini and Zhitomirsky 2002). While there are many similarities to electrodeposition, there are also some noticeable differences, which can form an advantage. Firstly, the deposition material leads to a porous barrier allowing solvent conductivity, so the material of the barrier does not need to be conductive itself and the solution is composed of NPs rather than ions. A polar solvent, such as water, induces an electric charge on the surface of the NP, which causes the formation of an electrostatic double layer surrounding the NP for electrostatic stabilization. The inner (Stern) layer consists of counterions strongly bound to the surface of the NP, outside of which is the diffuse (Gouy) layer comprising less tightly bound ions (Figure 20.3a) (Besra and Liu 2007). Electrophoresis occurs when an applied electric field causes the charged particles to move, along with the Stern and part of the diffuse layer. A slip plane exists, which is an imaginary boundary between the solvent molecules that move with the NP and those remaining with the surrounding bulk solution, and the electric potential at this location is called the zeta potential. The magnitude of the zeta potential loosely indicates the stability of a colloid solution (typically above 25 mV). Cao prepared a sol using titanium (IV) isopropoxide, glacial acetic acid, and water for the synthesis of titanium dioxide (TiO_2) nanorods via sol EPD (Cao 2004). The NWs were grown on a working electrode (Al sheets or carbon tape) with a platinum (Pt) counter electrode. Shown in Figure 20.3b and c, the NWs are polycrystalline with anatase phase identified from XRD analysis with no preferential orientation, but in further experiments with a different oxide material, vanadium pentoxide (V_2O_5), single-crystalline NWs were synthesized. Previously mentioned, the use of templates typically produces amorphous or polycrystalline structures, but in sol chemistry, vanadium oxide NPs can easily form ordered structures. The interactions between adjacent NPs are considerably weak, which allows for more rotation and migration to minimize the interface energy before aggregation, and the thermodynamically favorable

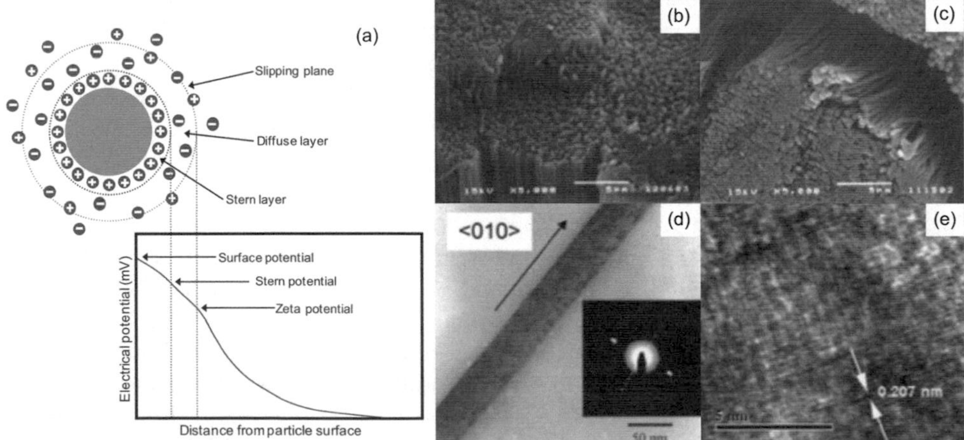

FIGURE 20.3 (a) Diagram showing the electrostatic double layer of a negatively charged Au NP with the corresponding graph of potential versus distance. (b and c) SEM images showing TiO_2 NWs grown via EPD in AAO, top and side view, respectively. (d) The low-resolution TEM image of V_2O_5 NWs; inset SAED pattern indexed to orthorhombic V_2O_5 on a [001] zone axis. (e) High-resolution TEM image showing the lattice fringes with a d-spacing of 0.207 nm and angle of 88.9° relative to the long axis of the nanorod. The d-spacing is similar to the spacing of (202) planes (0.204 nm), and the angle of the lattice fringes supports the proposed [010] growth direction of the NWs. ((b–e) Reproduced with permission from Cao 2004. Copyright 2004 American Chemical Society.)

state is for the NPs to epitaxially aggregate. Thus, the single-crystalline nature of the V_2O_5 NWs is thought to be due to the homoepitaxial aggregation of the NPs. The NWs were arranged parallel to each other and perpendicular to the substrate and indexed to orthorhombic V_2O_5 with a [010] growth direction as determined from transmission electron microscopy (TEM) images and selective area electron diffraction (SAED) patterns (Figure 20.3d and e).

20.2.3 Template Filling

A simple method which can nonetheless be effective is to simply fill the template directly using capillary action. Capillary forces drive the sol into the template when it is submerged in the colloidal liquid. After a predetermined time, the template is removed from the liquid, allowed to dry, and annealed at high temperatures to induce crystallinity in the residual colloids. While this technique is simple, its major shortcoming is, thus, that the templates are often not filled completely due to the high liquid-to-solid ratio, which can lead to shrinkage of the NW within the template pore. This can additionally lead to a competition between NW and nanotube formation, depending on the characteristics of the material; weak adhesion to walls or solidifying at the center, end, or uniformly leads to NW formation, whereas strong adhesion and solidifying inwards form a hollow structure. To increase the density of solid material during filling, Wen et al. used centrifugation to form metal oxide NW arrays of various materials (polymeric and colloidal silicon dioxide (SiO_2), TiO_2, and lead zirconate titanate ($Pb(Zr_{0.52}Ti_{0.48})O_3$, PZT) (Wen et al. 2005). First, the sols of the various materials were synthesized, and sol and template were centrifuged and annealed at varying temperatures. The TiO_2 and PZT both showed considerable shrinkage after annealing and formed the anatase and perovskite structures, respectively, whereas SiO_2 samples were amorphous and had a lower percentage of shrinkage. Ouyang et al. took this idea a step farther to make copper phthalocyanine (CuPc)-coated TiO_2 NW arrays (Ouyang et al. 2008). The AAO template was instead soaked in the TiO_2 solution, dried in air, and annealed in an oxygen environment, leaving a gap between the AAO template and the edge of the TiO_2 NW (Figure 20.4a–d). The EPD method was then utilized to form the CuPc coating using a $CHCl_3$/CuPc/CF_3OOH solution, where the TiO_2-infused AAO template (without Al substrate removal) formed the cathode and Pt the anode. On removal of the AAO template, the photoconductivity of the heterostructure NWs was tested and demonstrated a synergistic effect when compared to NW arrays of the separate constituents, likely because of the increased surface area between the two materials.

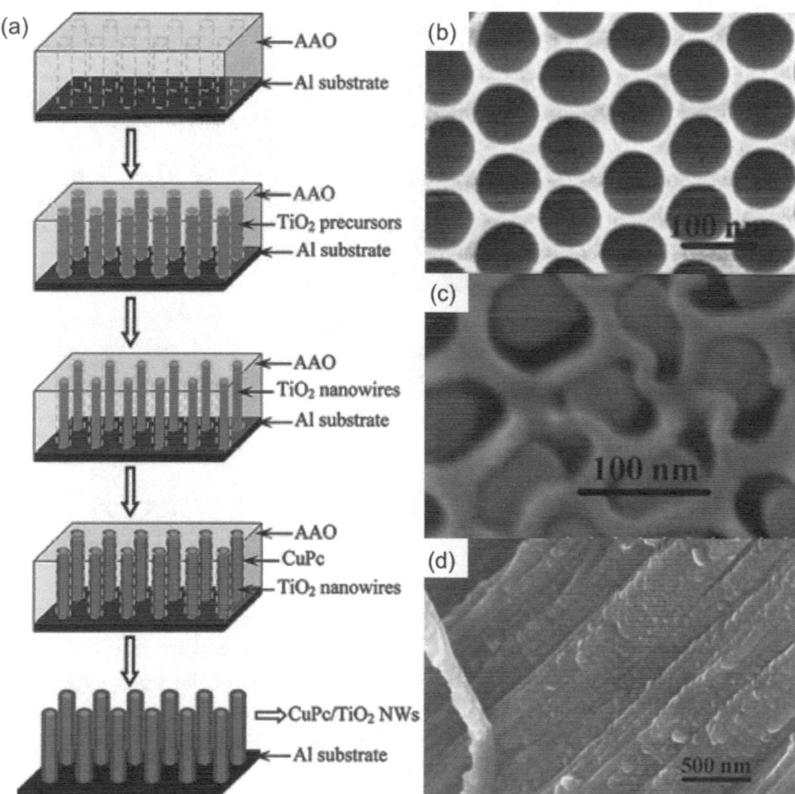

FIGURE 20.4 (a) Synthetic steps to form CuPc/TiO_2 NW arrays via template filling: prepared AAO template, TiO_2 filled template, annealed TiO_2 NWs showing diameter reduction, CuPc coating applied by EPD, and removal of the AAO template. SEM images of (b) top view of the AAO template showing uniform pores, (c) top view of the TiO_2 NWs in the AAO template, and (d) side view of the CuPc/TiO_2 NW arrays. (Reproduced with permission from Ouyang et al. 2008. Copyright 2008 American Chemical Society.)

20.2.4 Reactive Chemical Conversion

In a reactive chemical conversion, a material is deposited into a template and a chemical reaction is used to convert the precursor into the desired material. The template helps provide the support needed during the chemical reaction as well as the alignment for array formation. For metal oxide NWs, this method primarily involves oxidizing deposited metals, though more complex materials may be synthesized via multiple subsequent chemical reactions. Kolmakov et al. developed a method to convert tin NWs into stoichiometric oxides and used in situ XRD (with a slow and a fast oxidation procedure) to show the structural changes due to oxidation during heating (Figure 20.5a and b) (Kolmakov et al. 2003). In the slow heating, all phases (tin – Sn, tin oxide – SnO, and tin dioxide – SnO_2) were observed until the thermodynamically stable SnO_2 reached phase purity after annealing at 600°C for 2 h. In the fast oxidation, the results were similar during heating. On cooling, the low-intensity Sn metal peaks reappear, indicating a kinetically controlled oxidation mechanism for heat treatments, which do not allow for the full (slow) diffusion of oxygen into the metal. Perego et al. made arrays of the heterostructure Au/nickel oxide (NiO)/Au NWs by ECD in AAO to study electrical properties by conductive atomic force microscopy (AFM) (Figure 20.5c) (Perego et al. 2013). This study utilized an AAO template with 50 nm pores spaced 100 nm apart, with a 100 nm Au back contact added before NW growth. Heterojunctions were formed by alternating the growth between nickel and gold during ECD. The template was overfilled with Au to form a thin film, which was mechanically polished and chemically etched to remove any polishing artifacts and slightly reduce the thickness of AAO membrane to reveal the tip of the NWs still embedded in the template. The sample finally underwent a thermal oxidation process in a controlled oxygen environment to form polycrystalline NiO as seen in Figure 20.5d.

20.3 Lithography

Lithography is a process used to transfer patterns. In order to fully value how the methods work, why they are used, and the evolution of lithographic methods and their benefits in the alignment of NWs, one must take into consideration the historical developments of lithography and how they are currently incorporated in the semiconductor industry. In Greek, the literal translation of the word lithography is "writing on stone." (Venugopal and Kim 2013). Lithography originated in 1798 with an unintentional discovery by Alois Senefelder who noticed that a waxy coating on limestone resisted a washing with acid (Okoroanyanwu 2010). Many years later, in 1826, Joseph Nicéphore Niépce

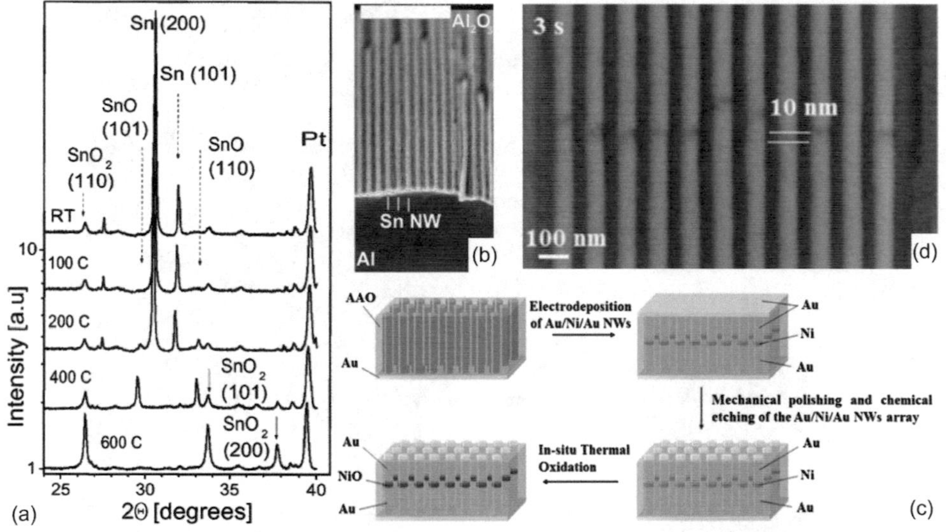

FIGURE 20.5 Reactive chemical conversion was used to convert metals into metal oxides utilizing a template to form ordered arrays. (a) XRD shows structural changes due to oxidation during the slow heating experiment. An array of NWs was removed from the AAO, placed on a Pt support, and the data were normalized to a Pt peak. At room temperature, the tin NWs are shown to preferentially grow along [100] direction with the existence of a thin SnO_2 film by the observation of the weak peaks (110), (101), and (200). As the temperature reaches near the melting point of tin (~232°C), the intensity of the Sn peaks decreases and the peaks for SnO appear. Annealing at 600°C for 2 h completes the conversion to SnO_2 NWs with no preferred orientation and a slight reduction in crystallinity as seen by peak broadening. (b) SEM image of as-grown Sn metal NWs in the AAO template. Scale 1,000 nm. (Reproduced with permission from Kolmakov et al. 2003. Copyright 2003 American Chemical Society.) (c) Schematic of grown heterostructure array formation by altering the electrolyte from cyanide based (for Au growth) to sulfate based (for Ni growth). The NWs were left in the AAO for electrical property measurements on single wires. (d) The correlation between Ni thickness and deposition time was studied in order to produce NW arrays with narrow (less than 100 nm) Ni segments. The cross-section SEM image shows an example of a 3 s deposition time to give a 10 nm Ni segment. (Reproduced with permission from Perego et al. 2013. Copyright 2013 IOP Publishing Ltd.)

used a photosensitive material and light to achieve a permanent image, thus, the beginning of photography and photolithography (Okoroanyanwu 2010). The invention of the transistor – a device that acts as a switch for electronic signals – and various other electrical components using photolithographic methods can be seen as the tipping point for all electronic devices. As such, the widespread incorporation of lithography in the semiconductor industry dates back to 1957 and is still the primary approach used today (Okoroanyanwu 2010).

20.3.1 Photolithography

In the most basic terms, optical lithography or photolithography uses light to transfer a pattern onto a substrate. Photolithography is the most commonly used technique to produce the silicon circuit boards in most electronic devices due to its high throughput, large-area capability, and cost-effectiveness. A polymeric material that can undergo changes in solubility when it interacts with light called a photoresist (PR) is spin coated on the substrate's surface. There are two kinds of PRs, positive and negative, that undergo divergent transformations when exposed to light. A positive PR causes the areas exposed to light to be more sensitive due to rupture of the polymer side chains (Figure 20.6a). Conversely, the areas exposed to light with a negative PR are stabilized by cross linking (Figure 20.6b). First, the substrate is pre-baked to remove any solvent and create good adhesion between the PR and the substrate. Typical examples of a positive PR and substrate materials are polyimide and silicon, respectively. Next, a mask is aligned, and the substrate is bombarded with high-intensity ultraviolet (UV) light. The PR is removed by placing the substrate in a developer, which removes the PR in the exposed areas of a positive PR, and the unexposed areas of a negative PR. A final hard bake is performed to make sure the remaining PR will protect the underlying surface. The areas not protected by the PR are etched to remove the uppermost layer of the substrate or other material. The PR is then completely removed by altering the chemical properties that allowed it to adhere to the substrate by soaking the substrate in a particular solvent. The completed process can be repeated in different directions and/or with a different mask to form a more complex pattern. There are three main types of optical lithography: projection, proximity, and contact lithography, in order of decreasing

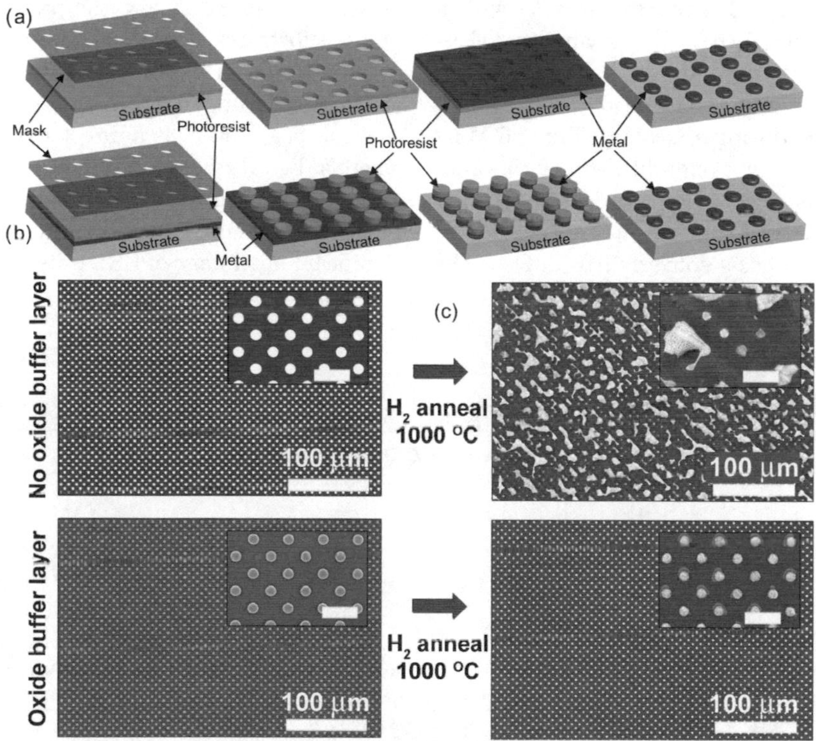

FIGURE 20.6 Schematic diagram showing photolithography with positive and negative PR for NP alignment on a substrate. (a) A substrate with a deposited positive PR (light gray) is irradiated with UV light through mask, the altered PR is removed by a developer, a metal (dark gray) film is deposited, and the PR is removed leaving behind metal NPs. (b) A substrate with a deposited metal film and negative PR is irradiated with UV light through mask, the unaltered PR is removed by a developer, the surface is etched removing any metal film not protected by the PR, and the PR is removed leaving behind metal NPs. (c) Photolithography was used to demonstrate the effectiveness of patterning the 300 nm oxide layer for preserving Au arrangement after annealing. When no oxide buffer layer was utilized, the deposited Au was able to migrate when annealed at high temperature. The addition of a thermally grown oxide buffer layer on the surface of the Si substrate reduced the migration of Au during annealing because the holes left in the patterned oxide layer confined the Au. ((c) Reproduced with permission from Kayes et al. 2007. Copyright 2007 AIP Publishing.)

distance between the photomask and the substrate. Contact printing achieves the best resolution but can cause defects in the pattern and deterioration of the mask. A 10–25 μm gap between the mask and substrate is used in proximity printing, which reduces the chance of mask damage but at a cost of resolution. In projection printing, since only a small portion of the mask is imaged by projecting it onto the wafer, the resolution is almost equal to contact lithography, but the mask may be held centimeters away, avoiding any chance of damage. Photolithography is a simple technique that can be used to pattern metal NPs for catalyst-assisted NW growth, to form trenches for lateral NW growth, or for the deposition of nucleation sites for NW growth in solution or in vapor phase. The major shortcoming of this technique is the resolution, in general, for proximity lithography, approaching 30 nm with the use of powerful excimer lasers with a wavelength of 193 nm (van Assenbergh et al. 2018).

Kayes et al. used photolithography to pattern a surface oxide layer to confine a growth catalyst (Au or copper) for VLS growth of NWs (Kayes et al. 2007). The thermally grown oxide layer was coated in a PR (S1813), which was then patterned by photolithography. The altered PR and underlying oxide were removed by submerging of the substrate in hydrofluoric acid leaving a substrate with a patterned oxide layer (Figure 20.6c). The catalyst was thermally deposited, and the PR was removed before annealing. The patterned oxide layer was crucial in order to contain the catalyst from diffusing during annealing. Tak and Yong (2005) patterned a substrate via photolithography in order to define specific regions of growth for dense vertically aligned ZnO NW arrays. A 40 nm zinc layer is added to the surface of a substrate and is converted to ZnO by the addition of zinc salt solution at 90°C for 6 h to improve the vertical alignment of nanorods and reduce chemical interactions with the developer. After the lithography process, the dried sample is added to a high concentration zinc salt solution and NWs are selectively grown in only the areas that have an exposed ZnO seed layer (Figure 20.7a–e). In a different approach by Shi et al., Figure 20.8a shows how photolithography was used to pattern as-grown vertically aligned ZnO NWs on a Si substrate by first completely coating them in PR (AZ1518) (Shi et al. 2016). Increasing the exposure time from 0 to 8.6 s, as seen in Figure 20.8b, acted to trim the length of the NWs from 12 to 7.8 μm. Additionally, Figure 20.8c shows a patterned mask with a range of different sized holes that were used to determine the smallest feature size to be about 5 μm, using 400 nm UV light.

20.3.2 Electron Beam Lithography (EBL)

EBL uses the same basic process as photolithography, but in this case, the PR is sensitive to electron beam exposure, rather than light. Here the exposure pattern does not require a photomask but instead a computer-aided design (CAD) drawing is loaded to an electron beam writer attached to an SEM, and the patterns are directly printed on the substrates. The chemistry, in this case, is also a little different; e-beam PR is typically polymethyl methacrylate (PMMA) and undergoes chain scission when exposed to the electron beam. EBL is a direct write (or maskless) method where a digital representation of a pattern is scanned across a resist-coated substrate with high precision but with lower processing speed compared to photolithography. The most significant advantage of EBL is its extremely high resolution,

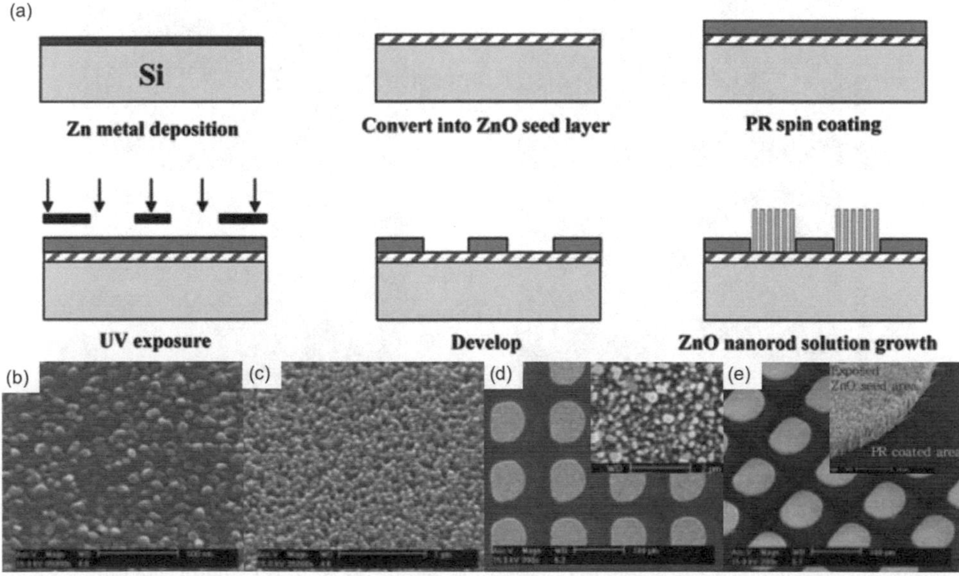

FIGURE 20.7 (a) Schematic diagram showing the photolithography process for the controlled growth of ZnO NW arrays. SEM images showing (b) Zn metal deposition (c) converted into ZnO seed layer and (d) top view of a ZnO nanorod array; inset higher magnification of a patterned NW section (e) angle view of a ZnO nanorod array; inset higher magnification showing the interface between the PR and the ZnO growth. (Reproduced with permission from Tak and Yong 2005. Copyright 2005 American Chemical Society.)

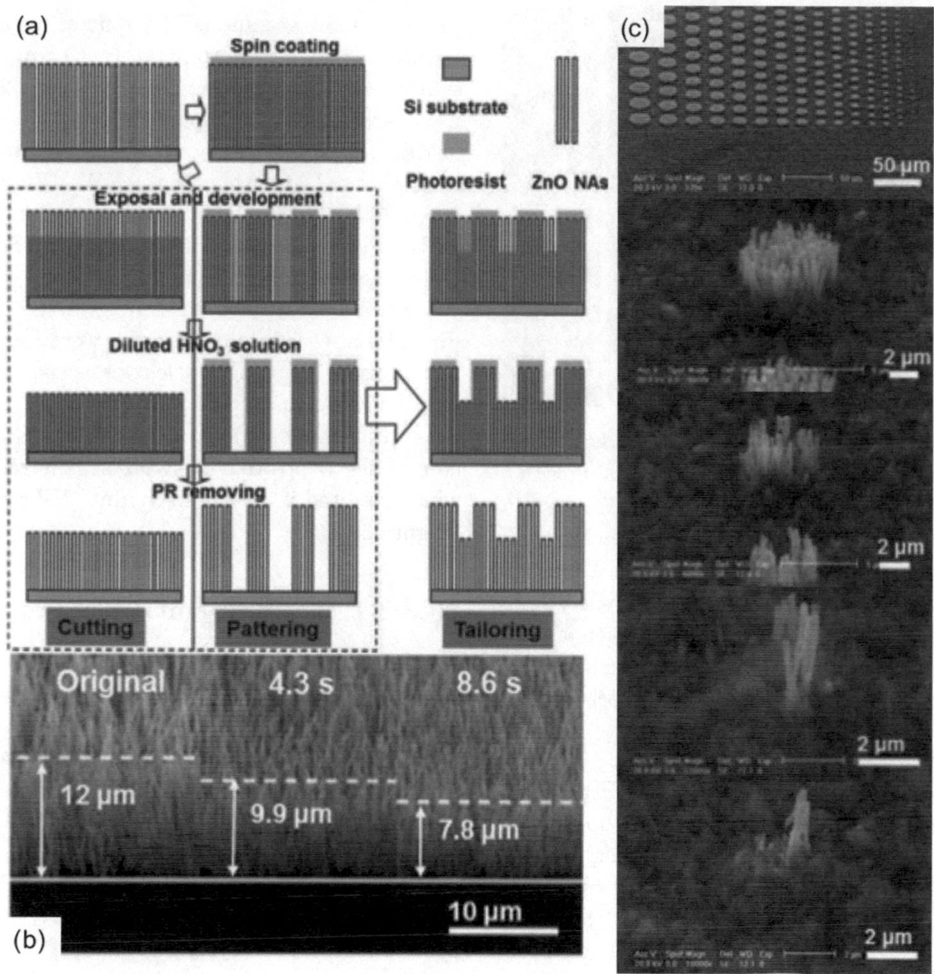

FIGURE 20.8 (a) Schematic of the fabrication process for cutting, pattering, and tailoring NW arrays using photolithography. (b) SEM images showing various UV exposure times and corresponding NW lengths. (c) SEM images showing a series of circular patterns of NWs ranging from 1 to 20 μm in diameter. The circular shape becomes less defined as the diameter decreases. (Reproduced with permission from Shi et al. 2016. Copyright 2016 Nature Publishing Group.)

which can reach as small a feature size as the 5 nm level, and the technique is commonly used to pattern masks for other lithographic processes (van Assenbergh et al. 2018). Xu et al. used EBL to pattern various sizes of lateral tracks with different orientations to the <0001> direction of the ZnO $[2\bar{1}\bar{1}0]$ substrate for subsequent hydrothermal decomposition of epitaxial ZnO NWs (Figure 20.9a–c) (Xu et al. 2009). The substrate was coated with PMMA as the PR with the addition of ESPACER, a conductive polymer, to avoid any artifacts from charging. The substrate was irradiated by the electron beam, washed with water (to remove ESPACER), and developed in a 1:3 mixture (by volume) of isopropyl alcohol and methyl isobutyl ketone to reveal the patterned PMMA. A combination of the confined pattern and growth parameters (concentration, temperature, and time) is used to tailor the dimensionality of the NWs. Once the growth exceeds the patterned PR array opening, there is significant lateral expansion; the width of the NW (800 nm) is roughly double the width of the patterned PR (400 nm) and the length is approximately four times the size. Ng et al. used controlled dwell time and set spacing between dwell points in order to increase the throughput over conventional EBL, which defines individual shapes from a set pattern often using several polygons to make a circle (Ng et al. 2018). Moreover, the diameter of the hole left in the resist is proportional to the dwell time, and utilizing a low beam dose means the uniformity is not compromised. A pattern of Au NPs was fabricated by exposing a PMMA film to different dwell times, evaporation of Au, and removal of the PR (Figure 20.10a and b).

20.3.3 Ion Beam Lithography (IBL)

There are numerous ways that ions can be beneficial for lithographic techniques and are broadly classified into three categories: focused ion beam (FIB), proton beam writing (*p*-beam writing), and ion projection lithography (IPL). The latter, IPL, is similar to traditional optical lithography utilizing conventional stencil masks for large-scale production, but instead of using photons for exposure, ions with

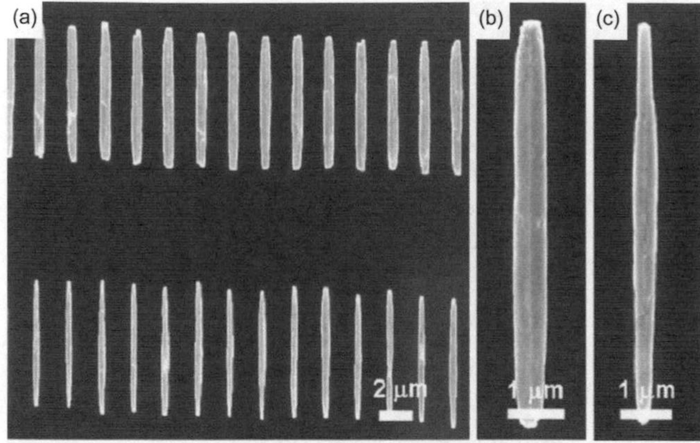

FIGURE 20.9 EBL was used to pattern lateral tracks in a ZnO substrate for low-temperature (<100°C) hydrothermal growth of ordered ZnO NWs. SEM images of horizontal ZnO NW arrays; (a) 2 µm by 400 nm (top) and 2 µm by 200 nm (bottom) etched tracks containing, 800 nm (top) and 400 nm (bottom) ZnO NWs. Length approximately 7.5 µm in both cases. The increase in width is due to lateral expansion during growth. Higher magnification of larger (b) and smaller (c) filled tracks. (Reproduced with permission from Xu et al. 2009. Copyright 2009 American Chemical Society.)

an energy of 100 keV are generally used. IBL (FIB and *p*-beam writing) is a direct write (or maskless) method. Moreover, the ion beam reduces the effects of scattering common to EBL due to the relatively much larger (with respect to electrons) ionic mass. Slow, heavy ions (e.g. Ga^+) with an energy of 30 keV and fast, light ions (e.g. H^+) with an energy of 1 MeV are used for FIB and *p*-beam writing, respectively. The smaller ions used in *p*-beam writing are able to penetrate deeper into the sample making this method faster than FIB. The FIB technology is highly commercialized due to its many capabilities: resist exposure, ion milling, ion-assisted etching, ion-induced deposition, and ion implantation and is therefore more common. The general resolution of IBL (not including IPL) is approximately 10 nm depending on the sample material (van Assenbergh et al. 2018). Zhou et al. used *p*-beam writing to pattern a PMMA layer on a gallium nitride (GaN)/sapphire substrate (Figure 20.11a) for the hydrothermal growth of ZnO NWs (Figure 20.11b) (Zhou et al. 2008). The *p*-beam writing method was chosen over EBL and FIB because of the deeper/straighter trench and the higher efficiency, respectively. Nam et al. used FIB to pattern a Pt catalyst for NW growth (Nam et al. 2005). First, features 1 µm in diameter were formed, but the large area caused very high nucleation density of NWs where Pt was deposited; no NWs were grown in the gallium milled holes (Figure 20.11c). To reduce the NW density, the diameter of the deposited Pt catalyst was reduced to 250 nm, which resulted in nearly individual NWs per nucleation site (Figure 20.11d).

20.3.4 Nanoimprint Lithography (NIL)

Compared to conventional lithography processes, the main difference with imprint lithography is that the chemical structure of the resist is not changed in the procedure. Instead, the resist is physically deformed by the forced pressure of a mold on the surface of the resist layer. The patterned resist can perform as the final product, simply utilizing the physical change, or it can be used as a pattern transfer for further etching procedures and eventual removal of the resist. This latter process has two steps: (i) An imprint resist such as polystyrene (PS) or PMMA is spin coated onto a cleaned substrate. A patterned mold is pressed on top of the resist-coated substrate and heated to soften the polymer resist. The sample is cooled, the mold is removed, and the pattern remains in the resist-coated substrate. (ii) A pattern transfer step is performed, where an etching process is used to remove the resist in the areas that were physically

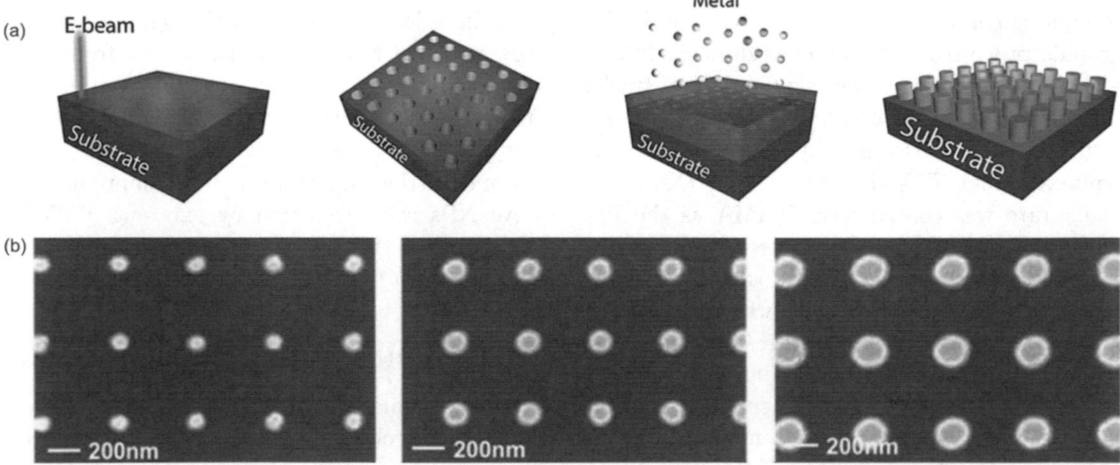

FIGURE 20.10 (a) The PMMA film was exposed to the electron beam and the PR was developed leaving holes defined by the dwell time. Evaporation of Au and removal of the PR leave a patterned array of Au NPs on the surface of the substrate. (b) SEM images of the Au NP array on Si substrate with a set distance of 500 nm between NPs. The dot sizes were 110 nm, 170 nm, and 210 nm. The dwell time ranged from 0.2 to 0.8 ms. (Reproduced with permission from Ng et al. 2018. Copyright 2018 Nature Publishing Group.)

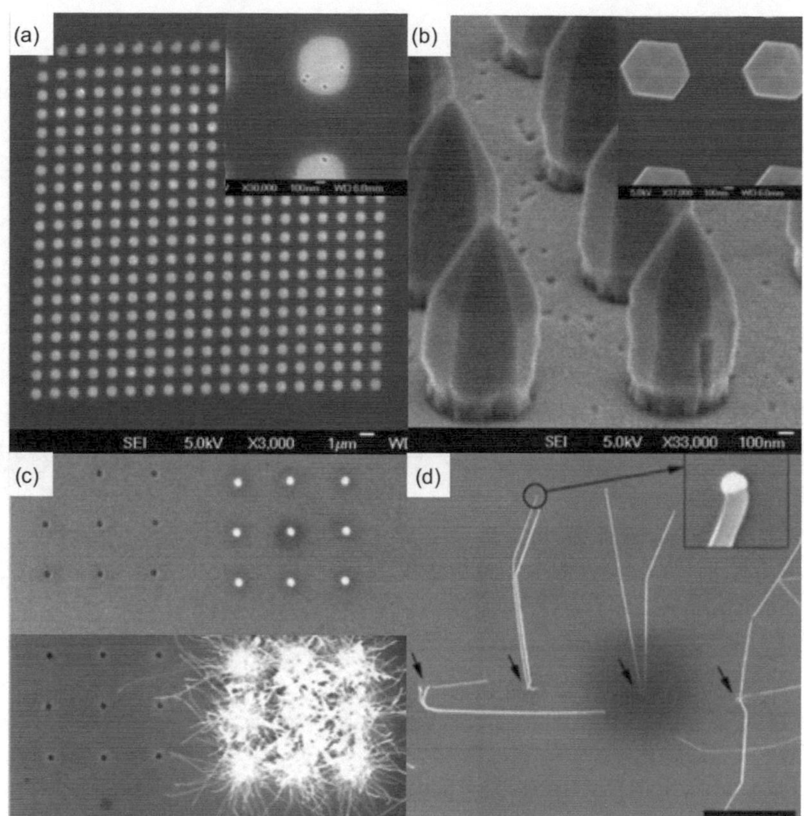

FIGURE 20.11 IBL was used to make ordered NW arrays. SEM images showing (a) templated PMMA array by *p*-beam writing; inset higher magnification. (b) Side view of ZnO arrays; inset top view. (Reproduced with permission from Zhou et al. 2008. Copyright 2008 American Chemical Society.) (c) A 1 μm diameter FIB milled hole in silicon with 200 nm thermal oxide (top left), Pt deposited with depth 250 nm (top right), no NW growth without Pt (bottom left), and Pt-catalyzed GaN NW growth (bottom right). (d) Pt diameter decreased; inset shows clear Pt tip. (Reproduced with permission from Nam et al. 2005. Copyright 2005 AIP Publishing.)

altered by compression. Since the resolution of patterns fabricated using NIL depends on the mold used, and the mold is created (typically from quartz) using EBL, the limit of feature size for NIL is typically 30 nm, but with the inclusion of UV-curable PR, the resolution can be improved to 10 nm (Schift 2008). Moreover, there are several derivatives of NIL such as UV-NIL, which involves light to cure the PR but still relies on direct contact of the mold. Jung and Lee used NIL with a UV-curable resist to pattern ZnO nanorods (Figure 20.12a) (Jung and Lee 2011). The UV-cured resists are more difficult to remove so a double layer resist technique is often utilized, and in this case, PMMA was used as the planar layer. The resist was cured by exposure to UV light, and reactive ion etching (RIE) was used to remove the PMMA to reveal the surface of the substrate in the compressed areas. On the substrate surface, a patterned 1-octadecyltrichlorosilane (OTS) self-assembled monolayer (SAM) was formed, the PR was removed by acetone, and a ZnO seed layer was spin coated on the surface. As shown in Figure 20.12b and d, the ZnO seed layer only deposited on the hydrophilic surface of the Si substrate that was revealed by the removal of the PR and not the hydrophobic areas patterned with the OTS. An oxygen plasma treatment was used to further increase the polarity of the Si surface and remove the OTS layer. The ZnO nanorods were grown hydrothermally from the patterned seed layer, as shown in Figure 20.12c and e. Mårtensson et al. (2004) used the NIL process to deposit Au NPs for NW growth by VLS. Figure 20.12f shows an image of the stamp made by EBL that was used for the NIL process. The process achieved a high throughput of uniform arrays with one NW grown per Au catalyst (Figure 20.12g and h). In this example, indium phosphide (InP) NWs with a diameter of 300 nm, a spacing of 1 μm, and hexagonal pattern were specifically chosen to study photonic crystals, but the technique could be modified to deposit smaller NPs and grow metal oxide NWs.

20.3.5 Nanosphere Lithography (NSL)

Nanosphere lithography (NSL) or colloidal lithography is a simple approach to replace the mask – whose creation can pose technical challenges – with an array of self-assembled spheres resulting in a regular pattern on the surface of a substrate. A layer of nanospheres (NS) is spin coated on the substrate, and these spheres self-assemble to form a close-packed layer. The spheres are typically made from silica or PS and range in diameter from 10 nm to several micrometers. The close-packed arrangement of

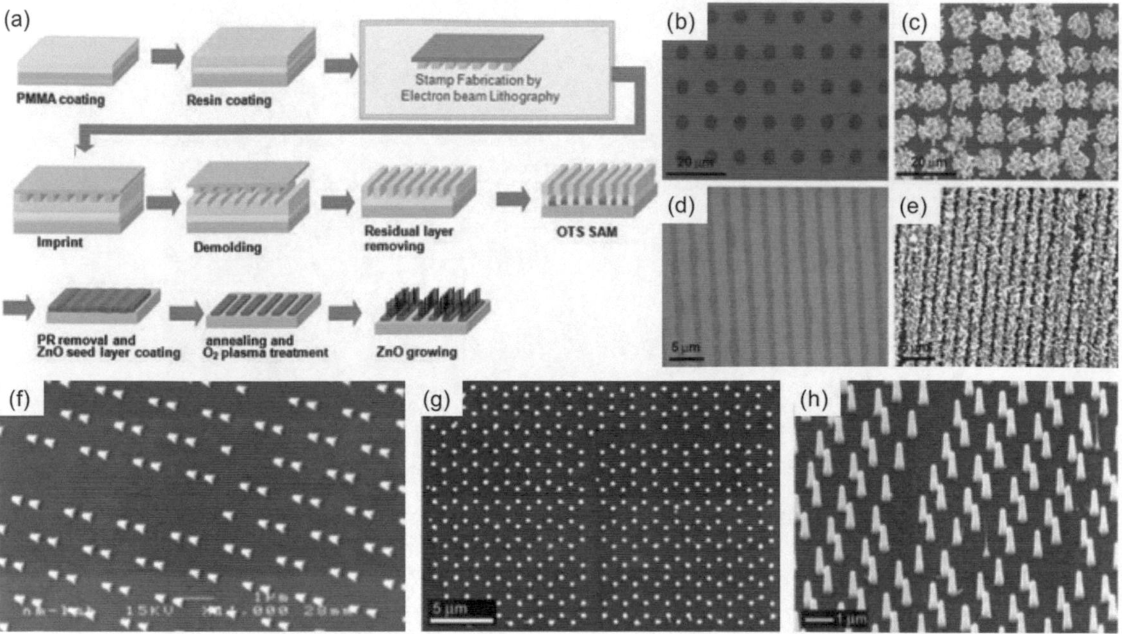

FIGURE 20.12 (a) Schematic diagram of a patterning process by NIL to form ZnO nanorod arrays. SEM images of (b and d) seed layer patterned with dots or lines, respectively. (c and e) Growth of ZnO nanorods from the patterned seed layer. (Reproduced with permission from Jung and Lee 2011. Copyright 2011 Springer.) SEM images of (f) the Si stamp made by EBL showing the hexagonal pattern with a diameter of 200 nm and height of 300 nm. Missing pattern array was intentional. (g) Top view of InP NW arrays with 290 nm diameter and (h) angled view (45°) showing a few inconsistent non-catalyzed NWs. (Reproduced with permission from Mårtensson et al. 2004. Copyright 2004 American Chemical Society.)

spheres leaves triangular gaps of equal spacing and size, of about 1/5th the area of the sphere diameter. A metal, which acts as a catalyst for NW growth, is coated over the SAM of spheres. Once the spheres are removed, the deposited metal remains only in these triangular-shaped areas with equal spacing, size, and depth. The SAM of spheres can also be used directly as a mask for NSL. NSL was used to synthesize ZnO NW arrays by, first, patterning a metal catalyst material and, second, using VLS growth to achieve an individual NW per site by Fan et al. (2006). Figure 20.13a shows the synthetic process where a mask transfer method was utilized to overcome the issue of using a hydrophobic substrate (GaN), which makes the direct adhesion of spheres more difficult. An initial Au film deposition helps to stabilize the SAM of PS spheres for transfer and further reduces the diameter of the Au NP arrays by reducing the gap between the spheres before the second Au deposition (Figure 20.13b). Figure 20.13c and d shows ZnO NWs grown by VLS in which NSL is used to pattern ZnO NW arrays on fluorine-doped tin oxide (FTO)-coated glass substrates by two routes: template growth (TG) and templated seeding (TS) (Colson et al. 2012). In the TG method, a ZnO seed layer is deposited followed by the addition of the NS, whereas in the TS method, the NS are deposited before the deposition of the seed layer and the NS are removed (Figure 20.13e). In both cases, the ZnO NWs were grown hydrothermally either between the NS (Figure 20.13f), which are subsequently removed after growth (Figure 20.13g) or only on the templated areas of the seed layer (Figure 20.13h and i).

20.3.6 Scanning Probe Lithography (SPL)

Scanning probe microscopes (SPMs) employ an atomically sharp tip, which is scanned across a substrate surface with a resolution in the 10 nm range. Tip–sample interactions can be used to gather information about the physical properties (electrical, optical, mechanical, etc.) of the sample. Since SPM techniques do not use light to directly image the surface and do not require expensive beam optics, SPM techniques are not diffraction limited. There are two main SPM systems: the scanning tunneling microscope (STM) and AFM. In an STM, a conducting tip is used to measure the tunneling current of electrons between the surface of a sample to the tip. In the constant current mode, feedback control is used to continuously adjust the Z position so as to maintain a constant tunneling current. Thus, the height of the tip as it is rastered over the substrate surface produces a topographic map and point spectroscopy may be used to determine the band structure of the sample in any position. Though the resolution of STM is excellent, this technique is limited to conducting samples. In AFM, a piezoelectric cantilever is employed, and the sample–tip force curve is measured. The deflection of the tip is measured by a laser spot reflected from the top of the cantilever into a camera. There are three main modes of operation: contact, noncontact, and tapping that can be used to study the topology and mechanical and physical properties of various materials, depending on the characteristics of the tip. Scanning probes can be used to make NW arrays in two main ways:

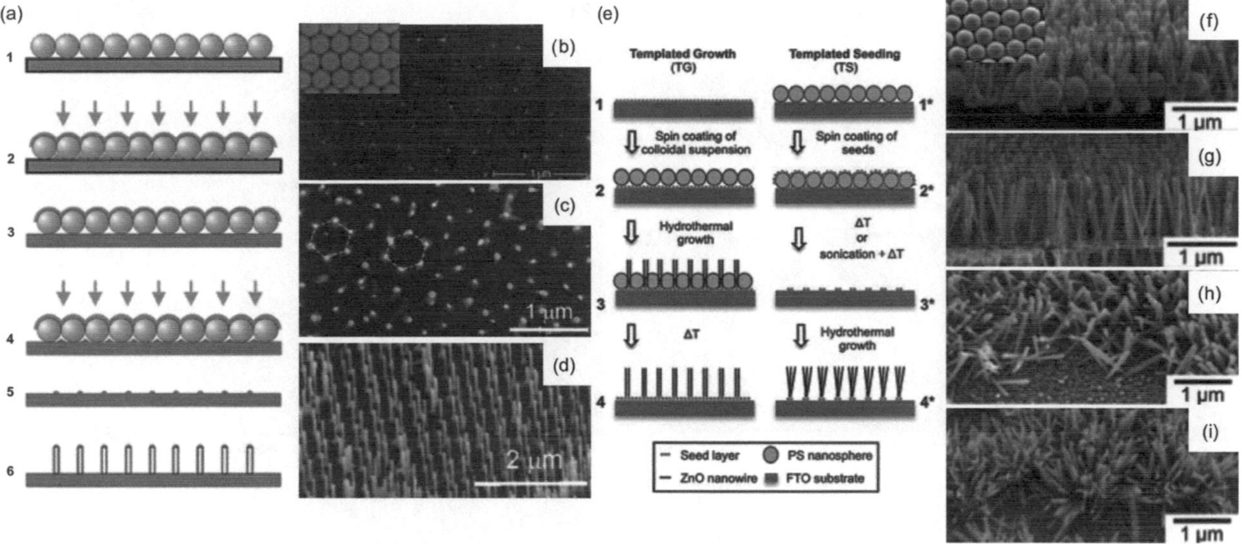

FIGURE 20.13 (a) Schematic diagram of the mask transfer NSL process. (1) NS deposition on glass, (2) Au film deposition for reinforcement, (3) transfer to a substrate, (4) Au film deposition for NP formation, (5) removal of PS spheres revealing patterned Au NPs, and (6) VLS growth ZnO NWs with an average diameter of 63 nm. SEM images showing the (b) deposition of the Au NPs; inset shows the initial NS monolayer, (c) top view image of the ZnO NW growth highlighting the hexagonal pattern formed by the close packing NPs, (d) side view of the free-standing NWs. (Reproduced with permission from Fan et al. 2006. Copyright 2006 Elsevier.) (e) Schematic diagram showing two methods of synthesizing ZnO nanorod arrays by NSL. (f–i) SEM images showing (f) the TG method after growth and before NS removal; inset monolayer of 490 nm diameter polystyrene spheres. (g) The TG method after NS removal. (h) The TS method with the removal of NS by calcination (i) the TS method with the removal of NS by sonication. (Reproduced with permission from Colson et al. 2012. Copyright 2012 Royal Society of Chemistry; permission conveyed through the Copyright Clearance Center, Inc.)

(i) The direct deposition of material by the tip. (ii) Using the tip to first form an etched template array for subsequent NW growth. Using the first method, Song et al. fabricated an array of gold dots using a gold-modified AFM tip, by voltage pulses across the tip–substrate gap (Song et al. 1998). In the latter approach, He et al. (2006) used AFM nanomachining to make a patterned arrangement of a metal catalyst for subsequent ZnO NW growth via the VLS method (Figure 20.14a–j). In this case, an AFM tip was used to remove polymer PR in a specific design, followed by the deposition of catalyst metal by sputtering. The PR was removed, and NWs were grown.

20.3.7 Other Lithographic Techniques

In addition to the most common techniques discussed above, there is an assortment of other lithographic techniques capable of producing NW arrays, and new technique development to improve the current limitations is a highly active research area. Some other common examples, not discussed here, are laser interference lithography (LIL), X-ray lithography (XRL), and extreme ultraviolet lithography (EUV) (Kim et al. 2007, Wei et al. 2010, Maldonado and Peckerar 2016, van Assenbergh et al. 2018). In general, the steps for LIL are similar to photolithography; the difference lies in how the sample undergoes exposure, wherein the PR records an interference pattern by splitting the coherent light into two (or more) beams, resulting in an elongated pathway between the source of light and the sample, consequently improving the resolution. For XRL, the short wavelength of X-rays (0.4–4 nm) with respect to visible light leads to a much smaller diffraction limit than that of photolithography, with a correspondingly smaller feature size. EUV also takes advantage of the smaller diffraction limit due to the short wavelength (13.5 nm), with correspondingly high resolution.

20.4 Post-Synthetic Nanowire Alignment

The previous sections have described various approaches to synthesizing metal oxide NWs such that they grow in an aligned array. An alternate approach to synthesizing NW arrays is to first grow free-standing NWs and subsequently align them into ordered arrays. Several such approaches are described in this section.

20.4.1 Langmuir–Blodgett Assembly

During the late eighteenth and early nineteenth centuries, Benjamin Franklin, Lord Rayleigh, and Agnes Pockels observed and studied the various behaviors of an oil film on water, laying the foundation for the experimental work by Irving Langmuir and Katharine Blodgett, which led to the awarding of the Nobel Prize in Chemistry to Langmuir in 1932 (Gaines 1983, Hussain 2009). The concept of forming a

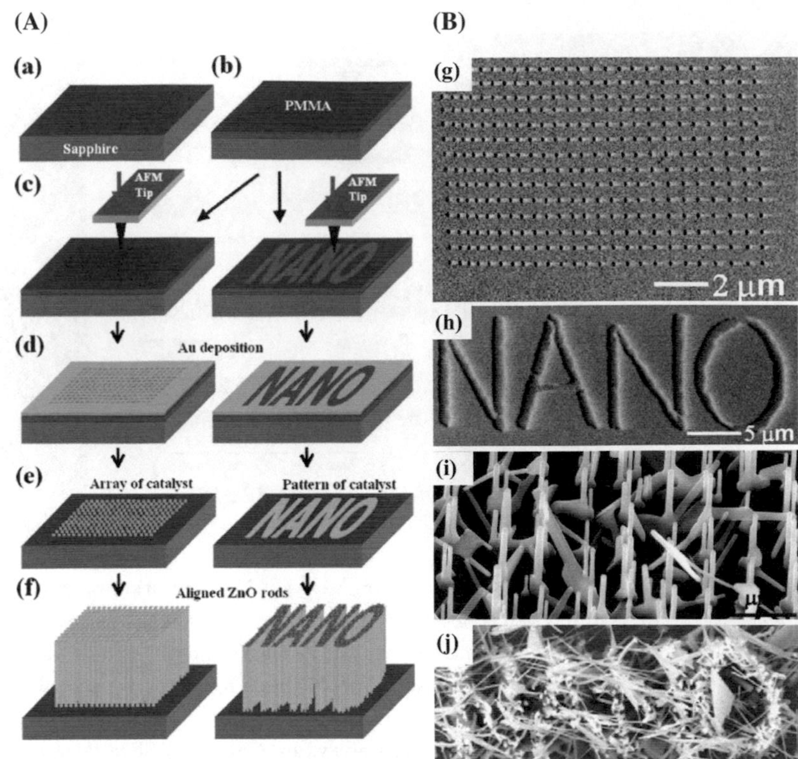

FIGURE 20.14 (A) Schematic diagram of the AFM nanolithography process. (a) A cleaned sapphire substrate, (b) spin-coated PR, (c) AFM tip used to pattern PR, (d) Au sputtered onto the surface, (e) PR is removed, (f) ZnO NWs grown. (B) SEM images showing (g) Au NP array, smallest diameter ∼70 nm, (h) continuous Au nanostructure with a complex pattern, (i) ZnO NWs growth from pattered Au NP array, the image was taken at 25° tilt, (j) ZnO NWs from continuous Au nanostructure. (Reproduced with permission from He et al. 2006. Copyright 2006 American Chemical Society.)

Langmuir monolayer is fairly simple: amphiphilic molecules that contain a hydrophobic "tail" and a hydrophilic "head," arranged in such a way as to minimize the interfacial energy between water and a molecular layer. The Langmuir–Blodgett (LB) assembly process leads to a monolayer of ordered amphiphilic molecules with all their heads in contact with the hydrophilic water. To extend this idea to create arrays of aligned NWs, surfactant-coated NWs are dissolved in a volatile organic solvent and the solution is carefully placed on the water's surface where the solvent evaporates spreading the NWs across the surface. As shown in Figure 20.15a, the trough consists of a movable barrier that can compress or expand the surface film against an electronic microbalance at the opposite end. Due to compression, the surface energy of the liquid is minimized when the NWs align perpendicular to the compression direction. The surface tension is directly proportional to the measurable force exerted by compression of the film. A graph of the surface pressure versus area per molecule known as an isotherm is used to identify the phase of the monolayer as it undergoes compression at a constant temperature. Similar to the three-dimensional phase changes of an ideal gas, the two-dimensional film undergoes phase changes from gas, to liquid, to solid, and if compressed past the solid state, the film buckles and the surface pressure abruptly decrease. The aligned NW film is deposited on a solid substrate by vertical or horizontal dipping of the substrate followed by removal at a precise rate, with spacing between wires controlled by the compression pressure and/or lifting speed of the substrate.

Mai et al. demonstrated this method using hydrothermally prepared vanadium (IV) oxide (VO_2) NWs with diameters of 30–60 nm, functionalized with stearic acid (SA) and cetyltrimethylammonium bromide (CTAB) to prevent aggregation and improve the quality of the NW film (Mai et al. 2009). They demonstrated that the LB assembly technique using functionalized NWs only showed $(00l)$ crystal planes in the XRD pattern of the films (Figure 20.15b – top) compared to the XRD of synthesized VO_2 NWs (Figure 20.15b – bottom). This indicated that the films had a precise orientation, which they attributed to the SA and CTAB preferentially coordinating to the (001) surface. Shown in Figure 20.15c, Huang et al. used the LB method to deposit lines of metallic NPs for the growth of NWs with controlled density and placement (Huang et al. 2006). Though this approach was applied to the growth of Si NW arrays, this would be an intriguing avenue to explore for the VLS growth of metal oxide NW arrays.

20.4.2 Blown Bubble Films

The blown bubble method is primarily used in the plastic industry to make highly efficient films for consumables such

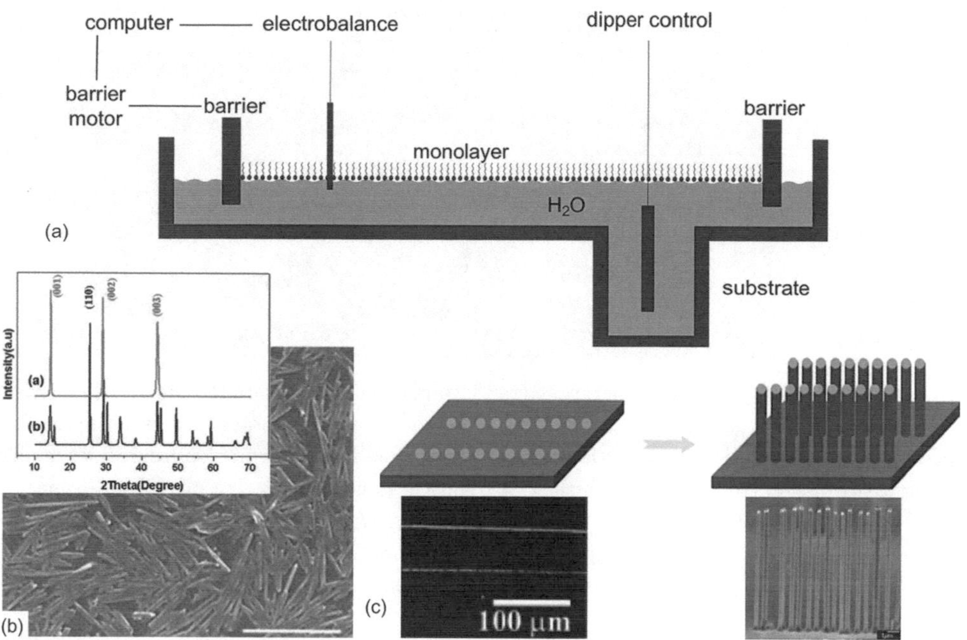

FIGURE 20.15 (a) Diagram of LB trough. (b) SEM image of VO_2 NWs with 40 mN/m surface pressure; inset XRD showing LB-assembled film (top) and the not LB-assembled VO_2 NWs (bottom). (Reproduced with permission from Mai et al. 2009. Copyright 2009 American Chemical Society.) (c) Schematic diagram showing Au NP array and NW formation with corresponding SEM images. (Reproduced with permission from Huang 2006. Copyright 2006 American Chemical Society.)

as its wide use for shopping bag manufacturing (Yu et al. 2008). The blown bubble process as applied to NW alignment occurs in several steps. First, the NWs are chemically functionalized to inhibit aggregation and are added to a polymer solution using a precise concentration of NWs. A crucial step in the process is the formation of a homogenous and stable suspension of the NW-loaded polymer solution with the ideal viscosity (dependent on materials used) to form a good bubble. A bubble is formed on a circular die with a highly controlled pressure and expansion rate (Figure 20.16a). Stabilization of the bubble is controlled by a motor-driven ring, which moves upwards during the expansion process. The film is transferred onto a substrate by contacting the bubble covering the entire wafer, as seen in Figure 20.16b (Yu et al. 2007). The NWs contained in the bubble align due to the shear force from the film expansion; the initial concentration of NWs controls the resultant density of the film. This technique allows for *meter length* arrays on any type of substrate and a wide variety of NWs. The disadvantage to this approach is that the NWs must be chemically functionalized and coated in a polymer matrix, which could inhibit their potential use in electronic devices. Additionally, the closed-packed distribution of NWs is challenging to achieve, because the expansion of the bubble spreads the NWs widely apart. Yu et al. pioneered the use of the blown bubble technique for nanocomposite films (Yu et al. 2007). The NWs were functionalized with 5,6-epoxyhexyltriethoxysilane in tetrahydrofuran (THF) solvent, removed from the substrate via sonication, and mixed with epoxy part A followed by epoxy part B. The viscosity of the suspension was closely monitored as the epoxy cured until the desired 15–25 Pa s was reached. A bubble with an average height of 50 cm and an average diameter of 25 cm was formed and silicon wafers were fixed, and the bubble was extended until the contact coated the whole substrate. Figure 20.16c shows the NW-coated wafer with SEM images from multiple areas indicating the reproducibility over a large area. A common polymer PMMA can be used as an alternative to an epoxy resin for the blown bubble method as shown by Wu et al. (2014). Furthermore, they demonstrated that it was possible to remove the PMMA for better device performance by magnetically clamping a clean wafer to the PMMA NW-coated substrate and suspending it in acetone to dissolve the PMMA.

20.4.3 Contact Printing Methods

Contact printing is a dry deposition strategy that occurs in two steps: first, NWs are synthesized using a nanocluster directed growth (Fan et al. 2008, 2009). Second, the NWs are transferred from the growth substrate to the device substrate that has been patterned by lithography. As seen in Figure 20.17a, the patterned substrate is fixed and the growth substrate containing the NWs is placed on top and slid across. The sliding motion, a shear force, causes the direction of the NWs to be parallel. Lubricants, such as mineral oil, can be used to reduce friction to avoid damage to the NWs during the transfer process. A subset of contact printing, differential roll printing (DRP), has the added benefit of being able to easily print over large areas with a continual replacement of wires contacting the surface.

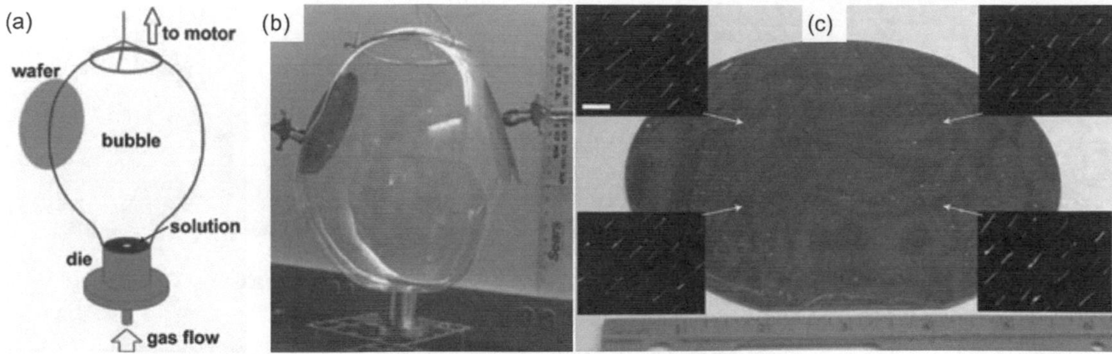

FIGURE 20.16 (a) Bubble formation equipment diagram shows a 50 mm circular die where NW-dissolved epoxy solution is deposited. The bubble expands upward by directed gas flow and a ring is attached to a controlled motor. The wafer(s) are fixed near the bubble and the bubble is expanded until contact is made with the wafer(s). (Reproduced with permission from Yu et al. 2008. Copyright 2008 Royal Society of Chemistry; permission conveyed through the Copyright Clearance Center, Inc.) (b) Optical image shows two 150 mm Si wafers in contact with the blown bubble on a 50 mm circular die. (c) Optical image showing coating of the entire 150 mm Si wafer with Si NW epoxy solution; insets show dark-field optical images of the uniform deposition and alignment of the Si NWs at different locations on the wafer. (Reproduced with permission from Yu et al. 2007. Copyright 2007 Nature Publishing Group.)

First, the NWs are grown on a cylindrical substrate, then the roller is attached to wheels, and the NWs are transferred to the patterned substrate by rolling the tube at a constant velocity (Figure 20.17b). Fan et al. used octane and mineral oil to reduce the friction between the sliding substrates to achieve better control of NW transfer and increase NW density (Figure 20.17c) (Fan et al. 2008). Additionally, the surface of the receiving substrate was modified with various chemical groups to observe the differences in NW density transfer; fluorinated surfaces decrease attachment, and nitrogen groups increase adhesion. Yerushalmi et al. (2007) grew randomly oriented NWs by VLS directly on a cylindrical substrate and deposited the NWs on a new flat substrate via DRP as shown in Figure 20.17d. Additionally, the velocity and pressure of the roller were studied; above a threshold rolling velocity (>20 mm/min) and pressure (>200 g/cm^2), the arrays were inconsistent.

20.4.4 Microfluidic Alignment

Another way to use shear force to help align NWs is to use a fluidic model; here, the NWs align parallel to the fluid flow to minimize the shear force, and the addition of micron-sized channels assists in their alignment. The channels are typically made from a poly(dimethylsiloxane) (PDMS) mold. Increasing the flow rate increases the shear force, causing a higher degree of alignment, and the flow duration may be tailored to optimize NW density. In this setup, the substrate is covered with a PDMS mold forming micron-sized channels. More complex structures are possible by changing the direction of the substrate. A solution of NWs is prepared and introduced to the start of the channel. In a simple approach, Messer et al. used a PDMS mold microchannel to align $[Mo_3Se_3^-]_\infty$ NWs by self-assembly (Messer et al. 2000). A droplet (0.1–10 μL) of $LiMo_3Se$ dissolved in DMSO or N-methylformamide was added to the channel opening and the channel was filled by capillary action (Figure 20.18a). As the solvent evaporates under vacuum, the meniscus withdraws into the corners of the channel providing a solvent convective flow that drives the NWs to the edge. The intermolecular forces involved, specifically the electrostatic interactions between the $[Mo_3Se_3^-]_\infty$ NWs and counter ions (Li), cause the NWs to aggregate and form bundles along the edge of the channel (Figure 20.18b). Huang et al. performed several experiments to identify the specific parameters to alter the alignment and separation of NWs using this approach (Huang et al. 2001). The results indicated that 80% of NWs were within 5° of the flow direction at the highest flow rate (9.40 mm/s). Alignment was further enhanced with the incorporation of specific chemical groups on the receiving substrate to encourage NW attachment. Increasing the duration of flow boosted the density of NWs, and the addition of NH_2 groups to the substrate's surface resulted in very fast NW deposition.

20.4.5 Chemically Driven Assembly

An alternate approach is to use various chemical modifications to the surface to preferentially assign the location of NWs. This relies on weak chemical bonding forces (instead of shear forces) such as hydrogen bonding, van der Waals forces, and electrostatic forces between the NW and the attached chemical group. While it is possible to use this technique as a standalone, it is often incorporated with another post-alignment technique to further enhance alignment. To enhance preferential attachment, biomolecules are used instead of organic molecules due to their explicit specificity. Myung et al. used lithography to pattern an SAM of the positively charged species, aminopropylethoxysilane (APTES), and OTS as the neutral species on Si substrates and the substrates were briefly submerged in an aqueous solution of V_2O_5 NWs (Myung et al. 2005). Due to electrostatic interactions, the negatively charged

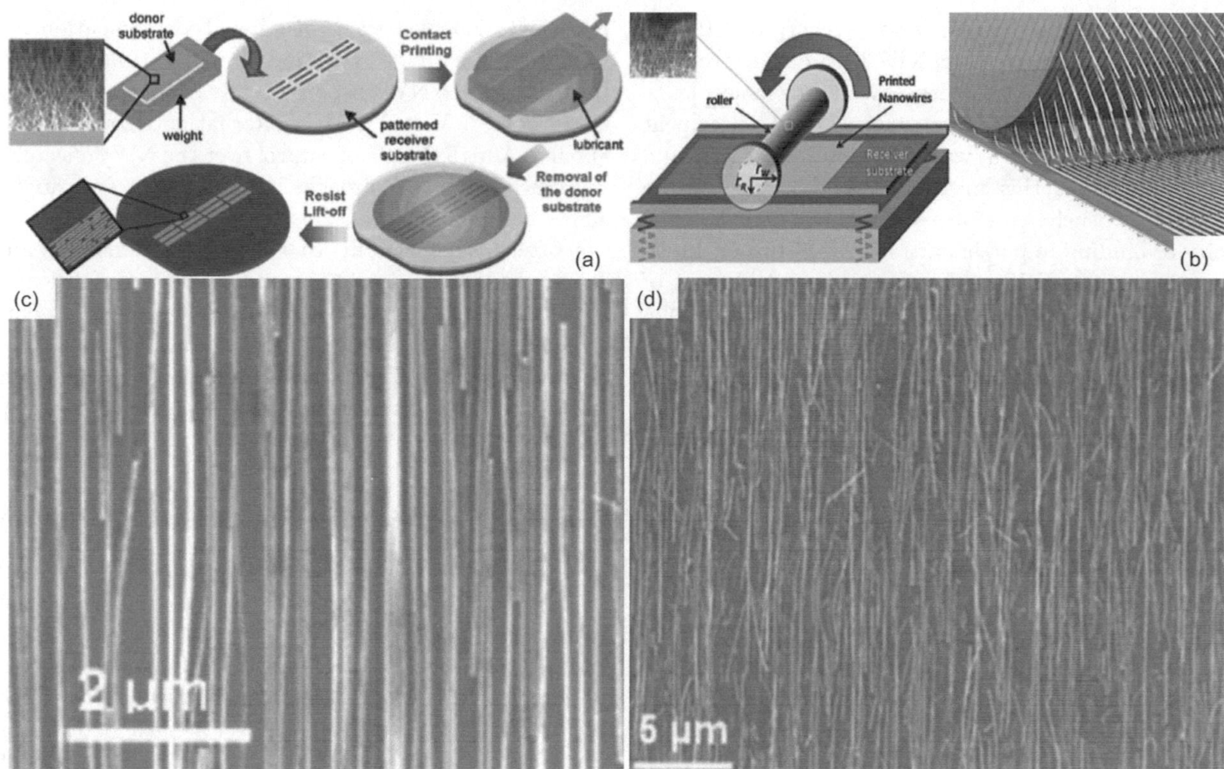

FIGURE 20.17 (a) Schematic diagram of the contact printing procedure. SEM image shows the donor substrate with randomly oriented Ge NWs. Patterned substrate receives NWs. (Reproduced with permission from Fan et al. 2008. Copyright 2008 American Chemical Society.) (b) Schematic diagram of the roll printing process. Cylindrical donor substrate containing randomly oriented Ge NWs deposits NWs which align on the receiver substrate. (Reproduced with permission from Fan et al. 2009. Copyright 2009 John Wiley and Sons.) (c) SEM images of 30 nm diameter NWs arranged in an aligned monolayer by contact printing. (Reproduced with permission from Fan et al. 2008. Copyright 2008 American Chemical Society.) (d) SEM images of 30 nm diameter NWs arranged by roll printing of Ge NWs on a Si substrate showing the dense (~6 NW/µm) parallel array. (Reproduced with permission from Yerushalmi et al. 2007. Copyright 2007 AIP Publishing.)

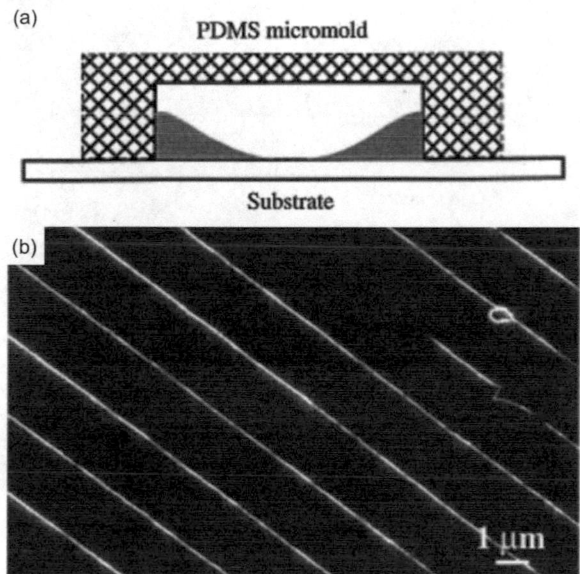

FIGURE 20.18 (a) Schematic diagram showing the liquid retreating to the corners of the channel. The channel has a length of 5–10 mm, a width of 1–10 µm, and a height of 1–4 µm. (b) SEM images of the $[Mo_3Se_3^-]_\infty$ NW arrays on Si substrate. The NWs may range in diameter from 10 to 200 nm and are made of bundles of smaller diameter NWs. The smaller NWs are made by decreasing the solution concentration and reducing microchannel volume. (Reproduced with permission from Messer et al. 2000. Copyright 2000 American Chemical Society.)

V_2O_5 NWs attach only to the positively charged, APTES, regions (Figure 20.19a). Further, NWs with distinctive fragments allow for functionality control by binding molecules to specific regions on the NWs. For example, a three-segment NW of Au/nickel (Ni)/Au was fabricated by Chen and Searson (2005) using this method. Furthermore, the use of biomolecules, such as biotin–avidin, can enhance the specificity of binding in particular regions. In this example, the Ni region of the NW was protected by attachment of palmitic acid to the native oxide layer, followed by thiol group attachment of biotin (or avidin) to the Au segments. A suspension of biotin-functionalized NWs is mixed with avidin-functionalized NWs, and due to the strong nature of the biotin–avidin bond, it is assumed that if a collision occurs between the two linkages, the NWs will align end to end (Figure 20.19b).

20.4.6 Electric Field-Assisted Alignment

Using an electric field to assist in the alignment of NWs is effective due to their inherent anisotropy. Utilizing dielectrophoresis (DEP), in which an external electric field dictates the motion of neutral particles, NWs may be aligned. In this procedure, voltage is applied to contact pads covered by a solution of NWs. The electric field strength required for effective alignment will depend on the dielectric constant of the solvent relative to the NW material; the higher the difference in the dielectric constant, the lower the required electric field strength. Lao et al. synthesized ZnO NWs and dispersed them in ethanol via sonication for the formation of a single NW diode by DEP (Lao et al. 2006). In this case, the NW suspension was dropped onto the electrodes, and 5 V and 1 MHz alternating current (AC) was applied. A high degree of control over the NW concentration was crucial for this approach, in order to achieve a single NW across the electrodes for this study; varying the concentration of such a solution in order to vary the density of aligned arrays would be an intriguing approach to the fabrication of metal oxide NW arrays. A disadvantage to this technique is that DEP may produce sufficient heat to cause electromigration of the electrodes, leading to device asymmetry and unpredictable function (Lao et al. 2006, Paunovic and Schlesinger 2006).

20.4.7 Magnetic Field-Assisted Alignment

Similar to electric fields, magnetic fields may also be used to align NWs in solution, ideally for ferromagnetic or superparamagnetic NWs. In order to minimize energy, the NWs interact with the external magnetic field and each other to assemble with a head-to-tail alignment. The degree of alignment is proportional to the strength of the applied magnetic field; higher fields lead to better alignment. In addition to using an external magnetic field, ferromagnetic electrodes can be added to the substrate forming a localized dipolar magnetic field enhancing the degree of alignment. Tanase et al. (2001) synthesized Ni NWs and functionalized them with highly fluorescent porphyrins to study

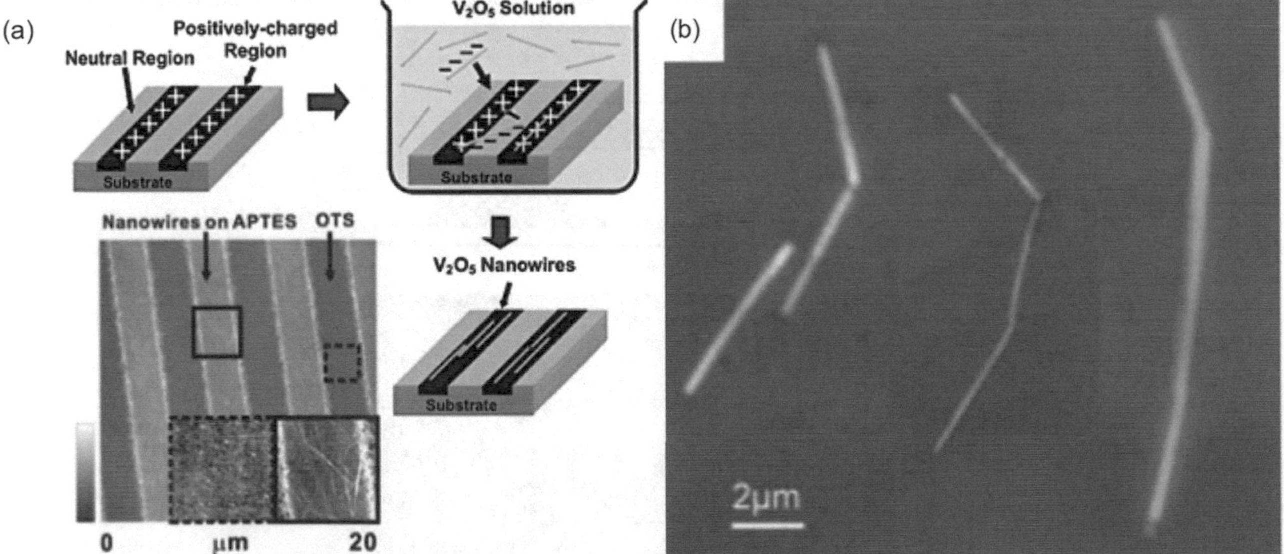

FIGURE 20.19 Chemically driven assembly using organic molecules (a) and biomolecules (b). (a) The schematic diagram shows the patterning of the substrate with positively charged and neutral groups. The assembly of NWs is directed by patterning of the surface with different types of molecular species. After rinsing with water, NWs are aligned. Corresponding AFM image showing V_2O_5 NWs only on the positively charged APTES region (solid box) and no NWs on neutral OTS region (dashed box). (Reproduced with permission from Myung et al. 2005. Copyright 2005 John Wiley and Sons.) (b) A chemically driven assembly utilizing biomolecules for enhanced preferential alignment. A light microscopy image showing Au/Ni/Au NWs attached end to end by functionalizing only the gold ends with biotin (or avidin) by preventing attachment to the Ni segment using palmitic acid, which selectively binds to the native oxide layer on nickel. (Reproduced with permission from Chen and Searson. Copyright 2005 John Wiley and Sons.)

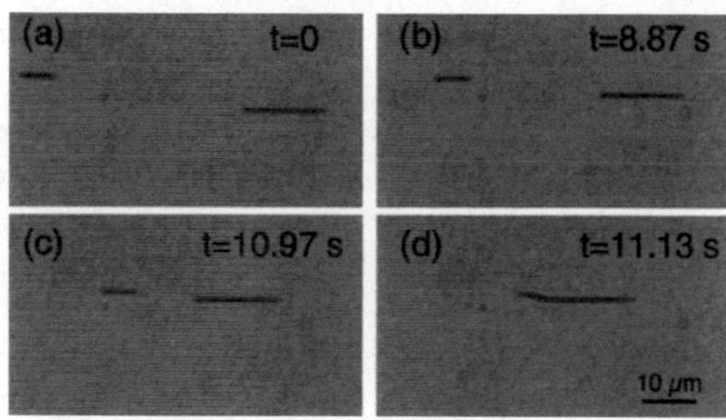

FIGURE 20.20 (a–d) Microscopy images from a video showing two NWs suspended in 35°C water attaching end to end as time is increased from 0 to 11.13 s due to the application of a small magnetic field (1 G). These wires align parallel to the magnetic field forming long chains (>100 μm) over time. (Reproduced with permission from Tanase et al. 2001. Copyright 2001 American Chemical Society.)

the assembly of these wires into arrays by magnetic fields. The results showed in Figure 20.20a–d that the NWs aligned end to end along the magnetic field due to Ni NWs having a large remnant magnetization and high aspect ratio and high viscosity solvents slowed the movement of the NWs.

20.5 Conclusion

In general, template-based methods allow for the production of well-aligned NWs of any material that can be chemically/physically deposited and have the capability of easily producing heterojunctions with different materials. Drawbacks to template-based approaches are that they are not generally able to produce complex patterns, that removing the template can damage the NWs and, in most cases, that the NWs grown within templates are polycrystalline.

Lithography involves transferring a pattern from one form onto another with the help of a resist that is chemically/physically changed by a chemical/physical means. In most cases, lithography is diffraction limited by the wavelength of light used, limiting the minimum feature size possible. Decreasing the wavelength by using electron, ion, or powerful excimer lasers over conventional UV light improves the resolution but also increases the cost associated with manufacturing. For NW arrays, lithography is commonly used to pattern NPs onto a surface or pattern holes to direct the growth of NWs to certain areas on a substrate. The formation of highly aligned NWs with few defects typically occurs at high temperatures and is limited to heteroepitaxial materials. Additionally, the use of PRs, developers, and etchants can damage the NWs.

In post-synthetic NW alignment, the NWs are first grown by an assortment of processes and subsequently aligned by a variety of methods. These methods utilize relatively simple and inexpensive procedures and in the case of blown bubble films are able to cover macroscopic length scales. Drawbacks to aligning NWs post-synthesis include limited accuracy to control the alignment and spacing (with respect to template-based or lithographic approaches). Moreover, for a majority of post-alignment techniques, the chemistry of the NWs is typically altered in some way, e.g. by dissolving them in a polymer solution, via chemical functionalization, or by introducing defects during mechanical manipulation.

In conclusion, there are a vast assortment of techniques to align NWs in specific orientations, and this is an exciting and currently active area of nanomaterials research. In a recent review regarding nanostructure fabrication by Assenbergh et al., the "performance-based pathway" is discussed, an approach in which the specific application or structure is first defined and is used to determine the optimal fabrication approach(es); such strategies will increasingly come to replace the previous processing-based approaches, in which the fabrication method determines the properties of the structure (van Assenbergh et al. 2018).

References

Besra, L. and Liu, M., 2007. A review on fundamentals and applications of electrophoretic deposition (EPD). *Progress in Materials Science*, 52 (1), 1–61.

Beyond the diffraction limit, 2009. *Nature Photonics*, 3, 361.

Boccaccini, A.R. and Zhitomirsky, I., 2002. Application of electrophoretic and electrolytic deposition techniques in ceramics processing. *Current Opinion in Solid State and Materials Science*, 6 (3), 251–260.

Cao, G., 2004. Growth of oxide nanorod arrays through sol electrophoretic deposition. *The Journal of Physical Chemistry B*, 108 (52), 19921–19931.

Cao, G. and Liu, D., 2008. Template-based synthesis of nanorod, nanowire, and nanotube arrays. *Advances in Colloid and Interface Science*, 136 (1–2), 45–64.

Chen, M. and Searson, P.C., 2005. The dynamics of nanowire self-assembly. *Advanced Materials*, 17 (22), 2765–2768.

Colson, P., Schrijnemakers, A., Vertruyen, B., Henrist C., and Cloots, R., 2012. Nanosphere lithography and hydrothermal growth: How to increase the surface area and control reversible wetting properties of ZnO nanowire arrays? *Journal of Materials Chemistry*, 22 (33), 17086.

Demazeau, G. and Largeteau, A., 2015. Hydrothermal/solvothermal crystal growth: An old but adaptable process: Hydrothermal/solvothermal crystal growth. *Zeitschrift für anorganische und allgemeine Chemie*, 641 (2), 159–163.

Fan, H.J., Fuhrmann, B., Scholz, R., Syrowatka, F., Dadgar, A., Krost, A., and Zacharias, M., 2006. Well-ordered ZnO nanowire arrays on GaN substrate fabricated via nanosphere lithography. *Journal of Crystal Growth*, 287 (1), 34–38.

Fan, Z., Ho, J.C., Jacobson, Z.A., Yerushalmi, R., Alley, R.L., Razavi, H., and Javey, A., 2008. Wafer-scale assembly of highly ordered semiconductor nanowire arrays by contact printing. *Nano Letters*, 8 (1), 20–25.

Fan, Z., Ho, J.C., Takahashi, T., Yerushalmi, R., Takei, K., Ford, A.C., Chueh, Y.-L., and Javey, A., 2009. Toward the development of printable nanowire electronics and sensors. *Advanced Materials*, 21 (37), 3730–3743.

Gaines, G.L., 1983. On the history of Langmuir-Blodgett films. *In*: *Langmuir–Blodgett Films, 1982*. Elsevier, New York, viii–xiii.

Gómez, H., Cantillana, S., Cataño, F.A., Altamirano, H., and Burgos, A., 2014. Template assisted electrodeposition of highly oriented ZnO nanowire arrays and their integration in dye sensitized solar cells. *Journal of the Chilean Chemical Society*, 59 (2), 2447–2450.

He, J.H., Hsu, J.H., Wang, C.W., Lin, H.N., Chen, L.J., and Wang, Z.L., 2006. Pattern and feature designed growth of ZnO nanowire arrays for vertical devices. *The Journal of Physical Chemistry B*, 110 (1), 50–53.

Huang, J., Tao, A.R., Connor, S., He, R., and Yang, P., 2006. A general method for assembling single colloidal particle lines. *Nano Letters*, 6 (3), 524–529.

Huang, Y., Duan, X., Wei, Q., and Lieber, C.M., 2001. Directed assembly of one-dimensional nanostructures into functional networks. *Science*, 291 (5504), 630–633.

Hussain, S.A., 2009. Langmuir-Blodgett Films a unique tool for molecular electronics. *Modern Physics Letters B* 23 (27), 9.

Jessensky, O., Müller, F., and Gösele, U., 1998. Self-organized formation of hexagonal pore arrays in anodic alumina. *Applied Physics Letters*, 72 (10), 1173–1175.

Jung, M.-H. and Lee, H., 2011. Selective patterning of ZnO nanorods on silicon substrates using nanoimprint lithography. *Nanoscale Research Letters*, 6 (1), 159.

Kayes, B.M., Filler, M.A., Putnam, M.C., Kelzenberg, M.D., Lewis, N.S., and Atwater, H.A., 2007. Growth of vertically aligned Si wire arrays over large areas (>1 cm^2) with Au and Cu catalysts. *Applied Physics Letters*, 91 (10), 103110.

Keller, F., Hunter, M.S., and Robinson, D.L., 1953. Structural features of oxide coatings on aluminum. *Journal of The Electrochemical Society*, 100 (9), 411.

Kim, D.S., Ji, R., Fan, H.J., Bertram, F., Scholz, R., Dadgar, A., Nielsch, K., Krost, A., Christen, J., Gösele, U., and Zacharias, M., 2007. Laser-interference lithography tailored for highly symmetrically arranged ZnO nanowire arrays. *Small*, 3 (1), 76–80.

Kolmakov, A., Zhang, Y., and Moskovits, M., 2003. Topotactic thermal oxidation of Sn nanowires: Intermediate suboxides and core-shell metastable structures. *Nano Letters*, 3 (8), 1125–1129.

Lao, C.S., Liu, J., Gao, P., Zhang, L., Davidovic, D., Tummala, R., and Wang, Z.L., 2006. ZnO Nanobelt/nanowire schottky diodes formed by dielectrophoresis alignment across Au electrodes. *Nano Letters*, 6 (2), 263–266.

Lee, W. and Park, S.-J., 2014. Porous anodic aluminum oxide: Anodization and templated synthesis of functional nanostructures. *Chemical Reviews*, 114 (15), 7487–7556.

Li, G., Sundararajan, A., Mouti, A., Chang, Y.-J., Lupini, A.R., Pennycook, S.J., Strachan, D.R., and Guiton, B.S., 2013. Synthesis and characterization of p–n homojunction-containing zinc oxide nanowires. *Nanoscale*, 5 (6), 2259.

Mai, L., Gu, Y., Han, C., Hu, B., Chen, W., Zhang, P., Xu, L., Guo, W., and Dai, Y., 2009. Orientated Langmuir–Blodgett assembly of VO_2 nanowires. *Nano Letters*, 9 (2), 826–830.

Maldonado, J.R. and Peckerar, M., 2016. X-ray lithography: Some history, current status and future prospects. *Microelectronic Engineering*, 161, 87–93.

Mankotia, D., Sharma, D.Y.C., and Sharma, D.S.K., 2014. Review of highly ordered anodic porous alumina membrane development. *International Journal of Recent Research Aspects* 1 (2), 6.

Mårtensson, T., Carlberg, P., Borgström, M., Montelius, L., Seifert, W., and Samuelson, L., 2004. Nanowire arrays defined by nanoimprint Lithography. *Nano Letters*, 4 (4), 699–702.

Masuda, H. and Fukuda, K., 1995. Ordered metal nanohole arrays made by a two-step replication of honeycomb structures of anodic alumina. *Science*, 268 (5216), 1466–1468.

Messer, B., Song, J.H., and Yang, P., 2000. Microchannel networks for nanowire patterning. *Journal of the American Chemical Society*, 122 (41), 10232–10233.

Moore, G.E., 1998. Cramming more components onto integrated circuits. *Proceedings of the IEEE*, 86 (1), 82–85.

Myung, S., Lee, M., Kim, G.T., Ha, J.S., and Hong, S., 2005. Large-scale "Surface-Programmed Assembly" of pristine vanadium oxide nanowire-based devices. *Advanced Materials*, 17 (19), 2361–2364.

Na, Y., Farva, U., Cho, S.M., and Park, C., 2009. Synthesis and optimization of porous anodic aluminum oxide nano-template for large area device applications. *Korean Journal of Chemical Engineering*, 26 (6), 1785–1789.

Nam, C.Y., Kim, J.Y., and Fischer, J.E., 2005. Focused-ion-beam platinum nanopatterning for GaN nanowires: Ohmic contacts and patterned growth. *Applied Physics Letters*, 86 (19), 193112.

Ng, W.H., Lu, Y., Liu, H., Carmalt, C.J., Parkin, I.P., and Kenyon, A.J., 2018. Controlling and modelling the wetting properties of III-V semiconductor surfaces using re-entrant nanostructures. *Scientific Reports*, 8 (1), 3544.

Okoroanyanwu, U., 2010. *Chemistry and Lithography*. Hoboken, NJ ; Bellingham, WA: Wiley; SPIE Press.

Ouyang, M., Bai, R., Chen, L., Yang, L., Wang, M., and Chen, H., 2008. Highly photoconductive copper phthalocyanine-coated titania nanoarrays via secondary deposition. *The Journal of Physical Chemistry C*, 112 (30), 11250–11256.

Paunovic, M. and Schlesinger, M., 2006. *Fundamentals of Electrochemical Deposition*. 2nd ed. Hoboken, NJ: Wiley-Interscience.

Perego, D., Franz, S., Bestetti, M., Cattaneo, L., Brivio, S., Tallarida, G., and Spiga, S., 2013. Engineered fabrication of ordered arrays of Au–NiO–Au nanowires. *Nanotechnology*, 24 (4), 045302.

Poinern, G.E.J., Ali, N., and Fawcett, D., 2011. Progress in nano-engineered anodic aluminum oxide membrane development. *Materials*, 4 (3), 487–526.

Sander, M.S. and Tan, L.-S., 2003. Nanoparticle arrays on surfaces fabricated using anodic alumina films as templates. *Advanced Functional Materials*, 13 (5), 393–397.

Schift, H., 2008. Nanoimprint lithography: An old story in modern times? A review. *Journal of Vacuum Science & Technology B: Microelectronics and Nanometer Structures*, 26 (2), 458.

Shi, R., Huang, C., Zhang, L., Amini, A., Liu, K., Shi, Y., Bao, S., Wang, N., and Cheng, C., 2016. Three dimensional sculpturing of vertical nanowire arrays by conventional photolithography. *Scientific Reports*, 6 (1), 18886.

Song, J., Liu, Z., Li, C., Chen, H., and He, H., 1998. SPM-based nanofabrication using a synchronization technique. *Applied Physics A: Materials Science & Processing*, 66 (7), S715–S717.

Tak, Y. and Yong, K., 2005. Controlled growth of well-aligned ZnO nanorod array using a novel solution method. *The Journal of Physical Chemistry B*, 109 (41), 19263–19269.

Tanase, M., Bauer, L.A., Hultgren, A., Silevitch, D.M., Sun, L., Reich, D.H., Searson, P.C., and Meyer, G.J., 2001. Magnetic alignment of fluorescent nanowires. *Nano Letters*, 1 (3), 155–158.

van Assenbergh, P., Meinders, E., Geraedts, J., and Dodou, D., 2018. Nanostructure and microstructure fabrication: From desired properties to suitable processes. *Small*, 14, 1703401.

Venugopal, G. and Kim, S.-J., 2013. Nanolithography. *In*: K. Takahata, ed. *Advances in Micro/Nano Electromechanical Systems and Fabrication Technologies*. InTech.

Wagner, R.S. and Ellis, W.C., 1964. Vapor-liquid-solid mechanism of single crystal growth. *Applied Physics Letters*, 4 (5), 89–90.

Waldrop, M.M., 2016. The chips are down for Moore's law. *Nature News*, 530 (7589), 144–147.

Wei, Y., Wu, W., Guo, R., Yuan, D., Das, S., and Wang, Z.L., 2010. Wafer-scale high-throughput ordered growth of vertically aligned ZnO nanowire arrays. *Nano Letters*, 10 (9), 3414–3419.

Wen, T., Zhang, J., Chou, T.P., Limmer, S.J., and Cao, G., 2005. Template-based growth of oxide nanorod arrays by centrifugation. *Journal of Sol-Gel Science and Technology*, 33 (2), 193–200.

Wu, S., Huang, K., Shi, E., Xu, W., Fang, Y., Yang, Y., and Cao, A., 2014. Soluble olymer-based, blown bubble assembly of single- and double-layer nanowires with shape control. *ACS Nano*, 8 (4), 3522–3530.

Xu, D. and Guo, G., 2003. Template assisted synthesis of semiconductor nanowires. In: Wang Z.L. (eds.) *Nanowires and Nanobelts*. Boston, MA: Springer.

Xu, S., Ding, Y., Wei, Y., Fang, H., Shen, Y., Sood, A.K., Polla, D.L., and Wang, Z.L., 2009. Patterned growth of horizontal ZnO nanowire arrays. *Journal of the American Chemical Society*, 131 (19), 6670–6671.

Yang, F., Huang, L., Guo, T., Wang, C., Wang, L., and Zhang, P., 2018. The precise preparation of anodic aluminum oxide template based on the current-controlled method. *Ferroelectrics*, 523 (1), 50–60.

Yerushalmi, R., Jacobson, Z.A., Ho, J.C., Fan, Z., and Javey, A., 2007. Large scale, highly ordered assembly of nanowire parallel arrays by differential roll printing. *Applied Physics Letters*, 91 (20), 203104.

Yu, G., Cao, A., and Lieber, C.M., 2007. Large-area blown bubble films of aligned nanowires and carbon nanotubes. *Nature Nanotechnology*, 2 (6), 372–377.

Yu, G., Li, X., Lieber, C.M., and Cao, A., 2008. Nanomaterial-incorporated blown bubble films for large-area, aligned nanostructures. *Journal of Materials Chemistry*, 18 (7), 728.

Zhang, Q., Yang, X., Li, P., Huang, G., Feng, S., Shen, C., Han, B., Zhang, X., Jin, F., Xu, F., and Lu, T.J., 2015. Bioinspired engineering of honeycomb structure – Using nature to inspire human innovation. *Progress in Materials Science*, 74, 332–400.

Zheludev, N.I., 2008. What diffraction limit? *Nature Materials*, 7 (6), 420–422.

Zhou, H.L., Shao, P.G., Chua, S.J., van Kan, J.A., Bettiol, A.A., Osipowicz, T., Ooi, K.F., Goh, G.K.L., and Watt, F., 2008. Selective growth of ZnO nanorod arrays on a GaN/sapphire substrate using a proton beam written mask. *Crystal Growth & Design*, 8 (12), 4445–4448.

21

Electrospinning and Electrospun Nanofibers

Dongyang Deng and
Lifeng Zhang
North Carolina Agricultural and Technical State University

21.1 Background

Fibrous materials have been accompanying human beings since the beginning of human civilization. Flax, wool, silk and cotton have been used for more than 5,000 years. Nowadays, fibrous materials are still very important in everyday life and widely employed in many fields such as textiles, automobile, aerospace, biomedical and orthopedic devices and advanced composites. The surface area per unit volume or mass (specific surface) is an important parameter for applications of fibrous materials. The specific surface increases with smaller fiber sizes. For example, the specific surface of polyacrylonitrile (PAN) fibers at 20 μm diameter is 0.17 m^2/g, and the specific surface of PAN fibers with a diameter of 200 nm is 100 times larger in theory. Higher specific surface area leads to smaller quantity of materials that are needed to reach desired surface functions, in other words, higher efficiency. Therefore, creating ultrathin fibers is very promising and attractive for applications that rely on specific surface area. It is well known that both natural and conventional spun fibers have diameters ranging from 10 to 200 μm. For instance, cotton fibers have diameters from 15 to 25 μm; silk fibers have diameters around 12 μm; synthetic fibers like polyester, nylon and polyolefin have diameters from 20 to 200 μm. Therefore, there is a need of spinning techniques today to produce ultrathin fibers with diameters in submicron range (<1 μm) or even nanometer range (<100 nm). At present, the most straightforward but efficient method to create submicron and/or nanometer size fibers is electrospinning. Because of simple setup, diverse materials suitable for use, and unique and interesting features especially the ultra-high specific surface area of the resultant fibrous material, electrospinning has been attracting people's attention for more than two decades.

The electrospinning setup is very straightforward (Figure 21.1). In a typical process of electrospinning, first polymer solution is placed in a capillary container with a small orifice at the tip. Next, a high electric field from a DC power generator is applied to the polymer solution. Then a polymer solution jet is ejected out of the orifice by the electrical driving force. Thereafter, the "whipping" jet and substantial solvent evaporation during the jet trip toward the grounded metal collector generate ultrathin polymer fibers. These fibers then deposit on the collector and form a kind of nonwoven fibrous membrane. This intrinsic product

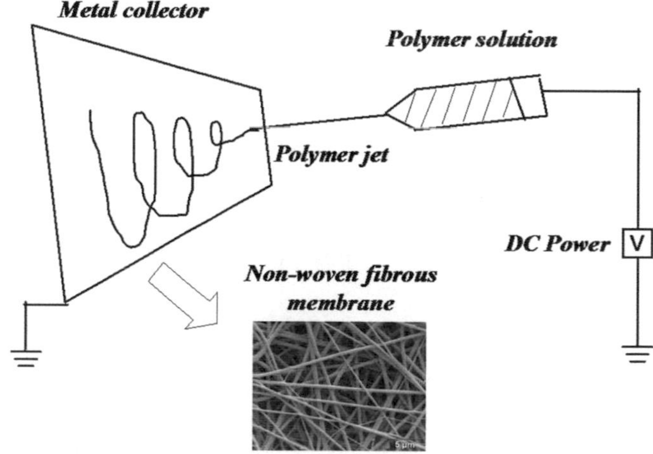

FIGURE 21.1 Schematic diagram of basic setup for electrospinning.

from electrospinning is normally termed as electrospun nanofibrous membrane, mat or felt.

The first patent on electrospinning was dated back to 1930s [1]. However, it did not generate the level of attention until the beginning of 1990s with the increasing interests in nanotechnology and needs for nanomaterials. Figure 21.2 shows the number of publications in the field of electrospinning between 1994 and 2017 and clearly illustrates the steady increase of attention on electrospinning after the year 2000.

In fact, the research booming on electrospinning was initiated by the creative work of Reneker and coworkers at the beginning of 1990s [2–9]. In their pioneer research, more than 20 polymers had been electrospun to fibers. These polymers included high-performance polymers such as polyamic acid and polyetherimide; liquid crystalline polymers such as polyaramid and poly(p-phenylene terephthalamide); copolymers such as nylon6-polyimide; textile fiber polymers such as PAN, poly(ethylene terephthalate) (PET) and nylon; conducting polymers such as polyaniline (PANi); biopolymers such as deoxyribonucleic acid (DNA) and polyhydroxybutyrate-valerate. Up to date, more than one hundred different polymers have been reported in electrospinning [10]. Most of them were electrospun from solutions and only a few were from melts. Electrospun fibrous materials are now widely used in many applications, such as high-efficiency filter media for submicron particles in separation industry; support materials for enzyme and catalyst; biomedical structural materials for tissue engineering, wound dressing, drug delivery, artificial organs and vascular grafts; nanocomposite and nanoceramics; protective shields on fabrics for special use; and elements for sensors, electronic or optical micro devices [10–18].

21.2 Electrospinning Mechanism

It is known that electrospun fibers with sub-micrometer diameters can be generated from an orifice with a diameter in millimeter scale. The understanding how electrospinning process transforms a millimeter diameter fluid stream into solid fibers with four or even five orders of magnitude smaller in diameter is both interesting and necessary in order to further control fiber properties. The exploration of electrospinning mechanism started with the booming of electrospinning [5]. The early mechanism supposed the following: At the tip of the capillary, the pendant hemispherical polymer solution drop takes a cone-like protrusion in the presence of an electric field. Charges in the solution move and concentrate on the part of the protruding surface. The accumulation of the charges causes the surface to protrude more. When the applied potential reaches a critical value, which is required to overcome the surface tension of the polymer solution, a jet of polymer solution is ejected from the cone tip and electrospinning begins. As the jet travels toward the collector, the forces from the external electric field accelerate and stretch the jet. The stretching force and the solvent evaporation cause the jet diameter to become thinner. As the radius of jet becomes smaller, the radial force from the charge can become large enough to overcome the cohesive force of the jet and cause it to split into two or more jets, which is called splay. This jet division process occurs several more times in rapid succession and produces a large number of small electrically charged fibers moving toward the collector. The divided jets repel each other because of the charge. Lateral velocities and chaotic trajectories give the jet a bush-like appearance beyond the first splay point. Splaying converts a single jet into many thinner jets. Thin fibers can also be created by elongating a

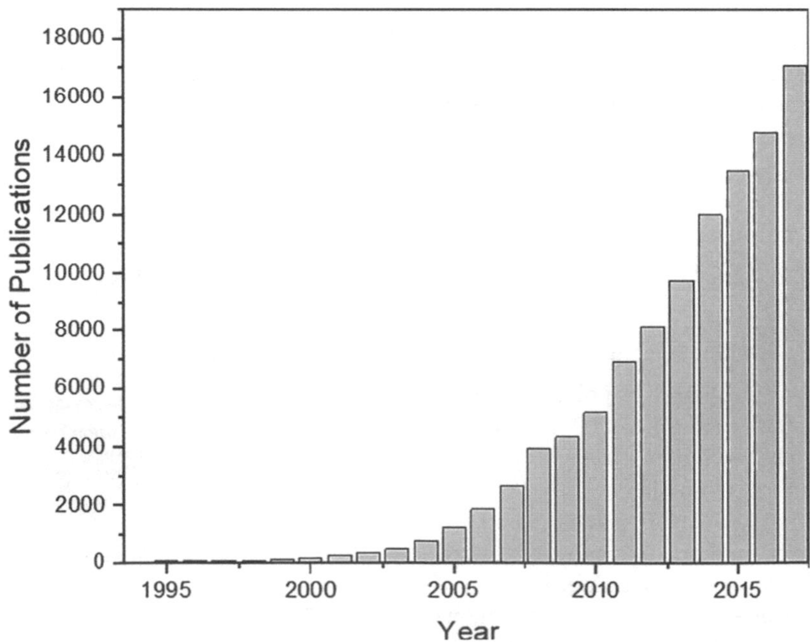

FIGURE 21.2 The annual number of publications in the field of electrospinning from 1994 to 2017, provided by Google Scholar.

single jet if splaying does not occur. Splaying and elongation appear to occur simultaneously in many cases.

Thanks to the high-speed photosystem (such as 2,000 frames per second) and capability of super short exposure time (such as 0.0125 ms), more accurate thinning mechanism was discovered later [19–23]. After a solution jet flows away from the solution droplet in a nearly straight line for some distance, it bends into a complex path. The high instability in this region is perceived to be crucial in the formation of submicron-scale fibers by electrospinning. When viewed with photos from exposure time down to millisecond range, the instability zone has the appearance of a conical region, which opens outward in the direction of flow and suggests an envelope of multiple jets (Figure 21.3a). However, when photos are taken using extreme short exposure time (such as 18 ns), the conical envelope in Figure 21.3a is actually a single, rapidly "whipping" jet (Figure 21.3b). The frequency of the whipping motion is so fast that conventional photography leads the appearance of splitting into multiple filaments. This is the reason why early theory believed that splaying was the thinning mechanism. More specifically, a solution jet from the orifice experiences an unstable bending back and forth with growing amplitude at the beginning and then the jet follows a bending, winding, spiraling and looping path in three dimensions. The solution jet in each loop grows longer and thinner with the increase in loop diameter and circumference. After some time, segments of a loop may suddenly develop a new bending instability, similar to the first but at a smaller scale. Each cycle of bending instability can be described in three steps:

1. A smooth segment of a solution jet that is straight or slightly curved suddenly develop to an array of bends
2. The segment of the jet in each bend elongates and the array of bends becomes a series of spiraling loops with growing diameters
3. With the increase in loops' perimeter, the jet becomes smaller and smaller in diameter.

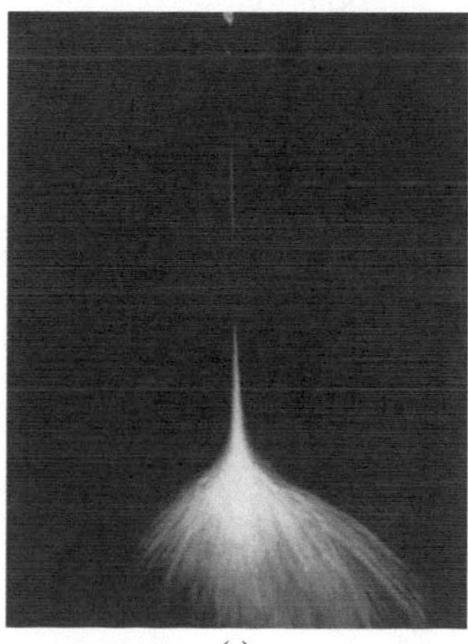

(a)

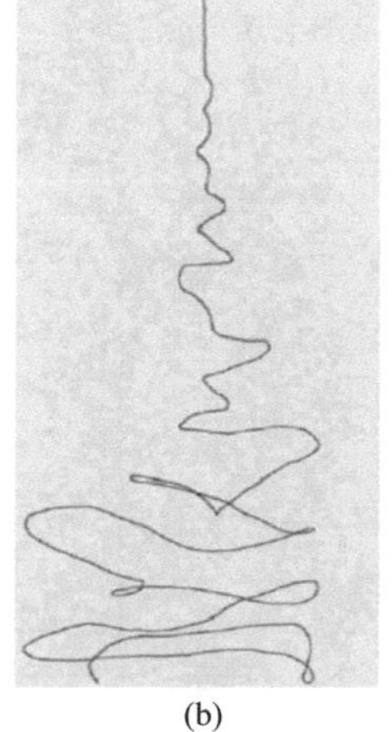

(b)

FIGURE 21.3 Instability region in an electrified polyethylene oxide (PEO)–water jet [22] Exposure time: (a) 1/250s; (b) 18 ns.

Step (1) may be re-established but with a smaller scale and then the next cycle of bending instability begins. It is inferred that more cycles can occur and the jet diameter can be reduced even more to nanoscale. After the second cycle, the axis of a particular segment jet may point in any direction. The fluid jet finally solidifies during the travel to the collector and nanofibers are then collected some distance away from the envelope cone. There were also some evidence of jet splaying [19]. A jet was observed to split into two jets with the axis of the thinner branch generally perpendicular to the axis of the primary jet. No bending instability was observed on the thinner branch jet. However, such events happened only once for every thousand times.

21.3 Electrospinning Parameters and Fiber Formation

The parameters which affect and/or control the process of electrospinning as well as subsequent fiber formation and morphology are known to be in two categories. One is solution parameters, such as solution concentration, viscosity, conductivity, surface tension, solvent volatility and polymer molecular weight; the other is processing parameters, such as applied voltage, distance between orifice and collector, electric field, solution flow rate, temperature and humidity. To obtain uniform fibers with desired diameters, all parameters should be carefully adjusted. For instance, increasing electrospinning voltage changed shape of the surface where the electrospinning jet originated in PEO aqueous solution [24]. This shape change decreased the stability of the initiating jet and increased the number of bead defects formed along the electrospun fibers. Diameter of the electrospun fibers was found to increase with solution concentration. In addition, surface tension, charge density carried by the solution jet and the viscoelasticity of PEO aqueous solution were found to be the key parameters in the process of forming beaded nanofibers [25]. For poly(D, L-lactic acid) (PDLLA) – N, N-dimethylformamide (DMF) solution, morphology of the electrospun polymer fibers depended on the strength of the electric field, the solution viscosity, the charge density of the solution and the solution feeding rate [26]. Addition of a small amount of salts such as KH_2PO_4, NaH_2PO_4 or NaCl greatly changed the morphology of electrospun PDLLA fibers from beads-on-fiber structure to uniform fiber structure with diameters in the range of 200–1,000 nm. An elastomeric poly(urethane-urea) copolymer – DMF solution was electrospun to fibers [27]. Morphology of the electrospun fibers was strongly correlated with solution viscosity, concentration and temperature. Low concentration solutions tend to form fibers with beads, whereas increased concentration favored the formation of curly fibers. Fibers spun at high temperature were uniform, unlike those fibers obtained at ambient temperature. Morphology and secondary structure (beaded, cylinder-shaped or ribbon-like) of the electrospun silk fibroin (SF) fibers were strongly influenced by solution concentration and processing voltage [28]. The conditions for preparing beadless circular cross-sectional fibers were found to be SF concentration 28 wt%, voltage 2 kV and working distance 12 cm.

However, not all the variables mentioned above are fundamental control parameters and they are not all independent of each other. For example, solution viscosity is a function of both concentration and polymer molecular weight; applied voltage and orifice–collector distance are interrelated as well as the orifice–collector distance and solvent volatility. The proper condition to get desired fibers usually is based on empirical experience. It always takes time to find the optimum condition. The theoretical guide, i.e. the correlation of independent parameters with fiber formation, is necessary. In fact, the polymer–polymer chain entanglement is reported to play an independent role on fiber formation in electrospinning [29–34]. Colby et al. revealed concentration dependence of solution viscosity for linear polymers in good solvents and identified four different concentration regimes in polymer solutions at different polymer concentration [35,36]. The four regimes, which are dilute, semidilute unentangled, semidilute entangled and concentrated regimes, are illustrated in Figure 21.4. In dilute regime, there is no chain entanglement at all. In semidilute unentangled regime, some kinds of chain overlap happen but not enough to cause any significant degree of entanglement. As the polymer concentration is further increased, chain entanglement shows up in the semidilute entangled regime and increases greatly in concentrated regime. The starting concentration for chain entanglement, C_e, is the boundary between the semidilute unentangled and semidilute entangled regimes and is defined as the point at which significant chain entanglements happen. C_e marks the distinct onset of polymer chain entanglement in solution.

McKee et al. studied electrospun fiber formation of a series of linear and branched poly(ethylene terephthalate-co-ethylene isophthalate) (PET-co-PEI) copolymer solutions in 70/30 w/w mixture of $CHCl_3$/DMF [30]. They analyzed the dependence of specific viscosity (η_{sp}) on solution concentration C. Specific viscosity was defined as

$$\eta_{sp} = \frac{\eta_0 - \eta_s}{\eta_s},$$

where η_0 is zero shear rate solution viscosity and η_s is solvent viscosity.

For neutral, linear polymers in a good solvent, η_{sp} is proportional to $C^{1.0}$, $C^{1.25}$, $C^{4.8}$ and $C^{3.6}$ in the dilute, semidilute unentangled, semidilute entangled and concentrated regimes, respectively. η_{sp} was plotted as a function of polymer concentration, and the changes in the slope marked the onset of the semidilute unentangled, semidilute entangled and concentrated regimes. The starting concentration for chain entanglement, C_e, was derived from the transition point between semidilute unentangled regime and semidilute entangled regime. For most of the copolymers studied, C_e was the minimum concentration required to obtain beaded nanofibers in electrospinning, while 2–2.5 times C_e was the minimum concentration required for uniform fibers. Shenoy et al. proposed a parameter called solution entanglement number $(n_e)_{soln}$ to predict fiber formation [32]. $(n_e)_{soln}$ is defined as the ratio of the polymer molecular weight to its solution entanglement molecular weight $(M_e)_{soln}$. For polydisperse systems, the weight–average molecular weight, M_w, is typically used as polymer molecular weight. In addition, relationship between the starting molecular weight for chain entanglement in solution $(M_e)_{soln}$ and the starting molecular weight for chain entanglement in melt M_e is $(M_e)_{soln} = M_e/\phi_p$, where ϕ_p is the polymer volume fraction. Thus,

$$(n_e)_{soln} = \frac{M_w}{(M_e)_{soln}} = \frac{(\phi_p M_w)}{M_e}$$

The results showed that for initiation of fibers in electrospinning, $(n_e)_{soln} = 2$ or $\phi_p M_w = 2M_e$. For complete fiber formation, $(n_e)_{soln} \geq 3.5$ or $\phi_p M_w \geq 3.5 M_e$. Therefore, with knowledge of the starting molecular weight for chain entanglement (M_e) and the weight-average molecular weight (M_w) for a given polymer, the concentration of polymer in solution can be predicted to get uniform fiber formation. They verified this relationship in the electrospinning of polystyrene (PS)/tetrahydrofuran (THF), poly(D,L-lactide) (PLA)/DMF, PEO/H_2O and poly(vinyl pyrrolidone) (PVP)/ethanol solutions. Gupta et al. used the ratio of polymer concentration (c) to critical chain overlap concentration (c*) to describe electrospun fiber

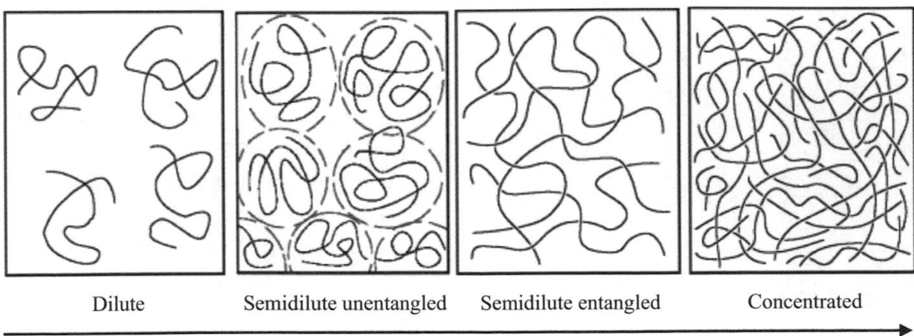

FIGURE 21.4 Schematic diagram of polymer molecules in solution.

formation of poly (methyl methacrylate) (PMMA) in DMF solution [33]. c* was determined by dynamic light scattering. The plot of zero shear viscosity vs. c/c* distinctly separated into different solution regimes: dilute (c/c* < 1), semidilute unentangled (1 < c/c* < 3) and semidilute entangled (c/c* > 3). Only polymer beads were formed when solutions were electrospun from the dilute concentration regime due to insufficient chain entanglement. With the increase in polymer concentration, beads and beaded fibers were observed in the semidilute unentangled regime, and beaded fibers and uniform fibers were observed in the semidilute entangled regime. Complete uniform fiber formation was observed at c/c* = ~6 for all the narrow molecular weight distribution (MWD) polymers (M_w = 12,470–205,800 g/mol, M_w/M_n = ~1.03–1.35), but complete uniform fibers were not formed until concentrations were higher (c/c* = ~10) for the relatively broad MWD polymers (M_w = 34,070, M_w/M_n = ~1.62 and M_w = 95,800 g/mol, M_w/M_n = ~2.12).

21.4 Applications of Electrospun Nanofibers

A brief discussion on some important applications of electrospun fibrous materials is given in the following part.

21.4.1 Filtration

Fibrous materials provide advantages of high filtration efficiency (FE) and low flow resistance when they are used for filters [37]. FE is much dependent on fiber fineness and is one of the most important concerns for the filter performance. The channels and structural elements of a filter must match the size of particles or droplets that are to be captured in the filter. The filter with smaller fiber sizes will have higher FE (Figure 21.5).

To effectively filter submicron-scale particles, a direct way is to use submicron or nanometer-sized fibers. In general, due to ultra-high specific surface, tiny particles with size <0.5 μm can be easily trapped in the filter made with electrospun fibrous materials and the FE is greatly improved [10]. Nanofibers, particularly, nonwoven nanofibrous mats or webs from electrospinning were already used in the field of filtration [38–40]. Experimental measurement and theoretical calculation showed that electrospun nanofibrous mat is extremely efficient at trapping aerosol particles [41]. The high FE is a direct result of the small size of electrospun fibers. Because the diameters of electrospun fibers are comparable to the mean free path of air molecules, the usual high-pressure drop penalty that happens to thin fiber filter materials in the continuum flow regime is much reduced with electrospun fibers. The research on filtration using electrospun fibers focused on higher selective efficiency by fiber surface functionalization and composite design. A highly effective sorption-filtering nonwoven polyvinyl alcohol (PVA) material was manufactured from electrospinning 11.8 wt% aqueous solution [42]. Introduction of Cr $(NO_3)_3$ to PVA fibers followed by thermal oxidation and pyrolysis increased the sorption – filtering efficiency. New fibrous materials for composite filtration membranes were prepared by means of electrospinning of a copolymer of tetrafluoroethylene with vinylidene fluoride (F-42) from 9% to 10% 1:1 acetone/DMF mixture or a copolymer of vinylidene fluoride with hexafluoropropylene (F-26) from 8% 1:0.1 acetone/DMF mixture [43]. The composite filtration membrane consisted of a sandwich structure: layers of thin fibers with diameters around 50 nm were sandwiched between layers of thick fibers with diameter around 1 μm. The thick fiber layers provided necessary mechanical strength while thin fiber layers determined filtration properties. A novel high-flux composite membrane for water filtration was developed based on a cross-linked electrospun PVA substrate coated with a hydrophilic polyether-b-polyamide copolymer or a cross-linked PVA hydrogel incorporated with surface-oxidized multiwalled carbon nanotubes (MWCNTs) [44]. The electrospun nanofibrous substrates provided good tensile strength, extremely light weight and interconnected porous structure with large specific surface area, making them excellent candidates for supporting scaffolds in filtration applications. Oil/water emulsion tests showed that this unique type of filtration media exhibited a high flux rate (up to 330 L/m^{2}h at the feed pressure of 100 psi) and an excellent total organic solute rejection rate (99.8%) without appreciable fouling. In another research, PS nanofibers with diameters around 600 nm were electrospun from recycled expanded PS [45]. These fibers were mixed with micro glass fibers and used as filter media to separate water from water-in-oil emulsion, which is important to petrochemical industries. The separation efficiency was improved from 68% to 88%.

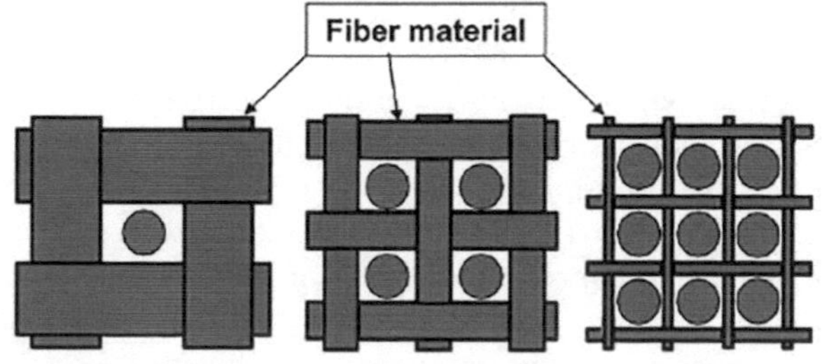

FIGURE 21.5 FE increases with smaller fiber sizes in fibrous materials [10].

21.4.2 Catalyst Support

Fibrous catalysts are flexible, versatile and easy to handle when they are compared with powder and monoliths [46]. Fibrous catalysts can adapt to any geometry with excellent mass transfer characteristics. The open structure inside fibrous catalysts can yield a low-pressure drop that competes with the monolithic catalysts. The pressure drop, usually one to two orders of magnitude smaller than that in packed beds, makes fibrous catalysts attractive for applications like three-phase operations where only low-pressure drops are allowed or high flow rates are required. Furthermore, small diameter fibers can essentially remove diffusion as a rate-controlling factor in liquid-phase operation. Therefore fibrous catalysts are good candidates for liquid-phase operations.

Nanofibers are ideal for use as catalyst support due to their super large surface area per unit mass and the feasibility for high loading of catalyst. When graphitic carbon nanofibers were used as catalyst support medium for Fe particles, the catalytic activity for hydrocarbon conversion was considerably higher than that of the Fe particles supported on either active carbon or γ-alumina under the same condition [47]. Enzyme is a category of highly efficient catalyst for bioreactions. Immobilization of enzyme on solid support makes easy separation, purification and recovery of enzyme. Electrospun fibers have been used as the support for enzyme immobilization [48–50]. Size reduction of carrier material can effectively improve the efficiency of immobilized enzymes. Electrospun PS was used as the immobilization support of a model enzyme, α-chymotrypsin [48]. The apparent hydrolytic activity of the nanofibrous enzyme in aqueous solution was over 65% of that of the native enzyme, much higher than other forms of immobilized enzymes. In addition, the nanofibrous enzyme exhibited a much-improved activity in organic solvents including hexane and isooctane than that of its native counterpart. The covalent binding also improved the enzyme's stability against structural denaturation. For instance, the half-life of the nanofibrous enzyme in methanol was 18-fold longer than that of the native enzyme. A lipase enzyme could be entrapped in sub-micrometer size PVA fibers via electrospinning [49]. The lipase in PVA/lipase electrospun nanofibrous membrane was 6 times more active than that in the cast membrane from the same solution. Lipase enzyme could also be grafted onto surface of electrospun cellulose nanofibers [50]. The reaction was performed with poly(ethylene glycol) (PEG) diacylchloride to simultaneously add amphiphilic spacers and reactive end groups for coupling lipase enzyme. The PEG cell-bound lipase showed much higher retention of catalytic activity after being exposed to cyclohexane, toluene and hexane than that of free lipase. Furthermore, the bound lipase exhibited significantly higher catalytic activity than free lipase at elevated temperatures (up to 8–10 times at 60°C and 70°C, respectively).

21.4.3 Nanocomposite and Nanoceramics

Due to much higher surface-to-volume ratio than conventional fibers, nanofibers may significantly increase the interaction between fibers and matrix material when they are used as reinforcing component in polymer composite materials. In addition, the porous structure as well as the hairiness of electrospun nanofibrous materials can take advantage of the mechanical interlocking mechanism for load transfer between the matrix and nanofibers; thus, they can significantly enhance performance of polymer composite materials than conventional fibers [12]. Furthermore, nanofiber-reinforced composites may have some additional functionalities, which cannot be shared by traditional micrometer-scale fiber-reinforced composites. For instance, if fiber and polymer matrix have different refractive indices, the resulting composite becomes opaque or nontransparent due to light scattering. This limitation can be overcome when fiber diameters become significantly smaller than the wavelength of visible light [13]. Electrospun nanofibers of polybenzimidazol (PBI) with average diameters of 300 nm were used to reinforce both epoxy matrix and styrene–butadiene rubber (SBR) [51]. A 15 wt% nanofiber-reinforced epoxy composite showed higher fracture toughness and modulus than 17 wt% fibrid-reinforced epoxy composite. A 10 wt% nanofiber-reinforced SBR composite showed 10 times Young's modulus and twice tear strength. Nylon-4,6 fibers were electrospun from formic acid solution with diameters in the range of 30–200 nm [52]. Because the fiber sizes are so small, epoxy composite containing the nylon-4,6 electrospun fibers could maintain optical transparency of the matrix. This transparent epoxy composite had significantly improved modulus and tensile strength. A nanocomposite fiber of nylon-6 with 7.5 wt% montmorillonite was produced by electrospinning from hexafluoroisopropanol solution, which illustrated the potential of using polymer nanocomposite as the foundation for fabricating nano- and meso-scopic structures [53].

In addition to reinforcement, if functional materials are included, the resultant nanocomposite material may have additional electrical or thermal properties. PMMA nanocomposite fibers containing well-dispersed MWCNTs were electrospun from DMF solution [54]. Diameters of the electrospun nanofibers were in the range of 120–710 nm. MWCNTs were found to be embedded in the polymer matrix and aligned along the fiber axis. The most interesting thing is that electrical conductivities of the PMMA/MWCNT nanofibrous membrane containing 1–5 wt% MWCNT were around 10^{-10}S/cm, much less than that of corresponding nanocomposite films, which was between 10^{-4} and 10^{-2} S/cm.

A new class of nanofibrous material, i.e. ceramic nanofibers, has attracted intensive attention [55–61]. This kind of material is from the combination of sol–gel method and electrospinning technique. In a typical procedure, a sol–gel precursor, such as a metal alkoxide, is mixed with polymer solution. The role of polymer herein is to enable fiber formation. After the solution is electrospun to thin fibers, the metal alkoxide immediately starts to hydrolyze by reacting with the moisture in air to generate a continuous gel network within the polymer matrix. As a result, inorganic/polymer composite nanofibers are produced. Ceramic nanofibers can be readily obtained by eliminating the polymer phase from the composite nanofibers via calcination at high temperature. In order to get ceramic nanofibers, three steps are generally involved:

1. Prepare a sol with suitable content of inorganic precursor and polymer and achieve right rheology for electrospinning;
2. Electrospin the sol to obtain polymer/inorganic composite nanofibers;
3. Calcinate the composite nanofibers to obtain final ceramic nanofibers.

For instance, in the work by Li and Xia [58], PVP and titanium tetraisopropoxide were mixed in ethanol solution and electrospun to composite nanofibers containing PVP and amorphous TiO_2. The composite fibers had lengths up to several centimeters and could be subsequently converted into anatase TiO_2 without changing their fibrous morphology via calcination in air at 500°C. The average diameter of these TiO_2 nanofibers could be controlled in the range of 20–200 nm by varying parameters like the concentration and composition of PVP and titanium tetraisopropoxide in solution, the strength of the electric field, and the feeding rate of the precursor solution. Besides PVP [58–61], PVA is another widely used polymer in the process of preparing ceramic nanofibers [56,62–66]. A number of oxides that include Al_2O_3, CuO, NiO, TiO_2-SiO_2, V_2O_5, ZnO, Co_3O_4, Nb_2O_5, MoO_3 and $MgTiO_3$ have been fabricated in fibrous form [13].

21.4.4 Biomedical Uses

The feasibility of electrospinning biopolymers, biocompatible polymers and biodegradable polymers makes electrospun nanofibers very useful in the biomedical field. Nonwoven mats of electrospun nanofibers are well known for their interconnected, three-dimensional porous structure and relatively large specific surface area. They provide a class of ideal material to mimic the natural extracellular matrix (ECM) required for tissue engineering. Poly(D,L-lactide-*co*-glycolide) (PLGA) was electrospun from THF/DMF mixture solution for tissue engineering applications [67]. The electrospun product contained PLGA electrospun fibers with diameters from 500 to 800 nm and had a wide range of pore diameter distribution, high porosity and effective mechanical properties, which represented a morphologic similarity to the ECM of natural tissue. Cell seeded on this structure tended to maintain phenotypic shape and grew according to nanofiber orientation. Mesenchymal stem cells derived from the bone marrow of neonatal rats were seeded and cultured on the poly(ε-caprolactone) (PCL) scaffolds, which came from electrospinning of 10 wt% chloroform solution [68]. The final cell–polymer constructs maintained size and shape of the original scaffolds. Electrospun collagen promoted cell growth and penetration into the matrix [69]. The structural properties of electrospun collagen varied with the tissue of origin (from which collagen was produced), the isotype and the concentration of the electrospinning solution. A mixture DMF solution of PLA with miscible PLGA random copolymers, poly(lactide-b-ethylene glycol-b-lactide) (PLA-b-PEG-b-PLA) tri-block copolymers and a lactide was electrospun to nonwoven scaffolds [70]. All the electrospun scaffolds showed fiber diameters in the range of 500 to 800 nm, 70%–75% porosity and 0.3–0.4 g/cm^3 density. The biodegradation rate, as well as the hydrophilicity of the electrospun scaffolds, could be finely tuned with different material compositions.

Electrospun fibers have also been applied to wound dressing because porous-structured membrane is able to exude fluid from the wound, does not build up fluid under the covering and does not cause desiccation because of good path for the transport of vapor. In addition, fibrous structures can protect the wound from bacterial penetration via the aerosol particle capturing mechanism [13]. Electrospun polyurethane (PU) membrane was used as a wound dressing [71]. This wound dressing showed controlled evaporative water loss, excellent oxygen permeability and promoted fluid drainage ability. There was no toxicity and no permeability to exogenous microorganism with the nanofibrous membrane. Epithelialization rate was found to be increased by histological examination. The exudates in the dermis were well controlled by covering the wound with the electrospun membrane. SF was electrospun to nanofiber nonwovens for wound dressing [72]. The nonwovens promoted cell adhesion and spreading of normal human keratinocytes and fibroblasts. With the aid of electric field, fine fibers of biodegradable polymers can be directly sprayed/spun onto the injured location of skin to form a fibrous mat dressing (Figure 21.6). This can heal wounds by encouraging normal skin growth and eliminating the formation of scar tissue which possibly occur in a traditional treatment [10].

Electrospinning can even be used to create biocompatible thin films with useful coating design and surface structure that can be deposited onto implantable devices in order to improve the biocompatibility of these devices with human body. A useful coating design would contain both functional and structural elements to encourage cellular ingrowths, reduce mismatch between the device and the tissue and

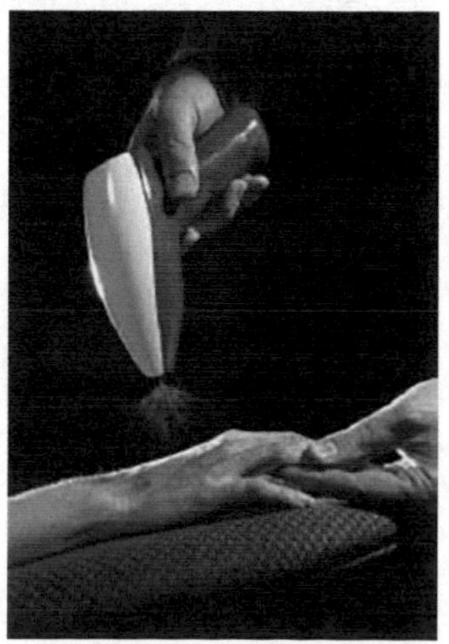

FIGURE 21.6 Wound dressing by direct spraying [10].

have a surface structure that could help integrate the device into the body. A silk-like polymer with fibronectin functionality (SLPF) was electrospun and deposited onto prosthetic devices designed to be implanted in the central nervous system [73]. The control of morphology and thickness of the coatings can be achieved by varying four processing parameters: solution concentration, applied field strength, deposition distance and deposition time.

21.5 Bicomponent and Multicomponent Electrospinning

21.5.1 Polymer–Polymer Phase Separation in the Process of Mixing

Phase separation is a common morphological characteristic of most polymer blends in the solid state [74,75]. When two polymers are mixed together at temperature T, the Gibbs free energy change $\Delta G_{\mathrm{m}} = \Delta H_{\mathrm{m}} - T\Delta S_{\mathrm{m}}$. If $\Delta G_{\mathrm{m}} < 0$, two polymers are completely miscible like a kind of special solution, which means the blend of two polymers has homogeneity at molecular level. If $\Delta G_{\mathrm{m}} > 0$, two polymers are not miscible and are phase separated in blend. Most polymers have small mixing entropy change due to its long chain molecules.

The quantitative analysis of entropy change during polymer–polymer mixing process could come from the Flory–Huggins lattice theory of polymer solutions, which can be applied to polymer mixtures because they can be considered as a kind of special polymer solution. The entropy change after mixing two polymers [76]

$$\begin{aligned}\Delta S_{\mathrm{m}} &= -R\left(n_{\mathrm{A}} \ln \phi_{\mathrm{A}} + n_{\mathrm{B}} \ln \phi_{\mathrm{B}}\right)\\ &= -R\left(\frac{\phi_{\mathrm{A}} V}{x_{\mathrm{A}} V_{\mathrm{A}}} \ln \phi_{\mathrm{A}} + \frac{\phi_{\mathrm{B}} V}{x_{\mathrm{B}} V_{\mathrm{B}}} \ln \phi_{\mathrm{B}}\right)\end{aligned}$$

R – gas constant; n_{A} – mole number of polymer A; n_{B} – mole number of polymer B
ϕ_{A} – volume fraction of polymer A; ϕ_{B} – volume fraction of polymer B
V – total volume of system; V_{A} – segment volume of polymer A
V_{B} – segment volume of polymer B; x_{A} – number of segments of polymer A
x_{B} – number of segments of polymer B.

It is known that high molecular weight polymer has larger number of segments, i.e. they have larger x_{A} and x_{B}. Since $\phi_{\mathrm{A}} + \phi_{\mathrm{B}} = 1$, $\ln\phi_{\mathrm{A}}$ and $\ln\phi_{\mathrm{B}} < 0$, smaller ΔS_{m} is obtained when high molecular weight polymers are mixed together, then the mixing Gibbs free energy change ΔG_{m} becomes larger and, therefore, phase separation becomes more significant. Thus, in theory, high molecular weight polymers have more distinct phase separation and larger phase domains than that of low molecular weight polymers in polymer blends.

21.5.2 Phase Separation Morphology in Polymer Blend

In a polymer blend, phase domain size and morphology may be controlled by varying the compatibility and composition of two polymers [76]. Suppose we have immiscible polymer A and polymer B, when we add B into A system, the phase separation morphology of A and B blend system can vary from B island in A sea, A–B co-continuous and A island in B sea with increasing the B content (Figure 21.7). For example, morphology of immiscible polymer blend of PS and poly (*tert*-butyl acrylate) (PtBA) in spin-coated thin films was characterized by x-ray photoelectron spectroscopy (XPS), Fourier-transform infrared spectroscopy (FTIR) and scanning force microscopy (SFM) combined with selective dissolution [77]. PS and PtBA are found to be phase separated laterally in the thin films after spin coating. The overall trend is when the PS:PtBA ratio is less than one, PS islands form and when the PS:PtBA ratio is larger than one, PtBA islands form.

21.5.3 Phase Separation in the Process of Electrospinning

In the electrospinning process, polymer solution jet solidifies during the travel from orifice to collector. If two or more polymers exist in solution, polymer–polymer phase separation occurs with the evaporation of solvent. Due to fast evaporation of solvent, polymer–polymer phase separation mechanism is not thermodynamic. There are two kinds of phase separation mechanism: nucleation and growth mechanism and spinodal decomposition mechanism. The former happens when the solution is in metastable state, i.e. the solution may stay for a long time and during this time period, the phase separation happens via nucleation and growth. The individual droplets of the minor phase that are formed in the early state of the process grow slowly. They are dispersed in the matrix of the corresponding coexisting phase and can become rather large in the end. The latter happens when the solution is in unstable state, i.e. phase separation takes place spontaneously and fast because any fluctuation in polymer concentration will inevitably lead to a reduction in the Gibbs free energy right away, which is a more stable state. For spinodal decomposition, the size of the coexisting phases is usually at least one order of magnitude smaller than the other mechanism and the phase-separation morphology tends to be co-continuous,

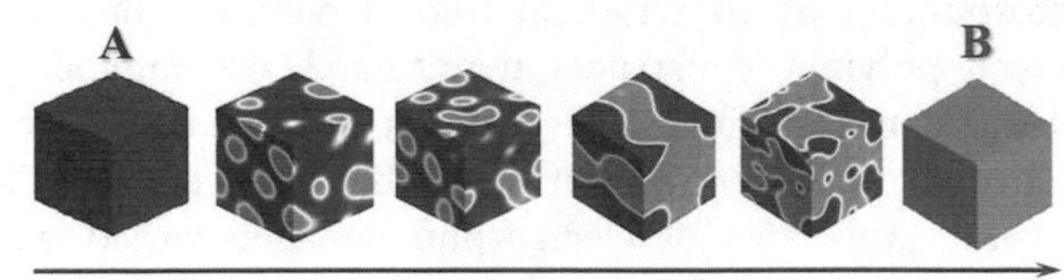

FIGURE 21.7 Schematic diagram of phase separation morphology.

i.e. for each phase, it is possible to find paths through the entire system. In the process of electrospinning, solvent evaporation is superfast and fibers solidify very quickly after ejecting from the electrospinning orifice. In this case, spinodal decomposition is preferred and phase-separated domains may not have enough time to become as large as they are in equilibrium state as described by Gibbs free energy [16,78,79]. Therefore, fine phase morphology should be observed.

Meanwhile, fast solvent evaporation of polymer jet during the trip from electrospinning orifice to collector could lead to lower temperature in electrospun fiber [79]. Low temperature is in favor of phase separation from the equation of ΔG_m. This thermally induced phase separation (TIPS) is a well-known process.

21.5.4 Advances of Bicomponent and Multicomponent Electrospinning

In the early stage of electrospinning, single-component polymer solution was mostly used and electrospun fibers usually show a solid interior and smooth surface. Nowadays, bicomponent or multicomponent electrospinning has attracted more attention. The advantages of bicomponent or multicomponent electrospinning include the following:

1. Multicomponent electrospun product may have combined properties;
2. This is an important approach to obtain ultrathin fibers of polymers or even inorganic materials which cannot be electrospun alone;
3. This makes special fiber structure and morphology available based on phase separation. These special surface topologies such as side-by-side, sheath/core, hollow and porous structure can greatly enhance specific surface and functionality, which are beneficial to fibers' wetting processes and adsorption behavior as well as applications such as micro-fluidics, photonics and energy storage.

PANi, a conductive polymer, was blended with a natural protein, gelatin, in 1,1,1,3,3,3-hexafluoro-2-propanol (HFP) and co-electrospun into nanofibers [80]. Experimental data demonstrated that electrospun PANi–gelatin bicomponent fibers were biocompatible and supported attachment, migration and proliferation of H9c2 rat cardiac myoblasts. At the same time, electrical stimulation can be applied to the cells with conductive scaffold.

Casein, a milk protein, cannot be electrospun by itself due to its strong intermolecular force and three-dimensional structure. However, by mixing with another polymer such as PEO and PVA, casein/PEO or casein/PVA polymer solutions can be easily and successfully electrospun [49]. PVA was also used to help chitosan fiber formation [81]; 82.5% deacetylated chitosan (M_v = 1,600 kDa) was mixed with PVA (M_w = 124–186 kDa) in 2% (v/v) aqueous acetic acid and successfully electrospun to nanofibers. With increasing PVA content, finer fibers, fewer beads and more efficient fiber formation were observed.

Poly(vinyl cinnamate)/poly(3-hydroxybutyrate-co-3-hydroxyvalerate) (PVCi/PHBV) bicomponent fibers were electrospun from chloroform solution [82]. PVCi and PHBV were immiscible and phase separation happened during the electrospinning process. After photocrosslinking PVCi, PHBV was removed from the bicomponent fibers upon chloroform treatment. A porous structure was observed from the remaining ultrafine PVCi fibers. In another research, the incompatibility between PAN and PVP induced phase separation, and microdomains of PVP were formed in the PAN/PVP electrospun fibers [83]. PVP microdomains were then leached out in water and porous PAN ultrafine fibers were obtained. Polyether imide (PEI)/PHBV was electrospun to bicomponent fibers, which had a relatively broad size distribution in the range of 2.6–15.1 μm [84]. PEI and PHBV were partially miscible and phase separation progressed fast during the electrospinning process. Porous PEI ultrafine fibers were prepared via selective thermal degradation of PHBV. PEI/PHBV(75/25) and PEI/PHBV(50/50) fibers showed highly porous surfaces after thermal degradation of PHBV at 210°C. PLA and PVP bicomponent fibers were electrospun from dichloromethane (DCM) solution [85]. Fibers with porous structure were obtained by selective removal of PVP in water or selective removal of PLA by heat.

Modifying electrospinning setup (spinneret) makes it easier to get bicomponent or multicomponent electrospun product with special structures. Gupta et al. simultaneously electrospun two polymer solutions in a side-by-side fashion with poly (vinyl chloride) (PVC)/segmented PU and PVC/poly (vinylidiene fluoride) (PVDF) [86]. In the electrospinning setup (Figure 21.8), two polymer solutions were separated and an electrode was immersed in each solution. The two polymer solutions flowed outwards through each of the Teflon needles until they came into contact at the ends of the needles. Some mixing of the two components was expected. In fact, when the tip–collector distance is 9 cm or larger, a single common Taylor cone was observed. At distance larger than 25 cm, the polymer jet was not

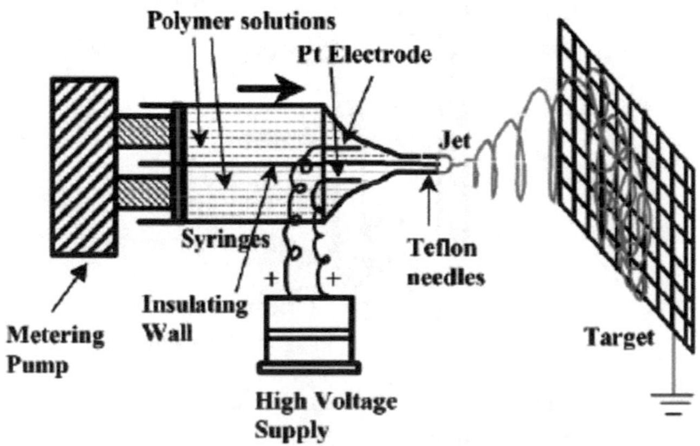

FIGURE 21.8 Side-by-side bicomponent electrospinning [86].

continuous. At distance <9 cm, two Taylor cones and two separated jets were observed from the two solutions. This led to two zones of fiber collection on collector, each corresponding to one of the two polymer components. In the case of one Taylor cone, it was identified that fibers were rich in one of the two polymer components, and the ratio of the major component and the minor component varied along fibers.

Sun et al. demonstrated a co-electrospinning setup where two coaxial capillaries were used instead of the normal single capillary (Figure 21.9) [87]. Two different polymer solutions were put into the inner and outer chambers, respectively. The combined solution droplet at the tip of the spinneret was electrospun to compound sheath–core fibers. The co-electrospinning process was actually fast enough to prevent the mixing of the core and sheath polymers. Co-electrospinning is of particular interest for those core materials that cannot form fibers alone via electrospinning. In this case, the sheath polymer serves as a template of fiber for the core material and results in a sheath–core structure. Nanofibers with sheath–core structures certainly encourage applications like in the field of microelectronics, optics and medicines. For example, PEO (sheath)/poly(dodecylthiophene) (PDT, core) nanofibers were obtained from co-electrospinning of 2 wt% PEO solution in chloroform and 1 wt% PDT solution in chloroform and PLA (sheath)/$Pd(OAc)_2$ (core) was obtained from co-electrospinning of 3 wt% PLA chloroform solution and 5 wt% solution of $Pd(OAc)_2$ in THF [87]. The non-electrospinnable materials such as PDT and Pd $(OAc)_2$ could be forced into fiber form and indicated that co-electrospinning method is versatile and may promote new material design. Another example is PCL (sheath)/gelatin (core) bicomponent fibers produced by co-electrospinning of their solutions in 2,2,2-trifluoroethanol (TFE) [88]. Li et al. demonstrated that nanofibers of conjugated polymers and their blends, such as poly(2-methoxy-5-(2-ethylhexloxy)-1,4-phenylenevinylene (MEH-PPV), could be fabricated by co-electrospinning their solutions with a spinnable polymer such as PVP as sheath polymer, followed by extraction of the sheath polymer [89].

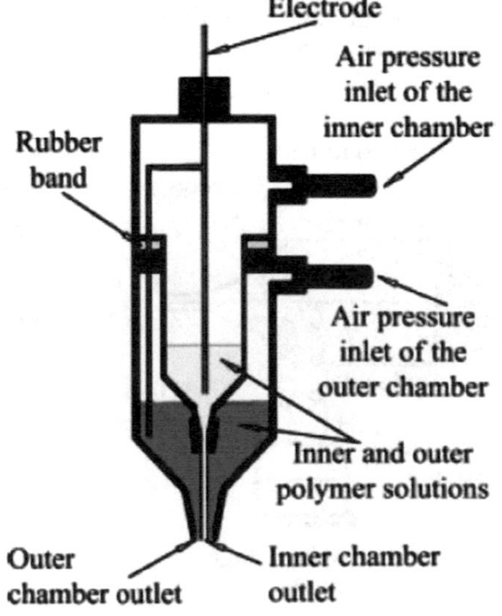

FIGURE 21.9 The co-electrospinning setup [87].

The sheath–core structure fibers also facilitated the manufacture of hollow fibers by selectively removing the core component [90,91]. Li and Xia combined the sol–gel process and co-electrospinning technique to produce hollow nanoceramic tubes. Two viscous but immiscible liquids, mineral oil and an ethanol solution containing PVP and $Ti(OiPr)_4$, were used as source materials for core and sheath in a co-electrospinning setup. The two liquids were ejected simultaneously through the inner and outer capillaries to form a stable compound jet. In situ formation of gel network in the sheath during electrospinning facilitated generating nanofibers with TiO_2/PVP sheath and mineral oil core. Selective removal of the oil phase by solvent extraction resulted in the formation of hollow fibers consisting of TiO_2/PVP composite walls. The sheath thickness and inner diameter of the nanofibers varied in the range of tens of nanometers to several hundred nanometers by controlling spinning conditions.

Li and Xia also generated porous ceramic nanofibers using co-electrospinning [92]. In this case, PS solution in DMF/THF mixture was used as core liquid and a PVP/$Ti(OiPr)_4$ solution in ethanol was used as sheath liquid. Although PS and PVP are immiscible, the solvents are miscible. It was found that these two liquids could be partially mixed in the electrospinning process because the mutual diffusion of the two solvents could bring part of PS and PVP/$Ti(OiPr)_4$ together. Two polymer phases were separated to generate nanoscale domain of PS embedded in a continuous TiO_2/PVP matrix after the solvents quickly evaporated. Once the PS phase had been removed by calcination along with the PVP phase, the resultant fibers became highly porous.

21.6 Conclusion

The top-down nanomanufacturing technique of electrospinning provides a universal approach to conveniently prepare fibers with typical diameters from a few tens to a few hundreds of nanometers (commonly termed as "electrospun nanofibers"). Unlike conventional fiber spinning techniques such as dry, wet or melt spinning, which is driven by mechanical force, electrospinning is driven by electrical force and follows a "bending instability" thinning mechanism. This unique thinning mechanism enables approximately two to three orders of magnitude thinner fibers than natural and conventional spun fibers. Compared to most of other 1D nanomaterials that are made from bottom-up methods such as nanotubes, nanowires and nanorods, electrospun nanofibers are collected intrinsically in the form of nonwoven mat and usually require no further purification. Electrospun nanofibers are inexpensive, continuous and relatively easy to be aligned, assembled and processed into many applications such as filtration, catalyst support, nanocomposite and nanoceramics and biomedical uses. It is noteworthy that electrospinning is currently the simplest and most cost-effective method to produce a variety of long and continuous nanofibers and it does have scalability for mass production. The rapidly growing research on electrospinning and electrospun nanofibers will pave the road toward new discoveries and realize brand-new applications.

References

1. Formhals, A. US Patent 1975504, 1934.
2. Doshi, J; Reneker, DH. *Journal of Electrostatics*, 1995, 35, 151–160.
3. Srinivasan, G; Reneker, DH. *Polymer International*, 1995, 36, 195–201.
4. Zachariades, AE; Porter, RS; Doshi, J; Srinivasan, G; Reneker, DH. *Polymer News*, 1995, 20, 206–207.
5. Reneker, DH; Chun I. *Nanotechnology*, 1996, 7, 216–223.
6. Fang, X; Reneker, DH. *Journal of Macromolecular Science, Physics*, 1997, B36, 169–173.
7. Fong, H; Reneker, DH. *Journal of Polymer Science, Part B: Polymer Physics*, 1999, 37, 3488–3493.
8. Kim, J-S; Reneker, DH. *Polymer Engineering and Science*, 1999, 39, 849–854.
9. Zachariades, AE; Porter, RS; Doshi, J; Srinivasan, G; Reneker, DH. *Polymer News*, 1995, 20, 206–207.
10. Huang, Z-M; Zhang, Y-Z; Kotaki, M; Ramakrishna, S. *Composites Science and Technology*, 2003, 63, 2223–2253.
11. Subbiah, T; Bhat, GS; Tock, RW; Parameswaran, S; Ramkumar, SS. *Journal of Applied Polymer Science*, 2005, 96, 557–569.
12. Chronakis, IS. *Journal of Materials Processing Technology*, 2005, 167, 283–293.
13. Li, D; Xia, Y. *Advanced Materials*, 2004, 14, 1151–1170.
14. Frenot, A; Chronakis, IS. *Current Opinion in Colloid and Interface Science*, 2003, 8, 64–75.
15. Smith, LA; Ma, PX. *Colloids and Surfaces B: Biointerfaces*, 2004, 39, 125–131.
16. Dersch, R; Steinhart, M; Boudriot, U; Greiner, A; Wendorff, JH. *Polymers for Advanced Technologies*, 2005, 16, 276–282.
17. Venugopal, J; Ramakrishna, S. *Applied Biochemistry and Biotechnology*, 2005, 125, 147–157.
18. Scheibel, T. *Current Opinion in Biotechnology*, 2005, 16, 427–433.
19. Reneker, DH; Yarin, AL; Fong, H; Koombhongse, S. *Journal of Applied Physics*, 2000, 87, 4531–4547.
20. Yarin, AL; Koombhongse, S; Reneker, DH. *Journal of Applied Physics*, 2001, 89, 3018–3026.
21. Yarin, AL; Koombhongse, S; Reneker, DH. *Journal of Applied Physics*, 2001, 90, 4836–4846.
22. Shin, YM; Hohman, MM; Brenner, MP; Rutledge, GC. *Polymer*, 2001, 42, 9955–9967.
23. Shin, YM; Hohman, MM; Brenner, MP; Rutledge, GC. *Applied Physics Letters*, 2001, 78, 1149–1151.
24. Deitzel, JM; Kleinmeyer, J; Harris, D; Beck Tan, NC. *Polymer*, 2001, 42, 261–272.
25. Fong, H; Chun, I; Reneker, DH. *Polymer*, 1999, 40, 4585–4592.
26. Zong, X; Kim, K; Fang, D; Ran, S; Hsiao, BS; Chu, B. *Polymer*, 2002, 43, 4403–4412.
27. Demir, MM; Yilgor, I; Yilgor, E; Erman, B. *Polymer*, 2002, 43, 3303–3309.
28. Wang, H; Zhang, Y; Shao, H; Hu, X. *Journal of Materials Science*, 2005, 40, 5359-5363.
29. Kenawy, E-R; Layman, JM; Watkins, JR; Bowlin, GL; Matthews, JA; Simpson, DG; Wnek, GE. *Biomaterials*, 2003, 24, 907–913.
30. McKee, MG; Wilkes, GL; Colby, RH; Long, TE. *Macromolecules*, 2004, 37, 1760–1767.
31. Woerdeman, DL; Ye, P; Shenoy, S; Parnas, RS; Wnek, GE; Trofimova, O. *Biomacromolecules*, 2005, 6, 707–712.
32. Shenoy, SL; Bates, WD; Frisch, HL; Wnek, GE. *Polymer*, 2005, 46, 3372–3384.
33. Gupta, P; Elkins, C; Long, TE; Wilkes, GL. *Polymer*, 2005, 46, 4799–4810.
34. Shenoy, SL; Bates, WD; Wnek, G. *Polymer*, 2005, 46, 8990–9004.
35. Colby, RH; Fetters, LJ; Funk, WG; Graessley, WW. *Macromolecules*, 1991, 24, 3873–3882.
36. Krause, WE; Bellomo, EG; Colby, RH. *Biomacromolecules*, 2001, 2, 65–69.
37. Tsaia, PP; Schreuder Gibson, H; Gibson, P. *Journal of Electrostatics*, 2002, 54, 333–341.
38. Madhavamoorthi, P. *Synthetic Fibres*, 2005, 34, 12–18.
39. Pawlowski, KJ; Barnes, CP; Boland, ED; Wnek, GE; Bowlin, GL. *Journal of Materials Education*, 2004, 26, 195–206.
40. Grafe, T; Graham, K. *International Nonwovens Journal*, 2003, 12, 51–55.
41. Gibson, P; Schreuder-Gibson, H; Rivin, D. *Colloids and Surfaces, A: Physicochemical and Engineering Aspects*, 2001, 187–188, 469–481.
42. Tovmash, AV; Polevov, VN; Mamagulashvili, VG; Chernyaeva, GA; Shepelev, AD. *Fibre Chemistry*, 2005, 37, 187–191.
43. Shutov, AA. *Technical Physics Letters*, 2005, 31, 1026–1028.
44. Wang, X; Chen, X; Yoon, K; Fang, D; Hsiao, BS; Chu, B. *Environmental Science and Technology*, 2005, 39, 7684–7691.
45. Shin, C; Chase, GG; Reneker, DH. *Colloids and Surfaces, A: Physicochemical and Engineering Aspects*, 2005, 262, 211–215.
46. Matatov-Meytal, Y; Sheintuch, M. *Applied Catalysis A: General*, 2002, 231, 1–16.
47. Rodriguez, NM; Kim, M-S; Baker RTK. *Journal of Physical Chemistry*, 1994, 98, 13108–13111.

48. Jia, H; Zhu, G; Vugrinovich, B; Kataphinan, W; Reneker, DH; Wang, P. *Biotechnology Progress*, 2002, 18, 1027–1032.
49. Xie, J; Hsieh, Y-L. *Journal of Materials Science*, 2003, 38, 2125–2133.
50. Wang, Y; Hsieh, Y-L. *Journal of Polymer Science: Part A: Polymer Chemistry*, 2004, 42, 4289–4299.
51. Kim, J-S; Reneker, DH. *Polymer Composties*, 1999, 20, 124–131.
52. Bergshoef, MM; Vancso, GJ. *Advanced Materials*, 1999, 11, 1362–1365.
53. Fong, H; Liu, W; Wang, C-S; Vaia, RA. *Polymer*, 2002, 43, 775–780.
54. Sung, JH; Kim, HS; Jin, H-J; Choi, HJ; Chin, I-J. *Macromolecules*, 2004, 37, 9899–9902.
55. Larsen, G; Velarde-Ortiz, R; Minchow, K; Barrero, A; Loscertales, IG. *Journal of the American Chemical Society*, 2003, 125, 1154–1155.
56. Guan, H; Shao, C; Wen, S; Chen, B; Gong, J; Yang, X. *Materials Chemistry and Physics* 2003, 82, 1002–1006.
57. Choi, S-S; Lee, SG. *Journal of Materials Science Letters*, 2003, 22, 891–893.
58. Li, D; Xia, Y. *Nano Letters*, 2003, 3, 555–560.
59. Li, D; Wang, Y; Xia, Y. *Nano Letters*, 2003, 3, 1167–1171.
60. Li, D; Wang, Y; Xia, Y. *Advanced Materials*, 2004, 16, 361–366.
61. Li, D; Herricks, T; Xia, Y. *Applied Physics Letters*, 2003, 83, 4586–4588.
62. Guan, H; Shao, C; Chen, B; Gong, J; Yang, X. *Inorganic Chemistry Communications*, 2003, 6, 1409–1411.
63. Guan, H; Shao, C; Wen, S; Chen, B; Gong, J; Yang, X. *Inorganic Chemistry Communications*, 2003, 6(10), 1302–1303.
64. Hong, Y; Wang, C; Yang, Q; Li, Z. *Polymeric Materials Science and Engineering*, 2003, 89, 468–469.
65. Dharmaraj, N; Park, HC; Lee, BM; Viswanathamurthi, P; Kim, HY; Lee, DR. *Inorganic Chemistry Communications*, 2004, 7, 431–433.
66. Yang, X; Shao, C; Guan, H; Li, X; Gong, J. *Inorganic Chemistry Communications*, 2004, 7, 176–178.
67. Li, W-J; Laurencin, CT; Caterson, EJ; Tuan, RS; Ko, FK. *Journal of Biomedical Materials Research*, 2002, 60, 613–621.
68. Yoshimoto, H; Shin, YM; Terai, H; Vacanti, JP. *Biomaterials*, 2003, 24, 2077–2082.
69. Matthews, JA; Wnek, GE; Simpson, DG; Bowlin, GL. *Biomacromolecules*, 2002, 3, 232–238.
70. Kim, K; Yu, M; Zong, X; Chiu, J; Fang, D; Seo, Y-S; Hsiao, BS; Chu, B; Hadjiargyrou, M. *Biomaterials*, 2003, 24, 4977–4985.
71. Khil, M-S; Cha, D-I; Kim, H-Y; Kim, I-S; Bhattarai, N. *Journal of Biomedical Materials Research Part B: Applied Biomaterials*, 2003, 67B, 675–679.
72. Min, B-M; Lee, G; Kim, SH; Nam, YS; Lee, TS; Park, WH. *Biomaterials*, 2004, 25, 1289–1297.
73. Buchko, CJ; Chen, LC; Shen, Y; Martin, DC. *Polymer*, 1999, 40, 7397–7407.
74. Olabisi, O; Robeson, LM; Shaw, MT. *Polymer-Polymer Miscibility*, New York: Academic Press, 1979.
75. Folkes, MJ; Hope, PS. *Polymer Blends and Alloys*, Glasgow: Blackie Academic & Professional, 1993.
76. He, MJ; Chen, WX; Dong, XX. *Polymer Physics*, Shanghai: Fudan University Press, 1990.
77. Wang, P; Koberstein, JT. *Macromolecules*, 2004, 37, 5671–5681.
78. Bognitzki, M; Czado, W; Frese, T; Schaper, A; Hellwig, M; Steinhart, M; Greiner, A; Wendorff, JH. *Advanced Materials*, 2001, 13, 70–72.
79. Megelsiki, S; Stephens, JS; Chase, DB; Rabolt, JF. *Macromolecules*, 2002, 35, 8456–8466.
80. Li, M; Guo, Y; Wei, Y; MacDiarmid, AG; Lelkes, PI. *Biomaterials*, 2006, 27, 2705–2715.
81. Li, L; Hsieh, Y-L. *Carbohydrate Research*, 2006, 341, 374–381.
82. Lyoo, WS; Youk, JH; Lee, SW; Park, WH. *Materials Letters*, 2005, 59, 3558–3562.
83. Li, X; Nie, G. *Chinese Science Bulletin*, 2004, 49, 2368–2371.
84. Han, SO; Son, WK; Cho, D; Youk, JH; Park, WH. *Polymer Degradation and Stability*, 2004, 86, 257–262.
85. Bognitzki, M; Frese, T; Steinhart, M; Greiner, A; Wendorff, JH; Schaper, A; Hellwig, M. *Polymer Engineering and Science*, 2001, 41, 982–989.
86. Gupta, P; Wilkes, GL. *Polymer*, 2003, 44, 6353–6359.
87. Sun, Z; Zussman, E; Yarin, AL; Wendorff, JH; Greiner A. *Advanced Materials*, 2003, 15, 1929–1932.
88. Zhang, Y; Huang, Z-M; Xu, X; Lim, CT; Ramakrishna, S. *Chemistry of Materials*, 2004, 16, 3406–3409.
89. Li, D; Babel, A; Jenekhe, SA; Xia, Y. *Advanced Materials*, 2004, 16, 2062–2066.
90. Loscertales, IG; Barrero, A; Marquez, M; Spretz, R; Velarde-Ortiz, R; Larsen, G. *Journal of the American Chemical Society*, 2004, 126, 5376–5377.
91. McCann, JT; Li, D; Xia, Y. *Journal of Materials Chemistry*, 2005, 15, 735–738.
92. Li, D; Xia, Y. *Nano Letters*, 2004, 4, 933–938.

22

Nanopore Structures and Their Applications

Sanghyeon Choi and
Gunuk Wang
Korea University

22.1 Introduction

Nanometric tools have been widely used for various applications, including nanoscale imaging[1] and electronics[2]. Recently, the solid-state nanopore structure enables to be utilized for resistive random access memory (RRAM),[3,4] a single-molecule detector, and for DNA (or RNA) sequencing.[5,6] Diverse methods including ion-beam (or electron-beam) sculpting,[5,7–11] ion-track etching,[12–15] and electrochemical anodization[16–19] had been suggested for producing the solid-state nanopore. In this chapter, we describe diverse examples for established nanotechnologies for the nanopore formations and summarize their advantages and challenges. We also introduce a nanometric tool, which was recently proposed by De Vreede et al.[3] This nanometric approach utilizes the annealing process of metal nanoparticles on a ceramic substrate. This allows the metal nanoparticle to perpendicularly penetrate the ceramic substrate while evaporating, eventually forming a single solid-state nanopore inside the ceramic substrate with low cost and safety. Then, we discuss their possible applications: RRAM and biomolecule sequencing (via single-molecule detection).

22.2 Methods for Conventional Nanopore Structures

22.2.1 Ion-Beam (Electron-Beam) Sculpting

Ion-beam (electron-beam) sculpting is a means of fabricating a solid-state nanopore structure in an insulating solid-state material, based on an argon ion-beam etching process after forming a cavity on the opposite side.[5,7–11] Li et al. used an Si_3N_4 insulating material to create the nanopore structure with an argon ion beam.[8] To form a bowl-shaped cavity, a 500 nm Si_3N_4 layer was deposited on a Si substrate using the chemical vapor deposition, followed by photolithography and wet etching of Si to construct a free-standing Si_3N_4 layer. The bowl-shaped cavity can be formed near the center of the Si_3N_4 layer by reactive ion-etching process. Next, the opposite side of the sample was bombarded with an Ar beam of 3 keV. This allowed the Si_3N_4 surface without a cavity to eventually reach the bottom of the bowl-shaped cavity, as atoms of the Si_3N_4 surface are removed by ion-beam sculpting (see Figure 22.1a). Note that the ion beam can be replaced with an electron beam (i.e., electron-beam sculpting).[10,11] When the sample was ion-sculpted to the intersection between the Si_3N_4 surface and the bottom of the bowl-shaped cavity, a solid-state nanopore structure in Si_3N_4 layer was formed. Note that an apparatus that can count the ions transmitted through the nanopore and suppress its bombardment should be used because the exposure time of ion beam should be controlled for the formation of the well-defined nanopore. This apparatus should also control other important parameters, including sample temperature, ion-beam duty cycle, and ion-beam flux. Note that the ion-beam duty cycle is defined as t_{on}/t_{all}, where t_{on} is the ion-beam exposing time during "on" and t_{all} is the total time for pulsed beams when "on" or "off".

Although the ion-beam sculpting can create a well-defined solid-state nanopore structure, it is hard to fabricate at the room temperature. In fact, the transition from an open to closed nanopore is frequently observed at room temperature during ion-beam process. Figure 22.1b–d shows the transition that has been investigated by counting ions transmitted through the nanopore and by transmission electron microscopy (TEM) before and after ion-beam exposure. Second sputtering induced by ion beam can remove

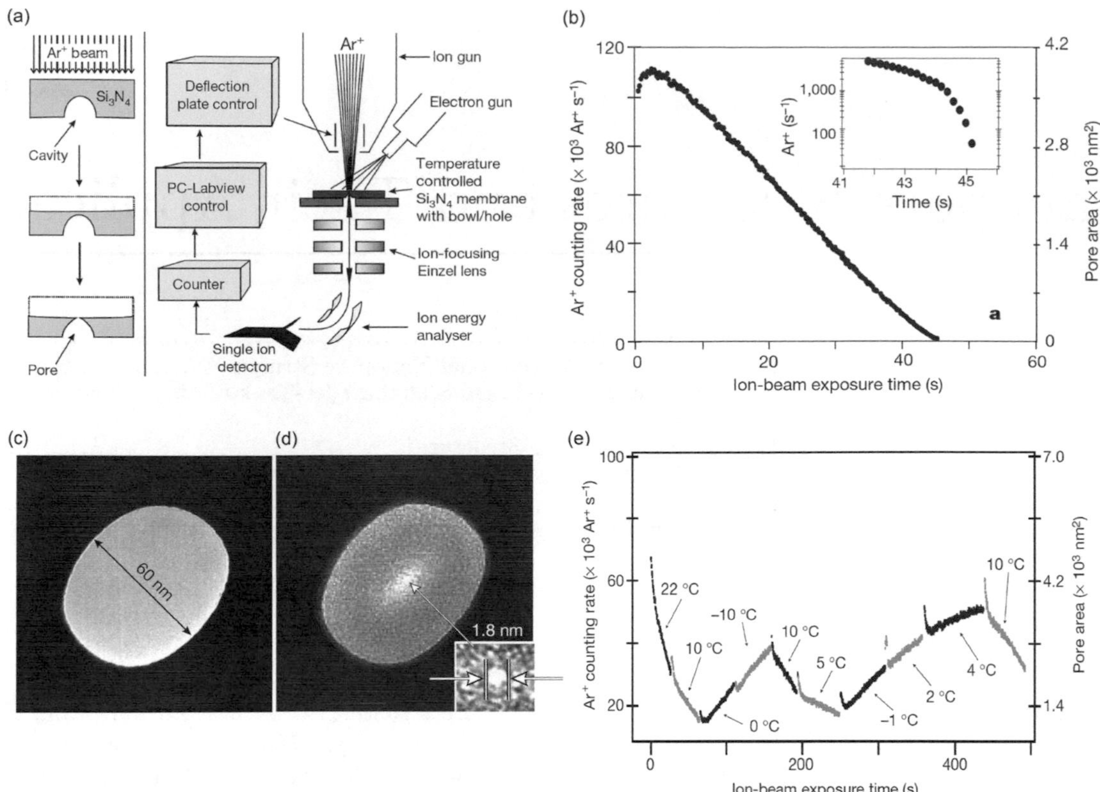

FIGURE 22.1 Ion-beam (or e-beam) sculpting. (a) Procedure for nanopore structure formation (left). An intersection between the even surface and the bowl-shaped cavity is formed by ion-beam bombardment, indicating the creation of a solid-state nanopore structure in the Si_3N_4 substrate. Equipment used for ion-beam sculpting through feedback control (right). (b) Ar^+ counting rate and pore area behavior as a function of ion-beam exposure time at 28°C. As the exposure time increases, both the ion-counting rate and pore area decrease. (c and d) TEM images of initial pore formed by focused ion beam (FIB) and closed pore after Ar^+ ion-beam exposure, respectively. (e) Transition behavior dependent on temperature. Opening and closing behaviors are determined by the temperature being higher or lower than around 5°C. (All figures are reproduced with permission from Li et al.[8] Copyright (2001) Springer Nature.)

surface atoms of the sample, and simultaneously, surface adatoms such as atoms or molecular clusters can diffuse to the pore, causing it to shrink. Note that this diffusion is generally explained by the adatoms diffusion model.[20] The transition to the shrunk pore is observed at around 5°C, implying that ion-beam sculpting requires lower temperature (see Figure 22.1e). In that case, the ion beam-induced electric field near the pore leads to accumulation of surface adatoms, as described by the modeling equation[8,21]:

$$X_m = \left(\frac{1}{D\tau_{\text{trap}}} + \frac{\sigma}{D}F \right)^{-\frac{1}{2}}$$

where X_m is the distance from the pore edge, D is the adatom diffusion coefficient, τ_{trap} is the trap lifetime, σ is the annihilation cross section, and F is the incident ion-beam flux. The surface adatoms within X_m are more likely to reach the pore than to be annihilated or trapped by ion beam and local surface traps, respectively. At very low temperature, it is believed that the diffusion effect on the sample can be reduced and X_m approaches zero value implying the suppression of adatom diffusion. As the temperature increases, however, the thermal activation can increase the adatom diffusion coefficient, so that X_m increases and the nanopore is stopped by diffusion. In addition, the ion-beam flux F affects the opening and closing of the nanopore owing to annihilation induced by the ion beam. Low F accelerates the adatom diffusion, leading to the nanopore closing.

To summarize, the solid-state nanopore structure can be fabricated on a sample with a prefabricated cavity using ion-beam sculpting. Counting the ions transmitted through the nanopore can affect the nanopore size and various parameters including temperature and ion-beam flux as well as the nanopore structure. Overall, ion-beam sculpting can create and manipulate a solid-state nanopore structure, but it includes some difficulties such as the need for sample preprocessing, complex apparatus, and the requirement for low temperature.

22.2.2 Ion-Track Etching

Ion-track etching, a type of selective etching, can be used to create nanopore structures in mainly soft materials such as polyethylene terephthalate (PET) and polyimide.[12–15] Initially, one side of a sample irradiated by heavy ions is in an etchant solution, whereas the opposite side of the sample is in a stopping solution such as potassium iodide (KI). When the side in the etchant solution has been etched by fragmentation of the area influenced by the heavy ions, and the sample has been completely penetrated, the stopping solution on the other side suppresses the etching process, creating diverse symmetric and asymmetric nanopore structures in the material, as shown in Figure 22.2a and b. In this way, the material is irradiated by heavy ions and then etched chemically. Note that the pore size and the number of pores can be controlled by varying the etching time and the number of ions per area, respectively.

Siwy et al. demonstrated the creation of a nanopore structure in polyimide using ion-track etching.[13] First, a sample of stacked polyimide foils (12.5 µm in thickness) was irradiated by heavy ions (uranium at 2,640 MeV) using a metallic mask to ensure the ions accessed only the desired area. Note that UV irradiation of the sample is performed before chemical etching to increase the sensitivity of the etching.[12] After irradiation, ion-track etching using sodium hypochlorite (NaOCl) was carried out at 50°C, where the specific temperature was selected to optimize the effectiveness of both ion-track etching and NaOCl decomposition. The NaOCl solution was at 12.6 pH and had an active chlorine (Cl) content of 13%. The ion-track etching was performed between two compartments filled with NaOCl and KI. The $2I^-$ in the stopping solution reacted with OCl^- ions in the etching solution, reducing OCl^- to Cl^- according to the reaction:

$$OCl^- + 2H^+ + 2I^- \rightarrow I_2 + Cl^- + H_2O \tag{22.1}$$

When the etchant had completely penetrated the samples, the chemical etching was suppressed through the above reaction. This suppression, occurring immediately upon penetration of the sample, results in a conical nanopore structure; otherwise, the size of the pore could increase above the nanometer scale, depending on the etching time. Detailed experimental results, equipment specifications, and purposes are described in the study by Siwy et al.[13]

As shown in Figure 22.2c, the nanopore structure can be formed in various shapes, both symmetric and asymmetric, depending on the chemical etching procedure. The conical nanopore structure resulting from the previously described procedure is an asymmetric shape. To construct a symmetrical nanopore (e.g., a cylinder), a sample is placed between two compartments, both filled with etchant solution. The chemical etching process is carried out bilaterally, resulting in a nanopore with a symmetric cylinder shape. Moreover, a symmetric hourglass shape can be obtained by etching the irradiated sample bilaterally and adding a stopping solution to slow down the subsequent chemical etching. Using such modifications of the chemical etching process, symmetric (cylindrical,[22,23] hourglass,[24,25] and cigar-like[26,27]) and asymmetric (conical[12–15] and bullet-like[28,29]) nanopore structures can be constructed in a material, as required for different research and practical applications.

In summary, ion-track etching is a method for creating solid-state nanopore structures in soft films. The first step exposes heavy ions to a sample such as PET and polyimide. This exposure forms an ion track in the sample, resulting in the exposed area being etched more extensively than the nonexposed area, owing to fragmentation. After the ion exposure, a suitable etchant solution is used to etch the ion-track area, which transforms it into the required nanopore structure. Furthermore, the nanopore structure can be manipulated to have various shapes, such

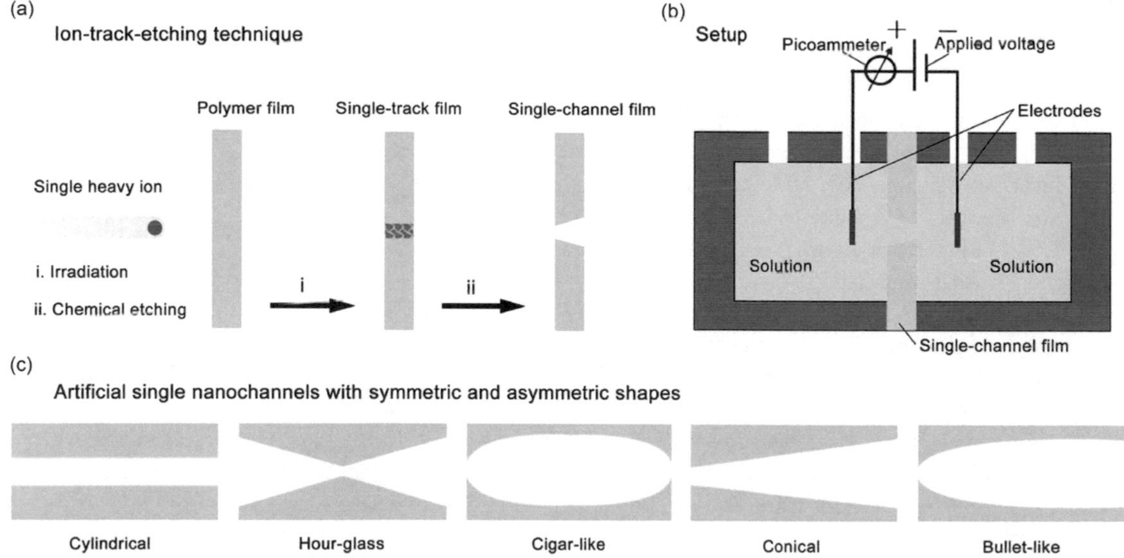

FIGURE 22.2 Ion-track etching. (a) Basic concept of ion-track etching. A substrate is irradiated by heavy ions, and then the fragile area induced by the irradiation is chemically etched. (b) Equipment used for chemical etching and control of ion-track etching. (c) Various nanopore structure shapes were produced with the ion-track etching technique. (All figures are from Zhang et al.[15] Reproduced by permission of The Royal Society of Chemistry. https://pubs.rsc.org/en/content/articlelanding/2013/cc/c3cc45526b#!divAbstract)

as cylindrical and conical, under suitable etching conditions. Despite these advantages, ion-track etching has some limitations, including high aspect ratio, cost, and complexity of the process. Note that the high aspect ratio of ion-track etching results from the thickness of the material used (e.g., the polymer film), which is of the order of 10 μm.

22.2.3 Electrochemical Anodization

Anodic oxides are used for electrochemical anodization, conducted with a two-electrode configuration. The anodic oxide is generally used as a protective or decorative layer on a metal surface. The oxide layer increases protection from corrosion, compared with a bare metal without any oxide layer, and also provides cosmetic effects such as dyes through thick porous coating. Recently, ordered and unordered nanopore structures have been fabricated using electrochemical anodization for various applications (e.g., RRAM,[16,17] sensors,[30,31] and drug delivery systems[32,33]).

The mechanism of the electrochemical anodization creating a solid-state nanopore structure is related to field-assisted transport and dissolution of the anodic oxide.[18,19] When a metal is anodized by application of a sufficient voltage in a galvanic cell, an oxidation reaction on the metal surface will initiate the following reactions (22.2–22.4)[18,19]:

$$M \rightarrow M^{z+} + ze^- \tag{22.2}$$

$$M + \frac{z}{2}H_2O \rightarrow TiO_{z/2} + zH^+ + ze^- \tag{22.3a}$$

$$M^{z+} + zH_2O \rightarrow M\left(OH\right)_z + zH^+ \tag{22.3b}$$

$$M\left(OH\right)_z \rightarrow MO_{z/2} + \frac{z}{2}H_2O \tag{22.4}$$

Simultaneously, hydrogen evolution occurs on the cathode, according to reaction (22.5):

$$zH_2O + ze^- \rightarrow \frac{z}{2}H_2 \uparrow + zOH^- \tag{22.5}$$

When an initial oxide layer has been formed on the metal surface by the anodization process, O^{2-} ions migrate toward the oxide/metal interface through the oxide layer and M^{z+} ions migrate toward the electrolyte/oxide interface through the oxide layer by field-assisted transport (see Figure 22.3a). This leads to further growth of the oxide layer either at the oxide/metal interface or at the electrolyte/oxide interface (in most cases, growth of the oxide layer at the oxide/metal interface is observed). According to the relation $V = Ed$, the electric field $E = V/d$ decreases with decreasing oxide thickness under a constant voltage V, indicating that migration of the ions gradually decreases with the depth of the oxide layer. Thus, the final thickness of oxide layer depends mainly on the anodization voltage, because the electric field is eventually lost, at which point migration of the ions no longer occurs.

Although a conventional anodic oxide (compact oxide) on a metal surface is formed through reactions (22.2–22.4) in the absence of fluoride ions (F^-), different oxide structures can be formed in the presence of F^- (Figure 22.3b). This is because F^- can etch the oxide layer to generate water-soluble species, and M^{z+} ions react with F^- to generate soluble species following reactions (22.6 and 22.7) (e.g., Ti metal):

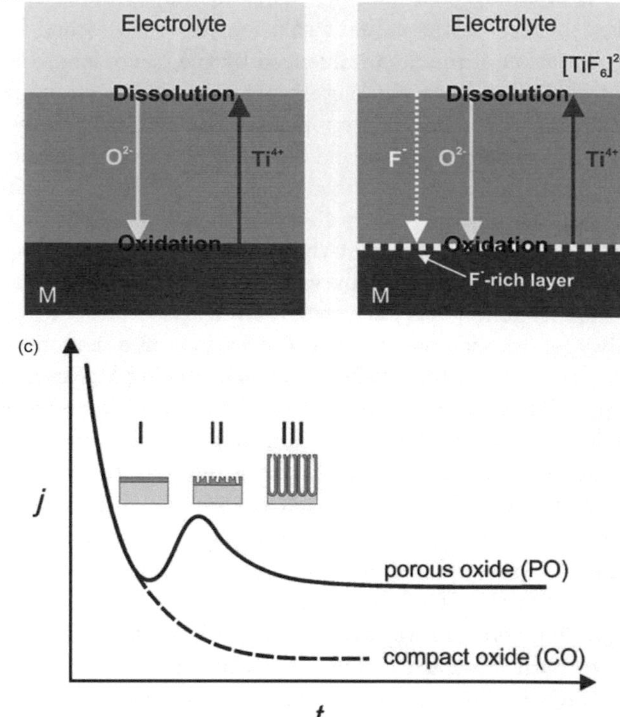

FIGURE 22.3 Electrochemical anodization. (a and b) Schematic diagrams of the growth of anodic oxide, influenced by field-assisted transport, and of the etching process with F^- ions due to formation of water-soluble species. (c) A typical current–time (I–t) curve for the electrochemical anodization, showing an exponential decay (step I), an increase (step II), and the quasi steady state (step III), according to F^- ion concentration. (All figures are reproduced with permission from Roy et al.[19] Copyright (2011) John Wiley and Sons.)

$$TiO_2 + 6F^- + 4H^+ \rightarrow \left[TiF_6\right]^{2-} + 2H_2O \tag{22.6}$$

$$Ti^{4+} + 6F^- \rightarrow \left[TiF_6\right]^{2-} \tag{22.7}$$

Depending on the F^- ion concentration, there are three cases: (i) free F^- ions, (ii) a high concentration of F^- ions, and (iii) an intermediate concentration of F^- ions.[34] In the case of free F^- ions, the F^- content of the electrolyte during anodization process is low, and a compact oxide will be formed on the metal surface. On the other hand, a high concentration of F^- ions results in all oxide and metal ions generating water-soluble species, so no anodic oxide is formed on the metal surface. In case of an intermediate concentration of F^- ions, there is competition between oxidation and dissolution, allowing a nanopore structure to be formed in the oxide layer.

Figure 22.3c shows a typical current–time (I–t) curve for electrochemical anodization with an intermediate concentration of F^- ions. The process consists of three steps: an exponential decay (step I), an increase (step II), and the quasi steady state (step III). Step I represents the formation of a compact oxide layer owing to the presence of free F^- ions, which decreases the electrical conductivity (an exponential drop). In step II, the initial compact oxide is etched according to reaction (22.6), and irregular nanopores are formed at the electrolyte/oxide interface because of defects such as impurities or roughness of the compact oxide. This etching (and thus reduction) of the oxide layer by F^- ions increases the current, which is reflected by the increase in the I–t curve. In step III, a regular nanopore or nanotube layer is formed, owing to initiation of competition between individual nanopores for the available currents; eventually, the nanopores equally share the available currents. This is because the electric field through the remaining oxide layer increases as the nanopores form. When the growth rate of oxidation is equal to the dissolution rate of F^- ions, a constant current is observed, indicating a quasi steady state. Note that the nanopore or nanotube layer forms a V-shape,[35] that is, a conical nanopore structure, owing to the less dense nature of the inner part of the oxide, which results in faster dissolution compared with the outer oxide part.[35] Furthermore, several factors such as diameter, structure, and geometry can be controlled by changing different parameters, including voltage, time, temperature, and electrolyte composition, as discussed in various review papers.[19,36,37]

In summary, electrochemical anodization is a nanometric tool that can be used to create a V-shaped (conical) nanopore or nanotube structure, based on field-assisted transport and dissolution. There are three cases, based on the F^- ion concentration: free, high concentration, and intermediate concentration. With an intermediate concentration of F^- ions, a compact oxide is initially formed on the metal surface, as in the case of free F^- ions. Subsequently, irregular nanopores are formed, leading eventually to a regular nanopore or nanotube structure in which the nanopores are V-shaped (conical) owing to the looser nature of the inner part of the oxide. However, electrochemical anodization has some limitations, including the requirement for hazardous solutions such as hydrofluoric acid (HF) and the difficulty of fabricating a single nanopore structure.

22.3 Conical Nanopore Structure

22.3.1 Fabrication

Figure 22.4a–d shows the fabrication procedure for the conical nanopore structure reported by Kwon et al.[4] Initially, SiO_x/Pt/Ta layers on an SiO_2/Si substrate were prepared. The SiO_2/Si substrate was thoroughly cleaned with acetone, isopropyl alcohol, and deionized water for 5 min. Pt and Ta metals were deposited using direct current (DC) sputtering, followed by deposition of an SiO_x layer (25 or 75 nm) on the Pt/Ta using a layer e-beam evaporator, under a pressure of 10^{-6} Torr and a deposition rate of 0.5 Å/s. To form a microscale circular gold (Au) patch, conventional photolithography was performed on the SiO_x/Pt/Ta substrate, and an Au patch of 1.5 μm diameter and 18 nm thickness on the photoresist (PR) patterned substrate was deposited by an e-beam evaporator under a pressure of 10^{-6} Torr and a deposition rate of 0.5 Å/s. Then, a lift-off process was performed to remove the PR and leave the circular Au patch. Using a programmable oven, the Au patch/SiO_x/Pt/Ta layer was heated in ambient air

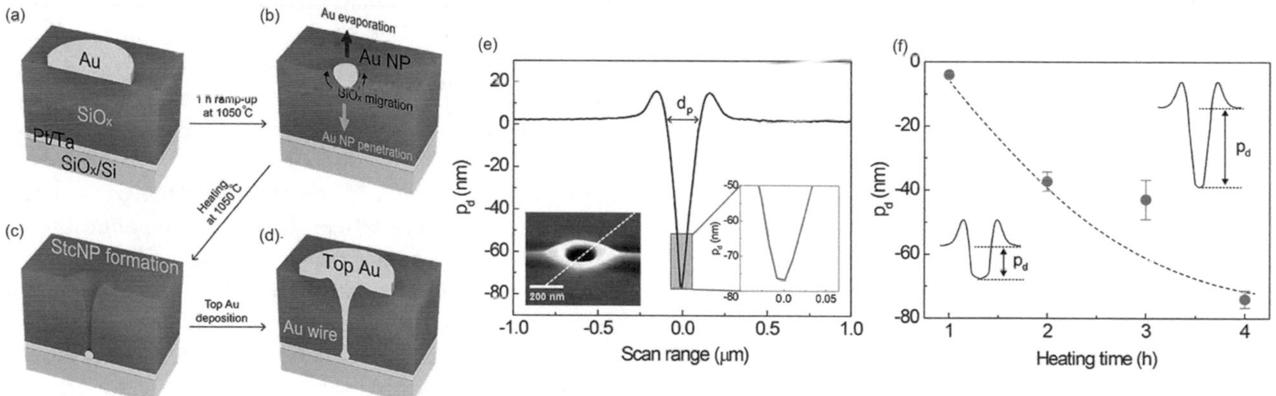

FIGURE 22.4 Fabrication procedure and investigation of nanopore structure using metal penetration. (a) Through conventional photolithography, an Au circular patch of 1.5 μm in diameter was patterned on an SiO_x/Pt/Ta substrate. (b) The Au circular patch/SiO_x/Pt/Ta layer was annealed in ambient air at 1,050°C with a 1 h ramp-up time. (c) After the temperature (1,050°C) had been maintained for a set time, the nanopore structure was formed. (d) Top Au deposition for SiO_x memory fabrication. Note that this process does not have to be performed when the nanopore structure is intended for applications other than RRAM. (e) Investigation of the nanopore structure fabricated by metal penetration through the noncontact mode of atomic force microscopy (AFM). The structure exhibited a conical nanopore structure that abruptly became narrow. (f) Change in pore depth (d_p) with heating time. As the heating time increased, the pore depth increased. (All figures are reproduced with permission from Kwon et al.[4] Copyright (2017) American Chemical Society.)

for different amounts of time at 1,050°C (the ramp-up time was 1 h). After heating for the desired length of time, the programmable oven was allowed to cool naturally to room temperature.

During the initial dewetting process in the oven, the patterned Au patch was cohered to form Au nanoparticles surrounded by a SiO_x ridge, after which the Au particles penetrated the SiO_x, creating a nanopore structure. The ridge was observed occasionally, as the heating temperature approached about half the ceramic melting point[3]; a detailed mechanism is given in Section 22.3.2. In brief, the Au particles move vertically into the SiO_x as the pore diameter decreases, and as the heating time increases, the Au particles further penetrate the SiO_x layer and bottom of the nanopore structure becomes narrower, owing to the thermally evaporated Au (see Figure 22.4e and f). Eventually, as the process continues, a vertical nanopore structure is created that becomes abruptly narrower with increasing depth, as a result of the annealing of the Au nanoparticle on the SiO_x ceramic substrate.

22.3.2 Mechanism

The melting temperature of Au is 1,064°C, which is close to the heating temperature used for the Au circular patch (1,050°C) in the fabrication process. At 1,050°C, some of the Au on the ceramic substrate forms nanocrystals that contain planes such as (111), while the remainder undergoes surface melting. Theoretical models and analyses predict that Au nanoparticles below a certain diameter will be melted at 1,050°C, whereas the Au nanoparticles suitably coated or embedded in a solid matrix will either melt or show superheating, forming a solid phase that persists in its metastable solid state under temperatures higher than the melting point, depending on the surface energy. A previous report presented some evidence for the latter,[38] the relation of surface (interfacial) energy between different parameters (e.g., cohesive energy, melting temperature, and vacancy formation energy for the Au nanocrystals on/in SiO_2). When the temperature is close to the melting temperature of the bulk Au, the melting behavior of the Au (111) can be divided into reconstructed and unreconstructed cases.[39] The reconstructed surface persists up to the bulk melting temperature, while several outer layers of the unreconstructed surface melt below the melting temperature, possibly owing to the boundary between the Au surface and air or SiO_2. The remaining outer layers retain some layers of movable atoms, according to the surface (interfacial) energy.[40]

When the annealing temperature is about half of SiO_2 melting temperature, a ridge encircling the Au nanoparticle on SiO_2 is formed. A metal nanoparticle encircled by the ridge on the ceramic substrate is usually observed when the annealing temperature increases to about the half the melting temperature of the ceramic substrate. This is due to ceramic migration along the surface of metal nanoparticle, resulting from surface diffusion toward the air/particle/ceramic triple line.[41–44] At the triple line, the equilibrium of the three interfaces is described by the Herring equation[43,45]:

$$\sum_{i=1}^{3}(\gamma_i\vec{t_i} + \frac{\partial\gamma_i}{\partial\vec{t_i}}) = 0$$

where γ_i is the interface energy; $\vec{t_i}$ is the vector in the plane of the ith interface, which is perpendicular to the triple line and points in an outward direction; and $\frac{\partial\gamma_i}{\partial t_i}$ is the torque, defined as the derivatives of interfacial energy corresponding to each vector. In the final equilibrium, material reconfiguration leads to the disappearance of the $\frac{\partial\gamma_i}{\partial t_i}$ term, and the Smith relation $\left(\frac{\gamma_{23}}{\sin\theta_1} = \frac{\gamma_{13}}{\sin\theta_2} = \frac{\gamma_{12}}{\sin\theta_3}\right)$ can explain the equilibrium,[46] where subscripts 1, 2, and 3 represent the respective phases at the three interfaces, and θ_1, θ_2, and θ_3 are the angles formed by the phases. However, because Au nanoparticles evaporate, decreasing the torque at the triple line, the Smith relation is insufficient to describe the equilibrium in these new nanotechnology systems. Thus, De Vreede et al. suggested repetitive remodeling of the SiO_2 ceramic substrate, taking into consideration the decreased torque at the triple line by Au evaporation. Note that the diffusional migration is fast, leading to swift reconfiguration owing to the small dimensions of the system. Furthermore, the pore configuration could be a kinetic rather than an equilibrium configuration, because only the SiO_2 close to the Au is movable, whereas that distant from the Au is essentially immovable.[47] Figure 22.5a–f shows the schematics of the suggested mechanism by De Vreede et al. The system tends toward an equilibrium state, where the angles are estimated to be $\theta_{Au} \sim 170°$, $\theta_{SiO_2} \sim 95°$, and $\theta_{air} \sim 95°$, given that the interface energies[48,49] at the three interfaces should satisfy the Smith relation. During the initial dewetting process, SiO_2 ceramic migration toward the triple line results in the formation of the SiO_2 ridge surrounding the Au nanoparticle, so that the center of the Au nanoparticle is on the SiO_2 ceramic substrate. This is followed by the migration of the SiO_2 around the Au nanoparticles above the substrate, as well as a slight bulge configuration of the Au nanoparticle. Note that the changed configuration of the Au nanoparticle follows the equilibrium angle $\theta_{air} \sim 95°$. As the heating time increases, the Au nanoparticle further penetrates into the SiO_2 ceramic substrate, leading to formation of the nanopore structure. This can be attributed to SiO_2 ceramic migration toward the triple line and the constantly evaporating Au nanoparticle. Also, the triple line gradually moves toward the top of the Au nanoparticle, in order for the system to satisfy the Smith relation. Note that given the equilibrium angle $\theta_{air} \sim 95°$, the nanopore structure will not be closed because $2\theta_{air} > 180°$. Assuming that the top side of the Au nanoparticle contains the refractory (111) plane[50] and θ_{SiO_2} is similar to the equilibrium value, the triple line will be "pinned" at the edge of the plane due to the energy increase induced by a decrease in the amount of Au containing the (111) plane that has low energy and by θ_{SiO_2} modulation. Thus, the

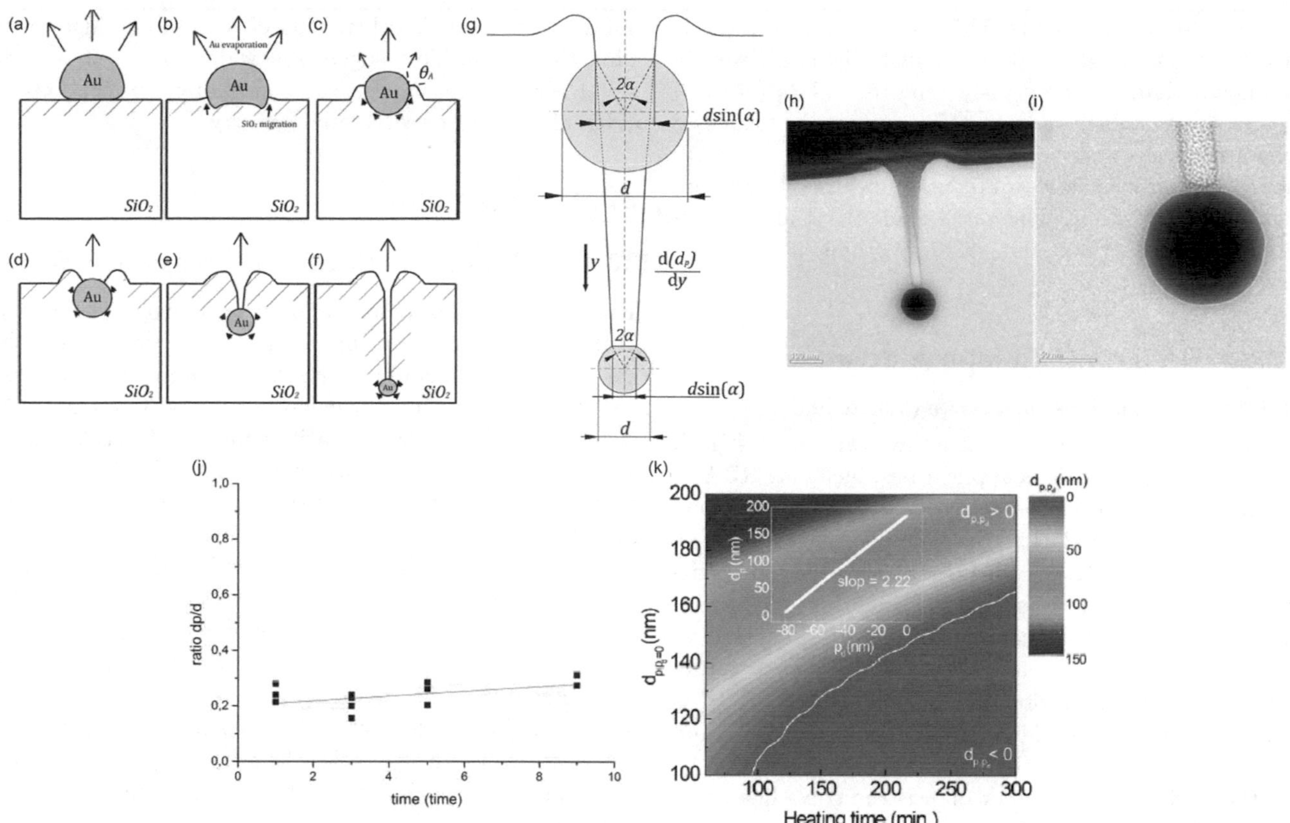

FIGURE 22.5 Schematic diagrams of mechanism and design of nanopore structure. (a–f) Schematic diagrams show individual processes from the initial dewetting to the formation of the nanopore. (a) After the initial dewetting process of Au, (b and c) an SiO_2 ridge surrounding the Au was formed. (d–f) SiO_2 migration toward the triple line and the constantly evaporated Au led to formation of the nanopore structure. (g) Modeling of the nanopore structure formed by metal penetration. The relationship between the Au nanoparticle and pore diameters is described by $d_{\text{pore}} = d_{\text{Au}}\sin(\alpha)$, while $d(d_{\text{pore}})/dy = -3/2\sin^3(\alpha)$ determines the pore-narrowing rate. Note that the d_{pore} and d_{Au} correspond to d_{p} and d in this figure, respectively. (h) Cross-sectional STEM image of nanopore structure after heating for a set time at 1,050°C. (i) Enlarged STEM image of the bottom of the nanopore structure. (j) $d_{\text{pore}}/d_{\text{Au}}$ values for different heating times. The ratio was almost constant at ~0.26. (All figures are reproduced with permission from de Vreede et al.[3] Copyright (2015) American Chemical Society.) (k) Contour plots for design guidelines based on the correlation between pore diameter (d_{p}) and pore depth (p_{d}). The inset shows the constant ratio of $d_{\text{p}}/p_{\text{d}}$. (The figure is reproduced with permission from Kwon et al.[4] Copyright (2017) American Chemical Society.)

angular diameter of the Au (111) influences the diameter of the nanopore structure. When the angular diameter is 2α, the diameter of the nanopore can be estimated as follows (see Figure 22.5g):

$$\sin\alpha = \frac{d_{\text{pore}}}{\frac{d_{\text{Au}}}{2}}$$

$$d_{\text{pore}} = \frac{d_{\text{Au}}}{2}\sin\alpha \times 2 = d_{\text{Au}}\sin\alpha.$$

In addition, the diameter of the Au nanoparticle is decreased by Au evaporation as the depth increases, so that the nanopore structure becomes narrow. Note that the top side of the Au nanoparticle will be not closed, owing to the pinned triple line. The relation between pore diameter and depth can be expressed by the following equations (see Figure 22.5g):

$$dV_{\text{Au}} = \text{d}\left(\frac{1}{6}\pi d^3\right) = -\frac{\pi(d\sin\alpha)^2}{4}dy$$

$$\frac{d\left(V_{\text{Au}}\right)}{dy} = -\frac{3}{2}\sin^2\alpha$$

$$\frac{d\left(d_{\text{pore}}\right)}{dy} = \frac{d\left(d\sin\alpha\right)}{dy} = -\frac{3}{2}\sin^3\alpha$$

$$\therefore \frac{d(d_{\text{pore}})}{dy} = \left(-\frac{3}{2}\sin^3(\alpha)\right)$$

where y is the coordinate perpendicular to the ceramic substrate. The pore-narrowing rate $(d(d_{pore})/dy)$ shows that the nanopore diameter depends linearly on the depth of the nanopore structure when the Au nanoparticle penetrates into the SiO_2 ceramic substrate. Based on the scanning transmission electron microscopy (STEM) cross-section images (Figure 22.5h and i), the results of the theoretical calculations were compared with the experimental values. The angular diameter was estimated to be $2\alpha = {\sim}33°$, based on the Au diameter (105 nm) and pore diameter (30 nm). Moreover, the pore-narrowing rate (−0.06) obtained from the images shown in Figure 22.5h and i gave an angular diameter of $2\alpha = \sim 35°$. Both angular diameters were a little higher than the previously reported values (25–30°) for (111) Au crystals in vacuum.[50] To support the suggested mechanism, the ratios of diameters ($d_{\text{pore}}/d_{\text{Au}}$) for 12 nanopores, obtained with different annealing times, are shown in Figure 22.5j. The ratio maintained an approximately constant value of ~0.26 at different annealin times.

22.3.3 Design of Nanopore Structure

Based on two equations, $d_{\text{pore}} = d_{\text{Au}} \sin\alpha$ and $\frac{d(d_{\text{pore}})}{dy} = (-\frac{3}{2}\sin^3(\alpha))$, the nanopore structure can be designed. This is desirable for various applications, including RRAM, single-molecule detection, and DNA (or RNA) sequencing, which require manipulation of diameter and depth. As discussed in Section 22.3.2, when the Au nanoparticle penetrates the SiO_2 ceramic substrate, the theoretical relations between the Au diameter, pore diameter, depth, and narrowing rate can be used to control the nanopore structure formed by metal penetration, as the pore diameter d_{pore} is dependent on the initial diameter of Au nanoparticle d_{i} and decreases as the depth of the pore increases (Figure 22.5). After the initial annealing, not only is the triple line gradually tied up at the (111) Au plane, but the pore structure also becomes narrow. Given this combination of materials, the maximum of pore depth (p_d) can be estimated by $p_{d,\text{max}}=2d_i/(3\sin^2(\alpha)) + p_{d,i} \simeq 940$ nm for $d_i = 100$ nm and $p_{d,i} = 200$ nm, where d_i is the initial diameter of the Au nanoparticle and $p_{d,i}$ is the initial depth of the pore.

Moreover, Kwon et al. suggested a method for designing nanopore structure in the same Au/SiO_x combination, based on the correlation between d_{pore} and p_d according to heating time.[4] First, the pore-narrowing ratio ($d(d_{\text{pore}})/dy$) was estimated to be ~−2.22, that is, d_{pore} decreased more abruptly for a given p_d than a reported rate of previous paper,[3] presumably owing to the effects of stoichiometry, morphology, and phase of ceramic substrate on various interface energies between Au and SiO_x. After estimation of the ratio, the correlation between pore diameter and heating time was investigated by polynomial fitting of Figure 22.4f, as shown in the contour plot in Figure 22.5k. This modeling provides a design guideline for physical nanopore fabrication using heating conditions, although for simplicity it disregards the possibility of modulating the narrowing ratio.

In brief, the nanotechnology discussed here is useful for creating strictly confined single nanopore structures and arrays, with potential applications in various fields including nanoelectronics and bio-applications. The metal penetration method utilizes the ceramic migration that occurs when the heating temperature approaches about half the metal's own melting point, owing to diffusion toward the air/particle/ceramic triple line. This simple, cheap, and safe technique could theoretically be applied to other metal/ceramic combinations, as long as the materials do not react with each other, and do exhibit a ridge encircling the metal. This suggests a wide variety of possible metal/ceramic materials, as well many applications in different areas of research and industry.

22.4 Applications

22.4.1 Resistive Random Access Memory

There is increasing demand for new nonvolatile forms of memory for use in electronic products and computing technology, as the commonly used flash memory reaches its physical limits.[51] One of the strongest candidates to replace flash memory is RRAM, owing to its simplicity,[52] scalability,[53] and fast and low-energy switching[3,16,17]. RRAM is usually referred to as a type of memristor that is able to memorize its resistance state based on the history of applied electrical stimulus.[54,55] The structure of RRAM consists of a metal oxide material sandwiched between two conducting electrodes, and it can alternate between a high- and low-resistance state according to the applied voltage and current. Generally, nanoscale conductive filaments (CFs) in oxide-based RRAM are thought to be responsible for this resistance switching; in principle, their formation and dissolution result from the dominance between an electric field and Joule heating.[55] Although nanoscale CFs potentially offer a way to overcome the scaling issues that limit the use of flash memory,[53,56,57] the randomness of their formation and their tendency to form multiple fragile CFs, instead of a single robust CF, lead to the degradation of device uniformity, switching performance, and device reliability.[58–60] Therefore, if they are to be of use in high-performance RRAM, tight control of CFs is required to overcome these problems. In this section, we introduce an example of fast and scalable SiO_x memory, based on a conical nanopore structure for control of CFs.

Kwon et al. suggested that the number, size, and length of CFs could be reliably controlled by a well-defined vertical single truncated conical nanopore (StcNP) SiO_x structure,[4] which could be fabricated by metal penetration into ceramic materials (see Section 22.3). Following the fabrication process in described in Section 22.3.1, an StcNP structure was constructed in the SiO_x active layer, and then an Au electrode was deposited for the formation of Au wire via the vertical StcNP using e-beam evaporation, as shown in Figure 22.4a–d. Using a single voltage sweep, the initial breakdown was performed to form a confined Si nanocrystal (Si-NC) CF in the StcNP structure where the current abruptly dropped, that is, by an electromigration process. Notably, depending on the electrical input, the semimetallic Si nanocrystals (Si-NC) in SiO_x increased and decreased, inducing a nonvolatile unipolar switching behavior of the SiO_x-based memory.[17] In case of planar SiO_x, similar breakdown phenomena have been reported,[61–63] and the Si-NC filament through nanosized gap following the breakdown has been investigated using real-time images obtained by in situ TEM[61] and electroluminescence spectrum profiles.[63]

During electromigration in the Au wire, the SiO_2 active layer undergoes an intrinsic post-breakdown process and a high electric field at the nanosized gap resulted from the collapsed junction, leading to the redox reaction from the SiO_x to Si phase (Figure 22.6a).[17,61–63] The electromigration around the nanosized gap results in a confined CF, demonstrating that randomness and formation of multiple fragile CFs can be sufficiently suppressed. The breakdown voltage (V_b) shown in Figure 22.6a can be deemed the electroforming voltage that results in the initial switchable-state device. As shown in Figure 22.6b, a statistical evaluation of V_b (across 18 devices) for StcNP SiO_x RRAM gave a V_b of 4.03 ± 0.22 V, which was half the corresponding deviation for multiple nanopore SiO_x memory (± 0.4 V).[17] This may have been due to the relatively uniform Au wire, in which individual devices undergo a similar electromigration process. After the initial device was set to a switchable state, the StcNP SiO_x device showed typical unipolar switching and a high ON/OFF ratio (>10^7). The current–voltage (I–V) curve showed that the current abruptly increased at ~4 V, which was defined as the set voltage V_{set} that could change the device to its ON state. The current then decreased and fluctuated at ~9 V, which was defined as the reset voltage V_{reset} that could revert the device to its OFF state. Depending on the programmed voltage, which ranged from 5 to 15 V with an interval of 2 V, the resistance at the read voltage 1.0 V (V_{read}) increased from 1.01×10^5 to 2.63×10^{11}, exhibiting multiple (six) states. This suggests that the StcNP SiO_x device has potential applications as high-capacity nonvolatile memory (Figure 22.6c). The elevated resistance results from the degree of contraction of the Si-NC by amorphization, as the applied voltage increases. In addition, the nonvolatile switching reliability of the device was evaluated to be acceptable in terms of retention (10^4 s) and endurance (10^3 times), as well as having a fast switching speed (up to 6 ns).

The switching performance was investigated according to the physical dimensions and number of Au wires (or pores), as shown in Figure 22.6d and e. Different lengths (25 and 75 nm) and numbers (1, 10, and 100) of Au wires were used, with p_d (#) denoting the number (#) of Au wires of length p_d. For example, "25 (1)" denotes a device with a single Au wire of ~25 nm length, and "75 (100) indicates that

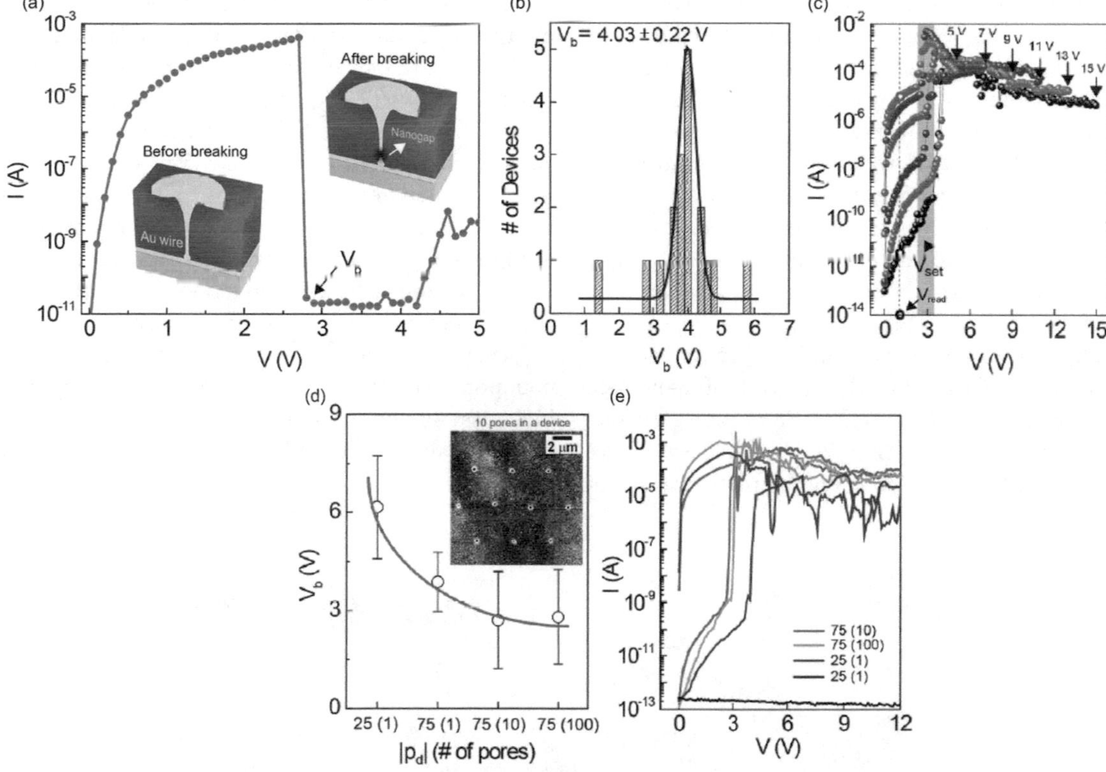

FIGURE 22.6 Basic concepts of RRAM devices and StcNP SiO_x memory based on metal penetration technique. (a) Breakdown (electroforming) by the electromigration process. The inset exhibits a nanogap formed by the breakdown process. The schematic diagrams correspond to before (left) and after (right) breaking, respectively. (b) Histogram and normal distribution of statistically evaluated V_b. (c) Multiple states of the StcNP SiO_x memory device depending on the applied voltage (from 5 to 15 V). (d) A plot of V_b according to the physical dimensions of the nanopore. Note that p_d (#) represents the number (#) of nanopores with pore depth (p_d) in a StcNP SiO_x memory device. The inset shows an AFM image of the 75 (10). (e) Representative current–voltage (I–V) switching behaviors of the StcNP SiO_x memory devices in cases 25 (1), 75 (10), and 75 (100). (All figures are reproduced with permission from Kwon et al.[4] Copyright (2017) American Chemical Society.)

the device has 100 Au wires of ~75 nm length. Figure 22.6d exhibits the different V_b values, which depend on the values of p_d, as a larger Au wire requires a higher V_b value during electromigration.[64–66] Note that in case of p_d = 75 nm, similar V_b values were obtained regardless of the number of Au wires in the device (Figure 22.6d). This showed that V_b was not dependent on the number of Au wires connected in parallel between the top and bottom electrodes. The 25 (1) devices mainly exhibited switching failure after the breakdown process, while the 75 (1), 75 (10), and 75 (100) devices mostly demonstrated switching performance, as shown in Figure 22.6e. However, as the number of Au wires in a device increases from 1 to 100, the chance that multiple Si-NC filaments are created in the StcNP structure probably increases. Therefore, a device with multiple Au wires tends to exhibit similar issues to multiple nanopore SiO_x memory (e.g., instable switching).[17] The switching failure and the degraded performance for the 25 (1) device might result from the creation of a larger gap than in the case of p_d = 75 nm after the electromigration process. Owing to the lower electric field at a given programmed voltage, there might be a reduced probability of Si phase transition, increasing the uncertainty of Si-NC creation. The 25 (1) device also has more opportunities to break during the repeated switching operation, leading to an increased chance of switching failure. In a similar context, depending on the physical dimensions of the Au wires, various nanogap sizes after the electromigration process were observed in the planar Au wires on a SiO_2 substrate.[4] Furthermore, Kwon et al. investigated the breakdown process that formed the nanogaps in Au wires, using a truncated conical structure and induced temperature during electromigration process. This analysis contributed to the literature on designing and controlling a single nanoscale CF; a detailed explanation is given in Kwon et al.[4]

Section 22.4.1 provides a good example of how a nanopore structure can be employed for the control of nanoscale CFs. The well-defined vertical StcNP SiO_x structure was used to confine the CF, resulting in improved switching performance and reduced variation. Therefore, the nanopore structure represents a useful means to improve memory devices through a filamentary mechanism.

22.4.2 Biomolecule Sequencing (Single-Molecule Detection)

There has been great interest in designing faster and lower-cost genome sequencing technologies for crucial applications in healthcare, genetics, and drug development, since the National Institutes of Health announced its genome plan in 2004. One of the most promising sequencing tools is the recently developed nanopore-based technique, which has the benefits of single-molecule detection and label-free sequencing.[5,6,67,68] The principle of sequencing based on nanopore structure can be briefly explained as follows. A membrane containing a nanopore structure is placed in a chamber divided into two compartments filled with conducting electrolytes. The two electrodes are placed across the nanopore structure. Then, using an apparatus for electrical measurements, an ionic current can be obtained as the charged ions migrate through the nanopore, depending on the applied voltage, as shown in Figure 22.7a. The nanopore-based architecture can sensitively detect the change in the ionic current induced by a molecule passing through the nanopore structure (Figure 22.7b). Based on electrical information, the structure and motion of molecules can be investigated even at a single-molecule level, enabling biomolecules to be sequenced. In other words, when a single strand of DNA passes through the nanopore structure, ionic currents that correspond to the four different bases (A, T, C, and G) will be observed; thus, the DNA can be sequenced by recording the ionic current. The nanopore structures used for biomolecule sequencing are generally classified into biological and solid-state nanopore structures. The solid-state nanopore structures have distinct advantages, such as stress tolerance,[67,68] stability,[69,70] and control of the nanopore structure[71–74] or array,[75] in comparison to the biological nanopore structures. For this reason, we focus here on fabrication and detection schemes for solid-state nanopores.

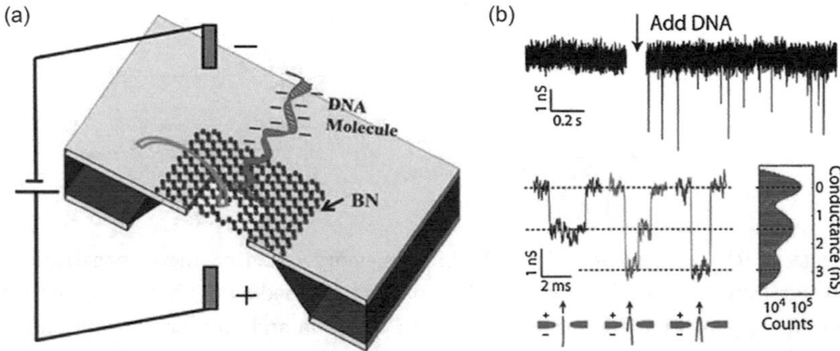

FIGURE 22.7 Nanopore-based technique for biomolecule sequencing. (a) A schematic diagram of nanopore-based sequencing. Reproduced with permission from Liu et al.[76] Copyright (2013) John Wiley and Sons.) (b) Detection of signals induced by biomolecules through the nanopore. According to the presence/absence of biomolecules, changes in the current can be observed and analyzed. (The figure is reproduced with permission from Schneider et al.[77] Copyright (2010) American Chemical Society.)

In addition to the ion-beam (electron-beam) sculpting, ion-track etching, electrochemical anodization, and metal penetration approaches that can be applied to produce nanopores for sequencing, a technique using a scanning helium ion microscope to create solid-state nanopores has been developed by Yang et al.[78,79] Fabrication using a helium ion microscope improves yield and lowers costs, because it is possible to form nanopore arrays of sufficiently small size in a relatively short time. Kwok et al. formed a single nanopore structure of less than 2 nm in size, with sub-nanometer controllability, based on the control of dielectric breakdown in the solution.[80] The nanometric tool uses a high electric field generated by the applied voltage, or the material's dielectric strength is locally controlled, to form a nanopore structure with monitoring of the induced leakage current, which reduces fabrication complexity and cost. Atomic layer deposition (ALD) was employed to control surface characteristics and deposit various materials for the formation of nanopore structures. Chen et al. modified the surface characteristics of nanopores using ALD and improved DNA translocation,[81] and Venkatesan et al. utilized aluminum oxide as a nanopore material to improve electrical performance and durability.[82]

Monitoring an ionic current through a nanopore is the most common strategy for biomolecule sequencing.[67,68] When a membrane containing the nanopore structure is located in a chamber with two compartments, a constant ionic current will be obtained according to the voltage applied across the membrane. Assuming that the chamber contains a dilute solution of nucleic acids that retains a relatively low potential, transient changes in the ionic current can be observed in the form of spikes due to individual nucleic acids. Thus, basic structural characteristics can be confirmed without labeling or amplification of the sample, based on analysis of the amplitude and duration of the ionic current pulse. For example, using ionic current measurements, Akeson et al. discriminated nucleic acids consisting of two homopolymers within single RNA molecules.[83] Moreover, a nanopore-based technique for sequencing has recently been developed using graphene,[84–86] which is a two-dimensional (2D) material with good electrical performance. The principle of the nanopore/graphene-based method involves obtaining the fluctuations of the ionic current,[87,88] tunneling current,[85,86,89] or nanoribbon conductance.[90–92] Some papers have reported ion transport-based detection of DNA translocation through nanopores in materials such as single- and multilayer graphene[93,94] and monolayer-coated graphene,[77] as well as in other 2D materials (e.g., boron nitride[76] or molybdenum sulfide[95]). However, the nanopore/graphene-based technologies suffer from a reduction of ion/DNA transport due to the hydrophobicity of the graphene.[68] Accordingly, the graphene surface needs to be treated with an appropriate agent to improve the ion/DNA transport. For example, Shankla et al. used simulations to investigate the control of the translocation velocity of single-stranded DNA resulting from the surface charge of a graphene sheet.[96]

In this section, we provided a basic introduction to biomolecule sequencing as an application of nanopore structures. Although biological nanopores were not dealt with, the concept of nanopore-based technologies for the sequencing was described. Based on electrical changes induced by biomolecules passing through the nanopore, biomolecules can be detected and sequenced. Therefore, nanopore structures represent a practical tool for use in biomolecule detection and sequencing, with low labor and cost requirements.[67,68]

References

1. Balasubramanian G, Chan I, Kolesov R, Al-Hmoud M, Tisler J, Shin C, Kim C, Wojcik A, Hemmer PR, Krueger A. 2008. Nanoscale imaging magnetometry with diamond spins under ambient conditions. *Nature*. 455(7213):648.
2. Ko SH, Park I, Pan H, Grigoropoulos CP, Pisano AP, Luscombe CK, Fréchet JM. 2007. Direct nanoimprinting of metal nanoparticles for nanoscale electronics fabrication. *Nano Letters*. 7(7):1869–1877.
3. de Vreede LJ, van den Berg A, Eijkel JC. 2015. Nanopore fabrication by heating Au particles on ceramic substrates. *Nano Letters*. 15(1):727–731.
4. Kwon S, Jang S, Choi J-W, Choi S, Jang S, Kim T-W, Wang G. 2017. Controllable switching filaments prepared via tunable and well-defined single truncated conical nanopore structures for fast and scalable SiO_x memory. *Nano Letters*. 17(12):7462–7470.
5. Li J, Gershow M, Stein D, Brandin E, Golovchenko JA. 2003. DNA molecules and configurations in a solid-state nanopore microscope. *Nature Materials*. 2(9):611.
6. Chang H, Kosari F, Andreadakis G, Alam M, Vasmatzis G, Bashir R. 2004. DNA-mediated fluctuations in ionic current through silicon oxide nanopore channels. *Nano Letters*. 4(8):1551–1556.
7. Kuan AT, Golovchenko JA. 2012. Nanometer-thin solid-state nanopores by cold ion beam sculpting. *Applied Physics Letters*. 100(21):213104.
8. Li J, Stein D, McMullan C, Branton D, Aziz MJ, Golovchenko JA. 2001. Ion-beam sculpting at nanometre length scales. *Nature*. 412(6843):166.
9. Mitsui T, Stein D, Kim Y-R, Hoogerheide D, Golovchenko JA. 2006. Nanoscale volcanoes: Accretion of matter at ion-sculpted nanopores. *Physical Review Letters*. 96(3):036102.
10. Apel PY, Korchev YE, Siwy Z, Spohr R, Yoshida M. 2001. Diode-like single-ion track membrane prepared by electro-stopping. *Nuclear Instruments and Methods in Physics Research Section B: Beam Interactions with Materials and Atoms*. 184(3): 337–346.

11. Spinney P, Howitt D, Smith R, Collins S. 2010. Nanopore formation by low-energy focused electron beam machining. *Nanotechnology.* 21(37): 375301.
12. Wu M-Y, Krapf D, Zandbergen M, Zandbergen H, Batson PE. 2005. Formation of nanopores in a SiN/SiO_2 membrane with an electron beam. *Applied Physics Letters.* 87(11):113106.
13. Siwy Z, Dobrev D, Neumann R, Trautmann C, Voss K. 2003. Electro-responsive asymmetric nanopores in polyimide with stable ion-current signal. *Applied Physics A.* 76(5):781–785.
14. Wang C, Wang L, Zhu X, Wang Y, Xue J. 2012. Low-voltage electroosmotic pumps fabricated from track-etched polymer membranes. *Lab on a Chip.* 12(9):1710–1716.
15. Zhang H, Tian Y, Jiang L. 2013. From symmetric to asymmetric design of bio-inspired smart single nanochannels. *Chemical Communications.* 49(86):10048–10063.
16. Wang G, Lee J-H, Yang Y, Ruan G, Kim ND, Ji Y, Tour JM. 2015. Three-dimensional networked nanoporous Ta_2O_5–x memory system for ultrahigh density storage. *Nano Letters.* 15(9):6009–6014.
17. Wang G, Yang Y, Lee J-H, Abramova V, Fei H, Ruan G, Thomas EL, Tour JM. 2014. Nanoporous silicon oxide memory. *Nano Letters.* 14(8): 4694–4699.
18. Huo K, Gao B, Fu J, Zhao L, Chu PK. 2014. Fabrication, modification, and biomedical applications of anodized TiO_2 nanotube arrays. *RSC Advances.* 4(33):17300–17324.
19. Roy P, Berger S, Schmuki P. 2011. TiO_2 nanotubes: Synthesis and applications. *Angewandte Chemie International Edition.* 50(13):2904–2939.
20. Cai Q, Ledden B, Krueger E, Golovchenko JA, Li J. 2006. Nanopore sculpting with noble gas ions. *Journal of Applied Physics.* 100(2):024914.
21. Hoogerheide DP, George HB, Golovchenko JA, Aziz MJ. 2011. Thermal activation and saturation of ion beam sculpting. *Journal of Applied Physics.* 109(7):074312.
22. Ali M, Ramirez P, Tahir MN, Mafe S, Siwy Z, Neumann R, Tremel W, Ensinger W. 2011. Biomolecular conjugation inside synthetic polymer nanopores via glycoprotein–lectin interactions. *Nanoscale.* 3(4):1894–1903.
23. Pevarnik M, Healy K, Toimil-Molares ME, Morrison A, Létant SE, Siwy ZS. 2012. Polystyrene particles reveal pore substructure as they translocate. *ACS Nano.* 6(8):7295–7302.
24. Hou X, Liu Y, Dong H, Yang F, Li L, Jiang L. 2010. A pH-gating ionic transport nanodevice: Asymmetric chemical modification of single nanochannels. *Advanced Materials.* 22(22):2440–2443.
25. Tian Y, Hou X, Jiang L. 2011. Biomimetic ionic rectifier systems: Asymmetric modification of single nanochannels by ion sputtering technology. *Journal of Electroanalytical Chemistry.* 656(1–2):231–236.
26. Ali M, Ramirez P, Nguyen HQ, Nasir S, Cervera J, Mafe S, Ensinger W. 2012. Single cigar-shaped nanopores functionalized with amphoteric amino acid chains: Experimental and theoretical characterization. *ACS Nano.* 6(4):3631–3640.
27. Zhang H, Hou X, Zeng L, Yang F, Li L, Yan D, Tian Y, Jiang L. 2013. Bioinspired artificial single ion pump. *Journal of the American Chemical Society.* 135(43):16102–16110.
28. Apel PY, Blonskaya IV, Orelovitch OL, Ramirez P, Sartowska BA. 2011. Effect of nanopore geometry on ion current rectification. *Nanotechnology.* 22(17):175302.
29. Apel PY, Blonskaya IV, Dmitriev SN, Orelovitch OL, Presz A, Sartowska BA. 2007. Fabrication of nanopores in polymer foils with surfactant-controlled longitudinal profiles. *Nanotechnology.* 18(30):305302.
30. Zhao R, Xu M, Wang J, Chen G. 2010. A pH sensor based on the TiO_2 nanotube array modified Ti electrode. *Electrochimica Acta.* 55(20):5647–5651.
31. Perillo P, Rodriguez D. 2014. A room temperature chloroform sensor using TiO_2 nanotubes. *Sensors and Actuators B: Chemical.* 193:263–266.
32. Gultepe E, Nagesha D, Sridhar S, Amiji M. 2010. Nanoporous inorganic membranes or coatings for sustained drug delivery in implantable devices. *Advanced Drug Delivery Reviews.* 62(3):305–315.
33. Simovic S, Losic D, Vasilev K. 2010. Controlled drug release from porous materials by plasma polymer deposition. *Chemical Communications.* 46(8): 1317–1319.
34. Beranek R, Hildebrand H, Schmuki P. 2003. Self-organized porous titanium oxide prepared in H_2SO_4/HF electrolytes. *Electrochemical and Solid-State Letters.* 6(3):B12–B14.
35. Albu SP, Ghicov A, Aldabergenova S, Drechsel P, LeClere D, Thompson GE, Macak JM, Schmuki P. 2008. Formation of double-walled TiO_2 nanotubes and robust anatase membranes. *Advanced Materials.* 20(21):4135–4139.
36. Huang J-Y, Zhang K-Q, Lai Y-K. 2013. Fabrication, modification, and emerging applications of TiO_2 nanotube arrays by electrochemical synthesis: A review. *International Journal of Photoenergy.* 2013, 19.
37. Paramasivam I, Jha H, Liu N, Schmuki P. 2012. A review of photocatalysis using self-organized TiO_2 nanotubes and other ordered oxide nanostructures. *Small.* 8(20):3073–3103.
38. Ruffino F, Grimaldi M, Giannazzo F, Roccaforte F, Raineri V. 2008. Thermodynamic properties of supported and embedded metallic nanocrystals: Gold on/in SiO_2. *Nanoscale Research Letters.* 3(11):454.

39. Carnevali P, Ercolessi F, Tosatti E. 1987. Melting and nonmelting behavior of the Au (111) surface. *Physical Review B.* 36(12):6701.
40. Pluis B, Frenkel D, Van der Veen J. 1990. Surface-induced melting and freezing II. A semi-empirical Landau-type model. *Surface Science.* 239(3):282–300.
41. Saiz E, Cannon RM, Tomsia AP. 2001. Reactive spreading in ceramic/metal systems. *Oil & Gas Science and Technology.* 56(1):89–96.
42. Saiz E, Tomsia A, Cannon R. 1998. Ridging effects on wetting and spreading of liquids on solids. *Acta Materialia.* 46(7):2349–2361.
43. Kaplan WD, Chatain D, Wynblatt P, Carter WC. 2013. A review of wetting versus adsorption, complexions, and related phenomena: The rosetta stone of wetting. *Journal of Materials Science.* 48(17):5681–5717.
44. Karakouz T, Tesler AB, Sannomiya T, Feldman Y, Vaskevich A, Rubinstein I. 2013. Mech anism of morphology transformation during annealing of nanostructured gold films on glass. *Physical Chemistry Chemical Physics.* 15(13): 4656–4665.
45. Herring, C. 1953. Surface tension as a motivation for sintering. In *The Physics of Powder Metallurgy*, ed. W. E. Kingston, 143–179, New York: McGraw-Hill.
46. Smith CS. 1948. Grains, phases, and interfaces: An introduction of microstructure. *Transactions of the Metallurgical Society AIME.* 175:15–51.
47. Ferreira Nascimento ML, Zanotto ED. 2007. Diffusion processes in vitreous silica revisited. *Physics and Chemistry of Glasses-European Journal of Glass Science and Technology Part B.* 48(4): 201–217.
48. Ricci E, Novakovic R. 2001. Wetting and surface tension measurements on gold alloys. *Gold Bulletin.* 34(2):41–49.
49. Brunauer S, Kantro D, Weise C. 1956. The surface energies of amorphous silica and hydrous amorphous silica. *Canadian Journal of Chemistry.* 34(10):1483–1496.
50. Heyraud J, Metois J. 1980. Equilibrium shape of gold crystallites on a graphite cleavage surface: Surface energies and interfacial energy. *Acta Metallurgica.* 28(12):1789–1797.
51. Chang T-C, Chang K-C, Tsai T-M, Chu T-J, Sze SM. 2016. Resistance random access memory. *Materials Today.* 19(5):254–264.
52. Li, Y.-T., Long, S.-B., Lv, H.-B., Liu, Q., Wang, Q., Wang, Y., Zhang, S., Lian, W.-T., Liu, S. Liu, M. 2010. A low-cost memristor based on titanium oxide. *2010 10th IEEE international conference on Solid-state and integrated circuit technology (ICSICT)*, IEEE.
53. Lee M-J, Lee CB, Lee D, Lee SR, Chang M, Hur JH, Kim Y-B, Kim C-J, Seo DH, Seo S. 2011. A fast, high-endurance and scalable non-volatile memory device made from asymmetric Ta_2O_{5-x}/TaO_{2-x} bilayer structures. *Nature Materials.* 10(8):625.
54. Strukov DB, Snider GS, Stewart DR, Williams RS. 2008. The missing memristor found. *Nature.* 453(7191):80.
55. Yang JJ, Strukov DB, Stewart DR. 2013. Memristive devices for computing. *Nature Nanotechnology.* 8(1):13.
56. Kwon D-H, Kim KM, Jang JH, Jeon JM, Lee MH, Kim GH, Li X-S, Park G-S, Lee B, Han S. 2010. Atomic structure of conducting nanofilaments in TiO_2 resistive switching memory. *Nature Nanotechnology.* 5(2):148.
57. Waser R. 2012. *Nanoelectronics and Information Technology.* New York: John Wiley & Sons.
58. Chen, A.; Lin, M.-R. 2011. Variability of resistive switching memories and its impact on crossbar array performance. *Reliability Physics Symposium (IRPS), 2011 IEEE International.* IEEE.
59. Guan X, Yu S, Wong H-SP. 2012. On the switching parameter variation of metal-oxide RRAM—Part I: Physical modeling and simulation methodology. *IEEE Transactions on Electron Devices.* 59(4):1172–1182.
60. Arita M, Takahashi A, Ohno Y, Nakane A, Tsurumaki-Fukuchi A, Takahashi Y. 2015. Switching operation and degradation of resistive random access memory composed of tungsten oxide and copper investigated using in-situ TEM. *Scientific Reports.* 5:17103.
61. Yao J, Zhong L, Natelson D, Tour JM. 2012. In situ imaging of the conducting filament in a silicon oxide resistive switch. *Scientific Reports.* 2:242.
62. Yao J, Sun Z, Zhong L, Natelson D, Tour JM. 2010. Resistive switches and memories from silicon oxide. *Nano Letters.* 10(10):4105–4110.
63. He C, Li J, Wu X, Chen P, Zhao J, Yin K, Cheng M, Yang W, Xie G, Wang D. 2013. Tunable electroluminescence in planar graphene/SiO_2 memristors. *Advanced Materials.* 25(39):5593–5598.
64. Bellisario DO, Ulissi Z, Strano MS. 2013. A quantitative and predictive model of electromigration-induced breakdown of metal nanowires. *The Journal of Physical Chemistry C.* 117(23): 12373–12378.
65. Durkan C, Schneider M, Welland M. 1999. Analysis of failure mechanisms in electrically stressed Au nanowires. *Journal of Applied Physics.* 86(3): 1280–1286.
66. Karim S, Maaz K, Ali G, Ensinger W. 2009. Diameter dependent failure current density of gold nanowires. *Journal of Physics D: Applied Physics.* 42(18):185403.
67. Li J, Yu D, Zhao Q. 2016. Solid-state nanopore-based DNA single molecule detection and sequencing. *Microchimica Acta.* 183(3):941–953.

68. Carson S, Wanunu M. 2015. Challenges in DNA motion control and sequence readout using nanopore devices. *Nanotechnology.* 26(7):074004.
69. Haque F, Li J, Wu H-C, Liang X-J, Guo P. 2013. Solid-state and biological nanopore for real-time sensing of single chemical and sequencing of DNA. *Nano Today.* 8(1):56–74.
70. Maitra RD, Kim J, Dunbar WB. 2012. Recent advances in nanopore sequencing. *Electrophoresis.* 33(23):3418–3428.
71. Liu S, Zhao Q, Li Q, Zhang H, You L, Zhang J, Yu D. 2011. Controlled deformation of Si_3N_4 nanopores using focused electron beam in a transmission electron microscope. *Nanotechnology.* 22(11):115302.
72. Zhang J, You L, Ye H, Yu D. 2007. Fabrication of ultrafine nanostructures with single-nanometre precision in a high-resolution transmission electron microscope. *Nanotechnology.* 18(15):155303.
73. Storm A, Chen J, Ling X, Zandbergen H, Dekker C. 2003. Fabrication of solid-state nanopores with single-nanometre precision. *Nature Materials.* 2(8):537.
74. Venkatesan BM, Dorvel B, Yemenicioglu S, Watkins N, Petrov I, Bashir R. 2009. Highly sensitive, mechanically stable nanopore sensors for DNA analysis. *Advanced Materials.* 21(27):2771–2776.
75. Kim MJ, Wanunu M, Bell DC, Meller A. 2006. Rapid fabrication of uniformly sized nanopores and nanopore arrays for parallel DNA analysis. *Advanced Materials.* 18(23):3149–3153.
76. Liu S, Lu B, Zhao Q, Li J, Gao T, Chen Y, Zhang Y, Liu Z, Fan Z, Yang F. 2013. Boron nitride nanopores: Highly sensitive DNA single-molecule detectors. *Advanced Materials.* 25(33):4549–4554.
77. Schneider GF, Kowalczyk SW, Calado VE, Pandraud G, Zandbergen HW, Vandersypen LM, Dekker C. 2010. DNA translocation through graphene nanopores. *Nano Letters.* 10(8):3163–3167.
78. Yang J, Ferranti DC, Stern LA, Sanford CA, Huang J, Ren Z, Qin L-C, Hall AR. 2011. Rapid and precise scanning helium ion microscope milling of solid-state nanopores for biomolecule detection. *Nanotechnology.* 22(28):285310.
79. Marshall MM, Yang J, Hall AR. 2012. Direct and transmission milling of suspended silicon nitride membranes with a focused helium ion beam. *Scanning.* 34(2):101–106.
80. Kwok H, Briggs K, Tabard-Cossa V. 2014. Nanopore fabrication by controlled dielectric breakdown. *PLoS One.* 9(3):e92880.
81. Chen P, Mitsui T, Farmer DB, Golovchenko J, Gordon RG, Branton D. 2004. Atomic layer deposition to fine-tune the surface properties and diameters of fabricated nanopores. *Nano Letters.* 4(7): 1333–1337.
82. Venkatesan BM, Shah AB, Zuo JM, Bashir R. 2010. DNA sensing using nanocrystalline surface-enhanced Al_2O_3 nanopore sensors. *Advanced Functional Materials.* 20(8):1266–1275.
83. Akeson M, Branton D, Kasianowicz JJ, Brandin E, Deamer DW. 1999. Microsecond time-scale discrimination among polycytidylic acid, polyadenylic acid, and polyuridylic acid as homopolymers or as segments within single RNA molecules. *Biophysical Journal.* 77(6):3227–3233.
84. Drndic M. 2014. Sequencing with graphene pores. *Nature Nanotechnology.* 9(10):743.
85. Avdoshenko SM, Nozaki D, Gomes da Rocha C, González JW, Lee MH, Gutierrez R, Cuniberti G. 2013. Dynamic and electronic transport properties of DNA translocation through graphene nanopores. *Nano Letters.* 13(5):1969–1976.
86. Postma HWC. 2010. Rapid sequencing of individual DNA molecules in graphene nanogaps. *Nano Letters.* 10(2):420–425.
87. Sathe C, Zou X, Leburton J-P, Schulten K. 2011. Computational investigation of DNA detection using graphene nanopores. *ACS Nano.* 5(11): 8842–8851.
88. Wells DB, Belkin M, Comer J, Aksimentiev A. 2012. Assessing graphene nanopores for sequencing DNA. *Nano Letters.* 12(8):4117–4123.
89. Prasongkit J, Grigoriev A, Pathak B, Ahuja R, Scheicher RH. 2011. Transverse conductance of DNA nucleotides in a graphene nanogap from first principles. *Nano Letters.* 11(5):1941–1945.
90. Girdhar A, Sathe C, Schulten K, Leburton J-P. 2013. Graphene quantum point contact transistor for DNA sensing. *Proceedings of the National Academy of Sciences.* 110(42):16748–16753.
91. Nelson T, Zhang B, Prezhdo OV. 2010. Detection of nucleic acids with graphene nanopores: Ab initio characterization of a novel sequencing device. *Nano Letters.* 10(9):3237–3242.
92. Saha KK, Drndic M, Nikolic BK. 2011. DNA base-specific modulation of microampere transverse edge currents through a metallic graphene nanoribbon with a nanopore. *Nano Letters.* 12(1):50–55.
93. Garaj S, Hubbard W, Reina A, Kong J, Branton D, Golovchenko J. 2010. Graphene as a subnanometre trans-electrode membrane. *Nature.* 467(7312):190.
94. Garaj S, Liu S, Golovchenko JA, Branton D. 2013. Molecule-hugging graphene nanopores. *Proceedings of the National Academy of Sciences.* 110(30):12192–12196.
95. Liu K, Feng J, Kis A, Radenovic A. 2014. Atomically thin molybdenum disulfide nanopores with high sensitivity for DNA translocation. *ACS Nano.* 8(3):2504–2511.
96. Shankla M, Aksimentiev A. 2014. Conformational transitions and stop-and-go nanopore transport of single-stranded DNA on charged graphene. *Nature Communications.* 5:5171.

23

Methane Storage in Nanoporous Carbons

Iván Cabria
Universidad de Valladolid

Fabián Suárez-García
Instituto Nacional del Carbón

Luis F. Mazadiego and
Marcelo F. Ortega
Universidad Politécnica de Madrid

23.1 Introduction: The Adsorbed Natural Gas Vehicle and the Nanoporous Carbons

Nowadays the most used fuels by road transport vehicles are gasoline and diesel. Both fuels contribute in important proportions to the air pollution and to the so-called greenhouse effect. At present, all the trademarks of vehicles have improved the use of catalysts, having considerably reduced the emission of carbon monoxide, CO, nitrogen oxides, NO_x, and solid pollutant particles. Greenhouse gas emissions (GHGs) in road transport are mainly dependent on two factors: the consumption and the type of fuel used by the vehicle. Europe has set a compulsory CO_2 emission target of 95 g CO_2/km for the vehicle fleet sold in 2020, which implies a decrease in GHG emissions from vehicles of approximately 30% compared to 2010. This trend is not isolated; countries such as Japan, China, India, Canada or USA have also set targets. In particular, USA has set targets for 2020 very similar to those in Europe. In addition, it is important to note that Europe wants to continue this trend in the face of 2030 by setting even more ambitious targets.

Natural gas vehicles (NGVs) are an alternative to the present gasoline and diesel vehicles. The use of these vehicles would allow to reach the European CO_2 emission targets. In the last two decades, the increase in the number of NGVs has been important, reaching more than 12 million natural gas (NG)-powered vehicles around the world. However, the rise of this fuel in the automotive sector has been heterogeneous. In some countries, the increase in the NGV sector has been very important (China, India, Pakistan, Iran, Argentina, Brazil, Ukraine and Italy) and in the rest has been modest.

The current storage methods of NG for vehicles are compressed natural gas (CNG), liquefied natural gas (LNG), and, more recently, adsorbed natural gas (ANG) and solidified natural gas (SNG). The main advantages of NG with respect to gasoline and diesel are:

1. Lower CO_2 emissions. The combustion of NG is more complete and pure than that of liquid fuels and produces less CO_2 than the combustion of other hydrocarbons to equal energy supply. The CO_2 emissions are between 15% and 25% lower.
2. Lower pollutant emissions. NO_x emissions are reduced by 80% and SO_x and solid particles are virtually non-existing.
3. To cover a distance of 100 km with NG is about 35% and 50% cheaper than using diesel and gasoline, respectively.
4. Longer life of the vehicle. Because NG does not produce carbon residue, it does not form sediments, nor does it wash the cylinder walls, and besides, it increases the life of the engine, spark plugs, filters and oil.
5. Suitability for the bivalent application (gas–gasoline or gas–diesel).
6. The reserves of NG cover a period clearly higher than those of crude oil (250 years compared to 60 years).
7. There are no volatilization losses when refueling.

CNG is handled at high pressures, 20–25 MPa; LNG requires cryogenic systems at very low temperature, 111 K, with significant additional costs to obtain these conditions and SNG operates at 273 K, has kinetic limitations related to the solubility of methane in water or ice and is only for stationary applications. For these reasons and in order to optimize the fuel of vehicles in terms of cost-safety-autonomy, research has been underway in the last two decades to use NG stored in porous materials (adsorbents) at moderate pressures: 3.5–4.0 MPa and also 5.0–6.5 MPa. These types of vehicles are called ANG vehicles. The power density provided by ANG at 3.4 MPa is equivalent to that of CNG at 16.5 MPa (Pfeifer, 2011; Mota, 2008). An 80% of the volumetric CNG energy density at 20.7 MPa can be obtained by ANG at 3.4 MPa (Nie et al., 2016; Mason et al., 2014; Menon and Komarneni, 1998). The use of moderate pressures with ANG, instead of high pressures with CNG, has several positive implications for ANG vehicles:

1. The use of lower pressure compressors, which are cheaper and more reliable than 20–25 MPa compressors and consume less energy to fill a tank. A domestic compressor of 5 MPa costs about 500 euros and a domestic compressor of 20–25 MPa costs about 3,000–4,000 euros (FuelMaker, 2018). A compressor of 5 MPa consumes approximately half of the electrical energy than a compressor of 25 MPa with the same power, to fill a tank of the same volume. In addition, the current shortage of NG filling stations in many countries could be solved with domestic compressors of 5 MPa.
2. The use of industrial compressors of lower pressures, which are cheaper and consume less energy to fill a tank and the use of cheaper NG filling stations. Industrial compressors are used in NG filling stations, not in households. Building a NG filling station of 5 MPa would cost about 450,000 euros and building a NG filling station of 20 MPa would cost about 800,000 euros (Gas & Go, 2018).
3. ANG tanks are made from materials much lighter than those of CNG tanks. Having to withstand much lower pressures, the tanks do not have to be made of heavy materials and can be made of lighter materials such as aluminum and reinforced plastic.
4. ANG tanks have a high degree of conformability (Pfeifer, 2011). These tanks do not have to be cylindrical and can be shaped like prisms or other forms, using about 20% more space of the vehicle for storage purposes than CNG tanks.

The United States Department of Energy (DOE) is funding the research program "Methane Opportunities for Vehicular Energy" (MOVE), where several methane targets for automotive applications have been set to meet in the coming years (Kumar et al., 2017; DOE, 2012; Burchell and Rogers, 2000; Wegrzyn and Gurevich, 1996). The main objective is to obtain a reasonable driving range in automotive applications based on methane. Other objectives of the MOVE program are oriented towards the integration of adsorbing materials into tanks to improve the conformability factor, storage density, etc.

At present, research in optimizing methane adsorption processes in nanoporous materials is considered essential for ANG vehicles. Porous solids in which the average pore diameter is <2 nm can be considered as candidates for adsorbing gas in quantities proportional to their pore volume (Mota, 2008). One of the most important lines of research in the coming years will be adsorbents. The development of new porous materials and the optimization of their methane storage capacity will probably define the upcoming research on this topic. Several types of porous adsorbents have been experimentally tested in storage conditions of NGVs: nanoporous carbons, zeolites, metal-organic frameworks (MOFs), covalent organic frameworks (COFs) and porous organic polymers (POPs), among others. Nanoporous carbons are the only type used as adsorbents in commercial ANG tanks (Cenergy Solutions, 2018).

Nanoporous carbons, which include activated carbons (ACs), high-surface activated carbons (HSACs), carbide-derived carbons (CDCs) and other porous carbon-based materials, are one of the main possible adsorbents for ANG vehicles. Their methane storage capacities are high. In addition, nanoporous carbons have other very relevant property, which make them one of the most versatile adsorbents: the possibility of regenerating them. This chapter will address in the next sections the methane adsorption mechanisms and storage properties of nanoporous carbons. A summarized comparison of the methane storage capacities of carbon and non-carbon-based porous materials is also included.

23.2 Types and Mechanisms of Methane Storage

NG is a fuel formed mainly by methane, although it also usually contains a variable proportion of N_2, CO_2, water and other heavier hydrocarbons such as ethane, propane, butane and others. NG is considered a very good fuel to replace gasoline in vehicles, thanks to different advantages, including a lower cost, a higher H/C ratio that makes it has the highest energy per unit mass of fuel (50 MJ/kg for methane and 46.4 MJ/kg for gasoline) and, therefore, lower emissions. Thus, comparing with a vehicle powered by gasoline or diesel, a vehicle that runs on NG emits less hydrocarbons and up to 70% less of CO, 87% less of NO_x and 20% less of CO_2 (CNG United, 2018).

Despite all these advantages, the main drawback of the use of NG as fuel in vehicles is its storage in the vehicular fuel tank. At standard temperature and pressure (STP) conditions, i.e., 298 K and 1 atm, the energy density of NG per unit volume is only 0.038 MJ/L, which corresponds to only the 0.11% of that of gasoline. Therefore, the autonomy of a vehicle running on NG would be very low. Hence, the use of NG as fuel on vehicles will depend on the ability to store an adequate amount in the onboard fuel tank.

There are different possibilities to store NG in a tank. NG is conventionally stored in heavy steel-made tanks at high pressures, 20–25 MPa. At these high pressures, CNG has an energy density of 9.2 MJ/L, which is almost 4 times lower than the energy density of gasoline, 34.2 MJ/L (DOE, 2012). Therefore, an onboard fuel tank of about four times more volume is needed to have an equivalent autonomy or mileage range. This is impractical, not only because of the large volume of the tank but also because of the heavy weight of a high-pressure tank made of steel.

Another method for storing NG is by LNG, by cooling it below its boiling temperature, 111 K. The energy density of LNG is 22.2 MJ/L, which is 65% of that of gasoline and 2.4 times greater than the energy density of compressed NG at 25 MPa. Despite of this high energy density, this storage method presents different technical problems related to the tank requirements. To avoid losses, the storage tank must be highly insulated, and therefore, it will be large, heavy and expensive. In addition, this method is not suitable for onboard NG storage in light-duty vehicle, i.e., domestic use vehicle, due to fuel boil-off during the inactivity periods of the vehicle (Kumar et al., 2017).

The third alternative for the storage of NG is by adsorption in porous solids. The study of the adsorption of NG in porous solids as a method for storing it dates back to several decades (Menon and Komarneni, 1998). Thus, in 1985, an ANG vehicular fuel storage system was patented and an ANG storage technology using carbon adsorbents, for fork lifts and welding equipment, was commercialized (Menon and Komarneni, 1998).

In the ANG method, the NG storage takes place by gas adsorption in the porosity of a solid. The adsorption phenomena occur as a consequence of the attractive forces established between the surface of a solid (the adsorbent) and the gas molecules (the adsorptive), which produce their accumulation on the surface of the solid after the thermodynamic equilibrium is achieved. In general, two types of adsorption, physical or chemical adsorption, are distinguished, which depend on the type of interaction established between the adsorbent and the adsorbate (the adsorptive actually adsorbed by the adsorbent is named adsorbate). In a chemisorption process, specific chemical interactions between adsorbent and adsorbate occur and the process is not reversible. On the other hand, in a physical adsorption, specific interactions are absent and the process is fully reversible (or almost reversible). In the case of NG, physisorption is the type of adsorption that takes place. Thus, the isosteric heat of adsorption at low coverage typically ranges between 9 and 20 kJ/mol in the case of MOFs (García Blanco et al., 2016; Furukawa and Yaghi, 2009) and between 16 and 30 kJ/mol for carbon-based adsorbents (Kumar et al., 2017; Menshchikov et al., 2017a,b; García Blanco et al., 2016; Yeon et al., 2009).

In order to achieve a comparable mileage to that obtained with CNG, the United States DOE has established very ambitious gravimetric and volumetric targets for the storage systems based on ANG in the MOVE program (DOE, 2012). These targets are established for a tank working at room temperature, i.e., 298 K, and pressures around 3.5 MPa. Thus, the volumetric DOE targets are 12.5 MJ/L (based on the sorbent volume) or 9.2 MJ/L (based on the inner tank volume), which correspond to 0.5 g CH_4/g sorbent or 0.4 g CH_4/g inner tank, respectively (DOE, 2012). These values correspond to volumetric capacities as high as 356 L (STP)/L or 263 L (STP)/L as working capacities, what implies methane densities >0.188 g/cm^3, which is the methane density at 25 MPa and 298 K. These values have not been achieved with any porous material so far and probably they could not be achieved.

Finally, another method to store NG, called natural gas hydrate (NGH) or solidified natural gas (SNG, has been reported recently and it is presented as a technology with high potential for this application. This method consists of the formation of methane hydrates confined in the nanopores of a material. Thus, it has been demonstrated that methane hydrates can be formed in the pores of different materials, including active carbons, at moderate pressures (<10 MPa) and temperatures between 256 and 275 K. This method consists of carrying out the adsorption of methane in a solid whose porosity has previously been saturated with water. Under these conditions, the formation of methane hydrates can be achieved and the methane storage capacity can be increased by 173% compared to the dry material (Borchardt et al., 2016). The main disadvantage of this method, in addition to having to operate at temperatures around 273 K, is that the methane hydrate formation has kinetic limitations associated with the solubility of methane in water or ice. Intensive research is being carried out in this field to resolve these limitations and demonstrate its practical viability. Detailed information on this topic can be found in different works (Veluswamy et al., 2018; Borchardt et al., 2018, 2016; Casco et al., 2015b).

23.3 Adsorption Properties of Nanoporous Carbons

In this section, some considerations and definitions about the adsorption at high pressures under supercritical conditions and about the experimental process to measure methane adsorption isotherms are given. These definitions can be applied to any adsorbent, not to nanoporous carbons alone.

As is known, the isotherm obtained experimentally is the excess adsorption isotherm or Gibbs isotherm, i.e., the amount of adsorbed gas whose density is greater than the density of the gas at the same pressure and temperature. Figure 23.1 is a schematic representation of the Gibbs adsorption, where the relation between excess adsorption, absolute adsorbed amount and gas amount compressed in the void space is shown.

The amount of CH_4 measured experimentally is the excess adsorption and corresponds to the region I in Figure 23.1. This amount is the value reported in the majority of the articles published. The absolute adsorption amount of a gas is the sum of regions I and II, and the relationship between

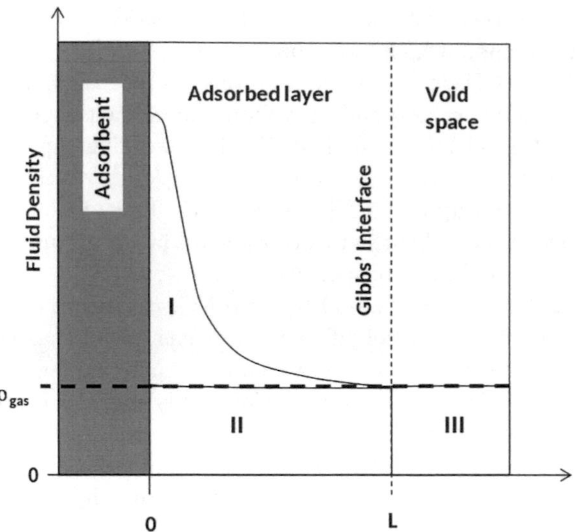

FIGURE 23.1 Schematic representation of the different regions in which the gas molecules can be located. Region I represents the excess adsorption. Region II is the compressed phase in the adsorbed layer and III is the compressed phase in the void space. The gas density in these last two regions is given by the P and T conditions.

the absolute and excess amounts is given by the following equation:

$$n_a = n_e + \rho_{\text{bulk}} V_{\text{ad}}, \tag{23.1}$$

where n_a is the absolute adsorption amount, n_e is the excess adsorption, i.e., the experimental isotherm, ρ_{bulk} is the bulk or free gas density at P and T and V_{ad} is the volume of the adsorbed phase. Both n_e and ρ_{bulk} are experimentally measurable parameters, while n_a and V_{ad} cannot be directly measured. Under subcritical conditions is established that the density of the adsorbed gas, ρ_{ad}, is constant and equal to the liquid or solid density of the adsorbate (Rouquerol et al., 1999). So, it is possible to obtain the absolute adsorption isotherm by introducing the following relation into Eq. 23.1:

$$V_{\text{ad}} = \frac{n_a}{\rho_{\text{ad}}}. \tag{23.2}$$

The absolute adsorption amount, n_a, can be obtained as:

$$n_a = \frac{n_e}{1 - \frac{\rho_{\text{bulk}}}{\rho_{\text{ad}}}}. \tag{23.3}$$

In the case of adsorption of NG at temperatures above 190.5 K, the critical temperature for methane is not possible to establish that the adsorbed phase density, ρ_{ad}, is equal to liquid or solid adsorbate density because NG cannot condense under supercritical conditions. There are different theoretical approaches to estimate the absolute adsorption isotherm, which can be classified into two groups (Murata et al., 2001): those which assume that ρ_{ad} is constant, as in the case of subcritical conditions (Eq. 23.3) and those which consider that V_{ad} is constant (Eq. 23.1). In some reports on the storage of NG, the authors estimate the absolute adsorption, so it should be taken into account when comparing these values with other works where excess adsorption is reported, since the values obtained in the first case will be highest.

Gravimetric and volumetric devices are the most extensively used for measuring the excess adsorption or Gibbs isotherms. In the gravimetric method, a degassed sample is introduced into a sample holder inside a chamber whose temperature remains constant and which can be pressurized. The chamber is pressurized with the gas, i.e., NG, and the amount adsorbed is determined when the thermodynamic equilibrium of adsorption is reached, i.e., the weight of the sample is constant. In this method, the isotherm is measured directly as the weight change of the sample. The main source error of this method is the correction of the buoyancy effect due to the gas displacement that is produced by the volume occupied by the components of the balance (sample hold and other balance components) and by the sample itself as well as the adsorbed phase. The buoyancy can be corrected by different ways (Sircar, 2001). After correcting the experimental data with the buoyancy due to the balance components, sample volume and gas adsorbed phase, the excess or Gibbs isotherm is obtained.

The volumetric method is probably the most widely used to measure adsorption isotherms. In this case, the amount of gas adsorbed by the sample is determined indirectly by measuring the pressure variation in a given volume and using a real gas equation. A typical device consists of two cells: the sample cells and the manifold. Both, sample cell and manifold volumes are known. In this method, a known amount of gas is expanded from the manifold to the sample cell and the pressure is measured when the equilibrium is achieved, i.e., the pressure is constant. There are different error sources in this method. One of them is the volume of the manifold. Since all the calculations to obtain the amount of gas adsorbed by the sample refer to this volume, is very important to calibrate perfectly this volume. One of the main error sources is leaks. Due to a little leak, it could be interpreted as gas adsorption if the equilibrium time is larger than the rate of dropped gas due to the leak. Finally, a real gas state equation must be used in the calculations. This is very important in the case of high pressure adsorption, where the density of the gas can vary considerably with respect to an ideal gas.

In addition to the considerations indicated above for the measurement of the excess adsorption isotherms, various properties of the adsorbent must be taken into account. Thus, the sample density is a fundamental parameter in this application, since it is used to determine the adsorption volumetric capacity, as well as for the determination of the total amount of methane stored, as will be indicated below. The excess adsorption capacity on volumetric basis can be obtained from the experimental isotherm by the following equation:

$$n_{\text{e,vol}} = n_{\text{e,grav}} \rho_{\text{mat}}. \tag{23.4}$$

Thus, the volumetric adsorption capacity, $n_{\text{e,vol}}$, is obtained by multiplying the experimental isotherm, $n_{\text{e,grav}}$, by the material density, ρ_{mat}. As can be deduced from Eq. 23.4, material density is a key factor in this application, and the higher the material density, the higher is the volumetric capacity.

Different kinds of material density are used in this field, but in some cases, they are erroneously used. Because of that, we want to introduce here the definitions of several samples densities to clarify what density should be used in each calculation.

True density or helium density, ρ_{He}, is defined as the solid mass divided by the volume occupied by the atoms of the sample. Thus, this density neither includes the interparticle space volume nor the sample porosity. True density is measured by helium pycnometry, and it does not depend on the degree of compaction of the solid. Bulk density is defined as the solid mass divided by total volume occupied by the solid. This density includes the volume occupied by the solid atoms, the sample pore volume and the interparticle space volume. Bulk density has different values depending on whether it is measured in the freely settled or compacted state (tapped density, ρ_{tap}). Bulk density is determined by putting a solid mass in a container, for example, in a burette and measuring the volume it occupies. The sample can be subjected to a specified compaction, usually involving vibration of the container, obtaining the tapped density. The sample compaction can be carried out by a compression process by applying axial pressure in order to decrease the interparticle space. In this case, the obtained bulk density is called packing density, ρ_{packing}. This density is usually reported and used in the gas storage topic. The final pressure applied for compacting the sample has a great effect on the density value obtained (in general, ρ_{packing} increases with pressure) (Alcañiz-Monge et al., 2009b). Therefore, when packing density is used, the pressure applied for measuring must be indicated. Furthermore, it should be checked that the sample holds the applied pressure as its porosity may be reduced when high pressures are used. For example, it was demonstrated that some MOFs do not support high compacting pressures and their porosity falls (Juan-Juan et al., 2010; Alcañiz-Monge et al., 2009b).

Finally, other type of density frequently used in works on the storage of gases using crystalline solids, specifically MOFs, as adsorbents, is the crystal density. Many authors use the crystallographic density of the material to calculate the volumetric amount of gas adsorbed. Crystal density refers to the mass of solid divided by the volume occupied by the particle. This density includes the volume occupied by the solid atoms and the particle's internal pore volume, but it does not include the interparticle space volume. Thus, as crystal density is not taking into account the interparticle space, this density is very high, and therefore, the values obtained for the amount of gas adsorbed on a volumetric basis are very high, although not real from a practical point of view (Casco et al., 2015a; Marco-Lozar et al., 2012). Some examples of the high values of methane adsorbed on a volumetric basis in MOFs using their crystallographic density can be found in the literature (Kumar et al., 2017; Li et al., 2016; Casco et al., 2015a; Wu et al., 2010).

As expected from the previous definitions, the density of the materials increases following the sequence: bulk density < tap density < packing density < crystal density < true density. Therefore, the density used will greatly influence the amount of NG adsorbed on volumetric basis. Special care should be taken when comparing the volumetric adsorption capacity of different types of adsorbents, taking into account which definition of the density of the material has been used.

The most interesting parameter, from a practical point of view, is the total amount of gas stored. This is the amount of NG that can be introduced in a tank filled with an adsorbent. The total amount of NG stored includes the gas adsorbed in the porosity of the material (excess or Gibbs isotherm, n_e) plus the compressed gas in the space not occupied by the atoms of the material, including the interparticle space and the porosity of the sample. Therefore, the total gas in a tank is the sum of regions I, II and III represented in Figure 23.1. Thus, to obtain the total amount of gas stored in a tank filled with an adsorbent, in addition to the excess isotherm, we need to know the space occupied by the skeleton of the material and how much material we can introduce inside the tank. The volume occupied by the atoms of the material, V_S, is:

$$V_S = \frac{1}{\rho_{\mathrm{He}}}, \tag{23.5}$$

and the amount of material, W, that we can put in the tank can be estimated from its bulk, tap or packing density:

$$W = V_{\mathrm{tank}}\rho_{\mathrm{mat}}, \tag{23.6}$$

where V_{tank} is the tank volume and ρ_{mat} is the bulk, tap or packing density. The free volume, V_f, i.e., regions II and III in Figure 23.1, is:

$$V_f = V_{\mathrm{tank}}\left(1 - \frac{\rho_{\mathrm{mat}}}{\rho_{\mathrm{He}}}\right). \tag{23.7}$$

Finally, the total amount of gas stored, n_s, per unit volume of the tank is:

$$n_s = n_e + \rho_{\mathrm{bulk}}\left(1 - \frac{\rho_{\mathrm{mat}}}{\rho_{\mathrm{He}}}\right). \tag{23.8}$$

With this simple equation and using only measurable parameter, the total amount of methane stored can be obtained from the experimental adsorption isotherms. The values obtained at the laboratory level, using Eq. 23.8, could be extrapolated to the total amount of NG stored in a real fuel tank, only if an appropriate and realistic material density is used, for example, its tap density or its packing density. From a practical point of view, in order to obtain high methane storage capacities, it is necessary to put as much material as possible in the tank. That is why materials with high densities are needed in this application. One way to increase the bulk density of the carbon materials is to reduce the interparticle space by compression, i.e., increasing their packing density or by conforming the carbon material into pieces or monoliths (Lozano-Castelló et al., 2002b,a).

23.4 Experimental Methane Storage Capacities of Nanoporous Carbons

The study of the storage of methane or natural gas by adsorption (ANG) on nanoporous carbons goes back to several decades. In fact, already at the end of the 90s, V. C. Menon and S. Komarneni reported the advances in this field in an interesting and complete review (Menon and Komarneni, 1998), where they indicated the key properties that nanoporous carbons must have to maximize the methane adsorption capacity, both on gravimetric and volumetric basis. Porosity of the materials is, of course, the main property that affects to the methane adsorption capacity. As expected, the greater the volume of micropores or the specific surface area (SSA) is, the greater the capacity of methane adsorption is. Figure 23.2 shows how the amount of methane adsorbed linearly increases with the SSA of the carbon materials. This realization has also been reported in other works, not only for carbon materials but also for zeolites, MOFs and other porous materials (Kumar et al., 2017; Policicchio et al., 2017; Li et al., 2016; Makal et al., 2012; Alcañiz-Monge et al., 2009a; Menon and Komarneni, 1998), and it will be commented in one of the last sections.

It can be concluded from these data that, for this application, materials with high SSA, i.e., high micropore volume, are needed. However, not only are high micropore volumes needed but also the pore size distribution (PSD) affects the amount of methane adsorbed. Thus, it has been shown that pore sizes around 0.8 nm are the most suitable (Kumar et al., 2017; Menon and Komarneni, 1998), although larger sizes have been also indicated to be necessary (Rodríguez-Reinoso et al., 2005). To maximize the density of the adsorbed methane, it is necessary to optimize the pore size of the adsorbents. A pore size of around 0.8 nm corresponds to the size where two molecules of methane can be adsorbed, and therefore, a high density of the adsorbed phase is expected. However, it must be taken into account that the optimum size will depend on the final pressure that is reached. As is well known, the process of physical adsorption begins in the smallest pores, i.e., in pores where the potential is greater due to the overlapping of the potentials of both pore walls, but as these pores are filled, i.e., when pressure increases, the adsorption in wider pores takes place. Thus, this pore size (0.8 nm) is the optimum pore size for the methane adsorption at moderate pressures, <2.5 MPa, but if the storage is carried out at higher pressures, wider micropores would be needed to improve the adsorption capacity (Rodríguez-Reinoso et al., 2005).

As indicated above, the methane adsorption capacity on a volumetric basis, as well as the total amount stored, directly depend on the bulk density of the material. The bulk (tap or packing) density of porous carbon materials decreases with increasing porosity of the sample (Jordá-Beneyto et al., 2008). Therefore, in order to reach a high adsorption capacity on a volumetric basis, it is necessary to reach a compromise between the development of the porosity and the bulk density of the material. Figures 23.3 and 23.4 show the methane adsorption capacities on volumetric basis for different carbon materials as a function of their SSA.

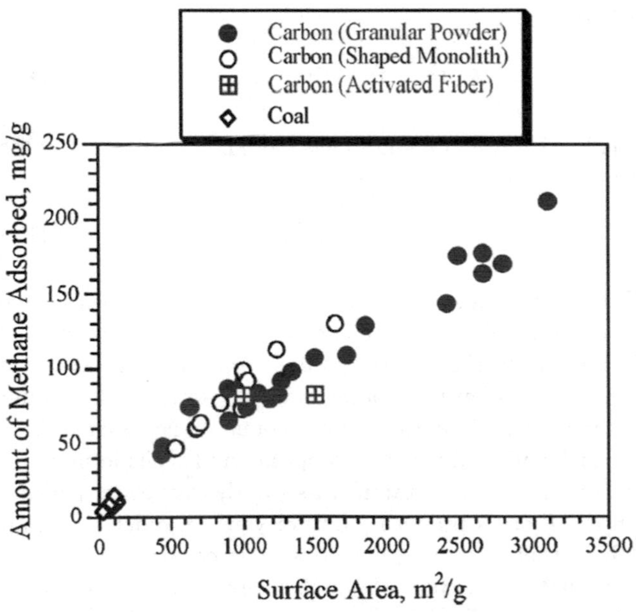

FIGURE 23.2 Variation in the experimental gravimetric methane adsorption capacity at 3.4 MPa and 298 K (373 K for AC fibers) with the SSA for different porous carbon materials. Reprinted with permission from Springer, J. Porous Mater. 5 (1998) 43–58 (Menon and Komarneni, 1998). Copyright 1998.

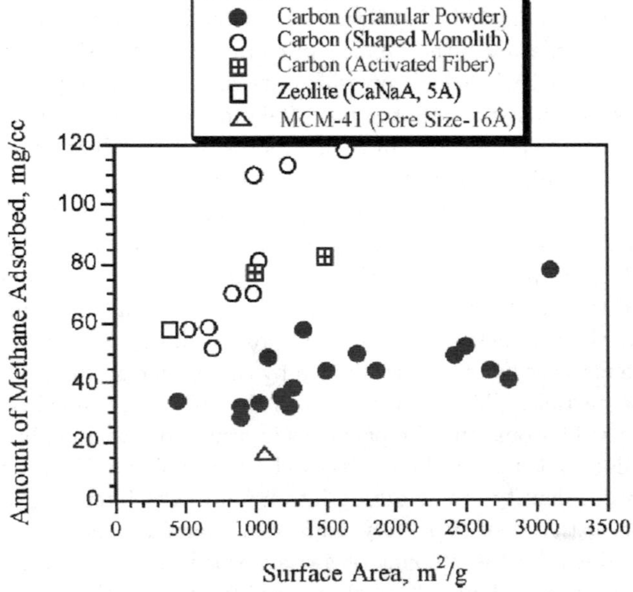

FIGURE 23.3 Experimental volumetric methane adsorption capacity of a zeolite, a microporous MCM-41 and several porous carbon materials at 3.4 MPa and 298 K, versus their specific surface area. Reprinted with permission from Springer, J. Porous Mater. 5 (1998) 43–58 (Menon and Komarneni, 1998). Copyright 1998.

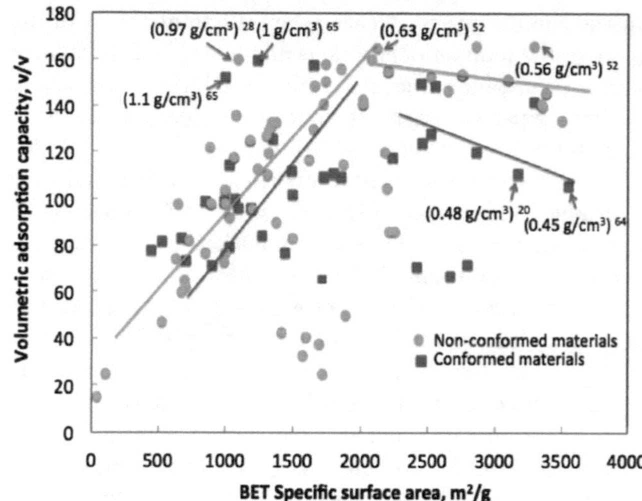

FIGURE 23.4 Volumetric adsorption capacity (v/v) versus the Brunauer—Emmett–Teller (BET) SSA of carbon structures (gray filled circles represent carbon materials corresponding to materials with slit-shaped pores, including ACs and carbon fibers). Adapted with permission from Chem. Rev. 117 (2017) 1796–1825 (Kumar et al., 2017). Copyright 2017 America Chemical Society.

The adsorption capacity on a volumetric basis does not increase linearly with the specific surface area, unlike what happened with the amount of methane adsorbed on a gravimetric basis (compare Figures 23.3 and 23.4 with Figure 23.2). In this case, the samples with higher density, such as carbon fibers or monoliths, have the highest volumetric adsorption capacity (Figures 23.3 and 23.4). As it is reflected in these Figures, the increase in the density of the material produces an increase in the amount of methane adsorbed. Thus, materials with higher density (achieved by decreasing the interparticle free space either by compacting under pressure or by conforming into pellets or monoliths) are those that reach higher values of methane adsorbed on volumetric basis. Volumetric adsorption capacities over 160 v/v have been reported by different authors (see Figure 23.4) (Giraldo et al., 2018; Kumar et al., 2017; Casco et al., 2015a; Marco-Lozar et al., 2012; Srinivas et al., 2012; Yeon et al., 2010, 2009; Celzard and Fierro, 2005; Lozano-Castelló et al., 2002a; Alcañiz-Monge et al., 1997).

Some of these results were reported for conformed carbon materials (monoliths), which probably are the highest methane adsorption capacities published to date, around 160 v/v (Marco-Lozar et al., 2012; Molina-Sabio et al., 2003). Although there are other higher values reported (for example, 195 v/v (Celzard and Fierro, 2005)), many of them correspond to fine powder carbons, where the high volumetric values are obtained after compaction of the powders under high pressures, in order to obtain high packing densities. The problem is that it is practically impossible to realize this method of compaction on a real scale, and these high packing densities could not be reached in a real onboard application. Therefore, the volumetric capacity at the laboratory scale for powder nanoporous carbons using their packing density could be considered as the practical limit for that particular adsorbent. In the case of monoliths, the piece density is used for calculating the volumetric capacity, so the value reported at laboratory scale will be closer to the real value.

23.5 Theoretical Methane Storage Capacities of Nanoporous Carbons

Theoretical simulations are useful to explain experimental results of the adsorption process at the molecular level, to find out the relationship between the adsorption at the molecular level and for the storage properties of a material at the macroscopic level, to design new materials for methane storage and to obtain information quickly. Experiments to obtain the same information would be longer and much more expensive. Many details reported by the simulations are difficult to obtain in experiments, for instance, the regions/surfaces of the material where the methane-surface interaction is more intense. The most used theoretical methodology to understand and predict the adsorption of gases into porous materials, such as nanoporous carbons, are grand canonical Monte Carlo (GCMC) simulations.

Nanoporous carbons are usually modeled as carbon slit pores in the GCMC simulations. According to experiments, many regions of the nanoporous carbons consist of flat graphitic-like surfaces parallel to each other and separated by a distance of a few nanometers (Park et al., 2015). Therefore, these carbons are usually simulated as carbon slit-shaped pores: Two infinite parallel graphitic sheets separated a distance w, called the pore size or pore width (See Figure 23.5). The width w of the slit pore is the distance between the centers of the two carbon atoms on opposite walls or sheets of the pore. Some authors use the effective pore width, w_e, and define it as the distance between the surfaces of the opposite walls of the pore. These two

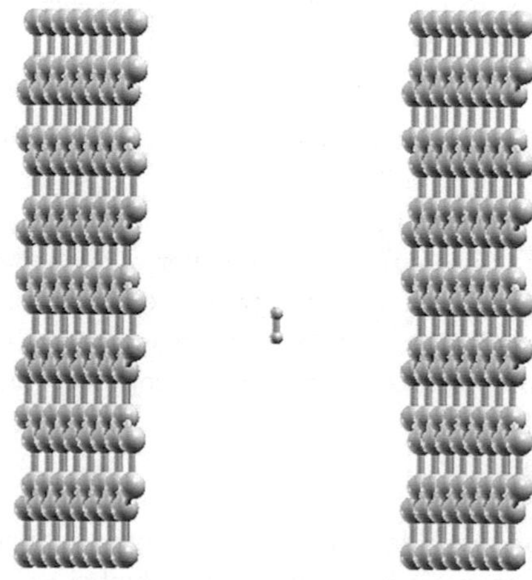

FIGURE 23.5 Slit-shaped pore with one hydrogen molecule inside. The two graphene sheets are flat and parallel.

definitions are related: $w_e = w - 2r_C$, where r_C is the radius of a carbon atom, whose value is approximately 1.42 Å. Nanoporous carbons consist of a set of slit pores of different sizes: They have a PSD.

GCMC simulations consider the μVT ensemble: The chemical potential μ, the volume and the temperature of the system are kept constant during the simulation. The number of methane molecules is allowed to change. The volume and size of the slit pore is kept constant. Each simulation is performed at a fixed temperature and pressure. The chemical potential is obtained from an equation of state, EOS, of methane, using the values of the fixed temperature and pressure. The most used EOS for methane are the Peng–Robinson (Peng and Robinson, 1976) and the Soave–Redlich–Kwong, SRK, (Soave, 1972) equations. The ideal gas EOS is also used by some authors at low pressures ($\leq$0.1 MPa). The details of the GCMC algorithm can be found at several sources in the scientific literature (Ohno et al., 1999; Frenkel and Smit, 1996; Allen and Tildesley, 1987), and a short description will be provided in the following paragraphs.

A GCMC simulation consists of millions of Monte Carlo, MC, iterations. Each MC iteration is an attempt to translate, insert or delete a molecule. Each attempt is assigned a probability, and this probability is kept fixed along all the simulation. The sum of the probabilities of the three types of attempts is the unity. Deletion and insertion should have the same probability to assure microscopic reversibility. Some authors assign the same probability to the translation, deletion and insertion attempts. Other authors assign the same probability to the deletion and insertion and a different and smaller probability to the translation, to speed up the convergence of the iterations towards the equilibrium. The attempt is chosen randomly: A random number between 0 and 1 is generated and compared with the probabilities of the three types of attempts. The three attempts are:

1. Translation: A randomly chosen molecule is moved a maximum distance d along a randomly chosen direction.
2. Insertion: A new methane molecule is inserted in a random location inside the pore.
3. Deletion: A randomly chosen molecule is deleted.

The attempts create a new or test configuration of the system. The Metropolis criterion is used to reject or accept this new configuration. The energy of the test configuration of the system is calculated. If the energy of the test configuration is lower than the energy of the last iteration, then the test configuration and energy are accepted. If not, then the Metropolis criterion is applied to reject or accept.

The set of iterations of a GCMC simulation has two parts: The first part are the iterations to reach the equilibrium, and the second part are the production or run iterations. Some authors consider that the equilibrium of the system is achieved after a fixed and large number of iterations, about 10^6–10^7. Other authors monitor the stability of the total energy or of the number of gas molecules and consider that the equilibrium is reached when the total energy of the system or the number of gas molecules is stable: It fluctuates around a constant value. After the equilibrium is reached, the three types of attempts continue on each production iteration but around the equilibrium.

The average values of the properties are calculated over the production iterations, spaced by 10^3–10^4 iterations to assure statistical independence and after the system has reached the equilibrium. The most important average calculated in a GCMC simulation is the average of the number of methane molecules, $< N >$, at the corresponding temperature, pressure (or chemical potential) and volume of the system (volume of the slit pore in this case). Many storage-related properties are calculated using $< N >$. Other important averages are the average energy of the system, $< E >$, and the average of the product energy and number of molecules, $< EN >$.

Some authors run preliminary simulation tests to obtain a value of d that minimizes the number of iterations to reach the equilibrium, and then they use that value in all the simulations. Other authors adjust the value of d during the first part of each GCMC simulation, the equilibrium part, such that the average acceptance of the attempted moves is 50%, which speeds up the convergence towards the equilibrium and then use the adjusted value in the production part of the simulation. To assure microscopic reversibility, the value of the maximum distance d must be kept constant during the production part of the simulation. Both strategies satisfy this requirement. A too small value of d implies usually very small fluctuations of the energy and number of molecules N and, hence, a slow convergence. A too large value implies large fluctuations of the energy and N and, again, a slow convergence. The most usual optimized values of d are in the range 0.1–0.5 Å.

The GCMC simulations are usually run to study carbon slit pores of different widths, between 6 Å and 50 Å, at pressures in the range 0.1–36 MPa and at room temperature, 298 or 300 K. Some authors also carried out simulations at temperatures between 273 and 300 K and at pressures between 0.001 and 0.1 MPa.

The interaction potentials between the methane molecules (methane–methane interaction) and between the methane molecules and the pore surfaces (or methane–carbon interaction) are key ingredients of an MC simulation. Lennard-Jones, LJ, potentials (Lennard-Jones, 1924) are usually employed for the methane–methane interactions and Steele potentials (Steele, 1974) for the methane–pore wall interaction.

The methane molecule, CH_4, is modeled as a one-center LJ interaction site. The LJ potential between two methane molecules is given by:

$$V_{LJ}(z) = 4\epsilon_{CH_4}\left[\left(\frac{\sigma_{CH_4}}{z}\right)^{12} - \left(\frac{\sigma_{CH_4}}{z}\right)^{6}\right], \qquad (23.9)$$

where z is the distance between the molecules and σ_{CH_4} and ϵ_{CH_4} are the LJ parameters of a methane molecule. Tabulated values of the parameters ϵ and σ for methane and other molecules and atoms can be found in reports of molecular parameters (Tee et al., 1966), in generic force fields, such as Universal Force Field, UFF, (Rappe et al., 1992) and Dreiding (Mayo et al., 1990) and in many works reporting GCMC and molecular dynamics, MD, simulations.

The Steele interaction potential between a methane molecule and a single infinite graphene layer is given by:

$$V_{10-4}(z) = 2\pi\epsilon_{gs}\rho_C\sigma_{gs}^2\Delta\left[\frac{2}{5}\left(\frac{\sigma_{gs}}{z}\right)^{10} - \left(\frac{\sigma_{gs}}{z}\right)^4\right]. \quad (23.10)$$

The Steele interaction potential between a methane molecule and a graphite slab made of n parallel graphene layers separated a distance, Δ, is given by the sum of the corresponding n V_{10-4} potentials:

$$V_{n,10-4}(z) = \sum_{i=0}^{n-1} V_{10-4}(z+i\Delta) = \sum_{i=0}^{n-1} 2\pi\epsilon_{gs}\rho_C\sigma_{gs}^2\Delta \times\left[\frac{2}{5}\left(\frac{\sigma_{gs}}{z+i\Delta}\right)^{10} - \left(\frac{\sigma_{gs}}{z+i\Delta}\right)^4\right]. \quad (23.11)$$

The Steele interaction potential V_{gs} is the interaction potential energy of a methane molecule with a graphite composed of an infinite number of graphene layers separated by a distance Δ, i.e., $V_{gs} = V_{n,10-4}$ with $n = \infty$. The infinite sum is approximated and the result is:

$$V_{gs}(z) = 2\pi\epsilon_{gs}\rho_C\sigma_{gs}^2\Delta \times\left[\frac{2}{5}\left(\frac{\sigma_{gs}}{z}\right)^{10} - \left(\frac{\sigma_{gs}}{z}\right)^4 - \frac{\sigma_{gs}^4}{3\Delta(z+0.61\Delta)^3}\right], \quad (23.12)$$

where Δ is the distance between the graphene layers in graphite, 3.35 Å, ρ_C is the number of carbon atoms per unit volume in graphite, 0.114 Å^{-3}, σ_{gs} and ϵ_{gs} are the LJ parameters of the interaction between a methane molecule (gas, g) and a carbon atom (solid, s) and z is the distance between the methane molecule and the first graphene layer. The parameters σ_{gs} and ϵ_{gs} are usually determined using the Lorentz–Berthelot combining rules: $\sigma_{gs} = (\sigma_C + \sigma_{CH_4})/2$ and $\epsilon_{gs} = \sqrt{\epsilon_C\epsilon_{CH_4}}$, where σ_C and ϵ_C are the LJ parameters of carbon and σ_{CH_4} and ϵ_{CH_4} are the LJ parameters of the methane molecule (Berthelot, 1898; Lorentz, 1881).

Slit-shaped pores have two walls. Therefore, the interaction potential of a single methane molecule with a slit pore of width w is given by:

$$V_{slit}(z; w) = V_{gs}(z) + V_{gs}(w - z), \quad (23.13)$$

where z is the distance between the molecule and one of the walls, w is the pore width and V_{gs} is the Steele potential. If the pore wall of the slit pore model is a single graphene layer, then the potential $V_{1,10-4}$ is used in Eq. 23.13, instead of V_{gs}. If the pore wall is composed of a slab of three parallel graphene layers, then $V_{3,10-4}$ is used.

The interaction potential energy $V_{slit}(z; w)$ of a methane molecule with a slit pore of $w = 12$ Å is plotted in Figure 23.6. Each wall pore is a single graphene layer. The two potential wells are located at 3.6 Å from the walls of the slit pore.

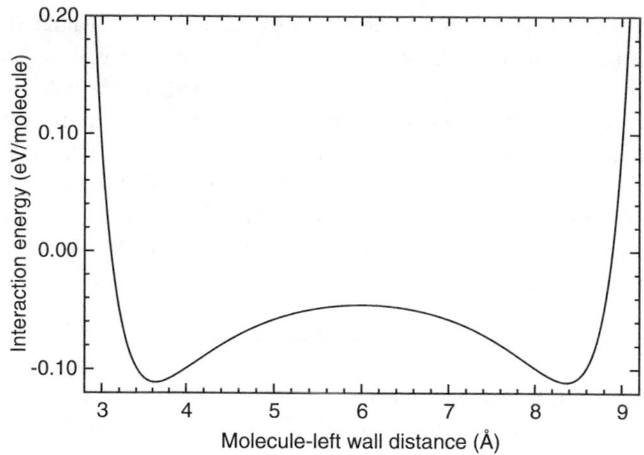

FIGURE 23.6 Interaction potential energy (in eV/molecule) of a methane molecule with a slit-shaped pore of 12 Å of width.

There are many instances of theoretical or simulated isotherms that predict or report storage capacities higher than the experimental ones. One reason of this discrepancy is that the model of carbon pores is perfect, infinite, without defects, impurities or solvent molecules used to synthesize the material. Another reason is that the LJ and Steele potentials reproduce approximately the methane–methane and the methane–wall interactions, respectively.

The main results of the GCMC simulations are the adsorption isotherms, the heat of adsorption, the adsorption sites, the density profiles or histograms of the gas inside the pores and the optimal pore size. Another result of the simulations is the potential energy surface, PES. This is a 3d map of the interaction energy of a methane molecule with the surface of the slit pore. This map shows the sites or regions where the interactions are higher, and hence, there will be more density of molecules adsorbed. In the case of slit pores, the 1d interaction potential energy $V_{slit}(z)$ is reported and analyzed, instead of the PES. The PES or the potential $V_{slit}(z)$ is related to the adsorption isotherms, the adsorption sites and the density profiles. A video of the "movement or molecular dynamics" of the methane gas molecules inside the pore can be also build with the snapshots of the molecular positions at each GCMC iteration to find out and understand the adsorption sites of the adsorbent material.

In the experiments, the excess adsorption isotherms n_e of methane on nanoporous carbons are measured, which are related to the number of excess molecules, N_{excess}. The direct result of the GCMC simulations is the number $N = N_{stored}$ of stored molecules, not N_{excess}. The number of stored molecules is the sum of the molecules in regions I, II and III in Figure 23.1 and the number of excess molecules is the number of molecules in region I in Figure 23.1. N_{excess} can be obtained from the simulations as follows. The number of excess molecules at P, T and for a pore width w is given by:

$$N_{excess}(P,T,w) = N_{stored}(P,T,w) - \rho_{bulk}(P,T)V, \quad (23.14)$$

where $\rho_{\text{bulk}}(P,T)$ is the density of bulk or free methane at P and T and V is the volume of the simulation cell, not the volume of the adsorbed layer. The pore width w is defined as the distance between the carbon nuclei at opposite walls, and therefore, the volume in Eq. 23.14 should be the volume of the simulation cell. The number of excess molecules $N_{\text{excess}}(P,T,w)$ is calculated from Eq. 23.14 and then the excess gravimetric isotherms are calculated using the following equation:

$$n_{\text{e,grav}}(P,T,w) = \frac{N_{\text{excess}}(P,T,w)m_{\text{CH}_4}}{M_{\text{adsorbent}}}, \tag{23.15}$$

where m_{CH_4} is the mass of a single methane molecule and $M_{\text{adsorbent}}$ is the mass of the adsorbent material. In the simulations of carbon slit pores, $M_{\text{adsorbent}} = N_C m_C$, where N_C is the number of carbon atoms of the slit pores and m_C is the mass of a single carbon atom.

The isosteric heat of adsorption is the amount of heat released upon methane adsorption. The typical value of the isosteric heat of commercial ACs is about 16 kJ/mol for methane at room temperature. The isosteric heat is related with fluctuations of the energy E and of the number of methane molecules N of the system and is calculated in the GCMC simulations as (Nicholson and Parsonage, 1982):

$$q_{st} = k_B T + \frac{< E >< N > - < EN >}{< N^2 > - < N >< N >}, \tag{23.16}$$

where k_B is the Boltzmann constant and T is the temperature of the system. The quantities $< E >$, $< N >$, $< N^2 >$ and $< EN >$ are the averages of the energy of the system, of the number of methane molecules N, of the square of N and of the product of the number of molecules and the energy of the system, respectively. The isosteric heat depends on the amount of methane adsorbed. According to GCMC simulations, the isosteric heat at room temperature increases linearly with the amount of methane adsorbed (Ortiz et al., 2016; Matranga et al., 1992).

Another interesting result of the GCMC simulations is the density profile. The density distribution or profile gives information about the most important adsorption sites. GCMC simulations reported by Ortiz et al. showed three regions on the methane density profiles inside carbon slit pores: Primary layers, secondary layers and the bulk region (Ortiz et al., 2016). Narrow pores ($w \leq 7$ Å) had only one primary or first layer of high methane density at all pressures, in the center of the pores. Larger pores (7 Å $< w <$ 15 Å) had two primary layers of high methane density at all pressures adsorbed on each wall of the slit pore. Pores of width $w \geq 15$ Å had two primary layers, and depending on pressure and size, they also had one or two secondary layers. They had also a bulk region in the center of the pore, which has the density of bulk, free or non-adsorbed methane gas, ρ_{bulk}. The secondary layers are located at about 4 Å of distance from the primary layers. The origin of these secondary layers is the weak van der Waals, vdW, attraction from the primary layers. Secondary layers can also enhance the methane density of the bulk region due to the vdW attraction, increasing that density above ρ_{bulk}.

Matranga et al. reported GCMC simulations of methane storage on slit pores (Matranga et al., 1992). They found that slit-shaped pores of width $w = 11.4$ Å had the highest excess methane storage capacity at 3.5 MPa and room temperature. The pore size $w = 11.4$ Å is the distance between the centers of the carbon atoms of opposite walls of the slit pore and corresponds to an effective pore size $w_e = 8.56$ Å, which is the distance between the surfaces of opposite walls. Pores of $w = 11.4$ Å contained two layers of methane, adsorbed on each sheet. Kumar et al. performed GCMC simulations using a slit-shaped pore whose walls were composed of three graphitic sheets, instead of a single sheet (Kumar et al., 2017). They obtained excess methane storage volumetric capacities of 129 v/v and 133 v/v at pressures of 3.5 and 5 MPa, respectively, at 298 K and for a pore width $w = 11.4$ Å. They also did simulations with slit pores made by single graphitic sheets, obtaining excess capacities of 282 v/v and 290 v/v at 3.5 and 5 MPa, respectively, for the same pore width and temperature. Heuchel et al. did GCMC simulations of methane inside a slit pore whose walls were composed by an infinite number of graphitic sheets. They did simulations at nine different pore widths w: 6.10, 7.62, 9.53, 11.43, 15.24, 22.86, 30.48, 38.10 and 47.63 Å and found an optimal pore width of $w = 11.43$ Å at 4 MPa and 293 K (Heuchel et al., 1999). Simulations at pore widths w between 10 and 12 Å should be carried out to obtain a more precise optimal pore size.

GCMC simulations and experimental data can be combined together to obtain the PSD of a nanoporous carbon. The GCMC-simulated isotherm curve due to a single slit pore of width w is $n(P,w;T)$, where T is the fixed temperature of the isotherm. The function $n(P,w;T)$ could be a gravimetric or a volumetric methane storage capacity. The experimental isotherm curve of the nanoporous carbon is $n(P;T)$. A nanoporous carbon can be modeled as a set of slit pores of different widths w and hence, the experimental isotherm curve of the nanoporous carbon can be assumed to be a sum of the isotherms of single pores of different widths w, weighted or multiplied by the relative distribution or "abundance" of the pores of width w in the nanoporous carbon:

$$n(P;T) = \int n(P,w;T)f(w)dw, \tag{23.17}$$

where $f(w)$ is the PSD function and the integral is over all the pore sizes of the nanoporous carbon. The function $f(w)$ is obtained by solving Eq. 23.17. The least square method is the most used method to solve this equation. This procedure to obtain the PSD can be found in detail in a variety of works in the scientific literature (Davies et al., 1999; Davies and Seaton, 1998; Samios et al., 1997; López-Ramón et al., 1997). It has been applied to experimental and GCMC-simulated isotherms of methane, hydrogen, nitrogen and carbon dioxide (Sriling et al., 2016; García Blanco et al., 2010; Heuchel et al., 1999). This procedure has been also applied to the isotherms of a couple of pure gases at the same time in some works, because then, the obtained PSD will probably be more appropriate for other gases or for mixtures of gases (Heuchel et al., 1999).

23.6 Comparison with the Methane Storage Capacities of Other Materials

Since the 1990s, many researchers have reported the methane storage properties of non-carbon-based porous materials. MOFs are the main group of non-carbon-based porous materials. They have emerged as an extensive class of crystalline materials with ultrahigh porosity (up to 90% free volume), enormous SSAs, extending beyond 6,000 m^2/g and tunable properties (He et al., 2018). These properties, together with the extraordinary degree of variability for both the organic and inorganic components of their structures, make MOFs of interest for potential applications in clean energy, most significantly as storage media for gases such as hydrogen and methane and as high-capacity adsorbents to meet various separation needs (Teo et al., 2017; Kayal et al., 2015; Zhou et al., 2012).

The number of MOFs that have been tested for methane storage is very small in comparison to the huge variety of MOFs that has been synthesized. Among porous MOFs reported for methane storage, those exhibiting high methane uptakes can be roughly grouped into the following categories: dicopper paddle wheel-based MOFs, Zn_4O-based MOFs, Zr-based MOFs, Al-based MOFs, melamine-based MOFs (MAFs) and flexible MOFs. By comparing their methane uptakes, some top-performing MOFs can be identified (He et al., 2018). Metal-organic frameworks of high SSA have made a significant impact on methane storage, but their great disadvantage is that they are not yet commercially viable because of their high fabrication cost (Liu et al., 2014).

Concerns over the stability and cost associated with the application of MOFs in methane adsorption have led to the evaluation of POPs as methane sorbents. Their tolerance to water and metal-free design makes POPs very attractive options in applications. Furthermore, many POPs exhibit exceptionally high SSAs and low framework density, which make them ideal in the gravimetric storage of gases (Makal et al., 2012).

A combined computational–experimental study has revealed that covalent organic frameworks (COFs) show great promise as methane sorbents (Makal et al., 2012). COFs are a class of crystalline porous polymers that allow the atomically precise integration of organic units to create predesigned skeletons and nanopores. They have recently emerged as a new molecular platform for designing promising organic materials for gas storage, catalysis, and optoelectronic applications (Feng et al., 2012). Finally, taking into account that large SSA and high micropore volume are the two most important factors to improve methane uptake capacity, many authors consider amorphous POPs as a promising material to store methane by adsorption. Amorphous POPs have received different names from different research groups: porous polymer networks (PPNs) (Makal et al., 2012; Yuan et al., 2011; Wood et al., 2008) and hypercrosslinked polymers (HCPs) (Wood et al., 2008), among others.

Figure 23.7 is a graphical comparison of the gravimetric capacities of carbon and non-carbon- based porous materials. MOFs and ACs have the highest gravimetric capacities, without significant differences between them, except that gravimetric capacities of MOFs have a larger dispersion of values, due to the diversity of MOFs. On the other hand, a graphical comparison of the volumetric capacities of MOFs and ACs is shown in Figure 23.8. In this case, there is a significant difference between the volumetric storage capacities: MOFs have higher volumetric capacities than ACs. The data represented in Figures 23.7, 23.8 and 23.9 were obtained in the experiments reported in (Alhasan et al., 2016; Beckner and Dailly, 2015; Liu et al., 2014; Makal et al., 2012; Lozano-Castelló et al., 2002c; Vasiliev et al., 2000).

According to the experimental results, the SSA of ACs is in the range 346.9–3,380 m^2/g, while the SSA of MOFs is in the range 824–6,240 m^2/g. The relationship between the gravimetric storage capacity and the SSA of ACs and MOFs can be noticed in Figure 23.9. There is an approximate linear relationship between SSA and gravimetric storage capacity, similar to the relationship found in Figure 23.2, in an earlier section.

Finally, in view of the collected results, it can be said that the storage capacities of the different types of materials are similar. In particular, MOFs and ACs have quite similar storage capacities; MOFs have slightly larger volumetric capacities. The main disadvantage of the MOFs

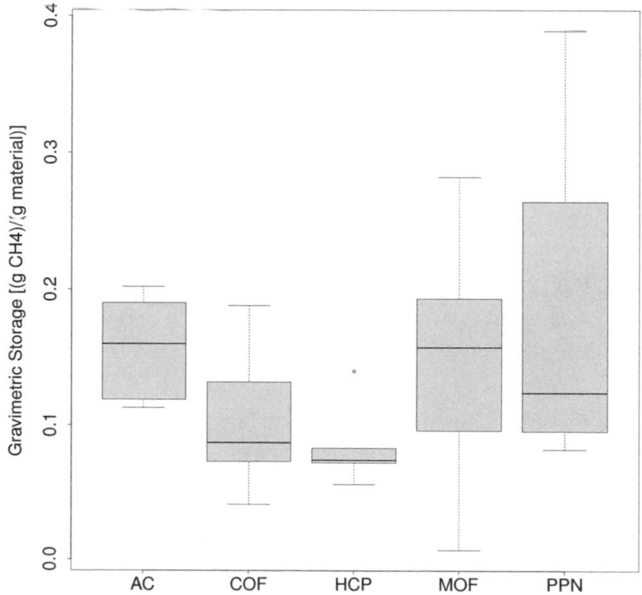

FIGURE 23.7 Comparison of the methane gravimetric capacities (in g CH_4/g adsorbent material) of different types of materials at 298 K and 3.5–4 MPa.

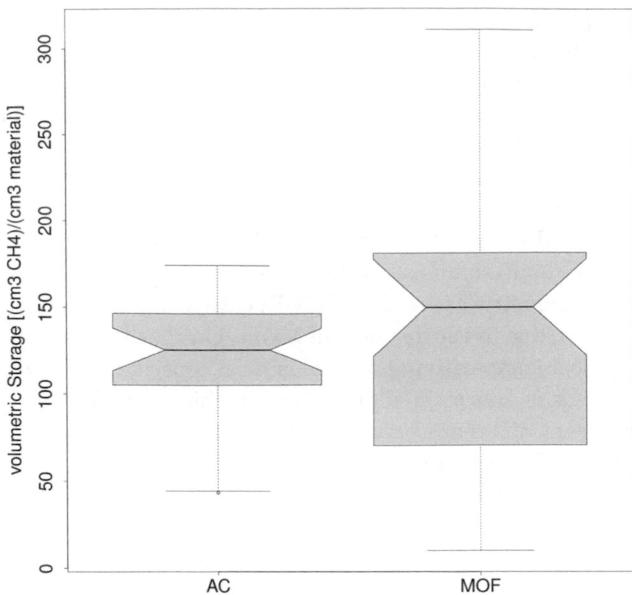

FIGURE 23.8 Comparison of the methane volumetric storage capacities (in cm^3 CH_4/cm^3 adsorbent material) of different types of materials at 298 K and 3.5–4 MPa.

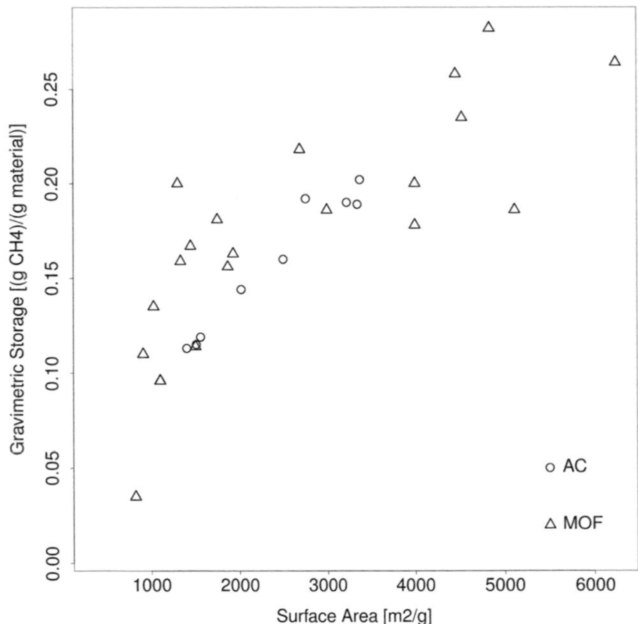

FIGURE 23.9 Methane gravimetric storage capacities (in g CH_4/g adsorbent material) of different types of materials at 298 K and 3.5–4 MPa versus the SSA (in m^2/g).

is, at present, the high manufacturing price, 1,000–19,000 euros/kg, in contrast to the low price of ACs, 1–30 euros/kg.

23.7 Adsorbed Natural Gas Tanks with Nanoporous Carbons

As explained at the beginning of this chapter, CNG and LNG are the most common technologies for the use of NG as a vehicle fuel. However, both technologies have drawbacks or points of weakness that need to be improved. Among these drawbacks can be cited the need for large containers to store NG as well as the costs derived from the processes of compression at high pressures, related to CNG, or liquefaction, related to LNG.

The adsorption of gas in porous materials at moderate pressures, 3.5–4 or 5–6.5 MPa, and room temperature allows to reach energy densities similar to those of CNG at 20–25 MPa and almost one- third of LNG, where both aspects, pressure and temperature, are advantageous for transport (Kumar et al., 2017). In short, the adsorption in nanoporous materials makes it possible to achieve energy densities similar to those of CNG but at substantially lower pressures, which has some advantages over CNG: Lighter tanks, conformable tanks and increased safety levels in the event of shocks.

In CNG vehicles, the gas is stored in tanks characterized by being manufactured normally with steel to withstand high pressures (up to 30 MPa). Because ANG needs lower pressures than CNG, the ANG tanks are made of materials much lighter than CNG tanks, such as aluminum and reinforced plastic. Besides, the thickness of the walls of the storage tanks, and therefore its economic cost, is reduced.

One of the main challenges in the research of ANG is the shape of the onboard storage tanks. While liquid fuels (gasoline, LNG and diesel) can be stored in tanks that are adjusted from the factory in the available space, without reducing the passenger space, this does not happen with the ANG tanks. In the case of CNG, the tanks are cylindrical, because this shape is more efficient to withstand the high pressures used in the CNG tanks. However, these cylinders leave unused parts in the spaces of the vehicle enabled for them. ANG tanks do not have to be cylindrical and can be shaped like prisms or other forms, with the aim of adapting the tank shape to the free space of the vehicles. This way, ANG tanks can use 20% more spacc than CNG tanks. They are said to be tanks with a high degree of "conformability". The factor of conformability is defined as the volume of the outer tank divided by the smaller box volume. This factor represents the efficiency of packing a tank inside a box. For instance, Zhengwei et al. explain that in rectangular plant tanks (ratio width/height equal to integer), cylinders, in which the gas is introduced, would only fill 75% of the total storage volume (Nie et al., 2016). Moreover, if that ratio was not equal to an integer, the percentage would be reduced to a scarce 50%. For this reason, one of the most active lines of research within ANG technology is aimed at maximizing the tanks. Pressurized conformable tanks are being developed for this purpose. Currently, non-cylindrical ANG tanks are being tested.

If the objective is to increase the use of ANG in vehicles and to be competitive with LNG or other fuels, storage and refueling costs should be reduced. ANG technology based on NG adsorption in porous materials at moderate pressures 3.5–4 MPa offers two main advantages: it allows good design flexibility in tank configuration and positioning and reduces the cost to compress NG at high pressures.

An example of research on this topic is the project "Advanced Natural Gas Fuel Tank" (Research and Division, 2016; Pfeifer et al., 2016; Xu and Lin, 2013). The main objectives are to design and build a light conformable tank for ANG to improve the adsorption capacity of AC and to develop a methodology to compact these ACs into monoliths.

Finally, the degree of reversibility of gas storage on ANG tanks should be considered. The storage capacities of the ACs inside the ANG tanks should be reversible, i.e., they should not depend on the number of cycles of filling-releasing gas. Interesting research is underway to study the storage capacities on ACs of pure methane and natural gas, which is a mixture of gases, mainly methane, after a number of cycles of filling-releasing (Romanos et al., 2017). For pure methane, the storage capacity presented no variation after a large number of cycles. For NG, the storage capacity decreased as the number of cycles increased. In these investigations, the gravimetric adsorption was found to be reduced by 33% after more than 100 cycles and its tendency remained decreasing until the 1,000th cycle. On the other hand, the volumetric storage capacity also decreased to 50% after the first 100 cycles and it remained constant in the following cycles. After analyzing the composition of the remaining gas at the end of a certain number of cycles (100, 500 and 1,000), the presence of gas molecules heavier than methane was observed, which could reduce the efficiency of the cycles of filling-releasing.

Acknowledgments

Financial support from both the Spanish Ministerio de Economía y Competitividad (MINECO) and the European Regional Development Fund (ERDF) through projects MAT2015-69844-R and MAT2014-54378-R, and from Junta de Castilla y León through project VA050U14 are gratefully acknowledged.

References

Alcañiz-Monge, J., Casa-Lillo, M., Cazorla-Amorós, D., et al., 1997. Methane storage in activated carbon fibres. *Carbon* 35, 291–297.

Alcañiz-Monge, J., Lozano-Castelló, D., Cazorla-Amorós, D., et al., 2009a. Fundamentals of methane adsorption in microporous carbons. *Microporous Mesoporous Mater.* 124, 110–116.

Alcañiz-Monge, J., Trautwein, G., Pérez-Cadenas, M., et al., 2009b. Effects of compression on the textural properties of porous solids. *Microporous Mesoporous Mater.* 126, 291–301.

Alhasan, S., Carriveau, R., Ting, D. S. K., 2016. A review of adsorbed natural gas storage technologies. *Int. J. Environ. Stud.* 73, 343–356.

Allen, M. P., Tildesley, D. J., 1987. *Computer Simulation of Liquids.* Oxford University Press, Oxford.

Beckner, M., Dailly, A., 2015. Adsorbed methane storage for vehicular applications. *Appl. Energy* 149, 69–74.

Berthelot, D., 1898. Sur le mélange des gaz. *Comptes rendus hebdomadaires des séances de l'Académie des Sciences* 126, 1703.

Borchardt, L., Casco, M. E., Silvestre-Albero, J., 2018. Methane hydrate in confined spaces - An alternative energy storage system. *ChemPhysChem* 19, 1–18.

Borchardt, L., Nickel, W., Casco, M., et al., 2016. Illuminating solid gas storage in confined spaces - Methane hydrate formation in porous model carbons. *Phys. Chem. Chem. Phys.* 18, 20607–20614.

Burchell, T., Rogers, M., 2000. Low pressure storage of natural gas for vehicular applications. SAE Technical Paper 2000-01-2205; SAE International: Warrendale, PA.

Casco, M. E., Martínez-Escandell, M., Gadea-Ramos, E., et al., 2015a. High-pressure methane storage in porous materials: Are carbon materials in the pole position? *Chem. Mater.* 27, 959–964.

Casco, M. E., Silvestre-Albero, J., Ramírez-Cuesta, A. J., et al., 2015b. Methane hydrate formation in confined nanospace can surpass nature. *Nat. Commun.* 6, 6432–6440.

Celzard, A., Fierro, V., 2005. Preparing a suitable material designed for methane storage: A comprehensive report. *Energy Fuels* 19, 573–583.

Cenergy Solutions, 2018. www.cenergysolutions.com, accessed April 30, 2018.

CNG United, 2018. How you can benefit from CNG conversion. www.cngunited.com/how-you-can-benefit-from-cng-conversion, accessed April 30, 2018.

Davies, G. M., Seaton, N. A., 1998. The effect of the choice of pore model on the characterization of the internal structure of microporous carbons using pore size distributions. *Carbon* 36, 1473–1490.

Davies, G. M., Seaton, N. A., Vassiliadis, V. S., 1999. Calculation of pore size distributions of activated carbons from adsorption isotherms. *Langmuir* 15, 8235–8245.

DOE, 2012. Methane Opportunities for Vehicular Energy (MOVE) Program Overview. http://arpa-e.energy.gov/sites/default/files/documents/files/MOVE_ProgramOverview.pdf, accessed April 30, 2018.

Feng, X., Ding, X., Jiang, D., 2012. Covalent organic frameworks. *Chem. Soc. Rev.* 41, 6010–6022.

Frenkel, D., Smit, B., 1996. *Understanding Molecular Simulation: From Algorithms to Applications.* Academic Press, San Diego, CA.

FuelMaker, 2018. Home-fueling CNG compressor. www.brcfuelmaker.com/phill-domestico-prodotto-brc-fuel-maker.aspx, accessed April 30, 2018.

Furukawa, H., Yaghi, O. M., 2009. Storage of hydrogen, methane, and carbon dioxide in highly porous covalent organic frameworks for clean energy applications. *J. Am. Chem. Soc.* 131, 8875–8883.

García Blanco, A. A., de Oliveira, J. C. A., López, R., et al., 2010. A study of the pore size distribution for activated carbon monoliths and their relationship with the storage of methane and hydrogen. *Colloid Surf. A-Physicochem. Eng. Asp.* 357, 74–83.

García Blanco, A. A., Vallone, A. F., Korili, S. A., et al., 2016. A comparative study of several microporous materials to store methane by adsorption. *Microporous Mesoporous Mater.* 224, 323–331.

Gas & Go, 2018. Private communication; http://gasngo.es/en.

Giraldo, L., Rodríguez-Estupiñán, P., Moreno-Pirajá́n, J., 2018. A microcalorimetric study of methane adsorption on activated carbons obtained from mangosteen peel at different conditions. *J. Therm. Anal. Calorim.* 132, 525–541.

He, Y., Zhou, W., Chen, B., 2018. Current status of porous metal - organic frameworks for methane storage. In *Metal-Organic Frameworks*, Wiley-VCH, Weinheim, Germany, Ch. 6, pp. 163–198.

Heuchel, M., Davies, G. M., Buss, E., et al., 1999. Adsorption of carbon dioxide and methane and their mixtures on an activated carbon: Simulation and experiment. *Langmuir* 15, 8695–8705.

Jordá-Beneyto, M., Lozano-Castelló, D., Suárez-García, F., et al., 2008. Advanced activated carbon monoliths and activated carbons for hydrogen storage. *Microporous Mesoporous Mater.* 112, 235–242.

Juan-Juan, J., Marco-Lozar, J. P., Suárez-García, F., et al., 2010. A comparison of hydrogen storage in activated carbons and a metal-organic framework (MOF-5). *Carbon* 48, 2906–2909.

Kayal, S., Sun, B., Chakraborty, A., 2015. Study of metal-organic framework MIL-101 (Cr) for natural gas (methane) storage and compare with other MOFs (metal-organic frameworks). *Energy* 91, 772–781.

Kumar, K. V., Preuss, K., Titirici, M., et al., 2017. Nanoporous materials for the onboard storage of natural gas. *Chem. Rev.* 117, 1796–1825.

Lennard-Jones, J. E., 1924. On the determination of molecular fields. *Proc. Roy. Soc. (London) A* 106, 463–477.

Li, B., Wen, H.-M., Zhou, W., et al., 2016. Porous metal-organic frameworks: Promising materials for methane storage. *Chem.* 1, 557–580.

Liu, B., Wang, W., Wang, N., et al., 2014. Preparation of activated carbon with high surface area for high-capacity methane storage. *J. Energy Chem.* 23, 662–668.

López-Ramón, M. V., Jagiello, J., Bandosz, T. J., et al., 1997. Determination of the pore size distribution and network connectivity in microporous solids by adsorption measurements and Monte Carlo simulation. *Langmuir* 13, 4435–4445.

Lorentz, H. A., 1881. Über die Anwendung des Satzes vom Virial in der kinetischen Theorie der Gase. *Ann. Physik* 248, 127–136.

Lozano-Castelló, D., Alcañiz-Monge, J., de la Casa-Lillo, M. A., et al., 2002a. Advances in the study of methane storage in porous carbonaceous materials. *Fuel* 81, 1777–1803.

Lozano-Castelló, D., Cazorla-Amorós, D., Linares-Solano, A., et al., 2002b. Activated carbon monoliths for methane storage: Influence of binder. *Carbon* 40, 2187–2825.

Lozano-Castelló, D., Cazorla-Amorós, D., Linares-Solano, A., et al., 2002c. Influence of pore size distribution on methane storage at relatively low pressure: Preparation of activated carbon with optimum pore size. *Carbon* 40, 989–1002.

Makal, T. A., Li, J.-R., Lu, W., et al., 2012. Methane storage in advanced porous materials. *Chem. Soc. Rev.* 41, 7761–7779.

Marco-Lozar, J. P., Kunowsky, M., Suárez-García, F., et al., 2012. Activated carbon monoliths for gas storage at room temperature. *Energy Environ. Sci.* 5, 9833–9842.

Mason, J. A., Veenstra, M., Long, J. R., 2014. Evaluating metal-organic frameworks for natural gas storage. *Chem. Sci.* 5, 32–51.

Matranga, K. R., Myers, A. L., Glandt, E. D., 1992. Storage of natural gas by adsorption on activated carbon. *Chem. Eng. Sci.* 47, 1569–1579.

Mayo, S. L., Olafson, B. D., Goddard III, W. A., 1990. DREIDING: A generic force field. *J. Phys. C: Solid State Phys.* 94, 8897–8909.

Menon, V. V., Komarneni, S., 1998. Porous adsorbents for vehicular natural gas storage: a review. *J. Porous Mat.* 5, 43–58.

Menshchikov, I., Fomkin, A. A., Shkolin, A. V., et al., 2017a. The energy of adsorption of methane on microporous carbon adsorbents. *Prot. Met. Phys. Chem. Surf.* 53, 780–785.

Menshchikov, I. E., Fomkin, A. A., Tsivadze, A. Y., et al., 2017b. Adsorption accumulation of natural gas based on microporous carbon adsorbents of different origin. *Adsoption* 23, 327–339.

Molina-Sabio, M., Almansa, C., Rodríguez-Reinoso, F., 2003. Phosphoric acid activated carbon discs for methane adsorption. *Carbon* 41, 2113–2119.

Mota, J. P., 2008. Adsorbed natural gas technology. In: Mota, J. P., Lyubchik, S. (Eds.), *Recent Advances in Adsorption Processes for Environmental Protection and Security.* NATO Science for Peace and Security Series C: Environmental Security. Springer Netherlands, Dordrecht, Vol. 1, pp. 177–192.

Murata, K., Merraoui, M. E., Kaneko, K., 2001. A new determination method of absolute adsorption isotherm of supercritical gases under high pressure with a special relevance to density-functional theory study. *J. Chem. Phys.* 114, 4196–4205.

Nicholson, D., Parsonage, N. G., 1982. *Computer Simulation and the Statistical Mechanics of Adsorption.* Academic Press, London.

Nie, Z., Lin, Y., Jin, X., 2016. Research on the theory and application of adsorbed natural gas used in new energy vehicles: A review. *Font. Mech. Eng.* 11, 258–274.

Ohno, K., Esfarjani, K., Kawazoe, Y., 1999. *Computational Materials Science: From ab initio to Monte Carlo Methods.* Springer-Verlag, Heidelberg.

Ortiz, L., Kuchta, B., Firley, L., et al., 2016. Methane adsorption in nanoporous carbon: The numerical estimation of optimal storage conditions. *Mater. Res. Express* 3, 055011.

Park, M.-S., Lee, S.-E., Kim, M. I., et al., 2015. CO_2 adsorption characteristics of slit-pore shaped activated carbon prepared from cokes with high crystallinity. *Carbon Lett.* 16, 45–50.

Peng, D. Y., Robinson, D. B., 1976. A new two-constant equation of state. *Ind. Eng. Chem. Fundam.* 15, 59–64.

Pfeifer, P., 2011. Advanced natural gas fuel tank project. In: Proceedings of Natural Gas Vehicle Technology Forum. San Francisco.

Pfeifer, P., Little, R., Rash, T., et al., 2016. Advanced natural gas fuel tank project. Publication No. CEC5002016038; California Energy Commission, Energy Research and Development Division: Sacramento, CA.

Policicchio, A., Filosa, R., Abate, S., et al., 2017. Activated carbon and metal organic framework as adsorbent for low-pressure methane storage applications: An overview. *J. Porous Mat.* 24, 905–922.

Rappe, A. K., Casewit, C. J., Colwell, K. S., et al., 1992. UFF, a full periodic table force field for molecular mechanics and molecular dynamics simulations. *J. Am. Chem. Soc.* 114, 10024–10035.

Research, E., Division, D., 2016. Advanced natural gas fuel tank project. University of Missouri Interim Final Project Report.

Rodríguez-Reinoso, F., Almansa, C., Molina-Sabio, M., 2005. Contribution to the evaluation of density of methane adsorbed on activated carbon. *J. Phys. Chem. B* 109, 20227–20231.

Romanos, J., Rash, T., Dargham, S. A., et al., 2017. Cycling and regeneration of adsorbed natural gas in mcroporous materials. *Energy Fuels* 31, 14332–14337.

Rouquerol, F., Rouquerol, J., Sing, K. S. W., 1999. *Adsorption by Powders & Porous Solids: Principles, Methodology and Applications.* Academic Press, New York.

Samios, S., Stubos, A. K., Kanellopoulos, N. K., et al., 1997. Determination of micropore size distribution from Grand Canonical Monte Carlo simulations and experimental CO_2 isotherm data. *Langmuir* 13, 2795–2802.

Sircar, S., 2001. Measurement of gibbsian surface excess. *AIChE J.* 47, 1169–1176.

Soave, G., 1972. Equilibrium constants from a modified Redlich-Kwong equation of state. *Chem. Eng. Sci.* 27, 1197–1203.

Sriling, P., Wongkoblap, A., Tangsathitkulchai, C., 2016. Computer simulation study for methane and hydrogen adsorption on activated carbon based catalyst. *Adsoption* 22, 707–715.

Srinivas, G., Burress, J., Yildirim, T., 2012. Graphene oxide derived carbons (GODCs): synthesis and gas adsorption properties. *Energy Environ. Sci.* 5, 6453–6459.

Steele, W. A., 1974. *The Interaction of Gases with Solid Surfaces.* Pergamon Press, Oxford.

Tee, L. S., Gotoh, S., Stewart, W. E., 1966. Molecular parameters for normal fluids. Lennard-Jones 12-6 potential. *Ind. Eng. Chem. Fundam.* 5, 356–363.

Teo, H. W. B., Chakraborty, A., Kayal, S., 2017. Evaluation of CH_4 and CO_2 adsorption on HKUST-1 and MIL-101 (Cr) MOFs employing Monte Carlo simulation and comparison with experimental data. *Appl. Therm. Eng.* 110, 891–900.

Vasiliev, L. L., Kanonchik, L., D. A. Mishkinis, M. I. R., 2000. Adsorbed natural gas storage and transportation vessels. *Int. J. Therm. Sci.* 39, 1047–1055.

Veluswamy, H., Kumar, A., Seo, Y., et al., 2018. A review of solidified natural gas (SNG) technology for gas storage via clathrate hydrates. *Appl. Energy* 216, 262–285.

Wegrzyn, J., Gurevich, M., 1996. Adsorbent storage of natural gas. *Appl. Energy* 55, 71–83.

Wood, C., Tan, B., Trewin, A., et al., 2008. Microporous organic polymers for methane storage. *Adv. Mater.* 20, 1916–1921.

Wu, H., Simmons, J., Liu, Y., et al., 2010. Metal-organic frameworks with exceptionally high methane uptake: Where and how is methane stored? *Chem. Eur. J.* 16, 5205–5214.

Xu, H., Lin, Y., 2013. Optimal design of conformable adsorbed natural gas tank. University of Missouri Project Report.

Yeon, S.-H., Knoke, I., Gogotsi, Y., et al., 2010. Enhanced volumetric hydrogen and methane storage capacity of monolithic carbide-derived carbon. *Microporous Mesoporous Mater.* 131, 423–428.

Yeon, S.-H., Osswald, S., Gogotsi, Y., et al., 2009. Enhanced methane storage of chemically and physically activated carbide-derived carbon. *J. Power Sources* 191, 560–567.

Yuan, D., Lu, W., Zhao, D., et al., 2011. Highly stable porous polymer networks with exceptionally high gas-uptake capacities. *Adv. Mater.* 23, 3723–3725.

Zhou, H.-C., Long, J., Yaghi, O., 2012. Introduction to metal-organic frameworks. *Chem. Rev.* 112, 673–674.

24

Metal Hydroxide and Oxide Nanocages

Jian Yu, Lidong Li, and Lin Guo
Beihang University

24.1 Introduction

Nanocages, defined as nanoscale architectures composed of clear inner cavities and outer shells [1,2]. They can be categorized into various morphologies, such as hollow nanospheres, nanoboxes, hollow octahedral nanocages, and so on [3–6]. On the basis of the number of layers of outer shell, they also can be termed as single-shelled [7,8], double-shelled [9–11], and multi-shelled (or walled) nanocages [12–14]. Over the past decades, metal hydroxide and oxide nanocages have intensively attracted a great deal of interest owing to their aesthetic beauty, unique structural features, high specific surface area, low density, high loading capacity, fascinating physicochemical properties [15–17], and widespread applications in drug delivery [18–26], energy storage and conversion [27–40], catalysis [41–49], sensors [50–60], water treatment [61–68], etc. Although they have great structural advantages, the controlled synthesis of nanocages remains challenging.

Since a pioneering report by Caruso and colleagues on the synthesis of hollow SiO_2 and inorganic-polymer hybrid hollow spheres in 1998 [69], many nanocages have been developed on the basis of the similar templating strategy [3,15,70,71]. These hard template strategies involve the removal of the template, which is generally based on some etching methods (such as acid etching method [72,73], redox etching method [74,75], coordinating etching method [76], etc.). Moreover, the templates have evolved from the classic SiO_2 nanospheres to complex metal-organic frameworks (MOFs) [77–83], all kinds of organic and inorganic nanomaterials [18], and various unconventional biological templates [84], while their morphologies have also been extended from the simple spheres to cubes, octahedron, dodecahedron, and so on. The final shape and void size of nanocage are determined by the shape and size of the hard template, while the shell thickness is mainly determined by the coating process [42]. Generally, the hard templates are removed thoroughly without any contribution to the composition of the final products. In contrast, the other soft template strategy is different from the hard template. It is unnecessary to remove the soft template, and a part of the template is deliberately left to become composition of the final product [2]. The soft templates are usually in the fluid form, such as vesicles/micelles, emulsion droplets, and gas bubbles, which provides more possibilities in tuning both the inner and the outer structures, producing more complex hollow nanostructures [15]. Interestingly, some "smarter" strategies, termed as self-templating strategies, have also been rapidly developed in recent years [1]. The self-templating strategy refers to the direct synthesis of metal hydroxide and oxide nanocages by various physical and/or chemical processes (such as Ostwald ripening [85], Kirkendall effect [86], surface-protected etching [87], and galvanic replacement method [7]) and without the need of additional templates [88]. It has many advantages such as simplified synthesis procedures, reduced production cost, and more efficient fabrication of various complex nanocages.

It is well known that the properties of materials are mainly determined by their structures, while the applications of materials are highly dependent on their properties [2]. Metal hydroxide and oxide nanocages have fascinating properties associated with the unique hollow structures. For example, nanocages have a larger specific surface area than solid materials due to the existence of the hollow cavities [16,17]. High surface area and open porous channels provide large accessible active sites for surface charge storage and catalytic reactions [41,42]. And the cavities within the porous shells can be used as nanoreactors for catalytically loading active species for catalytic reactions or nanocontainers for drug storage [19,89,90]. In short, as compared to their solid counterparts, metal hydroxide and oxide nanocages offer additional possibilities for structural and compositional tuning that can be well utilized for rational design of novel functional materials toward many desired applications [91,92].

This chapter focuses primarily on the developments in the various synthetic methods and potential applications of metal hydroxide and oxide nanocages, emphasizing the progress made in recent years. It is organized as follows: in the first section, the syntheses of metal hydroxide nanocages, including monometallic hydroxide nanocages and mixed metal hydroxide nanocages, are systematically discussed. In the second section, the representative synthetic strategies for monometallic oxide nanocages and mixed metal oxide nanocages are also comprehensively presented. The synthetic methods include hard templating strategies (such as acid/alkali etching method, redox etching method, coordinating etching method, and thermal decomposition method), soft templating strategies, and self-templating strategies (such as surface-protected etching, Ostwald ripening, Kirkendall effect, and galvanic replacement). For each method, we offer several critical comments based on our knowledge and related research experience. In the last section, the potential applications of both metal hydroxide and oxide nanocages in different fields are broadly reviewed, such as lithium-ion battery (LIB), lithium-sulfur (Li-S) batteries, supercapacitors, photocatalysis, and electrocatalysis. Some perspectives on the future research and development are also provided.

24.2 Syntheses of Metal Hydroxide Nanocages

Over the past decades, the research of metal hydroxide nanocages has been reported relatively less compared to metal oxide nanocages. There are two main reasons: (i) metal hydroxide nanocages have relatively weak stability owing to the hydroxyl prone to dehydration reaction into metal oxide nanocages at relatively high temperatures and (ii) probably because the potential application range of metal hydroxide nanocages is narrower than that of metal oxide nanocages. Nevertheless, researchers are still using various strategies (such as redox etching method, coordinating etching method, and assisted dissolution method) to fabricate monometallic hydroxide nanocages and mixed metal hydroxide nanocages.

24.2.1 Syntheses of Monometallic Hydroxide Nanocages

Monometallic hydroxide nanocages, as its name suggests, contain only one metal element in the hydroxides. In the following part, two typical synthesis strategies are mainly introduced.

Redox etching method: This is one of the hard template strategies. As shown in Figure 24.1, Wang et al. [75] demonstrated a facile approach for the preparation of uniform hollow $Fe(OH)_x$ nanocages with various shapes and dimensions by template-engaged redox etching

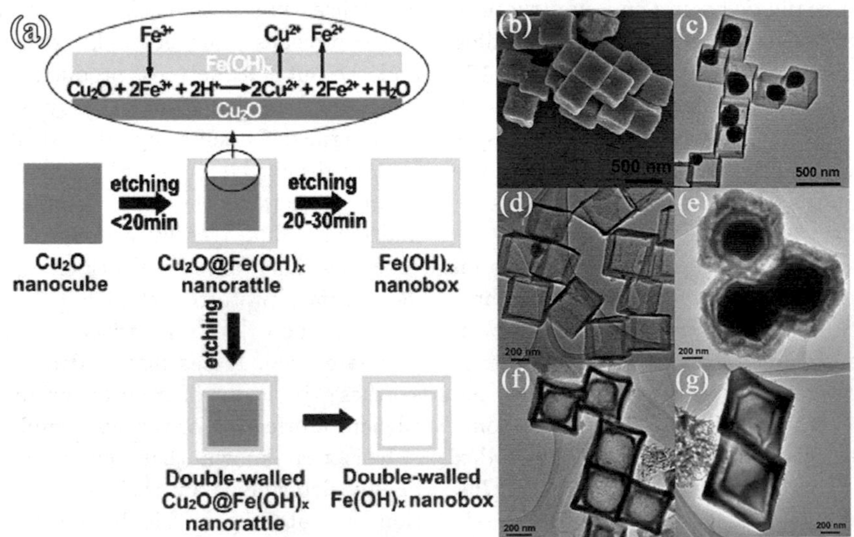

FIGURE 24.1 (a) Schematic illustration of the formation of single- or double-walled hollow structures by template-engaged redox etching of Cu_2O nanocubes. (b and c) SEM and TEM images of cubic Cu_2O@$Fe(OH)_x$ nanorattles; TEM images of $Fe(OH)_x$ nanoboxes (d), cubic double-walled Cu_2O@$Fe(OH)_x$ nanorattles (e), $Fe(OH)_x$ box-in-box (f), and octahedral box-in-box $Fe(OH)_x$ cages (g). (Reproduced with permission from Ref. [75]. Copyright 2010, American Chemical Society.)

of shape-controlled Cu_2O crystals. Because the standard reduction potential of Fe^{3+}/Fe^{2+} pair (0.77 V *vs* SHE) was higher than that of Cu^{2+}/Cu_2O (0.203 V *vs* SHE), Cu_2O crystals in suspension could be immediately oxidized by Fe(III) ions at room temperature according to the following redox reaction:

$$Cu_2O\,(s) + 2Fe^{3+}\,(aq) + 2H^+\,(aq) \rightarrow 2Cu^{2+}\,(aq) + 2Fe^{2+}\,(aq) + H_2O \quad (24.1)$$

Finally, the Fe(II) ions of template surface readily form hydroxides because of the depletion of H^+ ions. More remarkably, this strategy enables top-down engineering the interiors of hollow structures as demonstrated by the fabrication of double-walled nanorattles (Figure 24.1e) and nanoboxes (Figure 24.1f) and even box-in-box structures (Figure 24.1g).

Coordinating etching method: Inspired by Pearson's hard and soft acid–base principle, our group [93] developed a coordinating etching method for the fabrication of hollow $Ni(OH)_2$ nanocages. We selected soft base ligands ($S_2O_3{}^{2-}$) as the coordinating etchant because it prefers coordination with soft acid Cu^+. The amorphous $Ni(OH)_2$ nanoboxes formation is illustrated in Figure 24.2. Their chemical route could be described as follows:

$$Cu_2O + xS_2O_3^{2-} + H_2O \rightarrow \left[Cu_2\left(S_2O_3^{2-}\right)_x\right]^{2-2x} + 2OH^- \quad (24.2)$$

$$S_2O_3^{2-} + H_2O \rightleftharpoons HS_2O_3^{2-} + OH^- \quad (24.3)$$

$$Ni^{2+} + 2OH^- \rightarrow Ni\,(OH)_2 \downarrow \quad (24.4)$$

During this process, $S_2O_3{}^{2-}$ plays versatile roles: (i) coordinating etching of Cu_2O (Eq. 24.2) since the soft–soft interaction of Cu^+-$S_2O_3{}^{2-}$ was much stronger than the soft–hard interaction of Cu^+-O^{2-} within Cu_2O; (ii) Ni^{2+} was nearly free in the solution as the binding of borderline acid-soft base (Ni^{2+}-$S_2O_3{}^{2-}$) was unstable; and (iii) along with the OH^- released from the etching of Cu_2O (Eq. 24.2), those OH^- originated from hydrolysis of excess $S_2O_3{}^{2-}$ (Eq. 24.3) could facilitate the precipitation of Ni^{2+} (Eq. 24.4). More interestingly, based on this method, we have synthesized a series of different hydroxide nanocages such as $Mn(OH)_2/MnO(OH)$ (Figure 24.2a_{1-4}), $Fe(OH)_3$ (Figure 24.2b_{1-4}), $Co(OH)_2$ (Figure 24.2c_{1-4}), $Ni(OH)_2$ (Figure 24.2d_{1-4}), and $Zn(OH)_2$ (Figure 24.2d_{1-4}) nanocages [76].

24.2.2 Syntheses of Mixed Metal Hydroxide Nanocages

The mixed metal hydroxide nanocages contain two or more metal elements in the hydroxides. Lou and co-workers [94] have developed the synthesis of single-crystal $CoSn(OH)_6$ nanoboxes by a one-pot "pumpkin-carving" etching protocol (Figure 22.3a). Firstly, the $[Co(OH)_4]^{2-}$ and $[Sn(OH)_6]^{2-}$ could be generated when single-crystalline $CoSn(OH)_6$ solid nanocubes (Figure 24.3b) were dissolved in a concentrated alkaline solution. Meanwhile, $[Co(OH)_4]^{2-}$ ions would be oxidized by O_2 to form insoluble CoO(OH) species, which serve as passivation layers and protect the outer surface of nanocubes from further dissolution. Secondly, the

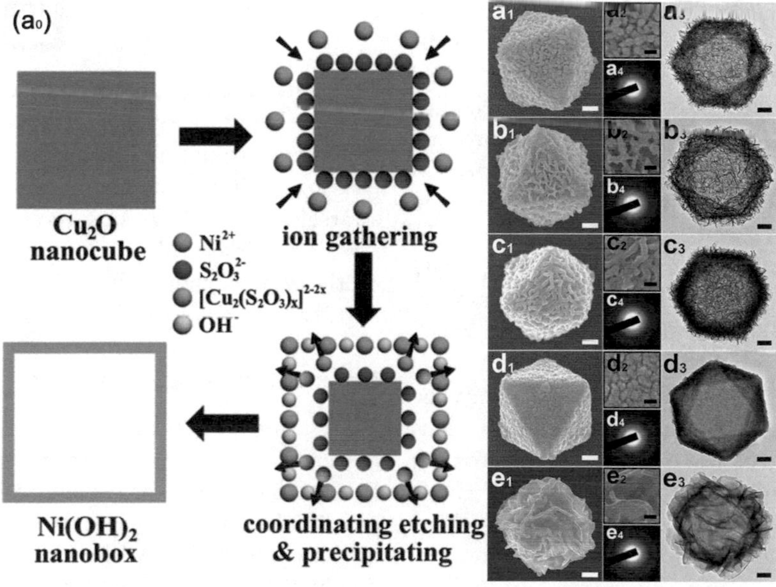

FIGURE 24.2 (a_0) Schematic illustration of the formation of $Ni(OH)_2$ nanoboxes by simultaneous coordinating etching of Cu_2O nanocubes. SEM, TEM, and SAED images of the (a_{1-4}) Mn, (b_{1-4}) Fe, (c_{1-4}) Co, (d_{1-4}) Ni, and (e_{1-4}) Zn hydroxide nanocages. Parts x_1 (x = a–e) and x_3 show a typical metal hydroxide cage image by SEM and TEM, respectively; part x_2 shows high-magnification images of the surface of the cage in part x_1; part x_4 is the selected area electron diffraction (SAED) pattern obtained from the whole cage in part x_3. The scale bars in parts x_1, x_2, and x_3 are 100, 20, and 100 nm, respectively. (Panel (a_0) reproduced with permission from Ref. [93]. Copyright 2013, Wiley-VCH Verlag GmbH & Co. KGaA, Weinheim. Panels (a_{1-4}–e_{1-4}) reproduced with permission from Ref. [76]. Copyright 2013, American Chemical Society.)

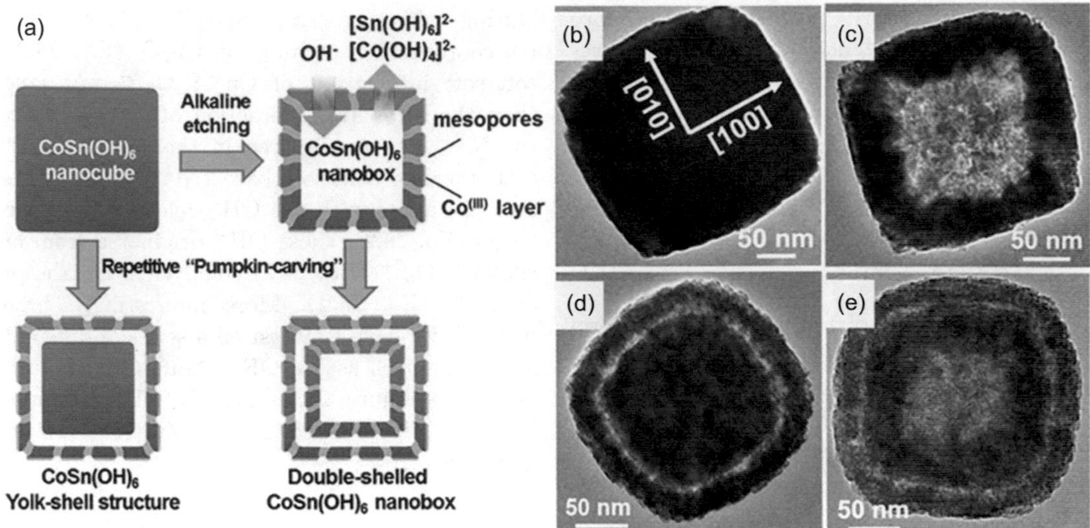

FIGURE 24.3 (a) Schematic illustration for the formation of $CoSn(OH)_6$ hollow nanostructures. TEM images of single $CoSn(OH)_6$ solid nanocube (b), single-shelled nanobox (c), yolk-shelled nanobox (d), and double-shelled nanobox (e). (Reproduced with permission from Ref. [94]. Copyright 2013, Nature Publishing Group.)

freshly exposed interiors were preferentially etched to form $CoSn(OH)_6$ nanoboxes (Figure 24.3c). Finally, $CoSn(OH)_6$ hollow particles with yolk-shelled (Figure 24.3d) or multi-shelled (Figure 24.3e) nanocages were obtained by repeating the template growth and etching processes.

In addition, Wang et al. [95] reported the synthesis of single-crystalline $ZnSn(OH)_6$ hollow cubes by an alkali-assisted dissolution process at room temperature. The formation of $ZnSn(OH)_6$ hollow cubes undergoes two steps. First, the $ZnSn(OH)_6$ solid cubes were formed owing to the coprecipitation of Sn (IV) and Zn (II) under basic conditions. Second, NaOH solution was added, and the solid cubes of $ZnSn(OH)_6$ as the self-templates were converted to hollow structures through an alkali-assisted dissolution process. $ZnSn(OH)_6$ hollow cubes were finally obtained successfully. Furthermore, our group [96] also reported a unique template-engaged approach for fabrication of amorphous Ni-Co double hydroxide nanocages. Briefly, we first used Cu_2O nanocrystals as the template. It could be gradually dissolved by $Na_2S_2O_3$ via a "coordinating etching" process, which could form a soluble complex $[Cu_2(S_2O_3)_x]_{2-2x}$. A large amount of OH^- ions were subsequently formed. Finally, the Ni^{2+} and Co^{2+} ions could coprecipitate with these OH^- ions at the etching interface to generate Ni-Co double hydroxides nanocages by perfectly inheriting the geometries of Cu_2O templates. More importantly, we could also flexibly adjust the Ni/Co ratio in Ni-Co double hydroxides nanocages.

24.3 Syntheses of Metal Oxide Nanocages

In the past few decades, metal oxide nanocages including transition metal and main group metal oxides have been intensively investigated owing to their intriguing structural advantages and attractive physical and chemical properties. The methods of synthesizing them have also been developed very richly, from the traditional hard templating strategies to the soft templating strategies and then to the smart self-templating strategies.

In this section, the various synthesis strategies of metal oxide nanocages will be summarized in detail.

24.3.1 Syntheses of Monometallic Oxide Nanocages

There are many methods such as hard templating strategies (including acid/alkali etching method, redox etching method, coordinating etching method, and thermal decomposition method), soft template strategies, and self-template strategies (including surface-protected etching, Ostwald ripening, Kirkendall effect, and galvanic replacement) for synthesizing monometallic oxide nanocages. In the following part, each method will be discussed in detail.

Hard Templating Strategies

Hard templating strategies for the synthesis of nanocages are simple and straightforward in concept. As shown in Figure 24.4, it involves three main steps: first, the synthesis of suitable hard template material; second, the deposition of the outer surface with a layer of desired material, and finally, the removal of the template to obtain the nanocage.

There are many compounds such as polymer-based materials (polystyrene or PS, formaldehyde resin, poly(methyl methacrylate), etc.), silica-based materials (solid SiO_2, mesoporous silica, and silica shell), carbon-based materials (solid carbon and mesoporous carbon), metal or metal oxide-based materials (Ag sphere, Au polyhedron, Cu_2O, Fe_2O_3, MgO, etc.), inorganic salt materials (NaCl, $CaCO_3$,

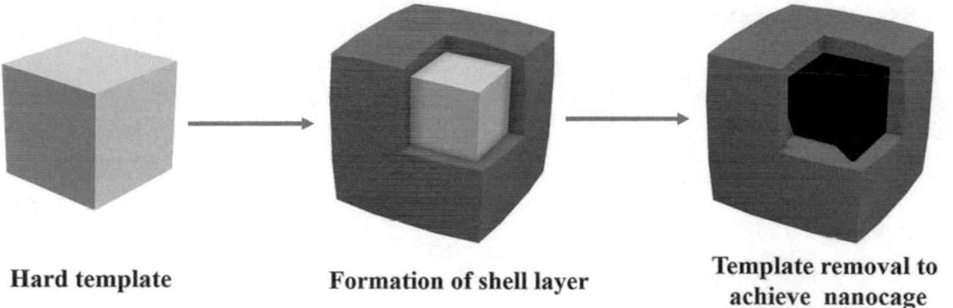

FIGURE 24.4 Schematic illustration of the hard templating synthesis process.

$MnCO_3$, etc.), and so forth, which can be chosen as hard template materials [2,3,15]. The final size and morphology of the metal oxide nanocages are also mainly dependent on the template materials. In order to successfully cover the target layer on the template, a surface modification step that can change the surface functionality, such as sol–gel process or hydro/solvothermal approach, is usually applied. The important selective removal of the hard template can be achieved through the acid/alkali etching method, redox etching method, coordinating etching method, and thermal decomposition method.

Acid/alkali etching method: Direct addition of acid or alkali to etch hard template is a simple and effective method of preparing the metal oxide nanocages. For example, Wang et al. [73] reported an additive-free, aqueous solution synthesis and interior functionalization of anisotropic TiO_2 nanocages by facile hydrothermal coating of anatase TiO_2 against Cu_2O polyhedra followed by acid etching removal of the Cu_2O core. As shown in Figure 24.5a, it mainly went through the following steps: (i) the highly uniform Cu_2O octahedral crystals were first synthesized (Figure 24.5b). (ii) In process i, a polycrystalline TiO_2 layer was preferentially deposited around the scaffold of Cu_2O polyhedra by the synergy of accelerated hydrolysis of TiF_4 and simultaneous slight etching of Cu_2O by HF released from the hydrolysis of TiF_4. Then by precise control of the initial concentration of TiF_4, uniform $Cu_2O@TiO_2$ core/shell structures were thus formed without the need for prior surface modification of the template. (iii) TiO_2 nanocages with entirely hollow interiors (Figure 24.5c–g) could be easily obtained by acidic etching removal of the $Cu_2O@TiO_2$ structures. In addition, using the same synthesis method, Cu_2O nanocages of different sizes could be precisely controlled by changing the size of the Cu_2O hard template (Figure 24.5h–j).

Furthermore, Lou and co-workers [83] fabricated $Fe(OH)_3$ hollow boxes by etching the Prussian blue ($Fe_4[Fe(CN)_6]_3$) cubic templates using an NaOH solution, followed by calcination to obtain Fe_2O_3 hollow boxes. For another example, Wang et al. [97] reported a way of wet chemical reduction for preparing Cu_2O nanocubes and subsequent conversion to Cu_2O nanoboxes by etching Cu_2O nanocubes with acetic acid solution at room temperature. The nanoboxes retained the size and external morphology of the Cu_2O templates. Similarly, Kuo et al. [72] reported a simple approach to prepare different types of cuprous oxide (Cu_2O) nanocages and nanoframes by using HCl etching method.

Redox etching method: This method obtains the nanocage structure by introducing the salt of the redox reaction with the template in the system. For example, Gu et al. [98] developed a convenient one-step approach of preparing various amorphous manganese oxides nanocages with different morphologies and uniform hollow structures by sacrificial template-engaged redox etching of the corresponding shape-controlled Cu_2O templates in $KMnO_4$ solution at room temperature. As shown in Figure 24.6a, different morphologies of Cu_2O nanocrystals were first

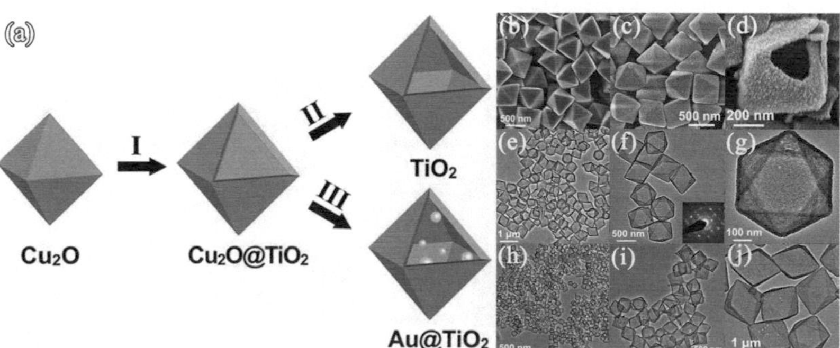

FIGURE 24.5 (a) Schematic illustration of the fabrication and interior functionalization of anisotropic TiO_2 nanocages. SEM images of Cu_2O octahedra (b) and octahedral TiO_2 nanocages (c and d); (e–g) TEM images of TiO_2 nanocages with an average edge length of 500 nm, the inset in (f) is an SAED pattern; (h–j) TEM images of octahedral TiO_2 nanocages with different edge lengths: (h) 100 nm, (i) 300 nm, and (j) 1,000 nm. (Reproduced with permission from Ref. [73]. Copyright 2012, Wiley-VCH Verlag GmbH & Co. KGaA, Weinheim.)

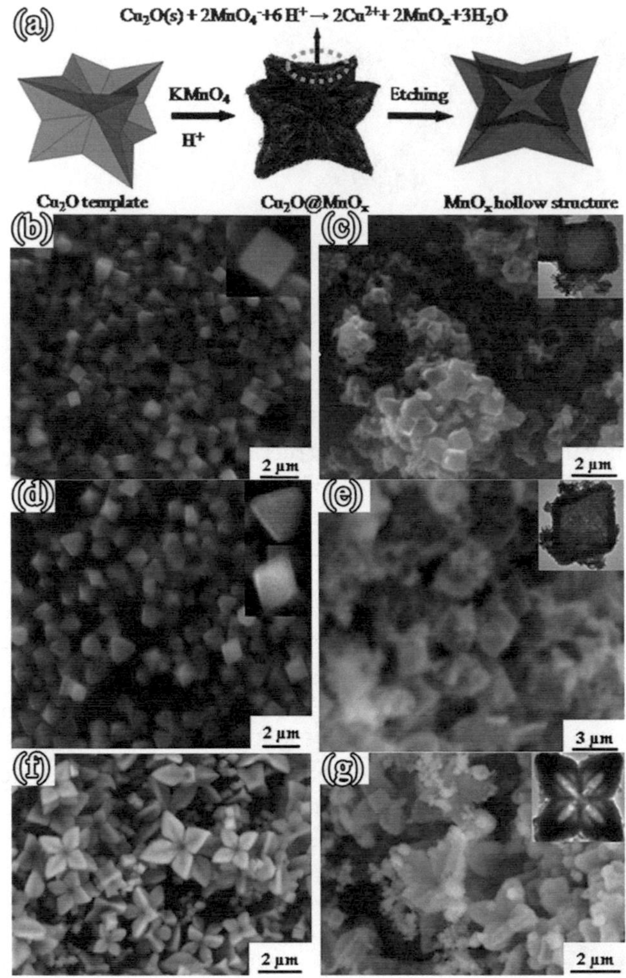

FIGURE 24.6 (a) A schematic formation process of the MnO_x hollow structure. SEM images of Cu_2O polyhedron templates with different morphologies: (b) cube, (d) octahedron, and (f) star; SEM images of manganese oxide hollow structures with different morphologies: (c) hollow cube, (e) hollow octahedron, and (g) hollow star. The inset of (c, e, g) shows the corresponding TEM images of the manganese oxide hollow structures with different morphologies. (Reproduced with permission from Ref. [98]. Copyright 2013, Elsevier B.V. All rights reserved.)

synthesized (Figure 24.6b, d, f). The redox reaction between Cu_2O templates and MnO_4^- ions was thermodynamically feasible owing to the Cu^{2+}/Cu_2O standard reduction potential (0.203 V *vs* SHE) which was much lower than that of $KMnO_4/MnO_2$ (1.679 V *vs* SHE). Cu_2O nanocrystals could be easily oxidized by MnO_4^- at room temperature according to the following redox reaction equation:

$$Cu_2O\,(s) + 2MnO_4^- + 6H^+ \rightarrow 2Cu^{2+} + 2MnO_x + 3H_2O \tag{24.5}$$

Finally, the manganese oxide hollow structures with different morphologies such as hollow cube (Figure 24.6c), hollow octahedron (Figure 24.6e), and hollow star (Figure 24.6g) were obtained by completely consuming Cu_2O templates.

In addition, as mentioned in Section 24.2.1, Wang et al. [75] reported that various morphologies and sizes of $Fe(OH)_x$ nanocages were successfully prepared by redox etching method. Based on the above method, the hollow cubic and octahedral Fe_2O_3 and Fe_3O_4 nanocages could be further obtained by calcination of $Fe(OH)_x$ nanocages under different conditions. Cleverly, Fei et al. [99] successfully prepared the spherical and cubic hollow MnO_2 nanostructures with controlled morphologies by combining redox etching and acid etching method. As shown in Figure 24.7, first of all, the $MnCO_3$ templates with different morphologies could be simply obtained by adding $(NH_4)_2SO_4$ solution in the reaction system. Then, a redox reaction that generated MnO_2 precipitate occurred between the $KMnO_4$ reactant and the $MnCO_3$ template. Finally, the HCl solution was used to selectively remove the $MnCO_3$ template to obtain the desired hollow nanostructures.

Coordinating etching method: The coordinating etching dissolution is usually used to dissolve many insoluble substances, such as silver halide or cuprous, which can coordinate with some ligands (e.g., Cl^-, $S_2O_3^{2-}$, CN^-, SCN^-, $NH_3{\cdot}H_2O$) to form soluble complexes in solution [76,100]. Therefore, by the use of coordinating etching method, it is possible to create the desired metal oxide nanocages. Wang et al. [100] reported that SnO_2 nanocages with uniform

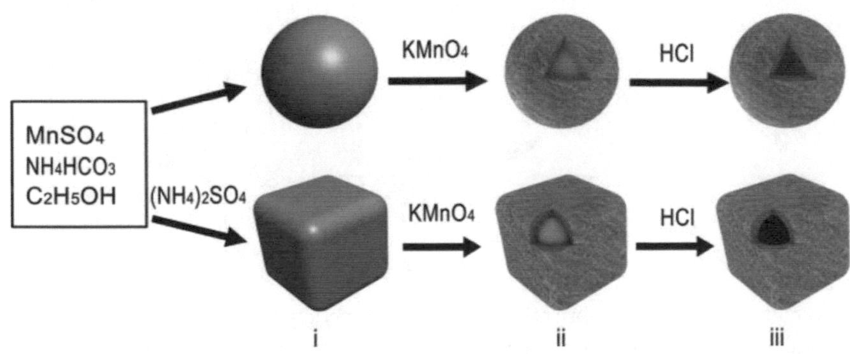

FIGURE 24.7 Schematic illustration of the controlled preparation of MnO_2 hierarchical hollow nanostructures: (i) intermediate $MnCO_3$ crystals with different morphologies, (ii) MnO_2 shell structures with $MnCO_3$ cores, and (iii) as-prepared MnO_2 hierarchical hollow nanostructures. (Reproduced with permission from Ref. [99]. Copyright 2008, Wiley-VCH Verlag GmbH & Co. KGaA, Weinheim.)

morphology, good structural stability, and an adjustable internal volume could be easily synthesized at room temperature through the template-engaged coordinating etching method. As shown in Figure 24.8a, driven by the interfacial reaction between the solid Cu_2O crystal and aqueous solution of $SnCl_4$, the SnO_2 precipitation layer was formed around the Cu_2O template by the accelerated hydrolysis reaction of Sn^{4+} (Eq. 24.6). During this period, an insoluble CuCl intermediate was formed, but it could be form soluble $[CuCl_x]^{1-x}$ by coordinating with Cl^- while being dissolved in an aqueous NaCl solution (Eq. 24.7). Due to the outward flow of $[CuCl_x]^{1-x}$ and the inward flow of Sn^{4+} and Cl^- ions, the void space in the SnO_2 shell was emerged. Finally, Cu_2O templates were completely consumed to form SnO_2 nanocages (Figure 24.8b–d). The overall chemical route for the fabrication of SnO_2 nanoboxes might be described as follows:

$$SnCl_4\,(aq) + xH_2O + 2Cu_2O\,(s) \rightarrow SnO_2 \cdot xH_2O\,(s) + 4CuCl\,(s) \tag{24.6}$$

$$CuCl\,(s) + (x-1)\,Cl^-\,(aq) \rightarrow [CuCl_x]_{1-x}\,(s) \tag{24.7}$$

As shown in the TEM images (Figure 24.8b–d, f), the SnO_2 nanoboxes well inherit the uniform dimensions of

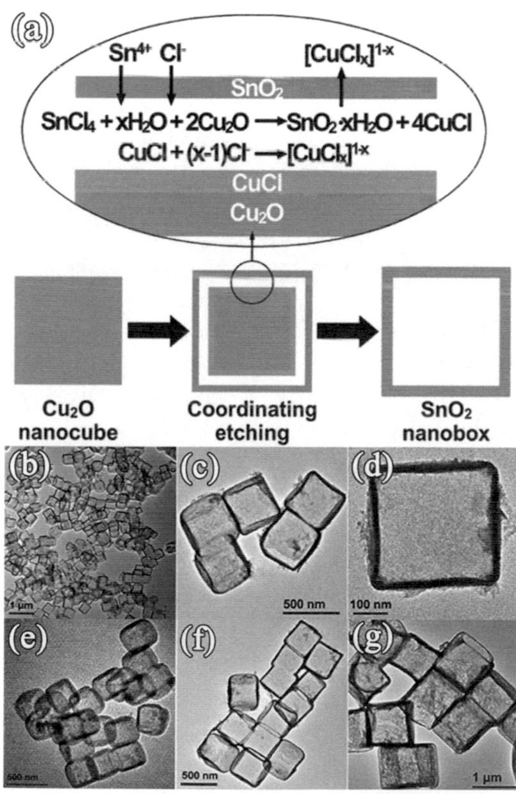

FIGURE 24.8 (a) Schematic illustration of the formation of SnO_2 nanoboxes. (b–d) TEM images of as-prepared SnO_2 nanoboxes; TEM images of SnO_2 nanoboxes with different edge lengths: (e) 200–250 nm, (f) 350–400 nm, and (g) 800–1,000 nm. (Reproduced with permission from Ref. [100]. Copyright 2011, American Chemical Society.)

Cu_2O nanocubes with edge length of around 350–400 nm. Moreover, the size and inner volume of the resulting SnO_2 nanocages could also be reasonably regulated using pregrown Cu_2O templates. For instance, SnO_2 nanoboxes with edge lengths of around 200–250 nm (Figure 24.8e) and 800–1000 nm (Figure 24.8g) could be synthesized using the same route by templating against Cu_2O nanocubes with corresponding sizes.

In addition, the similar coordinating etching method was also reported by our group [76]; as described above, we first synthesized a series of metal hydroxide nanocages by coordinated etching of Cu_2O with $S_2O_3{}^{2-}$ ions. And then, all kinds of metal oxide nanocages such as Mn_3O_4, Fe_2O_3, CoO, NiO, and ZnO nanocages with spherical, cubic, and octahedral structures were easily obtained by simple calcination in air.

Thermal decomposition method: This method involves a hard template that can be calcined into a gas. The commonly used hard templates include PS and carbon spheres, which result in the formation of metal oxide spherical nanocages. For example, Zhang et al. [101] reported a simple and cost-effective hard templating strategy to synthesize anatase TiO_2 hollow spheres composed of highly crystalline nanocrystals, in which carbonaceous spheres (C spheres) were chosen as the removable template. The formation of TiO_2 hollow spheres was simple and straightforward, as illustrated in Figure 24.9. First of all, the C spheres (Figure 24.9b) obtained by a hydrothermal method were well dispersed into anhydrous ethanol through sonication, followed by the addition of titanium tetrabutoxide as the Ti precursor. Next, ammonia solution was added to the reaction mixture to initiate hydrolysis of titanium tetrabutoxide in order to obtain a conformal and firm coating of

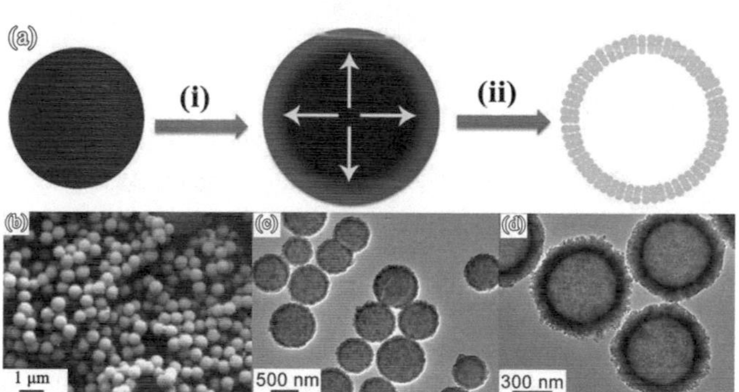

FIGURE 24.9 (a) Illustration of the formation of TiO_2 hollow spheres using carbon spheres as templates: (i) amorphous TiO_2 coating (dark gray) to form the C@TiO_2 core/shell structure and (ii) thermal annealing treatment to form polycrystalline TiO_2 hollow spheres (light gray). (b) SEM image of carbon spheres; TEM images of C@TiO_2 core/shell spheres (c) and crystallized TiO_2 hollow spheres (d). (Panel (b) reproduced with permission from Ref. [102]. Copyright 2004, Wiley-VCH Verlag GmbH & Co. KGaA, Weinheim. Panels (a, c, d) reproduced with permission from Ref. [101]. Copyright 2014, Wiley-VCH Verlag GmbH & Co. KGaA, Weinheim.)

an amorphous TiO_2 layer on the surface of the C spheres. A uniform and stable C@TiO_2 core/shell structure was formed after the reaction (Figure 24.9c), the formation of which was aided by the presence of many hydrophilic functional groups on the surface of the C spheres. Finally, polycrystalline TiO_2 hollow spheres composed of small nanocrystals could be obtained after a simple thermal annealing treatment in air (Figure 24.9d).

The same thermal decomposition method for the preparation of various spherical metal oxide nanocages was also reported by Thomas and co-workers [103]. It is worth mentioning that the carbon templates were synthesized during the hydrothermal process, and thus, it is a one-pot synthesis. As shown in Figure 24.10, various metal salts and D-glucose monohydrate (serve as the carbon precursor) were dissolved in water first, and the mixture was transferred into an autoclave and heated at 180°C for 24 h. During the hydrothermal treatment, carbon spheres were formed with metal ions incorporated into their hydrophilic shells. The removal of carbon via calcination yielded hollow metal oxide spheres. All kinds of metal oxide hollow spheres, such as Fe_2O_3, NiO, Co_3O_4, CeO_2, MgO, and CuO, could be created using this process.

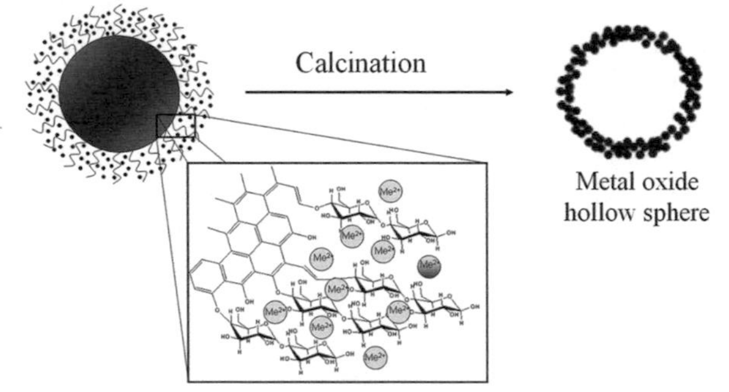

FIGURE 24.10 Schematic illustration of the synthesis of various metal oxide hollow spheres. (Reproduced with permission from Ref. [103]. Copyright 2006, American Chemical Society.)

Soft Templating Strategies

In this section, we will discuss soft template strategies for preparing various metal oxide nanocages. Unlike the above-mentioned hard templates, soft templates have a high degree of sensitivity, which are thermodynamically metastable and easily affected by various environmental and experimental factors such as pH, temperature, polarity of the solvent, ionic strength, and concentration of organic templates and inorganic additives [2,42,89]. Nevertheless, the soft template strategies also have its own advantages not only to tune the morphology but also to adjust the internal and external structures of hollow nanomaterials. More importantly, soft templates can generally be removed under mild conditions through a simple process such as washing or evaporation, reducing the chances for chemical or structural modification, or damage, of the synthesized hollow nanomaterial. There are many materials that can be selected as soft templates such as emulsion droplets (including oil-in-water (O/W) and water-in-oil (W/O) types), micelle/vesicle, polymer, gas bubble, and so on.

For example, some vesicles such as dioctadecyldimethylammonium bromide (DODAB) [105], poly(diallyldimethylammonium) chloride (PDADMAC) [106], and mixtures of dodecyltrimethylammonium bromide (DTAB) and sodium dodecylbenzenesulfonate (SDBS) [107] have been successfully used to synthesize single-shelled hollow nanomaterials (as shown in Figure 24.11a, b). Xu et al. [104] reported the vesicle templating synthesis of multi-shelled hollow Cu_2O spheres with a single-crystalline shell structure by using cetyltrimethylammonium bromide (CTAB) as surfactants. Theoretically, multi-shell hollow spheres can be obtained if all the layers of multilayer vesicle are used as growth sites (Figure 24.11c). They obtained different layers of vesicles by varying the concentration of CTAB from 0.1 to 0.15 M. Multilamellar vesicles could template the formation of Cu_2O hollow spheres with varying structures from single-, double-, triple- to quadruple-shelled (Figure 24.11d–g). Similarly, Chen and co-workers [108] prepared the cubic Co_3O_4 nanocages by a surfactant soft

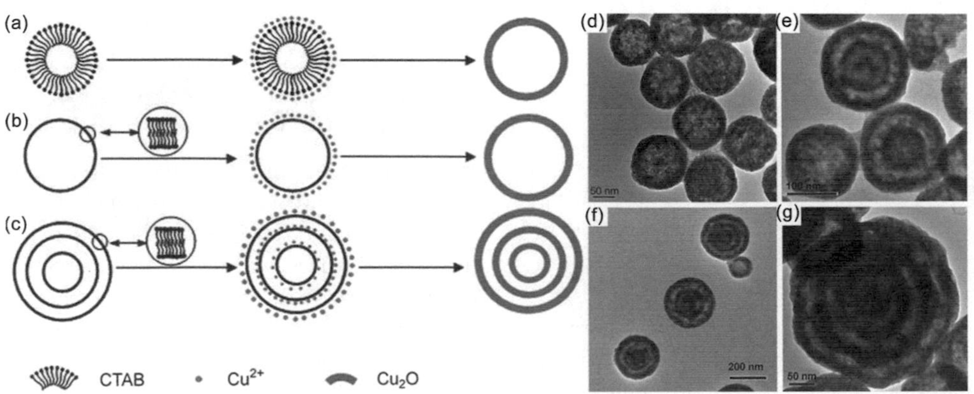

FIGURE 24.11 The formation of different Cu_2O hollow structures in the presence of a CTAB template: (a) micelle, (b) single-lamellar vesicle, (c) multilamellar vesicle. TEM images of Cu_2O hollow structures: (d) single-shelled, (e) double-shelled, (f) triple-shelled, and (g) quadruple-shelled. (Reproduced with permission from Ref. [104]. Copyright 2007, Wiley-VCH Verlag GmbH & Co. KGaA, Weinheim.)

template method. Briefly, $Co(NO_3)_2{\cdot}6H_2O$ and SDBS were first dispersed in absolute ethanol and heated at 90°C for 8 h and then at 180°C for 3 h, and finally, the organics were removed to achieve Co_3O_4 nanoboxes (Figure 24.12a–c).

Self-templating Strategies

In addition to the hard and soft template strategies, in recent years, many novel synthetic approaches directly converting templates into metal oxide nanocages have been developed. In order to distinguish the conventional template strategies, these approaches are also known as "template-free" or "template-engaged" methods in some articles [109–111]. Herein, we collectively term it as the "self-templating" strategies. Many types of self-templating strategies toward hollow nanostructures have been developed on the basis of different principles, including surface-protected etching, Kirkendall effect, Ostwald ripening, galvanic replacement, and so on. Generally, the self-templating strategies can be better described as a two-step synthesis: (i) the synthesis of required template materials and (ii) the transformation of templates into hollow structures. In this section, we will summarize the recent progress in the area of self-templating formation of monometallic oxide nanocages.

Surface-protected etching: Since 2008, Yin and co-workers [87] have reported a series of hollow mesoporous silica by a "surface-protected etching" strategy. Many metal oxide nanocages have also been prepared using similar strategies. A typical surface-protected etching process mainly involves the preparation of solid material as a template, pre-adsorption of solid templates with a protecting layer (such as polyvinylpyrrolidone or PVP, poly(acrylic acid), or sodium dodecyl sulfate), and subsequent preferential etching of interior material using an appropriate etching agent (such as NaOH or $NaBH_4$), as illustrated in Figure 24.13. It is worth mentioning that if a passivation layer is formed *in situ* on the surface of the template, the pre-coating process becomes unnecessary [94]. Yang et al. [112] further extended the surface-protected etching for the fabrication of various hollow metal oxides including CeO_2, Cu_2O, and ZnO. Polyol synthetic particles used as precursors, PVP as a protective agent, and low-concentration acids as etchants were the three prerequisites for constructing a protective etching system. Compared to previous methods, their current methods have several advantages, including compatibility of different metal oxides, one-pot reaction without multistep procedure, rapid synthesis from commonly used metal salts to porous products, crystalline porous framework, and a large surface area (up to 200 m^2/g).

Another example, Wang and co-workers [113] reported the preparation of porous amphoteric metal oxide hollow spheres of Al_2O_3 and ZnO by a facile self-template route. In this case, PVP was selected as the protection polymer and HCl (to hollow ZnO) or NaOH (to hollow Al_2O_3) was chosen as an etchant to selectively remove the interior of metal oxide solid spheres. Finally, the Al_2O_3 and ZnO hollow nanospheres were obtained.

Ostwald ripening: This is a classical physical phenomenon that was first systematically investigated by Wilhelm Ostwald around 1900 [114]. It is driven by chemical potential differences among different-sized particles. In 2007, IUPAC recommended that Ostwald ripening can be defined as "dissolution of small crystals or sol particles and the redeposition of the dissolved species on the surfaces of larger crystals or sol particles." [115] Since 2004, Yang and Zeng [85] have reported for the first time the preparation of TiO_2 hollow nanospheres under hydrothermal conditions through a simple "one-pot" approach based on the Ostwald ripening mechanism (Figure 24.14). Later, more and more research scholars synthesized various hollow metal oxide nanocages through Ostwald ripening mechanism. For example, Lou et al. [116] reported an inside-out Ostwald ripening mechanism for the synthesis of hollow SnO_2 nanostructures by hydrothermal treatment in an ethanol–water mixed solvent. First, the amorphous solid nanospheres were formed by the hydrolysis of the stannate. The small particles in the inner region may be looser than those in the outer layer, resulting in higher

FIGURE 24.12 TEM (a) and SEM (b and c) images of the Co_3O_4 nanoboxes: (b) a box without a top wall and (c) a complete box. (Reproduced with permission from Ref. [108]. Copyright 2006, Wiley-VCH Verlag GmbH & Co. KGaA, Weinheim.)

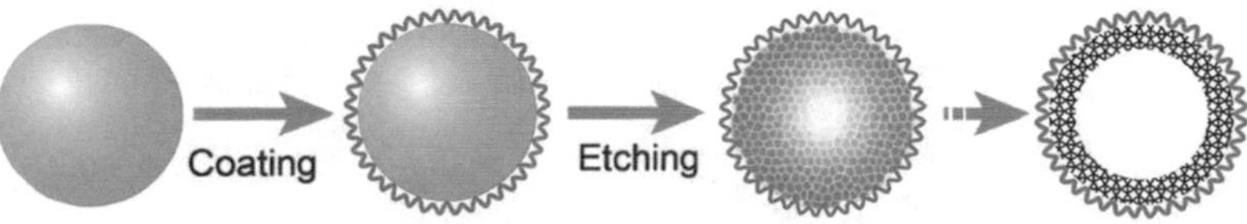

FIGURE 24.13 Schematic illustration of the concept of "surface-protected etching" for transforming solid spheres into hollow nanostructures. (Reproduced with permission from Ref. [87]. Copyright 2008, American Chemical Society.)

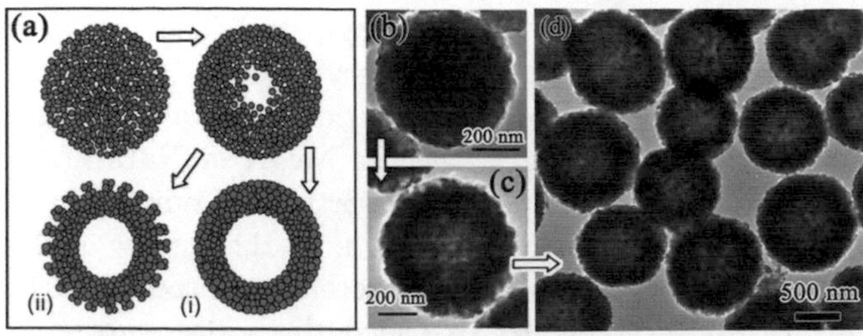

FIGURE 24.14 (a) Schematic illustration of the ripening process and two types (i and ii) of hollow structures. TEM images of TiO_2 nanospheres synthesized with different reaction times: (b) 2 h, (c) 20 h, and (d) 50 h. (Reproduced with permission from Ref. [85]. Copyright 2004, American Chemical Society.)

surface energy and a stronger tendency to dissolve. In addition, SnO_2 particles on the outer surface were relatively easy to crystallize because of the contact with a large amount of ambient solvent. With the prolongation of the hydrothermal treatment time, the spontaneous emptying of internal components will eventually produce SnO_2 hollow nanostructures.

Xue and co-workers [117] prepared porous hollow Fe_3O_4 beads via a facile solvothermal route on the basis of oriented assembly and Ostwald ripening mechanism. The as-prepared beads were spherical in morphology with an average size of 700 nm and had good monodispersion. Through a series of time-dependent experiments, they carefully observed the morphology and composition evolution of the product and proposed a four-step synthesis method for forming porous hollow Fe_3O_4 beads. In the first step, the spindle-shaped Fe_2O_3 nanoparticles were formed under the solvothermal condition. In the second step, the Fe_2O_3 nanoparticles were reduced to Fe_3O_4 nanoparticles by ethylene glycol (EG). In the third step, a well-defined Fe_3O_4 sphere was formed within 2 days under the combined action of a surfactant and water. In the fourth step, due to the Ostwald ripening process under solvothermal conditions, the Fe_3O_4 spheres gradually evolved into hollow and porous Fe_3O_4 beads. Besides these, various metal oxide hollow nanostructures, including flower-like NiO hollow nanospheres [118], Cu_2O hollow nanocubes [119], MnO_2 hollow spheres [120], and ZnO hollow spheres [121] have been fabricated by the Ostwald ripening process.

Kirkendall effect: This effect is a classical phenomenon in metallurgy, which describes the nonequilibrium mutual diffusion process of the boundary layer between two metals that occurs as a consequence of the difference in diffusion rates of the metal atoms. In 1942, Kirkendall for the first time suggested a difference in diffusion rates of two components based on interdiffusion experiments on copper and brass diffusion couple, which were welded together and subjected to an elevated temperature [122,123]. From the traditional point of view, the formation of Kirkendall voids in alloy and solder is not an ideal process for metallurgical manufacturing, because porosity reduces the mechanical properties of the interface or leads to the failure of welding line in integrated circuits. However, since the first report in 2004 on the preparation of hollow CoO and CoS nanocrystals by oxidation and sulfidation of Co nanocrystals, the Kirkendall effect has recently shown positive effects in the design and synthesis of various hollow nanomaterials [86].

As shown in Figure 24.15a, the nanoscale Kirkendall effect for forming nanocrystals with a hollow interior undergoes mainly two steps. The first step is to synthesize solid nanocrystals of at least one element of the final shell A. In the second step, a second element B in solution or gas phase reacts with A to produce an AB compound. During this time, the surface of the solid core nanocrystal A is usually first reacted with the reagent B to produce a shell material AB. The direct conversion of the core material into the shell material again is hindered by the layer, and further reactions will continue with the diffusion of atoms or ions through the barrier layer. If the diffusion rate of the core material is faster than the diffusion rate of the shell material, the preferential outdiffusion of atoms or ions from the core material to the shell material results in a net material flux across the nanocrystal interface and, at the same time, results in the flow of rapidly moving vacancies to the vicinity of the solid–liquid interface. Ultimately, hollow voids are formed by combining many vacancies based on the nanoscale Kirkendall effect [124]. Figure 24.15b and c shows the evolution of CoO hollow nanocrystals over time in response to a stream of O_2/Ar mixture (1:4 in volume ratio, 120 mL/min) being blown through a cobalt colloidal solution at 455 K. The formation of CoO hollow nanocrystals was completed by the Kirkendall effect after the flow of O_2/Ar for 210 min [86]. In addition, the Kirkendall effect has also been expanded to synthesize various hollow metal

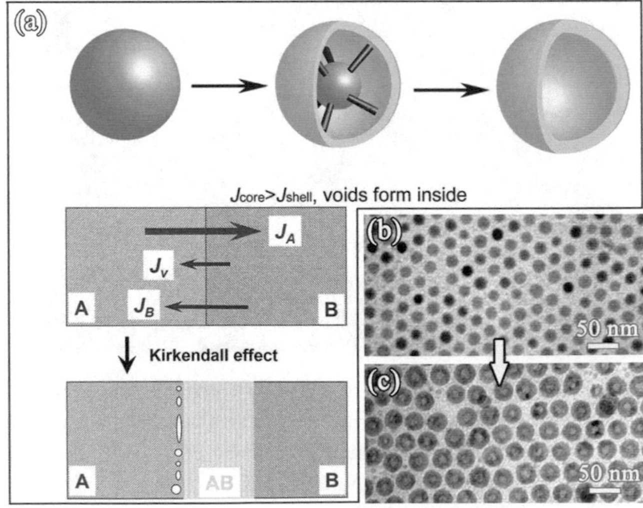

FIGURE 24.15 (a) Schematic illustration of the Kirkendall effect for the formation of hollow nanocrystals (J_A, J_B, and J_v are diffuse fluxes of metal A, B, and void, respectively). TEM images of (b) cobalt nanocrystals and (c) hollow CoO nanocrystals. (Panel (a) reproduced with permission from Ref. [124]. Copyright 2013, American Chemical Society. Panels (b and c) reproduced with permission from Ref. [86]. Copyright 2004, the American Association for the Advancement of Science.)

oxide nanostructures such as Co_3O_4 [125], TiO_2 [126], Fe_3O_4 [127], NiO [128], Al_2O_3 [129], In_2O_3 [130], Bi_2O_3 [131], and so on.

Galvanic replacement: This is an electrochemical redox process initiated by the difference in reduction potential between two species brought in direct contact. For example, one metal acts as the reducing agent (anode) and the salt of the other metal as the oxidizing agent (cathode). The anode metal nanostructures must first be synthesized and then contacted with the metal ions with higher reduction potential. Subsequently, the anode metal nanoparticles are oxidized and dissolved into the solution while the metal ions are reduced and plated onto the outer surface of the anode template. Finally, a hollow nanostructure is obtained, whose morphology is similar to that of the anode metal template. At present, galvanic replacement reaction has been regarded as a very effective self-templating strategy for the synthesis of various hollow nanostructures with tunable elemental composition, controlled shapes and sizes, and porous shells or walls. However, most of the prepared hollow nanomaterials are noble metal elements and few metal oxide nanocages. The application of galvanic replacement techniques for synthesis of hollow nanomaterials was first reported by Xia and co-workers in 2002 [132]. But the as-prepared hollow nanomaterials still belong to precious metals. Until 2013, various metal oxide nanocages (e.g., Mn_3O_4, Fe_2O_3, and Co_3O_4 nanoboxes) were prepared by Hyeon and co-workers through the galvanic replacement reactions [7]. This study breaks the limitation that galvanic replacement is only applicable to metal nanostructures. For comprehensive reviews on the synthesis of hollow materials via galvanic replacement routes, readers can refer to the review paper by Xia et al. [133]

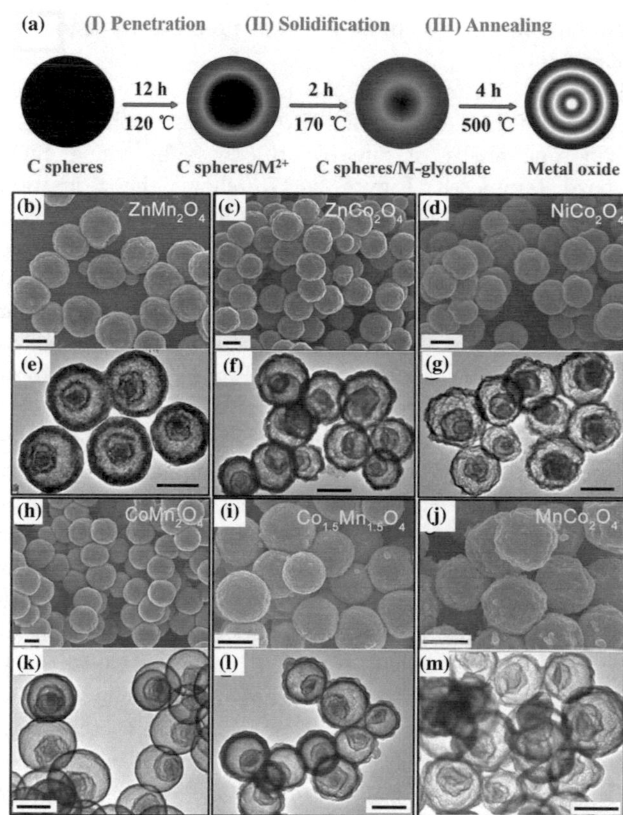

FIGURE 24.16 (a) Schematic illustration of the "penetration-solidification-annealing" process for the formation of complex metal oxide multi-shelled hollow spheres. (b–d, h–j) SEM and (e–g, k–m) TEM images of complex hollow spheres of mixed metal oxides: (b and e) $ZnMn_2O_4$; (c and f) $ZnCo_2O_4$; (d and g) $NiCo_2O_4$; (h and k) $CoMn_2O_4$; (i and l) $Co_{1.5}Mn_{1.5}O_4$; (j and m) $MnCo_2O_4$. All scale bars are 500 nm. (Reproduced with permission from Ref. [134]. Copyright 2014, Wiley-VCH Verlag GmbH & Co. KGaA, Weinheim.)

24.3.2 Syntheses of Mixed Metal Oxide Nanocages

Each metal oxide has its own intrinsic advantage, and the synergy between different metal oxides may further improve the material's performance. In recent years, in order to better meet the application requirements of nanomaterials, many mixed metal oxide nanocages have been synthesized mainly through the hard and self-templating strategies.

Hard Templating Strategies

The hard templating strategy is still a classical approach for the synthesis of mixed metal oxide nanocages. Zhang et al. [134] have developed a new "penetration-solidification-annealing" method which could realize the synthesis of various multi-shelled hollow spheres of mixed metal oxide, including $ZnMn_2O_4$, $ZnCo_2O_4$, $NiCo_2O_4$, $CoMn_2O_4$, $Co_{1.5}Mn_{1.5}O_4$, and $MnCo_2O_4$. The general "penetration-curing-annealing" process is shown in Figure 24.16a. First, carbonaceous spheres (C spheres) were obtained by a hydrothermal method. Second, in order to achieve "permeation" of metal ions deep into the C spheres, the reaction solution was heated in an oil bath to 120°C and stirred for 12 h at this temperature. Third, the "solidification" process was performed by further raising the temperature to 170°C and continuing the reflux for 2 h. During this process, the metal glycolate would be formed in the inner and outer surfaces of the C spheres. Finally, multi-layered hollow spheres could be easily formed by thermal annealing in air. In order to demonstrate the versatility of this method, multi-layered hollow spheres of three different mixed metal oxides including $ZnMn_2O_4$, $ZnCo_2O_4$, and $NiCo_2O_4$ were fabricated by simply using the respective precursors. Their morphological features were very similar in size and shape. Figure 24.16b–g shows the typical SEM and TEM images of $ZnMn_2O_4$ (Figure 24.16b and e), $ZnCo_2O_4$ (Figure 24.16c and f), and $NiCo_2O_4$ (Figure 24.16d and g) triple-shelled hollow spheres. And from the TEM images, we can clearly see that they all have a triple-layered shell structure. To further validate the generality of this method, the $Co_xMn_{3-x}O_4$ hollow spheres with controlled molar ratios of Mn and Co were also synthesized. As shown in Figure 24.16, the $CoMn_2O_4$ (Figure 24.16h and k), $Co_{1.5}Mn_{1.5}O_4$ (Figure 24.16i and l), and $MnCo_2O_4$ (Figure 24.16j and m) triple hollow spheres could be easily prepared by using the same method.

Recently, Yu et al. [135] reported the controllable synthesis of high-quality $Li_4Ti_5O_{12}$ hollow spheres with mesoporous shells of tunable thickness by using SiO_2 as hard templates. Figure 24.17a shows the formation process of the mesoporous $Li_4Ti_5O_{12}$ hollow spheres. First, the uniform coating of an amorphous TiO_2 shell on the silica sphere through a sol–gel process; second, chemical lithiation in an LiOH solution and subsequent annealing treatment to convert the core–shell sphere into the mesoporous $Li_4Ti_5O_{12}$ hollow sphere (Figure 24.17b). Moreover, Lou and co-workers [5] prepared a hierarchical nitrogen-doped carbon@$NiCo_2O_4$ (NC@$NiCo_2O_4$) double-shelled nanobox by using Fe_2O_3 hard templates. As shown in Figure 24.17c, the overall synthetic strategy involves three steps: (I) Fe_2O_3 nanocubes were used as templates (Figure 24.17d), and a layer of polydopamine (PDA) was coated on Fe_2O_3 templates by sol–gel method (Figure 24.17e); (II) the resulting Fe_2O_3@PDA core–shell nanocubes were then heated in an N_2 atmosphere to convert the PDA layer into an N-doped carbon (NC) shell, and the Fe_2O_3 cores were selectively dissolved by acid etching to obtain NC nanocages (Figure 24.17f); (III) a layer of $NiCo_2O_4$ nanosheets was grown on the NC nanobox by a hydrothermal reaction, followed by heat treatment to obtain hierarchical NC@$NiCo_2O_4$ double-shelled nanoboxes (Figure 24.17g).

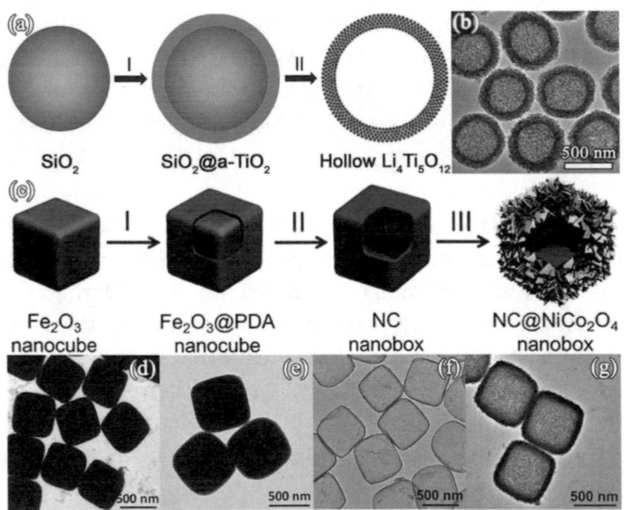

FIGURE 24.17 (a) Schematic illustration of the formation of mesoporous $Li_4Ti_5O_{12}$ hollow spheres through a templating approach. (b) TEM image of $Li_4Ti_5O_{12}$ hollow spheres. (c) Schematic illustration of the synthetic process of hierarchical NC@$NiCo_2O_4$ double-shelled nanoboxes. TEM images of Fe_2O_3 nanocube (d), Fe_2O_3@PDA core-shelled nanocube (e), NC nanoboxes (f), and hierarchical NC@$NiCo_2O_4$ double-shelled nanoboxes (g). (Panels (a and b) reproduced with permission from Ref. [135]. Copyright 2013, Wiley-VCH Verlag GmbH & Co. KGaA, Weinheim. Panels (c–g) reproduced with permission from Ref. [5]. Copyright 2018, Royal Society of Chemistry.)

Self-templating Strategies

As the composition and structure of metal oxide nanomaterials become more and more complicated, some novel synthesis methods have been developed recently. The smart self-templating methods have been always used to synthesize some intricate mixed metal oxide nanocages. For example, Hu et al. [136] reported a novel box-in-box nanocage with high complexity in shell architecture and composition, namely, $Co_3O_4/NiCo_2O_4$ double-shelled nanocage (DSNC). They used a self-template strategy to synthesize DSNCs ($Co_3O_4/NiCo_2O_4$), which involved the facile synthesis of zeolitic imidazolate framework-67/Ni-Co layered double hydroxides (ZIF-67/Ni-Co LDH) yolk-shelled structures and subsequent thermal annealing in air. Figure 24.18a shows the formation of $Co_3O_4/NiCo_2O_4$ DSNCs; it could be simply described as the following three steps. (I) Preparation of ZIF-67 solid templates: two methanol solutions of 2-methylimidazoleare and $Co(NO_3)_2$ mixed rapidly and aged for 24 h at room temperature to generate uniform ZIF-67 particles (Figure 24.18b). (II) Preparation of ZIF-67/Ni-Co LDH yolk-shelled structures (Figure 24.18c): the ZIF-67 solid templates were dispersed in the ethanol solution of $Ni(NO_3)_2$ and stirred for 30 min. In this process, protons generated by the hydrolysis of Ni^{2+} ions could gradually etch the ZIF-67 templates to release Co^{2+} ions, and the Co^{2+} ions were partially oxidized by the dissolved O_2 and NO_3^- ions in the solution. Then Co^{2+}/Co^{3+} ions coprecipitated with Ni^{2+} ions to form Ni-Co LDH shells. (III) ZIF-67/Ni-Co LDH yolk-shelled structures were calcined in air and converted to $Co_3O_4/NiCo_2O_4$ DSNCs (Figure 24.18d).

In addition to the DSNCs, mixed metal oxide nanocages with more complicated interior structures have also been successfully synthesized by Lou and co-worker [137]. They have developed a new and generally applicable strategy for the efficient fabrication of $NiCo_2O_4$ hollow spheres with complex interior structures. The formation of $NiCo_2O_4$ complex hollow structures is shown in Figure 24.19a. The uniform nickel–cobalt glycerate spheres (NiCo-glycerate

FIGURE 24.18 (a) Schematic illustration of the formation process of $Co_3O_4/NiCo_2O_4$ DSNCs. TEM images of ZIF-67 (b), ZIF-67/Ni-Co LDH yolk-shelled structures (c), and $Co_3O_4/NiCo_2O_4$ DSNCs (d). (Reproduced with permission from Ref. [136]. Copyright 2015, American Chemical Society.)

spheres) as the templates were first prepared by a facile solvothermal method, and the NiCo-glycerate spheres were converted into $NiCo_2O_4$ core-in-double shell hollow spheres by a nonequilibrium heat treatment process. In the stage I, the large temperature gradient (ΔT) present in the radial direction results in the rapid formation of $NiCo_2O_4$ shells on the surface of NiCo-glycerate spheres (the resulting core–shell structure is shown in Figure 24.19c). In the stage II, there were two actions in opposing directions (contraction F_c and adhesion action F_a) acting on the interface between the $NiCo_2O_4$ shell and the NiCo-glycerate core. When $F_c > F_a$, the inner core would shrink further inwards and separate from the preformed outer metal oxide shell (the resulting yolk-shell structure is shown in Figure 24.19d). With prolonged heating, in stage III, the heterogeneous contraction process continued to occur on the interior core. Finally, a unique three-layer core-in-double-shell hollow structure (Figure 24.19e) was formed by the same mechanism. Interestingly, this method could also be further extended to synthesize other mixed metal oxide hollow spheres with complex interior structures, including $ZnCo_2O_4$ and $CoMn_2O_4$.

Recently, Lu et al. [138] demonstrated a novel approach for the effective synthesis of multi-shelled hybrid nanoboxes by taking advantage of the unique reactivity of zeolitic imidazolate framework (ZIF-67) with the vanadium source of vanadium oxytriisopropoxide (VOT). The synthesis strategy is schematically illustrated in Figure 24.20a. The highly uniform ZIF-67 nanocubes were first prepared as templates. And these ZIF-67 nanocubes were dispersed in an absolute ethanol solution containing a certain amount of VOT, and stirring was continued for 20 minutes to form a clear purple solution for further use. In Step 1, a ZIF-67@amorphous-$Co_3V_2O_8$-50 (ZIF-67@a-$Co_3V_2O_8$-50) yolk-shell-structured precursor (Figure 24.20b) could be obtained by a solvothermal treatment of the above purple solution with the VOT concentration of 50 μL. During this period, the VOT-produced vanadate anion (VO_4^{3-}) gradually replaced the 2-methylimidazolium anion in ZIF-67 by ion exchange to form the amorphous-$Co_3V_2O_8$ (a-$Co_3V_2O_8$) shell. And the continual depletion of the ZIF-67 core results in a gap space between the newly formed a-$Co_3V_2O_8$ shell and the remaining ZIF-67 core. In Step 2, the as-prepared ZIF-67@a-$Co_3V_2O_8$-50 yolk-shell structure was thermally calcined in air, and the ZIF-67 core and the a-$Co_3V_2O_8$ shell were converted into Co_3O_4 and $Co_3V_2O_8$, respectively, namely, triple-shelled Co_3O_4@$Co_3V_2O_8$ nanoboxes (Figure 24.20c). Moreover, by precisely adjusting the VOT concentration, double-shelled Co_3O_4@$Co_3V_2O_8$ nanoboxes (Figure 24.20e) and single-shelled $Co_3V_2O_8$ nanoboxes (Figure 24.20g) could also be synthesized via the same mechanism.

Furthermore, Lou and co-workers [14] have developed a universal method for synthesizing various multi-shelled mixed metal oxide hollow structures (including Ni-Co oxide, Mn-Co oxide, Mn-Ni oxide, Zn-Mn oxide, and Mn-Co-Ni oxide multi-shelled structures) by using coordination polymers (CPs) as templates. Herein, the amorphous CPs play an important role as carboxylate ligands, which have been extensively demonstrated to form coordinate bonds with most transition metal ions. Using Ni-Co oxide as an example as shown in Figure 24.21a, Ni-Co CP spheres (CPSs)

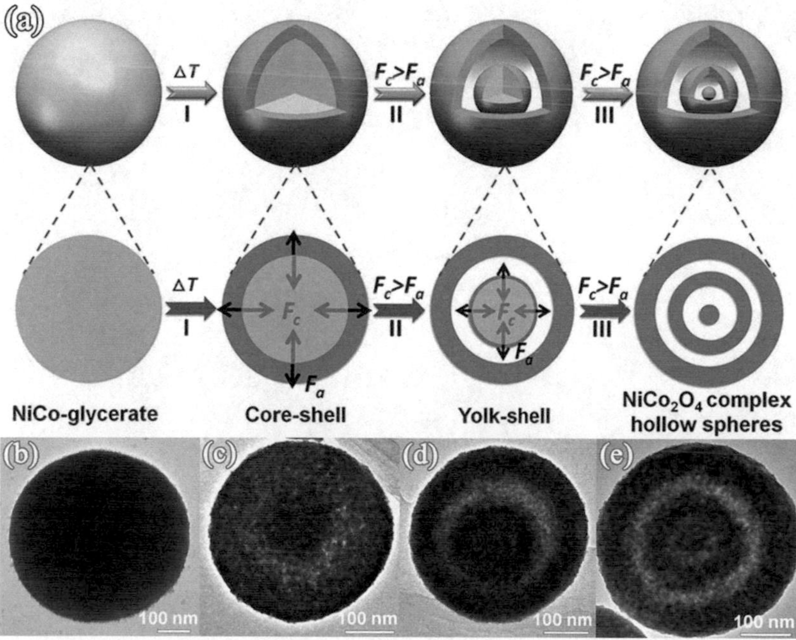

FIGURE 24.19 (a) Schematic illustration of the formation process of $NiCo_2O_4$ core-in-double-shell hollow spheres. TEM images of NiCo-glycerate sphere (b), NiCo-glycerate sphere@$NiCo_2O_4$ core–shell structure (c), NiCo-glycerate sphere@$NiCo_2O_4$ yolk-shell structure (d), and three-layer core-in-double shell $NiCo_2O_4$ hollow sphere (e). (Reproduced with permission from Ref. [137]. Copyright 2015, Wiley-VCH Verlag GmbH & Co. KGaA, Weinheim.)

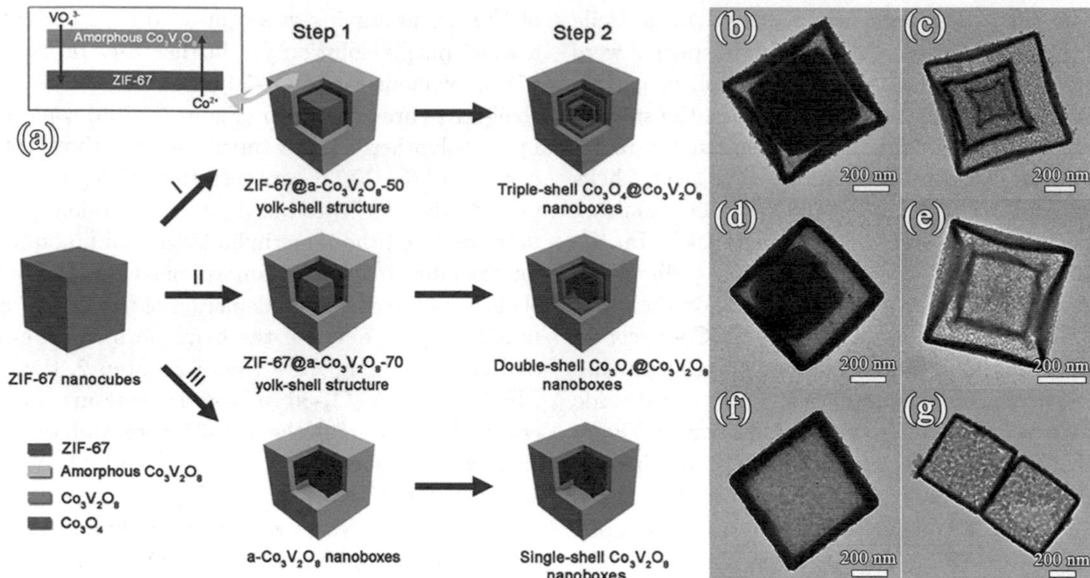

FIGURE 24.20 (a) Schematic illustration of the formation of complex $Co_3O_4@Co_3V_2O_8$ hollow structures. TEM images of ZIF-67@a-$Co_3V_2O_8$-50 yolk-shell structure (b), triple-shelled $Co_3O_4@Co_3V_2O_8$ nanoboxes (c), ZIF-67@a-$Co_3V_2O_8$-70 yolk-shell structure (d), double-shelled $Co_3O_4@Co_3V_2O_8$ nanoboxes (e), a-$Co_3V_2O_8$ nanoboxes (f), and single-shelled $Co_3V_2O_8$ nanoboxes (g). (Reproduced with permission from Ref. [138]. Copyright 2017, Wiley-VCH Verlag GmbH & Co. KGaA, Weinheim.)

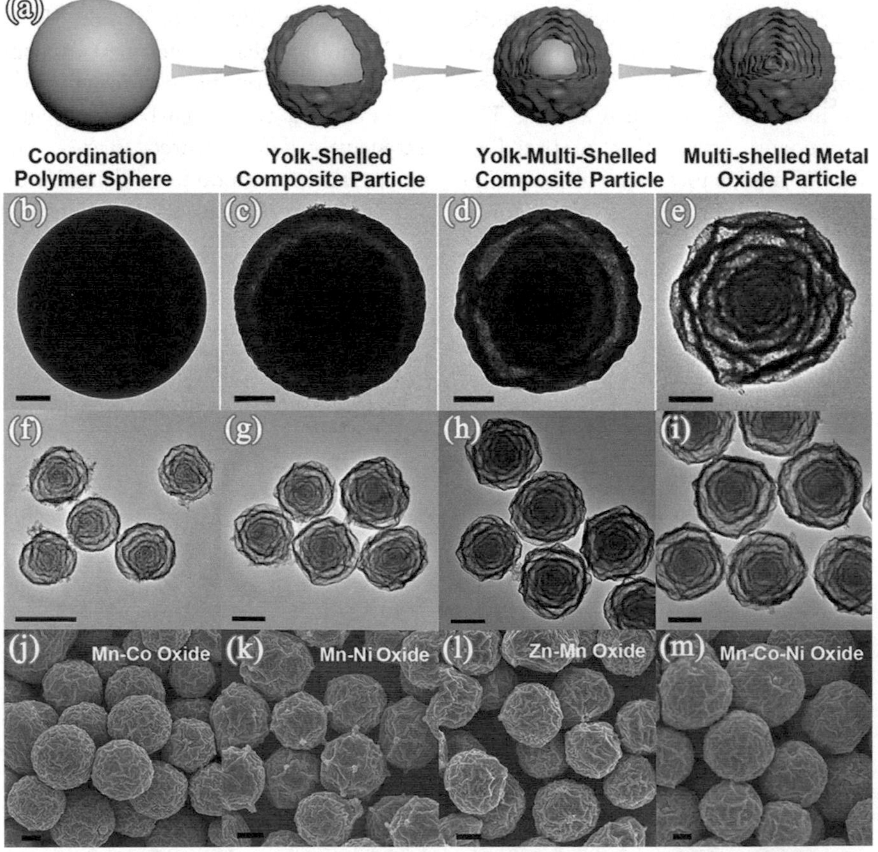

FIGURE 24.21 (a) Schematic illustration of the formation process of a multi-shelled structure. TEM images of (b) Ni-Co CPS and the products obtained after calcination of CPS to different temperatures: (c) 360°C, (d) 390°C, and (e) 500°C. The scale bars in panels (b–e) are 200 nm. TEM images of multi-shelled Ni-Co oxide particles with different sizes: (f) 500 nm, (g) 800 nm, (h) 1,000 nm, and (i) 1,600 nm. SEM images of (j) Mn-Co oxide, (k) Mn-Ni oxide, (l) Zn-Mn oxide, and (m) Mn-Co-Ni oxide multi-shelled structures. The scale bars in panels (f–m) are 500 nm. (Reproduced with permission from Ref. [14]. Copyright 2017, Wiley-VCH Verlag GmbH & Co. KGaA, Weinheim.)

(Figure 24.21b) were first synthesized by the coprecipitation of Ni^{2+} and Co^{2+} ions in the presence of organic ligand (isophthalic acid: H_2IPA). These Ni-Co CPSs could be easily oxidized in the air, while the oxide layer formed loose and peeled off, forming seven-layer multi-shelled Ni-Co oxide particles without any other template. Figure 24.21b–e shows the structure of the intermediates collected from different thermal decomposition stages (from solid CPS to multi-shelled Ni-Co oxide). The size of multi-shelled Ni-Co oxide could be easily tailored in the range of 500–1,600 nm by tuning the CPS diameter via simply adjusting the concentration of organic ligand and metal ions in the reaction medium (Figure 24.21f–i). Of note, four other multi-shelled mixed transition metal oxide particles including Mn-Co oxide (Figure 24.21j), Mn-Ni oxide (Figure 24.21k), Zn-Mn oxide (Figure 24.21l), and Mn-Co-Ni oxide (Figure 24.21m) could also be synthesized through the thermal oxidation of other mixed-metal CP precursors' base on the same mechanism.

24.4 Potential Applications of Metal Hydroxide and Oxide Nanocages

The inherent advantages of hollow structures, such as low density, large surface area, abundant inner void space, the specific compositions of the shell material, and reduced length for both mass and charge transport, make metal hydroxide and oxide nanocages good candidates for many applications. Besides, the nanocages can also be combined with other functional materials (such as noble metal, carbon materials, graphene, and polypyrrole) to form composite nanocages for further improvement of performance. In this section, we will mainly introduce some well-developed applications of metal hydroxide and oxide nanocages. According to their different application areas, five categories of the applications are reviewed: LIBs, Li-S batteries, supercapacitors, photocatalysis, and electrocatalysis.

24.4.1 Lithium-Ion Batteries

LIB technology has been widely used in portable consumer electronic devices such as laptops and cellular phones [139]. As we all know, the lithium-ion cell typically consists of two electrodes, a cathode and an anode, separated by an electrolyte and an electrolyte-permeable separator [140]. At present, the anode material used in commercial LIBs is mainly graphite (theoretical capacity is about 372 mAh/g), and the cathode material generally uses lithium metal oxides (such as $LiCoO_2$), lithium iron phosphate, and ternary $LiNi_xCo_yMn_{1-x-y}O_2$. Hence, much effort has been devoted in search of suitable nanostructured anode materials such as metal sulfides, metal oxides, and metal alloy. Although the new anode materials have significant advantages over traditional graphite, several drawbacks are still suffered: (i) pulverization, (ii) low cycling efficiency, and (iii) permanent capacity losses, due to large volume changes during lithiation and delithiation processes. Due to the above issues, the metal hydroxide and oxide nanocages are being used as new anode materials to replace the traditional carbonaceous materials.

SnO_2 is an important metal oxide, which has been widely used as anode materials for LIBs due to its low working potential (0.6 V *vs* Li/Li^+) and high theoretical capacity (790 mAh/g). Lithium storage in SnO_2 relies on a reversible alloy-dealloying reaction between metallic Li and Sn nanocrystals, which are produced by the initial irreversible reduction of SnO_2. Archer and co-workers [116] prepared SnO_2 hollow nanospheres (Figure 24.22a) with high capacity and improved cycling performance through an inside-out Ostwald ripening mechanism. The as-prepared SnO_2 hollow nanospheres have a large Brunauer–Emmett–Teller (BET) surface area of 110 m^2/g and large initial discharge capacity of 1,140 mAh/g, which is 75% larger than the pristine SnO_2 nanoparticles. In particular, although all materials will show a certain degree of decline over time, the capacity of as-prepared SnO_2 hollow nanospheres is still comparable to the theoretical capacity of SnO_2 after more than 30 cycles and much higher than the theoretical capacity of graphite after more than 40 cycles. The comprehensive performance of the LIB is expected to be further improved by increasing the number of shell layers. Archer and co-workers [141] have designed and synthesized double-shelled SnO_2 hollow spheres and SnO_2@carbon coaxial hollow nanospheres (Figure 24.22b and c). As expected, these SnO_2@carbon coaxial hollow spheres show excellent cycling performance compared with other SnO_2 anode materials, and the capacity gradually decays in the first 30 cycles and stabilizes around 460 mAh/g for more than 100 cycles. Besides, Lou and co-workers [100,142] have also successfully synthesized well-defined SnO_2 nanoboxes (Figure 24.22d) and SnO_2/nitrogen-doped carbon (SnO_2/NC) composite nanoboxes (Figure 24.22e). The as-synthesized SnO_2/NC nanoboxes have a size of about 400 nm, a shell thickness of 40 nm, and a high specific surface area of 125 m^2/g. Moreover, the SnO_2/NC nanoboxes show prominent electrochemical performance as an anode material for LIBs (Figure 24.22f and g). A high reversible capacity of 491 mAh/g can be retained after 100 cycles at a current density of 0.5 A/g.

In addition to SnO_2, many other metal oxide or mixed metal oxide nanocages (such as TiO_2, CoO, Fe_3O_4, and $CoSnO_3$) can also be used as excellent anode materials for LIBs. Recently, Yu et al. [143] reported hollow rutile TiO_2 nanoboxes with superior lithium storage properties (Figure 24.23a). The as-prepared rutile TiO_2 nanoboxes possess a relatively high BET surface area of 57 m^2/g and a very high initial discharge capacity of about 424 mAh/g (Figure 24.23e), which greatly exceeds the theoretical capacity of rutile TiO_2 (335 mAh/g). Moreover, the rutile TiO_2 nanoboxes can be cycled with high stability at higher current rates of 5 and 10 C, and a reversible capacity of 141 mAh/g at a current rate of 5 C after 500 cycles can still be retained with CE of almost 100% throughout the cycling

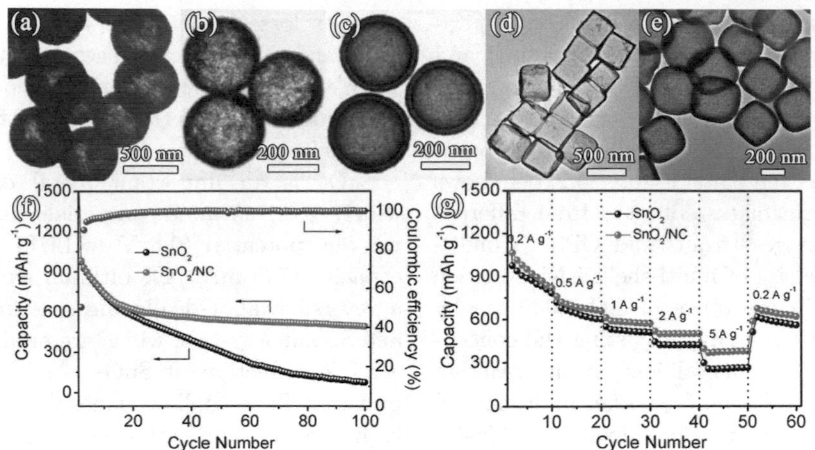

FIGURE 24.22 TEM images of (a) SnO_2 hollow nanospheres, (b) SnO_2@carbon coaxial hollow spheres, (c) double-shelled SnO_2 hollow spheres, (d) SnO_2 nanoboxes, and (e) SnO_2/NC nanoboxes. (f) Cycling performance of the SnO_2 nanoboxes and SnO_2/NC nanoboxes electrodes at the same current density of 0.5 A/g and the coulombic efficiency (CE). (g) Rate capability of the SnO_2 and SnO_2/NC nanoboxes electrodes. (Panel (a) reproduced with permission from Ref. [116]. Copyright 2006, Wiley-VCH Verlag GmbH & Co. KGaA, Weinheim. Panels (b, c) reproduced with permission from Ref. [141]. Copyright 2009, Wiley-VCH Verlag GmbH & Co. KGaA, Weinheim. Panel (d) reproduced with permission from Ref. [100]. Copyright 2011, American Chemical Society. Panels (e–g) reproduced with permission from Ref. [142]. Copyright 2016, Wiley-VCH Verlag GmbH & Co. KGaA, Weinheim.)

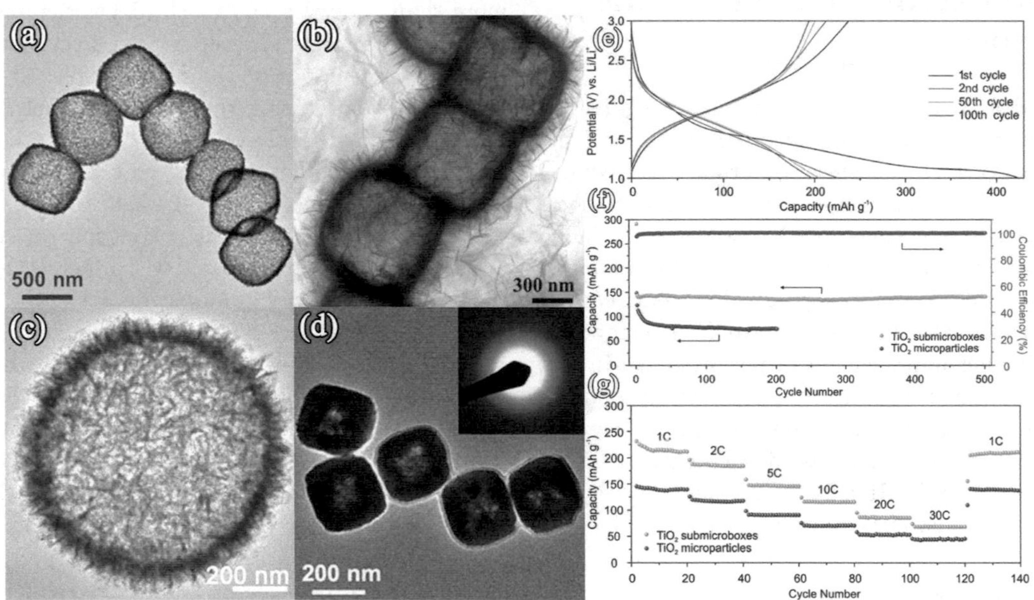

FIGURE 24.23 TEM images of (a) rutile TiO_2 nanoboxes, (b) CoO/RGO composite nanoboxes, (c) hierarchical Fe_3O_4 hollow spheres, and (d) amorphous $CoSnO_3$@C nanoboxes. (e) Charge-discharge voltage profiles of rutile TiO_2 nanoboxes at a current density of 1 C. (f) Cycling performance of rutile TiO_2 nanoboxes and TiO_2 particles at a current density of 5°C, and the corresponding CE of rutile TiO_2 nanoboxes. (g) Rate performance of rutile TiO_2 nanoboxes and TiO_2 particles at various current rates from 1 to 30 C (1 C = 170 mA/g). (Panels (a, e–g) reproduced with permission from Ref. [143]. Copyright 2015, Wiley-VCH Verlag GmbH & Co. KGaA, Weinheim. Panel (b) reproduced with permission from Ref. [144]. Copyright 2014, American Chemical Society. Panel (c) reproduced with permission from Ref. [145]. Copyright 2015, Wiley-VCH Verlag GmbH & Co. KGaA, Weinheim. Panel (d) reproduced with permission from Ref. [146]. Copyright 2013, Royal Society of Chemistry.)

(Figure 24.23f). Obviously, TiO_2 nanoboxes possess better cyclic stability and higher capacity than TiO_2 particles at each current rate (Figure 24.23g). These TiO_2 nanoboxes possess significantly improved lithium storage properties benefitting from hollow structures with large specific surface area, porous thin shells, and small primary nanoparticles.

Our group [144] also reported hollow CoO/reduced graphene oxide (RGO) composite nanoboxes (Figure 24.23b). The CoO/RGO composite nanoboxes demonstrate high lithium storage capacity, reaching 1,170 mAh/g at a current density of 150 mA/g, which is much higher than that of RGO-free hollow CoO nanocubes. Other iron group oxide, Fe_3O_4 hollow spheres, were prepared by Lou and co-workers [145] through a solvothermal approach. As shown in Figure 24.23c, in virtue of the structural advantages, the as-prepared hierarchical Fe_3O_4 hollow spheres manifest outstanding electrochemical properties as anode materials for LIBs in terms of high specific capacity, superior rate capability, and remarkable cyclability. Lou and co-workers [146] have also developed amorphous mixed metal oxide nanocages such as amorphous $CoSnO_3$@C nanoboxes, used as anode materials for LIBs. Benefiting from the unique structure, they exhibit exceptional long-term cycling stability over 400 cycles for highly reversible lithium storage.

24.4.2 Lithium-Sulfur Batteries

After more than two decades of development, LIBs based on intercalation compounds are approaching the limits of their energy density. Therefore, it is difficult to further meet the increasing demand for mobile electronic devices with increased power consumption. The rapidly developing market urgently needs new systems with higher energy density. In recent years, Li-S batteries have three major advantages: (i) high theoretical energy density, (ii) low cost, and (iii) environmental friendliness are considered to be very promising candidates for next-generation rechargeable batteries [147]. The Li-S battery can provide an average voltage of about 2.2 V and a high theoretical energy density of 2,600 Wh/kg through the coupled sulfur cathode (theoretical capacity is 1,675 mAh/g) and lithium anode (theoretical capacity up to 3,840 mAh/g), which can achieve a practical energy density that is 2–3 times higher than the most advanced commercial LIBs [148]. Despite the overwhelming advantages, the commercialization of lithium-sulfur batteries is still hampered by the following three main issues: (i) the poor electronic conductivity of sulfur and its end products of discharge (Li_2S/Li_2S_2) leads to a low specific capacity; (ii) the dissolution of intermediate lithium polysulfides (LiPSs) and their shuttle effect result in the loss of active materials, poor cycling stability, and low CE; (iii) large volumetric expansion in the lithiation process [149]. In view of these, tremendous efforts have been devoted to the development of advanced hollow-structured sulfur hosts to improve the electrochemical performance of Li-S batteries.

For example, Liang and Nazar [150] have reported that the hollow spherical MnO_2 shell could be formed as a bifunctional host *in situ* on the sulfur particles to provide physical limitations and chemical adsorption of LiPSs in Li-S batteries. As-prepared materials exhibit a very low capacity fade rate of 0.048% per cycle over 800 cycles at 2 C, providing a final usable capacity of 480 mAh/g. The manganese dioxide nanosheet-decorated hollow sulfur sphere nanocomposites [151] have been developed through a facile synthesis, exhibiting ultralong cycling stability over 1,500 cycles and high capacity for high-performance Li-S batteries. V_2O_5 hollow spheres [152] have also been proved to be effective LiPS mediators. Moreover, Lou and co-workers [153] have designed a sulfur host based on highly conductive polar TiO@C hollow nanospheres (TiO@C-HS) for Li-S batteries (Figure 24.24a). The host can effectively regulate the diffusion of LiPSs and at the same time enhance the redox reaction kinetics of sulfur species. Profiting from the good conductivity and the strong LiPSs adsorption capability of TiO@C-HSs, the TiO@C hollow nanosphere/sulfur (TiO@C-HS/S) composite (Figure 24.24b) cathode has a discharge capacity of up to 1,100 mA/g at 0.1 C (Figure 24.25g). At 0.2 and 0.5 C, the cycle life is stable for 500 cycles, and the capacity decay rate per cycle is 0.08% (Figure 24.24e). Furthermore, the TiO@C-HS/S composite electrode can also provide high areal capacities with good stability and high CE at various current densities when the areal loading of sulfur is increased to 4.0 mg/cm^2 (Figure 24.24h).

Compared with these simple metal oxide nanocages, mixed metal compounds with complex hollow structures may be better sulfur hosts. Recently, Zhang et al. [11] have designed and synthesized DSNCs with cobalt hydroxide inner shell and layered double hydroxides outer shell (CH@LDH) as a conceptually new sulfur host for Li-S batteries. As shown in Figure 24.24c, the hollow CH@LDH polyhedra provide enough self-functionalized surfaces for chemical bonding with polysulfide to inhibit its dissolution. When evaluated as cathode material for Li-S batteries, the CH@LDH/sulfur (CH@LDH/S) composites (Figure 24.24d) can load with a high content of sulfur (75 wt%), and it can maintain excellent cycling stability at both 0.1 and 0.5 C over 100 cycles (Figure 24.24f) and deliver high-rate capacities with relatively high sulfur loading of 3 mg/cm^2 (Figure 24.24i).

24.4.3 Supercapacitors

Supercapacitors, or electrochemical capacitors, are a unique class of electrochemical energy storage devices with high power density (10 kW/kg), superior rate capability, ultrafast charge/discharge rate (a few seconds), and long cycle life (100,000 cycles) [154]. The supercapacitor can make up for the gap between the traditional dielectric capacitor and the secondary battery and plays an important role in supplementing or replacing the secondary battery for energy storage [155]. According to the charge storage mechanism, supercapacitors can be classified into the following three categories: (i) electrical double-layer capacitors (EDLCs); (ii) pseudocapacitors or redox supercapacitors; and (iii) hybrid supercapacitors (HSCs), which integrate a battery-type electrode and a capacitive electrode in an electrochemical cell to achieve both high energy and high power densities [17,156].

In recent years, metal hydroxide and oxide nanocages have been selected as electrode materials of supercapacitors due to the following advantages: (i) the unique hollow structures can provide more accessible faradic reactive sites, resulting in a higher energy density in the real capacitive

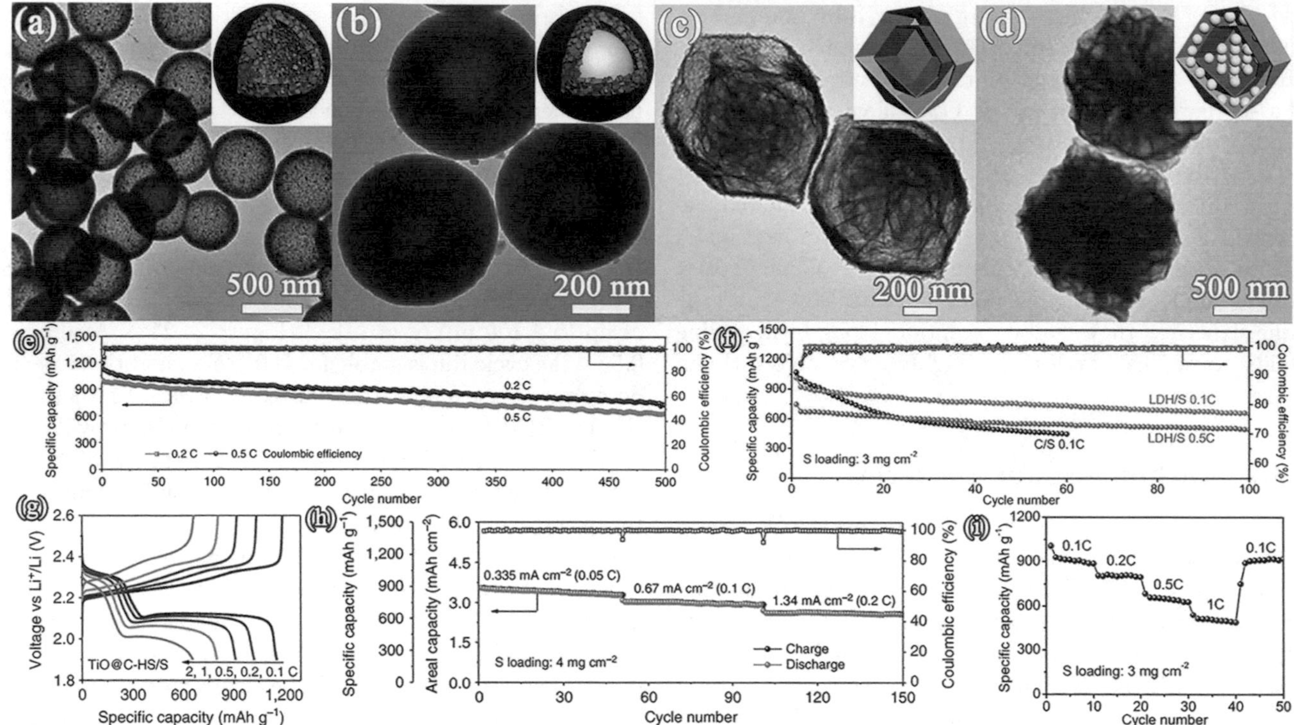

FIGURE 24.24 TEM images of (a) TiO@C-HS, (b) TiO@C-HS/S, (c) double-shelled CH@LDH, and (d) CH@LDH/S. (e) Cycle life and CE at 0.2 and 0.5 C of the TiO@C-HS/S electrode. (f) Cycle performance comparison between CH@LDH/S and mesoporous carbon/sulfur (C/S). (g) Voltage profiles at various current densities from 0.1 to 2 C of the TiO@C-HS/S electrode. (h) Areal capacities of the TiO@C-HS/S electrode with high sulfur mass loading of 4.0 mg/cm^2. (i) Discharge capacities of CH@LDH/S at various current densities from 0.1 to 1 C. (Panels (a, b, e, g, h) reproduced with permission from Ref. [153]. Copyright 2016, Nature Publishing Group. Panels (c, d, f, i) reproduced with permission from Ref. [11]. Copyright 2016, Wiley-VCH Verlag GmbH & Co. KGaA, Weinheim.)

process due to the enhanced surface-to-volume ratio or increased effective surface area; (ii) the porous outer shell of the nanocage can significantly increase the accessibility between the electrolyte and the active material, increase the conductivity, shorten the transmission length of ions and charges, and thus, increase the rate capability and power density; and (iii) for the complex multi-shell nanocages, the outer shell can protect the inner shell from electrochemical dissolution, resulting in better structural and electrochemical stability, thereby improving the cycle performance. Yang et al. [40] reported the preparation of NiO hollow nanospheres with controllable shell number including single-, double-, and triple-shelled hollow nanospheres (Figure 24.25a–c) through a facile layer-by-layer (LBL) self-assembly method. When three kinds of NiO hollow nanospheres are evaluated as the active electrode materials of the supercapacitors, the double-shelled NiO hollow nanosphere sample with largest surface area (92.99 m^2/g) shows the best performance in terms of the high capacitance of 612.5 F/g at 0.5 A/g and the superior long-term cyclic stability (over 90% specific capacitance retention after 1,000 cycles). This outstanding performance can be ascribed to the short diffusion path and large surface area of the unique hollow structure for bulk accessibility of faradaic reaction. Wang et al. [157] also reported Mn_2O_3 hollow spheres with a controllable shell number (including double, triple, quadruple shells). As shown in Figure 24.25d–f, three types of as-prepared Mn_2O_3 hollow spheres with thin porous shells all have a size of about 800 nm. Among these products, the triple-shelled Mn_2O_3 hollow spheres show an extremely high specific capacitance up to 1,651 F/g at a current density of 0.5 A/g and superior cycling stability with 92% retention after 2,000 consecutive cycles. Besides, the rate capability set a new record with a specific capacitance of 1,422 F/g at a high current density of 10 A/g.

Some mixed metal oxide nanocages can also be used as electrode materials in pseudocapacitors or HSCs. Lou and co-workers [136] reported the designed synthesis of complex hollow $Co_3O_4/NiCo_2O_4$ DSNCs by manipulating the ZIF-67 polyhedron-engaged reactions. When evaluated as electrodes for pseudocapacitors, the $Co_3O_4/NiCo_2O_4$ DSNCs show a large specific capacitance of 972 F/g at a current density of 5 A/g and remarkable stability with 92.5% capacitance retention after 12,000 cycles (Figure 24.25g,h). Guan et al. [14] have recently demonstrated an HSC device using $NiCo_2O_4$ seven-shelled hollow spheres as the battery-type electrode and graphene/multi-shelled mesoporous carbon spheres composites as the capacitive electrode (Figure 24.25i). The HSC device shows an excellent cycling stability with 91.3% of the initial capacitance

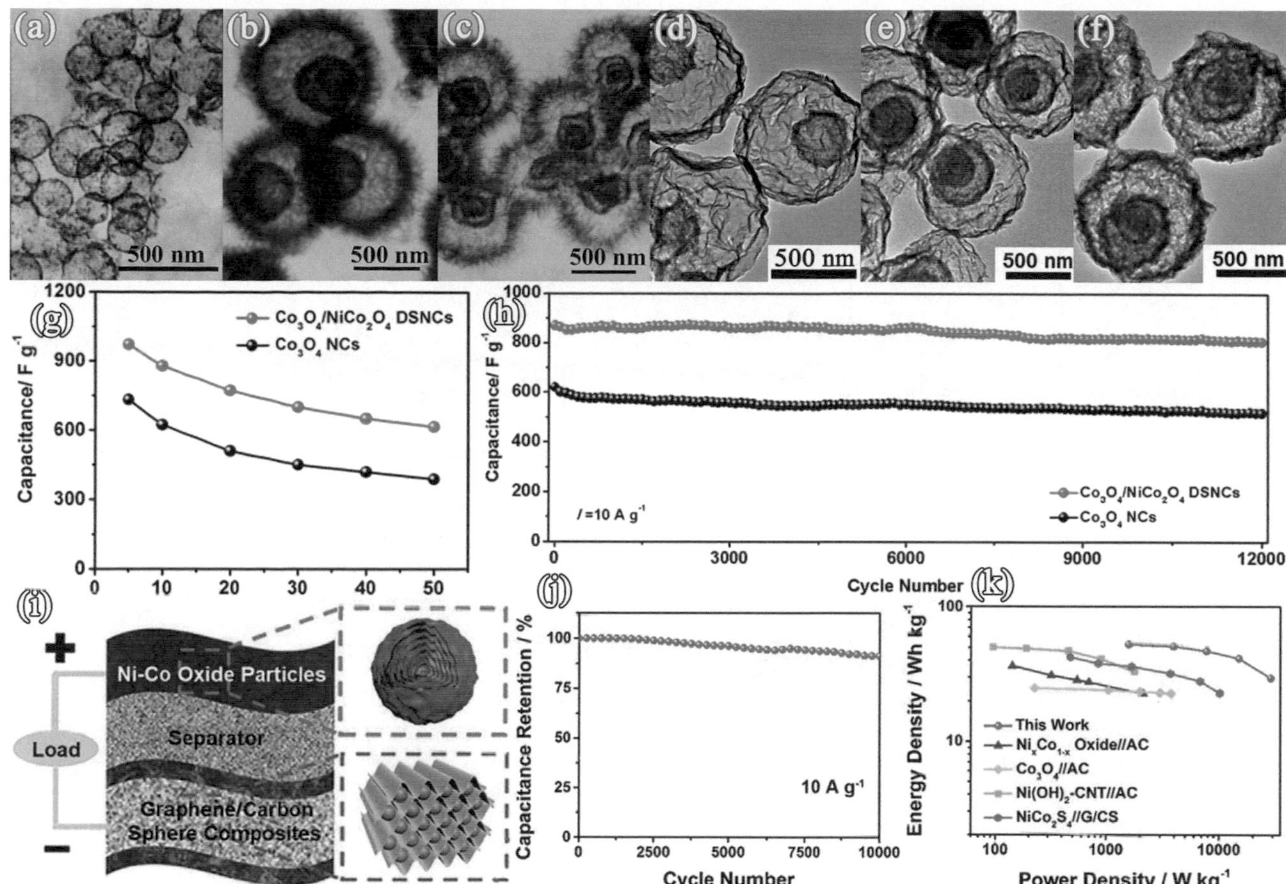

FIGURE 24.25 TEM images of (a) single-shelled NiO hollow spheres, (b) double-shelled NiO hollow spheres, (c) triple-shelled NiO hollow spheres, (d) double-shelled Mn_2O_3 hollow spheres, (e) triple-shelled Mn_2O_3 hollow spheres, and (f) quadruple-shelled Mn_2O_3 hollow spheres. Electrochemical characterizations of the $Co_3O_4/NiCo_2O_4$ DSNCs and Co_3O_4 nanocages (NCs): (g) specific capacitance as a function of current density and (h) cycling performance at a current density of 10 A/g. (i) Schematic illustration of the HSC device based on the multi-shelled Ni-Co oxide particles, (j) cycling performance of the HSC device at a current density of 10 A/g, and (k) the corresponding Ragone plot. (Panels (a–c) reproduced with permission from Ref. [40]. Copyright 2013, Elsevier B.V. All rights reserved. Panels (d–f) reproduced with permission from Ref. [157]. Copyright 2014, Wiley-VCH Verlag GmbH & Co. KGaA, Weinheim. Panels (g and h) reproduced with permission from Ref. [136]. Copyright 2015, American Chemical Society. Panels (i–k) reproduced with permission from Ref. [14]. Copyright 2017, Wiley-VCH Verlag GmbH & Co. KGaA, Weinheim.)

retained after 10,000 cycles (Figure 24.25j) and a high energy density of 52.6 Wh/kg at a power density of 1,604 W/kg (Figure 24.25k).

24.4.4 Photocatalysis

Since the first discovery of Fujishima and Honda on the photocatalytic splitting of water over TiO_2 electrodes [158], photocatalysis on a semiconductor has attracted continuous attention in many research fields [159]. The photocatalyst has a bandgap separating the conduction band (CB) and valence band (VB), as shown in Figure 24.26. When irradiated with light at an energy equal to or higher than the bandgap of the semiconductor (Eg), electrons (e^-) are excited from the VB to the CB while leaving holes (h^+) in the VB. The electrons and holes can then be migrated to the surface of the semiconductor to participate in the chemical reaction. Metal hydroxide and oxide nanocages as photocatalysts have several significant advantages over traditional solid photocatalysts: (i) a higher specific surface area and lower density; (ii) reduced transmission distance of mass and charge; (iii) more multiple reflections of the incident light enhance the light collection efficiency and further enhance their photocatalytic performance.

TiO_2 is a promising photocatalyst owing to its low toxicity, low cost, high chemical stability, and high photocatalytic activity. Yin and co-workers [160] prepared mesoporous hollow TiO_2 nanostructures with a high surface area (311 m^2/g) and a controlled crystallinity through a novel silica-protected calcination process. When used as photocatalysts for the degradation of Rhodamine B (RhB) under UV irradiation, the as-prepared mesoporous anatase TiO_2 shell shows significantly enhanced photocatalytic activity, and the sample calcined at higher temperature has greater enhancement due to its improved crystallinity. Zeng et al. [161] reported a simple hard template method for

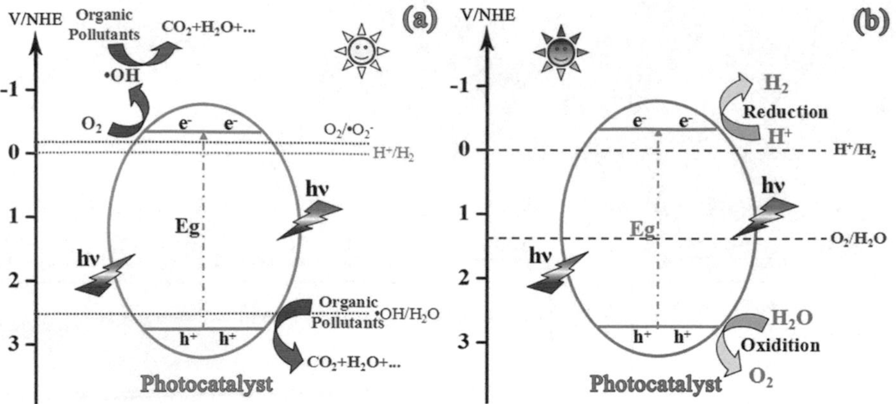

FIGURE 24.26 A schematic of the mechanism of photocatalytic degradation of organic compounds (a) and water splitting (b).

the preparation of multi-shell TiO_2 hollow spheres. As a photocatalyst, it exhibited stronger photocatalytic activity compared with sphere-in-sphere structures and nanoparticles in the experiment of RhB degradation. Some photocatalysts are also usually doped or supported with some cocatalysts such as noble metals and metal oxides to further increase photocatalytic activity and prolong the stability of the photocatalyst due to the synergistic effect of the cocatalysts (e.g., plasmon resonance effects and formation of heterojunction structures). Wang et al. [162] successfully prepared Ag/TiO_2 hollow spheres with highly uniform morphology and good structural stability by a one-pot hydrothermal method. The as-prepared Ag/TiO_2 composite hollow spheres as photocatalysts for the degradation of RhB and methyl orange (MO) under visible light illumination exhibit excellent photocatalytic activity, superior to commercial TiO_2 nanoparticles (P25). Its excellent photocatalytic activity is mainly attributed to an important optimal synergistic effect between the Ag and TiO_2 hollow spheres. Agrawal et al. [163] reported the fabrication of a mixed metal oxide (ZnO-TiO_2) hollow sphere via exploiting the template-assisted method. Compared to the pure ZnO and TiO_2, the as-prepared ZnO-TiO_2 hollow spheres exhibited higher photocatalytic activity for the degradation of Rhodamine 6G dye.

In addition to the photocatalytic degradation of organic dyes, solar water splitting directly to hydrogen and oxygen has become one of the most desirable methods for harvesting and conversion of solar energy into chemical energy. For example, a hollow nano-heterostructure comprising $ZnFe_2O_4$ and ZnO formed on a carbonaceous sacrificial support was reported [164]. Due to the unique structure of the hollow $ZnFe_2O_4/ZnO$, which gives enhanced separation capability of photo-generated carriers, the hydrogen generation rate of $ZnFe_2O_4/ZnO$ nano-heterostructures without cocatalysts is up to 2.15 mmol/h/g under visible light irradiation ($\lambda > 420$ nm), which is 45 times higher than the best yields ever reported for $ZnFe_2O_4$-based photocatalysts. Apart from the abovementioned studies, other metal oxide nanocages such as Bi_2O_3 [165], In_2O_3 [166], WO_3 [43], Ta_2O_5 [167], and CeO_2 [168] have also been reported for the photocatalytic degradation of organic dyes and water splitting.

24.4.5 Electrocatalysis

Electrocatalyst plays an important role in clean energy conversion technology. A series of electrochemical reactions such as the oxygen evolution reaction (OER), hydrogen evolution reaction (HER), and oxygen reduction reaction (ORR) are inseparable from the participation of electrocatalysts [169–171]. The most challenging issue is how to obtain highly efficient electrocatalysts that catalyze these reactions with high current density at low overpotential [16,171]. In this regard, the hollow structures exhibit various advantages as electrocatalysts, including increased contact area between electrolytes and catalysts, more exposed active sites, and facilitating the transmission of charges, making them promising candidates for these electrocatalytic reactions. Most metal hydroxide and oxide nanocages are commonly used as OER electrocatalysts. For example, our group explored the feasibility of the amorphous hollow nanomaterials for efficient electrochemical water oxidation [96]. The amorphous Ni-Co double hydroxides nanocages with well-defined shapes and structures were prepared by a "coordinating etching" process (Figure 24.27a). The Ni/Co molar ratio (including $Ni(OH)_2$, $Ni_2Co(OH)_x$, $NiCo(OH)_x$, $NiCo_{2.7}(OH)_x$, $NiCo_{5.7}(OH)_x$, $NiCo_{8.3}(OH)_x$, and $Co(OH)_2$) could be easily adjusted by adding the appropriate amount of reactants. When using them as electrocatalysts for OER, the amorphous $NiCo_{2.7}(OH)_x$ double hydroxide nanocages exhibit the best catalytic activity compared to other Ni-Co hydroxide nanocages in terms of the smaller overpotential, lower Tafel slope, and excellent stability (Figure 24.27d–f). The superior OER performance is attributed to its unique amorphous hollow structures, large specific surface area (137.8 m^2/g with an average pore size of 13.1 nm), and suitable metal element composition.

In addition to metal hydroxide nanocages, some mixed metal oxide nanocages are also used as excellent OER electrocatalysts. For example, Han et al. [172] successfully

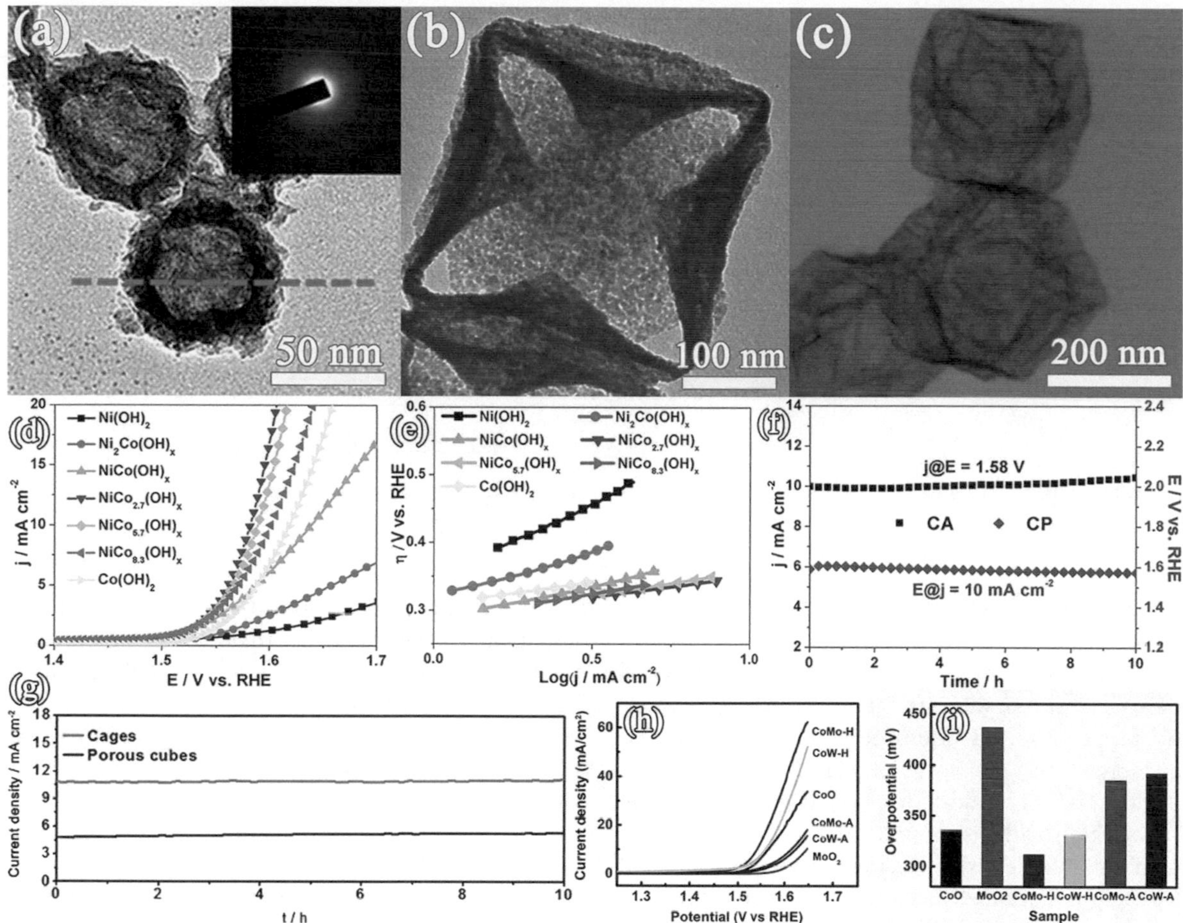

FIGURE 24.27 TEM images of (a) amorphous $NiCo_{2.7}(OH)_x$ nanocages, (b) Ni-Co mixed oxide nanocages, and (c) CoO-MoO_2 nanocages. (d) linear scan voltammetry (LSV) plots; (e) Tafel plots of the $Ni(OH)_2$, $Ni_2Co(OH)_x$, $NiCo(OH)_x$, $NiCo_{2.7}(OH)_x$, $NiCo_{5.7}(OH)_x$, $NiCo_{8.3}(OH)_x$, and $Co(OH)_2$ amorphous catalysts; (f) CA and CP plots of the $NiCo_{2.7}(OH)_x$ amorphous catalyst. (g) Chronoamperometry curves of the Ni-Co mixed oxide nanocages and porous cubes. (h) LSV curves and (i) overpotential at 10 mA/cm^2 for CoO, MoO_2, CoMo-H, CoW-H, CoMo-A, and CoW-A. Note: CoO-MoO_2 nanocages (denoted as CoMo-H) were obtained by calcining $CoMoO_4$-$Co(OH)_2$ nanocages under H_2/Ar; $CoMoO_4$-$Co(OH)_2$ calcined under air (denoted as CoMo-A); similarly, CoW-A and CoW-H are obtained. (Panels (a, d–f) reproduced with permission from Ref. [96]. Copyright 2015, Wiley-VCH Verlag GmbH & Co. KGaA, Weinheim. Panels (b, g) reproduced with permission from Ref. [172]. Copyright 2016, Wiley-VCH Verlag GmbH & Co. KGaA, Weinheim. Panels (c, h, i) reproduced with permission from Ref. [173]. Copyright 2017, Wiley-VCH Verlag GmbH & Co. KGaA, Weinheim.)

prepared Ni-Co mixed metal oxide nanocages (Figure 24.27b) by the thermal treatment of Ni-Co Prussian-blue-analog (PBA) cages in air. The resulting Ni-Co mixed oxide nanocages display the higher OER activity, a low overpotential of 0.38 V at a current density of 10 mA/cm^2, and excellent stability in an alkaline medium, compared to its counterpart Ni-Co mixed oxide cubes (Figure 24.27g). There are several main reasons for superior OER activity: (i) the complex 3D cage-like hollow porous nanocages give it structural advantages; (ii) the large electrode–electrolyte contact area is provided by the high surface area-to-volume ratio of the open structure; and (iii) building blocks composed of small nanoparticles have more electrochemically active sites.

Similarly, the electrocatalytic performance of mixed metal oxide nanocages for OER can be improved by controlling their architectures and compositions. Yin and co-workers [173] reported the preparation of CoO-MoO_2 nanocages (Figure 24.27c) by a self-templating route. The as-prepared CoO-MoO_2 nanocages delivered a lower overpotential of 312 mV at a current density of 10 mA/cm^2 and excellent stability toward OER. Compared to pure CoO and MoO_2, as well as other Co-Mo and Co-W nanocages, the CoMo-H ($CoMoO_4$-$Co(OH)_2$ nanocages were calcined under H_2/Ar to obtain CoO-MoO_2 nanocages, namely, CoMo-H) exhibits the best OER performance, as illustrated in Figure 24.27h and i. The improved OER activity of the CoO-MoO_2 nanocages may be attributed to functional CoO components with high-catalytic OER activity and excellent stability. Specifically, conductive MoO_2 components with reasonable OER activities promote charge transfer. Also the architectures of hollow porous nanocages provide multiple interface and reduced mass diffusion length.

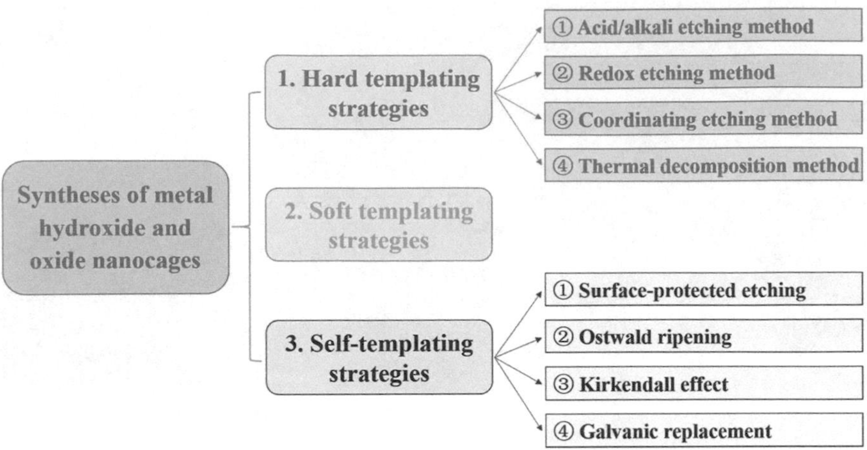

FIGURE 24.28 Classifcation of synthetic methodologies for metal hydroxide and oxide nanocages.

24.5 Conclusions and Outlook

Metal hydroxide and oxide nanocages have attracted rising research interest due to their unique structure-dependent properties and widespread potential applications. In this chapter, we have highlighted the synthetic methodologies and the most important applications of various metal hydroxide and oxide nanocages (including monometallic hydroxide/oxide nanocages and mixed metal hydroxide/oxide nanocages). So far, researchers commonly use the following three strategies to synthesize nanocages: (i) hard templating strategies; (ii) soft templating strategies; and (iii) self-templating strategies. Conceptually, the hard template strategy is the simplest way, that is, the surface of the pre-synthesized template is coated with a desired shell material, and the core of the template is selectively removed. Based on the selective removal of the template core, the hard templating strategies for nanocages are classified as (i) acid/alkali etching method, (ii) redox etching method, (iii) coordinating etching method, and (iv) thermal decomposition method. In principle, hard template strategies can be used to prepare any kind of metal hydroxide and oxide nanocages and, therefore, are also considered as the most popular methods. On the other hand, soft templating strategies generally use emulsion droplets, micelle/vesicle, polymer, and gas bubble as templates, which have the advantage of easier removal of templates. The newly developed self-templating strategies, as "smarter" strategies, have several advantages over the traditional templating strategies, including low cost, simplified synthesis procedures, and high product uniformity. Based on the formation mechanism of internal voids, it can be divided into (i) surface-protected etching, (ii) Ostwald ripening, (iii) Kirkendall effect, and (iv) galvanic replacement. The synthesis strategies for metal hydroxide and oxide nanocages are schematically illustrated in Figure 24.28.

The breakthroughs in the synthesis of metal hydroxide and oxide nanocages provide an opportunity to adjust their physicochemical properties and, thus, catalyze their exploration in various applications. However, the synthesis and application of metal hydroxide and oxide nanocages are still in its infancy. There are a considerable number of challenges that need to be overcome to promote their further development. For example, most syntheses of nanocages are at the laboratory level, the synthesis steps are cumbersome, the reproducibility is poor, it lacks uniformity and versatility, large-scale production is not feasible, and the cost is high. Therefore, it is important to develop a simple, universal, low-cost, and scalable approach to produce high-quality metal hydroxide/oxide nanocages with adjustable composition and precisely controlled structural parameters. Besides, a number of nanocages are synthesized by accidental windfalls or only on the basis of the experience of researchers, and the formation mechanism is still not very clear. Thus, deepening the understanding of the formation mechanism can not only help generalize these methods and promote industrial applications but also design and fabricate some fantastic hollow materials with complex structures and ideal performance that cannot be realized at present. Furthermore, although the properties and applications of metal hydroxide and oxide nanocages have been widely studied, better performance and new untapped properties are also strongly expected for fundamental research and practical applications by designing more complex hollow nanostructures, multiplex mixed metal hydroxide and oxide nanocages, and functionalized composites.

Finally, we hope that each of the examples selected in this chapter, including the preparation methods and applications, the synthesis principles, and our personal insights and perspectives can provide the reader with the necessary background knowledge and ideas to develop more in-depth expertise and skills. We are confident that more universal and more powerful synthesis strategies for metal hydroxide and oxide nanocages will be developed in the near future. We also believe that metal hydroxide and oxide nanocages with superior properties will break the bottleneck of current applications to achieve industrial production.

Acknowledgments

We are grateful for financial support from the National Natural Science Foundation of China (No. 51532001) and the National Basic Research Program of China (No. 2014CB931802).

References

1. Yu, L., Wu, H.B., and Lou, X.W. (2017) Self-templated formation of hollow structures for electrochemical energy applications. *Acc. Chem. Res.*, **50** (2), 293–301.
2. Wang, X., Feng, J., Bai, Y., Zhang, Q., and Yin, Y. (2016) Synthesis, properties, and applications of hollow micro-/nanostructures. *Chem. Rev.*, **116** (18), 10983–1060.
3. Hu, J., Chen, M., Fang, X., and Wu, L. (2011) Fabrication and application of inorganic hollow spheres. *Chem. Soc. Rev.*, **40** (11), 5472–5491.
4. Li, Y. and Shi, J. (2014) Hollow-structured mesoporous materials: Chemical synthesis, functionalization and applications. *Adv. Mater.*, **26** (20), 3176–3205.
5. Wang, S., Guan, B.Y., and Lou, X.W. (2018) Rationally designed hierarchical N-doped carbon@$NiCo_2O_4$ double-shelled nanoboxes for enhanced visible light CO_2 reduction. *Energy Environ. Sci.*, **11** (2), 306–310.
6. Sun, S. and Yang, Z. (2014) Cu_2O-templated strategy for synthesis of definable hollow architectures. *Chem. Commun.*, **50** (56), 7403–7415.
7. Oh, M.H., Yu, T., Yu, S.H., Lim, B., Ko, K.T., Willinger, M.G., Seo, D.H., Kim, B.H., Cho, M.G., Park, J.H., Kang, K., Sung, Y.E., Pinna, N., and Hyeon, T. (2013) Galvanic replacement reactions in metal oxide nanocrystals. *Science*, **340** (6135), 964–968.
8. Wu, Q., Yang, L., Wang, X., and Hu, Z. (2017) From carbon-based nanotubes to nanocages for advanced energy conversion and storage. *Acc. Chem. Res.*, **50** (2), 435–444.
9. Hu, M., Belik, A.A., Imura, M., and Yamauchi, Y. (2013) Tailored design of multiple nanoarchitectures in metal-cyanide hybrid coordination polymers. *J. Am. Chem. Soc.*, **135** (1), 384–391.
10. Zhang, P., Guan, B.Y., Yu, L., and Lou, X.W. (2017) Formation of double-shelled zinc-cobalt sulfide dodecahedral cages from bimetallic zeolitic imidazolate frameworks for hybrid supercapacitors. *Angew. Chem. Int. Ed.*, **56** (25), 7141–7145.
11. Zhang, J., Hu, H., Li, Z., and Lou, X.W. (2016) Double-shelled nanocages with cobalt hydroxide inner shell and layered double hydroxides outer shell as high-efficiency polysulfide mediator for lithium-sulfur batteries. *Angew. Chem. Int. Ed.*, **55** (12), 3982–3986.
12. Zhao, X., Yu, R., Tang, H., Mao, D., Qi, J., Wang, B., Zhang, Y., Zhao, H., Hu, W., and Wang, D. (2017) Formation of septuple-shelled $(Co_{2/3}Mn_{1/3})(Co_{5/6}Mn_{1/6})_2O_4$ hollow spheres as electrode material for alkaline rechargeable battery. *Adv. Mater.*, **29** (34), 1700550.
13. Zu, L., Su, Q., Zhu, F., Chen, B., Lu, H., Peng, C., He, T., Du, G., He, P., Chen, K., Yang, S., Yang, J., and Peng, H. (2017) Antipulverization electrode based on low-carbon triple-shelled superstructures for lithium-ion batteries. *Adv. Mater.*, **29** (34), 1701494.
14. Guan, B.Y., Kushima, A., Yu, L., Li, S., Li, J., and Lou, X.W. (2017) Coordination polymers derived general synthesis of multishelled mixed metal-oxide particles for hybrid supercapacitors. *Adv. Mater.*, **29** (17), 1605902.
15. Qi, J., Lai, X., Wang, J., Tang, H., Ren, H., Yang, Y., Jin, Q., Zhang, L., Yu, R., Ma, G., Su, Z., Zhao, H., and Wang, D. (2015) Multi-shelled hollow micro-/nanostructures. *Chem. Soc. Rev.*, **44** (19), 6749–6773.
16. Yu, L., Hu, H., Wu, H.B., and Lou, X.W. (2017) Complex hollow nanostructures: Synthesis and energy-related applications. *Adv. Mater.*, **29** (15), 1604563.
17. Zhou, L., Zhuang, Z., Zhao, H., Lin, M., Zhao, D., and Mai, L. (2017) Intricate hollow structures: Controlled synthesis and applications in energy storage and conversion. *Adv. Mater.*, **29** (20), 1602914.
18. An, K. and Hyeon, T. (2009) Synthesis and biomedical applications of hollow nanostructures. *Nano Today*, **4** (4), 359–373.
19. Yang, G., Xu, L., Chao, Y., Xu, J., Sun, X., Wu, Y., Peng, R., and Liu, Z. (2017) Hollow MnO_2 as a tumor-microenvironment-responsive biodegradable nano-platform for combination therapy favoring antitumor immune responses. *Nat. Commun.*, **8** (1), 902.
20. Zhu, Y.F., Shi, J.L., Shen, W.H., Dong, X.P., Feng, J.W., Ruan, M.L., and Li, Y.S. (2005) Stimuli-responsive controlled drug release from a hollow mesoporous silica sphere/polyelectrolyte multilayer core-shell structure. *Angew. Chem. Int. Ed.*, **44** (32), 5083–5087.
21. Shin, J., Anisur, R.M., Ko, M.K., Im, G.H., Lee, J.H., and Lee, I.S. (2009) Hollow manganese oxide nanoparticles as multifunctional agents for magnetic resonance imaging and drug delivery. *Angew. Chem. Int. Ed.*, **48** (2), 321–324.
22. Zhu, Y., Ikoma, T., Hanagata, N., and Kaskel, S. (2010) Rattle-type Fe_3O_4@SiO_2 hollow mesoporous spheres as carriers for drug delivery. *Small*, **6** (3), 471–478.
23. Ke, C.J., Su, T.Y., Chen, H.L., Liu, H.L., Chiang, W.L., Chu, P.C., Xia, Y., and Sung, H.W. (2011)

Smart multifunctional hollow microspheres for the quick release of drugs in intracellular lysosomal compartments. *Angew. Chem. Int. Ed.*, **50** (35), 8086–8089.

24. Gao, C., Zhang, H., Wu, M., Liu, Y., Wu, Y., Yang, X., and Feng, X. (2012) Polyethyleneimine functionalized polymer microsphere: A novel delivery vector for cells. *Polym. Chem.*, **3** (5), 1168–1173.
25. Chiang, W.L., Ke, C.J., Liao, Z.X., Chen, S.Y., Chen, F.R., Tsai, C.Y., Xia, Y., and Sung, H.W. (2012) Pulsatile drug release from plga hollow microspheres by controlling the permeability of their walls with a magnetic field. *Small*, **8** (23), 3584–3588.
26. Xing, R., Bhirde, A.A., Wang, S., Sun, X., Liu, G., Hou, Y., and Chen, X. (2013) Hollow iron oxide nanoparticles as multidrug resistant drug delivery and imaging vehicles. *Nano Res.*, **6** (1), 1–9.
27. Koo, H.J., Kim, Y.J., Lee, Y.H., Lee, W.I., Kim, K., and Park, N.G. (2008) Nano-embossed hollow spherical TiO_2 as bifunctional material for high-efficiency dye-sensitized solar cells. *Adv. Mater.*, **20** (1), 195–199.
28. Yang, S.C., Yang, D.J., Kim, J., Hong, J.M., Kim, H.G., Kim, I.D., and Lee, H. (2008) Hollow TiO_2 hemispheres obtained by colloidal templating for application in dye-sensitized solar cells. *Adv. Mater.*, **20** (5), 1059–1064.
29. Liu, J., Luo, T., Mouli, S.T., Meng, F., Sun, B., Li, M., and Liu, J. (2010) A novel coral-like porous SnO_2 hollow architecture: Biomimetic swallowing growth mechanism and enhanced photovoltaic property for dye-sensitized solar cell application. *Chem. Commun.*, **46** (3), 472–474.
30. Wu, X., Lu, G.Q., and Wang, L. (2011) Shell-in-shell TiO_2 hollow spheres synthesized by one-pot hydrothermal method for dye-sensitized solar cell application. *Energy Environ. Sci.*, **4** (9), 3565–3572.
31. Wang, H., Miyauchi, M., Ishikawa, Y., Pyatenko, A., Koshizaki, N., Li, Y., Li, L., Li, X., Bando, Y., and Golberg, D. (2011) Single-crystalline rutile TiO_2 hollow spheres: Room-temperature synthesis, tailored visible-light-extinction, and effective scattering layer for quantum dot-sensitized solar cells. *J. Am. Chem. Soc.*, **133** (47), 19102–19109.
32. Du, J., Qi, J., Wang, D., and Tang, Z. (2012) Facile synthesis of Au@TiO_2 core-shell hollow spheres for dye-sensitized solar cells with remarkably improved efficiency. *Energy Environ. Sci.*, **5** (5), 6914–6918.
33. Dong, Z., Lai, X., Halpert, J.E., Yang, N., Yi, L., Zhai, J., Wang, D., Tang, Z., and Jiang, L. (2012) Accurate control of multishelled ZnO hollow microspheres for dye-sensitized solar cells with high efficiency. *Adv. Mater.*, **24** (8), 1046–1049.
34. Wang, Z., Zhou, L., and Lou, X.W. (2012) Metal oxide hollow nanostructures for lithium-ion batteries. *Adv. Mater.*, **24** (14), 1903–1911.
35. Wang, J., Yang, N., Tang, H., Dong, Z., Jin, Q., Yang, M., Kisailus, D., Zhao, H., Tang, Z., and Wang, D. (2013) Accurate control of multishelled Co_3O_4 hollow microspheres as high-performance anode materials in lithium-ion batteries. *Angew. Chem. Int. Ed.*, **52** (25), 6417–6420.
36. Guan, C., Xia, X., Meng, N., Zeng, Z., Cao, X., Soci, C., Zhang, H., and Fan, H.J. (2012) Hollow core-shell nanostructure supercapacitor electrodes: Gap matters. *Energy Environ. Sci.*, **5** (10), 9085–9090.
37. Du, H., Jiao, L., Wang, Q., Yang, J., Guo, L., Si, Y., Wang, Y., and Yuan, H. (2013) Facile carbonaceous microsphere templated synthesis of Co_3O_4 hollow spheres and their electrochemical performance in supercapacitors. *Nano Res.*, **6** (2), 87–98.
38. Wu, X., Zeng, Y., Gao, H., Su, J., Liu, J., and Zhu, Z. (2013) Template synthesis of hollow fusiform $RuO_2{\cdot}xH_2O$ nanostructure and its supercapacitor performance. *J. Mater. Chem. A*, **1** (3), 469–472.
39. Wang, Y., Pan, A., Zhu, Q., Nie, Z., Zhang, Y., Tang, Y., Liang, S., and Cao, G. (2014) Facile synthesis of nanorod-assembled multi-shelled Co_3O_4 hollow microspheres for high-performance supercapacitors. *J. Power Sources*, **272**, 107–112.
40. Yang, Z., Xu, F., Zhang, W., Mei, Z., Pei, B., and Zhu, X. (2014) Controllable preparation of multishelled NiO hollow nanospheres via layer-by-layer self-assembly for supercapacitor application. *J. Power Sources*, **246**, 24–31.
41. Nguyen, C.C., Vu, N.N., and Do, T.O. (2015) Recent advances in the development of sunlight-driven hollow structure photocatalysts and their applications. *J. Mater. Chem. A*, **3** (36), 18345–18359.
42. Prieto, G., Tuysuz, H., Duyckaerts, N., Knossalla, J., Wang, G.H., and Schuth, F. (2016) Hollow nano- and microstructures as catalysts. *Chem. Rev.*, **116** (22), 14056–14119.
43. Xi, G., Yan, Y., Ma, Q., Li, J., Yang, H., Lu, X., and Wang, C. (2012) Synthesis of multiple-shell WO_3 hollow spheres by a binary carbonaceous template route and their applications in visible-light photocatalysis. *Chem. Eur. J.*, **18** (44), 13949–13953.
44. Mahmoud, M.A., Saira, F., and El-Sayed, M.A. (2010) Experimental evidence for the nanocage effect in catalysis with hollow nanoparticles. *Nano Lett.*, **10** (9), 3764–3769.
45. Zhong, J. and Cao, C. (2010) Nearly monodisperse hollow Fe_2O_3 nanoovals: Synthesis, magnetic property and applications in photocatalysis and gas sensors. *Sens. Actuators B*, **145** (2), 651–656.
46. Xu, X., Zhang, Z., and Wang, X. (2015) Well-defined metal-organic-framework hollow nanostructures for catalytic reactions involving gases. *Adv. Mater.*, **27** (36), 5365–5371.
47. Joo, J.B., Dahl, M., Li, N., Zaera, F., and Yin, Y. (2013) Tailored synthesis of mesoporous TiO_2

hollow nanostructures for catalytic applications. *Energy Environ. Sci.*, **6** (7), 2082–2092.

48. Joo, J.B., Lee, I., Dahl, M., Moon, G.D., Zaera, F., and Yin, Y. (2013) Controllable synthesis of mesoporous TiO_2 hollow shells: Toward an efficient photocatalyst. *Adv. Funct. Mater.*, **23** (34), 4246–4254.
49. Joo, J.B., Zhang, Q., Dahl, M., Zaera, F., and Yin, Y. (2013) Synthesis, crystallinity control, and photocatalysis of nanostructured titanium dioxide shells. *J. Mater. Res.*, **28** (3), 362–368.
50. Lee, J.H. (2009) Gas sensors using hierarchical and hollow oxide nanostructures: Overview. *Sens. Actuators B*, **140** (1), 319–336.
51. Liu, C., Zhao, L., Wang, B., Sun, P., Wang, Q., Gao, Y., Liang, X., Zhang, T., and Lu, G. (2017) Acetone gas sensor based on NiO/ZnO hollow spheres: Fast response and recovery, and low (ppb) detection limit. *J. Colloid Interface Sci.*, **495**, 207–215.
52. Liu, J., Wang, T., Wang, B., Sun, P., Yang, Q., Liang, X., Song, H., and Lu, G. (2017) Highly sensitive and low detection limit of ethanol gas sensor based on hollow ZnO/SnO_2 spheres composite material. *Sens. Actuators B*, **245**, 551–559.
53. Zhang, Z., Wen, Z., Ye, Z., and Zhu, L. (2017) Synthesis of Co_3O_4/Ta_2O_5 heterostructure hollow nanospheres for enhanced room temperature ethanol gas sensor. *J. Alloys Compd.*, **727**, 436–443.
54. Oh, K.H., Park, H.J., Kang, S.W., Park, J.C., and Nam, K.M. (2018) Synthesis of hollow iron oxide nanospheres and their application to gas sensors. *J. Nanosci. Nanotechno.*, **18** (2), 1356–1360.
55. Farbod, M., Joula, M.H., and Vaezi, M. (2016) Promoting effect of adding carbon nanotubes on sensing characteristics of ZnO hollow sphere-based gas sensors to detect volatile organic compounds. *Mater. Chem. Phys.*, **176**, 12–23.
56. Gan, T., Zhao, A., Wang, S., Lv, Z., and Sun, J. (2016) Hierarchical triple-shelled porous hollow zinc oxide spheres wrapped in graphene oxide as efficient sensor material for simultaneous electrochemical determination of synthetic antioxidants in vegetable oil. *Sens. Actuators B*, **235**, 707–716.
57. Wang, M., Jiang, X., Liu, J., Guo, H., and Liu, C. (2015) Highly sensitive H_2O_2 sensor based on Co_3O_4 hollow sphere prepared via a template-free method. *Electrochim. Acta*, **182**, 613–620.
58. Li, X.L., Lou, T.J., Sun, X.M., and Li, Y.D. (2004) Highly sensitive WO_3 hollow-sphere gas sensors. *Inorg. Chem.*, **43** (17), 5442–5449.
59. Wu, Z., Zhang, M., Yu, K., Zhang, S., and Xie, Y. (2008) Self-assembled double-shelled ferrihydrite hollow spheres with a tunable aperture. *Chem. Eur. J.*, **14** (17), 5346–5352.
60. Yang, H.M., Ma, S.Y., Jiao, H.Y., Chen, Q., Lu, Y., Jin, W.X., Li, W.Q., Wang, T.T., Jiang, X.H., Qiang, Z., and Chen, H. (2017) Synthesis of Zn_2SnO_4 hollow spheres by a template route for high-performance acetone gas sensor. *Sens. Actuators B*, **245**, 493–506.
61. Wei, Z., Xing, R., Zhang, X., Liu, S., Yu, H., and Li, P. (2013) Facile template-free fabrication of hollow nestlike alpha-Fe_2O_3 nanostructures for water treatment. *ACS Appl. Mater. Interfaces*, **5** (3), 598–604.
62. Zhang, D., Xu, D., Ni, Y., Lu, C., and Xu, Z. (2014) A facile one-pot synthesis of monodisperse ring-shaped hollow Fe_3O_4 nanospheres for waste water treatment. *Mater. Lett.*, **123**, 116–119.
63. Zhang, Y., Xu, S., Xia, H., and Zheng, F. (2016) Facile synthesis of Fe_3O_4@C hollow nanospheres and their application in polluted water treatment. *Solid State Sci.*, **61**, 16–23.
64. Zhu, D., Zhang, J., Song, J., Wang, H., Yu, Z., Shen, Y., and Xie, A. (2013) Efficient one-pot synthesis of hierarchical flower-like alpha-Fe_2O_3 hollow spheres with excellent adsorption performance for water treatment. *Appl. Surf. Sci.*, **284**, 855–861.
65. Cao, J., Mao, Q., Shi, L., and Qian, Y. (2011) Fabrication of gamma-MnO_2/alpha-MnO_2 hollow core/shell structures and their application to water treatment. *J. Mater. Chem.*, **21** (40), 16210–16215.
66. Cao, J., Zhu, Y., Shi, L., Zhu, L., Bao, K., Liu, S., and Qian, Y. (2010) Double-shelled Mn_2O_3 hollow spheres and their application in water treatment. *Eur. J. Inorg. Chem.*, **2010** (8), 1172–1176.
67. Shi, L. and Lin, H. (2010) Facile fabrication and optical property of hollow SnO_2 spheres and their application in water treatment. *Langmuir*, **26** (24), 18718–18722.
68. Wang, X., Zhong, Y., Zhai, T., Guo, Y., Chen, S., Ma, Y., Yao, J., Bando, Y., and Golberg, D. (2011) Multishelled Co_3O_4-Fe_3O_4 hollow spheres with even magnetic phase distribution: Synthesis, magnetic properties and their application in water treatment. *J. Mater. Chem.*, **21** (44), 17680–17687.
69. Caruso, F., Caruso, R.A., and Mohwald, H. (1998) Nanoengineering of inorganic and hybrid hollow spheres by colloidal templating. *Science*, **282** (6), 1111–1114.
70. Huang, C.C., Huang, W., and Yeh, C.S. (2011) Shell-by-shell synthesis of multi-shelled mesoporous silica nanospheres for optical imaging and drug delivery. *Biomaterials*, **32** (2), 556–564.
71. Wong, Y.J., Zhu, L., Teo, W.S., Tan, Y.W., Yang, Y., Wang, C., and Chen, H. (2011) Revisiting the stober method: In homogeneity in silica shells. *J. Am. Chem. Soc.*, **133** (30), 11422–11425.
72. Kuo, C.H. and Huang, M.H. (2008) Fabrication of truncated rhombic dodecahedral Cu_2O nanocages and nanoframes by particle aggregation and acidic etching. *J. Am. Chem. Soc.*, **130** (38), 12815–12820.
73. Wang, Z. and Lou, X.W. (2012) TiO_2 nanocages: Fast synthesis, interior functionalization and improved

lithium storage properties. *Adv. Mater.*, **24** (30), 4124–4129.

74. Liu, J., Xu, X., Hu, R., Yang, L., and Zhu, M. (2016) Uniform hierarchical Fe_3O_4@Polypyrrole nanocages for superior lithium ion battery anodes. *Adv. Energy Mater.*, **6** (13), 1600256.
75. Wang, Z., Luan, D., Li, C.M., Su, F., Madhavi, S., Boey, F.Y.C., and Lou, X.W. (2010) Engineering nonspherical hollow structures with complex interiors by template-engaged redox etching. *J. Am. Chem. Soc.*, **132** (45), 16271–16277.
76. Nai, J., Tian, Y., Guan, X., and Guo, L. (2013) Pearson's principle inspired generalized strategy for the fabrication of metal hydroxide and oxide nanocages. *J. Am. Chem. Soc.*, **135** (43), 16082–16091.
77. Li, B., Wen, H.M., Cui, Y., Zhou, W., Qian, G., and Chen, B. (2016) Emerging multifunctional metal-organic framework materials. *Adv. Mater.*, **28** (40), 8819–8860.
78. Cao, X., Tan, C., Sindoro, M., and Zhang, H. (2017) Hybrid micro-/nano-structures derived from metal-organic frameworks: Preparation and applications in energy storage and conversion. *Chem. Soc. Rev.*, **46** (10), 2660–2677.
79. Sun, C., Yang, J., Rui, X., Zhang, W., Yan, Q., Chen, P., Huo, F., Huang, W., and Dong, X. (2015) MOF-directed templating synthesis of a porous multi-component dodecahedron with hollow interiors for enhanced lithium-ion battery anodes. *J. Mater. Chem. A*, **3** (16), 8483–8488.
80. Chou, L.Y., Hu, P., Zhuang, J., Morabito, J.V., Ng, K.C., Kao, Y.C., Wang, S.C., Shieh, F.K., Kuo, C.H., and Tsung, C.K. (2015) Formation of hollow and mesoporous structures in single-crystalline microcrystals of metal-organic frameworks via double-solvent mediated overgrowth. *Nanoscale*, **7** (46), 19408–19412.
81. Sun, J.K. and Xu, Q. (2014) Functional materials derived from open framework templates/precursors: Synthesis and applications. *Energy Environ. Sci.*, **7** (7), 2071–2100.
82. Xia, B.Y., Yan, Y., Li, N., Wu, H.B., Lou, X.W., and Wang, X. (2016) A metal-organic framework-derived bifunctional oxygen electrocatalyst. *Nat. Energy*, **1** (1), 15006.
83. Zhang, L., Wu, H.B., and Lou, X.W. (2013) Metal-organic-frameworks-derived general formation of hollow structures with high complexity. *J. Am. Chem. Soc.*, **135** (29), 10664–10672.
84. Zhou, H., Fan, T., and Zhang, D. (2011) Biotemplated materials for sustainable energy and environment: Current status and challenges. *ChemSusChem*, **4** (10), 1344–1387.
85. Yang, H.G. and Zeng, H.C. (2004) Preparation of hollow anatase TiO_2 nanospheres via Ostwald ripening. *J. Phys. Chem. B*, **108** (11), 3492–3495.
86. Yin, Y., Rioux, R.M., Erdonmez, C.K., Hughes, S., Somorjai, G.A., and Alivisatos, A.P. (2004) Formation of hollow nanocrystals through the nanoscale Kirkendall effect. *Science*, **304** (5671), 711–714.
87. Zhang, Q., Zhang, T., Ge, J., and Yin, Y. (2008) Permeable silica shell through surface-protected etching. *Nano Lett.*, **8** (9), 2867–2871.
88. Zhang, Q., Wang, W., Goebl, J., and Yin, Y. (2009) Self-templated synthesis of hollow nanostructures. *Nano Today*, **4** (6), 494–507.
89. Lee, J., Kim, S.M., and Lee, I.S. (2014) Functionalization of hollow nanoparticles for nanoreactor applications. *Nano Today*, **9** (5), 631–667.
90. Fang, X., Zhao, X., Fang, W., Chen, C., and Zheng, N. (2013) Self-templating synthesis of hollow mesoporous silica and their applications in catalysis and drug delivery. *Nanoscale*, **5** (6), 2205–2218.
91. Wei, W., Wang, Z., Liu, Z., Liu, Y., He, L., Chen, D., Umar, A., Guo, L., and Li, J. (2013) Metal oxide hollow nanostructures: Fabrication and Li storage performance. *J. Power Sources*, **238**, 376–387.
92. Lou, X.W., Archer, L.A., and Yang, Z. (2008) Hollow micro-/nanostructures: Synthesis and applications. *Adv. Mater.*, **20** (21), 3987–4019.
93. Nai, J., Wang, S., Bai, Y., and Guo, L. (2013) Amorphous $Ni(OH)_2$ nanoboxes: Fast fabrication and enhanced sensing for glucose. *Small*, **9** (18), 3147–3152.
94. Wang, Z., Wang, Z., Wu, H., and Lou, X.W. (2013) Mesoporous single-crystal $CoSn(OH)_6$ hollow structures with multilevel interiors. *Sci. Rep.*, **3** (3), 1391.
95. Wang, L., Tang, K., Liu, Z., Wang, D., Sheng, J., and Cheng, W. (2011) Single-crystalline $ZnSn(OH)_6$ hollow cubes via self-templated synthesis at room temperature and their photocatalytic properties. *J. Mater. Chem.*, **21** (12), 4352–4357.
96. Nai, J., Yin, H., You, T., Zheng, L., Zhang, J., Wang, P., Jin, Z., Tian, Y., Liu, J., Tang, Z., and Guo, L. (2015) Efficient electrocatalytic water oxidation by using amorphous Ni-Co double hydroxides nanocages. *Adv. Energy Mater.*, **5** (10), 1401880.
97. Wang, Z., Chen, X., Liu, J., Mo, M., Yang, L., and Qian, Y. (2004) Room temperature synthesis of Cu_2O nanocubes and nanoboxes. *Solid State Commun.*, **130** (9), 585–589.
98. Gu, Y., Cai, J., He, M., Kang, L., Lei, Z., and Liu, Z.H. (2013) Preparation and capacitance behavior of manganese oxide hollow structures with different morphologies via template-engaged redox etching. *J. Power Sources*, **239**, 347–355.
99. Fei, J.B., Cui, Y., Yan, X.H., Qi, W., Yang, Y., Wang, K.W., He, Q., and Li, J.B. (2008) Controlled preparation of MnO_2 hierarchical hollow nanostructures and their application in water treatment. *Adv. Mater.*, **20** (3), 452–456.

100. Wang, Z., Luan, D., Boey, F.Y.C., and Lou, X.W. (2011) Fast formation of SnO_2 nanoboxes with enhanced lithium storage capability. *J. Am. Chem. Soc.*, **133** (13), 4738–4741.
101. Zhang, G., Wu, H.B., Song, T., Paik, U., and Lou, X.W. (2014) TiO_2 hollow spheres composed of highly crystalline nanocrystals exhibit superior lithium storage properties. *Angew. Chem. Int. Ed.*, **53** (46), 12590–12593.
102. Sun, X. and Li, Y. (2004) Colloidal carbon spheres and their core/shell structures with noble-metal nanoparticles. *Angew. Chem. Int. Ed.*, **43** (5), 597–601.
103. Titirici, M.M., Antonietti, M., and Thomas, A. (2006) A generalized synthesis of metal oxide hollow spheres using a hydrothermal approach. *Chem. Mater.*, **18** (18), 3808–3812.
104. Xu, H. and Wang, W. (2007) Template synthesis of multishelled Cu_2O hollow spheres with a single-crystalline shell wall. *Angew. Chem. Int. Ed.*, **119** (9), 1511–1514.
105. Hubert, D.H.W., Jung, M., and German, A.L. (2000) Vesicle templating. *Adv. Mater.*, **12** (17), 1291–1294.
106. Kepczynski, M., Ganachaud, F., and Hemery, P. (2004) Silicone nanocapsules from catanionic vesicle templates. *Adv. Mater.*, **16** (20), 1861–1863.
107. Caruso, F. (2000) Hollow capsule processing through colloidal templating and self-assembly. *Chem. Eur. J.*, **6** (3), 413–419.
108. He, T., Chen, D., Jiao, X., and Wang, Y. (2006) Co_3O_4 nanoboxes: Surfactant-templated fabrication and microstructure characterization. *Adv. Mater.*, **18** (8), 1078–1082.
109. Li, W.C., Qiao, X.J., Zheng, Q.Y., and Zhang, T.L. (2011) One-step synthesis of MFe_2O_4 (M = Fe, Co) hollow spheres by template-free solvothermal method. *J. Alloys Compd.*, **509** (21), 6206–6211.
110. Chu, X., Wang, H., Chi, Y., Wang, C., Lei, L., Zhang, W., and Yang, X. (2018) Hard-template-engaged formation of $Co_2V_2O_7$ hollow prisms for lithium ion batteries. *RSC Adv.*, **8** (4), 2072–2076.
111. Chen, G., Rosei, F., and Ma, D. (2015) Template engaged synthesis of hollow ceria-based composites. *Nanoscale*, **7** (13), 5578–5591.
112. Yang, N., Pang, F., and Ge, J. (2015) One-pot and general synthesis of crystalline mesoporous metal oxides nanoparticles by protective etching: Potential materials for catalytic applications. *J. Mater. Chem. A*, **3** (3), 1133–1141.
113. Pan, J.H., Bai, Y., and Wang, Q. (2015) Reconstruction of colloidal spheres by targeted etching: A generalized self-template route to porous amphoteric metal oxide hollow spheres. *Langmuir*, **31** (15), 4566–4572.
114. Ostwald, W.Z. (1900) Über die vermeintliche isomerie des roten und gelben quecksilberoxyds und die oberflächenspannung fester körper. *Phys. Chem.*, **34** (1), 495–503.
115. Aleman, J., Chadwick, A.V., He, J., Hess, M., Horie, K., Jones, R.G., Kratochvil, P., Meisel, I., Mita, I., Moad, G., Penczek, S., and Stepto, R.F.T. (2007) Definitions of terms relating to the structure and processing of sols, gels, networks, and inorganic-organic hybrid materials (IUPAC Recommendations 2007). *Pure Appl. Chem.*, **79** (10), 1801–1827.
116. Lou, X.W., Wang, Y., Yuan, C., Lee, J.Y., and Archer, L.A. (2006) Template-free synthesis of SnO_2 hollow nanostructures with high lithium storage capacity. *Adv. Mater.*, **18** (17), 2325–2329.
117. Chen, Y., Xia, H., Lu, L., and Xue, J. (2012) Synthesis of porous hollow Fe_3O_4 beads and their applications in lithium ion batteries. *J. Mater. Chem.*, **22** (11), 5006–5012.
118. Cao, C.Y., Guo, W., Cui, Z.M., Song, W.G., and Cai, W. (2011) Microwave assisted gas/liquid interfacial synthesis of flowerlike NiO hollow nanosphere precursors and their application as supercapacitor electrodes. *J. Mater. Chem.*, **21** (9), 3204–3209.
119. Teo, J.J., Chang, Y., and Zeng, H.C. (2006) Fabrications of hollow nanocubes of Cu_2O and Cu via reductive self-assembly of CuO nanocrystals. *Langmuir*, **22** (17), 7369–7377.
120. Xu, M., Kong, L., Zhou, W., and Li, H. (2007) Hydrothermal synthesis and pseudocapacitance properties of alpha-MnO_2 hollow spheres and hollow urchins. *J. Phys. Chem. C*, **111** (51), 19141–19147.
121. Wang, X., Liao, M., Zhong, Y., Zheng, J.Y., Tian, W., Zhai, T., Zhi, C., Ma, Y., Yao, J., Bando, Y., and Golberg, D. (2012) ZnO hollow spheres with double-yolk egg structure for high-performance photocatalysts and photodetectors. *Adv. Mater.*, **24** (25), 3421–3425.
122. Kirkendall, E., Thomassen, L., and Uethegrove, C. (1939) Rates of diffusion copper and zinc in alpha brass. *Trans. Amer. Inst. Mining Met. Eng.*, **133**, 186–203.
123. Kirkendall, E.O. (1942) Diffusion of zinc in alpha brass. *Trans. Amer. Inst. Mining Met. Eng.* , **147**, 104–110.
124. Wang, W., Dahl, M., and Yin, Y. (2013) Hollow nanocrystals through the nanoscale kirkendall effect. *Chem. Mater.*, **25** (8), 1179–1189.
125. Ha, D.H., Moreau, L.M., Honrao, S., Hennig, R.G., and Robinson, R.D. (2013) The oxidation of cobalt nanoparticles into Kirkendall-hollowed CoO and Co_3O_4: The diffusion mechanisms and atomic structural transformations. *J. Phys. Chem. C*, **117** (27), 14303–14312.
126. Yu, Y., Yin, X., Kvit, A., and Wang, X. (2014) Evolution of hollow TiO_2 nanostructures via the Kirkendall effect driven by cation exchange with

enhanced photoelectrochemical performance. *Nano Lett.*, **14** (5), 2528–2535.

127. Peng, S. and Sun, S. (2007) Synthesis and characterization of monodisperse hollow Fe_3O_4 nanoparticles. *Angew. Chem. Int. Ed.*, **46** (22), 4155–4158.
128. Railsback, J.G., Johnston Peck, A.C., Wang, J., and Tracy, J.B. (2010) Size-dependent nanoscale Kirkendall effect during the oxidation of nickel nanoparticles. *ACS Nano*, **4** (4), 1913–1920.
129. Nakamura, R., Tokozakura, D., Nakajima, H., Lee, J.G., and Mori, H. (2007) Hollow oxide formation by oxidation of Al and Cu nanoparticles. *J. Appl. Phys.*, **101** (7), 074303.
130. Jen La Plante, I. and Mokari, T. (2013) Harnessing thermal expansion mismatch to form hollow nanoparticles. *Small*, **9** (1), 56–60.
131. Niu, K.Y., Park, J., Zheng, H., and Aivisatos, A.P. (2013) Revealing bismuth oxide hollow nanoparticle formation by the Kirkendall effect. *Nano Lett.*, **13** (11), 5715–5719.
132. Sun, Y.G., Mayers, B.T., and Xia, Y.N. (2002) Template-engaged replacement reaction: A one-step approach to the large-scale synthesis of metal nanostructures with hollow interiors. *Nano Lett.*, **2** (5), 481–485.
133. Xia, X., Wang, Y., Ruditskiy, A., and Xia, Y. (2013) Galvanic replacement: A simple and versatile route to hollow nanostructures with tunable and well-controlled properties. *Adv. Mater.*, **25** (44), 6313–6333.
134. Zhang, G. and Lou, X.W. (2014) General synthesis of multi-shelled mixed metal oxide hollow spheres with superior lithium storage properties. *Angew. Chem. Int. Ed.*, **53** (34), 9041–9044.
135. Yu, L., Wu, H.B., and Lou, X.W. (2013) Mesoporous $Li_4Ti_5O_{12}$ hollow spheres with enhanced lithium storage capability. *Adv. Mater.*, **25** (16), 2296–2300.
136. Hu, H., Guan, B., Xia, B., and Lou, X.W. (2015) Designed formation of $Co_3O_4/NiCo_2O_4$ double-shelled nanocages with enhanced pseudocapacitive and electrocatalytic properties. *J. Am. Chem. Soc.*, **137** (16), 5590–5595.
137. Shen, L., Yu, L., Yu, X.Y., Zhang, X., and Lou, X.W. (2015) Self-templated formation of uniform $NiCo_2O_4$ hollow spheres with complex interior structures for lithium-ion batteries and supercapacitors. *Angew. Chem. Int. Ed.*, **54** (6), 1868–1872.
138. Lu, Y., Yu, L., Wu, M., Wang, Y., and Lou, X.W. (2018) Construction of complex Co_3O_4@$Co_3V_2O_8$ hollow structures from metal-organic frameworks with enhanced lithium storage properties. *Adv. Mater.*, **30** (1), 1702875.
139. Song, M.K., Park, S., Alamgir, F.M., Cho, J., and Liu, M. (2011) Nanostructured electrodes for lithium-ion and lithium-air batteries: The latest developments, challenges, and perspectives. *Mater. Sci. Eng. R*, **72** (11), 203–252.
140. Goodenough, J.B. and Park, K.S. (2013) The Li-ion rechargeable battery: A perspective. *J. Am. Chem. Soc.*, **135** (4), 1167–1176.
141. Lou, X.W., Li, C.M., and Archer, L.A. (2009) Designed synthesis of coaxial SnO_2@carbon hollow nanospheres for highly reversible lithium storage. *Adv. Mater.*, **21** (24), 2536–2539.
142. Zhou, X., Yu, L., and Lou, X.W. (2016) Formation of uniform N-doped carbon-coated SnO_2 submicroboxes with enhanced lithium storage properties. *Adv. Energy Mater.*, **6** (14), 1600451.
143. Yu, X.Y., Wu, H.B., Yu, L., Ma, F.X., and Lou, X.W. (2015) Rutile TiO_2 submicroboxes with superior lithium storage properties. *Angew. Chem. Int. Ed.*, **54** (13), 4001–4004.
144. Guan, X., Nai, J., Zhang, Y., Wang, P., Yang, J., Zheng, L., Zhang, J., and Guo, L. (2014) CoO hollow cube/reduced graphene oxide composites with enhanced lithium storage capability. *Chem. Mater.*, **26** (20), 5958–5964.
145. Ma, F.X., Hu, H., Wu, H.B., Xu, C.Y., Xu, Z., Zhen, L., and Lou, X.W. (2015) Formation of uniform Fe_3O_4 hollow spheres organized by ultrathin nanosheets and their excellent lithium storage properties. *Adv. Mater.*, **27** (27), 4097–4101.
146. Wang, Z., Wang, Z., Liu, W., Xiao, W., and Lou, X.W. (2013) Amorphous $CoSnO_3$@C nanoboxes with superior lithium storage capability. *Energy Environ. Sci.*, **6** (1), 87–91.
147. Manthiram, A., Fu, Y.Z., Chung, S.H., Zu, C.X., and Su, Y.S. (2014) Rechargeable lithium-sulfur batteries. *Chem. Rev.*, **114** (23), 11751–11787.
148. Manthiram, A., Fu, Y., and Su, Y.S. (2013) Challenges and prospects of lithium-sulfur batteries. *Acc. Chem. Res.*, **46** (5), 1125–1134.
149. Yin, Y.X., Xin, S., Guo, Y.G., and Wan, L.J. (2013) Lithium-sulfur batteries: Electrochemistry, materials, and prospects. *Angew. Chem. Int. Ed.*, **52** (50), 13186–13200.
150. Liang, X. and Nazar, L.F. (2016) *In situ* reactive assembly of scalable core-shell sulfur-MnO_2 composite cathodes. *ACS Nano*, **10** (4), 4192–4198.
151. Wang, X., Li, G., Li, J., Zhang, Y., Wook, A., Yu, A., and Chen, Z. (2016) Structural and chemical synergistic encapsulation of polysulfides enables ultralong-life lithium-sulfur batteries. *Energy Environ. Sci.*, **9** (8), 2533–2538.
152. Liang, X., Kwok, C.Y., Lodi-Marzano, F., Pang, Q., Cuisinier, M., Huang, H., Hart, C.J., Houtarde, D., Kaup, K., Sommer, H., Brezesinski, T., Janek, J., and Nazar, L.F. (2016) Tuning transition metal oxide-sulfur interactions for long life lithium sulfur batteries: The "goldilocks" principle. *Adv. Energy Mater.*, **6** (6), 1501636.
153. Li, Z., Zhang, J., Guan, B., Wang, D., Liu, L.M., and Lou, X.W. (2016) A sulfur host based on titanium monoxide@carbon hollow spheres for advanced

lithium-sulfur batteries. *Nat. Commun.*, **7**, 13065.
154. Simon, P. and Gogotsi, Y. (2008) Materials for electrochemical capacitors. *Nat. Mater.*, **7** (11), 845–854.
155. Yu, Z., Tetard, L., Zhai, L., and Thomas, J. (2015) Supercapacitor electrode materials: nanostructures from 0 to 3 dimensions. *Energy Environ. Sci.*, **8** (3), 702–730.
156. Choudhary, N., Li, C., Moore, J., Nagaiah, N., Zhai, L., Jung, Y., and Thomas, J. (2017) Asymmetric supercapacitor electrodes and devices. *Adv. Mater.*, **29** (21), 1605336.
157. Wang, J., Tang, H., Ren, H., Yu, R., Qi, J., Mao, D., Zhao, H., and Wang, D. (2014) pH-regulated synthesis of multi-shelled manganese oxide hollow microspheres as supercapacitor electrodes using carbonaceous microspheres as templates. *Adv. Sci.*, **1** (1), 1400011.
158. Fujishima, A. and Honda, K. (1972) Electrochemical photolysis of water at a semiconductor electrode. *Nature*, **238** (5358), 37–38.
159. Kudo, A. and Miseki, Y. (2009) Heterogeneous photocatalyst materials for water splitting. *Chem. Soc. Rev.*, **38** (1), 253–278.
160. Joo, J.B., Zhang, Q., Lee, I., Dahl, M., Zaera, F., and Yin, Y. (2012) Mesoporous anatase titania hollow nanostructures though silica-protected calcination. *Adv. Funct. Mater.*, **22** (1), 166–174.
161. Zeng, Y., Wang, X., Wang, H., Dong, Y., Ma, Y., and Yao, J. (2010) Multi-shelled titania hollow spheres fabricated by a hard template strategy: enhanced photocatalytic activity. *Chem. Commun.*, **46** (24), 4312–4314.
162. Wang, S., Qian, H., Hu, Y., Dai, W., Zhong, Y., Chen, J., and Hu, X. (2013) Facile one-pot synthesis of uniform TiO_2-Ag hybrid hollow spheres with enhanced photocatalytic activity. *Dalton Trans.*, **42** (4), 1122–1128.
163. Agrawal, M., Gupta, S., Pich, A., Zafeiropoulos, N.E., and Stamm, M. (2009) A facile approach to fabrication of ZnO-TiO_2 hollow spheres. *Chem. Mater.*, **21** (21), 5343–5348.
164. Song, H., Zhu, L., Li, Y., Lou, Z., Xiao, M., and Ye, Z. (2015) Preparation of $ZnFe_2O_4$ nanostructures and highly efficient visible-light-driven hydrogen generation with the assistance of nanoheterostructures. *J. Mater. Chem. A*, **3** (16), 8353–8360.
165. Qin, F., Zhao, H., Li, G., Yang, H., Li, J., Wang, R., Liu, Y., Hu, J., Sun, H., and Chen, R. (2014) Size-tunable fabrication of multifunctional Bi_2O_3 porous nanospheres for photocatalysis, bacteria inactivation and template-synthesis. *Nanoscale*, **6** (10), 5402–5409.
166. Tseng, W.J., Tseng, T.T., Wu, H.M., Her, Y.C., and Yang, T.J. (2013) Facile synthesis of monodispersed In_2O_3 hollow spheres and application in photocatalysis and gas sensing. *J. Am. Ceram. Soc.*, **96** (3), 719–725.
167. Lin, S., Shi, L., Yoshida, H., Li, M., and Zou, X. (2013) Synthesis of hollow spherical tantalum oxide nanoparticles and their photocatalytic activity for hydrogen production. *J. Solid State Chem.*, **199** (1), 15–20.
168. Deng, W., Chen, D., and Chen, L. (2015) Synthesis of monodisperse CeO_2 hollow spheres with enhanced photocatalytic activity. *Ceram. Int.*, **41** (9), 11570–11575.
169. Xia, W., Mahmood, A., Liang, Z.B., Zou, R.Q., and Guo, S.J. (2016) Earth-abundant nanomaterials for oxygen reduction. *Angew. Chem. Int. Ed.*, **55** (8), 2650–2676.
170. Wang, J.H., Cui, W., Liu, Q., Xing, Z.C., Asiri, A.M., and Sun, X.P. (2016) Recent progress in cobalt-based heterogeneous catalysts for electrochemical water splitting. *Adv. Mater.*, **28** (2), 215–230.
171. Jiao, Y., Zheng, Y., Jaroniec, M., and Qiao, S.Z. (2015) Design of electrocatalysts for oxygen- and hydrogen-involving energy conversion reactions. *Chem. Soc. Rev.*, **44** (8), 2060–2086.
172. Han, L., Yu, X.Y., and Lou, X.W. (2016) Formation of Prussian-blue-analog nanocages via a direct etching method and their conversion into Ni-Co-mixed oxide for enhanced oxygen evolution. *Adv. Mater.*, **28** (23), 4601–4605.
173. Lyu, F., Bai, Y., Li, Z., Xu, W., Wang, Q., Mao, J., Wang, L., Zhang, X., and Yin, Y. (2017) Self-templated fabrication of CoO-MoO_2 nanocages for enhanced oxygen evolution. *Adv. Funct. Mater.*, **27** (34), 1702324.

Index

D

E

F

G

H

I

J

K

L

M

N

O

T

U

V